Birkhäuser

Systems & Control: Foundations & Applications

For further volumes:
http://www.springer.com/series/4895

Ali Saberi • Anton A. Stoorvogel
• Peddapullaiah Sannuti

Internal and External Stabilization of Linear Systems with Constraints

Birkhäuser

Ali Saberi
School of Electrical Engineering
and Computer Science
Washington State University
Pullman, WA, USA

Anton A. Stoorvogel
Department of Electrical Engineering,
Mathematics, and Computer Science
University of Twente
Enschede, The Netherlands

Peddapullaiah Sannuti
Department of Electrical
and Computer Engineering
Rutgers University
Piscataway, NJ, USA

ISBN: 978-0-8176-4786-5 ISBN: 978-0-8176-4787-2 (eBook)
DOI: 10.1007/978-0-8176-4787-2
Springer New York Heidelberg Dordrecht London

Library of Congress Control Number: 2012939202

Mathematics Subject Classification (2010): 93-02, 93C10, 93D15

Printed on acid-free paper

Springer is part of Springer Science+Business Media (www.birkhauser-science.com)

This book is dedicated to:

Ann, Ingmar, Ula, Dmitri, and Mirabella	**(Ali Saberi)**
My parents	**(Anton A. Stoorvogel)**
Jayanth	**(Pedda Sannuti)**

and

to the glory of man

Contents

Preface

The Challenge of Constraints

Willst du ins Unendliche schreiten,
Geh nur im Endlichen nach allen seiten.
If to the Infinite you want to stride,
Just walk in the Finite to every side.
Gott, Gemüt und Welt
Johann Wolfgang von Goethe

Constraints are common and are everywhere. Hard time-domain constraints on actuators, sensors, and state variables of dynamic systems are the most ubiquitous nonlinearities in practical control systems. Their impacts on stability, control performance, and safety have been well recognized by both control engineers and control theorists for many decades. The challenge of such constraints in analysis as well as in design of control systems is intense and dauntingly formidable and familiar; it needs no elaboration and explanation.

The primary focus of this book is on the problem of achieving simultaneous internal and external stabilization of linear systems subject to constraints in both semi-global and global framework. Our intended audience includes practicing engineers, graduate students, and researchers in the field of systems and control. A vast majority of the contents of this book are drawn from the research of the authors, their coworkers, and students. Thus, it bears the signature of the authors and has a recognizable identity and a coherence of point of view which can be characterized as a structural view in both the analysis and design of dynamic systems.

No work of this magnitude and nature can be undertaken without many sacrifices. The deeds of this book absorbed our time infinitely more than the deeds of our households. We thank our families for their tolerance and understanding. Naturally, the debt of gratitude to our families is paid in some way by dedicating this book to them. The PhD thesis work of Mr. Xu Wang reflects in many places, needless to say that we are indebted to him enormously. Also, we are certainly indebted to our editor, Dr. Tamer Başar, and the editorial staff at Birkhäuser. Our special thanks go to the copy editor for a meticulous editing that improved the text. Ali spent countless number of hours brooding over the manuscript of this book at Bucer's; the great coffee house of Moscow, Idaho. Ali acknowledges the contribution of all the good people of Bucer's, special thanks go to Ms. Pat Greenfield.

Finally, we trust and hope that a proper study of this book leads to a bounty of applications of what we strived to develop here. We await to realize that this is no idle dream.

Ali Saberi
Washington State University, Pullman, Washington, U.S.A.

Anton A. Stoorvogel
University of Twente, Drienerlolaan 5, The Netherlands

Peddapullaiah Sannuti
Rutgers University, Piscataway, New Jersey, U.S.A.

1

Introduction

Constraints on inputs and other variables of a dynamic system are ubiquitous. Often they occur in the form of magnitude as well as rate saturation of a variable. Clearly, the capacity of every device is capped. Valves can only be operated between fully open and fully closed states, pumps and compressors have a finite throughput capacity, and tanks can only hold a certain volume. Force, torque, thrust, stroke, voltage, current, flow rate, and so on, are limited in their activation range in all physical systems. Servers can serve only so many consumers. In circuits, transistors and amplifiers are saturating components. Saturation and other physical limitations are dominant in maneuvering systems like aircrafts. Every physically conceivable actuator, sensor, or transducer has bounds on the magnitude as well as on the rate of change of its output. Thus, the *saturation* of a device presents a hard *constraint*.

The first period of intense research that was focused on constraints on inputs was in the middle of last century, mainly via optimal control theory and anti-windup compensation. The success of this period of research was somewhat limited and control under constraints remained for a long time as a challenge to control engineers. A second and intense period of research on constraints on inputs, primarily on control input magnitude and rate saturation, got started in the early 1990s. For the next decade or so, control of linear systems subject to actuator saturation both in magnitude and rate of change was the center of focus. Achieving internal (Lyapunov) stability as well as simultaneous internal and external (L_p or ℓ_p) stability of a closed-loop linear system subject to such constraints has been the impetus for much of the research performed. Internal as well as external stabilization was pursued in both global and semi-global setting. This phase of research has provided a rich variety of techniques for analysis and design of control systems subject to constraints. Tremendous strides have been and are being made to advance our understanding of such constraints. Both research monographs and special issues of control journals document the results available in this early phase of work.

Control of linear systems subject to constraints took another and important turn a decade or so ago when constraints not only on actuator and sensor saturation but also on state variables were imposed. This general type of constraints is modeled by introducing what are termed as *constrained outputs* which are

A. Saberi et al., *Internal and External Stabilization of Linear Systems with Constraints*, Systems & Control: Foundations & Applications, DOI 10.1007/978-0-8176-4787-2_1,

linear combinations of inputs and state variables of the controlled system. Internal stabilization of linear systems was sought with the constraint that the constrained outputs be confined to subsets of their respective spaces. To be specific, consider a linear system Σ commonly described by

$$\Sigma : \rho x = Ax + Bu \tag{1.1}$$

where $x \in \mathbb{R}^n$ is the state and $u \in \mathbb{R}^m$ is the control input and ρx denotes $\frac{dx}{dt}$ for continuous time and $x(k+1)$ for discrete time. Let us define next the *constrained output*

$$z = C_z x + D_z u, \quad z \in \mathbb{R}^p \tag{1.2}$$

with

$$z \in \mathcal{S} \quad \text{and} \quad \frac{dz}{dt} \in \mathcal{T} \text{ or } (z(k+1) - z(k)) \in \mathcal{T} \tag{1.3}$$

for all $t \geqslant 0$ or $k \geqslant 0$, where $\mathcal{S}$ and $\mathcal{T}$ are a priori given subsets of $\mathbb{R}^p$. Thus, the constrained output z captures constraints on both magnitude and rate of change on a part of input u as well as on a part of state x. We observe next that, based on the constrained output z, a taxonomy of constraints was also developed categorizing and delineating the constraints into different groups. Such a categorization of constraints paved distinct omnidirectional paths showing what can be achieved and what cannot be achieved.

More recently, the saga of research on constraints got elevated to another orbit when it embraced so-called sandwiched nonlinearities. Let us emphasize that most if not all systems encountered in practice indeed consist of an interconnection of components, or otherwise called subsystems, and some of these subsystems are well characterized as linear, whereas others are more distinctly nonlinear. Clearly, this results in a system configuration which is an interconnection of *separable* linear and nonlinear parts. In other words, a common paradigm of nonlinear systems is that they are indeed linear systems in which *nonlinear elements are sandwiched or embedded.* This provokes or motivates a thorough study of different types of nonlinear elements or constraints. As pointed out earlier, one of the ubiquitous static nonlinearities is saturation of a device. This implies that most often one encounters as a system model a collection of linear systems in which saturation nonlinearities are embedded or sandwiched. Thus, control of such systems for internal stabilization or for other performance requirements emerged as another immense focus of research.

In very recent years, based on a clear but very agonizing understanding of the complexity of achieving external stabilization in the presence of sustained disturbances, the focus of research activity has been directed toward identifying a class of sustained disturbances for which external stabilization of linear systems subject to constraints on control can be assured, while simultaneously assuring global internal stabilization of such systems in the absence of such disturbances.

The goal of this book is to lay brick by brick a foundation for a systematic analysis and design of linear systems subject to a variety of constraints while focussing mainly on internal stabilization as well as simultaneous internal and external stabilization. To elaborate, this book is an ardent story of mainly four topics:

- Internal stabilization of linear systems subject to constraints on control input magnitude and rate saturation set in both global and semi-global framework.
- Internal stabilization of linear systems subject to constraints on what we termed earlier as *constrained output* consisting of input and state variables of the system.
- Internal stabilization of linear systems sandwiched with static saturation nonlinearities set in both global and semi-global framework.
- Simultaneous internal and external stabilization of linear systems subject to constraints on control input saturation once again set in both global and semi-global framework.

Most of the results presented are of recent origin and are due to the authors or their coworkers and students. In this spirit, the book incorporates several published as well as yet unpublished results of the authors and their colleagues. As such, as can be expected, the exposition given is somewhat biased in the direction of the research work of the authors carried over a period of two decades or more.

It is appropriate now to preview briefly the contents of the book. Chapter 2 on preliminaries recalls several notations and notions of internal (Lyapunov) stability and external (L_p or ℓ_p) stability, while Chap. 3 presents a special coordinate basis (SCB) of linear systems and explores its properties. We emphasize that the SCB of the given system exhibits clearly its finite and infinite zero structure and thus plays a crucial role throughout the book in both analysis and design. Chapters 4–6 are devoted to internal stabilization of linear systems subject to control saturation, both magnitude as well as rate of change of it. Internal stabilization is sought here in both global and semi-global framework. Several control design methodologies such as low gain, low-and-high gain, scheduled low gain, scheduled low-and-high gain are developed by both direct as well as Riccati equation-based methods. Chapters 7–9 deal also with internal stabilization; however instead of actuator saturation, they consider input and state constraints. The constraints here are formulated in terms of the constrained output $z \in \mathbb{R}^p$ by imposing its magnitude and rate of change be confined respectively to certain a priori prescribed subsets $\mathcal{S}$ and $\mathcal{T}$ of $\mathbb{R}^p$ (see (1.2) and (1.3)). Chapter 10 continues further the theme of internal stabilization. Here we face a multiple of saturation nonlinearities each sandwiched between two linear systems, and hence, the scope of controller design is an intricate extension of design methodologies developed in Chaps. 4–6.

The goal of the rest of the chapters (except Chap. 17) is achieving simultaneous internal and external stabilization of linear systems subject to control saturation. This is done both in global and semi-global framework. Let us be explicit about this. Chapter 11 formulates precisely several internal and external stabilization problems. It is fitting to observe at this time that the external disturbance can enter a given system either additive to the control input or nonadditive to it. For the case of input additive external disturbance, Chap. 12 more or less resolves and develops appropriate control design methods to solve all the internal and external stabilization problems formulated in Chap. 11. However, for the case of non-input additive external disturbance, not all problems formulated in Chap. 11 are solvable for general linear systems subject to actuator saturation. Chapter 13 tackles and resolves some such problems by developing the needed controllers. Chapter 14 explores the intricacies involved in the case of non-input additive external disturbances by considering a canonical system, namely, a double integrator. Among many results developed here, one key result is that while external stabilization without finite gain is achievable for all L_p non-input additive disturbance signals with $p \in [1, \infty)$ (i.e., for all disturbances whose "energy" vanishes asymptotically), it is not achievable for sustained disturbances (L_∞ disturbances). It becomes imperative then to identify a class of sustained disturbances for which the states of controlled system are bounded. Chapter 14 for the canonical double integrator does identify a class of integral bounded non-input additive sustained disturbances for which L_∞ stabilization can be attained. This theme of identifying a class of sustained disturbances for which L_∞ stabilization can be attained is pursued for general linear systems subject to actuator saturation in Chap. 15 for continuous-time systems and in Chap. 16 for discrete-time systems.

Finally, as a prelude to future research, in Chap. 17, a *stochastic framework* is initiated laying out a road map for simultaneous internal and external stabilization of linear systems subject to constraints.

2

Preliminaries

In this chapter, we bring together the notations and acronyms used in this book as well as various definitions and facts related to matrices, linear spaces, linear operators, norms of deterministic as well as stochastic signals, norms of linear time- or shift-invariant systems, saturation functions, internal (Lyapunov) stability, and external stability.

2.1 A list of symbols

Throughout this book, we shall adopt the following conventions and notations:

$\mathbb{R}$	Set of real numbers
$\mathbb{R}^+$	Set of nonnegative real numbers
$\mathbb{Z}^+$	Set of nonnegative integers
$\mathbb{C}$	Entire complex plane
$\mathbb{C}^-$	Open left-half complex plane
$\mathbb{C}^0$	Imaginary axis
$\mathbb{C}^+$	Open right-half complex plane
$\mathbb{C}^{-0}$	Closed left-half complex plane
$\mathbb{C}^{+0}$	Closed right-half complex plane
$\mathbb{C}^{\ominus}$	Set of complex numbers inside the unit circle
$\mathbb{C}^{\bigcirc}$	Unit circle
$\mathbb{C}^{\oplus}$	Set of complex numbers outside the unit circle
$\mathbb{C}^{\otimes}$	Set of complex numbers inside and on the unit circle
$\mathbb{C}^{\odot}$	Set of complex numbers outside and on the unit circle
$\mathcal{B}(r)$	The set $\{\, x \in \mathbb{R}^n \mid \|x\| < r \,\}$
$\mathcal{B}(x_0, r)$	The set $\{\, x \in \mathbb{R}^n \mid \|x - x_0\| < r \,\}$

A. Saberi et al., *Internal and External Stabilization of Linear Systems with Constraints*, Systems & Control: Foundations & Applications, DOI 10.1007/978-0-8176-4787-2_2,

I	An identity matrix
I_k	Identity matrix of dimension $k \times k$
A'	Transpose of A
A^*	Complex conjugate transpose of A
$\lambda(A)$	Set of eigenvalues of A
$\sigma_{\max}(A)$	Maximum singular value of A
$\sigma_{\min}(A)$	Minimum singular value of A
$\rho(A)$	Spectral radius of A
trace A	Trace of A
ker A	The null space of A
im A	The range space of A
$\langle A \mid \operatorname{im} B \rangle$	The controllability subspace of the pair (A, B)
$\langle \ker C \mid A \rangle$	The unobservable subspace of the pair (A, C)
$\mathcal{V}^{\perp}$	Orthogonal complement of a subspace $\mathcal{V}$ in $\mathbb{R}^n$
$\boldsymbol{E}[\cdot]$	The expectation of a stochastic vector
$\mathbb{R}[s]$	Ring of polynomials with real coefficients
$\mathbb{R}^{n \times m}[s]$	Set of all $n \times m$ matrices with coefficients in $\mathbb{R}[s]$
$\mathbb{R}(s)$	Field of rational functions with real coefficients
$\mathbb{R}^{n \times m}(s)$	Set of all $n \times m$ matrices with coefficients in $\mathbb{R}(s)$

For any set $\mathcal{C} \subset \mathbb{R}^n$, $\operatorname{int} \mathcal{C}$ denotes the interior of set $\mathcal{C}$, $\partial \mathcal{C}$ the boundary of set $\mathcal{C}$, and $\overline{\mathcal{C}}$ the closure of set $\mathcal{C}$. For a dynamical system

$$\rho x = f(x, u),$$

the ρ denotes the time-derivative

$$\rho x = \tfrac{\mathrm{d}}{\mathrm{d}t} x$$

for continuous-time systems while it denotes the shift operator

$$(\rho x)(k) = x(k + 1)$$

for discrete-time systems.

2.2 Matrices, linear spaces, and linear operators

In this section, we recall certain fundamental facts and properties of matrices, linear spaces, and linear operators that are relevant to this book. We have done so for the ease of readers and to establish the related notations used throughout the book.

We say a matrix A is injective or surjective if A is of full column rank or full row rank, respectively. By $\operatorname{rank}_{\mathcal{K}}$, we denote the rank of a matrix over the field $\mathcal{K}$. We shall write *rank* only for the case when $\mathcal{K} = \mathbb{R}$ or $\mathcal{K} = \mathbb{C}$. Moreover, we use the term *normal rank* or *normrank* for $\operatorname{rank}_{\mathcal{K}}$ whenever $\mathcal{K} = \mathbb{R}(s)$. We note that if $A \in \mathbb{C}^{m \times n}$, we have that $\operatorname{im} A = \ker(A^*)^{\perp}$.

We recall next the classical concept of the Jordan form of a general matrix A and the concept of the multiplicity structure of an eigenvalue of a matrix A. Given any matrix A of dimension $n \times n$, we can always find a non-singular transformation matrix X (see [40]) such that

$$X^{-1}AX = J = \begin{pmatrix} J_1 & 0 & \cdots & 0 \\ 0 & J_2 & \ddots & \vdots \\ \vdots & \ddots & \ddots & 0 \\ 0 & \cdots & 0 & J_k \end{pmatrix}, \tag{2.1}$$

where $J_i, i = 1, \ldots, k$ are some $n_i \times n_i$ Jordan blocks,

$$J_i = \begin{pmatrix} \lambda_i & 1 & 0 & \cdots & 0 \\ 0 & \ddots & \ddots & \ddots & \vdots \\ \vdots & \ddots & \ddots & \ddots & 0 \\ \vdots & & \ddots & \ddots & 1 \\ 0 & \cdots & \cdots & 0 & \lambda_i \end{pmatrix}. \tag{2.2}$$

We note that

$$\sum_{i=1}^{k} n_i = n.$$

Then, the geometric multiplicity ν_λ of an eigenvalue $\lambda \in \lambda(A)$ is the number of Jordan blocks associated with λ in (2.1) as well as the number of linearly independent eigenvectors associated with λ. On the other hand, the algebraic multiplicity ρ_λ is the total number of repetitions of λ in $\lambda(A)$; equivalently, the algebraic multiplicity is equal to the sum of the number of rows of all Jordan blocks associated with λ.

We introduce next what is known as the multiplicity structure of an eigenvalue. For any given $\lambda \in \lambda(A)$, let there be ν_λ Jordan blocks of A associated with λ. Let

$$n_{\lambda,1} \geqslant n_{\lambda,2} \geqslant \cdots, n_{\lambda,\nu_\lambda}$$

be the dimensions of the corresponding Jordan blocks ordered in size. Then, λ is an eigenvalue of A with multiplicity structure S_λ^*,

$$S_\lambda^* = \{n_{\lambda,1}, n_{\lambda,2}, \ldots, n_{\lambda,\nu_\lambda}\}. \tag{2.3}$$

If $n_{\lambda,1} = n_{\lambda,2} = \cdots = n_{\lambda,\nu_\lambda} = 1$, then λ is called a semi-simple eigenvalue of A. Moreover, we call an eigenvalue a simple eigenvalue if $\nu_\lambda = 1$ and $n_{\lambda,1} = 1$ or equivalently if it has an algebraic multiplicity equal to 1.

The invariant factor $\Psi_i(s)$ of a matrix A is the monic polynomial of lowest degree such that for each eigenvalue λ with $\nu_\lambda \geqslant i$, $\Psi_i(s)$ has $n_{\lambda,i}$ zeros in λ. We note that algebraic multiplicity ρ_λ satisfies

$$\rho_\lambda = n_{\lambda,1} + n_{\lambda,2} + \cdots + n_{\lambda,\nu_\lambda}.$$

We recall next the following classic concepts of generalized eigenvectors and the eigenvector chain associated with an eigenvalue of a matrix. A vector x is said to be a generalized eigenvector of grade k associated with an eigenvalue λ of a matrix A if and only if

$$(A - \lambda I)^k x = 0 \quad \text{and} \quad (A - \lambda I)^{k-1} x \neq 0.$$

A generalized eigenvector of grade one (i.e., $k = 1$) is a standard eigenvector associated with an eigenvalue of a matrix. Let vector x be a generalized eigenvector of grade k associated with an eigenvalue λ of a matrix A. Let

$$\begin{aligned}
x_k &= x \\
x_{k-1} &= (A - \lambda I)V && = (A - \lambda I)x_k \\
x_{k-2} &= (A - \lambda I)^2 V && = (A - \lambda I)x_{k-1} \\
&\ \vdots && \quad \vdots \\
x_1 &= (A - \lambda I)^{k-1} V && = (A - \lambda I)x_2.
\end{aligned}$$

Such a set of vectors $\{x_1, x_2, \cdots, x_k\}$ is called a chain of generalized eigenvectors of length k associated with an eigenvalue λ.

For an eigenvalue λ with the multiplicity structure S_λ^* as given in (2.3), there are ν_λ chains of generalized eigenvectors with lengths $n_{\lambda,1}, n_{\lambda,2}, \cdots, n_{\lambda,\nu_\lambda}$. The total number of generalized eigenvectors in these chains equals the algebraic multiplicity ρ_λ. Moreover, these ρ_λ generalized eigenvectors are linearly independent.

If M is a subspace of $\mathbb{C}^n$, then we define the *orthogonal projection* P_M of $\mathbb{C}^n$ onto M by $P_M u = u$ if $u \in M$ and $P_M u = 0$ if $u \in M^\perp$. We note that $I - P_M = P_{M^\perp}$.

A matrix $U \in \mathbb{C}^{n\times n}$ is said to be a unitary matrix if $U^* = U^{-1}$. For a matrix $A \in \mathbb{C}^{m\times n}$, the generalized inverse of A (or Moore–Penrose inverse of A) is defined to be a unique matrix $A^\dagger$ in $\mathbb{C}^{n\times m}$ such that:

(a) $AA^\dagger$ is an orthogonal projection onto im A.

(b) $A^\dagger A$ is an orthogonal projection onto $(\ker A)^\perp = \operatorname{im} A^*$.

Another equivalent definition for a generalized inverse of $A \in \mathbb{C}^{m\times n}$ is a unique matrix $A^\dagger$ in $\mathbb{C}^{n\times m}$ such that:

(a) $AA^{\dagger}A = A$.

(b) $A^{\dagger}AA^{\dagger} = A^{\dagger}$.

(c) $AA^{\dagger}$ is a symmetric matrix.

(d) $A^{\dagger}A$ is a symmetric matrix.

Some basic properties of the generalized inverse of $A \in \mathbb{C}^{m \times n}$ are listed as follows:

- $(A^{\dagger})^{\dagger} = A$.
- $(A^{\dagger})^{*} = (A^{*})^{\dagger}$.
- If $\lambda \in \mathbb{C}$, $(\lambda A)^{\dagger} = \lambda^{\dagger}A^{\dagger}$, where $\lambda^{\dagger} = \frac{1}{\lambda}$ if $\lambda \neq 0$ and $\lambda^{\dagger} = 0$ if $\lambda = 0$.
- $A^{*} = A^{*}AA^{\dagger} = A^{\dagger}AA^{*}$.
- $(A^{*}A)^{\dagger} = A^{\dagger}(A^{*})^{\dagger}$.
- $A^{\dagger} = (A^{*}A)^{\dagger}A^{*} = A^{*}(AA^{*})^{\dagger}$.
- $(UAV)^{\dagger} = V^{*}A^{\dagger}U^{*}$, where U and V are unitary matrices.
- $\operatorname{im} A = \operatorname{im} AA^{\dagger} = \operatorname{im} AA^{*}$.
- $\operatorname{im} A^{\dagger} = \operatorname{im} A^{*} = \operatorname{im} A^{\dagger}A = \operatorname{im} A^{*}A$.
- $\operatorname{im}(I - AA^{\dagger}) = \ker AA^{\dagger} = \ker A^{*} = \ker A^{\dagger} = (\operatorname{im} A)^{\perp}$.
- $\operatorname{im}(I - A^{\dagger}A) = \ker A^{\dagger}A = \ker A = (\operatorname{im} A^{*})^{\perp}$.
- If $B \in \mathbb{C}^{n \times p}$, then $(AB)^{\dagger} = (P_{\operatorname{im} A^{*}} B)^{\dagger}(AP_{\operatorname{im} B})^{\dagger}$.
- If $A^{*}ABB^{*} = BB^{*}A^{*}A$, then $(AB)^{\dagger} = B^{\dagger}A^{\dagger}$.
- If $A = BC$, where $B \in \mathbb{C}^{m \times r}$ and $C \in \mathbb{C}^{r \times n}$, while $r = \operatorname{rank} A$, then $A^{\dagger} = C^{*}(CC^{*})^{-1}(B^{*}B)^{-1}B^{*}$.

The following necessary and sufficient conditions for a partitioned matrix to be *positive semi-definite* and *positive definite* are useful. Consider an arbitrarily partitioned Hermitian matrix Q:

$$Q = \begin{pmatrix} Q_{11} & Q_{12} \\ Q_{12}^{*} & Q_{22} \end{pmatrix}.$$

Then Q is *positive semi-definite* if and only if

$$\begin{cases} Q_{22} \geqslant 0 \\ Q_{12} = Q_{12}Q_{22}^{\dagger}Q_{22} \\ Q_{11} \geqslant Q_{12}Q_{22}^{\dagger}Q_{12}^{*}, \end{cases}$$

or, equivalently, Q is *positive semi-definite* if and only if

$$\begin{cases} Q_{11} \geqslant 0 \\ Q_{12} = Q_{11}Q_{11}^{\dagger}Q_{12} \\ Q_{22} \geqslant Q_{12}^{*}Q_{11}^{\dagger}Q_{12}. \end{cases}$$

Similarly, Q is *positive definite* if and only if

$$\begin{cases} Q_{22} > 0 \\ Q_{11} > Q_{12}Q_{22}^{-1}Q_{12}^{*} \end{cases}$$

or, equivalently,

$$\begin{cases} Q_{11} > 0 \\ Q_{22} > Q_{12}^{*}Q_{11}^{-1}Q_{12}. \end{cases}$$

Let us next discuss the addition of subspaces and the associated notations. Suppose $\mathcal{X}$, $\mathcal{Y}$, and $\mathcal{Z}$ are some subspaces of $\mathbb{R}^n$ or $\mathbb{C}^n$. Then,

$$\mathcal{X} + \mathcal{Y} = \{x + y \mid x \in \mathcal{X}, y \in \mathcal{Y}\}.$$

If $\mathcal{Z} = \mathcal{X} + \mathcal{Y}$ and $\mathcal{X} \cap \mathcal{Y} = \{0\}$, then $\mathcal{Z}$ is called the direct sum of $\mathcal{X}$ and $\mathcal{Y}$, and, in this case, $\mathcal{Z}$ is written as $\mathcal{X} \oplus \mathcal{Y}$. Consider a subspace $\mathcal{X}$ in $\mathbb{R}^n$. Then, the orthogonal complement $\mathcal{X}^{\perp}$ of the subspace $\mathcal{X}$ is defined as

$$\mathcal{X}^{\perp} = \{u \in \mathbb{R}^n \mid \langle u\,,\, v\rangle \;= 0 \text{ for every } v \in \mathcal{X}\}.$$

Let $\mathcal{X}$ and $\mathcal{Y}$ be two nontrivial subspaces of $\mathbb{R}^n$. If the inner product of x and y is zero for all $x \in \mathcal{X}$ and $y \in \mathcal{Y}$, then the two subspaces $\mathcal{X}$ and $\mathcal{Y}$ are said to be orthogonal, and this is denoted by $\mathcal{X} \perp \mathcal{Y}$.

Next, for a matrix $M \in \mathbb{R}^{m\times n}$, the linear transformation $M\mathcal{X}$ is defined as

$$M\mathcal{X} := \{Mx \mid x \in \mathcal{X}\}.$$

Also, for a matrix $N \in \mathbb{R}^{n\times m}$,

$$N^{-1}\mathcal{X} := \{z \in \mathbb{R}^m \mid Nz \in \mathcal{X}\}.$$

The following relations will be useful in algebraic manipulations regarding subspaces:

$$\begin{aligned}
\mathcal{X} \cap (\mathcal{Y} + \mathcal{Z}) &\supseteq (\mathcal{X} \cap \mathcal{Y}) + (\mathcal{X} \cap \mathcal{Z}) \\
\mathcal{X} + (\mathcal{Y} \cap \mathcal{Z}) &\subseteq (\mathcal{X} + \mathcal{Y}) \cap (\mathcal{X} + \mathcal{Z}) \\
(\mathcal{X}^{\perp})^{\perp} &= \mathcal{X} \\
(\mathcal{X} + \mathcal{Y})^{\perp} &= \mathcal{X}^{\perp} \cap \mathcal{Y}^{\perp} \\
(\mathcal{X} \cap \mathcal{Y})^{\perp} &= \mathcal{X}^{\perp} + \mathcal{Y}^{\perp} \\
M(\mathcal{X} \cap \mathcal{Y}) &\subseteq M\mathcal{X} \cap M\mathcal{Y} \\
M(\mathcal{X} + \mathcal{Y}) &= M\mathcal{X} + M\mathcal{Y} \\
N^{-1}(\mathcal{X} \cap \mathcal{Y}) &= N^{-1}\mathcal{X} \cap N^{-1}\mathcal{Y} \\
N^{-1}(\mathcal{X} + \mathcal{Y}) &\supseteq N^{-1}\mathcal{X} + N^{-1}\mathcal{Y}.
\end{aligned} \tag{2.4}$$

Also, let $\mathcal{V}$ be a subspace of dimension m. Then we have

$$M\mathcal{X} \subseteq \mathcal{V} \iff M'\mathcal{V}^{\perp} \subseteq \mathcal{X}^{\perp} \tag{2.5}$$

$$(M^{-1}\mathcal{V})^{\perp} = M'\mathcal{V}^{\perp}. \tag{2.6}$$

Let $A = \mathbb{R}^{n\times n}$. Then $\mathcal{T}$, a subspace of $\mathbb{R}^n$, is an A-invariant subspace if

$$A\mathcal{T} \subseteq \mathcal{T}.$$

The following properties of an A-invariant subspace are useful:

(a) A subspace $\mathcal{T}$ with T a matrix such that $\mathcal{T} = \operatorname{im} T$ is A-invariant if and only if a matrix X exists such that

$$AT = TX.$$

(b) Let $\mathcal{T}$ be an A-invariant subspace. Then a similarity transformation L exists such that

$$\tilde{A} := L^{-1}AL = \begin{pmatrix} \tilde{A}_{11} & \tilde{A}_{12} \\ 0 & \tilde{A}_{22} \end{pmatrix} \quad \text{and} \quad \mathcal{T} = \operatorname{im} L^{-1}\begin{pmatrix} I \\ 0 \end{pmatrix}$$

with $\tilde{A}_{11} \in \mathbb{R}^{h\times h}$, where $h := \dim \mathcal{T}$.

The proofs of the above relations are simple and can be found in standard books on vector spaces.

Consider a matrix $A \in \mathbb{R}^{n\times n}$ and an A-invariant subspace $\mathcal{T} \subseteq \mathbb{R}^{n\times n}$. Then the *restriction* of A to $\mathcal{T}$ is the linear map $A_{\mathcal{T}} : \mathcal{T} \to \mathcal{T}$ defined by

$$A_{\mathcal{T}}x = Ax \quad \text{for all } x \in \mathcal{T}.$$

The restriction of A to $\mathcal{T}$ is also often denoted by $A \mid \mathcal{T}$.

Next, we would like to recall some elementary concepts regarding modal subspaces. We first develop some notations used in continuous-time systems. Consider a matrix $A \in \mathbb{R}^{n\times n}$. Let $\alpha(s)$ denote the characteristic polynomial of A and factor it as $\alpha(s) = \alpha_-(s)\cdot\alpha_+(s)$, where $\alpha_-(s)$ has all its roots in the open left-half complex plane $\mathbb{C}^-$ and $\alpha_+(s)$ has all its roots in the closed right-half complex plane $\mathbb{C}^{+0}$. Then the stable and unstable modal subspaces of $\mathbb{R}^n$ related to A are

$$\mathcal{X}_-(A) = \ker \alpha_-(A),$$

$$\mathcal{X}_+(A) = \ker \alpha_+(A).$$

It is easy to show that $\mathcal{X}_-(A)$ is spanned by the real and the imaginary part of the generalized eigenvectors of A corresponding to the eigenvalues in $\mathbb{C}^-$. Similarly, $\mathcal{X}_+(A)$ is spanned by the real and imaginary parts of the generalized eigenvectors of A corresponding to the eigenvalues in $\mathbb{C}^{+0}$. These two modal subspaces are complementary; that is, they are independent and their sum is $\mathbb{R}^n$; thus,

$$\mathbb{R}^n = \mathcal{X}_-(A) \oplus \mathcal{X}_+(A).$$

Standard numerical linear algebra can be used to compute the bases for modal subspaces. For example, one can transform A via orthogonal transformation T to a real Schur form

$$T'AT = \begin{pmatrix} A_- & \star \\ 0 & A_+ \end{pmatrix}, \tag{2.7}$$

where the eigenvalues of A_- and A_+ are, respectively, located in $\mathbb{C}^-$ and $\mathbb{C}^{+0}$ and $\star$ denotes some matrix that is not necessarily zero. If we partition T in conformity with the partitioning on the right-hand side of (2.7),

$$T = \begin{pmatrix} T_1 & T_2 \end{pmatrix},$$

then it is obvious that the columns of T_1 form a basis for $\mathcal{X}_-(A)$. That is,

$$\mathcal{X}_-(A) = \operatorname{im} T_1.$$

Analogously, we develop some notations used in discrete-time systems. Consider a matrix $A \in \mathbb{R}^{n\times n}$. Let $\alpha(z)$ denote the characteristic polynomial of A and factor it as $\alpha(z) = \alpha_\ominus(z) \cdot \alpha_\odot(z)$, where $\alpha_\ominus(z)$ has all its roots within the unit circle $\mathbb{C}^\ominus$ in the complex plane and $\alpha_\odot(z)$ has all its roots on or outside the unit circle $\mathbb{C}^\odot$. Then the stable and unstable modal subspaces of $\mathbb{R}^n$ related to A are

$$\mathcal{X}_\ominus(A) = \ker \alpha_\ominus(A),$$
$$\mathcal{X}_\odot(A) = \ker \alpha_\odot(A).$$

It is easy to show that $\mathcal{X}_\ominus(A)$ is spanned by the generalized real eigenvectors of A corresponding to the eigenvalues in $\mathbb{C}^\ominus$. Similarly, $\mathcal{X}_\odot(A)$ is spanned by the generalized real eigenvectors of A corresponding to the eigenvalues in $\mathbb{C}^\odot$. These two modal subspaces are complementary; that is, they are independent and their sum is $\mathbb{R}^n$; thus,

$$\mathbb{R}^n = \mathcal{X}_\ominus(A) \oplus \mathcal{X}_\odot(A).$$

Again, as in the continuous-time case, standard numerical linear algebra can be used to compute the bases for modal subspaces.

2.3 Norms of deterministic signals

Many measures are used to describe the size of a signal. The measures of size are called norms. In this section, we recall some of the common norms for persistent or transient continuous-time (discrete-time) vector signals. We consider continuous-time vector signals $y : \mathbb{R}^+ \to \mathbb{R}^n$ and discrete-time vector signals $y : \mathbb{Z}^+ \to \mathbb{R}^n$.

Definition 2.1 *The* L_p ***space****, with* $p \in [1, \infty)$*, consists of all vector-valued continuous-time signals from* $\mathbb{R}^+$ *to* $\mathbb{R}^n$ *for which*

$$\int_0^\infty \sum_{i=1}^n |y_i(t)|^p dt$$

is well defined[1] *and finite. The space* L_∞ *consists of all vector-valued continuous-time signals for which*

$$\operatorname*{ess\,sup}_{t\in\mathbb{R}^+} \max_{1\le i\le n} |y_i(t)|$$

is finite.

The ℓ_p ***space****, with* $p \in [1, \infty)$*, consists of all vector-valued discrete-time signals from* $\mathbb{Z}^+$ *to* $\mathbb{R}^n$ *for which*

$$\sum_{k=0}^\infty \sum_{i=1}^n |y_i(k)|^p$$

is finite, and the space ℓ_∞ *consists of all vector-valued discrete-time signals for which*

$$\sup_{k\in\mathbb{Z}^+} \max_{1\le i\le n} |y_i(k)|$$

is finite.

Remark 2.2 *We will sometimes use* $L_p[t_0, \infty)$ *to refer to vector-valued signals from* $[t_0, \infty)$ *to* $\mathbb{R}^n$ *for which*

$$\int_{t_0}^\infty \sum_{i=1}^n |y_i(t)|^p dt$$

is well defined when $p \in [1, \infty)$ *or*

$$\operatorname*{ess\,sup}_{t\in[t_0,\infty)} \max_{1\le i\le n} |y_i(t)|$$

is finite in case $p = \infty$.

[1]This integral needs to be well defined in the sense of Lebesgue. A reader who has no prior acquaintance with the Lebesgue theory of measure and integration can simply think of all functions encountered here as piecewise-continuous functions and of all integrals as Riemann integrals. This would lead to no conceptual difficulties and no loss of insight except that occasionally some results from Lebesgue theory would have to be accepted on faith.

Similarly, we will sometimes use $\ell_p[k_0, \infty)$ to refer to vector-valued signals from $\{ k \in \mathbb{Z}^+ \mid k > k_0 \}$ to $\mathbb{R}^n$ for which

$$\sum_{k=k_0}^{\infty} \sum_{i=1}^{n} |y_i(k)|^p$$

is finite when $p \in [1, \infty)$ or

$$\sup_{k \in \mathbb{Z}^+, k \geq k_0} \max_{1 \leq i \leq n} |y_i(k)|$$

is finite when $p = \infty$.

However, we would like to note that L_p and ℓ_p will always refer to functions from $\mathbb{R}^+$ or $\mathbb{Z}^+$ to $\mathbb{R}^n$, respectively.

The spaces defined above are actually normed linear vector spaces if we define the appropriate norms.

Definition 2.3 *For a vector-valued continuous-time signal $y \in L_p$ with $p \in [1, \infty)$, the **L_p** norm is defined as*

$$\|y\|_p := \left(\int_0^{\infty} \sum_{i=1}^{n} |y_i(t)|^p dt \right)^{\frac{1}{p}},$$

*For a vector-valued continuous-time signal $y \in L_\infty$, the **L_∞ norm** is defined as*

$$\|y\|_\infty := \operatorname{ess\,sup}_{t \in \mathbb{R}^+} \max_{1 \leq i \leq n} |y_i(t)|.$$

*Analogously, for a vector-valued discrete-time signal $y \in \ell_p$ with $p \in [1, \infty)$, we define the **ℓ_p norm** as*

$$\|y\|_p := \left(\sum_{k=0}^{\infty} \sum_{i=1}^{n} |y_i(k)|^p \right)^{\frac{1}{p}}.$$

*Finally, for a vector-valued discrete-time signal $y \in \ell_\infty$, the **ℓ_∞ norm** is defined as*

$$\|y\|_\infty := \sup_{k \in \mathbb{Z}^+} \max_{1 \leq i \leq n} |y_i(k)|.$$

The following lemmas are useful in concluding attractivity in dealing with L_p stability to be defined shortly in a later section. These lemmas imply that if both a continuous-time signal and its derivative are in L_p for some $p \in [1, \infty)$, then it vanishes as time tends to infinity, and moreover, it is in L_∞

Lemma 2.4 *If $\phi : [0,\infty) \to \mathbb{R}$ is absolutely continuous, $\phi(t) \in L_{p_1}$ for some $p_1 \in [1,\infty)$, and its derivative $\dot{\phi}(t) \in L_{p_2}$ for some $p_2 \in [1,\infty)$, then* $\lim_{t\to\infty} \phi(t) = 0$.

Proof : Let $\alpha = p_1\left(1 - \frac{1}{p_2}\right) \geqslant 0$. Then, $\frac{\alpha}{p_1} + \frac{1}{p_2} = 1$. By Hölder's inequality,

$$\begin{aligned}
\frac{1}{\alpha+1}\left|\phi^{\alpha+1}(\tau) - \phi^{\alpha+1}(s)\right| &= \left|\int_s^\tau \phi^\alpha(t)\dot{\phi}(t)\mathrm{d}t\right| \\
&\leqslant \left\{\int_s^\tau |\phi^\alpha(t)|^{\frac{p_1}{\alpha}}\mathrm{d}t\right\}^{\frac{\alpha}{p_1}} \left\{\int_s^\tau |\dot{\phi}(t)|^{p_2}\mathrm{d}t\right\}^{\frac{1}{p_2}} \\
&= \left\{\int_s^\tau |\phi(t)|^{p_1}\mathrm{d}t\right\}^{\frac{\alpha}{p_1}} \left\{\int_s^\tau |\dot{\phi}(t)|^{p_2}\mathrm{d}t\right\}^{\frac{1}{p_2}}. \qquad (2.8)
\end{aligned}$$

Since $\phi \in L_{p_1}$ and $\dot{\phi}(t) \in L_{p_2}$, it is clear that $\{\phi^{\alpha+1}(t_k)\}_{k=1}^{\infty}$ is a Cauchy sequence for any sequence $t_k \to \infty$. Hence, we can assume that $\phi^{\alpha+1}(t_k) \to c$ as $k \to \infty$. That is, for all $\varepsilon > 0$, there exists a $K > 0$ such that

$$\left|\phi^{\alpha+1}(t_k) - c\right| < \frac{\varepsilon}{2}, \qquad \forall\, k \geqslant K. \qquad (2.9)$$

Also, by choosing t_K sufficiently large, we see from (2.8) that

$$\left|\phi^{\alpha+1}(t) - \phi^{\alpha+1}(t_K)\right| < \frac{\varepsilon}{2}, \qquad \forall\, t > t_K. \qquad (2.10)$$

Combining (2.9) and (2.10), we get

$$\left|\phi^{\alpha+1}(t) - c\right| \leqslant \left|\phi^{\alpha+1}(t) - \phi^{\alpha+1}(t_K)\right| + \left|\phi^{\alpha+1}(t_K) - c\right| < \varepsilon, \qquad \forall\, t > t_K.$$

Hence, $\lim_{t\to\infty} \phi^{\alpha+1}(t) = c$ or $\lim_{t\to\infty} \phi(t) = (c)^{\frac{1}{\alpha+1}}$. Since $\phi \in L_{p_1}$, it is obvious that $c = 0$. ■

Lemma 2.5 *For $\phi(t) : [0,\infty) \to \mathbb{R}$, if $\phi(t) \in L_{p_1}$ for some $p_1 \in [1,\infty)$ and its derivative $\dot{\phi}(t) \in L_{p_2}$ for some $p_2 \in [1,\infty)$, then $\phi \in L_\infty$.*

Proof : Let $\alpha = p_1(1-\frac{1}{p_2})$. Then $\frac{\alpha}{p_1}+\frac{1}{p_2}=1$. We have

$$\tfrac{1}{\alpha+1}|\phi^{\alpha+1}(t)-\phi^{\alpha+1}(0)| \leqslant \left|\int_0^t \phi^{\alpha}(s)\dot{\phi}(s)\mathrm{d}s\right| \tag{2.11}$$

$$\leqslant \left\{\int_0^t |\phi^{\alpha}(s)|^{\frac{p_1}{\alpha}}\right\}^{\frac{\alpha}{p_1}} \left\{\int_0^t |\dot{\phi}(s)|^{p_2}\right\}^{\frac{1}{p_1}}. \tag{2.12}$$

This implies that

$$|\phi^{\alpha+1}(t)| \leqslant |\phi^{\alpha+1}(0)| + (\alpha+1)\|\phi\|_{p_1}^{\alpha}\|\dot{\phi}\|_{p_2}.$$

This completes the proof. ∎

The square of the L_2 or ℓ_2 norm of a signal y is commonly termed as the total energy in the signal y. In many areas of engineering, the energy or square of the L_2 (ℓ_2) norm is used as a measure of the size of a transient signal y that decays to zero as time progresses toward infinity. By Parseval's theorem, $\|y\|_2$ can also be computed in the frequency domain as follows: for the continuous-time case,

$$\|y\|_2 = \Big(\frac{1}{2\pi}\int_{-\infty}^{\infty} Y(j\omega)^* Y(j\omega)\mathrm{d}\omega\Big)^{1/2},$$

where Y is the Fourier transform of y; similarly, for the discrete-time case,

$$\|y\|_2 = \Big(\frac{1}{2\pi}\int_{-\pi}^{\pi} Y(e^{j\omega})^* Y(e^{j\omega})\mathrm{d}\omega\Big)^{1/2},$$

where Y is the z-transform of y.

Definition 2.6 *A continuous-time signal y for which the following limit is well defined and finite:*

$$\lim_{T\to\infty}\frac{1}{T}\int_0^T y(t)'y(t)dt$$

is called an ***RMS (root mean square) or power signal****. The* ***RMS value*** *of such a continuous-time signal y is defined as*

$$\|y\|_{\text{RMS}} = \Big(\lim_{T\to\infty}\frac{1}{T}\int_0^T y(t)'y(t)dt\Big)^{1/2}. \tag{2.13}$$

Similarly, a discrete-time signal y for which the following limit is well defined and finite:

$$\lim_{T\to\infty} \frac{1}{T} \sum_{k=0}^{T} y(k)' y(k)$$

is called an ***RMS or power signal****. The* ***RMS value*** *of such a discrete-time signal y is defined as*

$$\|y\|_{\text{RMS}} = \left(\lim_{T\to\infty} \frac{1}{T} \sum_{k=0}^{T} y(k)' y(k)\right)^{1/2}. \tag{2.14}$$

Remark 2.7 *Note that sometimes, the RMS is defined by*

$$\|y\|_{\text{RMS}} = \left(\limsup_{T\to\infty} \frac{1}{T} \int_0^T y(t)' y(t) dt\right)^{1/2}$$

and

$$\limsup_{T\to\infty} \frac{1}{T} \sum_{k=0}^{T} y(k)' y(k),$$

respectively, for continuous- and discrete-time systems. This has the advantage that the class of signals for which the RMS is well defined and finite becomes a linear vectorspace. This additional structure can sometimes be convenient. Using limsup instead of the standard limit makes some of the derivations a bit more involved, but generally speaking, all properties carry over to this more general case.

The square of the RMS norm[2] of y is commonly termed as the average power of the signal y. Often, in engineering, the RMS norm or average power is used for signals y which are persistent. We note that the RMS norm is a steady-state measure of a signal and is not affected by any transients.

Remark 2.8 *It is obvious that an L_2 (ℓ_2) signal has a zero RMS value. Also, an L_1 signal does not necessarily have a finite or well-defined RMS value, whereas, in contrast, an ℓ_1 signal always has a zero RMS value. Finally, for an L_∞ (ℓ_∞) signal, the RMS value need not be well defined. However, if we use the generalized definition from Remark 2.7, then an L_∞ signal which is locally square Lebesgue integrable has a well-defined and finite RMS value which is less than its L_∞ norm.*

[2]We would like to remark that the RMS norm is a pseudo-norm because the RMS norm of any energy or transient signal is zero.

There exist some relationships among L_p spaces for different values of p. The L_p space can be visualized by a Venn diagram as clarified in the following remark.

Remark 2.9 *Consider a square in the Euclidean plane with vertices* $(0, 1)$, $(1, 0)$, $(0, -1)$, $(-1, 0)$; *that is, the vertices of square satisfy* $|x| + |y| = 1$. *Then, we can interpret the* L_∞ *space in a Venn diagram as the* $[-1, 1]$ *interval on the vertical axis and the* L_1 *space as the* $[-1, 1]$ *interval on the horizontal axis. Also, any* L_p *space can be interpreted as the rectangle whose vertices are given by* $(\pm 1/p, 0)$ *and* $(0, \pm 1 \mp 1/p)$; *that is,* $(1/p, 0), (-1/p, 0), (0, 1 - 1/p), (0, -1 + 1/p)$. *Figure 2.1 illustrates this.*

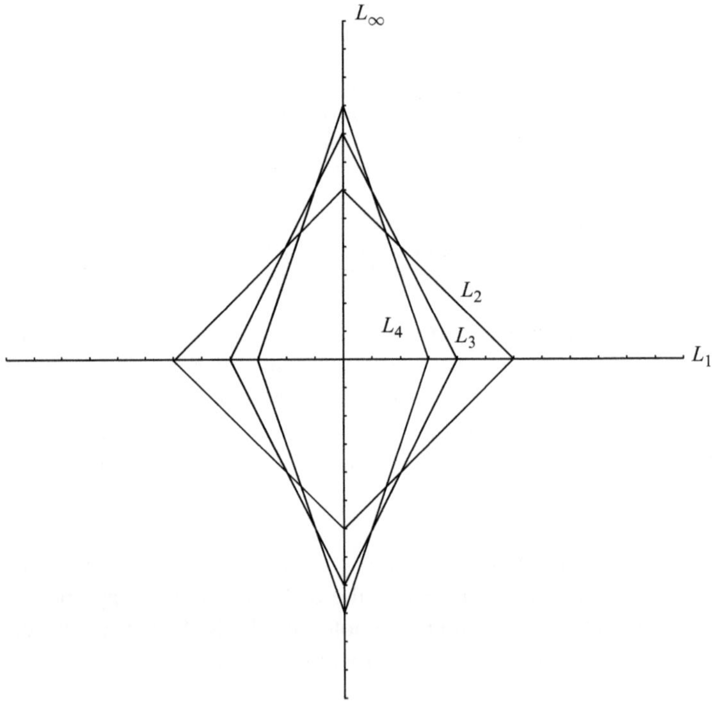

Figure 2.1: Venn diagram of L_p spaces

Note that L_p spaces for different values of p do not contain one another. Only the intersection of L_1 and L_∞ is contained in all L_p spaces. That is, the intersection of all L_p spaces is a single point in the Venn diagram which is equal to the intersection of L_1 and L_∞ spaces. In other words, we have

$$L_1 \cap L_\infty = \bigcap_{p=1}^{\infty} L_p.$$

Regarding the relationship among all ℓ_p spaces, we have for $1 < p < q < \infty$ that

$$\ell_1 \subset \ell_p \subset \ell_q \subset \ell_\infty.$$

To see this, note that $y \in \ell_p$ implies that $\|y(k)\| \leqslant \|y\|_{\ell_p}$ for any $k \geq 0$. Hence, for $p < q < \infty$,

$$\sum_{k=0}^{\infty} \|y(k)\|^q \leq \sum_{k=0}^{\infty} \|y(k)\|^p \|y\|_{\ell_p}^{q-p} = \|y\|_{\ell_p}^{q-p} \|y\|_{\ell_p}^{p} = \|y\|_{\ell_p}^{q} < \infty.$$

Obviously, unlike in the case of L_p space, each ℓ_p is a strict subset of ℓ_q whenever $p < q$.

2.4 Norms of stochastic signals

For a vector signal that is modeled as a wide-sense stationary or an asymptotically wide-sense stationary vector stochastic process (random sequence), the common measure of size is the RMS norm. We recall below the needed definition.

Definition 2.10 *For a wide-sense stationary vector stochastic process y with a bounded variance, we define the* ***stochastic RMS norm*** *as*

$$\|y\|_{\text{RMS}} = \left(\boldsymbol{E}\left[y(t)'y(t)\right]\right)^{1/2}. \tag{2.15}$$

Analogously, for a wide-sense stationary vector random sequence y with a bounded variance, we define the ***stochastic RMS norm*** *as*

$$\|y\|_{\text{RMS}} = \left(\boldsymbol{E}\left[y(k)'y(k)\right]\right)^{1/2}. \tag{2.16}$$

Here $\boldsymbol{E}[.]$ denotes the expectation. For stochastic processes (random sequences) that are only asymptotically wide-sense stationary as time goes to infinity [i.e., for asymptotically wide-sense stationary processes (random sequences)], (2.15) *and* (2.16) *need to be rewritten as*

$$\|y\|_{\text{RMS}} = \left(\lim_{t\to\infty} \boldsymbol{E}\left[y(t)'y(t)\right]\right)^{1/2} \tag{2.17}$$

and

$$\|y\|_{\text{RMS}} = \left(\lim_{k\to\infty} \boldsymbol{E}\left[y(k)'y(k)\right]\right)^{1/2}, \tag{2.18}$$

respectively.

Note that in (2.15) and (2.16), the result is independent of t or k because the stochastic process (random sequence) is wide-sense stationary.

We note that if y is an ergodic stochastic process (random sequence), then the deterministic RMS norm of any realization of the stochastic process (random sequence) y is equal to the stochastic RMS norm of y with probability one.

Also, we note that the RMS value of a wide-sense stationary process y can be expressed in terms of its autocorrelation matrix $R_y(\tau)$,

$$R_y(\tau) := \boldsymbol{E}[y(t)y'(t+\tau)],$$

or its power spectral density (PSD) $S_y(\omega)$,

$$S_y(\omega) := \int_{-\infty}^{\infty} R_y(\tau)e^{-j\omega\tau}\mathrm{d}\tau\,.$$

That is,

$$\|y\|_{\mathrm{RMS}} = \Big(\mathrm{trace}[R_y(0)]\Big)^{1/2} = \Big(\frac{1}{2\pi}\,\mathrm{trace}\Big[\int_{-\infty}^{\infty} S_y(\omega)\mathrm{d}\omega\Big]\Big)^{1/2}.$$

Similarly, for a wide-sense stationary random sequence y, let the autocorrelation matrix be

$$R_y(n) := \boldsymbol{E}[y(k)y'(k+n)]$$

and the power spectral density (PSD) be

$$S_y(\omega) := \sum_{n=-\infty}^{\infty} R_y(n)e^{-j\omega n}, \quad -\pi \le \omega \le \pi.$$

Then,

$$\|y\|_{\mathrm{RMS}} = \Big(\mathrm{trace}[R_y(0)]\Big)^{1/2} = \Big(\frac{1}{2\pi}\,\mathrm{trace}\Big[\int_{-\pi}^{\pi} S_y(\omega)\mathrm{d}\omega\Big]\Big)^{1/2}.$$

2.5 Norms of linear time- or shift-invariant systems

We recalled above the definitions of norms of signals. A notion related to the size of a signal is the gain of a transfer function of a linear time- or shift-invariant system. As in the case of a signal, once again, various norms are used to measure the size of a transfer function. In this section, we recall the definitions of certain such norms. Also, we recall methods of computing them.

Two well-known classic norms of linear time- or shift-invariant systems are the H_2 norm (which is the RMS value of the response of a system to white noise input of unit PSD) and the H_∞ norm (which is the RMS gain of the system). The definitions of these norms are recalled below.

Definition 2.11 *Consider a continuous-time system Σ having a $q\times\ell$ stable transfer function G. Then* **the H_2 norm of the continuous-time system Σ** *or, equivalently, of the transfer matrix G is defined as*

$$\|G\|_2 = \left(\frac{1}{2\pi}\,\mathrm{trace}\left[\int_{-\infty}^{\infty} G(j\omega)G^*(j\omega)d\omega\right]\right)^{1/2}. \tag{2.19}$$

We assign ∞ as the H_2 norm of an unstable continuous-time system.

We note that the H_2 norm is induced by an inner product; that is, we have

$$\|G\|_2 = \langle G, G\rangle^{1/2}$$

with the inner product defined by

$$\langle G_1, G_2\rangle = \frac{1}{2\pi}\,\mathrm{trace}\left[\int_{-\infty}^{\infty} G_1(j\omega)G_2^*(j\omega)d\omega\right].$$

By Parseval's theorem, the H_2 norm of the transfer matrix G can equivalently be defined as

$$\|G\|_2 = \left(\mathrm{trace}\left[\int_{0}^{\infty} g(t)g'(t)dt\right]\right)^{1/2}, \tag{2.20}$$

where $g(t)$ is the inverse Laplace transform of the transfer matrix or the unit impulse (Dirac distribution) response of the associated linear system. Thus, $\|G\|_2 = \|g\|_2$. It is also known that $\|G\|_2$ can be expressed in terms of the singular values of the matrix G at each frequency,

$$\|G\|_2 = \left(\frac{1}{2\pi}\int_{-\infty}^{\infty}\sum_{i=1}^{\min\{q,\ell\}}\sigma_i^2(G(j\omega))d\omega\right)^{1/2}. \tag{2.21}$$

where $\sigma_i(G(j\omega))$ is the ith singular value of $G(j\omega)$.

Remark 2.12 ***(Stochastic interpretation of the H_2 norm of a continuous-time system)*** *Let us consider a continuous-time system with a stable transfer function G. Let the input w to the system be a wide-sense stationary stochastic process. Let z be the corresponding output. It is well known that*

$$S_z(\omega) = G(j\omega)S_w(\omega)G^*(-j\omega), \tag{2.22}$$

where S_w and S_z are the PSDs of $w(t)$ and $z(t)$, respectively. Then, the H_2 norm of $G(s)$ can be interpreted as the *RMS* value of the output z when the given system is driven by zero mean white noise with unit PSD. *Note that formally, white noise with unit PSD does not exist, but the above can be formalized using Brownian motion and stochastic differential equations.*

Remark 2.13 *Note that the H_2 norm of a stable continuous-time system or transfer function $G(s)$ is finite if and only if it is strictly proper.*

Definition 2.14 *Consider a discrete-time system Σ having a $q \times \ell$ stable transfer function G. Then the* ***H_2 norm of the discrete-time system*** *Σ or, equivalently, of the transfer matrix G is defined as*

$$\|G\|_2 = \left(\frac{1}{2\pi}\operatorname{trace}\left[\int_{-\pi}^{\pi} G(e^{j\omega})G^*(e^{j\omega})d\omega\right]\right)^{1/2}. \tag{2.23}$$

We assign ∞ to the H_2 norm of an unstable discrete-time system.

Again, we note that the H_2 norm is induced by an inner product; that is, we have

$$\|G\|_2 = \langle G, G\rangle^{1/2}$$

with the inner product defined by

$$\langle G_1, G_2\rangle = \left(\frac{1}{2\pi}\operatorname{trace}\left[\int_{-\pi}^{\pi} G_1(e^{j\omega})G_2^*(e^{j\omega})d\omega\right]\right)^{1/2}.$$

Again, by Parseval's theorem, $\|G\|_2$ can equivalently be defined as

$$\|G\|_2 = \left(\operatorname{trace}\left[\sum_{k=0}^{\infty} g(k)g'(k)\right]\right)^{1/2} \tag{2.24}$$

where g is the inverse z-transform of the transfer matrix which is equal to the unit impulse response of the associated linear system. Thus, $\|G\|_2 = \|g\|_2$.

Remark 2.15 ***(Stochastic interpretation of the H_2 norm of a discrete-time system)*** *Let us consider a discrete-time system with a stable transfer function $G(z)$. Let the input $w(k)$ to the system be a wide-sense stationary random sequence. Let $z(k)$ be the corresponding output. Then, it is well known that*

$$S_z(\omega) = G(e^{j\omega})S_w(\omega)G^*(e^{-j\omega}), \quad -\pi \leq \omega \leq \pi, \tag{2.25}$$

where S_w and S_z are the PSDs of w and z, respectively. Then, once again, the H_2 norm of G can be interpreted as the RMS value of the output z when the given system is driven by a zero mean white noise random sequence having unit variance.

State-space method for computing the H_2 norm: We present here briefly some results on the computation of the H_2 norm of a transfer function matrix when its realization is given in a state-space form (for details, see [15]). Consider the transfer function G of a continuous-time system with realization (A, B, C, D) where A is Hurwitz stable. Let W^c_{obs} denote the observability grammian of the pair (A, C) and W^c_{con} the controllability grammian of (A, B). Note that D needs to be zero for a finite H_2 norm. We note that W^c_{obs} and W^c_{con} are the unique solutions of continuous-time Lyapunov equations:

$$\begin{aligned} A'W^c_{\text{obs}} + W^c_{\text{obs}}A + C'C &= 0, \\ AW^c_{\text{con}} + W^c_{\text{con}}A' + BB' &= 0. \end{aligned}$$

The H_2 norm of $G(s)$ can now be computed by

$$\begin{aligned} \|G\|_2 &= \left(\text{trace } B'W^c_{\text{obs}}B\right)^{1/2} \\ &= \left(\text{trace } CW^c_{\text{con}}C'\right)^{1/2}. \end{aligned}$$

The computation of the H_2 norm of a transfer function G of a discrete-time system with realization (A, B, C, D) where A is Schur stable, can be given along the same lines. That is, let W^d_{obs} and W^d_{con} be the unique solutions of discrete-time Lyapunov equations:

$$\begin{aligned} A'W^d_{\text{obs}}A - W^d_{\text{obs}} + C'C &= 0, \\ AW^d_{\text{con}}A' - W^d_{\text{con}} + BB' &= 0. \end{aligned}$$

The H_2 norm of G can now be computed by

$$\begin{aligned} \|G(z)\|_2 &= \left(\text{trace}[B'W^d_{\text{obs}}B + D'D]\right)^{1/2} \\ &= \left(\text{trace}[CW^d_{\text{con}}C' + DD']\right)^{1/2}. \end{aligned}$$

Definition 2.16 *Consider a continuous-time system having a $q \times \ell$ stable transfer function G. Then* ***the H_∞ norm*** *of G is defined as*

$$\|G\|_\infty := \sup_\omega \sigma_{\max}[G(j\omega)]. \tag{2.26}$$

Similarly, consider a discrete-time system having a $q \times \ell$ stable transfer function G. Then the H_∞ norm of G is defined as

$$\|G\|_\infty := \sup_{-\pi \le \omega \le \pi} \sigma_{\max}[G(e^{j\omega})]. \tag{2.27}$$

For a continuous-time system having a stable transfer function G, let w and z be energy signals that are, respectively, the input and the corresponding output of the given system. Similarly, for a discrete-time system having a stable transfer function G, let w and z be energy signals that are, respectively, the input and the corresponding output. Then it is easy to see that $\|G\|_\infty$ has the following interpretation for both continuous-time and discrete-time systems (where $\|\cdot\|_2$ denotes the L_2 and ℓ_2 norm, respectively):

$$\|G\|_\infty = \sup_{w \neq 0} \frac{\|z\|_2}{\|w\|_2}.$$

Also, when the input and the corresponding output (i.e., w and z) are power signals, the H_∞ norm of G turns out to coincide with its RMS gain, namely,

$$\|G\|_\infty = \|G\|_{\text{RMS gain}} = \sup_{\substack{w \\ \|w\|_{\text{RMS}} \neq 0}} \frac{\|z\|_{\text{RMS}}}{\|w\|_{\text{RMS}}}.$$

We have the following remarks.

Remark 2.17 *An important property of the H_∞ norm, for both continuous-time and discrete-time systems, is that it is submultiplicative. That is, for transfer matrices G_1 and G_2, we have*

$$\|G_1 G_2\|_\infty \leqslant \|G_1\|_\infty \|G_2\|_\infty.$$

Remark 2.18 *It is interesting to contrast the H_2 and H_∞ norms. Consider a transfer matrix H. Then the fact that $\|H\|_\infty < \alpha$ for some $\alpha > 0$ implies that*

$$\|Hu\|_{\text{RMS}} \leqslant \alpha \text{ for any input } u \text{ with } \|u\|_{\text{RMS}} \leqslant 1.$$

In contrast, the H_2 norm-bound specification $\|H\|_2 \leqslant \alpha$ implies that

$$\|Hu\|_{\text{RMS}} \leqslant \alpha \text{ when input } u \text{ is a white noise with unit intensity.}$$

State-space method for computing the continuous-time H_∞ norm:

Regarding the H_∞ norm computation for continuous time, there is a simple method to determine whether the inequality specification $\|G\|_\infty < \gamma$ is satisfied. To state this, given $\gamma > 0$, we define the matrix

$$M_\gamma = \begin{pmatrix} A + BR^{-1}D'C & \gamma^{-2}BR^{-1}B' \\ -C'(I + DR^{-1}D')C & -(A + BR^{-1}D'C)', \end{pmatrix}$$

where $R := \gamma^2 I - D'D > 0$. Then, we have

$$\|G\|_\infty < \gamma \iff M_\gamma \text{ has no imaginary eigenvalues and } \sigma_{\max}(D) < \gamma. \quad (2.28)$$

The above discussion provides a simple bisection algorithm that enables us to compute the H_∞ norm numerically to any degree of numerical accuracy. In the following algorithm, the first three steps represent initialization, whereas the last step represents the bisection principle:

(*i*) Set $i = 0$, and set $\gamma_\ell = \|D\|$.

(*ii*) Choose any $\gamma_0 > \gamma_\ell$.

(*iii*) Use (2.28) to test whether $\|G\|_\infty < \gamma_i$. If so, $\gamma_u = \gamma_i$ and continue with step (d). Otherwise, set $\gamma_{i+1} = 2\gamma_i$ and $i = i + 1$ and continue with step (*iii*).

(*iv*) Set $\gamma = (\gamma_u + \gamma_\ell)/2$. Use (2.28) to test whether $\|G\|_\infty < \gamma$. If so, set $\gamma_u = \gamma$ and otherwise set $\gamma_\ell = \gamma$ and then repeat step (*iv*).

We observe from step (d) that $\|G\|_\infty$ is within the interval $[\gamma_\ell, \gamma_u]$. After each iteration, the size of the interval divides itself into half. Hence, one can stop the iterations when the desired level of accuracy is reached.

For more details concerning the computation of the H_∞ norm, we refer to [14, 16].

State-space method for computing the discrete-time H_∞ norm:

Similar to the continuous time, for discrete time, there is a simple method to determine whether the inequality specification $\|G\|_\infty < \gamma$ is satisfied. To state this, given $\gamma > 0$, we define the matrix pencil

$$M_\gamma(z) = \begin{pmatrix} zI - A & -B & 0 \\ C'C & C'D & I - zA' \\ D'C & D'D - \gamma^2 I & -zB' \end{pmatrix}.$$

Then, we have

$$\|G\|_\infty < \gamma \iff M_\gamma(z) \text{ has no zeros on the unit circle and } \sigma_{\max}(D + C(I - A)^{-1}B) < \gamma.$$

The above theorem provides a simple bisection algorithm that enables us to compute the H_∞ norm numerically to any degree of numerical accuracy, which is completely similar to the continuous-time case.

The computation of the H_∞ norm of a transfer matrix of a discrete-time system through a bisection algorithm then follows similarly to the continuous-time case.

2.6 A class of saturation functions

As we said in Chap. 1, one of the most prevalent constraints is the one that arises from actuator saturation. We introduce below the class of saturation functions we consider throughout this book.

Definition 2.19 *The function $\sigma_1 : \mathbb{R}^m \to \mathbb{R}^m$ is called the* ***standard saturation function*** *if*

$$\sigma_1(s) = \begin{pmatrix} \text{sat}_1(s_1) \\ \text{sat}_1(s_2) \\ \vdots \\ \text{sat}_1(s_m) \end{pmatrix},$$

where

$$\text{sat}_1(s) = \text{sgn}(s)\ \min\{|s|, 1\}.$$

For convenience, we also introduce a scaled version of the standard saturation function:

$$\text{sat}_\Delta(s) = \Delta\ \text{sat}\left(\frac{s}{\Delta}\right)$$

and

$$\sigma_\Delta(s) = \begin{pmatrix} \text{sat}_\Delta(s_1) \\ \text{sat}_\Delta(s_2) \\ \vdots \\ \text{sat}_\Delta(s_m) \end{pmatrix}.$$

For ease of notation, we will often use $\sigma(s)$ as abbreviation for $\sigma_\Delta(s)$.

In reality, a saturation occurring in some device will never be equal to the above standard saturation function. Therefore, we introduce a class of saturation functions satisfying some minimum characteristics which are common for all saturation functions.

Definition 2.20 *A function* $\tilde{\sigma} : \mathbb{R}^m \to \mathbb{R}^m$ *is called a* ***saturation function*** *if:*

(i) $\tilde{\sigma}(u)$ *is decentralized, that is,*

$$\tilde{\sigma}(s) = \begin{pmatrix} \tilde{\sigma}_1(s_1) \\ \tilde{\sigma}_2(s_2) \\ \vdots \\ \tilde{\sigma}_m(s_m) \end{pmatrix}.$$

(ii) $\tilde{\sigma}_i$ *is globally Lipschitz, that is, for some* $\delta > 0$,

$$|\tilde{\sigma}_i(s_1) - \tilde{\sigma}_i(s_2)| \leqslant \delta |s_1 - s_2|.$$

(iii) $s\tilde{\sigma}_i(s) > 0$ *whenever* $s \neq 0$ *and* $\tilde{\sigma}_i(0) = 0$.

(iv) *The two limits*

$$\lim_{s \to 0+} \frac{\tilde{\sigma}_i(s)}{s}, \qquad \lim_{s \to 0-} \frac{\tilde{\sigma}_i(s)}{s}$$

both exist and are strictly positive.

(v) $\liminf_{|s| \to \infty} |\tilde{\sigma}_i(s)| > 0$.

Remark 2.21 *Note that the above definition for a saturation function does not enforce that a saturation function is bounded. Actually,* $\tilde{\sigma}(s) = s$ *satisfies all the properties above. Especially when establishing necessary conditions, it is sometimes useful to require the additional condition:*

(vi) *There exists a* $M > 0$ *such that* $|\tilde{\sigma}_i(s)| < M$ *for all* $s \in \mathbb{R}$.

Remark 2.22 *In some cases, we actually use the following condition:*

(vii) *There exist* $\theta > 0$ *and* $\psi > 0$ *such that*

$$|\tilde{\sigma}(s)| > \min\{\, \theta |s|, \psi \,\}$$

which is a consequence of Condition (iv), which guarantees that $|\sigma(s)|$ *is larger than* $\theta|s|$ *for some positive* θ *for small* s *Condition (v), which guarantees that* $\sigma(s)$ *is bounded away from zero for large* s *and Condition (iii), which guarantees that* $\sigma(s)$ *is never equal to zero for* $s \neq 0$.

2.7 Internal stability

In this section, we review various notions and definitions all pertaining to internal stability. Most of the definitions and results presented here are classical and can be found in many textbooks such as [58, 142, 189]. The most complete set of results can be found in [44]. We concentrate here only on continuous-time systems; however, all the following definitions can be easily modified for discrete-time systems.

We consider throughout this section a nonlinear ordinary differential equation of the form

$$\dot{x} = f(t, x), \quad x(t_0) = x_0, \tag{2.29}$$

where $x(t) \in \mathbb{R}^n$, $f : [0, \infty) \times \mathbb{R}^n \mapsto \mathbb{R}^n$ and

$$\mathcal{B}(r) = \{x \in \mathbb{R}^n \mid \|x\| < r\}.$$

Unless stated otherwise, we assume that the function f is such that, for all initial conditions in some open neighborhood of the equilibrium, the system (2.29) possesses a unique solution $x(t; t_0, x_0)$ for all $t_0 \geqslant 0$ and $t > t_0$.

The system (2.29) is referred to as a time-varying system. We also consider the case when the function f in (2.29) is not explicitly dependent on time t. In this case, the resulting system (2.29) is referred to as a time-invariant system which can be written as

$$\dot{x} = f(x), \quad x(t_0) = x_0. \tag{2.30}$$

We have the following definitions.

Definition 2.23 *A state x_e is said to be an* ***equilibrium state*** *of the system* (2.29) *if*

$$f(t, x_e) \equiv 0 \text{ for all } t \geqslant 0.$$

Definition 2.24 *The equilibrium state x_e of* (2.29) *is said to be an* ***isolated equilibrium state*** *if there exists a constant $\alpha > 0$ such that the system* (2.29) *does not contain any equilibrium other than x_e in the region*

$$\mathcal{B}(x_e, \alpha) := \{x \mid \|x - x_e\| < \alpha\} \subset \mathbb{R}^n.$$

Definition 2.25 *The equilibrium state x_e of* (2.29) *is said to be* ***stable*** *if for any $t_0 \geqslant 0$ and $\varepsilon > 0$ there exists a $\delta > 0$ such that for all $x_0 \in \mathbb{R}^n$ with $\|x_0 - x_e\| < \delta$, we have that $\|x(t; t_0, x_0) - x_e\| < \varepsilon$ for all $t \geq t_0$.*

Definition 2.26 *The equilibrium state x_e of* (2.29) *is said to be* ***unstable*** *if it is not stable.*

Definition 2.27 *The equilibrium state x_e of* (2.29) *is said to be* ***uniformly stable*** *if for any $\varepsilon > 0$, there exists a $\delta > 0$ such that for all $x_0 \in \mathbb{R}^n$ with $\|x_0 - x_e\| < \delta$, we have $\|x(t; t_0, x_0) - x_e\| < \varepsilon$ for all $t_0 \geqslant 0$ and for all $t \geq t_0$.*

Definition 2.28 *The equilibrium state x_e of* (2.29) *is said to be* ***asymptotically stable*** *if:*

(i) *It is stable.*

(ii) *For every $t_0 \geqslant 0$, there exists a $\delta > 0$ such that for all $x_0 \in \mathbb{R}^n$ with $\|x_0 - x_e\| < \delta$, we have $\lim_{t\to\infty} \|x(t; t_0, x_0) - x_e\| = 0$.*

Definition 2.29 *For any $t_0 \geqslant 0$, the set of all $x_0 \in \mathbb{R}^n$ such that $x(t; t_0, x_0) \to x_e$ as $t \to \infty$ is called the* ***region of attraction*** *at time t_0 of the equilibrium state x_e. If condition (ii) of Definition 2.28 is satisfied, then the equilibrium state x_e is said to be* ***attractive****.*

Definition 2.30 *The equilibrium state x_e of* (2.29) *is said to be* ***uniformly asymptotically stable*** *if*

(i) *It is uniformly stable.*

(ii) *For every $\varepsilon > 0$, there exist $T > 0$ and $\delta_0 > 0$ such that for all initial conditions $x_0 \in \mathbb{R}^n$ with $\|x_0 - x_e\| < \delta_0$, we have $\|x(t; t_0, x_0) - x_e\| < \varepsilon$ for all $t_0 \geqslant 0$ and for all $t \geq t_0 + T$.*

Definition 2.31 *The equilibrium state x_e of* (2.29) *is said to be* ***exponentially stable*** *if there exists an $\alpha > 0$ with the property that for every $\varepsilon > 0$ there exists a $\delta > 0$ such that for all initial conditions $x_0 \in \mathbb{R}^n$ for which $\|x_0 - x_e\| < \delta$, we have*

$$\|x(t; t_0, x_0) - x_e\| \leq \varepsilon e^{-\alpha(t-t_0)}$$

for all $t_0 \geqslant 0$ and for all $t \geq t_0$.

Definition 2.32 *A solution $x(t; t_0, x_0)$ for some $t_0 \geqslant 0$ and $x_0 \in \mathbb{R}^n$ of* (2.29) *is said to be* **bounded** *if there exists a $\beta > 0$ such that $\|x(t; t_0, x_0)\| < \beta$ for all $t \geq t_0$.*

Definition 2.33 *The solutions of* (2.29) *are said to be* **uniformly bounded** *if for any $\alpha > 0$, there exists a $\beta > 0$ such that for all initial conditions $x_0 \in \mathbb{R}^n$ for which $\|x_0\| < \alpha$, we have $\|x(t; t_0, x_0)\| < \beta$ for all $t_0 \geqslant 0$ and for all $t \geq t_0$.*

Definition 2.34 *The set of solutions of* (2.29) *is said to be* **uniformly ultimately bounded** *if there exists a $B > 0$ such that for any $\alpha > 0$, there exists a $T > 0$ such that for all initial conditions $x_0 \in \mathbb{R}^n$ for which $\|x_0\| < \alpha$, we have $\|x(t; t_0, x_0)\| < B$ for all $t_0 \geqslant 0$ and for all $t \geq t_0 + T$.*

Definition 2.35 *The equilibrium state x_e of* (2.29) *is said to be* **globally asymptotically stable** *if it is stable and every solution of* (2.29) *tends to x_e as $t \to \infty$ (i.e., the region of attraction of x_e is all of $\mathbb{R}^n$).*

Definition 2.36 *The equilibrium state x_e of* (2.29) *is said to be* **uniformly globally asymptotically stable** *if:*

(*i*) *It is uniformly stable.*

(*ii*) *The solutions of* (2.29) *are uniformly bounded.*

(*iii*) *For all $\alpha > 0$ and $\varepsilon > 0$, there exists a $T > 0$ such that for all initial conditions $x_0 \in \mathbb{R}^n$ satisfying $\|x_0 - x_e\| < \alpha$, we have $\|x(t; t_0, x_0) - x_e\| < \varepsilon$ for all for all $t_0 \geqslant 0$ and $t \geq t_0 + T$.*

Definition 2.37 *The equilibrium state x_e of* (2.29) *is said to be* **globally exponentially stable** *if there exists an $\alpha > 0$ with the property that for any $\beta > 0$, there exists a $k > 0$ such that for all initial conditions $x_0 \in \mathbb{R}^n$ for which $\|x_0 - x_e\| < \beta$, we have*

$$\|x(t; t_0, x_0) - x_e\| \leq k e^{-\alpha(t - t_0)}$$

for all $t_0 \geqslant 0$ and for all $t \geq t_0$.

Definition 2.38 *The trajectory $x(t; t_0, x_0)$ is said to be* ***stable (unstable, uniformly stable, asymptotically stable, uniformly asymptotically stable, exponentially stable)*** *if the equilibrium state $z_e = 0$ of the system*

$$\dot{z} = f(t, z + x(t; t_0, x_0)) - f(t, x(t; t_0, x_0))$$

is stable (unstable, uniformly stable, asymptotically stable, uniformly asymptotically stable, exponentially stable, respectively).

Definition 2.39 *The system (2.29) is said to be* ***internally L_p stable at t_0*** *if the trajectory $x(t; t_0, x_0)$ for any $x_0 \in \mathbb{R}^n$ belongs to $L_p[t_0, \infty)$.*

Definition 2.40 *The system (2.29) is said to be* ***uniformly internally L_p stable*** *if it is internally L_p stable at t_0 for all $t_0 \geqslant 0$.*

Remark 2.41 *For autonomous systems, internal L_p stability implies global attractivity of the origin when f is continuous at the origin (see Lemma 2.79).*

It is easy to see that in the case of autonomous system (2.30), all the references to the word "uniform" in the above definitions need not be evoked. That is, if the equilibrium state x_e is stable (asymptotically stable, exponentially stable, globally asymptotically stable), it is always uniformly stable (uniformly asymptotically stable, uniformly exponentially stable, uniformly globally asymptotically stable, respectively). Similarly, if the solution $x(t; t_0, x_0)$ is bounded, it is uniformly bounded as well.

2.7.1 Lyapunov's direct method

The stability properties of the equilibrium state x_e or the solution $x(t; t_0, x_0)$ of (2.29) can be verified by utilizing the well-known direct method of Lyapunov (also called as the second method of Lyapunov). The method seeks to answer various questions of stability by using the form of the function $f(t, x)$ in (2.29) rather than the explicit knowledge of the solutions. We need the following additional definitions before we introduce the method.

Definition 2.42 *A continuous function $\phi : [0, \infty) \mapsto \mathbb{R}^+$ is said to belong to class $\mathcal{K}$ (denoted by $\phi \in \mathcal{K}$) if:*

(*i*) $\phi(0) = 0$.

(*ii*) *ϕ is strictly increasing.*

Definition 2.43 *A continuous function $\phi : [0, \infty) \mapsto \mathbb{R}^+$ is said to belong to class $\mathcal{K}_\infty$ (denoted by $\phi \in \mathcal{K}_\infty$) if:*

(i) $\phi(0) = 0$.

(ii) ϕ *is strictly increasing.*

(iii) $\lim_{\tau \to \infty} \phi(\tau) = \infty$.

Definition 2.44 *Two functions $\phi_1, \phi_2 \in \mathcal{K}$ are said to be of the same order of magnitude if there exist positive constants k_1 and k_2 such that*

$$k_1\phi_1(r) \leq \phi_2(r) \leq k_2\phi_1(r),$$

for all $r \in [0, \infty)$.

Definition 2.45 *A function $V(t, x) : \mathbb{R}^+ \times \mathbb{R}^n \mapsto \mathbb{R}$ with $V(t, 0) = 0$ for all $t \in \mathbb{R}^+$ is said to be* **locally positive definite** *if there exists a subset $\mathcal{V}$ of $\mathbb{R}^n$ containing 0 in its interior and a continuous function $\phi \in \mathcal{K}$ such that $V(t, x) \geq \phi(\|x\|)$ for all $t \in \mathbb{R}^+$, $x \in \mathcal{V}$.*

$V(t, x)$ is called locally negative definite if $-V(t, x)$ is positive definite.

Definition 2.46 *A function $V(t, x) : \mathbb{R}^+ \times \mathcal{V} \mapsto \mathbb{R}$ with $V(t, 0) = 0$ for all $t \in \mathbb{R}^+$ is said to be* **locally positive semi-definite** *if if there exists a subset $\mathcal{V}$ of $\mathbb{R}^n$ containing 0 in its interior such that $V(t, x) \geq 0$ for all $t \in \mathbb{R}^+$, $x \in \mathcal{V}$.*

$V(t, x)$ is called negative semi-definite if $-V(t, x)$ is positive semi-definite.

Definition 2.47 *A function $V(t, x) : \mathbb{R}^+ \times \mathbb{R}^n \mapsto \mathbb{R}$ with $V(t, 0) = 0$ for all $t \in \mathbb{R}^+$ is said to be* **decrescent** *if there exists a subset $\mathcal{V}$ of $\mathbb{R}^n$ containing 0 in its interior and a continuous function $\phi \in \mathcal{K}$ such that $V(t, x) \leq \phi(\|x\|)$ for all $t \in \mathbb{R}^+$ and for all $x \in \mathcal{V}$.*

Definition 2.48 *A function $V(t, x) : \mathbb{R}^+ \times \mathbb{R}^n \mapsto \mathbb{R}$ with $V(t, 0) = 0$ for all $t \in \mathbb{R}^+$ is said to be* **positive definite** *if there exists a continuous function $\phi \in \mathcal{K}$ such that $V(t, x) \geq \phi(\|x\|)$ for all $t \in \mathbb{R}^+$ and $x \in \mathbb{R}^n$.*

Definition 2.49 *A function* $V(t,x) : \mathbb{R}^+ \times \mathbb{R}^n \mapsto \mathbb{R}$ *with* $V(t,0) = 0$ *for all* $t \in \mathbb{R}^+$ *is said to be a* **radially unbounded** *positive definite function if there exists a* $\phi \in \mathcal{K}_\infty$ *such that*

$$V(t,x) \geq \phi(\|x\|) \text{ for all } t \in \mathbb{R}^+ \text{ and } x \in \mathbb{R}^n.$$

In what follows, without loss of generality, we assume that $x_e = 0$ is an equilibrium point of (2.29). Also, we define $\dot{V}$ as the time derivative of the function $V(t,x)$ along the solution of (2.29), that is,

$$\dot{V} = \frac{\partial V}{\partial t} + (\nabla V) f(t,x),$$

where

$$\nabla V = \frac{\partial V}{\partial x} = \begin{pmatrix} \frac{\partial V}{\partial x_1} & \frac{\partial V}{\partial x_2} & \cdots & \frac{\partial V}{\partial x_n} \end{pmatrix}$$

is the gradient of V with respect to x.

The following theorem states the second method of Lyapunov.

Theorem 2.50 *Suppose that there exists a locally positive definite function*

$$V(t,x) : \mathbb{R}^+ \times \mathbb{R}^n \mapsto \mathbb{R}$$

for some $r > 0$ *with continuous first-order partial derivatives with respect to* t *and* x *and* $V(t,0) = 0$ *for all* $t \in \mathbb{R}^+$. *Then the following statements are true:*

(*i*) *If* $\dot{V}(t,x) \leq 0$ *for all* $t \in \mathbb{R}^+$ *and all* x *in some open neighborhood of* 0, *then* $x_e = 0$ *is stable.*

(*ii*) *If* V *is decrescent and* $\dot{V}(t,x) \leq 0$ *for all* $t \in \mathbb{R}^+$ *and all* x *in some open neighborhood of* 0, *then* $x_e = 0$ *is uniformly stable.*

(*iii*) *If* V *is decrescent and* $\dot{V}(t,x) < 0$ *for all* $t \in \mathbb{R}^+$ *and all* x *in some open neighborhood of* 0, *then* $x_e = 0$ *is uniformly asymptotically stable.*

(*iv*) *If* V *is decrescent and there exist functions* $\phi_1, \phi_2, \phi_3 \in \mathcal{K}$ *of the same order of magnitude such that*

$$\phi_1(\|x\|) \leq V(t,x) \leq \phi_2(\|x\|) \quad \text{and} \quad \dot{V}(t,x) \leq -\phi_3(\|x\|)$$

for all $t \in \mathbb{R}^+$ *and for all* x *in some open neighborhood of the origin, then* $x_e = 0$ *is exponentially stable.*

Let us examine statement (*ii*) of Theorem 2.50. If we remove the restriction of V being decrescent, we obtain statement (*i*). Therefore, one might be tempted to expect that by removing the condition of V being decrescent in statement (*iii*), we obtain a condition for asymptotic stabilty, that is, $\dot{V} < 0$ implies that $x_e = 0$ is asymptotic stability. Such an intuitive conclusion is not true, as demonstrated by a counterexample in [94], see also [44, Sect. 53], where a first-order differential equation and a positive definite, non-decrescent function $V(t,x)$ are used to show that $\dot{V} < 0$ alone does not imply that $x_e = 0$ is asymptotic stable.

The condition in statement (*iii*) of the above theorem, namely, $V(t,x)$, is decrescent, and $\dot{V}(t,x) < 0$ is also equivalent to the existence of functions $\phi_1, \phi_2, \phi_3 \in \mathcal{K}$, where ϕ_1, ϕ_2, ϕ_3 do not have to be of the same order of magnitude, such that

$$\phi_1(\|x\|) \le V(t,x) \le \phi_2(\|x\|) \text{ and } \dot{V}(t,x) \le -\phi_3(\|x\|)$$

for all $t \in \mathbb{R}^+$ and $x \in \mathcal{B}(r)$.

We recognize that, in the above theorem, the state x is restricted to a neighborhood of the origin. As such, the results (*i*)–(*iv*) of Theorem 2.50 are referred to as local results. The following theorems are concerned with the global results.

Theorem 2.51 *Assume that the solution of* (2.29) *exists and is unique for each* $x_0 \in \mathbb{R}^n$. *Suppose that there exists a decrescent, radially unbounded positive definite function* $V(t,x) : \mathbb{R}^+ \times \mathbb{R}^n \mapsto \mathbb{R}^+$ *with continuous first-order partial derivatives with respect to t and x and* $V(t,0) = 0$ *for all* $t \in \mathbb{R}^+$. *Then the following statements are true:*

(*i*) *If* $\dot{V}(t,x) < 0$ *for all* $t \in \mathbb{R}^+$ *and all* $x \in \mathbb{R}^n$, *then* $x_e = 0$ *is uniformly globally asymptotically stable.*

(*ii*) *If there exist functions* $\phi_1, \phi_2, \phi_3 \in \mathcal{K}_\infty$ *of the same order of magnitude such that*

$$\phi_1(\|x\|) \le V(t,x) \le \phi_2(\|x\|) \text{ and } \dot{V}(t,x) \le -\phi_3(\|x\|)$$

for all $t \in \mathbb{R}^+$ *and* $x \in \mathbb{R}^n$, *then* $x_e = 0$ *is globally exponentially stable.*

The condition in statement (*i*) of the above theorem, namely, $\dot{V}(t,x) < 0$, is also equivalent to the existence of functions $\phi_1, \phi_2 \in \mathcal{K}_\infty$ and $\phi_3 \in \mathcal{K}$ such that

$$\phi_1(\|x\|) \le V(t,x) \le \phi_2(\|x\|) \text{ and } \dot{V}(t,x) \le -\phi_3(\|x\|)$$

for all $t \in \mathbb{R}^+$ and $x \in \mathbb{R}^n$.

Theorem 2.52 *Let the solution of* (2.29) *be unique for each* $x_0 \in \mathbb{R}^n$. *Suppose that there exists a decrescent, positive definite function* $V(t,x) : \mathbb{R}^+ \times \mathbb{R}^n \mapsto \mathbb{R}^+$ *with continuous first-order partial derivatives with respect to t and x and* $V(t,0) = 0$ *for all* $t \in \mathbb{R}^+$. *Then the following statements are true:*

(*i*) *If there exists a* $R > 0$ *such that* $\dot{V}(t,x) \le 0$ *for all* $t \in R^+$ *and all* $x \in \mathbb{R}^n$ *with* $\|x\| > R$, *then* $x_e = 0$ *is uniformly bounded.*

(ii) *If there exists a* $R > 0$ *and* $\phi \in \mathcal{K}$ *such that* $\dot{V}(t,x) \leq -\phi(\|x\|)$ *for all* $t \in R^+$ *and all* $x \in \mathbb{R}^n$ *with* $\|x\| > R$*, then* $x_e = 0$ *is uniformly ultimately bounded.*

Theorems 2.50–2.52 also hold for the autonomous system (2.30) because it is a special case of (2.29)). In the case of (2.30), however, $V(t,x) = V(x)$, that is, it does not explicitly depend on time t, and all references to the words "decrescent" and "uniform" could be deleted. This is because $V(x)$ is always decrescent and the stability (respectively, asymptotic stability) of the equilibrium $x_e = 0$ of (2.30) implies uniform stability (respectively, uniform asymptotic stability). Also, for (2.30), we can obtain a stronger result than Theorem 2.51 for global asymptotic stability. Before we state this result, let us have the following definition.

Definition 2.53 *A set* $\Omega \in \mathbb{R}^n$ *is* ***invariant*** *with respect to* (2.30) *if every solution of* (2.30) *starting in* Ω *remains in* Ω *for all* t.

Theorem 2.54 *Let the solution of* (2.30) *be unique for each* $x_0 \in \mathbb{R}^n$*. Suppose that there exists a positive definite and radially unbounded function* $V(x) : \mathbb{R}^n \mapsto \mathbb{R}^+$ *with continuous first-order partial derivatives with respect to* x *and* $V(0) = 0$*. If*

(i) $\dot{V} \leq 0$ *for all* $x \in \mathbb{R}^n$.

(ii) *The origin* $x = 0$ *is the only invariant subset of the set*

$$\Omega = \{x \in \mathbb{R}^n \mid \dot{V} = 0\},$$

then the equilibrium $x_e = 0$ *of* (2.30) *is globally asymptotically stable.*

All the above theorems are referred to as Lyapunov-type theorems. The function $V(t,x)$ or $V(x)$ that satisfies any Lyapunov-type theorem is referred to as a Lyapunov function.

Lyapunov functions can also be used to predict the instability properties of an equilibrium point x_e. Several instability theorems based on the second method of Lyapunov are given in [44].

2.8 External stability

Securing the stability of the equilibrium point of a given system or physical process is central to any control system design. In this regard, the classical concept of internal stability or otherwise often known as Lyapunov stability of a given system, as discussed in Sect. 2.7, dwells on various notions of the stability of an equilibrium point. It is widely discussed in many text books. On the other hand, another classical notion well known in the context of linear systems is the concept of bounded input bounded output (BIBO) stability or input–output stability. It is rooted in the requirement that a "small" excitation should cause only a "small" response. Motivated by this, the notion of *external stability* is indeed an attempt to bring a notion similar to BIBO to nonlinear systems as well by defining an appropriate measure of "smallness". Most of the literature considers the L_p (or ℓ_p) norm as an appropriate measure of "smallness". Thus, external stability seeks the controlled output be in the L_p or ℓ_p space for $p \in [1, \infty]$ whenever the external input or disturbance of a system is in the L_p or ℓ_p space. Moreover, one can also define the notion of system gain as the induced norm of the mapping from the external input to the controlled output. Owing to the use of the L_p (or ℓ_p) norm as an appropriate measure, external stability is also known as L_p stability for continuous-time systems or ℓ_p stability for discrete-time systems. Thus, the notions of input–output stability, external stability, and L_p or ℓ_p stability tantamount to the same.

To distinguish internal stability, or otherwise called Lyapunov stability, from input–output stability, it is worth quoting here two paragraphs from J. C. Willem's book, *The Analysis of Feedback Systems* ([204], pp. 102–103), from which one may gain a historical view on how the Lyapunov stability and input–output stability are distinguished and get separated.

"Lyapunov stability considers stability as an internal property of a system, and inputs and outputs do not play a role. This formulation accounts for the early development and great historical importance of this type of stability. The study of systems without inputs and outputs is indeed basic to classical dynamics. The traditional question of the stability of the solar system, for example remains a long standing challenge and does not involve inputs in any way. It is thus more than natural that stability of control systems has been studied in this context; namely, as a condition on undriven classical *dynamical systems*. This is in spite of the fact that its founders, Lyapunov and Poincaré, were not primarily interested in control. It should be noted that this dynamical-system point of view is supported by the work of Maxwell and much of the subsequent work on regulators. Although Lyapunov stability remains important and very useful in many control applications, its basic philosophy can often be challenged and is somewhat out of line with the modern approach to systems, where inputs and outputs are the fundamental variables and the state is merely an auxiliary variable that essentially represents the contents of a memory bank. The development and success of input–output stability should thus come as no surprise. This does not exclude that for many applications Lyapunov stability does represent a very satisfactory type of stability, and thus its study will remain both important and fruitful."

"Input–output stability is, from an engineering point of view, a very significant and important type of stability. The informal definition of stability given for instance by Nyquist in his classic paper on *Regeneration Theory* is essentially that of input–output stability. It is intimately related to the idea of *stability under constant disturbances* and thus has some classical—although not system oriented—foundations. The concept of input–output stability stands in direct competition with the idea of stability in the sense of Lyapunov. Input–output stability considers the disturbance entering the system as a constantly acting input, where as stability in the sense of Lyapunov considers the initial conditions as the disturbance to the system. Which of these two types of stability is to be preferred clearly depends on the particular application. In a sense, input–output stability protects against noise disturbances, whereas Lyapunov stability protects against a single impulse-like disturbance."

Our concern in this section is to recall several definitions all pertaining to input–output stability or external stability or some other notions of stability related to it.

Throughout this section, we consider a system Σ of the form,

$$\begin{cases} \rho x = f(x,d), & x(0) = x_0 \\ z = g(x,d), \end{cases} \tag{2.31}$$

with $f(0,0) = 0$ and $g(0,0) = 0$, where, as before, ρ denotes the time derivative ($\rho x = \frac{\mathrm{d}}{\mathrm{d}t}x$) or the shift operator ($(\rho x)(k) = x(k+1)$) for continuous-time and discrete-time systems, respectively. Here x is the state, d is the input, and z is the controlled output.

At first, we start with various notions of L_p or ℓ_p stability where initial conditions are fixed; in fact as is customary, the initial conditions are fixed at the origin.

Definition 2.55 *Let $p \in [1,\infty]$. Consider a continuous-time system Σ as given in* (2.31)*. Then, it is said to be* ***L_p stable with fixed initial condition and without finite gain*** *if, given any input $d \in L_p$ and $x(0) = 0$, there exists a unique solution x such that the controlled output $z \in L_p$.*

Similarly, consider the discrete-time system Σ as given in (2.31)*. Then, it is said to be* ***ℓ_p stable with fixed initial condition and without finite gain*** *if, given any input $d \in \ell_p$ and $x(0) = 0$, the unique solution x is such that the controlled output $z \in \ell_p$.*

Remark 2.56 *We note that if the continuous-time system* (2.31) *has an L_p input, the classical condition of f being locally Lipschitz might not suffice to ensure the existence and uniqueness of solution. This can be shown in the following example. Consider*

$$\dot{x} = (1-x)d^3, \qquad x(0) = 0.$$

The above system has a right hand side which obviously is nicely differentiable. Let

$$d = \begin{cases} \left(\frac{1}{1-t}\right)^{\frac{1}{3}} & t < t_0, \\ 0 & t \geqslant t_0, \end{cases}$$

where $t_0 > 1$. *It is easy to verify that we have* $d \in L_1$. *One solution with this input is*

$$x(t) = \text{sat}_{t_0}(t).$$

Another solution is

$$x(t) = \text{sat}_1(t),$$

where sat_Δ *is a standard saturation function with saturation level* Δ *as defined in Definition 2.19. Both of these solutions are so-called weak solutions in the sense that*

$$x(t) = \int_0^t (1 - x(\tau))d^3(\tau)\, d\tau.$$

We should note that if we impose that $d \in L_p \cap L_\infty$, *then the solution is unique.*

Definition 2.57 *Let* $p \in [1, \infty]$. *Consider a continuous-time system* Σ *as given in* (2.31). *Then, it is said to be* ***L_p stable with fixed initial condition with finite gain and with zero bias*** *if, given any input* $d \in L_p$ *and* $x(0) = 0$, *there exists a unique solution* x *such that the controlled output* $z \in L_p$ *and, moreover, if there exists a positive constant* γ_p *such that the following holds:*

$$\|z\|_p \le \gamma_p \|d\|_p, \quad \textit{for all } d \in L_p.$$

Furthermore, the infimum over all such γ_p*'s is called the* L_p *gain of the system.*

Similarly, consider the discrete-time system Σ *as given in* (2.31). *Then, it is said to be* ***ℓ_p stable with fixed initial condition with finite gain and with zero bias*** *if, given any input* $d \in \ell_p$ *and* $x(0) = 0$, *the unique solution* x *is such that the controlled output* $z \in \ell_p$ *and, moreover, if there exists a positive constant* γ_p *such that the following holds:*

$$\|z\|_p \le \gamma_p \|d\|_p, \quad \textit{for all } d \in \ell_p.$$

Furthermore, the infimum of such γ_p*'s is called the* l_p *gain of the system.*

Definition 2.58 *Let* $p \in [1, \infty]$. *Consider a continuous-time system* Σ *as given in* (2.31). *Then, it is said to be* ***L_p stable with fixed initial condition with finite gain and with bias*** *if, given any input* $d \in L_p$ *and* $x(0) = 0$, *there exists a unique solution* x *such that the controlled output* $z \in L_p$ *and, moreover, if there exist positive constants* γ_p *and* b_p *such that the following holds:*

$$\|z\|_p \le \gamma_p \|d\|_p + b_p, \quad \textit{for all } d \in L_p.$$

Furthermore, the infimum over all such γ_p*'s is called the* L_p *gain of the system.*

Similarly, consider the discrete-time system Σ as given in (2.31). Then, it is said to be **ℓ_p stable with fixed initial condition with finite gain and with bias** *if, given any input $d \in \ell_p$ and $x(0) = 0$, the unique solution x is such that the controlled output $z \in \ell_p$, and moreover, if there exist positive constants γ_p and b_p such that the following holds:*

$$\|z\|_p \leq \gamma_p \|d\|_p + b_p, \quad \text{for all } d \in \ell_p.$$

Furthermore, the infimum of such γ_p's is called the ℓ_p-gain of the system.

Remark 2.59 *Definitions 2.57 and 2.58 are equivalent for linear systems as one implies the other and conversely.*

Remark 2.60 *In the literature, L_p or ℓ_p stability with fixed initial condition and without finite gain is often simply referred to as L_p or ℓ_p stability. Also, L_p or ℓ_p stability with fixed initial condition and with finite gain is often simply referred to as L_p or ℓ_p stability with finite gain. In this book, we do the same.*

In the definitions of L_p or ℓ_p stability given above, we assumed that the initial conditions of the given system are fixed at zero, that is, the system is at rest. Nevertheless, one can modify easily the above definitions by setting the initial conditions at any fixed nonzero point. This is done in the literature by Shi [149]. In this regard, we observe here one important aspect as pointed out by Shi and others (see [151]); that is, if a given system is L_p-stable in some sense for one fixed initial condition, it does not necessarily imply that it is L_p stable in the same sense for another fixed initial condition. The following example illustrates this.

Example 2.61 Consider the double integrator with a linear feedback control law and external input d:

$$\begin{cases} \dot{x}_1 = x_2 \\ \dot{x}_2 = \sigma(-x_1 - x_2) + \delta\sigma(d) \\ z = x_1, \end{cases} \tag{2.32}$$

where $\sigma(.)$ is a standard saturation function. This system is globally asymptotically stable and locally exponentially stable if $d = 0$. Hence, given the zero initial condition $x(0) = 0$, there exists a sufficiently small $\delta > 0$ such that for all $d \in L_p$, we have $x \in L_p$ for all $p \in [1, \infty]$. However, it is shown in Chap. 14 and in [169] that if $p > 2$, then, no matter how small δ is, there exist a $d^* \in L_p$ and an initial state (x_1^*, x_2^*) such that the state trajectory of the closed-loop system diverges to infinity. Thus, $z \notin L_p$.

To create an example for any $p \geqslant 1$ where L_p stability depends on the choice of the initial condition, we simply modify the above system. Consider the same double integrator but with the input passing through a nonlinear element $(\cdot)^{p/3}$ for some $p \geqslant 1$ and with a nonlinear output:

$$\begin{cases} \dot{x}_1 = x_2 \\ \dot{x}_2 = \sigma(-x_1 - x_2) + \delta\sigma(d^{p/3}) \\ z = |x_1|^{3/p}. \end{cases} \tag{2.33}$$

Consider the initial condition (x_1^*, x_2^*) and define $d = |d^*|^{3/p}\,\mathrm{sgn}(d^*)$ where $d^* \in L_3$ is an external signal for the system (2.32), which drives the state from initial condition (x_1^*, x_2^*) to infinity. Since $d^* \in L_3$, obviously $d \in L_p$. By construction, for initial condition (x_1^*, x_2^*), the state x of (2.33) diverges, and hence, $z \notin L_p$. This establishes that the system (2.33) is not L_p stable for initial condition (x_1^*, x_2^*).

On the other hand, we claim that the system (2.33) is L_p stable for zero initial conditions. After all, for any $d \in L_p$, we have $d^* = d^{p/3} \in L_3$. We know that system (2.32) with zero initial conditions and input d^* yields a state $x \in L_3$. But this clearly yields that the system (2.33) with zero initial conditions and input d yields a state $x \in L_3$ and an output $z \in L_p$. Hence, the system (2.33) is L_p stable for zero initial condition.

It is clear from the above discussion that the initial conditions of the given system play a dominant role in achieving or not achieving external stability (in one sense or other), and hence, any definition of external stability must take into account the initial conditions. Motivated by this, Shi [149] not only defines external stability by setting the initial conditions at a fixed point but also when initial conditions are arbitrary. That is, Shi [149] defines what is now known as external stability with arbitrary initial conditions. These definitions are recalled below.

Definition 2.62 *Let $p \in [1, \infty]$. Consider a continuous-time system Σ as given in* (2.31)*. Then, it is said to be* ***L_p stable with arbitrary initial conditions and without finite gain*** *if for any input $d \in L_p$ and for any arbitrary initial condition $x_0 \in \mathbb{R}^n$, there exists a unique solution x such that the controlled output $z \in L_p$.*

Similarly, consider the discrete-time system Σ as given in (2.31)*. Then, it is said to be* ***ℓ_p stable with arbitrary initial conditions and without finite gain*** *if for any input $d \in \ell_p$ and for any arbitrary initial condition $x_0 \in \mathbb{R}^n$, the unique solution x is such that the controlled output $z \in \ell_p$.*

Following the above definitions, a recent paper [126] defines external stability with arbitrary initial conditions with finite gain and with bias.

Definition 2.63 *Let $p \in [1, \infty]$. Consider a continuous-time system Σ as given in* (2.31)*. Then, it is said to be* ***L_p stable with arbitrary initial conditions with finite gain and with bias*** *if the following hold:*

(i) There exists a unique solution x for any $x(0) = x_0 \in \mathbb{R}^n$.

(ii) There exists a positive constant γ_p and a class $\mathcal{K}$-function b_p such that for any $x(0) = x_0 \in \mathbb{R}^n$, the following holds:

$$\|z\|_p \leq \gamma_p \|d\|_p + b_p(\|x_0\|), \quad \text{for all } d \in L_p.$$

Furthermore, the infimum of such γ_p's is called the L_p gain of the system.

Similarly, consider the discrete-time system Σ as in (2.31)*. Then, it is said to be* ***ℓ_p stable with arbitrary initial conditions with finite gain and with bias*** *if there exists a positive constant γ_p and a class $\mathcal{K}$-function b_p such that for any $x(0) = x_0 \in \mathbb{R}^n$, the following holds:*

$$\|z\|_p \leq \gamma_p \|d\|_p + b_p(\|x_0\|), \quad \text{for all } d \in \ell_p.$$

Furthermore, the infimum of such γ_p's is called the ℓ_p gain of the system.

More recently, some other notions of external stability are introduced. These notions are similar to the definitions of L_p or ℓ_p stability with arbitrary initial conditions as they incorporate within them in some sense or other the notion of internal stability in the absence of disturbance or external input d. One such definition is introduced in [156] and is called input-to-state stability (ISS). It makes an attempt to marry both the notions of internal Lyapunov stability and the L_∞ stability or ℓ_∞ stability. We first need to recall the definition of a $\mathcal{KL}$-function before we recall the definition of ISS.

Definition 2.64 *A continuous function $\psi : \mathbb{R}^+ \to \mathbb{R}^+$ is said to be of class $\mathcal{L}$ if it is monotonically decreasing and $\lim_{r\to\infty} \psi(r) = 0$. A function $\beta : \mathbb{R}^+ \times \mathbb{R}^+ \to \mathbb{R}^+$ is a class $\mathcal{KL}$-function if it is class $\mathcal{K}$ with respect to the first argument and class $\mathcal{L}$ with respect to the second argument.*

Definition 2.65 *Consider the continuous-time system Σ as given in* (2.31) *where $f : \mathbb{R}^n \times \mathbb{R}^s \to \mathbb{R}^n$. Then, Σ is said to be* ***input-to-state stable (ISS)*** *if there exist a class $\mathcal{KL}$-function $\beta : \mathbb{R}^+ \times \mathbb{R}^+ \to \mathbb{R}^+$ and a class $\mathcal{K}$-function α such that, for each input $d \in L_\infty$ and for each initial condition $x(0) = x_0$ with $x_0 \in \mathbb{R}^n$, there exists a unique solution $x(t; x_0, d)$ of Σ satisfying*

$$\|x(t; x_0, d)\| \leqslant \beta(\|x_0\|, t) + \alpha(\|d\|_{L_\infty}) \tag{2.34}$$

for each $t \geqslant 0$.

Definition 2.66 *Consider the discrete-time system Σ as given in* (2.31) *where $f : \mathbb{R}^n \times \mathbb{R}^s \to \mathbb{R}^n$. Then, Σ is said to be* **input-to-state stable (ISS)** *if there exist a class $\mathcal{KL}$-function $\beta : \mathbb{R}^+ \times \mathbb{R}^+ \to \mathbb{R}^+$ and a class $\mathcal{K}$-function α such that, for each input $d \in \ell_\infty$ and for each initial condition $x(0) = x_0$ with $x_0 \in \mathbb{R}^n$, the unique solution $x(k; x_0, d)$ of Σ satisfies*

$$\|x(k; x_0, d)\| \leqslant \beta(\|x_0\|, k) + \alpha(\|d\|_{\ell_\infty}) \tag{2.35}$$

for each integer $k \in \mathbb{Z}^+$.

Remark 2.67 *Note that, by causality, the same definition would result if one would replace* (2.35) *by*

$$\|x(k; x_0, d)\| \leqslant \beta(\|x_0\|, k) + \alpha(\|d_{[k-1]}\|_{\ell_\infty}),$$

where $k \geqslant 1$ and, for each $r \geqslant 0$, $d_{[r]}$ denotes the truncation of d at r; that is,

$$d_{[r]}(j) = \begin{cases} d(j) & \text{if } j \leqslant r, \\ 0 & \text{if } j > r. \end{cases}$$

Remark 2.68 *By definition, an immediate consequence of an ISS system is that, for any arbitrarily fixed initial state $x_0 \in \mathbb{R}^n$, any bounded input d must produce a bounded state. Moreover, when the input d is identically zero, the ISS implies the global asymptotic stability of the zero equilibrium point.*

Before we present another notion of external stability which is married in some sense or other to the notion of internal stability, let us next recall a well-known fact in linear system theory that global asymptotic stability implies that any vanishing external input produces a vanishing state and as such, external inputs that vanish as time progresses affect only the transient behavior of a given system. In general, such a behavior is not true for nonlinear systems. This is what is behind the notion of converging input converging state (CICS) stability as recalled shortly.

Let us first introduce some notation.

Definition 2.69 *The set of function $f : \mathbb{R}^+ \to \mathbb{R}^s$ with the property that*

$$\lim_{t\to\infty} f(t) = 0$$

is denoted by $\mathcal{C}_0$. Similarly, the set of function $f : \mathbb{Z}^+ \to \mathbb{R}^s$ with the property that

$$\lim_{k\to\infty} f(k) = 0$$

is denotd by c_0.

We have the following CICS definitions.

Definition 2.70 *Consider the continuous-time system Σ as given in* (2.31) *where $f : \mathbb{R}^n \times \mathbb{R}^s \to \mathbb{R}^n$. We say that the system Σ satisfies the CICS stability if for each input $d \in \mathcal{C}_0$ and for any initial condition $x(0) = x_0$ with $x_0 \in \mathbb{R}^n$, there exists a unique solution $x(t; x_0, d)$ that satisfies $x(\cdot; x_0, d) \in \mathcal{C}_0$.*

Definition 2.71 *Consider the discrete-time system Σ as given in* (2.31) *where $f : \mathbb{R}^n \times \mathbb{R}^s \to \mathbb{R}^n$. We say that the system Σ satisfies the CICS stability if for each input $d \in c_0$ and for any initial condition $x(0) = x_0$ with $x_0 \in \mathbb{R}^n$, the solution $x(k; x_0, d)$ of Σ satisfies $x(\cdot; x_0, d) \in c_0$.*

Remark 2.72 *We note that, in the absence of any disturbance d, CICS stability implies global attractivity of the origin.*

Remark 2.73 *Various concepts of external stability are formulated above. In these formulations, the space of external signals or disturbances has no restrictions, that is, it is the entire such possible space. In that sense, these results are global results even though we did not explicitly add "global" in the terminology. On the other hand, one can restrict the space of external signals to a certain proper subset of the entire possible space. Such a restriction can be useful to define semi-global rather than global external stability concepts, the study of which will be undertaken in detail in subsequent chapters. An example of a version of local external stability will be seen in the next section.*

2.9 Relationship between internal stability and external stability

Sections 2.7 and 2.8 deal respectively with various definitions of internal stability and external stability. One natural question that arises at this stage is whether there exists any relationship between these two notions of stability. We remark that a well-known result in linear system theory is that any asymptotically stable system has very good external stability properties. That is, for linear systems under some mild conditions, the notions of internal stability and external stability are very highly coupled and in fact simply coalesce. However, for nonlinear systems, as readers will observe in subsequent chapters, in general, just having internal stability in one sense or other does not necessarily imply external stability in some sense or another. One can generate several examples to illustrate this. To quote an example as given in Liu et al. [89, 90], consider a nonlinear system (a linear system subject to actuator saturation),

$$\dot{x}_1 = \sigma(-3x_1 + 7x_2 + d_1), \quad \dot{x}_2 = \sigma(-x_1 + 2x_2 + d_2),$$

where d_1 and d_2 are some inputs and $\sigma(.)$ is a standard saturation function with ± 1 as the saturation level. It is easy to see that in the absence of d_1 and d_2, such a system is locally asymtotically stable with the origin as its equilibrium point. On the other hand, one can find for some finite T some input functions $d_1(t)$ and $d_2(t)$ on the interval $[0, T]$ such that the origin $[0, 0]'$ of the considered system at $t = 0$ can be steered to $[1, 1]'$ at $t = T$. By defining $d_1 = 0$ and $d_2 = 0$ on (T, ∞), we have the solution of the considered system as $x_1 = t - T + 1$ and $x_2 = t - T + 1$. Thus, when we consider $d_1(t)$ and $d_2(t)$ as external input or disturbance signals, the considered system is not L_p stable for any $1 \le p \le \infty$ in the traditional sense of L_p stability with fixed initial conditions.

Then, another immediate query arises: Suppose the given system is externally stable in some sense. Does such an external stability have any consequences for internal stability of the given system in the absence of external input signals? In this regard, we have already remarked that having ISS implies global asymptotic stability of zero equilibrium point of a given nonlinear system in the absence of any external input signals. Additionally, if a nonlinear system is CICS stable, then it also has the property of global attractivity of the origin in the absence of any external input signals. However, it is not clear yet, whether having external stability in the classical sense of L_p stability with fixed initial conditions has any consequences for internal stability in some sense or other. It turns out that under some mild conditions on the given nonlinear system, external stability does imply certain properties of internal stability. We pursue such properties in this section based on the recent work of [191].

To be specific, in this section, for a nonlinear system that is L_p stable, our interest is to investigate the internal stability of the autonomous system (i.e, the given nonlinear system with the input zero). Our work here in this respect evolves along two main lines. The first line starts with L_p stability without finite gain. An important prior result in this direction is that in [89], which under a fairly restrictive condition on the structural property of the system, shows that L_p stability without finite gain implies global attractivity of the equilibrium point. Indeed, it turns out that this result of [89] can be obtained under weaker conditions. We show here that under mild conditions, global L_p stability without finite gain ensures attractivity of the equilibrium point in the absence of input and attractivity of the origin with any L_p input.

The other line emanates from L_p stability with finite gain. There is a large body of work in the literature in this direction; see, for instance, [25, 46, 89, 190]. Along this line of research, the objective here is to conclude local asymptotic stability of the equilibrium point based on L_p stability with finite gain. It was shown in [46] that under a uniform reachability condition, global L_p stability with finite gain implies local asymptotic stability of the equilibrium point. In [190], the notion of small-signal L_p stability with finite gain was introduced and its connection to attractivity of the equilibrium point was established. This concept of small-signal L_p stability was extended in [25] by so-called gain-over-set stability, and it was shown that finite-gain L_p stability over a set in L_p space yields local asymptotic stability of the equilibrium point. We prove here a result on the relationship between Lyapunov stability and local L_p stability with finite gain, which further extends, to some extent, the result in [25].

We consider a nonlinear system

$$\Sigma_1: \quad \dot{x} = f(x,d), \quad x(0) = x_0, \tag{2.36}$$

where $x \in \mathbb{R}^n$ and $d \in \mathbb{R}^m$. We assume that for all $x_0 \in \mathbb{R}^n$, Σ_1 has a unique solution defined on $[0,\infty)$, which is absolutely continuous on any compact interval. Moreover, we assume that $f(x,d)$ is continuous with respect to x. Let $x(t;t_0,x_0,d)$ denote the trajectory of Σ_1 initialized at time t_0 with input d and initial condition x_0. If $t_0 = 0$, we will use $x(t;x_0,d)$ instead of $x(t;t_0,x_0,d)$.

We shall investigate the internal stability of the unforced system

$$\Sigma_2: \quad \dot{x} = f(x,0), \quad x(0) = x_0, \tag{2.37}$$

under the assumption that Σ_1 is L_p stable in some sense.

We first recall some preliminaries. We defined several types of L_p stability earlier in Sect. 2.8. These were basically all global definitions even though, for brevity, we did not explicitly use the word "global." Here global is in the sense that $d \in L_p$ is not bounded in size. Below, we define a local version of L_p stability.

Definition 2.74 *The system Σ_1 is said to be locally L_p stable with fixed initial condition and with finite gain if there exists a δ and a γ such that for $x_0 = 0$ and any d with $\|d\|_{L_p} \leqslant \delta$, a unique solution exists and $\|x(t;0,d)\|_{L_p} \leqslant \gamma \|d\|_{L_p}$.*

The region or domain of attraction as defined in Definition 2.29 is denoted by $\mathcal{A}(\Sigma_2)$ for the system Σ_2, that is,

$$\mathcal{A}(\Sigma_2) = \{x_0 \in \mathbb{R}^n \mid x(t;x_0,0) \to 0 \text{ as } t \to \infty\}. \tag{2.38}$$

Definition 2.75 *A point $\xi \in \mathbb{R}^n$ is an L_p-reachable point of system Σ_1 if there exist a finite T, and an M, and a measurable input $d : [0,T] \to \mathbb{R}^m$ such that $x(T;0,d) = \xi$ and*

$$\int_0^T \|d(t)\|^p dt \leqslant M.$$

The set of all L_p-reachable points of Σ_1 is called the L_p-reachable set of Σ_1, which is denoted as $\mathcal{R}_p(\Sigma_1)$.

Remark 2.76 *The requirement in Definition 2.75 is a weak condition that ensures that the integral of* $\|d(t)\|^p$ *over the interval* $[0, T]$ *is finite. For example, any* x_0 *that is reachable via a signal* $d(t)$ *that is essentially bounded on* $[0, T]$ *is* L_p*-reachable for any* $p \in [1, \infty)$.

The following definition of small-signal local L_p reachability is adapted from [25].

Definition 2.77 *The system* Σ_1 *is said to be small-signal locally* L_p *reachable if for any* $\varepsilon > 0$*, there exists a* δ *such that for any* $\xi \in \mathbb{R}^n$ *with* $\|\xi\| \leqslant \delta$*, we can find a finite time* T *and a measurable input* $d : [0, T] \to \mathbb{R}^m$ *such that* $x(T; 0, d) = \xi$ *and* $\|d\|_p \leqslant \varepsilon$.

We have the following result.

Theorem 2.78 *Suppose the system* Σ_1 *is globally* L_p *stable with fixed initial condition without finite gain for some* $p \in [1, \infty)$*. Then* $\mathcal{A}(\Sigma_2) \supseteq \mathcal{R}_p(\Sigma_1)$.

In order to prove Theorem 2.78, we need the following lemma:

Lemma 2.79 *Consider the system* Σ_2*. If* $x(t; x_0, 0) \in L_p$ *for some* $p \in [1, \infty)$*, then* $x(t; x_0, 0) \to 0$.

Proof of Lemma 2.79 : For simplicity, we denote in this proof, $x(t; x_0, d)$ and $f(x(t), 0)$ by $x(t)$ and $f(x(t))$, respectively. Suppose, for the sake of establishing a contradiction, that $x(t) \to 0$ does not hold. Then there exists a $\delta > 0$ such that, for any arbitrarily large $T \geqslant 0$, there is a $\tau \geqslant T$ such that $\|x(\tau)\| \geqslant 2\delta$. Let m be a bound on $\|f(x)\|$ on the closed ball $\mathcal{B}(2\delta)$. This bound exists due to continuity of $f(x)$ with respect to x.

For some τ such that $\|x(\tau)\| \geqslant 2\delta$, let $t_2 > \tau$ be the smallest value such that $\|x(t_2)\| = \delta$, and let t_1 be the largest value such that $t_1 < t_2$ and $\|x(t_1)\| = 2\delta$. Such t_1 and t_2 exist because $x(t)$ is absolutely continuous and $x \in L_p$. Since $\|x(t)\| \in \mathcal{B}(2\delta)$ for all $t \in [t_1, t_2]$, we have, due to the absolute continuity of the solution,

$$\|x(t_1)\| - \|x(t_2)\| \leqslant \|x(t_2) - x(t_1)\| = \left\| \int_{t_1}^{t_2} f(x(\tau))\, \mathrm{d}\tau \right\|$$

$$\leqslant \int_{t_1}^{t_2} \|f(x(\tau))\|\, \mathrm{d}\tau \leqslant (t_2 - t_1)m.$$

Hence, $t_2 - t_1 \geqslant (\|x(t_1)\| - \|x(t_2)\|)/m = \delta/m$. Clearly, $\|x(t)\| \geqslant \delta$ for all $t \in [\tau, t_2]$, and furthermore, $t_2 - \tau \geqslant t_2 - t_1 \geqslant \delta/m$. It follows that for each τ such that $\|x(\tau)\| \geqslant 2\delta$, we have $\|x(t)\| \geqslant \delta$ for all $t \in [\tau, \tau + \delta/m]$.

Let T be chosen large enough that

$$\int_T^\infty \|x(t)\|^p \,\mathrm{d}\tau < \frac{\delta^{p+1}}{m}. \tag{2.39}$$

Such a T must exist, since $x(t) \in L_p$. Let $\tau \geqslant T$ be chosen such that $\|x(\tau)\| \geqslant 2\delta$. We have

$$\int_T^\infty \|x(t)\|^p \,\mathrm{d}\tau \geqslant \int_\tau^{\tau+\delta/m} \|x(t)\|^p \,\mathrm{d}\tau \geqslant \frac{\delta^{p+1}}{m}.$$

This contradicts (2.39), which proves that $x(t) \to 0$. ■

Proof of Theorem 2.78 : For any $x_0 \in \mathcal{R}_p(\Sigma_1)$, there exist finite T, M, and an input $d_0(t)$ for $t \in [0, T]$ such that $x(T; 0, d_0) = x_0$ and

$$\int_0^T \|d_0(t)\|^p \mathrm{d}t \leqslant M.$$

Define

$$d(t) = \begin{cases} d_0(t), & t \in [0, T], \\ 0, & t > T. \end{cases}$$

Clearly, $d \in L_p$. Since Σ_1 is globally L_p stable with fixed initial condition without finite gain, we have $x(\cdot; 0, d) \in L_p$. On the other hand, $d(t) = 0$ for $t > T$ implies that after T the system Σ_1 is equivalent with system Σ_2 initialized at x_0, that is, $x(t; 0, d, 0) = x(t - T; x_0, 0)$ with $t > T$. Therefore, $x(t; x_0, 0) \in L_p$ over $[0, \infty)$. It follows from Lemma 2.79 that $x(t; x_0, 0) \to 0$ as $t \to \infty$. This completes the proof. ■

Corollary 2.80 *If $\mathcal{R}_p(\Sigma_1) = \mathbb{R}^n$, then the origin of Σ_2 is globally attractive.*

The next theorem shows that under a certain condition on the structure of $f(x, d)$, the origin of Σ_1 is attractive for any input $d \in L_p$.

Theorem 2.81 *Suppose that Σ_1 is globally L_p stable with fixed initial condition without finite gain for some $p \in [1, \infty)$. If there exist δ, m_1, m_2, and $q \in [0, p)$ such that for any x with $\|x\| \leqslant \delta$ and for any d,*

$$\|f(x, d)\| \leqslant m_1 + m_2 \|d\|^q, \tag{2.40}$$

then for $x_0 = 0$ and any $d \in L_p$, $x(t, 0, d, 0) \to 0$ as $t \to \infty$.

Proof : Define a generalized saturation function $\bar{\sigma} : \mathbb{R}^n \to \mathbb{R}^n \in C^1$ as

$$\bar{\sigma}(x) = \begin{bmatrix} \bar{\sigma}_1(x_1) \\ \vdots \\ \bar{\sigma}_n(x_n) \end{bmatrix}, \quad \bar{\sigma}_i(x_i) = \begin{cases} -\frac{2\delta}{\pi}, & x_i < -\delta \\ \frac{2\delta}{\pi} \sin(\frac{\pi}{2\delta} x_i), & |x_i| \leqslant \delta \\ \frac{2\delta}{\pi}, & x_i > \delta. \end{cases}$$

Consider $\bar{x}(t) = \bar{\sigma}(x(t, 0, d, 0))$. Note that $\bar{x}(t)$ is still absolutely continuous on any compact interval. Let $\bar{x}_i$ and f_i denote the ith element of $\bar{x}$ and $f(x, d)$ respectively. We have

$$|\dot{\bar{x}}_i(t)| = \begin{cases} 0, & |x_i(t)| > \delta \\ |\cos(\frac{\pi}{2\delta} x_i) f_i(x(t), d(t))| \leqslant m_1 + m_2 \|d(t)\|^q, & |x_i(t)| \leqslant \delta, \end{cases}$$

Therefore, $\|\dot{\bar{x}}(t)\| \leqslant \sqrt{n}(m_1 + m_2\|d\|^q)$ for all $t > 0$. Note that $\|d(t)\|^q \leqslant 1 + \|d(t)\|^p$, and hence, $\|d\|^q$ is locally uniformly integrable. Then it follows from [182] that $\bar{x}(t) \to 0$ as $t \to 0$. This implies that $x(t, 0, d, 0) \to 0$ as $t \to 0$. ■

Remark 2.82 *In [89], in order to prove the same result as in Theorem 2.81, the following condition was imposed on $f(x, d)$: there exist δ_1, K_1, K_2, and $\alpha \in [0, p]$ such that for $x \in \mathbb{R}^n$ with $\|x\| \leqslant \delta_1$ and $d \in \mathbb{R}^m$,*

$$\|f(x, d)\| \leqslant K_1(\|x\| + \|d\|) + K_2(\|x\|^\alpha + \|d\|^\alpha).$$

Theorem 2.81 shows that the dependence on $\|x\|$ in the upper bound of the above condition is not necessary.

An immediate consequence of Theorem 2.81 is the next theorem.

Theorem 2.83 *Suppose that Σ_1 is globally L_p stable with fixed initial condition without finite gain and $\mathcal{R}_p(\Sigma_1) = \mathbb{R}^n$ for some $p \in [1, \infty)$. If there exist δ, m_1, m_2 and $q \in [0, p]$ such that for any x with $\|x\| \leqslant \delta$ and for any d,*

$$\|f(x, d)\| \leqslant m_1 + m_2 \|d\|^q,$$

then Σ_1 is globally L_p stable without finite gain with arbitrary initial condition. Moreover, for any $x_0 \in \mathbb{R}^n$ and any $d \in L_p$, $x(t; x_0, d) \to 0$ as $t \to \infty$.

Proof : Since $\mathscr{R}_p(\Sigma_1) = \mathbb{R}^n$, for any $x_0 \in \mathbb{R}^n$, there exist finite T, M, and a measurable input $d_0 : [0, T] \to \mathbb{R}^m$ such that $x(T; 0, d_0) = x_0$ and

$$\int_0^T \|d_0(t)\|^p \,\mathrm{d}t \leqslant M.$$

For any $d \in L_p$, define

$$\bar{d}(t) = \begin{cases} d_0(t), & t \in [0, T] \\ d(t - T), & t > T. \end{cases}$$

Then we have $x(t; x_0, d) = x(t + T, 0, \bar{d}, 0)$. Clearly, $\bar{d} \in L_p$. This implies that $x(\cdot; 0, \bar{d}) \in L_p$, and hence, $x(\cdot; x_0, d) \in L_p$. This proves L_p stability with arbitrary initial condition and it follows from Theorem 2.81 that $x(t; 0, \bar{d}) \to 0$ as $t \to \infty$ and therefore $x(t; x_0, d) \to 0$ as $t \to \infty$. ■

In what follows, we prove a theorem that is a slight generalization of results of [25].

Theorem 2.84 *Suppose that Σ_1 is locally L_p stable with fixed initial condition and with finite gain and small-signal locally L_p reachable. Then the origin of Σ_2 is locally asymptotically stable.*

Proof : Let ε be an arbitrary positive real number. We need to show that there exists a $\delta > 0$ such that $\|x_0\| \leqslant \delta$ implies that $\|x(t; x_0, 0)\| \leqslant \varepsilon$ for all $t \geqslant 0$. Toward this end, let $\delta \leqslant \frac{\varepsilon}{2}$ be chosen such that for any $x_0 \in \mathbb{R}^n$ with $\|x_0\| \leqslant \delta$, there exist a finite time T and a measurable input $d : [0, T] \to \mathbb{R}^m$ such that

$$x(T; 0, d) = x_0 \quad \text{and} \quad \|d\|_{L_p} < \tfrac{\varepsilon}{2\gamma} \left(\tfrac{\varepsilon}{2M(\varepsilon)}\right)^{\frac{1}{p}}.$$

This is possible due to L_p local reachability.

Set $d(t) = 0$ for $t > T$. Since Σ_1 is locally L_p stable with finite gain, from Definition 2.74, there exists a γ such that

$$\int_T^\infty \|x(t; 0, d)\|^p \,\mathrm{d}t \leqslant \int_0^\infty \|x(t; 0, d)\|^p \,\mathrm{d}t \leqslant \gamma^p \|d\|_{L_p}^p < \tfrac{\varepsilon^{p+1}}{2^{p+1} M(\varepsilon)}.$$

For $t > T$, the system Σ_1 is equivalent to Σ_2 initialized at $x(0) = x_0$, that is, $x(t;0,d) = x(t-T;x_0,0)$. Hence, we have

$$\int_0^\infty \|x(t;x_0,0)\|^p \,\mathrm{d}t < \frac{\varepsilon^{p+1}}{2^{p+1}M(\varepsilon)}. \tag{2.41}$$

It immediately follows from Lemma 2.79 that $x(t;x_0,0) \to 0$ as $t \to \infty$.

We proceed to show that $\|x(t;x_0,0)\| < \varepsilon$ for all $t \geqslant 0$. Suppose, for the sake of establishing a contradiction, that there exists a τ such that $\|x(\tau;x_0,0)\| \geqslant \varepsilon$. Let $t_1 < \tau$ be the largest value such that $\|x(t_1;x_0,0)\| = \varepsilon/2$, and let $t_2 \leqslant \tau$ be the smallest value such that $t_2 > t_1$ and $\|x(t_2;x_0,0)\| = \varepsilon$. Such t_1 and t_2 exist because $\|x_0\| \leqslant \frac{\varepsilon}{2}$. Then $\varepsilon/2 \leqslant \|x(t;x_0,0)\| \leqslant \varepsilon$ for all $t \in [t_1,t_2]$. Let $M(\varepsilon)$ be a bound on $f(x,0)$ for $\|x\| \leqslant \varepsilon$. We have, owing to the absolute continuity of $x(t;x_0,0)$,

$$\begin{aligned}\|x(t_2;x_0,0)\| - \|x(t_1;x_0,0)\| &\leqslant \|x(t_2;x_0,0) - x(t_1;x_0,0)\| \\ &\leqslant \left\| \int_{t_1}^{t_2} f(x(t),0)\,\mathrm{d}t \right\| \leqslant \int_{t_1}^{t_2} M(\varepsilon)\,\mathrm{d}t \leqslant M(\varepsilon)(t_2 - t_1).\end{aligned}$$

This gives that $t_2 - t_1 \geqslant \frac{\varepsilon}{2M(\varepsilon)}$ and, hence, that

$$\int_0^\infty \|x(t;x_0,0)\|^p \,\mathrm{d}t \geqslant \int_{t_1}^{t_2} \|x(t;x_0,0)\|^p \,\mathrm{d}t \geqslant \int_{t_1}^{t_2} \left(\frac{\varepsilon}{2}\right)^p \mathrm{d}t = \frac{\varepsilon^{p+1}}{2^{p+1}M(\varepsilon)},$$

which contradicts (2.41). Hence, $\|x(t;x_0,0)\| < \varepsilon$ for all $t \geqslant 0$, which completes the proof. ∎

Remark 2.85 *Compared with the result in [25], Theorem 2.84 only requires a finite gain within an arbitrary small neighborhood of the origin of L_p space.*

Remark 2.86 *We assume here that $f(x,d)$ is continuous with respect to x, which covers a large class of dynamical systems. In fact, it can be seen from the proof that we only need continuity of $f(x,d)$ with respect to x at $x = 0$.*

3

A special coordinate basis (SCB) of linear multivariable systems

3.1 Introduction

What is called the special coordinate basis (SCB) of a multivariable linear time-invariant system plays a dominant role throughout this book; hence a clear understanding of it is essential. The purpose of this chapter is to recall the SCB as well as its properties pertinent to this book. The SCB originated in [138, 140, 141] and was crystallized for strictly proper systems in [139] and for proper systems in [132]. Our presentation of SCB here omits all the proofs that can be found in the literature.

What is SCB? It is a fine-grained structural decomposition of a multivariable linear time-invariant system. It partitions a given system into separate but interconnected subsystems that reflect the architectural mapping of inner workings of the system. By this we mean that the SCB identifies all pertinent structural elements of a system and their functions, and most significantly, it also displays the interconnections among all such structural elements. In doing so, the SCB representation explicitly reveals the system's finite and infinite zero structure and invertibility properties. Since its introduction, the SCB has been used in a large body of research, on topics including loop transfer recovery, timescale assignment, disturbance rejection, H_2 control, and H_∞ control. It has also been used as a fundamental tool in the study of linear systems theory. For details on these topics, we refer to the books [18, 19, 123, 133, 136], all of which are based on the SCB, and the references therein. Other topics include multivariable root loci [138, 140], decoupling theory [139], factorization of linear systems [20], squaring down of nonsquare systems [130, 132, 139], and model order reduction [111]. As will be clear to the readers, the influence of the SCB will be felt amply throughout this book and often from different angles.

3.2 The SCB

We present the SCB in this section. For readers unfamiliar with the topic, the complexities of the SCB may initially appear overwhelming. This is only a reflection, however, of the inherent complexities that exist in general multivariable

A. Saberi et al., *Internal and External Stabilization of Linear Systems with Constraints*, Systems & Control: Foundations & Applications, DOI 10.1007/978-0-8176-4787-2_3,

linear time-invariant systems. In the following exposition, significant complexity is added to accommodate general non-strictly proper multivariable systems. To get an overview of the SCB of progressively complex systems, we recommend first reading the SCB of uniform rank systems [138], the SCB of invertible systems [140], the SCB of strictly proper systems [139], and then the SCB of general multivariable systems as presented shortly. Also, the complexities encountered can be dissipated by carefully following various notations used.

We consider a linear time-invariant system Σ_* characterized by a quadruple (A, B, C, D). Let the dynamic equations of Σ_* be

$$\Sigma_* : \begin{cases} \rho x = Ax + Bu \\ y = Cx + Du, \end{cases} \tag{3.1}$$

where ρ is an operator indicating the time derivative $\frac{d}{dt}$ for continuous-time systems and a forward unit time shift for discrete-time systems. Also, $x \in \mathbb{R}^n$ is the state, $u \in \mathbb{R}^m$ is the control, and $y \in \mathbb{R}^p$ is the output. Without loss of generality, we assume that $(B' \quad D')'$ and $(C \quad D)$ have full rank.

Next, it is simple to verify that nonsingular transformations $\widetilde{U}$ and $\widetilde{V}$ exist such that

$$\widetilde{U} D \widetilde{V} = \begin{pmatrix} I_{m_0} & 0 \\ 0 & 0 \end{pmatrix}, \tag{3.2}$$

where m_0 is the rank of matrix D. Hence, hereafter, without loss of generality, it is assumed that the matrix D has the form given on the right-hand side of (3.2). As such, without loss of generality, we can focus on a system Σ_* of the form

$$\Sigma_* : \begin{cases} \rho x = Ax + \begin{pmatrix} B_0 & \widehat{B}_1 \end{pmatrix} \begin{pmatrix} u_0 \\ \widehat{u}_1 \end{pmatrix} \\ \begin{pmatrix} y_0 \\ \widehat{y}_1 \end{pmatrix} = \begin{pmatrix} C_0 \\ \widehat{C}_1 \end{pmatrix} x + \begin{pmatrix} I_{m_0} & 0 \\ 0 & 0 \end{pmatrix} \begin{pmatrix} u_0 \\ \widehat{u}_1 \end{pmatrix}, \end{cases} \tag{3.3}$$

where the matrices B_0, $\widehat{B}_1$, C_0, and $\widehat{C}_1$ have appropriate dimensions.

By nonsingular transformation of the state, output, and input, the system (3.3) can be transformed to the SCB. We use the symbols $\widetilde{x}$, $\widetilde{y}$, and $\widetilde{u}$ to denote the state, output, and input of the system transformed to the SCB form. The transformations between the original system (3.3) and the SCB are called Γ_s, Γ_y, and Γ_u, and we write $x = \Gamma_s \widetilde{x}$, $y = \Gamma_y \widetilde{y}$, and $u = \Gamma_u \widetilde{u}$.

The state $\widetilde{x}$, output $\widetilde{y}$, and input $\widetilde{u}$ are partitioned as

$$\widetilde{x} = \begin{pmatrix} x_a \\ x_b \\ x_c \\ x_d \end{pmatrix}, \quad \widetilde{y} = \begin{pmatrix} y_0 \\ y_d \\ y_b \end{pmatrix}, \quad \widetilde{u} = \begin{pmatrix} u_0 \\ u_d \\ u_c \end{pmatrix}.$$

Each component of state $\widetilde{x}$ represents a particular subsystem described in the next section. In the partition of output $\widetilde{y}$, y_0 is the original part of output as given in (3.3), y_d is the output from the x_d subsystem, and y_b is the output from the x_b subsystem. Similarly, in the partition of input $\widetilde{u}$, u_0 is the original part of input as given in (3.3), u_d is the input to the x_d subsystem, and u_c is the input to the x_c subsystem. Because u_0 appears first in both u and $\widetilde{u}$, Γ_u is of the form $\text{diag}(I_{m_0}, \overline{\Gamma}_u)$, for some nonsingular $\overline{\Gamma}_u$.

3.2.1 Structure of the SCB

Consider first the case when the general system Σ_* as given in (3.1) is strictly proper, that is, the matrix $D = 0$, and consequently u_0 and y_0 do not exist in (3.3). The meaning of the four subsystems can be explained as follows:

- The x_a subsystem represents the zero dynamics. This part of the system is not directly affected by any inputs nor does it affect any outputs directly. It may be affected, however, by the outputs y_b and y_d from x_b and x_d subsystems.

  The state x_a can be partitioned further into three substates x_a^-, x_a^0, and x_a^+. These substates x_a^-, x_a^0, and x_a^+ are associated respectively with the dynamics of zeros which are in the open left-hand complex plane, on the imaginary axis, and in the open right-hand complex plane for continuous-time systems and are within the unit circle, on the unit circle, and outside the unit circle for discrete-time systems.

- The x_b subsystem has a direct effect on the output y_b, but it is not directly affected by any inputs. It may be affected, however, by the output y_d from the x_d subsystem. The x_b subsystem is observable from the output y_b. The existence of x_b subsystem renders the given system Σ_* non-right invertible. That is, Σ_* is right invertible whenever x_b subsystem is nonexistent.

- The x_c subsystem is directly affected by the input u_c, but it does not have a direct effect on any outputs. It may also be affected by the outputs y_b and y_d from the x_b and x_d subsystems, as well as the state x_a. However, the influence from x_a is matched with the input u_c (i.e., the influence from x_a is additive to that of the input u_c). The x_c subsystem is controllable from the input u_c. The existence of x_c subsystem renders the given system Σ_* non-left invertible. That is, Σ_* is left invertible whenever x_c subsystem is nonexistent.

- The x_d subsystem represents the infinite zero structure. This part of the system is directly affected by the input u_d, and it also affects the output y_d directly. The x_d subsystem can be further partitioned into m_d single-input single-output (SISO) subsystems with states x_i and outputs y_i for $i = 1, \ldots, m_d$. Each of these subsystems consists of a chain of integrators of length q_i, from the i'th element of u_d (denoted by u_i) to the i'th element of y_d (denoted by y_i). Each integrator chain may be affected at each stage by the output y_d from the x_d subsystem, and at the lowest level of the integrator chain (where the input appears), it may be affected by all the states of the system. The x_d subsystem is observable from y_d and controllable from u_d.

The structure of strictly proper SCB systems is summarized in Table 3.1. For non-strictly proper systems, the structure is the same, except for the existence of the direct-feedthrough output y_0, which is directly affected by the input u_0 and can be affected by any of the states of the system. It can also affect all the states of the system.

Table 3.1: Summary of strictly proper SCB structure

Subsystem	Input	Output	Interconnections
x_a	–	–	y_b, y_d
x_b	–	y_b	y_d
x_c	u_c	–	y_b, y_d, x_a^{a}
x_d	u_d	y_d	$x_a^{\text{a}}, x_b^{\text{a}}, x_c^{\text{a}}$

The *Interconnections* column indicates influences from other subsystems.
[a]Matched with input

A block diagram of the SCB is given in Figs. 3.1–3.4. Figure 3.1 expresses the zero dynamics. Figure 3.2 represents the dynamics that is present if and only if the system Σ_* is not right invertible. Figure 3.3 represents the dynamics that is present if and only if the system Σ_* is not left invertible. Finally, the dynamics in Fig. 3.4 is related to the infinite zero structure. In this last figure, a signal given by a double-edged arrow is some linear combination of outputs y_i, $i = 0$ to m_d, whereas the signal given by the double-edged arrow with a solid dot in it is some linear combination of all states.

3.2.2 SCB equations

The SCB representation of the system Σ_* as given in (3.3) is articulated by the following theorem.

Theorem 3.1 *For any given system Σ_* characterized by the matrix quadruple (A, B, C, D), there exist;*

(*i*) *Unique coordinate-free nonnegative integers:*
$n_{a-}(\Sigma_*)$, $n_{a\circ}(\Sigma_*)$, $n_{a+}(\Sigma_*)$, $n_b(\Sigma_*)$, $n_c(\Sigma_*)$, $n_d(\Sigma_*)$, $m_d \leqslant m - m_0$, *and* q_i, $i = 1, \ldots, m_d$.

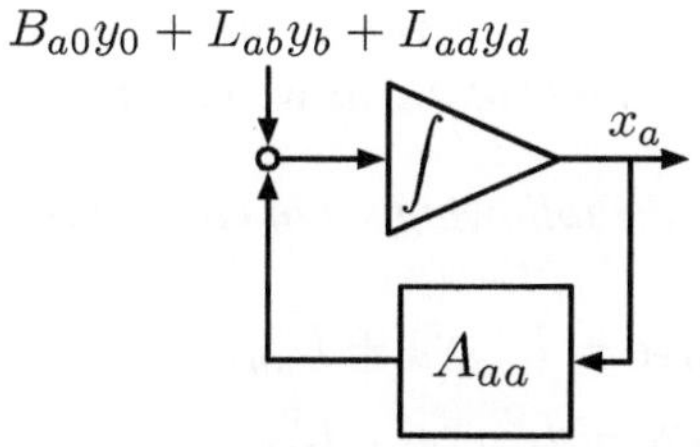

Figure 3.1: Zero dynamics

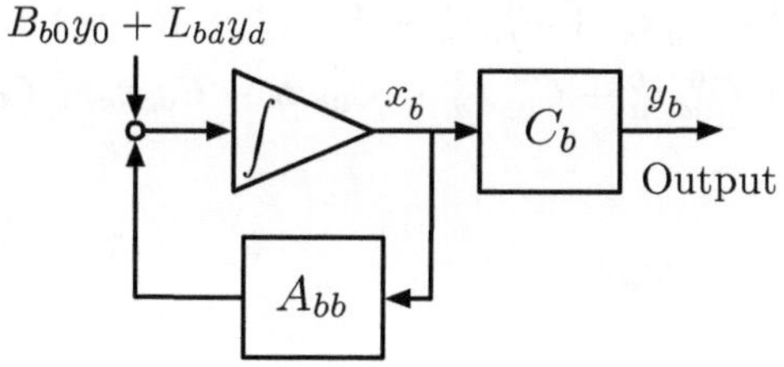

Figure 3.2: Non-right-invertible dynamics

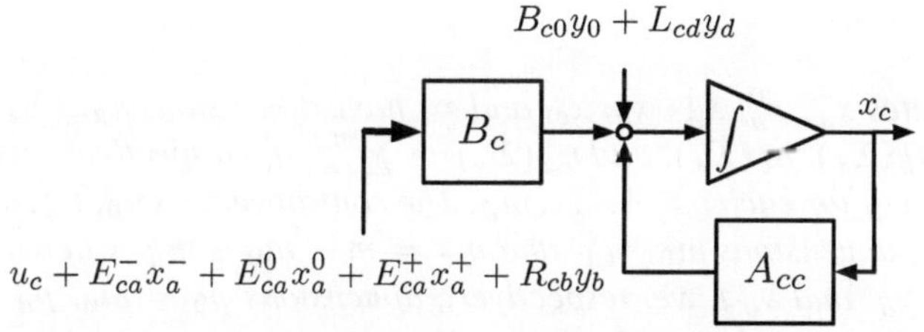

Figure 3.3: Non-left-invertible dynamics

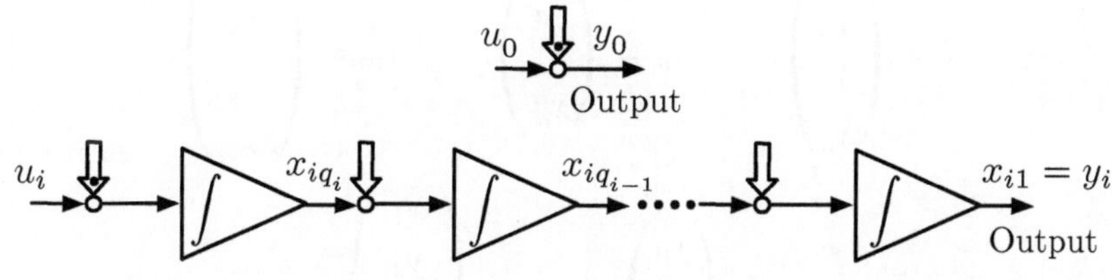

Figure 3.4: Infinite zero structure

(ii) Nonsingular state, output, and input transformations Γ_s, Γ_y, and Γ_u that take the given Σ_* into the SCB that displays explicitly both the finite and the infinite zero structures of Σ_* as well as its invertibility properties.

The SCB is described by the following set of equations:

$$\rho x_a^- = A_{aa}^- x_a^- + B_{a0}^- y_0 + L_{ad}^- y_d + L_{ab}^- y_b, \tag{3.4}$$

$$\rho x_a^0 = A_{aa}^0 x_a^0 + B_{a0}^0 y_0 + L_{ad}^0 y_d + L_{ab}^0 y_b, \tag{3.5}$$

$$\rho x_a^+ = A_{aa}^+ x_a^+ + B_{a0}^+ y_0 + L_{ad}^+ y_d + L_{ab}^+ y_b, \tag{3.6}$$

$$\rho x_b = A_{bb} x_b + B_{b0} y_0 + L_{bd} y_d, \tag{3.7}$$

$$\rho x_c = A_{cc} x_c + B_{c0} y_0 + L_{cd} y_d + B_c(E_{ca}^- x_a^- + E_{ca}^0 x_a^0 + E_{ca}^+ x_a^+ + R_{cb} y_b) + B_c u_c, \tag{3.8}$$

$$y_0 = C_{0a}^- x_a^- + C_{0a}^0 x_a^0 + C_{0a}^+ x_a^+ + C_{0b} x_b + C_{0c} x_c + C_{0d} x_d + u_0, \tag{3.9}$$

$$y_b = C_b x_b, \tag{3.10}$$

and for each $i = 1, \ldots, m_d$,

$$\rho x_i = A_{q_i} x_i + L_{i0} y_0 + L_{id} y_d + B_{q_i}\left(u_i + E_{ia} x_a + E_{ib} x_b + E_{ic} x_c + \sum_{j=1}^{m_d} E_{ij} x_j\right), \tag{3.11}$$

$$y_i = C_{q_i} x_i. \tag{3.12}$$

Here the states x_a^-, x_a^0, x_a^+, x_b, x_c, and x_d have dimensions $n_{a-}(\Sigma_*)$, $n_{a\circ}(\Sigma_*)$, $n_{a+}(\Sigma_*)$, $n_b(\Sigma_*)$, $n_c(\Sigma_*)$, and $n_d(\Sigma_*) = \sum_{i=1}^{m_d} q_i$, respectively, whereas x_i is of dimension q_i for each $i = 1, \ldots, m_d$. The control vectors u_0, u_d, and u_c have, respectively, dimensions m_0, m_d, and $m_c = m - m_0 - m_d$, whereas the output vectors y_0, y_d, and y_b have, respectively, dimensions $p_0 = m_0$, $p_d = m_d$, and $p_b = p - p_0 - p_d$. Also, we have

$$x = \Gamma_s \tilde{x}, \qquad y = \Gamma_y \tilde{y}, \qquad u = \Gamma_u \tilde{u},$$

$$\tilde{x} = \begin{pmatrix} x_a \\ x_b \\ x_c \\ x_d \end{pmatrix}, \quad x_a = \begin{pmatrix} x_a^- \\ x_a^0 \\ x_a^+ \end{pmatrix}, \quad x_d = \begin{pmatrix} x_1 \\ x_2 \\ \vdots \\ x_{m_d} \end{pmatrix},$$

$$\tilde{y} = \begin{pmatrix} y_0 \\ y_d \\ y_b \end{pmatrix}, \quad y_d = \begin{pmatrix} y_1 \\ y_2 \\ \vdots \\ y_{m_d} \end{pmatrix},$$

$$\widetilde{u} = \begin{pmatrix} u_0 \\ u_d \\ u_c \end{pmatrix}, \qquad u_d = \begin{pmatrix} u_1 \\ u_2 \\ \vdots \\ u_{m_d} \end{pmatrix}.$$

The x_i ($i = 1, \ldots, m_d$) together form x_d and, similarly, the y_i ($i = 1, \ldots, m_d$) together form y_d, and

$$y_d = C_d x_d, \quad \text{where} \quad C_d = \begin{pmatrix} C_{q_1} & 0 & \cdots & 0 \\ 0 & \ddots & \ddots & \vdots \\ \vdots & \ddots & \ddots & 0 \\ 0 & \cdots & 0 & C_{q_{m_d}} \end{pmatrix}. \tag{3.13}$$

The matrices A_{q_i}, B_{q_i}, and C_{q_i} have the following form:

$$A_{q_i} = \begin{pmatrix} 0 & I_{q_i-1} \\ 0 & 0 \end{pmatrix}, \quad B_{q_i} = \begin{pmatrix} 0 \\ 1 \end{pmatrix}, \quad C_{q_i} = \begin{pmatrix} 1 & 0 \end{pmatrix}. \tag{3.14}$$

(Obviously, for the case when $q_i = 1$, we have $A_{q_i} = 0$, $B_{q_i} = 1$, and $C_{q_i} = 1$.) Clearly, (A_{q_i}, B_{q_i}) and (C_{q_i}, A_{q_i}) form, respectively, controllable and observable pairs. This implies that all the states x_i are both controllable and observable. Assuming that the x_i are arranged such that $q_i \leqslant q_{i+1}$, the matrix L_{id} has the particular form

$$L_{id} = \begin{pmatrix} L_{i1} & L_{i2} & \cdots & L_{i\,i-1} & 0 & 0 & \cdots & 0 \end{pmatrix}.$$

The last row of each L_{id} is identically zero. Furthermore, the pair (A_{cc}, B_c) is controllable, and the pair (C_b, A_{bb}) is observable. Moreover, for continuous-time systems, we have $\lambda(A_{aa}^-) \in \mathbb{C}^-$, $\lambda(A_{aa}^0) \in \mathbb{C}^0$, $\lambda(A_{aa}^+) \in \mathbb{C}^+$, whereas for discrete-time systems, we have $\lambda(A_{aa}^-) \in \mathbb{C}^\ominus$, $\lambda(A_{aa}^0) \in \mathbb{C}^\circ$, and $\lambda(A_{aa}^+) \in \mathbb{C}^\oplus$.

3.2.3 A compact form

We can rewrite the SCB given by Theorem 3.1 in a more compact form as a system characterized by the quadruple $(\widetilde{A}, \widetilde{B}, \widetilde{C}, \widetilde{D})$:

$$\begin{cases} \rho\widetilde{x} = \widetilde{A}\widetilde{x} + \widetilde{B}\begin{pmatrix} y_0 \\ u_d \\ u_c \end{pmatrix} \\ \widetilde{y} = \widetilde{C}\widetilde{x} + \quad \widetilde{D}\widetilde{u}, \end{cases} \tag{3.15}$$

where

$$\widetilde{A} := \Gamma_s^{-1}(A - B_0C_0)\Gamma_s = \begin{pmatrix} A_{aa}^- & 0 & 0 & L_{ab}^-C_b & 0 & L_{ad}^-C_d \\ 0 & A_{aa}^0 & 0 & L_{ab}^0C_b & 0 & L_{ad}^0C_d \\ 0 & 0 & A_{aa}^+ & L_{ab}^+C_b & 0 & L_{ad}^+C_d \\ 0 & 0 & 0 & A_{bb} & 0 & L_{bd}C_d \\ B_cE_{ca}^- & B_cE_{ca}^0 & B_cE_{ca}^+ & B_cR_{cb}C_b & A_{cc} & L_{cd}C_d \\ B_dE_{da}^- & B_dE_{da}^0 & B_dE_{da}^+ & B_dE_{db} & B_dE_{dc} & A_{dd} \end{pmatrix}, \tag{3.16}$$

and where

$$A_{dd} = \begin{pmatrix} A_{q_1} & 0 & \cdots & 0 \\ 0 & \ddots & \ddots & \vdots \\ \vdots & \ddots & \ddots & 0 \\ 0 & \cdots & 0 & A_{q_{m_d}} \end{pmatrix} + L_{dd}C_d + B_dE_{dd}, \tag{3.17a}$$

$$B_d = \begin{pmatrix} B_{q_1} & 0 & \cdots & 0 \\ 0 & \ddots & \ddots & \vdots \\ \vdots & \ddots & \ddots & 0 \\ 0 & \cdots & 0 & B_{q_{m_d}} \end{pmatrix}, \tag{3.17b}$$

$$\widetilde{B} := \Gamma_s^{-1}\begin{pmatrix} B_0 & \widehat{B}_1 \end{pmatrix}\Gamma_u = \begin{pmatrix} B_{a0}^- & 0 & 0 \\ B_{a0}^0 & 0 & 0 \\ B_{a0}^+ & 0 & 0 \\ B_{b0} & 0 & 0 \\ B_{c0} & 0 & B_c \\ B_{d0} & B_d & 0 \end{pmatrix}, \tag{3.18}$$

$$\widetilde{C} := \Gamma_y^{-1}\begin{pmatrix} C_0 \\ \widehat{C}_1 \end{pmatrix}\Gamma_s = \begin{pmatrix} C_{0a}^- & C_{0a}^0 & C_{0a}^+ & C_{0b} & C_{0c} & C_{0d} \\ 0 & 0 & 0 & 0 & 0 & C_d \\ 0 & 0 & 0 & C_b & 0 & 0 \end{pmatrix}, \tag{3.19}$$

and

$$\widetilde{D} := \Gamma_y^{-1}D\Gamma_u = \begin{pmatrix} I_{m_0} & 0 & 0 \\ 0 & 0 & 0 \\ 0 & 0 & 0 \end{pmatrix}. \tag{3.20}$$

In the above equations, if one needs expanded expressions for the matrices E_{da}^-, E_{da}^0, E_{da}^+, E_{db}, E_{dc}, and E_{dd}, they can easily be obtained from (3.11). Note that we always have that (A_{cc}, B_c) and (A_{dd}, B_d) are controllable while (C_b, A_{bb}) is observable.

Admittedly, the SCB of Theorem 3.1 looks complicated with all its innate decompositions of state, output, and input variables. However, as will be evident throughout the book, the SCB of a linear system displays clearly the underlying structure of it. In fact, the proofs of several theorems, lemmas, and properties stated in later chapters will be hard to follow without having endured the intricacies of the SCB.

3.3 Properties of the SCB

The SCB is closely related to the canonical form of Morse [103], which is obtained through transformations of the state, input, and output spaces and the application of state feedback and output injection. A system in the canonical form of Morse consists of four decoupled subsystems that reflect essential geometric properties of the original system. The SCB form of a system largely reflects the same properties; however, the SCB is obtained through transformations of the state, input, and output spaces alone, without the application of state feedback and output injection. Thus, the SCB is merely a representation of the original system in a different coordinate basis, and it can therefore be used directly for design purposes.

Some properties of the SCB, which correspond directly to the properties of the canonical form of Morse, are first summarized as follows:

- The invariant zeros of the system Σ_* given in (3.1) are the eigenvalues of the matrix A_{aa}. Hence, the system is minimum-phase if and only if the eigenvalues of A_{aa} are located in the open left-half complex plane for continuous-time systems and within the unit circle for discrete-time systems.

- The system Σ_* given in (3.1) is right invertible if and only if the subsystem x_b is nonexistent.

- The system Σ_* given in (3.1) is left invertible if and only if the subsystem x_c is nonexistent.

- The system Σ_* given in (3.1) is invertible if and only if both the subsystem x_b and the subsystem x_c are nonexistent.

- The system Σ_* given in (3.1) has m_0 infinite zeros of order 0 and $i\overline{q}_i$ infinite zeros of order i, where $\overline{q}_i$ is the number integrator chains of length i in the x_d subsystem.

By studying the dynamics of the x_a subsystem and its connections to the rest of the system, one obtains a precise description of the invariant zero dynamics of the system and the classes of input signals that may be blocked by these zeros. The information thus obtained goes beyond what can be obtained through the notions of state and input *pseudo zero directions* (see [92, 122]).

The representation of the infinite zero structure through integrator chains in the x_d subsystem allows for the explicit construction of high-gain controllers and observers in a general multiple-input multiple-output setting (see, e.g., [131]). This removes unnecessary restrictions of square-invertibility and uniform relative degree that are found in much of the high-gain literature.

We next describe in detail the pertinent properties of the SCB as summarized above; each main property is stated in a subsection devoted to it. The properties discussed below are true for both continuous- and discrete-time systems. However, sometimes, for convenience of writing, we use the notations commonly used for continuous-time systems. The reader can easily decipher the corresponding notations for discrete-time systems. For clarity, whenever it is needed, we repeat our discussion for discrete-time systems.

3.3.1 Observability (detectability) and controllability (stabilizability)

In this subsection, we examine the issues related to observability, detectability, controllability, and stabilizability of a system via its SCB. Note that we simply use detectability and stabilizability, which for continuous-time systems refers to $\mathbb{C}^-$-detectability and $\mathbb{C}^-$-stabilizability, whereas for discrete-time systems, this refers to $\mathbb{C}^\ominus$-detectability and $\mathbb{C}^\ominus$-stabilizability.

We have the following property.

Property 3.2 *We note that (C_b, A_{bb}) and (C_{q_i}, A_{q_i}) form observable pairs. Unobservability can arise only in the variables x_a and x_c. In fact, the given system Σ_* is observable (detectable) if and only if $(C_{\rm obs}, A_{\rm obs})$ is observable (detectable), where*

$$A_{\rm obs} = \begin{pmatrix} A_{aa} & 0 \\ B_c E_{ca} & A_{cc} \end{pmatrix}, \qquad A_{aa} = \begin{pmatrix} A_{aa}^- & 0 & 0 \\ 0 & A_{aa}^0 & 0 \\ 0 & 0 & A_{aa}^+ \end{pmatrix},$$

$$C_{\rm obs} = \begin{pmatrix} C_{0a} & C_{0c} \\ B_d E_{da} & B_d E_{dc} \end{pmatrix}, \quad C_{0a} = \begin{pmatrix} C_{0a}^- & C_{0a}^0 & C_{0a}^+ \end{pmatrix},$$

$$E_{da} = \begin{pmatrix} E_{da}^- & E_{da}^0 & E_{da}^+ \end{pmatrix}, \quad E_{ca} = \begin{pmatrix} E_{ca}^- & E_{ca}^0 & E_{ca}^+ \end{pmatrix}.$$

Similarly, (A_{cc}, B_c) and (A_{q_i}, B_{q_i}) form controllable pairs. Basically, the variables x_a and x_b determine the controllability of the system. In fact, Σ_* is *controllable (stabilizable) if and only if* $(A_{\rm con}, B_{\rm con})$ *is controllable (stabilizable), where*

$$A_{\rm con} = \begin{pmatrix} A_{aa} & L_{ab}C_b \\ 0 & A_{bb} \end{pmatrix}, \quad B_{\rm con} = \begin{pmatrix} B_{a0} & L_{ad} \\ B_{b0} & L_{bd} \end{pmatrix},$$

$$B_{a0} = \begin{pmatrix} B_{a0}^- \\ B_{a0}^0 \\ B_{a0}^+ \end{pmatrix}, \quad L_{ab} = \begin{pmatrix} L_{ab}^- \\ L_{ab}^0 \\ L_{ab}^+ \end{pmatrix}, \quad L_{ad} = \begin{pmatrix} L_{ad}^- \\ L_{ad}^0 \\ L_{ad}^+ \end{pmatrix}.$$

3.3.2 Left and right invertibility

In this subsection, we examine the invertibility properties of Σ_* via its SCB. Let us first recall from [104] the definition of right and left invertibility.

Definition 3.3 *Consider a linear system* Σ_**:*

- *Let* u_1 *and* u_2 *be any inputs to the system* Σ_**, and let* y_1 *and* y_2 *be the corresponding outputs (for the same initial conditions). Then* Σ_* *is said to be* ***left invertible*** *if* $y_1(t) = y_2(t)$ *for all* $t \geq 0$ *implies that* $u_1(t) = u_2(t)$ *for all* $t \geq 0$.
- *The system* Σ_* *is said to be* ***right invertible*** *if, for any* $y_{ref}(t)$ *defined on* $[0, \infty)$*, an input* u *and a choice of* $x(0)$ *exist such that* $y(t) = y_{ref}(t)$ *for all* $t \in [0, \infty)$.
- *The system* Σ_* *is said to be invertible if the system is both left and right invertible.*

Remark 3.4 *One can easily deduce the following:*

(i) Σ_* *is right invertible if and only if its transfer function matrix is a surjective rational matrix.*

(ii) Σ_* *is right invertible if and only if the rank of* $P_{\Sigma_*}(s) = n + p$ *for all but finitely many* $s \in \mathbb{C}$*, where the polynomial matrix* $P_{\Sigma_*}(s)$ *is the Rosenbrock system matrix of* Σ_* *defined as*

$$P_{\Sigma_*}(s) := \begin{pmatrix} sI - A & -B \\ C & D \end{pmatrix}.$$

(iii) Σ_* *is left invertible if and only if its transfer function matrix is an injective rational matrix.*

(iv) Σ_* *is left invertible if and only if the rank of* $P_{\Sigma_*}(s) = n + m$ *for all but finitely many* $s \in \mathbb{C}$.

We have the following property connecting these properties to the SCB:

Property 3.5 *The given system Σ_* is right invertible if and only if x_b and hence y_b are nonexistent ($n_b = 0$, $p_b = 0$), left invertible if and only if x_c and hence u_c are nonexistent ($n_c = 0$, $m_c = 0$), and invertible if and only if both x_b and x_c are nonexistent. Moreover, Σ_* is called degenerate if it is neither left nor right invertible.*

3.3.3 Finite zero structure

In this subsection, we recall first the definition of invariant zeros of a system and their generalized associated right state and input zero direction chains, and then we discuss how SCB exhibits them in its structure.

The invariant zeros of a system Σ_* that is characterized by (A, B, C, D) are defined via the Smith canonical form of the Rosenbrock system matrix $P_{\Sigma_*}(s)$. Let us first briefly recall the Smith canonical form for any polynomial matrix $P(s) \in \mathbb{R}^{n\times m}[s]$. It is well known (see, e.g., [40]), that for any $P(s) \in \mathbb{R}^{n\times m}[s]$, there exist unimodular[1] matrices $U(s) \in \mathbb{R}^{n\times n}[s]$, $V(s) \in \mathbb{R}^{m\times m}[s]$ and $\Psi(s) \in \mathbb{R}^{n\times m}[s]$ with the latter of the form

$$\Psi(s) = \begin{pmatrix} \psi_1(s) & 0 & \cdots & \cdots & \cdots & 0 \\ 0 & \ddots & \ddots & & & \vdots \\ \vdots & \ddots & \psi_r(s) & \ddots & & \vdots \\ \vdots & & \ddots & 0 & \ddots & \vdots \\ \vdots & & & \ddots & \ddots & 0 \\ 0 & \cdots & \cdots & \cdots & 0 & 0 \end{pmatrix},$$

such that

$$P(s) = U(s)\Psi(s)V(s).$$

Here $\Psi(s)$ is called the Smith canonical form of $P(s)$ when the $\psi_i(s)$ are monic polynomials with the property that $\psi_i(s)$ divides $\psi_{i+1}(s)$ for $i = 1, \ldots, r-1$, and r is the normal rank of the matrix $P(s)$. The polynomials $\psi_i(s)$ are called the invariant factors of $P(s)$. Their product $\psi(s) = \psi_1(s)\psi_2(s)\cdots\psi_r(s)$ is called the zero polynomial of $P(s)$. Each invariant factor $\psi_i(s)$, $i = 1, 2, \ldots, r$, can be written as a product of linear factors

$$\psi_i(s) = (s-\lambda_{i1})^{\alpha_{i1}}(s-\lambda_{i2})^{\alpha_{i2}}\cdots(s-\lambda_{ik_i})^{\alpha_{ik_i}}, \quad i = 1, 2, \ldots, r,$$

[1] A polynomial matrix in $\mathbb{R}^{n\times m}[s]$ that is invertible with a polynomial inverse is called unimodular.

where $\lambda_{ik} \neq \lambda_{i\ell}$ $(k \neq \ell)$ are complex numbers and α_{ik} $(k, \ell \in \{1, \ldots, k_i\})$ are positive integers. Then the complete set of factors, $(s - \lambda_{ik})^{\alpha_{ik}}, k = 1, 2, \ldots, k_i$, and $i = 1, 2, \ldots, r$, is called the *elementary divisors* of the polynomial matrix $P(s)$.

Now we are ready to recall the definition of the invariant zeros [92] of Σ_*.

Definition 3.6 *The roots of the zero polynomial $\psi(s)$ of the (Rosenbrock) system matrix $P_{\Sigma_*}(s)$ are called the* ***invariant zeros*** *of Σ_*.*

Remark 3.7 *It is obvious from the above definition that an alternative way of defining an invariant zero of Σ_* is as follows: $\lambda \in \mathbb{C}$ is called an invariant zero of Σ_* if the rank of $P_{\Sigma_*}(\lambda)$ is strictly smaller than the normal rank of $P_{\Sigma_*}(s)$. Note that the normal rank is defined as the rank of a polynomial or rational matrix in all but finitely many $s \in \mathbb{C}$.*

The SCB of Theorem 3.1 shows explicitly the invariant zeros of the system. To be more specific, we have the following property.

Property 3.8 *Invariant zeros of Σ_* are the eigenvalues of A_{aa}. Moreover, for continuous-time systems, the invariant zeros in $\mathbb{C}^-$, $\mathbb{C}^0$, and $\mathbb{C}^+$ are the eigenvalues of A_{aa}^-, A_{aa}^0, and A_{aa}^+, respectively. Similarly, for discrete-time systems, the invariant zeros that are in $\mathbb{C}^{\ominus}$, $\mathbb{C}^{\circ}$, and $\mathbb{C}^{\oplus}$ are, respectively, the eigenvalues of A_{aa}^-, A_{aa}^0, and A_{aa}^+.*

For continuous-time systems, if all invariant zeros of a system Σ_* are in $\mathbb{C}^-$, then we say Σ_* is minimum phase; otherwise, Σ_* is said to be non-minimum phase. Those invariant zeros that are in $\mathbb{C}^-$ are called the stable invariant zeros. Also, those that are not in $\mathbb{C}^-$ are called the unstable invariant zeros. Analogously, for discrete-time systems, if all the invariant zeros of a system Σ_* are in $\mathbb{C}^{\ominus}$, then we say Σ_* is minimum phase; otherwise, Σ_* is said to be of non-minimum phase. Those invariant zeros that are in $\mathbb{C}^{\ominus}$ are called the stable invariant zeros. Also, those that are not in $\mathbb{C}^{\ominus}$ are called the unstable invariant zeros.

The following definition introduces the notions of algebraic and geometric multiplicities [57] of an invariant zero and its multiplicity structure.

Definition 3.9 *The* ***algebraic multiplicity*** *ρ_z of an invariant zero z is defined as the degree of the product of the elementary divisors of $P_{\Sigma_*}(s)$ corresponding to z. Likewise, the* ***geometric multiplicity*** *ν_z of an invariant zero z is defined as the number of the elementary divisors of $P_{\Sigma_*}(s)$ corresponding to z. Moreover, the invariant zero z is said to have a* ***semisimple structure*** *if its algebraic and* *geometric multiplicities are equal. Otherwise, it is referred to as an invariant zero with **nonsimple** structure.*

Given an invariant zero z, let $n_{z,i}$ be the degree of $(s - z)$ in the invariant factor $\Psi_i(s)$ of the Rosenbrock system matrix. Then the multiplicity structure of an invariant zero is defined as

$$S_z^* = \{n_{z,1}, n_{z,2}, \ldots, n_{z,\nu_z}\}. \tag{3.21}$$

If $n_{z,1} = n_{z,2} = \cdots = n_{z,\nu_z} = 1$, then we say z is a semisimple invariant zero of the given system Σ_*. It is called a simple invariant zero if $\nu_z = 1$ and $n_{z,1} = 1$.

We discuss next the invariant zeros together with their multiplicity structure of the system Σ_* as displayed by the SCB.

Property 3.10 *Consider the system Σ_* with its associated SCB. Then, z is an invariant zero of Σ_* with multiplicity structure S_z^* if and only if z is an eigenvalue of A_{aa} with multiplicity structure S_z^*.*

We need to recall next the notion of the right state and input zero directions and left state and input zero directions [57] associated with an invariant zero of a system. We focus first on the right state and input zero directions associated with an invariant zero for a left-invertible system Σ_* (left invertibility is discussed in Definition 3.3).

Definition 3.11 *Consider an invariant zero z with a semisimple structure of a left-invertible system Σ_*. Then the associated **right state and input zero directions**, $x_z \neq 0$ and u_z, of Σ_* are defined as those that satisfy the condition*

$$P_{\Sigma_*}(z)\begin{pmatrix} x_z \\ u_z \end{pmatrix} = \begin{pmatrix} zI - A & -B \\ C & D \end{pmatrix}\begin{pmatrix} x_z \\ u_z \end{pmatrix} = 0.$$

Some papers in the literature extend the above definition to non-left-invertible systems. This is incorrect as argued in [122].

Whenever an invariant zero has a nonsimple multiplicity structure, a concept of generalized right state and input zero direction[2] chain associated with that invariant zero exist. A proper definition of this is given in [122]. At this time, we would like to point out that although some researchers (e.g., see [92] and [157] among

[2]Generalized right state and input zero directions are also called pseudo-right state and input zero directions.

others) define the generalized right state and input zero direction chains as x_{R}^{j} and w_{R}^{j}, $j = 1, \cdots, \rho_Z - \sigma_Z$, satisfying

$$P_{\Sigma_*}(z)\begin{pmatrix} x_{\mathrm{R}}^{j} \\ w_{\mathrm{R}}^{j} \end{pmatrix} = -\begin{pmatrix} x_{\mathrm{R}}^{j-1} \\ 0 \end{pmatrix}, \quad j = 1, \ldots, \rho_Z - \sigma_Z. \tag{3.22}$$

However, this is also incorrect for non-left-invertible systems as argued in [122].

In what follows, we identify the right state and input zero direction chain associated with an invariant zero z of a general system whether it is left invertible or not, and whether z has a semisimple multiplicity structure or not. However, as discussed, in the absence of a precise definition that is not based on any SCB, we caution that Property 3.12 can be viewed either as a definition or as a property.

Let us start by defining the eigenvector chain associated with an eigenvalue of the matrix A_{aa}. Given an invariant zero z of the system Σ_* (i.e., the eigenvalue z of the matrix A_{aa}), for each $i = 1$ to ν_Z, a set of vectors in $\mathbb{R}^{n_a}$ that satisfies the following condition (3.23) is the eigenvector chain of A_{aa} associated with the invariant zero z:

$$A_{aa}x_{i,1}^{z} = zx_{i,1}^{z}, \quad \text{and} \quad (A_{aa} - zI_{n_a})x_{i,j+1}^{z} = x_{i,j}^{z}, \quad j = 1, \ldots, n_{z,i} - 1. \tag{3.23}$$

We have the following property regarding the right state and input zero direction chain associated with an invariant zero of a system.

Property 3.12

(i) For each $i \in \{1, \ldots, \nu_Z\}$, a set of vectors in $\mathbb{R}^n$ given in (3.24) *is the **generalized right state zero direction chain** of Σ_* associated with the invariant zero z*

$$x_{ij}^{z} = \Gamma_s \begin{pmatrix} x_{ij}^{z} \\ 0 \\ \vdots \\ 0 \end{pmatrix} \quad j \in \{1, \ldots, \nu_{z,i}\}. \tag{3.24}$$

Also, x_{i1}^{z} is the right state zero direction of Σ_* associated with z.

(ii) For each $i \in \{1, \ldots, \nu_Z\}$, a set of vectors w_{ij}^{z}, $j = 1$ to $n_{z,i}$, in $\mathbb{R}^m$ as given in (3.25) *is a **generalized right input zero direction chain** of Σ_* associated with the invariant zero z:*

$$w_{ij}^{z} = -\Gamma_y \begin{pmatrix} E_{da} \\ E_{dc} \end{pmatrix} x_{ij}^{z}, \tag{3.25}$$

where E_{da} is as defined in Property 3.2. Also, w_{i1}^{z} is said to be a right input zero direction of Σ_* associated with z.

The following property gives a dynamical interpretation of finite zero structure of a system. It is formulated for continuous-time systems. An analogous formulation is valid for discrete-time systems as well.

Property 3.13 ***(Dynamical interpretation of finite zero structure)*** *For a system Σ_* that is not necessarily left invertible, given that the initial condition, $x(0) = x^z_{i\alpha}$ for any $\alpha \le n_{z,i}$ and the input*

$$u = \sum_{j=1}^{\alpha} \frac{w^z_{ij} t^{\alpha-j} \exp(zt)}{(\alpha - j)!} \text{ for all } t \geqslant 0, \tag{3.26}$$

where z is any invariant zero of the system and $n_{z,i} \in S^*_z$, we have

$$y \equiv 0$$

and

$$x(t) = \sum_{j=1}^{\alpha} \frac{x^z_{ij} t^{\alpha-j} \exp(zt)}{(\alpha - j)!} \text{ for all } t \geqslant 0. \tag{3.27}$$

One can define the left state and input zero direction chain associated with an invariant zero of Σ_* as follows.

Definition 3.14 *The* ***left state and input zero direction chain*** *associated with each invariant zero of Σ_* are defined as the corresponding right state and input zero direction chain of the dual system Σ_{*d}.*

Remark 3.15 *We can connect the list of structural invariant indices due to Morse [103] to the SCB. In particular, the list $\mathfrak{I}_1$ of Morse is exactly equal to the invariant factors of A_{aa}.*

Next, we would like to recall the definition of the input decoupling zeros and the output decoupling zeros of a system.

Definition 3.16 *The zeros of the matrix pencil*

$$\begin{pmatrix} \lambda I - A & -B \end{pmatrix},$$

that is, the values of λ for which the above pencil loses rank, are called the ***input decoupling zeros*** *of Σ_*. They are also referred to as the input decoupling zeros of the pair (A, B).*

The zeros of the matrix pencil

$$\begin{pmatrix} \lambda I - A \\ C \end{pmatrix},$$

that is, the values of λ for which the above pencil loses rank, are called the output decoupling zeros of Σ_*. They are also referred to as the output decoupling zeros of the pair (C, A).

Remark 3.17 *In the literature, input decoupling zeros are also referred to as uncontrollable eigenvalues, whereas the output decoupling zeros are referred to as unobservable eigenvalues.*

Note that as we have done for invariant zeros, we can also associate a multiplicity structure with an input- or output-decoupling zero. The precise definition should be obvious from the above and hence is not included here.

The following property shows how input decoupling zeros and output decoupling zeros of Σ_* are displayed by the SCB.

Property 3.18 *Consider a system Σ_* with corresponding special coordinate basis. Define A_{con}, B_{con}, A_{obs}, and C_{obs} according to Property 3.2.*

(i) *The input decoupling zeros of Σ_* are the input decoupling eigenvalues of the pair $(A_{\mathrm{con}}, B_{\mathrm{con}})$. Also, input decoupling zeros of (A_{bb}, L_{bd}) are contained in the set of input decoupling zeros of Σ_*. Some of the input decoupling zeros of Σ_* could be contained among its invariant zeros.*

(ii) *The output decoupling zeros of Σ_* are the output decoupling eigenvalues of the pair $(C_{\mathrm{obs}}, A_{\mathrm{obs}})$. Also, output decoupling zeros of $(B_d E_{dc}, A_{cc})$ are contained in the set of output decoupling zeros of $(C_{\mathrm{obs}}, A_{\mathrm{obs}})$. Some of the output decoupling zeros of Σ_* could be contained among its invariant zeros.*

Remark 3.19 *As it is obvious from Property 3.18, it is crucial to realize that the input decoupling zeros and the output-decoupling zeros of a system need not be invariant zeros.*

3.3.4 *Infinite zero structure*

In this subsection, we examine the infinite zero structure of a system and how it is displayed by the SCB. Let us first recall some pertinent information from the literature. Infinite zeros are defined either in association with root-locus theory or as Smith–McMillan zeros of the transfer function at infinity.

Let us first view the infinite zeros from the viewpoint of root-locus theory. For this, consider a strictly proper system Σ_* subject to a high-gain feedback $u = \widetilde{\rho} y$ for a scalar gain $\widetilde{\rho}$. It can then be shown (see, e.g., Hung and MacFarlane [50]) that the unbounded closed-loop poles of the feedback system can be listed as

$$s_{j\ell}(\widetilde{\rho}) = \widetilde{\rho}^{1/\nu_j} \eta_{j\ell} + \zeta_{j\ell}(\widetilde{\rho}) \quad \text{for} \quad \ell = 1, \dots, \nu_j, \quad j = 1, \dots, m,$$

where

$$\lim_{\widetilde{\rho} \to \infty} \widetilde{\rho}^{-1/\nu_j} \zeta_{j\ell}(\widetilde{\rho}) = 0.$$

Here $s_{j\ell}$ is termed an infinite zero of order ν_j. Actually, until recently, the infinite zeros defined this way were considered to be fictitious objects introduced for the convenience of visualization.

Let us next consider the infinite zeros from the viewpoint of Smith–McMillan theory. To define the zero structure of the system Σ_* at infinity, one can use the familiar Smith–McMillan description of the zero structure at finite frequencies of the corresponding transfer matrix H, which need neither to be square nor strictly proper. A rational matrix $H(s)$ possesses an infinite zero of order k when $H(1/z)$ has a finite zero of precisely that order at $z = 0$; see [29, 115, 120, 188].

The number of zeros at infinity together with their orders indeed defines an infinite zero structure. It is important to note that for strictly proper transfer matrices, the above two definitions of the infinite zeros and their structure are consistent.

Owens [110] related the orders of the infinite zeros of the root loci of a square system with a nonsingular transfer function matrix to the $\mathfrak{C}^*$ structural invariant indices list $\mathfrak{I}_4$ of Morse [103]. This connection reveals that the *structure at infinity is in fact the topology of inherent integrations between the input and the output variables.* The SCB of Theorem 3.1 explicitly shows this topology of inherent integrations. The following property pinpoints this.

Property 3.20 *Let* $\overline{q}_0 = m_0$. *Let* $\overline{q}_j$ *be the integer such that exactly* $\overline{q}_j$ *elements of* $\{q_1, \dots, q_{m_d}\}$ *are equal to* j. *Also, let* σ *be the smallest integer such that* $\overline{q}_j = 0$ *for all* $j > \sigma$. *Then there are* $\overline{q}_0$ *infinite zeros of order* 0 *and* $j\overline{q}_j$ *number of infinite zeros of order* j, *for* $j = 1, \dots, \sigma$. *Moreover, the* $\mathfrak{C}^*$ *structural invariant indices list* $\mathfrak{I}_4$ *of Morse is given by*

$$\mathfrak{I}_4 = \{\overbrace{0, 0, \dots, 0}^{\overline{q}_0}, \overbrace{1, 1, \dots, 1}^{\overline{q}_1}, \dots, \overbrace{\sigma, \sigma, \dots, \sigma}^{\overline{q}_\sigma}\}.$$

Remark 3.21 *The state vector x_d in the SCB of a system is nonexistent if and only if the given system does not have infinite zeros of order greater than or equal to one.*

3.3.5 Geometric subspaces

Geometric theory is concerned with the study of subspaces of the state space with certain invariance properties, for example, A-invariant subspaces (which remain invariant under the unforced motion of the system), (A, B)-invariant subspaces (which can be made invariant by the proper application of state feedback), and (C, A) invariant subspaces (which can be made invariant by the proper application of output injection) (see, e.g., [184, 205]). Prominent examples of A-invariant subspaces are the controllable subspace (i.e., the image of the controllability matrix) and the unobservable subspace (the kernel of the observability matrix).

The development of geometric theory has in large part been motivated by the challenge of decoupling disturbance inputs from the outputs of a system, either exactly or approximately. Toward this end, a number of subspaces have been identified, which can be related to the different partitions in the SCB. Of particular importance in the context of control design for exact disturbance decoupling is the *weakly unobservable subspace*, which, by the proper selection of state feedback, can be made not to affect the outputs, and the *controllable weakly unobservable subspace*, which has the additional property that the dynamics restricted to this subspace is controllable. Of particular importance in the context of observer design for exact disturbance decoupling is the *strongly controllable subspace*, which, by the proper selection of output injection, is such that its quotient space can be rendered unaffected by the system inputs, and the *distributionally weakly unobservable subspace*, which has the additional property that the dynamics restricted to its quotient space is observable.

We proceed now to connect some classic subspaces from the geometric theory of linear systems to SCB. That is, in what follows, we show certain interconnections between the decomposition of the state space as done by the SCB and various invariant subspaces from the geometric theory. To do so, we recall the first two subspaces. The subspaces $\mathcal{V}_g(\Sigma_*)$ and $\mathcal{S}_g(\Sigma_*)$ are classic subspaces and are crucial elements of geometric theory of linear systems. Also, later on, we recall two more subspaces, $\mathcal{V}_\lambda(\Sigma_*)$ and $\mathcal{S}_\lambda(\Sigma_*)$, which are recently introduced in the context of H_∞ theory.

Definition 3.22 *Consider a linear system Σ_* characterized by the matrix quadruple (A, B, C, D). Then,*

(*i*) *The **$\mathbb{C}_g$-stabilizable weakly unobservable subspace $\mathcal{V}_g(\Sigma_*)$** is defined as the largest subspace of $\mathbb{R}^n$ for which a matrix F exists such that the subspace is $(A + BF)$ invariant, contained in $\ker(C + DF)$, whereas the eigenvalues of $(A + BF)|\mathcal{V}_g$ are contained in $\mathbb{C}_g \subseteq \mathbb{C}$.*

(ii) *The* $\mathbb{C}_g$*-detectable strongly controllable subspace* $\mathcal{S}_g(\Sigma_*)$ *is defined as the smallest subspace of* $\mathbb{R}^n$ *for which a matrix* K *exists such that the subspace is* $(A+KC)$ *invariant, contains* $\mathrm{im}(B+KD)$*, and is such that the eigenvalues of the map that is induced by* $(A+KC)$ *on the factor space* $\mathbb{R}^n/\mathcal{S}_g$ *are contained in* $\mathbb{C}_g \subseteq \mathbb{C}$.

For the case when $\mathbb{C}_g = \mathbb{C}$, $\mathcal{V}_g$ *and* $\mathcal{S}_g$ *are, respectively, denoted by* $\mathcal{V}^*$ *and* $\mathcal{S}^*$*; also, for the case when* $\mathbb{C}_g = \mathbb{C}^-$, $\mathcal{V}_g$ *and* $\mathcal{S}_g$ *are, respectively, denoted by* $\mathcal{V}^-$ *and* $\mathcal{S}^-$*, whereas for the case* $\mathbb{C}_g = \mathbb{C}^{-0}$, $\mathcal{V}_g$ *and* $\mathcal{S}_g$ *are, respectively, denoted by* $\mathcal{V}^{-0}$ *and* $\mathcal{S}^{-0}$*. Analogously, for the case when* $\mathbb{C}_g = \mathbb{C}^{\ominus}$, $\mathcal{V}_g$ *and* $\mathcal{S}_g$ *are, respectively, denoted by* $\mathcal{V}^{\ominus}$ *and* $\mathcal{S}^{\ominus}$*, whereas for the case* $\mathbb{C}_g = \mathbb{C}^{\otimes}$, $\mathcal{V}_g$ *and* $\mathcal{S}_g$ *are, respectively, denoted by* $\mathcal{V}^{\otimes}$ *and* $\mathcal{S}^{\otimes}$.

Finally, let a $\mathbb{C}_g$ *be chosen such that it has no common elements with the set of invariant zeros of* Σ_**. Then the corresponding* $\mathcal{V}_g(\Sigma_*)$*, which is always independent of the particular choice of such a* $\mathbb{C}_g$ *is referred to as the strongly controllable subspace* $\mathcal{R}^*(\Sigma_*)$.

Remark 3.23 *We note that* $\mathcal{V}_g(\Sigma_*)$ *and* $\mathcal{S}_g(\Sigma_*)$ *are dual in the sense that*

$$\mathcal{V}_g(\Sigma_{*d}) = \mathcal{S}_g(\Sigma_*)^{\perp},$$

where Σ_{*d} *is the dual system of* Σ_*.

Moreover, it can be shown that $\mathcal{R}^*(\Sigma_*)$ *equals* $\mathcal{V}^*(\Sigma_*) \cap \mathcal{S}^*(\Sigma_*)$.

Remark 3.24 *It is easy to observe that* $\mathcal{V}_g(\Sigma_*)$ *and* $\mathcal{S}_g(\Sigma_*)$ *are invariant under state feedback and output injection.*

Remark 3.25 *We should note that if* (A, B) *is* $\mathbb{C}_g$ *stabilizable, then for* $\mathcal{V}_g(\Sigma_*)$*, a matrix* F *exists that satisfies the conditions stated in Definition 3.22 and, moreover,* $A+BF$ *is* $\mathbb{C}_g$ *stable. An analogous comment can be made for* $\mathcal{S}_g(\Sigma_*)$.

Remark 3.26 *It is easily shown that the subspaces* $\mathcal{V}_g(\Sigma_*)$ *and* $\mathcal{S}_g(\Sigma_*)$ *satisfy the following:*

$$\begin{pmatrix} A \\ C \end{pmatrix} \mathcal{V}_g(\Sigma_*) \subseteq \big(\mathcal{V}_g(\Sigma_*) \oplus \{0\}\big) + \mathrm{im} \begin{pmatrix} B \\ D \end{pmatrix} \tag{3.28}$$

and

$$\ker \begin{pmatrix} C & D \end{pmatrix} \cap \begin{pmatrix} A & B \end{pmatrix} \big(\mathcal{S}_g(\Sigma_*) \oplus \mathbb{R}^m\big) \subseteq \mathcal{S}_g(\Sigma_*). \tag{3.29}$$

Remark 3.27 *In literature [101, 102], two geometric subspaces, namely, maximum uncontrollable subspace (MUCS) and maximum unobservable subspace (MUS) of Σ_*, are defined as follows: Consider the controllability matrix $Q_c(L)$ of Σ_* with output injection matrix L,*

$$Q_c(L) = \begin{pmatrix} B & (A + LC)B & \ldots & (A + LC)^{n-1}B \end{pmatrix}.$$

Let L be chosen such that the null space of $Q_c(L)'$ is of maximal order. Such a null space is called MUCS of Σ_*. Similarly, consider the observability matrix $Q_o(F)$ of Σ_* with state feedback gain F:

$$Q_o(F) = \begin{pmatrix} C \\ C(A + BF) \\ \vdots \\ C(A + BF)^{n-1} \end{pmatrix}.$$

Let F be chosen such that the null space of $Q_o(F)$ is of maximal order. Such a null space is called the MUS of Σ_*.

We observe that MUS equals $\mathcal{V}^*$ while MUCS is the orthogonal complement of $\mathcal{S}^*$.

We define below the notion of strong (also called perfect or ideal) controllability.

Definition 3.28 *A given system Σ_* is said to be* **strongly controllable** *if it is controllable under any arbitrary output injection matrix L, that is, the pair $(A + LC,\ B + LD)$ is controllable for every L of appropriate dimensions.*

Analogously, we define below the notion of strong (also called perfect or ideal) observability.

Definition 3.29 *A given system Σ_* is said to be* **strongly observable** *if it is observable under any arbitrary state feedback gain F, that is, the pair $(A+BF,\ C+DF)$ is observable for every F of appropriate dimensions.*

The above properties can be characterized in terms of the geometric subspaces defined above:

Theorem 3.30 *Consider a linear system Σ_* characterized by the matrix quadruple (A, B, C, D). Then,*

- *The system is strongly controllable if and only if* $\mathcal{V}(\Sigma_*) = \mathbb{R}^n$.
- *The system is strongly observable if and only if* $\mathcal{V}(\Sigma_*) = \{0\}$.

By now it is clear that the SCB decomposes the state space into several distinct parts. In fact, the state space $\mathcal{X}$ is decomposed as

$$\mathcal{X} = \mathcal{X}_a^- \oplus \mathcal{X}_a^0 \oplus \mathcal{X}_a^+ \oplus \mathcal{X}_b \oplus \mathcal{X}_c \oplus \mathcal{X}_d \,.$$

Here $\mathcal{X}_a^-$ is related to the stable invariant zeros, that is, to the eigenvalues of A_{aa}^-, which are the stable invariant zeros of Σ. Similarly, $\mathcal{X}_a^0 \oplus \mathcal{X}_a^+$ is related to the unstable invariant zeros of Σ. On the other hand, $\mathcal{X}_b$ is related to right invertibility, that is, the system is right invertible if and only if $\mathcal{X}_b = \{0\}$, whereas $\mathcal{X}_c$ is related to left invertibility, that is, the system is left invertible if and only if $\mathcal{X}_c = \{0\}$. The latter two equivalence are true provided that, as assumed before, $(B' \quad D')'$ and $(C \quad D)$ are of full rank. Finally, $\mathcal{X}_d$ is related to zeros of Σ at infinity.

We focus next on certain interrelationships between the SCB and some basic ingredients of the geometric control theory.

Property 3.31 *Consider a system Σ_* that has already been transformed in the SCB.*

(i) $\mathcal{X}_a^- \oplus \mathcal{X}_a^0 \oplus \mathcal{X}_a^+ \oplus \mathcal{X}_c$ *is equal to* $\mathcal{V}^*(\Sigma_*)$.

(ii) $\mathcal{X}_a^- \oplus \mathcal{X}_c$ *is equal to* $\mathcal{V}^-(\Sigma_*)$ *or* $\mathcal{V}^{\ominus}(\Sigma_*)$ *for continuous- and discrete-time systems, respectively.*

(iii) $\mathcal{X}_a^- \oplus \mathcal{X}_a^0 \oplus \mathcal{X}_c$ *is equal to* $\mathcal{V}^{-0}(\Sigma_*)$ *or* $\mathcal{V}^{\otimes}(\Sigma_*)$ *for continuous- and discrete-time systems, respectively.*

(iv) $\mathcal{X}_c \oplus \mathcal{X}_d$ *is equal to* $\mathcal{S}^*(\Sigma_*)$.

(v) $\mathcal{X}_a^+ \oplus \mathcal{X}_c \oplus \mathcal{X}_d$ *is equal to* $\mathcal{S}^{-0}(\Sigma_*)$ *or* $\mathcal{S}^{\otimes}(\Sigma_*)$ *for continuous- and discrete-time systems, respectively.*

(vi) $\mathcal{X}_a^0 \oplus \mathcal{X}_a^+ \oplus \mathcal{X}_c \oplus \mathcal{X}_d$ *is equal to* $\mathcal{S}^-(\Sigma_*)$ *or* $\mathcal{S}^{\ominus}(\Sigma_*)$ *for continuous- and discrete-time systems, respectively.*

(vii) $\mathcal{X}_c$ *is equal to* $\mathcal{R}^*(\Sigma_*)$.

Remark 3.32 *In view of Property 3.5, it is obvious that Σ_* is left invertible if and only if*

$$\mathcal{R}^*(\Sigma_*) = 0 \text{ and } \begin{pmatrix} B \\ D \end{pmatrix} \text{ is injective}$$

or equivalently

$$\mathcal{V}^*(\Sigma_*) \cap B \ker D = 0 \text{ and } \begin{pmatrix} B \\ D \end{pmatrix} \text{ is injective.}$$

Similarly, Σ_* is right invertible if and only if

$$\mathcal{V}^*(\Sigma_*) + \mathcal{S}^*(\Sigma_*) = \mathbb{R}^n \text{ and } \begin{pmatrix} C & D \end{pmatrix} \text{ is surjective}$$

or equivalently

$$\mathcal{S}^*(\Sigma_*) + C^{-1} \operatorname{im} D = \mathbb{R}^n \text{ and } \begin{pmatrix} C & D \end{pmatrix} \text{ is surjective.}$$

We recall now two more geometric subspaces $\mathcal{V}_\lambda(\Sigma_*)$ and $\mathcal{S}_\lambda(\Sigma_*)$ that were introduced in [144].

Definition 3.33 *For any $\lambda \in \mathbb{C}$, we define*

$$\mathcal{V}_\lambda(\Sigma_*) = \left\{ \zeta \in \mathbb{C}^n \;\middle|\; \exists\, \omega \in \mathbb{C}^m \; : \; 0 = \begin{pmatrix} A - \lambda I & B \\ C & D \end{pmatrix} \begin{pmatrix} \zeta \\ \omega \end{pmatrix} \right\} \tag{3.30}$$

and

$$\mathcal{S}_\lambda(\Sigma_*) = \left\{ \zeta \in \mathbb{C}^n \;\middle|\; \exists\, \omega \in \mathbb{C}^{n+m} \; : \; \begin{pmatrix} \zeta \\ 0 \end{pmatrix} = \begin{pmatrix} A - \lambda I & B \\ C & D \end{pmatrix} \omega \right\}. \tag{3.31}$$

We note that the geometric subspaces $\mathcal{V}_\lambda(\Sigma_*)$ and $\mathcal{S}_\lambda(\Sigma_*)$ are associated with the right state zero directions of Σ_* if λ is an invariant zero of Σ_*. These subspaces can also be displayed by the SCB of Σ_* as given in the following property.

Property 3.34 *We have*

$$\mathcal{V}_\lambda(\Sigma_*) = \operatorname{im} \left\{ \Gamma_s \begin{pmatrix} X_{a\lambda} & 0 \\ 0 & 0 \\ 0 & X_{c\lambda} \\ 0 & 0 \end{pmatrix} \right\}, \tag{3.32}$$

where $X_{a\lambda}$ is a matrix whose columns form a basis for the subspace,

$$\{\zeta_a \in \mathbb{C}^{n_a} \,|(\lambda I - A_{aa})\zeta_a = 0\} \tag{3.33}$$

and

$$X_{c\lambda} = (A_{cc} + B_c F_c - \lambda I)^{-1} B_c, \tag{3.34}$$

with F_c a matrix with suitable dimensions such that $A_{cc}+B_c F_c$ has no eigenvalue at λ. We note that the existence of such an F_c is guaranteed since (A_{cc}, B_c) is controllable.

Also, we have

$$\mathcal{S}_\lambda(\Sigma_*) = \operatorname{im}\left\{\Gamma_s \begin{pmatrix} \lambda I - A_{aa} & 0 & 0 & 0 \\ 0 & Y_{b\lambda} & 0 & 0 \\ 0 & 0 & I_{n_c} & 0 \\ 0 & 0 & 0 & I_{n_d} \end{pmatrix}\right\}, \tag{3.35}$$

where

$$\operatorname{im} Y_{b\lambda} = \ker C_b(A_{bb} + K_b C_b - \lambda I)^{-1}, \tag{3.36}$$

with K_b a matrix with suitable dimensions such that $A_{bb} + K_b C_b$ has no eigenvalue at λ. We note that the existence of such a K_b is guaranteed since (C_b, A_{bb}) is observable.

Clearly, if λ is not an eigenvalue of A_{aa}, then we have

$$\mathcal{V}_\lambda(\Sigma_*) \subseteq \mathcal{R}^*(\Sigma_*) \tag{3.37}$$

and

$$\mathcal{S}_\lambda(\Sigma_*) \supseteq \mathcal{V}^*(\Sigma_*) + \mathcal{S}^*(\Sigma_*). \tag{3.38}$$

Next, we would like to note that $\mathcal{V}_\lambda(\Sigma_*)$ and $\mathcal{S}_\lambda(\Sigma_*)$ are dual in the sense that $\mathcal{V}_\lambda(\Sigma_{*d}) = \mathcal{S}_\lambda(\Sigma_*)^\perp$. Also, $\mathcal{S}_\lambda(\Sigma_*) = \mathcal{V}_\lambda(\Sigma_{*d})^\perp$.

The subspaces $\mathcal{S}_\lambda(\Sigma_*)$ and $\mathcal{V}_\lambda(\Sigma_*)$ are the subspaces of $\mathbb{C}^n$ when we consider complex eigenvalues.

We have a precise characterization of this subspace. We factorize

$$X_a^0 = X_a^{01} \oplus X_a^{02}$$

such that X_a^{02} is defined by

$$X_a^{02} = \left\{v \in \mathbb{R}^{n_{ao}} \,\middle|\, \exists\lambda \in \mathbb{C}^0 \text{ such that } v' A_{aa}^0 = \lambda v'\right\}.$$

Then we obtain

$$\begin{aligned}\left(\mathcal{S}^*(\Sigma_*) + \mathcal{V}^{-0}(\Sigma_*)\right) \cap \{\cap_{\lambda\in\mathbb{C}^0} \mathcal{S}_\lambda(\Sigma_*)\} &= \mathcal{S}^*(\Sigma_*) + \mathcal{V}^-(\Sigma_*) + X_a^{01} \\ &= X_a^- \oplus X_a^{01} \oplus X_c \oplus X_d.\end{aligned}$$

Note that we will use later that the above structure results in a specific structure for A_{aa}^0,

$$A_{aa}^0 = \begin{pmatrix} A_{11} & A_{12} \\ 0 & A_{22} \end{pmatrix}, \quad B_{a0}^0 = \begin{pmatrix} B_1 \\ B_2 \end{pmatrix},$$

with

$$\left[\begin{pmatrix} A_{11} & A_{12} \\ 0 & A_{22} \end{pmatrix}, \begin{pmatrix} 0 \\ B_2 \end{pmatrix}\right]$$

controllable. The eigenvalues of A_{11} are contained in the set of eigenvalues of A_{22} with at least the same *geometric* multiplicity. Finally, A_{22} is diagonalizable.

3.3.6 Miscellaneous properties of the SCB

Several properties of linear multivariable time-invariant systems can trivially be visualized using the SCB of Theorem 3.1. We give below some of these properties.

Property 3.35 *The normal rank of Σ_* is equal to $m_d + m_0$. Moreover, it is easy to see the following:*

(*i*) normrank $P_{\Sigma_*}(s) = n + \text{normrank}[C(sI - A)^{-1}B + D]$.

(*ii*) *The normal rank of Σ_* is equal to p if and only if Σ_* is right invertible.*

(*iii*) *The normal rank of Σ_* is equal to m if and only if Σ_* is left invertible.*

Property 3.36

(*i*) *Σ_* is right invertible, and minimum phase $\Rightarrow$ Σ_* is stabilizable.*

(*ii*) *Σ_* is left invertible, and minimum phase $\Rightarrow$ Σ_* is detectable.*

(*iii*) *Σ_* is invertible, and minimum phase $\Rightarrow$ Σ_* is stabilizable and detectable.*

(*iv*) *Σ_* is right invertible, and the invariant zeros are disjoint from the eigenvalues (or unstable eigenvalues) of $A \Rightarrow \Sigma_*$ is controllable (stabilizable).*

(*v*) *Σ_* is left invertible, and its invariant zeros are disjoint from the eigenvalues (or unstable eigenvalues) of $A \Rightarrow \Sigma_*$ is observable (detectable).*

(*vi*) *Σ_* is invertible, and its invariant zeros are disjoint from the eigenvalues (or unstable eigenvalues) of $A \Rightarrow \Sigma_*$ is controllable and observable (stabilizable and detectable).*

(*vii*) *The feedthrough matrix D in Σ_* is injective $\Rightarrow$ Σ_* is left invertible and has no infinite zeros of order greater than or equal to one.*

(*viii*) *The feedthrough matrix D in Σ_* is surjective $\Rightarrow$ Σ_* is right invertible and has no infinite zeros of order greater than or equal to one.*

We connected in Sects. 3.3.3 and 3.3.4 the lists $\mathfrak{I}_1$ and $\mathfrak{I}_4$ of Morse [103] to SCB. The following property connects the lists $\mathfrak{I}_2$ and $\mathfrak{I}_3$ of Morse to SCB.

Property 3.37

The list $\mathfrak{I}_2$ of Morse = The controllability indices of the pair (A_{cc}, B_c).
The list $\mathfrak{I}_3$ of Morse = The observability indices of the pair (C_b, A_{bb}).

We have also the following remark.

Remark 3.38 *The integers*

$$n_{a-}(\Sigma_*), n_{a\circ}(\Sigma_*), n_{a+}(\Sigma_*), n_b(\Sigma_*), n_c(\Sigma_*), n_d(\Sigma_*), m_d, \quad \text{and } q_i \, (i = 1, \ldots, m_d)$$

are structurally invariant with respect to state feedback and output injection. Moreover, the integers $n_a(\Sigma_*) = n_{a-}(\Sigma_*) + n_{a\circ}(\Sigma_*) + n_{a+}(\Sigma_*)$, $n_b(\Sigma_*)$, $n_c(\Sigma_*)$, *and* $n_d(\Sigma_*)$ *are, respectively, equal to the number of elements in the lists* $\mathfrak{I}_1$, $\mathfrak{I}_2$, $\mathfrak{I}_3$, *and* $\mathfrak{I}_4$ *of Morse. For further details, one can refer to [103] and [129].*

3.3.7 Additional compact forms of the SCB

Finally, let us observe that, depending on some specific properties a given system satisfies, SCB can be written compactly in different formats. For convenience, we present below some such formats so that we can use them directly in later chapters as the need arises.

We will sometimes use the SCB in a more compact form where the special structure of x_d is not made explicit and where x_a^0 and x_a^+ are viewed together as x_a^{0+}. In this case, we get

$$\Gamma_s^{-1}(A - B_0 C_0)\Gamma_s = \begin{pmatrix} A_{aa}^- & 0 & L_{ab}^- C_b & 0 & L_{ad}^- C_d \\ 0 & A_{aa}^{0+} & L_{ab}^{0+} C_b & 0 & L_{ad}^{0+} C_d \\ 0 & 0 & A_{bb} & 0 & L_{bd} C_d \\ B_c E_{ca}^- & B_c E_{ca}^{0+} & B_c R_{cb} C_b & A_{cc} & L_{cd} C_d \\ B_d E_{da}^- & B_d E_{da}^{0+} & B_d E_{db} & B_d E_{dc} & A_{dd} \end{pmatrix}, \tag{3.39}$$

$$\Gamma_s^{-1}\begin{pmatrix} B_0 & \widehat{B}_1 \end{pmatrix}\Gamma_u = \begin{pmatrix} B_{a0}^- & 0 & 0 \\ B_{a0}^{0+} & 0 & 0 \\ B_{b0} & 0 & 0 \\ B_{c0} & 0 & B_c \\ B_{d0} & B_d & 0 \end{pmatrix}, \tag{3.40}$$

$$\Gamma_y^{-1}\begin{pmatrix} C_0 \\ \widehat{C}_1 \end{pmatrix}\Gamma_s = \begin{pmatrix} C_{0a}^- & C_{0a}^{0+} & C_{0b} & C_{0c} & C_{0d} \\ 0 & 0 & 0 & 0 & C_d \\ 0 & 0 & C_b & 0 & 0 \end{pmatrix}, \tag{3.41}$$

and

$$\Gamma_y^{-1}\begin{pmatrix} I_{m_0} & 0 \\ 0 & 0 \end{pmatrix}\Gamma_u = \begin{pmatrix} I_{m_0} & 0 & 0 \\ 0 & 0 & 0 \\ 0 & 0 & 0 \end{pmatrix}. \tag{3.42}$$

As discussed, the eigenvalues of A_{aa}^{-} and A_{aa}^{0+} are the invariant zeros of the given system Σ_*. Moreover, for continuous-time systems, the eigenvalues of A_{aa}^{-} are in the open left-half complex plane, whereas the eigenvalues of A_{aa}^{0+} are in the closed right-half complex plane. Similarly, for discrete-time systems, the eigenvalues of A_{aa}^{-} are within the unit circle, whereas the eigenvalues of A_{aa}^{0+} are on the unit circle or outside the unit circle.

If the given system Σ_* is left invertible, then the decomposition in (3.39)–(3.42) simplifies because x_c is no longer present (see Property 3.5), and we obtain the following structure:

$$\Gamma_s^{-1}(A - B_0 C_0)\Gamma_s = \begin{pmatrix} A_{aa}^{-} & 0 & L_{ab}^{-}C_b & L_{ad}^{-}C_d \\ 0 & A_{aa}^{0+} & L_{ab}^{0+}C_b & L_{ad}^{0+}C_d \\ 0 & 0 & A_{bb} & L_{bd}C_d \\ B_d E_{da}^{-} & B_d E_{da}^{0+} & B_d E_{db} & A_{dd} \end{pmatrix}, \tag{3.43}$$

$$\Gamma_s^{-1}\begin{pmatrix} B_0 & \widehat{B}_1 \end{pmatrix}\Gamma_u = \begin{pmatrix} B_{a0}^{-} & 0 \\ B_{a0}^{0+} & 0 \\ B_{b0} & 0 \\ B_{d0} & B_d \end{pmatrix}, \tag{3.44}$$

$$\Gamma_y^{-1}\begin{pmatrix} C_0 \\ \widehat{C}_1 \end{pmatrix}\Gamma_s = \begin{pmatrix} C_{0a}^{-} & C_{0a}^{0+} & C_{0b} & C_{0d} \\ 0 & 0 & 0 & C_d \\ 0 & 0 & C_b & 0 \end{pmatrix}, \tag{3.45}$$

and

$$\Gamma_y^{-1}\begin{pmatrix} I_{m_0} & 0 \\ 0 & 0 \end{pmatrix}\Gamma_u = \begin{pmatrix} I_{m_0} & 0 \\ 0 & 0 \\ 0 & 0 \end{pmatrix}. \tag{3.46}$$

Similarly, we will sometimes use the SCB in another more compact form where this time x_a^{-} and x_a^{0} are viewed together as x_a^{-0}. In this case, we get

$$\Gamma_s^{-1}(A - B_0 C_0)\Gamma_s = \begin{pmatrix} A_{aa}^{-0} & 0 & L_{ab}^{-0}C_b & 0 & L_{ad}^{-0}C_d \\ 0 & A_{aa}^{+} & L_{ab}^{+}C_b & 0 & L_{ad}^{+}C_d \\ 0 & 0 & A_{bb} & 0 & L_{bd}C_d \\ B_c E_{ca}^{-0} & B_c E_{ca}^{+} & B_c R_{cb} C_b & A_{cc} & L_{cd}C_d \\ B_d E_{da}^{-0} & B_d E_{da}^{+} & B_d E_{db} & B_d E_{dc} & A_{dd} \end{pmatrix}, \tag{3.47}$$

$$\Gamma_s^{-1}\begin{pmatrix} B_0 & \widehat{B}_1 \end{pmatrix}\Gamma_u = \begin{pmatrix} B_{a0}^{-0} & 0 & 0 \\ B_{a0}^{+} & 0 & 0 \\ B_{b0} & 0 & 0 \\ B_{c0} & 0 & B_c \\ B_{d0} & B_d & 0 \end{pmatrix}, \tag{3.48}$$

$$\Gamma_y^{-1}\begin{pmatrix} C_0 \\ \widehat{C}_1 \end{pmatrix}\Gamma_s = \begin{pmatrix} C_{0a}^{-0} & C_{0a}^{+} & C_{0b} & C_{0c} & C_{0d} \\ 0 & 0 & 0 & 0 & C_d \\ 0 & 0 & C_b & 0 & 0 \end{pmatrix}, \tag{3.49}$$

and

$$\Gamma_y^{-1}\begin{pmatrix} I_{m_0} & 0 \\ 0 & 0 \end{pmatrix}\Gamma_u = \begin{pmatrix} I_{m_0} & 0 & 0 \\ 0 & 0 & 0 \\ 0 & 0 & 0 \end{pmatrix}. \tag{3.50}$$

Once again, as discussed, the eigenvalues of A_{aa}^{-0} and A_{aa}^{0} are the invariant zeros of the given system Σ_*. Moreover, for continuous-time systems, the eigenvalues of A_{aa}^{-0} are in the closed left-half complex plane, whereas the eigenvalues of A_{aa}^{+} are in the open right-half complex plane. Similarly, for discrete-time systems, the eigenvalues of A_{aa}^{-0} are on the unit circle or within the unit circle, whereas the eigenvalues of A_{aa}^{+} are outside the unit circle.

If the given system Σ_* is left invertible, then the decomposition in (3.47)–(3.50) simplifies because x_c is no longer present (see Property 3.5), and we obtain the following structure:

$$\Gamma_s^{-1}(A - B_0C_0)\Gamma_s = \begin{pmatrix} A_{aa}^{-0} & 0 & L_{ab}^{-0}C_b & L_{ad}^{-0}C_d \\ 0 & A_{aa}^{+} & L_{ab}^{+}C_b & L_{ad}^{+}C_d \\ 0 & 0 & A_{bb} & L_{bd}C_d \\ B_dE_{da}^{-0} & B_dE_{da}^{+} & B_dE_{db} & A_{dd} \end{pmatrix}, \tag{3.51}$$

$$\Gamma_s^{-1}\begin{pmatrix} B_0 & \widehat{B}_1 \end{pmatrix}\Gamma_u = \begin{pmatrix} B_{a0}^{-0} & 0 \\ B_{a0}^{+} & 0 \\ B_{b0} & 0 \\ B_{d0} & B_d \end{pmatrix}, \tag{3.52}$$

$$\Gamma_y^{-1}\begin{pmatrix} C_0 \\ \widehat{C}_1 \end{pmatrix}\Gamma_s = \begin{pmatrix} C_{0a}^{-0} & C_{0a}^{+} & C_{0b} & C_{0d} \\ 0 & 0 & 0 & C_d \\ 0 & 0 & C_b & 0 \end{pmatrix}, \tag{3.53}$$

and

$$\Gamma_y^{-1}\begin{pmatrix} I_{m_0} & 0 \\ 0 & 0 \end{pmatrix}\Gamma_u = \begin{pmatrix} I_{m_0} & 0 \\ 0 & 0 \\ 0 & 0 \end{pmatrix}. \tag{3.54}$$

3.4 Software packages to generate SCB

While the SCB provides a fine-grained decomposition of multivariable linear time-invariant systems, transforming an arbitrary system to the SCB is a complex operation. A constructive algorithm for strictly proper systems is provided in

[139] based on a modified Silverman algorithm [152]. This algorithm is lengthy and involved and includes repeated rank operations and construction of nonunique transformations to divide the state space. Thus, the algorithm can realistically be executed by hand only for very simple systems.

To automate the process of finding transformations to the SCB, numerical algorithms were developed and implemented as part of the *Linear Systems Toolkit* for *Matlab*. The first among such work is by [80], and the resulting software was commercialized in early 1990s [79]. The Toolkit was revised subsequently twice each time with further improvements; the first revision was reported in [19, 28] and the second revision in [91]. Although these numerical algorithms are invaluable in practical applications, engineers also often operate on systems where some or all of the elements of the system matrices have a symbolic representation. There are obvious and definite advantages in being able to obtain symbolic representation of SCB without having to insert numerical values in place of symbolic variables. Furthermore, the numerical algorithms are based on inherently inaccurate floating-point operations that make them prone to numerical errors. Ideally, if the elements of the system matrices are represented by symbols and exact fractions, one would be able to obtain an exact SCB representation of that system, also represented by symbols and exact fractions. To address these issues, Grip and Saberi [42] developed recently a procedure for symbolic transformation of multivariable linear time-invariant systems to the SCB, using the commercial mathematics software suite "*Maple*." The procedure is based on the modified Silverman algorithm from [139], with a modification to achieve a later version of the SCB that includes an additional structural property (see, e.g., [122]) and an extension to SCB for non-strictly proper systems [132]. Symbolic transformations are useful complement to available numerical tools [19, 28]. Also, symbolic transformation to the SCB makes it possible to work directly on the SCB representation of a system without first inserting numerical values, thereby removing an obstacle to more widespread use of SCB such as squaring down of nonsquare systems and asymptotic timescale assignment and other topics where the SCB has previously been applied and to some other topics where the SCB has not yet been applied.

The "*Maple*" software code and its development are described in Sect. 3.A.

3.A "Maple" implementation

This Appendix in its entirety is the work of Grip and Saberi [42]. The intent here is to obtain, by utilizing the commercial mathematics software suite "*Maple*", various transformation matrices and dimensions of variables involved to transform a given multivariable linear time-invariant system to its SCB form as given in Theorem 3.1. Before we present the "*Maple*" code in detail, we describe first the concepts behind the algorithms used.

There are some notational changes here from those given in Sect. 3.2. In order to be transparent of these changes, we rewrite (3.3) that describes the given multivariable linear time-invariant system in the form given below where, for clarity, only the notation of continuous-time system is given:

$$\begin{aligned} \dot{\hat{x}} &= \hat{A}\hat{x} + \hat{B}\hat{u}, \\ \hat{y} &= \hat{C}\hat{x} + \hat{D}\hat{u}, \end{aligned} \tag{3.55}$$

where as usual $\hat{x} \in \mathbb{R}^n$ is the state, $\hat{u} \in \mathbb{R}^m$ is the input, and $\hat{y} \in \mathbb{R}^p$ is the output. We assume without loss of generality that the matrices $[\hat{B}', \hat{D}']'$ and $[\hat{C}, \hat{D}]$ are of full rank.

For simplicity in the non-strictly proper case (i.e., $\hat{D} \neq 0$), we assume here that the input and output are partitioned as

$$\hat{u} = \begin{pmatrix} u_0 \\ \hat{u}_1 \end{pmatrix} \text{ and } \hat{y} = \begin{pmatrix} y_0 \\ \hat{y}_1 \end{pmatrix},$$

where u_0 and y_0 are of dimension m_0, and furthermore that $\hat{D}$ has the form $\hat{D} = \operatorname{diag}(I_{m_0}, 0)$ as in (3.3). Then we may write

$$\hat{y} = \begin{pmatrix} y_0 \\ \hat{y}_1 \end{pmatrix} = \begin{pmatrix} C_0\hat{x} + u_0 \\ \hat{C}_1\hat{x} \end{pmatrix}, \tag{3.56}$$

where C_0 consists of the upper m_0 rows of $\hat{C}$, and $\hat{C}_1$ consists of the remaining rows of $\hat{C}$. The special form in (3.56) means that the input–output map is partitioned to separate the direct-feedthrough part from the rest: the output y_0 is directly affected by u_0, and the remainder of the output $\hat{y}_1$ is not directly affected by any input. Note that by substituting $u_0 = y_0 - C_0\hat{x}$, we can write the system (3.55) in the alternative form:

$$\begin{aligned} \dot{\hat{x}} &= (\hat{A} - B_0C_0)\hat{x} + \hat{B}\begin{pmatrix} y_0 \\ \hat{u}_1 \end{pmatrix}, \\ \hat{y} &= \hat{C}\hat{x} + \hat{D}\hat{u}, \end{aligned} \tag{3.57}$$

where B_0 consists of the left m_0 columns of $\hat{B}$. In the strictly proper case, B_0 and C_0 are nonexistent.

By nonsingular transformation of the state, output, and input, the system (3.55) can be transformed to the SCB. We use the symbols x, y, and u to denote the state, output, and input of the system transformed to SCB form (note that there is a change in notation; in the main text we used $\tilde{x}$, $\tilde{y}$, and $\tilde{u}$ instead). The transformations between the original system (3.55) and the SCB are called Γ_1, Γ_2, and Γ_3 (note again that there is a change in notation; in the main text we used Γ_s, Γ_y, and Γ_u instead), and we write $\hat{x} = \Gamma_1 x$, $\hat{y} = \Gamma_2 y$, and $\hat{u} = \Gamma_3 u$.

The state x is partitioned as $x = (x_a{}', x_b{}', x_c{}', x_d{}')'$, where each component represents a particular subsystem of SCB. The output is partitioned as $y = (y_0{}', y_d{}', y_b{}')'$, where y_0 is the original output y_0 from (3.55), y_d is the output from the x_d subsystem, and y_b is the output from the x_b subsystem. The input is partitioned as $u = (u_0', u_d', u_c')'$, where u_0 is the original input u_0 from (3.55), u_d is the input to the x_d subsystem, and u_c is the input to the x_c subsystem. Because u_0 appears first in both $\widehat{u}$ and u, Γ_3 is of the form $\operatorname{diag}(I_{m_0}, \overline{\Gamma}_3)$, for some nonsingular $\overline{\Gamma}_3$.

3.A.1 Pretransformation of non-strictly proper systems

We assumed in the above equations that the input and output vectors $\widehat{u}$ and $\widehat{y}$ have a special partitioning that separates the direct-feedthrough part from the rest, as shown in (3.56). A strictly proper system already has this form, but given a general non-strictly proper system, a pretransformation may have to be applied to put the system in the required form. Suppose that we initially have a system with input $\widetilde{u}$, output $\widetilde{y}$, input matrix $\widetilde{B}$, and output matrices $\widetilde{C}$ and $\widetilde{D}$. Then there are nonsingular transformations U and Y such that $\widetilde{u} = U\widehat{u}$ and $\widetilde{y} = Y\widehat{y}$, where $\widehat{u}$ and $\widehat{y}$ have the structure required in (3.56). The dimension m_0 of u_0 and y_0 is the rank of $\widetilde{D}$. The matrices $\widehat{B}$, $\widehat{C}$, and $\widehat{D}$ are obtained from $\widetilde{B}$, $\widetilde{C}$, and $\widetilde{D}$ by $\widehat{B} = \widetilde{B}U$, $\widehat{C} = Y^{-1}\widetilde{C}$, and $\widehat{D} = Y^{-1}\widetilde{D}U$ (we caution that the matrix $\widetilde{U}$ in (3.2) is denoted here as Y^{-1} and, similarly, the matrix $\widetilde{V}$ in (3.2) is denoted here as U). Our "*Maple*" procedure, in addition to returning the matrices A, B, C, and D of the SCB system (in the main text, the matrices of the SCB system are denoted by $\widetilde{A}$, $\widetilde{B}$, $\widetilde{C}$, and $\widetilde{D}$), the transformations Γ_1, Γ_2, and Γ_3 to transform (3.55) to SCB form, and the dimension of each subsystem, returns the transformations U and Y as well to take a general non-strictly proper system to the form required in (3.55).

3.A.2 "Maple" procedure

Our "*Maple*" procedure is invoked as follows:

A, B, C, D, G1, G2, G3, U, Y, dim := **scb**(Ai, Bi, Ci, Di)

The inputs Ai, Bi, Ci, and Di are system matrices describing a general multivariable linear time-invariant system. The outputs A, B, C, and D are the system matrices describing the corresponding SCB system. The outputs G1, G2, and G3 are the transformation matrices Γ_1, Γ_2, and Γ_3 between the system (3.55) and the SCB. The outputs U and Y are the pretransformations that must be applied to the system to put it in the form required of (3.55), as described earlier. Finally, the output dim is a list of four integers representing the dimensions of the x_a, x_b, x_c, and x_d subsystems, in that order.

The modified Silverman algorithm for transformation to the SCB is much too long to be presented here. For the details of the algorithm, we refer to [139]. In the following, we shall present a broad outline of the steps of the algorithm, and discuss issues that require particular attention in a symbolic implementation. Much of the algorithm consists of tedious but straightforward manipulation of matrices, which is not explicitly discussed here.

Throughout the algorithm, we identify a large number of variables that are linear transformations of the original state. We keep track of these by storing the matrices that transform the original state to the new variables. For example, the temporary variable y_{i0}, given by the expression $y_{i0} = C_i\hat{x}$, is represented internally by a "Matrix" data structure containing C_i. The procedure is not written to perform well on floating-point data. For this reason, all floating-point elements of the matrices passed to the procedure are converted to exact fractions before any other operations are performed, using "convert" function of "*Maple*". In many cases, we need to store a whole list of matrices, representing variables obtained during successive iterations of a particular part of the algorithm. To do this, we use the "*Maple*" data structures "Vector" and "Matrix" which can be used to store vectors or matrices whose elements are "Matrix" data structures.

Strictly proper case

We first consider the strictly proper case. The algorithm for this case is implemented as "**scbSP**." The first part of this algorithm identifies the two subsystems that directly influence the outputs, namely, the x_b and x_d subsystems, through a series of steps that are repeated until the outputs are exhausted. The algorithm works by identifying transformed input and output spaces such that each input channel is directly connected to one output channel by a specific number of inherent integrations.

Let the strictly proper system passed to the "**scbSP**" procedure be represented by the state equations $\dot{\hat{x}} = \hat{A}\hat{x} + \hat{B}\hat{u}$, $\hat{y} = \hat{C}\hat{x}$. In the first iteration, we start with the output $y_{10} = \hat{C}\hat{x}$ and determine whether its derivative $\dot{y}_{10} = \hat{C}\hat{A}\hat{x} + \hat{C}\hat{B}\hat{u}$ depends on any part of the input $\hat{u}$. If so, we use a transformation $\overline{S}_1$ to separate out a linear combination of outputs and inputs that are separated by one integration in a linearly independent manner. This will create an integrator chain of length one, as part of the x_d subsystem. A transformed part of the output derivative that is not directly influenced by the input is denoted $\widetilde{C}_1\hat{x}$ and is processed further. We use a transformation $\overline{\phi}_1$ to separate out any part of $\widetilde{C}_1\hat{x}$ that is linearly dependent on y_{10}. This will create states that are part of the x_b subsystem. After the linearly dependent components are separated out, the remaining part of the output derivative is given the name y_{20}. In the next iteration, we process y_{20} in the same fashion as y_{10} to identify integrator chains of length two and possibly further additions to the x_b subsystem. The algorithm continues in this fashion until the outputs are exhausted.

Constructing transformation matrices: strictly proper case

When implementing these steps in "*Maple*", the main part of each iteration consists of constructing transformation matrices $\overline{S}_i$ and $\overline{\phi}_i$. In particular, we are faced with the following problem at step i: given a matrix C_i of dimension $p_i \times n$ and a matrix $\overline{D}_{i-1}$ of dimension $\overline{q}_{i-1} \times m$ of maximal rank $\overline{q}_{i-1}$, let $\overline{q}_i$ be the rank of

$[\overline{D}'_{i-1}, (C_i\widehat{B})']'$, and let $\mathfrak{q}_i = \overline{\mathfrak{q}}_i - \overline{\mathfrak{q}}_{i-1}$. Find a nonsingular matrix $\overline{S}_i$ such that

$$\overline{S}_i \begin{pmatrix} \overline{D}_{i-1} \\ C_i\widehat{B} \end{pmatrix} = \begin{pmatrix} \overline{D}_{i-1} \\ \widehat{D}_i \\ 0 \end{pmatrix}, \quad \overline{S}_i = \begin{pmatrix} I_{\overline{\mathfrak{q}}_i-1} & 0 \\ S_{ia} & S_i \end{pmatrix}, \quad S_{ia} = \begin{pmatrix} 0 \\ S_{ib} \end{pmatrix}, \quad S_i = \begin{pmatrix} S_{i1} \\ S_{i2} \end{pmatrix},$$

where $\widehat{D}_i$ is a $\mathfrak{q}_i \times m$ matrix of maximal rank and where S_{i1}, S_{i2}, and S_{ib} are of dimensions $\mathfrak{q}_i \times p_i$, $(p_i - \mathfrak{q}_i) \times p_i$, and $(p_i - \mathfrak{q}_i) \times \overline{\mathfrak{q}}_{i-1}$. The meaning of the various dimensions are not important in this context. In general, $\overline{S}_i$ is not unique.

The rank of the matrix $[\overline{D}'_{i-1}, (C_i\widehat{B})']'$ can be obtained with the "Rank" function in the LinearAlgebra package. To construct the matrix $\overline{S}_i$, the first observation we make is that, since $S_{ib}\overline{D}_{i-1} + S_{i2}C_i\widehat{B} = 0$, the rows of the matrix $[S_{ib}, S_{i2}]$ must belong to the left null space of $[\overline{D}'_{i-1}, (C_i\widehat{B})']'$. If $[\overline{D}'_{i-1}, (C_i\widehat{B})']'$ has full rank $\overline{\mathfrak{q}}_{i-1} + p_i$, then S_{ib} and S_{i2} are empty matrices, and we may select $S_{ia} = 0$ and $S_{i1} = I_{p_i}$. Otherwise, we can obtain a set of linearly independent basis vectors for the left null space of $[\overline{D}'_{i-1}, (C_i\widehat{B})']'$ or, equivalently, for the right null space of its transpose, using the "NullSpace" function of the LinearAlgebra package. The transpose of the basis vectors can then be stacked to form the matrix $[S_{ib}, S_{i2}]$, which can be split up to form S_{ib} and S_{i2}. However, the null space basis is not unique and, moreover, the order in which the basis vectors are returned by "*Maple*" is not consistent. This may cause our procedure to produce different results on different executions with the same matrices, which is undesirable. To avoid this, we first stack the transpose of the basis vectors and then transform the resulting matrix to the unique reduced-row echelon form by using the "ReducedRowEchelonForm" function of the LinearAlgebra package. Since the transformation involves a finite number of row operations, the rows of the matrix in reduced-row echelon form remain in the left null space.

Since $\overline{S}_i$ should be a nonsingular matrix, the submatrix S_i must be nonsingular. This requires that S_{i2} has maximal rank, which is confirmed as follows: if any of the rows of S_{i2} are linearly dependent, a linear combination of rows in $[S_{ib}, S_{i2}]$ can be constructed to create a row vector v such that $v[\overline{D}'_{i-1}, (C_i\widehat{B})']' = 0$, where the rightmost p_i columns of v are zero. However, since the rows of $\overline{D}_{i-1}$ are linearly independent, this implies that $v = 0$, which in turn implies that $[S_{ib}, S_{i2}]$ must have linearly dependent rows. Since this is not the case, S_{i2} must have maximal rank.

We continue by constructing the matrix S_{i1}. Nonsingularity of S_i requires that the rows of S_{i1} must be linearly independent of the rows of S_{i2}. One way to produce S_{i1} is to choose its rows to be orthogonal to the rows of S_{i2}, which can be achieved by using a basis for the right null space of S_{i2}. However, since the matrix $\overline{S}_i$ will be used to transform the state of the original system, it is generally desirable for this matrix to have the simplest possible structure. This helps avoid unnecessary changes to the original states, and thus it generally produces more appealing solutions. We therefore construct S_{i1} by the following procedure: we start by initializing S_{i1} as the identity matrix of dimension $p_i \times p_i$. We then create a reduced-row echelon form of S_{i2} and iterate backward over the rows of this matrix. For each row, we search along the columns from the left until we reach the leading 1 on that row. We then delete the row in S_{i1} corresponding to the column with the leading 1. This ensures that $S_i = [S_{i1}', S_{i2}']'$ is nonsingular, with S_{i1} consisting of zeros except for a single element equal to 1 on each row. The construction of $\overline{S}_i$ is now easily completed.

At each step, we must also construct a nonsingular matrix $\overline{\phi}_i$. The problem of finding this matrix is analogous to the problem of finding $\overline{S}_i$, and we therefore use the same procedure. Finding the transformations $\overline{S}_i$ and $\overline{\phi}_i$ constitutes the most important part of finding the states x_b and x_d. After x_b and x_d are identified, finding the output transformation Γ_2 is straightforward, based on [139]. We also find an input transformation Γ_3' based on [139] and write $\hat{u} = \Gamma_3'[u_d', \overline{u}_c']'$, where $\overline{u}_c$ is a temporary input. Unlike [139], we shall apply a further transformation of $\overline{u}_c$ to achieve an input u_c that is matched with the influence from x_a on the right-hand side of the x_c equation.

Constructing the x_a and x_c states – strictly proper case

After finding the transformations from the original states to the x_b and x_d states, the next step is to find a transformation to a temporary state vector x_s that will be further decomposed into the states x_a and x_c. The requirements on x_s are that it must be linearly independent of the already identified states x_b and x_d, so that x_s, x_b, and x_d together span the entire state space, and that its derivative $\dot{x}_s$ must only depend on x_s itself, plus y_b, y_d, and $\overline{u}_c$, because those are the only quantities allowed in the derivatives of x_a and x_c in the strictly proper case.

Suppose that $(x_b', x_d')' = \Gamma_{bd}\hat{x}$. The procedure for finding x_s is to start with a temporary state vector $x_s^0 = \Gamma_s^0\hat{x}$ that is linearly independent of x_b and x_d. Hence, we select Γ_s^0 such that $[(\Gamma_s^0)', \Gamma_{bd}']'$ is nonsingular. To do so in our "*Maple*" procedure, we use the same technique as for finding S_{i1} based on S_{i2} as discussed earlier.

The derivative of x_s^0, written in terms of the states x_s^0, x_b, and x_d, and the inputs u_c and u_d, can be written as

$$\dot{x}_s^0 = A^0 \begin{pmatrix} x_s^0 \\ x_b \\ x_d \end{pmatrix} + B^0 \begin{pmatrix} u_d \\ \overline{u}_c \end{pmatrix} = A_s^0 x_s^0 + A_b^0 x_b + A_d^0 x_d + B_d^0 u_d + B_c^0 \overline{u}_c,$$

for some matrices $A^0 = [A_s^0, A_b^0, A_d^0]$ and $B^0 = [B_d^0, B_c^0]$. In our "*Maple*" procedure, we can easily calculate $A^0 = \Gamma_s^0 \hat{A}([(\Gamma_s^0)', \Gamma_{bd}']')^{-1}$ and $B^0 = \Gamma_s^0 \hat{B} \Gamma_3'$ and then extract the matrices A_s^0, A_b^0, A_c^0, B_d^0, and B_c^0. To do so, we use the "MatrixInverse" function of the LinearAlgebra package.

To conform with the SCB, we need to modify x_s^0 to eliminate the input u_d in $\dot{x}_s^0$. To eliminate u_d, we create a temporary state vector $x_{d0} = \Gamma_{d0}\hat{x}$, consisting of the lowermost level of each integrator chain in the x_d subsystem (i.e., the point where the input enters the integrator chain). According to Theorem 3.1, we then have $\dot{x}_{d0} = u_d + A_{d0}[x_s^{0'}, x_b', x_d']'$, for some matrix A_{d0}. Therefore, by defining a new temporary state $x_s^1 = x_s^0 - B_d^0 x_{d0}$, we have $\dot{x}_s^1 =$

$(A^0 - B_d^0 A_{d0})[x_s^{0\prime}, x_b', x_d'] + B_c^0 \bar{u}_c$. Hence, the derivative of the new temporary state vector x_s^1 is independent of u_d, bringing us one step closer to obtaining x_s. The elimination procedure is continued in a similar fashion, as described in [139], until we obtain a state x_s such that $\dot{x}_s$ depends only on x_s, y_b, y_d, and $\bar{u}_c$.

The final step is to decompose x_s into two subsystems, x_a and x_c, and to transform the input $\bar{u}_c$ into u_c in such a way that x_a is unaffected by u_c and x_c is controllable from u_c. Furthermore, the influence of x_a on x_c should be matched with u_c, as seen in Theorem 3.1. If $\bar{u}_c$ is nonexistent, then we simply set $x_a = x_s$. If $\bar{u}_c$ does exist, we proceed by first finding the derivative $\dot{x}_s = A_{ss}x_s + A_{sb}x_b + A_{sd}x_d + B_{sc}u_c$ for some matrices A_{ss}, A_{sb}, A_{sd}, and B_{sc}. We then obtain the proper transformations by calling **"scbSP"** recursively on the transposed system with system matrix A_{ss}', output matrix B_{sc}', and an empty input matrix. This recursive call returns a system consisting only of an x_a and an x_b subsystem. It is easily confirmed that, when transposed back again, this system has the desired structure. We therefore let $[x_a', x_c']' = \Gamma_1^{\star\prime} x_s$ and $u_c = \Gamma_2^{\star\prime}\bar{u}_c$, where $\Gamma_1^\star$ and $\Gamma_2^\star$ are the state and output transformations returned by the recursive call.

Non-strictly proper case

To handle the non-strictly proper case, the first step is to find the pretransformation matrices U and Y, described in Sect. 3.A.1. Suppose that the matrices passed to the procedure **"scb"** are $\widetilde{A}$, $\widetilde{B}$, $\widetilde{C}$, and $\widetilde{D}$. We need to find nonsingular U and Y such that, according to Sect. 3.A.1, $\widehat{B} = \widetilde{B}U$, $\widehat{C} = Y^{-1}\widetilde{C}$, and $\widehat{D} = Y^{-1}\widetilde{D}U$, where $\widehat{D}$ is of the form $\mathrm{diag}(I_{m_0}, 0)$. The rank m_0 of $\widetilde{D}$ is found using the "Rank" function.

Let $Y^{-1} = [Y_1', Y_2']'$, where Y_1 has m_0 rows. Then we have the equations

$$Y^{-1}\widetilde{D}U = \begin{pmatrix} Y_1\widetilde{D}U \\ Y_2\widetilde{D}U \end{pmatrix} = \begin{pmatrix} I_{m_0} & 0 \\ 0 & 0 \end{pmatrix}.$$

To solve these equations, we choose the rows of Y_2 from the left null space of $\widetilde{D}$, using the functions "NullSpace" and "ReducedRowEchelonForm" as before, and we select Y_1 such that $[Y_1', Y_2']'$ is nonsingular, using the same procedure as for finding S_{i1} given S_{i2} as discussed earlier for strictly proper case. This leaves us to solve the equation $Y_1\widetilde{D}U = [I_{m_0}, 0]$ with respect to some nonsingular U. Let $U^{-1} = [U_1', U_2']'$ such that U_1 has m_0 rows. We select $U_1 = Y_1\widetilde{D}$, and we select U_2 such that $[U_1', U_2']'$ is nonsingular, by the same procedure as before. It is then straightforward to confirm that $Y_1\widetilde{D}U = [I_{m_0}, 0]$. We can now calculate the matrices $\widehat{B}$, $\widehat{C}$, and $\widehat{D}$ that conform with the required structure of (3.55).

Let B_0 consist of the left m_0 columns of $\widehat{B}$, and let $\widehat{B}_1$ consist of the remaining columns of $\widehat{B}$. Similar to (3.57), we can write the system equations (3.55) as

$$\begin{aligned} \dot{\widehat{x}} &= (\widehat{A} - B_0C_0)\widehat{x} + B_0y_0 + \widehat{B}_1\widehat{u}_1, \\ y_0 &= C_0\widehat{x} + u_0, \\ \widehat{y}_1 &= \widehat{C}_1\widehat{x}. \end{aligned} \tag{3.58}$$

Suppose we obtain the SCB form of the strictly proper system described by the matrices $(\widehat{A} - B_0 C_0)$, $\widehat{B}_1$, and $\widehat{C}_1$ by invoking the procedure "**scbSP**" and suppose the transformation matrices returned for this system are $\bar{\Gamma}_1$, $\bar{\Gamma}_2$, and $\bar{\Gamma}_3$. Substituting $\hat{x} = \bar{\Gamma}_1 x$, $\hat{y}_1 = \bar{\Gamma}_2[y_d', y_b']'$, and $\hat{u}_1 = \bar{\Gamma}_3[u_d', u_c']'$ in (3.58) yields

$$\dot{x} = \bar{\Gamma}_1^{-1}(\widehat{A} - B_0 C_0)\bar{\Gamma}_1 x + \bar{\Gamma}_1^{-1} B_0 y_0 + \bar{\Gamma}_1^{-1}\widehat{B}_1\bar{\Gamma}_3 \begin{pmatrix} u_d \\ u_c \end{pmatrix},$$

$$y_0 = C_0 \bar{\Gamma}_1 x + u_0,$$

$$\begin{pmatrix} y_d \\ y_b \end{pmatrix} = \bar{\Gamma}_2^{-1}\widehat{C}_1\bar{\Gamma}_1 x.$$

It is easily confirmed that this system conforms to the SCB by defining $A = \bar{\Gamma}_1^{-1}(\widehat{A} - B_0 C_0)\bar{\Gamma}_1$, $B = \bar{\Gamma}_1^{-1}[B_0, \widehat{B}_1\bar{\Gamma}_3]$, $C = [C_0', (\bar{\Gamma}_2^{-1}\widehat{C}_1)']'\bar{\Gamma}_1$, and $D = \mathrm{diag}(I_{m_0}, 0)$. Defining the transformations for the non-strictly proper system as $\Gamma_1 = \bar{\Gamma}_1$, $\Gamma_2 = \mathrm{diag}(I_{m_0}, \bar{\Gamma}_2)$, and $\Gamma_3 = \mathrm{diag}(I_{m_0}, \bar{\Gamma}_3)$, we obtain $A = \Gamma_1^{-1}(\widehat{A} - B_0 C_0)\Gamma_1$, $B = \Gamma_1^{-1}\widehat{B}\Gamma_3$, $C = \Gamma_2^{-1}\widehat{C}\Gamma_1$, and $D = \Gamma_2^{-1}\widehat{D}\Gamma_3$, which are the proper expressions relating the matrices $\widehat{A}$, $\widehat{B}$, $\widehat{C}$, and $\widehat{D}$ to the SCB matrices.

3.A.3 Examples

In this section, we apply the SCB decomposition procedure to several example systems.

Example: linear single-track model

A widely used model for the lateral dynamics of a car is the linear single-track model (see, e.g., [60]). For a car on a horizontal surface, this model is described by the equations

$$\dot{v}_y = \frac{1}{m}(F_\mathrm{f} + F_\mathrm{r}) - r v_x,$$
$$\dot{r} = \frac{1}{J}(l_\mathrm{f} F_\mathrm{f} - l_\mathrm{r} F_\mathrm{r}),$$

where v_y is the lateral velocity at the center of gravity, r is the yaw rate (angular rate around the vertical axis), m is the mass, J is the moment of inertia around the vertical axis through the car's center of gravity, l_f and l_r are the longitudinal distances from the center of gravity to the front and rear axles, and F_f and F_r are the lateral road-tire friction forces on the front and rear axles. The longitudinal velocity v_x is assumed to be positive and to vary slowly enough compared to the lateral dynamics that it can be considered a constant. The friction forces can be modeled by the equations

$$\dot{F}_\mathrm{f} = \frac{c_\mathrm{f}}{T_\mathrm{r}}(\delta_\mathrm{f} - \frac{v_y}{v_x} - l_\mathrm{f}\frac{r}{v_x}) - \frac{1}{T_\mathrm{r}}F_\mathrm{f},$$
$$\dot{F}_\mathrm{r} = \frac{c_\mathrm{r}}{T_\mathrm{r}}(-\frac{v_y}{v_x} + l_\mathrm{r}\frac{r}{v_x}) - \frac{1}{T_\mathrm{r}}F_\mathrm{r},$$

where δ_f is the front-axle steering angle, c_f and c_r are the front- and rear-axle cornering stiffnesses, and T_r is a speed-dependent tire relaxation constant (see, e.g., [112]). In modern cars with electronic stability control, the main measurements that describe the lateral dynamics are the yaw rate r and the lateral acceleration $a_y = \frac{1}{m}(F_f + F_r)$. Considering δ_f as the input, the system is described by

$$\widehat{A} = \begin{pmatrix} 0 & -v_x & \frac{1}{m} & \frac{1}{m} \\ 0 & 0 & \frac{l_f}{J} & -\frac{l_r}{J} \\ -\frac{c_f}{T_r v_x} & -\frac{l_f c_f}{T_r v_x} & -\frac{1}{T_r} & 0 \\ -\frac{c_r}{T_r v_x} & \frac{l_r c_r}{T_r v_x} & 0 & -\frac{1}{T_r} \end{pmatrix}, \qquad \widehat{B} = \begin{pmatrix} 0 \\ 0 \\ \frac{c_f}{T_r} \\ 0 \end{pmatrix},$$

$$\widehat{C} = \begin{pmatrix} 0 & 1 & 0 & 0 \\ 0 & 0 & \frac{1}{m} & \frac{1}{m} \end{pmatrix}, \qquad \widehat{D} = \begin{pmatrix} 0 \\ 0 \end{pmatrix}.$$

If we pass these matrices to our "*Maple*" procedure, we obtain SCB system matrices

$$A = \begin{pmatrix} -\frac{1}{T_r} & 1 & 0 & \frac{T_r l_f m}{c_r(l_f+l_r)} \\ -\frac{l_r c_r(l_f+l_r)}{v_x T_r J} & 0 & 1 & \frac{l_f m}{c_r(l_f+l_r)} \\ -\frac{c_r(l_f+l_r)}{T_r J} & 0 & 0 & \frac{1}{v_x} \\ \frac{c_r(l_r c_r - l_f c_f)(l_f+l_r)}{m T_r^2 v_x J} & 0 & -\frac{c_f+c_r}{m T_r} & -\frac{1}{T_r} \end{pmatrix}, \qquad B = \begin{pmatrix} 0 \\ 0 \\ 0 \\ 1 \end{pmatrix},$$

$$C = \begin{pmatrix} 0 & 0 & 0 & 1 \\ 1 & 0 & 0 & 0 \end{pmatrix}, \qquad D = \begin{pmatrix} 0 \\ 0 \end{pmatrix},$$

and the transformations

$$\Gamma_1 = \begin{pmatrix} 0 & 0 & v_x & 0 \\ \frac{c_r(l_f+l_r)}{T_r J} & 0 & 0 & 0 \\ -\frac{c_r}{T_r^2} & \frac{c_r}{T_r} & 0 & m \\ \frac{c_r}{T_r^2} & -\frac{c_r}{T_r} & 0 & 0 \end{pmatrix}, \quad \Gamma_2 = \begin{pmatrix} 0 & \frac{c_r(l_f+c_r)}{T_r J} \\ 1 & 0 \end{pmatrix}, \quad \Gamma_3 = \frac{m T_r}{c_f}.$$

The dimension list "dim" returned by the procedure is 0, 3, 0, 1, meaning that the first three states belong to the x_b subsystem and the last state is an integrator chain of length 1 belonging to the x_d subsystem. Inspection of the SCB system immediately reveals that the system is observable, since both the x_b and x_d subsystems are always observable. The system is left invertible, since the state x_c is nonexistent, meaning that the steering angle can be identified from the outputs if the initial conditions are known. The system is not right invertible, since it has an x_b subsystem, reflecting the obvious fact that the yaw rate and lateral acceleration cannot be independently controlled from a single steering angle. There exists no state feedback that keeps the outputs identically zero, since the system has no zero dynamics subsystem x_a.

If we add rear-axle steering by augmenting the $\widehat{B}$ matrix with an additional column $[0, 0, 0, \frac{c_r}{T_r}]'$, the "*Maple*" procedure returns the SCB system matrices

$$A = \begin{pmatrix} 0 & 1 & -v_x & 0 \\ -\frac{c_f+c_r}{mT_r v_x} & -\frac{1}{T_r} & \frac{l_r c_r - l_f c_f}{mT_r v_x} & 0 \\ 0 & 0 & 0 & 1 \\ \frac{l_r c_r - l_f c_f}{JT_r v_x} & 0 & \frac{l_f^2 c_f + l_r^2 c_r}{JT_r v_x} & -\frac{1}{T_r} \end{pmatrix}, \qquad B = \begin{pmatrix} 0 & 0 \\ 1 & 0 \\ 0 & 0 \\ 0 & 1 \end{pmatrix},$$

$$C = \begin{pmatrix} 0 & 1 & 0 & 0 \\ 0 & 0 & 1 & 0 \end{pmatrix}, \qquad D = \begin{pmatrix} 0 & 0 \\ 0 & 0 \end{pmatrix},$$

and the transformations

$$\Gamma_1 = \begin{pmatrix} 1 & 0 & 0 & 0 \\ 0 & 0 & 1 & 0 \\ 0 & \frac{l_r m}{l_f+l_r} & 0 & \frac{J}{l_f+l_h} \\ 0 & \frac{l_f m}{l_f+l_r} & 0 & -\frac{J}{l_f+l_h} \end{pmatrix}, \quad \Gamma_2 = \begin{pmatrix} 0 & 1 \\ 1 & 0 \end{pmatrix}, \quad \Gamma_3 = \begin{pmatrix} \frac{l_r T_r m}{c_f(l_f+l_r)} & \frac{T_r J}{c_f(l_f+l_r)} \\ \frac{l_f T_r m}{c_r(l_f+l_r)} & -\frac{T_r J}{c_r(l_f+l_r)} \end{pmatrix},$$

with dimensions $1, 0, 0, 3$. This means that the first state of the system belongs to the zero dynamics x_a, and the remaining three states belong to the x_d subsystem. The x_d subsystem consists of two integrator chains: one of dimension one and one of dimension two. We conclude that the system is invertible due to the lack of x_b and x_c subsystems. The A_{aa} matrix is identically 0, meaning that the system has a zero at the origin. Hence, the relationship between the steering angle inputs and the yaw rate and lateral acceleration outputs is non-minimum-phase.

Referring back to our discussion of geometry theory, we see that the weakly unobservable subspace is spanned by the vector $[1, 0, 0, 0]'$. Transformed back to the original coordinate basis, this corresponds to the state v_y. We therefore know that a hypothetical disturbance occurring in $\dot{v}_y$ can be decoupled from the outputs a_y and r by state feedback (and the SCB representation tells us exactly how to do it). However, we also know that the resulting subsystem would not be asymptotically stable, since the non-minimum phase zero would become a pole of the closed-loop system.

Example – DC motor with friction

According to [32], a DC motor process can be described by the equations

$$\dot{\Omega} = \omega,$$
$$J\dot{\omega} = u - F,$$

where Ω is the shaft angular position, ω is the angular rate, u is the DC motor torque, F is a friction torque, and $J = 0.0023\,\text{kg m}^2$ is the motor and load inertia. The friction torque can be modeled by the dynamic LuGre friction model,

$$F = \sigma_0 z + \sigma_1 \dot{z} + \alpha_2 \omega,$$
$$\dot{z} = \omega - \frac{\sigma_0 z|\omega|}{\zeta(\omega)},$$

where $\zeta(\omega) = \alpha_0 + \alpha_1 \exp(-(\omega/\omega_0)^2)$. Numerical values for the friction parameters are $\sigma_0 = 260.0\,\mathrm{Nm/rad}$, $\sigma_1 = 0.6\,\mathrm{Nm\,s/rad}$, $\alpha_0 = 0.28\,\mathrm{Nm}$, $\alpha_1 = 0.05\,\mathrm{Nm}$, $\alpha_2 = 0.176\,\mathrm{Nm\,s/rad}$, and $\omega_0 = 0.01\,\mathrm{rad/s}$. The system can be viewed as consisting of a linear part with a nonlinear perturbation $\sigma_0 z|\omega|/\zeta(\omega)$. Assuming that only the shaft position Ω is measured, a nonlinear observer can be designed for this system by using the timescale assignment techniques from [131]. To do so, it is necessary to find the SCB form of the system, with the nonlinear perturbation $\sigma_0 z|\omega|/\zeta(\omega)$ considered as the sole input. The original system with the nonlinear perturbation as the input is described by the matrices

$$\widehat{A} = \begin{pmatrix} 0 & 1 & 0 \\ 0 & -\frac{1}{J}(\alpha_2 + \sigma_1) & -\frac{1}{J}\sigma_0 \\ 0 & 1 & 0 \end{pmatrix}, \qquad \widehat{B} = \begin{pmatrix} 0 \\ \frac{1}{J}\sigma_1 \\ -1 \end{pmatrix}, \tag{3.59a}$$

$$\widehat{C} = \begin{pmatrix} 1 & 0 & 0 \end{pmatrix}, \qquad \widehat{D} = 0. \tag{3.59b}$$

Inserting numerical values and using the Linear Systems Toolkit [91] yields the SCB matrices

$$A \approx \begin{pmatrix} -433.3 & -592.7 & 0 \\ 0 & 0 & 1 \\ -1.1\cdot 10^5 & -1.5\cdot 10^5 & 95.9 \end{pmatrix}, \qquad B = \begin{pmatrix} 0 \\ 0 \\ 1 \end{pmatrix},$$

$$C = \begin{pmatrix} 0 & 1 & 0 \end{pmatrix}, \qquad D = 0,$$

where the first state belongs to the zero dynamics subsystem x_a, and the remaining two states consist of an integrator chain of length two, in the x_d subsystem. As suggested by the large elements in the system matrices, the problem is poorly conditioned, and we find that we require very large gains to stabilize the system. Using our "*Maple*" procedure, we obtain the SCB matrices

$$A = \begin{pmatrix} -\frac{\sigma_0}{\sigma_1} & -\frac{\sigma_0(\sigma_0 J - \sigma_1\alpha_2)}{\sigma_1^3} & 0 \\ 0 & 0 & 1 \\ -\frac{\sigma_0}{J} & -\frac{\sigma_0(\sigma_0 J - \sigma_1\alpha_2)}{J\sigma_1^2} & \frac{\sigma_0 J - \sigma_1\alpha_2 - \sigma_1^2}{J\sigma_1} \end{pmatrix}, \qquad B = \begin{pmatrix} 0 \\ 0 \\ 1 \end{pmatrix},$$

$$C = \begin{pmatrix} 0 & 1 & 0 \end{pmatrix}, \qquad D = 0.$$

This reveals that a source of the conditioning problem is powers of the small parameter σ_1 appearing in the denominators, even though it does not appear in any denominators in (3.59). In particular, we see that σ_1 acts as a small regular perturbation that results in singularly perturbed zero dynamics, which happens when a regular perturbation reduces a system's relative degree [143]. Setting $\sigma_1 = 0$ results in a dramatically different structure, with the SCB consisting of a single integrator chain of length three, represented by the SCB matrices

$$A = \begin{pmatrix} 0 & 1 & 0 \\ 0 & 0 & 1 \\ 0 & -\frac{\sigma_0}{J} & -\frac{\alpha_2}{J} \end{pmatrix}, \qquad B = \begin{pmatrix} 0 \\ 0 \\ 1 \end{pmatrix}$$

$$C = \begin{pmatrix} 1 & 0 & 0 \end{pmatrix}, \qquad D = 0.$$

Proceeding with the observer gain selection based on this system, we obtain good results without using high gains.

Example – tenth-order system Our last example is a strictly proper, tenth-order system from [139]:

$$\widehat{A} = \begin{pmatrix} 0 & 0 & 0 & 0 & 1 & 1 & 0 & 0 & 0 & 0 \\ -1 & 0 & 0 & 0 & 1 & 0 & 1 & 0 & 0 & 0 \\ 1 & 1 & -1 & -1 & 0 & 0 & -1 & 0 & 0 & 0 \\ 0 & 1 & 1 & 1 & 0 & 0 & 0 & 0 & 0 & 0 \\ -1 & 2 & 0 & -1 & 2 & 0 & 1 & 0 & 0 & 0 \\ 0 & 0 & 0 & 0 & 1 & 0 & 0 & 0 & 0 & 0 \\ 0 & 0 & 0 & 0 & 0 & 1 & 0 & 0 & 0 & 0 \\ -1 & -1 & 0 & 0 & 1 & 0 & 1 & 0 & 0 & 1 \\ 0 & 0 & 0 & 0 & 0 & 0 & 0 & 0 & 0 & 0 \\ 0 & 0 & 0 & 0 & 0 & 0 & 0 & 0 & 1 & 0 \end{pmatrix},$$

$$\widehat{B} = \begin{pmatrix} 0 & 1 & 0 & 1 \\ 0 & 0 & 0 & 0 \\ 0 & 0 & 0 & 0 \\ 0 & 1 & 0 & 1 \\ 0 & 2 & 0 & 1 \\ 0 & 0 & 1 & 0 \\ 0 & 0 & 0 & 0 \\ 0 & 0 & 0 & 0 \\ 1 & 0 & 1 & 0 \\ 0 & 0 & 0 & 0 \end{pmatrix}, \quad \widehat{C}' = \begin{pmatrix} -1 & 0 & 0 & 0 \\ -1 & 0 & 0 & 0 \\ 0 & 0 & 1 & 0 \\ 0 & 0 & 0 & 0 \\ 1 & 0 & 0 & 0 \\ 0 & 0 & 0 & 0 \\ 2 & 1 & 0 & 0 \\ 0 & 1 & 0 & 1 \\ 0 & 0 & 0 & 0 \\ 0 & 0 & 0 & 0 \end{pmatrix}.$$

The "*Maple*" procedure gives the SCB system matrices

$$A = \begin{pmatrix} 1 & 0 & 0 & 0 & 1 & -1 & 0 & -1 & 0 & 0 \\ 0 & 0 & 1 & 0 & 0 & 0 & 0 & 0 & 0 & 0 \\ 0 & 0 & 0 & 0 & -1 & 1 & 0 & 1 & 0 & 0 \\ 2 & -12 & 0 & 2 & -8 & 8 & 0 & 8 & 0 & 0 \\ 2 & -4 & -2 & \frac{1}{2} & -2 & 1 & 1 & 1 & 1 & 0 \\ 0 & 0 & 0 & 0 & 0 & 0 & 1 & 0 & 0 & 0 \\ 2 & -2 & -2 & 0 & 0 & -1 & 1 & -1 & 1 & -1 \\ 0 & 0 & 0 & 0 & -1 & 0 & 0 & 0 & 1 & 0 \\ 0 & 0 & 0 & 0 & 0 & 0 & 0 & 0 & 0 & 1 \\ 0 & 2 & 0 & -\frac{1}{2} & 2 & -2 & 0 & -2 & 0 & 0 \end{pmatrix},$$

$$B = \begin{pmatrix} 0 & 0 & 0 & 0 \\ 0 & 0 & 0 & 0 \\ 0 & 0 & 0 & 0 \\ 0 & 0 & 0 & 1 \\ 1 & 0 & 0 & 0 \\ 0 & 0 & 0 & 0 \\ 0 & 1 & 0 & 0 \\ 0 & 0 & 0 & 0 \\ 0 & 0 & 0 & 0 \\ 0 & 0 & 0 & 1 \end{pmatrix}, \quad C = \begin{pmatrix} 0 & 0 & 0 & 0 \\ 0 & 0 & 0 & 1 \\ 0 & 0 & 0 & 0 \\ 0 & 0 & 0 & 0 \\ 1 & 0 & 0 & 0 \\ 0 & 1 & 0 & 0 \\ 0 & 0 & 0 & 0 \\ 0 & 0 & 1 & 0 \\ 0 & 0 & 0 & 0 \\ 0 & 0 & 0 & 0 \end{pmatrix},$$

and the dimensions $1, 2, 1, 6$. Hence, the first state belongs to the x_a subsystem, and we can therefore easily see that the system has a non-minimum-phase invariant zero at 1. The next two states belong to the x_b subsystem; thus, the system is not right invertible. The fourth state belongs to the x_c subsystem; thus, the system is not left invertible. Finally, the last six states consist of three integrator chains of length 1, 2, and 3, respectively, belonging to the x_d subsystem.

3.A.4 Numerical issues

The procedure described in this Appendix uses exact operations only; thus, there is no uncertainty in the results produced by the decomposition algorithm. The algorithm is primarily based on rank operations and the construction of bases for various subspaces. Rank operations are discontinuous in the sense that arbitrarily small perturbations to a matrix may alter its rank. This implies that, when a decomposition is carried out using exact operations, arbitrarily small perturbations to system matrices may fundamentally alter the identified structure of a system. This is in contrast to decompositions based on floating-point operations, which may be insensitive to small perturbations to the system matrices.

Whether exactness is desirable or not depends on the application. When the input data is exact or the system model is based on first principles, an exact decomposition may help to reveal fundamental structural properties of the system and how these properties are affected by various quantities in the system matrices. If, on the other hand, the system matrices have been derived based on experimental system identification, an exact decomposition may not be desirable, and it may even provide misleading information about the system structure. Thus, the exact procedure presented here is not a replacement for numerical tools developed for the same purpose.

Throughout the decomposition algorithm, a number of nonunique transformation matrices must be constructed. In the "*Maple*" procedure, these matrices are constructed with the goal of having a simple structure based on the assumption that fewer changes to the original states will result in the less complicated symbolic expressions in the computed SCB system. Depending on the structure and dimensions of the system, however, the procedure may still result in complicated expressions, and if the original system matrices contain complicated expressions, these will in general not be simplified.

A precise analysis of the computational complexity of the procedure is difficult due to the complex nature of the decomposition algorithm and the underlying "*Maple*" functions. However, it is possible to make some practical observations regarding this issue. Executed in "*Maple 12*" on an Intel Pentium processor with two 2 MHz cores, the total CPU time needed for decomposition of the single-track model was approximately 0.30 s for the single-input case and 0.21 s for the double-input case. For the DC motor example, the total CPU time was approximately 0.19 s, and for the tenth-order example, approximately 0.48 s. These execution times illustrate that an increase in the order of the system does not automatically result in a large increase in execution time; the structure of the system and the complexity of the expressions in the system matrices have a greater impact on execution time. For example, randomly generated, strictly proper systems with 20 states, 4 inputs, and 4 outputs, with the system matrices made up of integers between -10 and 10 with 25% density, are generally decomposed in less than 0.4 s. If, on the other hand, the number of inputs is reduced to 3, the decomposition generally takes around 50 s. The reason for this large difference is that, in the former case, the computed SCB systems generally consist of an x_a subsystem with 16 states and an x_d subsystem with four states, which requires only a single iteration of the algorithm for identifying x_b and x_d (described earlier when discussing strictly proper case). In the latter case, the computed SCB systems generally consist of an x_b subsystem with 17 states and an x_d subsystem with three states, which requires 17 increasingly complex iterations of the algorithm for identifying x_b and x_d.

3.A.5 "Maple" code

```
# Maple source code for structural decomposition of linear
# multivarible systems, version 0.2
```

```
#
# Copyright (C) 2009 Haavard Fjaer Grip and Ali Saberi
#
# Written and tested for Maple 12.0 with LinearAlgebra
# package
#
# Usage: A, B, C, D, G1, G2, G3, U, Y,
# dim := scb(Ai, Bi, Ci, Di);
#
# !!! Important: Ai, Bi, Ci, and Di must be of type Matrix
# (not scalars)
#
# This software is provided "as is", without warranty of any
# kind, either expressed or implied, including, without
# limitation, warranties of merchantability or fitness for a
# particular purpose.  The entire risk as to the quality and
# performance of the software is with you.
# Should the software prove defective in any respect,
# you assume the cost of any necessary servicing, repair, or
# correction.
#
# Under no circumstances will any of the copyright holders,
# developers, or any other party that modifies or conveys
# the software be liable to any person for  damages,
# including any general, indirect, special, incidental, or
# consequential damages of any kind, including, without
# limitation, damages for lost profits, loss of goodwill,
# work stoppage, computer failure or malfunction, even if
# such party has been informed of the possibility of such
# damages.

# Procedure to convert matrix M, possibly containing
# floating-point numbers, to matrix with only exact
# fractions:
# Use of floating-point elements in matrices is highly
# discouraged.
mconvert := proc (M)
    local n, m, i, j, Mr;
    uses LinearAlgebra:

    # Get matrix dimensions
    n, m := LinearAlgebra:-Dimension(M);

    # Create empty matrix
    Mr := Matrix(n, m);

    # Iterate over all rows
    for i to n do
        # Iterate over all columns
```

```
        for j to m do
            # Convert each element
            Mr[i, j] := convert(M[i, j], rational, exact);
        end do;
    end do;
    Mr;
end proc:

# Procedure to find matrix S1 such that [S2' S1']'
# is nonsingular, with S1 consisting of an identity
# matrix with rank(S2) missing rows:
nonsingular := proc(S2)
    local S2E, S1, n, m, i, j:
    uses LinearAlgebra:

    # Find dimensions of S2
    n, m := LinearAlgebra:-Dimension(S2):

    # Get reduced row echelon form of S2
    S2E := LinearAlgebra:-ReducedRowEchelonForm(S2):

    # Start by making S1 an m x m identity matrix
    S1 := LinearAlgebra:-IdentityMatrix(m):

    # Iterate from last row of echelon matrix
    for i from n by -1 to 1 do
        # On each row, iterate from leftmost column toward
        # the right for j from 1 by 1 to m do
            # If we encounter an element equal to 1,
            # then delete the row in S2 corresponding to
            # the column that we are currently at
            if S2E[i,j] = 1 then
                S1:= LinearAlgebra:-DeleteRow(S1,j):
                break:
            end if:
        end:
    end do:

    # Return S1
    S1:
end proc:

# Procedure for transformation to SCB for strictly proper
# systems
# Notation corresponds to Sannuti and Saberi (1987)
scbSP := proc(Ahi,Bhi,Chi)
    local p, n, m;
    local C, Cs, pp, pb, q, S, Dh, S1, S2, S1a, S1b, CCh, Ct, S01,
    S02, S0;
```

```
local phi, philb, phila, phi1, phi2, Db, PPsi, Cb, qb, r;
local alpha, beta, ggamma, xcfs, ycfs, xbcfs, xbiscfs, xifcfs;
local i, j, l, kern, kernmatrix, cfs;
local alphatmp, betatmp, gammatmp, qic, pic, K;
local x0, xf, xb, nf, nb, ns, na, nc;
local Gamma1, Gamma2, Gamma3;
local T, P0, P1;
local pf, rc, tmp, xs0, B0, Bu0, xs1, qc;
local X2, A2, R, As, Bs, Ctr, M, cspc, nspc, T1, T2, xs3, xa,
xc, Dt, xs2;
local A_scb, B_scb, C_scb, dim, Ah, Bh, Ch, xftmp;
local Gamma2inv, At, Bt, Gamma1t, Gamma2t, Gamma3t, dimt,
Gamma1inv, udim;
uses LinearAlgebra;

# Recast inputs as matrices, in case we received vectors
# or scalars
Ah := Matrix(Ahi):
Bh := Matrix(Bhi):
Ch := Matrix(Chi):

# Find system dimensions
p, n := LinearAlgebra:-Dimension(Ch):
n, m := LinearAlgebra:-Dimension(Bh):

# Create storage for various quantities used in the algorithm
C := Vector(n+1):
Cs := Vector(n+1): # C^*
pp := Vector(n+1): # p
pb := Vector(n+1): # \bar p
q := Vector(n):
S := Vector(n):
Dh := Vector(n): # \hat D
S1 := Vector(n): # S_i1
S2 := Vector(n): # S_i2
Sla := Vector(n): # S_ia
Slb := Vector(n): # S_ib
CCh := Vector(n): # \hat C
Ct := Vector(n): # \tilde C
phi := Vector(n):
philb := Vector(n): # \phi_ib
phila := Vector(n): # \phi_ia
phi1 := Vector(n): # \phi_i1
phi2 := Vector(n): # \phi_i2
Db := Vector(n): # \bar D
PPsi := Vector(n): # \Psi
Cb := Vector(n): # \bar C
qb := Vector(n): # \bar q
r := Vector(n):
```

```
alpha := Matrix(n, n):
beta := Matrix(n, n):
ggamma := Matrix(n, n):
xcfs := Matrix(n+1, n+1): # x_ij (coefficients)
(index+1)
ycfs := Matrix(n+2, n+2): # y_ij (coefficients)
(index+1)
xbcfs := Matrix(n+1, n+1): # x_bij (coefficients)
index+1)
xbiscfs := Vector(n): # x_bis (coefficients)
xifcfs := Vector(n): # xif (coefficients)

# Check that \hat B matrix and \hat C matrix are of
# maximal rank. If not, throw exception
if LinearAlgebra:-Rank(Ch) < p then
    error "Matrix C must be of maximal rank"
elif LinearAlgebra:-Rank(Bh) < m then
    error "Matrix B must be of maximal rank"
end if:

C[1] := Ch;
Cs[1] := Ch;
pp[1] := p;
pb[1] := p;
ycfs[1+1, 0+1] := C[1];

i := 1: # i counts the steps

# Step 1
if pp[1] > 0 then
    # Find rank q_1 of C_1*\hat B
    q[1] := LinearAlgebra:-Rank(C[1].Bh):

    #
    # S_12 must lie in left null space of C_1*\hat B.
    # Find basis for left null space and select S_12
    # as reduced-row echelon form of basis transpose
    kern := LinearAlgebra:-NullSpace(LinearAlgebra:-
    Transpose(C[1].Bh));
    kernmatrix := LinearAlgebra:-ReducedRowEchelon
    Form(Matrix(pp[1]-q[1], pp[1],
    LinearAlgebra:-Transpose(Matrix(pp[1], pp[1]-q[1],
    [kern[]]))));

    S2[1] := kernmatrix;

    # Get S_11 such that [S_11' S_12']' is nonsingular
    S1[1] := nonsingular(S2[1]);
```

```
# Define S_1 by stacking S_11 and S_12
S[1] := Matrix([[S1[1]], [S2[1]]]);

# Calculate \hat C_1, \tilde C_1, and \hat D_1
CCh[1] := S1[1].C[1].Ah:
Ct[1] := S2[1].C[1].Ah:
Dh[1] := S1[1].C[1].Bh;

# Find rank \bar p_2 of [Cs_1^*' \tilde C_1']' and
# calculate p_2 and r_1
pb[2] := LinearAlgebra:-Rank(Matrix([[Cs[1]], [Ct[1]]])):
pp[2] := pb[2]-pb[1]:
r[1] := pp[1]-q[1]-pp[2]:

# [\phi_1b \phi_12] must lie in the left null space of
# [C_1^*' \tilde C_1']'.

# Find basis for left null space of [C_1^*' \tilde C_1']'
# and find reduced-row echelon form of basis transpose
kern := LinearAlgebra:-NullSpace(LinearAlgebra:
-Transpose(Matrix([[Cs[1]], [Ct[1]]])));
kernmatrix := LinearAlgebra:-ReducedRowEchelonForm
(Matrix(r[1], pb[1]+pp[1]-q[1],
LinearAlgebra:-Transpose(Matrix(pb[1]+pp[1]-q[1],
r[1], [kern[]]))));

# Choose \phi_1b as left \bar p_1 columns
philb[1] := kernmatrix[1 .. r[1], 1 .. pb[1]];

# Choose \phi_2 as right p_1-q_1 columns
phi2[1] := kernmatrix[1 .. r[1], pb[1]+1 ..
pb[1]+pp[1]-q[1]];

# Define \phi_1a by stacking an appropriately sized
# zero matrix with \phi_1b
phila[1] := Matrix([[LinearAlgebra:
-ZeroMatrix(pp[2], pb[1])], [philb[1]]]);

# Get phi_11 such that [phi_11' phi_12']' is
# nonsingular
phi1[1] := nonsingular(phi2[1]);

# Define \phi_1 by stacking \phi_11 and \phi_12
phi[1] := Matrix([[phi1[1]], [phi2[1]]]);

# Calculate C_2 and transformation to y_20
C[2] := phi1[1].Ct[1]:
ycfs[2+1, 0+1] := C[2]:
```

```
# Calculate \Psi_1 and transformation to x_10,
# y_11, and x_b10
PPsi[1] := LinearAlgebra:-DiagonalMatrix
([<LinearAlgebra:-IdentityMatrix(q[1])>,
<phi[1]>]).S[1]:
cfs := PPsi[1].C[1]:
xcfs[1+1, 0+1] := cfs[1 .. q[1], 1 .. n]:
ycfs[1+1, 1+1] := cfs[q[1]+1 .. q[1]+pp[2], 1 .. n]:
xbcfs[1+1, 0+1] := cfs[q[1]+pp[2]+1 .. q[1]
+pp[2]+r[1], 1 .. n]:

# In preparation for next step, find C_2^s,
# \bar D_1, and \bar C_1
Cs[2] := Matrix([[Cs[1]], [C[2]]]):
Cb[1] := CCh[1]:
Db[1] := Dh[1]:

qb[1] := q[1];

i := 2:

# Step 2,3,...

# Repeat steps until p_i = 0
for i from 2 while pp[i] > 0 do
    # Find rank \bar q_i of [\bar D_{i-1}' C_i*\hat B']'
    # and calculate q_i
    qb[i] := LinearAlgebra:-Rank(Matrix([[Db[i-1]],
      [C[i].Bh]]));
    q[i] := qb[i]-qb[i-1];

    # [S_ib S_i2] must lie in the
    # left null space of [\bar D_{i-1}' C_i*\hat B']'.
    # Find reduced-row echelon form of basis
    transpose
    kern := LinearAlgebra:-NullSpace(LinearAlgebra:
    -Transpose(Matrix([[Db[i-1]], [C[i].Bh]])));
    kernmatrix := LinearAlgebra:
    -ReducedRowEchelonForm(Matrix(pp[i]-q[i], pp[i]+qb[i-1],
    LinearAlgebra:-Transpose(Matrix(pp[i]+qb[i-1],pp[i]-q[i],
    [kern[]]))));

    # Define S_ib as left \bar q_{i-1} columns
    S1b[i] := kernmatrix[1 .. pp[i]-q[i], 1 .. qb[i-1]];

    # Define S_i2 as right p[i] columns
    S2[i] := kernmatrix[1 .. pp[i]-q[i], qb[i-1]+1 ..
    pp[i]+qb[i-1]];
```

```
# Define S_ia by stacking appropriately sized zero
# matrix with S_ib
S1a[i] := Matrix([[LinearAlgebra:-ZeroMatrix(q[i],
qb[i-1])], [S1b[i]]]);

# Get S_i1 such that [S_i1' S_i2']' is nonsingular
S1[i] := nonsingular(S2[i]);

# Define S_i by stacking S_i1 and S_i2
S[i] := Matrix([[S1[i]], [S2[i]]]);

# Calculate \tilde C_i
Ct[i] := S1b[i].Cb[i-1]+S2[i].C[i].Ah;

# Find rank \bar p_{i+1} of [C_i^s' \tilde C_i']'
pb[i+1] := LinearAlgebra:
-Rank(Matrix([[Cs[i]], [Ct[i]]]));

# Calculate \hat C_i and \hat D_i
CCh[i] := S1[i].C[i].Ah;
Dh[i] := S1[i].C[i].Bh;

# Calculate p_{i+1} and r_i
pp[i+1] := pb[i+1]-pb[i];
r[i] := pp[i]-q[i]-pp[i+1];

# [\phi_ib \phi_i2] must lie in the left null
# space of [C_i^*' \tilde C_i']'.
# Find reduced-row echelon form of basis for
# left null space of [C_1^*' \tilde C_1']'
kern := LinearAlgebra:-NullSpace
(LinearAlgebra:-Transpose(Matrix([[Cs[i]],
[Ct[i]]])));
kernmatrix := LinearAlgebra:
-ReducedRowEchelonForm(Matrix(r[i],
pb[i]+pp[i]-q[i],
LinearAlgebra:
-Transpose(Matrix(pb[i]+pp[i]-q[i], r[i],
[kern[]]))));

# Choose \phi_ib as left \bar p_i columns
phi1b[i] := kernmatrix[1 .. r[i], 1 .. pb[i]];

# Choose \phi_i as right p_i-q_i columns of null
# space basis
phi2[i] := kernmatrix[1 .. r[i], pb[i]+1 ..
pb[i]+pp[i]-q[i]];

# Define \phi_ia by stacking an appropriately
```

```
#sized zero matrix with \phi_ib
phila[i] := Matrix([[LinearAlgebra:
-ZeroMatrix(pp[i+1], pb[i])], [philb[i]]]);

# Get phi_i1 such that [phi_i1' phi_i2']' is
# nonsingular
phi1[i] := nonsingular(phi2[i]);

# Define \phi_i by stacking \phi_i1 and \phi_i2
phi[i] := Matrix([[phi1[i]], [phi2[i]]]);

# Calculate C_{i+1} and transformation to y_{i+1 0}
C[i+1] := phi1[i].Ct[i];
ycfs[i+1+1, 0+1] := C[i+1];

# Find \Psi_i and store full alpha, beta, and gamma
  matrices
# temporarily (to be broken up)
PPsi[i] := LinearAlgebra:-DiagonalMatrix([
<LinearAlgebra:-IdentityMatrix(q[i])>, <phi[i]>]).S[i];
alphatmp := Matrix(pp[i+1], qb[i-1], phi1[i].S1b[i]);
betatmp := Matrix(r[i], qb[i-1], phi2[i].S1b[i]);
gammatmp := Matrix(r[i], pb[i], philb[i]);

# Iterate over j=1,..,i-1 and break out individual
  matrices of
# width q_j from alpha and beta matrices
qic := 0; # qic is the cumulative sum of q_k,
  k = 1,..,j-1
for j from 1 by 1 to i-1 do
    alpha[i, j] := Matrix(pp[i+1], q[j],
    alphatmp[1 .. pp[i+1], qic+1 .. qic+q[j]]);
    beta[i, j] := Matrix(r[i], q[j],
    betatmp[1 .. r[i], qic+1 .. qic+q[j]]);
    qic := qic+q[j];
end do;

# Iterate over j=1,..,i and break out individual matrices
  of
# width p_j from gamma matrix
pic := 0; # pic is the cumulative sum of p_k,
  k = 1,..,j-1
for j from 1 by 1 to i do
    ggamma[i, j] := Matrix(r[i], pp[j],
    gammatmp[1 .. r[i], pic+1 .. pic+pp[j]]);
    pic := pic+pp[j];
end do;

# Iterate over l=1,...,i-1 and compose transformations to
```

```
            # x_{i-l l}, y_{i-l l+1}, and x_{bi-l l}
            for l from 0 by 1 to i-1 do
                # First calculate the transformation to
                #[x_{i-l,l}' y_{i-l l+1}' x_{bi-l l}']', excluding
                  the
                # terms from alpha, beta, and gamma
                cfs := PPsi[i].ycfs[i-l+1, l+1];

                # Break the transformation into individual
                  transformations
                # for x_{i-l l}, y_{i-l l+1}, and  x_{bi-l l}
                # (still excluding the terms from alpha, beta,
                  and gamma)
                xcfs[i-l+1, l+1] := cfs[1 .. q[i], 1 .. n];
                ycfs[i-l+1, l+1+1] := cfs[q[i]+1 .. q[i]+
                pp[i+1], 1 .. n];
                xbcfs[i-l+1, l+1] := cfs[q[i]+pp[i+1]+1 .. q[i]+
                pp[i+1]+r[i], 1 .. n];

                # Add the terms from alpha and beta
                for j from l+1 by 1 to i-1 do
                    ycfs[i-l+1, l+1+1] :=
                    ycfs[i-l+1, l+1+1]+alpha[i, j].xcfs[j-l+1, l+1];
                    xbcfs[i-l+1, l+1] :=
                    xbcfs[i-l+1, l+1]+beta[i, j].xcfs[j-l+1, l+1]
                end do;

                # Add the terms from gamma
                for j from l+2 by 1 to i do
                    xbcfs[i-l+1, l+1] :=
                    xbcfs[i-l+1, l+1]+ggamma[i, j].ycfs[j-l-1+1,
                      l+1+1]
                end do
            end do;

            # In preparation for next step,
            # find C_{i+1}^s, \bar C_i, and \bar D_i
            Cs[i+1] := Matrix([[Cs[i]], [C[i+1]]]);
            Cb[i] := Matrix([[Cb[i-1]], [CCh[i]]]);
            Db[i] := Matrix([[Db[i-1]], [Dh[i]]]);
        end do:
    end if:

    # Define K as the number of steps taken
    K := i-1:

    # Find coefficients for x_f and x_b subsystems

    # Define matrix to hold the x_i0 coefficients
```

```
# (to be used later)
x0 := Matrix(0, n):

# Iterate over all chain lengths
for i from 1 to K do
    # Compose matrices of coefficients for
    # x_if and x_bis, for each i=1,...,K
    xifcfs[i] := Matrix(0, n);
    xbiscfs[i] := Matrix(0, n);
    for j to i do
        xifcfs[i] := Matrix([[xifcfs[i]],
        [xcfs[j+1, i-j+1]]]);
        xbiscfs[i] := Matrix([[xbiscfs[i]],
        [xbcfs[j+1, i-j+1]]])
    end do;

    # Add coefficients for x_i0 to x0 matrix
    x0 := Matrix([[x0], [xcfs[i+1, 0+1]]])
end do:

# Create matrices for x_f and x_b coefficients
# and add the coefficients
xf := Matrix(0, n):
xb := Matrix(0, n):
for i from 1 to K do
    xf := Matrix([[xf], [xifcfs[i]]]);
    xb := Matrix([[xb], [xbiscfs[i]]])
end do:

# Determine dimensions of x_f and x_b subsystems
nf, n := LinearAlgebra:-Dimension(xf):
nb, n := LinearAlgebra:-Dimension(xb):

# Find Gamma_2

# Different approach depending on whether K > 0 or not
if K > 0 then
    # If K > 0, define storage for matrices T_i, i=1,...,K
    T := Vector(K):

    # Set T_K as the identity matrix and
    # define T_i, i=1,...,K-1 recursively
    T[K] := LinearAlgebra:-IdentityMatrix(pp[K]):
    for i from K-1 by -1 to 1 do
        T[i] := LinearAlgebra:-DiagonalMatrix([
        <LinearAlgebra:-IdentityMatrix(q[i])>, <T[i+1].PPsi[i+1]>,
        <LinearAlgebra:-IdentityMatrix(r[i])>])
    end do:
```

```
    # Define cumulative counter p_f for the number of
    # measurements from the x_f chain
    pf := q[1]:

    # Define vector rc with each element i containing
    # the sum of r_j j=0,...,i-1, with r_0 defined as 0
    rc := Vector(K+1):

    # Initialize components 1 and 2 of rc
    rc[1] := 0:
    rc[2] := r[1]:

    # Create p_f and rc vector
    for i from 2 to K do
        pf := pf+q[i]:
        rc[i+1] := rc[i-1+1]+r[i]
    end do:

    # Use rc to create P_0 matrix, with identity
    #matrices of size r_i along the reversed diagonal
    P0 := Matrix(0, rc[K+1]):
    for i from K by -1 to 1 do
        tmp := Matrix([LinearAlgebra:-ZeroMatrix(r[i],
        rc[K+1]-rc[i+1]), LinearAlgebra:
        -IdentityMatrix(r[i]),
        LinearAlgebra:-ZeroMatrix(r[i], rc[i-1+1])]);
        P0 := Matrix([[tmp], [P0]])
    end do:

    #Define P_1 from P_0 and identity matrix
    P1 := LinearAlgebra:-DiagonalMatrix([
    <LinearAlgebra:-IdentityMatrix(pf)>, <P0>]):

    # Create Gamma_2
    Gamma2inv := P1.T[1].PPsi[1];
    Gamma2 := simplify(LinearAlgebra:
    -MatrixInverse(Gamma2inv)): else
    # If K = 0, meaning p = 0, define Gamma_2
    # as empty matrix
    Gamma2 := LinearAlgebra:-IdentityMatrix(p,p);
    Gamma2inv := Gamma2;
          end if:

        # Find Gamma_3

# Different approach depending on whether K > 0
# or not
if K > 0 then
    # \tilde D should be defined such that
```

```
    # [\bar D_K' \tilde D']' is nonsingular.
    # Hence, we may choose \tilde D' from the null
    # space of \bar D_K. Find basis for null space
    Dt := nonsingular(Db[K]);

    # Calculate Gamma_3 as inverse of [\bar D_K' Dt']'
    Gamma3 := simplify(LinearAlgebra:
    -MatrixInverse(Matrix([[Db[K]], [Dt]]))):
          else
    # If K = 0, define Gamma_3 as the identity matrix
    Gamma3 := LinearAlgebra:-IdentityMatrix(m);
end if:

# Different approach depending on whether K > 0
# or not
if K > 0 then
    # Define dimension n_s of x_a and x_c subsystem
    ns := n-pb[K]:

    # Find coefficients for x_s subsystem. Start by defining
    # coefficients for initial system x_s^0

    # The coefficients \tilde Gamma of x_s^0 should be
    # chosen so that [C_K^*' \tilde Gamma']' is nonsingular.
    # Hence, we may choose \tilde Gamma' from null space
    # of C_K^*. Find basis for null space
    xs0 := nonsingular(Cs[K]);

    # Find input matrix for x_s^0 system with respect to
    # new input vector [u' v']'
    B0 := xs0.Bh.Gamma3:

    # Split out input matrix for u input
    Bu0 := B0[1 .. ns, 1 .. qb[K]]: else
    # If K = 0, the entire system is part of the x_s subsystem.
    # Define the transformation to x_s^0 as identity matrix
    ns := n;
    xs0 := IdentityMatrix(n);

    # Find input matrix for x_s^0 system with respect to
    # new input vector [u' v']'
    B0 := xs0.Bh.Gamma3;

    # There is no u input
    Bu0 := Matrix(ns,0);
end if:

# Define new state x_s^1 by canceling nonzero
# occurences in Bu0, using x0
```

```
# (which describes x_i0 states)
xs1 := xs0-Bu0.x0:

# Define new state x_s^2 as x_s^1, to be
# recursively modified to remove occurences
# non-outputs from x_b and x_f subsystems
# in derivative
xs2 := xs1:

# We start by taking chains of length j = 2 and up,
# continue with 3 and up, and so forth,
# to eliminate remaining unwanted occurences
for j from 2 by 1 to K do
    # Initialize counters qc and rc, to represent the
    # number of states in levels below
    # the one currently processed
    qc := 0:
    rc := 0:

    # Add number of states in chains that are
    # shorter than j
    for i from 1 by 1 to j-1 do
        qc := qc+i*q[i];
        rc := rc+i*r[i];
    end do:

    # Iterate over all chains of length j and up
    for i from j by 1 to K do
        # Calculate coefficient matrix for current
        # version of x_s^2
        X2 := Matrix([[xs2], [xb], [xf]]);
        A2 := xs2.Ah.LinearAlgebra:-MatrixInverse(X2);

        # If there is an x_f chain of length i,
        # remove occurences of x_{i-j+2 j-2} by
        # subtracting linear combination of x_{i-j+1 j-1}
        if q[i] > 0 then
            R := Matrix(ns, q[i], A2[1 .. ns,
            ns+nb+qc+(i-j+1)*q[i]+1 .. ns+nb+qc+(i-j+2)*q[i]]):
            xs2 := xs2-R.xf[qc+(i-j)*q[i]+1 .. qc+
            (i-j+1)*q[i], 1 .. n]:
        end if:

        # If there is an x_b chain of length i,
        # remove occurences of x_{bi-j j-2} by
        # subtracting linear combination of x_{b i-j+1 j-1}
        if r[i] > 0 then
            R := A2[1 .. ns, ns+rc+(i-j+1)*r[i]+1 ..
              ns+rc+(i-j+2)*r[i]]:
```

```
                xs2 := xs2-R.xb[rc+(i-j)*r[i]+1 .. rc+(i-j+1)*r[i],
                  1 .. n]:
            end if:

            # Add to counters qc and rc
            qc := qc+i*q[i];
            rc := rc+i*r[i];
        end do;
end do:

# Change the order of the x_f system so that
# we have SISO integration chains

# Create new variable to be built up with new sorting
xftmp := Matrix(0, n):
qc := 0:
# qc counts the number of elements in chains already processed

# Iterate over all chains
for i from 1 to K do
    # Iterate over each level in integration chain i
    for j from 1 by 1 to q[i] do
        # Iterate over each component at level j of
        # integration chain i
        for l from 0 by 1 to i-1 do
            # Add component to xftmp
            xftmp := Matrix([[xftmp], [xf[qc+l*q[i]+j, 1 ..n]]]);
        end do:
    end do:
    # Add number of elements in chain i to cumulative counter
    qc := qc+i*q[i];
end do:
# Define x_f as xftmp
xf := xftmp;

# Transform x_s subsystem to Kalman controllable form

# Find coefficients of derivative of x_s in new coordinates
X2 := Matrix([[xs2], [xb], [xf]]):
A2 := xs2.Ah.LinearAlgebra:-MatrixInverse(X2);

# Separate out n_s\times n_s coefficient matrix for
#  x_s subsystem
As := A2[1 .. ns, 1 .. ns]:

# Find input matrix for x_s subsystem with respect
# to new [u' v']' inputs
Bs := xs2.Bh.Gamma3:
if K > 0 then
```

```
        udim := qb[K];
        Bs := Bs[1 .. ns, qb[K]+1 .. m]:
    else
        udim := 0;
    end if:

    if LinearAlgebra:-Rank(Bs) = 0 then
        xa := xs2;
        xc := Matrix(0,n);
        na := ns;
        nc := 0;
    else
        At, Bt, Ct, Gamma1t, Gamma2t, Gamma3t,
        dimt := scbSP(LinearAlgebra:-Transpose(As),
        Matrix(ns,0), LinearAlgebra:-Transpose(Bs));
        xs3 := LinearAlgebra:-Transpose(Gamma1t).xs2;
        na := dimt[1];
        nc := ns-na;
        xa := xs3[1 .. na, 1 .. n]:
        xc := xs3[na+1 .. ns, 1 .. n]:
        Gamma3 := Gamma3.LinearAlgebra:-DiagonalMatrix([
        <LinearAlgebra:-IdentityMatrix(udim)>,
        <LinearAlgebra:-MatrixInverse(LinearAlgebra:
        -Transpose(Gamma2t))>]);
    end if:

    # Calculate Gamma_1
    Gamma1inv := Matrix([[xa], [xb], [xc], [xf]]);
    Gamma1 := simplify(LinearAlgebra:
    -MatrixInverse(Gamma1inv)):

    # Calculate (simplified) system matrices for
    # new SCB system
    A_scb := simplify(Gamma1inv.Ah.Gamma1):
    B_scb := simplify(Gamma1inv.Bh.Gamma3):
    C_scb := simplify(Gamma2inv.Ch.Gamma1):

    # Create vector containing dimensions of
    # x_a, x_b, x_c, and x_f subsystems
    dim := [na, nb, nc, nf]:

    # Return SCB system matrices, state transformations,
    # and dimensions of subsystems
    A_scb, B_scb, C_scb, Gamma1, Gamma2, Gamma3, dim:
     end proc:

    # Procedure for finding SCB representation and state
    # transformation for general multivariable LTI systems
     scb := proc(Ai,Bi,Ci,Di)
```

```
local Ah, Bt, Ct, Dt, m0, C0, Ch1, U1, U2, Y1, Y2,
 kern, kernmatrix, p, n, m, Bh, B0, Bh1, D;
local U, Y, AsSP, BsSP, A, B, CsSP, C, G1, G2, G3, dim,
Gamma1SP, Gamma2SP, Gamma3SP;
uses LinearAlgebra;

# Recast inputs as matrices, in case we received vectors or
# scalars, and convert any floating-point elements to
# exact fractions
Ah := mconvert(Matrix(Ai)):
Bt := mconvert(Matrix(Bi)):
Ct := mconvert(Matrix(Ci)):
Dt := mconvert(Matrix(Di)):

# Find system dimensions
p, n := LinearAlgebra:-Dimension(Ct):
n, m := LinearAlgebra:-Dimension(Bt):

# Check that [\hat B' \hat D']' matrix and [\hat C \hat D]
# matrix are of maximal rank. If not, throw exception
if LinearAlgebra:-Rank(Matrix([Ct, Dt])) < p then
    error "Matrix [C, D]' must be of maximal rank"
elif LinearAlgebra:-Rank(Matrix([[Bt], [Dt]])) < m then
    error "Matrix [B', D']' must be of maximal rank"
end if:

# Start by treating non-strictly proper case

# Find rank of \tilde D
m0 := LinearAlgebra:-Rank(Dt):

# Y2 must lie in left null space of matrix \tilde D.
# Find basis for null space, and select Y2 as reduced-row
# echelon form of that basis
kern := LinearAlgebra:-NullSpace(LinearAlgebra:
-Transpose(Dt)):
kernmatrix := LinearAlgebra:-ReducedRowEchelonForm
(Matrix(p-m0, p, LinearAlgebra:-
Transpose(Matrix(p, p-m0, [kern[]])))):
Y2 := kernmatrix:

# Get Y1 such that [Y1' Y2']' is nonsingular
Y1 := nonsingular(Y2);

# Compute Y
Y := MatrixInverse(Matrix([[Y1], [Y2]]));

# Select U_1 as Y_1\tilde D
U1 := Y1.Dt;
```

```
# Compute U_2 such that [U_1' U_2']' is nonsingular
U2 := nonsingular(U1);

# Compute U
U := LinearAlgebra:
-MatrixInverse(Matrix([[U1], [U2]]));

# Find matrices C_0 and \hat C_1
C0 := Y1.Ct:
Ch1 := Y2.Ct:

# Find matrix \hat B and split it into B_0
# and \hat B_1
Bh := Bt.U;
B0 := Bh[1 .. n, 1 .. m0];
Bh1 := Bh[1 ..n, m0+1 .. m];

# Find transformations for strictly proper
# part of system
AsSP, BsSP, CsSP, Gamma1SP, Gamma2SP, Gamma3SP,
 dim := scbSP(Ah-B0.C0, Bh1, Ch1);

# Construct transformations matrices for
# non-strictly proper system
G1 := Gamma1SP;
G2 := simplify(LinearAlgebra:-DiagonalMatrix([
<IdentityMatrix(m0)>, <Gamma2SP>]));
G3 := simplify(LinearAlgebra:-DiagonalMatrix([
<IdentityMatrix(m0)>, <Gamma3SP>]));

# The SCB matrix for the non-strictly proper system
#  is the same as returned by scbSP
A := AsSP;

# Calculate the remaining SCB matrices using all the necessary
# transformations, including U and Y
B := simplify(LinearAlgebra:-MatrixInverse(G1).Bt.U.G3);
C := simplify(LinearAlgebra:-MatrixInverse(Y.G2).Ct.G1);
D := simplify(LinearAlgebra:-MatrixInverse(Y.G2).Dt.U.G3);

# Return SCB matrices, transformations and dimensions
A, B, C, D, G1, G2, G3, U, Y, dim:
      end proc:
```

4

Constraints on inputs: actuator saturation

4.1 Introduction

This chapter is concerned with designing controllers for linear systems subject to input saturation with the purpose of achieving internal stabilization. This is a prelude to most of the subsequent chapters, and presents basic problem statements of global and semi-global internal stabilization, necessary and sufficient conditions under which such a stabilization can be achieved, as well as control design methodologies that can be utilized for an appropriate design.

To start with, it is prudent to observe that control magnitude saturation or actuator saturation is pretty common and indeed is ubiquitous in engineering applications, and as such, it has been recognized ever since the beginnings of industrial revolution and automation. To exemplify this, let us note that the capacity of every device is capped. Valves can only be operated between fully open and fully closed states, pumps and compressors have a finite throughput capacity, and tanks can only hold a certain volume. Force, torque, thrust, stroke, voltage, current, flow rate, and so on, are limited in their activation range in all physical systems. Servers can serve only so many consumers. In circuits, transistors and amplifiers are saturating components. Saturation and other physical limitations are dominant in maneuvering systems like aircrafts. Every physically conceivable actuator, sensor, or transducer has bounds on the magnitude as well as on the rate of change of its output. Note that bounds on the rate of change and sensor saturation will be discussed in subsequent chapters. This chapter concentrates only on input magnitude saturation.

One of the foremost tasks of any control system design is to make sure that the given system under the designed control law is internally stable. In this regard, we recognize that most of the control theoretic concepts are developed for linear systems. Thus, in this framework, often, nonlinear elements are approximated as linear elements. Once the analysis or design of the resulting linear systems are complete, one tries to gauge the effect of nonlinearities on such an analysis or design. Such an ad hoc approach cannot always lead to satisfactory results. In fact, ignoring constraints can be detrimental to the stability and performance of control systems, and can lead to *catastrophic events*. A classical example for

A. Saberi et al., *Internal and External Stabilization of Linear Systems with Constraints*, Systems & Control: Foundations & Applications, DOI 10.1007/978-0-8176-4787-2_4,

the detrimental effect of neglecting *constraints* is the Chernobyl unit four nuclear power plant disaster in 1986. As such, direct methods of design are necessary to regulate and stabilize linear systems subject to actuator saturation. That is, the challenge is to take actuator saturation into account directly at the onset of design rather than resorting to ad hoc methods of design.

Although internal stabilization of linear systems subject to actuator saturation by direct methods rather than by ad hoc methods is an important problem, the arrows of history did not pierce through it for a very long time when Fuller in 1969 aimed an arrow and successfully hit it [38]. Fuller's seminal paper established that a chain of integrators with order higher than two cannot be globally asymptotically stabilized by any saturating linear static state feedback control law with only one input channel. Once that is done, history fell silent once more for about 20 years, and its arrows floated dimly overhead. The monumental work of Fuller, unfortunately, was not widely recognized in the following 20 years although researchers working in optimal control area were then making tremendous efforts to develop system theoretic criteria for global asymptotic stabilization of general linear time-varying systems by constrained controls, see [66, 145] among others.

In 1990, continuing the theme of Fuller, Sontag and Sussman [155] (see also [174]) established that, for continuous-time systems, in general, global asymptotic stabilization for linear systems with bounded inputs cannot be achieved using linear feedback laws, it is possible only by using *nonlinear* control laws. More precisely, they established that in general global asymptotic stabilization for linear systems with bounded inputs can be achieved using nonlinear feedback laws if and only if the system in the absence of saturation is stabilizable and critically unstable (equivalently, *asymptotically null controllable with bounded control* (ANCBC)). In the discrete-time setting, an analogous result is established by Yang [209, 210]. It states that a linear discrete-time system subject to input saturation can be globally asymptotically stabilized via nonlinear feedback if and only if it is stabilizable and all its poles are located inside or on the unit circle. A nonlinear globally stabilizing control law for such a system is also explicitly constructed in [209]. Let us emphasize that Sontag, Sussman, and Yang mostly point out the existence of controllers that would solve the global internal stabilization problem of linear systems with bounded controls under a set of necessary and sufficient conditions. However, methodologies to design appropriate controllers that achieve global internal stabilization are not clearly formulated.

The works of Sontag, Sussman, and Yang unleashed a flurry of activity in internally stabilizing linear systems subject to actuator saturation. Along one direction, Teel [179] proposed certain design methodologies to design appropriate controllers for global stabilization. Along another direction, Saberi and his students queried as to what can be achieved by utilizing linear feedback control laws. In this respect, they [74, 75, 77] proposed and emphasized a semi-global rather than a global framework for stabilization using bounded controls. Let us note that the concept of global asymptotic stabilization is well known. In contrast to global asymptotic stabilization, the possibility of semi-global exponential stabilization means that for any bounded set, we can find a controller which achieves local exponential stabilization of the system while the domain of attraction of the closed-loop system contains an a priori given bounded set. Relaxing the requirement of global stabilization to that of semi-global stabilization enables the utilization of linear feedback control laws. That is, Saberi et al. established that the semi-global exponential stabilization can be achieved by bounded controls, while utilizing linear feedback laws, under the same necessary and sufficient conditions required to achieve global asymptotic stabilization by nonlinear feedback laws. In other words, they established that for a linear system subject to actuator saturation there exists a family of linear feedback laws which achieves internal exponential stability if and only if it is stabilizable and has all its open-loop poles in the closed left-half plane for continuous-time systems and within or on the unit circle for discrete-time systems.

We observe that relaxing the requirement of global stabilization to that of semi-global stabilization not only enables the utilization of linear feedback control laws, but it also makes sense from an engineering point of view, since in general a plant's model is usually valid in some region of the state space. Moreover, it allows a stronger stability property for the closed-loop system, that is, the exponential stability of the closed-loop system, rather than mere asymptotic stability. In Sect. 4.7, we return to the question whether we can find linear controllers that achieve global stability.

Let us discuss next the available methods for designing controllers that can achieve semi-global stabilization. One method was proposed and detailed in [75] and [76]. It is based on the eigenstructure assignment and is referred to as a direct method of design for low-gain controllers. However, later on, algebraic Riccati equation (ARE)-based methods utilizing H_2 and H_∞ optimal control theory for designing low-gain controllers were also proposed independently in [88] and [181]. Also, the works of [74, 77, 78, 128] introduced yet another design technique, the so-called low-and-high-gain design technique. This design technique was basically conceived for semi-global control problems beyond stabilization and was related to the performance issues such as semi-global stabilization with enhanced utilization of the available control capacity of the system, semi-global disturbance rejection, and robustness of stability with respect to uncertainties. All these works have led to the development of low-gain and low-and-high-gain design methodologies now popular in this area (see [71, 135]). On the other hand, several nonlinear bounded feedback laws also were constructed explicitly for global asymptotic stabilization of linear systems with input saturation. Teel proposed a nested saturation design for a chain of integrators with bounded control [180]. This technique was later generalized by Sussman et al. to general linear systems [173]. Megretski came up with a scheduling-based nonlinear control law based on the H_∞ Riccati equation, which automatically increases the gain as the state approaches the origin [98] (such a control law also turns out to be crucial for external L_2 disturbance rejection). Two special issues published by the International Journal of Robust and Nonlinear Control have collected some contributions pertinent to this area [8, 134]. In particular, reference [7] has a comprehensive bibliography reflecting the earlier works on systems with saturating actuators.

The purpose of this chapter is to review this early phase of research dealing with global and semi-global internal stabilization of linear systems subject to saturation. In particular, we recall here the precise problem formulations of global and semi-global internal stabilization and various design methodologies to achieve such a stabilization and then review clearly the results obtained.

4.2 Problem statements and their solvability results

We consider a linear system subject to actuator saturation described by

$$\begin{aligned} \rho x &= Ax + B\sigma(u) \\ y &= Cx, \end{aligned} \tag{4.1}$$

where, as usual, $x \in \mathbb{R}^n$ is the state, $u \in \mathbb{R}^m$ is the control input to the saturator, and $y \in \mathbb{R}^p$ is the measurement output. Also, ρx denotes $\frac{dx}{dt}$ for continuous-time case and $x(k+1)$ for discrete time. Moreover, the function $\sigma(.) : \mathbb{R}^n \to \mathbb{R}^n$ is the saturation function defined in Definition 2.20.

There exist in the literature a number of saturation functions and some of the general ones are collected in Sect. 2.6. Most of the early results use the following standard saturation function:

$$\sigma(s) = \operatorname{sat}_\Delta(s), \tag{4.2a}$$

where

$$\operatorname{sat}_\Delta(s) = \Delta \operatorname{sat}\left(\frac{s}{\Delta}\right) \tag{4.2b}$$

and

$$\operatorname{sat}(s) = \begin{pmatrix} \operatorname{sat}_1(s_1) \\ \operatorname{sat}_1(s_2) \\ \vdots \\ \operatorname{sat}_1(s_m), \end{pmatrix} \tag{4.2c}$$

where $\Delta > 0$ is a constant and

$$\operatorname{sat}_1(s) = \begin{cases} -1 & \text{for } s < -1 \\ s & \text{for } |s| \leq 1 \\ 1 & \text{for } s > 1. \end{cases} \tag{4.2d}$$

Although early results are based on the above saturation function (4.2), most of the results which are obtained for the above saturation function can easily be re-obtained for a general saturation function as described in Sect. 2.6. This will be addressed in Chap. 5.

As pointed out in the introduction, although internally stabilizing a linear system subject to actuator saturation is an important problem, historically, no definite work existed until Fuller [38] established that a chain of integrators with order higher than two cannot be globally asymptotically stabilized by any saturating linear control law. In 1990, continuing the theme of Fuller, Sontag and Sussman [155] (see also [174]) formulated and solved the following problem:

Problem 4.1 Consider the system (4.1) in either continuous or discrete time. The **global asymptotic stabilization problem via state feedback** is to find, if possible, a state feedback (possibly nonlinear or time-varying) $u(t) = f(x(t), t)$ or $u(k) = f(x(k), k)$ such that the equilibrium point $x = 0$ of the closed-loop system is asymptotically stable.

Regarding the above problem, by now two facts (one positive, one negative) are well known, owing to the works of Sontag and Sussman [155] and Sussman and Yang [174]:

Fact 1: The global asymptotic stabilization problem via state feedback (Problem 4.1) is solvable if and only if the given linear system subject to magnitude saturation of the actuator is stabilizable and has all its open-loop poles in the closed left-half plane for continuous-time systems and within or on the unit circle for discrete-time systems.

Fact 2: The global asymptotic stabilization problem via state feedback (Problem 4.1) is, in general, not solvable by a linear state feedback but, whenever it is solvable, it can be done with a nonlinear, static, and time-invariant state feedback.

We will prove the necessity of Fact 1 at the end of this section. The sufficiency of Fact 1 and Fact 2 follows immediately from the construction of suitable controllers in the following sections. The fact that that we cannot always restrict attention to linear controllers for continuous-time systems is due to [38]. For discrete-time systems, this is a very strong conjecture.

It is worth pointing out here that in the literature a system is said to be asymptotically null controllable with bounded controls (ANCBC) if it is stabilizable, and has all its open-loop poles in the closed left-half plane for continuous-time systems and within or on the unit circle for discrete-time systems. This means that we can restate Fact 1 as follows: *The global asymptotic stabilization problem via state feedback (Problem 4.1) is solvable if and only if the given system is asymptotically null controllable with bounded controls.*

Remark 4.2 *For sample data systems with zero order hold and with sampling period T non-pathological,*[1] *if the given continuous-time system is ANCBC, so is the resulting sample data system.*

The case of measurement feedback instead of state feedback was not immediately obtained after the work of Sontag, Sussman, and Yang but was formulated and solved later on. The problem can be stated clearly as follows:

Problem 4.3 Consider the system (4.1) in either continuous or discrete time. The **global asymptotic stabilization via dynamic measurement feedback** is to find a dynamic measurement feedback control law (possibly nonlinear and time-varying) of the form

$$\begin{aligned}\dot{p}(t) &= \ell(p, y, t)\\ u(t) &= g(p, y, t),\end{aligned}$$

for continuous-time systems or of the form

$$\begin{aligned}p(k+1) &= l(p(k), y(k), k)\\ u(k) &= g(p(k), y(k), k),\end{aligned}$$

for discrete-time systems, such that the equilibrium point $(x, p) = (0, 0)$ of the closed-loop system is globally asymptotically stable.

Analogous to state feedback control problem, we can state the following facts regarding the measurement feedback control problem:

Fact 3: The global asymptotic stabilization problem via measurement feedback (Problem 4.3) is solvable if and only if the given linear system subject to magnitude saturation of the actuator is stabilizable, detectable, and has all its open-loop poles in the closed left-half plane for continuous-time systems and within or on the unit circle for discrete-time systems.

Fact 4: The global asymptotic stabilization problem via measurement feedback (Problem 4.3) cannot, in general, be solved by a linear measurement feedback of the form of a stabilizing state feedback combined with an observer.

Clearly, Fact 4 follows immediately from Fact 2. Fact 3 was introduced in [155] for continuous-time systems. We will prove the necessity of Fact 3 at the end of this section. The sufficiency follows immediately from the construction of suitable controllers in the following sections.

---

[1]Non-pathological sampling period T implies that there is no pair of distinct poles of the given continuous-time system, say λ_1 and λ_2, such that $|\mathrm{im}\,\lambda_1 - \mathrm{im}\,\lambda_2|$ equals $\frac{k2\pi}{T}$ for some integer k.

As we discussed in introduction, the work of Sontag, Sussman, and Yang unleashed a flood of activity. As pointed out above, by Facts 2 and 4, global asymptotic stabilization of linear system subject to magnitude saturation of the actuator requires *nonlinear* feedback control laws which are inherently cumbersome to implement. This implementation issue prompted Saberi and his students to ask a fundamental question of *how to weaken this requirement and be able to utilize only linear feedback control laws*. This leads to the formulation of semi-global internal stabilization problems as recalled below.

Problem 4.4 Consider the system (4.1) in either continuous or discrete time. The **semi-global exponential stabilization problem via state feedback** is to find, if possible, for any arbitrarily large bounded set $\mathcal{X} \subset \mathbb{R}^n$, a state feedback law $u(t) = f(x(t), t)$ or $u(k) = f(x(k), k)$, such that the equilibrium point $x = 0$ of the closed-loop system is locally exponentially stable with $\mathcal{X}$ contained in its domain of attraction.

Problem 4.5 Consider the system (4.1) in either continuous or discrete time. The **semi-global exponential stabilization via dynamic measurement feedback** is to find, if possible, an integer q such that, for any arbitrarily large bounded set $\mathcal{V} \subset \mathbb{R}^{n+q}$, there exists a controller of the form,

$$\begin{aligned} \dot{p}(t) &= \ell(p, y, t), \quad p(t) \in \mathbb{R}^q \\ u(t) &= g(p, y, t), \end{aligned}$$

for continuous-time systems or of the form,

$$\begin{aligned} p(k+1) &= l(p(k), y(k), k), \quad p(t) \in \mathbb{R}^q \\ u(k) \quad &= g(p(k), y(k), k), \end{aligned}$$

for discrete-time systems, such that the equilibrium point $(x, p) = (0, 0)$ of the closed-loop system is locally asymptotically stable and $\mathcal{V}$ is contained in its domain of attraction.

Note that the problem defined above considers nonlinear time-varying feedback controllers and arbitrary state dimension. However, from the work of Saberi and Lin [74, 75, 77], we can state the following:

Fact 5: The above state as well as measurement feedback semi-global stabilization problems (Problems 4.4 and 4.5) are solvable under the same conditions for which the respective global stabilization problems are solvable.

Fact 6: The above state as well as measurement feedback semi-global stabilization problems (Problems 4.4 and 4.5), if solvable, are always solvable by linear controllers, and, in the measurement feedback case, we can always choose $q = n$.

We will prove the necessity of Fact 5 at the end of this section. The sufficiency follows immediately from the construction of suitable controllers in the following sections. The above gives credence to a semi-global framework over a global framework as it utilizes only linear feedback while still yielding an arbitrary large domain of attraction. Moreover, we can achieve exponential stabilization rather than asymptotic stabilization.

In view of the above discussion, before we proceed further, we make the following standard assumptions on the triple (A, B, C) of the system 4.1:

Assumption 4.6 *The pair (A, B) is stabilizable.*

Assumption 4.7 *The eigenvalues of A are all located in the closed left-half plane for continuous-time systems and within or on the unit circle for discrete-time systems.*

In the literature, the above two assumptions are equivalently combined into one assumption as given by the following:

Assumption 4.8 *The pair (A, B) is asymptotically null controllable with bounded controls. That is:*

(i) All the eigenvalues of A are located in the closed left-half plane in the continuous time, while they are all located inside or on the unit circle in the discrete time.

(ii) The pair (A, B) is stabilizable.

Assumption 4.9 *The pair (A, C) is detectable.*

The rest of this chapter is devoted to discuss several design methodologies we employ throughout this book as well as their application for semi-global or global stabilization of linear systems subject to actuator saturation as described in the system (4.1). The primary design methodologies introduced early on into the literature are low-gain designs. Fundamentally, there are two different ways low-gain design can be accomplished:

- Direct eigenstructure assignment method as discussed in Sect. 4.3.
- Algebraic Riccati equation (ARE)-based methods as discussed in Sects. 4.4 and 4.8. This ARE-based method can utilize either the H_2 or H_∞ Riccati equation.

As discussed in detail later on, low-gain design methods underutilize the available control capacity. To rectify this and to utilize better the available control capacity, low-and-high-gain design methods were introduced [74, 77, 78, 128]. Both the low-gain design methods as well as low-and-high-gain design methods lead to linear controller designs. The low-gain design depends on what is called a *low-gain parameter* ε, while the low-and-high-gain design depends on ε as well as on a parameter α called a *high-gain parameter*. By utilizing the freedom in choosing the parameters ε and α appropriately, low-gain, and low-and-high-gain design methods have successfully been used, in connection with linear systems with saturating actuators, for internal stabilization [75, 76], simultaneous internal and external stabilization [86], robust stabilization [68, 78, 128], disturbance rejection [87], output regulation [88], etc.

Mainly, all of the above work is confined to a semi-global framework. In the next phase of research, the tuning parameters ε and α have been adapted or scheduled to depend on the state x or its estimate, and in so doing, nonlinear controllers have been designed [67, 70, 98, 126]. This is done to elevate the design framework from semi-global to global, both internally and externally.

We review in several subsequent sections the essence of low-gain, and low-and-high-gain design methods as well as scheduled-low-and-scheduled-high-gain design methods and their applications in early phase of research.

Proof of necessity of Facts 1–6 : To prove the necessity of Fact 1 in continuous time, it is immediate that (A, B) needs to be stabilizable. Next, assume that λ is an eigenvalue of A in the open right-half plane with corresponding left eigenvector p, i.e., $pA = \lambda p$. Then we have

$$\frac{\mathrm{d}}{\mathrm{d}t} px(t) = \lambda px(t) + v(t),$$

where

$$v := pB\sigma(u).$$

There clearly exists an $\widetilde{M} > 0$ such that $\|v(t)\| \leqslant \widetilde{M}$ for all $t > 0$ since $\sigma(u)$ is bounded. But then,

$$|px(t)| > \left|e^{\lambda t}\right| \left(|px(0)| - \frac{\widetilde{M}}{\operatorname{Re}\lambda} \right) + \frac{\widetilde{M}}{\operatorname{Re}\lambda},$$

which does not converge to zero since $\operatorname{Re}\lambda \geqslant 0$, provided the initial condition is such that

$$|px(0)| > \frac{\tilde{M}}{\operatorname{Re}\lambda}.$$

Note that this is valid for all controllers, and therefore, we can clearly not achieve global stability. In discrete time, the argument is similar. It is again immediate that (A, B) needs to be stabilizable. Next, assume that λ is an eigenvalue of A outside the closed unit disc, i.e., $|\lambda| > 1$, with corresponding left eigenvector p, i.e., $pA = \lambda p$. Then we have

$$px(k+1) = \lambda px(k) + v(k),$$

where

$$v := pB\sigma(u).$$

There clearly exists an $\tilde{M} > 0$ such that $\|v(k)\| \leqslant \tilde{M}$ for all $t > 0$ since $\sigma(u)$ is bounded. But then,

$$|px(k)| > \left|\lambda^k\right| \left(|px(0)| - \frac{\tilde{M}\lambda}{\lambda - 1} \right)$$

which does not converge to zero since $|\lambda| \geqslant 1$, provided the initial condition is such that

$$|px(0)| > \frac{\tilde{M}\lambda}{\lambda - 1}.$$

Note that this is again valid for all controllers, and therefore, we can clearly not achieve global stability.

The necessity of Fact 3 follows directly from the above since this argument is independent of the use of state or measurement feedback controllers. It also yield the necessity of the conditions for semi-global stabilization since the above argument identifies initial conditions that cannot be stabilized, and hence, semi-global stabilization is impossible as well. ■

4.3 Semi-global stabilization: direct eigenstructure assignment

In this section, we recall the low-gain design method by direct eigenstructure assignment that enables us to construct a family of state as well as measurement feedback control laws; semi-global stabilization can be achieved by utilizing a member of either of such a family of control laws. The low-gain design methodology was originally introduced by Lin and Saberi in [75]. Since this early work, there have been several variations of the method [76, 83, 84, 86–88].

4.3.1 Continuous-time systems

We first consider continuous-time systems and proceed to describe what is referred to as a direct method of design based on an eigenstructure assignment method.

Consider a linear system

$$\dot{x} = Ax + Bu, \quad x \in \mathbb{R}^n, \quad u \in \mathbb{R}^m, \tag{4.3}$$

where we assume that (A, B) is stabilizable and all the eigenvalues of A are in the closed left-half plane. Then, the low-gain design is carried out in the following three steps:

Step 1: Find a state transformation Γ_x and an input transformation Γ_u (a variation of the canonical forms presented in [56, Sect. 6.4.6]) such that $\Gamma_x^{-1} A \Gamma_x$ and $\Gamma_x^{-1} B \Gamma_u$ are in the following form:

$$\Gamma_x^{-1} A \Gamma_x = \begin{pmatrix} A_1 & A_{1,2} & \cdots & A_{1,q} & 0 \\ 0 & A_2 & \ddots & \vdots & \vdots \\ \vdots & \ddots & \ddots & A_{q-1,q} & \vdots \\ \vdots & & \ddots & A_q & 0 \\ 0 & \cdots & \cdots & 0 & A_{q+1} \end{pmatrix}, \tag{4.4}$$

$$\Gamma_x^{-1} B \Gamma_u = \begin{pmatrix} B_1 & 0 & \cdots & 0 & B_{1,q+1} \\ 0 & B_2 & \ddots & \vdots & \vdots \\ \vdots & \ddots & \ddots & 0 & \vdots \\ 0 & \cdots & 0 & B_q & B_{q,q+1} \\ B_{q+1,1} & \cdots & B_{q+1,q-1} & B_{q+1,q} & B_{q+1,q+1} \end{pmatrix}, \tag{4.5}$$

where q is an integer, and for $i = 1, 2, \ldots, q$,

$$A_i = \begin{pmatrix} 0 & 1 & 0 & \cdots & 0 \\ \vdots & \ddots & \ddots & \ddots & \vdots \\ \vdots & & \ddots & 1 & 0 \\ 0 & \cdots & \cdots & 0 & 1 \\ -a_{n_i}^i & \cdots & -a_3^i & -a_2^i & -a_1^i \end{pmatrix}, \quad B_i = \begin{pmatrix} 0 \\ \vdots \\ 0 \\ 0 \\ 1 \end{pmatrix}.$$

Clearly, (A_i, B_i) is controllable. The transformation Γ_x is such that all the eigenvalues of A_i are on the imaginary axis, and all the eigenvalues of A_{q+1} have strictly negative real parts.

Step 2: For each (A_i, B_i), let $F_{\varepsilon_i,i} \in \mathbb{R}^{1\times n_i}$ be the state feedback gain such that the eigenvalues of $A_i + B_i F_{\varepsilon_i,i}$ can be obtained from the eigenvalues of A_i by moving any eigenvalue λ_i on the imaginary axis to $\lambda_i - 2\varepsilon$ while all the eigenvalues in the open left-half plane remain at the same location.

We note here that such a gain $F_{\varepsilon_i,i}$ exists and is unique. Moreover, it can be obtained explicitly in terms of ε_i. The uniqueness follows since (A_i, B_i) is a single-input controllable pair.

Step 3 : The family of low-gain state feedback control laws parameterized in ε is defined by

$$u = F_\varepsilon x, \tag{4.6}$$

where the state feedback gain matrix F_ε is given by

$$F_\varepsilon = \Gamma_u \begin{pmatrix} F_{\varepsilon_1,1} & 0 & \cdots & \cdots & 0 \\ 0 & F_{\varepsilon_2,2} & \ddots & & \vdots \\ \vdots & \ddots & \ddots & \ddots & \vdots \\ \vdots & & \ddots & F_{\varepsilon_q,q} & 0 \\ 0 & \cdots & \cdots & 0 & 0 \end{pmatrix} \Gamma_x^{-1}, \tag{4.7}$$

where $\varepsilon_q = \varepsilon$, while for $i = 2, \ldots, q$, we have

$$\varepsilon_{i-1} = \varepsilon_i^{2+r_i},$$

where r_i is the largest algebraic multiplicity of the eigenvalues of A_i.

In order to show that the controller constructed above has certain desired properties, we first present and prove a crucial lemma:

Lemma 4.10 *Consider a linear single-input system in the controller canonical form:*

$$\dot{x} = Ax + Bu,$$

where

$$A = \begin{pmatrix} 0 & 1 & 0 & \cdots & 0 \\ \vdots & \ddots & \ddots & \ddots & \vdots \\ \vdots & & \ddots & 1 & 0 \\ 0 & \cdots & \cdots & 0 & 1 \\ -a_n & \cdots & -a_3 & -a_2 & -a_1 \end{pmatrix}, \quad B = \begin{pmatrix} 0 \\ \vdots \\ 0 \\ 0 \\ 1 \end{pmatrix},$$

with all eigenvalues of A in the closed left-half plane. Let F_ε be the unique matrix such that the eigenvalues of $A + BF_\varepsilon$ can be obtained from the eigenvalues of A by moving any eigenvalue λ_i on the imaginary axis to $\lambda_i - 2\varepsilon$ while all the eigenvalues in the open left-half plane remain at the same location.

Then, there exist β, γ, and ε^* such that for all $0 < \varepsilon \leqslant \varepsilon^*$, we have

$$\left\| F_\varepsilon e^{(A+BF_\varepsilon)t} \right\| \leqslant \varepsilon \beta e^{-\varepsilon t} \tag{4.8}$$

and

$$\left\| e^{(A+BF_\varepsilon)t} \right\| \leqslant \frac{\gamma}{\varepsilon^{r-1}} e^{-\varepsilon t} \tag{4.9}$$

for all $t > 0$, where r is the largest algebraic multiplicity among the eigenvalues of A.

Proof : Define $F_0 = 0$ and

$$p_\varepsilon(s) = \det(sI - A - BF_\varepsilon).$$

Since the state feedback moves all eigenvalues 2ε to the left, we note that $p_\varepsilon(s) = p_0(s + 2\varepsilon)$, where p_0 is obtained from p_ε by setting $\varepsilon = 0$. We define

$$q(s) = \begin{pmatrix} 1 \\ s \\ \vdots \\ s^{n-1} \end{pmatrix}.$$

Due to the structure of A and B, we have

$$(A + BF_\varepsilon)q(s) = sq(s) - Bp_\varepsilon(s) \tag{4.10}$$

for any $s \in \mathbb{C}$. If $\lambda_{\varepsilon,i}$ is an eigenvalue of $A + BF_\varepsilon$ with algebraic multiplicity m_i, then we can immediately obtain from (4.10) by differentiation,

$$\begin{aligned}
(A + BF_\varepsilon)q(\lambda_{\varepsilon,i}) &= \lambda_{\varepsilon,i}q(\lambda_{\varepsilon,i}) \\
(A + BF_\varepsilon)q^{(1)}(\lambda_{\varepsilon,i}) &= \lambda_{\varepsilon,i}q^{(1)}(\lambda_{\varepsilon,i}) + q(\lambda_{\varepsilon,i}) \\
\vdots \qquad & \qquad \vdots \\
(A + BF_\varepsilon)q^{(m_i-1)}(\lambda_{\varepsilon,i}) &= \lambda_{\varepsilon,i}q^{(m_i-1)}(\lambda_{\varepsilon,i}) + q^{(m_i-2)}(\lambda_{\varepsilon,i}).
\end{aligned}$$

If $\lambda_{\varepsilon,1}, \ldots, \lambda_{\varepsilon,k}$ are the eigenvalues of $A + BF_\varepsilon$ with multiplicity $m_1, \ldots, m_k$, then,

$$q(\lambda_{\varepsilon,1}), \ldots, q^{(m_1-1)}(\lambda_{\varepsilon,1}), \ldots, q(\lambda_{\varepsilon,k}), \ldots, q^{(m_k-1)}(\lambda_{\varepsilon,k}) \tag{4.11}$$

form a basis of $\mathbb{C}^n$. We note that

$$e^{(A+BF_\varepsilon)t} q^{(j)}(\lambda_{\varepsilon,i}) = e^{\lambda_{\varepsilon,i}t} \sum_{k=0}^{j} \frac{t^k}{k!} q^{j-k}(\lambda_{\varepsilon,i}).$$

Also, we note that

$$\left| e^{\lambda_{\varepsilon,i}t} \right| \leqslant e^{-2\varepsilon t},$$

and that there exists a $\widetilde{\rho} > 0$ such that

$$t^k e^{-\varepsilon t} \leqslant \frac{\widetilde{\rho}}{\varepsilon^k} \tag{4.12}$$

for $k = 0, \ldots, r-1$ and for all $t > 0$. Therefore, all the coefficients of the matrix $e^{(A+BF_\varepsilon)t}$ with respect to the basis (4.11) are bounded by

$$\frac{\mu}{\varepsilon^{r-1}} e^{-\varepsilon t}$$

for some constant $\mu > 0$ provided that $\varepsilon < 1$. The basis transformation associated with the basis (4.11) is continuous in ε and converges to the identity as ε converges to zero. Hence, there exists a γ such that (4.9) is satisfied.

Using (4.10), we find that

$$F_\varepsilon q(s) = (F_\varepsilon - F_0) q(s) = p_0(s) - p_\varepsilon(s) = p_0(s) - p_0(s + 2\varepsilon)$$

for all $s \in \mathbb{C}$. If $\lambda_{\varepsilon,i}$ is an eigenvalue of A of multiplicity m_i, then it is also a zero of p_0 of multiplicity m_i, and we find that there exists a $M > 0$ such that

$$\left| p_0^{(j)}(\lambda_{\varepsilon,i} - 2\varepsilon) - p_0^{(j)}(\lambda_{\varepsilon,i}) \right| \leqslant M \varepsilon^{m_i - j}$$

for all sufficiently small ε. But this implies that

$$\left\| F_\varepsilon q^{(j)}(\lambda_{\varepsilon,i}) \right\| \leqslant M \varepsilon^{m_i - j}. \tag{4.13}$$

We then have

$$F_\varepsilon e^{(A+BF_\varepsilon)t} q^{(j)}(\lambda_{\varepsilon,i}) = e^{\lambda_{\varepsilon,i}t} \sum_{k=0}^{j} \frac{t^k}{k!} F_\varepsilon q^{(j-k)}(\lambda_{\varepsilon,i}),$$

and, using (4.13), we find that

$$\left\| F_\varepsilon e^{(A+BF_\varepsilon)t} q^{(j)}(\lambda_{\varepsilon,i}) \right\| \leqslant e^{-2\varepsilon t} \sum_{k=0}^{j} \frac{t^k}{k!} M \varepsilon^{m_i - j + k}.$$

Using that $m_i - j > 0$, we find that

$$\left\| F_\varepsilon e^{(A+BF_\varepsilon)t} q^{(j)}(\lambda_{\varepsilon,i}) \right\| \leqslant e^{-2\varepsilon t} \sum_{k=0}^{j} \frac{t^k}{k!} M \varepsilon^{k+1}.$$

Using (4.12), this implies the existence of a $M_1 > 0$ such that, for all $t > 0$, we have

$$\left\| F_\varepsilon e^{(A+BF_\varepsilon)t} q^{(j)}(\lambda_{\varepsilon,i}) \right\| \leqslant M_1 \varepsilon e^{-\varepsilon t}.$$

The basis transformation associated with the basis (4.11) is continuous in ε and converges to the identity as ε converges to zero. The above bound then yields (4.8) for some suitably chosen constant β. ■

The parameterized state feedback gain F_ε as given by (4.7) is termed as a low-gain feedback, the name low-gain for F_ε is justified soon by (4.16). The low-gain F_ε has several inherent properties that enable its utilization to stabilize a linear system with constraints. These properties are addressed by the following theorem:

Theorem 4.11 *Consider the linear system as given by* (4.3). *Suppose that* (A, B) *is stabilizable and all the eigenvalues of* A *are in the closed left-half plane. Then we have the following properties:*

(*i*) *The closed-loop system matrix* $A + BF_\varepsilon$ *is Hurwitz stable for all* $\varepsilon > 0$.

(*ii*) *There exist constants* $M_1 > 0$, $M_2 > 0$, *and* $\varepsilon^* > 0$ *such that for all* $\varepsilon \in (0, \varepsilon^*]$ *and* $t \geq 0$, *we have*

$$\left\| e^{(A+BF_\varepsilon)t} \right\| \leqslant \frac{M_1}{\varepsilon_1^{r_1}} e^{-\varepsilon_1 t}, \tag{4.14}$$

$$\| F_\varepsilon e^{(A+BF_\varepsilon)t} \| \leqslant M_2 \varepsilon_q e^{-\varepsilon_1 t}. \tag{4.15}$$

We note that (4.15) implies that, for $t = 0$, we have

$$\| F_\varepsilon \| \leqslant M_2 \varepsilon_q. \tag{4.16}$$

Proof : From Lemma 4.10, it immediately follows that

$$\|x_q(t)\| = \left\| e^{(A_q+B_qF_{\varepsilon_q,q})t} x_q(0) \right\| \leqslant \frac{M_{1,q}}{\varepsilon_q^{r_q}} e^{-\varepsilon_q t} \|x(0)\|$$

and

$$\|F_{\varepsilon_q,q}x_q(t)\| = \left\| F_{\varepsilon_q,q} e^{(A_q+B_qF_{\varepsilon_q,q})t} x_q(0) \right\| \leqslant M_{2,q}\varepsilon_q e^{-\varepsilon_q t} \|x(0)\|.$$

Next, we will apply a recursion. Assume that for $k > i$, we have

$$\|x_k(t)\| \leqslant \frac{M_{1,k}}{\varepsilon_{i+1}^{r_{i+1}}} e^{-\varepsilon_{i+1}t} \|x(0)\|,$$

and

$$\|F_{\varepsilon_k,k}x_k(t)\| \leqslant M_{2,k}\varepsilon_q e^{-\varepsilon_{i+1}t} \|x(0)\|.$$

Next, we consider $x_i(t)$. We have

$$\dot{x}_i(t) = (A_i + B_iF_{\varepsilon_i,i})x_i(t) + d_i(t),$$

where $d_i(t)$ is a linear combination of $x_{i+1}(t), \ldots, x_q(t)$ and hence satisfies

$$\|d_i(t)\| \leqslant \frac{\widetilde{M}_{1,i}}{\varepsilon_{i+1}^{r_{i+1}}} e^{-\varepsilon_{i+1}t} \|x(0)\|. \tag{4.17}$$

We find that

$$x_i(t) = e^{(A_i+B_iF_{\varepsilon_i,i})t} x_i(0) + \int_0^t e^{(A_i+B_iF_{\varepsilon_i,i})(t-\tau)} d_i(\tau)\,\mathrm{d}\tau.$$

Combining Lemma 4.10 and (4.17) we get

$$\|x_i(t)\| \leqslant \frac{\gamma_i}{\varepsilon_i^{r_i-1}} e^{-\varepsilon_i t} \left[1 + \frac{\widetilde{M}_{1,i}}{\varepsilon_{i+1}^{r_{i+1}}} \frac{1}{\varepsilon_{i+1} - \varepsilon_i} \right] \|x(0)\|,$$

and, using that $\varepsilon_i = \varepsilon_{i+1}^{2+r_{i+1}}$, we get

$$\|x_i(t)\| \leqslant \frac{(1 + 3\widetilde{M}_{1,i})\gamma_i}{\varepsilon_i^{r_i}} e^{-\varepsilon_i t} \|x(0)\|. \tag{4.18}$$

Similarly, we find that

$$\|F_{\varepsilon_i,i}x_i(t)\| \leqslant \beta_i\varepsilon_i e^{-\varepsilon_i t}\left[1+\frac{\widetilde{M}_{1,i}}{\varepsilon_{i+1}^{r_{i+1}}}\frac{1}{\varepsilon_{i+1}-\varepsilon_i}\right]\|x(0)\|,$$

and hence,

$$\|F_{\varepsilon_i,i}x_i(t)\| \leqslant (1+3\widetilde{M}_{1,i})\beta_i\varepsilon_{i+1}e^{-\varepsilon_i t}\|x(0)\|. \tag{4.19}$$

Therefore, by recursion, we find that (4.18) and (4.19) hold for $i = 1,\dots,q$. It is then easy to show that we also have

$$\|x_{q+1}(t)\| \leqslant \widetilde{M}_{2,q+1}e^{-\varepsilon_1 t}\|x(0)\|.$$

The bounds we have obtained for the solution of the differential equation then immediately yield the two bounds presented in the theorem. ∎

Remark 4.12 *The above theorem brings into light important properties of low-gain matrix. Although such properties are obvious, to be explicit, we would like to describe them here in simple words. Justifying the name low-gain,* (4.16) *states that the magnitude or norm of the gain F_ε can be rendered as small as necessary by selecting the tuning parameter ε as small as needed. Similarly,* (4.14) *shows that the norm of the transition matrix of the closed-loop system under the low-gain control law* (4.6) *goes to zero exponentially as time tends to infinity but exhibits a peaking phenomenon for t small. However, even though the state might exhibit peaking,* (4.15) *shows that the norm of the feedback control can be rendered as small as necessary once again by appropriately selecting the parameter ε. These properties ensure that for an ε sufficiently small the actuator will not be driven to its saturation level.*

We can easily construct now the observer-based measurement feedback controllers based on the state feedback design (4.6). The family of low-gain measurement feedback control laws take the form,

$$\begin{cases} \dot{\widehat{x}} = A\widehat{x} + Bu + K(y - C\widehat{x}) \\ u = F_\varepsilon\widehat{x}, \end{cases} \tag{4.20}$$

where the state feedback gain F_ε is given by (4.7) and the observer gain K is any matrix such that $A - KC$ is Hurwitz stable. Note that another observer-based measurement feedback controller is also regularly used:

$$\begin{cases} \dot{\widehat{x}} = A\widehat{x} + B\sigma(u) + K(y - C\widehat{x}) \\ u = F_\varepsilon\widehat{x}. \end{cases} \tag{4.21}$$

The advantage of (4.20) is that the controller is completely linear. The advantage of (4.21) is that the observer is always such that

$$\dot{x} - \dot{\hat{x}} = (A - KC)(x - \hat{x}), \tag{4.22}$$

while in the case of (4.20), the observer error only satisfies (4.22) if we guarantee that the saturation does not get activated. Therefore, the analysis to prove, for instance, stability is often easier when using (4.21).

The following theorem shows that the family of state feedback laws given in (4.6) solves Problem 4.4, namely, the problem of semi-global exponential stabilization via linear static state feedback. Also, it shows that the family of measurement feedback laws given in (4.20) solves Problem 4.5, namely, the problem of semi-global exponential stabilization via linear dynamic measurement feedback.

Theorem 4.13 *Consider the continuous-time system given in* (4.1). *Assume that the pair* (A, B) *is stabilizable and the eigenvalues of* A *are all located in the closed left-half plane. Then the family of linear static state feedback laws given in* (4.6) *solves the problem of semi-global stabilization via state feedback as defined in Problem 4.4. More specifically, under the state feedback* (4.6), *for any given (arbitrarily large) bounded set* $\mathcal{X} \subset \mathbb{R}^n$, *there exists an* $\varepsilon^* \in (0, 1]$ *such that for all* $\varepsilon \in (0, \varepsilon^*]$, *the equilibrium point* $x = 0$ *of the closed-loop system is locally exponentially stable with* $\mathcal{X}$ *contained in its domain of attraction.*

Also, under the additional assumption that the pair (C, A) *is detectable, the family of linear dynamic measurement feedback laws given in* (4.20) *solves the problem of semi-global stabilization via measurement feedback, as defined in Problem 4.5. More specifically, under the measurement feedback law* (4.20), *for any given (arbitrarily large) bounded set* $\mathcal{X} \times \widehat{\mathcal{X}} \subset \mathbb{R}^{2n}$, *there exists an* $\varepsilon^* \in (0, 1]$ *such that for all* $\varepsilon \in (0, \varepsilon^*]$, *the equilibrium point* $(0, 0)$ *of the closed-loop system is locally exponentially stable with* $\mathcal{X} \times \widehat{\mathcal{X}}$ *contained in its domain of attraction.*

Proof : Given a bounded set $\mathcal{X} \subset \mathbb{R}^n$, choose ε^* such that for $\varepsilon < \varepsilon^*$ we have

$$M_2 \varepsilon_q \|x\| < 1$$

for all $x \in \mathcal{X}$, where M_2 is as defined in Theorem 4.11. Then for all initial conditions $x(0) \in \mathcal{X}$, we have

$$\left\| F_\varepsilon e^{(A+BF_\varepsilon)t} x(0) \right\| < 1,$$

and therefore, the saturation never gets activated. This implies that

$$x(t) = e^{(A+BF_\varepsilon)t} x(0)$$

for all $t > 0$, and hence, (4.14) implies that the state converges to zero exponentially as t tends to infinity.

In the case of measurement feedback, we will also establish that the saturation will never get activated. If the saturation does not get activated, we note that (4.22) is satisfied. Since $A - KC$ is stable, we find that

$$\|x(t) - \hat{x}(t)\| \leqslant M_3 e^{-\sigma t} \|x(0) - \hat{x}(0)\|$$

for some constants M_3 and $\sigma > 0$. Using

$$\|F_\varepsilon x(t)\| \leqslant \|F_\varepsilon e^{(A+BF_\varepsilon)t}\| \|x(0)\| + \int_0^t F_\varepsilon e^{(A+BF_\varepsilon)(t-\tau)} \|BF_\varepsilon\| \|x(t) - \hat{x}(\tau)\| \, d\tau,$$

we find that

$$\|F_\varepsilon x(t)\| \leqslant M_2 \varepsilon_q \left[\|x(0)\| + \frac{M_2 M_3 \varepsilon_q \|B\|}{\sigma - \varepsilon_1} \|x(0) - \hat{x}(0)\| \right].$$

Hence, for ε small enough, we have

$$\|F_\varepsilon \hat{x}(t)\| \leqslant \|F_\varepsilon x(t)\| + \|F_\varepsilon (x(t) - \hat{x}(t))\| < 1$$

for all $t > 0$. This implies that the saturation does not get activated for all initial conditions satisfying $x(0) \in \mathcal{X}$ and $\hat{x}(0) \in \hat{\mathcal{X}}$. This immediately yields the exponential convergence and the domain of attraction as presented in the theorem. ■

Remark 4.14 *In view of the above theorem, we can emphasize one aspect of the low-gain design. The state and measurement feedback laws* (4.6) *and* (4.20) *are parameterized in a single tuning parameter ε. One can include any a priori given (arbitrarily large) bounded set inside the basin of attraction of the closed-loop system equilibrium point (origin) by choosing the value of the tuning parameter ε sufficiently small. In this sense, the low-gain design described in this section is a "one shot" design as the design by itself does not depend on the a priori given set $\mathcal{X}$ or $(\mathcal{X}, \hat{\mathcal{X}})$.*

4.3.2 Discrete-time systems

We consider next discrete-time systems and proceed to describe a direct method of design based on an eigenstructure assignment method. Our development here parallels that for continuous-time systems.

Consider a linear system

$$x(k+1) = Ax(k) + Bu(k), \quad x \in \mathbb{R}^n, \, u \in \mathbb{R}^m, \tag{4.23}$$

where we assume that (A, B) is stabilizable and all the eigenvalues of A are in the closed unit disc. The low-gain design is carried out in three steps:

Step 1 : Find a state transformation Γ_x and an input transformation Γ_u (a variation of the canonical forms presented in [56, Sect. 6.4.6]) such that $\Gamma_x^{-1} A \Gamma_x$ and $\Gamma_x^{-1} B \Gamma_u$ are in the following form:

$$\Gamma_x^{-1} A \Gamma_x = \begin{pmatrix} A_1 & A_{1,2} & \cdots & A_{1,q} & 0 \\ 0 & A_2 & \ddots & \vdots & \vdots \\ \vdots & \ddots & \ddots & A_{q-1,q} & \vdots \\ \vdots & & \ddots & A_q & 0 \\ 0 & \cdots & \cdots & 0 & A_{q+1} \end{pmatrix},$$

$$\Gamma_x^{-1} B \Gamma_u = \begin{pmatrix} B_1 & 0 & \cdots & 0 & B_{1,q+1} \\ 0 & B_2 & \ddots & \vdots & \vdots \\ \vdots & \ddots & \ddots & 0 & \vdots \\ 0 & \cdots & 0 & B_q & B_{q,q+1} \\ B_{q+1,1} & \cdots & B_{q+1,q-1} & B_{q+1,q} & B_{q+1,q+1} \end{pmatrix},$$

where q is an integer, and for $i = 1, 2, \ldots, q$,

$$A_i = \begin{pmatrix} 0 & 1 & 0 & \cdots & 0 \\ \vdots & \ddots & \ddots & \ddots & \vdots \\ \vdots & & \ddots & 1 & 0 \\ 0 & \cdots & \cdots & 0 & 1 \\ -a_{n_i}^i & \cdots & -a_3^i & -a_2^i & -a_1^i \end{pmatrix}, \quad B_i = \begin{pmatrix} 0 \\ \vdots \\ 0 \\ 0 \\ 1 \end{pmatrix}.$$

Clearly, (A_i, B_i) is controllable. The transformation Γ_x is such that all the eigenvalues of A_i are on the unit circle, and all the eigenvalues of A_{q+1} are strictly inside the unit circle.

Step 2 : For each (A_i, B_i), let $F_{\varepsilon_i,i} \in \mathbb{R}^{1\times n_i}$ be the state feedback gain such that the eigenvalues of $A_i + B_i F_{\varepsilon_i,i}$ can be obtained from the eigenvalues of A_i by moving any eigenvalue λ_i on the unit circle to $(1-2\varepsilon)\lambda_i$ while all the eigenvalues in the open unit disc remain at the same location.

We note here that such a gain $F_{\varepsilon_i,i}$ exists and is unique. Moreover, it can be obtained explicitly in terms of ε_i. The uniqueness follows since (A_i, B_i) is a single-input controllable pair.

Step 3 : The family of low-gain state feedback control laws parameterized in ε is defined by

$$u = F_\varepsilon x, \tag{4.24}$$

where the state feedback gain matrix F_ε is given by

$$F_\varepsilon = \Gamma_u \begin{pmatrix} F_{\varepsilon_1,1} & 0 & \cdots & \cdots & 0 \\ 0 & F_{\varepsilon_2,2} & \ddots & & \vdots \\ \vdots & \ddots & \ddots & \ddots & \vdots \\ \vdots & & \ddots & F_{\varepsilon_q,q} & 0 \\ 0 & \cdots & \cdots & 0 & 0 \end{pmatrix} \Gamma_x^{-1}, \tag{4.25}$$

where $\varepsilon_q = \varepsilon$ while, for $i = 2, \dots, q$, we have

$$\varepsilon_{i-1} = \varepsilon_i^{2+r_i}, \tag{4.26}$$

where r_i is the largest algebraic multiplicity among the eigenvalues of A_i.

Lemma 4.15 *Consider a linear single-input system in the controller canonical form,*

$$x(k+1) = Ax(k) + Bu(k),$$

where

$$A = \begin{pmatrix} 0 & 1 & 0 & \cdots & 0 \\ \vdots & \ddots & \ddots & \ddots & \vdots \\ \vdots & & \ddots & 1 & 0 \\ 0 & \cdots & \cdots & 0 & 1 \\ -a_n & \cdots & -a_3 & -a_2 & -a_1 \end{pmatrix}, \quad B = \begin{pmatrix} 0 \\ \vdots \\ 0 \\ 0 \\ 1 \end{pmatrix},$$

with all eigenvalues of A in the closed unit disc. Let F_ε be the unique matrix such that the eigenvalues of $A + BF_\varepsilon$ can be obtained from the eigenvalues of A by moving any eigenvalue λ_i on the unit circle to $(1-2\varepsilon)\lambda_i$ while all the eigenvalues in the open unit disc remain at the same location. Then, there exist β, γ, and ε^* such that for all $0 < \varepsilon \leqslant \varepsilon^*$ we have

$$\left\| F_\varepsilon (A + BF_\varepsilon)^k \right\| \leqslant \varepsilon\beta(1-\varepsilon)^k, \tag{4.27}$$

and

$$\left\| (A + BF_\varepsilon)^k \right\| \leqslant \frac{\gamma}{\varepsilon^{r-1}}(1-\varepsilon)^k \tag{4.28}$$

for all $k > 0$, where r is the largest algebraic multiplicity among the eigenvalues of A.

Proof : Define $F_0 = 0$ and

$$p_\varepsilon(s) = \det(sI - A - BF_\varepsilon).$$

Note that

$$p_\varepsilon(s) = (1-2\varepsilon)^n p_0\left(\frac{s}{1-2\varepsilon}\right)$$

where p_0 is obtained from p_ε by setting $\varepsilon = 0$. We define

$$q(s) = \begin{pmatrix} 1 \\ s \\ \vdots \\ s^{n-1} \end{pmatrix}.$$

Due to the structure of A and B, we have

$$(A + BF_\varepsilon)q(s) = sq(s) - Bp_\varepsilon(s) \tag{4.29}$$

for any $s \in \mathbb{C}$. If $\lambda_{\varepsilon,i}$ is an eigenvalue of $A + BF_\varepsilon$ with algebraic multiplicity m_i, then we have,

$$\begin{aligned} (A + BF_\varepsilon)q(\lambda_{\varepsilon,i}) &= \lambda_{\varepsilon,i}q(\lambda_{\varepsilon,i}) \\ (A + BF_\varepsilon)q^{(1)}(\lambda_{\varepsilon,i}) &= \lambda_{\varepsilon,i}q^{(1)}(\lambda_{\varepsilon,i}) + q(\lambda_{\varepsilon,i}) \\ \vdots \qquad & \qquad \vdots \\ (A + BF_\varepsilon)q^{(m_i-1)}(\lambda_{\varepsilon,i}) &= \lambda_{\varepsilon,i}q^{(m_i-1)}(\lambda_{\varepsilon,i}) + q^{(m_i-2)}(\lambda_{\varepsilon,i}), \end{aligned}$$

which is immediately obtained from (4.29) by differentiation with respect to s and noting that $p_\varepsilon(s)$ has a zero in $\lambda_{\varepsilon,i}$ of order m_i. If $\lambda_{\varepsilon,1}, \ldots, \lambda_{\varepsilon,k}$ are the eigenvalues of $A + BF_\varepsilon$ with multiplicity $m_1, \ldots, m_k$, then

$$q(\lambda_{\varepsilon,1}), \ldots, q^{(m_1-1)}(\lambda_{\varepsilon,1}), \ldots, q(\lambda_{\varepsilon,k}), \ldots, q^{(m_k-1)}(\lambda_{\varepsilon,k}) \tag{4.30}$$

form a basis of $\mathbb{C}^n$. We note that

$$(A + BF_\varepsilon)^k q^{(j)}(\lambda_{\varepsilon,i}) = \sum_{\ell=0}^{\min(j,k)} \binom{k}{\ell} \lambda_{\varepsilon,i}^{k-\ell} q^{j-\ell}(\lambda_{\varepsilon,i}).$$

Also, we note that

$$\left|\lambda_{\varepsilon,i}^k\right| \leqslant (1-2\varepsilon)^k \leqslant (1-\varepsilon)^{2k}.$$

Moreover,

$$\sum_{\ell=0}^{k} \binom{k}{\ell} (1-\varepsilon)^{k-\ell}\varepsilon^\ell = ((1-\varepsilon)+\varepsilon)^k = 1,$$

and hence,

$$\binom{k}{\ell}(1-\varepsilon)^{k-\ell}\varepsilon^\ell \leqslant 1.$$

This implies that

$$\binom{k}{\ell}(1-\varepsilon)^{k-\ell} \leqslant \frac{1}{\varepsilon^\ell} \leqslant \frac{1}{\varepsilon^{r-1}} \tag{4.31}$$

for $\ell = 0, \ldots, r-1$ and for $k \geqslant \ell$. Therefore, all the coefficients of the matrix $(A + BF_\varepsilon)^k$ with respect to the basis (4.30) are bounded by

$$\frac{\mu}{\varepsilon^{r-1}}(1-\varepsilon)^k$$

for some constant $\mu > 0$ provided that $\varepsilon < 1$. The basis transformation associated with the basis (4.30) is continuous in ε and converges to the identity as ε converges to zero. Hence, there exists a γ such that (4.28) is satisfied.

Using (4.29), we find that

$$F_\varepsilon q(s) = (F_\varepsilon - F_0)q(s) = p_0(s) - p_\varepsilon(s) = p_0(s) - (1-2\varepsilon)^n p_0\left(\frac{s}{1-2\varepsilon}\right)$$

for all $s \in \mathbb{C}$. If λ_i is an eigenvalue of A of multiplicity m_i, then it is also a zero of p_0 of multiplicity m_i. We have

$$F_\varepsilon q(\lambda_{\varepsilon,i}) = p_0((1-2\varepsilon)\lambda_i).$$

Using that p_0 has a zero in λ_i of multiplicity m_i, we find that

$$\left| \frac{\mathrm{d}^j}{\mathrm{d}\varepsilon^j} p_0((1-2\varepsilon)\lambda_i) \right|_{\varepsilon=0} = 0$$

for $j < i$. We find that there exists a $M > 0$ such that

$$\left\| F_\varepsilon q^{(j)}(\lambda_{\varepsilon,i}) \right\| \leqslant M\varepsilon^{m_i - j} \tag{4.32}$$

for all sufficiently small ε. We have

$$F_\varepsilon (A + BF_\varepsilon)^k q^{(j)}(\lambda_{\varepsilon,i}) = \sum_{\ell=0}^{\min(j,k)} \binom{k}{\ell} \lambda_{\varepsilon,i}^{k-\ell} F_\varepsilon q^{j-\ell}(\lambda_{\varepsilon,i}),$$

and, using (4.32), we find that

$$\left\| F_\varepsilon (A + BF_\varepsilon)^k q^{(j)}(\lambda_{\varepsilon,i}) \right\| \leqslant \sum_{\ell=0}^{\min(j,k)} \binom{k}{\ell} (1-2\varepsilon)^{k-\ell} M\varepsilon^{m_i - j + \ell}.$$

Using that $m_i - j > 0$, we find that

$$\left\| F_\varepsilon (A + BF_\varepsilon)^k q^{(j)}(\lambda_{\varepsilon,i}) \right\| \leqslant M\varepsilon(1-\varepsilon)^k \sum_{\ell=0}^{\min(j,k)} \binom{k}{\ell} (1-\varepsilon)^{k-\ell} \varepsilon^{\ell},$$

which implies that

$$\left\| F_\varepsilon (A + BF_\varepsilon)^k q^{(j)}(\lambda_{\varepsilon,i}) \right\| \leqslant M\varepsilon(1-\varepsilon)^{k-j} \leqslant 2M\varepsilon(1-\varepsilon)^k$$

for ε sufficiently small. The basis transformation associated with the basis (4.30) is continuous in ε and converges to the identity as ε converges to zero. The above bound then yields (4.27) for some suitably chosen constant β. ■

The parameterized state feedback gain F_ε as given by (4.25) is termed as a low-gain feedback. It has several inherent properties that enable its utilization to stabilize a linear system with constraints. These properties are addressed by the following theorem:

Theorem 4.16 *Consider the linear system as given by* (4.3). *Suppose that* (A, B) *is stabilizable and all the eigenvalues of* A *are in the closed unit disc. Then we have the following properties:*

(*i*) *The closed-loop system matrix $A + BF_\varepsilon$ is Schur stable for all $\varepsilon > 0$.*

(*ii*) *There exist constants $M_1 > 0$, $M_2 > 0$ and $\varepsilon^* > 0$ such that for all $\varepsilon \in (0, \varepsilon^*]$ and $t \geq 0$, we have*

$$\left\|(A + BF_\varepsilon)^k\right\| \leqslant \frac{M_1}{\varepsilon_1^{r_1}}(1 - \varepsilon_1)^k \tag{4.33}$$

$$\|F_\varepsilon(A + BF_\varepsilon)^k\| \leqslant M_2\varepsilon_q(1 - \varepsilon_1)^k. \tag{4.34}$$

We note that (4.34) implies that, for $k = 0$,

$$\|F_\varepsilon\| \leqslant M_2\varepsilon_q. \tag{4.35}$$

Proof : From Lemma 4.15 it immediately follows that

$$\|x_q(t)\| = \left\|(A_q + B_q F_{\varepsilon_q,q})^k x_q(0)\right\| \leqslant \frac{M_{1,q}}{\varepsilon_q^{r_q}}(1 - \varepsilon_q)^k \|x(0)\|,$$

and

$$\|F_{\varepsilon_q,q}x_q(t)\| = \left\|F_{\varepsilon_q,q}(A_q + B_q F_{\varepsilon_q,q})^k x_q(0)\right\| \leqslant M_{2,q}\varepsilon_q(1 - \varepsilon_q)^k \|x(0)\|.$$

Next, we will apply a recursion. Assume that for $j > i$ we have

$$\|x_j(k)\| \leqslant \frac{M_{1,j}}{\varepsilon_{i+1}^{r_{i+1}}}(1 - \varepsilon_{i+1})^k \|x(0)\|,$$

and

$$\|F_{\varepsilon_j,j}x_j(k)\| \leqslant M_{2,j}\varepsilon_q(1 - \varepsilon_{i+1})^k \|x(0)\|.$$

Next, we consider $x_i(k)$. We have

$$x_i(k + 1) = (A_i + B_i F_{\varepsilon_i,i})x_i(k) + d_i(k),$$

where $d_i(k)$ is a linear combination of $x_{i+1}(k), \dots, x_q(k)$ and hence satisfies

$$\|d_i(k)\| \leqslant \frac{\widetilde{M}_{1,i}}{\varepsilon_{i+1}^{r_{i+1}}}(1 - \varepsilon_{i+1})^k \|x(0)\|. \tag{4.36}$$

We find that

$$x_i(k) = (A_i + B_i F_{\varepsilon_i,i})^k x_i(0) + \sum_{\ell=0}^{k} (A_i + B_i F_{\varepsilon_i,i})^{k-\ell} d_i(\ell).$$

Combining Lemma 4.15 and (4.36), we get

$$\|x_i(k)\| \leqslant \frac{\gamma_i}{\varepsilon_i^{r_i-1}}(1-\varepsilon_i)^k \left[1 + \frac{\widetilde{M}_{1,i}}{\varepsilon_{i+1}^{r_{i+1}}} \frac{1}{\varepsilon_{i+1}-\varepsilon_i}\right] \|x(0)\|,$$

and, using that $\varepsilon_i = \varepsilon_{i+1}^{2+r_{i+1}}$, we get

$$\|x_i(k)\| \leqslant \frac{(1+3\widetilde{M}_{1,i})\gamma_i}{\varepsilon_i^{r_i}}(1-\varepsilon_i)^k \|x(0)\|. \tag{4.37}$$

Similarly, we find that

$$\|F_{\varepsilon_i,i} x_i(k)\| \leqslant \beta_i \varepsilon_i (1-\varepsilon_i)^k \left[1 + \frac{\widetilde{M}_{1,i}}{\varepsilon_{i+1}^{r_{i+1}}} \frac{1}{\varepsilon_{i+1}-\varepsilon_i}\right] \|x(0)\|,$$

and hence,

$$\|F_{\varepsilon_i,i} x_i(k)\| \leqslant (1+3\widetilde{M}_{1,i})\beta_i \varepsilon_{i+1}(1-\varepsilon_i)^k \|x(0)\|. \tag{4.38}$$

Therefore, by recursion, we find that (4.37) and (4.38) hold for $i = 1, \ldots, q$. It is then easy to show that we also have

$$\|x_{q+1}(k)\| \leqslant \widetilde{M}_{2,q+1}(1-\varepsilon_1)^k \|x(0)\|.$$

The bounds we have obtained for the solution of the difference equation then immediately yield the two bounds presented in the theorem. ■

Remark 4.17 *We can essentially repeat Remark 4.12 which pertains to the direct method of design for continuous-time systems. To do so, all we need to do is to replace* (4.14), (4.15) *and* (4.16), *respectively, with* (4.33), (4.34), *and* (4.35).

We can easily construct now the observer-based measurement feedback controllers based on the state feedback design (4.24). The family of low-gain measurement feedback control laws take the form:

$$\begin{cases} \hat{x}(k+1) = A\hat{x}(k) + Bu(k) + K\left[y(k) - C\hat{x}(k)\right] \\ u(k) \quad\;\; = F_\varepsilon \hat{x}(k), \end{cases} \tag{4.39}$$

where the state feedback gain F_ε is given by (4.25) and the observer gain K is any matrix such that $A - KC$ is Schur stable. Note that another observer-based measurement feedback controller is also used often, and is given by

$$\begin{cases} \hat{x}(k+1) = A\hat{x}(k) + B\sigma(u(k)) + K[y(k) - C\hat{x}(k)] \\ u(k) \quad = F_\varepsilon \hat{x}(k). \end{cases} \tag{4.40}$$

Like in the continuous time, the advantage of (4.39) is that the controller is completely linear. The advantage of (4.40) is that the observer is always such that

$$x(k+1) - \hat{x}(k+1) = (A - KC)(x(k) - \hat{x}(k)), \tag{4.41}$$

while in the case of (4.39), the observer error only satisfies (4.41) if we guarantee that the saturation does not get activated. Therefore, the analysis to prove, for instance, stability is often easier when using (4.40).

The following theorem shows that the family of state feedback laws given in (4.24) solves Problem 4.4, namely, the problem of semi-global exponential stabilization via linear static state feedback. Also, it shows that the family of measurement feedback laws given in (4.39) solves Problem 4.5, namely, the problem of semi-global exponential stabilization via linear dynamic measurement feedback.

Theorem 4.18 *Consider the discrete-time system given in* (4.1). *Assume that the pair* (A, B) *is stabilizable and the eigenvalues of* A *are all located in the closed unit disc. Then the family of linear static state feedback laws given in* (4.24) *solves the problem of semi-global stabilization via state feedback as defined in Problem 4.4. More specifically, under the state feedback* (4.24), *for any given (arbitrarily large) bounded set* $\mathcal{X} \subset \mathbb{R}^n$, *there exists an* $\varepsilon^* \in (0, 1]$ *such that for all* $\varepsilon \in (0, \varepsilon^*]$, *the equilibrium point* $x = 0$ *of the closed-loop system is locally exponentially stable with* $\mathcal{X}$ *contained in its domain of attraction.*

Also, under the additional assumption that the pair (C, A) *is detectable, the family of linear dynamic measurement feedback laws given in* (4.39) *solves the problem of semi-global stabilization via measurement feedback, as defined in Problem 4.5. More specifically, under the measurement feedback law* (4.39), *for any given (arbitrarily large) bounded set* $\mathcal{X} \times \widehat{\mathcal{X}} \subset \mathbb{R}^{2n}$, *there exists an* $\varepsilon^* \in (0, 1]$ *such that for all* $\varepsilon \in (0, \varepsilon^*]$, *the equilibrium point* $(0, 0)$ *of the closed-loop system is locally exponentially stable with* $\mathcal{X} \times \widehat{\mathcal{X}}$ *contained in its domain of attraction.*

Proof : Given a bounded set $\mathcal{X} \subset \mathbb{R}^n$, choose ε^* such that for $\varepsilon < \varepsilon^*$ we have

$$M_2 \varepsilon_q \|x\| < 1$$

for all $x \in \mathcal{X}$, where M_2 is as defined in Theorem 4.16. Then, for all initial conditions $x(0) \in \mathcal{X}$, we have

$$\left\| F_\varepsilon(A + BF_\varepsilon)^k x(0) \right\| < 1,$$

and therefore, the saturation never gets activated. This implies that

$$x(k) = (A + BF_\varepsilon)^k x(0)$$

for all $k > 0$, and hence, (4.33) implies that the state converges to zero exponentially as t tends to infinity.

In the case of measurement feedback, we will also establish that the saturation will never get activated. If the saturation does not get activated, we note that (4.41) is satisfied. Since $A - KC$ is stable, we find that

$$\|x(k) - \hat{x}(k)\| \leqslant M_3(1-\sigma)^k \|x(0) - \hat{x}(0)\|$$

for some constants M_3 and $\sigma \in (0, 1]$. Using

$$\|F_\varepsilon x(k)\| \leqslant \|F_\varepsilon(A+BF_\varepsilon)^k\| \|x(0)\| + \sum_{\ell=0}^{k} F_\varepsilon(A+BF_\varepsilon)^{k-\ell} \|BF_\varepsilon\| \|x(\ell) - \hat{x}(\ell)\|,$$

we find that

$$\|F_\varepsilon x(k)\| \leqslant M_2\varepsilon_q \left[\|x(0)\| + \frac{M_2 M_3 \varepsilon_q \|B\|}{\sigma - \varepsilon_1} \|x(0) - \hat{x}(0)\| \right].$$

Hence, for ε small enough, we have

$$\|F_\varepsilon \hat{x}(k)\| \leqslant \|F_\varepsilon x(k)\| + \|F_\varepsilon (x(k) - \hat{x}(k))\| < 1$$

for all $k > 0$. This implies that the saturation does not get activated, and hence, we find exponential convergence to zero for all initial conditions satisfying $x(0) \in \mathcal{X}$ and $\hat{x}(0) \in \hat{X}$. ■

Remark 4.19 *In view of the above theorem, once again as in Remark 4.14, we can emphasize that the low-gain design described in this section is a "one shot" design as the design by itself does not depend on the a priori given set $\mathcal{X}$ or $(\mathcal{X}, \hat{\mathcal{X}})$.*

4.4 Semi-global stabilization: Riccati-based methods

As described in the previous section, direct method of designing low-gain feedback depends on directly assigning appropriate eigenstructure to the closed-loop system. To do so, it transforms the given system to a particular form (comprising of matrices in companion form) which reveals the innate structure of it. Although, the design method is straightforward and numerically efficient (see Remark 4.22 to follow), it is not very elegant to describe it. On the other hand, algebraic Riccati equation (ARE)-based methods of designing low-gain feedback can be described elegantly but lack numerical simplicity.

We describe in this section Riccati-based methods of designing low-gain feedback. As we said earlier, there exist in literature two types of Riccati-based methods, one method based on H_2 ARE and the other on H_∞ ARE. Although both the ARE-based methods are conceptually similar, they differ in details. Both of these methods are described below for both continuous- and discrete-time systems. The connections between the H_2- and H_∞-based low-gain feedbacks are further explored in Sect. 4.8.

4.4.1 H_2 ARE-based methods in continuous time

We consider an H_2 ARE-based method for continuous-time systems. We state the following lemma which plays an important role in the development to follow:

Lemma 4.20 *Let $Q_\varepsilon : (0, 1] \to \mathbb{R}^{n\times n}$ be a continuously differentiable matrix-valued function such that*

$$Q_\varepsilon > 0 \quad \text{and} \quad \frac{dQ_\varepsilon}{d\varepsilon} > 0$$

for any $\varepsilon \in (0, 1]$. Also, let

$$\lim_{\varepsilon\to 0} Q_\varepsilon = 0.$$

Assume that (A, B) is stabilizable and A has all its eigenvalues in the closed left-half plane. Then the H_2 continuous-time algebraic Riccati equation (CARE) defined as

$$PA + A'P - PBB'P + Q_\varepsilon = 0 \tag{4.42}$$

has a unique positive definite solution P_ε for any $\varepsilon \in (0, 1]$. Moreover, this positive definite solution P_ε has the following properties:

(*i*) *For any $\varepsilon \in (0, 1]$, the unique solution $P_\varepsilon > 0$ is such that $A - BB'P_\varepsilon$ is Hurwitz-stable.*

(*ii*) $\lim_{\varepsilon\to 0} P_\varepsilon = 0.$

(iii) P_ε *is continuously differentiable with respect to* ε *and*

$$\frac{dP_\varepsilon}{d\varepsilon} > 0, \quad \text{for any} \quad \varepsilon \in (0,1]. \tag{4.43}$$

Proof : The existence and uniqueness of a positive semi-definite solution P_ε is well known and follows, for instance, from [133]. It is also known (see [133]) that P_ε is the unique solution for which $A - BB'P_\varepsilon$ has all its eigenvalues in the closed left-half plane. For $\varepsilon = 0$, it is trivial to see that the positive semi-definite solution of the CARE (4.42) is equal to $P_0 = 0$ since, by assumption, $A - BB'P_0 = A$ has all its eigenvalues in the closed left-half complex plane. The fact that $\lim_{\varepsilon \to 0} P_\varepsilon = 0$ follows from standard continuity arguments. Note that for $\varepsilon > 0$ the solution is actually positive definite and is such that $A - BB'P_\varepsilon$ has all its eigenvalues in the open left-half plane.

Thus, we need to prove here only part *(iii)*. To do so, we observe that the continuous differentiability of P_ε for $\varepsilon > 0$ follows from the fact that the Hamiltonian matrix associated with the CARE (4.42) is a continuously differentiable function of ε and for $\varepsilon > 0$ the Hamiltonian matrix has no eigenvalues on the imaginary axis (see [64]). In order to show (4.43), we differentiate the CARE (4.42) to obtain the Lyapunov equation,

$$\frac{\mathrm{d}P_\varepsilon}{\mathrm{d}\varepsilon}(A - BB'P_\varepsilon) + (A - BB'P_\varepsilon)'\frac{\mathrm{d}P_\varepsilon}{\mathrm{d}\varepsilon} = -\frac{\mathrm{d}Q_\varepsilon}{\mathrm{d}\varepsilon}.$$

Now, (4.43) follows from the above equation since $A - BB'P_\varepsilon$ is asymptotically stable and $\frac{\mathrm{d}Q_\varepsilon}{\mathrm{d}\varepsilon} > 0$ for all $\varepsilon > 0$. ■

A family of low-gain state feedback control laws parameterized in ε is defined by

$$u = F_\varepsilon x \tag{4.44a}$$

where

$$F_\varepsilon := -B'P_\varepsilon, \quad \varepsilon \in (0,1], \tag{4.44b}$$

with P_ε being the positive definite solution of the algebraic Riccati equation (4.42). This family of state feedback gains (4.44) has the following property.

Theorem 4.21 *Consider the system* (4.45),

$$\dot{x} = Ax + Bu \tag{4.45}$$

where the state $x \in \mathbb{R}^n$ and the input $u \in \mathbb{R}^m$. Assume that (A, B) is stabilizable and A has all its eigenvalues in the closed left-half plane. Then, if we apply the state feedback law given by (4.44) to the system (4.45), the resulting closed-loop system,

$$\dot{x} = (A + BF_\varepsilon)x, \tag{4.46}$$

is asymptotically stable for all $\varepsilon > 0$. Moreover, there exist $\zeta_\varepsilon > 0$ and $\eta_\varepsilon > 0$ with $\zeta_\varepsilon \to 0$ as $\varepsilon \to 0$ such that for all $\varepsilon \in (0, 1]$,

$$\|F_\varepsilon e^{(A+BF_\varepsilon)t}\| \leqslant \zeta_\varepsilon e^{-\eta_\varepsilon t}. \tag{4.47}$$

Proof : The internal stability of the closed-loop system (4.46) follows trivially from Lemma 4.20. Next, we need to show (4.47). Using the CARE (4.42), we find that, for $\varepsilon \in (0, 1]$,

$$\begin{aligned} \frac{d}{dt}x'(t)P_\varepsilon x(t) &= -\|B'P_\varepsilon x(t)\|^2 - x'(t)Q_\varepsilon x(t) \\ &\leqslant -x'(t)Q_\varepsilon x(t) \\ &\leqslant -\eta_\varepsilon x'(t)P_\varepsilon x(t), \end{aligned}$$

where $\eta_\varepsilon = \lambda_{\min}(Q_\varepsilon)\|P_1\|^{-1}$. Hence

$$\|P_\varepsilon^{1/2}x(t)\| \leqslant e^{-\eta_\varepsilon t}\|P_\varepsilon^{1/2}x(0)\|. \tag{4.48}$$

Finally,

$$\begin{aligned} \|F_\varepsilon e^{(A+BF_\varepsilon)t}x(0)\| &= \|B'P_\varepsilon x(t)\| \\ &\leqslant \|B\|\|P_\varepsilon^{1/2}\|e^{-\eta_\varepsilon t}\|P_\varepsilon^{1/2}x(0)\|. \end{aligned} \tag{4.49}$$

Since (4.49) is true for all $x(0) \in \mathbb{R}^n$, it follows trivially that

$$\|F_\varepsilon e^{(A+BF_\varepsilon)t}\| \leq \|B\|\|P_\varepsilon^{1/2}\|^2 e^{-\eta_\varepsilon t} = \|B\|\|P_\varepsilon\|e^{-\eta_\varepsilon t}. \tag{4.50}$$

The proof is then completed by taking $\zeta_\varepsilon = \|B\|\|P_\varepsilon\|$. ■

For measurement feedback, the family of observer based low-gain measurement feedback control laws take the form,

$$\begin{cases} \dot{\hat{x}} = A\hat{x} + Bu + K(y - C\hat{x}) \\ u = -B'P_\varepsilon\hat{x}, \end{cases} \tag{4.51}$$

where, as before, P_ε is the positive definite solution of the continuous-time algebraic Riccati equation (4.42) and K is any matrix such that $A-KC$ is Hurwitz stable. As noted before, the alternative form,

$$\begin{cases} \dot{\hat{x}} = A\hat{x} + B\sigma(u) + K(y - C\hat{x}) \\ u = -B' P_\varepsilon \hat{x}, \end{cases} \tag{4.52}$$

is also often used in the literature since it is easier to analyze but for the moment we concentrate on the completely linear controller (4.51).

Remark 4.22 *We would like to emphasize here the numerical advantages of direct method in designing low-gain feedback compared to Riccati equation-based methods. As described earlier, in direct method, step 1 transforms the given system to a particular form. This obviously requires some numerical computations; however, such computations are independent of the parameter ε, and hence, they need to be done only once. Steps 2 and 3 where eigenstructure is assigned and the parameter ε is embedded are simple and straightforward, and thus do not involve much computational burden. In view of this, as the parameter ε varies, the design can easily be adapted. In contrast, in ARE-based methods, the pertinent ARE must be solved every time the parameter ε changes. This renders the design by ARE-based methods numerically expensive compared to the direct methods of eigenstructure assignment. Also, for a small value of the parameter ε, it is known that algebraic Riccati equations are numerically stiff to solve.*

In Theorem 4.13, we established that the low-gain feedback control laws based on the direct method solve the posed problems of semi-global stabilization via either linear state or measurement feedback. The following theorem presents the same result but relies on low-gain feedback laws designed via the H_2 algebraic Riccati equation:

Theorem 4.23 *Consider the continuous-time system given in* (4.1). *Assume that the pair (A, B) is stabilizable and the eigenvalues of A are all located in the closed left-half plane. Then the family of linear static state feedback laws given in* (4.44) *solves the problem of semi-global stabilization via state feedback as defined in Problem 4.4. More specifically, under the state feedback* (4.44a), *for any given (arbitrarily large) bounded set $\mathcal{X} \subset \mathbb{R}^n$, there exists an $\varepsilon^* \in (0, 1]$ such that for all $\varepsilon \in (0, \varepsilon^*]$, the equilibrium point $x = 0$ of the closed-loop system is locally exponentially stable with $\mathcal{X}$ contained in its domain of attraction.*

Also, under the additional assumption that the pair (C, A) is detectable, the family of linear dynamic measurement feedback laws given in (4.51) *solves the problem of semi-global stabilization via measurement feedback, as defined in Problem 4.5. More specifically, under the measurement feedback law* (4.51), *for any given (arbitrarily large) bounded set $\mathcal{X}\times\widehat{\mathcal{X}} \subset \mathbb{R}^{2n}$, there exists an $\varepsilon^* \in (0, 1]$*

such that for all $\varepsilon \in (0, \varepsilon^*]$*, the equilibrium point* $(0, 0)$ *of the closed-loop system is locally exponentially stable with* $\mathcal{X} \times \widehat{\mathcal{X}}$ *contained in its domain of attraction.*

Proof : Let us first consider the state feedback case. Under the state feedback law (4.44), the closed-loop system takes the following form:

$$\dot{x} = Ax + B\sigma(-B'P_\varepsilon x), \qquad \varepsilon \in (0, 1]. \tag{4.53}$$

Consider the Lyapunov function

$$V_\varepsilon(x) = x'P_\varepsilon x, \tag{4.54}$$

and let $c > 0$ be such that

$$c \geq \sup_{x \in \mathcal{X}, \varepsilon \in (0,1]} x'P_\varepsilon x. \tag{4.55}$$

Such a c exists because $\mathcal{X}$ is bounded and, by Lemma 4.20, $\lim_{\varepsilon \to 0} P_\varepsilon = 0$. Let ε^* be such that, for all $\varepsilon \in (0, \varepsilon^*]$,

$$x \in \mathcal{L}_V(c) = \{x \in \mathbb{R}^n : x'P_\varepsilon x \leq c\}$$

implies that $\|B'P_\varepsilon x\| \leq 1$, and hence, $\sigma(-B'P_\varepsilon x) = -B'P_\varepsilon x$. The existence of such an ε^* owes to the fact that $\lim_{\varepsilon \to 0} P_\varepsilon = 0$. It then follows that for all $\varepsilon \in (0, \varepsilon^*]$ and for $x \in \mathcal{L}_V(c)$, the closed-loop system (4.53) behaves linearly and can be written as

$$\dot{x} = (A - BB'P_\varepsilon)x. \tag{4.56}$$

The evaluation of $\dot{V}_\varepsilon$ along the trajectories of this closed-loop system gives

$$\dot{V}_\varepsilon = -x'Q_\varepsilon x - x'P_\varepsilon BB'P_\varepsilon x \leqslant -\lambda_{\min}(Q_\varepsilon)\|x\|^2 \tag{4.57}$$

for all $x \in \mathcal{L}_V(c)$. This shows that, for all $\varepsilon \in (0, \varepsilon^*]$, the equilibrium point $x = 0$ of the closed-loop system is locally exponentially stable with $\mathcal{L}_V(c)$ contained in its domain of attraction.

Finally, by observing that $\mathcal{X} \subset \mathcal{L}_V(c)$, we conclude our proof for the state feedback case.

Let us next consider the measurement feedback case. Under the output feedback law (4.51), the closed-loop system takes the form:

$$\begin{cases} \dot{x} = Ax + B\sigma(-B'P_\varepsilon \hat{x}) \\ \dot{\hat{x}} = A\hat{x} - BB'P_\varepsilon \hat{x} + K(y - C\hat{x}), \end{cases} \tag{4.58}$$

which, in the new coordinates (x, e) where $e = x - \widehat{x}$, becomes

$$\begin{cases} \dot{x} = Ax + B\sigma(-B'P_\varepsilon(x-e)) \\ \dot{e} = (A-KC)e + B[\sigma(-B'P_\varepsilon(x-e)) + B'P_\varepsilon(x-e)]. \end{cases} \tag{4.59}$$

Let $P_K > 0$ be the unique solution to the Lyapunov equation:

$$(A-KC)'P_K + P_K(A-KC) = -I. \tag{4.60}$$

The existence of such a P_K owes to the fact that $A - KC$ is Hurwitz stable.

Consider now the Lyapunov function,

$$V_\varepsilon(x,e) = x'P_\varepsilon x + e'P_K e, \tag{4.61}$$

and let $c > 0$ be such that

$$c \geq \sup_{(x,\widehat{x})\in\mathcal{X}\times\widehat{\mathcal{X}},\varepsilon\in(0,1]} \left[x'P_\varepsilon x + e'P_K e\right]. \tag{4.62}$$

Such a c exists because $\mathcal{X}$ and $\widehat{\mathcal{X}}$ are bounded and $\lim_{\varepsilon\to 0} P_\varepsilon = 0$. Let ε_1^* be such that, for all $\varepsilon \in (0, \varepsilon_1^*]$ and

$$(x,e) \in \mathcal{L}_V(c) = \{(x,e) \in \mathbb{R}^{2n} : x'P_\varepsilon x + e'P_K e \leqslant c\},$$

we have $\|B'P_\varepsilon(x-e)\| \leqslant \Delta$. The existence of such an ε_1^* again owes to the fact that $\lim_{\varepsilon\to 0} P_\varepsilon = 0$. The above immediately implies that

$$\sigma(-B'P_\varepsilon(x-e)) = -B'P_\varepsilon(x-e),$$

and hence, for all $\varepsilon \in (0, \varepsilon_1^*]$ and for all $(x,e) \in \mathcal{L}_V(c)$, the closed-loop system (4.59) behaves linearly and can be written as

$$\begin{cases} \dot{x} = (A - BB'P_\varepsilon)x + BB'P_\varepsilon e \\ \dot{e} = (A-KC)e. \end{cases} \tag{4.63}$$

The evaluation of $\dot{V}_\varepsilon$ along the trajectories of this closed-loop system gives, for all $(x,e) \in \mathcal{L}_V(c)$ and for all $\varepsilon \in (0, \varepsilon_1^*]$,

$$\begin{aligned} \dot{V}_\varepsilon &= -x'Q_\varepsilon x - x'P_\varepsilon BB'P_\varepsilon x + 2x'P_\varepsilon BB'P_\varepsilon e - e'e \\ &\leqslant -\lambda_{\min}(Q_\varepsilon)\|x\|^2 - (1 - \|B'P_\varepsilon\|^2)\|e\|^2. \end{aligned} \tag{4.64}$$

Let $\varepsilon^* \in (0, \varepsilon_1^*]$ be such that for all $\varepsilon \in (0, \varepsilon^*]$, $2\|B'P_\varepsilon\|^2 < 1$. Once again, the existence of such an ε^* owes to the fact that $\lim_{\varepsilon\to 0} P_\varepsilon = 0$. It then follows that,

$$\dot{V} \leq -\lambda_{\min}(Q_\varepsilon)\|x\|^2 - \tfrac{1}{2}\|e\|^2, \ \ \forall (x,e) \in \mathcal{L}_V(c) \text{ and } \forall \varepsilon \in (0, \varepsilon^*]. \tag{4.65}$$

This shows that, for all $\varepsilon \in (0, \varepsilon^*]$, the equilibrium point $(x, e) = (0, 0)$ of the closed-loop system is locally exponentially stable with $\mathcal{L}_V(c)$ contained in its domain of attraction.

Finally, we conclude our proof by observing that $(x, \hat{x}) \in \mathcal{X} \times \hat{\mathcal{X}}$ implies that $(x, e) \in \mathcal{L}_V(c)$. ■

4.4.2 H_2 ARE-based methods in discrete time

We consider here an H_2 ARE-based method for discrete-time systems. Our development here for discrete-time systems is conceptually analogous to the one presented in the previous section for continuous-time systems.

We first state the following lemma which plays an important role in the development to follow:

Lemma 4.24 *Let $Q_\varepsilon : (0, 1] \to \mathbb{R}^{n \times n}$ be a continuously differentiable matrix-valued function such that*

$$Q_\varepsilon > 0 \quad \text{and} \quad \frac{dQ_\varepsilon}{d\varepsilon} > 0$$

for any $\varepsilon \in (0, 1]$. Also, let

$$\lim_{\varepsilon \to 0} Q_\varepsilon = 0.$$

Assume that (A, B) is stabilizable and A has all its eigenvalues inside or on the unit circle. Then the H_2 discrete-time algebraic Riccati equation (DARE) defined as

$$P = A'PA + Q_\varepsilon - A'PB(B'PB + I)^{-1}B'PA \tag{4.66}$$

has a unique positive definite solution P_ε for any $\varepsilon \in (0, 1]$. Moreover, this positive definite solution P_ε has the following properties:

(i) For any $\varepsilon \in (0, 1]$, the unique matrix $P_\varepsilon > 0$ is such that

$$A - B(B'P_\varepsilon B + I)^{-1}B'P_\varepsilon A$$

is Schur stable.

(ii) $\lim_{\varepsilon \to 0} P_\varepsilon = 0.$

(iii) For any $\varepsilon \in (0, 1]$, we have

$$\|P_\varepsilon^{\frac{1}{2}} A P_\varepsilon^{-\frac{1}{2}}\| \leqslant \sqrt{2}.$$

(*iv*) *For any* $\varepsilon \in (0, 1]$,

$$\lambda_i(Q_\varepsilon) \leqslant \lambda_i(P_\varepsilon).$$

Moreover, strict inequality holds whenever A is non-singular.

(*v*) P_ε *is continuously differentiable with respect to* ε *and*

$$\frac{dP_\varepsilon}{d\varepsilon} > 0, \quad \text{for any } \varepsilon \in (0, 1].$$

Proof : The existence and uniqueness of a positive semi-definite solution P_ε is well known and follows from, for instance, [133]. It is also known (see [133]) that P_ε is the unique solution for which

$$A - B(B'P_\varepsilon B + I)^{-1}B'P_\varepsilon A$$

has all its eigenvalues on or inside the unit circle. For $\varepsilon = 0$, it is trivial to see that the DARE (4.66) has a solution $P_0 = 0$ since by assumption,

$$A - B(B'P_0 B + I)^{-1}B'P_0 = A$$

has all its eigenvalues inside or on the unit circle. The fact that $\lim_{\varepsilon\to 0} P_\varepsilon = 0$ follows from standard continuity arguments. Note that for $\varepsilon > 0$ the solution is actually positive definite and is such that $A - B(B'P_\varepsilon B + I)^{-1}B'P_\varepsilon A$ has all its eigenvalues inside the unit circle.

To show part (*iii*), we observe that by pre- and post-multiplying both sides of (4.66) with $P_\varepsilon^{-1/2}$, we obtain

$$V_\varepsilon \left[I - P_\varepsilon^{1/2} B(B'P_\varepsilon B + I)^{-1} B' P_\varepsilon^{1/2} \right] V_\varepsilon' = I - P_\varepsilon^{-1/2} Q_\varepsilon P_\varepsilon^{-1/2}, \quad (4.67)$$

where $V_\varepsilon = P_\varepsilon^{-1/2} A' P_\varepsilon^{1/2}$. Since $\lim_{\varepsilon\to 0} P_\varepsilon = 0$, it follows from the above equation that there exists an $\varepsilon^* > 0$ such that for all $\varepsilon \in (0, \varepsilon^*]$,

$$V_\varepsilon V_\varepsilon' \leqslant 2I - P_\varepsilon^{-1/2} Q_\varepsilon P_\varepsilon^{-1/2} \leqslant 2I.$$

This implies that

$$\|V_\varepsilon\| \le \sqrt{2}.$$

This completes the proof of part (*iii*).

Property (*iv*) follows from (4.67) since the left-hand side is clearly positive semi-definite for small ε, and therefore,

$$P_\varepsilon^{-1/2} Q_\varepsilon P_\varepsilon^{-1/2} \leqslant I$$

or, equivalently, $Q_\varepsilon \leqslant P_\varepsilon$. The proof of Property (*v*) follows along the same lines as that of its continuous-time counterpart given in the proof of Lemma 4.20. More specifically, we note that the fact that P_ε is continuously differentiable for $\varepsilon > 0$ follows from the fact that the Hamiltonian matrix associated with the DARE (4.66) is a continuously differentiable function of ε, and for $\varepsilon > 0$, the Hamiltonian matrix has no eigenvalues on the unit circle. Then, in order to prove Property (*v*) of Lemma 4.24, we need to show that $\frac{dP_\varepsilon}{d\varepsilon} > 0$. To show this, we observe that

$$\frac{\mathrm{d}(B'P_\varepsilon B + I)^{-1}}{\mathrm{d}\varepsilon} = -(B'P_\varepsilon B + I)^{-1} B' \frac{\mathrm{d}P_\varepsilon}{\mathrm{d}\varepsilon} B(B'P_\varepsilon B + I)^{-1}. \tag{4.68}$$

Now, differentiating (4.66) with respect to ε, and using equality (4.68), we get

$$\frac{\mathrm{d}P_\varepsilon}{\mathrm{d}\varepsilon} = (A' + F_\varepsilon' B')\frac{\mathrm{d}P_\varepsilon}{\mathrm{d}\varepsilon}(A + BF_\varepsilon) + \frac{\mathrm{d}Q_\varepsilon}{\mathrm{d}\varepsilon}, \tag{4.69}$$

where $F_\varepsilon = -(B'P_\varepsilon B + I)^{-1} B'P_\varepsilon A$. Noting that $A + BF_\varepsilon$ is Schur stable, and that $\frac{\mathrm{d}Q_\varepsilon}{\mathrm{d}\varepsilon} > 0$, we get $\frac{\mathrm{d}P_\varepsilon}{\mathrm{d}\varepsilon} > 0$. ■

The family of low-gain state feedback control laws parameterized in ε is defined by

$$u_\varepsilon = F_\varepsilon x, \tag{4.70a}$$

where

$$F_\varepsilon := -(B'P_\varepsilon B + I)^{-1} B'P_\varepsilon A, \quad \varepsilon \in (0, 1], \tag{4.70b}$$

and where P_ε is the positive definite solution of DARE (4.66). We refer to the control laws (4.44) and (4.70) as low-gain state feedback laws and ε as the *low-gain parameter* since, in view of Lemmas 4.20 and 4.24, one can make the norm of the feedback gain matrix F_ε arbitrarily small by choosing ε sufficiently small.

The family of state feedback gains (4.70) has the following property.

Theorem 4.25 *Consider the system*

$$x(k + 1) = Ax(k) + Bu(k), \tag{4.71}$$

where the state $x \in \mathbb{R}^n$ and the input $u \in \mathbb{R}^m$. Assume that (A, B) is stabilizable and A has all its eigenvalues within the closed unit circle. Then, if we apply the state feedback law given by (4.70) *to the system* (4.71), *the resulting closed-loop system,*

$$x(k + 1) = (A + BF_\varepsilon)x(k), \tag{4.72}$$

is asymptotically stable for all $\varepsilon > 0$*. Moreover, there exist positive-valued continuous functions* $\zeta_\varepsilon > 0$ *and* $0 < \eta_\varepsilon < 1$ *satisfying* $\lim_{\varepsilon \to 0} \zeta_\varepsilon = 0$ *such that*

$$\left\| F_\varepsilon [A + BF_\varepsilon]^k \right\| \leqslant \zeta_\varepsilon \eta_\varepsilon^k. \tag{4.73}$$

Proof : The internal stability of the closed-loop system (4.72) follows trivially from Lemma 4.24. Next, we need to show (4.73). Using the DARE (4.66), we find that, for $\varepsilon \in (0, 1]$,

$$\begin{aligned} x'(k+1) P_\varepsilon x(k+1) - x'(k) P_\varepsilon x(k) &= -\| F_\varepsilon x(k) \|^2 - x'(k) Q_\varepsilon x(k) \\ &\leqslant -x'(k) Q_\varepsilon x(k) \\ &\leqslant -\eta_\varepsilon x'(k) P_\varepsilon x(k), \end{aligned}$$

where $\eta_\varepsilon = \lambda_{\min}(Q_\varepsilon) \| P_1 \|^{-1}$. Hence,

$$\| P_\varepsilon^{1/2} x(k) \| \leqslant \eta_\varepsilon^k \| P_\varepsilon^{1/2} x(0) \|. \tag{4.74}$$

Finally,

$$\begin{aligned} \| F_\varepsilon (A + BF_\varepsilon)^k x(0) \| &\leqslant \left\| B' P_\varepsilon A x(k) \right\| \\ &\leqslant \| B' P_\varepsilon^{1/2} \| \| P_\varepsilon^{1/2} A P_\varepsilon^{-1/2} \| \| P_\varepsilon^{1/2} x(k) \| \\ &\leqslant \sqrt{2} \| B' P_\varepsilon^{1/2} \| \| P_\varepsilon^{1/2} x(0) \| \eta_\varepsilon^k. \end{aligned} \tag{4.75}$$

Since (4.75) is true for all $x(0) \in \mathbb{R}^n$, it follows trivially that

$$\| F_\varepsilon (A + BF_\varepsilon)^k \| \leq \sqrt{2} \| B \| \| P_\varepsilon^{1/2} \|^2 \eta_\varepsilon^k = \sqrt{2} \| B \| \| P_\varepsilon \| \eta_\varepsilon^k. \tag{4.76}$$

The proof is then completed by taking $\zeta_\varepsilon = \sqrt{2} \| B \| \| P_\varepsilon \|$. ■

The observer-based low-gain measurement feedback control laws take the form

$$\begin{cases} \hat{x}(k+1) = A\hat{x}(k) + Bu(k) + K(y(k) - C\hat{x}(k)) \\ \quad u(k) \quad = -(B' P_\varepsilon B + I)^{-1} B' P_\varepsilon A \hat{x}(k), \end{cases} \tag{4.77}$$

where, as before, P_ε is the positive definite solution of the DARE (4.66) and K is any matrix such that $A - KC$ is Schur stable. As noted before, the alternative form,

$$\begin{cases} \hat{x}(k+1) = A\hat{x}(k) + B\sigma(u(k)) + K(y(k) - C\hat{x}(k)) \\ \quad u(k) \quad = -(B' P_\varepsilon B + I)^{-1} B' P_\varepsilon A \hat{x}(k), \end{cases} \tag{4.78}$$

is also often used in the literature since it is easier to analyze, but for the moment, we concentrate on the completely linear controller (4.77).

Remark 4.26 *We can repeat here Remark 4.22 regarding the numerical advantages of direct method in designing low-gain feedback compared to the above ARE-based methods.*

The following theorem shows that the family of state feedback laws given in (4.70) solves Problem 4.4, namely, the problem of semi-global exponential stabilization via linear static state feedback. Also, it shows that the family of measurement feedback laws given in (4.77) solves Problem 4.5, namely, the problem of semi-global exponential stabilization via linear dynamic measurement feedback.

Theorem 4.27 *Consider the discrete-time system given in* (4.1). *Assume that the pair (A, B) is stabilizable and the eigenvalues of A are all located in the closed unit disc. Then the family of linear static state feedback laws given in* (4.70) *solves the problem of semi-global stabilization via state feedback as defined in Problem 4.4. More specifically, under the state feedback* (4.70a), *for any given (arbitrarily large) bounded set $\mathcal{X} \subset \mathbb{R}^n$, there exists an $\varepsilon^* \in (0, 1]$ such that for all $\varepsilon \in (0, \varepsilon^*]$, the equilibrium point $x = 0$ of the closed-loop system is locally exponentially stable with $\mathcal{X}$ contained in its domain of attraction.*

Also, under the additional assumption that the pair (C, A) is detectable, the family of linear dynamic measurement feedback laws given in (4.77) *solves the problem of semi-global stabilization via measurement feedback, as defined in Problem 4.5. More specifically, under the measurement feedback law* (4.77), *for any given (arbitrarily large) bounded set $\mathcal{X}\times\widehat{\mathcal{X}} \subset \mathbb{R}^{2n}$, there exists an $\varepsilon^* \in (0, 1]$ such that for all $\varepsilon \in (0, \varepsilon^*]$, the equilibrium point $(0, 0)$ of the closed-loop system is locally exponentially stable with $\mathcal{X} \times \widehat{\mathcal{X}}$ contained in its domain of attraction.*

Proof : The proof follows along the same lines as the proof of Theorem 4.23 that pertains to continuous-time systems with some subtle differences.

Let us first consider the state feedback case. Under the state feedback law (4.70), the closed-loop system takes the following form

$$x(k+1) = Ax(k) + B\sigma(-(B'P_\varepsilon B + I)^{-1}B'P_\varepsilon Ax(k)). \tag{4.79}$$

Consider the Lyapunov function

$$V_\varepsilon(x) = x'P_\varepsilon x, \tag{4.80}$$

and let $c > 0$ be such that

$$c \geq \sup_{x \in \mathcal{X}, \varepsilon \in (0,1]} x' P_\varepsilon x. \tag{4.81}$$

Such a c exists because $\mathcal{X}$ is bounded and, by Lemma 4.24, $\lim_{\varepsilon \to 0} P_\varepsilon = 0$. Let

$$x \in \mathcal{L}_V(c) = \{x \in \mathbb{R}^n : x' P_\varepsilon x \leq c\}.$$

We have

$$\left\|(B' P_\varepsilon B + I)^{-1} B' P_\varepsilon A x\right\| \leqslant \left\|B' P_\varepsilon A x\right\| \leqslant \|B' P_\varepsilon^{\frac{1}{2}}\| \|P_\varepsilon^{\frac{1}{2}} A P_\varepsilon^{-\frac{1}{2}}\| \|P_\varepsilon^{\frac{1}{2}} x\|.$$

Using (4.81) and Property (*iii*) of Lemma 4.24, we get

$$\left\|(B' P_\varepsilon B + I)^{-1} B' P_\varepsilon A x\right\| \leqslant \sqrt{2c} \|B' P_\varepsilon^{\frac{1}{2}}\|.$$

Let ε^* be such that, for all $\varepsilon \in (0, \varepsilon^*]$, we have $\sqrt{2c}\|B' P_\varepsilon^{\frac{1}{2}}\| \leqslant \Delta$. The existence of such an ε^* is guaranteed by the fact that $\lim_{\varepsilon \to 0} P_\varepsilon = 0$. It then follows that for all $\varepsilon \in (0, \varepsilon^*]$ and for all $x \in \mathcal{L}_V(c)$, the closed-loop system (4.79) behaves linearly and can be written as,

$$x(k+1) = (A - B(B' P_\varepsilon B + I)^{-1} B' P_\varepsilon A) x(k). \tag{4.82}$$

We get

$$V_\varepsilon(x(k+1)) - V_\varepsilon(x(k)) = -x' Q_\varepsilon x - x' F_\varepsilon' F_\varepsilon x \leqslant -\lambda_{\min}(Q_\varepsilon)\|x\|^2 \tag{4.83}$$

for all $x \in \mathcal{L}_V(c)$. This shows that, for all $\varepsilon \in (0, \varepsilon^*]$, the equilibrium point $x = 0$ of the closed-loop system is locally exponentially stable with $\mathcal{L}_V(c)$ contained in its domain of attraction.

Finally, by observing that $\mathcal{X} \subset \mathcal{L}_V(c)$, we conclude our proof for the state feedback case.

Let us next consider the measurement feedback case. Under the output feedback law (4.77), the closed-loop system takes the form,

$$\begin{cases} x(k+1) = Ax(k) + B\sigma(F_\varepsilon \hat{x}(k)) \\ \hat{x}(k+1) = A\hat{x}(k) + BF_\varepsilon \hat{x}(k) + K(y(k) - C\hat{x}(k)), \end{cases} \tag{4.84}$$

which, in the new coordinates (x, e) where $e = x - \hat{x}$, becomes

$$\begin{cases} x(k+1) = Ax(k) + B\sigma(F_\varepsilon(x(k) - e(k))) \\ e(k+1) = (A - KC)e(k) + B[\sigma(F_\varepsilon(x(k) - e(k))) - F_\varepsilon(x(k) - e(k))]. \end{cases} \tag{4.85}$$

Let $P_K > 0$ be the unique solution to the Lyapunov equation:

$$P_K = (A - KC)' P_K (A - KC) + I. \tag{4.86}$$

The existence of such a P_K owes to the fact that $A - KC$ is Schur stable.

Consider now the Lyapunov function,

$$V_\varepsilon(x, e) = x' P_\varepsilon x + e' P_K e, \tag{4.87}$$

and let $c > 0$ be such that

$$c \geq \sup_{(x,\hat{x}) \in \mathcal{X} \times \hat{\mathcal{X}}, \varepsilon \in (0,1]} \left[x' P_\varepsilon x + e' P_K e \right]. \tag{4.88}$$

Such a c exists because $\mathcal{X}$ and $\hat{\mathcal{X}}$ are bounded and $\lim_{\varepsilon \to 0} P_\varepsilon = 0$. Let ε_1^* be such that, for all $\varepsilon \in (0, \varepsilon_1^*]$,

$$(x, e) \in \mathcal{L}_V(c) = \{(x, e) \in \mathbb{R}^{2n} : x' P_\varepsilon x + e' P_K e \leqslant c\}.$$

We have $\| F_\varepsilon (x - e) \| \leqslant \Delta$, and hence,

$$\sigma(F_\varepsilon(x - e)) = F_\varepsilon(x - e).$$

The existence of such an ε_1^* again owes to the fact that $\lim_{\varepsilon \to 0} P_\varepsilon = 0$. Using (4.88) and Property *(iii)* of Lemma 4.24, we get

$$\left\| (B' P_\varepsilon B + I)^{-1} B' P_\varepsilon A (x - e) \right\| \leqslant M \| B' P_\varepsilon^{\frac{1}{2}} \|$$

for some constant M independent of ε, and hence, for ε small, we get $\| F_\varepsilon (x - e) \| \leqslant \Delta$ since $\lim_{\varepsilon \to 0} P_\varepsilon = 0$.

It then follows that for all $\varepsilon \in (0, \varepsilon_1^*]$ and for all $(x, e) \in \mathcal{L}_V(c)$, the closed-loop system (4.85) behaves linearly and can be written as

$$\begin{cases} x(k+1) = (A + BF_\varepsilon) x(k) - F_\varepsilon e(k) \\ e(k+1) = (A - KC) e(k). \end{cases} \tag{4.89}$$

The evaluation of V_ε along the trajectories of this closed-loop system gives, for all $(x, e) \in \mathcal{L}_V(c)$ and for all $\varepsilon \in (0, \varepsilon_1^*]$,

$$\begin{aligned} &V_\varepsilon(x(k+1), e(k+1)) - V_\varepsilon(x(k), e(k)) \\ &\quad = -x'(k) Q_\varepsilon x(k) - e'(k) e(k) - x'(k) F_\varepsilon' F_\varepsilon x(k) \\ &\qquad\quad + 2x'(k)(A + BF_\varepsilon)' P_\varepsilon B F_\varepsilon e(k) + e'(k) F_\varepsilon' B' P_\varepsilon B F_\varepsilon e(k) \\ &\quad \leqslant -x'(k) Q_\varepsilon x(k) - e'(k) e(k) + e'(k) F_\varepsilon' (I + B' P_\varepsilon B) F_\varepsilon e(k). \end{aligned}$$

Let $\varepsilon^* \in (0, \varepsilon_1^*]$ be such that for all $\varepsilon \in (0, \varepsilon^*]$,

$$\|F_\varepsilon'(I + B'P_\varepsilon B)F_\varepsilon\| < \tfrac{1}{2}.$$

Once again, the existence of such an ε^* owes to the fact that $\lim_{\varepsilon\to 0} P_\varepsilon = 0$. It then follows that,

$$V_\varepsilon(x(k+1), e(k+1)) - V_\varepsilon(x(k), e(k)) \leqslant -\lambda_{\min}(Q)\|x(k)\|^2 - \tfrac{1}{2}\|e(k)\|^2,$$

for all $(x, e) \in \mathcal{L}_V(c)$ and for all $\varepsilon \in (0, \varepsilon^*]$. This shows that, for all $\varepsilon \in (0, \varepsilon^*]$, the equilibrium point $(x, e) = (0, 0)$ of the closed-loop system is locally exponentially stable with $\mathcal{L}_V(c)$ contained in its domain of attraction.

Finally, we conclude our proof by observing that $(x, \hat{x}) \in \mathcal{X} \times \widehat{\mathcal{X}}$ implies that $(x, e) \in \mathcal{L}_V(c)$. ■

4.4.3 H_∞ ARE-based methods in continuous time

We consider here an H_∞ ARE-based method for continuous-time systems. We state the following lemma which plays an important role in the development to follow.

Lemma 4.28 *Let $Q_\varepsilon : (0, 1] \to \mathbb{R}^{n\times n}$ be a continuously differentiable matrix-valued function such that*

$$Q_\varepsilon > 0 \quad \text{and} \quad \frac{dQ_\varepsilon}{d\varepsilon} > 0$$

for any $\varepsilon \in (0, 1]$. Also, let

$$\lim_{\varepsilon\to 0} Q_\varepsilon = 0.$$

Assume that (A, B) is stabilizable and A has all its eigenvalues in the closed left-half plane. Then, there exists a positive real number γ^* such that the H_∞ continuous-time algebraic Riccati equation (CARE) defined as

$$PA + A'P - PBB'P + \gamma^{-2}PEE'P + Q_\varepsilon = 0, \tag{4.90}$$

has a unique positive definite solution $P_{\gamma,\varepsilon}$ such that

$$A - BB'P_{\gamma,\varepsilon} + \gamma^{-2}EE'P_{\gamma,\varepsilon} \tag{4.91}$$

is Hurwitz stable for any $\gamma > \gamma^*$ and $\varepsilon \in (0, 1]$. Moreover, this positive definite solution $P_{\gamma,\varepsilon}$ has the following properties:

(*i*) *For any $\varepsilon \in (0, 1]$, the unique solution $P_{\gamma,\varepsilon} > 0$ is such that $A - BB'P_{\gamma,\varepsilon}$ is also Hurwitz stable.*

(*ii*) $\lim_{\varepsilon \to 0} P_{\gamma,\varepsilon} = 0.$

(*iii*) *$P_{\gamma,\varepsilon}$ is continuously differentiable with respect to ε and*

$$\frac{dP_{\gamma,\varepsilon}}{d\varepsilon} > 0, \quad \textit{for any} \quad \varepsilon \in (0, 1]. \tag{4.92}$$

Remark 4.29 *Note that the matrix E can be arbitrarily chosen. In subsequent chapters, E is chosen appropriately to guarantee secondary goals such as disturbance rejection.*

Proof : Using the results of [34], since Q_1 is positive definite, there exists a γ^* such that, for each $\gamma > \gamma^*$, there exists a unique $P_{\gamma,1} > 0$ satisfying (4.90) for which (4.91) is asymptotically stable. Since $Q_\varepsilon \leqslant Q_1$ for $\varepsilon \in (0, 1]$, it implies that, for each $\gamma > \gamma^*$ and for $\varepsilon \subset (0, 1]$, there exists a unique $P_{\gamma,\varepsilon} > 0$ satisfying (4.90) for which (4.91) is asymptotically stable. We note that

$$\begin{aligned} P_{\gamma,\varepsilon}(A - BB'P_{\gamma,\varepsilon}) &+ (A - BB'P_{\gamma,\varepsilon})'P_{\gamma,\varepsilon} \\ &= -P_{\gamma,\varepsilon}BB'P_{\gamma,\varepsilon} - \gamma^{-2}P_{\gamma,\varepsilon}EE'P_{\gamma,\varepsilon} - Q_\varepsilon < 0, \end{aligned}$$

and then, standard properties of the Lyapunov equation guarantee that $A - BB'P_{\gamma,\varepsilon}$ is asymptotically stable.

To establish Property (*ii*), we first note that, by Lemma 4.20, there exists, for each $\varepsilon \in (0, 1]$ a unique positive-definite matrix $P_{\infty,\varepsilon}$ satisfying

$$P_{\infty,\varepsilon}A + A'P_{\infty,\varepsilon} - P_{\infty,\varepsilon}BB'P_{\infty,\varepsilon} + Q_\varepsilon = 0. \tag{4.93}$$

Moreover,

$$\lim_{\varepsilon \to 0} P_{\infty,\varepsilon} = 0. \tag{4.94}$$

It has been established in [161, Lemma 3.5] that

$$P_{\gamma,\varepsilon} \le \frac{\gamma^2}{\gamma^2 - (\gamma^*)^2} P_{\infty,\varepsilon}. \tag{4.95}$$

Property (*ii*) follows immediately from (4.94) and (4.95).

In order to show (4.92), we differentiate the CARE (4.90) to obtain the Lyapunov equation

$$\frac{\mathrm{d}P_{\gamma,\varepsilon}}{\mathrm{d}\varepsilon}(A - BB'P_{\gamma,\varepsilon} + \gamma^{-2}EE'P_{\gamma,\varepsilon}) + (A - BB'P_{\gamma,\varepsilon} + \gamma^{-2}EE'P_{\gamma,\varepsilon})'\frac{\mathrm{d}P_{\gamma,\varepsilon}}{\mathrm{d}\varepsilon} = -\frac{\mathrm{d}Q_\varepsilon}{\mathrm{d}\varepsilon}.$$

Now, (4.92) follows from the above equation since (4.91) is asymptotically stable and $\frac{\mathrm{d}Q_\varepsilon}{\mathrm{d}\varepsilon} > 0$ for all $\varepsilon > 0$. ■

A family of low-gain state feedback control laws parameterized in ε is defined by

$$u = F_{\gamma,\varepsilon}x, \tag{4.96a}$$

where

$$F_{\gamma,\varepsilon} := -B'P_{\gamma,\varepsilon}, \quad \varepsilon \in (0, 1], \tag{4.96b}$$

with $P_{\gamma,\varepsilon}$ the positive definite solution of the CARE (4.90). This family of state feedback gains (4.96) has the following property.

Theorem 4.30 *Consider the system*

$$\dot{x} = Ax + Bu, \tag{4.97}$$

where the state $x \in \mathbb{R}^n$ and the input $u \in \mathbb{R}^m$. Assume that (A, B) is stabilizable and A has all its eigenvalues in the closed left-half plane. Then, if we apply the state feedback law given by (4.96) *to the system* (4.97), *the resulting closed-loop system,*

$$\dot{x} = (A + BF_{\gamma,\varepsilon})x, \tag{4.98}$$

is asymptotically stable for all $\varepsilon > 0$. Moreover, there exist $\zeta_{\gamma,\varepsilon} > 0$ and $\eta_{\gamma,\varepsilon} > 0$ with $\zeta_{\gamma,\varepsilon} \to 0$ as $\varepsilon \to 0$ such that for all $\varepsilon \in (0, 1]$,

$$\|F_{\gamma,\varepsilon}e^{(A+BF_{\gamma,\varepsilon})t}\| \leqslant \zeta_{\gamma,\varepsilon}e^{-\eta_{\gamma,\varepsilon}t}. \tag{4.99}$$

Proof : The internal stability of the closed-loop system (4.98) follows trivially from Lemma 4.28. Next, we need to show (4.99). Using the CARE (4.90), we find that, for $\varepsilon \in (0, 1]$,

$$\begin{aligned}
\frac{d}{dt}x'(t)P_{\gamma,\varepsilon}x(t) &= -\|B'P_{\gamma,\varepsilon}x(t)\|^2 - \gamma^{-2}\|E'P_{\gamma,\varepsilon}x(t)\|^2 - x'(t)Q_\varepsilon x(t)\\
&\leqslant -x'(t)Q_\varepsilon x(t)\\
&\leqslant -\eta_{\gamma,\varepsilon}x'(t)P_{\gamma,\varepsilon}x(t),
\end{aligned}$$

where $\eta_{\gamma,\varepsilon} = \lambda_{\min}(Q_\varepsilon)\|P_{\gamma,1}\|^{-1}$. Hence,

$$\|P_{\gamma,\varepsilon}^{1/2}x(t)\| \leqslant e^{-\eta_{\gamma,\varepsilon}t}\|P_{\gamma,\varepsilon}^{1/2}x(0)\|. \tag{4.100}$$

Finally,

$$\begin{aligned}\|F_{\gamma,\varepsilon}e^{(A+BF_{\gamma,\varepsilon})t}x(0)\| &= \|B'P_{\gamma,\varepsilon}x(t)\| \\ &\leqslant \|B\|\|P_{\gamma,\varepsilon}^{1/2}\|\|P_{\gamma,\varepsilon}^{1/2}x(0)\|e^{-\eta_{\gamma,\varepsilon}t}.\end{aligned} \tag{4.101}$$

Since (4.101) is true for all $x(0) \in \mathbb{R}^n$, it follows trivially that

$$\|F_{\gamma,\varepsilon}e^{(A+BF_{\gamma,\varepsilon})t}\| \leq \|B\|\|P_{\gamma,\varepsilon}^{1/2}\|^2 e^{-\eta_{\gamma,\varepsilon}t} = \|B\|\|P_{\gamma,\varepsilon}\|e^{-\eta_{\gamma,\varepsilon}t}. \tag{4.102}$$

The proof is then completed by taking $\zeta_{\gamma,\varepsilon} = \|B\|\|P_{\gamma,\varepsilon}\|$. ■

For measurement feedback, the family of observer-based low-gain measurement feedback control laws take the form,

$$\begin{cases} \dot{\hat{x}} = A\hat{x} + Bu + K(y - C\hat{x}) \\ u = -B'P_{\gamma,\varepsilon}\hat{x}, \end{cases} \tag{4.103}$$

where, as before, $P_{\gamma,\varepsilon}$ is the positive definite solution of the CARE (4.90) and K is any matrix such that $A - KC$ is Hurwitz stable. As noted before, the alternative form,

$$\begin{cases} \dot{\hat{x}} = A\hat{x} + B\sigma(u) + K(y - C\hat{x}) \\ u = -B'P_{\gamma,\varepsilon}\hat{x}, \end{cases} \tag{4.104}$$

is also often used in the literature since it is easier to analyze, but for the moment, we concentrate on the completely linear controller (4.103).

The following theorem shows that the family of state feedback laws given in (4.96) solves Problem 4.4, namely, the problem of semi-global exponential stabilization via linear static state feedback. Also, it shows that the family of measurement feedback laws given in (4.103) solves Problem 4.5, namely, the problem of semi-global exponential stabilization via linear dynamic measurement feedback.

Theorem 4.31 *Consider the continuous-time system given in* (4.1). *Assume that the pair* (A, B) *is stabilizable and the eigenvalues of* A *are all located in the closed left-half plane. Then the family of linear static state feedback laws given in* (4.96) *solves the problem of semi-global stabilization via state feedback as defined in Problem 4.4. More specifically, under the state feedback* (4.96a), *for* *any given (arbitrarily large) bounded set* $\mathcal{X} \subset \mathbb{R}^n$*, there exists an* $\varepsilon^* \in (0, 1]$ *such that for all* $\varepsilon \in (0, \varepsilon^*]$*, the equilibrium point* $x = 0$ *of the closed-loop system is locally exponentially stable with* $\mathcal{X}$ *contained in its domain of attraction.*

Also, under the additional assumption that the pair (C, A) *is detectable, the family of linear dynamic measurement feedback laws given in* (4.103) *solves the problem of semi-global stabilization via measurement feedback as defined in Problem 4.5. More specifically, under the measurement feedback law* (4.103)*, for any given (arbitrarily large) bounded set* $\mathcal{X} \times \widehat{\mathcal{X}} \subset \mathbb{R}^{2n}$*, there exists an* $\varepsilon^* \in (0, 1]$ *such that for all* $\varepsilon \in (0, \varepsilon^*]$*, the equilibrium point* $(0, 0)$ *of the closed-loop system is locally exponentially stable with* $\mathcal{X} \times \widehat{\mathcal{X}}$ *contained in its domain of attraction.*

Proof : Let us first consider the state feedback case. Under the state feedback law (4.96), the closed-loop system takes the following form

$$\dot{x} = Ax + B\sigma(-B'P_{\gamma,\varepsilon}x), \qquad \varepsilon \in (0, 1]. \tag{4.105}$$

Consider the Lyapunov function

$$V_{\gamma,\varepsilon}(x) = x'P_{\gamma,\varepsilon}x, \tag{4.106}$$

and let $c > 0$ be such that

$$c \geq \sup_{x \in \mathcal{X}, \varepsilon \in (0,1]} x'P_{\gamma,\varepsilon}x. \tag{4.107}$$

Such a c exists because $\mathcal{X}$ is bounded and, by Lemma 4.28, $\lim_{\varepsilon \to 0} P_{\gamma,\varepsilon} = 0$. Let ε^* be such that, for all $\varepsilon \in (0, \varepsilon^*]$,

$$x \in \mathcal{L}_V(c) = \{x \in \mathbb{R}^n : x'P_{\gamma,\varepsilon}x \leq c\}$$

implies that $\|B'P_{\gamma,\varepsilon}x\| \leq 1$, and hence, $\sigma(-B'P_{\gamma,\varepsilon}x) = -B'P_{\gamma,\varepsilon}x$. The existence of such an ε^* owes to the fact that $\lim_{\varepsilon \to 0} P_{\gamma,\varepsilon} = 0$. It then follows that for all $\varepsilon \in (0, \varepsilon^*]$ and for $x \in \mathcal{L}_V(c)$, the closed-loop system (4.105) behaves linearly and can be written as,

$$\dot{x} = (A - BB'P_{\gamma,\varepsilon})x. \tag{4.108}$$

The evaluation of $\dot{V}_{\gamma,\varepsilon}$ along the trajectories of this closed-loop system gives,

$$\begin{aligned} \dot{V}_{\gamma,\varepsilon} &= -x'Q_\varepsilon x - x'P_{\gamma,\varepsilon}BB'P_{\gamma,\varepsilon}x - \gamma^{-2}x'P_{\gamma,\varepsilon}EE'P_{\gamma,\varepsilon}x \\ &\leqslant -\lambda_{\min}(Q_\varepsilon)\|x\|^2 \end{aligned}$$

for all $x \in \mathcal{L}_V(c)$. This shows that, for all $\varepsilon \in (0, \varepsilon^*]$, the equilibrium point $x = 0$ of the closed-loop system is locally exponentially stable with $\mathcal{L}_V(c)$ contained in its domain of attraction.

Finally, by observing that $\mathcal{X} \subset \mathcal{L}_V(c)$, we conclude our proof for the state feedback case.

Let us next consider the measurement feedback case. Under the output feedback law (4.103), the closed-loop system takes the form

$$\begin{cases} \dot{x} = Ax + B\sigma(-B'P_{\gamma,\varepsilon}\hat{x}) \\ \dot{\hat{x}} = A\hat{x} - BB'P_{\gamma,\varepsilon}\hat{x} + K(y - C\hat{x}), \end{cases} \tag{4.109}$$

which, in the new coordinates (x, e) where $e = x - \hat{x}$, becomes

$$\begin{cases} \dot{x} = Ax + B\sigma(-B'P_{\gamma,\varepsilon}(x-e)) \\ \dot{e} = (A - KC)e + B[\sigma(-B'P_{\gamma,\varepsilon}(x-e)) + B'P_{\gamma,\varepsilon}(x-e)]. \end{cases} \tag{4.110}$$

Let $P_K > 0$ be the unique solution to the Lyapunov equation

$$(A - KC)'P_K + P_K(A - KC) = -I. \tag{4.111}$$

The existence of such a P_K owes to the fact that $A - KC$ is Hurwitz stable.

Consider now the Lyapunov function,

$$V_{\gamma,\varepsilon}(x, e) = x'P_{\gamma,\varepsilon}x + e'P_K e, \tag{4.112}$$

and let $c > 0$ be such that

$$c \geq \sup_{(x,\hat{x}) \in \mathcal{X} \times \hat{\mathcal{X}}, \varepsilon \in (0,1]} \left[x'P_{\gamma,\varepsilon}x + e'P_K e\right]. \tag{4.113}$$

Such a c exists because $\mathcal{X}$ and $\hat{\mathcal{X}}$ are bounded and $\lim_{\varepsilon \to 0} P_{\gamma,\varepsilon} = 0$. Let ε_1^* be such that, for all $\varepsilon \in (0, \varepsilon_1^*]$,

$$(x, e) \in \mathcal{L}_V(c) = \{(x, e) \in \mathbb{R}^{2n} : x'P_{\gamma,\varepsilon}x + e'P_K e \leqslant c\},$$

we have $\|B'P_{\gamma,\varepsilon}(x - e)\| \leqslant \Delta$, and hence,

$$\sigma(-B'P_{\gamma,\varepsilon}(x - e)) = -B'P_{\gamma,\varepsilon}(x - e).$$

The existence of such an ε_1^* again owes to the fact that $\lim_{\varepsilon \to 0} P_{\gamma,\varepsilon} = 0$. It then follows that for all $\varepsilon \in (0, \varepsilon_1^*]$ and for all $(x, e) \in \mathcal{L}_V(c)$, the closed-loop system (4.110) behaves linearly and can be written as

$$\begin{cases} \dot{x} = (A - BB'P_{\gamma,\varepsilon})x + BB'P_{\gamma,\varepsilon}e \\ \dot{e} = (A - KC)e. \end{cases} \tag{4.114}$$

The evaluation of $\dot{V}_{\gamma,\varepsilon}$ along the trajectories of this closed-loop system gives, for all $(x,e) \in \mathcal{L}_V(c)$ and for all $\varepsilon \in (0, \varepsilon_1^*]$,

$$\begin{aligned}\dot{V}_{\gamma,\varepsilon} &= -x'Q_\varepsilon x - x'P_{\gamma,\varepsilon}BB'P_{\gamma,\varepsilon}x - \gamma^{-2}x'P_{\gamma,\varepsilon}EE'P_{\gamma,\varepsilon}x \\ &\quad + 2x'P_{\gamma,\varepsilon}BB'P_{\gamma,\varepsilon}e - e'e \\ &\leqslant -\lambda_{\min}(Q_\varepsilon)\|x\|^2 - (1 - \|B'P_{\gamma,\varepsilon}\|^2)\|e\|^2.\end{aligned}$$

Let $\varepsilon^* \in (0, \varepsilon_1^*]$ be such that for all $\varepsilon \in (0, \varepsilon^*]$, $2\|B'P_{\gamma,\varepsilon}\|^2 < 1$. Once again, the existence of such an ε^* owes to the fact that $\lim_{\varepsilon \to 0} P_{\gamma,\varepsilon} = 0$. It then follows that,

$$\dot{V} \leq -\lambda_{\min}(Q_\varepsilon)\|x\|^2 - \tfrac{1}{2}\|e\|^2, \quad \forall (x,e) \in \mathcal{L}_V(c) \text{ and } \forall \varepsilon \in (0, \varepsilon^*]. \tag{4.115}$$

This shows that, for all $\varepsilon \in (0, \varepsilon^*]$, the equilibrium point $(x,e) = (0,0)$ of the closed-loop system is locally exponentially stable with $\mathcal{L}_V(c)$ contained in its domain of attraction.

Finally, we conclude our proof by observing that $(x, \hat{x}) \in \mathcal{X} \times \hat{\mathcal{X}}$ implies that $(x,e) \in \mathcal{L}_V(c)$. ■

4.4.4 H_∞ ARE-based methods in discrete time

We consider here an H_∞ ARE-based method for discrete-time systems. Our development here for discrete-time systems is conceptually analogous to the one presented in the previous section for continuous-time systems.

We first state the following lemma which plays an important role in the development to follow:

Lemma 4.32 *Let $Q_\varepsilon : (0,1] \to \mathbb{R}^{n \times n}$ be a continuously differentiable matrix-valued function such that*

$$Q_\varepsilon > 0 \quad \text{and} \quad \frac{dQ_\varepsilon}{d\varepsilon} > 0$$

for any $\varepsilon \in (0,1]$. Also, let

$$\lim_{\varepsilon \to 0} Q_\varepsilon = 0.$$

Assume that (A, B) is stabilizable and A has all its eigenvalues inside or on the unit circle. Then, there exists a positive real number γ^* such that the H_∞ discrete-time algebraic Riccati equation (DARE),

$$P = A'PA + Q_\varepsilon - A'P\begin{pmatrix} B & E \end{pmatrix} G_\gamma(P)^{-1} \begin{pmatrix} B' \\ E' \end{pmatrix} PA, \tag{4.116a}$$

where

$$G_\gamma(P) = \begin{pmatrix} B'PB + I & B'PE \\ E'PB & E'PE - \gamma^2 I \end{pmatrix} \tag{4.116b}$$

with

$$R := \gamma^2 I - E'PE + E'PB(B'PB + I)^{-1}B'PE > 0, \tag{4.116c}$$

has a unique positive definite solution $P_{\gamma,\varepsilon}$ *for any* $\varepsilon \in (0, 1]$ *such that*

$$A - \begin{pmatrix} B & E \end{pmatrix} G_\gamma^{-1}(P_{\gamma,\varepsilon}) \begin{pmatrix} B' \\ E' \end{pmatrix} P_{\gamma,\varepsilon} A \tag{4.117}$$

is asymptotically stable. Moreover, this positive definite solution $P_{\gamma,\varepsilon}$ *has the following properties:*

(*i*) *For any* $\varepsilon \in (0, 1]$, *the unique matrix* $P_{\gamma,\varepsilon} > 0$ *is such that*

$$A - \begin{pmatrix} B & 0 \end{pmatrix} G_\gamma^{-1}(P_{\gamma,\varepsilon}) \begin{pmatrix} B' \\ E' \end{pmatrix} P_{\gamma,\varepsilon} A$$

is also Schur stable.

(*ii*) $\lim_{\varepsilon \to 0} P_{\gamma,\varepsilon} = 0.$

(*iii*) *For any* $\varepsilon \in (0, 1]$, *we have*

$$\| P_{\gamma,\varepsilon}^{\frac{1}{2}} A P_{\gamma,\varepsilon}^{-\frac{1}{2}} \| \leqslant \sqrt{2}.$$

(*iv*) *For any* $\varepsilon \in (0, 1]$,

$$\lambda_i(Q_\varepsilon) \leqslant \lambda_i(P_{\gamma,\varepsilon}).$$

Moreover, strict inequality holds whenever A *is non-singular.*

(*v*) $P_{\gamma,\varepsilon}$ *is continuously differentiable with respect to* ε *and*

$$\frac{dP_{\gamma,\varepsilon}}{d\varepsilon} > 0, \quad \textit{for any } \varepsilon \in (0, 1].$$

Remark 4.33 *Like in the continuous time, the matrix* E *can be arbitrarily chosen. In subsequent chapters,* E *is chosen appropriately to guarantee secondary goals such as disturbance rejection.*

Proof : Using the results of [160], since Q_1 is positive definite, there exists a γ^* such that, for each $\gamma > \gamma^*$, there exists a unique $P_{\gamma,1} > 0$ satisfying (4.116) with $\varepsilon = 1$ for which (4.117) is asymptotically stable. Since $Q_\varepsilon \leqslant Q_1$ for $\varepsilon \in (0,1]$, it implies that, for each $\gamma > \gamma^*$ and for $\varepsilon \in (0,1]$, there exists a unique $P_{\gamma,\varepsilon} > 0$ satisfying (4.116) for which (4.117) is asymptotically stable. Property (*i*) is a standard result which can be found in [160] as well.

To establish Property (*ii*), we first note that, by Lemma 4.20, there exists, for each $\varepsilon \in (0,1]$ a unique positive-definite matrix $P_{\infty,\varepsilon}$ satisfying

$$P_{\infty,\varepsilon} = A'P_{\infty,\varepsilon}A + Q_\varepsilon - A'P_{\infty,\varepsilon}B(B'P_{\infty,\varepsilon}B + I)^{-1}B'P_{\infty,\varepsilon}A. \tag{4.118}$$

Moreover,

$$\lim_{\varepsilon\to 0} P_{\infty,\varepsilon} = 0. \tag{4.119}$$

It has been established in [161, Lemma 9.7] that

$$P_{\gamma,\varepsilon} \leq \frac{\gamma^2}{\gamma^2 - (\gamma^*)^2} P_{\infty,\varepsilon}. \tag{4.120}$$

Property (*ii*) follows immediately from (4.119) and (4.120).

To show part (*iii*), we observe that by pre- and post-multiplying both sides of (4.116) with $P_\varepsilon^{-1/2}$, we obtain

$$V_{\gamma,\varepsilon}\left[I - P_{\gamma,\varepsilon}^{1/2}\begin{pmatrix} B & E\end{pmatrix} G_\gamma(P_{\gamma,\varepsilon})^{-1}\begin{pmatrix} B' \\ E'\end{pmatrix} P_{\gamma,\varepsilon}^{1/2}\right] V'_{\gamma,\varepsilon} = I - P_{\gamma,\varepsilon}^{-1/2} Q_\varepsilon P_{\gamma,\varepsilon}^{-1/2}, \tag{4.121}$$

where $V_{\gamma,\varepsilon} = P_{\gamma,\varepsilon}^{-1/2}A'P_{\gamma,\varepsilon}^{1/2}$. In view of Property (*ii*), it follows from the above equation that there exists an $\varepsilon^* > 0$ such that for all $\varepsilon \in (0,\varepsilon^*]$,

$$V_{\gamma,\varepsilon}V'_{\gamma,\varepsilon} \leqslant 2I - P_{\gamma,\varepsilon}^{-1/2} Q_\varepsilon P_{\gamma,\varepsilon}^{-1/2} \leqslant 2I.$$

This implies that

$$\|V_{\gamma,\varepsilon}\| \leq \sqrt{2}.$$

This completes the proof of part (*iii*).

Property (*iv*) follows from (4.121) since the left hand side is clearly positive semi-definite for small ε and therefore

$$P_{\gamma,\varepsilon}^{-1/2} Q_\varepsilon P_{\gamma,\varepsilon}^{-1/2} \leqslant I,$$

or, equivalently, $Q_\varepsilon \leqslant P_{\gamma,\varepsilon}$. The proof of Property (*v*) follows along the same lines as that of its continuous-time counterpart given in the proof of Lemma 4.28. More specifically, we note that the fact that $P_{\gamma,\varepsilon}$ is continuously differentiable for $\varepsilon > 0$ follows from the fact that the dependence on $\varepsilon > 0$ of the symplectic pencil associated with the DARE (4.116) is continuously differentiable, and for $\varepsilon > 0$, this symplectic pencil has no zeros on the unit circle. Then, in order to prove Property (v) of Lemma 4.32, we need to show that $\frac{\mathrm{d}P_{\gamma,\varepsilon}}{\mathrm{d}\varepsilon} > 0$. To show this, we observe that

$$\frac{\mathrm{d}G_\gamma^{-1}(P_{\gamma,\varepsilon})}{\mathrm{d}\varepsilon} = -G_\gamma^{-1}(P_{\gamma,\varepsilon})\begin{pmatrix} B' \\ E' \end{pmatrix}\frac{\mathrm{d}P_{\gamma,\varepsilon}}{\mathrm{d}\varepsilon}\begin{pmatrix} B & E \end{pmatrix}G_\gamma^{-1}(P_{\gamma,\varepsilon}). \tag{4.122}$$

Now, differentiating (4.116) with respect to ε, and using equality (4.122), we get

$$\frac{\mathrm{d}P_{\gamma,\varepsilon}}{\mathrm{d}\varepsilon} = A'_{\gamma,\varepsilon}\frac{\mathrm{d}P_{\gamma,\varepsilon}}{\mathrm{d}\varepsilon}A_{\gamma,\varepsilon} + \frac{\mathrm{d}Q_\varepsilon}{\mathrm{d}\varepsilon} \tag{4.123}$$

where

$$A_{\gamma,\varepsilon} = A - \begin{pmatrix} B & E \end{pmatrix}G_\gamma(P_{\gamma,\varepsilon})^{-1}\begin{pmatrix} B' \\ E' \end{pmatrix}P_{\gamma,\varepsilon}A.$$

Noting that $A_{\gamma,\varepsilon}$ is Schur stable, and that $\frac{\mathrm{d}Q_\varepsilon}{\mathrm{d}\varepsilon} > 0$, we get $\frac{\mathrm{d}P_{\gamma,\varepsilon}}{\mathrm{d}\varepsilon} > 0$. ■

The family of low-gain state feedback control laws parameterized in ε is defined by

$$u = F_{\gamma,\varepsilon}x, \tag{4.124a}$$

where

$$F_{\gamma,\varepsilon} := -(B'P_{\gamma,\varepsilon}B + I)^{-1}B'P_{\gamma,\varepsilon}A, \quad \varepsilon \in (0,1], \tag{4.124b}$$

and where $P_{\gamma,\varepsilon}$ is the positive definite solution of DARE (4.116). We refer to the control laws (4.96) and (4.124) as low-gain state feedback laws and ε as the *low-gain parameter* since, in view of Lemmas 4.28 and 4.32, one can make the norm of the feedback gain matrix F_ε arbitrarily small by choosing ε sufficiently small.

The family of state feedback gains (4.124) has the following property.

Theorem 4.34 *Consider the system*

$$x(k+1) = Ax(k) + Bu(k), \tag{4.125}$$

where the state $x \in \mathbb{R}^n$ and the input $u \in \mathbb{R}^m$. Assume that (A, B) is stabilizable and A has all its eigenvalues within the closed unit circle. Then, if we apply the state feedback law given by (4.124) *to the system* (4.125)*, the resulting closed-loop system,*

$$x(k+1) = (A + BF_{\gamma,\varepsilon})x(k), \tag{4.126}$$

is asymptotically stable for all $\varepsilon > 0$. *Moreover, there exist positive-valued continuous functions* $\zeta_{\gamma,\varepsilon} > 0$ *and* $0 < \eta_{\gamma,\varepsilon} < 1$ *satisfying* $\lim_{\varepsilon\to 0} \zeta_{\gamma,\varepsilon} = 0$ *such that*

$$\left\| F_{\gamma,\varepsilon}[A + BF_{\gamma,\varepsilon}]^k \right\| \leqslant \zeta_{\gamma,\varepsilon}\eta_{\gamma,\varepsilon}^k. \tag{4.127}$$

Proof : The internal stability of the closed-loop system (4.126) follows trivially from Lemma 4.32. Next, we need to show (4.127). Using the DARE (4.116), we find that, for $\varepsilon \in (0, 1]$,

$$\begin{aligned}
x'(k+1)&P_{\gamma,\varepsilon}x(k+1) - x'(k)P_{\gamma,\varepsilon}x(k) \\
&= -\|F_{\gamma,\varepsilon}x(k)\|^2 - x'(k)Q_\varepsilon x(k) \\
&\qquad - x'(k)A_F' P_{\gamma,\varepsilon}ER^{-1}E'P_{\gamma,\varepsilon}A_F x(k) \\
&\leqslant -x'(k)Q_\varepsilon x(k) \\
&\leqslant -\eta_{\gamma,\varepsilon}x'(k)P_{\gamma,\varepsilon}x(k),
\end{aligned}$$

where $A_F = A + BF_{\gamma,\varepsilon}$, R is defined by (4.116c), and $\eta_{\gamma,\varepsilon} = \lambda_{\min}(Q_\varepsilon)\|P_{\gamma,1}\|^{-1}$. Hence,

$$\|P_{\gamma,\varepsilon}^{1/2}x(k)\| \leqslant \eta_{\gamma,\varepsilon}^k\|P_{\gamma,\varepsilon}^{1/2}x(0)\|. \tag{4.128}$$

Finally,

$$\begin{aligned}
\|F_{\gamma,\varepsilon}(A + BF_{\gamma,\varepsilon})^k x(0)\| &\leqslant \left\|B'P_{\gamma,\varepsilon}Ax(k)\right\| \\
&\leqslant \|B'P_{\gamma,\varepsilon}^{1/2}\|\|P_{\gamma,\varepsilon}^{1/2}AP_{\gamma,\varepsilon}^{-1/2}\|\|P_{\gamma,\varepsilon}^{1/2}x(k)\| \\
&\leqslant \sqrt{2}\|B'P_{\gamma,\varepsilon}^{1/2}\|\|P_{\gamma,\varepsilon}^{1/2}x(0)\|\eta_{\gamma,\varepsilon}^k.
\end{aligned} \tag{4.129}$$

Since (4.129) is true for all $x(0) \in \mathbb{R}^n$, it follows trivially that

$$\|F_{\gamma,\varepsilon}(A + BF_{\gamma,\varepsilon})^k\| \leq \sqrt{2}\|B\|\|P_{\gamma,\varepsilon}^{1/2}\|^2\eta_{\gamma,\varepsilon}^k = \sqrt{2}\|B\|\|P_{\gamma,\varepsilon}\|\eta_{\gamma,\varepsilon}^k. \tag{4.130}$$

The proof is then completed by taking $\zeta_{\gamma,\varepsilon} = \sqrt{2}\|B\|\|P_{\gamma,\varepsilon}\|$. ■

The observer-based low-gain measurement feedback control laws the following form:

$$\begin{cases}
\hat{x}(k+1) = A\hat{x}(k) + Bu(k) + K(y(k) - C\hat{x}(k)) \\
u(k) \quad = -(B'P_{\gamma,\varepsilon}B + I)^{-1}B'P_{\gamma,\varepsilon}A\hat{x}(k),
\end{cases} \tag{4.131}$$

where, as before, $P_{\gamma,\varepsilon}$ is the positive definite solution of the DARE (4.116) and K is any matrix such that $A - KC$ is Schur stable. As noted before, the alternative form,

$$\begin{cases} \hat{x}(k+1) = A\hat{x}(k) + B\sigma(u(k)) + K(y(k) - C\hat{x}(k)) \\ \quad u(k) \quad = -(B'P_{\gamma,\varepsilon}B + I)^{-1}B'P_{\gamma,\varepsilon}A\hat{x}(k), \end{cases} \tag{4.132}$$

is also often used in the literature since it is easier to analyze, but for the moment, we concentrate on the completely linear controller (4.131).

Remark 4.35 *We can repeat here Remark 4.22 regarding the numerical advantages of direct method in designing low-gain feedback compared to the above ARE-based methods.*

The following theorem shows that the family of state feedback laws given in (4.124) solves Problem 4.4, namely, the problem of semi-global exponential stabilization via linear static state feedback. Also, it shows that the family of measurement feedback laws given in (4.131) solves Problem 4.5, namely, the problem of semi-global exponential stabilization via linear dynamic measurement feedback.

Theorem 4.36 *Consider the discrete-time system given in* (4.1). *Assume that the pair* (A, B) *is stabilizable and the eigenvalues of* A *are all located in the closed unit disc. Then the family of linear static state feedback laws given in* (4.124) *solves the problem of semi-global stabilization via state feedback as defined in Problem 4.4. More specifically, under the state feedback* (4.124a), *for any given (arbitrarily large) bounded set* $\mathcal{X} \subset \mathbb{R}^n$, *there exists an* $\varepsilon^* \in (0, 1]$ *such that for all* $\varepsilon \in (0, \varepsilon^*]$, *the equilibrium point* $x = 0$ *of the closed-loop system is locally exponentially stable with* $\mathcal{X}$ *contained in its domain of attraction.*

Also, under the additional assumption that the pair (C, A) *is detectable, the family of linear dynamic measurement feedback laws given in* (4.131) *solves the problem of semi-global stabilization via measurement feedback as defined in Problem 4.5. More specifically, under the measurement feedback law* (4.131), *for any given (arbitrarily large) bounded set* $\mathcal{X}\times\hat{\mathcal{X}} \subset \mathbb{R}^{2n}$, *there exists an* $\varepsilon^* \in (0, 1]$ *such that for all* $\varepsilon \in (0, \varepsilon^*]$, *the equilibrium point* $(0, 0)$ *of the closed-loop system is locally exponentially stable with* $\mathcal{X} \times \hat{\mathcal{X}}$ *contained in its domain of attraction.*

Proof : The proof follows along the same lines as the proof of Theorem 4.31 with some subtle differences.

Let us first consider the state feedback case. Under the state feedback law (4.124), the closed-loop system takes the following form:

$$x(k+1) = Ax(k) + B\sigma(F_{\gamma,\varepsilon}x(k)). \tag{4.133}$$

Consider the Lyapunov function

$$V_\varepsilon(x) = x' P_{\gamma,\varepsilon} x, \tag{4.134}$$

and let $c > 0$ be such that

$$c \geq \sup_{x \in \mathcal{X}, \varepsilon \in (0,1]} x' P_{\gamma,\varepsilon} x. \tag{4.135}$$

Such a c exists because $\mathcal{X}$ is bounded and, by Lemma 4.32, $\lim_{\varepsilon \to 0} P_{\gamma,\varepsilon} = 0$. Let

$$x \in \mathcal{L}_V(c) = \{x \in \mathbb{R}^n : x' P_{\gamma,\varepsilon} x \leq c\}.$$

We have

$$\begin{aligned}\|(B' P_{\gamma,\varepsilon} B + I)^{-1} B' P_{\gamma,\varepsilon} A x\| &\leqslant \|B' P_{\gamma,\varepsilon} A x\| \\ &\leqslant \|B' P_{\gamma,\varepsilon}^{\frac{1}{2}}\| \|P_{\gamma,\varepsilon}^{\frac{1}{2}} A P_{\gamma,\varepsilon}^{-\frac{1}{2}}\| \|P_{\gamma,\varepsilon}^{\frac{1}{2}} x\|.\end{aligned}$$

Using (4.135) and Property (*iii*) of Lemma 4.32, we get

$$\|(B' P_{\gamma,\varepsilon} B + I)^{-1} B' P_{\gamma,\varepsilon} A x\| \leqslant \sqrt{2c} \|B' P_{\gamma,\varepsilon}^{\frac{1}{2}}\|.$$

Let ε^* be such that, for all $\varepsilon \in (0, \varepsilon^*]$, we have $\sqrt{2c}\|B' P_{\gamma,\varepsilon}^{\frac{1}{2}}\| \leqslant \Delta$. The existence of such an ε^* is guaranteed by the fact that $\lim_{\varepsilon \to 0} P_{\gamma,\varepsilon} = 0$. It then follows that for all $\varepsilon \in (0, \varepsilon^*]$ and for all $x \in \mathcal{L}_V(c)$, the closed-loop system (4.133) behaves linearly and can be written as,

$$x(k+1) = (A - B(B' P_{\gamma,\varepsilon} B + I)^{-1} B' P_{\gamma,\varepsilon} A) x(k). \tag{4.136}$$

We get

$$\begin{aligned}&V_\varepsilon(x(k+1)) - V_\varepsilon(x(k)) \\ &\quad = -x'(k) Q_\varepsilon x(k) - x'(k) F_{\gamma,\varepsilon}' F_{\gamma,\varepsilon} x(k) \\ &\qquad - x'(k) A_F' P_{\gamma,\varepsilon} E R^{-1} E' P_{\gamma,\varepsilon} A_F x(k) \\ &\quad \leqslant -\lambda_{\min}(Q_\varepsilon)\|x\|^2\end{aligned}$$

for all $x \in \mathcal{L}_V(c)$. This shows that, for all $\varepsilon \in (0, \varepsilon^*]$, the equilibrium point $x = 0$ of the closed-loop system is locally exponentially stable with $\mathcal{L}_V(c)$ contained in its domain of attraction.

Finally, by observing that $\mathcal{X} \subset \mathcal{L}_V(c)$, we conclude our proof for the state feedback case.

Let us next consider the measurement feedback case. Under the output feedback law (4.131), the closed-loop system takes the form,

$$\begin{cases} x(k+1) = Ax(k) + B\sigma(F_{\gamma,\varepsilon}x(k)) \\ \widehat{x}(k+1) = A\widehat{x}(k) + BF_{\gamma,\varepsilon}\widehat{x}(k) + K(y(k) - C\widehat{x}(k)) \end{cases} \tag{4.137}$$

which, in the new coordinates (x, e) where $e = x - \widehat{x}$, becomes

$$\begin{cases} x(k+1) = Ax(k) + B\sigma(F_{\gamma,\varepsilon}(x(k) - e(k))) \\ e(k+1) = (A - KC)e(k) \\ \qquad\qquad +B[\sigma(F_{\gamma,\varepsilon}(x(k) - e(k))) - F_{\gamma,\varepsilon}(x(k) - e(k))]. \end{cases} \tag{4.138}$$

Let $P_K > 0$ be the unique solution to the Lyapunov equation,

$$P_K = (A - KC)' P_K (A - KC) + I. \tag{4.139}$$

The existence of such a P_K owes to the fact that $A - KC$ is Schur stable.

Consider now the Lyapunov function,

$$V_\varepsilon(x, e) = x' P_{\gamma,\varepsilon} x + e' P_K e, \tag{4.140}$$

and let $c > 0$ be such that

$$c \geq \sup_{(x,\widehat{x}) \in \mathcal{X} \times \widehat{\mathcal{X}}, \varepsilon \in (0,1]} \left[x' P_{\gamma,\varepsilon} x + e' P_K e \right]. \tag{4.141}$$

Such a c exists because $\mathcal{X}$ and $\widehat{\mathcal{X}}$ are bounded and $\lim_{\varepsilon \to 0} P_{\gamma,\varepsilon} = 0$. Let ε_1^* be such that, for all $\varepsilon \in (0, \varepsilon_1^*]$,

$$(x, e) \in \mathcal{L}_V(c) = \{(x, e) \in \mathbb{R}^{2n} : x' P_{\gamma,\varepsilon} x + e' P_K e \leqslant c\},$$

we have $\| F_{\gamma,\varepsilon}(x - e) \| \leqslant \Delta$, and hence,

$$\sigma(F_{\gamma,\varepsilon}(x - e)) = F_{\gamma,\varepsilon}(x - e).$$

The existence of such an ε_1^* again owes to the fact that $\lim_{\varepsilon \to 0} P_{\gamma,\varepsilon} = 0$. Using (4.141) and Property (*iii*) of Lemma 4.32, we get

$$\left\| (B' P_{\gamma,\varepsilon} B + I)^{-1} B' P_{\gamma,\varepsilon} A(x - e) \right\| \leqslant M \| B' P_{\gamma,\varepsilon}^{\frac{1}{2}} \|$$

for some constant M independent of ε, and hence, for ε small, we get $\| F_{\gamma,\varepsilon}(x - e) \| \leqslant \Delta$ since $\lim_{\varepsilon \to 0} P_{\gamma,\varepsilon} = 0$.

It then follows that for all $\varepsilon \in (0, \varepsilon_1^*]$ and for all $(x, e) \in \mathcal{L}_V(c)$, the closed-loop system (4.138) behaves linearly and can be written as

$$\begin{cases} x(k+1) = (A + BF_{\gamma,\varepsilon})x(k) - F_{\gamma,\varepsilon}e(k) \\ e(k+1) = (A - KC)e(k). \end{cases} \tag{4.142}$$

The evaluation of V_ε along the trajectories of this closed-loop system gives, for all $(x, e) \in \mathcal{L}_V(c)$ and for all $\varepsilon \in (0, \varepsilon_1^*]$,

$$\begin{aligned} &V_{\gamma,\varepsilon}(x(k+1), e(k+1)) - V_{\gamma,\varepsilon}(x(k), e(k)) \\ &\quad = -x'(k)Q_\varepsilon x(k) - e'(k)e(k) - x'(k)F'_{\gamma,\varepsilon}F_{\gamma,\varepsilon}x(k) \\ &\qquad - x'(k)A'_F P_{\gamma,\varepsilon}ER^{-1}E'P_{\gamma,\varepsilon}A_F x(k) \\ &\qquad + 2x'(k)(A + BF_{\gamma,\varepsilon})'P_{\gamma,\varepsilon}BF_{\gamma,\varepsilon}e(k) + e'(k)F'_{\gamma,\varepsilon}B'P_{\gamma,\varepsilon}BF_{\gamma,\varepsilon}e(k) \\ &\quad \leqslant -x'(k)Q_\varepsilon x(k) - e'(k)e(k) + e'(k)F'_{\gamma,\varepsilon}(I + B'P_{\gamma,\varepsilon}B)F_{\gamma,\varepsilon}e(k). \end{aligned}$$

Let $\varepsilon^* \in (0, \varepsilon_1^*]$ be such that for all $\varepsilon \in (0, \varepsilon^*]$,

$$\|F'_{\gamma,\varepsilon}(I + B'P_{\gamma,\varepsilon}B)F_{\gamma,\varepsilon}\| < \tfrac{1}{2}.$$

Once again, the existence of such an ε^* owes to the fact that $\lim_{\varepsilon \to 0} P_{\gamma,\varepsilon} = 0$. It then follows that

$$V_{\gamma,\varepsilon}(x(k+1), e(k+1)) - V_{\gamma,\varepsilon}(x(k), e(k)) \leqslant -\lambda_{\min}(Q_\varepsilon)\|x(k)\|^2 - \tfrac{1}{2}\|e(k)\|^2,$$

for all $(x, e) \in \mathcal{L}_V(c)$ and for all $\varepsilon \in (0, \varepsilon^*]$. This shows that, for all $\varepsilon \in (0, \varepsilon^*]$, the equilibrium point $(x, e) = (0, 0)$ of the closed-loop system is locally exponentially stable with $\mathcal{L}_V(c)$ contained in its domain of attraction.

Finally, we conclude our proof by observing that $(x, \hat{x}) \in \mathcal{X} \times \hat{\mathcal{X}}$ implies that $(x, e) \in \mathcal{L}_V(c)$. ■

4.5 Semi-global stabilization by low-and-high-gain design

The low-gain design method discussed in previous sections has one serious drawback. Although it helps to internally stabilize a given system, it does not make use of the available control capacity fully. Let us expand on this. The semi-globally stabilizing state or measurement feedback controllers are constructed by linear low-gain feedback in such a way that the control input does not saturate for a priori given arbitrarily large bounded set of initial conditions. As a result, the closed-loop system operates as a linear system. This leads to control capacity being not utilized fully. More specifically, in the low-gain design, for a large given set of initial conditions, the gain of the feedback is chosen sufficiently low so that the control magnitude will not saturate. Consequently, because of the linearity, whenever the state is close to the origin, the control magnitude will be far from its fully allowed value, and thus, the closed-loop system will be operating far from its full capacity. This is a critical issue in view of the fact that the control capacity of systems subject to input saturation is closely tied to performance issues such as fast response, disturbance rejection, and robustness in the face of uncertainties.

For continuous-time systems, the abovementioned drawback of low-gain feedback lead the researchers to develop an improved design method which can appropriately be termed as low-and-high-gain design method. Such a low-and-high-gain design method was basically conceived for semi-global control problems beyond stabilization and was related to the performance issues such as semi-global stabilization with enhanced utilization of the available control capacity, semi-global disturbance rejection, and robustness of stability with respect to uncertainties [74, 77, 78, 128]. The low-and-high-gain control laws are composite control laws. Namely, they are composed by adding a low-gain control law and a high-gain control law. The design is sequential. First, a low-gain control law is designed as in the previous sections while using a tuning parameter ε. Then, utilizing an appropriate Lyapunov function for the closed-loop system under such a low-gain control law, a high-gain control law is constructed by embedding another tuning parameter α. Thus, both the low-gain and high-gain controllers are equipped with tuning parameters. The role of the low-gain and that of the high-gain controller are completely separated. The role of the low-gain control law is to ensure, independent of the high-gain controller:

1. The asymptotic stability of the equilibrium of the closed-loop system.
2. That the basin of attraction of the closed-loop system contains an a priori given bounded set.

In fact, the tuning parameter ε in the low-gain controller can be tuned to enlarge the basin of attraction of the equilibrium of the closed-loop system such that it includes any a priori given (arbitrarily large) bounded set. On the other hand, the role of the high-gain controller is to achieve performance beyond stabilization such as disturbance rejection, robustness, and improved transient response. Again, this performance is achieved by appropriate choice of the tuning parameter α of the high-gain controller. We must mention that the low-gain parameter basically determines the domain of attraction. The high-gain parameter primary focus is to speed up the performance.

Obviously, one cannot have low-and-high-gain control laws for discrete-time systems in the same way as one does for continuous-time systems. This is so because one cannot arbitrarily increase gain for discrete-time systems. As mentioned above, the high-gain component of low-and-high-gain design is basically a "Lyapunov design". Recognizing this, one can produce the high-gain component for discrete-time systems although with profoundly different features than those obtained for the continuous-time case. This is what has been done in [84, 85] and will be reviewed shortly.

As in the case of designing low-gain control laws, there exist two methods of designing low-and-high-gain control laws, one based on direct eigenstructure assignment and the other ARE-based. We first describe the direct eigenstructure assignment method.

4.5.1 Direct eigenstructure assignment: continuous time

In Sect. 4.3, we designed low-gain feedback control laws that solve the semi-global stabilization problem. By reducing the parameter ε as close to zero as needed, we obtain as large a domain of attraction as needed. However, smaller ε also implies that the feedback gain becomes smaller, and hence, we have slow convergence to the equilibrium. We first present a preliminary lemma.

The following lemma gives a method to speed up the performance of the system by introducing a high-gain parameter:

Lemma 4.37 *Consider a linear system*

$$\dot{x} = Ax + Bu, \quad x(t) \in \mathbb{R}^n$$

with all eigenvalues of A in the closed left-half plane. Let $F_{L,\varepsilon}$ be a matrix such that there exists a $Q_\varepsilon > 0$ such that the solution $P_\varepsilon > 0$ of the Lyapunov equation,

$$(A + BF_{L,\varepsilon})' P_\varepsilon + P_\varepsilon(A + BF_{L,\varepsilon}) + Q_\varepsilon = 0,$$

has the property that

$$\lim_{\varepsilon \to 0} P_\varepsilon = 0, \tag{4.143}$$

and

$$P_\varepsilon \geqslant F_{L,\varepsilon}' F_{L,\varepsilon} \tag{4.144}$$

for all $\varepsilon > 0$, then the feedback controller

$$u = F_{L,\varepsilon} x + \alpha F_{H,\varepsilon} x, \tag{4.145}$$

where $F_{H,\varepsilon} = -B' P_\varepsilon$, Then, for any compact set $\mathcal{X}$ there exists an ε^* such that, for all $0 < \varepsilon < \varepsilon^*$ and all $\alpha \geqslant 0$, the system

$$\dot{x} = Ax + B\sigma(u), \quad x(t) \in \mathbb{R}^n$$

with controller (4.145) *is locally asymptotically stable with $\mathcal{X}$ contained in its domain of attraction.*

Proof : Consider the set

$$\mathcal{V}_\varepsilon := \left\{ x \in \mathbb{R}^n \mid x' P_\varepsilon x < 1 \right\}.$$

We choose ε sufficiently small such that $\mathcal{X} \subset \mathcal{V}_\varepsilon$. This is possible due to (4.143).

We note that $\mathcal{V}_\varepsilon$ is a positively invariant set in the sense that $x(t) \in \mathcal{V}_\varepsilon$ implies that $x(\tau) \in \mathcal{V}_\varepsilon$ for all $\tau > t$. In order to establish this, we first note that

$$\frac{\mathrm{d}}{\mathrm{d}t} x'(t) P_\varepsilon x(t) = -x'(t) Q_\varepsilon x(t) + 2x'(t) P_\varepsilon B v(t),$$

where

$$v(t) = \sigma\left(F_{L,\varepsilon} x(t) + \alpha F_{H,\varepsilon} x(t)\right) - F_{L,\varepsilon} x(t).$$

Since $x(t) \in \mathcal{V}_\varepsilon$, we have

$$x'(t) F'_{L,\varepsilon} F_{L,\varepsilon} x(t) \leqslant x'(t) P_\varepsilon x(t) < 1,$$

where we use (4.144), and hence, $\|F_{L,\varepsilon} x(t)\| < 1$. This implies that

$$\sigma((F_{L,\varepsilon} x(t) + \alpha F_{H,\varepsilon} x(t)) = F_{L,\varepsilon} x(t) + \begin{pmatrix} \mu_1 & 0 & \cdots & 0 \\ 0 & \ddots & \ddots & \vdots \\ \vdots & \ddots & \ddots & 0 \\ 0 & \cdots & 0 & \mu_m \end{pmatrix} \alpha F_{H,\varepsilon} x(t)$$

with $\mu_i \in [0, 1]$ for $i = 1, \ldots, m$. This implies that

$$\begin{aligned} \frac{\mathrm{d}}{\mathrm{d}t} x'(t) P_\varepsilon x(t) &= -x'(t) Q_\varepsilon x(t) \\ &\quad - \alpha 2x'(t) P_\varepsilon B \begin{pmatrix} \mu_1 & 0 & \cdots & 0 \\ 0 & \ddots & \ddots & \vdots \\ \vdots & \ddots & \ddots & 0 \\ 0 & \cdots & 0 & \mu_m \end{pmatrix} B' P_\varepsilon x(t), \end{aligned}$$

and hence,

$$\frac{\mathrm{d}}{\mathrm{d}t} x'(t) P_\varepsilon x(t) \leqslant -x'(t) Q_\varepsilon x(t) < 0.$$

Therefore, in the set $\mathcal{V}_\varepsilon$, we have $x'(t) P_\varepsilon x(t)$ exponentially decaying to zero, which implies that we do not leave $\mathcal{V}_\varepsilon$ and also that we have local exponential stability, and, by construction, the set $\mathcal{X}$ is contained in the domain of attraction.

■

Note that from the proof we can see that for α large, the input we choose approaches the input which minimizes

$$\frac{\mathrm{d}}{\mathrm{d}t} x'(t) P_\varepsilon x(t).$$

In other words, we maximize the decay of the Lyapunov function.

The above theorem does not make explicit how to find P_ε and Q_ε with the required properties. In the direct method to follow, our aim is to make the construction explicit and also make the dependency on ε explicit. This is the next phase of this section.

A design algorithm for a single-input system

We first consider single-input systems and then consider the general case. Consider a single input system (A, B) in the following controllable canonical form:

$$A = \begin{pmatrix} 0 & 1 & 0 & \cdots & 0 \\ \vdots & \ddots & \ddots & \ddots & \vdots \\ \vdots & & \ddots & 1 & 0 \\ 0 & \cdots & \cdots & 0 & 1 \\ -a_n & \cdots & -a_3 & -a_2 & -a_1 \end{pmatrix}, \quad B = \begin{pmatrix} 0 \\ \vdots \\ 0 \\ 0 \\ 1 \end{pmatrix}, \tag{4.146}$$

with all eigenvalues of A in the closed left-half plane. Let $F_{L,\varepsilon}$ be the matrix such that any eigenvalue λ_i on the imaginary axis is moved to $\lambda_i - 2\varepsilon$, while all the eigenvalues in the open left-half plane remain at the same location. We consider the equation

$$(A + BF_{L,\varepsilon})' P_\varepsilon + P_\varepsilon (A + BF_{L,\varepsilon}) + Q_\varepsilon = 0.$$

Borrowing the earlier construction from Sect. 4.3.1, we introduce the basis transformation Γ_ε associated with the basis (4.11) such that

$$\Gamma_\varepsilon^{-1}(A + BF_{L,\varepsilon})\Gamma_\varepsilon = \begin{pmatrix} J_{\varepsilon,1} & 0 & \cdots & 0 \\ 0 & \ddots & \ddots & \vdots \\ \vdots & \ddots & \ddots & 0 \\ 0 & \cdots & 0 & J_{\varepsilon,k} \end{pmatrix},$$

where $J_{\varepsilon,i}$ is a Jordan block associated with the eigenvalue $\lambda_i - 2\varepsilon$ of size $m_i \times m_i$. We introduce

$$\Gamma_\varepsilon' P_\varepsilon \Gamma_\varepsilon = \widetilde{P}_\varepsilon = \begin{pmatrix} \widetilde{P}_{\varepsilon,1} & 0 & \cdots & 0 \\ 0 & \ddots & \ddots & \vdots \\ \vdots & \ddots & \ddots & 0 \\ 0 & \cdots & 0 & \widetilde{P}_{\varepsilon,k} \end{pmatrix}$$

and

$$\Gamma_\varepsilon' Q_\varepsilon \Gamma_\varepsilon = \widetilde{Q}_\varepsilon = \begin{pmatrix} \widetilde{Q}_{\varepsilon,1} & 0 & \cdots & 0 \\ 0 & \ddots & \ddots & \vdots \\ \vdots & \ddots & \ddots & 0 \\ 0 & \cdots & 0 & \widetilde{Q}_{\varepsilon,k} \end{pmatrix}.$$

We get the equation

$$J_{\varepsilon,i}' \widetilde{P}_{\varepsilon,i} + \widetilde{P}_{\varepsilon,i} J_{\varepsilon,i} + \widetilde{Q}_{\varepsilon,i} = 0.$$

If $\operatorname{Re} \lambda_i < 0$, then we choose $\widetilde{Q}_{\varepsilon,i} = \varepsilon I$, and we obtain that $\widetilde{P}_{\varepsilon,i} = \varepsilon \overline{P}_i$ for all $\varepsilon > 0$ with $\overline{P}_i > 0$.

If $\operatorname{Re} \lambda_i = 0$, we observe that the Lyapunov equation is equal to

$$(\bar{J} - 2\varepsilon I)' \widetilde{P}_{\varepsilon,i} + \widetilde{P}_{\varepsilon,i} (\bar{J} - 2\varepsilon I) + \widetilde{Q}_{\varepsilon,i} = 0,$$

where $\bar{J}$ is a Jordan block associated with eigenvalue 0 of size $m_i \times m_i$. In other words, the imaginary part of λ_i does not play a role in the equation. In this case, we choose $\widetilde{Q}_{\varepsilon,i} = \varepsilon^{2m_i} I$, and we note that $\widetilde{P}_{\varepsilon,i}$ is polynomial in ε:

$$\widetilde{P}_{\varepsilon,i} = \sum_{j=1}^{2m_i-1} \varepsilon^j \, \overline{P}_{i,j}.$$

This can be established using the recursion

$$4\overline{P}_{i,j} = \bar{J}' \overline{P}_{i,j+1} + \overline{P}_{i,j+1} \bar{J}$$

for $j = 1, \ldots, 2m_i - 2$ with $\overline{P}_{i,2m_i-1} = \frac{1}{4} I$. The fact that $\overline{P}_{i,0} = 0$ follows from the fact that $\bar{J}$ is nilpotent and hence $\bar{J}^{m_i-1} = 0$.

In order to apply Lemma 4.37 to establish that we obtain in this way a proper low-and-high-gain design, we need to verify that $P_\varepsilon \to 0$ as $\varepsilon \to 0$ and that

$$P_\varepsilon \geqslant F_{L,\varepsilon}' F_{L,\varepsilon}. \tag{4.147}$$

The fact that $P_\varepsilon \to 0$ as $\varepsilon \to 0$ is a trivial consequence of our construction. In order to establish (4.147), we first note that

$$F_{L,\varepsilon}\Gamma_\varepsilon = \begin{pmatrix} \widetilde{F}_{\varepsilon,1} & \cdots & \widetilde{F}_{\varepsilon,k} \end{pmatrix},$$

and (4.13) implies that there exists a $M > 0$ such that

$$\|\widetilde{F}_{\varepsilon,i} e_\ell\| \leqslant M\varepsilon^{m_i-\ell+1}$$

for $i = 1, \ldots, k$ and $\ell = 1, \ldots, m_i$ where $e_1, \ldots, e_{m_i}$ is the standard basis for $\mathbb{R}^{m_i}$. Next, we note that our recursion implies that there exists a $m > 0$ such that

$$e_h' \bar{P}_j e_\ell = 0$$

for $h + \ell < 2m_i - j + 1$ and $e_h' \bar{P}_j e_\ell > m$ for $h + \ell = 2m_i - j + 1$. This implies that

$$e_h' \widetilde{P}_{\varepsilon,i} e_\ell > m\varepsilon^{2m_i+1-h-\ell}$$

for $h, \ell = 1, \ldots, m_i$ and ε sufficiently small. Combined with

$$e_h' \widetilde{F}_{\varepsilon,i}' \widetilde{F}_{\varepsilon,i} e_\ell < M^2 \varepsilon^{2m_i+2-h-\ell},$$

we find that for ε small enough, we have

$$\widetilde{F}_{\varepsilon,i}' \widetilde{F}_{\varepsilon,i} \leqslant \widetilde{P}_{\varepsilon,i}$$

and hence, (4.147) is satisfied. Therefore, Lemma 4.37 can be applied, and we can conclude that this design has the required properties.

A design algorithm for multi-input systems

We use the same initial state transformation as in Sect. 4.3.1, i.e., a state transformation Γ_x and an input transformation Γ_u such that the matrices $\Gamma_x^{-1} A \Gamma_x$ and $\Gamma_x^{-1} B \Gamma_u$ are in the following form:

$$\Gamma_x^{-1} A \Gamma_x = \begin{pmatrix} A_1 & A_{1,2} & \cdots & A_{1,q} & 0 \\ 0 & A_2 & \ddots & \vdots & \vdots \\ \vdots & \ddots & \ddots & A_{q-1,q} & \vdots \\ \vdots & & \ddots & A_q & 0 \\ 0 & \cdots & \cdots & 0 & A_{q+1} \end{pmatrix},$$

$$\Gamma_x^{-1} B \Gamma_u = \begin{pmatrix} B_1 & 0 & \cdots & 0 & B_{1,q+1} \\ 0 & B_2 & \ddots & \vdots & \vdots \\ \vdots & \ddots & \ddots & 0 & \vdots \\ 0 & \cdots & 0 & B_q & B_{q,q+1} \\ B_{q+1,1} & \cdots & B_{q+1,q-1} & B_{q+1,q} & B_{q+1,q+1} \end{pmatrix},$$

where q is an integer, and for $i = 1, 2, \ldots, q$,

$$A_i = \begin{pmatrix} 0 & 1 & 0 & \cdots & 0 \\ \vdots & \ddots & \ddots & \ddots & \vdots \\ \vdots & & \ddots & 1 & 0 \\ 0 & \cdots & \cdots & 0 & 1 \\ -a_{n_i}^i & \cdots & -a_3^i & -a_2^i & -a_1^i \end{pmatrix}, \quad B_i = \begin{pmatrix} 0 \\ \vdots \\ 0 \\ 0 \\ 1 \end{pmatrix}.$$

Clearly, (A_i, B_i) is controllable. The transformation Γ_x is such that all the eigenvalues of A_i are on the imaginary axis, and all the eigenvalues of A_{q+1} have strictly negative real parts.

Step 2: For each (A_i, B_i), let $F_{L,\varepsilon,i} \subset \mathbb{R}^{1\times n_i}$ be the state feedback gain such that the eigenvalues of $A_i + B_i F_{L,\varepsilon_i,i}$ can be obtained from the eigenvalues of A_i by moving any eigenvalue λ_i on the imaginary axis is to $\lambda_i - 2\varepsilon$ while all the eigenvalues in the open left-half plane remain at the same location.

We note here that such a gain $F_{L,\varepsilon,i}$ exists and is unique. Moreover, it can be obtained explicitly in terms of ε. The uniqueness follows since (A_i, B_i) is a single input controllable pair.

Step 3 : The family of low-gain state feedback control laws parameterized in ε is defined by

$$u = F_{L,\varepsilon} x, \tag{4.148}$$

where the state feedback gain matrix $F_{L,\varepsilon}$ is given by

$$F_{L,\varepsilon} = \Gamma_u \begin{pmatrix} F_{L,\varepsilon_1,1} & 0 & \cdots & \cdots & 0 \\ 0 & F_{L,\varepsilon_2,2} & \ddots & & \vdots \\ \vdots & \ddots & \ddots & \ddots & \vdots \\ \vdots & & \ddots & F_{L,\varepsilon_q,q} & 0 \\ 0 & \cdots & \cdots & 0 & 0 \end{pmatrix} \Gamma_x^{-1}, \tag{4.149}$$

where $\varepsilon_{q+1} = \varepsilon$ while for $i = 1, \ldots, q$, we have

$$\varepsilon_i = \varepsilon_{i+1}^{6r_{i+1}+1}, \tag{4.150}$$

where r_i is the largest algebraic multiplicity among the eigenvalues of A_i for $i = 1, \ldots, q$ while $r_{q+1} = 1$. Moreover, we construct $P_{\varepsilon,i}$ and $Q_{\varepsilon,i}$ such that

$$(A_i + B_i F_{\varepsilon_i,i})' P_{\varepsilon_i,i} + P_{\varepsilon_i,i}(A_i + B_i F_{\varepsilon_i,i}) + Q_{\varepsilon_i,i} = 0$$

with $P_{\varepsilon_i,i} \to 0$ as $\varepsilon_i \to 0$ and

$$P_{\varepsilon_i,i} \geqslant F'_{\varepsilon_i,i} F_{\varepsilon_i,i}$$

for $i = 1, \ldots, q$ such that there exist constants $\bar{m}_1, \bar{m}_2$, and $\bar{m}_3$ such that

$$\bar{m}_1 \varepsilon^{2r_i - 1} I \leqslant P_{\varepsilon_i,i} \leqslant \bar{m}_2 \varepsilon_i I \tag{4.151}$$

and

$$Q_{\varepsilon,i} \geqslant \bar{m}_3 \varepsilon^{2r_i} I. \tag{4.152}$$

The algorithm presented earlier for the single-input case tells us how to construct $P_{\varepsilon_i,i}$ and $Q_{\varepsilon_i,i}$. We define

$$F_{H,\varepsilon_i,i} = -B_i' P_{\varepsilon_i,i},$$

and the family of low- and high-gain state feedback control laws parameterized in ε is defined by

$$u = F_{L,\varepsilon} x + \alpha F_{H,\varepsilon} x, \tag{4.153}$$

where the state feedback gain matrix $F_{L,\varepsilon}$ is given by (4.149), and

$$F_{H,\varepsilon} = \Gamma_u \begin{pmatrix} F_{H,\varepsilon_1,1} & 0 & \cdots & \cdots & 0 \\ 0 & F_{H,\varepsilon_2,2} & \ddots & & \vdots \\ \vdots & \ddots & \ddots & \ddots & \vdots \\ \vdots & & \ddots & F_{H,\varepsilon_q,q} & 0 \\ 0 & \cdots & \cdots & 0 & 0 \end{pmatrix} \Gamma_x^{-1}. \tag{4.154}$$

We define $Q_{\varepsilon,q+1} = \varepsilon I$ and $P_{\varepsilon,q+1}$ is such that

$$A'_{q+1} P_{\varepsilon,q+1} + P_{\varepsilon,q+1} A_{q+1} + Q_{\varepsilon,q+1} = 0,$$

and define $r_{q+1} = 1$. We also define

$$P_\varepsilon = \Gamma_x \begin{pmatrix} P_{\varepsilon_1,1} & 0 & \cdots & 0 \\ 0 & P_{\varepsilon_2,2} & \ddots & \vdots \\ \vdots & \ddots & \ddots & 0 \\ 0 & \cdots & 0 & P_{\varepsilon_{q+1},q+1} \end{pmatrix} \Gamma_x^{-1}. \tag{4.155}$$

The initial conditions are in the compact set $\mathcal{X}$. This implies that for ε small enough, the initial conditions satisfy

$$x(0) \in \frac{1}{2}\mathcal{V}_\varepsilon = \left\{ x \in \mathbb{R}^n \mid x' P_\varepsilon x \leqslant \frac{1}{2} \right\}. \tag{4.156}$$

For $i = q + 1$, we have

$$x_i'(t) P_{\varepsilon_i,i} x_i(t) \leqslant e^{-\varepsilon_i^{2r_i} t} \tag{4.157}$$

given (4.156). Using (4.151), this yields

$$\|x_i(t)\| \leqslant \frac{\bar{m}_1}{\varepsilon_i^{r_i-1/2}} e^{-\varepsilon_i^{2r_i} t/2}. \tag{4.158}$$

We will prove (4.157) and (4.158) for $i = 1, \ldots, q$ through an induction argument. Assume that (4.157) and (4.158) are satisfied for $i = j+1, \ldots, q+1$. We will establish that this inequality is also satisfied for $i = j$.

To establish this, we consider the j'th subsystem

$$\dot{x}_j = A_j x_j + B_j u_j + w_j, \tag{4.159}$$

where

$$w_j = A_{j,j+1} x_{j+1} + \cdots + A_{j,q} x_q \tag{4.160}$$

while $w_q = 0$. Using (4.158) for $i = j+1, \ldots, q$, we obtain that there exists a M_j such that

$$\|w_j(t)\| \leqslant \frac{M_j}{\varepsilon_j^{r_{j+1}-1/2}} e^{-\varepsilon_{j+1}^{2r_{j+1}} t/2}$$

for all $t > 0$. We obtain in the set $\mathcal{V}_\varepsilon$

$$\frac{\mathrm{d}}{\mathrm{d}t} x_j'(t) P_{\varepsilon_j,j} x_j(t) = -x_j'(t) Q_{\varepsilon_j,j} x_j(t) + 2x_j'(t) P_{\varepsilon_j,j} B_j v_j(t) + 2x_j'(t) P_{\varepsilon_j,j} w_j(t),$$

where

$$v_j(t) = \sigma\left(F_{L,\varepsilon_j,j} x_j(t) + \alpha F_{H,\varepsilon_j,j} x_j(t)\right) - F_{L,\varepsilon_j,j} x_j(t).$$

Since by assumption $x(t) \in \mathcal{V}_\varepsilon$, we have

$$x_j'(t) F_{L,\varepsilon_j,j}' F_{L,\varepsilon_j,j} x(t) \leqslant x_j'(t) P_{\varepsilon,j} x_j(t) < 1,$$

where we use (4.144), and hence, $\| F_{L,\varepsilon_j,j} x_j(t) \| < 1$. This implies that

$$\sigma(\left(F_{L,\varepsilon_j,j} x_j(t) + \alpha F_{H,\varepsilon_j,j} x_j(t)\right) = F_{L,\varepsilon_j,j} x_j(t) + \mu\alpha F_{H,\varepsilon_j,j} x_j(t)$$

with $\mu \in [0, 1]$. This in turn implies that

$$\begin{aligned} \frac{\mathrm{d}}{\mathrm{d}t} x_j'(t) P_{\varepsilon_j,j} x_j(t) = & -x_j'(t) Q_{\varepsilon_j,j} x_j(t) + 2x_j'(t) P_{\varepsilon_j,j} w_j(t) \\ & - \mu\alpha 2 x_j'(t) P_{\varepsilon_j,j} B_j B_j' P_{\varepsilon_j,j} x_j(t), \end{aligned}$$

and hence,

$$\frac{\mathrm{d}}{\mathrm{d}t} x_j'(t) P_{\varepsilon_j,j} x_j(t) \leqslant -x_j'(t) Q_{\varepsilon_j,j} x_j(t) + 2x_j'(t) P_{\varepsilon_j,j} w_j(t). \tag{4.161}$$

By (4.151) and (4.152), we find that

$$x_j' Q_{\varepsilon_j,j} x_j \geqslant \frac{m_3}{m_2} \varepsilon^{2r_j - 1} x_j' P_{\varepsilon_j,j} x_j \geqslant 2\varepsilon^{2r_j} x_j' P_{\varepsilon_j,j} x_j,$$

where the last equality holds for sufficiently small $\varepsilon > 0$. Next, we note that

$$\begin{aligned} 2x_j' P_{\varepsilon_j,j} w_j & \leqslant 2\left(x_j' P_{\varepsilon_j,j} x_j\right)^{1/2} \sqrt{m_2} \sqrt{\varepsilon_j} \frac{M_j}{\varepsilon_{j+1}^{r_{j+1} - 1/2}} e^{-\varepsilon_{j+1}^{2r_{j+1}} t/2} \\ & \leqslant \left(x_j' P_{\varepsilon_j,j} x_j\right)^{1/2} \frac{2\sqrt{\varepsilon_j}}{\varepsilon_{j+1}^{r_{j+1}}} e^{-\varepsilon_{j+1}^{2r_{j+1}} t/2} \end{aligned}$$

for sufficiently small ε. We define

$$V_j(t) = x_j'(t) P_{\varepsilon_j,j} x_j(t),$$

and we obtain using the above two bounds and (4.161) that

$$V_j'(t) \leqslant -2\varepsilon_j^{2r_j} V(t) + \frac{2\sqrt{\varepsilon_j}}{\varepsilon_{j+1}^{r_{j+1}}} e^{-\varepsilon_{j+1}^{2r_{j+1}} t/2} \sqrt{V(t)}.$$

This yields

$$(V_j^{1/2})' \leqslant -\varepsilon_j^{2r_j} V_j^{1/2} + \frac{\sqrt{\varepsilon_j}}{\varepsilon_{j+1}^{r_{j+1}}} e^{-\varepsilon_{j+1}^{2r_{j+1}} t/2}.$$

Therefore,

$$V_j^{1/2}(t) \leqslant e^{-\varepsilon_j^{2r_j} t} \left[V_j^{1/2}(0) + \frac{4\sqrt{\varepsilon_j}}{\varepsilon_{j+1}^{3r_{j+1}}} \right] \leqslant e^{-\varepsilon_j^{2r_j} t}$$

given (4.150) and (4.156). This yields, for ε small enough, (4.157) and (4.158) for $i = j$, where we use for the latter (4.151). This recursively establishes (4.157) and (4.158) for $i = 1, \ldots, q+1$ given (4.156). Since $\mathcal{X} \subset \frac{1}{2}\mathcal{V}_\varepsilon$, we find that for all initial conditions in $\mathcal{X}$, we stay inside $\mathcal{V}_\varepsilon$ and converge to zero exponentially. This establishes the required properties of our design.

A design algorithm for measurement feedback

In case of measurement feedback we have,

$$\begin{aligned} \dot{x} &= Ax + B\sigma(u) \\ y &= Cx. \end{aligned} \tag{4.162}$$

We need to build an observer that can be used in conjunction with the low-and-high-gain state feedback designed before. To be more precise, we need a high-gain observer which yields fast performance. In addition to earlier assumptions that (A, B) is stabilizable and all eigenvalues of A being inside the closed left-half plane, we assume that the pair (C, A) is observable. For this purpose, we transform the system to the so-called Brunovski canonical form. That is, we choose a state transformation Γ_x, an output transformation Γ_y, and an output injection K such that $\Gamma_x^{-1}(A - KC)\Gamma_x$ and $\Gamma_x^{-1} B \Gamma_u$ are in the following form:

$$\Gamma_x^{-1}(A - KC)\Gamma_x = \bar{A} = \begin{pmatrix} A_1 & 0 & \cdots & 0 \\ 0 & A_2 & \ddots & \vdots \\ \vdots & \ddots & \ddots & 0 \\ 0 & \cdots & 0 & A_q \end{pmatrix},$$

$$\Gamma_y C \Gamma_x = \bar{C} = \begin{pmatrix} C_1 & 0 & \cdots & 0 \\ 0 & C_2 & \ddots & \vdots \\ \vdots & \ddots & \ddots & 0 \\ 0 & \cdots & 0 & C_q \\ C_{q+1,1} & \cdots & C_{q+1,q-1} & C_{q+1,q} \end{pmatrix},$$

where q is an integer, and for $i = 1, 2, \ldots, q$,

$$A_i = \begin{pmatrix} 0 & 1 & 0 & \cdots & 0 \\ \vdots & \ddots & \ddots & \ddots & \vdots \\ \vdots & & \ddots & 1 & 0 \\ 0 & \cdots & \cdots & 0 & 1 \\ 0 & \cdots & \cdots & 0 & 0 \end{pmatrix}$$

$$C_i = \begin{pmatrix} 1 & 0 & \cdots & \cdots & 0 & 0 \end{pmatrix}.$$

Clearly, (A_i, C_i) is observable. This transformation is closely related by duality to the transformation used to obtain (4.4) and (4.5). However, due to the output injection, we could obtain a block diagonal structure in the A matrix. In the state feedback case, when obtaining (4.4) and (4.5), we consider low-gain feedback, and we did not have the freedom to apply arbitrary preliminary state feedback laws. The consequence of this is that we could only obtain a block upper-diagonal matrix.

Choose K_i such that $A_i - K_i C_i$ is Hurwitz for $i = 1, \ldots q$ and let $\bar{K}$ be given by

$$\bar{K} = \begin{pmatrix} K_1 & 0 & \cdots & 0 & 0 \\ 0 & K_2 & \ddots & \vdots & \vdots \\ \vdots & \ddots & \ddots & 0 & 0 \\ 0 & \cdots & 0 & K_q & 0 \end{pmatrix}.$$

Next, we introduce

$$S_i(\ell) = \begin{pmatrix} \ell & 0 & \cdots & 0 \\ 0 & \ell^2 & \ddots & \vdots \\ \vdots & \ddots & \ddots & 0 \\ 0 & \cdots & 0 & \ell^{p_i} \end{pmatrix}$$

where p_i equals the number of columns of A_i and

$$S_\ell = \begin{pmatrix} S_1(\ell) & 0 & \cdots & 0 \\ 0 & S_2(\ell) & \ddots & \vdots \\ \vdots & \ddots & \ddots & 0 \\ 0 & \cdots & 0 & S_q(\ell) \end{pmatrix}.$$

In that case, it is easy to verify that

$$S_\ell^{-1}\left[\bar{A} - S_\ell \bar{K}\bar{C}\right] S_\ell = \ell(\bar{A} - \bar{K}\bar{C}),$$

and hence, the output injection

$$K_\ell = K + \Gamma_x S_\ell \bar{K} \Gamma_y$$

has the property that

$$S_\ell^{-1}\Gamma_x^{-1}\left[A - K_\ell C\right]\Gamma_x S_\ell = \ell(\bar{A} - \bar{K}\bar{C}).$$

We apply the measurement feedback controller

$$\begin{aligned} \dot{\hat{x}} &= A\hat{x} + B\sigma(u) + K_\ell(y - C\hat{x}) \\ u &= F_{L,\varepsilon}\hat{x} + \alpha F_{H,\varepsilon}\hat{x} \end{aligned} \tag{4.163}$$

to the system (4.162). We have the following result.

Theorem 4.38 *Consider the system* (4.162) *where* (A, B) *is stabilizable,* (C, A) *is observable and the eigenvalues of* A *are in the closed left-half plane. Then for any compact sets* $\mathcal{X}$ *and* $\widehat{\mathcal{X}}$*, there exists an* ε^* *such that for all* $\varepsilon < \varepsilon^*$ *and* $\alpha > 0$ *we can find* $\ell^*(\varepsilon, \alpha)$ *for which the controller* (4.163) *achieves local exponential stability and a domain of attraction containing* $\mathcal{X} \times \widehat{\mathcal{X}}$ *provided that* $\ell > \ell^*(\varepsilon, \alpha)$.

We establish here only a proof for the single-input case where a relatively simple Lyapunov-based proof can be given. In the general multi-input case, we need the same complex proof of the state feedback case establishing stability recursively for the subsystems (4.159), but in this case, the error term (4.160) contains an extra term due to measurement error.

Proof of the single-input case : In the single-input state feedback case, we have constructed P_ε and Q_ε such that

$$(A + BF_{L,\varepsilon})' P_\varepsilon + P_\varepsilon(A + BF_{L,\varepsilon}) + Q_\varepsilon = 0$$

while

$$P_\varepsilon \geqslant F_{L,\varepsilon}' F_{L,\varepsilon}, \qquad \lim_{\varepsilon \to 0} P_\varepsilon = 0$$

and

$$F_{H,\varepsilon} = -B' P_\varepsilon.$$

Under the output feedback law (4.163), the closed-loop system takes the form

$$\begin{cases} \dot{x} = Ax + B\sigma(F_{L,\varepsilon}\hat{x} - \alpha B'P_\varepsilon\hat{x}) \\ \dot{\hat{x}} = A\hat{x} + B\sigma(F_{L,\varepsilon}\hat{x} - \alpha B'P_\varepsilon\hat{x}) + K_\ell(y - C\hat{x}), \end{cases} \tag{4.164}$$

which, in the new coordinates (x, e) where $e = x - \hat{x}$, becomes

$$\begin{cases} \dot{x} = Ax + B\sigma((F_{L,\varepsilon} - \alpha B'P_\varepsilon)(x - e)) \\ \dot{e} = (A - K_\ell C)e. \end{cases} \tag{4.165}$$

Local exponential stability then immediately follows since $A - K_\ell C$ and $A + BF_{L,\varepsilon} - \alpha BB'P_\varepsilon$ are both Hurwitz. It remains to consider the domain of attraction. We define

$$z = \ell^n S_\ell^{-1} \Gamma_x^{-1} e.$$

Then, the closed-loop system can be rewritten as

$$\begin{cases} \dot{x} = Ax + B\sigma((F_{L,\varepsilon} - \alpha B'P_\varepsilon)(x - \ell^{-n} S_\ell \Gamma_x z)) \\ \dot{z} = \ell(\bar{A} - \bar{K}\bar{C})z. \end{cases} \tag{4.166}$$

Since σ is bounded and $x(0) \in \mathcal{X}$, there exists a bounded set $\widetilde{\mathcal{X}}$ such that $x(1) \in \widetilde{\mathcal{X}}$ for all initial conditions

$$(x(0), \hat{x}(0)) \in \mathcal{X} \times \hat{\mathcal{X}}$$

and independent of α and ℓ. We have

$$z(1) = e^{\ell(\bar{A} - \bar{K}\bar{C})} \ell^n S_\ell^{-1} \Gamma_x^{-1} e(0)$$

with $e(0)$ bounded. Hence, for an arbitrary compact set $\mathcal{Z}$, there exists an ℓ_1^* such that for $\ell > \ell_1^*$ we have $z(1) \in \mathcal{Z}$. After all, $e^{\ell(\bar{A} - \bar{K}\bar{C})}$ converges to zero exponentially as $\ell \to \infty$ while $\ell^n S_\ell^{-1}$ only grows polynomially in ℓ.

Consider now the Lyapunov function,

$$V_\varepsilon(x) = x'P_\varepsilon x + z'Rz, \tag{4.167}$$

where R is such that

$$(\bar{A} - \bar{K}\bar{C})R + R(\bar{A} - \bar{K}\bar{C}) + I = 0,$$

and let $c > 0$ be such that

$$c \geq \sup_{x \in \widetilde{\mathcal{X}}, z \in \mathcal{Z}, \varepsilon \in (0,1]} x'P_\varepsilon x + z'Rz. \tag{4.168}$$

Such a c exists because $\widetilde{\mathcal{X}}$ and $\mathcal{Z}$ are bounded. Moreover, P_ε is bounded by P_1 for $\varepsilon \in [0, 1]$. Let ε_1^* be such that, for all $\varepsilon \in (0, \varepsilon_1^*]$,

$$(x, z) \in \mathcal{L}_V(c) = \{x \in \mathbb{R}^n, z \in \mathbb{R}^n : x' P_\varepsilon x + z' R z \leqslant c\},$$

we have

$$\| F_{L,\varepsilon}(x - \ell^{-n} S_\ell \Gamma_x z) \| \leqslant \Delta$$

for all $\ell > \ell_1^*$. The existence of such an ε_1^* is due to the fact that $\ell^{-n} S_\ell$ is small for large ℓ while $\lim_{\varepsilon \to 0} F_{L,\varepsilon} = 0$.

We will establish that for all $t > 1$ we have $(x(t), z(t)) \in \mathcal{L}_V(c)$ with $x(t) \to 0$, $z(t) \to 0$ as $t \to \infty$. Note that for $(x, z) \in \mathcal{L}_V(c)$, we have

$$\sigma(F_{L,\varepsilon}(x - \ell^{-n} S_\ell \Gamma_x z)) = F_{L,\varepsilon}(x - \ell^{-n} S_\ell \Gamma_x z),$$

and hence,

$$\sigma((F_{L,\varepsilon} - \alpha B' P_\varepsilon)(x - \ell^{-n} S_\ell \Gamma_x z)) = (F_{L,\varepsilon} - \alpha \mu B' P_\varepsilon)(x - \ell^{-n} S_\ell \Gamma_x z))$$

for some $\mu \in [0, 1]$, provided that $(x, z) \in \mathcal{L}_V(c)$. It follows then that for all $\varepsilon \in (0, \varepsilon_1^*]$ and for all $(x, z) \in \mathcal{L}_V(c)$, the closed-loop system (4.166) can be written as

$$\begin{cases} \dot{x} = (A + B F_{L,\varepsilon} - \alpha \mu B B' P_\varepsilon) x + B(F_{L,\varepsilon} - \alpha \mu B' P_\varepsilon) \ell^{-n} S_\ell \Gamma_x z \\ \dot{z} = \ell(\bar{A} - \bar{K}\bar{C}) z. \end{cases} \tag{4.169}$$

The evaluation of $\dot{V}_\varepsilon$ along the trajectories of this closed-loop system gives, for all $(x, z) \in \mathcal{L}_V(c)$ and for all $\varepsilon \in (0, \varepsilon_1^*]$,

$$\begin{aligned} \dot{V}_\varepsilon &= -x' Q_\varepsilon x - 2\alpha\mu x' P_\varepsilon B B' P_\varepsilon x \\ &\qquad - 2x' P_\varepsilon B(F_{L,\varepsilon} - \alpha\mu B' P_\varepsilon) \ell^{-n} S_\ell \Gamma_x z - \ell z' z \\ &\leqslant -x' Q_\varepsilon x + \sqrt{\ell}\, x' Q_\varepsilon^{1/2} z - \ell z' z \\ &\leqslant -\tfrac{1}{2} x' Q_\varepsilon x - \tfrac{1}{2} \ell z' z, \end{aligned}$$

provided that ℓ is large enough such that

$$\| P_\varepsilon B(F_{L,\varepsilon} - \alpha\mu B' P_\varepsilon) \ell^{-n} S_\ell \Gamma_x \| \leqslant \frac{1}{2} \sqrt{\ell}\, \lambda_{\min}(Q_\varepsilon)^{1/2}.$$

The above can be clearly achieved since the left-hand converges to zero while the right-hand side converges to infinity as $\ell \to \infty$. This implies that (x, z) remains in the set $\mathcal{L}_V(c)$ for all $t > 0$ and that $V_\varepsilon(t) \to 0$ as $t \to \infty$ for ℓ sufficiently large. Therefore, we find that $x(t) \to 0$ as $t \to \infty$. This completes the proof of our result. ∎

The above observer, in general, has a large overshoot. This makes it sensitive to disturbances. But another example, presented in Chap. 5, is the case when the saturation function is not exactly known. The latter also causes the above high-gain observer to behave badly in some cases. If we assume that the given system is left invertible and of minimum phase, then these problems can be avoided.

We note that if the system $(A, B, C, 0)$ is left invertible and minimum phase, then there exists a basis transformation Γ_s such that

$$\Gamma_s^{-1} A \Gamma_s = \begin{pmatrix} A_0 & L_0 C_1 \\ B_1 E_1 & A_1 \end{pmatrix}, \quad \Gamma_s^{-1} B = \begin{pmatrix} 0 \\ B_1 \end{pmatrix}, \quad C \Gamma_s = \begin{pmatrix} 0 & C_1 \end{pmatrix}$$

with A_0 asymptotically stable. This follows from Chap. 3 where

$$\Gamma_s x = \begin{pmatrix} x_1 \\ x_2 \end{pmatrix}, \qquad x_1 = x_a, \qquad x_2 = \begin{pmatrix} x_b \\ x_d \end{pmatrix}$$

in the notation of SCB. Note that x_c is missing because the system is left invertible and A_0 is Hurwitz because the system is minimum phase.

Next, we note that the subsystem $(A_1, B_1, C_1, 0)$ based on the SCB has the following structure:

$$A_1 = \begin{pmatrix} A_{bb} & 0 & 0 & \cdots & 0 \\ B_{q_1} E_{1b} & A_{q_1} & 0 & \cdots & 0 \\ \vdots & 0 & \ddots & \ddots & \vdots \\ \vdots & \vdots & \ddots & \ddots & 0 \\ B_{q_{m_d}} E_{m_d b} & 0 & \cdots & 0 & A_{q_{m_d}} \end{pmatrix},$$

$$B_1 = \begin{pmatrix} 0 & 0 & \cdots & 0 \\ B_{q_1} & 0 & \cdots & 0 \\ 0 & \ddots & \ddots & \vdots \\ \vdots & \ddots & \ddots & 0 \\ 0 & \cdots & 0 & B_{q_{m_d}} \end{pmatrix},$$

$$C_1 = \begin{pmatrix} 0 & C_{q_1} & 0 & \cdots & 0 \\ 0 & 0 & C_{q_2} & \ddots & \vdots \\ \vdots & \vdots & \ddots & \ddots & 0 \\ 0 & 0 & \cdots & 0 & C_{q_{m_d}} \\ C_b & 0 & 0 & \cdots & 0 \end{pmatrix},$$

where q is an integer, and for $i = 1, 2, \ldots, m_d$,

$$A_{q_i} = \begin{pmatrix} 0 & 1 & 0 & \cdots & 0 \\ \vdots & \ddots & \ddots & \ddots & \vdots \\ \vdots & & \ddots & 1 & 0 \\ 0 & \cdots & \cdots & 0 & 1 \\ 0 & \cdots & \cdots & 0 & 0 \end{pmatrix} \qquad B_{q_i} = \begin{pmatrix} 0 \\ \vdots \\ 0 \\ 0 \\ 1 \end{pmatrix},$$

$$C_{q_i} = \begin{pmatrix} 1 & 0 & \cdots & \cdots & 0 & 0 \end{pmatrix}.$$

Choose K_{q_i} such that $A_{q_i} - K_{q_i} C_{q_i}$ is Hurwitz for $i = 1, \ldots m_d$ and let $K_{d,1}$ be given by

$$K_{d,1} = \begin{pmatrix} K_{q_1} & 0 & \cdots & 0 \\ 0 & K_{q_2} & \ddots & \vdots \\ \vdots & \ddots & \ddots & 0 \\ 0 & \cdots & 0 & K_{q_{m_d}} \end{pmatrix}.$$

Next, we introduce

$$S_i(\ell) = \ell^{-p_i} \begin{pmatrix} \ell & 0 & \cdots & 0 \\ 0 & \ell^2 & \ddots & \vdots \\ \vdots & \ddots & \ddots & 0 \\ 0 & \cdots & 0 & \ell^{p_i} \end{pmatrix},$$

where p_i equals the number of columns of A_{q_i} and

$$S_{d,\ell} = \begin{pmatrix} S_1(\ell) & 0 & \cdots & 0 \\ 0 & S_2(\ell) & \ddots & \vdots \\ \vdots & \ddots & \ddots & 0 \\ 0 & \cdots & 0 & S_{m_d}(\ell) \end{pmatrix}.$$

Finally, choose n_b distinct complex numbers $\lambda_1, \ldots, \lambda_{n_b}$ in the open left-half plane such that the set is symmetric with respect to the real axis. Choose $K_{b,\ell}$ such that $A_{bb} - K_{b,\ell} C_b$ has eigenvalues $\ell\lambda_1, \ldots \ell\lambda_{n_b}$ which is possible since (C_b, A_{bb}) is observable. Then there exists a $S_{b,\ell}$ such that

$$S_{b,\ell}^{-1}(A_{bb} - K_{b,\ell} C_b) S_{b,\ell} = \ell(A_{bb} - K_{b,1} C_b),$$

where, without loss of generality, we choose $S_{b,\ell}$ such that $\|S_{b,\ell}\| = 1$. Putting the above together, it is easy to verify that

$$S_{1,\ell}^{-1}\left[A_1 - K_{1,\ell}C_1\right]S_{1,\ell} = \ell(A_1 - \bar{K}_1 C_1),$$

where

$$S_{1,\ell} = \begin{pmatrix} S_{b,\ell} & 0 \\ 0 & S_{d,\ell} \end{pmatrix}, \qquad \bar{K}_1 = \begin{pmatrix} 0 & K_{b,1} \\ K_{d,1} & 0 \end{pmatrix},$$

and the output injection $K_{1,\ell}$ is given by

$$K_{1,\ell} = \begin{pmatrix} 0 & K_{b,\ell} \\ S_{1,\ell}K_{d,1} & 0 \end{pmatrix}.$$

Clearly, by construction, $S_{1,\ell}$ also has the property that

$$S_{1,\ell}^{-1}B_1 = B_1.$$

Moreover, there exists a M, independent of ℓ, such that $\|S_\ell\| < M$ for all $\ell \geqslant 1$.

In order to conclude, we define

$$K_\ell = \Gamma_s \begin{pmatrix} L_0 \\ K_{1,\ell} \end{pmatrix},$$

and we choose the feedback (4.163). With the new observer gain, the controller still has the properties of Theorem 4.38. This can be seen using the same arguments as in the proof of Theorem 4.38. The advantage of this new observer comes when we apply it to a system with input-additive disturbance d:

$$\begin{aligned} \dot{x} &= Ax + B\sigma(u) + Bd \\ y &= Cx. \end{aligned}$$

If we apply the controller (4.163) to this system, we obtain

$$x - \hat{x} = \Gamma_s \begin{pmatrix} I & 0 \\ 0 & S_{1,\ell} \end{pmatrix} \begin{pmatrix} \tilde{e}_1 \\ \tilde{e}_2 \end{pmatrix},$$

where

$$\dot{\tilde{e}}_1 = A_0 \tilde{e}_1 \tag{4.170}$$

$$\dot{\tilde{e}}_2 = \ell(A_1 - \bar{K}_1 C_1)\tilde{e}_2 + B_1(d + E_1\tilde{e}_1). \tag{4.171}$$

We note that increasing ℓ speeds up the convergence of $\tilde{e}_2$ to zero without increasing the effect of the disturbances. Since $S_{1,\ell}$ is bounded, we see also that $x - \hat{x}$ converges to zero faster without increasing the effect of the disturbances. Actually, by increasing ℓ, we see that the effect of the disturbances on the observer error can be made arbitrarily small. This is the advantage of this new observer which can only be obtained under the additional assumptions of left invertibility and minimum phase.

4.5.2 *Direct eigenstructure assignment: discrete time*

We note that one cannot have low-and-high-gain control laws for discrete-time systems in the same way as one does for continuous-time systems. This is because, for discrete-time systems, the closed-loop stability implies that the linear part of closed-loop system has poles within the unit circle rather than having arbitrarily large magnitude in the negative half plane as in the case of continuous-time systems. As before, we split the design into single- and multi-input systems.

A design algorithm for a single input system

The following lemma gives a method to speed up the performance of the system by introducing a high-gain parameter:

Lemma 4.39 *Consider a single-input linear system*

$$x(k+1) = Ax(k) + Bu(k), \quad x(k) \in \mathbb{R}^n$$

with all eigenvalues of A in the closed unit disc. Let $F_{L,\varepsilon}$ be a matrix such that the solution $P_\varepsilon > 0$ of the Lyapunov equation,

$$P_\varepsilon = (A + BF_{L,\varepsilon})' P_\varepsilon (A + BF_{L,\varepsilon}) + Q_\varepsilon,$$

has the property that

$$\lim_{\varepsilon \to 0} P_\varepsilon = 0, \tag{4.172}$$

and

$$P_\varepsilon \geqslant F'_{L,\varepsilon} F_{L,\varepsilon} \tag{4.173}$$

for all $\varepsilon > 0$, then the feedback controller

$$u = F_{L,\varepsilon} x + \alpha F_{H,\varepsilon} x \tag{4.174}$$

with $\alpha \in [0, 2]$ where

$$F_{H,\varepsilon} = -(B' P_\varepsilon B)^{-1} B' P_\varepsilon (A + BF_{L,\varepsilon})$$

is such that for any compact set $\mathcal{X}$ *there exists an* ε^* *such that for all* $0 < \varepsilon < \varepsilon^*$ *and all* $\alpha \in [0, 2]$, *the system*

$$x(k+1) = Ax(k) + B\sigma(u(k)), \quad x(k) \in \mathbb{R}^n$$

with controller (4.174) *is locally asymptotically stable with* $\mathcal{X}$ *contained in its domain of attraction.*

Proof : Consider the set

$$\mathcal{V}_\varepsilon := \left\{ x \in \mathbb{R}^n \mid x' P_\varepsilon x < 1 \right\}.$$

We choose ε sufficiently small such that $\mathcal{X} \subset \mathcal{V}_\varepsilon$. This is possible due to (4.172).

We note that $\mathcal{V}_\varepsilon$ is a positively invariant set in the sense that $x(k) \in \mathcal{V}_\varepsilon$ implies that $x(i) \in \mathcal{V}_\varepsilon$ for all $i > k$. In order to establish this, we first note that

$$\begin{aligned} x'(k+1)P_\varepsilon x(k+1) &= x'(k)P_\varepsilon x(k) - x'(k)Q_\varepsilon x(k) \\ &\quad + 2x'(k)(A + BF_{L,\varepsilon})' P_\varepsilon Bv(k) + v'(k)B'P_\varepsilon Bv(k), \end{aligned}$$

where

$$v(k) = \sigma\left(F_{L,\varepsilon}x(k) + \alpha F_{H,\varepsilon}x(k)\right) - F_{L,\varepsilon}x(k).$$

Since $x(k) \in \mathcal{V}_\varepsilon$, we have

$$x'(k)F_{L,\varepsilon}' F_{L,\varepsilon}x(k) \leqslant x'(k)P_\varepsilon x(k) < 1,$$

where we use (4.173), and hence, $\|F_{L,\varepsilon}x(k)\| < 1$. This implies that

$$\sigma((F_{L,\varepsilon}x(k) + \alpha F_{H,\varepsilon}x(k)) = F_{L,\varepsilon}x(k) + \mu\alpha F_{H,\varepsilon}x(k)$$

with $\mu \in [0, 1]$. This in turn implies that

$$\begin{aligned} x'(k+1)P_\varepsilon x(k+1) &= x'(k)P_\varepsilon x(k) - x'(k)Q_\varepsilon x(k) \\ &- (2\mu\alpha - \mu^2\alpha^2)x'(k)(A + BF_{L,\varepsilon})' P_\varepsilon B(B'PB)^{-1}B'P_\varepsilon(A + BF_{L,\varepsilon})x(k). \end{aligned}$$

Since $\mu \in [0, 1]$ and $\alpha \in [0, 2]$, we find $2\mu\alpha > \mu^2\alpha^2$, and hence,

$$x'(k+1)P_\varepsilon x(k+1) \leqslant x'(k)P_\varepsilon x(k) - x'(k)Q_\varepsilon x(k) < x'(k)P_\varepsilon x(k).$$

Therefore, in the set $\mathcal{V}_\varepsilon$, we have $x'(k)P_\varepsilon x(k)$ exponentially decaying to zero. This implies that we do not leave $\mathcal{V}_\varepsilon$, and also, we have local exponential stability, and, by construction, the set $\mathcal{X}$ is contained in the domain of attraction. ■

The above lemma does not make explicit how to find P_ε and Q_ε with the required properties. In the direct method to follow, our aim is to make the construction explicit and also make the dependency on ε explicit. This is the next phase of this section.

Consider a single input system (A, B) in the following controllable canonical form:

$$A = \begin{pmatrix} 0 & 1 & 0 & \cdots & 0 \\ \vdots & \ddots & \ddots & \ddots & \vdots \\ \vdots & & \ddots & 1 & 0 \\ 0 & \cdots & \cdots & 0 & 1 \\ -a_n & \cdots & -a_3 & -a_2 & -a_1 \end{pmatrix}, \quad B = \begin{pmatrix} 0 \\ \vdots \\ 0 \\ 0 \\ 1 \end{pmatrix}, \tag{4.175}$$

with all eigenvalues of A in the closed unit disc. Let $F_{L,\varepsilon}$ be the matrix such that

$$\lambda(A + BF_{L,\varepsilon}) = (1 - 2\varepsilon)\lambda(A) \subset \mathbb{C}^{\ominus}.$$

We consider the Lyapunov equation

$$P_\varepsilon = (A + BF_{L,\varepsilon})' P_\varepsilon (A + BF_{L,\varepsilon}) + Q_\varepsilon.$$

Borrowing the earlier construction from Sect. 4.3.2, we introduce a slightly modified basis transformation $\widetilde{\Gamma}_\varepsilon$ associated with the basis

$$q(\lambda_{\varepsilon,1}), \lambda_{\varepsilon,1} q^{(1)}(\lambda_{\varepsilon,1}), \ldots, \lambda_{\varepsilon,1}^{m_1-1} q^{(m_1-1)}(\lambda_{\varepsilon,1}), \ldots,$$
$$q(\lambda_{\varepsilon,k}), \lambda_{\varepsilon,k} q^{(1)}(\lambda_{\varepsilon,k}), \ldots, \lambda_{\varepsilon,1}^{m_k-1} q^{(m_k-1)}(\lambda_{\varepsilon,k}) \tag{4.176}$$

instead of the basis (4.30) such that

$$\widetilde{\Gamma}_\varepsilon^{-1}(A + BF_{L,\varepsilon})\widetilde{\Gamma}_\varepsilon = \begin{pmatrix} J_{\varepsilon,1} & 0 & \cdots & 0 \\ 0 & \ddots & \ddots & \vdots \\ \vdots & \ddots & \ddots & 0 \\ 0 & \cdots & 0 & J_{\varepsilon,k} \end{pmatrix},$$

where $J_{\varepsilon,i} = \lambda_{\varepsilon,i} \bar{J}_{m_i}$ where $\bar{J}_{m_i}$ is a Jordan block associated with the eigenvalue 1 of size $m_i \times m_i$. We introduce

$$\widetilde{\Gamma}_\varepsilon' P_\varepsilon \widetilde{\Gamma}_\varepsilon = \widetilde{P}_\varepsilon = \begin{pmatrix} \widetilde{P}_{\varepsilon,1} & 0 & \cdots & 0 \\ 0 & \ddots & \ddots & \vdots \\ \vdots & \ddots & \ddots & 0 \\ 0 & \cdots & 0 & \widetilde{P}_{\varepsilon,k} \end{pmatrix}$$

and

$$\widetilde{\Gamma}_\varepsilon' Q_\varepsilon \widetilde{\Gamma}_\varepsilon = \widetilde{Q}_\varepsilon = \begin{pmatrix} \widetilde{Q}_{\varepsilon,1} & 0 & \cdots & 0 \\ 0 & \ddots & \ddots & \vdots \\ \vdots & \ddots & \ddots & 0 \\ 0 & \cdots & 0 & \widetilde{Q}_{\varepsilon,k} \end{pmatrix}.$$

We get the equation

$$\widetilde{P}_{\varepsilon,i} = J_{\varepsilon,i}' \widetilde{P}_{\varepsilon,i} J_{\varepsilon,i} + \widetilde{Q}_{\varepsilon,i}.$$

If $|\lambda_i| < 1$, then we choose $\widetilde{Q}_{\varepsilon,i} = \varepsilon I$, and we obtain $\widetilde{P}_{\varepsilon,i} = \varepsilon \bar{P}_i$ for all $\varepsilon > 0$ with $\bar{P}_i > 0$.

If $|\lambda_i| = 1$, we note that the Lyapunov equation is equal to

$$\widetilde{P}_{\varepsilon,i} = (1-2\varepsilon)^2 \bar{J}_{m_i}' \widetilde{P}_{\varepsilon,i} \bar{J}_{m_i} + \widetilde{Q}_{\varepsilon,i}.$$

In other words, the phase angle of λ_i does not play a role in the equation. We define $\widetilde{J}_{m_i} = \bar{J}_{m_i} - I$ which is a Jordan block associated with the eigenvalue 0 and a nilpotent matrix. We get

$$(2\varepsilon - \varepsilon^2) \widetilde{P}_{\varepsilon,i} = (1-2\varepsilon)^2 \left(\widetilde{J}_{m_i}' \widetilde{P}_{\varepsilon,i} + \widetilde{P}_{\varepsilon,i} \widetilde{J}_{m_i} + \widetilde{J}_{m_i}' \widetilde{P}_{\varepsilon,i} \widetilde{J}_{m_i} \right) + \widetilde{Q}_{\varepsilon,i}.$$

Define

$$\delta = \frac{2\varepsilon - \varepsilon^2}{(1-2\varepsilon)^2}, \qquad \widetilde{Q}_{\varepsilon,i} = \frac{\delta^{2m_i}}{(1-2\varepsilon)^2} I,$$

and we get

$$\delta \widetilde{P}_{\varepsilon,i} = \left(\widetilde{J}_{m_i}' \widetilde{P}_{\varepsilon,i} + \widetilde{P}_{\varepsilon,i} \widetilde{J}_{m_i} + \widetilde{J}_{m_i}' \widetilde{P}_{\varepsilon,i} \widetilde{J}_{m_i} \right) + \delta^{2m_i} I.$$

In this case, we note that $\widetilde{P}_{\varepsilon,i}$ is a polynomial in δ:

$$\widetilde{P}_{\varepsilon,i} = \sum_{j=1}^{2m_i - 1} \delta^j \bar{P}_{i,j}.$$

This can be established using the recursion

$$\bar{P}_{i,j} = \widetilde{J}_{m_i}' \bar{P}_{i,j+1} + \bar{P}_{i,j+1} \widetilde{J}_{m_i} + \widetilde{J}_{m_i}' \bar{P}_{i,j+1} \widetilde{J}_{m_i}$$

for $j = 1, \ldots, 2m_i - 2$ with $\bar{P}_{i,2m_i-1} = I$. The fact that $\bar{P}_{i,0} = 0$ follows from the fact that $\widetilde{J}_{m_i}$ is nilpotent, and hence, $\widetilde{J}^{m_i-1} = 0$.

In order to apply Lemma 4.39 to establish that we obtain in this way a proper low-and-high-gain design, we need to verify that $P_\varepsilon \to 0$ as $\varepsilon \to 0$ and

$$P_\varepsilon \geqslant F'_{L,\varepsilon} F_{L,\varepsilon}. \tag{4.177}$$

The fact that $P_\varepsilon \to 0$ as $\varepsilon \to 0$ is a trivial consequence of our construction (note that $\varepsilon \to 0$ trivially implies that $\delta \to 0$). In order to establish (4.177), we first note that

$$F_{L,\varepsilon}\widetilde{\Gamma}_\varepsilon = \begin{pmatrix} \widetilde{F}_{\varepsilon,1} & \cdots & \widetilde{F}_{\varepsilon,k} \end{pmatrix},$$

and (4.32) implies that there exists a $M > 0$ such that

$$\| \widetilde{F}_{\varepsilon,i} e_\ell \| \leqslant M \varepsilon^{m_i - \ell + 1}$$

for $i = 1, \ldots, k$ and $\ell = 1, \ldots, m_i$ where $e_1, \ldots, e_{m_i}$ is the standard basis for $\mathbb{R}^{m_i}$. Next, we note that our recursion implies that there exists a $m > 0$ such that

$$e'_h \bar{P}_j e_\ell = 0$$

for $h + \ell < 2m_i - j + 1$ and $e'_h \bar{P}_j e_\ell > m$ for $h + \ell = 2m_i - j + 1$. This implies that

$$e'_h \widetilde{P}_{\varepsilon,i} e_\ell > m \varepsilon^{2m_i + 1 - h - \ell}$$

for $h, \ell = 1, \ldots, m_i$ and ε sufficiently small. Combined with

$$e'_h \widetilde{F}'_{\varepsilon,i} \widetilde{F}_{\varepsilon,i} e_\ell < M^2 \varepsilon^{2m_i + 2 - h - \ell},$$

we find that for ε small enough, we have

$$\widetilde{F}'_{\varepsilon,i} \widetilde{F}_{\varepsilon,i} \leqslant \widetilde{P}_{\varepsilon,i},$$

and hence, (4.177) is satisfied. Therefore, Lemma 4.39 can be applied and we can conclude that this design has the required properties.

A design algorithm for multi-input systems

We use the same initial state transformation as in Sect. 4.3.2, i.e., a state transformation Γ_x and an input transformation Γ_u such that the matrices $\Gamma_x^{-1} A \Gamma_x$ and $\Gamma_x^{-1} B \Gamma_u$ are in the following form:

$$\Gamma_x^{-1} A \Gamma_x = \begin{pmatrix} A_1 & A_{1,2} & \cdots & A_{1,q} & 0 \\ 0 & A_2 & \ddots & \vdots & \vdots \\ \vdots & \ddots & \ddots & A_{q-1,q} & \vdots \\ \vdots & & \ddots & A_q & 0 \\ 0 & \cdots & \cdots & 0 & A_{q+1} \end{pmatrix},$$

$$\Gamma_x^{-1} B \Gamma_u = \begin{pmatrix} B_1 & 0 & \cdots & 0 & B_{1,q+1} \\ 0 & B_2 & \ddots & \vdots & \vdots \\ \vdots & \ddots & \ddots & 0 & \vdots \\ 0 & \cdots & 0 & B_q & B_{q,q+1} \\ B_{q+1,1} & \cdots & B_{q+1,q-1} & B_{q+1,q} & B_{q+1,q+1} \end{pmatrix},$$

where q is an integer, and for $i = 1, 2, \ldots, q$,

$$A_i = \begin{pmatrix} 0 & 1 & 0 & \cdots & 0 \\ \vdots & \ddots & \ddots & \ddots & \vdots \\ \vdots & & \ddots & 1 & 0 \\ 0 & \cdots & \cdots & 0 & 1 \\ -a_{n_i}^i & \cdots & -a_3^i & -a_2^i & -a_1^i \end{pmatrix}, \quad B_i = \begin{pmatrix} 0 \\ \vdots \\ 0 \\ 0 \\ 1 \end{pmatrix}.$$

Clearly, (A_i, B_i) is controllable. The transformation Γ_x is such that all the eigenvalues of A_i are on the unit circle, and all the eigenvalues of A_{q+1} are strictly inside the unit circle.

Step 2: For each (A_i, B_i), let $F_{L,\varepsilon,i} \in \mathbb{R}^{1 \times n_i}$ be the state feedback gain such that

$$\lambda(A_i + B_i F_{L,\varepsilon_i,i}) = (1 - 2\varepsilon_i)\lambda(A_i) \subset \mathbb{C}^{\ominus}.$$

We note here that such a gain $F_{L,\varepsilon_i,i}$ exists and is unique. Moreover, it can be obtained explicitly in terms of ε. The uniqueness follows since (A_i, B_i) is a single input controllable pair.

Step 3 : The family of low-gain state feedback control laws parameterized in ε is defined by

$$u = F_{L,\varepsilon} x, \tag{4.178}$$

where the state feedback gain matrix $F_{L,\varepsilon}$ is given by

$$F_{L,\varepsilon} = \Gamma_u \begin{pmatrix} F_{L,\varepsilon_1,1} & 0 & \cdots & \cdots & 0 \\ 0 & F_{L,\varepsilon_2,2} & \ddots & & \vdots \\ \vdots & \ddots & \ddots & \ddots & \vdots \\ \vdots & & \ddots & F_{L,\varepsilon_q,q} & 0 \\ 0 & \cdots & \cdots & 0 & 0 \end{pmatrix} \Gamma_x^{-1}. \tag{4.179}$$

Here, $\varepsilon_{q+1} = \varepsilon$, and for $i = 1, \ldots, q$ we have

$$\varepsilon_i = \varepsilon_{i+1}^{6r_{i+1}+1}, \tag{4.180}$$

where r_i is the largest algebraic multiplicity among the eigenvalues of A_i for $i = 1, \ldots, q$ while $r_{q+1} = 1$. Moreover, we construct $P_{\varepsilon_i,i}$ and $Q_{\varepsilon_i,i}$ such that

$$P_{\varepsilon_i,i} = (A_i + B_i F_{\varepsilon_i,i})' P_{\varepsilon_i,i} (A_i + B_i F_{\varepsilon_i,i}) + Q_{\varepsilon_i,i}$$

with $P_{\varepsilon_i,i} \to 0$ as $\varepsilon \to 0$ and

$$P_{\varepsilon_i,i} \geqslant F'_{\varepsilon_i,i} F_{\varepsilon_i,i}$$

for $i = 1, \ldots, q$ such that there exist $\bar{m}_1, \bar{m}_2$, and $\bar{m}_3$ for which

$$\bar{m}_1 \varepsilon^{2r_i - 1} I \leqslant P_{\varepsilon,i} \leqslant \bar{m}_2 \varepsilon I \tag{4.181}$$

and

$$Q_{\varepsilon_i,i} \geqslant \bar{m}_3 \varepsilon^{2r_i} I. \tag{4.182}$$

The algorithm presented earlier for the single-input case tells us how to construct $P_{\varepsilon_i,i}$ and $Q_{\varepsilon_i,i}$. We define

$$F_{H,\varepsilon_i,i} = -(B_i' P_{\varepsilon_i,i} B_i)^{-1} B_i' P_{\varepsilon_i,i} = (A_i + B_i F_{\varepsilon_i,i}),$$

and the family of low-and-high-gain state feedback control laws parameterized in ε is defined by

$$u = F_{L,\varepsilon} x + \alpha F_{H,\varepsilon} x, \tag{4.183}$$

where the state feedback gain matrix $F_{L,\varepsilon}$ is given by (4.179), and

$$F_{H,\varepsilon} = \Gamma_u \begin{pmatrix} F_{H,\varepsilon_1,1} & 0 & \cdots & \cdots & 0 \\ 0 & F_{H,\varepsilon_2,2} & \ddots & & \vdots \\ \vdots & \ddots & \ddots & \ddots & \vdots \\ \vdots & & \ddots & F_{H11,\varepsilon_q,q} & 0 \\ 0 & \cdots & \cdots & 0 & 0 \end{pmatrix} \Gamma_x^{-1}. \tag{4.184}$$

We define $Q_{\varepsilon,q+1} = \varepsilon I$ and $P_{\varepsilon,q+1}$ such that

$$P_{\varepsilon,q+1} = A'_{q+1} P_{\varepsilon,q+1} A_{q+1} + Q_{\varepsilon,q+1}$$

and define $r_{q+1} = 1$. We also define

$$P_\varepsilon = \Gamma_x \begin{pmatrix} P_{\varepsilon_1,1} & 0 & \cdots & 0 \\ 0 & P_{\varepsilon_2,2} & \ddots & \vdots \\ \vdots & \ddots & \ddots & 0 \\ 0 & \cdots & 0 & P_{\varepsilon_{q+1},q+1} \end{pmatrix} \Gamma_x^{-1}. \tag{4.185}$$

The initial conditions are in the compact set $\mathcal{X}$. This implies that for ε small enough, the initial conditions satisfy

$$x(0) \in \frac{1}{2}\mathcal{V}_\varepsilon = \left\{ x \in \mathbb{R}^n \mid x' P_\varepsilon x \leqslant \frac{1}{2} \right\}. \tag{4.186}$$

For $i = q + 1$, we have

$$x_i'(k) P_{\varepsilon_i,i} x_i(k) \leqslant (1 - \varepsilon_i^{2r_i})^k \tag{4.187}$$

given (4.186). This yields

$$\|x_i(k)\| \leqslant \frac{\bar{m}_1}{\varepsilon_i^{r_i - 1/2}} (1 - \varepsilon_i^{2r_i})^{k/2} \tag{4.188}$$

using (4.181). We will prove (4.187) and (4.188) for $i = 1, \ldots, q$ through an induction argument. Assume (4.187) and (4.188) are satisfied for $i = j+1, \ldots, q+1$. We will establish that this inequality is also satisfied for $i = j$.

In order to establish this, we consider the j'th subsystem

$$x_j(k+1) = A_j x_j(k) + B_j u_j(k) + w_j(k),$$

where

$$w_j = A_{j,j+1} x_{j+1} + \cdots + A_{j,q} x_q$$

while $w_q = 0$. Using (4.188) for $i = j + 1, \ldots, q$, we obtain that there exists a M_j such that

$$\|w_j(k)\| \leqslant \frac{M_j}{\varepsilon_j^{r_{j+1} - 1/2}} (1 - \varepsilon_{j+1}^{2r_{j+1}})^{k/2}$$

for all $k > 0$. We obtain in the set $\mathcal{V}_\varepsilon$

$$\begin{aligned} x_j'(k+1) P_{\varepsilon_j,j} x_j(k+1) &= x_j'(k) P_{\varepsilon_j,j} x_j(k) - x_j'(k) Q_{\varepsilon_j,j} x_j(k) \\ &\quad + 2x_j'(k)(A_j + B_j F_{L,\varepsilon_j,j})' P_{\varepsilon_j,j} B_j v_j(k) + v_j'(k) B_j' P_{\varepsilon_j,j} B_j v_j(k) \\ &\quad + 2x_j'(k)(A_j + B_j F_{L,\varepsilon_j,j})' P_{\varepsilon_j,j} w_j(k) + w_j'(k) P_{\varepsilon_j,j} w_j(k) \\ &\quad + 2v_j'(k) B_j' P_{\varepsilon_j,j} w_j(k), \end{aligned}$$

where

$$v_j(k) = \sigma\left(F_{L,\varepsilon_j,j}x_j(k) + \alpha F_{H,\varepsilon_j,j}x_j(k)\right) - F_{L,\varepsilon_j,j}x_j(k).$$

Since by assumption $x(k) \in \mathcal{V}_\varepsilon$, we have

$$x_j'(k)F_{L,\varepsilon_j,j}'F_{L,\varepsilon_j,j}x_j(k) \leqslant x_j'(k)P_{\varepsilon_j,j}x + j(k) < 1,$$

where we use (4.173), and hence, $\|F_{L,\varepsilon_j,j}x_j(k)\| < 1$. This implies that

$$\sigma(\left(F_{L,\varepsilon_j,j}x_j(k) + \alpha F_{H,\varepsilon_j,j}x_j(k)\right) = F_{L,\varepsilon_j,j}x_j(k) + \mu_j\alpha F_{H,\varepsilon_j,j}x_j(k)$$

with $\mu_j \in [0,1]$, and hence,

$$\begin{aligned} x_j'(k+1)P_{\varepsilon_j,j}x_j(k+1) &\leqslant x_j'(k)P_{\varepsilon_j,j}x_j(k) - x_j'(k)Q_{\varepsilon_j,j}x_j(k) \\ &\quad + 2x_j'(k)(A_j + B_jF_{L,\varepsilon_j,j})'P_{\varepsilon_j,j}w_j(k) + w_j'(k)P_{\varepsilon_j,j}w_j(k) \\ &\quad + 2v_j'(k)B_j'P_{\varepsilon_j,j}w_j(k). \end{aligned} \tag{4.189}$$

By (4.181) and (4.182), we find that

$$x_j'Q_{\varepsilon_j,j}x_j \geqslant \frac{m_3}{m_2}\varepsilon^{2r_j-1}x_j'P_{\varepsilon_j,j}x_j \geqslant \varepsilon^{2r_j}x_j'P_{\varepsilon_j,j}x_j,$$

where the last equality holds for sufficiently small $\varepsilon > 0$. Using the Lyapunov equation, we note that

$$x_j'P_{\varepsilon_j,j}x_j \geqslant x_j'(k)(A_j + B_jF_{L,\varepsilon_j,j})'P_{\varepsilon_j,j}(A_j + B_jF_{L,\varepsilon_j,j})x_j(k).$$

Next, we note that

$$\begin{aligned} &2x_j'(A_j + B_jF_{L,\varepsilon_j,j})'P_{\varepsilon_j,j}w_j \\ &\qquad \leqslant 2\left(x_j'P_{\varepsilon_j,j}x_j\right)^{1/2}\sqrt{m_2}\sqrt{\varepsilon_j}\frac{M_j}{\varepsilon_{j+1}^{r_{j+1}-1/2}}(1-\varepsilon_{j+1}^{2r_j+1})^{k/2} \\ &\qquad \leqslant \frac{\sqrt{\varepsilon_j}}{3\varepsilon_{j+1}^{r_{j+1}}}(1-\varepsilon_{j+1}^{2r_j+1})^{k/2}, \end{aligned}$$

where we use the fact that, by assumption, $x \in \mathcal{V}_\varepsilon$. The last inequality follows for sufficiently small ε. We also note that

$$w_j'P_{\varepsilon_j,j}w_j \leqslant m_2\varepsilon_j\frac{M_j^2}{\varepsilon_{j+1}^{2r_{j+1}-1}}(1-\varepsilon_{j+1}^{2r_j+1})^k$$

and

$$\begin{aligned}2v_j'(k)B_j'P_{\varepsilon_j,j}w_j(k) &\leqslant \mu_j \alpha x_j' F_{H,\varepsilon_j,j} B_j' P_{\varepsilon_j,j} w_j\\ &\leqslant 2\|w_j\|\|P_{\varepsilon_j,j}^{1/2}\|\|P_{\varepsilon_j,j}(A_j + B_j F_{L,\varepsilon_j,j})x_j\|\\ &\leqslant 2\|w_j\|\|P_{\varepsilon_j,j}^{1/2}\|\|P_{\varepsilon_j,j}x_j\|\\ &\leqslant 2\sqrt{m_2}\sqrt{\varepsilon_j}\frac{M_j}{\varepsilon_{j+1}^{r_{j+1}-1/2}}(1-\varepsilon_{j+1}^{2r_{j+1}})^{k/2}\\ &\leqslant \frac{\sqrt{\varepsilon_j}}{3\varepsilon_{j+1}^{r_{j+1}}}(1-\varepsilon_{j+1}^{2r_{j+1}})^{k/2},\end{aligned}$$

where we again use that $x \in \mathcal{V}_\varepsilon$, and for the last inequality, we have to make sure that ε is sufficiently small. We define

$$V_j(k) = x_j'(k)P_{\varepsilon_j,j}x_j(k),$$

and we obtain using the above bounds and (4.189) that

$$V_j(k+1) \leqslant (1-\varepsilon_j^{2r_j})V(k) + \frac{\sqrt{\varepsilon_j}}{\varepsilon_{j+1}^{r_{j+1}}}(1-\varepsilon_{j+1}^{2r_{j+1}})^{k/2}.$$

Therefore,

$$V_j(k) \leqslant (1-\varepsilon_j^{2r_j})^k\left[V_j(0) + \sum_{i=0}^{\infty}\frac{\sqrt{\varepsilon_j}}{\varepsilon_{j+1}^{r_{j+1}}}\left(\frac{\sqrt{1-\varepsilon_{j+1}^{2r_{j+1}}}}{1-\varepsilon_j^{2r_j}}\right)^i\right],$$

which gives us

$$V_j(k) \leqslant (1-\varepsilon_j^{2r_j})^k\left[V_j(0) + \frac{2\sqrt{\varepsilon_j}}{\varepsilon_{j+1}^{3r_{j+1}}}\right].$$

This yields, for ε small enough, (4.187) and (4.188) for $i = j$, where we use (4.180) and (4.181). This establishes recursively (4.187) and (4.188) for $i = 1, \ldots, q+1$ given (4.186). Since $\mathcal{X} \subset \frac{1}{2}\mathcal{V}_\varepsilon$, we find that for all initial conditions in $\mathcal{X}$, we stay inside $\mathcal{V}_\varepsilon$ and converge to zero exponentially. This establishes the required properties of our design.

A design algorithm for measurement feedback

In case of measurement feedback, we have the system,

$$\begin{aligned}x(k+1) &= Ax(k) + B\sigma(u(k))\\ y(k) &= Cx(k).\end{aligned}$$

We need to build an observer that can be used in conjunction with the low-and-high-gain state feedback designed before. In addition to earlier assumptions that (A, B) is stabilizable and all eigenvalues of A being inside or on the unit circle, we assume that the pair (C, A) is observable. We construct the observer-based controller as

$$\begin{aligned} \hat{x}(k+1) &= A\hat{x}(k) + B\sigma(u(k)) + K(y(k) - C\hat{x}(k)) \\ u(k) &= F_{L,\varepsilon}\hat{x}(k) + \alpha F_{H,\varepsilon}\hat{x}(k), \end{aligned}$$

where K is such that the eigenvalues of $A - KC$ are all in the origin. This implies that $x(n) = \hat{x}(n)$. Given a bounded set $\mathcal{X}$ of initial conditions for $x(0)$, we know that $x(n)$ is contained in some compact set $\widetilde{\mathcal{X}}$ given the fact that $\sigma(u(k))$ is bounded for all k. Then we choose ε small enough to guarantee that $\widetilde{\mathcal{X}}$ is contained in the domain of attraction. Finally, we choose $\alpha \in [0, 2]$, and it is clear that this controller will achieve the required semi-global stability.

So far, we pursued low-and-high-gain feedback design based on direct eigenstructure assignment method. We present next another low-and-high-gain design method which, however, is based on solving a parameterized H_2 ARE.

4.5.3 ARE-based methods: continuous time

As in the previous subsections, a low-and-high-gain feedback law is a composite control law. It is composed by adding together a low-gain feedback control and a high-gain feedback control. For continuous-time systems, the low-gain state feedback gain $F_{L,\varepsilon}$ is defined as

$$F_{L,\varepsilon} := -B'P_\varepsilon, \tag{4.190}$$

where, as in (4.44b), P_ε is the positive definite solution of the following CARE:

$$PA + A'P - PBB'P + Q_\varepsilon = 0, \tag{4.191}$$

where $Q_\varepsilon > 0$ for all $\varepsilon > 0$ and $Q_\varepsilon \to 0$ as $\varepsilon \to 0$. The high-gain state feedback gain $F_{H,\varepsilon}$ is simply defined as

$$F_{H,\varepsilon} := -\alpha B'P_\varepsilon, \tag{4.192}$$

where, as in (4.190), P_ε is the positive definite solution of the CARE (4.191). As in the previous section, we use the qualifier *high* to the above state feedback gain because of the presence of the parameter α which can possibly take high values. For the same reason, α is called the *high-gain parameter*.

The family of parameterized low-and-high-gain state feedback laws is composed of families of low-gain and high-gain state feedback laws, and is defined by

$$u = F_{LH,\varepsilon,\alpha}x, \tag{4.193a}$$

where

$$F_{LH,\varepsilon,\alpha} := F_{L,\varepsilon} + F_{H,\varepsilon} = -(1+\alpha)B'P_\varepsilon. \tag{4.193b}$$

Remark 4.40 *The CARE-based low-and-high-gain state feedback design formulated above is actually an optimal design for the linear system (A, B) in the absence of input saturation with appropriately chosen Q_n and R_n. More specifically, choosing $R_n = I/(1+\alpha)$ and $Q_n = Q + \alpha PBB'P$, it is easy to verify that P is the solution to the new CARE:*

$$A'P + PA - PBR_n^{-1}B'P + Q_n = 0,$$

and hence, $u = -(1+\alpha)B'Px = -R_n^{-1}B'Px$ is an optimal control for each fixed $\alpha \geqslant 0$.

Once again, as in the previous subsections, we emphasize the roles played by the low-gain and high-gain parameters ε and α. We note from [74, 77, 78, 128] that, in the context of semi-global internal stabilization of linear systems with saturating actuators, the parameter α does not play any role in influencing the domain of attraction which, indeed, is basically controlled by the low-gain parameter ε. However, the parameter α plays a crucial role in the context of issues other than the internal stabilization, such as external stabilization, robust stabilization, disturbance rejection, and other performance-related issues.

Theorem 4.41 *Consider the continuous-time system* (4.1). *Let Assumption 4.8 hold, i.e., the pair (A, B) is stabilizable and the eigenvalues of A are in the closed left-half plane. Then the family of linear low-and-high-gain state feedback laws given by* (4.193) *solves Problem 4.4. More specifically, under the state feedback* (4.193), *for any given (arbitrarily large) bounded set $\mathcal{X} \subset \mathbb{R}^n$, there exists an $\varepsilon^* \in (0, 1]$ such that for all $\varepsilon \in (0, \varepsilon^*]$ and for all $\alpha \geq 0$, the equilibrium point $x = 0$ of the closed-loop system is locally exponentially stable with $\mathcal{X}$ contained in its domain of attraction.*

Proof : Consider the Lyapunov function

$$V_\varepsilon(x) = x'P_\varepsilon x \tag{4.194}$$

and let $c > 0$ be such that

$$c \geq \sup_{x\in\mathcal{X},\varepsilon\in(0,1]} x'P_\varepsilon x. \tag{4.195}$$

Such a c exists because $\mathcal{X}$ is bounded and, by Lemma 4.20, $\lim_{\varepsilon\to 0} P_\varepsilon = 0$. Let ε^* be such that, for all $\varepsilon \in (0, \varepsilon^*]$, having

$$x \in \mathcal{L}_V(c) = \{x \in \mathbb{R}^n : x'P_\varepsilon x \leq c\}$$

implies that $\|B'P_\varepsilon x\| \leq \Delta$, and hence,

$$\sigma(-B'P_\varepsilon x) = -B'P_\varepsilon x, \tag{4.196}$$

where Δ is the actuator saturation level. We note that $\mathcal{L}_V(c)$ is a positively invariant set in the sense that $x(t) \in \mathcal{V}_\varepsilon$ implies that $x(\tau) \in \mathcal{V}_\varepsilon$ for all $\tau > t$. In order to establish this, we first note that

$$\frac{\mathrm{d}}{\mathrm{d}t} x'(t) P_\varepsilon x(t) = -x'(t) Q_\varepsilon x(t) + 2x'(t) P_\varepsilon B v(t),$$

where

$$v(t) = \sigma\left(F_{L,\varepsilon}x(t) + \alpha F_{H,\varepsilon}x(t)\right) - F_{L,\varepsilon}x(t).$$

Since $x(t) \in \mathcal{L}_V(c)$, we have (4.196), and hence, $\|F_{L,\varepsilon}x(t)\| < \Delta$. This implies that

$$\sigma((F_{L,\varepsilon}x(t) + \alpha F_{H,\varepsilon}x(t)) = F_{L,\varepsilon}x(t) + \begin{pmatrix} \mu_1 & 0 & \cdots & 0 \\ 0 & \ddots & \ddots & \vdots \\ \vdots & \ddots & \ddots & 0 \\ 0 & \cdots & 0 & \mu_m \end{pmatrix} \alpha F_{H,\varepsilon}x(t)$$

with $\mu_i \in [0, 1]$ for $i = 1, \ldots, m$. Therefore,

$$\begin{aligned} \frac{\mathrm{d}}{\mathrm{d}t} x'(t) P_\varepsilon x(t) &= -x'(t) Q_\varepsilon x(t) \\ &\quad - \alpha 2x'(t) P_\varepsilon B \begin{pmatrix} \mu_1 & 0 & \cdots & 0 \\ 0 & \ddots & \ddots & \vdots \\ \vdots & \ddots & \ddots & 0 \\ 0 & \cdots & 0 & \mu_m \end{pmatrix} B' P_\varepsilon x(t), \end{aligned}$$

and hence,

$$\frac{\mathrm{d}}{\mathrm{d}t} x'(t) P_\varepsilon x(t) \leqslant -x'(t) Q_\varepsilon x(t) < 0.$$

The above yields that in the set $\mathcal{L}_V(c)$ we have $x'(t)P_\varepsilon x(t)$ exponentially decaying to zero, which implies that we do not leave $\mathcal{L}_V(c)$, and also that we have local exponential stability and, by construction, the set $\mathcal{X}$ is contained in the domain of attraction. ■

We proceed now to construct a family of parameterized low-and-high-gain measurement feedback control laws. We first present a preliminary lemma.

Lemma 4.42 *Let the pair (A, C) be observable. Also, let Q_ℓ be the solution of the following dual algebraic Riccati equation:*

$$0 = (A + \ell I)Q + Q(A + \ell I)' - QC'CQ + I. \tag{4.197}$$

Then, we have for sufficiently large ℓ

$$\left\| e^{(A-Q_\ell C'C)t} \right\| \leqslant e^{-\ell t/2} \tag{4.198}$$

for all $t > 1$.

Proof : Standard properties of the CARE (4.197) imply that

$$A + \ell I - Q_\ell C'C$$

is asymptotically stable, and hence, the eigenvalues of $A - Q_\ell C'C$ have real part strictly less than $-\ell$. It is also easy to verify that Q_ℓ is increasing in ℓ since differentiating the Riccati equation with respect to ℓ yields

$$(A - QC'C)\dot{Q} + \dot{Q}(A - QC'C)' + 2Q = 0.$$

To show (4.198), we note that Q_ℓ has an interpretation as the optimal cost of the following optimal control problem:

$$\dot{x} = A'x + C'u, \quad x(0) = \xi$$

and

$$\xi' Q_\ell \xi = \inf_u \int_0^t \|e^{\ell s} x(s)\|^2 + \|e^{\ell s} u(s)\|^2 \mathrm{d}s + e^{2\ell t} x(t)' Q_\ell x(t).$$

There exists a $M > 0$ such that for any ξ there exists an input u such that $x(1/2) = 0$, $\|u\|_2 < M\|\xi\|$ and $\|x\|_2 < M\|\xi\|$. When we choose this suboptimal input in the above optimization problem, we find that

$$\xi' Q_\ell \xi \leqslant 2e^{\ell/2} M^2 \|\xi\|^2.$$

On the other hand, for the optimal input $u = -CQ_\ell x$, we get

$$x(t) = e^{(A'-C'CQ_\ell)t}\xi,$$

and therefore,

$$\begin{aligned} e^{2\ell t} \xi' e^{(A-Q_\ell C'C)t} Q_\ell e^{(A'-C'CQ_\ell)t} \xi = e^{2\ell t} x(t)' Q_\ell x(t) \leqslant \xi' Q_\ell \xi \\ \leqslant 2e^{\ell/2} M^2 \|\xi\|^2. \end{aligned}$$

Hence, there exists a $N > 0$ such that

$$\begin{aligned} e^{(A-Q_\ell C'C)t}e^{(A'-C'CQ_\ell)t} &\leqslant Ne^{(A-Q_\ell C'C)t}Q_1e^{(A'-C'CQ_\ell)t} \\ &\leqslant Ne^{(A-Q_\ell C'C)t}Q_\ell e^{(A'-C'CQ_\ell)t} \\ &\leqslant 2e^{-3\ell t/2}NM^2 \end{aligned}$$

since $t > 1$ and Q_ℓ is increasing in ℓ. Therefore, we find for sufficiently large ℓ that (4.198) is satisfied. ■

A family of parameterized high-gain observer based low-and-high-gain measurement feedback control laws take the form

$$\begin{cases} \dot{\hat{x}} = A\hat{x} + B\sigma(u) + K_\ell(y - C\hat{x}) \\ u = F_{LH,\varepsilon,\alpha}\hat{x}, \end{cases} \tag{4.199}$$

where the state feedback gain $F_{LH,\varepsilon,\alpha}$ is as given by (4.193b) and the observer gain $K_\ell = Q_\ell C'$ where Q_ℓ is the solution of the algebraic Riccati equation (4.197).

We have the following theorem pertaining to semi-global stabilization:

Theorem 4.43 *Consider the continuous-time system* (4.1). *Let Assumption 4.8 hold, i.e., assume that the pair (A, B) is stabilizable and the eigenvalues of A are in the closed left-half plane. Moreover, assume that (C, A) is observable.*

In that case, the family of linear dynamic measurement feedback laws given in (4.199) *solves Problem 4.5. More specifically, under the measurement feedback law* (4.199), *for any given (arbitrarily large) bounded set $(\mathcal{X}, \widehat{\mathcal{X}}) \subset \mathbb{R}^{2n}$, there exists an $\varepsilon^* \in (0, 1]$ such that, for all $\alpha \geq 0$ and for all $\varepsilon \in (0, \varepsilon^*]$, there exists an ℓ^* such that for $\ell > \ell^*$ the equilibrium point $(0, 0)$ of the closed-loop system is locally exponentially stable with $(\mathcal{X}, \widehat{\mathcal{X}})$ contained in its domain of attraction.*

Proof : Under the output feedback law (4.51), the closed-loop system takes the form,

$$\begin{cases} \dot{x} = Ax + B\sigma(-(1+\alpha)B'P_\varepsilon\hat{x}) \\ \dot{\hat{x}} = A\hat{x} + B\sigma(-(1+\alpha)B'P_\varepsilon\hat{x}) + K_\ell(y - C\hat{x}) \end{cases} \tag{4.200}$$

which, in the new coordinates (x, e) where $e = x - \hat{x}$, becomes

$$\begin{cases} \dot{x} = Ax + B\sigma(-(1+\alpha)B'P_\varepsilon(x - e)) \\ \dot{e} = (A - K_\ell C)e. \end{cases} \tag{4.201}$$

Local exponential stability then immediately follows since $A - K_\ell C$ and $A-(1+\alpha)BB'P_\varepsilon$ are both Hurwitz. It remains to consider the domain of attraction.

Since σ is bounded and $x(0) \in \mathcal{X}$, there exists a bounded set $\widetilde{\mathcal{X}}$ such that $x(1) \in \widetilde{\mathcal{X}}$ for all initial conditions

$$(x(0), \hat{x}(0)) \in \mathcal{X} \times \widehat{\mathcal{X}}$$

and independent of α and ℓ. Next, we note from Lemma 4.42 that

$$\|e(t)\| \leqslant \widetilde{M} e^{-\ell t/2}$$

for all $t > 1$ where $\widetilde{M}$ only depends on $\mathcal{X}$ and $\widehat{\mathcal{X}}$ and is independent of ℓ and α. Consider now the Lyapunov function,

$$V_\varepsilon(x) = x' P_\varepsilon x, \tag{4.202}$$

and let $c > 0$ be such that

$$c \geq \sup_{x \in \widetilde{\mathcal{X}}, \varepsilon \in (0,1]} x' P_\varepsilon x. \tag{4.203}$$

Such a c exists because $\widetilde{\mathcal{X}}$ is bounded. Moreover, P_ε is bounded by P_1 for $\varepsilon \in [0, 1]$. Let ε_1^* be such that, for all $\varepsilon \in (0, \varepsilon_1^*]$,

$$x \in \mathcal{L}_V(c) = \{x \in \mathbb{R}^n : x' P_\varepsilon x \leqslant c\}$$

and e with $\|e\| < \widetilde{M}$ we have $2\|B' P_\varepsilon(x - e)\| \leqslant \Delta$. The existence of such an ε_1^* again owes to the fact that $\lim_{\varepsilon \to 0} P_\varepsilon = 0$.

We will establish that for all $t > 1$, we have $x(t) \in 2\mathcal{L}_V(c)$ with $x(t) \to 0$ as $t \to \infty$. Note that for $x \in 2\mathcal{L}_V(c)$ and $\|e\| < \widetilde{M}$, we have $\sigma(B' P_\varepsilon(x - e)) = B' P_\varepsilon(x - e)$, and hence,

$$\sigma(-B' P_\varepsilon(x - e)) = -B' P_\varepsilon(x - e) - \alpha \begin{pmatrix} \mu_1 & 0 & \cdots & 0 \\ 0 & \ddots & \ddots & \vdots \\ \vdots & \ddots & \ddots & 0 \\ 0 & \cdots & 0 & \mu_m \end{pmatrix} B' P_\varepsilon(x - e)$$

with $\mu_i(t) \in [0, 1]$ for $i = 1, \ldots, m$ and for all $t > 0$, provided that $x \in \mathcal{L}_V(c)$ and e satisfies $\|e\| < \widetilde{M}$. Define

$$M = \begin{pmatrix} \mu_1 & 0 & \cdots & 0 \\ 0 & \ddots & \ddots & \vdots \\ \vdots & \ddots & \ddots & 0 \\ 0 & \cdots & 0 & \mu_m \end{pmatrix}.$$

It then follows that for all $\varepsilon \in (0, \varepsilon_1^*]$ and for all $x \in 2\mathcal{L}_V(c)$ and $\|e\| < \tilde{M}$, the closed-loop system (4.201) can be written as

$$\begin{cases} \dot{x} = (A - BB'P_\varepsilon - \alpha BMB'P_\varepsilon)x + BB'P_\varepsilon e + \alpha BMB'P_\varepsilon e \\ \dot{e} = (A - K_\ell C)e. \end{cases} \tag{4.204}$$

The evaluation of $\dot{V}_\varepsilon$ along the trajectories of this closed-loop system gives, for all $(x, e) \in \mathcal{L}_V(c)$ and for all $\varepsilon \in (0, \varepsilon_1^*]$,

$$\begin{aligned} \dot{V}_\varepsilon &= -x'Q_\varepsilon x - x'P_\varepsilon B(I + 2\alpha M)B'P_\varepsilon x - 2x'P_\varepsilon B(I + \alpha M)B'P_\varepsilon e \\ &\leqslant -\delta V_\varepsilon - x'P_\varepsilon B(I + \alpha M)B'P_\varepsilon x - 2x'P_\varepsilon B(I + \alpha M)B'P_\varepsilon e \\ &\leqslant -\delta V_\varepsilon + e'P_\varepsilon B(I + \alpha M)B'P_\varepsilon e \end{aligned}$$

for some $\delta > 0$. Using (4.198), we find given ε and α there exists a $\bar{M} > 0$ such that

$$e(t)'P_\varepsilon B(I + \alpha M)B'P_\varepsilon e(t) \leqslant \bar{M}e^{-\ell t}$$

for $t > 1$. Moreover, $V_\varepsilon(1) \leqslant c$. But then

$$\dot{V}_\varepsilon \leqslant -\delta V_\varepsilon + \bar{M}e^{-\ell t}$$

implies that

$$V_\varepsilon(t) \leqslant e^{-\delta t}V_\varepsilon(1) + \frac{\bar{M}}{\ell - \delta}\left(e^{-\delta t} - e^{-\ell t}\right)$$

for $t > 1$. This yields that $V_\varepsilon(t) < 2c$ and $V_\varepsilon(t) \to 0$ as $t \to \infty$ for ℓ sufficiently large. Therefore, we find that $x(t) \to 0$ as $t \to \infty$, and combined with the fact that $e(t) \to 0$ as $t \to \infty$, we obtain that $\mathcal{X} \times \widehat{\mathcal{X}}$ is contained in the domain of attraction. This completes the proof of our result. ■

4.5.4 ARE-based methods: discrete time

As we said earlier, one cannot have low-and-high-gain control laws for discrete-time systems in the same way as one does for continuous-time systems. As in Sect. 4.5.2, however, by utilizing the ARE-based method, we develop here a family of improved low-gain control laws. An improved low-gain control law is a composite control law consisting of a low-gain control law and another part akin to what we termed as high-gain control law in the case of continuous-time systems.

For discrete-time systems, a family of low-gain state feedback control laws by H_2 ARE-based method is developed in Sect. 4.4.2. It is given by (4.70) and repeated below as

$$u_L = F_{L,\varepsilon}x, \tag{4.205a}$$

where

$$F_{L,\varepsilon} := -(B'P_\varepsilon B + I)^{-1}B'P_\varepsilon A, \quad \varepsilon \in (0,1], \tag{4.205b}$$

and where P_ε is the positive definite solution of DARE:

$$P = A'PA + Q_\varepsilon - A'PB(B'PB + I)^{-1}B'PA. \tag{4.206}$$

In order to improve the above low-gain control law, we add to it another part and formulate a family of composite state feedback control laws as

$$u = [F_{L,\varepsilon} + \alpha \bar{F}_{H,\varepsilon}]x, \quad \alpha \in \left[0, \frac{2}{\|B'P_\varepsilon B\|}\right], \tag{4.207}$$

where

$$\bar{F}_{H,\varepsilon} = F_{L,\varepsilon} = -(B'P_\varepsilon B + I)^{-1}B'P_\varepsilon A. \tag{4.208}$$

Note that we can rewrite this controller for the case of single-input systems in the form

$$u = [F_{L,\varepsilon} + \tilde{\alpha}\tilde{F}_{H,\varepsilon}]x,$$

with $\tilde{\alpha} \in [0,2]$ and where

$$\tilde{F}_{H,\varepsilon} = -(B'P_\varepsilon B)^{-1}B'P_\varepsilon A_c$$

with $A_c = A + BF_{L,\varepsilon}$. This follows immediately from the fact that for single-input systems,

$$\frac{1}{\|B'P_\varepsilon B\|} = (B'P_\varepsilon B)^{-1}.$$

We should note that in our earlier work elsewhere we have used

$$u = [F_{L,\varepsilon} + \tilde{\alpha}\kappa(x)\tilde{F}_{H,\varepsilon}]x, \tag{4.209}$$

where $\kappa : \mathbb{R}^n \to \mathbb{R}^+$ is defined as

$$\kappa(x) = \max_{z\in[0,1]} \{z : \|[F_{L,\varepsilon} + \alpha z F_{H,\varepsilon}]x\|_\infty \leqslant \Delta\}. \tag{4.210}$$

If, in the above maximization, there exists no z for which the inequality is satisfied, then z is chosen equal to 0. Note that (4.209) has the disadvantage over (4.207) that the resulting controller is nonlinear. However, this nonlinear feedback does have a slightly larger range for the high-gain term.

We emphasize that, unlike in continuous-time systems, for discrete-time systems, the parameter α (or in the single-input case $\tilde{\alpha}$) is limited in its range.

In the direct design, our method decouples the different inputs by using different timescales. Therefore, the issue with the nonlinear term κ does not show up. In the current design, we do not have these different timescales and then the nonlinear term κ is needed in the multi-input case.

Remark 4.44 *Clearly, when* $\alpha = 0$, *the new composite state feedback laws as given in* (4.207) *reduce to the low-gain-based linear state feedback laws as given in* (4.205). *In connection with output regulation problems, it is demonstrated in [84, 85] that the choice of* $F_{H,\varepsilon}$ *as given above and a value of*

$$\alpha \in \left[0, \frac{2}{\|B'P_\varepsilon B\|}\right]$$

represents fuller utilization of the control capacity and leads to an improved closed-loop transient performance.

Theorem 4.45 *Consider the discrete-time system* (4.1). *Let Assumption 4.8 hold, i.e., assume that the pair* (A, B) *is stabilizable and the eigenvalues of* A *are within or on the unit circle. Then the family of linear static state feedback laws given in* (4.207) *solves Problem 4.4. More specifically, under the state feedback* (4.207), *for any given (arbitrarily large) bounded set* $\mathcal{X} \subset \mathbb{R}^n$, *there exists an* $\varepsilon^* \in (0, 1]$ *such that, for all*

$$\varepsilon \in (0, \varepsilon^*] \text{ and } \alpha \in \left[0, \frac{2}{\|B'P_\varepsilon B\|}\right],$$

the equilibrium point $x = 0$ *of the closed-loop system is locally exponentially stable with* $\mathcal{X}$ *contained in its domain of attraction.*

Proof : Consider the closed-loop system under the state feedback law (4.207):

$$x(k+1) = A_c x + B[\sigma(F_{L,\varepsilon}x + \alpha \bar{F}_{H,\varepsilon}x) - F_{L,\varepsilon}x]. \tag{4.211}$$

Also, consider the Lyapunov function

$$V_1(x) = x'P_\varepsilon x, \tag{4.212}$$

and let $c_1 > 0$ be such that

$$c_1 \geq \sup_{x\in\mathcal{X},\varepsilon\in(0,1]} x'P_\varepsilon x. \tag{4.213}$$

Such a c_1 exists since $\lim_{\varepsilon\to 0} P_\varepsilon = 0$ by Lemma 4.24 and $\mathcal{X}$ is bounded. We note here that such a c_1 guarantees that $\mathcal{X} \subset L_{V_1}(c_1)$, $\forall \varepsilon \in (0, 1]$, where the level set $L_{V_1}(c_1)$ is defined as $L_{V_1}(c_1) = \{x \in \mathbb{R}^n : V_1(x) \leq c_1\}$. Let ε_1^* be such that for all $\varepsilon \in (0, \varepsilon_1^*]$, $x \in L_{V_1}(c_1)$ implies that $\|F_{L,\varepsilon}x\|_\infty \leq \Delta$. Such an ε_1^* exists because of Lemma 4.24 and the fact that $\lim_{\varepsilon\to 0} P_\varepsilon = 0$. Note also that from the DARE (4.206), it follows that,

$$A_c' P_\varepsilon A_c - P_\varepsilon = -Q_\varepsilon - F_{L,\varepsilon}' F_{L,\varepsilon}. \tag{4.214}$$

The evaluation of the difference $V_1(\rho x) - V_1(x)$ along the trajectories of this linear closed-loop system shows that, for $x \in L_{V_1}(c_1)$,

$$\begin{aligned} V_1(\rho x) - V_1(x) &= -x' Q_\varepsilon x - \sigma(u)'\sigma(u) \\ &\quad + [\sigma(u) - F_{L,\varepsilon}x]' (I + B' P_\varepsilon B) [\sigma(u) - F_{L,\varepsilon}x], \end{aligned} \tag{4.215}$$

where, as usual, $\rho x(k)$ means $x(k+1)$. Defining $\delta = \|B' P_\varepsilon B\|$, we get

$$\begin{aligned} &V_1(\rho x) - V_1(x) \\ &\quad \leqslant -x' Q_\varepsilon x - \sigma(u)'\sigma(u) + (1+\delta)\,[\sigma(u) - F_{L,\varepsilon}x]'\,[\sigma(u) - F_{L,\varepsilon}x] \\ &\quad = -x' Q_\varepsilon x + \delta \left\|\sigma(u) - \tfrac{1+\delta}{\delta} F_{L,\varepsilon}x\right\|^2 - \tfrac{1+\delta}{\delta}\|F_{L,\varepsilon}x\|^2. \end{aligned} \tag{4.216}$$

Since

$$|F_{L,\varepsilon}x| \leqslant |\sigma(u)| \leqslant \left(\tfrac{2}{\delta} + 1\right) |F_{L,\varepsilon}x|$$

componentwise, and since $\sigma(u)$ and $F_{L,\varepsilon}x$, also componentwise, have the same sign, it follows that

$$\|\sigma(u) - \tfrac{1+\delta}{\delta} F_{L,\varepsilon}x\| \leqslant \tfrac{1}{\delta}\|F_{L,\varepsilon}x\|.$$

Using this in (4.216) yields,

$$V_1(\rho x) - V_1(x) \leqslant -x' Q_\varepsilon x - \|F_{L,\varepsilon}x\|^2,$$

for all $x \in L_{V_1}(c_1)$ which implies that the closed-loop system (4.211) is locally exponentially stable with $\mathcal{X}$ contained in its basin of attraction. We note here that the choice of α determines the decay rate of $V_1(x(k+1)) - V_1(x(k))$, and hence, the freedom in choosing α can be utilized to ensure fast convergence. ∎

The family of observer-based measurement feedback control laws take the form,

$$\begin{cases} \hat{x}(k+1) = A\hat{x}(k) + B\sigma(u(k)) + K(y(k) - C\hat{x}(k)) \\ \quad u(k) \quad = [F_{L,\varepsilon} + \alpha \bar{F}_{H,\varepsilon}]\hat{x}(k), \end{cases} \tag{4.217}$$

where $F_{L,\varepsilon}$ and $\bar{F}_{H,\varepsilon}$ are as in (4.208) and K is a matrix such that all the eigenvalues of $\bar{A} = A - KC$ are at the origin.

The following theorem shows that the family of measurement feedback laws given in (4.217) solves Problem 4.5, namely, the problem of semi-global exponential stabilization via linear dynamic measurement feedback:

Theorem 4.46 *Consider the discrete-time system* (4.1)*. Let Assumption 4.8 hold, i.e., assume that the pair (A, B) is stabilizable and the eigenvalues of A are within or on the unit circle. Moreover, assume that the pair (C, A) is observable.*

In that case, the family of linear dynamic measurement feedback laws given in (4.217) *solves Problem 4.5. More specifically, under the measurement feedback law* (4.217)*, for any given (arbitrarily large) bounded set $(\mathcal{X}, \widehat{\mathcal{X}}) \subset \mathbb{R}^{2n}$, there exists an $\varepsilon^* \in (0, 1]$ such that for all*

$$\varepsilon \in (0, \varepsilon^*], \qquad \alpha \in \left[0, \frac{2}{\|B' P_\varepsilon B\|}\right],$$

the equilibrium point $(0, 0)$ of the closed-loop system is locally exponentially stable with $(\mathcal{X}, \widehat{\mathcal{X}})$ contained in its domain of attraction.

Proof : With the family of measurement feedback laws as given by (4.217), the closed-loop system is given by

$$\begin{cases} x(k+1) = Ax(k) + B\sigma(u(k)) \\ \hat{x}(k+1) = A\hat{x}(k) + B\sigma(u(k)) + K(y(k) - C\hat{x}(k)) \\ \quad u(k) \quad = (F_{L,\varepsilon} + \alpha F_{H,\varepsilon})\hat{x}(k). \end{cases} \tag{4.218}$$

We then adopt the invertible change of state variables,

$$\tilde{x} = x - \hat{x},$$

and rewrite the closed-loop system (4.218) as

$$\begin{cases} x(k+1) = Ax(k) + B\sigma((F_{L,\varepsilon} + \alpha F_{H,\varepsilon})\hat{x}(k) \\ \tilde{x}(k+1) = (A - KC)\tilde{x}(k). \end{cases} \tag{4.219}$$

Since all eigenvalues of $\bar{A} = A - KC$ are at the origin, it is easy to verify that for time $k \geq n$, $\tilde{x}(k) \equiv 0$. As a result, for $k \geq n$, $\hat{x}(k) \equiv x(k)$. On the other hand, for all $x(0) \in \mathcal{X}$ and $\hat{x}(0) \in \widehat{\mathcal{X}}$, $x(n)$ belongs to a bounded set $\widetilde{\mathcal{X}}$ since $\sigma(u)$ is bounded. Hence, the rest of the proof becomes the same as the proof for state feedback case as long as we guarantee that $\widetilde{\mathcal{X}}$ is in the domain of attraction. ■

4.6 Global stabilization

Let us summarize briefly what has been done so far in this chapter. As discussed earlier, low-gain design methodology can be utilized successfully to achieve semi-global stabilization of linear systems subject to actuator saturation. The low-and-high-gain design has been put forward to improve transient performance. However, all these techniques only yield semi-global stabilization. In this section, our intent is to address global stabilization.

In the seminal work of Fuller [38], as discussed earlier, it was established that global stabilization *in general* requires nonlinear feedback control laws. In this section, we first present a result for neutral systems which form a class of systems for which linear feedback control laws can still achieve global stabilization. Note that we cannot limit ourselves to linear feedback control laws if we want to address global stabilization for more general classes of systems. Since our design methodologies of low-gain and low-and-high-gain feedback yield linear feedback controllers, it might seem that these techniques are unsuitable to achieve global stabilization. However, it is transparent that both the low-gain and high-gain parameters ε and α have an asymptotic nature, and as such, for readers familiar with the adaptive control literature, it is clear that one can possibly adapt or schedule these parameters to depend on the state x or its estimate. This has been done in literature. The low-gain parameter ε is adapted (scheduled) to achieve global internal stabilization [98]. In subsequent chapters, we will also address control problems where it is useful to adapt also the high-gain parameter. In view of the above discussion, we review in this section a certain method of adapting the low-gain parameter ε to depend on the state x. This is done to construct globally stabilizing controllers. Obviously, whenever ε is adapted to depend on x, the resulting controllers are nonlinear.

4.6.1 Linear feedback controllers for neutral systems

As mentioned before, it was already established in [38] that *in general* nonlinear feedback control laws are needed for global stabilization. In the literature, some researchers study, for what class of systems with input saturation, global stabilization can be achieved via linear feedback controllers. In this subsection, we look at neutral systems for which it has been known for a long time that global stabilization can be achieved by linear feedback controllers. Even though we restrict attention to neutral systems, it should be noted that global stabilization can be achieved with linear feedback control even for more general classes of systems such as continuous-time systems for which all imaginary-axis poles are semi-simple (Jordan blocks of size at most 1) except for the origin where Jordan blocks of size 2 are allowed (see Sect. 4.7.1).

Let us first mention that the linear feedback control laws we use in this subsection for global stabilization of neutral systems fall into the class of low-gain and low-and-high-gain feedback control laws after we *generalize* them. Before we proceed to the said generalization, let us first briefly recall the ARE-based design we introduced earlier. For continuous-time systems, we introduced earlier the CARE,

$$A'P_\varepsilon + P_\varepsilon A - P_\varepsilon B'BP_\varepsilon + Q_\varepsilon = 0, \tag{4.220}$$

and then introduced low-gain feedback control laws of the form

$$u = -B'P_\varepsilon x. \tag{4.221}$$

Also, we introduced low-and-high-gain feedback control laws of the form

$$u = -(1+\alpha)B'P_\varepsilon x, \tag{4.222}$$

where P_ε is the solution of CARE (4.220) and where Q_ε is a positive-definite matrix for all $\varepsilon > 0$ with the property that $Q_\varepsilon \to 0$ as $\varepsilon \to 0$.

We observe that the above ARE-based low-gain or low-and-high-gain design methodology can be generalized to allow Q_ε to be positive semi-definite, provided that (Q_ε, A) is detectable and $Q_\varepsilon \to 0$ as $\varepsilon \to 0$. How does such a generalization of allowing Q_ε to be positive semi-definite help us? As we shall see shortly, such a generalization leads us to a generalization of low-and-high-gain feedback control laws of the type (4.222). Then, such generalized control laws, although they are linear feedback control laws, can be used to globally asymptotically stabilize neutral systems.

To proceed with our development, let us first consider a continuous-time linear system subject to actuator saturation:

$$\dot{x} = Ax + B\sigma(u). \tag{4.223}$$

Neutral systems are linear systems which are already stable but not yet asymptotically stable. That is, the system matrix A has all its eigenvalues in the closed left-half plane, but those eigenvalues on the imaginary axis are semi-simple (Jordan blocks of size at most 1). It is well known that this implies that there exists a positive definite matrix P such that

$$A'P + PA \leqslant 0. \tag{4.224}$$

In order to develop our new generalized design methodology, let us first choose,

$$Q_\varepsilon = -\varepsilon(A'P + PA) + \varepsilon^2 PBB'P, \tag{4.225}$$

where P is a positive definite solution of (4.224). We note that the above chosen Q_ε satisfies the properties: (1) Q_ε is positive semi-definite, (2) (Q_ε, A) is detectable, and (3) $Q_\varepsilon \to 0$ as $\varepsilon \to 0$. With Q_ε as chosen above, it is easy to verify that the positive definite solution P_ε of CARE (4.220) is given by

$$P_\varepsilon = \varepsilon P. \tag{4.226}$$

With such a solution of P_ε, we can rewrite the low-and-high-gain feedback control law (4.222) as

$$u = -(1+\alpha)\varepsilon B' P x. \tag{4.227}$$

The above control law can in turn be written as

$$u = -\lambda B' P x, \tag{4.228}$$

in which the low-gain parameter ε and high-gain parameter α are merged into a new parameter $\lambda = (1+\alpha)\varepsilon$. Note that this merging has occurred crucially because of the fact that the selection of positive semi-definite Q_ε as in (4.225) resulted in the solution P_ε of CARE (4.220) as εP; consequently, ε can be easily merged with $(1+\alpha)$ in order to create a new parameter λ.

We emphasize that the feedback control law given in (4.228) is obtained indeed from a generalization of ARE-based low-and-high-gain feedback design methodology in which the matrix Q_ε is allowed to be positive semi-definite. However, low-gain parameter ε and high-gain parameter α collapsed into a single parameter λ which can take any positive value. As such, we refer to the feedback control law given in (4.228) as a generalized low-and-high-gain linear state feedback law, and simply treat λ as a design parameter. Note that such a merging obviously obliterates the upper bound on low-gain parameter that dictates the domain of attraction of the closed-loop system. As such, this generalized linear state feedback law (4.228) leads to global stabilization in the case of neutrally stable systems. This is formalized below in Theorem 4.47. In fact, the following two theorems show that, for neutral systems of the form (4.223) which are subject to actuator saturation, the linear state feedback controller (4.228) is globally stabilizing for any $\lambda > 0$ while, when combined with a suitable observer, it results in a globally stabilizing linear measurement feedback controller:

Theorem 4.47 *Consider the system* (4.223) *with* (A, B) *stabilizable, while* A *has all its eigenvalues in the closed left-half plane, and those eigenvalues on the imaginary axis are semi-simple (Jordan blocks of size at most* 1*). In that case, the state feedback controller* (4.228)*, where* P *satisfies* (4.224) *and* $\lambda > 0$*, achieves global asymptotic stability.*

Proof : We consider the candidate Lyapunov function,

$$V(x) = x' P x,$$

and we find that

$$\frac{\mathrm{d}}{\mathrm{d}t} V(x(t)) = x'(A'P + PA)x - 2\lambda x' P B \sigma(B' P x) \leqslant 0$$

for all x. In order to prove global asymptotic stability, we apply LaSalle's invariance principle. Consider the closed-loop system

$$\dot{x} = Ax + B\sigma(-\lambda B' P x). \tag{4.229}$$

If this system satisfies

$$\frac{\mathrm{d}}{\mathrm{d}t} V(x(t)) = 0$$

for all t, then we find that $B'Px = 0$ and $(A'P + PA)x = 0$. Therefore, $x(t)$ is a solution of the differential equation

$$\dot{x} = Ax.$$

In view of the above equation and noting that $PAx = -A'Px$ and $F'B'Px = 0$ for any matrix F, we have

$$P\dot{x} = PAx = -(A'Px + F'B'Px) = -(A' + F'B')Px. \tag{4.230}$$

Let F be any matrix such that $A + BF$ is asymptotically stable. Since $-(A' + F'B')$ has all its eigenvalues in the open right-half plane, (4.230) implies that Px is exponentially growing which yields a contradiction since $x'Px$ is bounded and $P > 0$. Hence, LaSalle's invariance principle yields that the closed-loop system (4.229) is globally asymptotically stable. ∎

Theorem 4.48 *Consider the system*

$$\begin{aligned} \dot{x} &= Ax + B\sigma(u) \\ y &= Cx, \end{aligned}$$

where (A, B) is stabilizable, (C, A) is detectable, while A has all its eigenvalues in the closed left-half plane, and those eigenvalues on the imaginary axis are semi-simple (Jordan blocks of size at most 1*). Consider the dynamic measurement feedback controller*

$$\begin{cases} \dot{\widehat{x}} = A\widehat{x} + B\sigma(u) + K(y - C\widehat{x}) \\ u = -\lambda B' P \widehat{x}, \end{cases} \tag{4.231}$$

where $A - KC$ is asymptotically stable, P satisfies (4.224)*, and $\lambda > 0$. Then, the controller* (4.231) *achieves global asymptotic stability.*

Proof : The closed-loop system can be written as

$$\begin{aligned} \dot{\widehat{x}} &= A\widehat{x} - B\sigma(\lambda B' P \widehat{x}) + KCe \\ \dot{e} &= (A - KC)e, \end{aligned}$$

where $e = x - \hat{x}$. The feedback is clearly locally asymptotically stable, and hence, there exist $\delta_1, \delta_2 > 0$ such that if at time t_0 we have

$$\hat{x}(t_0)' P \hat{x}(t_0) \leqslant \delta_1^2, \qquad \|e(t_0)\| \leqslant \delta_2, \tag{4.232}$$

then $\hat{x}(t) \to 0$ and $e(t) \to 0$ as $t \to \infty$. In order to prove global asymptotic stability, we only need to establish that for any initial condition, there exists a $t_0 > 0$ for which (4.232) is satisfied.

Consider arbitrary initial conditions. Choose T such that

$$\int_T^\infty \|e(t)\| \, \mathrm{d}t < \frac{\delta_1}{2\tilde{\alpha}}, \qquad \|e(t)\| \leqslant \delta_2 \text{ for } t > T,$$

where $\tilde{\alpha} = 2\|P^{1/2} K C\|$. Consider x_1 and x_2 with $x_1(T) = x_2(T) = \hat{x}(T)$ while

$$\begin{aligned}
\dot{x}_1 &= A x_1 - B\sigma(\lambda B' P x_1) + KCe \\
\dot{x}_2 &= A x_2 - B\sigma(\lambda B' P x_2),
\end{aligned}$$

where e is the error signal defined before. Clearly, this implies that $x_1 = \hat{x}$. We find that $V = (x_1 - x_2)' P (x_1 - x_2)$ satisfies

$$\begin{aligned}
\dot{V} &= (x_1 - x_2)'(A'P + PA)(x_1 - x_2) \\
&\qquad - 2(x_1 - x_2) PB \left[\sigma(\lambda B' P x_1) - \sigma(\lambda B' P x_2)\right] + 2(x_1 - x_2)' PKCe \\
&\leqslant 2(x_1 - x_2)' PKCe \\
&\leqslant \tilde{\alpha} V^{1/2} \|e\|.
\end{aligned}$$

This yields

$$\frac{\mathrm{d}}{\mathrm{d}t} V(t)^{1/2} \leqslant \tilde{\alpha} \|e(t)\|.$$

Recalling that $x_1 = \hat{x}$, we find that

$$\begin{aligned}
\hat{x}(t)' P \hat{x}(t) &\leqslant 2(\hat{x}(t) - x_2(t))'(\hat{x}(t) - x_2(t)) + 2 x_2(t)' P x_2(t) \\
&\leqslant \frac{\delta_1^2}{2} + 2 x_2(t)' P x_2(t)
\end{aligned}$$

for all $t > T$. Since $x_2(t) \to 0$ as $t \to \infty$, we find that there exists a $t_0 > T$ such that (4.232) is satisfied, and hence, we are inside the domain of attraction, and therefore, $\hat{x}(t) \to 0$ and $e(t) \to 0$ as $t \to \infty$. Since this holds for all initial conditions, we obtain global asymptotic stability. ∎

We proceed now to address neutral systems for discrete time. Our development parallels the one for continuous-time systems. Consider the linear system subject to actuator saturation:

$$x(k+1) = Ax(k) + B\sigma(u(k)). \tag{4.233}$$

As said earlier, neutral systems are linear systems which are already stable but not yet asymptotically stable. That is, for discrete-time systems, the system matrix A has all its eigenvalues in the closed unit disc, but those eigenvalues on the unit circle are semi-simple (Jordan blocks of size at most 1). It is well known that this implies that there exists a positive definite matrix P such that

$$A'PA \leqslant P. \tag{4.234}$$

We will then consider the following state feedback controller:

$$u = -\lambda B'PAx, \tag{4.235}$$

where

$$\lambda \in \left[0, \frac{2}{\|B'PB\|}\right]. \tag{4.236}$$

The above design methodology, as in the case of continuous-time systems, can be related to low-and-high-gain design methodology based on AREs as developed in the previous sections by an appropriate choice of Q and R matrices in DARE (4.206) (we note that, in DARE (4.206), we had chosen R as identity matrix, but it can be chosen differently as long as the matrix remains positive definite). For example, for simplicity of presentation and without loss of generality, let us assume that all the eigenvalues of A in (4.233) are simple and are on the unit circle, that is, $A'A = I$. It is shown in [5, 26] that the linear feedback controller

$$u = -\kappa B'Ax$$

achieves global stabilization if

$$\kappa B'B \leqslant 2I. \tag{4.237}$$

Choose

$$R_\varepsilon = (2 + \varepsilon\|B'B\|)I - \varepsilon B'B, \quad Q_\varepsilon = \tfrac{\varepsilon^2}{2+\varepsilon\|B'B\|} A'BB'A.$$

Then the DARE,

$$P_\varepsilon = A'P_\varepsilon A + Q_\varepsilon - A'P_\varepsilon B(B'P_\varepsilon B + R_\varepsilon)^{-1}B'P_\varepsilon A,$$

has a unique positive definite solution $P_\varepsilon = \varepsilon I$. Following the design in Sect. 4.5.4, we can construct a low-and-high-gain feedback for discrete-time neutrally stable system as

$$u = -(1+\alpha)(B'P_\varepsilon B + R_\varepsilon)^{-1}B'P_\varepsilon Ax = -(1+\alpha)\tfrac{\varepsilon}{2+\varepsilon\|B'B\|} B'Ax,$$

where

$$\alpha \in \left[0, \tfrac{2}{\varepsilon\|B'B\|}\right].$$

Define

$$\kappa = (1+\alpha)\tfrac{\varepsilon}{2+\varepsilon\|B'B\|},$$

and hence,

$$u = -\kappa B'Ax.$$

As in continuous-time case, the low-gain and high-gain parameters merge into one parameter κ which satisfies

$$\kappa = (1+\alpha)\tfrac{\varepsilon}{2+\varepsilon\|B'B\|} \leqslant \tfrac{2}{\|B'B\|}. \tag{4.238}$$

This is exactly the condition (4.237).

The following two theorems show that the linear state feedback controller given in (4.235) is globally stabilizing for neutral systems subject to actuator saturation, while, when combined with a suitable observer, it results in a globally stabilizing linear measurement feedback controller:

Theorem 4.49 *Consider the system* (4.233) *with* (A, B) *stabilizable, while* A *has all its eigenvalues in the closed unit disc, and those eigenvalues on the unit circle are semi-simple (Jordan blocks of size at most* 1*). In that case, the state feedback controller* (4.235) *where* P *satisfies* (4.234) *and* λ *satisfies* (4.236) *achieves global asymptotic stability.*

Proof : We consider the candidate Lyapunov function,

$$V(x) = x'Px,$$

and we find that

$$\begin{aligned}
V(x(k+1)) - V(x(k)) &= x'(A'PA-P)x - 2x'A'PB\sigma(\lambda B'PAx) \\
&\qquad + \sigma(\lambda B'PAx)'B'PB\sigma(\lambda B'PAx) \\
&\leqslant x'(A'PA-P)x \\
&\qquad - \sigma(\lambda B'PAx)'\left[\frac{2}{\lambda} - B'PB\right]\sigma(\lambda B'PAx) \\
&\leqslant 0
\end{aligned}$$

where, in the above, x is an abbreviation for $x(k)$. In order to prove global asymptotic stability, as in the case of continuous-time systems, we apply LaSalle's invariance principle. Consider the closed-loop system,

$$x(k+1) = Ax(k) + B\sigma(-\lambda B'PAx(k)). \tag{4.239}$$

If this system satisfies

$$V(x(k+1)) - V(x(k)) = 0$$

for all k, then we find that $B'PAx = 0$ and $A'PAx = Px$. Therefore, $x(k)$ is a solution of the difference equation

$$x(k+1) = Ax(k).$$

In view of the above equation and noting that $A'PAx = Px$ and $F'B'PAx = 0$ for any matrix F, we have

$$(A + BF)'PAx(k+1) = A'PAx(k+1) = Px(k+1) = PAx(k). \quad (4.240)$$

Let F be any matrix such that $A + BF$ is asymptotically stable. Since $A + BF$ has all its eigenvalues inside the unit circle, (4.240) implies that PAx is exponentially growing which yields a contradiction since $x'Px$ is bounded and $P > 0$. Hence, LaSalle's invariance principle yields that the closed-loop system (4.239) is globally asymptotically stable. ■

Theorem 4.50 *Consider the system*

$$\begin{aligned} x(k+1) &= Ax(k) + B\sigma(u(k)),\\ y(k) &= Cx(k), \end{aligned}$$

where (A, B) is stabilizable, (C, A) is detectable, while A has all its eigenvalues in the closed unit disc, and those eigenvalues on the unit circle are semi-simple (Jordan blocks of size at most 1). Consider the dynamic measurement feedback controller

$$\begin{cases} \hat{x}(k+1) = A\hat{x}(k) + B\sigma(u(k)) + K(y(k) - C\hat{x}(k)) \\ u(k) \quad\;\; = -\lambda B'PA\hat{x}(k), \end{cases} \quad (4.241)$$

where $A - KC$ is asymptotically stable, P satisfies (4.234)*, and $\lambda > 0$ satisfies the bound* (4.236)*. Then, the controller* (4.241) *achieves global asymptotic stability.*

Proof : The closed-loop system can be written as

$$\begin{aligned} \hat{x}(k+1) &= A\hat{x}(k) - B\sigma(\lambda B'PA\hat{x}(k)) + KCe(k)\\ e(k+1) &= (A - KC)e(k), \end{aligned}$$

where $e = x - \hat{x}$. The feedback is clearly locally asymptotically stable, and hence, there exist $\delta_1 > 0$ and $\delta_2 \in (0, 1)$ such that if at time k_0 we have

$$\hat{x}(k_0)' P \hat{x}(k_0) \leqslant \delta_1^2, \qquad \|e(k_0)\| \leqslant \delta_2, \tag{4.242}$$

then $\hat{x}(k) \to 0$ and $e(k) \to 0$ as $k \to \infty$. In order to prove global asymptotic stability, we only need to establish that for any initial condition, there exists a $k_0 > 0$ for which (4.242) is satisfied.

Consider arbitrary initial conditions. For any $k_1 > 0$, consider x_1 and x_2 with $x_1(k_1) = x_2(k_1) = \hat{x}(k_1)$ while

$$\begin{aligned} x_1(k+1) &= Ax_1(k) - B\sigma(\lambda B' P A x_1(k)) + KCe(k) \\ x_2(k+1) &= Ax_2(k) - B\sigma(\lambda B' P A x_2(k)). \end{aligned}$$

Clearly, we have $x_1 = \hat{x}$. We find that $V = (x_1 - x_2)' P (x_1 - x_2)$ satisfies

$$\begin{aligned} V(k+1) - V(k) \leqslant{}& (x_1 - x_2)'(A'PA - P)(x_1 - x_2) \\ &+ 2(x_1 - x_2)' A' P K C e + e' C' K' P K C e \\ &- 2e' C' K' P B \left[\sigma(\lambda B' P A x_1) - \sigma(\lambda B' P A x_2)\right] \\ \leqslant{}& 2\tilde{\alpha} \|e\| V(k)^{1/2} + \tilde{\alpha}^2 \|e\|^2 \end{aligned}$$

for some suitably chosen $\tilde{\alpha}$. Here, we use that

$$\begin{aligned} &\|e' C' K' P B \left[\sigma(\lambda B' P A x_1) - \sigma(\lambda B' P A x_2)\right]\| \\ &\quad\leqslant \|e\| \|C' K' P^{1/2}\| \|P^{1/2} B\| \|\lambda B' P A (x_1 - x_2)\| \\ &\quad\leqslant \lambda \|e\| \|C' K' P^{1/2}\| \|B' P B\| \|P^{1/2} A (x_1 - x_2)\| \\ &\quad\leqslant \lambda \|e\| \|C' K' P^{1/2}\| \|B' P B\| \left[(x_1 - x_2)' A' P A (x_1 - x_2)\right]^{1/2} \\ &\quad\leqslant \lambda \|e\| \|C' K' P^{1/2}\| \|B' P B\| V(k)^{1/2} \\ &\quad\leqslant 2 \|e\| \|C' K' P^{1/2}\| V(k)^{1/2}. \end{aligned}$$

This yields

$$V(k+1) \leqslant \left(V(k)^{1/2} + \tilde{\alpha}\|e\|\right)^2,$$

and hence,

$$V(k+1)^{1/2} - V(k)^{1/2} \leqslant \tilde{\alpha} \|e(k)\|.$$

The above derivation was independent of k_1 and, in particular, we could choose $\widetilde{\alpha}$ independent of k_1. However, we now choose k_1 such that

$$\sum_{k=k_1}^{\infty} \|e(k)\| < \frac{\delta_1}{2\widetilde{\alpha}}, \qquad \|e(k)\| \leqslant \delta_2 \text{ for } k > k_1.$$

and, as before, $x_1(k_1) = x_2(k_1) = \widehat{x}(k_1)$. Recalling that $x_1 = \widehat{x}$, we find that

$$\begin{aligned}\widehat{x}(k)'P\widehat{x}(k) &\leqslant 2(\widehat{x}(k) - x_2(k))'(\widehat{x}(k) - x_2(k)) + 2x_2(k)'Px_2(k)\\ &\leqslant \frac{\delta_1^2}{2} + 2x_2(k)'Px_2(k)\end{aligned}$$

for all $k > k_1$. Since $x_2(k) \to 0$ as $k \to \infty$, we find that there exists a $k_0 > K$ such that (4.242) is satisfied, and hence, we are inside the domain of attraction, and therefore, $\widehat{x}(k) \to 0$ and $e(k) \to 0$ as $k \to \infty$. Since this holds for all initial conditions, we obtain global asymptotic stability. ∎

4.6.2 *Nonlinear feedback controllers based on adaptive-low-gain design methodology*

In the previous subsection, we used linear feedback controllers to achieve global asymptotic stability for neutral systems. However, this is not always possible for general linear systems subject to actuator saturation. In this subsection, we consider nonlinear feedback controllers obtained by modifying the low-gain feedback design methodology we studied in earlier sections. This is done by a suitable adaptation (scheduling) of the low-gain parameter. In what follows, by a scheduled parameter or hereafter called an adaptive-low-gain parameter, we mean a function $\varepsilon_a(x)$, from $\mathbb{R}^n \to (0,1]$. As discussed above, in order to elevate the low-gain design from a semi-global framework to a global framework, for continuous-time systems, one should look for an adaptive-low-gain parameter $\varepsilon_a(x)$ to have the following properties:

(*i*) $\varepsilon_a(x) \in C^1$.

(*ii*) $\varepsilon_a(x) = 1$ for all x in an open neighborhood of the origin.

(*iii*) For any $x \in \mathbb{R}^n$, we have

$$\|B'P_{\varepsilon_a(x)}x\|_\infty \leqslant \Delta.$$

(*iv*) $\varepsilon_a(x) \to 0$ as $\|x\|_\infty \to \infty$.

(*v*) $\{\, x \in \mathbb{R}^n \mid x'P_{\varepsilon_a(x)}x \leqslant c \,\}$ is a bounded set for all $c > 0$.

(*vi*) For any $x_1, x_2 \in \mathbb{R}^n$,

$$x_1' P_{\varepsilon_a(x_1)} x_1 \leqslant x_2' P_{\varepsilon_a(x_2)} x_2$$

implies $\varepsilon_a(x_1) \geqslant \varepsilon_a(x_2)$.

In item (*iii*), P_ε is the positive definite solution of the CARE given in (4.42). Here, as usual, Δ is the saturation level.

A particular choice[2] for the adaptive-low-gain parameter $\varepsilon_a(x)$ having all the above properties was introduced by Megretski [98] for continuous-time systems and it is given by

$$\varepsilon_a(x) = \max\{\, r \in (0, 1] \; : \; (x' P_r x) \operatorname{trace} B' P_r B \leqslant \Delta^2 \,\}. \tag{4.243}$$

Note that this is a nontrivial choice since Property (*iii*) requires that $\varepsilon(x)$ converges to 0 fast enough yet Properties (*v*) and (*vi*) restrict the speed with which ε can converge to 0. An option is to replace trace $B' P_r B$ by $\|B' P_r B\|$ which has been used in the literature. However, using this latter choice, we run into difficulties deriving a key technical bound given in Lemma 12.54.

For discrete-time systems, one should look for an adaptive-low-gain parameter $\varepsilon_a(x)$ to have the following properties:

(*i*) $\varepsilon_a(x) \in C^1$.

(*ii*) $\varepsilon_a(x) = 1$ for all x in an open neighborhood of the origin.

(*iii*) For any $x \in \mathbb{R}^n$, we have

$$\|(I + B' P_{\varepsilon_a(x)} B)^{-1} B' P_{\varepsilon_a(x)} A x\|_\infty \leqslant \Delta.$$

(*iv*) $\varepsilon_a(x) \to 0$ as $\|x\|_\infty \to \infty$.

(*v*) $\{\, x \in \mathbb{R}^n \mid x' P_{\varepsilon_a(x)} x \leqslant c \,\}$ is a bounded set for all $c > 0$.

(*vi*) For any $x_1, x_2 \in \mathbb{R}^n$,

$$x_1' P_{\varepsilon_a(x_1)} x_1 \leqslant x_2' P_{\varepsilon_a(x_2)} x_2$$

implies $\varepsilon_a(x_1) \geqslant \varepsilon_a(x_2)$.

In item (*iii*), P_ε is the positive definite solution of DARE given in (4.66). For discrete-time systems, one can also give an example of adapting the low-gain parameter $\varepsilon_a(x)$ satisfying the above properties. This is given by

$$\varepsilon_a(x) = \max\{\, r \in (0, 1] \; : \; x' P_r x \operatorname{trace} B' P_r B \leqslant \tfrac{1}{2}\Delta^2 \,\}. \tag{4.244}$$

---

[2] In several papers, the same scheduling is used with the trace of $B' P_r B$ instead of the norm. This alternative has the same properties as the scheduling presented in (4.243).

In order to establish Property (*iii*) for this adaptation, we need to use Property (*iii*) of Lemma 4.24. This can be seen from the following analysis:

$$\begin{aligned}\|(I + B'P_{\varepsilon_a}B)^{-1}B'P_{\varepsilon_a}Ax\|_\infty^2 &\leqslant \|(I + B'P_{\varepsilon_a}B)^{-1}B'P_{\varepsilon_a}Ax\|^2 \\ &\leqslant \|B'P_{\varepsilon_a}Ax\|^2 \\ &\leqslant \|B'P_{\varepsilon_a}^{1/2}\|^2\|P_{\varepsilon_a}^{1/2}AP_{\varepsilon_a}^{-1/2}\|^2\|P_{\varepsilon_a}^{1/2}x\|^2 \\ &\leqslant 2\|B'P_{\varepsilon_a}B\|x'P_{\varepsilon_a}x \leqslant \Delta^2.\end{aligned}$$

There are other possibilities for the scheduling. An option is to replace the term trace $B'P_rB$ by $\|B'P_rB\|$ in (4.244) which has been used in the literature. However, using this latter choice, we run into difficulties deriving a key technical bound (12.61) in Chap. 12.

State feedback:

Having introduced the adaptive-low-gain parameter $\varepsilon_a(x)$, we can now formulate new state feedback control laws by an appropriate modification of the low-gain state feedback control law (4.44). Thus for continuous-time systems, we have an adaptive-low-gain feedback control law given by

$$u := -B'P_{\varepsilon_a(x)}x, \tag{4.245}$$

where $P_{\varepsilon_a(x)}$ is the positive definite solution of the CARE (4.42) when ε is replaced by $\varepsilon_a(x)$ as defined in (4.243).

Similarly, for discrete-time systems, we can introduce what can be termed as an adaptive-low-gain feedback control law as

$$u := -(B'P_{\varepsilon_a(x)}B + I)^{-1}B'P_{\varepsilon_a(x)}Ax, \tag{4.246}$$

where P_ε is the positive definite solution of the DARE (4.66) when ε is replaced by $\varepsilon_a(x)$ as defined in (4.244).

We emphasize that both the above adaptive-low-gain feedback control laws are nonlinear state feedback control laws. Utilization of these control laws lead to the following results.

Theorem 4.51 *Consider the continuous-time linear system subject to input saturation*

$$\dot{x} = Ax + B\sigma(u),$$

where the pair (A, B) *is stabilizable and the eigenvalues of* A *are in the closed left-half plane. Let* $\varepsilon_a(x)$ *be as defined in* (4.243)*. Then, the zero equilibrium point of the closed-loop system with the nonlinear control law*

$$u(x) = -B'P_{\varepsilon_a(x)}x \tag{4.247}$$

is globally asymptotically stable and locally exponentially stable.

Proof : We first note that if $\varepsilon_a(x(t))$ is increasing, then we have

$$x'(s) P_{\varepsilon_a(x(s))} x(s) \operatorname{trace} B' P_{\varepsilon_a(x(s))} B = \Delta^2 \tag{4.248}$$

for all s in a sufficiently small open neighborhood of t and, since $\varepsilon_a(x(t))$ increasing implies that $P_{\varepsilon_a(x(t))}$ is increasing, we get from (4.248) that

$$x'(t) P_{\varepsilon_a(x(t))} x(t) \tag{4.249}$$

is decreasing in t. If we reverse this implication, we find that if (4.249) is nondecreasing, i.e., constant or increasing, then $\varepsilon_a(x(t))$ is nonincreasing. If we apply the feedback control (4.247), then we get

$$\begin{aligned}\frac{\mathrm{d}}{\mathrm{d}t} x'(t) P_{\varepsilon_a(x(t))} x(t) &= -x'(t) Q_{\varepsilon_a(x(t))} x(t) \\ &\quad - x'(t) P_{\varepsilon_a(x(t))} B B' P_{\varepsilon_a(x(t))} x(t) + x'(t) \left[\frac{\mathrm{d}}{\mathrm{d}t} P_{\varepsilon_a(x(t))} \right] x(t).\end{aligned}$$

As argued above, if (4.249) is nondecreasing, then $\varepsilon_a(x(t))$ is nonincreasing, and hence, also $P_{\varepsilon_a(x(t))}$ is nonincreasing. This implies that

$$\frac{\mathrm{d}}{\mathrm{d}t} x'(t) P_{\varepsilon_a(x(t))} x(t) \leqslant -x'(t) Q_{\varepsilon_a(x(t))} x(t) < 0,$$

which yields a contradiction. We conclude that (4.249) is decreasing. This immediately implies asymptotic stability. Let us observe the fact that, in a neighborhood of the origin, $\varepsilon_a(x) = 1$ and the system is locally linear and exponentially stable. This yields local exponential stability. ■

Theorem 4.52 *Consider the discrete-time linear system subject to input saturation*

$$x(k+1) = Ax(k) + B\sigma(u(k)),$$

where the pair (A, B) is stabilizable and the eigenvalues of A are in the closed unit disc. Let $\varepsilon_a(x)$ be as defined in (4.244). Then, the zero equilibrium point of the closed-loop system with the nonlinear control law

$$u(x) = F_{L,\varepsilon_a(x)} x = -\left(B' P_{\varepsilon_a(x)} B + I\right)^{-1} B' P_{\varepsilon_a(x)} A x, \tag{4.250}$$

is globally asymptotically stable and locally exponentially stable.

Proof : We first note that if $\varepsilon_a(x(k))$ is increasing, then $P_{\varepsilon_a(x(k))}$ is increasing. We then find that

$$\begin{aligned}2x'(k)P_{\varepsilon_a(x(k))}x(k)\,\mathrm{trace}\,B'P_{\varepsilon_a(x(k))}B &= \Delta^2\\ &\geqslant 2x'(k+1)P_{\varepsilon_a(x(k+1))}x(k+1)\,\mathrm{trace}\,B'P_{\varepsilon_a(x(k+1))}B\\ &\geqslant 2x'(k+1)P_{\varepsilon_a(x(k+1))}x(k+1)\,\mathrm{trace}\,B'P_{\varepsilon_a(x(k))}B,\end{aligned}$$

which yields

$$x'(k)P_{\varepsilon_a(x(k))}x(k) \geqslant x'(k+1)P_{\varepsilon_a(x(k+1))}x(k+1).$$

If we reverse this implication, we find that if $x'(k)P_{\varepsilon_a(x(k))}x(k)$ is nondecreasing then $\varepsilon_a(x(k))$ is nonincreasing. If we apply the feedback control (4.250), then we get

$$\begin{aligned}x'(k+1)P_{\varepsilon_a(x(k+1))}x(k+1) - x'(k)P_{\varepsilon_a(x(k))}x(k) &\leqslant -x'(k)Q_{\varepsilon_a(x(k))}x(k)\\ &\quad + x'(k)\left[P_{\varepsilon_a(x(k+1))} - P_{\varepsilon_a(x(k))}\right]x(k).\end{aligned}$$

As argued above if $x'(k)P_{\varepsilon_a(x(k))}x(k)$ is nondecreasing, then $\varepsilon_a(x(k))$ is nonincreasing, and hence, also $P_{\varepsilon_a(x(k))}$ is nonincreasing. This implies that

$$x'(k+1)P_{\varepsilon_a(x(k+1))}x(k+1) - x'(k)P_{\varepsilon_a(x(k))}x(k) \leqslant -x'(k)Q_{\varepsilon_a(x(k))}x(k) < 0,$$

which yields a contradiction. We conclude that $x'(k)P_{\varepsilon_a(x(k))}x(k)$ is decreasing. This immediately implies asymptotic stability. Let us observe the fact that, in a neighborhood of the origin, $\varepsilon_a(x) = 1$ and the system is locally linear and exponentially stable. This yields local exponential stability. ■

Measurement feedback:

Theorems 4.51 and 4.52 pertain to state feedback control. However, in many cases, we are restricted to measurement feedback. In this case, we clearly need to introduce an observer. However, it turns out that it is better to modify the low-gain state feedback in such a way that the observer error does not cause any problems. We will use here H_∞ ARE-based scheduled or adaptive-low-gain feedback controls (adaptive versions of those introduced in Sects. 4.4.3 and 4.4.4 for continuous-time and discrete-time systems, respectively) instead of the H_2 ARE-based adaptive-low-gain feedback control laws used in Theorems 4.51 and 4.52. In the H_∞ control problem, we deal with worst-case disturbances, while in the H_2 control problem, we only deal with white-noise disturbances. The measurement error can be handled easily in the H_∞ control context since it does not have white-noise characteristics.

We will use the same way of scheduling or adapting the low-gain parameter $\varepsilon_a(x)$ as before. Thus, for continuous-time systems, we have an adaptive-low-gain observer-based measurement feedback control law as

$$u = F_{L,\varepsilon_a(\widehat{x})}\widehat{x} = -B' P_{\gamma,\varepsilon_a(\widehat{x})}\widehat{x}, \tag{4.251}$$

where to start with $P_{\gamma,\varepsilon}$ is the positive definite solution of CARE (4.90) with $E = I$, such that (4.91) is asymptotically stable. Here γ is chosen such that (4.90) has a unique positive definite solution for which (4.91) is asymptotically stable for all $\varepsilon \in (0, 1]$. The existence of such a γ is discussed in Lemma 4.28. Next, in order to obtain $P_{\gamma,\varepsilon_a(\widehat{x})}$, at first ε is replaced by $\varepsilon_a(x)$ of (4.243) where P_r is replaced by $P_{\gamma,r}$, and then $\varepsilon_a(x)$ is replaced by $\varepsilon_a(\widehat{x})$. We connect this low-gain feedback with an observer,

$$\dot{\widehat{x}} = A\widehat{x} + B\sigma(u) + K(y - C\widehat{x}), \tag{4.252}$$

where K is such that $A - KC$ is asymptotically stable.

Similarly, for discrete-time systems, we can introduce an adaptive-low-gain observer-based measurement feedback control law as

$$u = F_{L,\varepsilon_a(\widehat{x})}\widehat{x} = -(B' P_{\gamma,\varepsilon_a(\widehat{x})} B + I)^{-1} B' P_{\gamma,\varepsilon_a(\widehat{x})} A\widehat{x}, \tag{4.253}$$

where to start with $P_{\gamma,\varepsilon}$ is the positive definite solution of DARE (4.116) with $E = I$ such that (4.117) is asymptotically stable. Here, γ is chosen such that (4.116) has a unique positive definite solution for which (4.117) is asymptotically stable for all $\varepsilon \in (0, 1]$. The existence of such a γ is discussed in Lemma 4.32. Next, in order to obtain $P_{\gamma,\varepsilon_a(\widehat{x})}$, at first, ε is replaced by $\varepsilon_a(x)$ of (4.244) where P_r is replaced by $P_{\gamma,r}$ and then $\varepsilon_a(x)$ is replaced by $\varepsilon_a(\widehat{x})$. We connect this low-gain feedback with an observer

$$\widehat{x}(k+1) = A\widehat{x}(k) + B\sigma(u(k)) + K(y(k) - C\widehat{x}(k)), \tag{4.254}$$

where K is such that $A - KC$ is asymptotically stable.

Theorem 4.53 *Consider the continuous-time linear system subject to input saturation,*

$$\begin{aligned} \dot{x} &= Ax + B\sigma(u) \\ y &= Cx, \end{aligned}$$

where the pair (A, B) is stabilizable, (C, A) is detectable, and the eigenvalues of A are in the closed left-half plane. Let γ be chosen such that (4.90) *with $E = I$ has a unique positive definite solution such that* (4.91) *is asymptotically stable for all $\varepsilon \in (0, 1]$, and define $\varepsilon_a(x)$ by*

$$\varepsilon_a(x) = \max\left\{\varepsilon \in (0, 1] : x' P_{\gamma,\varepsilon} x \ \left\| B' P_{\gamma,\varepsilon} B \right\| \leqslant \Delta^2\right\}. \tag{4.255}$$

Then, the zero equilibrium point of the closed-loop system with the nonlinear control law given by (4.251) *and* (4.252) *is globally asymptotically stable and locally exponentially stable.*

Proof : The closed-loop system is given by

$$\begin{aligned}\dot{e} &= (A - KC)e\\ \dot{\hat{x}} &= A\hat{x} - B\sigma(B'P_{\gamma,\varepsilon_a(\hat{x})}\hat{x}) + KCe,\end{aligned}$$

where $e = x - \hat{x}$. We consider the candidate Lyapunov function,

$$V(x, \hat{x}) = \hat{x}'P_{\gamma,\varepsilon_a(\hat{x})}\hat{x} + \beta e'P_K e,$$

where P_K is such that

$$(A - KC)'P_K + P_K(A - KC) + I = 0.$$

As in the state feedback case, we distinguish two cases. If $\varepsilon_a(\hat{x})$ is increasing, then $\hat{x}'P_{\gamma,\varepsilon_a(\hat{x})}\hat{x}$ is decreasing. Moreover, it is trivially verified that $e'P_K e$ is decreasing as well. This implies that $V(x, \hat{x})$ is decreasing.

On the other hand, if $\varepsilon_a(\hat{x})$ is nonincreasing, then $P_{\gamma,\varepsilon_a(\hat{x})}$ is nonincreasing and hence,

$$\begin{aligned}\frac{\mathrm{d}}{\mathrm{d}t}V(x, \hat{x}) \leqslant{}& -\hat{x}'Q_{\varepsilon_a(\hat{x})}\hat{x} - \hat{x}'P_{\gamma,\varepsilon_a(\hat{x})}BB'P_{\gamma,\varepsilon_a(\hat{x})}\hat{x}\\ &- \frac{1}{\gamma^2}\hat{x}'P^2_{\gamma,\varepsilon_a(\hat{x})}\hat{x} + 2\hat{x}'P_{\gamma,\varepsilon_a(\hat{x})}KCe - \beta e'e. \qquad (4.256)\end{aligned}$$

We have

$$\begin{aligned}2\hat{x}'P_{\gamma,\varepsilon_a(\hat{x})}KCe &\leqslant \frac{1}{\gamma^2}\hat{x}'P^2_{\gamma,\varepsilon_a(\hat{x})}\hat{x} + \gamma^2 e'C'K'KCe\\ &\leqslant \frac{1}{\gamma^2}\hat{x}'P^2_{\gamma,\varepsilon_a(\hat{x})}\hat{x} + \frac{\beta}{2}e'e,\end{aligned}$$

provided that we choose $\beta > 2\gamma^2\|KC\|^2$. Using this in (4.256), we get

$$\frac{\mathrm{d}}{\mathrm{d}t}V(x, \hat{x}) \leqslant -\hat{x}'Q_{\varepsilon_a(\hat{x})}\hat{x} - \frac{\beta}{2}e'e,$$

which is clearly negative. We find that the function V is always decreasing and hence is a Lyapunov function for the system proving global asymptotic stability. We observe that in an open neighborhood of the origin, we have $\varepsilon_a(\hat{x}) = 1$, and hence, the feedback is locally linear. Then, local exponential stability immediately follows. ■

Theorem 4.54 *Consider the discrete-time linear system subject to input saturation*

$$\begin{aligned} x(k+1) &= Ax(k) + B\sigma(u(k)) \\ y(k) &= Cx(k), \end{aligned}$$

where the pair (A, B) *is stabilizable,* (C, A) *is detectable, and the eigenvalues of* A *are in the closed unit disc. Let* γ *be chosen such that* (4.116) *has a unique positive definite solution for* $E = I$ *for which* (4.117) *is asymptotically stable for all* $\varepsilon \in (0, 1]$*, and define* $\varepsilon_a(x)$ *by*

$$\varepsilon_a(x) = \max\left\{\varepsilon \in (0,1] : 2x' P_{\gamma,\varepsilon} x \left\| B' P_{\gamma,\varepsilon} B \right\| \leqslant \Delta^2\right\}. \tag{4.257}$$

Then, the zero equilibrium point of the closed-loop system with the nonlinear control law given by (4.253) *and* (4.254) *is globally asymptotically stable and locally exponentially stable.*

Proof : The closed-loop system is given by

$$\begin{aligned} e(k+1) &= (A - KC)e(k) \\ \widehat{x}(k+1) &= A\widehat{x}(k) + B\sigma\left(F_{L,\varepsilon_a(\widehat{x}(k))}\widehat{x}(k)\right) + KCe(k), \end{aligned}$$

where $e = x - \widehat{x}$. We consider the candidate Lyapunov function,

$$V(x, \widehat{x}) = \widehat{x}' P_{\gamma,\varepsilon_a(\widehat{x})}\widehat{x} + \beta e' P_K e,$$

where P_K is such that

$$P_K = (A - KC)' P_K (A - KC) + I.$$

As in the state feedback case, we distinguish two cases. If $\varepsilon_a(\widehat{x})$ is increasing, then $\widehat{x}' P_{\gamma,\varepsilon_a(\widehat{x})}\widehat{x}$ is decreasing. Moreover, it is trivially verified that $e' P_K e$ is decreasing as well. This implies that $V(x, \widehat{x})$ is decreasing.

On the other hand, if $\varepsilon_a(\widehat{x})$ is nonincreasing, then $P_{\gamma,\varepsilon_a(\widehat{x})}$ is nonincreasing, and hence,

$$\begin{aligned} V_{k+1} - V_k \leqslant{} & -\widehat{x}' Q_{\varepsilon_a(\widehat{x})}\widehat{x} - \|F_{\gamma,\varepsilon_a(\widehat{x})}\widehat{x}\|^2 + 2\widehat{x}' A_F' P_{\gamma,\varepsilon_a(\widehat{x})} KCe \\ & - \widehat{x}' A_F' P_{\gamma,\varepsilon_a(\widehat{x})} R_{\gamma,\varepsilon_a(\widehat{x})}^{-1} P_{\gamma,\varepsilon_a(\widehat{x})} A_F \widehat{x} - \beta e'e, \end{aligned} \tag{4.258}$$

where we abbreviate $\widehat{x}(k)$ and $e(k)$ by $\widehat{x}$ and e, respectively, while

$$\begin{aligned} V_k &= V(x(k), \widehat{x}(k)) \\ A_F &= A - B(B' P_{\gamma,\varepsilon_a(\widehat{x})} B + I)^{-1} B' P_{\gamma,\varepsilon_a(\widehat{x})} A \\ R_{\gamma,\varepsilon} &= \gamma^2 I - P_{\gamma,\varepsilon} + P_{\gamma,\varepsilon} B(B' P_{\gamma,\varepsilon} B + I)^{-1} B' P_{\gamma,\varepsilon} > 0. \end{aligned}$$

Since $R_{\gamma,\varepsilon}$ is bounded away from 0, we can find μ independent of $\varepsilon \in [0,1]$ such that

$$R_{\gamma,\varepsilon} \leqslant \mu I$$

for all $\varepsilon \in [0,1]$. We have

$$\begin{aligned}2\hat{x}' A_F P_{\gamma,\varepsilon_a(\hat{x})} KCe &\leqslant \frac{1}{\mu}\hat{x}' A_F' P^2_{\gamma,\varepsilon_a(\hat{x})} A_F \hat{x} + \mu e' C' K' K C e \\ &\leqslant \hat{x}' A_F' P_{\gamma,\varepsilon_a(\hat{x})} R^{-1}_{\gamma,\varepsilon_a(\hat{x})} P_{\gamma,\varepsilon_a(\hat{x})} A_F \hat{x} + \frac{\beta}{2} e' e,\end{aligned}$$

provided that we choose $\beta > 2\mu \|KC\|^2$. Using this in (4.258), we get

$$\frac{\mathrm{d}}{\mathrm{d}t} V(x,\hat{x}) \leqslant -\hat{x}' Q_{\varepsilon_a(\hat{x})} \hat{x} - \frac{\beta}{2} e' e,$$

which is clearly negative. We find that the function V is always decreasing and hence is a Lyapunov function for the system proving global asymptotic stability. We observe that in an open neighborhood of the origin, we have $\varepsilon_a(\hat{x}) = 1$, and hence, the feedback is locally linear. Then, local exponential stability immediately follows. ■

4.6.3 *Adaptive-low-gain and high-gain design methodology*

As we discussed in earlier sections, use of high-gain enhances the utilization of available control capacity, and in so doing renders the transient response to die faster. In this regard, we utilize below the adaptive-low-gain feedback as constructed in the previous section, however, with an additional high-gain part. This leads to control laws referred to as "adaptive-low-gain and high-gain" control laws. As in the previous sections, the high-gain part does not interfere with global stabilization, its purpose as demonstrated in subsequent chapters is to achieve fast transient response and for disturbance quenching.

State feedback:

For continuous-time systems, the adaptive-low-gain feedback control law introduced in (4.245) can be modified by adding a high-gain component. The resulting adaptive-low-gain and high-gain control law is given by

$$u := -(1+\alpha) B' P_{\varepsilon_a(x)} x, \tag{4.259}$$

where, as in (4.245), $P_{\varepsilon_a(x)}$ is the positive definite solution of the CARE (4.42) when ε is replaced by $\varepsilon_a(x)$ of (4.243), while α is a high-gain parameter.

Similarly, for discrete-time systems, the adaptive-low-gain feedback control law introduced in (4.246) can be modified by adding a high-gain component. The resulting adaptive-low-gain and high-gain control law is given by

$$u := -(1+\alpha)(B'P_{\varepsilon_a(x)}B + I)^{-1}B'P_{\varepsilon_a(x)}Ax, \tag{4.260}$$

where, as in (4.246), P_ε is the positive definite solution of the DARE (4.66) when ε is replaced by $\varepsilon_a(x)$ of (4.244), while α is a high-gain parameter satisfying

$$\alpha \in \left[0, \frac{2}{\|B'P_1B\|}\right]. \tag{4.261}$$

We emphasize once again that both the above control laws are nonlinear state feedback control laws. The following results show that the addition of a high-gain component does not interfere with global stabilization as long as the high-gain parameter $\alpha \geqslant 0$ for discrete-time systems satisfies the bound in (4.261).

Theorem 4.55 *Consider the continuous-time linear system subject to actuator saturation*

$$\dot{x} = Ax + B\sigma(u),$$

where the pair (A, B) is stabilizable and the eigenvalues of A are in the closed left-half plane. The zero equilibrium point of the closed-loop system with the nonlinear control law (4.259) *is globally asymptotically stable and locally exponentially stable.*

Proof : As in the proof of Theorem 4.51, we note that if $\varepsilon_a(x(t))$ is increasing, then

$$x'(t)P_{\varepsilon_a(x(t))}x(t) \tag{4.262}$$

is decreasing in t. If we reverse this implication, we find that if (4.262) is nondecreasing, i.e., constant or increasing, then $\varepsilon_a(x(t))$ is nonincreasing. If we apply the feedback (4.259), then we get

$$\frac{\mathrm{d}}{\mathrm{d}t}x'(t)P_{\varepsilon_a(x(t))}x(t) = -x'(t)Q_{\varepsilon_a(x(t))}x(t) + x'(t)\left[\frac{\mathrm{d}}{\mathrm{d}t}P_{\varepsilon_a(x(t))}\right]x(t)$$

using similar arguments as in the proof of Theorem 4.41. As argued above, if (4.262) is nondecreasing, then $\varepsilon_a(x(t))$ is nonincreasing, and hence, also $P_{\varepsilon_a(x(t))}$ is nonincreasing. This implies that

$$\frac{\mathrm{d}}{\mathrm{d}t}x'(t)P_{\varepsilon_a(x(t))}x(t) \leqslant -x'(t)Q_{\varepsilon_a(x(t))}x(t) < 0,$$

which yields a contradiction. We conclude that (4.262) is decreasing. This immediately implies asymptotic stability. Let us observe the fact that, in a neighborhood of the origin, $\varepsilon_a(x) = 1$ and the closed-loop system is locally linear and exponentially stable. This yields local exponential stability. ■

Theorem 4.56 *Consider the discrete-time linear system subject to actuator saturation*

$$x(k+1) = Ax(k) + B\sigma(u(k)),$$

where the pair (A, B) is stabilizable and the eigenvalues of A are in the closed unit disc. Let $\varepsilon_a(x)$ be as defined in (4.244)*. Then, the zero equilibrium point of the closed-loop system with the nonlinear control law* (4.260) *is globally asymptotically stable and locally exponentially stable.*

Proof : As in the proof of Theorem 4.52, we note that if $\varepsilon_a(x(k))$ is increasing, then

$$x'(k)P_{\varepsilon_a(x(k))}x(k) \geqslant x'(k+1)P_{\varepsilon_a(x(k+1))}x(k+1).$$

If we reverse this implication, we find that if $x'(k)P_{\varepsilon_a(x(k))}x(k)$ is nondecreasing then $\varepsilon_a(x(k))$ is nonincreasing. If we apply the feedback (4.244), then we get

$$\begin{aligned} x'(k+1)P_{\varepsilon_a(x(k+1))}x(k+1) &- x'(k)P_{\varepsilon_a(x(k))}x(k) \\ &\leqslant -x'(k)Q_{\varepsilon_a(x(k))}x(k) + x'(k)\left[P_{\varepsilon_a(x(k+1))} - P_{\varepsilon_a(x(k))}\right]x(k). \end{aligned}$$

As argued above, if $x'(k)P_{\varepsilon_a(x(k))}x(k)$ is nondecreasing, then $\varepsilon_a(x(k))$ is nonincreasing, and hence, also $P_{\varepsilon_a(x(k))}$ is nonincreasing. This implies that

$$x'(k+1)P_{\varepsilon_a(x(k+1))}x(k+1) - x'(k)P_{\varepsilon_a(x(k))}x(k) \leqslant -x'(k)Q_{\varepsilon_a(x(k))}x(k) < 0$$

which yields a contradiction. We conclude that $x'(k)P_{\varepsilon_a(x(k))}x(k)$ is decreasing. This immediately implies asymptotic stability. Let us observe the fact that, in a neighborhood of the origin, $\varepsilon_a(x) = 1$ and the closed-loop system is locally linear and exponentially stable. This yields local exponential stability. ■

Measurement feedback:

Theorems 4.55 and 4.56 pertain to state feedback control. However, in many cases, we are restricted to measurement feedback. In this case, we clearly need to introduce an observer. However, it turns out that it is better to modify the low-gain state feedback in such a way that the observer error does not cause any problems. We will use here H_∞ ARE-based adaptive-low-gain feedback control laws (as in the previous subsection).

We will use the same way of adapting the low-gain parameter $\varepsilon_a(x)$ as before. However, this time we add a high-gain parameter. Thus, for continuous-time systems, we have an adaptive-low-gain and high-gain observer-based measurement feedback control law as

$$u = (1+\alpha)F_{L,\varepsilon_a(\hat{x})}\hat{x} = -(1+\alpha)B'P_{\gamma,\varepsilon_a(\hat{x})}\hat{x}, \tag{4.263}$$

where to start with $P_{\gamma,\varepsilon}$ is the positive definite solution of the CARE (4.90) with $E = I$ such that (4.91) is asymptotically stable. Here, γ is chosen such that (4.90) has a unique positive definite solution for which (4.91) is asymptotically stable for all $\varepsilon \in (0, 1]$. The existence of such a γ is discussed in Lemma 4.28. Next, in order to obtain $P_{\gamma,\varepsilon_a(\hat{x})}$, at first, ε is replaced by $\varepsilon_a(x)$ of (4.243) where P_r is replaced by $P_{\gamma,r}$ and then $\varepsilon_a(x)$ is replaced by $\varepsilon_a(\hat{x})$. We connect this low-and-high-gain feedback with an observer:

$$\dot{\hat{x}} = A\hat{x} + B\sigma(u) + K(y - C\hat{x}), \tag{4.264}$$

where K is such that $A - KC$ is asymptotically stable.

Similarly, for discrete-time systems, we can introduce an adaptive-low-gain and high-gain feedback control law as

$$u = (1+\alpha)F_{L,\varepsilon_a(\hat{x})}\hat{x} = -(1+\alpha)(B'P_{\gamma,\varepsilon_a(\hat{x})}B + I)^{-1}B'P_{\gamma,\varepsilon_a(\hat{x})}A\hat{x}, \tag{4.265}$$

where to start with $P_{\gamma,\varepsilon}$ is the positive definite solution of the DARE (4.116) with $E = I$ such that (4.117) is asymptotically stable. Here, γ is chosen such that (4.116) has a unique positive definite solution for which (4.117) is asymptotically stable for all $\varepsilon \in (0, 1]$. The existence of such a γ is discussed in Lemma 4.32. Next, in order to obtain $P_{\gamma,\varepsilon_a(\hat{x})}$, at first, ε is replaced by $\varepsilon_a(x)$ of (4.244) where P_r is replaced by $P_{\gamma,r}$, and then $\varepsilon_a(x)$ is replaced by $\varepsilon_a(\hat{x})$. Finally,

$$\alpha \in \left[0, \frac{2}{\|B'P_{\gamma,1}B\|}\right].$$

We connect this low-and-high-gain feedback with an observer,

$$\hat{x}(k+1) = A\hat{x}(k) + B\sigma(u(k)) + K(y(k) - C\hat{x}(k)), \tag{4.266}$$

where K is such that $A - KC$ is asymptotically stable.

Theorem 4.57 *Consider the continuous-time linear system subject to input saturation*

$$\begin{aligned}\dot{x} &= Ax + B\sigma(u)\\ y &= Cx,\end{aligned}$$

where the pair (A, B) *is stabilizable,* (C, A) *is detectable, and the eigenvalues of* A *are in the closed left-half plane. Let* γ *be chosen such that* (4.90) *with* $E = I$ *has a unique positive definite solution such that* (4.91) *is asymptotically stable for all* $\varepsilon \in (0, 1]$ *and define* $\varepsilon_a(x)$ *by*

$$\varepsilon_a(x) = \max\left\{\varepsilon \in (0,1] : x' P_{\gamma,\varepsilon} x \left\| B' P_{\gamma,\varepsilon} B \right\| \leqslant \Delta^2 \right\}. \tag{4.267}$$

Then, the zero equilibrium point of the closed-loop system with the nonlinear control law given by (4.263) *and* (4.264) *is globally asymptotically stable and locally exponentially stable.*

Proof : The closed-loop system is given by

$$\begin{aligned} \dot{e} &= (A - KC)e \\ \dot{\widehat{x}} &= A\widehat{x} - B\sigma((1+\alpha)B' P_{\gamma,\varepsilon_a(\widehat{x})}\widehat{x}) + KCe, \end{aligned}$$

where $e = x - \widehat{x}$. We consider the candidate Lyapunov function,

$$V(x, \widehat{x}) = \widehat{x}' P_{\gamma,\varepsilon_a(\widehat{x})} \widehat{x} + \beta e' P_K e,$$

where P_K is such that

$$(A - KC)' P_K + P_K (A - KC) + I = 0.$$

As in the state feedback case, we distinguish two cases. If $\varepsilon_a(\widehat{x})$ is increasing, then $\widehat{x}' P_{\gamma,\varepsilon_a(\widehat{x})} \widehat{x}$ is decreasing. Moreover, it is trivially verified that $e' P_K e$ is decreasing as well. This implies that $V(x, \widehat{x})$ is decreasing.

On the other hand, if $\varepsilon_a(\widehat{x})$ is nonincreasing, then $P_{\gamma,\varepsilon_a(\widehat{x})}$ is nonincreasing and hence,

$$\begin{aligned} \frac{\mathrm{d}}{\mathrm{d}t} V(x, \widehat{x}) \leqslant & -\widehat{x}' Q_{\varepsilon_a(\widehat{x})} \widehat{x} - \widehat{x}' P_{\gamma,\varepsilon_a(\widehat{x})} B B' P_{\gamma,\varepsilon_a(\widehat{x})} \widehat{x} \\ & - \frac{1}{\gamma^2} \widehat{x}' P^2_{\gamma,\varepsilon_a(\widehat{x})} \widehat{x} + 2\widehat{x}' P_{\gamma,\varepsilon_a(\widehat{x})} KCe - \beta e' e. \end{aligned} \tag{4.268}$$

We have

$$\begin{aligned} 2\widehat{x}' P_{\gamma,\varepsilon_a(\widehat{x})} KCe &\leqslant \frac{1}{\gamma^2} \widehat{x}' P^2_{\gamma,\varepsilon_a(\widehat{x})} \widehat{x} + \gamma^2 e' C' K' KCe \\ &\leqslant \frac{1}{\gamma^2} \widehat{x}' P^2_{\gamma,\varepsilon_a(\widehat{x})} \widehat{x} + \frac{\beta}{2} e' e, \end{aligned}$$

provided that we choose $\beta > 2\gamma^2 \| KC \|^2$. Using this in (4.268), we get

$$\frac{\mathrm{d}}{\mathrm{d}t} V(x, \widehat{x}) \leqslant -\widehat{x}' Q_{\varepsilon_a(\widehat{x})} \widehat{x} - \frac{\beta}{2} e' e,$$

which is clearly negative. We find that the function V is always decreasing, and hence is a Lyapunov function for the system proving global asymptotic stability. We observe that in an open neighborhood of the origin, we have $\varepsilon_a(\widehat{x}) = 1$, and hence, the feedback is locally linear. Then, local exponential stability immediately follows. ■

Theorem 4.58 *Consider the discrete-time linear system subject to input saturation*

$$\begin{aligned} x(k+1) &= Ax(k) + B\sigma(u(k)) \\ y(k) &= Cx(k), \end{aligned}$$

where the pair (A, B) *is stabilizable,* (C, A) *is detectable, and the eigenvalues of* A *are in the closed unit disc. Let* γ *be chosen such that* (4.116) *has a unique positive definite solution for* $E = I$ *for which* (4.117) *is asymptotically stable for all* $\varepsilon \in (0, 1]$ *and define* $\varepsilon_a(x)$ *by*

$$\varepsilon_a(x) = \max \left\{ \varepsilon \in (0, 1] : 2x' P_{\gamma,\varepsilon} x \left\| B' P_{\gamma,\varepsilon} B \right\| \leqslant \Delta^2 \right\}. \tag{4.269}$$

Then, the zero equilibrium point of the closed-loop system with the nonlinear control law given by (4.265) *and* (4.266) *is globally asymptotically stable and locally exponentially stable.*

Proof : The closed-loop system is given by

$$\begin{aligned} e(k+1) &= (A - KC)e(k) \\ \widehat{x}(k+1) &= A\widehat{x}(k) + B\sigma\left(F_{L,\varepsilon_a(\widehat{x}(k))}\widehat{x}(k)\right) + KCe(k), \end{aligned}$$

where $e = x - \widehat{x}$. We consider the candidate Lyapunov function,

$$V(x, \widehat{x}) = \widehat{x}' P_{\gamma,\varepsilon_a(\widehat{x})} \widehat{x} + \beta e' P_K e,$$

where P_K is such that

$$P_K = (A - KC)' P_K (A - KC) + I.$$

As in the state feedback case, we distinguish two cases. If $\varepsilon_a(\widehat{x})$ is increasing, then $\widehat{x}' P_{\gamma,\varepsilon_a(\widehat{x})} \widehat{x}$ is decreasing. Moreover, it is trivially verified that $e' P_K e$ is decreasing as well. This implies that $V(x, \widehat{x})$ is decreasing.

On the other hand, if $\varepsilon_a(\hat{x})$ is nonincreasing, then $P_{\gamma,\varepsilon_a(\hat{x})}$ is nonincreasing, and hence,

$$V_{k+1} - V_k \leqslant -\hat{x}' Q_{\varepsilon_a(\hat{x})}\hat{x} - \|F_{\gamma,\varepsilon_a(\hat{x})}\hat{x}\|^2 + 2\hat{x}' A_F' P_{\gamma,\varepsilon_a(\hat{x})} KCe - \hat{x}' A_F' P_{\gamma,\varepsilon_a(\hat{x})} R^{-1}_{\gamma,\varepsilon_a(\hat{x})} P_{\gamma,\varepsilon_a(\hat{x})} A_F \hat{x} - \beta e'e, \quad (4.270)$$

where we abbreviate $\hat{x}(k)$ and $e(k)$ by $\hat{x}$ and e, respectively, while

$$\begin{aligned} V_k &= V(x(k), \hat{x}(k)) \\ A_F &= A - B(B' P_{\gamma,\varepsilon_a(\hat{x})} B + I)^{-1} B' P_{\gamma,\varepsilon_a(\hat{x})} A \\ R_{\gamma,\varepsilon} &= \gamma^2 I - P_{\gamma,\varepsilon} + P_{\gamma,\varepsilon} B(B' P_{\gamma,\varepsilon} B + I)^{-1} B' P_{\gamma,\varepsilon} > 0. \end{aligned}$$

Since $R_{\gamma,\varepsilon}$ is bounded away from 0, we can find μ independent of $\varepsilon \in [0,1]$ such that

$$R_{\gamma,\varepsilon} \leqslant \mu I$$

for all $\varepsilon \in [0,1]$. We have

$$\begin{aligned} 2\hat{x}' A_F P_{\gamma,\varepsilon_a(\hat{x})} KCe &\leqslant \frac{1}{\mu}\hat{x}' A_F' P^2_{\gamma,\varepsilon_a(\hat{x})} A_F \hat{x} + \mu e' C' K' KCe \\ &\leqslant \hat{x}' A_F' P_{\gamma,\varepsilon_a(\hat{x})} R^{-1}_{\gamma,\varepsilon_a(\hat{x})} P_{\gamma,\varepsilon_a(\hat{x})} A_F \hat{x} + \frac{\beta}{2} e'e, \end{aligned}$$

provided that we choose $\beta > 2\mu\|KC\|^2$. Using this in (4.270), we get

$$\frac{\mathrm{d}}{\mathrm{d}t} V(x, \hat{x}) \leqslant -\hat{x}' Q_{\varepsilon_a(\hat{x})}\hat{x} - \frac{\beta}{2} e'e,$$

which is clearly negative. We find that the function V is always decreasing and hence is a Lyapunov function for the system proving global asymptotic stability. We observe that in an open neighborhood of the origin, we have $\varepsilon_a(\hat{x}) = 1$, and hence, the feedback is locally linear. Then, local exponential stability immediately follows. ■

4.7 Issues on global stabilization of linear systems subject to actuator saturation

There are still unresolved issues on global stabilization of linear systems subject to actuator saturation. One specific issue is whether such a stabilization can be accomplished utilizing linear feedbacks or requires nonlinear feedbacks. Let us briefly recollect what has been done. In the introductory section of this chapter (Sect. 4.1), we recalled a result of [38] which says that a chain of integrators with order greater or equal to three cannot be globally stabilized by any saturating linear static state feedback control law with only one input channel. Also, we recalled a result of [155] which states that, global stabilization of continuous-time linear systems with bounded input can be achieved if and only if the linear system in the absence of actuator saturation is stabilizable and critically unstable (that is, *asymptotically null controllable with bounded control* (ANCBC)). In general, this requires nonlinear feedback control laws.

In view of the results of [38, 155], one can enquire whether linear globally stabilizing feedback control laws exist for any class of linear systems subject to actuator saturation. For certain cases, global stabilization can indeed be achieved by linear feedback control laws. In fact, in Sect. 4.6.1, both for continuous- and discrete-time linear *neutral* systems, we constructed bounded linear static state feedbacks as well as bounded observer-based linear measurement feedback control laws that achieve global asymptotic stabilization. Moreover, for continuous-time systems which are ANCBC, the results of [186] state, but without a full proof, that global asymptotic stabilization can be achieved by linear static state feedback control laws if all nonzero eigenvalues on the imaginary axis are semi-simple (Jordan blocks of size at most 1) while zero is allowed to be an eigenvalue whose Jordan blocks can be at most of size 2×2 (which are associated with double integrators). In this section, we will present for continuous-time systems which are ANCBC, the recent work of [206] which goes beyond the work of [186], and presents constructive globally stabilizing saturating linear static state feedback control laws for linear systems which consist of a mix of *single integrators, double integrators, and neutrally stable dynamics.*

We also observe that, in the literature, there is this general belief for continuous-time systems that if there are Jordan blocks of size greater or equal to three associated with an eigenvalue at zero, then one needs nonlinear feedback control laws to globally stabilize such linear systems subject to actuator saturation. This is a misconception. Such a misconception is possibly based on a misreading of the result of [38]. One should emphasize that the beautiful result of Fuller does not claim or consider anything beyond static state feedbacks for a chain of integrators with a single input. We will present in this section our recent work [206] which reexamines this classical issue and resolves the general misconception that the size of Jordan block associated with a zero eigenvalue determines whether linear control laws or nonlinear control laws are needed for globally stabilizing a linear system subject to actuator saturation.

Global asymptotic stabilization of discrete-time systems subject to actuator saturation turns out to be much more complex than that of continuous-time systems. A discrete-time version of the classical result of Fuller is still not available. There is a large class of linear feedback control laws which achieve local asymptotic stability for the discrete-time equivalent of the double integrator but fail to achieve global asymptotic stability. This result is indeed in direct contrast with continuous-time case where local asymptotic stability for the double integrator always implies global asymptotic stability. We will present this recent work from [207] which completely classifies which linear feedback laws result in global asymptotic stability for the discrete-time equivalent of the double integrator. As far as we know today, the discrete-time equivalent of the double integrator and the class of linear neutral systems (Sect. 4.6.1) are the only discrete-time systems which can be rendered globally asymptotically stable by bounded linear feedback controllers.

As of this writing, it is fair to say that there still exist two general unresolved open problems:

- Under what conditions one can utilize a linear *static* state feedback control law to globally stabilize a linear system subject to actuator saturation?
- Under what conditions one can utilize a linear *dynamic* state feedback control law to globally stabilize a linear system subject to actuator saturation?

4.7.1 Mixed case of single integrators, double integrators, and neutrally stable dynamics

In continuous-time case, it is clear that both double integrator and neutrally stable systems subject to actuator saturation can be globally stabilized by linear static state feedback control laws. However, the mixture of these cases is not well studied. As recalled above, for the mixed case, [186] gave a sufficient condition that guarantees global stability of the closed-loop system by using linear static state feedback control laws, but this result is not studied from a design point of view. In this subsection, we present a linear static state feedback control design methodology to globally stabilize the mixed system subject to actuator saturation based on the very recent work of [206].

To start with, we present an algorithm which gives us a methodology for designing a linear static state feedback control laws for so-called *mixed* systems, namely, systems with double integrators, single integrators, and neutrally stable dynamics subject to actuator saturation. We then prove via a Lyapunov argument that such a control law globally stabilizes a mixed system subject to actuator saturation.

Consider a continuous-time linear system called *mixed* system subject to actuator saturation:

$$\dot{\tilde{x}} = \tilde{A}\tilde{x} + \tilde{B}\sigma(\tilde{u}), \tag{4.271}$$

where $\tilde{x} \in \mathbb{R}^n$, $\tilde{u} \in \mathbb{R}^m$ and σ is the standard saturation function defined in Definition 2.19.

As in the previous sections, we assume here as well that the pair $(\tilde{A}, \tilde{B})$ is stabilizable, and the eigenvalues of $\tilde{A}$ are all located in the closed left-half complex plane (i.e., the pair $(\tilde{A}, \tilde{B})$ is ANCBC). Furthermore, we assume that $\tilde{A}$ has eigenvalue zero with geometric multiplicity m and algebraic multiplicity $m + q$ with no Jordan blocks of size larger than 2 while the remaining eigenvalues are simple purely imaginary eigenvalues. Obviously, for such systems, stabilizability of the pair $(\tilde{A}, \tilde{B})$ is equivalent to controllability of the pair $(\tilde{A}, \tilde{B})$.

The algorithm for designing a linear static state feedback control law to globally stabilize the system described in (4.271) is carried out in three steps:
Step 1: We can obviously find a basis transformation Γ_x such that

$$A = \Gamma_x^{-1} \widetilde{A} \Gamma_x = \begin{pmatrix} A_d & 0 & 0 \\ 0 & A_s & 0 \\ 0 & 0 & A_\omega \end{pmatrix}$$

with

$$A_d = \begin{pmatrix} 0 & I \\ 0 & 0 \end{pmatrix},$$

while $A_s = 0$ and A_ω satisfies $A_\omega + A_\omega' = 0$. With respect to this basis transformation, we obtain

$$B = \Gamma_x^{-1} \widetilde{B} = \begin{pmatrix} B_d \\ B_s \\ B_\omega \end{pmatrix},$$

with

$$B_d = \begin{pmatrix} B_{d,1} \\ B_{d,2} \end{pmatrix}$$

compatible with the structure of A_d. The system in the new coordinates is given by

$$\dot{x} = Ax + B\sigma(u). \tag{4.272}$$

Step 2: Design K such that

$$K = \begin{pmatrix} K_1 \\ K_2 \\ \vdots \\ K_m \end{pmatrix}$$

satisfies

$$\begin{aligned} KA + B'\Lambda &= 0 \\ KB + (KB)' &< 0, \end{aligned}$$

where Λ is a diagonal matrix such that

$$\Lambda = \begin{pmatrix} \Lambda_1 & 0 & 0 \\ 0 & 0 & 0 \\ 0 & 0 & I \end{pmatrix},$$

with

$$\Lambda_1 = \begin{pmatrix} 0 & 0 \\ 0 & I \end{pmatrix}$$

compatible with the structure of A_d. The existence of such a K is shown in the proof of Theorem 4.61.

Step 3: Construct the linear static state feedback control law

$$u = \widetilde{K}\widetilde{x}, \tag{4.273}$$

where $\widetilde{K} = K\Gamma_x^{-1}$. The control law (4.273) globally stabilizes the system described in (4.271).

To prove that the above algorithm generates a $\widetilde{K}$ that globally stabilizes the system described in (4.271), we need the following lemmas. These two lemmas are very well known and can be found in [63] and [58], respectively.

Lemma 4.59 *Given two matrices X and Y, there exists a matrix Z such that*

$$ZX = Y$$

if and only if

$$\ker X \subset \ker Y,$$

where $\ker A$ *is the null space of a matrix* $A \in \mathbb{R}^{m\times n}$ *defined as*

$$\ker A := \{x \in \mathbb{R}^n \mid Ax = 0\}. \tag{4.274}$$

We present next a special case of LaSalle's invariance principle, where $V(x)$ is positive definite, which is also known as Krasovskii Theorem.

Lemma 4.60 *Consider the system*

$$\dot{x} = f(x),$$

where $f : \mathbb{R}^n \to \mathbb{R}^n$. Let $x = 0$ be an equilibrium point. Let $V : \mathbb{R}^n \to \mathbb{R}$ be a continuously differentiable, radially unbounded, positive definite function such that $\dot{V}(x) \leqslant 0$ for all $x \in \mathbb{R}^n$. Let $\mathcal{S} = \{x \in \mathbb{R}^n \mid \dot{V}(x) = 0\}$ and suppose that no solution can stay in $\mathcal{S}$ for all $t \geqslant 0$ other than the trivial solution $x(t) = 0$ for all $t \geqslant 0$. Then, the origin is globally asymptotically stable.

Next, we prove that the design given in the above algorithm has the desired properties.

Theorem 4.61 *Consider a linear system as given in* (4.271) *with input* $u(t) \in \mathbb{R}^m$. *Assume that the pair* $(\widetilde{A}, \widetilde{B})$ *is controllable. Moreover, we assume that* $\widetilde{A}$ *has eigenvalue zero with geometric multiplicity* m *and algebraic multiplicity* $m+q$ *with no Jordan blocks of size larger than* 2 *while the remaining eigenvalues are simple purely imaginary eigenvalues. The linear state feedback control law* $u = \widetilde{K}\widetilde{x}$ *given in* (4.273) *can globally stabilize the system* (4.271).

Proof : Through step 1 of the algorithm, we can transfer the system (4.271) into (4.272) as

$$\dot{x} = Ax + B\sigma(u).$$

The state vector has a decomposition

$$x = \begin{pmatrix} x_d \\ x_s \\ x_\omega \end{pmatrix}$$

compatible to the decomposition of A. Moreover,

$$x_d = \begin{pmatrix} x_{d,1} \\ x_{d,2} \end{pmatrix}.$$

We prove the theorem via Lyapunov argument. Consider a Lyapunov candidate

$$V(x) = \frac{1}{2}x'_\omega x_\omega + \frac{1}{2}x'_{d,2}x_{d,2} + \sum_{i=1}^{m} \int_0^{K_i x} \sigma_1(y)\mathrm{d}y. \tag{4.275}$$

The evaluation of $\dot{V}(x)$ along the trajectories of the closed-loop system yields,

$$\dot{V}(x) = x'_\omega \dot{x}_\omega + x'_{d,2}\dot{x}_{d,2} + \sigma'(Kx)K\dot{x}.$$

With some algebra, we can write the above equation in the matrix form as

$$\dot{V}(x) = \sigma'(Kx)(KAx + B'\Lambda x) + \sigma'(Kx)KB\sigma(Kx). \tag{4.276}$$

In order to make $\dot{V}(x)$ nonpositive, it is sufficient to guarantee that the gain matrix K satisfies

$$\begin{aligned} KA + B'\Lambda &= 0 \\ KB + (KB)' &< 0. \end{aligned} \tag{4.277}$$

Let us write (4.277) in matrix equality form,

$$K\begin{pmatrix} A & B \end{pmatrix} = \begin{pmatrix} -B'\Lambda & S \end{pmatrix}, \tag{4.278}$$

where S is any matrix satisfying $S + S' < 0$. We get

$$KB + (KB)' = S + S' < 0.$$

Now, let us show that a K which satisfies (4.278) exists. From Lemma 4.59, we see that the solvability of (4.278) is equivalent to showing

$$\begin{pmatrix} -B'\Lambda & S \end{pmatrix}\begin{pmatrix} x \\ u \end{pmatrix} = 0 \tag{4.279}$$

given

$$\begin{pmatrix} A & B \end{pmatrix}\begin{pmatrix} x \\ u \end{pmatrix} = 0. \tag{4.280}$$

Since the pair (A, B) is controllable, from the Hautus test [45], we know that

$$\operatorname{rank}\begin{pmatrix} A & B \end{pmatrix} = n.$$

Moreover, from the structure of matrices A and B, we know that $\operatorname{rank} A = n - m$ and $\operatorname{rank} B = m$. Thus,

$$\operatorname{im} A \cap \operatorname{im} B = 0,$$

where $\operatorname{im} Z$ is the range space of a matrix $Z \in \mathbb{R}^{m \times n}$ defined as

$$\operatorname{im} Z := \{Zx \mid x \in \mathbb{R}^n\}.$$

Therefore, (4.280) implies that $Ax = 0$ and $Bu = 0$. Clearly, $Ax = 0$ implies that $x_{d,2} = 0$ and $x_\omega = 0$ which yields $\Lambda x = 0$. Hence,

$$-B'\Lambda x = 0.$$

Moreover, $\operatorname{rank} B = m$ while B has m columns. This implies that B is injective. Therefore, $Bu = 0$ implies $u = 0$, and thus, $Su = 0$. Hence, (4.279) is satisfied, and we have shown that (4.278) is solvable. Since $(A \quad B)$ is surjective, for any given S, we have a unique solution K for (4.278) such that

$$\dot{V}(x) = \sigma'(Kx)KB\sigma(Kx) \leqslant 0,$$

provided that $S + S' < 0$. In order to prove asymptotic stability, we apply Lemma 4.60. Clearly, our Lyapunov candidate function $V(x)$ given in (4.275) is continuously differentiable, radially unbounded, and is positive definite.

Next, we note that the $\dot{V}(x) = 0$ if and only if $Kx = 0$. When $Kx = 0$, the dynamics obviously becomes $\dot{x} = Ax$. We need to show that there exists no initial condition $x(0) = x_0 \neq 0$ such that $Kx(t) = 0$ for all $t > 0$ while $\dot{x}(t) = Ax(t)$. We have

$$x(t) = \begin{pmatrix} x_{d,1}(0) + t x_{d,2}(0) \\ x_{d,2}(0) \\ x_s(0) \\ x_\omega(t) \end{pmatrix}.$$

Since $x_\omega(t)$ is only related to nonzero eigenvalues, we get from $Kx(t) = 0$ for all $t \geqslant 0$ that

$$K \begin{pmatrix} x_{d,2}(0) \\ 0 \\ 0 \\ 0 \end{pmatrix} = 0, \quad K \begin{pmatrix} x_{d,1}(0) \\ x_{d,2}(0) \\ x_s(0) \\ 0 \end{pmatrix} = 0. \tag{4.281}$$

The first equality in (4.281) implies that

$$0 = K \begin{pmatrix} x_{d,2}(0) \\ 0 \\ 0 \\ 0 \end{pmatrix} = KA \begin{pmatrix} 0 \\ x_{d,2}(0) \\ 0 \\ 0 \end{pmatrix} = -B' \begin{pmatrix} 0 \\ x_{d,2}(0) \\ 0 \\ 0 \end{pmatrix},$$

which yields $B'_{d,2} x_{d,2}(0) = 0$. Controllability of the pair (A, B) implies that $B_{d,2}$ must be surjective. Hence, $B'_{d,2}$ is injective, and we obtain $x_{d,2}(0) = 0$. Next, we note that the second equality in (4.281) yields

$$\begin{pmatrix} x_{d,1}(0) \\ 0 \\ x_s(0) \\ 0 \end{pmatrix} = Ax + Bu \text{ where } x = \begin{pmatrix} x_1 \\ x_2 \\ x_3 \\ x_4 \end{pmatrix}$$

for suitably chosen x and u since $(A \quad B)$ is surjective (because of controllability). Obviously, this implies that $0 = B_{d,2} u$ while x_4 satisfies

$$x_4 = -A_\omega^{-1} B_\omega u.$$

We find that

$$
\begin{aligned}
0 = u'K \begin{pmatrix} x_{d,1}(0) \\ 0 \\ x_s(0) \\ 0 \end{pmatrix} &= u'K\,[Ax + Bu] \\
&= -u'B'\Lambda x + u'Su \\
&= -u'B'_{d,2}x_2 - u'B'_\omega x_4 + u'Su \\
&= u'B'_\omega A_\omega^{-1} B_\omega u + u'Su \\
&= u'Su,
\end{aligned}
$$

where we used that $B_{d,2}u = 0$ and the fact that A_ω^{-1} is skew-symmetric. Since $S + S' < 0$, we find $u = 0$. But this immediately yields that $x_s(0) = 0$. Using that $x_s(0) = 0$ and $x_{d,2}(0) = 0$, we get from the second equality in (4.281) that

$$
0 = K \begin{pmatrix} x_{d,1}(0) \\ 0 \\ 0 \\ 0 \end{pmatrix} = KA \begin{pmatrix} 0 \\ x_{d,1}(0) \\ 0 \\ 0 \end{pmatrix} = -B' \begin{pmatrix} 0 \\ x_{d,1}(0) \\ 0 \\ 0 \end{pmatrix},
$$

which yields $B'_{d,2}x_{d,1}(0) = 0$. As noted before, $B'_{d,2}$ is injective, and therefore, $x_{d,1}(0) = 0$.

It remains to show that $x_\omega(0) = 0$. We note that

$$
K \begin{pmatrix} 0 \\ 0 \\ 0 \\ x_\omega(0) \end{pmatrix} = K_\omega x_\omega(0) = 0,
$$

where K_ω is the gain matrix associated with neutrally stable dynamics. We know that $x(t)$ remains in the kernel of K with $u(t) = 0$. Hence,

$$
KA \begin{pmatrix} 0 \\ 0 \\ 0 \\ x_\omega(0) \end{pmatrix} = K_\omega A_\omega x_\omega(0) = 0.
$$

But since $KA = -B'\Lambda$, this yields

$$
B'_\omega x_\omega(0) = 0.
$$

Hence, if $x_\omega(0) \neq 0$, we have a nontrivial A_ω-invariant subspace which is contained in $\ker B'_\omega$. Using the skew symmetry of A_ω, we find that this subspace is also A'_ω-invariant. However, the existence of a nontrivial A'_ω invariant subspace contained in ker B'_ω yields a contradiction with the observability of the pair (B'_ω, A'_ω) or, equivalently, a contradiction with the controllability of the pair (A_ω, B_ω). Therefore, $x_\omega(0) = 0$.

Hence, the origin is the only solution within the subset of $\mathbb{R}^n$ for which $\dot{V}(x) = 0$. Hence, the global asymptotic stability of the closed-loop system follows from Lemma 4.60. ∎

Let us consider an example, which contains two double integrators, one single integrator, and neutrally stable dynamics with A and B as follows:

$$A = \begin{pmatrix} 0 & 0 & 1 & 0 & 0 & 0 & 0 \\ 0 & 0 & 0 & 1 & 0 & 0 & 0 \\ 0 & 0 & 0 & 0 & 0 & 0 & 0 \\ 0 & 0 & 0 & 0 & 0 & 0 & 0 \\ 0 & 0 & 0 & 0 & 0 & 0 & 0 \\ 0 & 0 & 0 & 0 & 0 & 0 & 1 \\ 0 & 0 & 0 & 0 & 0 & -1 & 0 \end{pmatrix}, \quad B = \begin{pmatrix} 0 & 1 & 3 \\ 0 & 0 & 5 \\ 1 & 2 & 4 \\ 0 & 1 & 6 \\ 0 & 0 & 1 \\ 1 & 1 & 0 \\ 1 & 0 & 1 \end{pmatrix}.$$

It is easy to check that the pair (A, B) is controllable. Also, we observe that since S is arbitrary such that $S + S' < 0$, the solution for (4.278) is **not** unique. Therefore, the linear static state feedback control laws which can globally stabilize the closed-loop system are **not** unique either. However, for a given S, we have a unique solution for (4.278); therefore, we have an associated unique linear static state feedback control law which can globally stabilize the closed-loop system.

For this example, we choose

$$S = \begin{pmatrix} -1 & -1 & 1 \\ 1 & -3 & 1 \\ -1 & -1 & -53 \end{pmatrix}.$$

Then, the unique possible globally stabilizing linear static state feedback control law is $u = Kx$, where

$$K = \begin{pmatrix} K_1 \\ K_2 \\ K_3 \end{pmatrix}$$

with K_1, K_2, and K_3 as given below:

$$K_1 = \begin{pmatrix} -1 & 0 & -1 & 3 & -11 & -1 & 1 \end{pmatrix},$$
$$K_2 = \begin{pmatrix} -2 & -1 & 0 & -1 & 17 & 0 & 1 \end{pmatrix},$$
$$\text{and} \quad K_3 = \begin{pmatrix} -4 & -6 & 0 & 4 & -35 & -1 & 0 \end{pmatrix}.$$

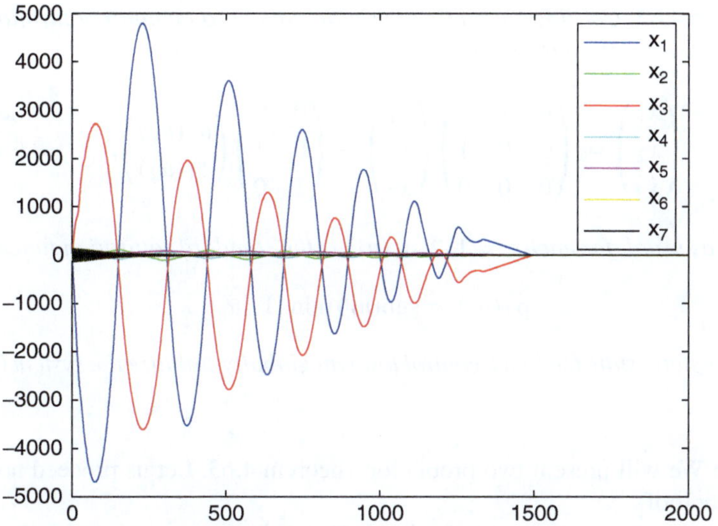

Figure 4.1: Global stabilization via a linear static state feedback

For the initial condition

$$x_0 = \begin{pmatrix} 100 & -100 & -100 & 100 & 100 & -100 & 100 \end{pmatrix}',$$

the dynamics are shown in Fig. 4.1, which clearly is consistent with our result that the closed-loop system is asymptotically stable.

4.7.2 *Triple integrator with multiple inputs*

As pointed out earlier, there exists a general belief that for a system which has an eigenvalue at zero with associated Jordan block of size greater or equal to three, there does not exist a saturating linear static state feedback control law which can globally stabilize the system. We claim that this is a misconception. More precisely, in simple words, whether a saturating linear static state feedback control law exists or not does not merely depend on the size of Jordan block. We prove here that there exists a saturating linear static state feedback control law which can globally stabilize a triple-integrator system with two inputs.

We present first a useful Lemma 4.62 from [155] (see also [172]).

Lemma 4.62 *Assume that $\dot{\zeta} = f(\zeta, v)$ has a globally Lipschitz right-hand side, and that the origin is a globally asymptotically stable state for $\dot{\zeta} = f(\zeta, 0)$. Then there exists some $\lambda > 0$ such that every solution of $\dot{\zeta} = f(\zeta, v)$ converges to zero, for every v such that $\|v(t)\| \leqslant \kappa e^{-\lambda t}$.*

We present next a theorem pertaining to a triple integrator with two inputs.

Theorem 4.63 *Consider a triple integrator subject to actuator saturation, with two input channels, described by*

$$\begin{pmatrix} \dot{x}_1 \\ \dot{x}_2 \\ \dot{x}_3 \end{pmatrix} = \begin{pmatrix} 0 & 1 & 0 \\ 0 & 0 & 1 \\ 0 & 0 & 0 \end{pmatrix} \begin{pmatrix} x_1 \\ x_2 \\ x_3 \end{pmatrix} + \begin{pmatrix} 0 & 0 \\ 0 & 1 \\ 1 & 0 \end{pmatrix} \begin{pmatrix} \sigma_1(u_1) \\ \sigma_1(u_2) \end{pmatrix}, \tag{4.282}$$

where, as usual, for each $i = 1, 2$, $\sigma_1(u_i)$ *is the standard saturation function:*

$$\sigma_1(u_i) = \operatorname{sgn}(u_i) \min\{1, |u_i|\}.$$

A linear static state feedback control law can globally stabilize the system (4.282).

Proof : We will present two proofs for Theorem 4.63. Let us proceed now with the first proof.

Consider a linear static state feedback control law

$$\begin{aligned} u_1 &= -\gamma x_3 \\ u_2 &= -\tilde{\alpha} x_1 - \beta x_2, \end{aligned}$$

where $\tilde{\alpha} > 0$, $\beta > 0$, and $\gamma >> 0$. By applying this particular state feedback control law, we get the closed-loop system as

$$\dot{x}_1 = x_2 \tag{4.283a}$$

$$\dot{x}_2 = x_3 + \sigma_1(-\tilde{\alpha} x_1 - \beta x_2) \tag{4.283b}$$

$$\dot{x}_3 = \sigma_1(-\gamma x_3). \tag{4.284}$$

The asymptotic stability of the closed-loop system follows from the fact that the poles of the linearized system are in the open left-half plane. In order to prove global asymptotic stability of the closed-loop system, we need to show that the closed-loop system is globally attractive.

We can view the system comprising of (4.283) and (4.284) as two subsystems, where the dynamics of subsystem 2 (4.284) consisting of $x_3(t)$ is decoupled from the dynamics of subsystem 1 (4.283) which consists of $x_1(t)$ and $x_2(t)$ and in which the dynamics of $x_3(t)$ is viewed as a disturbance.

In order to prove the global attractivity of the whole system, we use Lemma 4.62. To do so, let us first check all the conditions of the Lemma 4.62. Consider the subsystem 1. Let us define

$$f(\zeta, v) = \begin{pmatrix} x_2 \\ x_3 + \sigma_1(-\tilde{\alpha} x_1 - \beta x_2) \end{pmatrix}, \quad \zeta = \begin{pmatrix} x_1 \\ x_2 \end{pmatrix} \quad \text{and} \quad v = x_3.$$

We then have the dynamics of subsystem 1 as $\dot{\zeta} = f(\zeta, v)$.

Clearly, the origin is globally asymptotically stable for $\dot{\zeta} = f(\zeta, 0)$, since for $v = 0$, it becomes a double-integrator subject to actuator saturation with arbitrary negative linear state feedback control law, for which it is well known that the closed-loop system is globally asymptotically stable. It is also easily verified that f is globally Lipschitz.

In order to apply Lemma 4.62, we must guarantee $\|x_3(t)\| \leqslant \kappa e^{-\lambda t}$ for some λ determined by the dynamics of subsystem 1. First, choose $\gamma > \lambda$.

Obviously, we see that for big initial condition $\|x_3(0)\|$, the subsystem 2 given by (4.284) is subject to actuator saturation at the beginning, $\|x_3(t)\|$ decays linearly up to certain point; once it gets out of saturation region, $\|x_3(t)\|$ decays exponentially fast to zero, with a rate $\gamma > \lambda$. Thus, we can choose κ sufficiently large such that $\|x_3(t)\| \leqslant \kappa e^{-\lambda t}$. We automatically see that all the conditions of Lemma 4.62 are satisfied; therefore, every solution of the first subsystem converges to zero. Thus, the closed-loop system is globally attractive. Hence, we have proved that the closed-loop system is globally asymptotically stable. This completes our first proof.

We proceed next to present the second proof by constructing a Lyapunov function. The Lyapunov approach followed here guarantees stability for all $\tilde{\alpha}, \beta, \gamma > 0$ while the first proof given above proves stability only for γ sufficiently large. This proof demonstrates also the fact that searching for a Lyapunov function even for a simple (low order) linear system subject to actuator saturation is very complicated.

Let us partition $\mathbb{R}^3$ into 4 regions:

$$\begin{aligned}
R_1 &= \left\{(x_1, x_2, x_3) \in \mathbb{R}^3 | x_2 x_3 > 0, |\gamma x_3| > 1\right\}, \\
R_2 &= \left\{(x_1, x_2, x_3) \in \mathbb{R}^3 | x_2 x_3 < 0, |\gamma x_3| > 1\right\}, \\
R_3 &= \left\{(x_1, x_2, x_3) \in \mathbb{R}^3 | x_2 x_3 > 0, |\gamma x_3| < 1\right\}, \\
R_4 &= \left\{(x_1, x_2, x_3) \in \mathbb{R}^3 | x_2 x_3 < 0, |\gamma x_3| < 1\right\}.
\end{aligned}$$

Consider a Lyapunov candidate

$$\begin{aligned}
V(x) = &\int_0^{\tilde{\alpha} x_1 + \beta x_2} \sigma_1(y) dy + \frac{\tilde{\alpha}}{2} x_2^2 \\
&+ \frac{\tilde{\alpha}}{\gamma} \max\{0, x_2 x_3, x_2 x_3 |\gamma x_3|\} + r \max\left\{(\gamma x_3)^2, (\gamma x_3)^4\right\}, \quad (4.285)
\end{aligned}$$

where $r > 0$ is to be suitably selected shortly in subsequent development. We want to show that the Lyapunov candidate shown in (4.285) is indeed a Lyapunov function; thus, the global asymptotic stability of the closed-loop system follows. First, it is easy to see that $V(x)$ is continuous and positive definite. Also, $V(x)$ is radially unbounded.

In order to show global asymptotic stability of the closed-loop system, we need to show that $\dot{V}(x)$ in each region is negative. In regions R_1 and R_2, the system is described by

$$\begin{aligned}
\dot{x}_1 &= x_2\\
\dot{x}_2 &= x_3 + \sigma_1(-\tilde{\alpha}x_1 - \beta x_2)\\
\dot{x}_3 &= -\operatorname{sgn}(\gamma x_3).
\end{aligned}$$

For the region R_1, the evaluation of $\dot{V}(x)$ along the trajectories of the closed-loop system yields

$$\begin{aligned}
\dot{V} &= \sigma_1(\tilde{\alpha}x_1 + \beta x_2)[\tilde{\alpha}x_2 + \beta(x_3 - \sigma_1(\tilde{\alpha}x_1 + \beta x_2))]\\
&\quad + \tilde{\alpha}x_2[x_3 - \sigma_1(\tilde{\alpha}x_1 + \beta x_2)]\\
&\quad + \tilde{\alpha}[x_3 - \sigma_1(\tilde{\alpha}x_1 + \beta x_2)]x_3|x_3|\\
&\quad - 2\tilde{\alpha}x_2x_3 - 4r\gamma|\gamma x_3|^3\\
&= \beta x_3\sigma_1(\tilde{\alpha}x_1 + \beta x_2) - \beta\sigma_1^2(\tilde{\alpha}x_1 + \beta x_2) - \tilde{\alpha}x_2x_3\\
&\quad + \tilde{\alpha}|x_3|^3 - \tilde{\alpha}\sigma_1(\tilde{\alpha}x_1 + \beta x_2)x_3|x_3| - 4r\gamma|\gamma x_3|^3.
\end{aligned}$$

For $|\gamma x_3| > 1$, the following hold:

$$\begin{aligned}
x_3\sigma(\tilde{\alpha}x_1 + \beta x_2) &\leqslant |x_3| \leqslant \frac{1}{\gamma}|\gamma x_3| \leqslant \frac{1}{\gamma}|\gamma x_3|^3,\\
-\sigma_1(\tilde{\alpha}x_1 + \beta x_2)x_3|x_3| &\leqslant |x_3|^2 = \frac{1}{\gamma^2}|\gamma x_3|^2 \leqslant \frac{1}{\gamma^2}|\gamma x_3|^3.
\end{aligned}$$

We then get

$$\dot{V} \leqslant -\beta\sigma_1^2(\tilde{\alpha}x_1 + \beta x_2) - \tilde{\alpha}x_2x_3 - (4r\gamma - \frac{\tilde{\alpha}}{\gamma^2} - \frac{\tilde{\alpha}}{\gamma^3} - \frac{\beta}{\gamma})|\gamma x_3|^3.$$

Choosing r such that

$$4r\gamma^4 > \tilde{\alpha}\gamma + \tilde{\alpha} + \beta\gamma^2, \tag{4.286}$$

we get $\dot{V} < 0$.

For the region R_2, the evaluation of $\dot{V}_2$ along the trajectories of the closed-loop system yields

$$\begin{aligned}
\dot{V} &= \sigma_1(\tilde{\alpha}x_1 + \beta x_2)[\tilde{\alpha}x_2 + \beta(x_3 - \sigma_1(\tilde{\alpha}x_1 + \beta x_2))]\\
&\quad + \tilde{\alpha}x_2[x_3 - \sigma_1(\tilde{\alpha}x_1 + \beta x_2)] - 4r\gamma|\gamma x_3|^3\\
&= \beta x_3\sigma_1(\tilde{\alpha}x_1 + \beta x_2) - \beta\sigma_1^2(\tilde{\alpha}x_1 + \beta x_2) + \tilde{\alpha}x_2x_3\\
&\quad - 4r\gamma|\gamma x_3|^3\\
&\leqslant -\beta\sigma_1^2(\tilde{\alpha}x_1 + \beta x_2) + \tilde{\alpha}x_2x_3 - (4r\gamma - \frac{\beta}{\gamma})|\gamma x_3|^3.
\end{aligned}$$

Choosing r such that

$$4r\gamma^2 > \beta, \tag{4.287}$$

we get $\dot{V} < 0$.

In regions R_3 and R_4, the system is described by

$$\begin{aligned}\dot{x}_1 &= x_2\\ \dot{x}_2 &= x_3 + \sigma_1(\tilde{\alpha}x_1 - \beta x_2)\\ \dot{x}_3 &= -\gamma x_3.\end{aligned}$$

For the region R_3, the evaluation of $\dot{V}_3$ along the trajectories of the closed-loop system yields

$$\begin{aligned}\dot{V} &= \sigma_1(\tilde{\alpha}x_1 + \beta x_2)[\tilde{\alpha}x_2 + \beta(x_3 - \sigma_1(\tilde{\alpha}x_1 + \beta x_2))]\\ &\quad + \tilde{\alpha}x_2[x_3 - \sigma_1(\tilde{\alpha}x_1 + \beta x_2)]\\ &\quad + \frac{\tilde{\alpha}}{\gamma}[x_3 - \sigma_1(\tilde{\alpha}x_1 + \beta x_2)]x_3\\ &\quad - \tilde{\alpha}x_2x_3 - 2r\gamma^3x_3^2\\ &= (\beta - \frac{\tilde{\alpha}}{\gamma})x_3\sigma_1(\tilde{\alpha}x_1 + \beta x_2) - \beta\sigma_1^2(\tilde{\alpha}x_1 + \beta x_2)\\ &\quad + (\frac{\tilde{\alpha}}{\gamma} - 2r\gamma^3)x_3^2.\end{aligned}$$

In this case, let us choose an ϵ small enough such that

$$\frac{\epsilon}{2}|\beta - \frac{\tilde{\alpha}}{\gamma}| \leqslant \frac{\beta}{2}.$$

Next, using

$$|x_3\sigma_1(\tilde{\alpha}x_1 + \beta x_2)| \leqslant \frac{\epsilon}{2}\sigma_1^2(\tilde{\alpha}x_1 + \beta x_2) + \frac{1}{2\epsilon}x_3^2,$$

we get

$$\dot{V} \leqslant -\frac{\beta}{2}\sigma_1^2(\tilde{\alpha}x_1 + \beta x_2) + \left(\frac{1}{2\epsilon}|\beta - \frac{\tilde{\alpha}}{\gamma}| + \frac{\tilde{\alpha}}{\gamma} - 2r\gamma^3\right)x_3^2.$$

Choosing r such that

$$2r\gamma^3 > \frac{1}{2\epsilon}|\beta - \frac{\tilde{\alpha}}{\gamma}| + \frac{\tilde{\alpha}}{\gamma}, \tag{4.288}$$

we get $\dot{V} < 0$.

For the region R_4, the evaluation of $\dot{V}_4$ along the trajectories of the closed-loop system yields

$$\begin{aligned}
\dot{V} &= \sigma_1(\tilde{\alpha}x_1 + \beta x_2)[\tilde{\alpha}x_2 + \beta(x_3 - \sigma_1(\tilde{\alpha}x_1 + \beta x_2))] \\
&\quad + \tilde{\alpha}x_2[x_3 - \sigma_1(\tilde{\alpha}x_1 + \beta x_2)] - 2r\gamma^3 x_3^2 \\
&= \beta x_3 \sigma_1(\tilde{\alpha}_1 + \beta x_2) - \beta\sigma_1^2(\tilde{\alpha}x_1 + \beta x_2) \\
&\quad + \tilde{\alpha}x_2x_3 - 2r\gamma^3 x_3^2 \\
&\leqslant \frac{\beta}{2}[x_3^2 + \sigma_1^2(\tilde{\alpha}x_1 + \beta x_2)] - \beta\sigma_1^2(\tilde{\alpha}x_1 + \beta x_2) \\
&\quad + \tilde{\alpha}x_2x_3 - 2r\gamma^3 x_3^2 \\
&= -\frac{\beta}{2}\sigma_1^2(\tilde{\alpha}x_1 + \beta x_2) + \tilde{\alpha}x_2x_3 - (2r\gamma^3 - \frac{\beta}{2})x_3^2.
\end{aligned}$$

Choosing r such that

$$4r\gamma^3 > \beta, \tag{4.289}$$

we get $\dot{V} < 0$.

Thus, by choosing r sufficiently large such that all the inequalities (4.286)–(4.289) are satisfied, we get $\dot{V}(x) < 0$ for all four different regions. Therefore, $\dot{V}(x)$ is negative for the entire region $\mathbb{R}^3 - \{\mathbf{0}\}$. Thus, the closed-loop system is globally asymptotically stable. Hence, the triple-integrator subject to actuator saturation can be globally asymptotically stabilized via linear static state feedback control laws. ■

4.7.3 Discrete-time equivalent of the double integrator

There exists a vast difference between a discrete-time double integrator subject to actuator saturation and its counterpart in continuous time. In continuous time, any linear state feedback control law which locally stabilizes the double integrator, also globally stabilizes the double integrator in the presence of actuator saturation. In discrete time, the equivalent of the double integrator has intrinsically different behavior. Some linear feedbacks which locally stabilize the double integrator, also globally stabilize the double integrator in the presence of actuator saturation. However, some other linear feedbacks which locally stabilize the double integrator, do not globally stabilize the double integrator in the presence of actuator saturation. In this subsection, we classify a large class of linear feedback laws which achieve global stability. But, on the other hand, we also classify a class of linear feedback laws which achieve local stability but yield nonzero periodic solutions in the presence of actuator saturation and therefore cannot achieve global asymptotic stability. This subsection is based on the recent work of [207, 208].

Consider a discrete-time double integrator subject to actuator saturation described by

$$\begin{cases} x_1(k+1) = x_1(k) + x_2(k) \\ x_2(k+1) = x_2(k) + \sigma(u(k)), \end{cases} \tag{4.290}$$

where, as usual, $\sigma(u(k))$ is the standard saturation function

$$\sigma(u(k)) = \operatorname{sgn}(u(k)) \min\{1, |u(k)|\},$$

and

$$u(k) = f_1 x_1(k) + f_2 x_2(k) \tag{4.291}$$

with f_1 and f_2 as feedback gains.

Let us first consider the system (4.290) with the feedback control law (4.291) in the absence of actuator saturation. From Jury's test, we see that any feedback control law (4.291) with f_1 and f_2 which satisfy the following conditions:

$$\begin{aligned} f_1 &< 0 \\ f_1 - 2f_2 &< 4 \\ |f_2 - f_1 + 1| &< 1 \\ 1 - (f_2 - f_1 + 1)^2 &> |(f_2 + 2)(f_1 - f_2)| \end{aligned}$$

or, equivalently, which satisfy the condition

$$\frac{1}{2} f_1 - 2 < f_2 < f_1 < 0, \tag{4.292}$$

stabilizes the system (4.290) in the absence of actuator saturation or, in other words, achieves local asymptotic stability for the closed-loop system.

The question is whether such feedback control laws also globally stabilize the system (4.290) subject to actuator saturation. The motivation for studying the discrete-time double integrator subject to actuator saturation comes from the well-known result on its counterpart in continuous-time setting (see previous subsection). For contrast, let us first clearly state the result for the continuous-time double integrator subject to actuator saturation.

Theorem 4.64 *Consider the system*

$$\begin{cases} \dot{x}_1 = x_2 \\ \dot{x}_2 = \sigma(f_1 x_1 + f_2 x_2), \end{cases} \tag{4.293}$$

where f_1 and f_2 are constants. For any $f_1, f_2 < 0$, the system (4.293) *is globally asymptotically stable.*

From the above theorem, we see that an arbitrary linear state feedback control law which stabilizes a continuous-time double integrator in the absence of actuator saturation (which is equivalent to the conditions $f_1, f_2 < 0$) also globally stabilizes the system subject to actuator saturation. Then, an important question that arises is the following: Is this also true for a discrete-time double integrator subject to actuator saturation? As shown in this subsection, the answer to this question is indeed negative. In this regard, we first show that the system (4.290) is not globally stabilized for all feedback control laws (4.291) which locally stabilize the system (4.290).

Theorem 4.65 *If f_1 and f_2 satisfy Jury's condition* (4.292) *plus the following condition:*

$$f_2 > \tfrac{3}{2} f_1, \tag{4.294}$$

then the closed-loop system exhibits nonzero limit cycles, that is, there exist initial conditions that yield nonzero periodic solutions; hence, the closed-loop system is not globally asymptotically stable.

The above theorem considers the case $f_2 > \frac{3}{2} f_1$. An obvious question that arises next is "what can we expect if $f_2 < \frac{3}{2} f_1$?". The following theorem answers this question:

Theorem 4.66 *If f_1 and f_2 satisfy Jury's condition* (4.292) *plus the following condition:*

$$f_2 < \tfrac{3}{2} f_1, \tag{4.295}$$

then the closed-loop system is globally asymptotically stable.

The above results can be illustrated by Fig. 4.2. Note that in Fig. 4.2, line AB is $f_2 = f_1$, line BC is $f_2 = \frac{1}{2} f_1 - 2$, line AD is $f_2 = \frac{3}{2} f_1$, and line AC is $f_1 = 0$. Jury's test establishes that whenever f_1 and f_2 take their values within the triangle ABC, the closed-loop system is locally asymptotically stable, otherwise unstable. The triangle ABC can be bisected into two regions, triangle ABD (Region II) and triangle ADC (Region III). As shown by Theorem 4.65, whenever f_1 and f_2 take their values within the triangle ABD, there exist initial conditions that lead to nonzero periodic solutions, and hence, global asymptotic stability of the closed-loop system is impossible. On the other hand, as shown by Theorem 4.66, whenever f_1 and f_2 take their values within the triangle ADC, the closed-loop system is globally asymptotically stable. Theorem 4.65 can be illustrated by the following example:

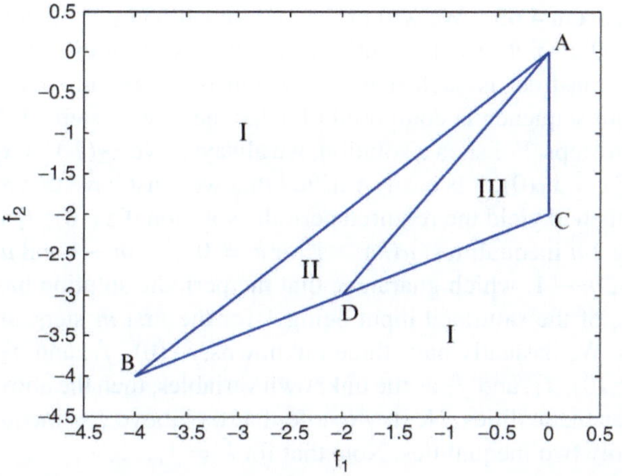

Figure 4.2: Stability characteristics as a function of f_1 and f_2

Example 4.67 Consider the system (4.290) with a state feedback control law (4.291) with gain parameters $f_1 = -1$ and $f_2 = -1.2$. The system has a periodic solution of period $T = 56$ for the following initial conditions: $3.6 \leqslant x_1(0) \leqslant 10.4$ and $x_2(0) = -14$. The state trajectory for $x_1(0) = 10.4$ and $x_2(0) = -14$ is given in Fig. 4.3, where we clearly see the symmetric periodic orbit. Note that the state trajectory moves clockwise along the periodic orbit shown in Fig. 4.3.

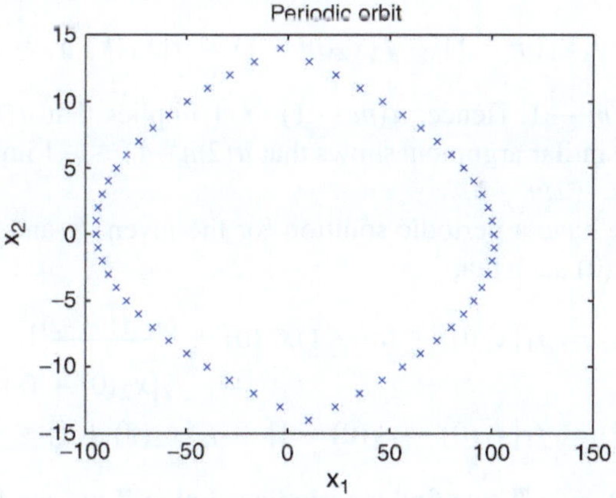

Figure 4.3: Periodic orbit of period 56

The proofs of the above theorems are considered next with the proofs of two technical lemma diverted to an appendix.

Proof of Theorem 4.65 : We will prove Theorem 4.65 by explicitly constructing nonzero periodic solutions. The periodic solution with an even period $T = 2m$ that we will construct is such that the system is always in saturation, and the saturated input sequence is composed of 1 for the first m steps, followed by -1 for the next m steps. For such a solution, we always have $x_2(T) = x_2(0)$. In order to have $x_1(T) = x_1(0)$, it is easily verified that we must have that $x_2(0) = -\frac{m}{2}$.

Clearly, this will yield the required periodic solution if $x_1(0)$, f_1 and f_2 satisfy the following $2m$ inequalities: $u(k) \geqslant 1$ for $k = 0, \dots, m-1$ and $u(k) \leqslant -1$ for $k = m, \dots, 2m-1$, which guarantee that the periodic solution has the required characteristic of the saturated input being 1 for the first m steps and -1 for the next m steps. We basically have three unknowns, $x_1(0)$, f_1, and f_2. However, if we view $f_1x_1(0)$, f_1, and f_2 as the unknown variables, then the above inequalities become linear inequalities. Next, we note that the above $2m$ inequalities can be reduced to only two inequalities. Note that for $k = 1, \dots, m$,

$$\begin{aligned} x_1(k) &= x_1(0) + kx_2(0) + \tfrac{k(k-1)}{2} \\ x_2(k) &= x_2(0) + k \end{aligned}$$

since by construction the input saturate to 1 for the first m steps. We then note that Jury's conditions (4.292) imply that $f_2 < f_1$ and $f_1 < 0$. This implies that $f_2 < f_1 - \frac{1}{2}kf_1$ for $k = 0, \dots, m-1$. This yields $f_1x_2(0) + f_2 < \frac{1}{2}f_1(-m-k+2)$ since $x_2(0) = -\frac{m}{2}$. We know $m-k-1 \geqslant 0$ and multiplying the above inequality on both sides with $m-k-1$ yields

$$f_1(m-k-1)x_2(0) + f_2(m-k-1) \leqslant f_1\left[\tfrac{k(k-1)}{2} - \tfrac{(m-1)(m-2)}{2}\right].$$

This is equivalent to

$$u(m-1) = f_1x_1(m-1) + f_2x_2(m-1) \leqslant f_1x_1(k) + f_2x_2(k) = u(k)$$

for $k = 0, \dots, m-1$. Hence, $u(m-1) \geqslant 1$ implies that $u(k) \geqslant 1$ for $k = 0, \dots, m-1$. A similar argument shows that $u(2m-1) \leqslant -1$ implies that $u(k) \leqslant -1$ for $k = m, \dots, 2m-1$.

Therefore, we have a periodic solution for the given f_1 and f_2 provided that there exists a $x_1(0)$ such that

$$\begin{aligned} u(m-1) &= f_1[x_1(0) + (m-1)x_2(0) + \tfrac{(m-1)(m-2)}{2}] \\ &\qquad + f_2[x_2(0) + (m-1)] \geqslant 1 \\ u(2m-1) &= f_1[x_1(0) - x_2(0) - 1] + f_2[x_2(0) + 1] \leqslant -1. \end{aligned}$$

Using that $x_2(0) = -\frac{m}{2}$, we find a periodic solution if we can find a $x_1(0)$ such that the following two inequalities are satisfied:

$$\begin{aligned} f_1[x_1(0) - (m-1)] + f_2[\tfrac{m}{2} - 1] &\geqslant 1 \\ f_1[x_1(0) + \tfrac{m}{2} - 1] + f_2[-\tfrac{m}{2} + 1] &\leqslant -1, \end{aligned}$$

which is equivalent to

$$1 + f_1(m-1) - f_2[\tfrac{m}{2}-1] \leqslant f_1 x_1(0) \leqslant -1 - f_1[\tfrac{m}{2}-1] + f_2[\tfrac{m}{2}-1]. \quad (4.296)$$

Clearly, a suitable $x_1(0)$ exists if and only if

$$1 + f_1(m-1) - f_2[\tfrac{m}{2}-1] \leqslant -1 - f_1[\tfrac{m}{2}-1] + f_2[\tfrac{m}{2}-1].$$

This implies that

$$f_1(3m-4) - f_2(2m-4) \leqslant -4,$$

which, for $m > 2$, is equivalent to

$$\tfrac{3m-4}{2m-4} f_1 + \tfrac{2}{m-2} \leqslant f_2. \quad (4.297)$$

From (4.294), it is clear that

$$\lim_{m\to\infty} (\tfrac{3m-4}{2m-4} f_1 + \tfrac{2}{m-2}) = \tfrac{3}{2} f_1 \leqslant f_2.$$

Therefore, for any f_1, f_2 which satisfy Jury's condition (4.292) and the additional condition (4.294), there exists a m sufficiently large such that (4.297) is satisfied. But in the above, we have seen that this implies that the system (4.290) with a feedback control law (4.291) exhibits periodic behavior for certain initial conditions with period $2m$. Hence, the system (4.290) can never be globally asymptotically stabilized by the feedback control law (4.291) if f_1 and f_2 satisfy (4.292) and (4.294). ∎

In order to prove Theorem 4.66, we need to establish asymptotic stability in the region III depicted in Fig. 4.2. A basis transformation turns out to be useful for establishing this result. We define $y_1(k) = u(k)$ and $y_2(k) = f_1 x_2(k)$. The closed-loop system is then given by

$$\begin{cases} y_1(k+1) = y_1(k) + y_2(k) + f_2\sigma(y_1(k)) \\ y_2(k+1) = y_2(k) + f_1\sigma(y_1(k)). \end{cases} \quad (4.298)$$

We sometimes denote

$$y(k) = \begin{pmatrix} y_1(k) \\ y_2(k) \end{pmatrix},$$

and y, y_1, or y_2 without explicitly indicating time will refer to $y(k)$, $y_1(k)$, or $y_2(k)$, respectively. We first establish asymptotic stability for the region IV, as depicted in Fig. 4.4, by constructing a classical Lyapunov function.

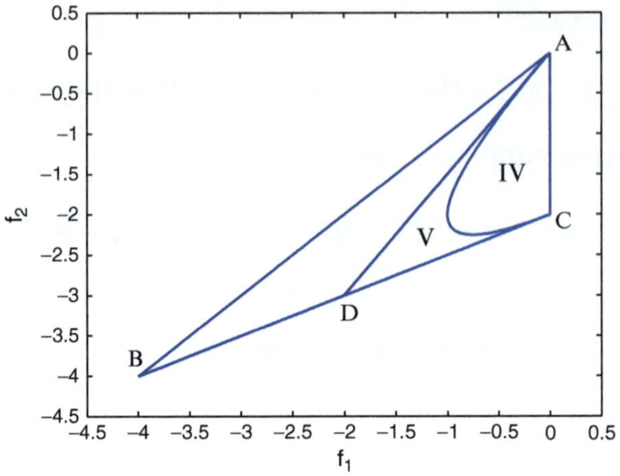

Figure 4.4: Stability characteristics as a function of f_1 and f_2

Lemma 4.68 *Consider system* (4.290) *with a feedback control law* (4.291)*. If feedback gains* f_1 *and* f_2 *satisfy Jury's condition* (4.292) *and the following condition:*

$$(f_2 - f_1 + 1)^2 - 1 < f_1, \tag{4.299}$$

then the closed-loop system is globally asymptotically stable.

Proof : Without loss of generality, we write the closed-loop system as (4.298). Next, we consider the following Lyapunov candidate:

$$V_k = V(y(k)) = 2y_1(k)\sigma(y_1(k)) - \sigma^2(y_1(k)) - 2\sigma(y_1(k))y_2(k) - \tfrac{1}{f_1}y_2^2(k).$$

Then with some algebra, we get

$$\begin{aligned} V_{k+1} = 2y_1\sigma(\tilde{y}_1) + 2(f_2 - f_1)\sigma(y_1)\sigma(\tilde{y}_1) - \sigma(\tilde{y}_1)^2 - \tfrac{1}{f_1}y_2^2 \\ - 2y_2\sigma(y_1) - f_1\sigma^2(y_1), \end{aligned}$$

where to simplify notation we have used $\tilde{y}_1 = y_1(k+1)$, while $y_1(k)$ and $y_2(k)$ are abbreviated to y_1 and y_2, respectively. We find that

$$\begin{aligned} \Delta V = V_{k+1} - V_k = 2y_1\sigma(\tilde{y}_1) + 2(f_2 - f_1)\sigma(y_1)\sigma(\tilde{y}_1) - \sigma(\tilde{y}_1)^2 \\ - 2y_1\sigma(y_1) - (f_1 - 1)\sigma^2(y_1). \end{aligned}$$

Next, we show that $\Delta V \leqslant 0$ by considering three different cases: case I: $y_1 \geqslant 1$, case II: $y_1 \leqslant -1$, and case III: $|y_1| < 1$.

In case I, using condition (4.299), we get

$$\begin{aligned}
\Delta V &= 2y_1\sigma(\tilde{y}_1) + 2(f_2 - f_1)\sigma(\tilde{y}_1) - \sigma(\tilde{y}_1)^2 - 2y_1 - (f_1 - 1) \\
&= 2(y_1 - 1)(\sigma(\tilde{y}_1) - 1) - (\sigma(\tilde{y}_1) - f_2 + f_1 - 1)^2 + (f_2 - f_1 + 1)^2 \\
&\qquad - (f_1 + 1) \\
&\leqslant (f_2 - f_1 + 1)^2 - (f_1 + 1) < 0.
\end{aligned}$$

In case II, using condition (4.299), we get:

$$\begin{aligned}
\Delta V &= 2y_1\sigma(\tilde{y}_1) - 2(f_2 - f_1)\sigma(\tilde{y}_1) - \sigma(\tilde{y}_1)^2 + 2y_1 - (f_1 - 1) \\
&= 2(y_1 + 1)(\sigma(\tilde{y}_1) + 1) - (\sigma(\tilde{y}_1) + f_2 - f_1 + 1)^2 + (f_2 - f_1 + 1)^2 \\
&\qquad - (f_1 + 1) \\
&\leqslant (f_2 - f_1 + 1)^2 - (f_1 + 1) < 0.
\end{aligned}$$

Finally, in case III, using (4.299), we get

$$\begin{aligned}
\Delta V &= 2(f_2 - f_1 + 1)y_1\sigma(\tilde{y}_1) - \sigma(\tilde{y}_1)^2 - (f_1 + 1)y_1^2 \\
&= -[\sigma(\tilde{y}_1) - (f_2 - f_1 + 1)y_1]^2 + [(f_2 - f_1 + 1)^2 - (f_1 + 1)]y_1^2 \leqslant 0.
\end{aligned}$$

We also see that equality holds only if $y_1 = 0$ and $\tilde{y}_1 = 0$ which implies that $x_1 = x_2 = 0$. Hence, the global asymptotic stability of the closed-loop system follows. ■

In order to prove Theorem 4.66, it remains to show that the closed-loop system is globally asymptotically stable in region V as depicted in Fig. 4.4. As before we assume the closed-loop system is given by (4.298). Next, let us consider a Lyapunov candidate in the presence of saturation, which is based on the linearized system as follows:

$$\begin{aligned}
V_k = V(y(k)) &= 2y_1(k)\sigma(y_1(k)) - \sigma^2(y_1(k)) \\
&\qquad + 2b\sigma(y_1(k))y_2(k) - \tfrac{1}{f_1}y_2^2(k). \qquad (4.300)
\end{aligned}$$

We choose

$$b = \begin{cases} \frac{2}{f_2} & f_2^2 + 4f_1 \geqslant 0 \\ -\frac{f_2}{2f_1} & f_2^2 + 4f_1 < 0. \end{cases} \qquad (4.301)$$

It is easily verified that in the triangle ADC of Fig. 4.4, we have $b \in [-1, -0.5)$ while for $b = -1$ we get the Lyapunov function used in Lemma 4.68, and in region V of Fig. 4.4, we have $b \in (-1, -0.5)$. We sometimes refer to the first case, when $f_2^2 + 4f_1 \geqslant 0$, as the real case since in that case the linearized system has real eigenvalues while the second case, when $f_2^2 + 4f_1 < 0$, is referred to as the complex case since in that case the linearized system has complex eigenvalues.

It is easy to see that the Lyapunov candidate (4.300) works for the linearized closed-loop system. In order to be a valid Lyapunov function, it is necessary that it must work when $\sigma(y_1)$ stays at 1 or at -1 in two consecutive time instants. It is easy to verify that in that case,

$$\Delta V = (2b-1)f_1 + 2f_2, \tag{4.302}$$

where $(\Delta V)(k) = V_{k+1} - V_k$, while $V_k = V(y(k))$. Thus, $\Delta V = (f_2^2 + 4f_1)/f_2 + (f_2 - f_1) < 0$ in the real case while $\Delta V = f_2 - f_1 < 0$ in the complex case.

Therefore, the Lyapunov candidate (4.300) has the required properties when $\sigma(y_1)$ is in saturation for two consecutive time instants or is out of saturation for two consecutive time instants. Note that for a continuous-time problem, we would be done since y_1 is continuous. However, for discrete-time systems, y_1 obviously jumps from one time to the other, and hence, if $\sigma(y_1(k))$ saturates, then it might well be that $\sigma(y_1(k+1))$ is out of saturation or conversely. This is intrinsically different from the continuous-time case. Thus, we have to show that the Lyapunov candidate (4.300) also decreases when y_1 jumps. The traditional Lyapunov argument is to show that $V_{k+1} - V_k < 0$ for all initial conditions. However, this approach does not work here. For the real case, if $f_2 < -2$, there exist initial conditions such that $V_{k+1} - V_k > 0$. A similar problem can arise in the complex case. Thus, we need a different technique. The main idea is to show that V decreases over a specifically chosen number of time steps, and V is bounded in the interim. In order to proceed with this idea, we first choose suitable time instants k_i. The formal definition of k_i is given next.

Definition 4.69 *$k_0 = 0$, and k_i is the smallest integer larger than k_{i-1}, such that either*

- $|y_1(k_i)| < 1$*; or*
- $y_1(k_i)y_1(k_i+1) < 0$ *and* $|y_1(k_i+1)| \geq 1$.

In other words, k_i is defined as the first time instant $k > k_{i-1}$ where $y_1(k)$ either gets out of saturation or where $y_1(k)$ switches the sign. It is easily seen that k_i is well defined given k_{i-1} since the only way k_i would not be well defined is if $y_1(k) > 1$ for all $k > k_{i-1}$ or if $y_1(k) < -1$ for all $k > k_{i-1}$. It is easily seen from the dynamics (4.298) that this is not possible. Instead of a classical Lyapunov design, we will study whether V_{k_i} is decreasing as a function of i. Before we formally prove Theorem 4.66, we present two crucial lemmas.

Lemma 4.70 *Let the Lyapunov candidate V be defined in* (4.300) *and assume that the feedback gains f_1 and f_2 are in region V of Fig. 4.4, that is,* (4.292) *and* (4.295) *are satisfied and*

$$(f_2 - f_1 + 1)^2 - 1 > f_1. \tag{4.303}$$

In that case, if $|y_1(k_i)| \leqslant 1$ and $V_{k_{i-1}} \neq 0$, then

$$V_{k_i} - V_{k_{i-1}} < 0.$$

Proof : Since the proof is very lengthy, for the readability, we give the proof in Appendix 4.A. ■

Lemma 4.71 *Let the Lyapunov candidate V be defined in* (4.300) *and assume that the feedback gains f_1 and f_2 are in region V of Fig. 4.4, that is,* (4.292) *and* (4.295) *are satisfied and*

$$(f_2 - f_1 + 1)^2 - 1 > f_1.$$

In that case, if $|y_1(k_i)| \geqslant 1$ and $V_{k_{i-1}} \neq 0$, then

$$V_{k_i} - V_{k_{i-1}} < 0 \quad \textit{or} \quad V_{k_{i+1}} - V_{k_{i-1}} < 0.$$

Proof : Since the proof is very lengthy, for the readability, we give the proof in Appendix 4.B. ■

Remark 4.72 *Note that if the feedback gains f_1 and f_2 take their values within the triangle ABD (region II) in Fig. 4.2, there actually exist initial conditions for which $V_{k_{i+1}} - V_{k_{i-1}} = 0$ since $k_{i+1} - k_{i-1}$ is precisely the period of the periodic behavior as constructed in the proof of Theorem 4.65.*

Proof of Theorem 4.66 : We already know that the system is locally asymptotically stable from Jury's test. It remains to show global attractivity of the origin. If (4.299) is satisfied, Lemma 4.68 guarantees global asymptotic stability. Therefore, we only need to consider the case where (4.303) is satisfied in addition to (4.292) and (4.295).

We first note that (4.300) can be rewritten as

$$V(y) = 2\sigma(y_1)\left[y_1 - \sigma(y_1)\right] + \left[\sigma(y_1) + by_2\right]^2 - \left(b^2 + \tfrac{1}{f_1}\right) y_2^2.$$

It is easy to show that for $f_2^2 + 4f_1 \neq 0$, we have $b^2 + \frac{1}{f_1} > 0$. This immediately implies $V(y) > 0$ if $y \neq 0$. Lemmas 4.70 and 4.71 imply that either $V_{k_{i+1}} - V_{k_i} < 0$ or $V_{k_{i+2}} - V_{k_i} < 0$ if $y(k_i) \neq 0$. This results in a sequence $\{\bar{k}_i\}$ such that $V_{\bar{k}_{i+1}} < V_{\bar{k}_i}$ for all i. This clearly implies that $V_{\bar{k}_i}$ is bounded, and hence, $\bar{k}_{i+1} - \bar{k}_i$ is bounded as well. This implies that $V_{\bar{k}_i} \to 0$ as $i \to \infty$. Local asymptotic stability implies that if $V_{\bar{k}_i}$ is small enough for some i, then $y(k) \to 0$ as $k \to \infty$, and therefore, we have global attractivity.

For the case that $b^2 + \frac{1}{f_1} = 0$, we have $V(y) \geqslant 0$, and $V(y) = 0$ implies $y_1 + by_2 = 0$ and $y_1 \in [-1, 1]$. Lemmas 4.70 and 4.71 imply that either $V_{k_{i+1}} - V_{k_i} < 0$ or $V_{k_{i+2}} - V_{k_i} < 0$ if $V_{k_i} \neq 0$. Similar to as before, we can show that $V(k) \to 0$ as $k \to \infty$. Let $\mathcal{X}$ denote the compact set of $y \in \mathbb{R}^2$ for which $V(y) = 0$. Then, it is easily verified that $y(k) \in \mathcal{X}$ implies that $y(k+1) \in \mathcal{X}$ and, moreover, $y(k_0) \in \mathcal{X}$ implies that $y(k) \to 0$ as $k \to \infty$. Then a minor variation of the classical LaSalle argument implies that the system is globally attractive. ■

4.8 H_2 and H_∞ low-gain theory

We understand from previous sections that, whenever it is feasible, semi-global stabilization of linear systems subject to input saturation can be achieved by an appropriate low-gain state feedback. So far, we presented in this chapter three distinct methods of low-gain design, namely, direct eigenstructure assignment method and continuous-time H_2 and H_∞ algebraic Riccati equation (ARE)-based methods. Recently, another method of low-gain design was developed based on a parametric Lyapunov equation [215, 216]. Although these four methods of low-gain design were proposed and developed independently in literature, it turns out that they are all rooted in and can be unified under two fundamental control theories, H_2 and H_∞ theory. Our intension here is to show this. Also, all the above four methods consider only the case where low-gains are demanded by all the input channels. Hence, they all require the asymptotic null controllability with bounded input (ANCBC) of the given system. Here, in this section, we introduce the concept of H_2 and H_∞ low gains in a general setting where only some or all the input channels are engaged with low gain. Then, we provide explicit existence conditions and design methods which yield the classical ANCBC condition and the aforementioned four design methods as special cases.

Let us emphasize that the aforementioned unification via H_2 and H_∞ theory not only brings the existing low-gain methods together but also reveals the interconnections between them. Also, by making explicit the connection between the proposed low-gain design methodologies and stabilization of linear systems subject to saturation, it also gives the necessary and sufficient conditions for semi-global stabilization of linear systems when only some but not necessarily all the input channels are subject to saturation; this aspect has never been considered in the literature. This section follows the recent work of [202] and [201].

Throughout this section, we consider the following linear time invariant system:

$$\Sigma : \begin{cases} \rho x = Ax + Bu \\ z = Du, \end{cases} \tag{4.304}$$

where $x \in \mathbb{R}^n$, $u \in \mathbb{R}^m$ and $z \in \mathbb{R}^{m_0}$. Here, the variable z represents a desired variable that can be constrained as required. Without loss of generality, we assume that

$$D = \begin{pmatrix} I_{m_0} & 0 \end{pmatrix}.$$

This is because for a general D, there always exist non-singular matrices U and V such that $UDV = (I_{m_0} \quad 0)$.

In what follows, a state feedback gain such as F_ε parameterized in a parameter ε is called a gain sequence since as ε changes, one obtains a sequence of gains. We define below formally what we mean by H_2 and H_∞ low-gain sequences.

Definition 4.73 *For the system Σ in* (4.304)*, the **H_2 low-gain sequence** is a sequence of parameterized static state feedback gains F_ε for which there exists an ε^* such that for any $\varepsilon \in (0, \varepsilon^*]$ the following properties hold:*

(i) For continuous-time case, there exists a M such that $\|F_\varepsilon\| \leqslant M$; no such bound is needed for discrete-time case.

(ii) $A + BF_\varepsilon$ is Hurwitz stable for continuous time or Schur stable for discrete time.

(iii) For any $x(0) \in \mathbb{R}^n$, the closed-loop system with $u = F_\varepsilon x$ satisfies

$$\lim_{\varepsilon \to 0} \|z\|_2 = 0.$$

The H_∞ low-gain sequence will depend on an a priori given data γ; hence, we define it as a γ-level H_∞ low-gain sequence to explicitly indicate such a dependence. When we refer to H_∞ low-gain sequence, we always imply γ-level H_∞ low-gain sequence.

Definition 4.74 *For the system Σ in* (4.304) *and an arbitrary given $E \in \mathbb{R}^{n \times p}$, define an auxiliary system,*

$$\Sigma_\infty : \begin{cases} \rho x = Ax + Bu + E\omega \\ z = Du, \end{cases} \tag{4.305}$$

and the infimum

$$\gamma^* = \inf_F \left\{ \|DF(sI - A - BF)^{-1}E\|_\infty \mid \lambda(A + BF) \in C^- \right\} \tag{4.306}$$

for continuous time, or the infimum

$$\gamma^* = \inf_F \{\|DF(zI - A - BF)^{-1}E\|_\infty \mid \lambda(A + BF) \in \mathbb{C}^\ominus\} \tag{4.307}$$

for discrete-time case.

*For the system Σ in (4.304), and for a $\gamma > \gamma^*$, the **γ-level H_∞ low-gain sequence** is a sequence of parameterized static state feedback gains $F_\varepsilon(E,\gamma)$ for which there exists an ε^* such that for any $\varepsilon \in (0, \varepsilon^*]$ the following properties hold:*

(i) *For continuous-time case, there exists a M such that $\|F_\varepsilon(E,\gamma)\| \leqslant M$; no such bound is needed for discrete-time case.*

(ii) *$A + BF_\varepsilon(E,\gamma)$ is Hurwitz stable for continuous time or Schur stable for discrete time.*

(iii) *Consider a signal ω in L_2 space (continuous time) or in ℓ_2 space (discrete time). For the system Σ_∞ with any $x(0) \in \mathbb{R}^n$,*

$$\lim_{\varepsilon \to 0} \left\{ \sup_\omega (\|z\|_2^2 - \gamma\|\omega\|_2^2) \right\} = 0.$$

Remark 4.75 *Unlike continuous time, for discrete-time systems, we do not require boundedness of F_ε or $F_\varepsilon(E,\gamma)$ as a function of ε. However, from the perspective of applications, a bounded F_ε or $F_\varepsilon(E,\gamma)$ is desirable and in fact can always be constructed.*

4.8.1 *Properties of H_2 and H_∞ low-gain sequences*

The first theorem given below shows the relationship between the H_2 and H_∞-γ-level low-gain sequences.

Theorem 4.76 *For the system Σ in (4.304) with a given $E \in \mathbb{R}^{n\times p}$ and a $\gamma > \gamma^*$ where γ^* is as defined in (4.306) for continuous-time case or in (4.307) for discrete-time case, a sequence of feedback gains $F_\varepsilon(E,\gamma)$ is a γ-level H_∞ low-gain sequence only if it is an H_2 low-gain sequence.*

Proof : By setting $\omega = 0$ in the definition of $H_\infty\gamma$-level low-gain sequence, we immediately conclude this result. ■

Remark 4.77 *The inverse of Theorem 4.76 is not true. For any given E, we can always construct a γ_1-level H_∞ low-gain sequence with $\gamma_1 > \gamma$ which, according to Theorem 4.76, is an H_2 low-gain sequence but not a γ-level H_∞ low-gain sequence.*

Theorem 4.76 can be visualized by the Venn diagram in Fig. 4.5. The next theorem shows that for the closed-loop consisting of system Σ in (4.304) and either an H_2 low-gain controller $u = F_\varepsilon x$ or an H_∞ low-gain controller $u = F_\varepsilon(E, \gamma)x$, the magnitude of z and DF_ε or $DF_\varepsilon(E, \gamma)$ can be made arbitrarily small.

Theorem 4.78 *The closed-loop system comprising* (4.304) *and either* $u = F_\varepsilon x$ *or* $u = F_\varepsilon(E, \gamma)x$ *satisfies the following properties:*

(i) $\lim_{\varepsilon\to 0} \|z\|_\infty = 0$.

(ii) $\lim_{\varepsilon\to 0} DF_\varepsilon = 0$ *and* $\lim_{\varepsilon\to 0} DF_\varepsilon(E, \gamma) = 0$.

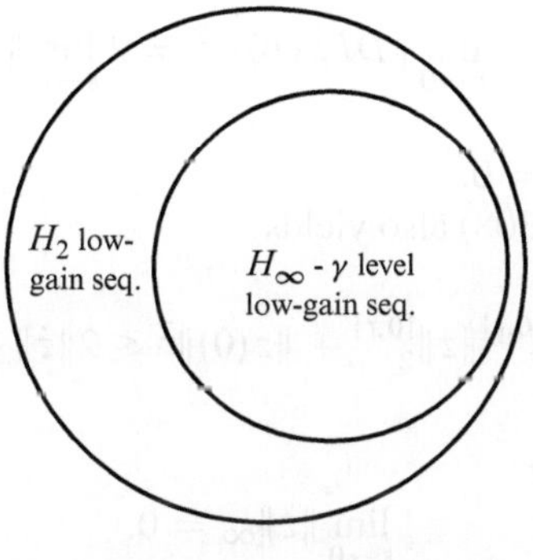

Figure 4.5: Venn diagram

Proof : Owing to Theorem 4.76, we only need to prove these two properties for an H_2 low-gain sequence.

Let us first consider the continuous-time case. The fact that $\|z\|_2 \to 0$ as $\varepsilon \to 0$ for any $x(0)$ implies that

$$\lim_{\varepsilon\to 0} \| F_\varepsilon e^{(A+BF_\varepsilon)t} \| = 0.$$

Since $\|F_\varepsilon\|$ is bounded for all $\varepsilon \in (0, \varepsilon^*]$, $\|A + BF_\varepsilon\|$ is also bounded for all $\varepsilon \in (0, \varepsilon^*]$. We have

$$\lim_{\varepsilon\to 0} \| F_\varepsilon(A + BF_\varepsilon)e^{(A+BF_\varepsilon)t} \| = 0.$$

This implies that $\dot{z} \in L_2$, and moreover, $\lim_{\varepsilon\to 0} \|\dot{z}\|_2 = 0$. Applying Cauchy-Schwartz inequality as

$$\begin{aligned} \left| \|z(t)\|^2 - \|z(0)\|^2 \right| &= 2\left| \int_0^t \dot{z}(\tau)' z(\tau) \mathrm{d}\tau \right| \\ &\leqslant 2\left(\int_0^t \|\dot{z}(\tau)\|^2 \mathrm{d}\tau \right)^{1/2} \left(\int_0^t \|z(\tau)\|^2 \mathrm{d}\tau \right)^{1/2}, \end{aligned} \tag{4.308}$$

we find that

$$\|z(0)\|^2 \leqslant 2\|\dot{z}\|_2^{[0,t]} \|z\|_2^{[0,t]} + \|z(t)\|_2.$$

Let ε be fixed and $t \to \infty$. Since $A + BF_\varepsilon$ is Hurwitz, $\|z(t)\| \to 0$. We then have

$$\|z(0)\|^2 \leqslant 2\|\dot{z}\|_2 \|z\|_2.$$

Then let $\varepsilon \to 0$. We conclude that for any $x(0) \in \mathbb{R}^n$,

$$\lim_{\varepsilon\to 0} \|z(0)\|^2 = \lim_{\varepsilon\to 0} \|DF_\varepsilon x(0)\|^2 = 2 \lim_{\varepsilon\to 0} \|\dot{z}\|_2 \|z\|_2 = 0,$$

and hence, $\lim_{\varepsilon\to 0} DF_\varepsilon = 0$.

On the other hand, (4.308) also yields

$$\|z(t)\|^2 \leqslant 2\|\dot{z}\|_2^{[0,t]} \|z\|_2^{[0,t]} + \|z(0)\|^2 \leqslant 2\|\dot{z}\|_2 \|z\|_2 + \|z(0)\|^2.$$

Therefore,

$$\lim_{\varepsilon\to 0} \|z\|_\infty = 0.$$

Let us next consider the discrete-time case. Since

$$\|z\|_\infty \leqslant \|z\|_2,$$

the fact that $\|z\|_2 \to 0$ as $\varepsilon \to 0$ for any $x(0)$ immediately yields (i). Moreover,

$$\|z(0)\| = \|DF_\varepsilon x(0)\| \leqslant \|z\|_\infty.$$

Therefore, $\|z\|_\infty \to 0$ as $\varepsilon \to 0$ for any $x(0)$ implies that $\|DF_\varepsilon\| \to 0$ as $\varepsilon \to 0$. ■

Remark 4.79 *If F_ε is not bounded for continuous time, the above theorem is not true in general. Consider the example of a system*

$$\begin{cases} \begin{pmatrix} \dot{x}_1 \\ \dot{x}_2 \end{pmatrix} = \begin{pmatrix} 0 & 1 \\ 0 & 0 \end{pmatrix} \begin{pmatrix} x_1 \\ x_2 \end{pmatrix} + \begin{pmatrix} 0 \\ 1 \end{pmatrix} u_1 + \begin{pmatrix} 1 \\ 0 \end{pmatrix} u_0 \\ z \quad = u_0. \end{cases}$$

Choosing $u_0 = \varepsilon x_2$ yields

$$\begin{cases} \begin{pmatrix} \dot{x}_1 \\ \dot{x}_2 \end{pmatrix} = \begin{pmatrix} 0 & 1+\varepsilon \\ 0 & 0 \end{pmatrix} \begin{pmatrix} x_1 \\ x_2 \end{pmatrix} + \begin{pmatrix} 0 \\ 1 \end{pmatrix} u_1 \\ z \quad = \varepsilon x_2. \end{cases}$$

Choose $u_1 = -\frac{1}{(1+\varepsilon)\varepsilon^2} x_1 - \frac{1}{\varepsilon} x_2$. Define

$$y_1(\tilde{t}) = \varepsilon x_1(t), \quad y_2(\tilde{t}) = (1+\varepsilon)\varepsilon^2 x_2(t) \text{ and } \tilde{t} = \tfrac{1}{\varepsilon}.$$

The closed-loop system in the new coordinates and in the new time scale is given by

$$\begin{pmatrix} \dot{y}_1 \\ \dot{y}_2 \end{pmatrix} = \begin{pmatrix} 0 & 1 \\ -1 & -1 \end{pmatrix} \begin{pmatrix} y_1 \\ y_2 \end{pmatrix}. \tag{4.309}$$

We first verify that the controller (u_0, u_1) is an H_2 low-gain controller. Note that

$$\|y(0)\| \leqslant \varepsilon \|x(0)\|,$$

provided that ε is small. There exists a γ_2 independent of ε such that

$$\|y_2\|_2 \leqslant \gamma_2 \|y(0)\| \leqslant \varepsilon \gamma_2 \|x(0)\|.$$

Then

$$\|z\|_2^2 = \varepsilon^2 \|x_2\|_2^2 = \varepsilon^2 \int_0^\infty \frac{y_2^2(\tilde{t})}{(1+\varepsilon)^2 \varepsilon^4} \varepsilon \, d\tilde{t} \leqslant \tfrac{1}{\varepsilon} \|y_2\|_2^2 \leqslant \varepsilon \gamma_2 \|x(0)\|.$$

Therefore, for any $x(0)$, we have $\lim_{\varepsilon \to 0} \|z\|_2 = 0$. However, if we fix initial condition at $(x_1(0), x_2(0)) = (1, 0)$, we get

$$(y_1(0), y_2(0)) = (\varepsilon, 0) \text{ and } \|y_2\|_\infty = \varepsilon \gamma_\infty,$$

where

$$\gamma_\infty := \|y_2\|_\infty \text{ with } \begin{pmatrix} y_1(0) \\ y_2(0) \end{pmatrix} = \begin{pmatrix} 1 \\ 0 \end{pmatrix}.$$

Note that γ_∞ is independent of ε. Then

$$\|z\|_\infty = \varepsilon \frac{1}{(1+\varepsilon)\varepsilon^2} \varepsilon \gamma_\infty = \frac{\gamma_\infty}{1+\varepsilon},$$

and $\|z\|_\infty$ cannot be reduced to zero.

Theorem 4.78 enables us to connect to the literature and explain why the H_2 and γ-level H_∞ sequences as defined in Definitions 4.73 and 4.74 are termed as "*low-gain*" sequences. As we alluded to in the beginning of this chapter, the name *low-gain* sequence arose or has roots in one of the classical problems, namely, the problem of semi-globally stabilizing a linear system subject to actuator saturation. To be precise, let us consider a linear system

$$\rho \bar{x} = \bar{A}\bar{x} + \bar{B}\sigma(\bar{u}), \tag{4.310}$$

where the function $\sigma(\cdot)$ denotes a standard saturation. Let the pair $(\bar{A}, \bar{B})$ be stabilizable and $\bar{A}$ has all its eigenvalues in the closed left-half plane for continuous-time systems or within or on the unit disc for discrete-time systems. Consider a state feedback controller

$$\bar{u} = \bar{F}_\varepsilon \bar{x}, \tag{4.311}$$

where $\bar{F}_\varepsilon$ is a parameterized sequence with the parameter as ε. If the feedback sequence $\bar{F}_\varepsilon$ satisfies all the three conditions posed in Theorem 4.11 (continuous time) or in Theorem 4.16 (discrete time), it is known as a "*low-gain*" feedback in the context of stabilization of linear systems subject to saturation. In fact, the state feedback controller $\bar{u} = \bar{F}_\varepsilon \bar{x}$, where $\bar{F}_\varepsilon$ is such a *low-gain* sequence, semi-globally stabilizes (4.310) for a small enough value of ε. That is, there exists an ε^* such that for all $\varepsilon \in (0, \varepsilon^*)$, the closed-loop system comprising (4.310) and (4.311) is semi-globally stable with a priori given (arbitrarily large) bounded set Ω being in the region of attraction, and moreover, the smaller the value of ε the larger can be the *a priori* prescribed set Ω.

Having recalled above the classical semi-global stabilization problem of a linear system with saturating linear feedbacks, we can now emphasize its connection to Theorem 4.78. As is done in classical semi-global stabilization problem, let us first assume that all the control channels are subject to saturation. Then, to see the connection between such a semi-global stabilization problem and Theorem 4.78, set $D = I_m$ and thus take $z = u$ as the constrained variable subject to saturation. Then, Theorem 4.78 shows that the H_2 and γ-level H_∞ sequences as defined in Definitions 4.73 and 4.74 satisfy all the three conditions posed in Theorems 4.11 and 4.16, and hence, they can appropriately be termed as *low-gain* sequences. Furthermore, as is evident from Theorem 4.78, they can readily achieve semi-global stabilization of a linear system where all control inputs are subject to saturation whenever it is achievable.

We now proceed with the general setting, where we assume without loss of generality $D = (I_{m_0} \;\; 0)$ for some $m_0 < m$. This means, in the scenario of a linear system subject to input saturation, all the input channels are not necessarily constrained, that is, some are constrained and others are not. To be precise, we can assume the following system configuration:

$$\rho\xi = A\xi + B_0\sigma(u_0) + B_1u_1, \tag{4.312}$$

where $\xi \in \mathbb{R}^n$, $u_0 \in \mathbb{R}^{m_0}$, $u_1 \in \mathbb{R}^{m-m_0}$, and $B = (B_0 \quad B_1)$. Some inputs as represented by u_0 are subject to saturation. In other words, we have the constrained variable $z = Du = u_0$. In this case, properties of Theorem 4.78 imply that for an initial condition x_0 in a given set and a prespecified saturation level Δ, there exists an ε^* such that for all $\varepsilon \in (0, \varepsilon^*)$ the closed-loop system satisfies the following:

$$\|z(t)\| = \|u_0(t)\| = \|DF_\varepsilon e^{(A+BF_\varepsilon)t}x_0\| \leqslant \Delta$$

for all $t \geqslant 0$ for continuous-time systems and

$$\|z(k)\| = \|u_0(k)\| = \|DF_\varepsilon(A + BF_\varepsilon)^k x_0\| \leqslant \Delta$$

for all $k \geqslant 0$ for discrete-time systems. This implies that the saturation can be made inactive for all the time, and hence, the closed-loop system can in fact be linear. Therefore, the stability of the closed-loop system directly follows from Definitions 4.73 and 4.74.

4.8.2 Existence of H_2 and H_∞ low-gain sequences

We have the following theorem regarding the existence of H_2 and H_∞ low-gain sequences.

Theorem 4.80 *For the system Σ in* (4.304) *with an arbitrarily given $E \in \mathbb{R}^{n\times p}$ and $\gamma > \gamma^*$ where γ^* is defined in* (4.306) *for continuous-time systems or in* (4.307) *for discrete-time systems, the H_2 and γ-level H_∞ low-gain sequences exist if and only if*

(i) (A, B) is stabilizable.

(ii) $(A, B, 0, D)$ is at most weakly non-minimum phase.

Remark 4.81 *In the special case of $D = I_m$, the invariant zeros of $(A, B, 0, I)$ coincide with the eigenvalues of A. Hence, condition (ii) requires that all the eigenvalues of A are in the closed left-half plane for continuous time and within the closed unit disc for discrete time. In this case, a system that satisfies conditions (i) and (ii) is indeed asymptotically null controllable with bounded control (ANCBC).*

Proof : Consider the case of H_2 low-gain sequence. For continuous time, define $\gamma_2^* = \sqrt{\text{trace}(P)}$ where P is the unique semi-stabilizing solution to the continuous-time linear matrix inequality (CLMI)[3]

$$\begin{pmatrix} A'P + PA & PB \\ B'P & D'D \end{pmatrix} \geqslant 0. \tag{4.313}$$

For discrete time, define $\gamma_2^* = \sqrt{\text{trace}(P)}$ where P is the unique semi-stabilizing solution to the discrete-time linear matrix inequality (DLMI)

$$\begin{pmatrix} A'PA - P & A'PB \\ B'PA & D'D + B'PB \end{pmatrix} \geqslant 0. \tag{4.314}$$

It was shown in [133] that H_2 low-gain sequence exists if and only if $\gamma_2^* = 0$, i.e., $P = 0$. This is equivalent to the conditions that (A, B) is stabilizable and

$$\text{rank} \begin{pmatrix} sI - A & -B \\ 0 & D \end{pmatrix} = \text{normrank} \begin{pmatrix} sI - A & -B \\ 0 & D \end{pmatrix}$$

for any $s \in \mathbb{C}^+$ for continuous time or for any $s \in \mathbb{C}^\oplus$ for discrete time, i.e., $(A, B, 0, D)$ is at most weakly non-minimum phase.

Consider next the case of γ-level H_∞ low-gain sequence. For continuous-time case, it is shown in [161] that given $\gamma > \gamma^*$, the γ-level H_∞-low-gain sequences exist if and only if, $P = 0$ is a semi-stabilizing solution to the continuous-time quadratic matrix inequality (CQMI)

$$\begin{pmatrix} A'P + PA + \gamma^{-2}PEE'P & PB \\ B'P & D'D \end{pmatrix} \geqslant 0.$$

Similarly, for discrete-time case, following [162], we can easily verify that given $\gamma > \gamma^*$, the discrete-time γ-level H_∞-low-gain sequences exist if and only if $P = 0$ is a semi-stabilizing solution to the discrete algebraic Riccati equation (DARE)

$$P = A'PA - \begin{pmatrix} B'PA \\ E'PA \end{pmatrix}' G(P)^\dagger \begin{pmatrix} B'PA \\ E'PA, \end{pmatrix}$$

where

$$G(P) = \begin{pmatrix} D'D & 0 \\ 0 & -\gamma^{-2}I \end{pmatrix} + \begin{pmatrix} B' \\ E' \end{pmatrix} P \begin{pmatrix} B & E \end{pmatrix}.$$

[3] The definition of a semi-stabilizing solution of a CLMI or a DLMI is given in [133].

The above properties are equivalent to the conditions that (A, B) is stabilizable and that the matrix pencil

$$\begin{pmatrix} sI - A & -B \\ 0 & D \end{pmatrix}$$

does not have any zeros in $\mathbb{C}^+$ for continuous time or in $\mathbb{C}^\oplus$ for discrete time, i.e., the system is at most weakly non-minimum phase. ∎

Remark 4.82 *As shown in the foregoing discussion, the low-gain sequences achieve semi-global stabilization of linear systems subject to input saturation. Consider the system* (4.310)*. In order to design a low-gain sequence for this system, one can choose $D = I_m$ in* (4.304)*. The above theorem then shows that the necessary and sufficient conditions for semi-global stabilization are that (A, B) is stabilizable and all the invariant zeros of $(A, B, 0, I_m)$ are in the closed left-half plane (continuous time) or in the closed unit disc (discrete time). It is known that the invariant zeros of $(A, B, 0, I_m)$ coincide with eigenvalues of A. Hence, condition (ii) implies that all the eigenvalues of A are in the closed left-half plane (continuous time) or in the closed unit disc (discrete time). Note that in this particular case of $D = I_m$, conditions (i) and (ii) are well known as classical ANCBC conditions as discussed in previous sections.*

However, in general, all the system inputs may not have to be subject to saturation as given in (4.312)*. To design a low-gain feedback sequence for this type of system, we can choose $D = (I_{m_0} \;\; 0)$ in* (4.304)*. Then the necessary and sufficient conditions as required in Theorem 4.80 are that (A, B) is stabilizable and the invariant zeros of $(A, B, 0, D)$ are in the closed left-half plane (continuous time) or in the closed unit disc (discrete time). It can be shown that the invariant zeros of $(A, B, 0, D)$ in this case are a subset of eigenvalues of A (see [139]). Therefore, only some eigenvalues of A have to be constrained while the others can be completely free. Moreover, Theorem 4.80 identifies those eigenvalues that need to be restricted. This can be illustrated by the following example. Consider a linear system with some components of input subject to saturation:*

$$\begin{pmatrix} \rho x_1 \\ \rho x_2 \\ \rho x_3 \\ \rho x_4 \end{pmatrix} = \begin{pmatrix} 0 & 1 & 0 & 0 \\ -1 & 0 & 0 & 0 \\ 0 & 0 & 2 & 1 \\ 1 & 2 & 0 & 3 \end{pmatrix} \begin{pmatrix} x_1 \\ x_2 \\ x_3 \\ x_4 \end{pmatrix} + \begin{pmatrix} 0 \\ 1 \\ 1 \\ 0 \end{pmatrix} \sigma(u_0) + \begin{pmatrix} 0 \\ 0 \\ 0 \\ 1 \end{pmatrix} u_1. \tag{4.315}$$

Clearly, (A, B) is stabilizable. Matrix A has eigenvalues $(j, -j, 2, 3)$. It can be identified that $(j, -j)$ are the invariant zeros of $(A, B, 0, D)$, which are obviously on the imaginary axis (continuous time) and are also on the unit circle *(discrete time). Hence, the two conditions in Theorem 4.80 are satisfied while the other two eigenvalues* $(2, 4)$ *are clearly in the right-half plane (continuous time) and are also outside the unit circle (discrete time).*

4.8.3 Design of H_2 low-gain sequences

We consider here design methods of H_2 low-gain sequences in this subsection while the next subsection considers the design of H_∞ low-gain sequences. We emphasize that for each design, we present different alternate procedures. Thus, the designer has a choice of choosing one method or the other. The design procedures we develop here yield the classical low-gain design methods as special cases.

From the definition, designing an H_2 low-gain sequence for the system Σ in (4.304) is equivalent to designing a bounded H_2 suboptimal control for the following auxiliary system:

$$\Sigma_2 \begin{cases} \rho x = Ax + Bu + \omega \\ z = Du. \end{cases}$$

Such an H_2 suboptimal controller for Σ_2 can be constructed using either direct eigenstructure assignment method or perturbation method (see [82, 133]).

Direct eigenstructure assignment method:
The design basically follows the "H_2 suboptimal state feedback gain sequence" (H_2-SOSFGS) algorithm developed in [81, 82] (see also Sect. 4.3). There exists a non-singular state transformation $(x_a', x_c')' = T_1 x$ such that the system Σ_2 can be transformed into a compact special coordinate basis (SCB) form:

$$\bar{\Sigma}_2 : \begin{cases} \begin{pmatrix} \rho x_a \\ \rho x_c \end{pmatrix} = \begin{pmatrix} A_a & 0 \\ \star & A_c \end{pmatrix} \begin{pmatrix} x_a \\ x_c \end{pmatrix} + \begin{pmatrix} 0 \\ B_c \end{pmatrix} u_1 + \begin{pmatrix} B_a \\ B_{ac} \end{pmatrix} u_0 + T_1 \omega \\ z = u_0, \end{cases} \tag{4.316}$$

where $x_a \in \mathbb{R}^{n_a}$, $x_c \in \mathbb{R}^{n_c}$, $u_0 \in \mathbb{R}^{m_0}$, $u_c \in \mathbb{R}^{m_c}$, $n_a + n_c = n$, and $m_0 + m_c = m$ and $\star$ denotes matrix of not much interest. The eigenvalues of A_a are the invariant zeros of the system Σ. When the given system is in the form of SCB, Theorem 4.80 implies that (A_a, B_a) is stabilizable and A_a has all its eigenvalues in the closed left-half plane (continuous time) or in the closed unit disc (discrete time). Moreover, it follows directly from the properties of the SCB that (A_c, B_c) is controllable (see Chap. 3).

In order to use the eigenstructure assignment method, we need to perform another transformation $(\bar{x}_a', x_c')' = T_2 (x_a', x_c')'$ such that the system can be transformed further into

$$\widetilde{\Sigma}_2 : \begin{cases} \begin{pmatrix} \rho\bar{x}_a \\ \rho x_c \end{pmatrix} = \begin{pmatrix} \bar{A}_a & 0 \\ \star & A_c \end{pmatrix} \begin{pmatrix} \bar{x}_a \\ x_c \end{pmatrix} + \begin{pmatrix} 0 \\ B_c \end{pmatrix} u_1 + \begin{pmatrix} \bar{B}_a \\ B_{ac} \end{pmatrix} u_0 + T\omega \\ \qquad z = u_0, \end{cases}$$

where $T = T_2 T_1$, $\bar{A}_a$ and $\bar{B}_a$ are in the following form:

$$\bar{A}_a = \begin{pmatrix} A_1 & A_{12} & \cdots & A_{1\ell} & 0 \\ 0 & A_2 & \cdots & A_{2\ell} & 0 \\ \vdots & \vdots & \ddots & \vdots & \vdots \\ 0 & 0 & \cdots & A_\ell & 0 \\ 0 & 0 & 0 & 0 & A_o \end{pmatrix},$$

$$\bar{B}_a = \begin{pmatrix} B_1 & 0 & \cdots & 0 & B_{1,o} \\ 0 & B_2 & \cdots & 0 & B_{2,o} \\ \vdots & \vdots & \ddots & \vdots & \vdots \\ 0 & 0 & \cdots & B_\ell & B_{\ell,o} \\ B_{o,1} & B_{o,2} & \cdots & B_{o,\ell} & B_o \end{pmatrix}, \tag{4.317}$$

and where A_o is Hurwitz stable for continuous time or Schur stable for discrete time, (A_i, B_i) is controllable, and A_i has all its eigenvalues on the imaginary axis for continuous time or on the unit circle for discrete time. Moreover, (A_i, B_i) is in the controllability canonical form (which has the same structure for continuous- and discrete-time systems) as given by

$$A_i = \begin{pmatrix} 0 & 1 & \cdots & 0 & 0 \\ 0 & 0 & \ddots & \vdots & 0 \\ \vdots & \vdots & \ddots & 1 & \vdots \\ 0 & 0 & \cdots & 0 & 1 \\ -\alpha_{i,0} & -\alpha_{i,1} & \cdots & -\alpha_{i,n_i-2} & -\alpha_{i,n_i-1} \end{pmatrix}, \quad B_i = \begin{pmatrix} 0 \\ 0 \\ \vdots \\ 0 \\ 1 \end{pmatrix}. \tag{4.318}$$

For each pair (A_i, B_i), let the feedback gain $F_i(\varepsilon)$ be such that

$$\lambda(A_i + B_i F_i(\varepsilon)) = -\varepsilon - \lambda(A_i)$$

for continuous time, or

$$\lambda(A_i + B_i F_i(\varepsilon)) = (1 - \varepsilon)\lambda(A_i)$$

for discrete time.

Define

$$F_{a,\varepsilon} = \begin{pmatrix} F_1(\varepsilon_1) & 0 & \cdots & 0 & 0 \\ 0 & F_2(\varepsilon_2) & \ddots & \vdots & \vdots \\ \vdots & \ddots & \ddots & 0 & 0 \\ 0 & \cdots & 0 & F_\ell(\varepsilon_\ell) & 0 \\ 0 & \cdots & 0 & 0 & 0 \end{pmatrix},$$

where $\varepsilon_i = \varepsilon^{2^{\ell-1}(r_{i+1}+1)\cdots(r_\ell+1)}$ for $i = 1, \ldots, \ell-1$ and $\varepsilon_\ell = \varepsilon$ where r_i is the largest algebraic multiplicity of eigenvalues of A_i.

Since (A_c, B_c) is controllable, we can choose a bounded F_c such that $A_c + B_c F_c$ is Hurwitz stable for continuous time or Schur stable for discrete time, and has a desired set of eigenvalues.

The sequence of feedback gains for the system Σ_2 can then be constructed as

$$F_\varepsilon = \begin{pmatrix} F_{a,\varepsilon} & 0 \\ 0 & F_c \end{pmatrix} T_2 T_1.$$

Clearly, F_ε is bounded and $A + BF_\varepsilon$ is Hurwitz stable for continuous time or Schur stable for discrete time. It follows from Sect. 4.3 (see also [82]) that F_ε also satisfies Property *(iii)* in Definition 4.73, namely, for any $x(0) \in \mathbb{R}^n$,

$$\lim_{\varepsilon \to 0} \|z\|_2 = 0.$$

Therefore, F_ε is an H_2 low-gain sequence.

Remark 4.83 *In the special case of $D = I_m$, the above design procedure recovers the direct eigenstructure assignment method in the classical low-gain design developed in Sect. 4.3 for linear systems subject to input saturation.*

To highlight the explicit nature of the above method and to illustrate the constructive procedure, we design an H_2 low-gain sequence for the discrete-time example given in (4.315). Note that for this system, A and B are already in the form of (4.317) and (4.318) where

$$A_1 = \begin{pmatrix} 0 & 1 \\ -1 & 0 \end{pmatrix}, \quad A_c = \begin{pmatrix} 2 & 1 \\ 0 & 3 \end{pmatrix}, \quad B_1 = \begin{pmatrix} 0 \\ 1 \end{pmatrix}, \quad B_c = \begin{pmatrix} 0 \\ 1 \end{pmatrix}.$$

With a bit of algebra, we find

$$F_1(\varepsilon) = \begin{pmatrix} 1 - (1-\varepsilon)^2 & 0 \end{pmatrix}.$$

It is easy to verify that $A_1 + B_1 F_1(\varepsilon)$ has eigenvalues at $((1-\varepsilon)j, -(1-\varepsilon)j)$. Choose $F_c = (-3.75 \quad -5)$ so that $A_c + B_c F_c$ has eigenvalues at $(0.5, -0.5)$. The discrete H_2 low-gain sequence can then be constructed as

$$F_\varepsilon = \begin{pmatrix} 1-(1-\varepsilon)^2 & 0 & 0 & 0 \\ 0 & 0 & -3.75 & -5 \end{pmatrix}.$$

Perturbation methods

There exists a classical perturbation method that has long been used in H_2 suboptimal controller design (see, for instance, [133]). The philosophy of perturbation methods used in H_2 low-gain design is the same as for H_2 suboptimal controller design, that is, to perturb the data of the system so that an H_2 optimal controller exists for the perturbed system, and then, based on continuity argument, we can obtain a sequence of H_2 low-gains for the original system utilizing H_2 optimal control design techniques developed in [133].

For a given quadruple (A, B, C, D), let a sequence of perturbed data $(A_\varepsilon, B_\varepsilon, C_\varepsilon, D_\varepsilon)$ be such that $A_\varepsilon \to A$, $B_\varepsilon \to B$, $\bar{Q}_\varepsilon \to \bar{Q}_0$ as $\varepsilon \to 0$ and $\bar{Q}_\varepsilon$ is continuous at $\varepsilon = 0$ where

$$\bar{Q}_0 = \begin{pmatrix} C & D \end{pmatrix}' \begin{pmatrix} C & D \end{pmatrix}, \quad \bar{Q}_\varepsilon = \begin{pmatrix} C_\varepsilon & D_\varepsilon \end{pmatrix}' \begin{pmatrix} C_\varepsilon & D_\varepsilon \end{pmatrix}. \tag{4.319}$$

In order for this perturbation $(A_\varepsilon, B_\varepsilon, C_\varepsilon, D_\varepsilon)$ to be admissible for H_2 low-gain design, it has to satisfy the following conditions:

(*i*) For continuous-time systems, the solution $P_{\varepsilon,2}$ to the CLMI,

$$\begin{pmatrix} A_\varepsilon' P_{\varepsilon,2} + P_{\varepsilon,2} A_\varepsilon & P_{\varepsilon,2} B_\varepsilon + C_\varepsilon' D_\varepsilon \\ B_\varepsilon' P_{\varepsilon,2} + D_\varepsilon' C_\varepsilon & D_\varepsilon' D_\varepsilon \end{pmatrix} \geqslant 0, \tag{4.320}$$

converges to 0. Similarly, for discrete-time systems, the smallest positive semi-definite semi-stabilizing solution $P_{\varepsilon,2}$ to the DLMI,

$$\begin{pmatrix} C_\varepsilon' C_\varepsilon + A_\varepsilon' P_{\varepsilon,2} A_\varepsilon - P_{\varepsilon,2} & A_\varepsilon' P_{\varepsilon,2} B_\varepsilon + C_\varepsilon' D_\varepsilon \\ B_\varepsilon' P_{\varepsilon,2} A_\varepsilon + D_\varepsilon' C_\varepsilon & D_\varepsilon' D_\varepsilon + B_\varepsilon' P_{\varepsilon,2} B_\varepsilon \end{pmatrix} \geqslant 0, \tag{4.321}$$

converges to 0.

(*ii*) An H_2 optimal state feedback controller of a static type F_ε exists for the perturbed system characterized by $(A_\varepsilon, B_\varepsilon, C_\varepsilon, D_\varepsilon, I)$. One can construct the F_ε using, for instance, the $(COGFMDZ)$ or $(COGFMDZ)_{nli}$ algorithms for left-invertible and non-left-invertible continuous-time systems, respectively. Similarly, we have the $(DOGFMDZ)$ or $(DOGFMDZ)_{nli}$ algorithm for discrete-time systems. Further details are in Chap. 7 of [133].

Moreover, the obtained F_ε should satisfy the next three conditions:

(*iii*) F_ε is bounded for continuous-time systems; although it is desirable, it need not be bounded for discrete-time systems.

(*iv*) F_ε is such that $A + BF_\varepsilon$ is Hurwitz stable for continuous-time systems or Schur stable for discrete-time systems.

(*v*) F_ε satisfies that $\|(C + DF_\varepsilon)(sI - A - BF_\varepsilon)^{-1}\|_2 \to 0$ as $\varepsilon \to 0$.

If $(A_\varepsilon, B_\varepsilon, C_\varepsilon, D_\varepsilon)$ and one of the correspondingly constructed F_ε satisfy all the 5 conditions stated above, such an F_ε is an H_2 low-gain sequence.

Remark 4.84 *Since F_ε is obtained from H_2 optimal controller design, we immediately see that $A_\varepsilon + B_\varepsilon F_\varepsilon$ is Hurwitz stable for continuous-time systems or Schur stable for discrete-time systems, and $\|(C_\varepsilon + D_\varepsilon F_\varepsilon)(sI - A_\varepsilon - B_\varepsilon F_\varepsilon)^{-1}\|_2 \to 0$. But these do not necessarily imply that $A + BF_\varepsilon$ is Hurwitz stable for continuous-time systems or Schur stable for discrete-time systems, and that $\|(C + DF_\varepsilon)(sI - A - BF_\varepsilon)^{-1}\|_2 \to 0$ as $\varepsilon \to 0$ even though the continuity is satisfied.*

Specifically in our problem, for the system Σ in (4.304) characterized by (A, B, C, D) with $C = 0$, we can use two perturbation methods to design an H_2 low-gain sequence.

Perturbation method I: The classical perturbations used in H_2 suboptimal control are of the form $(A, B, C_\varepsilon, D_\varepsilon)$ where C_ε and D_ε are such that $(A, B, C_\varepsilon, D_\varepsilon)$ has no zero structure (that is, neither invariant zeros nor infinite zeros), $\bar{Q}_\varepsilon \to \bar{Q}_0$ as $\varepsilon \to 0$, and there exists a β such that

$$\bar{Q}_{\varepsilon_1} \leqslant \bar{Q}_{\varepsilon_2}, \quad \text{for} \quad 0 \leqslant \varepsilon_1 \leqslant \varepsilon_2 \leqslant \beta. \tag{4.322}$$

This leads to a perturbed system

$$\Sigma_2^\varepsilon : \begin{cases} \rho x &= Ax + Bu + w \\ z_\varepsilon &= C_\varepsilon x + D_\varepsilon u. \end{cases}$$

For this perturbation, we have:

- Since C_ε and D_ε satisfy (4.322), condition (*i*) follows from Theorems 4.103 and 4.105 in Appendix 4.D.
- Since the quadruple $(A, B, C_\varepsilon, D_\varepsilon)$ does not have zero structures (that is, neither invariant zeros nor infinite zeros), condition (*ii*) follows from Theorem 4.101 in Appendix 4.D.
- Since we do not perturb A and B, condition (*iv*) is obvious.

- Since $u = F_\varepsilon x$ is an H_2 optimal state feedback for the perturbed system and $P_{\varepsilon,2} \to 0$, we have $\|(C_\varepsilon + D_\varepsilon F_\varepsilon)(sI - A - BF_\varepsilon)^{-1}\|_2 \to 0$. Then, (4.322) implies that

$$\|(C + DF_\varepsilon)(sI - A - BF_\varepsilon)^{-1}\|_2 \leqslant \|(C_\varepsilon + D_\varepsilon F_\varepsilon)(sI - A - BF_\varepsilon)^{-1}\|_2.$$

Therefore, $\|(C + DF_\varepsilon)(sI - A - BF_\varepsilon)^{-1}\|_2 \to 0$ as $\varepsilon \to 0$.

We find that conditions (*i*), (*ii*), (*iv*), and (*v*) are always satisfied by this type of perturbation. It remains to verify condition (*iii*). We note that since we can select $C = 0$ in our problem, we can always find a $(C_\varepsilon, D_\varepsilon)$ such that a bounded F_ε can be constructed following $(COGFMDZ)$ or $(COGFMDZ)_{nli}$ algorithm for continuous-time systems or $(DOGFMDZ)$ or $(DOGFMDZ)_{nli}$ algorithm for discrete-time systems [133]. In what follows, we give two examples for this type of perturbation which recover, in the special case of $D = I_m$, the standard H_2-ARE low-gain design for linear systems subject to input saturation.

Example 1. Consider the perturbed system Σ_2^ε defined earlier where

$$C_\varepsilon = \begin{pmatrix} 0 \\ 0 \\ \sqrt{Q_\varepsilon} \end{pmatrix}, \quad D_\varepsilon = \begin{pmatrix} D \\ \varepsilon I \\ 0 \end{pmatrix},$$

and $Q_\varepsilon \in \mathbb{R}^{n\times n}$ is such that

$$Q_\varepsilon > 0 \text{ and } \lim_{\varepsilon\to 0} Q_\varepsilon = 0. \tag{4.323}$$

Clearly, $(A, B, C_\varepsilon, D_\varepsilon)$ does not have any zero structure (that is, neither invariant zeros nor infinite zeros), and $(C_\varepsilon, D_\varepsilon)$ satisfies (4.322). Hence, we only need to check condition (*iii*) for continuous time systems. For such systems, define the H_2 optimal static state feedback for the perturbed system as

$$F_\varepsilon = -(D_\varepsilon' D_\varepsilon)^{-1} B' X_{\varepsilon,2},$$

where $X_{\varepsilon,2}$ is the positive definite solution of H_2 CARE:

$$A' X_{\varepsilon,2} + X_{\varepsilon,2} A + Q_\varepsilon - X_{\varepsilon,2} B' (D_\varepsilon' D_\varepsilon)^{-1} B X_{\varepsilon,2} = 0. \tag{4.324}$$

Although we do not need boundedness of F_ε for discrete-time systems, we can similarly define the H_2 optimal static state feedback for the perturbed system as

$$F_\varepsilon = -(B' X_{\varepsilon,2} B + D_\varepsilon' D_\varepsilon)^{-1} B' X_{\varepsilon,2} A,$$

where $X_{\varepsilon,2}$ is the positive definite solution of H_2 DARE:

$$X_{\varepsilon,2} = A' X_{\varepsilon,2} A + Q_\varepsilon - A' B X_{\varepsilon,2} (B' X_{\varepsilon,2} B + D_\varepsilon' D_\varepsilon)^{-1} X_{\varepsilon,2} B' A. \tag{4.325}$$

When $m_0 = m$, that is, $D = I_m$, F_ε is bounded for $\varepsilon \in [0, 1]$ and hence is an H_2 low-gain sequence. Moreover, it recovers the standard H_2-ARE-based low-gain design for linear systems subject to input saturation as discussed in

Sect. 4.4. However, when $m_0 < m$, the boundedness of F_ε needs to be proved for continuous-time systems. In the next example, we present an alternative perturbation of $(C_\varepsilon, D_\varepsilon)$ which automatically generates a bounded F_ε for any $m_0 \leqslant m$.

Example 2. First, we can transform the system into the form (4.316) with transformation $(x_a', x_c')' = T_1 x$. Then consider a perturbed system based on (4.316) as

$$\bar{\Sigma}^{\varepsilon}_{2,I} : \begin{cases} \begin{pmatrix} \rho x_a \\ \rho x_c \end{pmatrix} = \begin{pmatrix} A_a & 0 \\ \star & A_c \end{pmatrix} \begin{pmatrix} x_a \\ x_c \end{pmatrix} + \begin{pmatrix} 0 \\ B_c \end{pmatrix} u_c + \begin{pmatrix} B_a \\ B_{ac} \end{pmatrix} u_0 + T_1 \omega \\[2ex] \begin{pmatrix} z_0 \\ z_{\varepsilon,1} \end{pmatrix} = \begin{pmatrix} 0 & 0 \\ \sqrt{Q_\varepsilon} & 0 \end{pmatrix} \begin{pmatrix} x_a \\ x_c \end{pmatrix} + \begin{pmatrix} I_{m_0} & 0 \\ 0 & 0 \end{pmatrix} \begin{pmatrix} u_0 \\ u_c \end{pmatrix}, \end{cases}$$

where Q_ε satisfies (4.323). In this case,

$$C_\varepsilon = \begin{pmatrix} 0 & 0 \\ \sqrt{Q_\varepsilon} & 0 \end{pmatrix}, \quad D_\varepsilon = \begin{pmatrix} I_{m_0} & 0 \\ 0 & 0 \end{pmatrix}.$$

The perturbed system does not have zero structure (that is, neither invariant zeros nor infinite zeros) and $(C_\varepsilon, D_\varepsilon)$ satisfies (4.322). We proceed to check condition (*iii*).

For continuous-time systems, define the H_2 optimal static state feedback for the perturbed system as

$$F_\varepsilon = \begin{pmatrix} -B_a' X_{\varepsilon,2} & 0 \\ 0 & F_c \end{pmatrix} T_1,$$

where $X_{\varepsilon,2}$ is the positive definite solution of H_2 CARE:

$$A_a' X_{\varepsilon,2} + X_{\varepsilon,2} A_a + Q_\varepsilon - X_{\varepsilon,1} B_a B_a' X_{\varepsilon,2} = 0,$$

and where F_c is bounded such that $A_c + B_c F_c$ is Hurwitz stable. Although we do not need boundedness of F_ε for discrete-time systems, we can similarly define the H_2 optimal static state feedback for the perturbed system as

$$F_\varepsilon = \begin{pmatrix} -(I + B_a' X_{\varepsilon,2} B_a)^{-1} B_a' X_{\varepsilon,2} A_a & 0 \\ 0 & F_c \end{pmatrix} T_1,$$

where $X_{\varepsilon,2}$ is the positive definite solution of H_2 DARE:

$$X_{\varepsilon,2} = A_a' X_{\varepsilon,2} A_a + Q_\varepsilon - A_a' X_{\varepsilon,2} B_a (I + B_a' X_{\varepsilon,2} B_a)^{-1} B_a' X_{\varepsilon,2} A_a,$$

and where F_c is bounded such that $A_c + B_c F_c$ is Schur stable.

We note that F_ε selected above is bounded for any $m_0 \leqslant m$ and $\varepsilon \in [0,1]$. Therefore, it is an H_2 low-gain sequence.

When $m_0 = m$, that is when $D = I_m$, the above perturbation method also recovers the standard H_2-ARE low-gain design developed in Sect. 4.4.

Perturbation method II for continuous-time systems: In perturbation method I, we utilize fictitious outputs to remove zero dynamics completely. However, we can also directly perturb system dynamics to move those invariant zeros on the imaginary axis without adding fictitious outputs. Consider a perturbation (A_ε, B_ε, 0, D_ε) which leads to the following perturbed system:

$$\bar{\Sigma}^{\varepsilon}_{2,II} : \begin{cases} \rho\bar{x} = A_\varepsilon\bar{x} + B_\varepsilon u + \omega \\ \bar{z} = D_\varepsilon u, \end{cases} \tag{4.326}$$

where

$$A_\varepsilon = (1+\varepsilon)A, \quad B_\varepsilon = (1+\varepsilon)B, \quad D_\varepsilon = (1+\varepsilon)D,$$

and ε small enough such that $((1+\varepsilon)A, (1+\varepsilon)B)$ is stabilizable. For the sake of clarity, we focus on this particular choice of perturbation. The conditions required for perturbation can be verified as follows:

- Since $(A_\varepsilon, B, 0, D)$ always has the same normal rank as that of $(A, B, 0, D)$, condition (*i*) follows from Theorem 4.102 in Appendix 4.D.
- Since $(A_\varepsilon, B, 0, D)$ does not have any invariant zeros on the imaginary axis and has no infinite zeros, condition (*ii*) follows from Theorem 4.101 in Appendix 4.C.
- Note that

$$DF_\varepsilon e^{(A+BF_\varepsilon+\frac{\varepsilon}{2}I)t} = e^{\frac{\varepsilon}{2}t} DF_\varepsilon e^{(A+BF_\varepsilon)t}.$$

  By the definition of H_2 norm, this implies that

$$\|DF_\varepsilon(sI - A - BF_\varepsilon)\|_2 \leqslant \|DF_\varepsilon(sI - A - \tfrac{\varepsilon}{2}I - BF_\varepsilon)\|_2.$$

  Therefore, $\|DF_\varepsilon(sI - A - BF_\varepsilon)\|_2 \to 0$ if $\|DF_\varepsilon(sI - A - \frac{\varepsilon}{2}I - BF_\varepsilon)\|_2 \to 0$. We find that condition (*v*) is satisfied.
- Obviously, $A + BF$ is Hurwitz stable if $A + BF + \frac{\varepsilon}{2}I$ is Hurwitz stable. Therefore, condition (*iv*) is satisfied.

Therefore, the conditions (*i*), (*ii*), (*iv*), and (*v*) can be satisfied. For this perturbation, we can always construct a bounded H_2 optimal controller following $(COGFMDZ)_{nli}$ algorithm. This can be done as follows:

We first find a non-singular state transformation independent of ε

$$\begin{pmatrix} x_a^- \\ x_a^\circ \\ x_c \end{pmatrix} = T_2 x$$

such that the perturbed system can be transformed into its SCB form

$$\bar{\Sigma}_{2,II}^{\varepsilon}: \begin{cases} \begin{pmatrix} \dot{x}_a^- \\ \dot{x}_a^\circ \\ \dot{x}_c \end{pmatrix} = \begin{pmatrix} A_a^- + \frac{\varepsilon}{2}I & 0 & 0 \\ 0 & A_a^\circ + \frac{\varepsilon}{2}I & 0 \\ \star & \star & A_c + \frac{\varepsilon}{2}I \end{pmatrix} \begin{pmatrix} x_a^- \\ x_a^\circ \\ x_c \end{pmatrix} \\ \qquad + \begin{pmatrix} 0 \\ 0 \\ B_c \end{pmatrix} u_c + \begin{pmatrix} B_a^- \\ B_a^\circ \\ B_{ac} \end{pmatrix} u_0 + E\omega \\ z \quad = u_0, \end{cases}$$

where A_a^- is Hurwitz stable, the pairs (A_a°, B_a°) and (A_c, B_c) are controllable, and the eigenvalues of A_a° are on the imaginary axis. The eigenvalues of $(1+\varepsilon)A_a^\circ$ and $(1+\varepsilon)A_a^-$ are the invariant zeros of the perturbed system. For a small ε, $(1+\varepsilon)A_a^-$ is also Hurwitz stable. Let $X_{\varepsilon,2}$ be the positive definite solution of CARE:

$$(A_a^\circ + \tfrac{\varepsilon}{2}I)'X_{\varepsilon,2} + X_{\varepsilon,2}(A_a^\circ + \tfrac{\varepsilon}{2}I) - X_{\varepsilon,2}B_a^\circ B_a^{\circ\prime} X_{\varepsilon,2} = 0. \tag{4.327}$$

It is shown in [215] that $X_{\varepsilon,2}$ strictly decreases to zero as ε goes to zero, and hence, X_ε is bounded. Choose a bounded F_c such that $A_c + B_c F_c$ is Hurwitz. The H_2 low-gain sequence F_ε can then be constructed as

$$F_\varepsilon = \begin{pmatrix} 0 & -B_a^{\circ\prime} X_{\varepsilon,2} & 0 \\ 0 & 0 & F_c \end{pmatrix} T_2.$$

Obviously, the bounded condition (*iii*) on F_ε is satisfied.

Remark 4.85 *In the special case when $D = I_m$, the above perturbation method recovers the parametric Lyapunov approach to low-gain design developed in [215] for linear systems subject to input saturation.*

Perturbation method II for discrete-time systems: As in the case of continuous-time systems, for clarity, we focus on a particular perturbation as given in the system (4.326). The conditions required for perturbation can then be verified:

- Since $((1+\varepsilon)A, (1+\varepsilon)B, 0, (1+\varepsilon)D)$ has the same normal rank as that of $(A, B, 0, D)$, condition (*i*) follows from Theorems 4.102 and 4.104 in Appendix.

- Since $((1+\varepsilon)A, (1+\varepsilon)B, 0, (1+\varepsilon)D)$ does not have any invariant zeros on the unit circle and has no infinite zeros, condition (*ii*) follows from Theorem 4.101 in Appendix.
- Obviously, any F_ε for which $(1+\varepsilon)A + (1+\varepsilon)BF_\varepsilon$ is Schur stable also yields that $A + BF_\varepsilon$ is Schur stable. Therefore, condition (*iv*) is satisfied.
- Note that

$$(1+\varepsilon)DF_\varepsilon((1+\varepsilon)A + (1+\varepsilon)BF_\varepsilon)^k = (1+\varepsilon)^{k+1}DF_\varepsilon(A+BF_\varepsilon)^k.$$

By the definition of the H_2 norm, this implies that

$$\|DF_\varepsilon(zI - A - BF_\varepsilon)\|_2 \leqslant \|(1+\varepsilon)DF_\varepsilon(zI - (1+\varepsilon)A - (1+\varepsilon)BF_\varepsilon)\|_2.$$

Therefore, $\|DF_\varepsilon(zI - A - BF_\varepsilon)^{-1}\|_2 \to 0$ if

$$\|(1+\varepsilon)DF_\varepsilon(zI - (1+\varepsilon)A - (1+\varepsilon)BF_\varepsilon)^{-1}\|_2 \to 0.$$

We find that condition (v) is satisfied.

Hence, there exists an ε^* such that for any $\varepsilon \in (0, \varepsilon^*]$, conditions (*i*), (*ii*), (*iv*), and (v) are satisfied. For this specific perturbation, following the $(DOGFMDZ)_{nli}$ algorithm, we can directly construct a discrete-time H_2 optimal sequence controller. This can be done as follows: the perturbed system can be transformed into its compact SCB form using a non-singular state transformation. Let

$$\begin{pmatrix} x_a^{\ominus\prime} & x_a^{\bigcirc\prime} & x_c' \end{pmatrix}' = T_3 x$$

be such that

$$\bar{\Sigma}_{2,II}^{\varepsilon}: \begin{cases} \begin{pmatrix} \rho\bar{x}_a^{\ominus} \\ \rho\bar{x}_a^{\bigcirc} \\ \rho\bar{x}_c \end{pmatrix} = (1+\varepsilon)\begin{pmatrix} A_a^{\ominus} & 0 & 0 \\ 0 & A_a^{\bigcirc} & 0 \\ \star & \star & A_c \end{pmatrix}\begin{pmatrix} \bar{x}_a^{\ominus} \\ \bar{x}_a^{\bigcirc} \\ \bar{x}_c \end{pmatrix} + T_3\omega \\ \qquad + (1+\varepsilon)\begin{pmatrix} 0 \\ 0 \\ B_c \end{pmatrix} u_c + (1+\varepsilon)\begin{pmatrix} B_a^{\ominus} \\ B_a^{\bigcirc} \\ B_{ac} \end{pmatrix} u_0 \\ \qquad \bar{z} = (1+\varepsilon)u_0, \end{cases} \tag{4.328}$$

where $A_a^{\ominus}$ is Schur stable, the pairs $(A_a^{\bigcirc}, B_a^{\bigcirc})$ and (A_c, B_c) are controllable, and the eigenvalues of $A_a^{\bigcirc}$ are on the unit circle. The eigenvalues of $(1+\varepsilon)A_a^{\bigcirc}$ and $(1+\varepsilon)A_a^{\ominus}$ are the invariant zeros of the perturbed system. For a small ε, $(1+\varepsilon)A_a^{\ominus}$ is also Schur stable. Moreover, T_3 is independent of ε. Let $X_{\varepsilon,2}$ be the positive definite solution of DARE,

$$\tfrac{1}{(1+\varepsilon)^2}X_{\varepsilon,2} = A_a^{\bigcirc\prime}X_{\varepsilon,2}A_a^{\bigcirc} - A_a^{\bigcirc\prime}X_{\varepsilon,2}B_a^{\bigcirc}(I + B_a^{\bigcirc\prime}X_\varepsilon B_a^{\bigcirc})^{-1}B_a^{\bigcirc\prime}X_{\varepsilon,2}A_a^{\bigcirc},$$

and choose a bounded F_c such that $A_c + B_c F_c$ is Schur stable. The discrete-time H_2 suboptimal controller sequence F_ε can then be constructed as

$$F_\varepsilon = \begin{pmatrix} 0 & -(I + B_a^{\bigcirc\prime} X_{\varepsilon,2} B_a^{\bigcirc})^{-1} B_a^{\bigcirc\prime} X_{\varepsilon,2} A_a^{\bigcirc} & 0 \\ 0 & 0 & F_c \end{pmatrix} T_3.$$

Clearly, F_ε is bounded for any $\varepsilon \in (0, \varepsilon^*]$; therefore, condition *(iii)* is satisfied. We conclude that F_ε is an H_2 low-gain sequence.

Remark 4.86 *In the special case when $D = I_m$, the above perturbation methods recover the parametric Lyapunov approach to low-gain design developed in [215] for linear systems subject to input saturation.*

4.8.4 Design of H_∞ low-gain sequences

We consider here design of γ-level H_∞ low-gain sequences. As in the preceding subsection, we give here different alternate procedures. The design procedures we develop here recover the classical H_∞ ARE low-gain design methods discussed in Sect. 4.4 as special cases.

The direct eigenstructure assignment method of γ-level H_∞ low-gain design can be found in [18]. We focus here on designing γ-level H_∞ low-gain sequences using perturbation methods.

Perturbation methods

The philosophy of perturbation methods is similar to that in H_2 low-gain design. However, the conditions imposed on perturbations are more restrictive. For the auxiliary system Σ_∞ of (4.305), consider a sequence of perturbations (A_ε, B_ε, C_ε, D_ε, E_ε) which leads to the following perturbed system. Let

$$\Sigma_\infty^\varepsilon : \begin{cases} \rho x &= A_\varepsilon x + B_\varepsilon u + E_\varepsilon w \\ z_\varepsilon &= C_\varepsilon x + D_\varepsilon u \end{cases} \tag{4.329}$$

be such that $A_\varepsilon \to A$, $B_\varepsilon \to B$, $E_\varepsilon \to E$, and $\bar{Q}_\varepsilon \to \bar{Q}_0$, where $\bar{Q}_\varepsilon$ and $\bar{Q}_0$ are defined in (4.319). The quintuple $(A_\varepsilon, B_\varepsilon, C_\varepsilon, D_\varepsilon, E_\varepsilon)$ is admissible for γ-level H_∞ low-gain design if it satisfies the following conditions:

(*i*) For continuous-time systems, define

$$\gamma_\varepsilon^* = \inf_F \{\|(C_\varepsilon + D_\varepsilon F)(sI - A_\varepsilon - B_\varepsilon F)^{-1} E_\varepsilon\|_\infty \mid \lambda(A_\varepsilon + B_\varepsilon F) \in \mathbb{C}^-\}.$$

Similarly, for discrete-time systems, define

$$\gamma_\varepsilon^* = \inf_F \{\|(C_\varepsilon + D_\varepsilon F)(zI - A_\varepsilon - B_\varepsilon F)^{-1} E_\varepsilon\|_\infty \mid \lambda(A_\varepsilon + B_\varepsilon F) \in \mathbb{C}^{\odot}\}.$$

Given $\gamma > \gamma^*$ where γ^* is defined in (4.306) or (4.307), for a sufficiently small ε, we have $\gamma_\varepsilon^* < \gamma$.

(*ii*) Let $\gamma > \gamma_\varepsilon^*$. For continuous-time systems, the solution $P_{\varepsilon,\infty}$ of the CQMI,

$$\begin{pmatrix} A_\varepsilon' P_{\varepsilon,\infty} + P_{\varepsilon,\infty} A_\varepsilon + C_\varepsilon' C_\varepsilon + \gamma^{-2} P_{\varepsilon,\infty} E_\varepsilon E_\varepsilon' P_{\varepsilon,\infty} & P_{\varepsilon,\infty} B_\varepsilon + C_\varepsilon' D_\varepsilon \\ B_\varepsilon' P_{\varepsilon,\infty} + D_\varepsilon' C_\varepsilon & D_\varepsilon' D_\varepsilon \end{pmatrix} \geqslant 0,$$

satisfies $P_{\varepsilon,\infty} \to 0$ as $\varepsilon \to 0$.

For discrete-time systems, consider the H_∞ DARE:

$$P_\varepsilon = A_\varepsilon' P_\varepsilon A_\varepsilon + C_\varepsilon' C_\varepsilon - \begin{pmatrix} B_\varepsilon' P_\varepsilon A_\varepsilon + D_\varepsilon' C_\varepsilon \\ E_\varepsilon' P_\varepsilon A_\varepsilon \end{pmatrix}' G(P_\varepsilon)^\dagger \begin{pmatrix} B_\varepsilon' P_\varepsilon A_\varepsilon + D_\varepsilon' C_\varepsilon \\ E_\varepsilon' P_\varepsilon A_\varepsilon \end{pmatrix}$$

with

$$G(P_\varepsilon) = \begin{pmatrix} D_\varepsilon' D_\varepsilon & 0 \\ 0 & -\gamma^{-2} I \end{pmatrix} + \begin{pmatrix} B_\varepsilon' \\ E_\varepsilon' \end{pmatrix} P_\varepsilon \begin{pmatrix} B_\varepsilon & E_\varepsilon \end{pmatrix}$$

$$E_\varepsilon' P_\varepsilon E_\varepsilon + E_\varepsilon' P_\varepsilon B_\varepsilon (D_\varepsilon' D_\varepsilon + B_\varepsilon' P B_\varepsilon)^\dagger B_\varepsilon' P_\varepsilon E_\varepsilon < \gamma^2 I.$$

The smallest positive semi-definite semi-stabilizing solution P_ε satisfies the property that $P_\varepsilon \to 0$ as $\varepsilon \to 0$.

(*iii*) $(A_\varepsilon, B_\varepsilon, C_\varepsilon, D_\varepsilon)$ does not have invariant zeros on the imaginary axis (continuous time) or on the unit circle (discrete time). Under this condition, a γ-level H_∞ suboptimal state feedback $F_\varepsilon(E, \gamma)$ with $\gamma > \gamma^*(\varepsilon)$ for the perturbed system can be easily constructed using the techniques developed in [161]. Moreover, such an $F_\varepsilon(E, \gamma)$ should satisfy the next three conditions:

(*iv*) The $F_\varepsilon(E, \gamma)$ is bounded for continuous-time systems; although it is desirable, it need not be bounded for discrete-time systems.

(*v*) The $F_\varepsilon(E, \gamma)$ is such that $A + BF_\varepsilon(E, \gamma)$ is Hurwitz stable (continuous-time case) or Schur stable (discrete-time case).

(*vi*) Let $w \in L_2$ for continuous-time systems or $w \in \ell_2$ for discrete-time systems. Then, the closed-loop system comprising Σ_∞ and $u = F_\varepsilon(E,\gamma)x$ satisfies

$$\lim_{\varepsilon\to 0}\left\{\sup_w(\|z\|_2^2 - \gamma^2\|w\|_2^2)\right\} = 0.$$

If $(A_\varepsilon, B_\varepsilon, C_\varepsilon, D_\varepsilon, E_\varepsilon)$ and one of the correspondingly constructed $F_\varepsilon(E,\gamma)$ satisfy all six conditions, then this $F_\varepsilon(E,\gamma)$ is a γ-level H_∞ low-gain sequence.

In our problem, for a given quintuple (A, B, C, D, E) with $C = 0$ and the given $\gamma > 0$ satisfying $\gamma > \gamma^*$, two perturbation methods can be used for γ-level H_∞ low-gain design.

Perturbation method I for continuous-time systems:

Similar to that in H_2 low-gain design, the first perturbation is in the form of $(A, B, C_\varepsilon, D_\varepsilon, E)$ where C_ε and D_ε satisfy (4.322). We give two examples.

Example I. Consider a sequence of perturbations $(A, B, C_\varepsilon, D_\varepsilon, E)$ where

$$C_\varepsilon = \begin{pmatrix} 0 \\ 0 \\ \sqrt{Q_\varepsilon} \end{pmatrix}, \quad D_\varepsilon = \begin{pmatrix} D \\ \varepsilon I \\ 0 \end{pmatrix},$$

where Q_ε satisfies (4.323). We first verify below that this perturbation is admissible for H_∞ low-gain design.

- Suppose we apply any bounded F to the system (4.304) characterized by $(A, B, 0, D, E)$ such that $A + BF$ is Hurwitz stable. Also, let

$$\gamma_F = \|DF(sI - A - BF)^{-1}E\|_\infty.$$

  We then have

$$(C_\varepsilon + D_\varepsilon F)(sI - A - BF)^{-1}E = \begin{pmatrix} DF(sI - A - BF)^{-1}E \\ \varepsilon F(sI - A - BF)^{-1}E \\ \sqrt{Q_\varepsilon}(sI - A - BF)^{-1}E \end{pmatrix}.$$

  Since $A + BF$ is Hurwitz stable, F is bounded, there exists a M such that

$$\gamma_F \leqslant \|(C_\varepsilon + D_\varepsilon F)(sI - A - BF)^{-1}E\|_\infty \leqslant \gamma_F + \max\{\lambda_{max}(Q_\varepsilon), \varepsilon\}M.$$

  This together with (4.323) implies that for a given γ, there exists an ε^* such that for $\varepsilon \in (0, \varepsilon^*]$, conditions (*i*) are satisfied.

- Clearly, $(A, B, C_\varepsilon, D_\varepsilon)$ does not have any zero structure (that is, neither invariant zeros nor infinite zeros). One can then design a γ-level H_∞ sub-optimal feedback $F_\varepsilon(E,\gamma)$ using the techniques developed in [161].

- It is easy to see that C_ε and D_ε satisfy (4.322). Then, condition (*ii*) follows from Theorem 4.103 in Appendix 4.D.

- Since we only perturb C and D and $F_\varepsilon(E,\gamma)$ is obtained using H_∞ optimal control techniques, condition (*v*) is obvious.

Therefore, for $\varepsilon \in (0, \varepsilon^*]$, conditions (*i*), (*ii*), (*iii*), (*v*), and (*vi*) are all satisfied. Next, we construct a γ-level H_∞ suboptimal controller using the techniques developed in [161].

In order to construct $F_\varepsilon(E,\gamma)$, let $X_{\varepsilon,\infty}$ be the positive definite solution of H_∞ CARE:

$$A'X_{\varepsilon,\infty} + X_{\varepsilon,\infty}A + C_\varepsilon'C_\varepsilon - X_{\varepsilon,\infty}B'(D_\varepsilon'D_\varepsilon)^{-1}BX_{\varepsilon,\infty} + \gamma^{-2}X_{\varepsilon,\infty}EE'X_{\varepsilon,\infty} = 0.$$

Then a γ-level H_∞ suboptimal static state feedback can be constructed as

$$F_\varepsilon(E,\gamma) = -(D_\varepsilon'D_\varepsilon)^{-1}B'X_{\varepsilon,\infty}.$$

When $D = I_m$, $F_\varepsilon(E,\gamma)$ constructed above is bounded for $\varepsilon \in (0,\varepsilon^*]$. Therefore, the condition (*iv*) is satisfied, and $F_\varepsilon(E,\gamma)$ is a γ-level H_∞ low-gain sequence. Moreover, it recovers the H_∞ ARE-based low-gain design for semi-global stabilization of linear systems subject to input saturation as discussed in Sect. 4.4. When $D = (I_{m_0} \;\; 0)$ with some $m_0 < m$, the boundedness of F_ε needs to be proved. However, we present shortly an alternative perturbation $(C_\varepsilon, D_\varepsilon)$ which automatically yields a bounded $F_\varepsilon(E,\gamma)$.

Example 2. First, we can transform the system into the form (4.316) with transformation $(x_a', x_c')' = T_1 x$. Then consider a perturbed system based on (4.316) as

$$\Sigma_{\infty,I}^{\varepsilon}: \begin{cases} \begin{pmatrix} \rho x_a \\ \rho x_c \end{pmatrix} = \begin{pmatrix} A_a & 0 \\ \star & A_c \end{pmatrix}\begin{pmatrix} x_a \\ x_c \end{pmatrix} + \begin{pmatrix} 0 \\ B_c \end{pmatrix} u_c + \begin{pmatrix} B_a \\ B_{ac} \end{pmatrix} u_0 + \begin{pmatrix} E_a \\ E_c \end{pmatrix}\omega \\ z_\varepsilon = \begin{pmatrix} z_0 \\ z_{\varepsilon,1} \end{pmatrix} = \begin{pmatrix} 0 & 0 \\ \sqrt{Q_\varepsilon} & 0 \end{pmatrix}\begin{pmatrix} x_a \\ x_c \end{pmatrix} + \begin{pmatrix} I_{m_0} & 0 \\ 0 & 0 \end{pmatrix}\begin{pmatrix} u_0 \\ u_c \end{pmatrix}, \end{cases}$$

where Q_ε satisfies (4.323). For the same reasons as argued in the previous example, there exists an ε^* such that for any $\varepsilon \in (0,\varepsilon^*]$, conditions (*i*), (*ii*), (*iii*), (*v*), and (*vi*) are satisfied. It remains to check condition (*iv*).

Next, we construct a γ-level H_∞ suboptimal feedback F_ε for the perturbed system following the design procedure in [161]. Let $P_{\varepsilon,\infty}$ be the positive definite solution of H_∞ CARE,

$$A_a'P_{\varepsilon,\infty} + P_{\varepsilon,\infty}A_a + Q_\varepsilon - P_{\varepsilon,\infty}B_aB_a'P_{\varepsilon,\infty} + \gamma^{-2}P_{\varepsilon,\infty}E_aE_a'P_{\varepsilon,\infty} = 0,$$

and choose a bounded F_c such that $A_c + B_c F_c$ is Hurwitz. Then, $F_\varepsilon(E,\gamma)$ can be constructed as

$$F_\varepsilon(E,\gamma) = \begin{pmatrix} -B_a' P_{\varepsilon,\infty} & 0 \\ 0 & F_c \end{pmatrix} T_1.$$

Clearly, $F_\varepsilon(E,\gamma)$ is bounded for any $\varepsilon \in (0,\varepsilon^*]$. Therefore, $F_\varepsilon(E,\gamma)$ is a γ-level low-gain sequence.

When $D = I_m$, the above method recovers the H_∞ ARE-based low-gain design for semi-global stabilization of linear systems subject to input saturation [181].

Perturbation method II for continuous-time systems: Consider the perturbation $(A + \frac{\varepsilon}{2}I, B, 0, D)$ with ε small enough such that $(A + \frac{\varepsilon}{2}I, B)$ is stabilizable.

- Given $A + \frac{\varepsilon}{2}I + BF$ Hurwitz stable, we have

  $$\|DF(sI - A - BF)^{-1}E\|_\infty \leqslant \|DF(sI - A - \tfrac{\varepsilon}{2}I - BF)^{-1}E\|_\infty.$$

  This implies that conditions (*i*) and (*vi*) are satisfied.

- Since $(A + \frac{\varepsilon}{2}I, B, 0, D)$ always have the same normal rank as that of $(A, B, 0, D)$, condition (*ii*) follows from Theorem 4.102 in Appendix 4.C.

- Since $(A + \frac{\varepsilon}{2}I, B, 0, D)$ does not have any invariant zeros on the imaginary axis, condition (*iii*) is satisfied.

- $A + BF$ is Hurwitz if $A + \frac{\varepsilon}{2}I + BF$ is Hurwitz.

Therefore, conditions (*i*), (*ii*), (*iii*), (*v*), and (*vi*) are satisfied for a sufficiently small ε. Moreover, one can always design a bounded γ-level H_∞ state feedback according to [161] as follows.

There exists a non-singular state transformation independent of ε

$$\begin{pmatrix} x_a^- \\ x_a^0 \\ x_c \end{pmatrix} = T_2 x$$

such that the transformed system is in the SCB form,

$$\bar{\Sigma}_{\infty,II}^{\varepsilon} : \begin{cases} \begin{pmatrix} \dot{x}_a^- \\ \dot{x}_a^0 \\ \dot{x}_c \end{pmatrix} = \begin{pmatrix} A_a^- + \frac{\varepsilon}{2}I & 0 & 0 \\ 0 & A_a^0 + \frac{\varepsilon}{2}I & 0 \\ \star & \star & A_c + \frac{\varepsilon}{2}I \end{pmatrix} \begin{pmatrix} x_a^- \\ x_a^0 \\ x_c \end{pmatrix} \\ \qquad + \begin{pmatrix} 0 \\ 0 \\ B_c \end{pmatrix} u_c + \begin{pmatrix} B_a^- \\ B_a^0 \\ B_{ac} \end{pmatrix} u_0 + \begin{pmatrix} E_a^- \\ E_a^0 \\ E_c \end{pmatrix} \omega \\ z = u_0, \end{cases}$$

where A_a^- is Hurwitz, (A_c, B_c) is controllable, and $(A_a^0, {B_a}^0)$ is controllable. For a sufficiently small ε, $A_a^- + \frac{\varepsilon}{2}I$ is Hurwitz as well. Let $X_{\varepsilon,\infty}$ be the positive definite solution of H_∞ CARE:

$$(A_a^0 + \tfrac{\varepsilon}{2}I)'X_{\varepsilon,\infty} + X_{\varepsilon,\infty}(A_a^0 + \tfrac{\varepsilon}{2}I) - X_{\varepsilon,\infty}B_a^0(B_a^0)'X_{\varepsilon,\infty} \\ + \gamma^{-2}X_{\varepsilon,\infty}E_a^0(E_a^0)'X_{\varepsilon,\infty} = 0.$$

Let F_c be bounded and such that $A_c + B_cF_c$ is Hurwitz, and the γ-level H_∞ suboptimal controller is given by

$$F_\varepsilon(E, \gamma) = \begin{pmatrix} 0 & -(B_a^0)'X_{\varepsilon,\infty} & 0 \\ 0 & 0 & F_c \end{pmatrix} T_2.$$

Since $X_{\varepsilon,\infty}$ is bounded, $F_\varepsilon(E, \gamma)$ is bounded. Therefore, $F_\varepsilon(E, \gamma)$ is a γ-level H_∞ low-gain sequence.

Perturbation method I for discrete-time systems: Similar to the discrete-time H_2 low-gain design, the first class of perturbations for system Σ_∞ in (4.305) is in the form of $(A, B, C_\varepsilon, D_\varepsilon, E)$ where C_ε and D_ε satisfy (4.322). We give two examples.

Example 1:

One classical perturbation for system Σ_∞ which is widely used in the literature is $(A, B, C_\varepsilon, D_\varepsilon, E)$, where

$$C_\varepsilon = \begin{pmatrix} 0 \\ 0 \\ \sqrt{Q_\varepsilon} \end{pmatrix}, \quad D_\varepsilon = \begin{pmatrix} D \\ \varepsilon I \\ 0 \end{pmatrix},$$

and Q_ε satisfies (4.323). We first verify below that this perturbation is admissible for H_∞ low-gain design:

- Condition (*i*) is proved in the proof of Theorem 4.105 in Appendix.
- It is easy to see that C_ε and D_ε satisfy (4.322). Then, the condition (*ii*) follows from Theorem 4.105 in Appendix.
- Clearly, $(A, B, C_\varepsilon, D_\varepsilon)$ does not have invariant zeros. One can then design a discrete-time γ-level H_∞ suboptimal feedback $F_\varepsilon(E, \gamma)$ using the techniques developed in [161].
- Since we only perturb C and D and $F_\varepsilon(E, \gamma)$ is obtained using H_∞ optimal control techniques, condition (vi) is obvious.

Therefore, for $\varepsilon \in (0, \varepsilon^*]$, conditions (*i*), (*ii*), (*iii*), and (*v*) are all satisfied. Next, we construct a discrete-time γ-level H_∞ suboptimal controller using the techniques developed in [161]. Let P_ε be the unique positive semi-definite semi stabilizing solution of H_∞ DARE,

$$P_\varepsilon = A'P_\varepsilon A + Q_\varepsilon - \begin{pmatrix} B'P_\varepsilon A \\ E'P_\varepsilon A \end{pmatrix}' G(P_\varepsilon)^{-1} \begin{pmatrix} B'P_\varepsilon A \\ E'P_\varepsilon A, \end{pmatrix} \tag{4.330}$$

where

$$G(P_\varepsilon) = \begin{pmatrix} D_\varepsilon' D_\varepsilon & 0 \\ 0 & -\gamma^{-2} I \end{pmatrix} + \begin{pmatrix} B' \\ E' \end{pmatrix} P_\varepsilon \begin{pmatrix} B & E \end{pmatrix} \tag{4.331}$$

subject to

$$E_\varepsilon' P_\varepsilon E_\varepsilon + E_\varepsilon' P_\varepsilon B_\varepsilon (D_\varepsilon' D_\varepsilon + B_\varepsilon' P_\varepsilon B_\varepsilon)^{-1} B_\varepsilon' P_\varepsilon E_\varepsilon < \gamma^2 I.$$

Then, a discrete-time γ-level H_∞ suboptimal static state feedback can be constructed as

$$\begin{aligned} F_\varepsilon(E,\gamma) = {} & (B'P_\varepsilon B + D_\varepsilon' D_\varepsilon + B'P_\varepsilon E(\gamma^2 I - E'P_\varepsilon E)^{-1} E'P_\varepsilon B)^{-1} \\ & (B'P_\varepsilon A + B'P_\varepsilon E(\gamma^2 I - E'P_\varepsilon E)^{-1} E'P_\varepsilon A). \end{aligned}$$

If we apply this $u = F_\varepsilon(E,\gamma)x$ to the original system Σ_∞ and the perturbed system $\Sigma_\infty^\varepsilon$ with this class of perturbation, since our perturbation satisfies (4.322), we have $\|z\|_2 \leqslant \|z_\varepsilon\|_2$ for the same initial condition x_0 and w. This implies that

$$\sup_{w \in \ell_2} (\|z\|_2^2 - \gamma^2 \|w\|_2^2) \leqslant \sup_{w \in \ell_2} (\|z_\varepsilon\|_2^2 - \gamma^2 \|w\|_2^2) = x_0' P_\varepsilon x_0.$$

The last equality follows from [161]. Since $P_\varepsilon \to 0$ as $\varepsilon \to 0$ according to Theorem 4.105 of the Appendix, condition (*vi*) is satisfied. Therefore, $F_\varepsilon(E,\gamma)$ is a discrete-time γ-level H_∞ low-gain sequence. Moreover, it recovers the H_∞ ARE-based low-gain design for semi-global stabilization of linear system subject to input saturation introduced in [181].

Example 2:

Similar to that in H_2 low-gain sequence design, we can first transform the system into its SCB form with transformation $(x_a', x_c')' = T_1 x$:

$$\Sigma_{\infty,I} : \begin{cases} \begin{pmatrix} \rho x_a \\ \rho x_c \end{pmatrix} = \begin{pmatrix} A_a & 0 \\ \star & A_c \end{pmatrix} \begin{pmatrix} x_a \\ x_c \end{pmatrix} + \begin{pmatrix} 0 \\ B_c \end{pmatrix} u_c + \begin{pmatrix} B_a \\ B_{ac} \end{pmatrix} u_0 + \begin{pmatrix} E_a \\ E_c \end{pmatrix} w \\ \quad z = u_0. \end{cases}$$

Then, we perturb the above transformed system. After doing so, we get

$$\Sigma^\varepsilon_{\infty,I}: \begin{cases} \begin{pmatrix} \rho x_a \\ \rho x_c \end{pmatrix} = \begin{pmatrix} A_a & 0 \\ \star & A_c \end{pmatrix}\begin{pmatrix} x_a \\ x_c \end{pmatrix} + \begin{pmatrix} 0 \\ B_c \end{pmatrix} u_c + \begin{pmatrix} B_a \\ B_{ac} \end{pmatrix} u_0 + \begin{pmatrix} E_a \\ E_c \end{pmatrix} w \\ \begin{pmatrix} z \\ z_{\varepsilon,1} \end{pmatrix} = \begin{pmatrix} 0 & 0 \\ \sqrt{Q_\varepsilon} & 0 \end{pmatrix}\begin{pmatrix} x_a \\ x_c \end{pmatrix} + \begin{pmatrix} I_{m0} & 0 \\ 0 & 0 \end{pmatrix}\begin{pmatrix} u_0 \\ u_c \end{pmatrix}, \end{cases}$$

where Q_ε satisfies (4.323). For the same reasons as argued in the previous example, there exists an ε^* such that for any $\varepsilon \in (0, \varepsilon^*]$, conditions (*i*), (*ii*), (*iii*), and (*v*) are all satisfied.

Next we construct a discrete-time γ-level H_∞ suboptimal feedback F_ε for the perturbed system following the design procedure in [161]. Let P_ε be the positive semi-definite semi-stabilizing solution of H_∞ DARE

$$P_\varepsilon = A_a' P_\varepsilon A_a + Q_\varepsilon - \begin{pmatrix} B_a' P_\varepsilon A_a \\ E_a' P_\varepsilon A_a \end{pmatrix}' G(P_\varepsilon)^{-1} \begin{pmatrix} B_a' P_\varepsilon A_a \\ E_a' P_\varepsilon A_a, \end{pmatrix}$$

where

$$G(P_\varepsilon) = \begin{pmatrix} I & 0 \\ 0 & -\gamma^{-2} I \end{pmatrix} + \begin{pmatrix} B_a' \\ E_a' \end{pmatrix} P_\varepsilon \begin{pmatrix} B_a & E_a \end{pmatrix}$$

subject to

$$E_a' P_\varepsilon E_a + E_a' P_\varepsilon B_a (I + B_a' P_\varepsilon B_a)^{-1} B_a' P_\varepsilon E_a < \gamma^2 I,$$

and choose F_c such that $A_c + B_c F_c$ is Schur stable. The $F_\varepsilon(E, \gamma)$ can be constructed as

$$F_\varepsilon(E, \gamma) = \begin{pmatrix} \bar{F}_\varepsilon & 0 \\ 0 & F_c \end{pmatrix} T_1,$$

where

$$\bar{F}_\varepsilon = (B_a' P_\varepsilon B_a + I + B_a' P_\varepsilon E_a (\gamma^2 I - E_a' P_\varepsilon E_a)^{-1} E_a' P_\varepsilon B_a)^{-1} \\ (B_a' P_\varepsilon A + B_a' P_\varepsilon E_a (\gamma^2 I - E_a' P_\varepsilon E_a)^{-1} E_a' P_\varepsilon A_a).$$

If we apply this constructed feedback $u = F_\varepsilon(E, \gamma) x$ to the original system Σ_∞ and perturbed system $\Sigma^\varepsilon_\infty$ with this class of perturbation, since our perturbation satisfies (4.322), we have $\|z\|_{\ell_2} \leqslant \|z_\varepsilon\|_{\ell_2}$ for the same initial condition x_0 and w. This implies that

$$\sup_{w \in \ell_2} (\|z\|^2_{\ell_2} - \gamma^2 \|w\|^2_{\ell_2}) \leqslant \sup_{w \in \ell_2} (\|z_\varepsilon\|^2_{\ell_2} - \gamma^2 \|w\|^2_{\ell_2}) = x_a(0)' P_\varepsilon x_a(0),$$

where $x_a(0)$ is the initial condition of x_a dynamics. The last equality follows from [161]. Since $P_\varepsilon \to 0$ as $\varepsilon \to 0$, according to Theorem 4.105 of the Appendix, we find that condition (vi) is satisfied. Therefore, $F_\varepsilon(E,\gamma)$ is a γ-level low-gain sequence.

Perturbation method II for discrete-time systems: We can directly perturb the system dynamics to move those invariant zeros on the unit circle away from it. Consider the perturbation $(A_\varepsilon, B_\varepsilon, 0, D, E_\varepsilon)$ where

$$A_\varepsilon = (1+\varepsilon)A,\ B_\varepsilon = (1+\varepsilon)B,\ E_\varepsilon = (1+\varepsilon)E,$$

and ε small enough such that $((1+\varepsilon)A, (1+\varepsilon)B)$ is stabilizable and $(1+\varepsilon)A$ does not have eigenvalues on the unit circle. We focus on this particular choice of perturbation:

- Given $(1+\varepsilon)A + (1+\varepsilon)BF$ Schur stable, we note that $\|DF(zI - A - BF)^{-1}E\|_\infty \leqslant \|(1+\varepsilon)DF(zI-(1+\varepsilon)A-(1+\varepsilon)BF)^{-1}E\|_\infty$. This implies that condition (*i*) is satisfied.
- Since $((1+\varepsilon)A, (1+\varepsilon)B, 0, D)$ always have the same normal rank as that of $(A, B, 0, D)$, the condition (*ii*) follows from Theorem 4.104 in Appendix.
- Since $((1+\varepsilon)A, (1+\varepsilon)B, 0, D)$ does not have any invariant zeros on the unit circle, the condition (*iii*) is satisfied.
- $A + BF$ is Schur stable if $(1+\varepsilon)A + (1+\varepsilon)BF$ is Schur stable.

Therefore, conditions (*i*), (*ii*), (*iii*), and (*v*) are satisfied for a sufficiently small ε. Moreover, one can always design a bounded discrete-time γ-level H_∞ state feedback according to [161] as follows:

The perturbed system can be transformed into its compact SCB form using a non-singular state transformation. Let $\begin{pmatrix} x_a^{\ominus\prime} & x_a^{\bigcirc\prime} & x_c' \end{pmatrix}' = T_3 x$ is such that

$$\begin{cases} \begin{pmatrix} \rho\bar{x}_a^{\ominus} \\ \rho\bar{x}_a^{\bigcirc} \\ \rho\bar{x}_c \end{pmatrix} = (1+\varepsilon)\begin{pmatrix} A_a^{\ominus} & 0 & 0 \\ 0 & A_a^{\bigcirc} & 0 \\ \star & \star & A_c \end{pmatrix}\begin{pmatrix} \bar{x}_a^{\ominus} \\ \bar{x}_a^{\bigcirc} \\ \bar{x}_c \end{pmatrix} + (1+\varepsilon)\begin{pmatrix} 0 \\ 0 \\ B_c \end{pmatrix} u_c \\ \qquad\qquad +(1+\varepsilon)\begin{pmatrix} B_a^{\ominus} \\ B_a^{\bigcirc} \\ B_{ac} \end{pmatrix} u_0 + (1+\varepsilon)\begin{pmatrix} E_a^{\ominus} \\ E_a^{\bigcirc} \\ E_c \end{pmatrix} w \\ \quad \bar{z} \quad = u_0, \end{cases}$$

where $A_a^{\ominus}$ is Schur stable, (A_c, B_c) is controllable, $(A_a^{\bigcirc}, B_a^{\bigcirc})$ is controllable, and $A_a^{\bigcirc}$ has all its eigenvalues on the unit circle. The eigenvalues of $(1+\varepsilon)A_a^{\bigcirc}$ and $(1+\varepsilon)A_a^{\ominus}$ are the invariant zeros of the perturbed system. For a sufficiently small

ε, $(1+\varepsilon)A_a^{\ominus}$ is also Schur stable. Moreover, T_3 is independent of ε. Let $P_{\varepsilon,\infty}$ be the positive semi-definite semi-stabilizing solution of H_∞ DARE,

$$\frac{1}{(1+\varepsilon)^2}P_{\varepsilon,\infty} = A_a^{\bigcirc\prime} P_{\varepsilon,\infty} A_a^{\bigcirc} - \begin{pmatrix} B_a^{\bigcirc\prime} P_{\varepsilon,\infty} A_a^{\bigcirc} \\ E_a^{\bigcirc\prime} P_{\varepsilon,\infty} A_a^{\bigcirc} \end{pmatrix}' G(P_{\varepsilon,\infty})^{-1} \begin{pmatrix} B_a^{\bigcirc\prime} P_{\varepsilon,\infty} A_a^{\bigcirc} \\ E_a^{\bigcirc\prime} P_{\varepsilon,\infty} A_a^{\bigcirc} \end{pmatrix}, \tag{4.332}$$

where

$$G(P_{\varepsilon,\infty}) = \begin{pmatrix} \frac{1}{(1+\varepsilon)^2} I & 0 \\ 0 & -\frac{\gamma^{-2}}{(1+\varepsilon)^2} I \end{pmatrix} + \begin{pmatrix} B_a^{\bigcirc\prime} \\ E_a^{\bigcirc\prime} \end{pmatrix} P_{\varepsilon,\infty} \begin{pmatrix} B_a^{\bigcirc} & E_a^{\bigcirc} \end{pmatrix}.$$

Let F_c be bounded and such that $A_c + B_c F_c$ is Schur stable, and the γ-level H_∞ suboptimal controller is given by

$$F_\varepsilon(E,\gamma) = \begin{pmatrix} 0 & \bar{F}_\varepsilon & 0 \\ 0 & 0 & F_c \end{pmatrix} T_3,$$

where

$$\bar{F}_\varepsilon = H_\varepsilon^{-1}\left[B_a^{\bigcirc\prime} P_{\varepsilon,\infty} A_a^{\bigcirc} + B_a^{\bigcirc\prime} P_{\varepsilon,\infty} E_a^{\bigcirc} (\gamma^2 I - E_a^{\bigcirc\prime} P_{\varepsilon,\infty} E_a^{\bigcirc})^{-1} E_a^{\bigcirc\prime} P_{\varepsilon,\infty} A_a^{\bigcirc} \right]$$

and

$$H_\varepsilon = B_a^{\bigcirc\prime} P_{\varepsilon,\infty} B_a^{\bigcirc} + I + B_a^{\bigcirc\prime} P_{\varepsilon,\infty} E_a^{\bigcirc} (\gamma^2 I - E_a^{\bigcirc\prime} P_{\varepsilon,\infty} E_a^{\bigcirc})^{-1} E_a^{\bigcirc\prime} P_{\varepsilon,\infty} B_a^{\bigcirc}.$$

Since $P_{\varepsilon,\infty}$ is bounded, $F_\varepsilon(E,\gamma)$ is bounded. Therefore, $F_\varepsilon(E,\gamma)$ is a discrete-time γ-level H_∞ low-gain sequence as discussed in detail below.

Note that if we apply $u = F_\varepsilon(E,\gamma)$ to the original system Σ_∞ and perturbed system $\Sigma_\infty^\varepsilon$ with this perturbation data, we have, for the same initial condition and w, $z_\varepsilon(k) = (1+\varepsilon)^k z(k)$, and hence, $\|z\|_2 \leqslant \|z_\varepsilon\|_2$. Therefore,

$$\sup_{w\in\ell_2} (\|z\|_2^2 - \gamma^2\|w\|_2^2) \leqslant \sup_{w\in\ell_2} (\|\bar{z}\|_2^2 - \gamma^2\|w\|_2^2) = x_a^{\bigcirc\prime}(0) P_\varepsilon x_a^{\bigcirc}(0).$$

The last inequality follows from [161]. Since $P_\varepsilon \to 0$ as $\varepsilon \to 0$, according to Theorem 4.102 of the Appendix, we find that condition (*vi*) is satisfied. Therefore, $F_\varepsilon(E,\gamma)$ is a discrete-time γ-level H_∞ low-gain sequence.

4.8.5 Low-gain and delay

Low-gain methodology can also be used to enhance the tolerance of delay in feedback loop even in the presence of saturation. This is what is pursued here. We note that, in recent years, time-delayed systems have been greeted with great enthusiasm from researchers in recognition of their theoretical and applied importance [118]. Many control problems have been extensively studied, among which stability and stabilization are of particular interest (see, for instance, [24, 37, 43, 59, 106, 107] and reference therein). When both actuator saturation and input time-delay are present, controller design can be challenging. What is worse, the precise knowledge of delay is not available in most circumstances while only an approximation, usually an upper bound, is known. In this case, [97] studied the global asymptotic stabilization for chains of integrators using nested-saturation-type controller originally developed by [180]. This result was later on extended to a class of nonlinear feedforward systems by [96]. Chains of integrators were also studied by [100]. A linear low-gain state feedback was constructed to achieve the semi-global stabilization for integrator chains with input saturation and unknown input delay that has a known upper bound which can be arbitrarily large. A different low-gain design based on the parametric Lyapunov equation was used by [217] to prove a similar result for a broader class of critically unstable systems with eigenvalues on the imaginary axis being zero. Both state and measurement feedback were developed. However, in the measurement feedback case, delays have to be known by the observer.

We investigate here the stabilization of general linear critically unstable systems with multiple unknown constant input delays as well as stabilization of such systems subject to actuator saturation. We give nonconservative upper bounds on the delays which are inversely proportional to the maximal magnitude of the open-loop eigenvalues on the imaginary axis. This makes sense because when a delay is unknown, a system with highly oscillatory behavior is obviously more difficult to stabilize than a system with dynamics that do not change "direction" so frequently. As the eigenvalues on imaginary axis move toward the origin, the upper bounds on delay turn to infinity. For unknown input delays satisfying these bounds, a linear low-gain state or finite dimensional measurement feedback controller can be designed to achieve semi-global stabilization. The design here relies only on the upper bounds.

The development in this subsection is based on our recent work [194].

Throughout this subsection, we denote a diagonal matrix as

$$\operatorname{diag}\{A_i\}_{i=1}^m = \begin{pmatrix} A_1 & 0 & \cdots & 0 \\ 0 & \ddots & \ddots & \vdots \\ \vdots & \ddots & \ddots & 0 \\ 0 & \cdots & 0 & A_m \end{pmatrix}.$$

Also, $\mathcal{C}_\tau^n := C([-\tau, 0], \mathbb{R}^n)$ denotes the Banach space of all continuous functions from $[-\tau, 0] \to \mathbb{R}^n$ with norm $\|x\|_C = \sup_{s\in[-\tau,0]} \|x(s)\|$.

Consider the following system:

$$\begin{cases} \dot{x} &= Ax + \sum_{i=1}^m B_i \sigma[u_i(t-\tau_i)], \\ y &= Cx, \\ x(\theta) &= \phi(\theta), \theta \in [-\bar{\tau}, 0], \end{cases} \tag{4.333}$$

where $x \in \mathbb{R}^n$, $u_i \in \mathbb{R}$, $y \in \mathbb{R}^p$, $\phi \in \mathcal{C}^n_{\overline{\tau}}$. Each input u_i has a delay $\tau_i \in [0, \overline{\tau}_i]$ and $\overline{\tau} = \max \overline{\tau}_i$.

We formulate next formally two semi-global stabilization problems, one with state feedback and another with observer-based measurement feedback:

Problem 4.87 The semi-global asymptotic stabilization via state feedback problem for system (4.333) is to find, if possible, a set of $\overline{\tau}_i > 0$ and in addition to find, for any a priori given bounded set of initial conditions $\mathcal{W} \subset \mathcal{C}^n_{\overline{\tau}}$ with $\overline{\tau} = \max\{\overline{\tau}_i\}$, a linear state feedback controller $u = Fx$ independent of specific delay such that the zero solution of the closed-loop system is locally asymptotically stable for any $\tau_i \in [0, \overline{\tau}_i]$ with $\mathcal{W}$ contained in its domain of attraction, i.e., the following properties hold for all $\tau_i \in [0, \overline{\tau}_i]$, $i = 1, ..., m$:

(*i*) For all $\varepsilon > 0$, there exists a $\delta > 0$ such that for all $\phi \in \mathcal{C}^n_{\overline{\tau}}$ which satisfy $\|\phi\|_C \leqslant \delta$ we have $\|x(t)\| \leqslant \varepsilon$ for all $t \geqslant 0$.

(*ii*) For all $\phi \in \mathcal{W}$, we have that $x(t) \to 0$ as $t \to \infty$.

Problem 4.88 The semi-global asymptotic stabilization via measurement feedback problem for system (4.337) is to find, if possible, a set of $\overline{\tau}_i > 0$, a positive integer $q > 0$, and in addition, for any a priori given bounded set $\mathcal{W} \subset \mathcal{C}^{n+q}_{\overline{\tau}}$ with $\overline{\tau} = \max\{\overline{\tau}_i\}$, to find a dynamic measurement feedback controller independent of delay,

$$\begin{cases} \dot{\chi} &= A_k \chi + B_k y \\ u &= C_k \chi + D_k y \\ \chi(\theta) &= \psi(\theta), \ \forall \theta \in [-\overline{\tau}, 0], \end{cases}$$

where $\chi(t) \in \mathbb{R}^q$ and $\psi \in \mathcal{C}^q_{\overline{\tau}}$ such that the zero solution of the closed-loop system is locally asymptotically stable for all $\tau_i \in [0, \tau_i]$ with $\mathcal{W}$ contained in its domain of attraction, i.e., the following properties hold for all $\tau_i \in [0, \overline{\tau}_i]$:

(*i*) For all $\varepsilon > 0$, there exists a $\delta > 0$ such that for all $\phi \in \mathcal{C}^n_{\overline{\tau}}$ and $\psi \in \mathcal{C}^q_{\overline{\tau}}$ which satisfy $\|\phi\|_C \leqslant \delta$ and $\|\psi\|_C \leqslant \delta$, we have $\|x(t)\| \leqslant \varepsilon$ for all $t \geqslant 0$.

(*ii*) For all $(\phi, \psi) \in \mathcal{W}$, we have that $x(t) \to 0$ and $\chi(t) \to 0$ as $t \to \infty$.

The above two problems are cast in a semi-global setting where the bounded set of initial conditions $\mathcal{W}$ is arbitrarily prescribed a priori. The same problems can be set in the global setting by considering the set $\mathcal{W}$ as the entire Banach space $\mathcal{C}^n_{\overline{\tau}}$ for Problem 4.87 and $\mathcal{C}^{n+q}_{\overline{\tau}}$ for Problem 4.88.

If $\tau_i = 0$, $i = 1, ..., m$, it is well known that the semi-global stabilization problem is solvable only if system (4.337) is asymptotically null controllable with bounded control (ANCBC), i.e., the following assumption holds:

Assumption 4.89 *(A, B) is stabilizable with $B = [B_1 \cdots B_m]$ and A has all its eigenvalues in the closed left-half plane.*

Moreover, for stabilization via measurement feedback, the next assumption is also necessary.

Assumption 4.90 *(A, C) is detectable.*

We assume throughout that the above assumptions hold.

Low-gain design:
Here, we utilize H_2 ARE-based design as discussed in Sect. 4.4.1. Assume that (A, B) is stabilizable and A has all its eigenvalues in the closed left-half plane. For $\varepsilon \in (0, 1]$, let P_ε be the solution of algebraic Riccati equation

$$A'P_\varepsilon + P_\varepsilon A - P_\varepsilon BB'P_\varepsilon + \varepsilon I = 0. \tag{4.334}$$

The low-gain state feedback can be constructed as

$$u = F_\varepsilon = -B'P_\varepsilon x. \tag{4.335}$$

An observer-based low-gain feedback controller which we refer to as a low-gain compensator can be realized as follows:

$$\begin{cases} \dot{\chi} = A\chi + BF_\varepsilon\chi + K(y - C\chi) \\ u = F_\varepsilon\chi, \end{cases} \tag{4.336}$$

where K is chosen such that $A - KC$ is Hurwitz stable. As before, the ε is called a low-gain parameter. With a properly chosen ε, the low-gain feedback (4.335) and low-gain compensator (4.336) solve Problems 4.87 and 4.88, respectively. To prove this, we will proceed in two steps: First, we will show that our controllers globally asymptotically stabilize (4.333) without saturation and provide us with a nonconservative input-delay tolerance. Then, we will prove that our controllers semi-globally asymptotically stabilize the system (4.333) where saturation is present by selecting the low-gain parameter differently.

As said above, we first consider the case when input saturation is not present. That is, we consider the system

$$\begin{cases} \dot{x} &= Ax + \sum_{i=1}^m B_i u_i(t - \tau_i) \\ y &= Cx \\ x(\theta) &= \phi(\theta),\ \forall \theta \in [-\bar{\tau}, 0]. \end{cases} \tag{4.337}$$

Since the system is linear, it is possible to solve the global asymptotic stabilization problems for (4.337) using the low-gain feedback (4.335) and the compensator (4.336).

In order to present our result, we need the following notation. For each input u_i $i = 1, ..., m$, define the *maximal controllable frequency* as

$$\omega_{\max}^i := \max\{\omega \in \mathbb{R} \mid \exists v \in \mathbb{C}^n, \text{ s.t. } A'v = j\omega v \text{ and } B_i'v \neq 0\}. \tag{4.338}$$

It is clear that $j\omega_{\max}^i$ is the eigenvalue of A on the imaginary axis with the maximal magnitude which is at least partially controllable via input channel u_i.

Before we present our main result, we present two lemmas which are instrumental in proving our results. The first lemma is classical and can be found in [31, 214].

Lemma 4.91 *The system* (4.341) *is asymptotically stable if and only if*

$$\det[I - G_\varepsilon(j\omega)\Delta(j\omega)] \neq 0, \ \forall \omega, \ \forall \tau_i \in [0, \overline{\tau}_i], \tag{4.339}$$

where

$$\Delta(s) = \operatorname{diag}\{e^{-\tau_i s} - 1\}_{i=1}^m.$$

Assume that A has r eigenvalues on the imaginary axis which are denoted by $j\omega_k, k = 1, \ldots, r$. Given

$$\overline{\tau}_i < \frac{\pi}{3\omega_{\max}^i}, \quad i = 1, \ldots, m, \tag{4.340}$$

there exists a $\delta > 0$ such that:

(*i*) The neighborhoods $\mathcal{E}_k := (\omega_k - \delta, \omega_k + \delta)$, $k = 1, \ldots, r$ around these eigenfrequencies are mutually disjoint.

(*ii*) For any $i = 1, \ldots, m$ we have that $\omega\overline{\tau}_i < \frac{\pi}{3}$ for $\omega \in \mathcal{E}_k$ for any k for which ω_k is at least partially controllable through input i.

We have the following lemma whose proof is presented in Appendix 4.E:

Lemma 4.92 *The following properties hold:*

(*i*) *If $j\omega_k$ is not controllable via input u_i for some i, then*

$$\lim_{\varepsilon \downarrow 0} F_\varepsilon(j\omega I - A - BF_\varepsilon)^{-1} B e_i = 0,$$

uniformly in ω for $\omega \in \mathcal{E}_k$, where e_i is the standard basis of $\mathbb{R}^m$ and F_ε is given by (4.335).

(*ii*) *There exists an ε^* such that for $\varepsilon \in (0, \varepsilon^*]$,*

$$\|F_\varepsilon(j\omega I - A - BF_\varepsilon)^{-1} B\| \leqslant \tfrac{1}{3}, \ \forall \omega \in \Omega := \mathbb{R} \setminus \cup_{k=1}^r \mathcal{E}_k.$$

Now, we are ready to present the following theorem:

Theorem 4.93 *Consider the system* (4.337) *and assume* (4.340) *is satisfied. For any, there exists an* ε^* *such that for any* $\varepsilon \in (0, \varepsilon^*]$, *the closed-loop of* (4.337) *and the low-gain feedback* (4.335) *is globally asymptotically stable for any* $\tau_i \in [0, \overline{\tau}_i]$, $i = 1, \ldots, m$.

Proof : Consider the closed-loop system

$$\dot{x} = Ax + \sum_{i=1}^{m} B_i F_i x(t - \tau_i). \tag{4.341}$$

Define

$$G_\varepsilon(s) = F_\varepsilon(sI - A - BF_\varepsilon)^{-1} B.$$

Note that in general (4.339) has to be satisfied for all $\omega \in \mathbb{R}$. However, due to the merit of low-gain feedback, we are only concerned with those ω's that are in a finite number of small intervals.

Thanks to Lemma 4.92, we find that there exists an ε_1 such that (4.339) is satisfied if for all $k = 1, \ldots, r$,

$$\det\left[I - G_\varepsilon(j\omega)\widetilde{\Delta}_k(j\omega)\right] \neq 0, \quad \forall \omega \in \mathcal{E}_k, \forall \tau_i \in [0, \overline{\tau}_i], \tag{4.342}$$

provided that $\varepsilon \leqslant \varepsilon_1$ where $\widetilde{\Delta}_k(s)$ is $\Delta(s)$ with $\tau_i = 0$ for all i's such that the eigenvalue $j\omega_k$ is completely uncontrollable via the input channel i.

Let us consider (4.342):

$$\begin{aligned}
&I - G_\varepsilon(j\omega)(\widetilde{D}_k(j\omega) - I)\\
=&I - (I + G_\varepsilon(j\omega))(\widetilde{D}_k(j\omega) - I) + (\widetilde{D}_k(j\omega) - I)\\
=&\widetilde{D}_k(j\omega) - (I + G_\varepsilon(j\omega))(\widetilde{D}_k(j\omega) - I).
\end{aligned}$$

First of all, we know that for all $\varepsilon > 0$ [2, see Sect. 5.4, p.122],

$$\sigma_{\min}[I - F_\varepsilon(j\omega I - A)^{-1} B] \geqslant 1, \quad \forall \omega,$$

and this implies that

$$\sigma_{\max}[I + G_\varepsilon(j\omega)] \leqslant 1, \quad \forall \omega. \tag{4.343}$$

Since $\widetilde{\Delta}_k(j\omega) + I$ is unitary, it is easy to see that given (4.343),

$$(\widetilde{\Delta}_k(j\omega) + I) - (I + G_\varepsilon(j\omega))\widetilde{\Delta}_k(j\omega)$$

is non-singular if $\sigma_{\max}\widetilde{\Delta}_k(j\omega) < 1$. Therefore, we have the condition (4.342) holding for $\varepsilon \leqslant \varepsilon_1$ if for all $k = 1, \ldots, r$,

$$\sigma_{\max}\widetilde{\Delta}_k(j\omega) < 1, \quad \forall \omega \in \mathcal{E}_k, \forall \tau_i \in [0, \overline{\tau}_i]. \tag{4.344}$$

This is guaranteed by our choice of δ and $\mathcal{E}_k$. ∎

In a special case where A has all its eigenvalues at the origin, the low-gain feedback can tolerate any bounded delay that can be arbitrarily large.

Corollary 4.94 *Suppose A has only zero eigenvalues. For any $\overline{\tau}_i > 0$, $i = 1, \ldots, m$, there exists an ε^* such that for $\varepsilon \in (0, \varepsilon^*]$, the closed-loop system of* (4.337) *and* (4.335) *is asymptotically stable for any $\tau_i \in [0, \overline{\tau}_i]$, $i = 1, \ldots, m$.*

Before we present our results for the measurement feedback case, we present the following technical lemma whose proof is presented in Appendix 4.E:

Lemma 4.95 *Let $G_\varepsilon(s) = F_\varepsilon(sI - A - BF_\varepsilon)^{-1}B$. Then*

$$\lim_{\varepsilon \downarrow 0} G_\varepsilon^m(j\omega) = G_\varepsilon(j\omega)$$

uniformly in ω.

The next theorem concerns with the stabilization of (4.337) via measurement feedback.

Theorem 4.96 *Consider the system* (4.337). *If*

$$\overline{\tau}_i < \frac{\pi}{3\omega_{\max}^i}, \quad i = 1, \ldots, m,$$

then there exists an ε^* such that for $\varepsilon \in (0, \varepsilon^*]$, the closed-loop system of (4.337) *and low-gain compensator* (4.336) *is asymptotically stable for $\tau_i \in [0, \overline{\tau}_i]$.*

Proof : The closed-loop system is given by

$$\left\{ \begin{array}{rcl} \dot{x} & = & Ax + \sum_{i=1}^m B_i F_i \chi(t - \tau_i) \\ \dot{\chi} & = & (A + BF_\varepsilon + KC)\chi + KCx \\ x(\theta) = \phi(\theta), & & \forall \theta \in [-\overline{\tau}, 0] \\ \chi(\theta) = \psi(\theta), & & \forall \theta \in [-\overline{\tau}, 0]. \end{array} \right. \tag{4.345}$$

Define

$$G_\varepsilon^m(s) = F_\varepsilon(sI - A - BF_\varepsilon)^{-1}KC(sI - A + KC)^{-1}B.$$

Obviously, $G_\varepsilon^m(s)$ is stable.

It follows from Lemma 4.91 that the closed-loop system of (4.337) and (4.336) is global asymptotically stable if and only if

$$\det[I - G_\varepsilon^m(j\omega)\,(D(j\omega) - I)] \neq 0, \forall \omega \in \mathbb{R}, \forall \tau_i \in [0, \overline{\tau}_i]. \tag{4.346}$$

If, by Theorem 4.93, there exists an $\varepsilon_2 \leqslant \varepsilon_1$ such that for all $\varepsilon \in (0, \varepsilon_2]$ we have (4.339) satisfied with $G_\varepsilon(j\omega)$, then we can find, with the help of Lemma 4.95, an $\varepsilon_3 \leqslant \varepsilon_2$ such that (4.346) holds for all $\varepsilon \in (0, \varepsilon_3]$. ■

We proceed next to extend the above results pertaining to linear systems without actuator saturation to the case where actuator saturation is present, and solve the semi-global stabilization problems as formulated in Problems 4.87 and 4.88.

We have the following result pertaining to Problem 4.87.

Theorem 4.97 *Consider the system* (4.333). *The semi-global asymptotic stabilization via state feedback problem can be solved by a low-gain feedback of the form* (4.335). *Specifically, for a set of positive real numbers*

$$\overline{\tau}_i < \frac{\pi}{3\omega^i_{\max}}, \quad i = 1, \dots, m, \tag{4.347}$$

and for any a priori given compact set of initial conditions $\mathcal{W} \subset \mathcal{C}^n_{\overline{\tau}}$, there exists an ε^* such that for any $\varepsilon \in (0, \varepsilon^*]$, the low-gain feedback (4.335) *achieves local asymptotic stability of the closed-loop system with the domain of attraction containing $\mathcal{W}$ for any $\tau_i \in [0, \overline{\tau}_i]$, $i = 1, \dots, m$.*

Proof : The closed-loop system can be written as

$$\begin{cases} \dot{x} &= Ax + \sum_{i=1}^m B_i \sigma(F_i x(t - \tau_i)) \\ x(\theta) &= \phi(\theta), \ \forall \theta \in [-\overline{\tau}, 0]. \end{cases} \tag{4.348}$$

Suppose $\overline{\tau}_i$'s satisfy the bound (4.347). Let ε_1 be such that the closed-loop system in the absence of saturation, i.e., (4.341), is asymptotically stable. Then, the local stability of (4.348) for $\varepsilon \leqslant \varepsilon_1$ follows.

It remains to show the attractivity. It is sufficient to prove that for the system (4.348), given $\mathcal{W}$, there exists an $\varepsilon^* \leqslant \varepsilon_1$ such that for $\varepsilon \in (0, \varepsilon^*]$, we have

$$\|F_\varepsilon x(t - \overline{\tau})\| \leqslant 1, \forall t \geqslant 0.$$

Then, we can avoid saturation for all $t \geqslant 0$. The closed-loop system becomes linear, and the attractivity of zero solution is therefore guaranteed with $\varepsilon \leqslant \varepsilon_1$.

Let us define two linear time invariant operators g_ε and δ with the following transfer matrices:

$$\begin{aligned} G_\varepsilon(s) &= F_\varepsilon(sI - A - BF_\varepsilon)^{-1}B \\ \Delta(s) &= \operatorname{diag}\{e^{-\tau_i s} - 1\}_{1=1}^m. \end{aligned}$$

Note that the operators g_ε and δ have zero initial conditions. If all the delays satisfy the bound in Theorem 4.93, there exists an ε_1 such that

$$\sigma_{\min}(I - G_\varepsilon(j\omega)\Delta(j\omega)) > \mu, \forall \omega \in \mathbb{R}, \forall \tau_i \in [0, \overline{\tau}_i]$$

for all $\varepsilon \leqslant \varepsilon_1$ and some $\mu > 0$ where μ only depends on $\overline{\tau}_i$, provided that $\varepsilon \leqslant \varepsilon_1$. This implies that

$$\|(I - G_\varepsilon(s)\Delta(s))^{-1}\|_\infty \leqslant \tfrac{1}{\mu}.$$

Moreover, we already have in (4.343)

$$\sigma_{\max}(I + G_\varepsilon(j\omega)) \leqslant I,\ \forall\omega \in \mathbb{R},$$

which implies that $\|G_\varepsilon(s)\|_\infty \leqslant 2$.

Note that for $t \geqslant 0$,

$$\dot{x} = (A + BF_\varepsilon)x + B\delta(F_\varepsilon x) + Bv_\varepsilon,$$

where

$$v_\varepsilon(t) = \begin{pmatrix} v_1(t) \\ \vdots \\ v_m(t) \end{pmatrix}, \quad v_i(t) = \begin{cases} F_i\phi(t - \tau_i), & t < \tau_i, \\ 0, & t \geqslant \tau_i. \end{cases}$$

Since $v_\varepsilon(t)$ vanishes for $t \geqslant \overline{\tau}$, $\phi \in \mathcal{W}$ and $F_\varepsilon \to 0$, we have for any $\phi \in \mathcal{W}$, $\|v_\varepsilon\|_\infty \to 0$ and $\|v_\varepsilon\|_2 \to 0$ as $\varepsilon \to 0$.

We have

$$F_\varepsilon x(t) = F_\varepsilon e^{(A+BF_\varepsilon)t}x(0) + (g_\varepsilon \circ \delta)(F_\varepsilon x)(t) + g_\varepsilon(v_\varepsilon)(t),$$

and hence,

$$F_\varepsilon x(t) = (1 - g_\varepsilon \circ \delta)^{-1}\left[F_\varepsilon e^{(A+BF_\varepsilon)t}x(0) + g_\varepsilon(v_\varepsilon)(t)\right]. \tag{4.349}$$

Let $w_\varepsilon(t) = g_\varepsilon(v_\varepsilon)(t)$. By the definition of g_ε, we have

$$\begin{cases} \dot{\xi} &= (A + BF_\varepsilon)\xi + Bv_\varepsilon, \quad \xi(0) = 0 \\ w_\varepsilon &= F_\varepsilon\xi. \end{cases}$$

Clearly,

$$\|w_\varepsilon\|_2 \leqslant \|G(s)\|_\infty\|v_\varepsilon\|_2 \leqslant 2\|v_\varepsilon\|_2.$$

Hence, for any given initial condition ϕ, $\|w_\varepsilon\|_2 \to 0$ as $\varepsilon \to 0$.

For $t \in [0, \overline{\tau}]$,

$$\begin{aligned} \dot{w}_\varepsilon(t) &= F_\varepsilon(A + BF_\varepsilon)\xi(t) + F_\varepsilon Bv_\varepsilon(t) \\ &= F_\varepsilon(A + BF_\varepsilon)\int_0^t e^{(A+BF_\varepsilon)(t-s)}Bv_\varepsilon(s)ds + F_\varepsilon Bv_\varepsilon(t). \end{aligned}$$

Since $A + BF_\varepsilon$ is bounded for all $\varepsilon \in [0, 1]$ and $s \in [0, \overline{\tau}]$ and $\|v_\varepsilon\|_\infty \to 0$ as $\varepsilon \to 0$, we will have

$$\sup_{t \in [0,\overline{\tau}]} \|\dot{w}_\varepsilon(t)\| \to 0 \text{ as } \varepsilon \to 0. \tag{4.350}$$

This also implies that

$$\int_0^{\overline{\tau}} \|\dot{w}(t)\|^2 dt \to 0 \text{ as } \varepsilon \to 0. \tag{4.351}$$

From $\overline{\tau}$ onward, $v_\varepsilon(t)$ vanishes and

$$\dot{w}(t) = F_\varepsilon e^{(A+BF_\varepsilon)t}(A + BF_\varepsilon)\xi(\overline{\tau}).$$

It is shown by [200] that

$$\int_{\overline{\tau}}^{\infty} \|\dot{w}(t)\|^2 dt \to 0 \text{ as } \varepsilon \to 0, \tag{4.352}$$

provided that $\xi(\overline{\tau})$ is bounded which is obvious by noticing that

$$\xi(\overline{\tau}) = \int_0^{\overline{\tau}} e^{(A+BF)(\overline{\tau}-t)} Bv_\varepsilon(t) dt v$$

and $\|v_\varepsilon\|_\infty \to 0$ as $\varepsilon \to 0$. Combining (4.351) and (4.352), we have shown that for any given $\phi \in \mathcal{W}$, $\|\dot{w}\|_2 \to 0$ as $\varepsilon \to 0$.

Now let us go back to (4.349). We get

$$\begin{aligned}
\|F_\varepsilon x\|_2 \leqslant& \|(1 - G_\varepsilon(s)D(s))^{-1}\|_\infty \|F_\varepsilon e^{(A+BF_\varepsilon)t} x(0)\|_2 \\
&+ \|(1 - G_\varepsilon(s)D(s))^{-1}\|_\infty \|w_\varepsilon\|_2 \\
\leqslant& \tfrac{1}{\mu} \|F_\varepsilon e^{(A+BF_\varepsilon)t} x(0)\|_2 + \tfrac{1}{\mu}\|w_\varepsilon\|_2.
\end{aligned}$$

Since for any ϕ, $\|F_\varepsilon e^{(A+BF_\varepsilon)t} x(0)\|_2 \to 0$ [200] and $v_\varepsilon \to 0$ as $\varepsilon \to 0$ and μ is independent of ε (provided that ε is smaller than ε_1), there exists an ε_3 such that for $\varepsilon \in (0, \varepsilon_3]$, we get

$$\|F_\varepsilon x\|_2 \leqslant \tfrac{1}{2}, \ \forall \phi \in \mathcal{W}. \tag{4.353}$$

Note that (4.349) also yields

$$F\dot{x}(t) = (1 - g_\varepsilon \circ \delta)^{-1} \left[F_\varepsilon e^{(A+BF_\varepsilon)t}(A + BF_\varepsilon)x(0) + \dot{w}_\varepsilon(t) \right],$$

and thus,

$$\begin{aligned}\|F_\varepsilon \dot{x}\|_2 \leqslant & \|(1-G_\varepsilon(s)D(s))^{-1}\|_\infty \|F_\varepsilon e^{(A+BF_\varepsilon)t}\widetilde{x}\|_2 \\ & + \|(1-G_\varepsilon(s)D(s))^{-1}\|_\infty \|\dot{w}_\varepsilon\|_2 \\ \leqslant & \tfrac{1}{\mu}\|F_\varepsilon e^{(A+BF_\varepsilon)t}\widetilde{x}\| + \tfrac{1}{\mu}\|\dot{w}_\varepsilon\|_2\end{aligned}$$

with $\widetilde{x} = (A+BF_\varepsilon)x(0)$. There exists an ε_4 such that for $\varepsilon \in (0, \varepsilon_4]$, we have

$$\|F_\varepsilon \dot{x}\|_2 \leqslant \tfrac{1}{2}, \ \forall \phi \in \mathcal{W}. \tag{4.354}$$

Applying Cauchy-Schwartz inequality, we can prove that for any $t \geqslant 0$,

$$\left| \|F_\varepsilon x(t)\|^2 - \|F_\varepsilon x(0)\|^2 \right| \leqslant 2\|F_\varepsilon \dot{x}\|_2 \|F_\varepsilon x\|_2,$$

and

$$\|F_\varepsilon x(t)\|^2 \leqslant \|F_\varepsilon x(0)\|^2 + 2\|F_\varepsilon \dot{x}\|_2 \|F_\varepsilon x\|_2. \tag{4.355}$$

Finally, there exists an ε_5 such that for $\varepsilon \in (0, \varepsilon_5]$,

$$\|F_\varepsilon x(0)\|^2 \leqslant \|F_\varepsilon \phi\|_C^2 \leqslant \tfrac{1}{2}, \quad \phi \in \mathcal{W}. \tag{4.356}$$

Let

$$\varepsilon^* = \min\{\varepsilon_1, \ldots, \varepsilon_5\}.$$

We conclude from (4.353) to (4.356) that for $\varepsilon \in (0, \varepsilon^*]$,

$$\|F_\varepsilon x(t-\overline{\tau})\| \leqslant 1, \forall t \geqslant 0.$$

■

The next theorem solves Problem 4.88. However, we first present a required lemma, the proof of which is presented in Appendix 4.E.

Lemma 4.98 *For any* $\xi \in \mathbb{R}^{2n}$,

$$\lim_{\varepsilon \downarrow 0} \int_0^\infty \|\mathcal{F}_\varepsilon e^{(\mathcal{A}+\mathcal{B}\mathcal{F}_\varepsilon)t}\xi\|^2 dt = 0,$$

where

$$\mathcal{A} = \begin{pmatrix} A & BF_\varepsilon \\ KC & A+BF_\varepsilon - KC \end{pmatrix}, \ \mathcal{B} = \begin{pmatrix} B \\ 0 \end{pmatrix}, \ \mathcal{F} = \begin{pmatrix} 0 & F_\varepsilon \end{pmatrix}.$$

Theorem 4.99 *Consider the system* (4.333)*. The semi-global asymptotic stabilization via measurement feedback problem can be solved by the low-gain compensator* (4.336)*. Specifically, for any a priori given compact set of initial conditions $\mathcal{W} \subset \mathcal{C}^{2n}_{\overline{\tau}}$ and a set of positive real numbers*

$$\overline{\tau}_i < \frac{\pi}{3\omega^i_{\max}}, \quad i = 1, \ldots, m, \tag{4.357}$$

there exists an ε^* such that for any $\varepsilon \in (0, \varepsilon^*]$, the low-gain feedback (4.336) *achieves local asymptotic stability of the closed-loop system for any $\tau_i \in [0, \overline{\tau}_i]$, $i = 1, \ldots, m$ with the domain of attraction containing $\mathcal{W}$.*

Proof : The closed-loop system can be written as

$$\begin{cases} \dot{x} &= Ax + \sum_{i=1}^m B_i \sigma(F_i \chi(t - \tau_i)) \\ \dot{\chi} &= (A + BF_\varepsilon - KC)\chi + KCx \\ x(\theta) &= \phi(\theta), \ \forall \theta \in [-\overline{\tau}, 0] \\ \chi(\theta) &= \psi(\theta), \ \forall \theta \in [-\overline{\tau}, 0]. \end{cases} \tag{4.358}$$

Assume that $\overline{\tau}_i$'s satisfy the bound (4.357). Let ε_1 be given by Theorem 4.96 such that the closed-loop system without saturation is asymptotically stable. Then, the local stability of (4.358) for $\varepsilon \leqslant \varepsilon_1$ follows.

Define two linear time invariant operators g^m_ε and δ with Laplace transforms as

$$\begin{aligned} G^m_\varepsilon(s) &= F_\varepsilon(sI - A - BF_\varepsilon)^{-1} KC(sI - A + KC)^{-1} B \\ \Delta(s) &= \operatorname{diag}\{e^{-\tau_i s} - 1\}^m_{i=1}. \end{aligned}$$

From the proof of Theorem 4.96, we know that (4.346) holds for $\varepsilon \leqslant \varepsilon_1$. There exists a $\mu > 0$ such that

$$\sigma_{\min}(I - G^m_\varepsilon(j\omega)\Delta(j\omega)) > \mu, \forall \omega \in \mathbb{R}, \forall \tau_i \in [0, \overline{\tau}_i], \tag{4.359}$$

where μ is independent of ε, provided that $\varepsilon \leqslant \varepsilon_1$. It follows from Lemma 4.95 that $G^m_\varepsilon(j\omega) \to G_\varepsilon(j\omega)$ uniformly in ω, where $G_\varepsilon(s) = F_\varepsilon(sI - A - BF_\varepsilon)^{-1}B$. Hence, given $\sigma_{\max}(G_\varepsilon(j\omega)) \leqslant 2$ for any $\varepsilon > 0$ and $\omega \in \mathbb{R}$, there exists an ε_2 such that

$$\sigma_{\max}(G^m_\varepsilon(j\omega)) \leqslant 3, \ \forall \omega \in \mathbb{R}. \tag{4.360}$$

Given (4.359), (4.360), and Lemma 4.98, we can use exactly the same argument as in the proof of Theorem 4.97 to prove that there exists an $\varepsilon^* \leqslant \varepsilon_1$ such that for $\varepsilon \in (0, \varepsilon^*]$,

$$\|F_\varepsilon \chi(t - \overline{\tau})\| \leqslant 1, \ \forall t \geqslant 0, \ (\phi, \psi) \in \mathcal{W}.$$

■

Example 4.100 Consider the following example:

$$\begin{pmatrix}\dot{x}_1\\ \dot{x}_2\\ \dot{x}_3\\ \dot{x}_4\end{pmatrix} = \begin{pmatrix}0 & 1 & 1 & 0\\ 0 & 0 & 0 & 1\\ 0 & 0 & 0 & 1\\ 0 & 0 & -1 & 0\end{pmatrix}\begin{pmatrix}x_1\\ x_2\\ x_3\\ x_4\end{pmatrix} + \begin{pmatrix}0 & 0\\ 1 & 0\\ 0 & 0\\ 0 & 1\end{pmatrix}\begin{pmatrix}u_1(t-\tau_1)\\ u_2(t-\tau_2)\end{pmatrix}$$

$$y_1 = x_1, \quad y_2 = x_2.$$

First, we have

$$\omega_{\max}^1 = 0, \quad \omega_{\max}^2 = 1.$$

The upper bounds on delay are given by

$$\overline{\tau}_1 < \infty, \quad \overline{\tau}_2 < \tfrac{\pi}{3}.$$

In this example, we choose $\overline{\tau}_1 = 1$ and $\overline{\tau}_2 = \frac{\pi}{4}$. The initial condition is given by

$$x(\theta) = \phi(\theta) = \begin{pmatrix}2\\ -2\\ 2\\ -2\end{pmatrix}, \quad \forall\theta \in [-1, 0].$$

We first consider state feedback. Choose $\varepsilon = 0.001$. The low-gain state feedback can be constructed according to (4.335), and this is given by

$$F_\varepsilon = \begin{pmatrix}-0.0281 & -0.2319 & 0.2262 & -0.0587\\ -0.0145 & -0.0587 & 0.0512 & -0.1120\end{pmatrix}.$$

The simulation data is shown in Figs. 4.6 and 4.7.

Let us next consider measurement feedback. The low-gain compensator can be constructed as in (4.336) with

$$K = \begin{pmatrix}7 & 2\\ -1 & 7\\ 11 & 7\\ -7 & 9\end{pmatrix},$$

and the initial condition of the compensator is given by

$$\chi(\theta) = \psi(\theta) = \begin{pmatrix}2\\ 2\\ 2\\ 2\end{pmatrix}, \quad \forall\theta \in [-1, 0].$$

In this case, ε is chosen to be 0.0001. Simulation data is shown in Figs. 4.8 and 4.9.

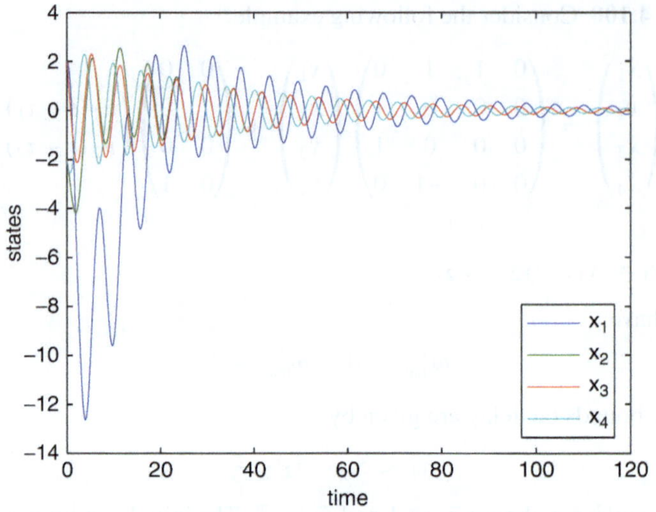

Figure 4.6: Evolution of states

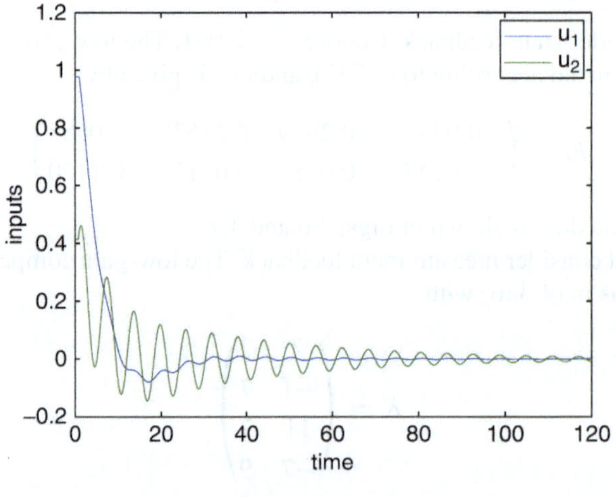

Figure 4.7: Inputs to the system

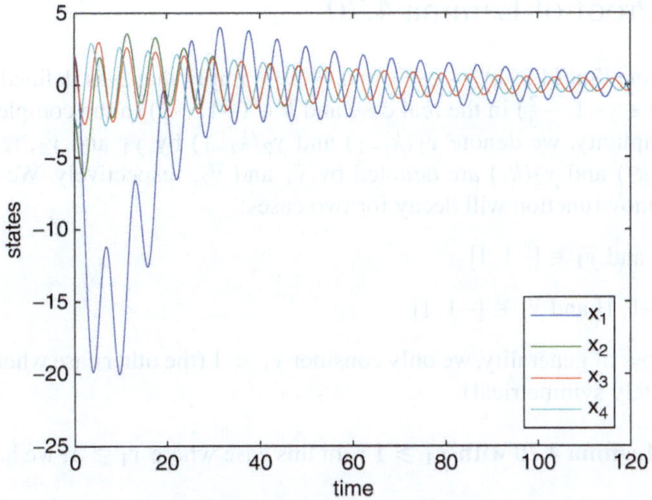

Figure 4.8: Evolution of states

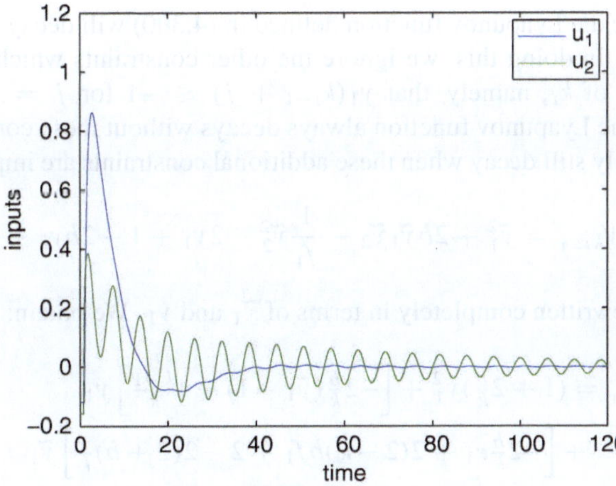

Figure 4.9: Inputs to the system

4.A Proof of Lemma 4.70

We first note that (4.292), (4.295), and (4.303) imply that b as defined in (4.301) satisfies $b \in (-1, -\frac{2}{3})$ in the real case and $b \in (-1, -\frac{3}{4})$ in the complex case.

For simplicity, we denote $y_1(k_{i-1})$ and $y_2(k_{i-1})$ by y_1 and y_2, respectively, while $y_1(k_i)$ and $y_2(k_i)$ are denoted by $\widetilde{y}_1$ and $\widetilde{y}_2$, respectively. We will prove the Lyapunov function will decay for two cases:

- $y_1 \geqslant 1$ and $\widetilde{y}_1 \in [-1, 1]$
- $y_1 \in [-1, 1]$ and $\widetilde{y}_1 \in [-1, 1]$

Without loss of generality, we only consider $y_1 \geqslant 1$ (the other case where $y_1 \leqslant -1$ is completely symmetrical).

Proof of Lemma 4.70 with $y_1 \geqslant 1$: In this case where $y_1 \geqslant 1$, we have

$$\begin{aligned} \widetilde{y}_1 &= y_1 + k y_2 + e_1 \\ \widetilde{y}_2 &= y_2 - (k-2) f_1, \end{aligned}$$

where we denote $k = k_i - k_{i-1}$ while

$$e_1 = f_2 + (k-1)(f_1 - f_2) - \tfrac{f_1}{2}(k-1)(k-2). \tag{4.361}$$

We will prove the Lyapunov function defined in (4.300) will decay if $y_1 \geqslant 1$ and $\widetilde{y}_1 \in [-1, 1]$. In doing this, we ignore the other constraints which follow from the definition of k_i, namely, that $y_1(k_{i-1} + j) \leqslant -1$ for $j = 1, \ldots, k-1$. However, if the Lyapunov function always decays without these constraints, then it will definitely still decay when these additional constraints are imposed. We get

$$V_{k_i} - V_{k_{i-1}} = \widetilde{y}_1^2 + 2b\widetilde{y}_1\widetilde{y}_2 - \frac{1}{f_1}\widetilde{y}_2^2 - 2y_1 + 1 - 2by_2 + \frac{1}{f_1}y_2^2.$$

This can be rewritten completely in terms of $\widetilde{y}_1$ and y_1. We obtain:

$$\begin{aligned} V_{k_i} - V_{k_{i-1}} = {} & (1 + 2\tfrac{b}{k})\widetilde{y}_1^2 + \left[-2\tfrac{b}{k}(\widetilde{y}_1 - 1) - 4\tfrac{k-1}{k}\right] y_1 \\ & + \left[-2\tfrac{b}{k}e_1 + 2(2-k)bf_1 + 2 - 2(2+b)\tfrac{1}{k}\right]\widetilde{y}_1 \\ & - 2e_1 + 2(2+b)\tfrac{1}{k}e_1 - (k-2)^2 f_1 + 1. \end{aligned}$$

We need to show this is negative for all $y_1 \geqslant 1$ and all $\widetilde{y}_1 \in [-1, 1]$. However, this is a linear function of y_1 whose coefficient is negative, and hence, $V_{k_i} - V_{k_{i-1}}$ is maximal for $y_1 = 1$. We find

$$\begin{aligned} V_{k_i} - V_{k_{i-1}} \leqslant {} & (1 + 2\tfrac{b}{k})\widetilde{y}_1^2 + \left[-2\tfrac{b}{k}e_1 + 2(2-k)bf_1 + 2 - 4(1+b)\tfrac{1}{k}\right]\widetilde{y}_1 \\ & - 2e_1 + 2(2+b)\tfrac{1}{k}e_1 + 2\tfrac{b}{k} - 4\tfrac{k-1}{k} - (k-2)^2 f_1 + 1. \end{aligned} \tag{4.362}$$

The upper bound is a quadratic function which we need to maximize. Clearly, the sign of the quadratic term is crucial here. For $k = 1$, the coefficient of the quadratic term is negative, and the maximum is obtained by setting the derivative equal to zero (if we ignore that $\tilde{y}_1 \in [-1, 1]$). We obtain for $k = 1$:

$$\begin{aligned} V_{k_i} - V_{k_{i-1}} \leqslant (1+2b)\tilde{y}_1^2 + [2b(f_1 - f_2) - 2(1+2b)]\tilde{y}_1 \\ +2b + 1 - f_1 + 2(1+b) f_2. \end{aligned} \tag{4.363}$$

For the real case ($f_2^2 + 4f_1 > 0$), since $b = 2f_2^{-1}$ and using (4.292), we obtain from (4.363) that

$$V_{k_i} - V_{k_{i-1}} \leqslant (f_2^2 + 4f_1)(4 + 2f_2 - f_1) f_2^{-1}/(f_2 + 4) < 0.$$

For the complex case ($f_2^2 + 4f_1 < 0$), since $b = -f_2 f_1^{-1}/2$ and again using (4.292), we obtain from (4.363) that

$$V_{k_i} - V_{k_{i-1}} \leqslant (f_2 - f_1)(f_2^2 + 4f_1) f_1^{-1}/4 < 0.$$

Finally, if $f_2^2 + 4f_1 = 0$, then it is easily verified from (4.363) that $V_{k_i} - V_{k_{i-1}} < 0$ unless $y_1 = 1$ and $\tilde{y}_1 = 1 + f_2/2$. However, it can be seen that in that case, $V_{k_{i-1}} = 0$.

Next, we return to the case where $k > 1$. In that case, the upper bound (4.362) has a quadratic term with a positive coefficient. Therefore, the maximum is attained on the boundary, i.e., $\tilde{y}_1 = 1$ or $\tilde{y}_1 = -1$. For $\tilde{y}_1 = 1$ we obtain

$$\begin{aligned} V_{k_i} - V_{k_{i-1}} &\leqslant (k-2)\left[-2bf_1 - \tfrac{4}{k}(f_2 - f_1) - 3f_1 + 2f_2\right] \\ &\leqslant (k-2)\left[-2bf_1 - 2(f_2 - f_1) - 3f_1 + 2f_2\right] \\ &\leqslant (2-k)(2b+1) f_1 \\ &\leqslant 0, \end{aligned}$$

where we used $k \geqslant 2$ and $b < -0.5$. Note that we have the upper bound negative unless $k = 2$ in which case it is easily verified that the decay equals zero only if $\tilde{y}_1 = 1$ and $y_1 = 1$. The latter is inconsistent with $k = 2$ since we then get

$$y(k_{i-1} + 1) = y_1 + y_2 + f_2 = 1 - \tfrac{1}{2} f_1 + f_2 \geqslant -1,$$

where we used $f_2 > \frac{1}{2} f_1 - 2$ for the inequality, while we should have $y(k_{i-1}+1) \leqslant -1$. Next, we need to investigate the other boundary where $\tilde{y}_1 = -1$. We get

$$V_{k_i} - V_{k_{i-1}} \leqslant [2b - (k-2)]\left[(3f_1 - 2f_2) - \tfrac{4}{k}(f_1 - f_2 - 1)\right] < 0. \tag{4.364}$$

The first inequality is a simple rewriting of our upper bound for $\tilde{y}_1 = -1$. The second inequality is more subtle. It is easy to see that $2b - (k-2) < 0$.

If $f_1 - f_2 - 1 \leqslant 0$, we immediately find that the expression is negative since we know from (4.295) that $3f_1 - 2f_2 > 0$. On the other hand, if $f_1 - f_2 - 1 > 0$, then we find that

$$(3f_1 - 2f_2) - \tfrac{4}{k}(f_1 - f_2 + 1) \geqslant (3f_1 - 2f_2) - 2(f_1 - f_2 - 1) = f_1 + 2 > 0,$$

and the inequality (4.364) is satisfied. The fact that $f_1 > -2$ follows from (4.292) and (4.295). ■

Next, we need to study case 2 where

$$y_1 \in [-1, 1] \quad \text{and} \quad \tilde{y}_1 \in [-1, 1]. \tag{4.365}$$

The proof is split into two cases: the real case where $f_2^2 + 4f_1 \geqslant 0$ and the complex case where $f_2^2 + 4f_1 \leqslant 0$.

Proof of Lemma 4.70 with $y_1 \in [-1, 1]$ and $f_2^2 + 4f_1 \geqslant 0$: In this case, we have

$$\tilde{y}_1 = y_1(k_i) = d_4 y_1 + k y_2 + e_4 \tag{4.366}$$
$$\tilde{y}_2 = y_2(k_i) = f_1 y_1 + y_2 - (k-1) f_1, \tag{4.367}$$

where we denote $k = k_i - k_{i-1}$ and

$$d_4 = 1 + f_2 + (k-1) f_1 \tag{4.368}$$
$$e_4 = -(k-1)\left(f_2 - f_1 + \tfrac{1}{2} f_1 k\right). \tag{4.369}$$

Given (4.365), we find that

$$V_{k_i} - V_{k_{i-1}} = \tilde{y}_1^2 + 2b\tilde{y}_1\tilde{y}_2 - \tfrac{1}{f_1}\tilde{y}_2^2 - y_1^2 - 2b y_1 y_2 + \tfrac{1}{f_1} y_2^2.$$

We can eliminate $\tilde{y}_2$ and y_2 from the above expression by using (4.366) and (4.367):

$$\begin{aligned}V_{k_i} - V_{k_{i-1}} &= \tilde{y}_1^2 + 2b\tilde{y}_1\left[f_1 y_1 + \tfrac{1}{k}(\tilde{y}_1 - d_4 y_1 - e_4) - (k-1) f_1\right] \\ &\quad - \tfrac{1}{f_1}\left[f_1 y_1 + \tfrac{1}{k}(\tilde{y}_1 - d_4 y_1 - e_4) - (k-1) f_1\right]^2 - y_1^2 - 2\tfrac{b}{k} y_1\left[\tilde{y}_1 - d_4 y_1 - e_4\right] \\ &\quad + \tfrac{1}{k^2 f_1}\left[\tilde{y}_1 - d_4 y_1 - e_4\right]^2.\end{aligned} \tag{4.370}$$

Our objective is now to prove that (4.370) is negative. We first note that for $k = 1$, we only need to study the unsaturated linear system, and it is easily verified that we have

$$V_{k_i} - V_{k_{i-1}} < 0 \tag{4.371}$$

provided that $V_{k_{i-1}} \neq 0$. For $k = 2$, we will show that (4.370) is negative for all

$$-1 \leqslant y_1 \leqslant 1, \qquad -1 \leqslant \tilde{y}_1 \leqslant -1 - (1 + f_2 - f_1)(y_1 + 1) - f_1, \tag{4.372}$$

where the upper bound for $\tilde{y}_1$ follows from the constraint that $y_1(k_i - 1) \leqslant -1$. For $k > 2$, we consider all

$$-1 \leqslant y_1 \leqslant 1, \qquad -1 \leqslant \tilde{y}_1 \leqslant -1, \tag{4.373}$$

and we ignore all other constraints which follow from the definition of k_i, namely, that $y_1(k_{i-1} + j) \leqslant -1$ for $j = 1, \dots, k - 1$.

The quadratic term in $\tilde{y}_1$ in (4.370) is equal to

$$1 + 2b\tfrac{1}{k}$$

which is positive for $k \geqslant 2$ since $b > -1$. Therefore, we know (4.370) is maximal in a boundary point, i.e.,

$$\tilde{y}_1 = -1 \text{ or } \tilde{y}_1 = -1 - (1 + f_2 - f_1)(y_1 + 1) - f_1$$

for $k = 2$ and

$$\tilde{y}_1 = -1 \quad \text{or} \quad \tilde{y}_1 = 1$$

for $k > 2$. For the lower bound $\tilde{y}_1 = -1$, we do not need to distinguish between $k = 2$ and $k > 2$ and we obtain that

$$\begin{aligned} V_{k_i} - V_{k_{i-1}} &= 1 - 2bf_1\left[y_1 - (k-1)\right] - \tfrac{2b}{k}(1 + y_1)\left[-1 - d_4 y_1 - e_4\right] \\ &\quad - \tfrac{2}{k}\left[y_1 - (k-1)\right]\left[-1 - d_4 y_1 - e_4\right] - f_1\left[y_1 - (k-1)\right]^2 - y_1^2, \end{aligned} \tag{4.374}$$

and we need to show this expression is negative. The expression has the form,

$$\bar{a}y_1^2 + \bar{b}y_1 + \bar{c}. \tag{4.375}$$

Here, we have

$$\begin{aligned} \bar{a} &= (2b+1)f_1 - 1 + \tfrac{2}{k}(b+1)(1 + f_2 - f_1) \\ \bar{b} &= -(1+b)f_1 k + \left[(1+b)(3f_1 - 2f_2) - 2(1 + f_2 - f_1)\right] \\ &\qquad + \tfrac{4}{k}(1+b)(1 + f_2 - f_1) \\ \bar{c} &= \left[(1+b)f_1 - (3f_1 - 2f_2)\right]k + 1 - (2b+1)f_1 + (b+1)(3f_1 - 2f_2) \\ &\qquad - 2(1 + f_2 - f_1) + \tfrac{2}{k}(1+b)(1 + f_2 - f_1). \end{aligned} \tag{4.376}$$

We note that $\bar{a} < 0$. After all, if $1 + f_2 - f_1 \leqslant 0$, we have

$$\bar{a} \leqslant (2b+1)f_1 - 1 \leqslant (2b+1)(-\tfrac{1}{4}f_2^2) - 1 = -(\tfrac{1}{2}f_2 + 1)^2 < 0,$$

where we used that $2b + 1 < 0$, the fact that in the real case $f_2^2 + 4f_1 \geqslant 0$, and the definition of b. If $1 + f_2 - f_1 > 0$, then we obtain that $\overline{a}$ is maximal for $k = 2$ and we obtain,

$$\overline{a} \leqslant bf_1 + (1+b)f_2 + b \leqslant b(-\tfrac{1}{4}f_2^2) + (1+b)f_2 + b = \tfrac{1}{2f_2}(f_2+2)^2 < 0.$$

We need to verify that (4.375) is negative for all $y_1 \in [-1, 1]$. We first verify it is negative in the boundary points. We get for $y_1 = -1$ that (4.375) equals

$$\overline{a} - \overline{b} + \overline{c} = [2(1+b)f_1 - (3f_1 - 2f_2)]\,k < 0.$$

In the proof of Lemma 4.70 with $y_1 \geqslant 1$, we already established that for $y_1 = 1$, we have

$$V_{k_i} - V_{k_{i-1}} < 0.$$

Finally, (4.375) may attain its maximum in the interior where

$$y_1 = -\frac{\overline{b}}{2\overline{a}} \quad \text{with} \quad \left|\frac{\overline{b}}{2\overline{a}}\right| < 1,$$

but then, the maximum is less than $\overline{c} - \overline{a}$, and we get

$$\overline{c} - \overline{a} = [(1+b)f_1 - (3f_1 - 2f_2)]\,k + (3-b)f_1 - (2b+4)f_2.$$

Note that the above expression is a linear function of k whose coefficient is negative since $b > -1$, $f_1 < 0$, and $3f_1 - 2f_2 > 0$, and hence, it is maximal for all $k \geqslant 2$ when $k = 2$, and we get

$$\begin{aligned}\overline{c} - \overline{a} &\leqslant (b-1)f_1 - 4 = \tfrac{1}{f_2}\,[2f_1 - f_2(f_1+4)] \\ &\leqslant \tfrac{1}{f_2}\,[2f_1 + 2(f_1+4)] = \tfrac{4}{f_2}(f_1+2) < 0.\end{aligned}$$

In other words, for $\tilde{y}_1 = -1$, we have (4.370) negative.

Remains to check whether (4.370) is negative for the upper bound for $\tilde{y}_1$. Unfortunately, here, we have to distinguish between $k = 2$ and $k > 2$. For $k = 2$, we have

$$\tilde{y}_1 = -1 - (1 + f_2 - f_1)(y_1 + 1) - f_1$$

for the upper bound. We obtain that

$$V_{k_i} - V_{k_{i-1}} = \hat{a}y_1^2 + \hat{b}y_1 + \hat{c}. \tag{4.377}$$

Here, we have

$$\begin{aligned}\hat{a} &= (1+2b)(f_1 - f_2 - 1)^2 - (f_1+1) + 2(1+b)(f_2+1) \\ \hat{b} &= -2(1+b)(f_1 - f_2 - 1)(f_2+2) - 2b(f_1 - f_2 - 1)(1+f_1) \\ &\qquad + 2(1+b) + 2(f_1 - f_2 - 1) \\ \hat{c} &= (f_2+2)^2 + 2b(f_2+2)(f_1+1) + 2(f_1 - f_2 - 1) + 2f_2 - 3f_1.\end{aligned}$$

Using $-4f_1 \leqslant f_2^2$, we get

$$\widehat{a} \leqslant (1+2b)(f_1 - f_2 - 1)^2 + \tfrac{1}{4f_2}(f_2+4)(f_2+2)^2 < 0$$

since $1+2b<0$ and $-3<f_2<-2$. Therefore, the maximum is attained on the boundary or in the interior. We show that $-\frac{\widehat{b}}{2\widehat{a}} \geqslant 1$. Since $\widehat{a}<0$, it is equivalent to show that $\widehat{b}+2\widehat{a}>0$ and, with some algebra, we get

$$\widehat{b} + 2\widehat{a} = 2(2f_2 - f_1 + 4)[-(f_1 - f_2 - 1) + 1](1 + \tfrac{2}{f_2}) > 0.$$

Thus, for case $k=2$, $V_{k_i} - V_{k_{i-1}}$ (4.375) is maximal for all $y_1 \in [-1,1]$ when $y_1 = 1$, and the maximum is $\widehat{a}+\widehat{b}+\widehat{c}$. Next, we show that this is indeed negative. We get

$$\begin{aligned}\widehat{a}+\widehat{b}+\widehat{c} &= 4(f_2 - \tfrac{1}{2}f_1 + 2)(f_2 + \tfrac{4}{f_2} - \tfrac{1}{2}f_1 + 3)\\ &< 4(f_2 - \tfrac{1}{2}f_1 + 2)(-\tfrac{1}{2}f_1 - 1) < 0.\end{aligned}$$

The following step is to check the upper bound $\widetilde{y}=1$ for $k>2$. We obtain that

$$\begin{aligned}V_{k_i} - V_{k_{i-1}} &= 1 + 2bf_1\,[y_1 - (k-1)] + \tfrac{2b}{k}(1-y_1)\,[1 - d_4y_1 - e_4]\\ &\quad - \tfrac{2}{k}\,[y_1 - (k-1)]\,[1 - d_4y_1 - e_4] - f_1\,[y_1 - (k-1)]^2 - y_1^2,\end{aligned} \tag{4.378}$$

and we need to show this expression is negative. The expression has the form

$$\widetilde{a}y_1^2 + \widetilde{b}y_1 + \widetilde{c}. \tag{4.379}$$

Here, we have

$$\begin{aligned}\widetilde{a} &= (2b+1)f_1 - 1 + \tfrac{2}{k}(b+1)(1+f_2-f_1)\\ \widetilde{b} &= -(b+1)kf_1 + b(3f_1 - 2f_2) - 2 - 4f_2 + 5f_1 + \tfrac{1}{k}(-4b - 4f_1 + 4f_2)\\ \widetilde{c} &= k(-bf_1 + 2f_2 - 2f_1) + 3 + 4f_1 - 4f_2 + b(2f_2 - f_1)\\ &\quad + \tfrac{2}{k}(b-1)(1 - f_2 + f_1).\end{aligned}$$

Since $\widetilde{a}=\bar{a}$ and we already showed that $\bar{a}$ is negative, $\widetilde{a}$ is negative. In the proof of Lemma 4.70 with $y_1 \geqslant 1$, we already established that for $y_1 = 1$, we have

$$V_{k_i} - V_{k_{i-1}} < 0.$$

On the other hand for $y_1 = -1$ we have

$$\widetilde{a} - \widetilde{b} + \widetilde{c} = k(2f_2 - f_1) + 2b(2f_2 - f_1) + 4 + \frac{8b}{k}.$$

For $k=3$, we get

$$6f_2 - 3f_1 + 12 - \tfrac{4}{f_2}f_1 + \tfrac{16}{3f_2} = 6f_2 - (3 + \tfrac{4}{f_2})f_1 + 12 + \tfrac{16}{3f_2} < 6f_2 - 2f_1 + 12 + \tfrac{16}{3f_2}.$$

This upper bound equals

$$\tfrac{2}{3}(2f_2 - 3f_1) + \tfrac{1}{3f_2}(f_2 + 2)(14f_2 + 8) < 0.$$

For $k > 3$, we have

$$\begin{aligned}\widetilde{a} - \widetilde{b} + \widetilde{c} &< k(2f_2 - f_1) + 2b(2f_2 - f_1) + 4 \\ &\leqslant 4(2f_2 - f_1) + 2b(2f_2 - f_1) + 4 = 8f_2 - 4(1 + \tfrac{1}{f_2})f_1 + 12 \\ &< 8f_2 - \tfrac{8}{3}f_1 + 12 < 0.\end{aligned}$$

Remains to show that if the maximum of (4.379) is attained in the interior, i.e., $y_1 \in (-1, 1)$, the maximum of (4.379) is also negative. As before, we note that the maximum is less than $\widetilde{c} - \widetilde{a}$, and we get

$$\widetilde{c} - \widetilde{a} = k[-(b+2)f_1 + 2f_2] + 3(1-b)f_1 + 2(b-2)f_2 + 4 + \tfrac{4}{k}[b(f_1 - f_2) - 1]. \tag{4.380}$$

For $k = 3$, we get

$$\widetilde{c} - \widetilde{a} = -(\tfrac{28}{3f_2} + 3)f_1 + 2f_2 + 4 < \tfrac{1}{9}f_1 + 2f_2 + 4 < 0,$$

while for $k > 3$, we get

$$\begin{aligned}\widetilde{c} - \widetilde{a} &< 4[-(b+2)f_1 + 2f_2] + 3(1-b)f_1 + 2(b-2)f_2 + 4 \\ &= -(7b+5)f_1 + 4(f_2 + 2).\end{aligned}$$

If $7b + 5 < 0$, then this expression is negative since $f_1 < 0$ and $f_2 < -2$. On the other hand, if $7b + 5 > 0$, then

$$\begin{aligned}-(7b+5)f_1 + 4f_2 + 8 &< \tfrac{28}{f_2} + 4f_2 + 18 \\ &< \tfrac{22}{f_2} + 4f_2 + 16 = \tfrac{2}{f_2}\left(2(f_2 + 2)^2 + 3\right) < 0,\end{aligned}$$

since $f_1 \in (-2, 0)$ and $f_2 \in (-3, -2)$. This completes the proof of Lemma 4.70 with $y_1 \in [-1, 1]$ and $f_2^2 + 4f_1 \geqslant 0$. ■

Next, we study the complex case:

Proof of Lemma 4.70 with $y_1 \in [-1, 1]$ and $f_2^2 + 4f_1 \leqslant 0$: We again want to establish that $V_{k_i} - V_{k_{i-1}} < 0$. However, in this case, it is not sufficient to consider the case $\widetilde{y}_1 \in [-1, 1]$ and $y_1 \in [-1, 1]$ since in that case the result is simply not true for f_1 and f_2 sufficiently small. But recall that we ignored the constraints that $y_1(k_{i-1} + j) \leqslant -1$ for $j = 1, \ldots, k-1$. In this case we, actually ignore the constraint that $\widetilde{y}_1 < 1$ and replace it by the constraint that $y_1(k_i - 1) < -1$. Note that

$$y_1(k_i - 1) = d_3 y_1 + (k-1)y_2 + e_3 \leqslant -1, \tag{4.381}$$

where

$$\begin{aligned} d_3 &= 1 + f_2 + (k-2) f_1 \\ e_3 &= -(k-2)[f_2 - f_1 + \tfrac{1}{2} f_1 (k-1)]. \end{aligned}$$

We also recall (4.366) and (4.367), and we again obtain (4.370). However, this time, we want to prove that (4.370) is negative for all y_1 and $\widetilde{y}_1$ for which $y_1 \in [-1, 1]$ and

$$-1 \leqslant \widetilde{y}_1 \leqslant -1 - (1 + f_2 - f_1)\tfrac{1}{k-1}(y_1 + 1) - \tfrac{1}{2} f_1 k,$$

where the upper bound for $\widetilde{y}_1$ follows from (4.381). We note that (4.370) is a quadratic function in $\widetilde{y}_1$, and the quadratic term has coefficient $1 + 2b\frac{1}{k}$ which is positive for $k \geqslant 2$ (note that, like in the real case, for $k = 1$, we have a linear system without saturation, and hence, we can trivially verify $V_{k_i} - V_{k_{i-1}} < 0$). Hence, for $k \geqslant 2$, (4.370) attains its maximum on the boundary where either $\widetilde{y}_1 = -1$ or

$$\widetilde{y}_1 \leqslant -1 - (1 + f_2 - f_1)\tfrac{1}{k-1}(y_1 + 1) - \tfrac{1}{2} f_1 k.$$

On the boundary $\widetilde{y}_1 = -1$, we have (4.370) equal to

$$\widetilde{a}_1 y_1^2 + \widetilde{b}_1 y_1 + \widetilde{c}_1, \tag{4.382}$$

where

$$\begin{aligned} \widetilde{a}_1 &= (2b+1) f_1 - 1 + \tfrac{2}{k}(b+1)(1 + f_2 - f_1) \qquad (4.383)\\ \widetilde{b}_1 &= -k f_1 (1+b) - 2(1 - f_1 + f_2) + (b+1)(3 f_1 - 2 f_2) \\ &\quad + \tfrac{4}{k}(b+1)(1 + f_2 - f_1) \\ \widetilde{c}_1 &= 2k(f_2 - f_1) + k b f_1 - 1 + 4 f_1 - 4 f_2 + b(f_1 - 2 f_2) \\ &\quad + \tfrac{2}{k}(1+b)(1 + f_2 - f_1). \end{aligned}$$

We have

$$\widetilde{a}_1 \leqslant (1+b) f_2 + b f_1 + b = \tfrac{f_2}{2 f_1}(f_1 - f_2 - 1) < 0,$$

where we used that $\widetilde{a}_1$ is maximal for $k = 2$ and that in the complex case, where $f_2^2 + 4 f_1 \leqslant 0$, we have $2 b f_1 = -f_2$ and

$$1 + f_2 - f_1 > 0. \tag{4.384}$$

We note that

$$\widetilde{b}_1 - 2\widetilde{a}_1 = -k f_1 (1+b) + (1+b)(3 f_1 - 2 f_2) > 0,$$

since $b > -1$, $f_1 < 0$, $k > 0$, and $3 f_1 - 2 f_2 > 0$. This implies that

$$-\tfrac{\widetilde{b}_1}{2\widetilde{a}_1} > -1,$$

which implies that (4.382) attains its maximum for $y_1 > -1$ (recall that $y_1 \in [-1, 1]$). Next, we assume that the maximum is attained for $y_1 \in (-1, 1)$ which implies that $y_1 = -\frac{\widetilde{b}_1}{2\widetilde{a}_1}$. The maximum is then equals

$$\widetilde{c}_1 - \frac{\widetilde{b}_1^2}{4\widetilde{a}_1} < \widetilde{c}_1 - \widetilde{a}_1,$$

where we used that $\widetilde{a}_1 < 0$ and $|\frac{\widetilde{b}_1}{2\widetilde{a}_1}| < 1$. In that case, we obtain,

$$\widetilde{c}_1 - \widetilde{a}_1 = k[(1+b)f_1 - (3f_1 - 2f_2)] + [-1 + 3f_1 - 4f_2 - b(f_1 + 2f_2)],$$

which is maximal for $k = 2$, and hence, we obtain

$$\widetilde{c}_1 - \widetilde{a}_1 \leqslant -1 - f_1 + b(f_1 - 2f_2) = -1 - f_1 - \tfrac{1}{2} f_2 + \tfrac{f_2}{f_1} f_2 < -1 - f_1 + f_2 < 0,$$

where we used that $2f_2 < 3f_1$. It remains to show that (4.382) is negative if the maximum is attained for $y_1 = 1$. In that case, the maximum equals

$$\begin{aligned}\widetilde{a}_1 + \widetilde{b}_1 + \widetilde{c}_1 = k(2f_2 - 3f_1) - 4 + 10f_1 - 8f_2 + 2b(3f_1 - 2f_2) \\ + \tfrac{8}{k}(1+b)(1 + f_2 - f_1),\end{aligned}$$

which is maximal for $k = 2$ and hence less than or equal to

$$b(4 + 2f_1),$$

which is negative. The above establishes that $V_{k_i} - V_{k_{i-1}} < 0$ if it attains its maximum on the boundary where $\widetilde{y}_1 = -1$. Remains to show that $V_{k_i} - V_{k_{i-1}} < 0$ if it attains its maximum on the other boundary where

$$\widetilde{y}_1 = -1 - (1 + f_2 - f_1)\tfrac{1}{k-1}(y_1 + 1) - \tfrac{1}{2} f_1 k.$$

In that case, we get that $V_{k_i} - V_{k_{i-1}}$ as given in (4.370) is equals

$$\widetilde{a}_2(y_1 + 1)^2 + \widetilde{b}_2(y_1 + 1) + \widetilde{c}_2, \tag{4.385}$$

where

$$\begin{aligned}
\widetilde{a}_2 &= (1 + 2b)\left[\tfrac{1}{k-1}(1 + f_2 - f_1)\right]^2 + \tfrac{2}{k-1}(b+1)(1 + f_2 - f_1) \\
&\qquad - (1 + f_2 - f_1) \\
\widetilde{b}_2 &= \tfrac{1}{k-1}(1 + f_2 - f_1)\left[(1+b)f_1 + b(3f_1 - 2f_2)\right] + (b+1)(3f_1 - 2f_2) \\
&\qquad + (1 + f_2 - f_1)f_1(1 + 2b) - (1+b)f_1(k-1) \\
\widetilde{c}_2 &= (f_2 - f_1)\left[-\tfrac{1}{4}k^2 f_1 + k(1 + \tfrac{1}{2} f_2)\right].
\end{aligned}$$

It is easily verified that in the region of interest, we have that $\widetilde{c}_2 < 0$ since

$$-\tfrac{1}{4}k^2 f_1 + k(1 + \tfrac{1}{2} f_2) > -\tfrac{1}{4}k^2 f_1 + k(1 + f_1) = -\tfrac{1}{4}(k-2)^2 f_1 + k + f_1 > 0.$$

Secondly, for $\widetilde{b}_2$, it is easily verified that the coefficient of $1/(k-1)$ is negative while the coefficient of $(k-1)$ is positive which implies that $\widetilde{b}_2$ is increasing in k and attains its maximum for $k = 2$. Moreover, for $k = 2$, we find that $\widetilde{b}_2$ is equals

$$2(b+1)(f_1 - f_2)(2 + f_2 - f_1) > 0.$$

We therefore note that $\widetilde{b}_2 > 0$ for $k \geqslant 2$. Finally, $\widetilde{a}_2 < 0$ in the region of interest since $1 + 2b < 0$ while $b + 1 > 0$ and $1 + f_2 - f_1 > 0$ imply

$$\tfrac{2}{k-1}(b+1)(1+f_2-f_1)-(1+f_2-f_1) < 2(b+1)(1+f_2-f_1)-(1+f_2-f_1) < 0.$$

We need to show that (4.385) is negative for all $y_1 \in [-1, 1]$. We use two bounds:

$$\begin{aligned}
\widetilde{a}_2 < \bar{a}_2 &= -\tfrac{1}{2}\left[\tfrac{1}{k-1}(1 + f_2 - f_1)\right]^2 - \tfrac{1}{2}(1 + f_2 - f_1)\\
\widetilde{b}_2 < \bar{b}_2 &= (1 + f_2 - f_1)\tfrac{1}{k-1}(f_1 - \tfrac{1}{2} f_2) + (1 + f_2 - f_1)(f_1 - f_2)\\
&\quad - 2f_2 + 3f_1 - (f_1 - \tfrac{1}{2} f_2)(k-1),
\end{aligned}$$

where we used that $2bf_1 = -f_2$ and we note that $2f_1 - f_2 < 0$ in our region of interest.

We get that (4.385) is negative for $y_1 = -1$ since $\widetilde{c}_2 < 0$. For $k = 2$, it is also negative for $y_1 = 1$ since

$$4\widetilde{a}_2 + 2\widetilde{b}_2 + \widetilde{c}_2 = \tfrac{1}{f_1}(f_1 - f_2)(f_1 - f_2 - 2)(f_1 - 2f_2 - 4) < 0.$$

For $y_1 = 1$ and $k = 3$, we get

$$\begin{aligned}
4\widetilde{a}_2 + 2\widetilde{b}_2 + \widetilde{c}_2 &= \tfrac{1}{4}(4 - 2f_2^2 + f_2 f_1 + f_1^2) - \tfrac{1}{f_1}(3f_2 + f_2^2)\\
&< \tfrac{1}{4}(-14 - 6f_2 - 2f_2^2 + f_2 f_1 + f_1^2)\\
&< \tfrac{1}{4}(-14 - 8f_2 - 2f_2^2 + f_1^2)\\
&= \tfrac{1}{4}(f_1^2 - 4) - \tfrac{1}{2}(1 + (f_2 + 2)^2) < 0,
\end{aligned}$$

where in the first inequality we used that $3f_1 > 2f_2$ and $f_2 > -3$ while for $k > 3$ we have

$$4\widetilde{a}_2 + 2\widetilde{b}_2 + \widetilde{c}_2 \leqslant 4\bar{a}_2 + 2\bar{b}_2 + \widetilde{c}_2,$$

and

$$\begin{aligned}4\bar{a}_2 + 2\bar{b}_2 + \tilde{c}_2 = &-2\left[\tfrac{1}{k-1}(1+f_2-f_1)\right]^2 - \tfrac{1}{4}k(f_2-f_1)(kf_1-2f_2))\\ &+ (k-2)(2f_2-3f_1+2(f_1-\tfrac{1}{2}f_2)\\ &+ 2(1+f_2-f_1)\left[\tfrac{1}{k-1}(f_1-\tfrac{1}{2}f_2)-(1+f_2-f_1)\right] < 0.\end{aligned}$$

Remains to prove that the maximum of (4.385) is also negative if (4.385) attains its maximum in the interior of the interval $(-1,1)$. We get

$$\tilde{a}_2(y_1+1)^2 + \tilde{b}_2(y_1+1) + \tilde{c}_2 \leqslant \tilde{c}_2 - \frac{\tilde{b}_2^2}{4\tilde{a}_2} < \tilde{c}_2 + \tilde{b}_2 < \tilde{c}_2 + \bar{b}_2,$$

where we used that the maximum is attained in the interior and hence

$$-\frac{\tilde{b}_2}{2\tilde{a}_2} < 2.$$

We find that

$$\begin{aligned}\tilde{c}_2 + \bar{b}_2 = &-\tfrac{1}{4}(f_2-f_1)(k^2f_1-2kf_2) - 2k(f_1-\tfrac{3}{4}f_2)\\ &+ (1+f_2-f_1)\tfrac{1}{k-1}(f_1-\tfrac{1}{2}f_2) - (f_1-f_2)^2 + (5f_1-\tfrac{7}{2}f_2).\end{aligned}$$

The first term is decreasing in k for $k \geqslant 2$ since $2f_1 - f_2 < 0$ and $f_2 - f_1 < 0$. It is then easy to verify that this complete upper bound is decreasing in k and therefore is maximal for $k = 2$, and we get

$$\tilde{c}_2 + \bar{b}_2 \leqslant (f_1 - \tfrac{1}{2}f_2)(2 - f_1 + f_2) < 0.$$

This completes the proof. ■

4.B Proof of Lemma 4.71

Due to the symmetry, we only need to consider the case that $y_1(k_i) \leqslant -1$. The case where $y_1(k_i) \geqslant 1$ then follows trivially. Note that the case when $y_1(k_i) \in [-1,1]$ has already been treated by Lemma 4.70. As before, we define

$$k = k_i - k_{i-1}.$$

By definition, we have $k \geqslant 1$. On the other hand, $k = 1$ would imply

$$y_1(k_i - 1) \geqslant -1, \quad y_1(k_i) \leqslant -1, \quad y_1(k_i+1) \geqslant 1,$$

and it is easily verified, given the system dynamics (4.298), that this can only happen if $f_1 > 2f_2 + 4$, which contradicts Jury's conditions (4.292). Therefore, we only need to address the case where $y_1(k_{i-1}+j) = -1$ for $j = 1,\ldots,k$ with $k \geqslant 2$.

Proof of Lemma 4.71 for $f_1 > -1.6$ with $y_1(k_{i-1}) \in [-1, 1]$ and $f_2^2 + 4f_1 \geqslant 0$: From the system equations (4.298), we have

$$y_1(k_i + 1) = y_1(k_{i-1} + k + 1) = d_5 y_1 + (k + 1)y_2 + e_5,$$

where $d_5 = 1 + f_2 + kf_1$ and $e_5 = -k(f_2 + \frac{1}{2}(k-1)f_1)$. Since $y_1(k_i + 1) \geqslant 1$, we get

$$y_2 \geqslant \tfrac{1}{k+1}(1 - e_5 - d_5 y_1). \tag{4.386}$$

As argued before, our Lyapunov candidate has a constant decay, which is given in (4.302), if we are in -1 for two consecutive time instants; thus, we analyze $V_{k_i} - V_{k_{i-1}}$ as

$$\begin{aligned} V_{k_i} - V_{k_{i-1}} &= V_{k_{i-1}+1} - V_{k_{i-1}} + (k-1)[(2b-1)f_1 + 2f_2] \\ &= -2(1 + bf_1 + f_2)y_1 - 2(1+b)(1+y_1)y_2 - 1 - (f_1+1)y_1^2 \\ &\quad + (k-1)[(2b-1)f_1 + 2f_2]. \end{aligned}$$

Since $b > -1$ and $-1 < y_1 < 1$, the term $-2(1+b)(1+y_1)y_2$ is maximal for minimal y_2, and using the bound (4.386), we get

$$\begin{aligned} V_{k_i} - V_{k_{i-1}} &\leqslant -2(1 + bf_1 + f_2)y_1 - 2(1+b)(1+y_1)\tfrac{1}{k+1}(1 - e_5 - d_5 y_1) \\ &\quad - 1 - (f_1+1)y_1^2 + (k-1)[(2b-1)f_1 + 2f_2] \\ &= \bar{a}_3 y_1^2 + \bar{b}_3 y_1 + \bar{c}_3, \end{aligned} \tag{4.387}$$

where

$$\begin{aligned} \bar{a}_3 &= [(2b+1)f_1 - 1] + \tfrac{2}{k+1}(1+b)(f_2 - f_1 + 1) \\ \bar{b}_3 &= -(1+b)f_1 k + (1+b)(4f_1 - 2f_2) - 2(1 + bf_1 + f_2) \\ &\quad + \tfrac{4}{k+1}(1+b)(f_2 - f_1) \\ \bar{c}_3 &= [(1+b)f_1 + 2f_2 - 3f_1]k + 2(1+b)(f_1 - f_2) - (2b-1)f_1 - 2f_2 - 1 \\ &\quad + \tfrac{2}{k+1}(1+b)(f_2 - f_1 - 1). \end{aligned}$$

Note that $\bar{a}_3 < 0$ since it is the same as $\bar{a}$ given in (4.376) (with k replaced by $k+1$), which has been shown to be negative. Next, let us show that $V_{k_i} - V_{k_{i-1}} < 0$.

We show that $-\frac{\bar{b}_3}{2\bar{a}_3} > 1$ for all $k \geqslant 1$, which implies that our bound for $V_{k_i} - V_{k_{i-1}}$ for all $k \geqslant 1$ is maximal at $y_1 = 1$. Since $\bar{a}_3 < 0$, this implies that we need to show that $\bar{b}_3 + 2\bar{a}_3 > 0$. We get

$$\begin{aligned} \bar{b}_3 + 2\bar{a}_3 &= -(1+b)f_1 k + (1+b)(4f_1 - 2f_2) - 2(1 + bf_1 + f_2) \\ &\quad + 2[(1+2b)f_1 - 1] + \frac{4}{k+1}(1+b)(2f_2 - 2f_1 + 1). \end{aligned}$$

Since $2f_2 - 2f_1 + 1 < 0$ and $1 + b > 0$ in the region of interest, we find

$$\begin{aligned}\bar{b}_3 + 2\bar{a}_3 &> -(1+b)f_1 + (1+b)(4f_1 - 2f_2) - 2(1 + bf_1 + f_2) \\ &\qquad + 2[(1+2b)f_1 - 1] + 2(1+b)(2f_2 - 2f_1 + 1) \\ &= (1+b)(2 + f_1) > 0,\end{aligned}$$

where we used that $bf_2 = 2$ and $1 + b > 0$ and $f_1 > -2$. We find that our bound for $V_{k_i} - V_{k_{i-1}}$ for all $k \geqslant 1$ is maximal at $y_1 = 1$, and hence,

$$V_{k_i} - V_{k_{i-1}} \leqslant \bar{a}_3 + \bar{b}_3 + \bar{c}_3,$$

and we find

$$\begin{aligned}\bar{a}_3 + \bar{b}_3 + \bar{c}_3 = k(2f_2 - 3f_1) + 4(2f_1 - 2f_2 - 1) + 4b(f_1 - f_2) \\ - \frac{8}{k+1}(1+b)(f_1 - f_2). \qquad (4.388)\end{aligned}$$

Unfortunately, this is not always negative (choose $k = 3$), and we conclude that we cannot guarantee that $V_{k_i} - V_{k_{i-1}} < 0$. However, it can be verified that it is negative, provided that $f_1 > -1.6$. ■

Proof of Lemma 4.71 for $f_1 > -1.6$ with $y_1(k_{i-1}) \in [-1, 1]$ and $f_2^2 + 4f_1 \leqslant 0$: Similar to the proof for the real case, we obtain (4.387) with the same expressions for $\bar{a}_3, \bar{b}_3$, and $\bar{c}_3$. This time the fact that $\bar{a}_3 < 0$ follows from the fact that it is the same as $\tilde{a}_1$ as defined in (4.383) (with k replaced by $k + 1$) which has been shown to be negative.

We first show that the bound in (4.387) is negative for both $y_1 = -1$ and $y_1 = 1$. We get for $y_1 = -1$,

$$\bar{a}_3 - \bar{b}_3 + \bar{c}_3 = (f_2 - f_1)k < 0.$$

On the other hand, for $y_1 = 1$, we get the same expression (4.388) as in the real case. Also, in the complex case, this expression is not guaranteed to be negative (choose $k = 4$). However, it can be verified that it is negative, provided that $f_1 > -1.6$, since (4.387) is less than

$$3(2f_2 - 3f_1) + 4(2f_1 - 2f_2 - 1) + 4b(f_1 - f_2) = 2f_2 - f_1 - 4 + 4b(f_1 - f_2),$$

which is negative for $f_1 > -1.6$. Remains to check that, if $f_1 > -1.6$, then (4.387) is also negative if the maximum is attained in the interior. Using the same arguments as before and the fact that $\bar{a}_3 < 0$, we find that the maximum is less than

$$\bar{c}_3 - \bar{a}_3 = k(\tfrac{3}{2}f_2 - 2f_1) + 2f_1 - 3f_2 - 2bf_2 - \frac{4}{k+1}(1+b),$$

and we find that

$$\bar{c}_3 - \bar{a}_3 < (\tfrac{3}{2} f_2 - 2f_1) + 2f_1 - 3f_2 - 2bf_2 = b(3f_1 - 2f_2) < 0. \qquad \blacksquare$$

$\blacksquare$

Proof of Lemma 4.71 for $f_1 > -1.6$ with $y_1(k_{i-1}) \geqslant 1$ and $f_2^2 + 4f_1 \geqslant 0$: From the system equations (4.298), we have

$$y_1(k_i + 1) = y_1(k_{i-1} + k + 1) = y_1 + (k+1)y_2 + e_2,$$

where $e_2 = f_2 + k(f_1 - f_2) - \frac{f_1}{2}k(k-1)$, and, for ease of presentation, we denote $y_1(k_{i-1}) = y_1$ and $y_2(k_{i-1}) = y_2$. Since $y_1(k_i + 1) \geqslant 1$, we get

$$y_2 \geqslant \tfrac{1}{k+1}(1 - y_1 - e_2). \tag{4.389}$$

As noted before, if $\sigma(y_1)$ stays at 1 for two consecutive time instants, our Lyapunov candidate actually has a constant decay, which is given in (4.302). Therefore, we obtain

$$\begin{aligned} V_{k_i} - V_{k_{i-1}} &= V_{k_{i-1}+1} - V_{k_{i-1}} + (k-1)[(2b-1)f_1 + 2f_2] \\ &= -4y_1 - 4(1+b)y_2 - (1+2b)f_1 - 2f_2 \\ &\qquad\qquad + (k-1)[(2b-1)f_1 + 2f_2]. \end{aligned}$$

Since $b > -1$, the term $-4(1+b)y_2$ is maximal for minimal y_2, i.e., (4.389); thus, we get

$$\begin{aligned} V_{k_i} - V_{k_{i-1}} \leqslant{}& [-4 + 4(1+b)\tfrac{1}{k+1}]y_1 - 4(1+b)\tfrac{1}{k+1}(1 - c) \\ &- (1+2b)f_1 - 2f_2 + (k-1)[(2b-1)f_1 + 2f_2]. \end{aligned}$$

Since $b < 0$, we have $-4 + 4(1+b)\frac{1}{k+1} < 0$ for all $k \geqslant 2$, and hence, the upper bound is maximal for minimal y_1, i.e., $y_1 = 1$. With some algebra, we get

$$V_{k_i} - V_{k_{i-1}} \leqslant \bar{a}_4 k + \bar{b}_4 + \bar{c}_4 \tfrac{1}{k+1}, \tag{4.390}$$

where

$$\begin{aligned} \bar{a}_4 &= 2f_2 - 3f_1, \\ \bar{b}_4 &= -4 + 4(b+2)(f_1 - f_2) \\ \bar{c}_4 &= 8(1+b)(f_2 - f_1). \end{aligned}$$

This is equal to the upper bound found in (4.388), and it can be verified that, in the real case, this bound can become positive for $k = 3$ for certain f_1 and f_2 in the region of interest. On the other hand, as already noted earlier, this expression is negative, provided that $f_1 > -1.6$. $\blacksquare$

Proof of Lemma 4.71 for $f_1 > -1.6$ with $y_1(k_{i-1}) \geqslant 1$ and $f_2^2 + 4f_1 \leqslant 0$: We get the same expressions as in the proof for the case $f_2^2 + 4f_1 \geqslant 0$ resulting in (4.390). This is equal to the upper bound found in (4.388), and it can be verified that, in the complex case, this bound can become positive for $k = 4$ for certain f_1 and f_2 in the region of interest while, as already noted earlier, this expression is negative, provided that $f_1 > -1.6$. ∎

The problem we have is that the Lyapunov candidate given in (4.300) increases for some initial conditions y_1 and y_2, i.e.,

$$V_{k_i} - V_{k_{i-1}} = -4y_1 - 4(1+b)y_2 - (1+2b)f_1 - 2f_2 + (k-1)[(2b-1)f_1 + 2f_2] > 0 \quad (4.391)$$

when $y_1(k_{i-1}) > 1$, or

$$V_{k_i} - V_{k_{i-1}} = -2(1+bf_1+f_2)y_1 - 2(1+b)(1+y_1)y_2 - 1 - (f_1+1)y_1^2 + (k-1)[(2b-1)f_1 + 2f_2] > 0 \quad (4.392)$$

when $y_1(k_{i-1}) \in [-1, 1]$, where $k = k_i - k_{i-1}$. We have verified before that this can only happen when $f_1 < -1.6$. We will show that if (4.391) is positive, then $V(k_{i+1}) - V(k_{i-1}) < 0$. We proceed to show this by considering several scenarios depending on whether or not $y_1(k_{i-1})$ or $y_1(k_{i+1})$ is saturated, and we will use the notation that $\ell = k_{i+1} - k_i$.

Proof of Lemma 4.71 for $f_1 < -1.6$ with $y_1(k_{i-1})$ and $y_1(k_{i+1})$ both saturated : Due to the symmetry, we only need to consider the case where $y_1(k_{i-1}) \geqslant 1$, and then, $\sigma(y_1)$ switches from $+1$ to -1, and stays at -1 for k steps, after which it switches to $+1$ and stays at $+1$ for ℓ steps, and finally $y_1(k_{i+1}+1) \leqslant -1$. Clearly, for this case, we need $k \geqslant 2$ and $\ell \geqslant 2$.

We obtain

$$V_{k_{i+1}} - V_{k_i} = 4y_1(k_i) + 4(1+b)y_2(k_i) - (1+2b)f_1 - 2f_2 + (\ell-1)[(2b-1)f_1 + 2f_2]. \quad (4.393)$$

Combining (4.391) and (4.393), we get

$$V_{k_{i+1}} - V_{k_{i-1}} = 4[ky_2 + e_1] - 4(1+b)(k-2)f_1 - 2(1+2b)f_1 - 4f_2 + (k+\ell-2)[(2b-1)f_1 + 2f_2], \quad (4.394)$$

where e_1 is defined in (4.361). We need to show that $V_{k_{i+1}} - V_{k_{i-1}} < 0$ given (4.391) and the following constraints on y_1 and y_2:

$$\begin{aligned}
4y_1(k_{i-1}) &= y_1 \geqslant 1 && (4.395)\\
y_1(k_{i-1}+1) &= y_1 + y_2 + f_2 \leqslant -1 && (4.396)\\
\vdots \quad &\quad \vdots \\
y_1(k_i) &= y_1 + ky_2 + e_1 \leqslant -1 && (4.397)\\
y_1(k_i+1) &= y_1 + (k+1)y_2 + e_5 \geqslant 1 && (4.398)\\
\vdots \quad &\quad \vdots \\
y_1(k_{i+1}) &= y_1 + (k+\ell)y_2 + e_1 + e_6 - (k-2)\ell f_1 \geqslant 1 && (4.399)\\
y_1(k_{i+1}+1) &= y_1 + (k+\ell+1)y_2 + e_1 + e_7 - (k-1)(\ell+1)f_1 \leqslant -1, && (4.400)
\end{aligned}$$

where $k = k_i - k_{i-1}$ and $\ell = k_{i+1} - k_i$ and

$$\begin{aligned}
e_5 &= f_2 + k(f_1 - f_2) - \tfrac{f_1}{2}k(k-1),\\
e_6 &= -f_2 - (\ell-1)(f_1 - f_2) + \tfrac{f_1}{2}(\ell-1)(\ell-2),\\
e_7 &= -f_2 - \ell(f_1 - f_2) + \tfrac{f_1}{2}\ell(\ell-1).
\end{aligned} \tag{4.401}$$

We first note that if $k = 2$, we get

$$y_1(k_{i-1}) = y_1 \geqslant 1, \qquad y_1(k_i+2) = y_1 + 4y_2 \geqslant 1,$$

and then

$$-1 \geqslant y_1(k_i) = y_1 + 2y_2 + f_1 > 1 + f_1,$$

which yields a contradiction with $f_1 > -2$. Therefore, we have $k \geqslant 3$. We claim that $\ell \geqslant k - 4$. Since $\ell \geqslant 2$, we only need to prove this property for $k \geqslant 6$. We have:

$$\begin{aligned}
y_1(k_i+j) &= y_1 + (k+j)y_2 + (k-j)(f_1 - f_2) - \tfrac{f_1}{2}(k-1)(k-2)\\
&\quad + \tfrac{f_1}{2}(j-1)(j-2) - (k-2)jf_1\\
&= \tfrac{k+j-1}{k}\left(y_1 + (k+1)y_2\right) - \tfrac{j-1}{k}\left(y_1 + y_2\right) + (k-j)(f_1 - f_2)\\
&\quad - \tfrac{f_1}{2}(k-1)(k-2) + \tfrac{f_1}{2}(j-1)(j-2) - (k-2)jf_1.
\end{aligned}$$

Using the inequalities (4.396) and (4.398) in the above, we get

$$\begin{aligned}
y_1(k_i+j) &\geqslant \tfrac{k+j-1}{k}(1 - e_5) + \tfrac{j-1}{k}(1 + f_2) + (k-j)(f_1 - f_2)\\
&\quad - \tfrac{f_1}{2}(k-1)(k-2) + \tfrac{f_1}{2}(j-1)(j-2) - (k-2)jf_1\\
&= \tfrac{k+2j-2}{k} - f_2 - (2j-1)(f_1 - f_2) + \tfrac{1}{2}f_1k + \tfrac{1}{2}f_1j^2\\
&\quad - \tfrac{1}{2}f_1kj + \tfrac{1}{2}f_1.
\end{aligned}$$

Note that this lower bound is a concave function in j. Therefore, if this larger bound is larger than or equal to 1 for $j = 1$ and $j = k - 4$, then it is larger than for all j satisfying $1 \leqslant j \leqslant k - 4$, and this implies that $\ell \geqslant k - 4$. For $j = 1$, the lower bound is actually equal to 1, while for $j = k - 4$, we find

$$y_1(k_i + k - 4) \geqslant \tfrac{3k-10}{k} + (k-5)(2f_2 - \tfrac{7}{2}f_1).$$

For $f_1 < -1.6$, we have $2f_2 - \frac{7}{2}f_1 > 0$, and hence, this expression is increasing in k. The minimum is achieved for $k = 6$, and we get

$$y_1(k_i + k - 4) \geqslant \tfrac{4}{3} + 2f_2 - \tfrac{7}{2}f_1$$

which is larger than 1 in the critical region. This completes the proof that $\ell \geqslant k - 4$.

Using (4.395) and (4.397) in (4.394), we get

$$\begin{aligned} V_{k_{i+1}} - V_{k_{i-1}} \leqslant\ & -8 - 4(1+b)(k-2)f_1 - 2(1+2b)f_1 - 4f_2 \\ & + (k + \ell - 2)[(2b-1)f_1 + 2f_2]. \end{aligned}$$

If we use $\ell \geqslant k$, we get

$$\begin{aligned} V_{k_{i+1}} - V_{k_{i-1}} &\leqslant -8 - 4(1+b)(k-2)f_1 - 2(1+2b)f_1 - 4f_2 \\ &\quad + (2k-2)[(2b-1)f_1 + 2f_2]. \\ &= -8 + 8f_1 - 8f_2 + k(-6f_1 + 4f_2) \\ &< -8 - 4f_1 \\ &< 0 \end{aligned}$$

for $k \geqslant 2$. Therefore, it only remains $h = k - \ell \in \{1,2,3,4\}$. Inequality (4.400) combined with inequality (4.395) yields

$$(k + \ell + 1)y_2 \leqslant -2 + (k-1)(\ell+1)f_1 - (e_1 + e_7),$$

and working this out using the definitions of e_1 and e_7, we get

$$(2k - h + 1)y_2 \leqslant -2 + \left[k^2 - k + 1 - \tfrac{1}{2}h(h+1)\right] f_1 + (h-1)f_2.$$

Using this bound in (4.394), we get

$$\begin{aligned} V_{k_{i+1}} - V_{k_{i-1}} \leqslant\ & -2k^2 f_1 + 4kf_1 + hf_1 - 2hf_2 - 2hbf_1 \\ & + \tfrac{4k}{2k-h+1}\left[-2 + \left[k^2 - k + 1 - \tfrac{1}{2}h(h+1)\right] f_1 + (h-1)f_2\right]. \end{aligned}$$

Rewriting this equation, we obtain

$$\begin{aligned} V_{k_{i+1}} - V_{k_{i-1}} \leqslant\ & -4 + 2kf_1 + (2-h^2)f_1 - 2f_2 - 2hbf_1 \\ & + \tfrac{2h-2}{2k-h+1}\left[-2 + \left[k^2 - k + 1 - \tfrac{1}{2}h(h+1)\right] f_1 + (h-1)f_2\right]. \end{aligned}$$

It is easily verified that this upper bound is negative in the specified region for $h = 1$. For $h = \{2, 3, 4\}$, we want to show this upper bound is decreasing in k, and therefore, we differentiate the upper bound with respect to k. This results in

$$\tfrac{2(h-1)}{(2k-h+1)^2}\Big[\big(\tfrac{4}{h-1}+2\big)k^2 f_1 - (2+2h)kf_1 + 4 + (2-2h)f_2 + (h^2+3h-4)f_1\Big].$$

This is clearly negative, provided that

$$\big(\tfrac{4}{h-1}+2\big)k^2 f_1 - (2+2h)kf_1 + 4 + (2-2h)f_2 + (h^2+3h-4)f_1 \tag{4.402}$$

is negative. This is a simple quadratic function in k which achieves its maximum for $k = (h-1)/2 < 3$. In our case, we know $k = \ell + h \geqslant 2 + h$. Using that in (4.402), we get

$$4 - 2(h-1)f_2 + \left[(h+4)(h+5) + \frac{36}{h-1}\right] f_1,$$

which is easily checked to be negative in the specified region for $h = 2, 3, 4$. This proves that our upper bound for $V_{k_{i+1}} - V_{k_{i-1}}$ is decreasing in k, and hence, we only need to verify the worst case when $k = h + 2$. In that case, we find

$$V_{k_{i+1}} - V_{k_{i-1}} \leqslant \frac{1}{h+5}\Big[-8h - 16 + (h^2+17h+24)f_1 + (2h^2-6h-8)f_2 - 2(h^2+5h)f_1 b\Big].$$

It is then straightforward to verify that for the three remaining cases $h = 2, 3, 4$, we have $V_{k_{i+1}} - V_{k_{i-1}}$ negative. This completes the proof. ■

Proof of Lemma 4.71 for $f_1 < -1.6$ with $y_1(k_{i-1})$ saturated and $y_1(k_{i+1})$ unsaturated : Again, due to the symmetry, we only need to consider the case where $y_1(k_{i-1}) \geqslant 1$, and then, $\sigma(y_1)$ switches from $+1$ to -1 and stays at -1 for k steps, then it switches to $+1$ and stays at $+1$ for $\ell - 1$ steps, and finally $y_1(k_{i+1}) \in [-1, 1]$. Clearly, for this case, we need $k \geqslant 2$, and $\ell \geqslant 1$.

The arguments used in the case where $y_1(k_{i-1})$ and $y_1(k_{i+1})$ are both saturated immediately imply that we have $\ell \geqslant k - 3$ for $k \geqslant 4$. Note that $\ell = k - 4$ in our case is not possible since in the earlier argument it was shown that $y_1(k_i+k-4) > 1$ while we currently consider the case that $y_1(k_{i+1})$ is unsaturated. We claim that we also have $\ell \leqslant k$. We first note that the bounds (4.395)–(4.398) are still valid when $y(k_{i+1})$ is unsaturated. However, (4.399) and (4.400) no longer hold, and instead, we have

$$y_1(k_{i+1}-1) = y_1 + (k+\ell-1)y_2 + e_1 + e_8 - (k-2)(\ell-1)f_1 \geqslant 1$$

$$y_1(k_{i+1}) = y_1 + (k+\ell)y_2 + e_1 + e_6 - (k-2)\ell f_1 \geqslant -1 \tag{4.403}$$

$$y_1(k_{i+1}) = y_1 + (k+\ell)y_2 + e_1 + e_6 - (k-2)\ell f_1 \leqslant 1, \tag{4.404}$$

where

$$e_8 = -f_2 - (\ell - 2)(f_1 - f_2) + \tfrac{f_1}{2}(\ell - 2)(\ell - 3).$$

Now, if we assume that $\ell > k$, then we have

$$y_1(k_i) = y_1 + ky_2 + e_1 \leqslant -1 \tag{4.405}$$

$$y_1(k_i + k) = y_1 + 2ky_2 - (k-2)kf_1 \geqslant 1 \tag{4.406}$$

together with (4.395). From (4.406), we obtain

$$ky_2 \geqslant \tfrac{1}{2}(1 - y_1 + (k-2)kf_1).$$

Using this combined with (4.395) in (4.405), we get

$$1 + \tfrac{1}{2}(k-2)kf_1 + e_1 \leqslant -1,$$

which yields

$$f_1 + \tfrac{1}{2}(k-2)(3f_1 - 2f_2) \leqslant -2.$$

We obtain a contradiction since $k \geqslant 2$, $f_1 > -2$ and $3f_1 - 2f_2 > 0$, and hence, we must have $\ell \leqslant k$. We obtain

$$\begin{aligned} V_{k_{i+1}} - V_{k_i} &= (y_1(k_i) + \ell y_2(k_i) + e_6)^2 \\ &\quad + 2b(y_1(k_i) + \ell y_2(k_i) + e_6)[y_2(k_i) + (\ell - 2) f_1] \\ &\quad - 2(\ell - 2) y_2(k_i) - (\ell - 2)^2 f_1 + 2y_1(k_i) + 1 + 2by_2(k_i), \end{aligned}$$

which yields

$$\begin{aligned} V_{k_{i+1}} - V_{k_{i-1}} &= \{y_1 + ky_2 + e_1 + \ell[y_2 - (k-2) f_1] + e_6\}^2 \\ &\quad + 2b\,\{y_1 + ky_2 + e_1 + \ell[y_2 - (k-2) f_1] + e_6\}\,[y_2 + (\ell - k) f_1] \\ &\quad - 2(\ell - 2)[y_2 - (k-2) f_1] - (\ell - 2)^2 f_1 \\ &\quad + 2[y_1 + ky_2 + e_1] + 1 + 2b[y_2 - (k-2) f_1] \\ &\quad - 4y_1 - 4(1+b) y_2 - (1+2b) f_1 - 2f_2 \\ &\quad + (k-1)[(2b-1) f_1 + 2f_2]. \end{aligned} \tag{4.407}$$

We will show that $V_{k_{i+1}} - V_{k_{i-1}} < 0$ for all y_1 and y_2 satisfying (4.395), (4.403), and (4.404). By ignoring some of the constraints, we actually prove that $V_{k_{i+1}} - V_{k_{i-1}} < 0$ for a larger class of y_1 and y_2.

Note that the coefficient of y_2^2 term in (4.407) is $(k + \ell)^2 + 2b(k + \ell)$ which is positive since $b > -1$ and $k + \ell \geqslant 3$. Thus, $V_{k_{i+1}} - V_{k_{i-1}}$ is maximal as a function of y_2 if y_2 takes a boundary value. Recall that we ignore all constraints on y_1 and y_2 except (4.395), (4.403), and (4.404). Hence, a boundary value for y_2 implies that either (4.403) or (4.404) is an equality.

In case (4.403) is an equality, we get

$$(2k-h)y_2 = -1 + (-\tfrac{1}{2}h^2 - \tfrac{1}{2}h + k^2 - 2k)f_1 + hf_2 - y_1,$$

where $\ell = k - h$ with $h \in \{0, 1, 2, 3\}$. This yields

$$V_{k_{i+1}} - V_{k_{i-1}} = \tfrac{2h-4b}{2k-h}\left[-1 + (-\tfrac{1}{2}h^2 - \tfrac{1}{2}h + k^2 - 2k)f_1 + hf_2 - y_1\right] + 2 - 2y_1 + 2hbf_1 - h^2 f_1.$$

We note that this expression is linear in y_1 with a negative coefficient, and hence, it is maximal for $y_1 = 1$ given (4.395). We obtain

$$\begin{aligned} V_{k_{i+1}} - V_{k_{i-1}} &\leqslant \tfrac{2h-4b}{2k-h}\left[-2 + (-\tfrac{1}{2}h^2 - \tfrac{1}{2}h + k^2 - 2k)f_1 + hf_2\right] - (h-2b)if_1 \\ &\leqslant \tfrac{h-2b}{2k-h}\left[-4 + (2k^2 - 2(2+h)k - h)f_1 + 2hf_2\right]. \end{aligned}$$

The sign of the above upper bound is determined by the sign of

$$-4 + (2k^2 - 2(2+h)k - h)f_1 + 2hf_2.$$

This expression is decreasing in k given that $k \geqslant 2$ and $k \geqslant h + 1$, and hence, the maximum is obtained for $k = \max\{2, h + 1\}$, and it is then easily verified that this expression is negative in the region of interest for $h \in \{0, 1, 2, 3\}$ which establishes that

$$V_{k_{i+1}} - V_{k_{i-1}} < 0 \tag{4.408}$$

if (4.403) is an equality. The only other possible alternative is that (4.404) is an equality. In that case, we obtain

$$(2k-h)y_2 = 1 + (-\tfrac{1}{2}h^2 - \tfrac{1}{2}h + k^2 - 2k)f_1 + hf_2 - y_1,$$

where $\ell = k - h$ with $h \in \{0, 1, 2, 3\}$. This yields

$$V_{k_{i+1}} - V_{k_{i-1}} = 2 - 2y_1 + \tfrac{2h}{2k-h}\left[1 + (-\tfrac{1}{2}h^2 - \tfrac{1}{2}h + k^2 - 2k)f_1 + hf_2 - y_1\right] - 2hbf_1 - h^2 f_1.$$

We note that this expression is linear in y_1 with a negative coefficient, and hence, it is maximal for $y_1 = 1$ given (4.395). We obtain

$$\begin{aligned} V_{k_{i+1}} - V_{k_{i-1}} &\leqslant \tfrac{2h}{2k-h}\left[(-\tfrac{1}{2}h^2 - \tfrac{1}{2}h + k^2 - 2k)f_1 + hf_2\right] - 2hbf_1 - h^2 f_1 \\ &\leqslant \tfrac{h}{2k-h}\left[(2k^2 - 2(2b + h + 2)k + (2b - 1)h)f_1 + 2hf_2\right]. \end{aligned} \tag{4.409}$$

For $h = 0$, this establishes that

$$V_{k_{i+1}} - V_{k_{i-1}} \leqslant 0. \tag{4.410}$$

An equality would imply $y_1 = 1$ and $2y_2 = (k-2)f_1$ in which case we obtain

$$y_1(k_i) = 2(f_2 - f_1 + 1) - 1 + \tfrac{k}{2}(3f_1 - 2f_2) \geqslant f_1 + 1 > -1,$$

where in the first inequality, we used that $3f_1 - 2f_2 > 0$ and $k \geqslant 2$. This yields a contradiction with (4.397), and hence, we must have a strict inequality in (4.410) for $h = 0$.

The sign of the upper bound in (4.409) for $h \in \{1, 2, 3\}$ is determined by the sign of

$$(2k^2 - 2(2b + h + 2)k + (2b - 1)h)f_1 + 2hf_2.$$

This expression is decreasing in k given that $k \geqslant 2$ and $k \geqslant h + 1$, and hence, the maximum is obtained for $k = h + 1$, and it is then easily verified that for $h \in \{1, 2, 3\}$, the choice $k = h + 1$ yields $2f_2 - (5 + 6b)f_1$, $4f_2 - (8 + 8b)f_1$, and $6f_2 - (11 + 10b)f_1$, respectively, which are all negative in the area of interest. This establishes that

$$V_{k_{i+1}} - V_{k_{i-1}} < 0,$$

and this completes the proof. ■

Proof of Lemma 4.71 for $f_1 < -1.6$ with $y_1(k_{i-1})$ unsaturated and $y_1(k_{i+1})$ saturated : Clearly, in this case, we have $k \geqslant 2$ and $\ell \geqslant 2$. The following constraints are satisfied:

$$y_1(k_{i-1}) = y_1 \in (-1, 1) \tag{4.411}$$
$$y_1(k_{i-1} + 1) = (1 + f_2)y_1 + y_2 \leqslant -1 \tag{4.412}$$
$$\vdots \qquad \vdots$$
$$y_1(k_i) = d_4 y_1 + k y_2 + e_4 \leqslant -1 \tag{4.413}$$
$$y_1(k_i + 1) = (d_4 + f_1)y_1 + (k + 1)y_2 + e_4 - (k - 1)f_1 - f_2 \geqslant 1 \tag{4.414}$$
$$\vdots \qquad \vdots$$
$$y_1(k_{i+1}) = y_1(k_i) + \ell y_2(k_i) + e_6 \geqslant 1 \tag{4.415}$$
$$y_2(k_{i+1}) = y_2(k_i) + (\ell - 2)f_1 \tag{4.416}$$
$$y_1(k_{i+1} + 1) = y_1(k_i) + \ell y_2(k_i) + e_6 + y_2(k_i) + (\ell - 2)f_1 - f_2 \leqslant -1. \tag{4.417}$$

Notice that $y_1(k_i)$ and $y_2(k_i)$ are given in (4.366) and (4.367), respectively, d_4 and e_4 are defined in (4.368) and (4.369), while e_6 is defined in (4.401).

We will show that ℓ satisfies $k - 4 \leqslant \ell < k$. We first establish that $\ell \geqslant k - 4$. Since $\ell \geqslant 2$, we only need to show this property for $k \geqslant 6$.

Using (4.414) to obtain a lower bound for y_2, we get

$$\begin{aligned} y_1(k_i + j) &= y_1(k_i) + j y_2(k_i) - f_2 - (j-1)(f_1 - f_2) + \tfrac{f_1}{2}(j-1)(j-2) \\ &\geqslant \tfrac{(f_1-f_2-1)(j-1)}{k+1} y_1 + (k+j)\left(\tfrac{f_1-f_2+1}{k+1} + f_2 - f_1 + \tfrac{f_1}{2}k\right) + e_4 \\ &\quad - j(k-1)f_1 - f_2 - (j-1)(f_1 - f_2) + \tfrac{f_1}{2}(j-1)(j-2). \end{aligned}$$

Note that this lower bound is a concave function in j. Therefore, if this lower bound is larger than or equal to 1 for $j = 1$ and $j = k - 4$, then it is larger than 1 for all j satisfying $1 \leqslant j \leqslant k - 4$, and this implies that $\ell \geqslant k - 4$. For $j = 1$, the lower bound is actually equal to 1, while for $j = k - 4$, we find using that $y_1 \in (-1, 1)$ that

$$y_1(k_i + j) \geqslant -\left|\tfrac{(f_1-f_2-1)(k-5)}{k+1}\right| + \tfrac{2k-4}{k+1}(f_1 - f_2 + 1) - f_1(4k-19) + f_2(2k-9).$$

If $f_1 - f_2 - 1 > 0$, we get

$$\begin{aligned} y_1(k_i + j) &\geqslant 2(f_2 - 2f_1)k + 20f_1 - 10f_2 + 3 - \tfrac{12}{k+1} \\ &\geqslant 2(f_2 - 2f_1) + \tfrac{9}{7} > 1, \end{aligned}$$

where we have used that $f_2 > 2f_1$ and that $k \geqslant 6$. On the other hand, if $f_1 - f_2 - 1 < 0$, we get

$$\begin{aligned} y_1(k_i + j) &\geqslant 2(f_2 - 2f_1)k + 22f_1 - 12f_2 + 1 - \tfrac{12(f_1-f_2)}{k+1} \\ &\geqslant \tfrac{12}{7}(f_2 - 2f_1) - \tfrac{2}{7}f_1 + 1 > 1, \end{aligned}$$

where we have used that $2f_1 < f_2 < f_1$ and that $k \geqslant 6$. Next, we establish that $\ell \leqslant k$. We show this by contradiction. Assume that $\ell \geqslant k$, then we have

$$y_1(k_i + k) = y_1(k_i) + k y_2(k_i) - f_2 - (k-1)(f_1 - f_2) + \tfrac{f_1}{2}(k-1)(k-2) \geqslant 1.$$

Using this, we obtain

$$2ky_2 \geqslant 1 - (d_4 + kf_1)y_1 - e_4 + k(k-1)f_1 + f_2 + (k-1)(f_1 - f_2) - \tfrac{f_1}{2}(k-1)(k-2).$$

Applying this lower bound in (4.413), we get

$$(1 + f_2 - f_1)y_1 + 3(f_2 - f_1) + 1 - k(2f_2 - 3f_1) \leqslant -2. \tag{4.418}$$

Now, let us show that we obtain a contradiction from the above equation. Let us first consider the case $1 + f_2 - f_1 < 0$. Since $y_1 \leqslant 1$, we obtain

$$\begin{aligned} &(1 + f_2 - f_1)y_1 + 3(f_2 - f_1) + 1 - k(2f_2 - 3f_1) \\ &\qquad \geqslant 2(f_1 + 1) + (k-2)(3f_1 - 2f_2) > -2, \end{aligned}$$

where we used that $k \geqslant 2$, $f_1 > -2$, and $3f_1 - 2f_2 > 0$, and we obtain a contradiction with (4.418). Next, let us consider the case $1 + f_2 - f_1 > 0$. Using that $y_1 \geqslant -1$, we obtain

$$(1+f_2-f_1)y_1+3(f_2-f_1)+1-k(2f_2-3f_1) \geqslant f_1+(k-1)(3f_1-2f_2) > -2$$

because $k \geqslant 2$, $f_1 > -2$, and $3f_1 - 2f_2 > 0$, and we again obtain a contradiction with (4.418). Therefore, we can conclude that $\ell < k$.

Returning to our Lyapunov candidate, we note that, for this case, we have

$$\begin{aligned} V_{k_{i+1}} - V_{k_i} &= 2[y_1(k_i) + \ell y_2(k_i) + e_6] + 2b[y_2(k_i) + (\ell-2) f_1] \\ &\quad - 2(\ell-2)y_2(k_i) - (\ell-2)^2 f_1 + 2y_1(k_i) + 2by_2(k_i). \end{aligned}$$

Therefore, together with (4.392), we have

$$\begin{aligned} V_{k_{i+1}} - V_{k_{i-1}} &= 2[y_1(k_i) + \ell y_2(k_i) + e_6] + 2b[y_2(k_i) + (\ell-2) f_1] \\ &\quad - 2(\ell-2)y_2(k_i) - (\ell-2)^2 f_1 + 2y_1(k_i) + 2by_2(k_i) \\ &\quad - 2(1 + bf_1 + f_2)y_1 - 2(1+b)(1+y_1)y_2 - 1 \\ &\quad - (f_1+1)y_1^2 + (k-1)[(2b-1)f_1 + 2f_2]. \end{aligned}$$

Note that the coefficient of y_2 term is

$$2(k+\ell)+2b-2(\ell-2)+2k+2b-2(1+b)(1+y_1) = 4k+2(1+b)(1-y_1) > 0,$$

where we have used that $b > -1$ and $y_1 < 1$. Therefore, $V_{k_{i+1}} - V_{k_{i-1}}$ is maximal for the maximum value of y_2.

For the upper bound for y_2, we use (4.417) while ignoring all other constraints except for $y_1 \in [-1, 1]$. We note that (4.417) implies

$$\begin{aligned} y_2 \leqslant -f_1 y_1 + \tfrac{1}{2(2k-h+1)}\Big[&-2(1+f_2-f_1)y_1 - 2 + (2k^2-h-h^2) f_1 \\ &+ (4+2h) f_2\Big], \end{aligned}$$

where $h = k - \ell \in \{1, 2, 3, 4\}$.

Since we know $V_{k_{i+1}} - V_{k_{i-1}}$ is maximal for the maximum value of y_2, we can replace y_2 by its upper bound to obtain

$$V_{k_{i+1}} - V_{k_{i-1}} \leqslant a_5 y_1^2 + b_5 y_1 + c_5, \tag{4.419}$$

where

$$\begin{aligned} a_5 &= (1+2b)f_1 - 1 + \tfrac{2}{2k-h+1}(b+1)(1+f_2-f_1) \\ b_5 &= \tfrac{2(h-1)}{2k-h+1}(f_1-f_2-1) \\ &\quad + \tfrac{1+b}{2k-h+1}\left[2(f_1-f_2)-(2k^2-h-h^2)f_1-(4+2h)f_2\right] \\ c_5 &= f_1[1-2b(h+1)-h^2]+2f_2-3 \\ &\quad + \tfrac{b+h}{2k-h+1}[-2+(2k^2-h-h^2)f_1+(4+2h)f_2]. \end{aligned}$$

We first note that a_5 is equal to $\overline{a}$ defined in (4.376) with k replaced by $2k - h + 1 \geqslant 3$. Therefore, the earlier argument also implies that $a_5 < 0$. The upper bound (4.419) is a quadratic function in y_1. We will show that $2a_5 + b_5 > 0$. This implies that the upper bound given $y_1 \in [-1, 1]$ takes it maximum for $y_1 = 1$. On the other hand, we have already shown that $V_{k_{i+1}} - V_{k_{i-1}}$ only subject to $y_1(k_{i+1}+1) \leqslant -1$ and $y_1 \geqslant 1$ is negative in an earlier part of the proof of Lemma (4.71) for the case $f_1 < -1.6$ with $y_1(k_{i-1})$ saturated and $y_1(k_{i+1})$ unsaturated. Remains to establish that $2a_5 + b_5 > 0$. We have

$$\begin{aligned} 2a_5 + b_5 = \tfrac{1}{2k-h+1}\Big[&-2(1+b)f_1(k-1)^2 + 4(bf_1-1)k \\ &+ f_1\big((1+b)(h^2-3h+4)+4(h-1)\big) \\ &- f_2\big(2(1+b)(1+h)-2(1-h)\big)+4(1+b)\Big]. \end{aligned}$$

Clearly, the factor $1/(2k - h + 1)$ is irrelevant for the sign of $2a_5 + b_5$. Remains to establish that

$$\begin{aligned} -2(1+b)f_1(k-1)^2 + 4(bf_1-1)k + f_1\big((1+b)(h^2-3h+4)+4(h-1)\big) \\ - f_2\big(2(1+b)(1+h)-2(1-h)\big)+4(1+b) > 0. \qquad (4.420) \end{aligned}$$

We note that $k = h + \ell \geqslant h + 2$. By taking the derivative of (4.420) with respect to k, we obtain

$$-4(1+b)f_1(k-1)+4(bf_1-1) > 0$$

since $1 + b > 0$, $f_1 < 0$, $k > 1$, and

$$bf_1 - 1 > -\tfrac{3}{4}f_1 - 1 = -\tfrac{3}{4}(f_1 + \tfrac{4}{3}) > 0,$$

where we have used that $b \geqslant -\frac{3}{4}$ and $f_1 < -1.6$ in the region of interest. This implies that (4.420) is minimal for the smallest possible k, i.e., $k = h+2$. Setting $k = h + 2$ in (4.420), we get

$$\begin{aligned} f_1\big((1+b)(-h^2-3h+10)-12\big) - f_2\big(2(1+b)(1+h)-2(1-h)\big) \\ - 4h - 8 + 4(1+b) > 0. \end{aligned}$$

Next, we note that the derivative with respect to h equals

$$-3f_1(1+b) - 2(f_2+2) - 2hf_1(1+b) - 2(1+b)f_2 > 0,$$

where we have used that $f_1 < 0$, $f_2 < -2$, $1+b > 0$, and $h > 0$. Therefore, the expression is minimal for $h = 1$, and we obtain

$$6(b-1)f_1 - 4(1+b)f_2 + 4(b-2),$$

which is positive in the region of interest. Therefore, we conclude that $b_5 + 2a_5 > 0$. As noted before, this yields that $V_{k_{i+1}} - V_{k_{i-1}}$ is negative, and the proof is complete. ■

Proof of Lemma 4.71 for $f_1 < -1.6$ with both $y_1(k_{i-1})$ and $y_1(k_{i+1})$ unsaturated : Clearly, in this case, we have $k \geqslant 2$ and $\ell \geqslant 1$. The following constraints are satisfied:

$$y_1(k_{i-1}) = y_1 \in [-1, 1] \tag{4.421}$$
$$y_1(k_{i-1}+1) = (1+f_2)y_1 + y_2 \leqslant -1 \tag{4.422}$$
$$\vdots \qquad \vdots$$
$$y_1(k_i) = d_4 y_1 + k y_2 + e_4 \leqslant -1 \tag{4.423}$$
$$y_1(k_i+1) = (d_4+f_1)y_1 + (k+1)y_2 + e_4 - (k-1)f_1 - f_2 \geqslant 1 \tag{4.424}$$
$$\vdots \qquad \vdots$$
$$y_1(k_{i+1}) = y_1(k_i) + \ell y_2(k_i) + e_6 \geqslant -1 \tag{4.425}$$
$$y_1(k_{i+1}) = y_1(k_i) + \ell y_2(k_i) + e_6 \leqslant 1. \tag{4.426}$$

Notice that $y_1(k_i)$ and $y_2(k_i)$ are given in (4.366) and (4.367), respectively, d_4 and e_4 are defined in (4.368) and (4.369), while e_6 is defined in (4.401). Finally,

$$y_2(k_{i+1}) = y_2(k_i) + (\ell-2)f_1.$$

The argument used in the case where $y_1(k_{i-1})$ is unsaturated and $y_1(k_{i+1})$ is saturated immediately implies that we have $\ell \geqslant k-3$ for $k \geqslant 4$, and $\ell \leqslant k$. We will show that

$$V_{k_{i+1}} - V_{k_{i-1}} < 0.$$

Let us define

$$\begin{aligned} \hat{y}_1 &= y_1(k_{i+1}) = (d_4 + \ell f_1)y_1 + (k+\ell)y_2 + e_4 + e_6 - \ell(k-1)f_1 \\ \hat{y}_2 &= y_2(k_{i+1}) = f_1 y_1 + y_2 + (\ell - k - 1)f_1. \end{aligned} \tag{4.427}$$

Then, we have

$$V_{k_{i+1}} - V_{k_i} = \widehat{y}_1^2 + 2b\widehat{y}_1\widehat{y}_2 - \tfrac{1}{f_1}\widehat{y}_2^2 + 2y_1(k_i) + 1 + 2by_2(k_i) + \tfrac{1}{f_1}y_2^2(k_i).$$

Combining this with (4.392) and eliminating $y_1(k_i)$ and $y_2(k_i)$ by using (4.366) and (4.367) and eliminating $\widehat{y}_2$ using (4.427), we obtain

$$\begin{aligned} V_{k_{i+1}} - V_{k_{i-1}} &= \widehat{y}_1^2 + 2b\widehat{y}_1\left[f_1y_1 + y_2 + (\ell-k-1)f_1\right] - (\ell-2)^2 f_1 \\ &\quad -2(\ell-2)\left[f_1y_1 + y_2 - (k-1)f_1\right] + 2(d_4y_1 + ky_2 + e_4) - 2(b+1)y_1y_2 \\ &\quad - 2(1+f_2)y_1 - 2y_2 - (f_1+1)y_1^2 + (k-1)\left[2f_2 - f_1\right]. \end{aligned}$$

Let us write this expression in terms of y_1 and $\widehat{y}_1$ by eliminating y_2 using

$$(k+\ell)y_2 = \widehat{y}_1 - (d_4 + \ell f_1)y_1 + \ell(k-1)f_1 - e_4 - e_6.$$

We get

$$\begin{aligned} V_{k_{i+1}} - V_{k_{i-1}} &= \widehat{y}_1^2 + 2b\widehat{y}_1\left[f_1y_1 + (\ell-k-1)f_1\right] - (\ell-2)^2 f_1 \\ &\qquad + 2(d_4y_1 + e_4) - (f_1+1)y_1^2 \\ &\quad - 2(\ell-2)f_1\left[y_1 - (k-1)\right] - 2(1+f_2)y_1 + (k-1)\left[2f_2 - f_1\right] \\ &\qquad + \tfrac{1}{k+\ell}\left[2b\widehat{y}_1 + 2(k-\ell+1) - 2(1+b)y_1\right] \\ &\qquad \times \left[\widehat{y}_1 - (d_4 + \ell f_1)y_1 + \ell(k-1)f_1 - e_4 - e_6\right]. \end{aligned} \tag{4.428}$$

Our objective is now to prove that this expression is always negative. We only consider the constraints (4.421), (4.425), and (4.426), that is,

$$-1 \leqslant y_1 \leqslant 1, \quad -1 \leqslant \widehat{y}_1 \leqslant 1, \tag{4.429}$$

while we ignore all other constraints. Note that the coefficient of the term $\widehat{y}_1^2$ is equal to

$$1 + \tfrac{2b}{k+\ell},$$

which is positive for $k+\ell \geqslant 2$ and $b > -1$. Therefore, we know (4.428) is maximal at the boundary of $\widehat{y}_1$, that is, either $\widehat{y}_1 = -1$ or $\widehat{y}_1 = 1$. Define $h = k - \ell = \{0, 1, 2, 3\}$.

Let us first consider the boundary $\widehat{y}_1 = -1$. We obtain that

$$\begin{aligned} V_{k_{i+1}} - V_{k_{i-1}} &= 1 + 2b(h+1)f_1 - (h+1)^2 f_1 + \left[(1+2b)f_1 - 1\right]y_1^2 \\ &\qquad - \tfrac{1}{2k-h}\left[(1+b)(y_1+1) - (h+2)\right] \times \\ &\quad \times \left[-2 - 2(1+f_2-f_1)y_1 + 2f_2(h+1) + f_1(2k^2 - h^2 - 3h - 2)\right]. \end{aligned} \tag{4.430}$$

Note that the coefficient of the term y_1^2 is

$$a_6 = (1+2b)f_1 - 1 + \tfrac{2}{2k-h}(b+1)(1+f_2-f_1),$$

which is negative as it is the same as $\bar{a}$ given in (4.376), with k replaced by $k+\ell$. Let us now derive the coefficient of the term y_1:

$$b_6 = -\tfrac{2(1+f_2-f_1)}{2k-h}(h+2) - \tfrac{1+b}{2k-h}\left[h(2f_2-3f_1) + f_1(2k^2-h^2) - 4\right].$$

We will show that $2a_6 + b_6 > 0$. This implies that the upper bound given the constraints (4.429) takes it maximum for $y_1 = 1$ and $\widehat{y}_1 = -1$. On the other hand, we have already shown that $V_{k_{i+1}} - V_{k_{i-1}}$ only subject to $y_1(k_{i+1}) \in [-1,1]$ and $y_1 \geqslant 1$ is negative in an earlier part of the proof of Lemma (4.71) for the case $f_1 < -1.6$ with $y_1(k_{i-1})$ saturated and $y_1(k_{i+1})$ unsaturated.

Therefore, we have

$$\begin{aligned} b_6 + 2a_6 = \tfrac{1}{2k-h}\Big[&-2(1+f_2-f_1)(h+2) - (1+b)[h(2f_2-3f_1) \\ &+ f_1(2k^2-h^2) - 4\Big] + 2(2b+1)f_1(2k-h) \\ &- 2(2k-h) + 4(1+b)(1+f_2-f_1)]. \end{aligned}$$

Note that the term $\frac{1}{2k-h}$ does not affect the sign of $b_6 + 2a_6$. Therefore, to show $b_6 + 2a_6 > 0$ is equivalent to showing that

$$\begin{aligned} &-2(1+f_2-f_1)(h+2) - (1+b)[h(2f_2-3f_1) + f_1(2k^2-h^2) - 4] \\ &+ 2(2b+1)f_1(2k-h) - 2(2k-h) + 4(1+b)(1+f_2-f_1) > 0. \end{aligned} \tag{4.431}$$

We first show that the left-hand side of the above inequality is increasing in k, and therefore, we differentiate the left-hand side with respect to k.This results in

$$-4(1+b)f_1(k-1) + 4(bf_1-1),$$

which is positive since $1+b > 0$, $f_1 < 0$, $k > 1$, and $bf_1 - 1 > 0$. Thus, the left-hand side of (4.431) achieves its minimum for

$$k = 1 + \tfrac{bf_1-1}{(1+b)f_1} < 1,$$

where we have used that $1+b > 0$, $f_1 < 0$, and $bf_1 - 1 > 0$. In our case, we know that $k = \ell + h \geqslant h+1$ and $k \geqslant 2$. Therefore, we have $k \geqslant \max\{2, h+1\}$. Thus, the left-hand side of (4.431) achieves its minimal for $k = 2$ when $h = 0$, while for $h = 1,2,3$, it achieves it minimal when $k = h+1$.

For the case where $k = 2$ and $h = 0$, the left-hand side of (4.431) is

$$4bf_2 + 4bf_1 + 4(2b-1) = 4b(f_2+2) + 4(bf_1-1) > 0,$$

where we have used that $b < 0$, $f_2 < -2$, and $bf_1 - 1 > 0$.

For the case where $k = h+1$ and $h = 1, 2, 3$, using $k = h+1$ in the left-hand sider of (4.431) yields

$$\begin{aligned} &-2(1+f_2-f_1)(h+2)-(1+b)[h(2f_2-3f_1)+f_1(h^2+4h+2)-4] \\ &\quad +2(2b+1)f_1(h+2)-2(h+2)+4(1+b)(1+f_2-f_1). \end{aligned} \tag{4.432}$$

We first show that this is increasing in h, and therefore, we differentiate with respect to h. This results in

$$-4+3(1+b)f_1-2(b+2)f_2-2(1+b)f_1h > (1+b)(3f_1-2f_2)-2(f_2+2) > 0,$$

where for the first inequality we have used that $1+b > 0$, $f_1 < 0$, and $h > 0$, while for the second inequality we have used that $1+b > 0$, $3f_1 - 2f_2 > 0$, and $f_2 < -2$. Therefore, (4.432) is minimal for the minimum value of h, that is, $h = 1$. When $h = 1$, we have

$$-4(1-2b)+4(1+b)f_1+2(b-2)f_2$$

which is positive in the region of interest. Therefore, we conclude that $b_6 + 2a_6 > 0$, which as argued before implies that $V_{k_{i+1}} - V_{k_{i-1}} < 0$.

The only other possible alternative is that $\widehat{y}_1 = 1$. In that case, we obtain

$$\begin{aligned} V_{k_{i+1}} - V_{k_{i-1}} &= 1 - 2b(h+1)f_1 - (h+1)^2 f_1 + [(1+2b)f_1 - 1]\, y_1^2 \\ &\quad + \tfrac{1}{2k-h}\Big[(1+b)(1-y_1)+h\Big]\Big[2-2(1+f_2-f_1)y_1 \\ &\quad + 2f_2(h+1) + f_1(2k^2-h^2-3h-2)\Big]. \end{aligned} \tag{4.433}$$

Note that the coefficient of the term y_1^2 is

$$a_7 = (1+2b)f_1 - 1 + \tfrac{2}{2k-h}(b+1)(1+f_2-f_1),$$

which is equal to a_6, and thus, it is negative. Let us now derive the coefficient of the term y_1:

$$\begin{aligned} b_7 = \tfrac{1}{2k-h}\Big[&-2(1+f_2-f_1)(h+2+2b) - (1+b)h(2f_2-3f_1) \\ &- (1+b)f_1(2k^2-h^2)\Big]. \end{aligned}$$

We will show that $2a_7 + b_7 > 0$. This implies that the upper bound given the constraints (4.429) takes it maximum for $y_1 = 1$ and $\widehat{y}_1 = 1$. As argued before, we have already shown that $V_{k_{i+1}} - V_{k_{i-1}}$ only subject to $y_1(k_{i+1}) \in [-1, 1]$ and $y_1 \geqslant 1$ is negative in an earlier part of the proof of Lemma (4.71) for the case $f_1 < -1.6$ with $y_1(k_{i-1})$ saturated and $y_1(k_{i+1})$ unsaturated.

With just a little bit algebra, we obtain that

$$b_7 + 2a_7 = \tfrac{1}{2k-h}[(2f_1(1+2b)-2)(2k-h) \\ -2(1+f_2-f_1)h-(1+b)h(2f_2-3f_1)-(1+b)f_1(2k^2-h^2)].$$

Note that the term $\frac{1}{2k-h}$ does not affect the sign of b_7+2a_7. Therefore, showing that $b_7+2a_7>0$ is equivalent to showing that

$$[2f_1(1+2b)-2](2k-h)-2(1+f_2-f_1)h-(1+b)h(2f_2-3f_1) \\ -(1+b)f_1(2k^2-h^2)>0. \quad (4.434)$$

We first show that the left-hand side of the above inequality is increasing in k, and therefore, we differentiate the left-hand side with respect to k. This results in

$$-4(1+b)f_1(k-1)+4(bf_1-1),$$

which is positive since $1+b>0$, $f_1<0$, $k>1$, and $bf_1-1>0$. Thus, the left-hand side of (4.434) achieves its minimum for

$$k = 1+\tfrac{bf_1-1}{(1+b)f_1}<1,$$

where we have used that $1+b>0$, $f_1<0$, and $bf_1-1>0$. In our case, we know that $k=\ell+h\geqslant h+1$ and $k\geqslant 2$. Therefore, we have $k\geqslant \max\{2,h+1\}$. Thus, the left-hand side of (4.431) achieves its minimal for $k=2$ when $h=0$, while for $h=1,2,3$, it achieves it minimal when $k=h+1$.

For the case where $k=2$ and $h=0$, the left-hand side of (4.434) is

$$8(bf_1-1)>0,$$

where we have used that $bf_1-1>0$.

For the case where $k=h+1$ and $h=1,2,3$, using $k=h+1$ in the left-hand side of (4.434) yields

$$[2f_1(1+2b)-2](h+2)-2(1+f_2-f_1)h-(1+b)h(2f_2-3f_1) \\ -(1+b)f_1(h^2+4h+2). \quad (4.435)$$

We first show that this is increasing in h, and therefore, we differentiate with respect to h. This results in

$$-4+3(1+b)f_1-2(b+2)f_2-2(1+b)f_1h>(1+b)(3f_1-2f_2)-2(f_2+2)>0,$$

where for the first inequality we have used that $1+b>0$, $f_1<0$, and $h>0$, while for the second inequality we have used that $1+b>0$, $3f_1-2f_2>0$, and $f_2 < -2$. Therefore, (4.435) is minimal for the minimum value of h, that is, $h = 1$. When $h = 1$, we have

$$-8 + 4(1 + 2b) f_1 - 2(b + 2) f_2,$$

which is positive in the region of interest. Therefore, we conclude that $b_7 + 2a_7 > 0$, which as argued before implies that $V_{k_{i+1}} - V_{k_{i-1}} < 0$. This completes our proof. ■

4.C Existence of H_2 optimal controller

Consider the system,

$$\Sigma_a : \begin{cases} \rho x = Ax + Bu + I\omega \\ z = Cx + Du. \end{cases} \tag{4.436}$$

We recall the following existence conditions of H_2 optimal static state feedback controller for the system Σ_a:

Theorem 4.101 *For the system Σ_a in (4.436), H_2 optimal state feedback controller of a static type exists if and only if*

(i) *(A, B) is stabilizable.*

(ii) *Σ_a does not have any invariant zero on the imaginary axis (continuous time) or on the unit circle (discrete time).*

(iii) *Σ_a does not have any infinite zero of order greater or equal to 1 (continuous time).*

Proof: The results follow from [133] (see Lemmas 5.6.3 and 5.4.1 for continuous-time case and Lemmas 6.6.5 and 6.6.1 for discrete-time case). ■

4.D Continuity of solution of CQMI and DARE

Theorem 4.102 given below recalls from [163] the continuity properties of the solution of the CQMI associated with the quintuple (A, B, C, D, E) and $\gamma > \gamma^*$:

$$F_\gamma(P) = \begin{pmatrix} A'P + PA + C'C + \gamma^{-2} PEE'P & PB + C'D \\ B'P + D'C & D'D \end{pmatrix} \geqslant 0, \tag{4.437}$$

where

$$\gamma^* := \inf_F \left\{ \|(C + DF)(sI - A - BF)^{-1}E\|_\infty \mid \lambda(A + BF) \in \mathbb{C}^- \right\}. \quad (4.438)$$

Theorem 4.102 *Consider a quintuple (A, B, C, D, E). Suppose (A, B) is stabilizable, (A, B, C, D) does not have any invariant zeros in the open right-half plane, and $\gamma > \gamma^*$ where γ^* is defined in* (4.438)*. Let a sequence of perturbed data $(A_\varepsilon, B_\varepsilon, C_\varepsilon, D_\varepsilon, E_\varepsilon)$ converges to (A, B, C, D, E). Moreover, assume that the normal rank of $C_\varepsilon(sI - A_\varepsilon)^{-1}B_\varepsilon + D_\varepsilon$ is equal to the normal rank of $C(sI - A)^{-1}B + D$ for all ε. Then, the smallest positive semi-definite semi-stabilizing solution of CQMI* (4.437) *associated with $(A_\varepsilon, B_\varepsilon, C_\varepsilon, D_\varepsilon, E_\varepsilon)$ converges to the solution of CQMI associated with (A, B, C, D, E).*

In the perturbation method I of both H_2 and H_∞ low-gain design, we use perturbations which do not necessarily preserve the normal rank. In this case, we need the following result.

Theorem 4.103 *Consider a quintuple (A, B, C, D, E) and $\gamma > \gamma^*$. Suppose a sequence of perturbations $(C_\varepsilon, D_\varepsilon)$ converges to (C, D), and satisfies the following conditions:*

(i) $\bar{Q}_\varepsilon$ is continuous at $\varepsilon = 0$, where $\bar{Q}_\varepsilon$ is defined as in (4.319).

(ii) There exists a β such that for $0 \leqslant \varepsilon_1 \leqslant \varepsilon_2 \leqslant \beta$, we have $\bar{Q}_{\varepsilon_1} \leqslant \bar{Q}_{\varepsilon_2}$.

Then the solution to CQMI (4.437) *associated with $(A, B, C_\varepsilon, D_\varepsilon, E)$ converges to the solution of CQMI* (4.437) *associated with (A, B, C, D, E).*

Proof : The case $\gamma = \infty$ was proved in [183]. We shall prove this result for a finite γ.

First, we need to show that given $\gamma > \gamma^*$, for sufficiently small ε, we have $\gamma > \gamma_\varepsilon^*$ where γ_ε^* is defined in (4.438) with (A, B, C, D, E) replaced by $(A_\varepsilon, B_\varepsilon, C_\varepsilon, D_\varepsilon, E_\varepsilon)$. This follows from the fact that there exists a stabilizing state feedback $u = Fx$ such that the H_∞ norm from w to z equals $\gamma_0 < \gamma$. The transfer matrix $G_{\varepsilon,\mathrm{cl}}$ from w to z_ε satisfies,

$$G_{\varepsilon,\mathrm{cl}}(s)' G_{\varepsilon,\mathrm{cl}}(s) = G_{\mathrm{cl}}(s)' G_{\mathrm{cl}}(s) + G_0(s)'(\bar{Q}_\varepsilon - \bar{Q}_0) G_0(s),$$

where G_{cl} is the transfer matrix from w to z while G_0 is defined by

$$G_0(s) = \begin{pmatrix} I \\ F \end{pmatrix} (sI - A - BF)^{-1} E.$$

Since G_0 has a finite H_∞ norm and $\bar{Q}_\varepsilon \to \bar{Q}_0$, we find that

$$\lim_{\varepsilon \downarrow 0} \|G_{\varepsilon,\mathrm{cl}}\|_\infty \to \|G_{\mathrm{cl}}\|_\infty = \gamma_0 < \gamma.$$

Next, we investigate

$$x_0' P_\varepsilon x_0 = \sup_{w \in L_2} \inf_{u \in L_2} \left\{ \|z_\varepsilon\|_2^2 - \gamma^2 \|w\|_2^2 \mid x \in L_2 \right\},$$

where $x(0) = x_0$. Since $\bar{Q}_\varepsilon > \bar{Q}_0$ for small ε, we find that

$$x_0' P_\varepsilon x_0 \geqslant x_0' P_0 x_0$$

for small ε. If we choose $u = Fx$, we obtain

$$0 \leqslant x_0' P_\varepsilon x_0 \leqslant \left\{ \sup_{w \in L_2} \|z_\varepsilon\|_2^2 - \gamma^2 \|w\|_2^2 \mid u = Fx \right\}.$$

We always have for any η,

$$\|a + b\|^2 \leqslant (1 + \tfrac{1}{\eta})\|a\|^2 + (1 + \eta)\|b\|^2.$$

Let γ_1 be such that $\gamma_0 < \gamma_1 < \gamma$ such that for all sufficiently small ε, we know the H_∞ norm from w to z_ε is less than γ_1 for the feedback $u = Fx$. With u fixed by $u = Fx$, we can write $z_\varepsilon = z_{x_0} + z_w$ where z_{x_0} is the output for initial condition x_0 and $w = 0$ and z_w is the output for initial condition $x_0 = 0$ and disturbance w. Let L be such that

$$\|x_{x_0}\|_2^2 = x_0' L x_0,$$

where x_{x_0} is the state for initial condition x_0 and $w = 0$. Choose

$$\eta = \frac{\gamma^2 - \gamma_1^2}{2\gamma_1^2}.$$

We find

$$\|z_\varepsilon\|_2^2 \leqslant \tfrac{\gamma^2+\gamma_1^2}{\gamma^2-\gamma_1^2} \|\bar{Q}_\varepsilon\| x_0' L x_0 + \tfrac{\gamma^2+\gamma_1^2}{2} \|w\|_2^2.$$

But then if w is such that

$$\|w\|^2 > \beta x_0' L x_0, \tag{4.439}$$

where $\beta > \beta_\varepsilon$ for all sufficiently small ε with

$$\beta_\varepsilon = \frac{2(\gamma^2 + \gamma_1^2)}{2(\gamma^2 - \gamma_1^2)^2} \|\bar{Q}_\varepsilon\|,$$

we have

$$\|z_\varepsilon\|_2^2 - \gamma^2 \|w\|_2^2 < 0$$

for $u = Fx$. We find that for w for which (4.439) is satisfied, we obtain for a suboptimal u already a negative cost. Since

$$\sup_{w\in L_2} \inf_{u\in L_2} \left\{ \|z_\varepsilon\|_2^2 - \gamma^2 \|w\|_2^2 \mid x \in L_2 \right\} > 0,$$

we can without loss of generality assume that w satisfies

$$\|w\|^2 < \beta x_0' L x_0, \tag{4.440}$$

provided that ε is small enough. By setting $u = Fx + v$, the above inf–sup problem is equivalent to

$$\sup_{w\in L_2} \inf_{v\in L_2} \left\{ \|z_\varepsilon\|_2^2 - \gamma^2 \|w\|_2^2 \mid u = Fx + v \right\} > 0.$$

Since we showed that we can assume, without loss of generality, that w is bounded, it is clear that in the above optimization for $\varepsilon = 0$, we can also assume without loss of generality that v is bounded as well, i.e.,

$$\|v\|_2 \leqslant N \|x_0\|.$$

If the system is left invertible from v to z then as v gets sufficiently large in L_2 norm, then the cost can be made arbitrarily large. If the system is not left invertible we can split the input in to one part which has no effect on the output and another part which has a left-invertible effect on the output. The latter has to be bounded for a bounded cost. The first can be set to zero without loss of generality. But with v and w bounded, we can find an M such that

$$\|x\|_2 \leqslant M \|x_0\|^2,$$

but then,

$$\begin{aligned}
&\sup_{w\in L_2} \inf_{v\in L_2} \left\{ \|z_\varepsilon\|_2^2 - \gamma^2 \|w\|_2^2 \right\} \\
&\quad \leqslant \sup_{w\in L_2} \inf_{v\in L_2} \left\{ \|z_\varepsilon\|_2^2 - \gamma^2 \|w\|_2^2 \mid \|v\|_2 \leqslant N \|x_0\| \right\} \\
&\quad \leqslant \sup_{w\in L_2} \inf_{v\in L_2} \left\{ \|z_0\|_2^2 - \gamma^2 \|w\|_2^2 \mid \|v\|_2 \leqslant N \|x_0\| \right\} + \|\bar{Q}_\varepsilon - \bar{Q}_0\| M \|x_0\|^2 \\
&\quad = \sup_{w\in L_2} \inf_{v\in L_2} \left\{ \|z_0\|_2^2 - \gamma^2 \|w\|_2^2 \right\} + \|\bar{Q}_\varepsilon - \bar{Q}_0\| M \|x_0\|^2 \\
&\quad = x_0' P_0 x_0 + \|\bar{Q}_\varepsilon - \bar{Q}_0\| M \|x_0\|^2,
\end{aligned}$$

where in each case $u = Fx + v$. In conclusion, we find

$$x_0' P_0 x_0 \leqslant x_0' P_\varepsilon x_0 \leqslant x_0' P_0 x_0 + \|\bar{Q}_\varepsilon - \bar{Q}_0\| M \|x_0\|^2,$$

which implies that $P_\varepsilon \to P_0$ as $\varepsilon \downarrow 0$. ∎

Our concern next is with the continuity of semi-stabilizing solution of the following DARE associated with the quintuple (A, B, C, D, E) and $\gamma > \gamma^*$:

$$P = A'PA + C'C - \begin{pmatrix} B'PA + D'C \\ E'PA \end{pmatrix}' G(P)^\dagger \begin{pmatrix} B'PA + D'C \\ E'PA \end{pmatrix}, \tag{4.441}$$

where

$$G(P) = \begin{pmatrix} D'D & 0 \\ 0 & -\gamma^{-2} I \end{pmatrix} + \begin{pmatrix} B' \\ E' \end{pmatrix} P \begin{pmatrix} B & E \end{pmatrix}$$

and

$$\gamma^* = \inf_F \left\{ \|(C + DF)(zI - A - BF)^{-1} E\|_\infty \mid \lambda(A + BF) \in \mathbb{C}^{\odot} \right\}. \tag{4.442}$$

We recall the following theorem from [163]:

Theorem 4.104 *Consider a quintuple (A, B, C, D, E). Suppose (A, B) is stabilizable, (A, B, C, D) does not have any invariant zeros in $\mathbb{C}^{\otimes}$ and $\gamma > \gamma^*$, where γ^* is defined in* (4.442)*. Let a sequence of perturbed data $(A_\varepsilon, B_\varepsilon, C_\varepsilon, D_\varepsilon, E_\varepsilon)$ converges to (A, B, C, D, E). Moreover, assume that the normal rank of $C_\varepsilon(zI - A_\varepsilon)^{-1} B_\varepsilon + D_\varepsilon$ is equal to the normal rank of $C(zI - A)^{-1} B + D$ for all ε. Then, the smallest positive semi-definite semi-stabilizing solution of DARE* (4.441) *associated with $(A_\varepsilon, B_\varepsilon, C_\varepsilon, D_\varepsilon, E_\varepsilon)$ converges to the smallest positive semi-definite semi-stabilizing solution of DARE associated with (A, B, C, D, E).*

In the perturbation method I of both H_2 and H_∞ low-gain design, we use perturbations which do not necessarily preserve the normal rank. In this case, we need the following result.

Theorem 4.105 *Consider a quintuple (A, B, C, D, E) and $\gamma > \gamma^*$ where γ^* is defined in* (4.442)*. Suppose a sequence of perturbations $(C_\varepsilon, D_\varepsilon)$ converges to (C, D) and satisfies the following conditions:*

(i) $\bar{Q}_\varepsilon$ is continuous at $\varepsilon = 0$.

(ii) There exists a β such that for $0 \leqslant \varepsilon_1 \leqslant \varepsilon_2 \leqslant \beta$, we have $\bar{Q}_{\varepsilon_1} \leqslant \bar{Q}_{\varepsilon_2}$,

where $\bar{Q}_\varepsilon$ is defined in (4.319)*. Then the semi-stabilizing solution of the DARE* (4.441) *associated with $(A, B, C_\varepsilon, D_\varepsilon, E)$ converges to the semi-stabilizing solution of DARE* (4.441) *associated with (A, B, C, D, E).*

Proof : The case $\gamma = \infty$ was proved in [183]. We shall prove this result for a finite γ. Our proof here follows more or less along the same lines as the proof of Theorem 4.103.

First, we need to show that given $\gamma > \gamma^*$, for sufficiently small ε, we have $\gamma > \gamma_\varepsilon^*$ where γ_ε^* is defined in (4.438) with (A, B, C, D, E) replaced by (A_ε, B_ε, C_ε, D_ε, E_ε). This follows from the fact that there exists a stabilizing state feedback $u = Fx$ such that the H_∞ norm from w to z equals $\gamma_0 < \gamma$. The transfer matrix $G_{\varepsilon,\text{cl}}$ from w to z_ε satisfies,

$$G_{\varepsilon,\text{cl}}(z)' G_{\varepsilon,\text{cl}}(z) = G_{\text{cl}}(z)' G_{\text{cl}}(z) + G_0(z)'(\bar{Q}_\varepsilon - \bar{Q}_0) G_0(z),$$

where G_{cl} is the transfer matrix from w to z while G_0 is defined by

$$G_0(z) = \begin{pmatrix} I \\ F \end{pmatrix} (zI - A - BF)^{-1} E.$$

Since G_0 has a finite H_∞ norm and $\bar{Q}_\varepsilon \to \bar{Q}_0$, we find that

$$\lim_{\varepsilon \downarrow 0} \|G_{\varepsilon,\text{cl}}\|_\infty \to \|G_{\text{cl}}\|_\infty = \gamma_0 < \gamma.$$

Next, we investigate

$$x_0' P_\varepsilon x_0 = \sup_{w \in \ell_2} \inf_{u \in L_2} \left\{ \|z_\varepsilon\|_2^2 - \gamma^2 \|w\|_2^2 \mid x \in \ell_2 \right\},$$

where $x(0) = x_0$. Since $\bar{Q}_\varepsilon > \bar{Q}_0$ for small ε, we find that

$$x_0' P_\varepsilon x_0 \geqslant x_0' P_0 x_0$$

for a small ε. If we choose $u = Fx$, we obtain

$$0 \leqslant x_0' P_\varepsilon x_0 \leqslant \left\{ \sup_{w \in \ell_2} \|z_\varepsilon\|_2^2 - \gamma^2 \|w\|_2^2 \mid u = Fx \right\}.$$

We always have for any η,

$$\|a + b\|^2 \leqslant (1 + \tfrac{1}{\eta})\|a\|^2 + (1 + \eta)\|b\|^2.$$

Let γ_1 be such that $\gamma_0 < \gamma_1 < \gamma$ and that, for all sufficiently small ε, H_∞ norm from w to z_ε be less than γ_1 for the feedback $u = Fx$. With u fixed by $u = Fx$, we can write $z_\varepsilon = z_{x_0} + z_w$ where z_{x_0} is the output for initial condition x_0 and $w = 0$ and z_w is the output for initial condition $x_0 = 0$ and disturbance w. Let L be such that

$$\|x_{x_0}\|_2^2 = x_0' L x_0,$$

where x_{x_0} is the state for initial condition x_0 and $w = 0$. Choose

$$\eta = \frac{\gamma^2 - \gamma_1^2}{2\gamma_1^2}.$$

We find

$$\|z_\varepsilon\|_2^2 \leqslant \tfrac{\gamma^2+\gamma_1^2}{\gamma^2-\gamma_1^2}\|\bar{Q}_\varepsilon\| x_0' L x_0 + \tfrac{\gamma^2+\gamma_1^2}{2}\|w\|_2^2.$$

But then if w is such that

$$\|w\|^2 > \beta x_0' L x_0, \tag{4.443}$$

where $\beta > \beta_\varepsilon$ for all sufficiently small ε with

$$\beta_\varepsilon = \frac{2(\gamma^2 + \gamma_1^2)}{2(\gamma^2 - \gamma_1^2)^2}\|\bar{Q}_\varepsilon\|,$$

we have

$$\|z_\varepsilon\|_2^2 - \gamma^2\|w\|_2^2 < 0$$

for $u = Fx$. We find that for w for which (4.443) is satisfied, we obtain for a suboptimal u already a negative cost. Since

$$\sup_{w\in\ell_2}\inf_{u\in\ell_2}\left\{ \|z_\varepsilon\|_2^2 - \gamma^2\|w\|_2^2 \mid x \in \ell_2 \right\} > 0,$$

we can without loss of generality assume that w satisfies

$$\|w\|^2 < \beta x_0' L x_0, \tag{4.444}$$

provided that ε is small enough. By setting $u = Fx+v$ the above inf–sup problem is equivalent to

$$\sup_{w\in\ell_2}\inf_{v\in\ell_2}\left\{ \|z_\varepsilon\|_2^2 - \gamma^2\|w\|_2^2 \mid u = Fx + v \right\} > 0.$$

Since we showed that we can assume, without loss of generality, that w is bounded, it is clear that in the above optimization for $\varepsilon = 0$, we can also assume without loss of generality that v is bounded as well, i.e.,

$$\|v\|_2 \leqslant N\|x_0\|.$$

If the system is left invertible from v to z, then as v gets sufficiently large in ℓ_2 norm, the cost can be made arbitrarily large. If the system is not left invertible, we can split the input in to one part which has no effect on the output and another part which has a left-invertible effect on the output. The latter has to be bounded for a bounded cost. The first can be set to zero without loss of generality. But with v and w bounded, we can find an M such that

$$\|x\|_2 \leqslant M\|x_0\|^2,$$

but then,

$$\begin{aligned}
&\sup_{w\in\ell_2}\inf_{v\in\ell_2}\left\{\|z_\varepsilon\|_2^2-\gamma^2\|w\|_2^2\right\}\\
&\quad\leqslant \sup_{w\in\ell_2}\inf_{v\in\ell_2}\left\{\|z_\varepsilon\|_2^2-\gamma^2\|w\|_2^2 \mid \|v\|_2\leqslant N\|x_0\|\right\}\\
&\quad\leqslant \sup_{w\in\ell_2}\inf_{v\in\ell_2}\left\{\|z_0\|_2^2-\gamma^2\|w\|_2^2 \mid \|v\|_2\leqslant N\|x_0\|\right\}+\|\bar{Q}_\varepsilon-\bar{Q}_0\|M\|x_0\|^2\\
&\quad= \sup_{w\in\ell_2}\inf_{v\in\ell_2}\left\{\|z_0\|_2^2-\gamma^2\|w\|_2^2\right\}+\|\bar{Q}_\varepsilon-\bar{Q}_0\|M\|x_0\|^2\\
&\quad= x_0'P_0x_0+\|\bar{Q}_\varepsilon-\bar{Q}_0\|M\|x_0\|^2,
\end{aligned}$$

where in each case $u = Fx + v$. In conclusion, we find that

$$x_0'P_0x_0 \leqslant x_0'P_\varepsilon x_0 \leqslant x_0'P_0x_0+\|\bar{Q}_\varepsilon-\bar{Q}_0\|M\|x_0\|^2,$$

which implies that $P_\varepsilon \to P_0$ as $\varepsilon \downarrow 0$. ∎

4.E Proofs of Lemmas 4.92, 4.95, and 4.98

Proof of Lemma 4.92 : To prove item (*i*), we first note that

$$\begin{aligned}
&F_\varepsilon(j\omega I-A-BF_\varepsilon)^{-1}Be_i\\
&=F_\varepsilon(I-(j\omega I-A)^{-1}BF_\varepsilon)^{-1}(j\omega I-A)^{-1}Be_i\\
&=(I-F_\varepsilon(j\omega I-A)^{-1}B)^{-1}F_\varepsilon(j\omega I-A)^{-1}Be_i.
\end{aligned}$$

Next, we note that [2, see Sect. 5.4, p.122]

$$\sigma_{\max}(I-F_\varepsilon(j\omega I-A)^{-1}B)^{-1}\leqslant 1,\ \forall\omega\in\mathbb{R}.$$

Moreover, $\forall\omega\in\mathcal{E}_k$, $(j\omega I-A)^{-1}Be_i$ has no pole, and therefore,

$$\|(j\omega I-A)^{-1}Be_i\|\leqslant M, \forall\omega\in\mathcal{E}_k$$

for $M > 0$ independent of ω.

But then,

$$\|F_\varepsilon(j\omega I-A-BF_\varepsilon)^{-1}Be_i\|\leqslant M\|F_\varepsilon\|, \forall\omega\in\mathcal{E}_k,$$

and since F_ε converges to zero, we get

$$\|F_\varepsilon(j\omega I - A - BF_\varepsilon)^{-1} Be_i\| \to 0$$

as $\varepsilon \to 0$ uniformly in $\mathcal{E}_k$.

It remains to show item (*ii*). By definition, $\det(j\omega I - A) \neq 0$ for all $\omega \in \Omega$. There exists a μ such that

$$\sigma_{\min}(j\omega I - A) > \mu, \ \forall \omega \in \Omega.$$

After all, assume this is not the case. Then there exists a sequence $\omega^i \in \Omega$ such that

$$\sigma_{\min}(j\omega^i I - A) \to 0$$

as $i \to \infty$. We can ensure that this sequence ω^i is bounded since for ω satisfying $|\omega| > \|A\| + 1$, we have

$$\sigma_{\min}(j\omega I - A) > |\omega| - \|A\| > 1.$$

But a bounded sequence ω^i has a convergent subsequence whose limit, denoted by $\bar{\omega}$, is in Ω (since Ω is closed). The limit $\bar{\omega}$ would have the property

$$\sigma_{\min}(j\bar{\omega} I - A) = 0.$$

This implies that $\bar{\omega}$ is an eigenvalue of A which is in contradiction with the definition of Ω.

Choose ε^* such that $\|F_\varepsilon\| \leqslant \frac{\mu}{4}\|B\|^{-1}$ for $\varepsilon \leqslant \varepsilon^*$. In that case,

$$\sigma_{\min}(\omega I - A - BF) > \mu - \|B\|\|F_\varepsilon\| > \tfrac{3\mu}{4}, \forall \omega \in \Omega,$$

and hence,

$$\|(j\omega I - A - BF_\varepsilon)^{-1}\| < \tfrac{4}{3\mu}, \forall \omega \in \Omega,$$

but then,

$$\|F_\varepsilon(j\omega I - A - BF_\varepsilon)^{-1} B\| \leqslant \|F_\varepsilon\|\|(j\omega I - A - BF_\varepsilon)^{-1}\|\|B\| \leqslant \tfrac{1}{3}$$

for all $\omega \in \Omega$. ■

Proof of Lemma 4.95 : The difference between $G_\varepsilon^m(s)$ and $G_\varepsilon(s)$ equals

$$\begin{aligned} G_\varepsilon(s) - G_\varepsilon^m(s) &= \left[I + F_\varepsilon(sI - A - BF_\varepsilon)^{-1} B\right] F_\varepsilon(sI - A + KC)^{-1} B \\ &= [I + G_\varepsilon(s)]\, F_\varepsilon(sI - A + KC)^{-1} B. \end{aligned}$$

We obtain from (4.343)

$$\sigma_{\max}(I + G_\varepsilon(j\omega)) \leqslant 1,\ \forall \varepsilon > 0,\ \omega \in \mathbb{R}.$$

Moreover,

$$\|F_\varepsilon(sI - A + KC)^{-1}B\|_\infty \leqslant \|F_\varepsilon\|\|(sI - A + KC)^{-1}B\|_\infty.$$

Since $F_\varepsilon \to 0$ as $\varepsilon \to 0$, we immediately have

$$\lim_{\varepsilon \downarrow 0} G_\varepsilon^m(j\omega) - G_\varepsilon(j\omega) = 0$$

uniformly in ω. ■

Proof of Lemma 4.98 : Define a system as

$$\begin{cases} \dot{x}_1 = Ax_1 + BF_\varepsilon x_2 \\ \dot{x}_2 = (A + BF_\varepsilon - KC)x_2 + KCx_1 \ , \\ \ z = F_\varepsilon x_2 \end{cases} \qquad \begin{pmatrix} x_1(0) \\ x_2(0) \end{pmatrix} = \xi.$$

It is obvious that for any ξ,

$$\|z\|_2 = \int_0^\infty \|\mathcal{F}_\varepsilon e^{(\mathcal{A}+\mathcal{B}\mathcal{F})t}\xi\|^2 dt.$$

Let $e = x_1 - x_2$. In the new coordinates of (x_1, e), the above system can be written as

$$\begin{cases} \dot{x}_1 = (A + BF_\varepsilon) - BF_\varepsilon e \\ \dot{e} = (A - KC)e \\ \dot{z} = F_\varepsilon(x_1 - e), \end{cases}$$

with $e_1(0) = x_1(0) - x_2(0)$. We get $\|z\|_2 \leqslant \|F_\varepsilon e\|_2 + \|F_\varepsilon x_1\|_2$.

Since $A - KC$ is Hurwitz, there exists a γ such that $\|e\|_2 \leqslant \gamma\|e(0)\|$ for any $e(0) \in \mathbb{R}^n$. Then

$$\|F_\varepsilon e\|_2 \leqslant \gamma\|F_\varepsilon\|\|e(0)\| \to 0 \text{ as } \varepsilon \to 0.$$

But for x_1, we have

$$\begin{aligned} \|F_\varepsilon x_1\|_2 &\leqslant \|G_\varepsilon(s)\|_\infty\|F_\varepsilon e\|_2 + \int_0^\infty \|F_\varepsilon e^{(A+BF_\varepsilon)t}x_1(0)\|^2 dt \\ &\leqslant 2\gamma\|F_\varepsilon\|\|e(0)\| + \int_0^\infty \|F_\varepsilon e^{(A+BF_\varepsilon)t}x_1(0)\|^2 dt, \end{aligned}$$

where $G_\varepsilon(s) = F_\varepsilon(sI - A - BF_\varepsilon)^{-1}B$. It was shown in [200] that

$$\lim_{\varepsilon \downarrow 0} \int_0^\infty \|F_\varepsilon e^{(A+BF_\varepsilon)t} x_1(0)\|^2 dt = 0,$$

and thus, $\lim_{\varepsilon \downarrow 0} \|F_\varepsilon x_1\|_2 = 0$. We conclude that $\lim_{\varepsilon \downarrow 0} \|z\|_2 = 0$. ∎

5

Robust semi-global internal stabilization

5.1 Introduction

In the previous chapter, we discussed semi-global internal stabilization of linear systems subject to control magnitude saturation. However, all of this is obtained for an ideal saturation element $\sigma(u)$. In reality, we are often faced with a saturation element which differs from the ideal saturation but still satisfies some of the basic properties as outlined in Sect. 2.6. One of the objectives of this chapter is to show in which respect the design methodologies such as low-gain and low-and-high-gain, which were described in detail in the previous chapter, still apply in case the saturation function has a different shape.

However, this is only one aspect of robustness for our design methodologics. We also like to see how uncertainties due to more general modeling errors effect our designs. In this case, we often can no longer rely just on low-gain design, and the low-and-high-gain controllers need to be used. Model uncertainty can effect the system through input-additive disturbances of the form

$$\dot{x} = Ax + B\sigma(u + g(x,t))$$

or through matched uncertainty

$$\dot{x} = Ax + B\sigma(u) + Bg(x,t).$$

The unmatched case

$$\dot{x} = Ax + B\sigma(u) + Eg(x,t),$$

where $B \neq E$ is in general very hard, and only limited results are available in the literature.

5.2 Generalized saturation functions: continuous time

In Sect. 2.6, we defined, besides a standard saturation function, a so-called generalized saturation function satisfying the following properties:

A. Saberi et al., *Internal and External Stabilization of Linear Systems with Constraints*, Systems & Control: Foundations & Applications, DOI 10.1007/978-0-8176-4787-2_5,

(*i*) $\widetilde{\sigma}(u)$ is decentralized, i.e.

$$\widetilde{\sigma}(s) = \begin{pmatrix} \widetilde{\sigma}_1(s_1) \\ \widetilde{\sigma}_2(s_2) \\ \vdots \\ \widetilde{\sigma}_m(s_m) \end{pmatrix}.$$

(*ii*) $\widetilde{\sigma}_i$ is globally Lipschitz, i.e. for some $\delta > 0$,

$$|\widetilde{\sigma}_i(s_1) - \widetilde{\sigma}_i(s_2)| \leqslant \delta |s_1 - s_2|.$$

(*iii*) $s\widetilde{\sigma}_i(s) > 0$ whenever $s \neq 0$ and $\widetilde{\sigma}_i(0) = 0$.

(*iv*) The two limits

$$\lim_{s \to 0+} \frac{\widetilde{\sigma}_i(s)}{s}, \quad \lim_{s \to 0-} \frac{\widetilde{\sigma}_i(s)}{s}$$

both exist and are strictly positive.

(*v*) $\liminf_{|s| \to \infty} |\widetilde{\sigma}_i(s)| > 0$.

(*vi*) There exists an $M > 0$ such that $|\widetilde{\sigma}_i(s)| < M$ for all $s \in \mathbb{R}$.

We note that the above conditions imply that there exists positive constants θ and ψ such that

$$|\widetilde{\sigma}(s)| > \min\{\, \theta |s|, \psi \,\} \tag{5.1}$$

componentwise.

In the previous chapter, we investigated the semi-global stabilization problem for a system of the form:

$$\begin{aligned} \dot{x} &= Ax + B\sigma(u) \\ y &= Cx. \end{aligned} \tag{5.2}$$

The two basic design methodologies presented in the previous chapter are the low-gain method and the low-and-high-gain method where the feedback gain is computed either via the direct-eigenstructure assignment or via Riccati equations. One of the basic result obtained in the previous chapter for continuous-time systems is as follows:

Theorem 5.1 *Consider the system* (5.2) *with* (A, B) *stabilizable and with all the eigenvalues of* A *in the closed left-half plane. Consider the low-gain state feedbacks* F_ε *designed either through the direct-eigenstructure method satisfying:*

- $\left\| F_\varepsilon e^{(A+BF_\varepsilon)t} \right\| \leqslant \varepsilon \beta e^{-\varepsilon t}$

- $\left\| e^{(A+BF_\varepsilon)t} \right\| \leqslant \frac{\gamma}{\varepsilon^{r-1}} e^{-\varepsilon t}$

for suitably chosen $\beta > 0$ and $\gamma > 0$ with r an integer less than or equal to n, or designed via the Riccati equation

$$A'P_\varepsilon + P_\varepsilon A - P_\varepsilon BB'P_\varepsilon + Q_\varepsilon = 0$$

with $F_\varepsilon = -B'P_\varepsilon$ and $Q_\varepsilon \to 0$ as $\varepsilon \to 0$. Both of these designs have the property that for an arbitrary compact set $\mathcal{X} \subset \mathbb{R}^n$, there exists an ε^* such that for all ε with $0 < \varepsilon < \varepsilon^*$, the feedback $u = F_\varepsilon x$ renders the closed-loop system asymptotically stable with $\mathcal{X}$ contained in its domain of attraction.

In the case of a generalized saturation function, we have the system

$$\begin{aligned} \dot{x} &= Ax + B\tilde{\sigma}(u) \\ y &= Cx \end{aligned} \tag{5.3}$$

where $\tilde{\sigma}$ satisfies properties (*i*)–(*vi*) described above. The next theorem establishes that the Riccati based design is robust in the sense that the low-gain design still works for the generalized saturation function.

Theorem 5.2 *Consider the system* (5.3) *with (A, B) stabilizable and with all the eigenvalues of A in the closed left half plane. Assume that $\tilde{\sigma}$ satisfies properties (i)–(vi) on page 340 which implies that there exist θ and ψ such that* (5.1) *is satisfied. Consider the low-gain state feedbacks F_ε designed via the Riccati equation*

$$A'P_\varepsilon + P_\varepsilon A - P_\varepsilon BB'P_\varepsilon + Q_\varepsilon = 0$$

with $F_\varepsilon = -B'P_\varepsilon$ and $Q_\varepsilon \to 0$ as $\varepsilon \to 0$. The system (5.3) *has the property that for an arbitrary compact set $\mathcal{X} \subset \mathbb{R}^n$, there exists an ε^* such that for all ε with $0 < \varepsilon < \varepsilon^*$, the feedback $u = \theta^{-1}F_\varepsilon x$ results in a closed-loop system which is asymptotically stable with $\mathcal{X}$ contained in its domain of attraction.*

Proof : The basic properties which yield the above result are the gain and phase margin properties of the ARE-based state feedback. We define

$$\mathcal{V}_\varepsilon = \left\{ x \in \mathbb{R}^n \mid x'P_\varepsilon x < 1 \right\}.$$

It is then easy to verify that for ε small enough, we have $\mathcal{X} \subset \mathcal{V}_\varepsilon$, and for all $x \in \mathcal{V}_\varepsilon$,

$$|\tilde{\sigma}(\theta^{-1}F_\varepsilon x)| > |F_\varepsilon x|$$

componentwise, while $\tilde{\sigma}(\theta^{-1}F_\varepsilon x)$ and $F_\varepsilon x$ have, again componentwise, the same sign.

Next, we consider the system (5.3). Let ε be chosen satisfying the earlier properties. It is then straightforward to verify that $x' P_\varepsilon x$ is a suitable Lyapunov function in the region $\mathcal{V}_\varepsilon$ for the system (5.3) and that the domain of attraction will contain the set $\mathcal{X}$. ■

The low-gain design studied in the above two theorems in general might yield slow transient response. For this purpose, in the previous chapter, the low-and-high-gain design was introduced to have a fast enough transient performance. Later on in this chapter, it also becomes obvious that the high-gain parameter plays an important role in reducing the effect of model uncertainties. In the next chapters, it will become obvious that the low-and-high-gain design also plays a crucial role in the rejection of disturbances.

Theorem 5.3 *Consider the system* (5.2) *with* (A, B) *stabilizable and with all the eigenvalues of* A *in the closed left-half plane. Let* $P_\varepsilon > 0$ *be the solution of the algebraic Riccati equation and*

$$A' P_\varepsilon + P_\varepsilon A - P_\varepsilon B B' P_\varepsilon + Q_\varepsilon = 0$$

with $Q_\varepsilon \to 0$ *as* $\varepsilon \to 0$. *In that case, for an arbitrary compact set* $\mathcal{X} \subset \mathbb{R}^n$, *there exists an* ε^* *such that for all* ε *with* $0 < \varepsilon < \varepsilon^*$ *and* $\alpha \geqslant 0$, *the feedback*

$$u = -B' P_\varepsilon x - \alpha B' P_\varepsilon x$$

has the property that the closed-loop system is asymptotically stable with $\mathcal{X}$ *contained in its domain of attraction.*

Remark 5.4 *In the above, the role of the high-gain parameter is not clarified. Basically, decreasing* ε *will increase the domain of attraction. On the other hand, the high-gain parameter* α *does not affect the domain of attraction, but it plays a crucial role in improving the transient performance and, as we will see later, in rejecting model uncertainty and disturbances.*

Theorem 5.5 *Consider the system* (5.3) *with* (A, B) *stabilizable and with all the eigenvalues of* A *in the closed left-half plane. Assume that* $\tilde{\sigma}$ *satisfies properties (i)–(vi) on page 340 which implies that there exist* θ *and* ψ *such that* (5.1) *is satisfied. Let* $P_\varepsilon > 0$ *be the solution of the algebraic Riccati equation and*

$$A' P_\varepsilon + P_\varepsilon A - P_\varepsilon B B' P_\varepsilon + Q_\varepsilon = 0$$

with $Q_\varepsilon \to 0$ *as* $\varepsilon \to 0$. *In that case, for an arbitrary compact set* $\mathcal{X} \subset \mathbb{R}^n$, *there exists an* ε^* *such that for all* ε *with* $0 < \varepsilon < \varepsilon^*$ *and* $\alpha \geqslant 0$, *the feedback*

$$u = -\theta^{-1} B' P_\varepsilon x - \alpha B' P_\varepsilon x$$

has the property that the closed-loop system is asymptotically stable with $\mathcal{X}$ *contained in its domain of attraction.*

Proof : Property (5.1) guarantees that the generalized saturation function is bounded away from zero. Let ε be such that

$$|B'P_\varepsilon x| < \psi$$

componentwise. Due to this property and (5.1), we know that

$$|\widetilde{\sigma}(\theta^{-1}B'P_\varepsilon x + \alpha B'P_\varepsilon x)| > |B'P_\varepsilon x|$$

componentwise for an ε sufficiently small. Moreover, $\widetilde{\sigma}(\theta^{-1}B'P_\varepsilon x + \alpha B'P_\varepsilon x)$ and $\sigma(B'P_\varepsilon x)$ also have, again componentwise, the same sign. It is then straightforward to verify that $x'P_\varepsilon x$ is a suitable Lyapunov function for the system and that the domain of attraction will contain the set $\mathcal{X}$ for a sufficiently small ε. ■

5.3 Generalized saturation functions: discrete time

In the previous section, we compared a so-called generalized saturation function with the standard saturation function for continuous-time systems. In this section, we make this same comparison for discrete-time systems. We consider a system of the form,

$$\begin{aligned} \rho x &= Ax + B\sigma(u) \\ y &= Cx \end{aligned} \tag{5.4}$$

with the standard saturation function. The two basic design methodologies presented in the previous chapter are the low-gain method and the low-and-high-gain methods where the feedback gain is computed either via the direct-eigenstructure assignment or via Riccati equations. The basic result obtained in the previous chapter for discrete-time systems is as follows:

Theorem 5.6 *Consider the system* (5.4) *with* (A, B) *stabilizable and with all the eigenvalues of* A *in the closed unit disc. Consider the low-gain state feedbacks* F_ε *designed either through the direct-eigenstructure method satisfying:*

- $\left\| F_\varepsilon(A + BF_\varepsilon)^k \right\| \leqslant \varepsilon\beta(1-\varepsilon)^k$
- $\left\| (A + BF_\varepsilon)^k \right\| \leqslant \dfrac{\gamma}{\varepsilon^{r-1}}(1-\varepsilon)^k$

for suitably chosen $\beta > 0$ *and* $\gamma > 0$ *with* r *an integer less than or equal to* n, *or designed via the Riccati equation*

$$P_\varepsilon = A'P_\varepsilon A - A'P_\varepsilon B(I + B'P_\varepsilon B)^{-1}B'P_\varepsilon A + Q_\varepsilon \tag{5.5}$$

with

$$F_\varepsilon = -(I + B'P_\varepsilon B)^{-1}B'P_\varepsilon A \tag{5.6}$$

and $Q_\varepsilon \to 0$ *as* $\varepsilon \to 0$. *Both of these designs have the property that for an arbitrary compact set* $\mathcal{X} \subset \mathbb{R}^n$, *there exists an* ε^* *such that for all* ε *with* $0 < \varepsilon < \varepsilon^*$, *the feedback* $u = F_\varepsilon x$ *renders the closed-loop system asymptotically stable with* $\mathcal{X}$ *contained in its domain of attraction.*

In the case of a generalized saturation function, we have the system

$$\begin{aligned} \rho x &= Ax + B\widetilde{\sigma}(u) \\ y &= Cx, \end{aligned} \tag{5.7}$$

with $\widetilde{\sigma}$ satisfies properties (*i*)–(*vi*) described in the previous section. The next theorem establishes that the Riccati-based design is robust in the sense that the low-gain design still works for the generalized saturation function.

Theorem 5.7 *Consider the system* (5.7) *with* (A, B) *stabilizable and with all the eigenvalues of A in the closed unit disc. Assume that* $\widetilde{\sigma}$ *satisfies properties (i)–(vi) on page 340 which implies that there exist* θ *and* ψ *such that* (5.1) *is satisfied. Consider the low-gain state feedbacks* F_ε *designed via the Riccati equation* (5.5) *and* (5.6) *with* $Q_\varepsilon \to 0$ *as* $\varepsilon \to 0$. *The system has the property that for an arbitrary compact set* $\mathcal{X} \subset \mathbb{R}^n$, *there exists an* ε^* *such that for all* ε *with* $0 < \varepsilon < \varepsilon^*$, *the feedback* $u = \theta^{-1} F_\varepsilon x$ *has the property that the closed-loop system is asymptotically stable with* $\mathcal{X}$ *contained in its domain of attraction.*

Proof : We define

$$\mathcal{V}_\varepsilon = \left\{ x \in \mathbb{R}^n \mid x' P_\varepsilon x < 1 \right\}.$$

It is then easy to verify that for ε small enough, $\mathcal{X} \subset \mathcal{V}_\varepsilon$, and

$$|\widetilde{\sigma}(\theta^{-1} F_\varepsilon x)| > |F_\varepsilon x| \tag{5.8}$$

componentwise, while $\widetilde{\sigma}(\theta^{-1} F_\varepsilon x)$ and $F_\varepsilon x$ have, again componentwise, the same sign for all $x \in \mathcal{V}_\varepsilon$.

Next, we consider the system (5.3). Given the global Lipschitz bound δ of $\widetilde{\sigma}$ of property (*ii*), we can also guarantee that

$$\frac{\delta}{\theta} < 1 + \frac{2}{\| B' P_\varepsilon B \|} \tag{5.9}$$

for a sufficiently small ε. Define

$$\begin{aligned} u &= \theta^{-1} F_\varepsilon x \\ v &= F_\varepsilon x. \end{aligned}$$

Then (5.8) and (5.9) imply that

$$|v| \leqslant \widetilde{\sigma}(u) \leqslant \left(1 + \frac{2}{\|B'P_\varepsilon B\|}\right)|v| \tag{5.10}$$

componentwise, and v and $\widetilde{\sigma}(u)$ have, again componentwise, the same sign. We have:

$$\begin{aligned}
&(\rho x)'P_\varepsilon(\rho x) - x'P_\varepsilon x \\
&\quad = v'(I + B'P_\varepsilon B)v - x'Q_\varepsilon x - 2v'(I + B'P_\varepsilon B)\widetilde{\sigma}(u) \\
&\qquad\qquad + \widetilde{\sigma}(u)'B'P_\varepsilon B\widetilde{\sigma}(u) \\
&\quad = [v - \widetilde{\sigma}(u)]'(I + B'P_\varepsilon B)[v - \widetilde{\sigma}(u)] - x'Q_\varepsilon x - \widetilde{\sigma}(u)'\widetilde{\sigma}(u) \\
&\quad \leqslant (1 + \mu)[v - \widetilde{\sigma}(u)]'[v - \widetilde{\sigma}(u)] - x'Q_\varepsilon x - \widetilde{\sigma}(u)'\widetilde{\sigma}(u) \\
&\quad = \mu\widetilde{\sigma}(u)'\widetilde{\sigma}(u) - 2(1 + \mu)\widetilde{\sigma}(u)'v + (1 + \mu)v'v - x'Q_\varepsilon x \\
&\quad \leqslant \mu\widetilde{\sigma}(u)'\widetilde{\sigma}(u) - 2(1 + \mu)\widetilde{\sigma}(u)'v + (2 + \mu)v'v - x'Q_\varepsilon x \\
&\quad = \mu\left[\widetilde{\sigma}(u) - \tfrac{1+\mu}{\mu}v\right]'\left[\widetilde{\sigma}(u) - \tfrac{1+\mu}{\mu}v\right] - \tfrac{1}{\mu}v'v - x'Q_\varepsilon x \\
&\quad \leqslant -x'Q_\varepsilon x.
\end{aligned}$$

where $\mu = \|B'P_\varepsilon B\|$. The last inequality follows from the fact that (5.10) implies that

$$-\tfrac{1}{\mu}v \leqslant \widetilde{\sigma}(u) - \tfrac{1+\mu}{\mu}v \leqslant \tfrac{1}{\mu}v,$$

and hence,

$$\mu\left[\widetilde{\sigma}(u) - \tfrac{1+\mu}{\mu}v\right]'\left[\widetilde{\sigma}(u) - \tfrac{1+\mu}{\mu}v\right] - \tfrac{1}{\mu}v'v \leqslant 0.$$

It is then straightforward to verify that $x'P_\varepsilon x$ is a suitable Lyapunov function in the region $\mathcal{V}_\varepsilon$ for the system (5.3) and that the domain of attraction will contain the set $\mathcal{X}$. ■

The low-gain design studied in the above two theorems in general might yield slow transient response. For this purpose, in the previous chapter, the low-and-high-gain design was introduced to render the transient performance faster. As in the case of continuous-time systems, later on in this chapter, it also becomes obvious that the high-gain parameter plays an important role in mitigating the effect of model uncertainties. Also, in subsequent chapters, it will become obvious that the low-and-high-gain design also plays a crucial role in the rejection of disturbances.

Theorem 5.8 *Consider the system* (5.2) *with* (A, B) *stabilizable and with all the eigenvalues of* A *in the closed unit disc. Let* $P_\varepsilon > 0$ *be the solution of the algebraic Riccati equation* (5.5) *and let* F_ε *be defined by* (5.6)*, where* $Q_\varepsilon \to 0$ *as* $\varepsilon \to 0$.

In that case, for an arbitrary compact set $\mathcal{X} \subset \mathbb{R}^n$, there exists an ε^* such that for all ε with $0 < \varepsilon < \varepsilon^*$ and

$$0 < \alpha < \frac{2}{\|B' P_\varepsilon B\|}, \tag{5.11}$$

the feedback

$$u = F_\varepsilon x + \alpha F_\varepsilon x$$

has the property that the closed-loop system is asymptotically stable with $\mathcal{X}$ contained in its domain of attraction.

Remark 5.9 *In the above, the role of the high-gain parameter is not clarified. Basically, descreasing ε will increase the domain of attraction. On the other hand, the high-gain parameter α does not affect the domain of attraction, but it is important to improve the transient performance and, as we shall see later on, to reject model uncertainty and disturbances.*

Theorem 5.10 *Consider the system* (5.3) *with (A, B) stabilizable and with all the eigenvalues of A in the closed unit disc. Assume that $\tilde{\sigma}$ satisfies properties (i)–(vi) on page 340 which implies that there exist θ and ψ such that* (5.1) *is satisfied. Let $P_\varepsilon > 0$ be the solution of the algebraic Riccati equation* (5.5)*, and let F_ε be defined by* (5.6)*, where $Q_\varepsilon \to 0$ as $\varepsilon \to 0$. In that case, for an arbitrary compact set $\mathcal{X} \subset \mathbb{R}^n$, there exists an ε^* such that for all ε with $0 < \varepsilon < \varepsilon^*$ and*

$$0 < \alpha < \frac{1}{\delta}\left[1 + \frac{2}{\|B' P_\varepsilon B\|} - \frac{\delta}{\theta}\right], \tag{5.12}$$

the feedback

$$u = \theta^{-1} F_\varepsilon x + \alpha F_\varepsilon x$$

has the property that the closed-loop system is asymptotically stable with $\mathcal{X}$ contained in its domain of attraction.

Remark 5.11 *Note that $P_\varepsilon \to 0$ as $\varepsilon \to 0$. Hence, for small ε, we can choose α very large and hence speed up the convergence considerably. Also note that if $\tilde{\sigma} = \sigma$, then $\delta = \theta = 1$, and we re-obtain the result from Theorem 5.8.*

Proof : We define

$$\mathcal{V}_\varepsilon = \left\{ x \in \mathbb{R}^n \mid x' P_\varepsilon x < 1 \right\}.$$

It is then easy to verify that for ε small enough $\mathcal{X} \subset \mathcal{V}_\varepsilon$ and

$$|F_\varepsilon x| < \psi$$

componentwise. Due to this property and (5.1), we know that

$$|\widetilde{\sigma}(\theta^{-1}F_\varepsilon x + \alpha F_\varepsilon x)| > |F_\varepsilon x| \tag{5.13}$$

componentwise for a sufficiently small ε. Moreover, $\widetilde{\sigma}(\theta^{-1}B'P_\varepsilon x + \alpha B'P_\varepsilon x)$ and $\sigma(B'P_\varepsilon x)$ also have, again componentwise, the same sign. Using the Lipschitz property of $\widetilde{\sigma}$, we find that

$$|\widetilde{\sigma}(\theta^{-1}F_\varepsilon x + \alpha F_\varepsilon x)| < \left(1 + \frac{2}{\|B'P_\varepsilon B\|}\right)|F_\varepsilon x| \tag{5.14}$$

componentwise provided α satisfies (5.12). Moreover, $\widetilde{\sigma}(\theta^{-1}B'P_\varepsilon x + \alpha B'P_\varepsilon x)$ and $B'P_\varepsilon x$ have, again componentwise, the same sign.

Define

$$\begin{aligned} u &= \theta^{-1}F_\varepsilon x + \alpha F_\varepsilon x \\ v &= F_\varepsilon x. \end{aligned}$$

Then (5.13) and (5.14) imply that

$$|v| \leqslant \widetilde{\sigma}(u) \leqslant \left(1 + \frac{2}{\|B'P_\varepsilon B\|}\right)|v| \tag{5.15}$$

componentwise, and v and $\widetilde{\sigma}(u)$ have, again componentwise, the same sign. We obtain, similar to that in the proof of Theorem 5.7,

$$(\rho x)'P_\varepsilon(\rho x) - x'P_\varepsilon x \leqslant -x'Q_\varepsilon x.$$

It is then straightforward to verify that $x'P_\varepsilon x$ is a suitable Lyapunov function in the region $\mathcal{V}_\varepsilon$ for the system (5.7) and that the domain of attraction will contain the set $\mathcal{X}$. ■

5.4 Systems with saturation and input-additive uncertainty: continuous time

We consider a class of nonlinear systems which are obtained by cascading linear systems with memory-free input nonlinearities of saturation type

$$\Sigma_{ud}: \begin{cases} \dot{x}(t) = Ax(t) + B\widetilde{\sigma}(u(t) + g(x,t)) \\ y(t) = Cx(t), \end{cases} \tag{5.16}$$

where $x \in \mathbb{R}^n$ is the state, $u \in \mathbb{R}^m$ is the control input, and $y \in \mathbb{R}^p$ is the measurement output. As before, we assume that $\widetilde{\sigma}$ satisfies properties *(i)*–*(vi)* on page 340. The uncertain element $g: \mathbb{R}^n \times \mathbb{R}^+ \to \mathbb{R}^m$ represents both the uncertainties and the disturbances. We impose only one requirement that we know an upper bound on its norm. More specifically, we make the following assumption:

Assumption 5.12 *The uncertain element $g(x,t)$ is piecewise continuous in t, locally Lipschitz in x, and its norm is bounded by a known function*

$$\|g(x,t)\| \leq g_0(\|x\|) + D_0, \quad \forall (t,x) \in \mathbb{R}^+ \times \mathbb{R}^n. \tag{5.17}$$

where D_0 is a known positive constant, and the known function $g_0(x) : \mathbb{R}^+ \to \mathbb{R}^+$ is locally Lipschitz and satisfies

$$g_0(0) = 0\,. \tag{5.18}$$

We are interested in finding controllers that achieve semi-global stabilization, independent of the specific saturation function $\tilde{\sigma}$ and independent of the precise g that satisfies Assumption 5.12.

The main state feedback problem we solve in this section is the following:

Problem 5.13 Consider the system (5.16), where $\tilde{\sigma}$ satisfies properties (*i*)–(*vi*) on page 340 and g satisfies the properties in Assumption 5.12 with $D_0 = 0$. The objective is to find, for any compact set $\mathcal{X}$, a feedback gain matrix F such that, for all functions g satisfying Assumption 5.12 with $D_0 = 0$, the closed-loop system Σ_{ud} is asymptotically stable and contains $\mathcal{X}$ in its domain of attraction.

If $D_0 \neq 0$, then it is not possible to achieve asymptotic stability without more precise information about the function g. However, in that case, we can still achieve practical stability:

Problem 5.14 Consider the system (5.16), where $\tilde{\sigma}$ satisfies properties (*i*)–(*vi*) on page 340 and g satisfies the properties in Assumption 5.12. The objective is to find, for any compact sets $\mathcal{X}_1$ and $\mathcal{X}_2$ containing 0 in their interior and with $\mathcal{X}_1 \supset \mathcal{X}_2$, a feedback gain matrix F such that, for all functions g satisfying Assumption 5.12 and initial conditions inside $\mathcal{X}_1$, the closed-loop system Σ_{ud} is such that the state enters and remains in $\mathcal{X}_2$ after some finite amount of time.

Remark 5.15 *For the case when $g_0 \equiv 0$ and $D_0 = 0$, as discussed in the previous chapter, the above problem reduces to the semi-global stabilization problem by state feedback control. When $g_0 \not\equiv 0$, but $D_0 = 0$, it can be called the robust semi-global stabilization problem by state feedback control. When $g_0 \not\equiv 0$ and $D_0 > 0$, it can be called the robust semi-global practical stabilization problem by state feedback control.*

The main measurement feedback problem we solve in this section is the following:

Problem 5.16 Consider the system (5.16), where $\tilde{\sigma}$ satisfies properties (*i*)–(*vi*) on page 340 and g satisfies the properties in Assumption 5.12 with $D_0 = 0$. The objective is to find, for any compact set $\mathcal{X} \subset \mathbb{R}^{2n}$, a dynamic feedback of the form

$$\begin{aligned} \dot{x}_c &= A_c x_c + B_c y, \quad x_c \in \mathbb{R}^n \\ u &= C_c x_c, \end{aligned} \tag{5.19}$$

such that, for all functions g satisfying Assumption 5.12 with $D_0 = 0$, the closed-loop system Σ_{ud} is asymptotically stable and contains $\mathcal{X}$ in its domain of attraction.

If $D_0 \neq 0$, then it is again not possible to achieve asymptotic stability without more precise information about the function g. However, in that case, we can still achieve practical stability.

Problem 5.17 Consider the system (5.16), where $\tilde{\sigma}$ satisfies properties (*i*)–(*vi*) on page 340 and g satisfies the properties in Assumption 5.12. The objective is to find, for any pair of compact sets $\mathcal{X}_1, \mathcal{X}_2 \subset \mathbb{R}^{2n}$ containing 0 in their interior and with $\mathcal{X}_1 \supset \mathcal{X}_2$, a dynamic feedback of the form (5.19) such that for all functions g satisfying Assumption 5.12 and initial conditions inside $\mathcal{X}_1$, the closed-loop system Σ_{ud} is such that the state enters and remains in $\mathcal{X}_2$ after some finite amount of time.

We note that the requirement that all the eigenvalues have nonpositive parts is necessary for semi-global stabilization as discussed in the previous chapter. However, without such an assumption on the open-loop eigenvalues, interesting local results are still possible and will be pointed out later.

5.4.1 State feedback results

Before we discuss the solvability of the above-defined state feedback control problem, we recall below the family of parameterized low-and-high-gain state feedback laws, denoted by

$$u = \theta^{-1} F_\varepsilon x + \alpha F_\varepsilon x, \tag{5.20}$$

where $F_\varepsilon = -B' P_\varepsilon$, and P_ε is the positive definite solution of

$$A' P_\varepsilon + P_\varepsilon A - P_\varepsilon B B' P_\varepsilon + Q_\varepsilon = 0$$

with $Q_\varepsilon > 0$ such that $Q_\varepsilon \to 0$ as $\varepsilon \to 0$. As we discussed in Chap. 4, the parameter ε is the low-gain tuning parameter, whereas α is the high-gain tuning parameter.

Theorem 5.18 *Consider the system* (5.16) *and the robust semi-global stabilization problem formulated in Problem 5.13. Assume that* (A, B) *is stabilizable and all the eigenvalues of* A *are in the closed left-half plane. Also, assume that* $\tilde{\sigma}$ *satisfies properties (i)–(vi) on page 340 which implies that there exist* θ *and* ψ *such that* (5.1) *is satisfied. In that case, there exist, for any compact set* $\mathcal{X}$*, an* ε^* *and a function* $\alpha^* : (0, \varepsilon^*) \to \mathbb{R}^+$ *such that for all* $\varepsilon \in (0, \varepsilon^*)$ *and* $\alpha > \alpha^*(\varepsilon)$*, the feedback* (5.20) *has the property that for all functions* g *satisfying Assumption 5.12 with* $D_0 = 0$*, the resulting closed-loop system is asymptotically stable and contains* $\mathcal{X}$ *in its domain of attraction.*

Proof : We first choose ε^* such that

$$\mathcal{X} \subset \mathcal{V}_\varepsilon := \{ x \in \mathbb{R}^n \mid x' P_\varepsilon x < 1 \}$$

and

$$|F_\varepsilon x| < \psi$$

componentwise for all $x \in \mathcal{V}_\varepsilon$ and for all $\varepsilon \in (0, \varepsilon^*)$. The last condition implies that

$$|\tilde{\sigma}(\gamma F_\varepsilon x)| > \min\{ \theta |\gamma F_\varepsilon x|, \psi \} > |F_\varepsilon x| \tag{5.21}$$

for $\gamma > \theta^{-1}$.

For all $x \in \mathcal{V}_\varepsilon$, there exist M_1 and M_2 such that

$$\|g(x,t)\|^2 \leqslant g_0(\|x\|)^2 \leqslant M_1 \|x\|^2 \leqslant M_2 x' Q_\varepsilon x, \tag{5.22}$$

where the first inequality follows from Assumption 5.12, while the existence of an M_1 such that the second inequality is satisfied is a consequence of the fact that g_0 is locally Lipschitz and x is in a bounded set. The existence of M_2 such that the final inequality is satisfied follows from the fact that Q_ε is positive definite.

Next, we look at the candidate Lyapunov function

$$V_\varepsilon(t) := x(t)' P_\varepsilon x(t).$$

We obtain

$$\dot{V}_\varepsilon = -x' Q_\varepsilon x - v'v - 2v' \left[\tilde{\sigma}(\theta^{-1} v + \alpha v + g) - v \right], \tag{5.23}$$

where

$$v = -B' P_\varepsilon x.$$

To analyze further the derivative of our candidate Lyapunov function, we consider two cases. Let v_i and g_i denote the ith component of v and g, respectively. If

$$|\alpha v_i| \geqslant |g_i|$$

is satisfied, we then obtain

$$\theta^{-1} v_i + \alpha v_i + g_i = \gamma v_i$$

with $\gamma > \theta^{-1}$. Using (5.21), we then find that

$$v_i \left[\tilde{\sigma}(\theta^{-1} v_i + \alpha v_i + g_i) - v_i\right] > 0.$$

On the other hand, if

$$|\alpha v_i| < |g_i| \tag{5.24}$$

is satisfied, then we get

$$|\tilde{\sigma}(\theta^{-1} v_i + \alpha v_i + g_i) - v_i| \leqslant \delta|\theta^{-1} v_i + \alpha v_i + g_i| + |v_i| \leqslant M_3|g_i|$$

for some constant $M_3 > 0$ (independent of α) using (5.24) and the fact that $\tilde{\sigma}$ is globally Lipschitz with Lipschitz constant δ. Here we used, without loss of generality, that $\alpha > 1$. This implies that

$$v_i \left[\tilde{\sigma}(\theta^{-1} v_i + \alpha v_i + g_i) - v_i\right] \geqslant -M_3|g_i||v_i| \geqslant -\frac{M_3}{\alpha} g_i^2$$

where we again used (5.24). If we bring together the above two cases, we find that

$$2v' \left[\tilde{\sigma}(\theta^{-1} v + \alpha v + g) - v\right] \geqslant -\frac{M_3}{\alpha}\|g\|^2 \geqslant -\frac{M_3 M_2}{\alpha} x' Q_\varepsilon x,$$

where we used (5.22). Using this in (5.23), we get

$$\dot{V}_\varepsilon \leqslant -x' Q_\varepsilon x - v'v + \frac{M_3 M_2}{\alpha} x' Q_\varepsilon x < -\frac{1}{2} x' Q_\varepsilon x,$$

provided $\alpha > 2M_3M_2$. Note that the lower bound for α depends on ε since the bound M_2 depends on ε. We conclude that in $\mathcal{V}_\varepsilon$ the Lyapunov function V_ε is strictly decaying which implies asymptotic stability and, moreover, that $\mathcal{V}_\varepsilon$ is positively invariant and contained in the domain of attraction. ■

We consider next the semi-global practical stabilization problem.

Theorem 5.19 *Consider the system* (5.16) *and the robust semi-global practical stabilization problem formulated in Problem 5.14. Assume that* (A, B) *is stabilizable, and all the eigenvalues of* A *are in the closed left-half plane. Also, assume that* $\tilde{\sigma}$ *satisfies properties (i)–(vi) on page 340 which implies that there exist* θ *and* ψ *such that* (5.1) *is satisfied. In that case, there exist, for any pair of compact sets* $\mathcal{X}_1$ *and* $\mathcal{X}_2$ *containing* 0 *in their interior and with* $\mathcal{X}_1 \supset \mathcal{X}_2$, *an* ε^* *and a function* $\alpha^* : (0, \varepsilon^*) \to \mathbb{R}^+$ *such that for all* $\varepsilon \in (0, \varepsilon^*)$ *and* $\alpha > \alpha^*(\varepsilon)$, *the feedback* (5.20) *has the property that for all functions* g *satisfying Assumption 5.12, the resulting closed-loop system is such that for all initial conditions in* $\mathcal{X}_1$ *the state enters and remains in the set* $\mathcal{X}_2$ *within a finite amount of time.*

Proof : We first choose an ε^* such that

$$\mathcal{X}_1 \subset \mathcal{V}_\varepsilon := \{\, x \in \mathbb{R}^n \mid x' P_\varepsilon x < 1 \,\}$$

and

$$|F_\varepsilon x| < \psi$$

componentwise for all $x \in \mathcal{V}_\varepsilon$ and for all $\varepsilon \in (0, \varepsilon^*)$. The last condition implies that

$$|\tilde{\sigma}(\gamma F_\varepsilon x)| > \min\{\, \theta |\gamma F_\varepsilon x|, \psi \,\} > |F_\varepsilon x| \tag{5.25}$$

for $\gamma > \theta^{-1}$.

For all $x \in \mathcal{V}_\varepsilon$, there exist M_1 and M_2 such that

$$\begin{aligned} \|g(x,t)\|^2 &\leqslant [g_0(\|x\|) + D_0]^2 \\ &\leqslant M_1 \|x\|^2 + 2D_0^2 \\ &\leqslant M_2 x' Q_\varepsilon x + 2D_0^2, \end{aligned} \tag{5.26}$$

where the first inequality follows from Assumption 5.12, while the existence of an M_1 such that the second inequality is satisfied is a consequence of the fact that g_0 is locally Lipschitz and x is in a bounded set. The existence of M_2 such that the final inequality is satisfied follows from the fact that Q_ε is positive definite.

Next we look at the candidate Lyapunov function,

$$V_\varepsilon(t) := x(t)' P_\varepsilon x(t).$$

We obtain

$$\dot{V}_\varepsilon = -x' Q_\varepsilon x - v'v - 2v' \left[\tilde{\sigma}(\theta^{-1} v + \alpha v + g) - v \right], \tag{5.27}$$

where

$$v = -B' P_\varepsilon x.$$

To analyze further the derivative of our candidate Lyapunov function, we consider two cases. Let v_i and g_i denote the ith component of v and g, respectively. If

$$|\alpha v_i| \geqslant |g_i|$$

is satisfied, we then obtain,

$$\theta^{-1} v_i + \alpha v_i + g_i = \gamma v_i$$

with $\gamma > \theta^{-1}$. Using (5.21), we find that

$$v_i \left[\tilde{\sigma}(\theta^{-1} v_i + \alpha v_i + g_i) - v_i \right] > 0.$$

On the other hand, if

$$|\alpha v_i| < |g_i| \tag{5.28}$$

is satisfied, we get

$$|\tilde{\sigma}(\theta^{-1}v_i + \alpha v_i + g_i) - v_i| \leqslant \delta|\theta^{-1}v_i + \alpha v_i + g_i| + |v_i| \leqslant M_3|g_i|$$

for some constant $M_3 > 0$ (independent of α) using (5.28) and the fact that $\tilde{\sigma}$ is globally Lipschitz with Lipschitz constant δ. Here we used, without loss of generality, that $\alpha > 1$. This implies that

$$v_i\left[\tilde{\sigma}(\theta^{-1}v_i + \alpha v_i + g_i) - v_i\right] \geqslant -M_3|g_i||v_i| \geqslant -\frac{M_3}{\alpha}g_i^2,$$

where we again used (5.28). If we bring together the above two cases, we find that

$$2v'\left[\tilde{\sigma}(\theta^{-1}v + \alpha v + g) - v\right] \geqslant -\frac{M_3}{\alpha}\|g\|^2 \geqslant -\frac{M_3M_2}{\alpha}x'Q_\varepsilon x - \frac{2M_3}{\alpha}D_0^2,$$

where we used (5.26). Using this in (5.27) we get,

$$\begin{aligned}\dot{V}_\varepsilon &\leqslant -x'Q_\varepsilon x - v'v + \frac{M_3M_2}{\alpha}x'Q_\varepsilon x + \frac{2M_3}{\alpha}D_0^2\\ &< -\frac{1}{2}x'Q_\varepsilon x + \frac{2M_3}{\alpha}D_0^2,\end{aligned} \tag{5.29}$$

provided $\alpha > 2M_3M_2$. Moreover, let μ_1, μ_2 be such that

$$\{x \in \mathbb{R}^n \mid x'Q_\varepsilon x < \mu_1\} \subset \mathcal{W}_\mu = \{x \in \mathbb{R}^n \mid x'P_\varepsilon x < \mu_2\} \subset \mathcal{X}_2.$$

Choose $\alpha > 2M_3M_2$ such that

$$\frac{4M_3}{\alpha}D_0^2 < \mu_1.$$

Then it is obvious from (5.29) that inside $\mathcal{V}_\varepsilon$, the Lyapunov function V_ε is decreasing outside $\mathcal{W}_\mu$. This implies that the solution will enter and stay in $\mathcal{W}_\mu$ within a finite amount of time. By construction, this implies that the solution enters and stays in the set $\mathcal{X}_2$ after a finite amount of time for all initial conditions in the set $\mathcal{X}_1$. ■

5.4.2 *Measurement feedback results*

In the measurement feedback case, we use the low-and-high-gain state feedback (5.20) as studied in the previous subsection and introduced in Chap. 4. We combine this with a high-gain observer, i.e.,

$$\begin{aligned}\dot{\hat{x}} &= (A + BF_\varepsilon)\hat{x} + K_\ell(y - C\hat{x})\\ u &= \theta^{-1}F_\varepsilon\hat{x} + \alpha F_\varepsilon\hat{x},\end{aligned} \tag{5.30}$$

where $F_\varepsilon = -B'P_\varepsilon$, and P_ε is the positive definite solution of

$$A'P_\varepsilon + P_\varepsilon A - P_\varepsilon BB'P_\varepsilon + Q_\varepsilon = 0$$

with $Q_\varepsilon > 0$ such that $Q_\varepsilon \to 0$ as $\varepsilon \to 0$.

In order to construct a suitable observer, we note that if the system $(A, B, C, 0)$ is left invertible and minimum phase then there exists a basis transformation Γ_s such that

$$\Gamma_s^{-1} A \Gamma_s = \begin{pmatrix} A_0 & L_0 C_1 \\ B_1 E_1 & A_1 \end{pmatrix}, \quad \Gamma_s^{-1} B = \begin{pmatrix} 0 \\ B_1 \end{pmatrix}, \quad C \Gamma_s = \begin{pmatrix} 0 & C_1 \end{pmatrix}$$

with A_0 asymptotically stable. This follows from Chap. 3 where

$$\Gamma_s x = \begin{pmatrix} x_1 \\ x_2 \end{pmatrix}, \qquad x_1 = x_a, \qquad x_2 = \begin{pmatrix} x_b \\ x_d \end{pmatrix}$$

in the notation of SCB. Note that x_c is missing because the system is left-invertible, and A_0 is asymptotically stable because the system is minimum-phase. Since the subsystem $(A_1, B_1, C_1, 0)$ is observable and its transfer matrix has a polynomial inverse, it has been established in Sect. 4.5.1 of Chap. 4 that there exist $K_{1,\ell}$ and $S_{1,\ell}$ such that

$$S_{1,\ell}^{-1}(A_1 - K_{1,\ell} C_1) S_{1,\ell} = \ell \widetilde{A}_1, \qquad S_{1,\ell}^{-1} B_1 = B_1,$$

where there exists an M, independent of ℓ, such that $\|S_{1,\ell}\| < M$ for all $\ell \geqslant 1$. We define,

$$K_\ell = \Gamma_s \begin{pmatrix} L_0 \\ K_{1,\ell} \end{pmatrix}.$$

The claim is that the above dynamic observer-based controller can be used to solve Problems 5.16 and 5.17 for ε sufficiently small and α and ℓ sufficiently large. The following two theorems make this result more precise.

Theorem 5.20 *Consider the system* (5.16) *and the robust semi-global stabilization problem formulated in Problem 5.16. Assume that* (A, B) *is stabilizable, all the eigenvalues of* A *are in the closed left-half plane, and the system* $(A, B, C, 0)$ *is left invertible and minimum phase. Also, assume that* $\widetilde{\sigma}$ *satisfies properties* (*i*)–(*vi*) *on page 340 which implies that there exist* θ *and* ψ *such that* (5.1) *is satisfied. In that case, there exist, for any compact set* $\mathcal{X}_{cl} \subset \mathbb{R}^{2n}$, *an* ε^*, *a function* $\alpha^* : (0, \varepsilon^*) \to \mathbb{R}^+$, *and a function* $\ell^* : (0, \varepsilon^*) \times \mathbb{R}^+ \to \mathbb{R}^+$ *such that for all* $\varepsilon \in (0, \varepsilon^*)$, $\alpha > \alpha^*(\varepsilon)$, *and* $\ell > \ell^*(\varepsilon, \alpha)$, *the feedback* (5.30) *has the property that, for all functions* g *satisfying Assumption 5.12 with* $D_0 = 0$, *the resulting closed-loop system is asymptotically stable and contains* $\mathcal{X}_{cl}$ *in its domain of attraction.*

Proof : We decompose $\hat{x}$ compatible with the decomposition of x, and we define e_1 and e_2 as

$$\Gamma_s \hat{x} = \begin{pmatrix} \hat{x}_1 \\ \hat{x}_2 \end{pmatrix}, \quad e_1 = x_1 - \hat{x}_1, \quad e_2 = S_{1,\ell}^{-1}(x_2 - \hat{x}_2). \tag{5.31}$$

The closed-loop system can then be written in terms of these new variables as

$$\begin{aligned}
\dot{\widehat{x}}_1 &= A_0\widehat{x}_1 + L_0C_1x_2\\
\dot{x}_2 &= A_1x_2 + B_1(\widetilde{\sigma}(u+g) + E_1(\widehat{x}_1 + e_1))\\
\dot{e}_1 &= A_0e_1\\
\dot{e}_2 &= \ell\widetilde{A}_1e_2 + B_1(\widetilde{\sigma}(u+g) + E_1(\widehat{x}_1 + e_1))\\
u &= \left(\theta^{-1}+\alpha\right) F_\varepsilon \Gamma_s \begin{pmatrix} \widehat{x}_1 \\ x_2 - S_{1,\ell}e_2 \end{pmatrix}.
\end{aligned} \tag{5.32}$$

For ease of presentation, we will define $F_{\varepsilon,1}$ and $F_{\varepsilon,2}$ such that

$$u = \left(\theta^{-1}+\alpha\right)\left[F_{\varepsilon,1}\widehat{x}_1 + F_{\varepsilon,2}(x_2 - S_{1,\ell}e_2)\right]. \tag{5.33}$$

Moreover, we define

$$\widetilde{P}_\varepsilon = \Gamma_s' P_\varepsilon \Gamma_s, \qquad \widetilde{Q}_\varepsilon = \Gamma_s' Q_\varepsilon \Gamma_s, \qquad \widetilde{x} = \begin{pmatrix} \widehat{x}_1 \\ x_2 \end{pmatrix}, \qquad \widetilde{e} = \begin{pmatrix} e_1 \\ e_2 \end{pmatrix}. \tag{5.34}$$

Clearly, we can construct compact sets $\mathcal{X}_1, \mathcal{X}_2, \mathcal{E}_1, \mathcal{E}_2$ such that $(x, \widehat{x}) \in \mathcal{X}_{\text{cl}}$ implies that $\widehat{x}_1 \in \mathcal{X}_1$, $x_2 \in \mathcal{X}_2$, $e_1 \in \mathcal{E}_1$, and $e_2 \in \mathcal{E}_2$.

First, we choose β such that

$$\beta \geqslant 2\|E_1\|^2 + 1. \tag{5.35}$$

Next, let c be such that

$$\beta\widetilde{e}' R\widetilde{e} < \frac{c}{2}$$

for all $\widetilde{e} \in \mathcal{E}_1 \times \mathcal{E}_2$ and

$$R = \begin{pmatrix} R_0 & 0 \\ 0 & R_1 \end{pmatrix} \tag{5.36}$$

where

$$A_0'R_0 + R_0A_0 + I = 0, \qquad \widetilde{A}_1'R_1 + R_1\widetilde{A}_1 + I = 0.$$

Finally, choose ε^* such that for any $\varepsilon \in (0, \varepsilon^*)$, we have

$$\widetilde{x}'\widetilde{P}_\varepsilon\widetilde{x} < \frac{c}{2}$$

for all $\widetilde{x} \in \mathcal{X}_1 \times \mathcal{X}_2$ and

$$|F_{\varepsilon,1}\widehat{x}_1 + F_{\varepsilon,2}x_2| < \psi \tag{5.37}$$

componentwise for all $(\widehat{x}_1, x_2)$ with $\widetilde{x}'\widetilde{P}_\varepsilon\widetilde{x} < c$. Next, we look at the candidate Lyapunov function

$$L_\varepsilon(t) = V_\varepsilon(t) + \beta W(t),$$

where

$$V_\varepsilon(t) := \tilde{x}(t)'\tilde{P}_\varepsilon\tilde{x}(t), \qquad W(t) := \tilde{e}(t)'R\tilde{e}(t).$$

Clearly, for all initial conditions in $\mathcal{X}_{\rm cl}$, we have $L_\varepsilon(0) \leqslant 1$, and (5.37) is satisfied for all $t \geqslant 0$ as long as $L_\varepsilon(0) \leqslant 1$.

Before we formally prove that L_ε is a suitable Lyapunov function, we note that, similar to the proof of Theorem 5.18, (5.37) implies that

$$|\tilde{\sigma}(\gamma F_{\varepsilon,1}\hat{x}_1 + \gamma F_{\varepsilon,2}x_2)| > |F_{\varepsilon,1}\hat{x}_1 + F_{\varepsilon,2}x_2| \tag{5.38}$$

componentwise for $\gamma > \theta^{-1}$. Given our assumptions on the function g, there exist M_1, M_2, and M_3 such that

$$\begin{aligned}\|g(x,t)\|^2 &\leqslant [g_0(\|x\|)]^2\\ &\leqslant M_1\|x\|^2\\ &\leqslant M_2\tilde{x}'\tilde{Q}_\varepsilon\tilde{x} + M_3\|e_1\|^2.\end{aligned} \tag{5.39}$$

We obtain

$$\dot{V}_\varepsilon = -\tilde{x}'\tilde{Q}_\varepsilon\tilde{x} - v'v - 2v'\left[\tilde{\sigma}(\theta^{-1}v + \alpha v + h + g) - v + E_1e_1\right] \tag{5.40}$$

where

$$v = F_{\varepsilon,1}\hat{x}_1 + F_{\varepsilon,2}x_2, \qquad h = -(\theta^{-1}+\alpha)F_{\varepsilon,2}S_{1,\ell}e_2.$$

Let v_i, g_i, and h_i denote the ith component of v, g, and h, respectively. If

$$|\alpha v_i| \geqslant |g_i + h_i|$$

is satisfied, we then obtain

$$\theta^{-1}v_i + \alpha v_i + g_i + h_i = \gamma v_i$$

with $\gamma > \theta^{-1}$. Using (5.38), we get

$$v_i\left[\tilde{\sigma}(\theta^{-1}v_i + \alpha v_i + h_i + g_i) - v_i\right] > 0.$$

On the other hand, if

$$|\alpha v_i| < |g_i + h_i| \tag{5.41}$$

is satisfied, we then use

$$\begin{aligned}|\tilde{\sigma}(\theta^{-1}v_i + \alpha v_i + h_i + g_i) - v_i| &\leqslant \delta|\theta^{-1}v_i + \alpha v_i + h_i + g_i| + |v_i|\\ &\leqslant M_4|g_i + h_i|\end{aligned}$$

for some constant $M_4 > 0$ (independent of α) using (5.24) and the fact that $\tilde{\sigma}$ is globally Lipschitz with Lipschitz constant δ. Here we used, without loss of generality, that $\alpha > 1$. This implies that

$$\begin{aligned}v_i\left[\tilde{\sigma}(\theta^{-1}v_i + \alpha v_i + h_i + g_i) - v_i\right] &\geqslant -M_4|g_i + h_i||v_i|\\ &\geqslant -\frac{2M_4}{\alpha}(g_i^2 + h_i^2),\end{aligned}$$

where we again used (5.41). If we bring together the above two cases, we find

$$\begin{aligned}
2v'\left[\tilde{\sigma}(\theta^{-1}v + \alpha v + h + g) - v\right] \\
&\geqslant -\frac{4M_4}{\alpha}\left(\|g\|^2 + \|h\|^2\right) \\
&\geqslant -\frac{4M_4M_2}{\alpha}\tilde{x}'\widetilde{Q}_\varepsilon\tilde{x} - \frac{4M_4M_3}{\alpha}\|e_1\|^2 - \frac{4M_4}{\alpha}\|h\|^2, \qquad (5.42)
\end{aligned}$$

where we used (5.39). Using (5.40) and (5.42), we get

$$\begin{aligned}
\dot{V}_\varepsilon &\leqslant -\tilde{x}'\widetilde{Q}_\varepsilon\tilde{x} - v'v - 2v'E_1e_1 + \frac{4M_4M_2}{\alpha}\tilde{x}'\widetilde{Q}_\varepsilon\tilde{x} \\
&\qquad + \frac{4M_4M_3}{\alpha}\|e_1\|^2 + \frac{4M_4}{\alpha}\|h\|^2 \\
&< -\frac{2}{3}\tilde{x}'\widetilde{Q}_\varepsilon\tilde{x} + \frac{3\beta}{4}e_1'e_1 + M_5e_2'e_2, \qquad (5.43)
\end{aligned}$$

provided $\alpha > 12M_3M_2$ and $\alpha > \frac{16M_4M_3}{\beta}$, and where M_5 is a suitable constant depending on our choice of α but independent of ℓ. For the latter, we need to use that the norm of $S_{1,\ell}$ is bounded as a function of ℓ. We also used that, based on (5.35), we have

$$-v'v - 2v'E_1e_1 \leqslant \|E_1e_1\|^2 \leqslant \frac{\beta}{2}\|e_1\|^2. \qquad (5.44)$$

Next, we consider W. We obtain

$$\dot{W} \leqslant -e_1'e_1 - \ell e_2'e_2 + 2e_2'R_1B_1[\tilde{u} + E_1(\hat{x}_1 + e_1)], \qquad (5.45)$$

where

$$\tilde{u} = \tilde{\sigma}(\theta^{-1}v + \alpha v + h + g).$$

We obtain

$$\dot{W} \leqslant -e_1'e_1 - \frac{\ell}{2}e_2'e_2 + \frac{M_6}{\ell}\|\tilde{u} + E_1(\hat{x}_1 + e_1)\|^2,$$

for a suitable constant M_6 independent of ℓ and α. We note that

$$\|\tilde{u} + E_1(\hat{x}_1 + e_1)\|^2 \leqslant M_7\tilde{x}'\widetilde{Q}_\varepsilon\tilde{x} + M_8\tilde{e}'\tilde{e}$$

for suitable constants M_7 and M_8 which depend on α but which are independent of ℓ and we obtain

$$\dot{W} \leqslant -e_1'e_1 - \frac{\ell}{2}e_2'e_2 + \frac{M_6M_7}{\ell}\tilde{x}'\widetilde{Q}_\varepsilon\tilde{x} + \frac{M_6M_8}{\ell}\tilde{e}'\tilde{e}.$$

Combining this with (5.43), we obtain

$$\begin{aligned}
\dot{L}_\varepsilon &\leqslant -\frac{2}{3}\tilde{x}'\widetilde{Q}_\varepsilon\tilde{x} + M_5e_2'e_2 - \frac{\beta}{4}e_1'e_1 - \frac{\beta\ell}{2}e_2'e_2 \\
&\qquad + \frac{\beta M_6M_7}{\ell}\tilde{x}'\widetilde{Q}_\varepsilon\tilde{x} + \frac{\beta M_6M_8}{\ell}\tilde{e}'\tilde{e} \\
&\leqslant -\frac{1}{3}\tilde{x}'\widetilde{Q}_\varepsilon\tilde{x} - \frac{\beta}{4}\tilde{e}'\tilde{e}
\end{aligned}$$

for a sufficiently large ℓ. This immediately implies the required asymptotic stability and the required domain of attraction. ∎

Theorem 5.21 *Consider the system* (5.16) *and the robust semi-global stabilization problem formulated in Problem 5.16. Assume that* (A, B) *is stabilizable, all the eigenvalues of* A *are in the closed left-half plane, and the system* $(A, B, C, 0)$ *is left invertible and minimum phase. Also, assume that* $\widetilde{\sigma}$ *satisfies properties (i)–(vi) on page 340 which implies that there exist* θ *and* ψ *such that* (5.1) *is satisfied. In that case, there exist, for any pair of compact sets* $\mathcal{X}_{cl,1}$ *and* $\mathcal{X}_{cl,2}$ *in* $\mathbb{R}^{2n}$ *containing* 0 *in their interior and with* $\mathcal{X}_{cl,1} \supset \mathcal{X}_{cl,2}$*, an* ε^**, a function* $\alpha^* : (0, \varepsilon^*) \to \mathbb{R}^+$*, and a function* $\ell^* : (0, \varepsilon^*) \times \mathbb{R}^+ \to \mathbb{R}^+$ *such that for all* $\varepsilon \in (0, \varepsilon^*)$*,* $\alpha > \alpha^*(\varepsilon)$*, and* $\ell > \ell^*(\varepsilon, \alpha)$*, the feedback* (5.30) *has the property that, for all functions g satisfying Assumption 5.12, the resulting closed-loop system is such that for all initial conditions in* $\mathcal{X}_{cl,1}$ *the state enters and remains in the set* $\mathcal{X}_{cl,2}$ *within a finite amount of time.*

Proof : Following the proof of Theorem 5.20, we decompose $\hat{x}$ compatible with the decomposition of x and we define e_1 and e_2 as in (5.31), and in terms of these new variables, we obtain the closed-loop system (5.32). We again define $F_{\varepsilon,1}$ and $F_{\varepsilon,2}$ such that (5.33) is satisfied and use the definition in (5.34). Clearly, we can construct compact sets $\mathcal{X}_1, \mathcal{X}_2, \mathcal{E}_1, \mathcal{E}_2$ such that $(x, \hat{x}) \in \mathcal{X}_{\text{cl},2}$ implies that $\hat{x}_1 \in \mathcal{X}_1$, $x_2 \in \mathcal{X}_2$, $e_1 \in \mathcal{E}_1$, and $e_2 \in \mathcal{E}_2$.

First, we choose β such that (5.35) is satisfied, and let c be such that

$$\beta \widetilde{e}' R \widetilde{e} < \frac{c}{2}$$

for all $\widetilde{e} \in \mathcal{E}_1 \times \mathcal{E}_2$, where R as defined in (5.36). Finally, choose ε^* such that for any $\varepsilon \in (0, \varepsilon^*)$, we have

$$\widetilde{x}' \widetilde{P}_\varepsilon \widetilde{x} < \frac{c}{2}$$

for all $\widetilde{x} \in \mathcal{X}_1 \times \mathcal{X}_2$ and (5.37) is satisfied componentwise for all $(\hat{x}_1, x_2)$ with $\widetilde{x}' \widetilde{P}_\varepsilon \widetilde{x} < c$. Next, we look at the candidate Lyapunov function,

$$L_\varepsilon(t) = V_\varepsilon(t) + \beta W(t),$$

where

$$V_\varepsilon(t) := \widetilde{x}(t)' \widetilde{P}_\varepsilon \widetilde{x}(t), \qquad W(t) := \widetilde{e}(t)' R \widetilde{e}(t).$$

Clearly, for all initial conditions in $\mathcal{X}_{\text{cl},2}$, we have $L_\varepsilon(0) \leqslant 1$, and (5.37) is satisfied for all $t \geqslant 0$ as long as $L_\varepsilon(0) \leqslant 1$.

Let $\widetilde{\mathcal{X}}_{\text{cl},1}$ be such that $(x, \hat{x}) \in \mathcal{X}_{\text{cl},1}$ if and only if $(\widetilde{x}, \widetilde{e}) \in \widetilde{\mathcal{X}}_{\text{cl},1}$. Next, we choose δ such that

$$\widetilde{x}' \widetilde{Q}_\varepsilon \widetilde{x} + \frac{\beta}{4} \widetilde{e}' \widetilde{e} \leqslant \delta D_0^2$$

for all $(\widetilde{x}, \widetilde{e}) \in \widetilde{\mathcal{X}}_{\text{cl},1}$.

We first obtain as before (5.40). Given our assumptions on the function g, there exist M_1, M_2, and M_3 such that

$$\begin{aligned}\|g(x,t)\|^2 &\leqslant [g_0(\|x\|) + D_0]^2\\ &\leqslant M_1\|x\|^2 + 2D_0^2\\ &\leqslant M_2\tilde{x}'\tilde{Q}_\varepsilon\tilde{x} + M_3\|e_1\|^2 + 2D_0^2. \end{aligned} \tag{5.46}$$

We note that, similar to the proof of Theorem 5.20, we obtain

$$\begin{aligned}&2v'\left[\tilde{\sigma}(\theta^{-1}v + \alpha v + h + g) - v\right]\\ &\quad\geqslant -\frac{4M_4}{\alpha}\left(\|g\|^2 + \|h\|^2\right)\\ &\quad\geqslant -\frac{4M_4M_2}{\alpha}\tilde{x}'\tilde{Q}_\varepsilon\tilde{x} - \frac{4M_4M_3}{\alpha}\|e_1\|^2 - \frac{4M_4}{\alpha}\|h\|^2 - \frac{8M_4}{\alpha}D_0^2, \end{aligned} \tag{5.47}$$

where we used in the last inequality (5.46) instead of (5.39). Using (5.40) and (5.47), we get

$$\begin{aligned}\dot{V}_\varepsilon &\leqslant -\tilde{x}'\tilde{Q}_\varepsilon\tilde{x} - v'v - 2v'E_1e_1 + \frac{4M_4M_2}{\alpha}\tilde{x}'\tilde{Q}_\varepsilon\tilde{x}\\ &\qquad + \frac{4M_4M_3}{\alpha}\|e_1\|^2 + \frac{4M_4}{\alpha}\|h\|^2 + \frac{8M_4}{\alpha}D_0^2\\ &< -\frac{2}{3}\tilde{x}'\tilde{Q}_\varepsilon\tilde{x} + \frac{3\beta}{4}e_1'e_1 + M_5e_2'e_2 + \frac{\delta}{2}D_0^2, \end{aligned} \tag{5.48}$$

provided $\alpha > 12M_3M_2$, $\alpha > \frac{16M_4M_3}{\beta}$, and $\alpha > \frac{16M_4}{\delta}$, where M_5 is a suitable constant depending on our choice of α but independent of ℓ. For the latter, we need to use that the norm of $S_{1,\ell}$ is bounded as a function of ℓ. We also used (5.44). Next, we consider W. We obtain

$$\dot{W} \leqslant -e_1'e_1 - \ell e_2'e_2 + 2e_2'R_1B_1[\tilde{u} + E_1(\hat{x}_1 + e_1)], \tag{5.49}$$

where

$$\tilde{u} = \tilde{\sigma}(\theta^{-1}v + \alpha v + h + g).$$

We obtain

$$\dot{W} \leqslant -e_1'e_1 - \frac{\ell}{2}e_2'e_2 + \frac{M_6}{\ell}\|\tilde{u} + E_1(\hat{x}_1 + e_1)\|^2,$$

for a suitable constant M_6 independent of ℓ and α. We note that

$$\|\tilde{u} + E_1(\hat{x}_1 + e_1)\|^2 \leqslant M_7\tilde{x}'\tilde{Q}_\varepsilon\tilde{x} + M_8\tilde{e}'\tilde{e} + M_9D_0^2$$

for suitable constants M_7, M_8, and M_9 which depend on α but which are independent of ℓ, and we obtain

$$\dot{W} \leqslant -e_1'e_1 - \frac{\ell}{2}e_2'e_2 + \frac{M_6M_7}{\ell}\tilde{x}'\tilde{Q}_\varepsilon\tilde{x} + \frac{M_6M_8}{\ell}\tilde{e}'\tilde{e} + \frac{M_6M_9}{\ell}D_0^2.$$

Combining this with (5.48), we obtain

$$\begin{aligned}\dot{L}_\varepsilon &\leqslant -\frac{2}{3}\widetilde{x}'\widetilde{Q}_\varepsilon\widetilde{x} + M_5 e_2' e_2 - \frac{\beta}{4} e_1' e_1 - \frac{\beta\ell}{2} e_2' e_2 \\ &\qquad + \frac{\beta M_6 M_7}{\ell}\widetilde{x}'\widetilde{Q}_\varepsilon\widetilde{x} + \frac{\beta M_6 M_8}{\ell}\widetilde{e}'\widetilde{e} + \left(\frac{\delta}{2} + \frac{M_6 M_9}{\ell}\right) D_0^2 \\ &\leqslant -\frac{1}{3}\widetilde{x}'\widetilde{Q}_\varepsilon\widetilde{x} - \frac{\beta}{4}\widetilde{e}'\widetilde{e} + \delta D_0^2\end{aligned}$$

for a sufficiently large ℓ. This immediately implies that L_ε will be decaying for $(x, \widehat{x}) \notin \mathcal{X}_{\mathrm{cl},2}$. This in turn implies that the solution of the closed-loop system will enter the set $\mathcal{X}_{\mathrm{cl},2}$ after a finite time which completes the proof of the theorem. ■

5.5 Systems with saturation and matched uncertainty: continuous time

We consider a class of nonlinear systems which are obtained by cascading linear systems with memory-free input nonlinearities of saturation type

$$\Sigma_{ud} : \begin{cases} \dot{x}(t) = Ax(t) + B\widetilde{\sigma}(u(t)) + Bg(x,t) \\ y(t) = Cx(t), \end{cases} \tag{5.50}$$

where $x \in \mathbb{R}^n$ is the state, $u \in \mathbb{R}^m$ is the control input, and $y \in \mathbb{R}^p$ is the measurement output. As before, we assume that $\widetilde{\sigma}$ satisfies properties (*i*)–(*vi*) on page 340. The uncertain element $g : \mathbb{R}^n \times \mathbb{R}^+ \to \mathbb{R}^m$ represents both the uncertainties and the disturbances. We impose only one requirement that we know an upper bound on its norm. More specifically, we make the following assumptions:

Assumption 5.22 *The uncertain element $g(x,t)$ is piecewise continuous in t, locally Lipschitz in x, and there exists a $\chi > 0$ such that its norm is bounded,*

$$\|g(x,t)\| \leqslant \psi - \chi, \quad \forall (t,x) \in \mathbb{R}^+ \times \mathbb{R}^n, \tag{5.51}$$

where ψ is such that (5.1) *is satisfied for appropriate θ.*

In order to obtain asymptotic stability, we need stronger assumptions on the disturbance g as formulated in the following assumption:

Assumption 5.23 *Assume $\widetilde{\sigma}$ and g satisfy Assumption 5.22. Additionally, assume that g is such that*

$$\|g(x,t)\| \leqslant g_0(\|x\|), \quad \forall (t,x) \in \mathbb{R}^+ \times \mathbb{R}^n,$$

where the known function $g_0(x) : \mathbb{R}^+ \to \mathbb{R}^+$ is locally Lipschitz and satisfies

$$g_0(0) = 0. \tag{5.52}$$

Note that Assumption 5.22 guarantees that $\tilde{\sigma}$ can be made larger than the uncertain element g. If this is not the case, then there exist uncertain elements which cancel the stabilizing effect of u and cause instability (if the nominal system is not yet stable).

We are interested in finding controllers that achieve semi-global results independent of the specific saturation function $\tilde{\sigma}$ and independent of the precise g that satisfies Assumption 5.22.

The main state feedback problems we solve in this section are the following:

Problem 5.24 Consider the system (5.50), where $\tilde{\sigma}$ satisfies properties *(i)*–*(vi)* on page 340 and $\tilde{\sigma}$ and g satisfy the properties in Assumption 5.23. The objective is to find, for any compact set $\mathcal{X}$, a feedback gain matrix F such that, for all functions g satisfying Assumption 5.23, the closed-loop system Σ_{ud} is asymptotically stable and contains $\mathcal{X}$ in its domain of attraction.

If g and $\tilde{\sigma}$ do not satisfy Assumption 5.23 but only Assumption 5.22, then it is not possible to achieve asymptotic stability without more precise information about the function g. However, in that case, we can still achieve practical stability:

Problem 5.25 Consider the system (5.50), where $\tilde{\sigma}$ satisfies properties *(i)*–*(vi)* on page 340 and g satisfies the properties in Assumption 5.22. The objective is to find, for any compact set $\mathcal{X}_1$ and $\mathcal{X}_2$ containing 0 in their interior and with $\mathcal{X}_1 \supset \mathcal{X}_2$, a feedback gain matrix F such that, for all functions g satisfying Assumption 5.12 and initial conditions inside $\mathcal{X}_1$, the closed-loop system Σ_{ud} is such that the state enters and remains in $\mathcal{X}_2$ after some finite amount of time.

The main measurement feedback problem we solve in this section is the following:

Problem 5.26 Consider the system (5.50), where $\tilde{\sigma}$ satisfies properties *(i)*–*(vi)* on page 340 and g satisfies the properties in Assumption 5.23. The objective is to find, for any compact set $\mathcal{X} \subset \mathbb{R}^{2n}$, a dynamic feedback of the form

$$\begin{aligned} \dot{x}_c &= A_c x_c + B_c y, \quad x_c \in \mathbb{R}^n \\ u &= C_c x_c, \end{aligned} \tag{5.53}$$

such that for all functions g satisfying Assumption 5.23, the closed-loop system Σ_{ud} is asymptotically stable and contains $\mathcal{X}$ in its domain of attraction.

If g and $\widetilde{\sigma}$ do not satisfy Assumption 5.23 but only Assumption 5.22, then it is again not possible to achieve asymptotic stability without more precise information about the function g. However, in that case we can still achieve practical stability:

Problem 5.27 Consider the system (5.50) where $\widetilde{\sigma}$ satisfies properties (*i*)–(*vi*) on page 340 and g satisfies the properties in Assumption 5.22. The objective is to find, for any pair of compact sets $\mathcal{X}_1, \mathcal{X}_2 \subset \mathbb{R}^{2n}$ containing 0 in their interior and with $\mathcal{X}_1 \supset \mathcal{X}_2$, a dynamic feedback of the form (5.53) such that, for all functions g satisfying Assumption 5.22 and initial conditions inside $\mathcal{X}_1$, the closed-loop system Σ_{ud} is such that the state enters and remains in $\mathcal{X}_2$ after some finite amount of time.

We note that the requirement that all the eigenvalues have nonpositive parts is necessary for semi-global stabilization as discussed in the previous chapter. However, without such an assumption on the open-loop eigenvalues, interesting local results are still possible and will be pointed out later.

5.5.1 State feedback results

Before we discuss the solvability of the above-defined state feedback control problem, we recall below the family of parameterized low-and-high-gain state feedback laws, denoted by

$$u = \theta^{-1} F_\varepsilon x + \alpha F_\varepsilon x, \tag{5.54}$$

where $F_\varepsilon = -B' P_\varepsilon$, and P_ε is the positive definite solution of

$$A' P_\varepsilon + P_\varepsilon A - P_\varepsilon B B' P_\varepsilon + Q_\varepsilon = 0$$

with $Q_\varepsilon > 0$ such that $Q_\varepsilon \to 0$ as $\varepsilon \to 0$. As we discussed in Chap. 4, the parameter ε is the low-gain tuning parameter, whereas α is the high-gain tuning parameter.

Theorem 5.28 *Consider the system* (5.50) *and the robust semi-global stabilization problem formulated in Problem 5.24. Assume that* (A, B) *is stabilizable, all the eigenvalues of* A *are in the closed left-half plane, and Assumption 5.23 is satisfied.*

In that case, there exist, for any compact set $\mathcal{X}$*, an* ε^* *and a function* $\alpha^* : (0, \varepsilon^*) \to \mathbb{R}^+$ *such that for all* $\varepsilon \in (0, \varepsilon^*)$ *and* $\alpha > \alpha^*(\varepsilon)$*, the feedback* (5.54) *has the property that, for all functions* g *satisfying Assumption 5.23, the resulting closed-loop system is asymptotically stable and contains* $\mathcal{X}$ *in its domain of attraction.*

Proof : We first choose an ε^* such that

$$\mathcal{X} \subset \mathcal{V}_\varepsilon := \{ x \in \mathbb{R}^n \mid x' P_\varepsilon x < 1 \},$$

and

$$|F_\varepsilon x| < \chi$$

componentwise for all $x \in \mathcal{V}_\varepsilon$ and for all $\varepsilon \in (0, \varepsilon^*)$, where χ is defined in Assumption 5.23. For all $x \in \mathcal{V}_\varepsilon$, there exist M_1 and M_2 such that

$$\|g(x,t)\|^2 \leqslant g_0(\|x\|)^2 \leqslant M_1 \|x\|^2 \leqslant M_2 x' Q_\varepsilon x, \tag{5.55}$$

where the first inequality follows from Assumption 5.12, while the existence of an M_1 such that the second inequality is satisfied is a consequence of the fact that g_0 is locally Lipschitz and x is in a bounded set. The existence of M_2 such that the final inequality is satisfied follows from the fact that Q_ε is positive definite.

Next, we look at the candidate Lyapunov function

$$V_\varepsilon(t) := x(t)' P_\varepsilon x(t).$$

We obtain

$$\dot{V}_\varepsilon = -x' Q_\varepsilon x - v'v - 2v' \left[\widetilde{\sigma}(\theta^{-1} v + \alpha v) + g - v \right] \tag{5.56}$$

where

$$v = -B' P_\varepsilon x.$$

To analyze further the derivative of our candidate Lyapunov function, we consider two cases. Let v_i and g_i denote the ith component of v and g, respectively. If

$$|\theta \alpha v_i| \geqslant |g_i|$$

is satisfied, we then obtain

$$\widetilde{\sigma}(\theta^{-1} v_i + \alpha v_i) \geqslant \min\{(1 + \theta\alpha)|v_i|, \psi\} \geqslant |g_i| + |v_i|$$

with $\gamma > \theta^{-1}$. We then find that

$$v_i \left[\widetilde{\sigma}(\theta^{-1} v_i + \alpha v_i) + g_i - v_i \right] > 0.$$

On the other hand, if

$$|\theta \alpha v_i| < |g_i| \tag{5.57}$$

is satisfied, then we get

$$|\widetilde{\sigma}(\theta^{-1} v_i + \alpha v_i) + g_i - v_i| \leqslant \delta |\theta^{-1} v_i + \alpha v_i| + |g_i| + |v_i| \leqslant M_3 |g_i|$$

for some constant $M_3 > 0$ (independent of α) using (5.57) and the fact that $\widetilde{\sigma}$ is globally Lipschitz with Lipschitz constant δ. Here we used, without loss of generality, that $\alpha > 1$. This implies that

$$v_i \left[\widetilde{\sigma}(\theta^{-1} v_i + \alpha v_i) + g_i - v_i \right] \geqslant -M_3 |g_i||v_i| \geqslant -\frac{M_3}{\alpha} g_i^2,$$

where we again used (5.57). If we bring together the above two cases, we find that

$$2v'\left[\tilde{\sigma}(\theta^{-1}v+\alpha v)+g-v\right] \geqslant -\frac{M_3}{\alpha}\|g\|^2 \geqslant -\frac{M_3M_2}{\alpha}x'Q_\varepsilon x,$$

where we used (5.55). Using this in (5.56), we get

$$\dot{V}_\varepsilon \leqslant -x'Q_\varepsilon x - v'v + \frac{M_3M_2}{\alpha}x'Q_\varepsilon x < -\frac{1}{2}x'Q_\varepsilon x,$$

provided $\alpha > 2M_3M_2$. Note that the lower bound for α depends on ε since the bound M_2 depends on ε. We conclude that in $\mathcal{V}_\varepsilon$, the Lyapunov function V_ε is strictly decaying which implies asymptotic stability and, moreover, that $\mathcal{V}_\varepsilon$ is positively invariant and contained in the domain of attraction. ■

Theorem 5.29 *Consider the system* (5.50) *and the robust semi-global practical stabilization problem formulated in Problem 5.25. Assume that* (A, B) *is stabilizable, all the eigenvalues of* A *are in the closed left-half plane, and Assumption 5.22 is satisfied.*

In that case, there exist, for any pair of compact sets $\mathcal{X}_1$ *and* $\mathcal{X}_2$ *containing* 0 *in their interior and with* $\mathcal{X}_1 \supset \mathcal{X}_2$*, an* ε^* *and a function* $\alpha^* : (0, \varepsilon^*) \to \mathbb{R}^+$ *such that for all* $\varepsilon \in (0, \varepsilon^*)$ *and* $\alpha > \alpha^*(\varepsilon)$*, the feedback* (5.54) *has the property that, for all functions* g *satisfying Assumption 5.22, the resulting closed-loop system is such that, for all initial conditions in* $\mathcal{X}_1$*, the state enters and remains in the set* $\mathcal{X}_2$ *within a finite amount of time.*

Proof : We first choose an ε^* such that

$$\mathcal{X}_1 \subset \mathcal{V}_\varepsilon := \{\, x \in \mathbb{R}^n \mid x'P_\varepsilon x < 1 \,\},$$

and

$$|F_\varepsilon x| < \psi$$

componentwise for all $x \in \mathcal{V}_\varepsilon$ and for all $\varepsilon \in (0, \varepsilon^*)$. The last condition implies that

$$|\tilde{\sigma}(\gamma F_\varepsilon x)| > \min\{\, \theta|\gamma F_\varepsilon x|, \psi \,\} > |F_\varepsilon x| \tag{5.58}$$

for $\gamma > \theta^{-1}$.

For all $x \in \mathcal{V}_\varepsilon$, there exist M_1 and M_2 such that

$$\begin{aligned}\|g(x,t)\|^2 &\leqslant [g_0(\|x\|) + D_0]^2 \\ &\leqslant M_1\|x\|^2 + 2D_0^2 \\ &\leqslant M_2 x'Q_\varepsilon x + 2D_0^2,\end{aligned} \tag{5.59}$$

where the first inequality follows from Assumption 5.12, while the existence of an M_1 such that the second inequality is satisfied is a consequence of the fact that g_0 is locally Lipschitz and x is in a bounded set. The existence of M_2 such that the final inequality is satisfied follows from the fact that Q_ε is positive definite.

Next, we look at the candidate Lyapunov function,

$$V_\varepsilon(t) := x(t)' P_\varepsilon x(t).$$

We obtain

$$\dot{V}_\varepsilon = -x' Q_\varepsilon x - v'v - 2v' \left[\tilde{\sigma}(\theta^{-1} v + \alpha v) + g - v\right] \tag{5.60}$$

where

$$v = -B' P_\varepsilon x.$$

To analyze further the derivative of our candidate Lyapunov function, we consider two cases. Let v_i and g_i denote the ith component of v and g, respectively. If

$$|\theta \alpha v_i| \geqslant |g_i|$$

is satisfied, we then obtain,

$$\tilde{\sigma}(\theta^{-1} v_i + \alpha v_i) \geqslant |v_i| + |g_i|.$$

We then find that

$$v_i \left[\tilde{\sigma}(\theta^{-1} v_i + \alpha v_i) + g_i - v_i\right] > 0.$$

On the other hand, if

$$|\theta \alpha v_i| < |g_i| \tag{5.61}$$

is satisfied, then we get

$$|\tilde{\sigma}(\theta^{-1} v_i + \alpha v_i) + g_i - v_i| \leqslant \delta |\theta^{-1} v_i + \alpha v_i| + |g_i| + |v_i| \leqslant M_3 |g_i|$$

for some constant $M_3 > 0$ (independent of α) using (5.61) and the fact that $\tilde{\sigma}$ is globally Lipschitz with Lipschitz constant δ. Here we used, without loss of generality, that $\alpha > 1$. This implies that

$$v_i \left[\tilde{\sigma}(\theta^{-1} v_i + \alpha v_i) + g_i - v_i\right] \geqslant -M_3 |g_i||v_i| \geqslant -\frac{M_3}{\alpha} g_i^2,$$

where we again used (5.61). If we bring together the above two cases, we find that

$$2v' \left[\tilde{\sigma}(\theta^{-1} v + \alpha v) + g - v\right] \geqslant -\frac{M_3}{\alpha} \|g\|^2 \geqslant -\frac{2M_3}{\alpha} \chi^2,$$

where we used (5.59). Using this in (5.60), we get

$$\begin{aligned} \dot{V}_\varepsilon &\leqslant -x' Q_\varepsilon x - v'v + \frac{2M_3}{\alpha} \chi^2 \\ &< -x' Q_\varepsilon x + \frac{2M_3}{\alpha} \chi^2. \end{aligned} \tag{5.62}$$

Moreover, let μ_1, μ_2 be such that

$$\{ x \in \mathbb{R}^n \mid x' Q_\varepsilon x < \mu_1 \} \subset \mathcal{W}_\mu = \{ x \in \mathbb{R}^n \mid x' P_\varepsilon x < \mu_2 \} \subset \mathcal{X}_2.$$

Choose α such that

$$\frac{2M_3}{\alpha} \chi^2 < \mu_1.$$

Then it is obvious from (5.62) that inside $\mathcal{V}_\varepsilon$, the Lyapunov function V_ε is decreasing outside $\mathcal{W}_\mu$. This implies that the solution will enter and stay in $\mathcal{W}_\mu$ within a finite amount of time. By construction, this implies that the solution enters and stays in the set $\mathcal{X}_2$ after a finite amount of time for all initial conditions in the set $\mathcal{X}_1$. ∎

5.5.2 *Measurement feedback results*

In the measurement feedback case, we use the low-and-high-gain state feedback (5.54) as studied in the previous subsection and introduced in Chap. 4. We combine this with a high-gain observer, i.e.

$$\begin{aligned} \dot{\widehat{x}} &= (A + BF_\varepsilon)\widehat{x} + K_\ell(y - C\widehat{x}) \\ u &= \theta^{-1} F_\varepsilon \widehat{x} + \alpha F_\varepsilon \widehat{x}, \end{aligned} \tag{5.63}$$

where $F_\varepsilon = -B' P_\varepsilon$, and P_ε is the positive definite solution of

$$A' P_\varepsilon + P_\varepsilon A - P_\varepsilon B B' P_\varepsilon + Q_\varepsilon = 0$$

with $Q_\varepsilon > 0$ such that $Q_\varepsilon \to 0$ as $\varepsilon \to 0$.

In order to construct a suitable observer, we note that if the system $(A, B, C, 0)$ is left invertible and minimum phase then there exists a basis transformation Γ_s such that

$$\Gamma_s^{-1} A \Gamma_s = \begin{pmatrix} A_0 & L_0 C_1 \\ B_1 E_1 & A_1 \end{pmatrix}, \quad \Gamma_s^{-1} B = \begin{pmatrix} 0 \\ B_1 \end{pmatrix}, \quad C\Gamma_s = \begin{pmatrix} 0 & C_1 \end{pmatrix}$$

with A_0 asymptotically stable. This follows from Chap. 3 where

$$\Gamma_s x = \begin{pmatrix} x_1 \\ x_2 \end{pmatrix}, \qquad x_1 = x_a, \qquad x_2 = \begin{pmatrix} x_b \\ x_d \end{pmatrix}$$

in the notation of SCB. Note that x_c is missing because the system is left invertible and A_0 is asymptotically stable because the system is minimum phase. Since the subsystem $(A_1, B_1, C_1, 0)$ is observable and its transfer matrix has a polynomial inverse, it has been established in Sect. 4.5.1 of Chap. 4 that there exist $K_{1,\ell}$ and $S_{1,\ell}$ such that

$$S_{1,\ell}^{-1}(A_1 - K_{1,\ell} C_1) S_{1,\ell} = \ell \widetilde{A}_1, \qquad S_{1,\ell}^{-1} B_1 = B_1,$$

where there exists an M, independent of ℓ, such that $\|S_{1,\ell}\| < M$ for all $\ell \geqslant 1$. We define

$$K_\ell = \Gamma_s \begin{pmatrix} L_0 \\ K_{1,\ell} \end{pmatrix}.$$

The claim is that the above dynamic observer-based controller can be used to solve Problems 5.26 and 5.27 for ε sufficiently small and α and ℓ sufficiently large. The following two theorems make this result more precise:

Theorem 5.30 *Consider the system* (5.50) *and the robust semi-global stabilization problem formulated in Problem 5.26. Assume that (A, B) is stabilizable, all the eigenvalues of A are in the closed left-half plane, and the system $(A, B, C, 0)$ is left invertible and minimum phase. Also, assume that $\widetilde{\sigma}$ and g satisfy Assumption 5.23, in addition to the standard properties (i)–(vi) on page 340.*

In that case, there exist, for any compact set $\mathcal{X}_{cl} \subset \mathbb{R}^{2n}$, an ε^*, a function $\alpha^* : (0, \varepsilon^*) \to \mathbb{R}^+$, and a function $\ell^* : (0, \varepsilon^*) \times \mathbb{R}^+ \to \mathbb{R}^+$ such that, for all $\varepsilon \in (0, \varepsilon^*)$, $\alpha > \alpha^*(\varepsilon)$, and $\ell > \ell^*(\varepsilon, \alpha)$, the feedback (5.63) *has the property that, for all functions g satisfying Assumption 5.23, the resulting closed-loop system is asymptotically stable and contains $\mathcal{X}_{cl}$ in its domain of attraction.*

Proof : We decompose $\widehat{x}$ compatible with the decomposition of x, and we define e_1 and e_2,

$$\Gamma_s \widehat{x} = \begin{pmatrix} \widehat{x}_1 \\ \widehat{x}_2 \end{pmatrix}, \quad e_1 = x_1 - \widehat{x}_1, \quad e_2 = S_{1,\ell}^{-1}(x_2 - \widehat{x}_2). \tag{5.64}$$

The closed-loop system can then be written in terms of these new variables:

$$\begin{aligned} \dot{\widehat{x}}_1 &= A_0 \widehat{x}_1 + L_0 C_1 x_2 \\ \dot{x}_2 &= A_1 x_2 + B_1(\widetilde{\sigma}(u) + g + E_1(\widehat{x}_1 + e_1)) \\ \dot{e}_1 &= A_0 e_1 \\ \dot{e}_2 &= \ell \widetilde{A}_1 e_2 + B_1(\widetilde{\sigma}(u) + g + E_1(\widehat{x}_1 + e_1)) \\ u &= \left(\theta^{-1} + \alpha\right) F_\varepsilon \Gamma_s \begin{pmatrix} \widehat{x}_1 \\ x_2 - S_{1,\ell} e_2 \end{pmatrix}. \end{aligned} \tag{5.65}$$

For ease of presentation, we will define $F_{\varepsilon,1}$ and $F_{\varepsilon,2}$ such that

$$u = \left(\theta^{-1} + \alpha\right) \left[F_{\varepsilon,1} \widehat{x}_1 + F_{\varepsilon,2}(x_2 - S_{1,\ell} e_2) \right]. \tag{5.66}$$

Moreover, we define

$$\widetilde{P}_\varepsilon = \Gamma_s' P_\varepsilon \Gamma_s, \qquad \widetilde{Q}_\varepsilon = \Gamma_s' Q_\varepsilon \Gamma_s, \qquad \widetilde{x} = \begin{pmatrix} \widehat{x}_1 \\ x_2 \end{pmatrix}, \qquad \widetilde{e} = \begin{pmatrix} e_1 \\ e_2 \end{pmatrix}. \tag{5.67}$$

Clearly, we can construct compact sets $\mathcal{X}_1, \mathcal{X}_2, \mathcal{E}_1, \mathcal{E}_2$ such that $(x, \widehat{x}) \in \mathcal{X}_{\text{cl}}$ implies that $\widehat{x}_1 \in \mathcal{X}_1$, $x_2 \in \mathcal{X}_2$, $e_1 \in \mathcal{E}_1$, and $e_2 \in \mathcal{E}_2$.

First, we choose β such that

$$\beta \geqslant 2\|E_1\|^2 + 1. \tag{5.68}$$

Next, let c be such that

$$\beta \widetilde{e}' R \widetilde{e} < \frac{c}{2}$$

for all $\widetilde{e} \in \mathcal{E}_1 \times \mathcal{E}_2$ and

$$R = \begin{pmatrix} R_0 & 0 \\ 0 & R_1 \end{pmatrix} \tag{5.69}$$

where

$$A_0' R_0 + R_0 A_0 + I = 0, \qquad \widetilde{A}_1' R_1 + R_1 \widetilde{A}_1 + I = 0.$$

Finally, choose ε^* such that for any $\varepsilon \in (0, \varepsilon^*)$, we have

$$\widetilde{x}' \widetilde{P}_\varepsilon \widetilde{x} < \frac{c}{2}$$

for all $\widetilde{x} \in \mathcal{X}_1 \times \mathcal{X}_2$, and

$$|F_{\varepsilon,1}\widehat{x}_1 + F_{\varepsilon,2}x_2| < \chi \tag{5.70}$$

componentwise for all $(\widehat{x}_1, x_2)$ with $\widetilde{x}' \widetilde{P}_\varepsilon \widetilde{x} < c$. Next, we look at the candidate Lyapunov function,

$$L_\varepsilon(t) = V_\varepsilon(t) + \beta W(t),$$

where

$$V_\varepsilon(t) := \widetilde{x}(t)' \widetilde{P}_\varepsilon \widetilde{x}(t), \qquad W(t) := \widetilde{e}(t)' R \widetilde{e}(t).$$

Clearly, for all initial conditions in $\mathcal{X}_{\text{cl}}$, we have $L_\varepsilon(0) \leqslant 1$, and (5.70) is satisfied for all $t \geqslant 0$ as long as $L_\varepsilon(0) \leqslant 1$.

Given our assumptions on the function g, there exist M_1, M_2, and M_3 such that

$$\begin{aligned} \|g(x,t)\|^2 &\leqslant [g_0(\|x\|)]^2 \\ &\leqslant M_1 \|x\|^2 \\ &\leqslant M_2 \widetilde{x}' \widetilde{Q}_\varepsilon \widetilde{x} + M_3 \|e_1\|^2. \end{aligned} \tag{5.71}$$

We obtain

$$\dot{V}_\varepsilon = -\widetilde{x}' \widetilde{Q}_\varepsilon \widetilde{x} - v'v - 2v' \left[\widetilde{\sigma}(\theta^{-1} v + \alpha v + h) + g - v + E_1 e_1 \right], \tag{5.72}$$

where

$$v = F_{\varepsilon,1}\widehat{x}_1 + F_{\varepsilon,2}x_2, \qquad h = -(\theta^{-1} + \alpha) F_{\varepsilon,2} S_{1,\ell} e_2.$$

Let v_i, g_i, and h_i denote the ith component of v, g, and h, respectively. Assume that

$$|\theta\alpha v_i| \geqslant |g_i| + |h_i|$$

is satisfied, then there exists an α_1 such that

$$|\theta\alpha_1 v_i \geqslant |g_i| \text{ and } |\theta(\alpha - \alpha_1)v_i \geqslant |h_i|.$$

Using this, we obtain

$$\widetilde{\sigma}(\theta^{-1}v_i + \alpha v_i + h_i) \geqslant \min\{(1+\theta\alpha_1)|v_i|, \psi\} \geqslant |g_i| + |v_i|.$$

We then find that

$$v_i\left[\widetilde{\sigma}(\theta^{-1}v_i + \alpha v_i + h_i) + g_i - v_i\right] > 0.$$

On the other hand, if

$$|\theta\alpha v_i| < |g_i| + |h_i| \tag{5.73}$$

is satisfied, then we get

$$\begin{aligned}|\widetilde{\sigma}(\theta^{-1}v_i + \alpha v_i + h_i) + g_i - v_i| &\leqslant \delta|\theta^{-1}v_i + \alpha v_i + h_i| + |g_i| + |v_i| \\ &\leqslant M_4\left(|g_i| + |h_i|\right)\end{aligned}$$

for some constant $M_4 > 0$ (independent of α) using (5.73) and the fact that $\widetilde{\sigma}$ is globally Lipschitz with Lipschitz constant δ. Here we used, without loss of generality, that $\alpha > 1$. This implies that

$$\begin{aligned}v_i\left[\widetilde{\sigma}(\theta^{-1}v_i + \alpha v_i + h_i) + g_i - v_i\right] &\geqslant -M_4\left(|g_i| + |h_i|\right)|v_i| \\ &\geqslant -\frac{2M_4}{\alpha}(g_i^2 + h_i^2),\end{aligned}$$

where we again used (5.73). If we bring together the above two cases, we find that

$$\begin{aligned}&2v'\left[\widetilde{\sigma}(\theta^{-1}v + \alpha v + h) + g - v\right] \\ &\quad\geqslant -\frac{4M_4}{\alpha}\left(\|g\|^2 + \|h\|^2\right) \\ &\quad\geqslant -\frac{4M_4M_2}{\alpha}\widetilde{x}'\widetilde{Q}_\varepsilon\widetilde{x} - \frac{4M_4M_3}{\alpha}\|e_1\|^2 - \frac{4M_4}{\alpha}\|h\|^2,\end{aligned} \tag{5.74}$$

where we used (5.71). Using (5.72) and (5.74), we get

$$\begin{aligned}\dot{V}_\varepsilon &\leqslant -\widetilde{x}'\widetilde{Q}_\varepsilon\widetilde{x} - v'v - 2v'E_1e_1 + \frac{4M_4M_2}{\alpha}\widetilde{x}'\widetilde{Q}_\varepsilon\widetilde{x} \\ &\qquad + \frac{4M_4M_3}{\alpha}\|e_1\|^2 + \frac{4M_4}{\alpha}\|h\|^2 \\ &< -\frac{2}{3}\widetilde{x}'\widetilde{Q}_\varepsilon\widetilde{x} + \frac{3\beta}{4}e_1'e_1 + M_5e_2'e_2,\end{aligned} \tag{5.75}$$

provided $\alpha > 12M_3M_2$ and $\alpha > \frac{16M_4M_3}{\beta}$, and where M_5 is a suitable constant depending on our choice of α but independent of ℓ. For the latter, we need to use that the norm of $S_{1,\ell}$ is bounded as a function of ℓ. We also used that, based on (5.68), we have

$$-v'v - 2v'E_1e_1 \leqslant \|E_1e_1\|^2 \leqslant \frac{\beta}{2}\|e_1\|^2. \tag{5.76}$$

Next, we consider W. We obtain

$$\dot{W} \leqslant -e_1'e_1 - \ell e_2'e_2 + 2e_2'R_1B_1[\widetilde{u} + g + E_1(\widehat{x}_1 + e_1)], \tag{5.77}$$

where

$$\widetilde{u} = \widetilde{\sigma}(\theta^{-1}v + \alpha v + h).$$

We obtain

$$\dot{W} \leqslant -e_1'e_1 - \frac{\ell}{2}e_2'e_2 + \frac{M_6}{\ell}\|\widetilde{u} + g + E_1(\widehat{x}_1 + e_1)\|^2$$

for a suitable constant M_6 independent of ℓ and α. We note that

$$\|\widetilde{u} + g + E_1(\widehat{x}_1 + e_1)\|^2 \leqslant M_7\widetilde{x}'\widetilde{Q}_\varepsilon\widetilde{x} + M_8\widetilde{e}'\widetilde{e}$$

for suitable constants M_7 and M_8 which depend on α but which are independent of ℓ, and we obtain

$$\dot{W} \leqslant -e_1'e_1 - \frac{\ell}{2}e_2'e_2 + \frac{M_6M_7}{\ell}\widetilde{x}'\widetilde{Q}_\varepsilon\widetilde{x} + \frac{M_6M_8}{\ell}\widetilde{e}'\widetilde{e}.$$

Combining this with (5.75), we obtain

$$\begin{aligned}\dot{L}_\varepsilon &\leqslant -\frac{2}{3}\widetilde{x}'\widetilde{Q}_\varepsilon\widetilde{x} + M_5e_2'e_2 - \frac{\beta}{4}e_1'e_1 - \frac{\beta\ell}{2}e_2'e_2 \\ &\qquad + \frac{\beta M_6M_7}{\ell}\widetilde{x}'\widetilde{Q}_\varepsilon\widetilde{x} + \frac{\beta M_6M_8}{\ell}\widetilde{e}'\widetilde{e} \\ &\leqslant -\frac{1}{3}\widetilde{x}'\widetilde{Q}_\varepsilon\widetilde{x} - \frac{\beta}{4}\widetilde{e}'\widetilde{e}\end{aligned}$$

for a sufficiently large ℓ. This immediately implies the required asymptotic stability and the required domain of attraction. ■

Theorem 5.31 *Consider the system* (5.50) *and the robust semi-global stabilization problem formulated in Problem 5.26. Assume that* (A, B) *is stabilizable, all the eigenvalues of* A *are in the closed left-half plane, and the system* $(A, B, C, 0)$ *is left invertible and minimum phase. Also, assume that* $\widetilde{\sigma}$ *and* g *satisfy Assumption 5.22, in addition to the standard properties (i)–(vi) on page 340.*

In that case, there exist, for any pair of compact sets $\mathcal{X}_{cl,1}$ and $\mathcal{X}_{cl,2}$ in $\mathbb{R}^{2n}$ containing 0 in their interior and with $\mathcal{X}_{cl,1} \supset \mathcal{X}_{cl,2}$, an ε^*, a function $\alpha^* : (0,\varepsilon^*) \to \mathbb{R}^+$, and a function $\ell^* : (0,\varepsilon^*)\times\mathbb{R}^+ \to \mathbb{R}^+$ such that for all $\varepsilon \in (0,\varepsilon^*)$, $\alpha > \alpha^*(\varepsilon)$, and $\ell > \ell^*(\varepsilon,\alpha)$, the feedback (5.63) has the property that, for all functions g satisfying Assumption 5.22, the resulting closed-loop system is such that, for all initial conditions in $\mathcal{X}_{cl,1}$, the state enters and remains in the set $\mathcal{X}_{cl,2}$ within a finite amount of time.

Proof : Following the proof of Theorem 5.30, we decompose $\hat{x}$ compatible with the decomposition of x and we define e_1 and e_2 as in (5.64), and in terms of these new variables, we obtain the closed-loop system (5.65). We again define $F_{\varepsilon,1}$ and $F_{\varepsilon,2}$ such that (5.66) is satisfied and use the definition in (5.67). Clearly, we can construct compact sets $\mathcal{X}_1, \mathcal{X}_2, \mathcal{E}_1, \mathcal{E}_2$ such that $(x,\hat{x}) \in \mathcal{X}_{cl,2}$ implies that $\hat{x}_1 \in \mathcal{X}_1$, $x_2 \in \mathcal{X}_2$, $e_1 \in \mathcal{E}_1$, and $e_2 \in \mathcal{E}_2$.

First, we choose β such that (5.68) is satisfied, and let c be such that

$$\beta\tilde{e}' R\tilde{e} < \frac{c}{2}$$

for all $\tilde{e} \in \mathcal{E}_1 \times \mathcal{E}_2$, where R as defined in (5.69). Finally, choose ε^* such that for any $\varepsilon \in (0,\varepsilon^*)$, we have

$$\tilde{x}'\tilde{P}_\varepsilon\tilde{x} < \frac{c}{2}$$

for all $\tilde{x} \in \mathcal{X}_1 \times \mathcal{X}_2$, and (5.70) is satisfied componentwise for all $(\hat{x}_1, x_2)$ with $\tilde{x}'\tilde{P}_\varepsilon\tilde{x} < c$. Next, we look at the candidate Lyapunov function,

$$L_\varepsilon(t) = V_\varepsilon(t) + \beta W(t),$$

where

$$V_\varepsilon(t) := \tilde{x}(t)'\tilde{P}_\varepsilon\tilde{x}(t), \qquad W(t) := \tilde{e}(t)'R\tilde{e}(t).$$

Clearly, for all initial conditions in $\mathcal{X}_{cl,2}$, we have $L_\varepsilon(0) \leqslant 1$, and (5.70) is satisfied for all $t \geqslant 0$ as long as $L_\varepsilon(0) \leqslant 1$.

Let $\tilde{\mathcal{X}}_{cl,1}$ be such that $(x,\hat{x}) \in \mathcal{X}_{cl,1}$ if and only if $(\tilde{x},\tilde{e}) \in \tilde{\mathcal{X}}_{cl,1}$. Next, we choose δ such that

$$\tilde{x}'\tilde{Q}_\varepsilon\tilde{x} + \frac{\beta}{4}\tilde{e}'\tilde{e} \leqslant \delta\chi^2$$

for all $(\tilde{x},\tilde{e}) \in \tilde{\mathcal{X}}_{cl,1}$.

As before, we first obtain (5.72). Given our assumptions on the function g, we note that

$$\|g(x,t)\|^2 \leqslant \chi^2. \tag{5.78}$$

Similar to the proof of Theorem 5.20, we obtain

$$\begin{aligned} 2v'\left[\tilde{\sigma}(\theta^{-1}v + \alpha v + h) + g - v\right] &\geqslant -\frac{4M_4}{\alpha}\left(\|g\|^2 + \|h\|^2\right) \\ &\geqslant -\frac{4M_4}{\alpha}\|h\|^2 - \frac{4M_4}{\alpha}\chi^2, \end{aligned} \tag{5.79}$$

where we used in the last inequality (5.78) instead of (5.71). Using (5.72) and (5.79), we get

$$\begin{aligned} \dot{V}_\varepsilon &\leqslant -\tilde{x}'\tilde{Q}_\varepsilon\tilde{x} - v'v - 2v'E_1e_1 + \frac{4M_4}{\alpha}\|h\|^2 + \frac{4M_4}{\alpha}\chi^2 \\ &< -\frac{2}{3}\tilde{x}'\tilde{Q}_\varepsilon\tilde{x} + \frac{3\beta}{4}e_1'e_1 + M_5e_2'e_2 + \frac{\delta}{2}\chi^2 \end{aligned} \tag{5.80}$$

provided $\alpha > \frac{8M_4}{\delta}$, where M_5 is a suitable constant depending on our choice of α but independent of ℓ. For the latter, we need to use that the norm of $S_{1,\ell}$ is bounded as a function of ℓ. We also used (5.76). Next, we consider W. We obtain

$$\dot{W} \leqslant -e_1'e_1 - \ell e_2'e_2 + 2e_2'R_1B_1[\tilde{u} + g + E_1(\hat{x}_1 + e_1)], \tag{5.81}$$

where

$$\tilde{u} = \tilde{\sigma}(\theta^{-1}v + \alpha v + h).$$

We obtain

$$\dot{W} \leqslant -e_1'e_1 - \frac{\ell}{2}e_2'e_2 + \frac{M_6}{\ell}\|\tilde{u} + g + E_1(\hat{x}_1 + e_1)\|^2$$

for a suitable constant M_6 independent of ℓ and α. We note that

$$\|\tilde{u} + g + E_1(\hat{x}_1 + e_1)\|^2 \leqslant M_7\tilde{x}'\tilde{Q}_\varepsilon\tilde{x} + M_8\tilde{e}'\tilde{e} + M_9\chi^2$$

for suitable constants M_7, M_8, and M_9 which depend on α but which are independent of ℓ, and we obtain

$$\dot{W} \leqslant -e_1'e_1 - \frac{\ell}{2}e_2'e_2 + \frac{M_6M_7}{\ell}\tilde{x}'\tilde{Q}_\varepsilon\tilde{x} + \frac{M_6M_8}{\ell}\tilde{e}'\tilde{e} + \frac{M_6M_9}{\ell}\chi^2.$$

Combining this with (5.80), we obtain

$$\begin{aligned} \dot{L}_\varepsilon \leqslant{}& -\frac{2}{3}\tilde{x}'\tilde{Q}_\varepsilon\tilde{x} + M_5e_2'e_2 - \frac{\beta}{4}e_1'e_1 - \frac{\beta\ell}{2}e_2'e_2 \\ &+ \frac{\beta M_6M_7}{\ell}\tilde{x}'\tilde{Q}_\varepsilon\tilde{x} + \frac{\beta M_6M_8}{\ell}\tilde{e}'\tilde{e} + \left(\frac{\delta}{2} + \frac{M_6M_9}{\ell}\right)\chi^2 \\ \leqslant{}& -\frac{1}{3}\tilde{x}'\tilde{Q}_\varepsilon\tilde{x} - \frac{\beta}{4}\tilde{e}'\tilde{e} + \delta\chi^2 \end{aligned}$$

for a sufficiently large ℓ. This immediately shows that L_ε will be decaying for $(x, \hat{x}) \notin \mathcal{X}_{\mathrm{cl},2}$. This implies that the solution of the closed-loop system will enter the set $\mathcal{X}_{\mathrm{cl},2}$ after a finite time which completes the proof of the theorem. ■

5.6 Systems with saturation and uncertainty: discrete time

In the previous two sections, we analyzed both input additive and matched uncertainty for continuous-time systems. The main design tool is a high-gain feedback generated by the low-and-high-gain design methodology. This design methodology is also available for discrete-time systems. However, there is a crucial restriction on the high-gain parameter. This immediately implies that the problems as formulated in the previous sections are not solvable in the discrete-time case. We would like to illustrate this by a simple example.

Let us first have a look at practical stabilization:

Example 5.32 Consider the system

$$x(k+1) = x(k) + \sigma(u(k) + g(x,k)),$$

where $g(x,k)$ is locally Lipschitz in x and its norm is bounded by a known function

$$\|g(x,k)\| \leqslant g_0(\|x\|) + D_0, \quad \forall (k,x) \in \mathbb{Z}^+ \times \mathbb{R}, \tag{5.82}$$

where D_0 is a known positive constant and the known function $g_0(x) : \mathbb{R}^+ \to \mathbb{R}^+$ is locally Lipschitz and satisfies $g(0) = 0$.

Then there exists no feedback $u = f(x)$ which achieves practical stability in the sense that we cannot guarantee that, for any compact set $\mathcal{X}_1, \mathcal{X}_2$ containing 0 in their interior, there exists a feedback $u = f(x)$ such that the closed-loop system is such that for all initial conditions in $\mathcal{X}_1$ the state enters and remains in $\mathcal{X}_2$ after some finite amount of time.

This is easily verified. The fact that we need to remain in an arbitrarily small region around 0 requires that $f(0) = 0$ which is easily seen by choosing $g(x,k) = 0$. But then, for initial condition 0, we have to remain in $\mathcal{X}_2$. However, by choosing $x(0) = 0$ and $g(x,j) = 0$ for $j < k$ and $g(x,k) = D_0$ yields $x(k+1) = \sigma(D_0)$, which will not be in $\mathcal{X}_2$ if $\mathcal{X}_2$ has been chosen sufficiently small.

Next, we show that even if $D_0 = 0$ in (5.82), we will not be able to achieve stabilization when g_0 is too large.

Example 5.33 Consider the system

$$x(k+1) = x(k) + \sigma(u(k) + g(x,k)),$$

where $g(x,k)$ is locally Lipschitz in x and its norm is bounded by a known function:

$$\|g(x,k)\| \leqslant g_0(\|x\|), \quad \forall (k,x) \in \mathbb{Z}^+ \times \mathbb{R},$$

where the known function $g_0(x) : \mathbb{R}^+ \to \mathbb{R}^+$ is locally Lipschitz and satisfies $g(0) = 0$.

Then for g_0 sufficiently large, there exists no feedback $u = f(x)$ which achieves practical stability in the sense that we cannot guarantee that for any compact set $\mathcal{X}$, there exists a feedback $u = f(x)$ such the closed-loop system is asymptotically stable with $\mathcal{X}$ contained in its domain of attraction.

This is easily verified. Choose $g_0(x)$ such that $g_0(\|x\|) \geqslant 2\|x\|$. We choose $g(x,k) = 0$ if $f(x)$ and x have the same sign and choose $g(x,k) = -2x$ otherwise. Then it is easily verified that for any $x(k)$ with $|x(k)| < 1/2$, we have $|x(k+1)| \geqslant |x(k)|$, and the origin is therefore clearly not asymptotically stable.

5.7 Saturation with deadzone

The saturation function studied in this chapter satisfies the conditions *(i)*–*(vi)* on page 340 which implies that the saturation function is bounded away from zero outside the origin as formulated in condition (5.1). However, in practice, we do encounter saturation functions which include a deadzone and hence do not satisfy property *(iv)* on page 340 and also do not satisfy (5.1). We will study in this section saturation functions with deadzone, denoted by $\tilde{\sigma}_d$, which satisfy the following properties:

(i) $\tilde{\sigma}_d(u)$ is decentralized, i.e.,

$$\tilde{\sigma}_d(s) = \begin{pmatrix} \tilde{\sigma}_{d,1}(s_1) \\ \tilde{\sigma}_{d,2}(s_2) \\ \vdots \\ \tilde{\sigma}_{d,m}(s_m). \end{pmatrix}.$$

(ii) $\tilde{\sigma}_{d,i}$ is globally Lipschitz, i.e. for some $\delta > 0$,

$$|\tilde{\sigma}_{d,i}(s_1) - \tilde{\sigma}_{d,i}(s_2)| \leqslant \delta |s_1 - s_2|.$$

(iii) $s\tilde{\sigma}_{d,i}(s) \geqslant 0$ and $\tilde{\sigma}_i(0) = 0$.

(iv) $\liminf_{|s|\to\infty} |\tilde{\sigma}_{d,i}(s)| > 0$.

(v) There exists an $M > 0$ such that $|\tilde{\sigma}_{d,i}(s)| < M$ for all $s \in \mathbb{R}$.

The above conditions imply that there exist positive constants θ, ψ, and a such that

$$|\tilde{\sigma}_d(s)| > \min\{\, \theta(|s| - a), \psi \,\}. \tag{5.83}$$

Graphically, the above conditions imply that the graph of the saturation function is in the hatched area of Fig. 5.1.

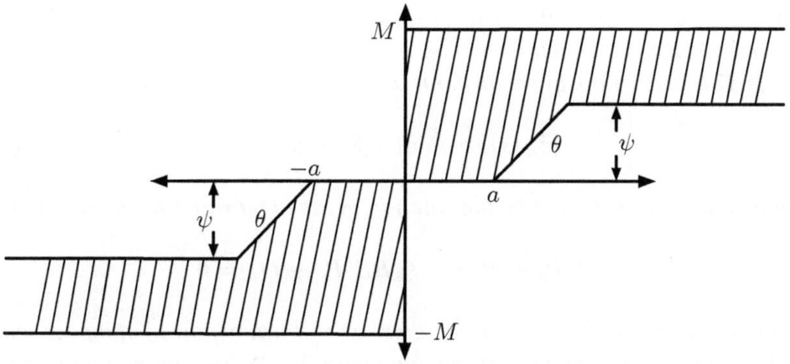

Figure 5.1: Saturation with deadzone

5.7.1 Continuous time

One approach for controlling a system with a deadzone is using dithering. Consider the system

$$\begin{aligned} \dot{x} &= Ax + B\tilde{\sigma}_d(u) \\ y &= Cx. \end{aligned} \tag{5.84}$$

We apply a high-frequency sinusoid as a preliminary feedback,

$$u(t) = v(t) + r(t) = v(t) + A\sin(\omega t).$$

Then the system can be approximated by

$$\begin{aligned} \dot{x} &= Ax + B\tilde{\sigma}(v) \\ y &= Cx \end{aligned} \tag{5.85}$$

for some $\tilde{\sigma}$ satisfying the classical properties (*i*)–(*vi*) on page 340. This approximation does require that $A > a$ and the approximation gets arbitrary accurate as $\omega \to \infty$. We will not pursue this approach in the book. We do note that this approach was introduced in [212, 213], although the use of dither signals has a much longer history going back to the work of Schuchman [146]. For recent extensions, we refer to [51, 52].

Our approach will be closer to [68] using low-and-high-gain state feedbacks as already exploited earlier in this chapter. We note that if $\tilde{\sigma}_d$ satisfies properties (*i*)–(*v*) above, then

$$\tilde{\sigma}(s) = \tilde{\sigma}_d(s - g_d(s)), \quad \text{where} \quad g_d(s) = -a\,\mathrm{sgn}(s) \tag{5.86}$$

satisfies properties (*i*)–(*vi*) on page 340. Using this relationship, we can derive the following theorems:

Theorem 5.34 *Consider the system* (5.84) *with* (A, B) *stabilizable and all eigenvalues of* A *in the closed left-half plane. Assume that* $\tilde{\sigma}_d$ *satisfies properties* (i)–(v) *on page 374, and let positive constants* a, θ, *and* ψ *be such that*

$$|\tilde{\sigma}_d(s)| > \min\{\, \theta(|s| - a), \psi \,\} \tag{5.87}$$

componentwise. Let $P_\varepsilon > 0$ *be the solution of the algebraic Riccati equation*

$$A'P_\varepsilon + P_\varepsilon A - P_\varepsilon BB'P_\varepsilon + Q_\varepsilon = 0$$

with $Q_\varepsilon \to 0$ *as* $\varepsilon \to 0$. *In that case, there exist, for any pair of compact sets* $\mathcal{X}_1$ *and* $\mathcal{X}_2$ *containing* 0 *in their interior and with* $\mathcal{X}_1 \supset \mathcal{X}_2$, *an* ε^* *and a function* $\alpha^* : (0, \varepsilon^*) \to \mathbb{R}^+$ *such that for all* $\varepsilon \in (0, \varepsilon^*)$ *and* $\alpha > \alpha^*(\varepsilon)$, *the feedback*

$$u = -\theta^{-1}B'P_\varepsilon x - \alpha B'P_\varepsilon x \tag{5.88}$$

has the property that the closed-loop system is such that for all initial conditions in $\mathcal{X}_1$ *the state enters and remains in the set* $\mathcal{X}_2$ *within a finite amount of time.*

Remark 5.35 *Note that the above theorem only yields practical stability. It is easily verified that any continuous feedback can only achieve practical stability of the equilibrium point in zero. After all, a continuous feedback* $u = f(x)$ *will result either in a nonzero equilibrium or in a saturated input which is equal to zero in a neighborhood of the origin. The latter implies that the feedback will only yield asymptotic stability if the system is already asymptotically stable.*

Proof : Let ε^* and the function $\alpha^* : (0, \varepsilon^*) \to \mathbb{R}^+$ be such that the controller (5.88) solves the robust semi-global practical stabilization problem for the system (5.16) for all $\varepsilon \in (0, \varepsilon^*)$, and $\alpha > \alpha^*(\varepsilon)$ with respect to the compact sets $\mathcal{X}_1$ and $\mathcal{X}_2$, where $\tilde{\sigma}$ is given by (5.86), while g satisfies the properties in Assumption 5.12 with $D_0 > a$. Then the same controller applied to the system (5.84) has the property that the closed-loop system is such that for all initial conditions in $\mathcal{X}_1$ the state enters and remains in the set $\mathcal{X}_2$ within a finite amount of time. This follows from the fact that

$$g = g_d(u) = g_d\left(-\theta^{-1}B'P_\varepsilon x - \alpha B'P_\varepsilon x\right)$$

satisfies the conditions of Theorem 5.12, and after setting $g = g_d$, we note that the system (5.16) with controller (5.88) yields the same closed-loop system as the system (5.84) with the same controller (5.88). The rest is then a direct application of Theorem 5.19. ■

Theorem 5.36 *Consider the system*

$$\Sigma_{ud} : \begin{cases} \dot{x} = Ax + B\widetilde{\sigma}_d(u) \\ y = Cx. \end{cases} \tag{5.89}$$

Assume that (A, B) is stabilizable, all the eigenvalues of A are in the closed left-half plane, and the system $(A, B, C, 0)$ is left invertible and minimum phase. Also, assume that $\widetilde{\sigma}_d$ satisfies the standard properties (i)–(v) on page 374, which guarantees the existence of positive constants a, θ, and ψ such that (5.87) *is satisfied componentwise. In that case, there exist, for any pair of compact sets $\mathcal{X}_{cl,1}$ and $\mathcal{X}_{cl,2}$ in $\mathbb{R}^{2n}$ containing 0 in their interior and with $\mathcal{X}_{cl,1} \supset \mathcal{X}_{cl,2}$, an ε^*, a function $\alpha^* : (0, \varepsilon^*) \to \mathbb{R}^+$, and a function $\ell^* : (0, \varepsilon^*) \times \mathbb{R}^+ \to \mathbb{R}^+$ such that for all $\varepsilon \in (0, \varepsilon^*)$, $\alpha > \alpha^*(\varepsilon)$, and $\ell > \ell^*(\varepsilon, \alpha)$, the feedback* (5.30) *has the property that the resulting closed-loop system is such that for all initial conditions in $\mathcal{X}_{cl,1}$ the state enters and remains in the set $\mathcal{X}_{cl,2}$ within a finite amount of time.*

Proof : Let ε^* and the functions $\alpha^* : (0, \varepsilon^*) \to \mathbb{R}^+$ and $\ell > \ell^*(\varepsilon, \alpha)$ be such that the controller (5.30) solves the robust semi-global practical stabilization problem for the system (5.16) for all $\varepsilon \in (0, \varepsilon^*)$, $\alpha > \alpha^*(\varepsilon)$ and $\ell > \ell^*(\varepsilon, \alpha)$ with respect to the compact sets $\mathcal{X}_1$ and $\mathcal{X}_2$, where $\widetilde{\sigma}$ is given by (5.86), while g satisfies the properties in Assumption 5.12 with $D_0 > a$. Then the same controller applied to the system (5.84) has the property that the closed-loop system is such that for all initial conditions in $\mathcal{X}_1$ the state enters and remains in the set $\mathcal{X}_2$ within a finite amount of time. This follows from the fact that

$$g = g_d(u) = g_d\left(-\theta^{-1} B' P_\varepsilon \hat{x} - \alpha B' P_\varepsilon \hat{x}\right)$$

satisfies the conditions of Assumption 5.12, and after setting $g = g_d$, we note that the system (5.89) with controller (5.30) yields the same closed-loop system as the system (5.84) with the same controller (5.30). The rest is then a direct application of Theorem 5.21. ■

The above two theorems only consider the case of stabilization in case of a saturation with a deadzone. If we have model uncertainty as studied before in Sect. 5.4, then the above two theorems can be trivially expanded. Consider the system,

$$\Sigma_{ud} : \begin{cases} \dot{x}(t) = Ax(t) + B\widetilde{\sigma}_d(u(t) + \widetilde{g}(x,t)) \\ y(t) = Cx(t), \end{cases} \tag{5.90}$$

where $\widetilde{g}$ satisfies the following assumption:

Assumption 5.37 *The uncertain element $\widetilde{g}(x,t)$ is piecewise continuous in t, locally Lipschitz in x and its norm is bounded by a known function*

$$\|\widetilde{g}(x,t)\| \leq g_0(\|x\|) + \widetilde{D}_0, \quad \forall (t,x) \in \mathbb{R}^+ \times \mathbb{R}^n, \tag{5.91}$$

where $\widetilde{D}_0$ is a known positive constant, and the known function $g_0(x) : \mathbb{R}^+ \to \mathbb{R}^+$ is locally Lipschitz and satisfies

$$g_0(0) = 0. \tag{5.92}$$

We then obtain the following two theorems:

Theorem 5.38 *Consider the system* (5.90) *with (A, B) stabilizable and all eigenvalues of A in the closed left-half plane. Assume that $\widetilde{\sigma}_d$ satisfies properties (i)–(v) on page 374, and let positive constants a, θ, and ψ be such that*

$$|\widetilde{\sigma}_d(s)| > \min\{\, \theta(|s| - a), \psi \,\} \tag{5.93}$$

componentwise, while $\widetilde{g}$ satisfies the conditions of Assumption 5.37.
Let $P_\varepsilon > 0$ be the solution of the algebraic Riccati equation and

$$A'P_\varepsilon + P_\varepsilon A - P_\varepsilon BB'P_\varepsilon + Q_\varepsilon = 0$$

with $Q_\varepsilon \to 0$ as $\varepsilon \to 0$. In that case, there exist, for any pair of compact sets $\mathcal{X}_1$ and $\mathcal{X}_2$ containing 0 in their interior and with $\mathcal{X}_1 \supset \mathcal{X}_2$, an ε^* and a function $\alpha^* : (0, \varepsilon^*) \to \mathbb{R}^+$ such that for all $\varepsilon \in (0, \varepsilon^*)$ and $\alpha > \alpha^*(\varepsilon)$, the feedback

$$u = -\theta^{-1}B'P_\varepsilon x - \alpha B'P_\varepsilon x$$

has the property that the closed-loop system is such that for all initial conditions in $\mathcal{X}_1$ and all possible functions $\widetilde{g}$ satisfying Assumption 5.37, the state enters and remains in the set $\mathcal{X}_2$ within a finite amount of time.

Proof : Let ε^* and the function $\alpha^* : (0, \varepsilon^*) \to \mathbb{R}^+$ be such that the controller (5.88) solves the robust semi-global practical stabilization problem for the system (5.16) for all $\varepsilon \in (0, \varepsilon^*)$ and $\alpha > \alpha^*(\varepsilon)$ with respect to the compact sets $\mathcal{X}_1$ and $\mathcal{X}_2$, where $\widetilde{\sigma}$ is given by (5.86), while g satisfies the properties in Assumption 5.12 with $D_0 > \widetilde{D}_0 + a$. Then the same controller applied to the system (5.84) has the property that the closed-loop system is such that for all initial conditions in $\mathcal{X}_1$ the state enters and remains in the set $\mathcal{X}_2$ within a finite amount of time. This follows from the fact that

$$g = \widetilde{g} + g_d(u) = \widetilde{g} + g_d\left(-\theta^{-1}B'P_\varepsilon x - \alpha B'P_\varepsilon x\right)$$

satisfies the conditions of Assumption 5.12, and after setting $g = \widetilde{g} + g_d$, we note that the system (5.16) with controller (5.88) yields the same closed-loop system as the system (5.84) with the same controller (5.88). The rest is then a direct application of Theorem 5.19. ■

Theorem 5.39 *Consider the system* (5.90) *with* (A, B) *stabilizable, with all the eigenvalues of* A *in the closed left-half plane, and with the system* $(A, B, C, 0)$ *left invertible and minimum phase. Assume that* $\widetilde{\sigma}_d$ *satisfies the standard properties (i)–(v) on page 374, which guarantees the existence of positive constants* a, θ, *and* ψ *such that* (5.93) *is satisfied componentwise, while* $\widetilde{g}$ *satisfies the conditions of Assumption 5.37.*

In that case, there exist, for any pair of compact sets $\mathcal{X}_{cl,1}$ *and* $\mathcal{X}_{cl,2}$ *in* $\mathbb{R}^{2n}$ *containing 0 in their interior and with* $\mathcal{X}_{cl,1} \supset \mathcal{X}_{cl,2}$, *an* ε^*, *a function* $\alpha^* : (0, \varepsilon^*) \to \mathbb{R}^+$, *and a function* $\ell^* : (0, \varepsilon^*) \times \mathbb{R}^+ \to \mathbb{R}^+$ *such that for all* $\varepsilon \in (0, \varepsilon^*)$, $\alpha > \alpha^*(\varepsilon)$, *and* $\ell > \ell^*(\varepsilon, \alpha)$, *the feedback* (5.30) *has the property that the resulting closed-loop system is such that for all initial conditions in* $\mathcal{X}_{cl,1}$ *and all possible functions* $\widetilde{g}$ *satisfying Assumption 5.37, the state enters and remains in the set* $\mathcal{X}_{cl,2}$ *within a finite amount of time.*

Proof : Let ε^* and the functions $\alpha^* : (0, \varepsilon^*) \to \mathbb{R}^+$ and $\ell > \ell^*(\varepsilon, \alpha)$ be such that the controller (5.30) solves the robust semi-global practical stabilization problem for the system (5.16) for all $\varepsilon \in (0, \varepsilon^*)$, $\alpha > \alpha^*(\varepsilon)$ and $\ell > \ell^*(\varepsilon, \alpha)$ with respect to the compact sets $\mathcal{X}_1$ and $\mathcal{X}_2$ where $\widetilde{\sigma}$ is given by (5.86) while g satisfies the properties in Assumption 5.12 with $D_0 > \widetilde{D}_0 + a$. Then the same controller applied to the system (5.84) has the property that the closed-loop system is such that for all initial conditions in $\mathcal{X}_1$ the state enters and remains in the set $\mathcal{X}_2$ within a finite amount of time. This follows from the fact that

$$g = \widetilde{g} + g_d(u) = \widetilde{g} + g_d\left(-\theta^{-1} B' P_\varepsilon \widehat{x} - \alpha B' P_\varepsilon \widehat{x}\right)$$

satisfies the conditions of Theorem 5.12, and after setting $g = \widetilde{g} + g_d$, we note that the system (5.89) with controller (5.30) yields the same closed-loop system as the system (5.84) with the same controller (5.30). The rest is then a direct application of Theorem 5.21. ∎

5.7.2 Discrete time

Consider the system

$$\begin{aligned} x(k+1) &= Ax(k) + B\widetilde{\sigma}_d(u(k)) \\ y(k) &= Cx(k). \end{aligned} \tag{5.94}$$

The approach based on dithering mentioned in the previous subsection clearly does not work for discrete-time systems since we are intrinsically limited in our use of high-frequency signals by the sampling rate of the system.

But also practical stabilization is simply not achievable in the presence of a deadzone. This is illustrated in the following example which has the same structure as the examples in Sect. 5.6.

Example 5.40 Consider the system,

$$x(k+1) = x(k) + \sigma(u(k) + g(x,k)).$$

There exists saturation functions $\tilde{\sigma}_d$ with a deadzone satisfying properties (*i*)–(*v*) on page 374 for which there exists no feedback $u = f(x)$ which achieves practical stability in the sense that we cannot guarantee that, for any compact set $\mathcal{X}_1, \mathcal{X}_2$ containing 0 in their interior, there exists a continuous feedback $u = f(x)$ such that the closed-loop system is such that for all initial conditions in $\mathcal{X}_1$ the state enters and remains in $\mathcal{X}_2$ after some finite amount of time.

This can be seen by choosing $\mathcal{X}_2$ such that $x \in \mathcal{X}_2$ implies $|x| \leqslant \alpha/2$. Moreover, choose $\tilde{\sigma}_d$ such that $\tilde{\sigma}_d(x) = 0$ when x satisfies $\|x\| < \alpha$. This is clearly consistent with properties (*i*)–(*v*). In that case, we find that for all $x(k) \notin \mathcal{X}_2$ with $\|x(k)\| < \alpha$, we have that $x(k)$ first has to increase in amplitude before it can enter $\mathcal{X}_2$. This structure to achieve attractivity of $\mathcal{X}_2$ is however impossible to achieve by a continuous feedback.

6
Control magnitude and rate saturation

6.1 Introduction

Chapter 4 considered global and semi-global stabilization of linear systems with actuators subject to magnitude saturation alone. In this chapter, we revisit the same internal stabilization, however, with actuators subject to both magnitude and rate saturation. Rate saturation refers to the case when actuator outputs cannot change faster than a certain value.

In contrast with the amount of literature that exists when there are only magnitude bounds, not much literature exists when both magnitude and rate saturation are present. If we have only rate saturation, we can approach the study of the given system for stabilization by viewing the derivative of the input signal as a new fictitious input signal which is then bounded only in magnitude, and hence, we can readily apply the development given in the previous chapter. However, if we have simultaneous bounds on the rate as well as on the magnitude, then such a simple approach does not work out. We would like to point out right at the outset that, unlike magnitude saturation which is a static nonlinearity, the rate saturation is a dynamic nonlinearity.

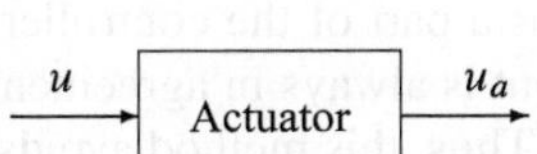

Figure 6.1: Actuator

In modeling magnitude and rate saturation for actuators, one can basically choose one of the following two approaches:

- A natural way is to write down first the linear dynamic equations of an actuator (see block diagram of Fig. 6.1) and then impose on such equations both the magnitude and rate saturation as shown in the following equation for continuous-time systems:

$$
\begin{aligned}
\dot{x}_a &= \sigma_{\Delta_2}(A_a x_a + B_a u) \\
u_a &= \sigma_{\Delta_1}(C_a x_a),
\end{aligned}
\tag{6.1}
$$

A. Saberi et al., *Internal and External Stabilization of Linear Systems with Constraints*, Systems & Control: Foundations & Applications, DOI 10.1007/978-0-8176-4787-2_6,

where $\sigma_{\Delta_1}(\cdot)$ and $\sigma_{\Delta_2}(\cdot)$ are standard saturation functions as defined in Definition (2.19). Also, $u \in \mathbb{R}^m$ is the input signal, $x_a \in \mathbb{R}^{na}$ is the state of the actuator, and $u_a \in \mathbb{R}^m$ is the output of the actuator which is applied to the plant.

We would like to point out that it is rather difficult to incorporate the "nice" external stability behavior of actuators into a state space model characterized by the matrix triple (A_a, B_a, C_a). Anyway, having modeled the actuator as in (6.1), the next obvious step is to augment (6.1) with the plant model in order to obtain the model for both the actuator and the plant. It is well known that it is hard to analyze and design dynamical systems which are modeled as such. However, if one imposes a certain strong mathematical structure on the dynamics of the actuator given in (6.1), one can avoid the complexity in analysis and design. The required mathematical structure is typically to assume that A_a, B_a, and C_a are diagonal matrices. Essentially, such a mathematical structure implies that the actuator dynamics for each component of the input is a first-order or scalar dynamics. This approach is taken in [6, 69, 72, 73]. We note that [6, 73] deal with rate saturation while [69, 72] deal with magnitude as well as rate saturation. It is obvious to see that in [73] with the imposed structure of diagonality of the matrices A_a, B_a, and C_a, the control of linear systems with rate saturation reduces to the control of the augmented plant with only input magnitude saturation. However, we would like to point out two interesting but undesirable aspects of this approach. First, it yields the necessity of using the state of the actuator for control feedback. Secondly if the system is in rate saturation for a long time, x_a might become very large. But that implies that u_a is saturated in amplitude as well, **but** we cannot get u_a out of saturation for a long time since we first have to make x_a small which, due to the rate saturation, cannot be achieved quickly.

- The second approach is to model the constraints in such a way that they can be incorporated as a part of the controller and then to design the controller so that its output is always in agreement with the constraints as dictated by the actuator. Thus, this method avoids overloading of the actuator. By incorporating the actuator constraints in the design of controllers, this method essentially sidesteps the shortcomings of the first method. The work of [1] takes this approach to model the rate saturation. The block diagram in Fig. 6.2 depicts the philosophy of the method.

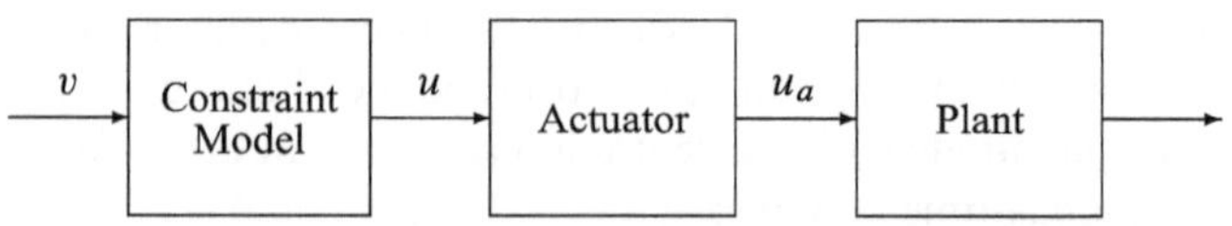

Figure 6.2: Constraint model, actuator, and plant

We take the second approach here, and model the constraints by introducing a nonlinear operator that captures both magnitude and rate constraints. We refer to the new nonlinear operator as a standard magnitude+rate operator (see Fig. 6.3). Such an operator has a "deadbeat stability" property. With such a property being valid, we study and examine the state space realizations of this operator. It turns out that, although one could obtain a useful state space realization in discrete-time systems, one cannot do so in continuous-time systems. This indicates that a framework that includes functional differential equations for modeling the plant and magnitude as well as rate saturation constraints of the actuator is indeed a natural framework.

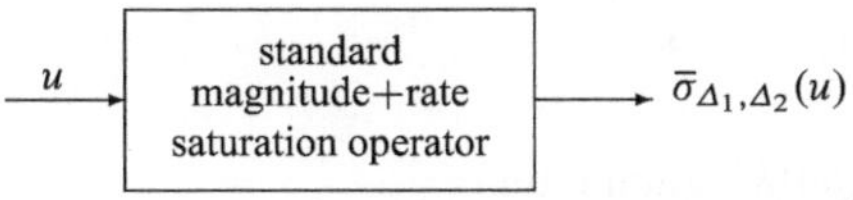

Figure 6.3: Amplitude+rate saturation operator

Utilizing this functional differential framework and knowing that it satisfies a "deadbeat stability" property, we redefine the notions of semi-global stabilization. We then proceed to show that the same low-gain design methodology that was successfully used in the previous chapter to design controllers for linear systems having only input magnitude saturation can also successfully be used for linear systems with both magnitude and rate saturation on the control input. In fact, we present here explicit controller design methods via low-gain design methodology in order to semi-globally stabilize linear systems with both input magnitude and rate saturation. It will be clarified why we cannot expect in case of rate and amplitude constraints that scheduling the low-gain parameter will result in global stability. As illustrated in the previous chapter, controllers designed via low-gain design methodology exhibit unacceptably slow transient behavior. For that reason, we also utilize low-and-high-gain controllers where the high-gain component does not affect the stability and its associated domain of attraction. On the other hand, the high-gain component greatly improves the transient behavior. However, this low-and-high-gain controller cannot be obtained in the same way as in the previous chapter owing to the specific dynamic structure of a rate limiter.

We consider both continuous- and discrete-time systems together in this chapter. This chapter is based completely on the work of the authors in [164].

6.2 Modeling issues: standard magnitude + rate saturation operator

In this section, we present modeling aspects of actuators with both magnitude and rate saturation. This will first be done for discrete-time systems, and then we will present the continuous-time result which contains some more subtleties.

6.2.1 Discrete time

We will first consider a discrete-time system of the form,

$$x(k+1) = Ax(k) + B\sigma_{\Delta_1,\Delta_2}(u)(k). \tag{6.2}$$

Here $\sigma_{\Delta_1,\Delta_2}$ is a diagonal operator with identical nonlinear elements on the diagonal given by $\overline{\sigma}_{\Delta_1,\Delta_2}$ which is uniquely defined by the following properties:

- We have

  $$|\overline{\sigma}_{\Delta_1,\Delta_2}(u)(k)| \leqslant \Delta_1, \text{ and } |\overline{\sigma}_{\Delta_1,\Delta_2}(u)(k+1) - \overline{\sigma}_{\Delta_1,\Delta_2}(u)(k)| \leqslant \Delta_2$$

  for all k.
- If $u(k) > \overline{\sigma}_{\Delta_1,\Delta_2}(u)(k)$, then either
  - ⋆ $u(k) > \Delta_1$ and $\overline{\sigma}_{\Delta_1,\Delta_2}(u)(k) = \Delta_1$ or
  - ⋆ $\overline{\sigma}_{\Delta_1,\Delta_2}(u)(k) = \overline{\sigma}_{\Delta_1,\Delta_2}(u)(k-1) + \Delta_2$.
- If $u(k) < \overline{\sigma}_{\Delta_1,\Delta_2}(u)(k)$, then either
  - ⋆ $u(k) < -\Delta_1$ and $\overline{\sigma}_{\Delta_1,\Delta_2}(u)(k) = -\Delta_1$, or
  - ⋆ $\overline{\sigma}_{\Delta_1,\Delta_2}(u)(k) = \overline{\sigma}_{\Delta_1,\Delta_2}(u)(k-1) - \Delta_2$.

Obviously, $\overline{\sigma}_{\Delta_1,\Delta_2}$ is a dynamic nonlinearity. We can also describe this operator by a state space model

$$x_\infty(k+1) = \Delta_2 \operatorname{sgn}(\sigma_{\Delta_1}(u(k+1)) - x_\infty(k)), \quad x_\infty(0) = \sigma_{\Delta_1}(u(0)), \tag{6.3}$$

where $\overline{\sigma}_{\Delta_1,\Delta_2}(u) = x_\infty$. This is a good model for our magnitude+rate operator which is consistent with our earlier description.

It is easy to see that $\overline{\sigma}_{\Delta_1,\Delta_2}$ is dynamic and has initial conditions. Also, it is not difficult to see that $\overline{\sigma}_{\Delta_1,\Delta_2}(u)(k)$ can be viewed as the state of this system at time k. For our purpose, it will in general be sufficient to note that $\overline{\sigma}_{\Delta_1,\Delta_2}$ is deadbeat. We even have the stronger property that if u is such that

$$\|u(k)\|_\infty < \Delta_1 \text{ and } \|u(k) - u(k-1)\|_\infty < \Delta_2 \text{ for } k > K,$$

then

$$\overline{\sigma}_{\Delta_1,\Delta_2}(u)(k) = u(k) \text{ for } k > K + \Delta_1/\Delta_2.$$

If we refer to arbitrary initial conditions of $\overline{\sigma}_{\Delta_1,\Delta_2}$ at time 0, then we mean an arbitrary input signal u in the interval $[-\Delta_1/\Delta_2, 0]$. In our definitions, we will refer to the initial conditions of $\overline{\sigma}_{\Delta_1,\Delta_2}$ as $\overline{x}_s$, and we will refer to the space of all possible initial conditions for the operator $\overline{\sigma}_{\Delta_1,\Delta_2}$, namely, the space of all signals defined on the interval $[-\Delta_1/\Delta_2, 0]$, as $\overline{\mathcal{X}}_s$. Similarly, for the operator $\sigma_{\Delta_1,\Delta_2}$, we denote the initial conditions by x_s and the space of all initial signals as $\mathcal{X}_s$.

6.2.2 Continuous time

We consider a continuous-time system of the form

$$\dot{x}(t) \quad = \quad Ax(t) + B\sigma_{\Delta_1,\Delta_2}(u)(t). \tag{6.4}$$

Here again, $\sigma_{\Delta_1,\Delta_2}$ is a diagonal operator with identical nonlinear elements on the diagonal given by $\overline{\sigma}_{\Delta_1,\Delta_2}$. We seek an operator $\overline{\sigma}_{\Delta_1,\Delta_2}$ with the following properties:

- For any continuously differentiable u, $\overline{\sigma}_{\Delta_1,\Delta_2}(u)$ is differentiable and

$$|\overline{\sigma}_{\Delta_1,\Delta_2}(u)(t)| \leqslant \Delta_1, \text{ and } |\tfrac{\mathrm{d}}{\mathrm{d}t}\overline{\sigma}_{\Delta_1,\Delta_2}(u)(t)| \leqslant \Delta_2$$

  for all t.

- If $u(t) > \overline{\sigma}_{\Delta_1,\Delta_2}(u)(t)$, then either
  - ⋆ $u(t) > \Delta_1$ and $\overline{\sigma}_{\Delta_1,\Delta_2}(u)(t) = \Delta_1$, or
  - ⋆ $\frac{\mathrm{d}}{\mathrm{d}t}\overline{\sigma}_{\Delta_1,\Delta_2}(u)(t) = \Delta_2$.
- If $u(t) < \overline{\sigma}_{\Delta_1,\Delta_2}(u)(t)$, then either
  - ⋆ $u(t) < -\Delta_1$ and $\overline{\sigma}_{\Delta_1,\Delta_2}(u)(t) = -\Delta_1$, or
  - ⋆ $\frac{\mathrm{d}}{\mathrm{d}t}\overline{\sigma}_{\Delta_1,\Delta_2}(u)(t) = -\Delta_2$.

However, in the above case of continuous-time systems, it is not clear whether this uniquely determines the operator $\overline{\sigma}_{\Delta_1,\Delta_2}$. Moreover, for instance due to measurement noise, we might have an input signal which is not smooth but might only be piecewise continuous, and for this general class of signals, the above definition is clearly not sufficient.

We now consider a different way of looking at $\overline{\sigma}_{\Delta_1,\Delta_2}$. Consider the class of models,

$$\dot{x}_\lambda = \sigma_{\Delta_2}(\lambda(\sigma_{\Delta_1}(u) - x_\lambda)), \quad x_\lambda(0) = \sigma_{\Delta_1}(u(0)). \tag{6.5}$$

It is well known that this differential equation has a unique solution for any measurable input signal u. Next, we define $\overline{\sigma}_{\Delta_1,\Delta_2}$ by

$$\overline{\sigma}_{\Delta_1,\Delta_2}(u) = \lim_{\lambda\to\infty} x_\lambda. \tag{6.6}$$

The following lemma shows that the operator as defined by (6.6) has all the required properties.

Lemma 6.1 *For any piecewise continuous function u, the limit in* (6.6) *exists in L_∞, and the limit $\overline{\sigma}_{\Delta_1,\Delta_2}(u)$ has an L_∞ norm less than Δ_1 and is Lipschitz continuous with Lipschitz constant Δ_2.*

Proof : Let $\lambda^* > 0$. For any $\lambda_1, \lambda_2 > \lambda^*$, we have $x_{\lambda_1}(0) - x_{\lambda_2}(0) = 0$. Moreover, if $x_{\lambda_1}(t) - x_{\lambda_2}(t) > 2\Delta_2/\lambda^*$, we have $\dot{x}_{\lambda_1}(t) - \dot{x}_{\lambda_2}(t) \leqslant 0$. After all, we have only two possibilities:

- $x_{\lambda_1}(t) > \sigma_{\Delta_1}(u)(t) + \Delta_2/\lambda^*$ in which case $\dot{x}_{\lambda_1}(t) = -\Delta_2$ and $\dot{x}_{\lambda_2}(t) \geqslant -\Delta_2$. Therefore, $\dot{x}_{\lambda_1}(t) - \dot{x}_{\lambda_2}(t) \leqslant 0$.
- $x_{\lambda_2}(t) < \sigma_{\Delta_1}(u)(t) - \Delta_2/\lambda^*$ in which case $\dot{x}_{\lambda_1}(t) \leqslant \Delta_2$ and $\dot{x}_{\lambda_2}(t) = \Delta_2$. Therefore, $\dot{x}_{\lambda_1}(t) - \dot{x}_{\lambda_2}(t) \leqslant 0$.

Similarly, if $x_{\lambda_1}(t) - x_{\lambda_2}(t) < -2\Delta_2/\lambda^*$, we have $\dot{x}_{\lambda_1}(t) - \dot{x}_{\lambda_2}(t) \geqslant 0$. This shows that for all $\lambda_1, \lambda_2 > \lambda^*$, we have

$$\|x_{\lambda_1} - x_{\lambda_2}\|_\infty \leqslant \tfrac{2\Delta_2}{\lambda^*}.$$

Therefore, by definition, $\{x_\lambda\}$ is a Cauchy sequence, and it has a limit which we call $x_\infty \in L_\infty$.

We know that $x_\lambda(t) \geqslant \Delta_1$ implies that $\dot{x}_\lambda(t) \leqslant 0$, and $x_\lambda(t) \leqslant -\Delta_1$ implies that $\dot{x}_\lambda(t) \geqslant 0$. Combined with $\|x_\lambda(0)\| \leqslant \Delta_1$, we find then that $\|x_\lambda\|_\infty \leqslant \Delta_1$. This obviously implies that

$$\|x_\infty\| = \lim_{\lambda\to\infty} \|x_\lambda\| \leqslant \Delta_1.$$

Finally, we have $\|x_\lambda(t_2) - x_\lambda(t_1)\| \leqslant \Delta_2|t_2 - t_1|$ for any $t_1, t_2 > 0$. By letting $\lambda \to \infty$, we find that

$$\|x_\infty(t_2) - x_\infty(t_1)\| \leqslant \Delta_2|t_2 - t_1| \qquad \text{for all } t_1, t_2 > 0,$$

and hence, x_∞ is Lipschitz continuous with Lipschitz constant Δ_2. ■

Note that a Lipschitz continuous function is absolutely continuous, and hence, it is easy to see that $\overline{\sigma}_{\Delta_1,\Delta_2}(u)$ is differentiable almost everywhere, and there exists an L_∞ function w with L_∞ norm less than Δ_2 such that

$$\overline{\sigma}_{\Delta_1,\Delta_2}(u)(t) = \overline{\sigma}_{\Delta_1,\Delta_2}(u)(0) + \int_0^t w(t)\,\mathrm{d}t.$$

However, $\overline{\sigma}_{\Delta_1,\Delta_2}(u)$ need not be differentiable everywhere. An example is given by the function,

$$u(t) = \begin{cases} 0 & t = 0 \\ t\sin\left(\frac{1}{t}\right) & \text{elsewhere,} \end{cases}$$

for which $\overline{\sigma}_{\Delta_1,\Delta_2}(u)$ is not differentiable in 0. Obviously, with the more precise definition given in (6.6), $\overline{\sigma}_{\Delta_1,\Delta_2}$ is uniquely determined. Moreover, as soon as u is sufficiently smooth, it is easy to verify that the mathematically precise definition given in (6.6) is consistent with our intuitive definition given initially.

Note that we might define the state model for $\overline{\sigma}_{\Delta_1,\Delta_2}$ as,

$$\dot{x}_\infty = \Delta_2 \operatorname{sgn}(\sigma_{\Delta_1}(u) - x_\infty), \quad x_\infty(0) = \sigma_{\Delta_1}(u(0)), \tag{6.7}$$

with $\overline{\sigma}_{\Delta_1,\Delta_2}(u) = x_\infty$. This is consistent with our intuitive description, and it looks like the appropriate model given the state space models for x_λ. However, note that if u satisfies the rate and saturation bounds, then we expect that $u = x_\infty$ (this can also be formally shown), but then the above differential equation shows that $\dot{x}_\infty = 0$ which obviously need not be the case. Therefore, the model given in (6.7) is **incorrect**.

Like in discrete time, we see that $\overline{\sigma}_{\Delta_1,\Delta_2}$ is dynamic and has initial conditions. We note that again $\overline{\sigma}_{\Delta_1,\Delta_2}(u)(t)$ can be viewed as the state of this system at time t. For our purpose, it will in general be sufficient to note that $\overline{\sigma}_{\Delta_1,\Delta_2}$ is deadbeat. We again have the stronger property that if u is such that

$$\|u(t)\|_\infty < \Delta_1 \text{ and } \|\dot{u}(t)\|_\infty < \Delta_2 \text{ for } t > t_1,$$

then

$$\overline{\sigma}_{\Delta_1,\Delta_2}(u)(t) = u(t) \text{ for } t > t_1 + \Delta_1/\Delta_2.$$

If we refer to arbitrary initial conditions of $\overline{\sigma}_{\Delta_1,\Delta_2}$ at time 0, then we mean an arbitrary input signal u in the interval $[-\Delta_1/\Delta_2, 0]$. As in discrete-time systems, we will refer to the initial conditions of $\overline{\sigma}_{\Delta_1,\Delta_2}$ as $\overline{x}_s$, and we will refer to the space of all possible initial conditions for the operator $\overline{\sigma}_{\underline{\Delta}_1,\Delta_2}$, namely, the space of all signals defined on the interval $[-\Delta_1/\Delta_2, 0]$, as $\mathcal{X}_s$. Similarly, for the operator $\sigma_{\Delta_1,\Delta_2}$, we denote the initial conditions by x_s, and the space of all initial signals as $\mathcal{X}_s$.

In the literature, there have been many different models for a rate limit in combination with a saturation. All other models use a form of modeling of the form

$$\dot{x}_r = f(x_r, u), \quad x_r \in \mathbb{R}^k.$$

We do not model the actuator with its rate and magnitude limits. We model these limits and constraints and view them as part of the controller. Namely, we use an operator as part of the controller, which guarantees that the control signal satisfies the bounds in the actuator and avoids overloading the actuator. This is the important difference, but also the fact that our operator only has a state space model in discrete time and can only be approximated by state space models in continuous time yields differences in the analysis, while leading to the fact that its state space equals $[-\Delta_1, \Delta_1]$. Our approach is really different but, as we will see, very powerful since all results we have obtained in the previous chapter for control magnitude saturation alone can be easily extended to the case with rate limits including the low-and-high-gain design which is difficult to analyze for a state space model with $x_r \in \mathbb{R}^k$.

In summary, in view of the above discussions, the functional differential equations (6.2) and (6.4) along with the definition for the constraint operator $\sigma_{\Delta_1,\Delta_2}$ given in (6.3), (6.5) and (6.6) are the appropriate models for discrete- and continuous-time systems, respectively, whenever the actuators are constrained by both magnitude and rate saturation.

6.3 Preliminaries and problem statements

In view of the modeling aspects of actuators with both magnitude and rate saturation as discussed in Sect. 6.2, the given system is modeled by functional difference equations (6.2) in discrete time and by functional differential equation (6.4) in continuous time. Clearly, this includes the definition for the constraint operator $\sigma_{\Delta_1,\Delta_2}$ given in (6.3) (discrete time) and in (6.5) and (6.6) (continuous time). Thus, we consider a dynamic system of the form

$$\Sigma : \begin{cases} \rho x = Ax + B\sigma_{\Delta_1,\Delta_2}(u) \\ y \;= C_y x + D_y u, \end{cases} \tag{6.8}$$

where $x \in \mathbb{R}^n, u \in \mathbb{R}^m, and y \in \mathbb{R}^r$ are respectively state, input, and measurement output. Also, ρx denotes $\frac{dx}{dt}$ for continuous-time case and $x(k+1)$ for discrete time. Moreover, $\sigma_{\Delta_1,\Delta_2}$ is a diagonal operator with identical nonlinear elements on the diagonal given by $\bar{\sigma}_{\Delta_1,\Delta_2}$ which is uniquely defined as discussed in Sect. 6.2. The space of all possible initial conditions for the operator $\sigma_{\Delta_1,\Delta_2}$, namely, the space of all functions from $[-\Delta_1/\Delta_2, 0]$ to $\mathbb{R}^m$, is referred as $\mathcal{X}_s$.

We give below a precise definition of the concept of semi-global stabilization for the above system, at first via linear state feedback laws and then via measurement feedback laws.

Problem 6.2 Consider a continuous- or discrete-time system of the form (6.8) along with the definition for the constraint operator $\sigma_{\Delta_1,\Delta_2}$ given in (6.3) (discrete) or (6.5) and (6.6) (continuous). The problem of **semi-global stabilization via linear state feedback** is to find, if possible, a family of feedback gains F_ε parameterized by $\varepsilon > 0$ such that for any a priori given (arbitrarily large) bounded set $\mathcal{X}_0 \subset \mathbb{R}^n$, there exists an ε^* such that for all $\varepsilon < \varepsilon^*$, the linear static feedback law $u = F_\varepsilon x$ is such that the equilibrium $x = 0, x_s = 0$ of the continuous-time system

$$\dot{x}(t) = Ax(t) + B\sigma_{\Delta_1,\Delta_2}(Fx)(t) \tag{6.9}$$

or the discrete-time system

$$x(k+1) = Ax(k) + B\sigma_{\Delta_1,\Delta_2}(Fx)(k) \tag{6.10}$$

is locally exponentially stable with $\mathcal{X}_0 \times \mathcal{X}_s$ contained in its basin of attraction.

Problem 6.3 Consider a continuous- or discrete-time system of the form (6.8) along with the definition for the constraint operator $\sigma_{\Delta_1,\Delta_2}$ given in (6.3) (discrete) or (6.5) and (6.6) (continuous). The problem of **semi-global stabilization via observer-based measurement feedback** is defined as follows. For any a priori given (arbitrarily large) bounded sets $\mathcal{X}_0 \subset \mathbb{R}^n$ and $\mathcal{Z}_0 \subset \mathbb{R}^n$, find, if possible, a measurement feedback law of the form

$$\begin{aligned} \rho\widehat{x} &= A\widehat{x} + B\sigma_{\Delta_1,\Delta_2}(u) + K\left(y - C_y\widehat{x} - D_y u\right) \\ u &= F\widehat{x} \end{aligned} \tag{6.11}$$

such that the equilibrium $(x, x_s, \widehat{x}) = (0, 0, 0)$ of

$$\begin{aligned} \rho x &= Ax + B\sigma_{\Delta_1,\Delta_2}(F\widehat{x}) \\ \rho\widehat{x} &= A\widehat{x} + B\sigma_{\Delta_1,\Delta_2}(F\widehat{x}) + KC_y(x - \widehat{x}) \end{aligned}$$

is asymptotically stable with $\mathcal{X}_0 \times \mathcal{X}_s \times \mathcal{Z}_0$ contained in its basin of attraction.

Remark 6.4 *We would like to emphasize that our definitions of the above semi-global stabilization problems do not view the set of initial conditions of the plant and the initial conditions of the controller dynamics as given data. The given data is simply the model of the plant. Therefore, the solvability conditions* must *be independent of the set of initial conditions of the plant $\mathcal{X}_0 \times \mathcal{X}_s$, and the set of initial conditions for the controller dynamics, $\mathcal{Z}_0$.*

The standard approach to address the above problems is the use of a low-gain feedback. The low-gain parameter ε guarantees that

$$F_\varepsilon x(k+1) - F_\varepsilon x(k)$$

is sufficiently small when ε is small enough and hence satisfies the rate limitations (as well as the amplitude constraints) on the input. Let us rewrite the above when the low-gain parameter is scheduled:

$$\left[F_{\varepsilon(x(k+1))} - F_{\varepsilon(x(k))}\right] x(k+1) + F_{\varepsilon(x(k))}\left[x(k+1) - x(k)\right].$$

The second term can be made sufficiently small to satisfy the rate limitations. However, the first term might cause violation of the rate limitations, and hence, it is quite hard to establish whether such design would result in global stability in the presence of rate and amplitude constraints. Except for these scheduled low-gain approaches, there are basically no techniques available to design a globally stabilizing static state feedback in case of rate and amplitude constraints. Although not presented here in detail, we should note that the dynamic state feedbacks of the form (6.30) (even for $\alpha = 0$) studied in Sect. 6.5 do allow for a scheduling to yield global stabilization for systems subject to rate and amplitude saturation.

6.4 Semi-global stabilization via low-gain feedback

We have the following necessary and sufficient conditions for the solvability of the semi-global stabilization problem via linear state feedback laws utilizing low gain. We first present the continuous-time result.

Theorem 6.5 *Consider the continuous-time system* (6.8) *along with the definition for the constraint operator* $\sigma_{\Delta_1,\Delta_2}$ *given in* (6.5) *and* (6.6)*. The semi-global stabilization Problem 6.2 via linear state feedback laws is solvable if and only if* (A, B) *is stabilizable and the eigenvalues of* A *are in the closed left half plane.*

Moreover, in that case, the semi-global stabilization problem via linear state feedback laws is solved by the family of feedback laws

$$u = F_\varepsilon x = -B' P_\varepsilon x$$

where

$$0 = A' P_\varepsilon + P_\varepsilon A - P_\varepsilon B B' P_\varepsilon + Q_\varepsilon, \tag{6.12}$$

and where Q_ε *is a continuously differentiable matrix-valued function such that* $Q_\varepsilon > 0$, $\frac{dQ_\varepsilon}{d\varepsilon} > 0$ *for any* $\varepsilon \in (0, 1]$, *and* $\lim_{\varepsilon\to 0} Q_\varepsilon = 0$.

Proof : As we have seen in Chap. 4, the conditions that, (A, B) is stabilizable and the eigenvalues of A are in the closed left half plane, are necessary for semi-global stabilization even if we only have magnitude saturation but not rate saturation. Therefore, obviously, they are still necessary when we have magnitude saturation and rate limits.

To prove that these conditions are also sufficient, it is obviously sufficient to verify that the given family of feedback laws has the desired properties. In other words, while utilizing the given family of feedback laws, we need to show that for each given set $\mathcal{X}_0 \times \mathcal{X}_s$, there exists an $\varepsilon^* > 0$ such that, for all $\varepsilon \in (0, \varepsilon^*]$, we have local asymptotic stability with $\mathcal{X}_0 \times \mathcal{X}_s$ contained in its domain of attraction. In view of Theorem 4.21, we have

$$\|F_\varepsilon e^{(A+BF_\varepsilon)t}\| \leqslant \zeta_\varepsilon e^{-\eta_\varepsilon t}. \tag{6.13}$$

Here $F_\varepsilon = -B' P_\varepsilon$ where ζ and η depend continuously on ε and are positive for $\varepsilon > 0$. Moreover $\zeta_0 = \eta_0 = 0$. Note that we also find

$$\|F_\varepsilon (A + BF_\varepsilon) e^{(A+BF_\varepsilon)t}\|_\infty \leqslant \bar{\zeta}_\varepsilon e^{-\eta_\varepsilon t}, \tag{6.14}$$

where $\bar{\zeta}$ also depends continuously on ε, is positive for $\varepsilon > 0$, and satisfies $\bar{\zeta}_0 = 0$. The latter follows from the fact that

$$\begin{aligned}\|F_\varepsilon (A + BF_\varepsilon) e^{(A+BF_\varepsilon)t}\|_\infty &= \|F_\varepsilon e^{(A+BF_\varepsilon)t}(A + BF_\varepsilon)\|_\infty \\ &\leqslant \|(A + BF_\varepsilon)\| \zeta_\varepsilon e^{-\eta_\varepsilon t}.\end{aligned}$$

Also, we can rewrite (6.9) as

$$\dot{x} = Ax + B\sigma_{\Delta_1,\Delta_2}(F_\varepsilon x). \tag{6.15}$$

In the absence of saturation elements, the above system takes the form,

$$\dot{x} = (A + BF_\varepsilon)x. \tag{6.16}$$

It then follows from (6.13) and (6.14) that there exists an $\varepsilon_1^* > 0$ such that for all $\varepsilon \in (0, \varepsilon_1^*]$, we have

$$\|F_\varepsilon x\|_\infty \leqslant \Delta_1, \quad \|F_\varepsilon \dot{x}\|_\infty \leqslant \Delta_2 \quad \text{for all } x(0) \in \mathcal{X}_0 \text{ and } x_s \in \mathcal{X}_s.$$

This shows that for all $\varepsilon \in (0, \varepsilon_1^*]$ and for all $x(0) \in \mathcal{X}_0$ and $x_s \in \mathcal{X}_s$, system (6.15) operates in the linear regions of saturation elements, and hence, we can conclude that the equilibrium $x = 0$ and $x_s = 0$ of the system (6.15) is asymptotically stable with $\mathcal{X}_0 \times \mathcal{X}_s$ contained in its basin of attraction. ■

Next, we present the discrete-time version of the above result.

Theorem 6.6 *Consider the discrete-time system* (6.8) *along with the definition for the constraint operator $\sigma_{\Delta_1,\Delta_2}$ given in* (6.3). *The semi-global stabilization Problem 6.2 via linear state feedback laws is solvable if and only if (A, B) is stabilizable and the eigenvalues of A are in the closed unit disc.*

Moreover, in that case, the semi-global stabilization problem via linear state feedback is solved by the family of feedback laws

$$u = F_\varepsilon x = -(B'P_\varepsilon B + I)^{-1}B'P_\varepsilon Ax$$

where

$$P_\varepsilon = A'P_\varepsilon A - A'P_\varepsilon B(B'P_\varepsilon B + I)^{-1}B'P_\varepsilon A + Q_\varepsilon, \tag{6.17}$$

and where Q_ε is a continuously differentiable matrix-valued function such that $Q_\varepsilon > 0$, $\frac{dQ_\varepsilon}{d\varepsilon} > 0$ for any $\varepsilon \in (0, 1]$, and $\lim_{\varepsilon \to 0} Q_\varepsilon = 0$.

Proof : It follows directly using the same arguments as in the proof of Theorem 6.5 and is based on Theorem 4.25. ■

The solvability conditions for the semi-global stabilization problem via measurement feedback are given in the following theorem.

Theorem 6.7 *Consider a continuous- or discrete-time system* (6.8) *along with the definition for the constraint operator $\sigma_{\Delta_1,\Delta_2}$ given in* (6.3), (6.5), *and* (6.6).

The semi-global stabilization Problem 6.3 via linear measurement feedback laws is solvable if and only if (A, B) is stabilizable, A has all its eigenvalues in the closed left half plane (continuous time) or in the closed unit disc (discrete time), and the pair (C_y, A) is detectable.

Moreover, in that case, a suitable family of linear static measurement feedback laws is given by

$$\begin{aligned} \rho\hat{x} &= A\hat{x} + B\sigma_{\Delta_1,\Delta_2}(u) + KC_y(x - \hat{x}) \\ u &= F_\varepsilon \hat{x}, \end{aligned} \tag{6.18}$$

where

$$F_\varepsilon = -B' P_\varepsilon$$

with P_ε defined by (6.12) *in continuous time, while in discrete time*

$$F_\varepsilon = -(B' P_\varepsilon B + I)^{-1} B' P_\varepsilon A$$

with P_ε defined by (6.17)*. The gain K is chosen such that the matrix $A - KC_y$ is Hurwitz stable for continuous-time systems and Schur stable for discrete-time systems.*

Proof : We prove this theorem only for continuous time. The discrete-time version can be derived similarly. Consider the family of feedback laws given in (6.18). We know that (6.13) and (6.14) are satisfied for $F_\varepsilon = -B' P_\varepsilon$ where $\zeta_\varepsilon, \overline{\zeta}_\varepsilon$ and η_ε are continuous, positive-valued function which are equal to zero for $\varepsilon = 0$.

The closed-loop system consisting of the given system (6.8) and the given family of feedback laws can be written as

$$\begin{aligned} \dot{x} &= Ax + B\sigma_{\Delta_1,\Delta_2}(F_\varepsilon \hat{x}) \\ \dot{\hat{x}} &= A\hat{x} + B\sigma_{\Delta_1,\Delta_2}(F_\varepsilon \hat{x}) + KC_y(x - \hat{x}). \end{aligned} \tag{6.19}$$

As usual, we then adopt the invertible change of variable $e = x - \hat{x}$ and then rewrite the closed-loop system (6.19) as

$$\begin{aligned} \dot{x} &= Ax + B\sigma_{\Delta_1,\Delta_2}(F_\varepsilon x - F_\varepsilon e) \\ \dot{e} &= (A - KC_y)e. \end{aligned} \tag{6.20}$$

Recalling that the matrix $A - KC_y$ is Hurwitz stable, it readily follows that there exists a $T_1 \geqslant 0$ such that, for all possible initial conditions $e(0)$,

$$\|F_\varepsilon e\|_{\infty,\mathrm{T}_1} \leqslant \frac{\Delta_1}{4}, \ \|F_\varepsilon \dot{e}\|_{\infty,\mathrm{T}_1} \leqslant \frac{\Delta_2}{4}, \tag{6.21}$$

for all $\varepsilon \in (0, 1]$. We next consider the first equation of (6.20). Note that $x(T_1)$ belongs to a bounded set independent of ε since $x(0)$ is bounded and since x

is determined via a linear differential equation with bounded input $\sigma_{\Delta_1,\Delta_2}(u)$. Hence, there exists an M_1 such that for all possible initial conditions,

$$\|x(T_1)\| \leqslant M_1, \quad \text{for all } \varepsilon \in (0,1]. \tag{6.22}$$

Let us now assume that, from time T_1 onward, the saturation elements are non-existent. In this case, the first equation of (6.20) can be written as

$$\dot{x} = (A + BF_\varepsilon)x - BF_\varepsilon e. \tag{6.23}$$

Since $e \to 0$ exponentially with a decay rate independent of ε as $t \to \infty$, it follows trivially from (6.13) and (6.14) that there exist an $\varepsilon_1^* > 0$ and an $M_2 > 0$ such that for any possible initial condition $e(0)$,

$$\int_{T_1}^{\infty} \|e^{\eta_\varepsilon \tau} BF_\varepsilon e(\tau)\| \mathrm{d}\tau \leqslant M_2. \tag{6.24}$$

This in turn shows that for $t \geqslant T_1$,

$$\begin{aligned}
\|F_\varepsilon x(t)\| &= \left\| F_\varepsilon e^{(A+BF_\varepsilon)t} x(T_1) - \int_{T_1}^{t} F_\varepsilon e^{(A+BF_\varepsilon)(t-\tau)} BF_\varepsilon e(\tau) \mathrm{d}\tau \right\| \\
&\leqslant \zeta_\varepsilon M_1 + \zeta_\varepsilon \int_{T_1}^{\infty} \|e^{\eta_\varepsilon \tau} BF_\varepsilon e(\tau)\| \mathrm{d}\tau \\
&\leqslant \zeta_\varepsilon (M_1 + M_2).
\end{aligned}$$

Choose $\varepsilon_2^* \in (0, \varepsilon_1^*]$ such that for all $\varepsilon \in (0, \varepsilon_2^*]$,

$$\|F_\varepsilon x\|_{\infty,T_1} \leqslant \frac{\Delta_1}{4}. \tag{6.25}$$

Similarly, we can show that there exists an $\varepsilon_3^* \in (0, \varepsilon_2^*]$ such that for all $\varepsilon \in (0, \varepsilon_3^*]$,

$$\|F_\varepsilon \dot{x}\|_{\infty,T_1} \leqslant \frac{\Delta_2}{4}. \tag{6.26}$$

These two bounds, together with (6.21), show that the system (6.20) will operate linearly after time T_1 and local exponential stability of this linear system follows from the separation principle.

In summary, we have shown that there exists an $\varepsilon_3^* > 0$ such that, for all $\varepsilon \in (0, \varepsilon_3^*]$, the equilibrium point $(0,0,0)$ of the system (6.20) is asymptotically stable, with $(\mathcal{X}_0, \mathcal{X}_s, \mathcal{Z}_0)$ contained in its basin of attraction. This completes our proof. ■

As in the previous chapter, the low-gain technique presented in this section basically avoids saturation by squeezing the gain. This of course results in a very slow transient response. We present next an additional high gain that greatly improves the transient response.

6.5 Semi-global stabilization via low-and-high-gain feedback

As already seen in the previous chapter, low-gain-based designs underutilize the available control capacity, and the resulting convergence of the state to zero is very slow. Clearly, in the case of continuous time, the state feedback law $u = -B'P_\varepsilon x$ utilizes a low gain since P_ε as given by (6.12) converges to zero as ε becomes small. The same holds true for the measurement feedback designs which are based on the same low-gain state feedback. Similar discussion holds also for discrete time. Our next goal is to recall another design methodology which incorporates a significant improvement to the low-gain design method and leads to a better utilization of the available control capacity and hence better closed-loop performance.

The improved design utilizes the concepts of low-and-high-gain feedback as presented in the previous two chapters. Let us consider first continuous-time systems. Clearly, as mentioned before, the low-gain feedback $u = -B'P_\varepsilon x$ for the continuous-time system (6.8) achieves stability, and the resulting domain of attraction is arbitrarily large for ε small enough. When only magnitude saturation is present, the modified feedback $u = -(\alpha + 1)B'P_\varepsilon x$ was shown to achieve stability and the same domain of attraction for any $\alpha > 0$ and, moreover, has improved transient performance. However, such a modified feedback cannot be applied here because it is easy to construct examples to show that, in the case of rate limits, for α large, the domain of attraction can become arbitrarily small. The main problem is the fact that the rate limiter has memory. As such, if u was large for some time, then it takes a while before the input can become negative again and this delay causes the instability for large α. Therefore, at first, present here a different low-gain design which is more suited for this low-and-high-gain methodology.

Consider the continuous-time system (6.8). We introduce the following modified system:

$$\begin{aligned} \dot{x}(t) &= Ax(t) + Bu(t) \\ \dot{u}(t) &= \sigma_{\Delta_2}(v(t)). \end{aligned} \tag{6.27}$$

Then, for this system, we derive a low-gain state feedback that solves the semi-global stabilization problem. Let $P_{\varepsilon,\zeta}$ be the solution of the following continuous-time algebraic Riccati equation (CARE):

$$0 = \begin{pmatrix} A & B \\ 0 & 0 \end{pmatrix}' P + P \begin{pmatrix} A & B \\ 0 & 0 \end{pmatrix} - P \begin{pmatrix} 0 & 0 \\ 0 & I \end{pmatrix} P + \begin{pmatrix} Q_\varepsilon & 0 \\ 0 & \zeta^2 I \end{pmatrix}. \tag{6.28}$$

Then, the feedback law

$$v = -(\alpha + 1)\begin{pmatrix} 0 & I \end{pmatrix} P_{\varepsilon,\zeta} \begin{pmatrix} x \\ u \end{pmatrix} \tag{6.29}$$

for any fixed ζ solves the semi-global stabilization problem for the system (6.27) in the sense that the closed-loop system is asymptotically stable, and for ε small enough, the domain of attraction can be chosen arbitrarily large. Hence, by choosing α large, we can improve the transient performance without affecting the domain of attraction. Then, we can apply the following dynamic state feedback to the original system (6.8):

$$\dot{u} = \sigma_{\Delta_2}\left(-(\alpha + 1)\begin{pmatrix} 0 & I \end{pmatrix} P_{\varepsilon,\zeta} \begin{pmatrix} x \\ u \end{pmatrix} \right). \tag{6.30}$$

Obviously, this interconnection works beautifully if we only have rate constraints and no magnitude constraints because the interconnection of (6.8) and (6.30) for $\Delta_1 = \infty$ is equal to the interconnection of (6.27) and (6.29). In general, this is not the case because of the magnitude saturation. However, we will show that for a suitable choice for ε and ζ, the above feedback (6.30) has the desired properties when applied to the given system (6.8).

We have the following result.

Theorem 6.8 *Consider the continuous-time system* (6.8) *along with the definition for the constraint operator* $\sigma_{\Delta_1,\Delta_2}$ *given in* (6.5) *and* (6.6). *Let the solvability conditions for semi-global stabilization as in Theorem 6.5 prevail, i.e., let* (A, B) *be stabilizable and the eigenvalues of* A *be in the closed left half plane. Consider the family of controllers* (6.30). *Then, for any a priori given (arbitrarily large) bounded set* $\mathcal{X}_0 \subset \mathbb{R}^n$ *and any* $\zeta \in (0, 1)$, *there exists an* $\varepsilon^* > 0$ *such that for each* $\varepsilon \in (0, \varepsilon^*]$ *and for each* $\alpha \geqslant 0$, *the controller* (6.30) *solves the semi-global stabilization Problem 6.2, i.e., the interconnection of the controller* (6.30) *and* (6.8) *is locally exponentially stable with* $\mathcal{X}_0 \times \mathcal{X}_s \times [-\Delta_1, \Delta_1]^m$ *contained in its basin of attraction.*

Remark 6.9 *Note that for* $\alpha = 0$, $\zeta \to 0$, *and for a fixed* ε, *the controller converges to the low-gain feedback as presented in Theorem 6.5.*

Proof : There exists a compact set $\mathcal{X}_1$ such that $x(t) \in \mathcal{X}_1$ for all $t \leqslant \Delta_1/\Delta_2$, any $x(0) \in \mathcal{X}_0$, any input u, and any initial condition for the rate limiter (because the input to the system is bounded).

Choose the Lyapunov function

$$V_\varepsilon(x,u) = \begin{pmatrix} x' & u' \end{pmatrix} P_{\varepsilon,\zeta} \begin{pmatrix} x \\ u \end{pmatrix}$$

and let $c \geqslant 0$ be such that

$$\sup\{\, V_\varepsilon(x,u) \mid x \in \mathcal{X}_1, u \in [-\Delta_1, \Delta_1]^m, \varepsilon \in (0,1] \,\} \leqslant c.$$

Next, we note that there exists an $\varepsilon_1^* \in (0,1]$ such that

$$\left\| \begin{pmatrix} 0 & I \end{pmatrix} P_{\varepsilon,\zeta} \begin{pmatrix} x \\ u \end{pmatrix} \right\|_\infty < \Delta_1$$

for all x, u such that $V(x,u) < c$ and

$$\left[\begin{pmatrix} 0 & I \end{pmatrix} P_{\varepsilon,\zeta} \begin{pmatrix} x \\ u \end{pmatrix} \right]_i > 0$$

for any $i = 1, \dots, m$ and all x, u such that $V(x,u) < c$ and such that $u_i \geqslant \Delta_1$ where $[\cdot]_i$ denotes the ith element of a vector. Because of symmetry, we then also have,

$$\left[\begin{pmatrix} 0 & I \end{pmatrix} P_{\varepsilon,\zeta} \begin{pmatrix} x \\ u \end{pmatrix} \right]_i < 0$$

for any $i = 1, \dots, m$ and all x, u such that $V(x,u) < c$ and such that $u_i \leqslant -\Delta_1$.

The existence of ε_1^* is guaranteed by the fact that

$$P_{\varepsilon,\zeta} \to \begin{pmatrix} 0 & 0 \\ 0 & \zeta I \end{pmatrix} \tag{6.31}$$

as $\varepsilon \to 0$.

The above properties imply that $u(0) \in [-\Delta_1, \Delta_1]^m$ and $V(x(t), u(t)) < c$ for all $t \in [0, T]$ guarantee that $u(T) \in [-\Delta_1, \Delta_1]^m$ for any $T > 0$. After all, if the ith coefficient of u becomes Δ_1, then the derivative is negative, and if the ith coefficient of u becomes $-\Delta_1$, then the derivative is positive.

Since $\|u(t)\|_\infty < \Delta_1$ and $\|\dot{u}(t)\|_\infty < \Delta_2$ for all $t \in [0, \Delta_1/\Delta_2]$, we find from the characteristics of the rate limiter that $(\sigma_{\Delta_1,\Delta_2}(u))(\Delta_1/\Delta_2) = u(\Delta_1/\Delta_2)$ independent of the initial conditions for u and the rate limiter. We also know that $x(\Delta_1/\Delta_2) \in \mathcal{X}_1$.

Consider the interconnection of (6.27) and (6.29) with the same initial conditions as the interconnection of (6.8) and (6.30) at time $t = \Delta_1/\Delta_2$. We can easily prove that the interconnection of (6.27) and (6.29) is stable while $V(x(t), u(t)) < c$ for all $t \in [\Delta_1/\Delta_2, \infty)$. Therefore, the interconnection is such that $\|u(t)\|_\infty$ is bounded by Δ_1 for all $t > \Delta_1/\Delta_2$. The latter then implies that the solution of the interconnection of (6.27) and (6.29) is equal to the solution of the stable interconnection of (6.8) and (6.30) for all $t > \Delta_1/\Delta_2$. This clearly implies stability and the required domain of attraction. ■

Next, we focus on discrete-time systems. Consider the discrete-time system (6.8). We introduce the following modified system:

$$\begin{aligned} x(k+1) &= Ax(k) + Bu(k) \\ u(k+1) &= u(k) + \sigma_{\Delta_2}(v(k)). \end{aligned} \tag{6.32}$$

Then for this system, we derive a low-gain state feedback law that solves the semi-global stabilization problem. Let $P_{\varepsilon,\zeta}$ be the solution of the discrete-time algebraic Riccati equation (DARE)

$$\begin{aligned} P = &\begin{pmatrix} A & B \\ 0 & I \end{pmatrix}' P \begin{pmatrix} A & B \\ 0 & I \end{pmatrix} + \begin{pmatrix} Q_\varepsilon & 0 \\ 0 & \zeta^2 I \end{pmatrix} \\ &- \begin{pmatrix} A & B \\ 0 & I \end{pmatrix}' P \begin{pmatrix} 0 \\ I \end{pmatrix} \left[\begin{pmatrix} 0 & I \end{pmatrix} P \begin{pmatrix} 0 \\ I \end{pmatrix} + I \right]^{-1} \begin{pmatrix} 0 \\ I \end{pmatrix}' P \begin{pmatrix} A & B \\ 0 & I \end{pmatrix}. \end{aligned} \tag{6.33}$$

Then the feedback law,

$$v = -(1+\alpha) F_{\varepsilon,\zeta} \begin{pmatrix} x \\ u \end{pmatrix}, \tag{6.34}$$

for a fixed ζ, where

$$F_{\varepsilon,\zeta} = \left[\begin{pmatrix} 0 & I \end{pmatrix} P_{\varepsilon,\zeta} \begin{pmatrix} 0 \\ I \end{pmatrix} + I \right]^{-1} \begin{pmatrix} 0 \\ I \end{pmatrix}' P_{\varepsilon,\zeta} \begin{pmatrix} A & B \\ 0 & I \end{pmatrix},$$

and where

$$\alpha \in \left[0, 2 \left\| \begin{pmatrix} 0 & I \end{pmatrix} P_{\varepsilon,\zeta} \begin{pmatrix} 0 \\ I \end{pmatrix} \right\|^{-1} \right] \tag{6.35}$$

solves the semi-global stabilization problem for the system (6.32) in the sense that the closed-loop system is asymptotically stable, while for ε small enough, the domain of attraction can be chosen arbitrarily large. Moreover, by choosing α, we can improve the transient performance without affecting the domain of attraction. Then, we can apply the following dynamic state feedback to the original system (6.8):

$$\rho u = u + \sigma_{\Delta_2}\left(-(F_{\varepsilon,\zeta} + \alpha K_{\varepsilon,\zeta}) \begin{pmatrix} x \\ u \end{pmatrix} \right). \tag{6.36}$$

Obviously, this interconnection works beautifully if we only have rate constraints and no magnitude constraints because the interconnection of (6.8) and (6.36) for $\Delta_1 = \infty$ is equal to the interconnection of (6.32) and (6.34). In general, this is not the case because of the magnitude saturation. However, we will show that for a suitable choice for ζ, the above feedback (6.36) has the desired properties when applied to the system (6.8).

We have the following result.

Theorem 6.10 *Consider the discrete-time system* (6.8) *along with the definition for the constraint operator* $\sigma_{\Delta_1,\Delta_2}$ *given in* (6.3)*. Assume that* B *is injective. Let the solvability conditions for semi-global stabilization as in Theorem 6.6 prevail, i.e., let* (A, B) *be stabilizable and the eigenvalues of* A *be in the closed unit disc. Consider the family of controllers* (6.36)*. Then, for any a priori given (arbitrarily large) bounded set* $\mathcal{X}_0 \subset \mathbb{R}^n$ *and any* $\zeta \in (0,1)$*, there exists an* $\varepsilon^* > 0$ *such that for each* $\varepsilon \in (0, \varepsilon^*]$ *and for each* α *satisfying* (6.35)*, the controller* (6.36) *solves the semi-global stabilization Problem 6.2, i.e., the interconnection of the controller* (6.36) *and* (6.8) *is locally exponentially stable with* $\mathcal{X}_0 \times \mathcal{X}_s \times [-\Delta_1, \Delta_1]^m$ *contained in its basin of attraction.*

Remark 6.11 *Note that for* $\alpha = 0$*,* $\zeta \to 0$*, and for a fixed* ε*, the controller converges to the low-gain feedback as presented in Theorem 6.6.*

Proof : The proof of this theorem can be obtained using the same kind of arguments as in the proof of Theorem 6.8. ■

We now proceed to discuss measurement feedback controllers while using low-and-high-gain feedback. In connection with continuous-time systems, it turns out that we can apply the same argument as before to improve the performance of measurement feedback controllers by combining the observer used in (6.18) with the low-and-high-gain state feedback controller presented in (6.30).

To start with, we define the following family of linear dynamic measurement feedback laws:

$$\begin{aligned}\dot{\hat{x}} &= A\hat{x} + B\sigma_{\Delta_1,\Delta_2}(u) + K(y - C_y\hat{x} - D_y u)\\ \dot{u} &= \sigma_{\Delta_2}\left(-(\alpha+1)\begin{pmatrix}0 & I\end{pmatrix}P_{\varepsilon,\zeta}\begin{pmatrix}\hat{x}\\ u\end{pmatrix}\right).\end{aligned} \tag{6.37}$$

However, we will see that we need stronger conditions on the observer. It is no longer sufficient to choose a fixed observer such that $A - KC_y$ is stable. Therefore, we choose the observer gain parameterized by ℓ.

Let R_ℓ be the solution of the dual algebraic Riccati equation,

$$0 = (A + \ell I)R_\ell + R_\ell(A + \ell I)' - R_\ell C_y' C_y R_\ell + I. \tag{6.38}$$

This Riccati equation has been used before in this book, and a desirable property of this Riccati equation has been obtained in Lemma 4.42. We choose the following observer gain:

$$K_\ell = R_\ell C_y'.$$

We have the following result.

Theorem 6.12 *Consider the continuous-time system* (6.8) *along with the definition for the constraint operator $\sigma_{\Delta_1,\Delta_2}$ given in* (6.5) *and* (6.6)*. Under the solvability conditions of Theorem 6.7, there exists, for any a priori given (arbitrarily large) bounded sets $\mathcal{X}_0 \subset \mathbb{R}^n$ and $\mathcal{Z}_0 \subset \mathbb{R}^n$ and any $\zeta \in (0,1)$, an $\varepsilon^* > 0$ such that for all $\alpha \geq 0$ and for all $\varepsilon \in (0,\varepsilon^*]$, there exists an ℓ^* such that for $\ell > \ell^*$, the interconnection of* (6.8) *and* (6.37) *with $K = K_\ell$ is locally exponentially stable with $\mathcal{X}_0 \times \mathcal{Z}_0 \times \mathcal{X}_s \times [-\Delta_1, \Delta_1]^m$ contained in its basin of attraction.*

Proof : The proof basically uses the same arguments as the proof of Theorem 6.8 in combination with the measurement feedback arguments of Theorem 4.43. ■

In connection with discrete-time systems, we use the combination of the observer used in (6.18) with the low-and-high-gain state feedback controller presented in (6.36).

We define the following family of linear dynamic measurement feedback laws:

$$\begin{aligned} \rho \hat{x} &= A\hat{x} + B\sigma_{\Delta_1,\Delta_2}(u) + K\left(y - C_y\hat{x} - D_y u\right) \\ \rho u &= u + \sigma_{\Delta_2}\left(-(1+\alpha)F_{\varepsilon,\zeta}\begin{pmatrix}\hat{x}\\ u\end{pmatrix}\right). \end{aligned} \tag{6.39}$$

We choose the observer gain K such that $A - KC_y$ has all its eigenvalues in the origin. We have the following result.

Theorem 6.13 *Consider the discrete-time system* (6.8) *with B injective, along with the definition for the constraint operator $\sigma_{\Delta_1,\Delta_2}$ given in* (6.3)*. Under the solvability conditions of Theorem 6.7 , there exists, for any a priori given (arbitrarily large) bounded sets $\mathcal{X}_0 \subset \mathbb{R}^n$ and $\mathcal{Z}_0 \subset \mathbb{R}^n$ and any $\zeta \in (0,1)$, an $\varepsilon^* > 0$ such that for each $\varepsilon \in (0,\varepsilon^*]$ and α satisfying* (6.35)*, the interconnection of* (6.8) *and* (6.39) *is locally exponentially stable with $\mathcal{X}_0 \times \mathcal{Z}_0 \times \mathcal{X}_s \times [-\Delta_1, \Delta_1]^m$ contained in its basin of attraction.*

Proof : The proof basically uses the same arguments as the proof of Theorem 6.10 in combination with the measurement feedback arguments of Theorem 4.46. ■

7

State and input constraints: Semi-global and global stabilization in admissible set

7.1 Introduction

Chapter 4 considers internal stabilization of linear systems subject to control magnitude constraints, while Chap. 6 considers the same, however, with both control magnitude and rate constraints. Although such constraints on control variables occur prominently, magnitude and rate constraints on state variables are also of a major concern in many plants. Nearly every application imposes constraints on state as well as control variables. We observe that dynamic models of physical systems are often nonlinear. Linear approximations of such nonlinear systems are obviously valid only in certain constraint regions of state and control spaces. In process control, state and control constraints arise from economic necessity of operating the plants near the boundaries of feasible regions. In connection with safety issues, state and control constraints are a major concern in many plants. In certain possibly hazardous systems, such as a nuclear power plant, safety limits on some variables are often imposed. The violations of such predetermined safety measures may cause system malfunction or even damage. This implies that magnitude constraints or bounds on states must be taken as integral parts of any control system design.

In the literature on internal stabilization, there have been some efforts to deal with state and input constraints utilizing the concept of positive invariant sets. Blanchini [10] gives a good overview of these efforts. The available tools presented in this line of work, however, are computationally very demanding and yield highly complex controllers. Model predictive control, which is a popular design technique for industrial processes [17, 93], also has been used to deal with constraints on states as well as inputs [95]. However, this technique is also intrinsically computationally intensive and therefore not suitable for systems with fast dynamics. Secondly, it is fundamentally a numerical tool and gives only limited insight into the structural properties and effects of constraints on a system.

In recent years, we and our students focused on stabilization of linear systems with both state and control magnitude and rate constraints [124, 125, 137, 167]. The emphasis in our work has been on identifying the structural properties of linear plants under which the semi-global and global stabilization problems are solvable. Whenever the required structural properties are satisfied, design

A. Saberi et al., *Internal and External Stabilization of Linear Systems with Constraints*, Systems & Control: Foundations & Applications, DOI 10.1007/978-0-8176-4787-2_7,

methodologies for semi-global and global stabilization follow from the constructive methods of proving the obtained results. These aspects of our work distinguish us from other works dealing with state and input constraints.

In this chapter, we impose constraints on both the input and state variables. Such constraints are modeled in terms of what is called a constrained output with its magnitude and rate of change (or increment in discrete-time case) required to be in some prescribed constraint sets. It follows then that the initial state of the system must also obviously be restricted since we cannot satisfy the constraints if the initial state of the system is arbitrary. For this reason, we then define what can be called as an *admissible set* of initial conditions which is the set of initial conditions that do not violate the constraints at the initial time. This leads us to formulate precisely both the semi-global and global stabilization problems in the admissible set.

It is logical to expect that the solvability conditions as well as the design of appropriate controllers to achieve semi-global and global stabilization in the admissible set depend very much on certain innate structural properties of the given linear system along with the constrained output. Indeed, certain structural properties of the mapping from the input vector to the constrained output vector play dominant roles in dictating what kind of stabilization is feasible and what is not feasible under what conditions. This leads us to develop here a *taxonomy of constraints* that categorizes and as such delineates the structural properties in different directions. Such a categorization of constraints paves the architecture of our development as given in subsequent chapters.

7.2 Problem formulations

In this section, we consider a general linear time-invariant system of the form,

$$\Sigma : \begin{cases} \rho x = Ax + Bu \\ y \;\;= C_y x + D_y u \\ z \;\;= C_z x + D_z u, \end{cases} \tag{7.1}$$

where ρx denotes $\frac{dx}{dt}$ for continuous-time case and $x(k+1)$ for discrete-time. We discuss next the nature of constraints we impose on state and control variables and then formulate precisely both the semi-global and global stabilization problems in the admissible set.

As usual, the general linear time-invariant system we deal with has $x \in \mathbb{R}^n$ as its state, $u \in \mathbb{R}^m$ as its control input, and $y \in \mathbb{R}^r$ as its measured output. Besides these variables, we introduce here another new variable $z \in \mathbb{R}^p$ termed as *constrained output*. This constrained output has as its components certain control inputs and certain state variables. That is, all the control inputs and state variables that need to be constrained are collected together to form the constrained output z. We model all the prescribed constraints by imposing that the magnitude of the constrained output z lies in some prescribed constraint set $\mathcal{S}$, and the rate of change (or increment in discrete-time case) of z lies in some other prescribed constraint set $\mathcal{T}$. In this way, the constrained output z and the constraint sets $\mathcal{S}$ and $\mathcal{T}$, as their names imply, replicate or reflect all the constraints.

Thus, we consider a linear time-invariant system as depicted in the block diagram of Fig. 7.1. Without loss of generality, we assume that the matrices,

$$\begin{pmatrix} C_z & D_z \end{pmatrix} \text{ and } \begin{pmatrix} B \\ D_z \end{pmatrix},$$

are surjective and injective, respectively.

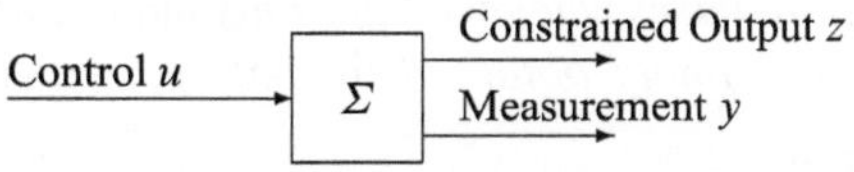

Figure 7.1: Linear system with control, measurement, and constrained output shown

For the system Σ of (7.1), our interest here and in subsequent chapters is on stabilization via state as well as measurement feedback. Namely, for two a priori given sets $\mathcal{S} \subset \mathbb{R}^p$ and $\mathcal{T} \subset \mathbb{R}^p$ which we alluded to as constraint sets, we are interested in the stabilization of the plant Σ subject to the requirement that the constrained output z remains in the set $\mathcal{S}$, while the derivative of the constrained output z (that is, $\dot{z}$) remains in the set $\mathcal{T}$ for continuous time or the increment $z(k+1) - z(k)$ remains in the set $\mathcal{T}$ for $k \geqslant 0$ for discrete time. In particular, we will be first interested in obtaining necessary and sufficient conditions for the existence of a feedback controller that achieves such a stabilization, and then designing appropriate controllers that achieve such a stabilization.

We discuss next the nature of the constraint sets $\mathcal{S} \subset \mathbb{R}^p$ and $\mathcal{T} \subset \mathbb{R}^p$. In general, the constraint sets $\mathcal{S}$ and $\mathcal{T}$ when discussed in the literature are nearly always bounded and convex. In fact, typically they are hypercubes. In this book as well, we essentially make the same assumption. However, one can slightly generalize the constraint sets to a broader class of convex sets which are not necessarily bounded. This generalization is possible in the light of an observation that any convex set can be uniquely projected to a bounded convex set. Thus, we assume that $\mathcal{S}$ and $\mathcal{T}$ are only convex sets, and we make the following fundamental assumptions on the nature of these constraint sets:

Assumption 7.1 *The following conditions on $\mathcal{S}$ and $\mathcal{T}$ are satisfied:*

(*i*) *The sets $\mathcal{S}$ and $\mathcal{T}$ are closed, convex, and contain* 0 *as an interior point.*

(*ii*) *$\mathcal{S} \cap \mathcal{T}$ is bounded.*

(*iii*) *We have* $C_z' D_z = 0$ *and*

$$\mathcal{S} = (\mathcal{S} \cap \operatorname{im} C_z) \oplus (\mathcal{S} \cap \operatorname{im} D_z)$$
$$\mathcal{T} = (\mathcal{T} \cap \operatorname{im} C_z) \oplus (\mathcal{T} \cap \operatorname{im} D_z)\,.$$

Remark 7.2 *The decomposition of* $\mathcal{S}$ *and* $\mathcal{T}$ *as required in* (*iii*) *implies simply that we have constraints on states and/or inputs, and we have no mixed constraints where allowable inputs depend on the current state and conversely. As such, we observe that* $\operatorname{im} C_z$ *reflects the state constraints while* $\operatorname{im} D_z$ *reflects the input constraints.*

Remark 7.3 *The special cases when we have only magnitude constraints or only rate constraints can be simply obtained by setting, respectively* $\mathcal{T} = \mathbb{R}^p$ *or* $\mathcal{S} = \mathbb{R}^p$.

Whenever $\mathcal{T} = \mathbb{R}^p$, Assumption 7.1 simplifies as given below.

Assumption 7.4 *The following conditions on* $\mathcal{S}$ *are satisfied:*

(*i*) *The set* $\mathcal{S}$ *is compact, convex, and contains* 0 *as an interior point.*

(*ii*) $\mathcal{S}$ *is bounded.*

(*iii*) *We have* $C_z' D_z = 0$ *and* $\mathcal{S} = (\mathcal{S} \cap \operatorname{im} C_z) \oplus (\mathcal{S} \cap \operatorname{im} D_z)$.

It is clear that the initial state of the system must obviously be restricted since we cannot satisfy the constraints if the initial state of the system is arbitrary. For this reason, we define an admissible set of initial conditions which is the set of initial conditions that do not violate the constraints at the initial time. It is straightforward to show that if the initial state does not belong to this set, then we can never satisfy our constraint requirements.

Definition 7.5 *Consider the system* (7.1) *along with the constraint sets* $\mathcal{S} \subset \mathbb{R}^p$ *and* $\mathcal{T} \subset \mathbb{R}^p$. *We define*

$$\mathcal{V}(\mathcal{S},\mathcal{T}) := \{\, x_0 \in \mathbb{R}^n \mid \exists\, u_0 \text{ such that } C_z x_0 + D_z u_0 \in \mathcal{S}, \text{ and } C_z(Ax_0 + Bu_0) \in \mathcal{T}\}$$

as the ***admissible set of initial conditions*** *for continuous-time case. Similarly, we define*

$$\mathcal{V}(\mathcal{S},\mathcal{T}) := \{\, x \in \mathbb{R}^n \mid \exists\, u_0 \text{ such that } C_z x_0 + D_z u_0 \in \mathcal{S}, \text{ and } [C_z(Ax_0 + Bu_0) - C_z x_0] \in \mathcal{T}\}$$

as the ***admissible set of initial conditions*** *for the discrete-time case.*

Remark 7.6 *For continuous-time case, in the derivative at time 0, we might expect a term $D_z\dot{u}(0)$ since the derivative of the input affects the derivative of the output z. However, we can omit this term because part (iii) of Assumption 7.1 implies that $C_z(Ax_0 + Bu_0) + D_z\dot{u}(0) \in \mathcal{T}$ if and only if $C_z(Ax_0 + Bu_0) \in \mathcal{T}$ and $D_z\dot{u}(0) \in \mathcal{T}$. However, at time 0, we do not impose rate constraints. Due to continuity, we still need to have $C_z(Ax_0 + Bu_0) \in \mathcal{T}$, but we do not need to impose a condition on the derivative of u since that need not be continuous.*

Similarly, for discrete-time case, in the increment at time 0, we might expect a term $D_z(u(1) - u(0))$ since the increment of the input affects the increment of the output z. However, we can omit this term because part (iii) of Assumption 7.1 implies that $C_z(Ax_0 + Bu_0) + D_z(u(1) - u(0)) \in \mathcal{T}$ if and only if $C_z(Ax_0 + Bu_0) \in \mathcal{T}$ and $D_z(u(1) - u(0)) \in \mathcal{T}$. Since we can assign the input at time 0, we can trivially guarantee $D_z(u(1) - u(0)) \in \mathcal{T}$ and it can be omitted from the above definition.

Remark 7.7 *Consider the case when there are no rate constraints, that is, when $\mathcal{T} = \mathbb{R}^p$. Then, in view of Assumption 7.1, the admissible set of initial conditions $\mathcal{V}(\mathcal{S}, \mathbb{R}^p)$ can be rewritten as*

$$\mathcal{V}(\mathcal{S}, \mathbb{R}^p) := \{ x_0 \in \mathbb{R}^n \mid C_z x_0 \in \mathcal{S} \}.$$

We proceed now to formulate precisely the internal stabilization problems either in semi-global or global setting. In a semi-global setting, we assume that the initial conditions are in some arbitrary compact set contained in the interior of the set of admissible initial conditions, and in the global setting, we consider arbitrary initial conditions in the set of admissible initial conditions. We consider both state and measurement feedback.

The following two problems pertain to state feedback controllers.

Problem 7.8 Consider the system (7.1) along with the constraint sets $\mathcal{S} \subset \mathbb{R}^p$ and $\mathcal{T} \subset \mathbb{R}^p$. The **semi-global stabilization in the admissible set via state feedback** is to find, if possible, for any a priori given compact set $\mathcal{W}$ contained in the interior of $\mathcal{V}(\mathcal{S}, \mathcal{T})$, a state feedback (possibly nonlinear and time varying) $u(t) = f(x(t), t)$ or $u(k) = f(x(k), k)$ such that the following conditions hold:

(*i*) The equilibrium point $x = 0$ of the closed-loop system is asymptotically stable with $\mathcal{W}$ contained in its basin of attraction.

(*ii*) In the case of continuous-time systems, for any $x_0 \in \mathcal{W}$, we have $z(t) \in \mathcal{S}$ for all $t \geq 0$ and $\dot{z}(t) \in \mathcal{T}$ for all $t > 0$. In the case of discrete-time systems, for any $x_0 \in \mathcal{W}$, we have $z(k) \in \mathcal{S}$ for all $k \geqslant 0$ and $(z(k+1) - z(k)) \in \mathcal{T}$ for all $k \geqslant 0$.

Problem 7.9 Consider the system (7.1) along with the constraint sets $\mathcal{S} \subset \mathbb{R}^p$ and $\mathcal{T} \subset \mathbb{R}^p$. The **global stabilization in the admissible set via state feedback** is to find, if possible, a state feedback (possibly nonlinear and time varying) $u(t) = f(x(t), t)$ or $u(k) = f(x(k), k)$ such that the following conditions hold:

(*i*) The equilibrium point $x = 0$ of the closed-loop system is asymptotically stable with $\mathcal{V}(\mathcal{S}, \mathcal{T})$ contained in its basin of attraction.

(*ii*) In the case of continuous-time systems, for any $x_0 \in \mathcal{V}(\mathcal{S}, \mathcal{T})$, we have $z(t) \in \mathcal{S}$ for all $t \geq 0$ and $\dot{z}(t) \in \mathcal{T}$ for all $t > 0$. In the case of discrete-time systems, for any $x_0 \in \mathcal{V}(\mathcal{S}, \mathcal{T})$, we have $z(k) \in \mathcal{S}$ for all $k \geqslant 0$ and $(z(k+1) - z(k)) \in \mathcal{T}$ for all $k \geqslant 0$.

The following two problems pertain to measurement feedback controllers.

Problem 7.10 Consider the system (7.1) along with the constraint sets $\mathcal{S} \subset \mathbb{R}^p$ and $\mathcal{T} \subset \mathbb{R}^p$. The **semi-global stabilization in the admissible set via measurement feedback** is to find, if possible, an integer q and, for any (arbitrarily large) a priori given compact sets $\mathcal{W}$ contained in the interior of the set $\mathcal{V}(\mathcal{S}, \mathcal{T}) \times \mathbb{R}^q$, a measurement feedback controller of the form,

$$\begin{aligned} \dot{p}(t) &= \ell(p(t), y(t), t), \quad p(t) \in \mathbb{R}^q \\ u(t) &= g(p(t), y(t), t) \end{aligned} \tag{7.2}$$

for continuous-time case or of the form,

$$\begin{aligned} p(k+1) &= \ell(p(k), y(k), k), \quad p(k) \in \mathbb{R}^q \\ u(k) &= g(p(k), y(k), k) \end{aligned} \tag{7.3}$$

for discrete-time case, such that the following conditions hold:

(*i*) The equilibrium point $(x, p) = (0, 0)$ of the closed-loop system comprising of the given system (7.1) and either the controller (7.2) (for continuous-time case) or the controller (7.3) (for discrete-time case) is asymptotically stable with $\mathcal{W}$ contained in its basin of attraction.

(*ii*) For any $(x_0, p_0) \in \mathcal{W}$, we have $z(t) \in \mathcal{S}$ for all $t \geqslant 0$ and $\dot{z}(t) \in \mathcal{T}$ for all $t > 0$ (for continuous-time case), or for any $(x_0, p_0) \in \mathcal{W}$, we have $z(k) \in \mathcal{S}$ for all $k \geqslant 0$ and $(z(k+1) - z(k)) \in \mathcal{T}$ for all $k \geqslant 0$ (for discrete-time case).

Problem 7.11 Consider the system (7.1) along with the constraint sets $\mathcal{S} \subset \mathbb{R}^p$ and $\mathcal{T} \subset \mathbb{R}^p$. The **global stabilization in the admissible set via measurement**

feedback is to find, if possible, an integer q and a measurement feedback controller of the form

$$\begin{aligned} \dot{p}(t) &= \ell(p(t), y(t), t), \quad p(t) \in \mathbb{R}^q \\ u(t) &= g(p(t), y(t), t) \end{aligned} \tag{7.4}$$

for continuous-time case or of the form

$$\begin{aligned} p(k+1) &= l(p(k), y(k), k), \quad p(k) \in \mathbb{R}^q \\ u(k) &= g(p(k), y(k), k) \end{aligned} \tag{7.5}$$

for discrete-time case, such that the following conditions hold:

(*i*) The equilibrium point $(x, p) = (0, 0)$ of the closed-loop system comprising of the given system (7.1) and either the controller (7.4) (for continuous-time case) or the controller (7.5) (for discrete-time case) is asymptotically stable with $\mathcal{V}(\mathcal{S}, \mathcal{T}) \times \mathbb{R}^q$ contained in its basin of attraction.

(*ii*) For any $(x_0, p_0) \in \mathcal{V}(\mathcal{S}, \mathcal{T}) \times \mathbb{R}^q$, we have $z(t) \in \mathcal{S}$ for all $t \geqslant 0$ and $\dot{z}(t) \in \mathcal{T}$ for all $t > 0$ (for continuous-time case), or for any $(x_0, p_0) \in \mathcal{V}(\mathcal{S}, \mathcal{T}) \times \mathbb{R}^q$, we have $z(k) \in \mathcal{S}$ for all $k \geqslant 0$ and $(z(k+1) - z(k)) \in \mathcal{T}$ for all $k \geqslant 0$ (for discrete-time case).

Remark 7.12 *In* (7.1)*, if* $C_z = 0$*, the above problems are referred to as* input-constrained stabilization problems*, while if* $D_z = 0$*, the above problems are referred to as* state-constrained stabilization problems.

7.3 A taxonomy of constraints

The taxonomy of constraints described in this section emerged and evolved over a number of years from various efforts and attempts we made to solve the semi-global and global stabilization problems for linear systems with both input and state constraints. We learned that the solvability and controller design for the above posed stabilization problems in the admissible set depend very much on the structural properties of the mapping from input u to the constrained output z of the given system (7.1). The mapping from input u to the constrained output z (denoted by Σ_{uz}) is characterized by the quadruple (A, B, C_z, D_z). The taxonomy of constraints developed in this section categorizes and as such delineates the structural properties of Σ_{uz} in three different directions, and it forms a basis in subsequent chapters for stating clearly the solvability conditions and accordingly designing appropriate stabilizing controllers.

The first category in the taxonomy of constraints is based on whether the system Σ_{uz} is right invertible or not.

Definition 7.13 *The constraints are said to be:*

- ***Right invertible constraints*** *if the system Σ_{uz} is right invertible.*
- ***Non-right-invertible constraints*** *if the system Σ_{uz} is non-right invertible.*

The following remark points out one important special case of right-invertible constraints, namely, the case of having **constraints only on actuators**.

Remark 7.14 *Let us assume that we have both magnitude and rate constraints only on actuators, i.e.,* ***constraints only on the control variable*** *u. For this special case, we have $z = u$, implying that $C_z = 0$ and $D_z = I_m$. Since Σ characterized by $(A, B, 0, I_m)$ can easily be verified to be right invertible, we note that the magnitude and rate constraints only on actuators are indeed right-invertible constraints.*

Remark 7.15 *In view of the above remark, it is clear that non-right-invertible constraints arise inherently due to magnitude and rate constraints on the state x.*

As we shall see in subsequent chapters, in order to study the posed semi-global and global stabilization problems, a further elaboration on the non-right-invertible constraints is needed.

Definition 7.16 *The constraints are said to be* ***weakly non-right invertible*** *if the constraints are non-right invertible and are such that*

$$C_z^{-1} \operatorname{im} D_z \subset \mathcal{V}^*(\Sigma_{uz}) + \mathcal{S}^*(\Sigma_{uz}), \tag{7.6}$$

where $\mathcal{V}^*(\Sigma_{uz})$ and $\mathcal{S}^*(\Sigma_{uz})$ are respectively the weakly unobservable subspace and the strongly controllable subspace as defined in Definition 3.22 with $\mathbb{C}_g$ being the entire complex plane $\mathbb{C}$. If the condition (7.6) does not hold, then the constraints are said to be ***strongly non-right invertible****.*

Remark 7.17 *In the SCB of Σ_{uz} (see Theorem 3.1), the condition (7.6) is equivalent to the matrix C_b being injective. Knowing this fact will be very helpful in understanding the design methodology to be developed later on. Also, note that for the case of right-invertible constraints, we always have*

$$\mathcal{V}^*(\Sigma_{uz}) + \mathcal{S}^*(\Sigma_{uz}) = \mathbb{R}^n,$$

and as such, (7.6) is always satisfied.

The second category in the taxonomy of constraints is based on the location of the invariant zeros of the system Σ_{uz}. Because of its importance, we specifically label the invariant zeros of the system Σ_{uz} as the *constraint invariant zeros* of the plant.

Definition 7.18 *The invariant zeros of the system Σ_{uz} are called the* ***constraint invariant zeros*** *of the plant associated with the constrained output z.*

Definition 7.19 *The constraints are said to be:*

- ***Minimum-phase constraints*** *if all the constraint invariant zeros are in the open left-half complex plane $\mathbb{C}^-$ for continuous-time systems or within the unit circle $\mathbb{C}^{\ominus}$ for discrete-time systems.*
- ***Weakly minimum-phase constraints*** *if all the constraint invariant zeros are in the closed left-half complex plane $\mathbb{C}^{-0}$ for continuous-time systems or within as well as on the unit circle $\mathbb{C}^{\otimes}$ for discrete-time systems with one restriction that at least one such constraint invariant zero is on the imaginary axis (continuous time) or on the unit circle (discrete time) and any such constraint invariant zero is simple.*
- ***Weakly non-minimum-phase constraints*** *if all the constraint invariant zeros are in the closed left half complex plane $\mathbb{C}^{-0}$ for continuous-time systems or within as well as on the unit circle $\mathbb{C}^{\otimes}$ for discrete-time systems, and at least one constraint invariant zero which is on the imaginary axis (continuous time) or on the unit circle (discrete time) is not simple.*
- ***Strongly non-minimum-phase constraints*** *if one or more of the constraint invariant zeros are in the right hand complex plane $\mathbb{C}^+$ for continuous-time systems or outside the unit circle $\mathbb{C}^{\oplus}$ for discrete-time systems.*

Remark 7.20 *Whenever we say that the constraints are* ***at most weakly non-minimum phase****, we mean that either the constraints are minimum phase, or weakly minimum phase, or weakly non-minimum phase, but not strongly non-minimum phase.*

The third categorization is based on the order of the infinite zeros of the system Σ_{uz}. Because of its importance, we specifically label the infinite zeros of the system Σ_{uz} as the constraint infinite zeros of the plant.

Definition 7.21 *The infinite zeros of the subsystem Σ_{uz} are called the* ***constraint infinite zeros*** *of the plant associated with the constrained output z.*

Definition 7.22 *The constraints are said to be* ***type one constraints*** *if the order of all constraint infinite zeros is less than or equal to one.*

The impact of the taxonomy of constraints as developed above in solving the posed semi-global and global stabilization problems in the admissible set is explored in subsequent chapters which clearly display what is feasible under (1) right- or non-right-invertible constraints, (2) minimum- and non-minimum-phase constraints, and (3) the type of infinite zero order.

8

Solvability conditions and design for semi-global and global stabilization in the admissible set

8.1 Introduction

In Chap. 7, we formulated two important problems, (1) the semi-global stabilization problem in the admissible set and (2) the global stabilization problem in the admissible set. Moreover, based on the structural properties of the mapping from the control input to the constrained output, a taxonomy of constraints was developed there. In view of such a taxonomy, this chapter concentrates on semi-global as well as global stabilization problems in the admissible set. The nature and solvability of these stabilization problems as well as appropriate design of controllers differ profoundly for the two different cases of right and non-right-invertible constraints. Because of this, we consider here these two cases separately. In particular, we consider the case of right-invertible and non-right-invertible constraints, respectively, in Sects. 8.2 and 8.3 for continuous-time systems. Similarly, we consider the same in Sects. 8.4 and 8.5 for discrete-time systems. This chapter is primarily based on our work in [124, 125, 137].

8.2 Semi-global and global stabilization in admissible set for right-invertible constraints: continuous time

In this section, for continuous-time linear system Σ as described in (7.1), we study in detail both the semi-global and global stabilization in the admissible set as formulated in Problems 7.8 and 7.9 for state feedback and in Problems 7.10 and 7.11 for measurement feedback. We provide here the necessary and sufficient conditions for the solvability of Problems 7.8 and 7.9 and sufficient conditions for the solvability of Problem 7.10 whenever the constraints are right invertible, as defined in Definition 7.13. The global measurement feedback problem 7.11 will not be discussed in this book because the solvability conditions are extremely restrictive. Once the solvability conditions are formulated, we also develop here constructive methods of designing appropriate controllers that achieve semi-global or global stabilization as required.

A. Saberi et al., *Internal and External Stabilization of Linear Systems with Constraints*, Systems & Control: Foundations & Applications, DOI 10.1007/978-0-8176-4787-2_8,

We have the following theorem which is concerned with the semi-global stabilization Problem 7.8 via state feedback.

Theorem 8.1 *Consider the continuous-time system* Σ *as given by* (7.1) *and constraint sets* $\mathcal{S}$ *and* $\mathcal{T}$ *that satisfy Assumption 7.1. Assume that the constraints are right invertible. Then, the semi-global stabilization problem in the admissible set via state feedback as defined in Problem 7.8 is solvable if and only if the following conditions hold:*

(*i*) (A, B) *is stabilizable.*

(*ii*) *The constraint invariant zeros of the given system* Σ *are all in the closed left-half complex plane, i.e., the system* Σ *has at most weakly non-minimum-phase constraints.*

We have the following theorem which is concerned with the global stabilization Problem 7.9 via state feedback.

Theorem 8.2 *Consider the continuous-time system* Σ *as given by* (7.1) *and constraint sets* $\mathcal{S}$ *and* $\mathcal{T}$ *that satisfy Assumption 7.1. Assume that we only have amplitude constraints, i.e.,* $\mathcal{T} = \mathbb{R}^p$ *and that the constraints are right invertible. Then the global stabilization problem in the admissible set via state feedback as defined in Problem 7.9 is solvable if and only if the following conditions hold:*

(*i*) (A, B) *is stabilizable.*

(*ii*) *The constraint invariant zeros of the given system* Σ *are all in the closed left-half complex plane, i.e., the system* Σ *has at most weakly non-minimum-phase constraints.*

(*iii*) *The constraints are of type one, i.e., the given system* Σ *has no constraint infinite zeros of order greater than one.*

Remark 8.3 *In case of rate constraints, the above problem is an open problem. This should not be surprising since even the global stabilization problem by static state feedback with amplitude and rate constraints on the input is basically open.*

We now move on to the case of measurement feedback that concerns with the semi-global stabilization. We have the following theorem.

Theorem 8.4 *Consider the continuous-time system* Σ *as given by* (7.1) *and constraint sets* $\mathcal{S}$ *and* $\mathcal{T}$ *that satisfy Assumption 7.1. Assume that the constraints are* *right invertible. Then, the semi-global stabilization problem in the admissible set via measurement feedback as defined in Problem 7.10 is solvable if the following conditions hold:*

(i) (A, B) *is stabilizable.*

(ii) *The constraint invariant zeros of the given system* Σ *are all in the closed left-half plane, i.e., the system* Σ *has at most weakly non-minimum-phase constraints.*

(iii) *The pair* (C_y, A) *is observable.*

Moreover, conditions (i) and (ii) are necessary for the solvability of the semi-global stabilization problem in the admissible set via measurement feedback.

The following remark addresses the need for condition (*iii*).

Remark 8.5 *Some discussion regarding the condition (iii) in Theorem 8.4 is in order. Clearly, condition (iii) is sufficient but not necessary. Also, obviously, the detectability of the pair* (C_y, A) *is necessary to solve the posed problem. On the other hand, if the pair* (C_y, A) *is not observable, we can have a problem with unobservable dynamics which we cannot observe but which might result in violation of our constraints. Relaxing the condition (iii) and developing a set of necessary and sufficient conditions are not very difficult but highly technical and not that interesting.*

The following remark points out that the specific features of the given constraint sets $\mathcal{S}$ and $\mathcal{T}$ do not have any role in the solvability of semi-global and global stabilization in the admissible set for right-invertible constraints.

Remark 8.6 *We emphasize here a fundamental aspect of solvability conditions as given by Theorems 8.1, 8.2, and 8.4, namely, that they are independent of any specific features of the given constraint sets* $\mathcal{S}$ *and* $\mathcal{T}$*. That is, for the case of right-invertible constraints, if the semi-global and global stabilization problems in the admissible set are solvable for some given constraint sets* $\mathcal{S}$ *and* $\mathcal{T}$ *satisfying Assumption 7.1, then such semi-global and global stabilization problems are also solvable for all constraint sets satisfying Assumption 7.1.*

Remark 7.14 dictates that the constraints are right invertible whenever we have constraints only on actuators. The following remark exemplifies several aspects of having constraints only on actuators.

Remark 8.7 *Consider the case when we have constraints only on actuator magnitude and rate, i.e., the case when* $C_z = 0$. *In other words, there are no state constraints, and only a subset of the input channels is subject to magnitude and rate constraints. Then, as Remark 7.14 points out, the constraints are right invertible. Also, in this case, it is straightforward to show that the constraint invariant zeros of* Σ *(i.e., the invariant zeros of the system* Σ *characterized by* $(A, B, 0, D_z)$*) coincide with a subset of the eigenvalues of* A. *This observation implies that the requirement of at most weakly non-minimum-phase constraints in Theorems 8.1, 8.2, and 8.4 is equivalent to requiring that a particular subset of eigenvalues of* A *lies in the closed left-half plane. Obviously, such a condition is always satisfied if we are dealing with critically unstable systems. Let us next assume that the given system* Σ *is controllable via the unconstrained input channels, it is then straightforward to see that the system characterized by* $(A, B, 0, D_z)$ *does not have any invariant zeros, i.e.,* Σ *does not have any constraint invariant zeros. Thus, for this special case, obviously, there will not be any constraints on the eigenvalues of* A *as can be expected.*

It is worthwhile to consider another special case of the above when all the input channels are subject to magnitude and rate constraints which is the case considered in Chap. 6. In this case, $C_z = 0$ *and* $D_z = I_m$. *For this special case, we observe that the admissible set of initial conditions* $\mathcal{V}(\mathcal{S}, \mathcal{T})$ *is indeed* $\mathbb{R}^n$. *Moreover, then the constraint invariant zeros of* Σ *coincide with all the eigenvalues of* A. *As such, the requirement of at most weakly non-minimum-phase constraints in Theorems 8.1, 8.2, and 8.4 is equivalent to requiring that the given system be critically unstable. Furthermore, for this special case, it is easy to see that there are no constraint infinite zeros of order greater than 1, and hence, the condition (iii) of Theorem 8.2 is automatically satisfied.*

In view of the above observations, for this special case when $C_z = 0$ *and* $D_z = I_m$, *the conditions of Theorems 8.1, 8.2, and 8.4 can be restated as follows: Under the assumptions of Theorem 8.1, the necessary and sufficient conditions for semi-global stabilization via state feedback are indeed (1)* (A, B) *is stabilizable and (2) all the eigenvalues of* A *lie in the closed left-half plane (i.e., the given system is critically unstable). Under the assumptions of Theorem 8.4, for semi-global stabilization via measurement feedback, the sufficient conditions are (1)* (A, B) *is stabilizable, (2) all the eigenvalues of* A *lie in the closed left-half plane (i.e., the given system is critically unstable), and (3) the pair* (C_y, A) *is observable. All the above results coincide with what has been reported earlier in Chaps. 4 and 6.*

Let us next recall one of the fundamental and important facts that emerged in Chap. 4. Namely, for systems with only input saturation, in general global stabilization requires nonlinear feedback laws, while semi-global stabilization can be achieved whenever it can be done by utilizing simply linear time-invariant feedback laws. Then, a question that arises naturally is whether an analogous result is valid under a broad framework of state as well as input constraints that are being considered in this chapter. The following theorem answers this question:

Theorem 8.8 *Consider the continuous-time system Σ as given by* (7.1) *and constraint sets $\mathcal{S}$ and $\mathcal{T}$ that satisfy Assumption 7.1. Assume that the constraints are right invertible. Then the following hold:*

(i) *Under the condition that* $\operatorname{im} C_z \subset \mathcal{T}$ *(i.e., no rate constraints on states), if a semi-global stabilization problem in the admissible set via state feedback as defined in Problem 7.8 is solvable, then it is also solvable via a linear time-invariant state feedback law.*

(ii) *If* $\operatorname{im} C_z \not\subset \mathcal{T}$ *(i.e., rate constraints on states are present), whenever a semi-global stabilization problem in the admissible set via state feedback as defined in Problem 7.8 is solvable, in general it might not be solvable via a linear time-invariant state feedback law. That is, there exist a system Σ as given by* (7.1) *and constraint sets $\mathcal{S}$ and $\mathcal{T} \not\supset \operatorname{im} C_z$ that satisfy Assumption 7.1 for which the semi-global stabilization problem is solvable via a nonlinear feedback law but for which there exists no linear feedback law that solves the problem.*

8.2.1 Proofs and construction of controllers

The proofs of all theorems as well as construction of appropriate controllers rely on *one* specific decomposition of the system. This decomposition is nothing else than the decomposition related to the special coordinate basis (SCB) as presented in Chap. 3 (see Theorem 3.1). However, for the presentation of the results here, a compact form of SCB suffices and is given below.

For the right-invertible constraints considered in this section, by considering the SCB for the system Σ_{uz} characterized by the quadruple (A, B, C_z, D_z), one can find suitable transformation matrices Γ_s and Γ_u such that (7.1) can be rewritten in a compact form as

$$
\begin{aligned}
\begin{pmatrix} \dot{x}_1 \\ \dot{x}_2 \end{pmatrix} &= \begin{pmatrix} A_{11} & 0 \\ A_{21} & A_{22} \end{pmatrix}\begin{pmatrix} x_1 \\ x_2 \end{pmatrix} + \begin{pmatrix} 0 \\ B_2 \end{pmatrix}\tilde{u} + \begin{pmatrix} K_1 \\ K_2 \end{pmatrix} z \\
z &= \begin{pmatrix} 0 & C_{z,2} \end{pmatrix}\begin{pmatrix} x_1 \\ x_2 \end{pmatrix} + \tilde{D}_z \tilde{u},
\end{aligned}
\tag{8.1}
$$

where

$$
\begin{pmatrix} x_1 \\ x_2 \end{pmatrix} = \tilde{x} = \Gamma_s^{-1} x \quad \text{and} \quad \tilde{u} = \Gamma_u^{-1} u.
$$

Here $x_1 \in \mathbb{R}^{n_1}$ and $x_2 \in \mathbb{R}^{n_2}$ with $n_1 + n_2 = n$. Note that no transformation is done on the constrained output z in order to preserve the constraint sets $\mathcal{S}$ and $\mathcal{T}$. We observe also that x_a in the SCB is renamed here as x_1, x_b is nonexistent for systems with right-invertible constraints, while the rest of the states x_c and x_d form x_2.

We emphasize that many submatrices given in (8.1) have definite structure which we judiciously point out as the need arises. At this time, we point out that the subsystem characterized by the quadruple $(A_{22}, B_2, C_{z,2}, \widetilde{D}_z)$ is strongly controllable[1] and has no constraint invariant zeros. Also, there exists a matrix H such that $A_{21} = B_2 H$ and $\widetilde{D}_z H = 0$. Moreover, the eigenvalues of A_{11} are equal to the constraint invariant zeros of the system Σ.

We can extract from (8.1) two subsystems. The first subsystem is given by

$$\dot{x}_1 = A_{11}x_1 + K_1 z, \quad x_1 \in \mathbb{R}^{n_1}. \tag{8.2}$$

The second subsystem extracted from (8.1) is given by

$$\begin{aligned} \dot{x}_2 &= A_{22}x_2 + B_2(\widetilde{u} + Hx_1) + K_2 z, \qquad x_2 \in \mathbb{R}^{n_2}; \\ z &= C_{z,2}x_2 + \widetilde{D}_z \widetilde{u}. \end{aligned} \tag{8.3}$$

The above two subsystems (8.2) and (8.3) form the system Σ_{uz}. However, in view of (8.1), it is important to point out that the state and input constraints exist only on the second subsystem (8.3) and there are no constraints whatsoever on subsystem (8.2). As such, the admissible set of initial conditions $\mathcal{V}(\mathcal{S}, \mathcal{T})$ for the given system Σ can simply be redefined as $\mathcal{V}_2(\mathcal{S}, \mathcal{T})$ for the subsystem (8.3):

$$\begin{aligned} \mathcal{V}_2(\mathcal{S}, \mathcal{T}) = \{\, x_{2,0} \in \mathbb{R}^{n_2} \mid \exists\, u_0 \text{ such that } z_0 \in \mathcal{S} \\ \text{and } C_{z,2}(A_{22}x_{2,0} + B_2 u_0 + K_2 z_0) \in \mathcal{T} \}, \end{aligned} \tag{8.4}$$

where $z_0 = C_{z,2}x_{2,0} + \widetilde{D}_z u_0$. Note that x_1 has no effect on the class of admissible initial conditions for the second subsystem since, in view of $\widetilde{D}_z H$ being zero, Hx_1 can be exactly canceled by a suitable choice of u_0.

We emphasize that the subsystem (8.2) represents the zero dynamics of the system Σ_{uz} characterized by the quadruple (A, B, C_z, D_z). Moreover, as mentioned above, the eigenvalues of A_{11} are equal to the constraint invariant zeros of the given system Σ.

Both for construction of controllers and the proofs of the stated results, we use the decomposition in the two subsystems (8.2) and (8.3) as defined above. We observe clearly that we can control the first subsystem (8.2) only through z. Also, from the SCB decomposition (see Theorem 3.1), as we alluded to earlier, it follows that the second subsystem characterized by the quadruple (A_{22}, B_2, $C_{z,2}$, $\widetilde{D}_z$) has no finite invariant zeros and is right invertible. This implies that we can guarantee by a suitable choice of $\widetilde{u}$ that z is arbitrarily close to any desired signal as will be evident soon. Therefore, we basically design a controller in two phases:

---

[1] Definition 3.28 discusses strong controllability.

- First, design a desired feedback for the first subsystem (8.2) using z as the (constrained) input signal such that the first subsystem exhibits a desired closed-loop behavior.
- Second, design a feedback for the second subsystem (8.3) with state x_2, input $\tilde{u}$, and output z such that
  - *(i)* The output z is close to the desired feedback for the first subsystem.
  - *(ii)* The output satisfies the constraints.
  - *(iii)* The state x_2 of the second subsystem exhibits a desirable behavior.

All feedback designs here are constructed in accordance with this two-phase design.

We need to discuss next what kind of initial conditions can be considered for the first subsystem (8.2). In fact, since we have no state constraints on this subsystem, we can have arbitrary initial conditions for it. Hence, we consider arbitrary initial conditions in $\mathcal{W}_1 = \mathbb{R}^{n_1}$ in the global case, while in the semi-global case we consider initial conditions in some arbitrary compact set $\mathcal{W}_1$.

Similarly, the initial conditions for the second subsystem must be in some set $\mathcal{W}_2$. In the global case, we have $\mathcal{W}_2 = \mathcal{V}_2(\mathcal{S}, \mathcal{T})$ using the definition in (8.4), while in the semi-global case we have $\mathcal{W}_2$ as an arbitrary compact set contained in the interior of $\mathcal{V}_2(\mathcal{S}, \mathcal{T})$.

Proof of Theorem 8.1

In view of the decomposition given in (8.2) and (8.3), the following lemma is obvious:

Lemma 8.9 *Let the system* (7.1) *and constraint sets* $\mathcal{S}$ *and* $\mathcal{T}$ *be given. There exists a state feedback that solves the semi-global stabilization problem for the system* (7.1) *only if the system* (8.2) *is semi-globally stabilizable by a state feedback, i.e., for any compact set* $\mathcal{W}_1$*, there exists a state feedback* $z = f(x_1)$ *such that*

(i) The equilibrium point $x_1 = 0$ *of the closed-loop system is asymptotically stable with* $\mathcal{W}_1$ *contained in its basin of attraction.*

(ii) For any $x_1(0) \in \mathcal{W}_1$*, we have* $z(t) \in \mathcal{S}$ *and* $\dot{z}(t) \in \mathcal{T}$ *for all* $t > 0$.

Note that (8.2) is semi-globally stabilizable by a state feedback only if A_{11} has all its eigenvalues in the closed left-half plane. Secondly, if the given system Σ has right-invertible constraints (the case we are considering in this section), then the eigenvalues of A_{11} are the constraint invariant zeros of the system, and hence the above proves necessity of Theorem 8.1. We can hence assume from now on that A_{11} has all its eigenvalues in the closed left-half plane. It remains to prove sufficiency of the conditions in Theorem 8.1. We prove sufficiency by an explicit design of a suitable controller. We begin with the design of a state feedback controller as needed for Theorem 8.1.

The basic philosophy of our controller design as said earlier is as follows. We first design a suitable stabilizing controller $z = f_1(x_1)$ for the subsystem (8.2). Next, we consider the subsystem (8.3). We need to design an input u such that the output z tracks the desired feedback for the first subsystem while avoiding constraint violation and while guaranteeing stability of the second subsystem.

State feedback controller design for semi-global stabilization

For the semi-global design, we need to guarantee stability for all initial conditions for the first subsystem in some compact subset $\mathcal{W}_1 \subset \mathbb{R}^{n_1}$ and for all initial conditions of the second subsystem in a compact subset $\mathcal{W}_2$ where $\mathcal{W}_2$ is such that there exists a $\zeta \in (0, 0.5)$ such that $\mathcal{W}_2 \subset (1 - 2\zeta)\mathcal{V}_2(\mathcal{S}, \mathcal{T})$. Our design has a larger domain of attraction for smaller ζ but at the expense of the need for a higher feedback gain in the second subsystem. Obviously, there exists a compact set $\tilde{\mathcal{S}} \subset \mathcal{S}$ such that $\mathcal{W}_2 \subset (1-\zeta)\mathcal{V}_2(\tilde{\mathcal{S}}, \mathcal{T})$. We assume that $\tilde{\mathcal{S}}$ and $\mathcal{T}$ still satisfy Assumption 7.1 (this is always possible). We actually design a feedback such that $z(t) \in \tilde{\mathcal{S}}$ for all $t > 0$. Because we are proving sufficiency, this restriction is without loss of generality and enables a simplification in the proof.

Assumption 7.1 enables us to decompose $\tilde{\mathcal{S}}$ and $\mathcal{T}$ as

$$\tilde{\mathcal{S}} \cap \operatorname{im} C_z = \tilde{\mathcal{S}}_1, \quad \tilde{\mathcal{S}} \cap \operatorname{im} D_z = \tilde{\mathcal{S}}_2,$$
$$\mathcal{T} \cap \operatorname{im} C_z = \mathcal{T}_1, \quad \mathcal{T} \cap \operatorname{im} D_z = \mathcal{T}_2.$$

Step 1 (Controller design for the zero dynamics)

We now focus our design for the first subsystem (8.2) while viewing z as an input variable. At first, we let $z = z_0 + v$ and rewrite subsystem (8.2) as

$$\dot{x}_1 = A_{11}x_1 + K_1 z_0 + K_1 v. \tag{8.5}$$

We note that the conditions of the theorem imply that all the eigenvalues of A_{11} are in the closed left-half plane. As will become transparent in the design for our second subsystem, we need to choose a δ_2 such that

$$\delta_2 \bar{\mathcal{S}}_1 \subset \frac{\zeta}{3}\mathcal{T}_1, \tag{8.6}$$

where

$$\bar{\mathcal{S}}_1 = \{\, z_1 - z_2 \mid z_1 \in \tilde{\mathcal{S}}_1, z_2 \in \tilde{\mathcal{S}}_1 \,\}. \tag{8.7}$$

Clearly, such a δ_2 exists since $\bar{\mathcal{S}}_1$ is bounded.

Our objective is to design a stabilizing feedback $z_0 = f(x_1)$ such that the equilibrium point of the closed-loop system of (8.5) and $z_0 = f(x_1)$ with $v = 0$ is asymptotically stable. Moreover, for all v satisfying the bound

$$\|v(t)\| \leqslant Me^{-\delta_2 t} \tag{8.8}$$

for all $t > 0$, and for all initial conditions in some arbitrarily large but compact subset $\mathcal{W}_1 \subset \mathbb{R}^{n_1}$, we satisfy $x_1(t) \to 0$ as $t \to \infty$. Finally, we need to guarantee also that

$$z_0(t) \in (1-\zeta)\widetilde{\mathcal{S}}_1, \qquad \dot{z}_0(t) \in \frac{\zeta\delta_2}{2e^{-1}}\widetilde{\mathcal{S}}_1 \cap \frac{\zeta}{3}\mathcal{T}_1 \tag{8.9}$$

for all $t > 0$. Again, it will become clear in our design for our second subsystem why it is desirable to guarantee that z_0 satisfies these bounds. Following the method of design given in Chap. 6, we can design such a suitable feedback law $f(x_1)$ as a linear state feedback law as described below.

Let P_μ be the solution of the continuous-time algebraic Riccati equation,

$$A_{11}'P_\mu + P_\mu A_{11} - P_\mu K_1 K_1' P_\mu + \mu^2 I = 0.$$

As discussed in Chap. 4,

$$\lim_{\mu\to 0} P_\mu = 0.$$

Also, for any compact subset $\mathcal{W}_1$, there exists a μ^* such that for all $\mu \in (0, \mu^*]$, we have

(*i*) $z_0 = -K_1'P_\mu x_1$ is a stabilizing controller for the system (8.5) with $\mathcal{W}_1$ contained in the domain of attraction.

(*ii*) (8.9) is satisfied for all t and for all v satisfying (8.8).

Hence we can choose

$$f(x_1) = F_\mu x_1 = -K_1'P_\mu x_1$$

for some $\mu \in (0, \mu^*]$ to obtain a suitable feedback for this first subsystem. In particular, (8.8) implies that

$$w_\mu(t) := \int_0^t e^{(A_{11}+K_1F_\mu)(t-\tau)} K_1 v(\tau)\,d\tau$$

is uniformly bounded for μ small enough. But this implies that

$$z_0(t) = f(x_1(t)) = F_\mu x_1(t) = F_\mu e^{(A_{11}+K_1F_\mu)(t-\tau)} + F_\mu w_\mu(t)$$

will respect our constraints (8.9) for μ sufficiently small. In the rest of the proof, we assume in fact that f is a linear function of the state as presented above.

Step 2 (Controller design for the second subsystem)

Our next design objective is to find a suitable input $\widetilde{u}$ to the second subsystem (8.3) such that (8.8) is satisfied where

$$v(t) = z(t) - f(x_1(t))$$

for all $t > 0$ and where $z_0(t) = f(x_1(t))$ has been designed to satisfy (8.9) provided v satisfies (8.8). Obviously, we must guarantee also that $z(t) \in \widetilde{\mathcal{S}}$ for all $t \geq 0$ and $\dot{z}(t) \in \mathcal{T}$ for all $t > 0$ while assuring the stability of the resulting closed-loop system with the desired domain of attraction.

At this stage, in order to proceed further with our design, we need to reveal from SCB certain finer structure of the matrices A_{22}, B_2, and $C_{z,2}$. Indeed, we have:

$$A_{22} = \widetilde{A}_{22} + \widetilde{B}_2 G, \; B_2 H = \widetilde{B}_2 H_1 + \widetilde{B}_3 H_2, \; C_{z,2} = \begin{pmatrix} C_1 \\ 0 \end{pmatrix} \text{ and } \widetilde{D}_z = \begin{pmatrix} 0 \\ D_1 \end{pmatrix}$$

for some compatible matrices G, H_1, and H_2 while

$$\widetilde{A}_{22} = \begin{pmatrix} A_1 & 0 & \cdots & 0 \\ 0 & \ddots & \ddots & \vdots \\ \vdots & \ddots & A_s & 0 \\ 0 & \cdots & 0 & A_c \end{pmatrix}, \quad \widetilde{B}_2 = \begin{pmatrix} B_{2,1} & 0 & \cdots & 0 \\ 0 & \ddots & \ddots & \vdots \\ \vdots & \ddots & B_{2,s} & 0 \\ 0 & \cdots & 0 & 0 \end{pmatrix},$$

$$\widetilde{B}_3 = \begin{pmatrix} 0 \\ \vdots \\ 0 \\ B_c \end{pmatrix}, \quad C_1 = \begin{pmatrix} C_{11} & 0 & \cdots & 0 & 0 \\ 0 & \ddots & \ddots & \vdots & \vdots \\ \vdots & \ddots & \ddots & 0 & 0 \\ 0 & \cdots & 0 & C_{1s} & 0 \end{pmatrix},$$

and for $i = 1, 2, \cdots, s$,

$$A_i = \begin{pmatrix} 0 & 1 & 0 & \cdots & 0 \\ 0 & \ddots & \ddots & \ddots & \vdots \\ 0 & \ddots & \ddots & \ddots & 0 \\ \vdots & \ddots & \ddots & \ddots & 1 \\ 0 & \cdots & 0 & 0 & 0 \end{pmatrix} \in \mathbb{R}^{k_i \times k_i}, \quad B_{2,i} = \begin{pmatrix} 0 \\ \vdots \\ 0 \\ 0 \\ 1 \end{pmatrix},$$

$$C_{1i} = \begin{pmatrix} 1 & 0 & \cdots & 0 \end{pmatrix}.$$

Here A_c and B_c are matrices of appropriate dimension such that the pair (A_c, B_c) is controllable.

In view of the above and in view of (8.21), the system of equations given in (8.3) can be rewritten as

$$\begin{aligned} \dot{x}_2 &= \widetilde{A}_{22} x_2 + \widetilde{B}_2(\widetilde{u}_2 + H_1 x_1 + G x_2) + \widetilde{B}_3(H_2 x_1 + \widetilde{u}_3) + K_2 z, \\ z &= \begin{pmatrix} z_1 \\ z_2 \end{pmatrix} = \begin{pmatrix} C_1 x_2 \\ D_1 \widetilde{u}_1 \end{pmatrix} \end{aligned} \tag{8.10}$$

with D_1 invertible. Next, we partition K_2, x_2, z_1, $\widetilde{u}_2$, and $f(x_1)$ in conformity with the partitioning of the other matrices. We have

$$K_2 = \begin{pmatrix} K_{2,1} \\ \vdots \\ K_{2,s} \\ K_{2,c} \end{pmatrix}, \quad x_2 = \begin{pmatrix} x_{2,1} \\ x_{2,2} \\ \vdots \\ x_{2,s} \\ x_c \end{pmatrix}, \quad z_1 = \begin{pmatrix} z_{1,1} \\ z_{1,2} \\ \vdots \\ z_{1,s} \end{pmatrix},$$

$$\widetilde{u}_2 = \begin{pmatrix} u_{2,1} \\ u_{2,2} \\ \vdots \\ u_{2,s} \end{pmatrix}, \quad f(x_1) = \begin{pmatrix} f_1(x_1) \\ f_2(x_1) \end{pmatrix},$$

$$f_1(x_1) = \begin{pmatrix} f_{1,1}(x_1) \\ f_{1,2}(x_1) \\ \vdots \\ f_{1,s}(x_1) \end{pmatrix}, \quad H_1 = \begin{pmatrix} H_{1,1} \\ \vdots \\ H_{1,s} \end{pmatrix}, \quad G = \begin{pmatrix} G_1 \\ \vdots \\ G_s \end{pmatrix}.$$

We are now ready to design $\widetilde{u}$. We will first focus on $\widetilde{u}_1$ and $\widetilde{u}_3$. We choose $\widetilde{u}_1 = D_1^{-1} f_2(x_1)$. To get $\widetilde{u}_3$, we choose first a matrix F_c such that the eigenvalues of $A_{cc} + B_c F_c$ are all at desired locations in the open left-half plane. Such a selection of F_c is possible since (A_{cc}, B_c) is controllable. We then choose $\widetilde{u}_3 = F_c x_c - H_2 x_1$.

We need to choose $\widetilde{u}_2$. To do so, let us study the system (8.10). By substituting for $\widetilde{u}_1$ and $\widetilde{u}_3$ as chosen, we can rewrite (8.10) as

$$\begin{aligned} \dot{x}_{2,i} &= A_i x_{2,i} + B_{2,i}(u_{2,i} + H_{1,i} x_1 + G_i x_2) + K_{2,i} z \\ z_{1,i} &= C_{1i} x_{2,i}, \end{aligned} \tag{8.11}$$

for $i = 1, 2, \cdots, s$, and

$$\dot{x}_c = (A_{cc} + B_c F_c) x_c + K_{2,c} z.$$

Now our objective in designing $\widetilde{u}_2$ is to guarantee that $z_{1,i} - f_{1,i}(x_1)$ converges to zero exponentially while making sure that the state constraints are satisfied. To proceed further, we define functions $m_{i,j}(x)$, $i = 1, \cdots, s,\ j = 1, \cdots, k_i + 1$ as follows:

$$m_{i,1}(x) := f_{1,i}(x_1),$$

and for $i = 1, \ldots, s,\ j = 2, \ldots, k_i + 1$,

$$m_{i,j}(x) := -K_{2,i,j-1} z - \delta_j \left(x_{2,i,j-1} - m_{i,j-1}(x) \right) + \frac{d}{dt} m_{i,j-1}(x),$$

where the parameter δ_j is such that δ_2 is as chosen before while

$$\delta_3 > 0, \ldots, \delta_{k_i+1} > 0$$

are to be chosen subsequently. We would like to point out that $m_{i,j}(x)$ as defined above are linear functions of x. We define next certain variables, $\varepsilon_{i,j}$, $i = 1, \cdots, s$, $j = 2, \cdots, k_i$ such that

$$\varepsilon_{i,j}(t) := x_{2,i,j} - m_{i,j}(x).$$

We are now ready to choose the components of $\widetilde{u}_2$, namely, $u_{2,i}$, $i = 1, \cdots, s$. If $k_i > 1$, we choose

$$u_{2,i} = -H_{1,i}x_1 - G_i x_2 + m_{i,k_i+1}(x). \tag{8.12}$$

If $k_i = 1$ for some i, say $i = \alpha$, then $u_{2,\alpha}$ is chosen as

$$u_{2,\alpha} = -H_{1,\alpha}x_1 - G_\alpha x_2 + m_{\alpha,2}(x) + \varepsilon_{\alpha,2}(t), \tag{8.13}$$

where $\varepsilon_{\alpha,2}(t)$ is to be chosen soon. We note that if $k_i \neq 1$ for any $i = 1, \cdots, s$, obviously, the system (8.11) with the choice of $u_{2,i}$ as chosen in (8.12) results in $x_{2,i,j}(t) - m_{i,j}(t) \to 0$ as $t \to \infty$ for $j = 1, \ldots, k_i$. We define next

$$\varepsilon_2(t) := \begin{pmatrix} \varepsilon_{1,2}(t) \\ \vdots \\ \varepsilon_{s,2}(t) \end{pmatrix}.$$

Let us next focus on the behavior of the constraints under the feedback laws chosen above. We observe that z_2 satisfies the constraints due to the choice of $\widetilde{u}_1$. Hence we focus on z_1. We have

$$\begin{aligned} z_1 = C_1 x_2(t) &= e^{-\delta_2 t} C_1 x_2(0) + f_1(x_1(t)) \\ &\quad - e^{-\delta_2 t} f_1(x_1(0)) + \int_0^t e^{-\delta_2(t-\tau)} \varepsilon_2(\tau) d\tau \\ &= e^{-\delta_2 t} C_1 x_2(0) + \left(1 - e^{-\delta_2 t}\right) f_1(x_1(t)) \\ &\quad + e^{-\delta_2 t} \int_0^t \tfrac{d}{dt} f_1(x_1(\tau)) d\tau + \int_0^t e^{-\delta_2(t-\tau)} \varepsilon_2(\tau) d\tau. \end{aligned}$$

Since $z_0 = f_1(x_1)$ satisfies (8.9), we get

$$e^{-\delta_2 t} \int_0^t \tfrac{d}{dt} f_1(x_1(\tau)) d\tau \in \tfrac{\zeta}{2} \widetilde{\mathcal{S}}_1.$$

Moreover, $C_1 x_2(0) \in (1-\zeta)\widetilde{\mathcal{S}}_1$ and $f_1(x_1(t)) \in (1-\zeta)\widetilde{\mathcal{S}}_1$. Therefore, if we guarantee that

$$\int_0^t e^{-\delta_2(t-\tau)}\varepsilon_2(\tau)d\tau \in \tfrac{\zeta}{2}\widetilde{\mathcal{S}}_1, \tag{8.14}$$

then we obtain $C_1 x_2(t) \in \widetilde{\mathcal{S}}_1$ as required.

Next, we need to consider the rate constraint on z_1. We have

$$\dot{z}_1 = \frac{d}{dt}C_1 x_2 = -\delta_2(C_1 x_2 - f_1(x_1)) + \frac{d}{dt}f_1(x_1) + \varepsilon_2(t). \tag{8.15}$$

We know from (8.9) that $\frac{d}{dt}f_1(x_1) \in \frac{\zeta}{3}\mathcal{T}_1$. Since δ_2 satisfies (8.6), combined with the fact that $(C_1 x_2 - f_1(x_1)) \in \overline{\mathcal{S}}_1$, we find that $\delta_2(C_1 x_2 - f_1(x_1)) \in \frac{\zeta}{3}\mathcal{T}_1$. Hence, we obtain that $\frac{d}{dt}C_1 x_2 \in \mathcal{T}_1$ if we guarantee that

$$\varepsilon_2(t) \in \left(1 - \tfrac{2\zeta}{3}\right)\mathcal{T}_1. \tag{8.16}$$

Therefore, if we can guarantee (8.14) and (8.16), then we satisfy our constraints. We still need to show that the difference between z and $f(x_1)$, which is equal to the disturbance v in the first subsystem, satisfies (8.8). We have

$$v = \begin{pmatrix} v_1 \\ v_2 \end{pmatrix},$$

where $v_2 = 0$, and rewriting (8.15), we get

$$\dot{v}_1(t) = -\delta_2 v_1(t) + \varepsilon_2(t)$$

with $v_1(0) \in \overline{\mathcal{S}}_1$. Therefore, v satisfies (8.8) if ε_2 satisfies

$$\int_0^t e^{\delta_2 \tau}\varepsilon_2(\tau)d\tau \in \widetilde{\mathcal{S}}_1. \tag{8.17}$$

We will next consider how we can guarantee that $\varepsilon_2(t)$ satisfies (8.14), (8.16), and (8.17). It is easy to see that these three conditions are satisfied if for all $t > 0$ we have

$$\varepsilon_2(t) \in e^{-\delta_3 t}\left(1 - \tfrac{2\zeta}{3}\right)\mathcal{T}_1 \tag{8.18}$$

for δ_3 large enough.

For ease of notation, we define

$$\mathcal{A} = \{\, i \mid i = 1,\ldots,s,\ k_i = 1 \,\},$$
$$\mathcal{A}^c = \{\, i \mid i = 1,\ldots,s,\ k_i > 1 \,\}.$$

With $i \in \mathcal{A}^c$, we have

$$\varepsilon_{i,2}(t) = e^{-\delta_3 t}\varepsilon_{i,2}(0) + \int_0^t e^{-\delta_3(t-\tau)}\varepsilon_{i,3}(\tau)d\tau$$

with $\varepsilon_{i,3}(t) = 0$ if $k_i = 2$. For $i \in \mathcal{A}$, we can obtain arbitrary $\varepsilon_{i,2}$ by choosing $u_{2,i}$ as in (8.13). Assume that for $i \in \mathcal{A}$, we choose

$$\varepsilon_{i,2}(t) = e^{-\delta_3 t}\varepsilon_{i,2}(0) \tag{8.19}$$

with $\varepsilon_{i,2}(0)$ still to be chosen. If we guarantee:

(*i*) $\varepsilon_2(0) \in (1-\zeta)\mathcal{T}_1$,

(*ii*) $\int_0^t e^{\delta_3 t}\varepsilon_{i,3}(\tau)d\tau$ small enough for i with $k_i \geqslant 3$,

then (8.18) is satisfied. We will see later how, by choosing $\delta_4, \delta_5, \cdots$, we can guarantee item (*ii*). Let us next consider item (*i*). We have

$$\varepsilon_2(0) = \dot{z}_1(0) + \delta_2\left(z_1(0) - f_1(x_1(0))\right) - \tfrac{d}{dt} f_1(x_1(0)).$$

We have

$$\begin{gathered}\delta_2\left(z_1(0) - f_1(x_1(0))\right) \in \delta_2\bar{\mathcal{S}}_1 \subset \tfrac{\zeta}{3}\mathcal{T}_1\\ \tfrac{d}{dt} f_1(x_1(0)) \in \tfrac{\zeta}{3}\mathcal{T}_1,\end{gathered}$$

and hence, as soon as we guarantee that $\dot{z}_1(0) \in (1-2\zeta)\mathcal{T}_1$, we know that item (*i*) is satisfied.

However, because $x_2(0) \in (1-2\zeta)\mathcal{V}_2(\mathcal{S},\mathcal{T})$, there exists a $u(0)$ such that $\dot{z}_1(0) \in (1-2\zeta)\mathcal{T}_1$. It is easily verified that the only components of $u(0)$ which affect $\dot{z}_1(0)$ are exactly the $u_{2,i}(0)$ with $i \in \mathcal{A}$ which, according to (8.13), is equivalent to choosing $\varepsilon_{i,2}(0)$ with $i \in \mathcal{A}$ appropriately.

This yields a system with the desired properties, but the feedback is partially determined in open loop due to our choice in (8.19) and therefore not acceptable. Choose instead for each t, $\varepsilon_{i,2}(t)$ with $i \in \mathcal{A}$ to minimize the criterion:

$$\min\{\, \gamma \mid \varepsilon_2(t) \in \gamma\mathcal{T}_1 \,\}. \tag{8.20}$$

Note that the existence of $\varepsilon_{i,2}(t)$ with $i \in \mathcal{A}$ that minimize this criterion is a consequence of the fact that the set $\mathcal{T}_1$ is bounded. Clearly, the optimal $\varepsilon_{i,2}(t)$ with $i \in \mathcal{A}$ becomes a function of $\varepsilon_{i,2}(t)$ with $i \in \mathcal{A}^c$. But for $i \in \mathcal{A}^c$, the $\varepsilon_{i,2}(t)$ are a function of the state, and hence the $\varepsilon_{i,2}(t)$ with $i \in \mathcal{A}$ are determined according to a state feedback. Note that (8.19) is a suboptimal choice for the optimization in (8.20) yielding (8.18), and therefore we have

$$\gamma \leqslant e^{-\delta_3 t}\left(1 - \tfrac{2\zeta}{3}\right),$$

and hence, the choice for $\varepsilon_{i,2}(t)$ according to the optimization is a state feedback which also satisfies (8.18). Note that in general, the dependence of $\varepsilon_{i,2}$ with $i \in \mathcal{A}$ on the $\varepsilon_{i,2}$ with $i \in \mathcal{A}^c$ is nonlinear. There are a few instances where we can guarantee a linear feedback. Clearly, if either the set $\mathcal{A}$ or the set $\mathcal{A}^c$ is empty, then this mapping is automatically linear since either its domain or its range is zero dimensional. Moreover, if $\operatorname{im} C_z \subset \mathcal{T}$, then $\mathcal{T}_1$ is equal to the whole space, and we get an optimal value $\gamma = 0$ by choosing $\varepsilon_{i,2}(t) = 0$ for $i \in \mathcal{A}$ which clearly also yields a linear feedback.

Finally, we still need to choose $\delta_4, \delta_5, \ldots$. We note that we have the following structure when $k_i > 2$:

$$\begin{cases} \dot{\varepsilon}_{i,j} = -\delta_j \varepsilon_{i,j} + \varepsilon_{i,j+1} & \text{for } j = 1, \ldots, k_i - 1, \\ \dot{\varepsilon}_{i,j} = -\delta_j \varepsilon_{i,j} & \text{for } j = k_i. \end{cases}$$

From the above structure, it should be obvious that we can make the $\varepsilon_{i,j}$ small by a suitable design of the δ_j. We have to make sure that

$$\int_0^t e^{\delta_3 t} \varepsilon_{i,3}(\tau) d\tau$$

is small enough for those i with $k_i \geqslant 3$. By making δ_4 large enough, we can make this arbitrarily small provided that

$$\int_0^t e^{\delta_4 t} \varepsilon_{i,4}(\tau) d\tau$$

is small enough. If $k_i = 4$, this is actually equal to zero, and otherwise, we can use a similar argument to make this small enough by choosing δ_5 large enough. In this way, we can recursively determine $\delta_4, \delta_5, \ldots, \delta_{k_i+1}$.

Finally, note that all the $\varepsilon_{i,j}(t)$ converge to zero exponentially, and therefore, for $i = 2$, this implies that $\varepsilon_{i,2}(t)$ converges to zero exponentially, and hence, the difference between $x_{i,1}$ and $f_{1,i}(x_1)$ converges to zero exponentially. Since $f_1(x_1)$ also converges to zero exponentially, we find that $x_{i,1}$ converges to zero exponentially. This also implies that z converges to zero exponentially. Moreover, it implies that $m_{i,2}(x)$ converges to zero exponentially. Similarly, since $\varepsilon_{i,3}(t)$ converges to zero exponentially, we have the difference between $x_{i,1}$ and $m_{i,2}(x)$ converging to zero exponentially. Hence, if $k_i > 1$, $m_{i,2}(x)$ converges to zero exponentially, and we find that $x_{i,2}$ converges to zero exponentially. As before, this also implies that $m_{i,3}(x_1)$ converges to zero exponentially. Continuing with this recursive argument, we find that all states converge to zero exponentially, and therefore, the constructed feedback has the desired attractivity as well as stability. ■

Proof of Theorem 8.2

We first show that the conditions of Theorem 8.2 are necessary. The necessity of condition (*i*) of Theorem 8.2 is trivial. As in the semi-global case, the necessity of condition (*ii*) of Theorem 8.2 is a consequence of the results obtained in Chap. 4 and the following lemma which is a direct consequence of the decomposition given in (8.2) and (8.3).

Lemma 8.10 *Let the system* (7.1) *and constraint sets* $\mathcal{S}$ *and* $\mathcal{T} = \mathbb{R}^p$ *be given. There exists a state feedback that solves the global stabilization problem for the system* (7.1) *only if the system* (8.2) *is globally stabilizable by a state feedback* $z = f(x_1)$, *i.e.,*

(*i*) *The equilibrium point* $x_1 = 0$ *of the closed-loop system is globally asymptotically stable.*

(*ii*) *For any initial condition, we have* $z(t) \in \mathcal{S}$ *for all* $t > 0$.

In order to complete the proof that the conditions of Theorem 8.2 are necessary, we need to prove the following lemma which states that condition (*iii*) is also a necessary condition for the solvability of Problem 7.9.

Lemma 8.11 *Consider the system* Σ *as given by* (7.1). *Let the assumptions of Theorem 8.2 be satisfied. Then, the global stabilization problem as defined in Problem 7.9 is solvable only if the given system has no constraint infinite zeros of order greater than one.*

Proof : First, note that, since the system is right invertible, having no infinite zeros of order greater than one is equivalent to $(C_z B \quad D_z)$ being surjective. Therefore, if the system has infinite zeros of order greater than one, then there exists a vector $c \neq 0$ such that $c' D_z = 0$ and $c' C_z B = 0$. Let $z_0 \in \mathcal{S}$ be such that $\langle z, c\rangle \leqslant \langle z_0, c\rangle$ for all $z \in \mathcal{S}$. Since $\mathcal{S}$ is convex as well as compact, such a z_0 always exists. Next, because the given system has right-invertible constraints, there exist initial condition x_0 and input function $\bar{u}$ such that the output z satisfies $z(0) = z_0$ and $\dot{z}(0) = c$, i.e.,

$$z_0 = C_z x_0 + D_z \bar{u}(0) \in \mathcal{S}.$$

Clearly, $x_0 \in \mathcal{V}(\mathcal{S}, \mathbb{R}^p)$.

But if we start at time 0 in x_0, then we have for any input signal u

$$\langle c, z(0)\rangle = \langle c, C_z x_0\rangle = \langle c, z_0\rangle,$$
$$\tfrac{d}{dt}\langle c, z(t)\rangle \,|_{t=0} = \langle c, C_z A x_0\rangle = \langle c, c\rangle > 0.$$

Therefore, $\langle c, z(t)\rangle > \langle c, z_0\rangle$ for small $t > 0$ and for any input u. By definition of z_0, this implies that $z(t) \notin \mathcal{S}$ for small $t > 0$ and for any input u. Therefore, there exist initial conditions in $\mathcal{V}(\mathcal{S}, \mathbb{R}^p)$ which cannot be stabilized without violating our constraints which yields the required contradiction. ∎

This establishes the necessity of our conditions. The next step is to prove sufficiency by explicitly designing a suitable feedback. Before we do so, in view of SCB, we need to recall a finer structure of (8.1), namely,

$$C_{z,2} = \begin{pmatrix} C_1 \\ 0 \end{pmatrix}, \quad \widetilde{D}_z = \begin{pmatrix} 0 & 0 & 0 \\ D_1 & 0 & 0 \end{pmatrix}, \quad B_2 = \begin{pmatrix} 0 & \widetilde{B}_2 & \widetilde{B}_3 \end{pmatrix}, \tag{8.21}$$

with D_1 invertible. Since we have no constraint infinite zeros of order greater than 1, we have the additional structure that $C_1 \widetilde{B}_2$ is invertible. Also, we decompose $\widetilde{u}$ and z to be compatible with the above:

$$z = \begin{pmatrix} z_1 \\ z_2 \end{pmatrix}, \quad \text{and} \quad \widetilde{u} = \begin{pmatrix} \widetilde{u}_1 \\ \widetilde{u}_2 \\ \widetilde{u}_3 \end{pmatrix}. \tag{8.22}$$

We observe that the assumptions on the set $\mathcal{S}$ guarantees that we can decompose the set $\mathcal{S}$ compatible with the decomposition of z:

$$\mathcal{S} = \mathcal{S}_1 \times \mathcal{S}_2, \tag{8.23}$$

such that $z \in \mathcal{S}$ if and only if $z_1 \in \mathcal{S}_1$ and $z_2 \in \mathcal{S}_2$.

State feedback controller design for global stabilization

For the design for the first subsystem, we choose an input z which satisfies the constraints such that $z(t) \in (1-\widetilde{\rho})\mathcal{S}$ where the parameter $\widetilde{\rho} \in (0,1)$. By choosing $\widetilde{\rho}$ close to 1, we have limited control effort for the first subsystem, but in our design for $\widetilde{u}$ to track the desired output z, we have more flexibility. Conversely, choosing $\widetilde{\rho}$ small gives us more control effort for the first subsystem, but in our design for $\widetilde{u}$, we need to track the desired output z quite accurately. Based on these arguments, we fix the parameter $\widetilde{\rho} \in (0,1)$. We now focus our design for the first subsystem (8.2) while viewing z as an input variable. At first, we let $z = z_0 + v$ and rewrite subsystem (8.2) as

$$\dot{x}_1 = A_{11}x_1 + K_1 z_0 + K_1 v. \tag{8.24}$$

We note that the conditions of the theorem imply that all the eigenvalues of A_{11} are in the closed left-half plane. Next, we would like to construct a state feedback law $z_0 = f(x_1)$ such that it satisfies the constraints $z_0(t) \in (1-\widetilde{\rho})\mathcal{S}$ for all $t > 0$ while rendering the zero equilibrium point of the closed-loop system of (8.24) and $z_0 = f(x_1)$ globally attractive (i.e., $x_1(t) \to 0$ as $t \to \infty$) in the presence of signals v satisfying the bound

$$\|v(t)\| \leqslant M e^{-\delta t} \tag{8.25}$$

for some $M > 0$. Moreover, the feedback law $z_0 = f(x_1)$ should render the zero equilibrium point of the closed-loop system when $v = 0$ locally exponentially stable. Such a nonlinear feedback law $z_0 = f(x_1)$ can be obtained from the adaptive-low-gain design based on the H_∞ ARE as discussed in Chap. 4. Note that the effect of the exponentially decaying signal v can then be handled in the same way as the effect of the exponentially decaying observer error e in the proof of Theorem 4.53.

Next, we consider the second subsystem, namely, (8.3). The main design objective is to find a suitable input $\widetilde{u}$ to the second subsystem (8.3) such that for any initial condition in $\mathcal{W}_2$ of the first subsystem (8.3) and for any initial condition of the second subsystem (8.2), we have

$$\|z(t) - f(x_1(t))\| \leqslant Me^{-\delta t} \tag{8.26}$$

for all $t > 0$ and for any function f satisfying

$$f(x_1(t)) \in (1 - \widetilde{\rho})\mathcal{S}$$

for all $t > 0$. Obviously, we must also guarantee that $z(t) \in \mathcal{S}$ for all $t > 0$.

To proceed further, let us next partition $f(x_1)$ to be compatible with the partitioning of z:

$$f(x_1) = \begin{pmatrix} f_1(x_1) \\ f_2(x_1) \end{pmatrix}.$$

We are now ready to construct the required feedback laws for $\widetilde{u}$. Our objective in designing it is to guarantee that $v := \widetilde{u} - f(x_1)$ satisfies (8.25) while z satisfies the constraints. Knowing the properties of SCB, it can be shown that one can choose $\widetilde{u}_3 = Fx_2$ such that the system (8.1) with inputs $\widetilde{u}_1$ and $\widetilde{u}_2$ and output z is invertible and moreover, the additional invariant zeros introduced by the feedback $\widetilde{u}_3 = Fx_2$ are placed in a desired location in the open left-half plane. With this choice of $\widetilde{u}_3$, we obtain

$$\begin{aligned} z_1 &= C_1 x_2, \\ z_2 &= D_1 \widetilde{u}_1, \\ \dot{z}_1 &= C_1(B_2 H x_1 + \widetilde{B}_2 G x_2 + \widetilde{B}_3 F x_2 + K_2 z) + C_1 \widetilde{B}_2 \widetilde{u}_2. \end{aligned}$$

Then choose the feedback laws,

$$\begin{aligned} \widetilde{u}_1 &= D_1^{-1} f_2(x_1), \\ \widetilde{u}_2 &= (C_1 \widetilde{B}_2)^{-1} \Big(-C_1(B_2 H x_1 + \widetilde{B}_2 G x_2 + \widetilde{B}_3 F x_2 + K_2 z) \\ &\qquad - \delta (z_1 - f_1(x_1)) + (1 - e^{-\delta t}) \frac{d}{dt} f_1(x_1(t)) \Big), \end{aligned}$$

where $\delta > 0$ is a positive constant. We emphasize that the above feedback laws are time-varying nonlinear state feedback laws. These feedback laws guarantee that $z_2 = f_2(x_1)$ and that $z_1(t) \to f_1(x_1(t))$ as $t \to \infty$ for all initial conditions in the set of admissible set of initial conditions $\mathcal{V}(\mathcal{S}, \mathcal{T})$. We show next that z_1

and z_2 with the above feedback laws satisfy all the constraints. We observe first that $f(x_1) \in \mathcal{S}$ and $\frac{d}{dt} f(x_1) \in \widetilde{\mathcal{T}}$ which guarantees that $z_2 = f_2(x_1) \in \mathcal{S}_2$ and $\dot{z}_2 = \frac{d}{dt} f_2(x_1) \in \widetilde{\mathcal{T}}_2$. This implies that z_2 satisfies all the constraints. We focus next on showing that z_1 satisfies all the constraints, i.e., $z_1(t) \in \mathcal{S}_1$ and $\dot{z}_1(t) \in \widetilde{\mathcal{T}}_1$ for all $t > 0$. We have

$$\dot{z}_1(t) = -\delta\,(z_1(t) - f_1(x_1(t))) + (1 - e^{-\delta t})\frac{d}{dt} f_1(x_1(t)). \tag{8.27}$$

Integrating this equation, we obtain,

$$z_1(t) = e^{-\delta t} z_1(0) + (1 - e^{-\delta t}) f_1(x_1(t)). \tag{8.28}$$

Since $f_1(x_1) \in \mathcal{S}_1$ and $z_1(0) \in \mathcal{S}_1$, we find, using the convexity of $\mathcal{S}_1$, that $z_1(t) \in \mathcal{S}_1$. Note that $v_1(t) = z_1(t) - f_1(x_1(t))$ satisfies according to (8.28):

$$v_1(t) = e^{-\delta t} z_1(0) - e^{-\delta t} f_1(x_1(t)) \in e^{-\delta t}\,\overline{\mathcal{S}}_1$$

for any $t > 0$, and hence, the error signal $v = z - f(x_1)$ satisfies

$$\|v(t)\| \leqslant M e^{-\delta t}$$

for all $t > 0$ where M is some positive constant. This immediately shows that, for all initial conditions in $\mathcal{V}(\mathcal{S}, \mathbb{R}^p)$, $x_1(t) \to 0$ as $t \to \infty$ which in turn guarantees that $z(t) \to 0$ as $t \to \infty$. Moreover, it is straightforward to show that $x_2(t) \to 0$ as $t \to \infty$ for all initial conditions in $\mathcal{V}(\mathcal{S}, \mathbb{R}^p)$. Thus, we conclude that the zero equilibrium point of the closed-loop system is attractive for all initial conditions in $\mathcal{V}(\mathcal{S}, \mathbb{R}^p)$. Finally, it is obvious that the zero equilibrium point of the closed-loop system is locally asymptotically stable. This concludes our proof. ∎

Proof of Theorem 8.8 : The first statement of Theorem 8.8 follows from Theorem 8.1. The second statement follows from the following counterexample. Consider the system

$$\dot{x} = \begin{pmatrix} 0 & 0 & 0 & 0 & 0 & 0 & 0 \\ 0 & 0 & 0 & 0 & 0 & 0 & 0 \\ 0 & 0 & 0 & 0 & 0 & 0 & 0 \\ 1 & 0 & 0 & 0 & 3 & 0 & 0 \\ 0 & 1 & 0 & -3/2 & 3/2 & -3 & 0 \\ 0 & 0 & 1 & -3 & 0 & 0 & 0 \\ 0 & 0 & 0 & 0 & 0 & -3 & 0 \end{pmatrix} x + \begin{pmatrix} 1 & 0 & 0 & 0 \\ 0 & 1 & 0 & 0 \\ 0 & 0 & 1 & 0 \\ 0 & 0 & 0 & 1 \\ 0 & 0 & 0 & 1 \\ 0 & 0 & 0 & 1 \\ 0 & 0 & 0 & 1 \end{pmatrix} u,$$

$$z = \begin{pmatrix} 0 & 0 & 0 & 1 & 0 & 0 & 0 \\ 0 & 0 & 0 & 0 & 1 & 0 & 0 \\ 0 & 0 & 0 & 0 & 0 & 1 & 0 \\ 0 & 0 & 0 & 0 & 0 & 0 & 1 \end{pmatrix} x$$

with no magnitude constraints; however, the rate constraint $\mathcal{T}$ is given by

$$\mathcal{T} = \{ z \in \mathbb{R}^4 \mid \|z\|_\infty < 3 \}.$$

Consider the following four possible initial conditions:

$$x_1 = \begin{pmatrix} 0 \\ 0 \\ 0 \\ 1 \\ 1 \\ 1 \\ 0 \end{pmatrix}, \quad x_2 = \begin{pmatrix} 0 \\ 0 \\ 0 \\ 1 \\ 1 \\ -1 \\ 0 \end{pmatrix}, \quad x_3 = \begin{pmatrix} 0 \\ 0 \\ 0 \\ -1 \\ 1 \\ 1 \\ 0 \end{pmatrix}, \quad x_4 = \begin{pmatrix} 0 \\ 0 \\ 0 \\ 1 \\ -1 \\ 1 \\ 0 \end{pmatrix}.$$

These initial conditions are all feasible. But in order to satisfy the rate constraint at time 0, we see that $u_4(0)$ must be equal to $0, 0, 0$, and 3 for $x(0) = x_1, x(0) = x_2, x(0) = x_3$, and $x(0) = x_4$, respectively. One can show easily that this can, however, never be done via a linear feedback. Note that this also excludes solvability of the semi-global problem by linear feedback because that still requires that for initial conditions close to x_1, we must have $u(0)$ close to 0, and similarly for the other three initial conditions. This is not possible with arbitrary accuracy with a linear feedback. It remains to show that we can solve the semi-global stabilization problem for this system. This is, however, a direct consequence of Theorem 8.1. Note that the constructed system has *no* finite constraint invariant zeros.

Finally, that we can find a linear feedback in case rate constraints are not present is clear from the construction of a controller in the proof of Theorem 8.1. ■

Proof of Theorem 8.4 : It remains to prove Theorem 8.4. This is straightforward given the solution of the state feedback problem. We have a solution for the measurement feedback case for input rate and amplitude saturation in Chap. 6. Basically by exploiting the fact that we have a bounded input, we can find $t_0 > 0$ small such that $x(t_0)$ is bounded away from the boundary of $\mathcal{V}(\mathcal{S}, \mathcal{T})$ since we start at time 0 in the interior of $\mathcal{V}(\mathcal{S}, \mathcal{T})$. This t_0 is independent of the input signal and only exploits some upper bound on the size of the input. Next, since the system is observable, we can find for any arbitrarily small ν and arbitrarily large η an observer such that

$$\|\hat{x}(t) - x(t)\| < \nu e^{-\eta t}$$

for all $t > t_0$. If we then start our design at time t_0 with the same state feedback as before but with x replaced by $\hat{x}$, then it is easy to check that this will yield stability provided ν is small enough and η is large enough. ■

8.3 Semi-global and global stabilization in admissible set for non-right-invertible constraints: continuous time

For continuous-time systems subject to right-invertible constraints, Sect. 8.2 considers both semi-global and global stabilization in the admissible set. In this section, we consider the same, however, for non-right-invertible constraints. It is worth to recall here that non-right-invertible constraints arise inherently due to state constraints as Remark 7.15 points out.

It is clearly evident from Sect. 8.2 that, for right-invertible constraints, if a stabilization problem is solvable for a pair of constraint sets $\mathcal{S}$ and $\mathcal{T}$ satisfying Assumption 7.1, the same stabilization problem is solvable for any pair of constraint sets $\mathcal{S}$ and $\mathcal{T}$ satisfying Assumption 7.1 irrespective of their shape. In general this is not so for non-right-invertible constraints. This adds a layer of complexity and renders the case of non-right-invertible constraints profoundly different from the case of right-invertible constraints. Because of such complexity, we need to examine carefully the posed semi-global and global stabilization problems based on the type of constraint sets $\mathcal{S}$ and $\mathcal{T}$ that are prescribed, that is, whether we need stabilization results for fixed sets $\mathcal{S}$ and $\mathcal{T}$ or for all sets $\mathcal{S}$ and $\mathcal{T}$, etc.

In this section, at first, for a given pair of constraint sets $\mathcal{S}$ and $\mathcal{T}$ that satisfy Assumption 7.1, we present the necessary and sufficient conditions under which semi-global stabilization is solvable for non-right-invertible constraints. Such conditions depend on the shape of the *given* pair of constraint sets $\mathcal{S}$ and $\mathcal{T}$. Such a dependence gives rise to two important questions that need to be answered.

Suppose we consider strengthened versions of Problems 7.8–7.11 in which *all* possible pairs of constraint sets $\mathcal{S}$ and $\mathcal{T}$ that satisfy Assumption 7.1 are considered. Then, the first question that needs to be answered is this. Given a system Σ with non-right-invertible constraints, for such a Σ, what are the solvability conditions for such enhanced stabilization Problems 7.8–7.11 in which all possible pairs of constraint sets $\mathcal{S}$ and $\mathcal{T}$ that satisfy Assumption 7.1 are considered instead of just one given pair of constraint sets $\mathcal{S}$ and $\mathcal{T}$?

The second important question that needs to be answered is this. Given a system Σ with non-right-invertible constraints, for such a Σ, does there exist a pair of constraint sets $\mathcal{S}$ and $\mathcal{T}$ that satisfy Assumption 7.1 for which semi-global and global stabilization problems as formally formulated in Problems 7.8–7.11 are solvable?

Answers to both of the above questions will definitely shed light on the complexities involved with non-right-invertible constraints. Section 8.3.1 answers these questions. In this section, we often focus only on amplitude constraints on the constrained output as expressed by $z(t) \in \mathcal{S}$ for all $t \geqslant 0$.

To proceed with our development, we need to recall the SCB of the system Σ_{uz} characterized by the quadruple (A, B, C_z, D_z). Consider the state, input, and constrained output transformation matrices, Γ_s, Γ_u, and Γ_z, and let

$$x = \Gamma_s \bar{x}, \quad u = \Gamma_u \bar{u}, \quad \text{and} \quad z = \Gamma_z \bar{z}$$

so that Σ_{uz} is in its SCB as given by Theorem 3.1. That is,

$$\bar{\Sigma} : \begin{cases} \dot{x}_a &= A_{aa} x_a + K_a \bar{z} \\ \dot{x}_b &= A_{bb} x_b + K_b \bar{z} \\ \dot{x}_c &= A_{cc} x_c + B_c (u_c + H_a x_a) + K_c \bar{z} \\ \dot{x}_d &= A_{dd} x_d + B_d (u_d + G\bar{x}) + K_d \bar{z} \\ y &= C_y \bar{x} + D_y \bar{u} \\ \bar{z} &= \begin{pmatrix} z_b \\ z_0 \\ z_d \end{pmatrix} = \begin{pmatrix} C_b x_b \\ u_0 \\ C_d x_d \end{pmatrix}, \end{cases} \tag{8.29}$$

where

$$\bar{x} = \begin{pmatrix} x_a \\ x_b \\ x_c \\ x_d \end{pmatrix}, \quad \bar{u} = \begin{pmatrix} u_0 \\ u_c \\ u_d \end{pmatrix}, \quad \bar{z} = \begin{pmatrix} z_0 \\ z_b \\ z_d \end{pmatrix},$$

and $G\bar{x} = G_a x_a + G_b x_b + G_c x_c + G_d x_d$. We define the constraint set $\bar{\mathcal{S}} = T_z^{-1} \mathcal{S}$ so that under the SCB for Σ_{uz}, we have $\bar{z}(t) \in \bar{\mathcal{S}}, \ \forall t \geqslant 0$. Because of its importance, we extract a subsystem from $\bar{\Sigma}$ consisting of the state variables x_a and x_b:

$$\Sigma_1 : \begin{cases} \dot{x}_a &= A_{aa} x_a + K_{ab} C_b x_b + K_{a2} \zeta \\ \dot{x}_b &= (A_{bb} + K_{bb} C_b) x_b + K_{b2} \zeta \\ \bar{z} &= \begin{pmatrix} C_b \\ 0 \end{pmatrix} x_b + \begin{pmatrix} 0 \\ I \end{pmatrix} \zeta, \end{cases} \tag{8.30}$$

where in the first two equations we rewrote $\bar{z}$ in terms of $C_b x_b$ and ζ. One can view ζ as the input and $\bar{z}$ as the output of this subsystem. Apparently, both the input and output are constrained. We let $x_{ab} = (x_a', \ x_b')' \in \mathbb{R}^{n_1}$ and $x_{cd} = (x_c', \ x_d')' \in \mathbb{R}^{n-n_1}$, where $n_1 = n_a + n_b$. The admissible set for Σ_1 is defined by

$$\mathcal{V}_{ab}(\bar{\mathcal{S}}) = \mathbb{R}^{n_a} \oplus \mathcal{V}_b(\bar{\mathcal{S}}),$$

where

$$\mathcal{V}_b(\bar{\mathcal{S}}) := \left\{ x_b \in \mathbb{R}^{n_b} \mid \exists \zeta \text{ such that } \begin{pmatrix} C_b x_b \\ \zeta \end{pmatrix} \in \bar{\mathcal{S}} \right\}. \tag{8.31}$$

Here $\mathcal{V}_b(\bar{\mathcal{S}})$ basically represents the admissible set of the subsystem of Σ_1 obtained by leaving out the zero dynamics.

For semi-global stabilization with non-right-invertible constraints, a characterization of the boundary of the admissible set $\mathcal{V}_b(\bar{\mathcal{S}})$ is crucial. For this purpose, we define for any $\xi \in \partial \mathcal{V}_b(\overline{\mathcal{S}})$ (the boundary of $\mathcal{V}_b(\overline{\mathcal{S}})$), the set of normals to this convex set (for further properties of this set, see [119]). Let

$$N(\xi) := \left\{ \eta \in \mathbb{R}^{n_b} \mid \|\eta\| = 1 \text{ and } \langle \xi' - \xi, \eta \rangle \leqslant 0,\ \forall\, \xi' \in \mathcal{V}_b(\overline{\mathcal{S}}) \right\}, \tag{8.32}$$

and

$$T(\xi) := \Bigg\{ \mu \in \mathbb{R}^{n_b} \mid \mu = \widetilde{A}_b \xi + K_{b2}\zeta \text{ with} \begin{pmatrix} C_b \xi \\ \zeta \end{pmatrix} \in \overline{\mathcal{S}} \\ \text{and } \langle \mu, \eta \rangle \leqslant 0,\ \forall\, \eta \in N(\xi) \Bigg\}, \tag{8.33}$$

where $\widetilde{A}_b = A_{bb} + K_{bb}C_b$. Note that $N(\xi)$ is the normalized normal vectors at the boundary point ξ of $\mathcal{V}_b(\overline{\mathcal{S}})$. By corollary 11.6.1 in [119], this set is guaranteed to be nonempty. On the other hand, $T(\xi)$ is the collection of directions at the boundary point ξ along which the admissible set $\mathcal{V}_b(\overline{\mathcal{S}})$ remains invariant under the dynamics of Σ_1.

We have the following theorem which is concerned with the semi-global stabilization Problem 7.8 via state feedback.

Theorem 8.12 *Consider the continuous-time system Σ as given by (7.1) and constraint set $\mathcal{S}$ that satisfies Assumption 7.1. Suppose the system Σ_{uz} is represented in its SCB. Then, the semi-global stabilization problem in the admissible set via state feedback as defined in Problem 7.8 is solvable if and only if the following conditions hold:*

(i) *(A, B) is stabilizable.*

(ii) *The constraint invariant zeros of the given system Σ are all in the closed left-half complex plane, i.e., the system Σ has at most weakly non-minimum-phase constraints.*

(iii) *The constraints are either right invertible or weakly non-right invertible.*[2]

(iv) *If the constraints are weakly non-right invertible, then the set $T(\xi)$ is non-empty for all $\xi \in \partial \mathcal{V}_b(\overline{\mathcal{S}})$.*

We note that the constraint sets for most systems are polyhedral. In this case, the necessary and sufficient conditions stated in Theorem 8.12 can be simplified, that is, we only need to check a finite number of corner points for the condition (*iv*) of Theorem 8.12.

[2]Weakly non-right-invertible constraints are defined in Definition 7.16.

Theorem 8.13 *Let the assumptions stated in Theorem 8.12 hold and moreover assume that the constraint set $\mathcal{S}$ is polyhedral. Then we have:*

(i) The set $\mathcal{V}_b(\overline{\mathcal{S}})$ is also polyhedral; moreover, it has no more corner points than the set $\overline{\mathcal{S}}$.

(ii) $T(\xi)$ is nonempty for all $\xi \in \partial\mathcal{V}_b(\overline{\mathcal{S}})$ provided that $T(\xi)$ is nonempty for all corner points of $\mathcal{V}_b(\overline{\mathcal{S}})$.

We now move on to the case of measurement feedback that concerns with the semi-global stabilization. We have the following theorem.

Theorem 8.14 *Let the assumptions stated in Theorem 8.12 hold. Then, the semi-global stabilization via measurement feedback problem is solvable if the following conditions hold:*

(i) (A, B) is stabilizable.

(ii) The constraint invariant zeros of the given system Σ are all in the closed left-half complex plane, i.e., the system Σ has at most weakly non-minimum-phase constraints.

(iii) The constraints are either right invertible or weakly non-right invertible.

(iv) If the constraints are not right invertible, then the set $T(\xi)$ is nonempty for all $\xi \in \partial\mathcal{V}_b(\overline{\mathcal{S}})$.

(v) The pair (C_y, A) is observable.

Moreover, the first four conditions are necessary.

Remark 8.15 *We observe that for semi-global stabilization, the conditions for the case of right-invertible constraints remain necessary even for the case of non-right-invertible constraints. However, these conditions by themselves are no longer sufficient. In fact, as seen above, an intricate set of additional conditions are required.*

Proofs and construction of controllers

We proceed now to prove Theorems 8.12–8.14 and to construct appropriate controllers. We first need some definitions. We will be dealing extensively with recoverable regions or sets and their computation in Chap. 9; however, at this time, the following definition will enhance our presentation.

Definition 8.16 *Let a system Σ as in* (7.1) *and a constraint set $\mathcal{S}$ that satisfies Assumption 7.1 be given. Then, the* recoverable region $\mathcal{R}(\Sigma, \mathcal{S})$ *is defined to be the set of all initial states $x(0)$ in the set of admissible initial conditions $\mathcal{V}(\mathcal{S})$ (where $\mathcal{V}(\mathcal{S}) = \mathcal{V}(\mathcal{S}, \mathbb{R}^p)$) for which there exists a control input u such that $x(t) \to 0$ as $t \to \infty$ while $z(t) \in \mathcal{S}$ for all $t \geqslant 0$.*

For brevity, it is beneficial to define the domain of attraction $\mathcal{R}^f(\Sigma, \mathcal{S})$ of a given system Σ with a feedback control law $u = f(x,t)$.

Definition 8.17 *Let a system Σ as in* (7.1) *and a constraint set $\mathcal{S}$ that satisfies Assumption 7.1 be given. Assume $u = f(x,t)$ is a control law for the system Σ. The* domain of attraction $\mathcal{R}^f(\Sigma, \mathcal{S})$ *is defined to be the set of all $x(0) \in \mathbb{R}^n$ such that the state trajectory of the closed-loop system satisfies $x(t) \to 0$ as $t \to \infty$ while $z(t) \in \mathcal{S}$ for all $t > 0$. In this case, we say that the controller f achieves the domain of attraction $\mathcal{R}^f(\Sigma, \mathcal{S})$.*

We present next a reduction procedure that reduces the posed stabilization of the given system Σ to that of a lower order system having a special structure. Then, we establish semi-global stabilization for the reduced system.

Reduction of the stabilization problem

We first establish a relationship between the semi-global stabilization of the original system Σ and that of the subsystem Σ_1 in (8.30). Let Π_{ab} be the projection operator from $\mathbb{R}^n$ to $\mathbb{R}^{n_1}$ containing states x_a and x_b.

We have the following result.

Theorem 8.18 *Let the given system Σ as in* (7.1) *satisfy Assumption 7.1. Suppose the system Σ_{uz} is represented in its SCB. Then we have:*

(i) *If $\bar{u} = f(\bar{x})$ is a stabilizing control law for the system $\bar{\Sigma}$ given in* (8.29) *with the domain of attraction $\mathcal{R}^f(\bar{\Sigma}, \bar{\mathcal{S}})$, then, for any $\tilde{\rho} \in [0,1)$, there is a control law $\zeta = f_1(x_{ab})$ for the system Σ_1 such that*

$$\Pi_{ab}\left\{\tilde{\rho}\mathcal{R}^f(\bar{\Sigma}, \bar{\mathcal{S}})\right\} \subset \mathcal{R}^{f_1}(\Sigma_1, \bar{\mathcal{S}}). \tag{8.34}$$

(ii) *If $\zeta = f_1(x_{ab})$ is a stabilizing law for the system Σ_1 with the domain of attraction $\mathcal{R}^{f_1}(\Sigma_1, \bar{\mathcal{S}})$, then for any $\tilde{\rho} \in [0,1)$, there is a stabilizing law $\bar{u} = f(\bar{x})$ for the system $\bar{\Sigma}$ such that*

$$\tilde{\rho}\mathcal{R}^{f_1}(\Sigma_1, \bar{\mathcal{S}}) \subset \Pi_{ab}\left\{\mathcal{R}^f(\bar{\Sigma}, \bar{\mathcal{S}})\right\}. \tag{8.35}$$

For the proof of this theorem, we need to recall the following important result from [167] which is a consequence of Theorem 9.18 which will be presented in the next chapter.

Lemma 8.19 *Let the given system Σ in (7.1) satisfy Assumption 7.1. Assume that there exists a state feedback f (discontinuous or dynamic) with a domain of attraction $\mathcal{R}^f(\Sigma, \mathcal{S})$. Then, for any $\widetilde{\rho} \in [0, 1)$, there exists a globally Lipschitz static controller g such that*

$$\widetilde{\rho}\mathcal{R}^f(\Sigma, \mathcal{S}) \subset \mathcal{R}^g(\Sigma, \mathcal{S})$$

while guaranteeing uniform exponential convergence to the origin for all initial conditions in the set $\widetilde{\rho}\mathcal{R}^f(\Sigma, \mathcal{S})$.

Proof of Theorem 8.18 : The proof is carried out assuming that Σ_{uz} is in its SCB. Let $\overline{u} = f(\overline{x})$ be a stabilizing controller for $\overline{\Sigma}$ with a domain of attraction $\mathcal{R}^f(\overline{\Sigma}, \overline{\mathcal{S}})$. If we view the following dynamics

$$\Sigma_F : \begin{cases} \dot{x}_c &= A_{cc}x_c + B_c(u_c + H_a x_a) + K_c\overline{z} \\ \dot{x}_d &= A_{dd}x_d + B_d(u_d + G\overline{x}) + K_d\overline{z} \\ \zeta &:= \begin{pmatrix} z_0 \\ z_d \end{pmatrix} = \begin{pmatrix} 0 \\ C_d \end{pmatrix} x_d + \begin{pmatrix} I \\ 0 \end{pmatrix} u_0 \end{cases} \tag{8.36}$$

with input x_a and x_b and output ζ as a controller for the subsystem Σ_1, it is obvious that this controller is stabilizing and achieves a domain of attraction $\mathcal{R}^f(\overline{\Sigma}, \overline{\mathcal{S}})$. By Lemma 8.19, there is a static controller f_1 which achieves a domain of attraction satisfying (8.34). This proves part (*i*).

Conversely, let a controller for Σ_1 with a domain of attraction $\mathcal{R}^{f_1}(\Sigma_1, \overline{\mathcal{S}})$ be given. Then, according to Lemma 8.19, we can get a globally Lipschitz controller for Σ_1 that achieves exponential stability and achieves a domain of attraction containing $\widetilde{\rho}_1\mathcal{R}^{f_1}(\Sigma_1, \overline{\mathcal{S}})$ for any $\widetilde{\rho}_1 \in (\widetilde{\rho}, 1)$. If we add a disturbance to the system,

$$\Sigma_1^d : \begin{cases} \dot{x}_a = A_{aa}x_a + K_{ab}C_b x_b + K_{a2}(\zeta + d) \\ \dot{x}_b = (A_{bb} + K_{bb}C_b)\, x_b + K_{b2}(\zeta + d) \\ \overline{z} = \begin{pmatrix} C_b \\ 0 \end{pmatrix} x_b + \begin{pmatrix} 0 \\ I \end{pmatrix} \zeta, \end{cases} \tag{8.37}$$

with d satisfying $\|d(t)\| \leqslant Me^{-\gamma t}$ for all $t > 0$, then the fact that the controller achieves exponential stability guarantees that for γ large enough we have, for all initial conditions in $\widetilde{\rho}_2\mathcal{R}^{f_1}(\Sigma_1, \overline{\mathcal{S}})$ with $\widetilde{\rho}_2 \in (\widetilde{\rho}, \widetilde{\rho}_1)$, that the constraints are satisfied for all time and the state converges to zero as time tends to infinity.

Given a controller for Σ_1 which can handle exponentially decaying disturbances and achieves a domain of attraction $\widetilde{\rho}_2\mathcal{R}^{f_1}(\Sigma_1, \overline{\mathcal{S}})$, the technique used in Step 2 on page 419 gives an explicit design for the full system with a domain of attraction satisfying (8.35). ■

Theorem 8.18 leads to the following corollary immediately.

Corollary 8.20 *Let the given system Σ in* (7.1) *satisfy Assumption 7.1. Suppose the system Σ_{uz} is represented in its SCB. Then the system Σ is semi-globally stabilizable if and only if the system Σ_1 is semi-globally stabilizable.*

In view of the above corollary, the semi-global stabilization problem of system Σ can be studied by that of subsystem Σ_1. We pursue below a further reduction of subsystem Σ_1 under the condition that the constraints are at most weakly non-minimum phase. For this purpose, we extract the following subsystem from Σ_1,

$$\Sigma_b : \begin{cases} \dot{x}_b = (A_{bb} + K_{bb}C_b)\, x_b + K_{b2}\zeta, \\ \overline{z} = \begin{pmatrix} C_b \\ 0 \end{pmatrix} x_b + \begin{pmatrix} 0 \\ I \end{pmatrix} \zeta \end{cases} \tag{8.38}$$

which contains only the dynamics of x_b. Analogously, we can establish a connection between the semi-global stabilization of the subsystems Σ_1 and Σ_b. Let Π_b be the projection operator from $\mathbb{R}^{n_1}$ to $\mathbb{R}^{n_b}$ containing only state x_b. We have the following result.

Theorem 8.21 *Let the given system Σ in* (7.1) *satisfy Assumption 7.1 and the system Σ_{uz} be represented in its SCB. Assume that the constraints are non-right invertible and at most weakly non-minimum phase. Then*

(*i*) *If $\zeta = f(x_{ab})$ is a stabilizing controller for the system Σ_1 with a domain of attraction $\mathcal{R}^f(\Sigma_1, \overline{\mathcal{S}})$, then for any $\widetilde{\rho} \in [0, 1)$, there is a stabilizing controller $\zeta = f_b(x_b)$ for the system Σ_b such that*

$$\Pi_b \left\{ \widetilde{\rho}\mathcal{R}^f(\Sigma_1, \overline{\mathcal{S}}) \right\} \subset \mathcal{R}^{f_b}(\Sigma_b, \overline{\mathcal{S}}). \tag{8.39}$$

(*ii*) *If $\zeta = f_b(x_b)$ is a stabilizing controller for the system Σ_b with a domain of attraction $\mathcal{R}^{f_b}(\Sigma_b, \overline{\mathcal{S}})$, then for any $\widetilde{\rho} \in [0, 1)$ and any compact set $\mathcal{K}_a \subset \mathbb{R}^{n_a}$, there is a stabilizing controller $\zeta = f(x_{ab})$ for the system Σ_1 such that*

$$\mathcal{K}_a \times \widetilde{\rho}\mathcal{R}^{f_b}(\Sigma_b, \overline{\mathcal{S}}) \subset \mathcal{R}^f(\Sigma_1, \overline{\mathcal{S}}). \tag{8.40}$$

Proof : The proof of part (*i*) is similar to that of part (*i*) of Theorem 8.18. Given a stabilizing controller $\zeta = f(x_{ab})$ for Σ_1 which achieves a domain of attraction $\mathcal{R}^f(\Sigma_1, \overline{\mathcal{S}})$, the following dynamics

$$\Sigma_a : \left\{ \dot{x}_a = A_{aa}x_a + K_{ab}C_b x_b + K_{a2}\zeta, \right. \tag{8.41}$$

where $\zeta = f(x_{ab})$, can be viewed as a dynamic controller for Σ_b. This controller is stabilizing and achieves a domain of attraction $\mathcal{R}^f(\Sigma_1, \bar{\mathcal{S}})$. We can find a static controller to achieve a domain of attraction satisfying (8.39) according to Lemma 8.19.

Conversely, given a controller $\zeta = f_b(x_b)$ for Σ_b which achieves a domain of attraction $\mathcal{R}^{f_b}(\Sigma_b, \bar{\mathcal{S}})$, according to Lemma 8.19, we can get a globally Lipschitz controller $\zeta = \bar{f}_b(x_b)$ for Σ_b that achieves exponential stability with a domain of attraction containing $\tilde{\rho}\mathcal{R}^{f_b}(\Sigma_b, \bar{\mathcal{S}})$. Therefore, we have

$$\|x_b(t)\| \leqslant Me^{-\varepsilon t} \|x_b(0)\| . \tag{8.42}$$

Next, let P_0 be the semi-stabilizing solution of the algebraic Riccati equation,

$$A_0' P_0 + P_0 A_0 - P_0 B_0 B_0' P_0 + C_0' C_0 = 0,$$

where

$$A_0 = \begin{pmatrix} A_{aa} & K_{ab} C_b \\ 0 & \tilde{A}_b \end{pmatrix}, \quad B_0 = \begin{pmatrix} K_{a2} \\ K_{b2} \end{pmatrix}, \quad C_0 = \begin{pmatrix} 0 & C_b \end{pmatrix}.$$

It is clear to see that

$$P_0 \begin{pmatrix} x_a \\ 0 \end{pmatrix} = 0 \tag{8.43}$$

for all $x_a \in \mathbb{R}^{n_a}$. Let $c > 0$ be such that for all $\xi \in \mathcal{V}_0(c) := \{\xi \in \mathbb{R}^{n_1} \mid \xi' P_0 \xi \leqslant c\}$,

$$\begin{pmatrix} C_0 \\ -B_0' P_0 \end{pmatrix} \xi \in \bar{\mathcal{S}}/3.$$

Equations (8.42) and (8.43) together imply that with the feedback $\zeta = \bar{f}_b(x_b)$ there exists a $T > 0$ such that for all initial conditions in the set $\tilde{\rho}\mathcal{R}^{f_1}(\Sigma_1, \bar{\mathcal{S}})$ we have

$$\begin{pmatrix} x_a(T) \\ x_b(T) \end{pmatrix} \in \mathcal{V}_0(c). \tag{8.44}$$

Furthermore, for all initial conditions

$$\begin{pmatrix} x_a(0) \\ x_b(0) \end{pmatrix} \in \mathcal{K}_a \times \tilde{\rho}\mathcal{R}^{f_b}(\Sigma_b, \bar{\mathcal{S}}),$$

there exists a compact set $\mathcal{M} \supset \mathcal{K}_a$ such that $x_a(T) \in \mathcal{M}$.

Define P_ε as the stabilizing solution of the algebraic Riccati equation,

$$A_0' P_\varepsilon + P_\varepsilon A_0 - P_\varepsilon B_0 B_0' P_\varepsilon + C_0' C_0 + \varepsilon I = 0.$$

We have $P_\varepsilon \to P_0$ as $\varepsilon \to 0$. Let $\mathcal{V}_\varepsilon(c) := \{\xi \in \mathbb{R}^{n_1} \mid \xi' P_\varepsilon \xi \leqslant c\}$. Then (8.44) and $x_a(T) \in \mathcal{M}$ with $\mathcal{M}$ compact implies that there exists an $\varepsilon > 0$ such that the state is in the set $2\mathcal{V}_\varepsilon(c)$ at time T for any initial condition in $\mathcal{K}_a \times \tilde{\rho}\mathcal{R}^{f_b}(\Sigma_b, \bar{\mathcal{S}})$. Moreover, for ε small enough, it holds that

$$\begin{pmatrix} C_0 \\ -B_0' P_\varepsilon \end{pmatrix} \xi \in \bar{\mathcal{S}}$$

for any initial condition $\xi \in 2\mathcal{V}_\varepsilon(c)$. Let $\zeta = -B_0' P_\varepsilon \xi$ be the feedback for $\xi \in 2\mathcal{V}_\varepsilon(c)$, which is an asymptotically stabilizing law for the system Σ_1. Moreover, this feedback makes the set $2\mathcal{V}_\varepsilon(c)$ invariant thus, the set $2\mathcal{V}_\varepsilon(c)$ is in the domain of attraction.

Consequently, we have constructed a composite controller,

$$\zeta = f(x_{ab}) := \begin{cases} \bar{f}(x_b), & x_{ab} \notin 2\mathcal{V}_\varepsilon(c), \\ -B_0' P_\varepsilon x_{ab}, & x_{ab} \in 2\mathcal{V}_\varepsilon(c), \end{cases}$$

with its domain of attraction satisfying (8.40). ■

The following corollaries are direct consequences of Theorems 8.18 and 8.21.

Corollary 8.22 *Let the given system Σ in* (7.1) *satisfy Assumption 7.1. Suppose the system Σ_{uz} is represented in its SCB. Denote by $\bar{\Pi}_b$ the projection from $\bar{x}$ to x_b. Assume that the constraints are at most weakly non-minimum phase and non-right invertible. Then*

(i) *If $\zeta = f(\bar{x})$ is a stabilizing controller for the system $\bar{\Sigma}$ with a domain of attraction $\mathcal{R}^f(\bar{\Sigma}, \bar{\mathcal{S}})$, then for any $\tilde{\rho} \in [0, 1)$, there is a stabilizing controller $\zeta = f_b(x_b)$ for the system Σ_b such that*

$$\bar{\Pi}_b \left\{ \tilde{\rho}\mathcal{R}^f(\bar{\Sigma}, \bar{\mathcal{S}}) \right\} \subset \mathcal{R}^{f_b}(\Sigma_b, \bar{\mathcal{S}}). \tag{8.45}$$

(ii) *If $\zeta = f_b(x_b)$ is a stabilizing controller for the system Σ_b with a domain of attraction $\mathcal{R}^{f_b}(\Sigma_b, \bar{\mathcal{S}})$, then for any $\tilde{\rho} \in [0, 1)$, there is a stabilizing controller $\zeta = f(\bar{x})$ for the system $\bar{\Sigma}$ such that*

$$\tilde{\rho}\mathcal{R}^{f_b}(\Sigma_b, \bar{\mathcal{S}}) \subset \bar{\Pi}_b \left\{ \mathcal{R}^f(\bar{\Sigma}, \bar{\mathcal{S}}) \right\}. \tag{8.46}$$

Corollary 8.23 *Let the given system Σ in* (7.1) *satisfy Assumption 7.1. Suppose the system Σ_{uz} is represented in its SCB. Then the system Σ is semi-globally stabilizable if and only if the system Σ_b is semi-globally stabilizable.*

In summary, if the constraints are at most weakly non-minimum phase and non-right invertible, whether or not the semi-global stabilization of Σ is possible is determined by the subsystem Σ_b.

Semi-global stabilization of Σ_b

We proceed now to semi-globally stabilize Σ_b. We first establish an auxiliary lemma.

Lemma 8.24 *Consider a strictly proper linear time-invariant system*

$$\begin{cases} \dot{x} = Ax + Bu \\ z = Cx \end{cases}$$

which is left invertible with no invariant zeros in $\mathbb{C}^- \cup \mathbb{C}^0$ and subject to the constraint

$$\begin{pmatrix} z \\ u \end{pmatrix} \in \mathcal{B}$$

for all $t > 0$ with $\mathcal{B}$ compact, convex, and containing a neighborhood of 0. *Then the recoverable set for this system is bounded.*

Proof : If all the eigenvalues of A are in the open right-half plane, then due to the fact that the input is bounded, the recoverable set is bounded. Otherwise, in a suitable basis, the system can be written as

$$\begin{aligned} \dot{\bar{x}} &= \begin{pmatrix} A_u & 0 \\ 0 & A_s \end{pmatrix} \begin{pmatrix} x_u \\ x_s \end{pmatrix} + \begin{pmatrix} B_u \\ B_s \end{pmatrix} u \\ z &= \begin{pmatrix} C_u & C_s \end{pmatrix} \begin{pmatrix} x_u \\ x_s \end{pmatrix}, \end{aligned}$$

where the eigenvalues of A_u are antistable and the eigenvalues of A_s are in the closed left-half plane. Since the system is assumed left invertible, all unobservable eigenvalues are invariant zeros. Since there are no stable invariant zeros, we find that (C_s, A_s) is observable. Obviously, x_u must be bounded by the same reason stated at the beginning. On the other hand, x_s has to be bounded as well. This follows from

$$z(t) = C_u x_u(t) + C_s e^{A_s t} x_{s0} + \int_0^t C_s e^{A_s(t-\tau)} B_s u(\tau) \mathrm{d}\tau.$$

If x_{s0} is unbounded, then there exists a x_{s0} such that the above equation fails to hold because all other terms $z(t)$, $x_u(t)$ and the integral are bounded and (C_s, A_s) is observable. ■

It follows from Lemma 8.24 that the recoverable set of Σ_b is bounded because Σ_b is left invertible and has no invariant zeros in the closed left-half plane.

A necessary and sufficient condition for semi-global stabilization is that the closure of the recoverable set is equal to the closure of the admissible set. Hence, a necessary condition for semi-global stabilization of Σ_b is that the admissible set is bounded. This requires that C_b be injective since all the components of x_b have to be constrained. We obtain the following lemma.

Lemma 8.25 *Let the given system Σ in* (7.1) *satisfy Assumption 7.1. Suppose the system Σ_{uz} is represented in its SCB. If the system Σ_b is semi-globally stabilizable, then C_b is injective, i.e.,* (7.6) *holds.*

The next lemma gives a necessary and sufficient condition for the semi-global stabilization of the system Σ_b.

Lemma 8.26 *Semi-global stabilization for the system Σ_b given by* (8.38) *is possible only if C_b is injective and for all $\xi \in \partial \mathcal{V}_b(\bar{\mathcal{S}})$, the set $T(\xi)$ defined in* (8.33) *is nonempty.*

On the other hand, semi-global stabilization for the system Σ_b given by (8.38) *is possible if additionally $\bar{\mathcal{S}}$ is polytopic and defined by*

$$\bar{\mathcal{S}} = \{\, x_b \mid Lx_b \leqslant c \text{ componentwise} \,\},$$

where L has rows L_i and c has elements such that if for any row L_i we have $L_i K_{b2} = 0$, then we have

$$L_i \tilde{A}_b x_b < 0 \text{ for all } x_b \in \mathcal{V}_b(\bar{\mathcal{S}}) \text{ with } L_i x_b = c_i$$

Remark 8.27 *The fact that $T(\xi)$ is nonempty on the boundary of $\mathcal{V}_b(\bar{\mathcal{S}})$ implies that, if for any row L_i we have $L_i K_{b2} = 0$, then we have*

$$L_i \tilde{A}_b x_b \leqslant 0 \text{ for all } x_b \in \mathcal{V}_b(\bar{\mathcal{S}}) \text{ with } L_i x_b = c_i\,.$$

In other words, the sufficient conditions are only slightly stronger since they only replace an inequality by a strict inequality for those rows satisfying the specific property that $L_i K_{b2} = 0$.

Proof : The proof of this lemma uses the set-valued mapping techniques which will be presented in Chap. 9.

(Necessity) If semi-global stabilization is possible, the closure of the admissible set is equal to the closure of the recoverable set. By Lemma 9.21, for any initial condition in $\partial \mathcal{V}_b(\bar{\mathcal{S}})$, there exists an input ζ such that the trajectory stays in the closure of the admissible set. Hence $T_1(\xi)$ is nonempty.

(Sufficiency) Let $\Sigma_{b,\kappa}$ denote the system Σ_b defined in (8.38) with $\widetilde{A}_b$ replaced by $\widetilde{A}_b + \kappa I$ where $\kappa \geqslant 0$. Let $N_{\kappa,\rho}(\xi)$ and $T_{\kappa,\rho}(\xi)$ be the sets defined as $N(\xi)$ and $T(\xi)$ in (8.32) and (8.33) but with $\widetilde{A}_b$ replaced by $\widetilde{A}_b + \kappa I$ and $\mathcal{V}_b(\overline{\mathcal{S}})$ replaced by the recoverable set $\rho\mathcal{V}_b(\overline{\mathcal{S}})$.

Due to the extra condition, it should be noted that not only $T_{0,\rho}(\xi)$ is nonempty for all $\xi \in \rho\partial\mathcal{V}_b(\overline{\mathcal{S}})$ when $\rho < 1$ but also that we can actually choose ζ such that the derivative actually points inside $\mathcal{V}_b(\rho\overline{\mathcal{S}})$ (i.e., we have $\langle \mu, \eta \rangle < 0$ for all $\eta \in N(\xi)$ where μ as defined in the definition of $T(\xi)$). This implies that $T_{0,\rho}(\xi)$ is lower semicontinuous and also that the set-valued mapping $T_{\kappa,\rho}(\xi)$ is also nonempty for all $\xi \in \rho\partial\mathcal{V}_b(\overline{\mathcal{S}})$ provided κ is sufficiently small. Then, by Theorem 9.33, there exists a continuous mapping $\zeta = f(x_b)$ defined on $\rho\partial\mathcal{V}_b(\overline{\mathcal{S}})$ such that for all $x_b \in \widetilde{\rho}\partial\mathcal{V}_b(\overline{\mathcal{S}})$, we have $\dot{x}_b \in N_{\widetilde{\rho}}(x_b)$ (using $\widetilde{A}_b + \kappa I$ instead of $\widetilde{A}_b$). Moreover, since $\dot{x}_b$ can be actually chosen to point inside and not just along the boundary of $\rho\mathcal{V}_b(\overline{\mathcal{S}})$, we can modify this feedback slightly without affecting the property that $\dot{x}_b \in N_{\widetilde{\rho}}(x_b)$ to ensure that the feedback becomes Lipschitz continuous. This feedback can be extended to the whole state space by imposing that the feedback be homogeneous (which uniquely determines the feedback). Then, by Theorem 9.31, we conclude that this feedback makes the set $\rho\mathcal{V}_b(\overline{\mathcal{S}})$ invariant using $\widetilde{A}_b + \kappa I$ instead of $\widetilde{A}_b$. Then it is easily verified that for the original system we have $\rho\mathcal{V}_b(\overline{\mathcal{S}})$ invariant but additionally the state converging to zero exponentially for all initial conditions inside $\rho\mathcal{V}_b(\overline{\mathcal{S}})$. Since we can choose ρ arbitrarily close to 1, we note that we can in this way achieve semi-global stabilization in the admissible set. ∎

Proof of Theorem 8.12 : This theorem is a completion of Theorem 8.1 where the right-invertible constraints are considered. If the constraints are weakly non-right invertible, the condition *(iv)* follows from Lemma 8.26. ∎

Proof of Theorem 8.13 : Part *(i)* can be easily checked from the definition of the set $\mathcal{V}_b(\overline{\mathcal{S}})$ after noting that a linear transformation does not increase the number of corner points. Regarding part *(ii)*, we note that the crucial aspect is that if $T(\xi_1)$ and $T(\xi_2)$ are both nonempty, then $T\ (\lambda\xi_1 + (1-\lambda)\xi_2)$ is also nonempty for any $\lambda \in [0, 1]$. Since a polyhedral set is the convex hull of its corner points, the result thus follows. ∎

In order to prove Theorem 8.14, we need two lemmas. Lemma 4.62 has already been presented earlier in the book. This lemma is the first preparatory step toward an observer design. However, if we adopt a fast observer to make an extremely accurate estimation of the state, there is an unavoidable annoying phenomenon called *peaking* associated with high-gain observer design. The peaking phenomenon must be taken care of seriously, for we are facing a control system with constraints. Fortunately, we have another lemma which provides us a mechanism to avoid the negative effect of peaking. As the second step, we recall the lemma here because it is instrumental to the high-gain observer design.

Lemma 8.28 *Consider the system*

$$\dot{\eta} = (A - LC)\eta \tag{8.47}$$

where $A \in \mathbb{R}^{n\times n}$ and (C, A) is observable. Then, for any $N > 0$, $\varepsilon > 0$, $r > 0$, and $\tau > 0$, there exists a matrix L such that $A - LC + rI$ is Hurwitz stable and

$$\|\eta(t)\| \leqslant \varepsilon e^{-rt} \tag{8.48}$$

for all $t \geqslant \tau$ and for all initial conditions $\eta(0) \in \mathbb{R}^n$ satisfying $\|\eta(0)\| \leqslant N$.

Remark 8.29 *Note that the high-gain observers designed in Chap. 4 based on either the direct method or on Riccati equations will satisfy the above property for ℓ sufficiently large.*

Proof : This lemma follows from the results in Izmaĭlov [53, 54] and Theorem 8.2 in Sussmann and Kokotovic [172]. Since (C, A) is observable, for any $\rho > r$, there exists a matrix L such that $A - LC + 2\rho I$ is Hurwitz stable. Let ν be the largest observability index of the pair (C, A). Then according to Theorem 8.2 in [172], the state η has a peaking exponent $\nu - 1$. In other words, there exists a constant $\alpha > 0$ independent of ρ such that

$$\|\eta(t)\| \leqslant \alpha\|\eta(0)\|(2\rho)^{\nu-1}e^{-2\rho t}$$

for all $t \geqslant 0$. For any given $\varepsilon > 0$, $\tau > 0$, and $r > 0$, we choose $\rho > r$ large enough so that $\alpha\|\eta(0)\|(2\rho)^{\nu-1}e^{-\rho t} \leqslant \varepsilon$ for $t \geqslant \tau$. Then this choice of L which guarantees Hurwitz stability of $A - LC + 2\rho I$ leads to η satisfying (8.48) for all $t \geqslant \tau$ and for any initial condition $\eta(0) \in \mathbb{R}^n$ satisfying $\|\eta(0)\| \leqslant N$. ■

So far, we have prepared enough tools for the observer design. As the last step in the observer design, we construct the observer-based controller in the proof of Theorem 8.14.

Proof of Theorem 8.14 : The necessity of the first four conditions is a consequence of the state feedback design. The sufficiency of the conditions are proven by an explicit design as presented below.

We need a high-gain observer to estimate the state. The observer takes the standard form

$$\dot{\widehat{x}} = A\widehat{x} + Bu + L(y - C_y\widehat{x}).$$

By the assumption that (C_y, A) is observable, we can choose a gain matrix L such that the eigenvalues of the matrix $A_{\text{obs}} := A - LC_y$ can be assigned anywhere in the left-half complex plane. The estimation error $e := \widehat{x} - x$ satisfies

$$\dot{e} = A_{\text{obs}}e. \tag{8.49}$$

Our goal is to devise a measurement feedback such that the set $\mathcal{X} \times \mathcal{V}$ is contained in the domain of attraction; meanwhile, for all initial states in this set, the constraints are satisfied. Due to the possible peaking of the state estimate caused by the high-gain observer, the state estimate during the short period at the beginning of time is not useful. To insure that the constraints are not violated, we need to borrow an idea from Esfandiari and Khalil [35], and saturate the control so that the peaking signal does not enter the plant. The appropriate level of saturation is to be specified below. In this way the control law always generates bounded control signal, regardless of the peaking. Then, the design objective is to guarantee that the control law is functioning as closely as the state feedback law after the peaking is over. But during the short period of peaking, the state starting from $\mathcal{X}$ may drift to a larger set, say $\widehat{\mathcal{X}}$, because we are not controlling the plant, although the input is kept bounded. For this reason, we need two components in our design. One is to design from the beginning a state feedback $u = Fx$ for a larger set of initial condition, say set $\widehat{\mathcal{X}}$ which satisfies $\mathcal{X} \subset \operatorname{int} \widehat{\mathcal{X}}$, and make sure that $\widehat{\mathcal{X}}$ is contained in the domain of attraction and for all initial conditions in the set $\widehat{\mathcal{X}}$ the output $z(t) \in \rho\mathcal{S}$ for some $\rho \in (0, 1)$. One can use the design technique provided in [124] for this task. The other component is a saturation element, the level of which is specified below.

Consider the following system:

$$\dot{x} = (A + BF)x + BFe,$$

where e is the estimation error. It follows from Lemma 4.62 that if $e(t)$ satisfies

$$\|e(t)\| \leqslant \varepsilon e^{-rt} \tag{8.50}$$

for certain $\varepsilon \in (0, 1)$, $r > 0$ and for all $t > 0$, then for all initial conditions in the set $\widehat{\mathcal{X}}$, the constraints on z are satisfied; meanwhile, the state trajectory remains in a compact set $\Omega(\widehat{\mathcal{X}}) \subset \operatorname{int} \mathcal{A}(\mathcal{S})$ and the state converges to zero. Hence, there exists an $M_1 > 0$ such that $\|Fx\|_\infty \leqslant M_1$ for all $x(0) \in \widehat{\mathcal{X}}$ and for all e satisfying (8.50), where

$$M_1 = \sup_{x \in \Omega(\widehat{\mathcal{X}})} \|Fx\|.$$

Define $M_2 = \varepsilon\|F\|$. Let $\tau > 0$ be such that

$$\dot{x} = Ax + Bu$$

satisfies $x(t) \in \widehat{\mathcal{X}}$ for all $t \in [0, \tau]$ for all u satisfying $\|u(t)\| \leqslant M_1 + M_2$ and for all $x(0) \in \mathcal{X} \subset \operatorname{int} \widehat{\mathcal{X}}$. Then, we can choose the observer gain matrix L so that the error-bound (8.50) holds for $t \geqslant \tau$, where τ, ε, and r are as specified above. Consequently, the combination of the observer and the state feedback

$$u = \operatorname{sat}_{M_1+M_2}(F\hat{x}),$$

where $\text{sat}_M(\cdot)$ is a standard saturation function with saturation level M, has the following properties: For any given initial state $x(0) \in \mathcal{X}$, we have $x(t) \in \widehat{\mathcal{X}}$ for all $t \in [0, \tau]$, and for $t \geqslant \tau$, we have

$$u(t) = \text{sat}_{M_1+M_2}(F\hat{x}(t)) = F\hat{x}(t) = Fx(t) + Fe(t).$$

Then stabilization follows from Lemma 8.28. ∎

8.3.1 *Exploration of complexity of non-right-invertible constraints*

In this subsection, we proceed to answer the questions we posed earlier in order to shed some light on the complexities inherent in dealing with systems having non-right-invertible constraints. The two questions are as follows:

(*i*) Given a system Σ with non-right-invertible constraints, for such a Σ, what are the solvability conditions for the Problems 7.8–7.11 when they are enhanced in the sense that they consider all possible pairs of constraint sets $\mathcal{S}$ and $\mathcal{T}$ that satisfy Assumption 7.1 instead of just one given pair of constraint sets $\mathcal{S}$ and $\mathcal{T}$?

(*ii*) Given a system Σ with non-right-invertible constraints, for such a Σ, does there exist a pair of constraint sets $\mathcal{S}$ and $\mathcal{T}$ that satisfy Assumption 7.1 for which semi-global and global stabilization problems as formally formulated in Problems 7.8–7.11 are solvable?

At first, we proceed to answer the first question. As in the previous section, we need to rewrite Σ in a particular form. In fact, the SCB given in (8.29) can be rewritten by using a preliminary state feedback in the form

$$\begin{aligned}
\begin{pmatrix} \dot{x}_{ab} \\ \dot{x}_{cd} \end{pmatrix} &= \begin{pmatrix} A_{11} & 0 \\ A_{21} & A_{22} \end{pmatrix} \begin{pmatrix} x_{ab} \\ x_{cd} \end{pmatrix} + \begin{pmatrix} 0 \\ B_2 \end{pmatrix} \widetilde{u} + \begin{pmatrix} K_1 \\ K_2 \end{pmatrix} z, \\
z &= \begin{pmatrix} C_{z,1} & C_{z,2} \end{pmatrix} \begin{pmatrix} x_{ab} \\ x_{cd} \end{pmatrix} + \widetilde{D}_z \widetilde{u},
\end{aligned} \tag{8.51}$$

where x_{ab} consists of x_a and x_b while x_{cd} consists of x_c and x_d, respectively. Also note that in the above,

$$\widetilde{u} = \begin{pmatrix} u_c + H_a x_a \\ u_d + G\overline{x} \end{pmatrix}.$$

Also, we can decompose the constrained output $\overline{z}$ as

$$\overline{z} = \begin{pmatrix} z_b \\ z_d \\ z_0 \end{pmatrix} = \begin{pmatrix} C_{z,12} \\ 0 \\ 0 \end{pmatrix} x_{ab} + \begin{pmatrix} 0 \\ \overline{C}_1 \\ 0 \end{pmatrix} x_{cd} + \begin{pmatrix} 0 \\ 0 \\ D_1 \end{pmatrix} u_0. \tag{8.52}$$

Note that choosing a basis in the constrained output space affects the constraint set $\mathcal{S}$ since we moved from output z to output $\bar{z} = \Gamma_z z$, and therefore, we obtain new constraint set $\widetilde{\mathcal{S}}$.

Consider (8.51) together with the $\bar{z}$ as in (8.52). By decomposing x_{ab} into x_a and x_b and only considering z_b as output, we obtain the following subsystem:

$$\begin{aligned}\begin{pmatrix}\dot{x}_a\\ \dot{x}_b\end{pmatrix} &= \begin{pmatrix}\widetilde{A}_1^1 & \widetilde{A}_1^2\\ 0 & \widetilde{A}_1^3\end{pmatrix}\begin{pmatrix}\widetilde{x}_1^1\\ \widetilde{x}_1^2\end{pmatrix} + \begin{pmatrix}\widetilde{B}_1^1\\ \widetilde{B}_1^2\end{pmatrix}\bar{z},\\ z_b &= \begin{pmatrix}0 & \widetilde{C}_1^2\end{pmatrix}\begin{pmatrix}\widetilde{x}_1^1\\ \widetilde{x}_1^2\end{pmatrix},\end{aligned} \tag{8.53}$$

where x_a represent the zero dynamics. Using this decomposition explicitly, we get

$$\begin{aligned}\begin{pmatrix}\dot{\widetilde{x}}_1^1\\ \dot{\widetilde{x}}_1^2\end{pmatrix} &= \begin{pmatrix}\widetilde{A}_1^1 & \widetilde{A}_1^2\\ 0 & \widetilde{A}_1^3\end{pmatrix}\begin{pmatrix}x_a\\ x_b\end{pmatrix}\\ &\quad + \begin{pmatrix}\widetilde{B}_1^{11} & \widetilde{B}_1^{12} & \widetilde{B}_1^{13}\\ \widetilde{B}_1^{21} & \widetilde{B}_1^{22} & \widetilde{B}_1^{23}\end{pmatrix}\begin{pmatrix}z_b\\ z_d\\ z_0\end{pmatrix}\\ z_b &= \begin{pmatrix}0 & \widetilde{C}_1^2\end{pmatrix}\begin{pmatrix}x_a\\ x_b\end{pmatrix}.\end{aligned}$$

We can eliminate the z_b from the state equation by substituting the output equation, and we obtain the following system:

$$\begin{aligned}\begin{pmatrix}\dot{x}_a\\ \dot{x}_b\end{pmatrix} &= \begin{pmatrix}\widetilde{A}_1^1 & \widetilde{A}_1^2 + \widetilde{B}_1^{11}\widetilde{C}_1^2\\ 0 & \widetilde{A}_1^3 + \widetilde{B}_1^{21}\widetilde{C}_1^2\end{pmatrix}\begin{pmatrix}x_a\\ x_b\end{pmatrix}\\ &\quad + \begin{pmatrix}\widetilde{B}_1^{12} & \widetilde{B}_1^{13}\\ \widetilde{B}_1^{22} & \widetilde{B}_1^{23}\end{pmatrix}\begin{pmatrix}z_d\\ z_0\end{pmatrix},\\ z_b &= \begin{pmatrix}0 & \widetilde{C}_1^2\end{pmatrix}\begin{pmatrix}x_a\\ x_b\end{pmatrix},\end{aligned}$$

which we can write down compactly as

$$\bar{\Sigma}_1 : \begin{cases}\begin{pmatrix}\dot{x}_a\\ \dot{x}_b\end{pmatrix} = \begin{pmatrix}\widetilde{A}_1^1 & \bar{A}_1^2\\ 0 & \bar{A}_1^3\end{pmatrix}\begin{pmatrix}x_a\\ x_b\end{pmatrix} + \begin{pmatrix}\bar{B}_1^1\\ \bar{B}_1^2\end{pmatrix}\bar{v}_1\\ z_b = \begin{pmatrix}0 & \widetilde{C}_1^2\end{pmatrix}\begin{pmatrix}\widetilde{x}_1^1\\ \widetilde{x}_1^2\end{pmatrix}\end{cases} \tag{8.54}$$

using the obvious definitions. Note that if we impose amplitude constraints on $\bar{v}_1$ and z_b, then we can always translate these back to constraints on the original output z provided the constraint set $\bar{\mathcal{S}}$ for $\bar{v}_1$ decomposes as $\bar{\mathcal{S}} = \bar{\mathcal{S}}_1 \times \bar{\mathcal{S}}_2$ compatible with the decomposition of $\bar{v}_1$ into z_d and z_0. A similar statement can be made with the rate constraint set $\bar{\mathcal{T}}$.

We have the following result.

Theorem 8.30 *Consider the system Σ given by* (7.1) *with* (A, B) *stabilizable. Then the following two statements are equivalent:*

(*i*) *Semi-global stabilization is possible for all constraint sets $\mathcal{S}$ and $\mathcal{T}$ that satisfy Assumption 7.1.*

(*ii*) *The system Σ has at most weakly non-minimum-phase constraint invariant zeros. Moreover, if we construct $\bar{\Sigma}_1$ of the form* (8.54)*, then we have $\bar{A}_1^3 = \alpha I$ with $\alpha \leqslant 0$ and $\widetilde{C}_1^2$ injective.*

Proof : In order to show that item (*i*) implies item (*ii*), we first use the necessary conditions of the previous subsection, to obtain that the system Σ has at most weakly non-minimum-phase constraint invariant zeros. Secondly, if we construct the system $\bar{\Sigma}_1$ of the form (8.54) according to arguments given before, then it is obvious from our earlier results that this system must be semi-globally stabilizable for all possible amplitude and rate constraint sets for the input $\bar{v}_1$ and the output z_b.

The necessary conditions of Theorem 8.32 (which we will present later) yield that we must have

$$\ker \begin{pmatrix} 0 & \widetilde{C}_1^2 \end{pmatrix} \subset \ker \begin{pmatrix} 0 & \widetilde{C}_1^2 \end{pmatrix} \begin{pmatrix} \bar{A}_1^1 & \bar{A}_1^2 \\ 0 & \bar{A}_1^3 \end{pmatrix},$$

which is equivalent to $\ker \widetilde{C}_1^2 \subset \ker \widetilde{C}_1^2 \bar{A}_1^3$. Therefore, $\ker \widetilde{C}_1^2$ is part of the zero dynamics and the partitioning of our system guarantees that $\widetilde{x}_1^2$ does not contain zero dynamics. Hence, we obtain that $\widetilde{C}_1^2$ must be injective.

Now, we can consider any initial condition for $x_b(0)$. If $\bar{A}_1^3 x_b(0) = 0$, obviously $x_b(0)$ is an eigenvector of $\bar{A}_1^3$.

Otherwise, for some time interval $(0, T)$ and any $t \in (0, T)$, we have

$$x_b(t) = x_b(0) + t \left(\bar{A}_1^3 \widetilde{x}_1^2(\tau) + \bar{B}_1^2 \bar{v}_1(\tau)\right)$$

for some $\tau \in (0, t)$. Choosing our amplitude constraint set such that $x_b(t)$ is constrained to be confined to a very small neighborhood of the line between 0 and $x_b(0)$, we can force that $x_b(t) - (1 - \mu)x_b(0)$ is arbitrarily small for some $\mu \in [0, 1]$. By also choosing the constraint set for $\bar{v}_1(t)$ to be very small, we can force that

$$\bar{A}_1^3 \widetilde{x}_1^2(\tau) - \lambda x_b(0)$$

is arbitrarily small for $\lambda = -\mu/t \leqslant 0$. By continuity of $x_b(\tau)$, we find that we must be able to make

$$\bar{A}_1^3 x_b(0) - \lambda x_b(0) \tag{8.55}$$

arbitrarily small for some choice of $\lambda \leqslant 0$. But since the expression in (8.55) is fixed (and independent of the input), we must have that it is equal to zero for some λ. In other words, we have $\overline{A}_1^3 x_b(0) = \lambda x_b(0)$, i.e., $x_b(0)$ is an eigenvector of $\overline{A}_1^3$. Since any vector we choose as initial condition is an eigenvector of $\overline{A}_1^3$ with a real eigenvalue less than or equal to zero, we must have $\overline{A}_1^3 = \alpha I$ for some $\alpha \leqslant 0$.

Conversely, we need to show that item (*ii*) implies item (*i*). We again construct $\bar{\Sigma}_1$ of the form (8.54) according to arguments given before. From the controller design presented in the previous subsection, it should be obvious that we only need to show that we can achieve semi-global stabilization of $\bar{\Sigma}_1$ for any amplitude and rate constraint sets for the input $\overline{v}_1$ and the output $\widetilde{z}_1$. If $\alpha < 0$, then it is not difficult to see that

$$\|z_b\|_\infty \leqslant \|\widetilde{z}_1(0)\| + \frac{1}{\alpha}\|\widetilde{C}_1^2\|\|\overline{B}_1^2\overline{v}_1\|_\infty,$$

and hence, we satisfy the constraints on $\widetilde{z}_1$ if we make sure that the amplitude of $\overline{v}_1$ is small enough. Therefore, we basically only have input constraints and then we can obviously always design a suitable controller. The situation for $\alpha = 0$ is a bit more subtle. We first apply a preliminary feedback

$$\overline{v}_1 = -\varepsilon(\overline{B}_1^2)^R x_b + \widehat{v}_1,$$

where $(\overline{B}_1^2)^R$ denotes a right inverse of $\overline{B}_1^2$ which must exist because of stabilizability of $(\overline{A}_1^3, \overline{B}_1^2)$ with $\overline{A}_1^3 = 0$. It is then easy to see that if we constrain $\widehat{v}_1$ to a small enough, neighborhood and choose ε small enough then we will automatically satisfy the constraints. But this preliminary feedback basically changes the system from $\overline{A}_1^3 = 0$ to a new system with $\overline{A}_1^3 = -\varepsilon I$. Hence, we can use our earlier arguments for the case $\alpha < 0$ to derive the existence of a stabilizing feedback satisfying our constraints on $\widehat{v}_1$. It is easy to see that the above technique can also be used to satisfy possible rate constraints. ∎

For global constrained stabilization, we can use similar arguments to obtain the following result.

Theorem 8.31 *Consider the system* Σ *given by* (7.1) *with* (A, B) *stabilizable. Then the following two statements are equivalent:*

(*i*) *Global stabilization is possible for all constraint sets* $\mathcal{S}$ *and* $\mathcal{T} = \mathbb{R}^n$ *that satisfy Assumption 7.1.*

(*ii*) *The system* Σ *has at most weakly non-minimum-phase constraint invariant zeros and has no infinite zeros of order greater than* 1. *Moreover, if we construct* $\bar{\Sigma}_1$ *of the form* (8.54), *then we have* $\overline{A}_1^3 = \alpha I$ *with* $\alpha \leqslant 0$ *and* $\widetilde{C}_1^2$ *injective.*

We proceed next to answer the second question posed in the beginning of this subsection. To start with, we recall the subsystem $\widetilde{\Sigma}_1$ as formulated in (8.54), and the way it was formulated starting with the SCB representation of Σ. We obtained the following system for $i = 1$ and $\widetilde{z}_1 = z_b$:

$$\widetilde{\Sigma}_i : \begin{cases} \dot{\widetilde{x}}_i = \widetilde{A}_i \widetilde{x}_i + \widetilde{B}_i \widetilde{v}_i \\ \widetilde{z}_i = \widetilde{C}_i \widetilde{x}_i . \end{cases} \tag{8.56}$$

Note that in constructing $\widetilde{\Sigma}_1$ we deleted the dynamics associated to the infinite zeros and the non-left invertibility and concentrate on the non-right-invertible output. Consider $\widetilde{\Sigma}_1$ with associated amplitude and rate constraints on $\widetilde{z}_1$. We can now bring $\widetilde{\Sigma}_1$ into SCB and again delete (as before) the dynamics associated with the infinite zeros and the non-left invertibility and concentrate on the non-right-invertible output. In this way, we obtain a system $\widetilde{\Sigma}_2$ of the form (8.56) with $i = 2$. By bringing this system in to SCB and again deleting the dynamics associated with the infinite zeros and the non-left invertibility and concentrating on the non-right-invertible output, we obtain $\widetilde{\Sigma}_3$. In this way, we can recursively define a sequence of systems of the form (8.56). Note that this sequence of systems is closely related to the chain of systems as studied in [129].

At each step of developing $\widetilde{\Sigma}_i$, we make sure that the matrix $\widetilde{B}_i$ has full column rank and the matrix $\widetilde{C}_i$ has full row rank to proceed with the next step of developing $\widetilde{\Sigma}_{i+1}$. This can of course be done without loss of generality. This chain of construction ends if we obtain a subsystem $\widetilde{\Sigma}_i$ which is right invertible in the sense that $\widetilde{\Sigma}_{i+1}$ satisfies $\widetilde{C}_{i+1} = 0$. Another possibility for termination is that after some steps we get $\widetilde{B}_i = 0$ which obviously implies that we can end the chain. We know that (it can be shown easily) if the pair (A, B) of the given system Σ is stabilizable, then all the systems $\widetilde{\Sigma}_i$ as defined in (8.56) are stabilizable.

We have the following result for the case of amplitude constraints:

Theorem 8.32 *Consider the plant Σ as given by* (7.1) *with constraint sets $\mathcal{S}$ and $\mathcal{T} = \mathbb{R}^p$ satisfying Assumption 7.1. Let the chain of systems $\widetilde{\Sigma}_i$ $(i = 1, \dots s)$ be as described above. Then the semi-global stabilization problem as formulated in Problem 7.8 is solvable only if the following conditions are satisfied:*

(i) (A, B) is stabilizable.

(ii) The constraints are at most weakly non-minimum phase.

(iii) All the systems $\widetilde{\Sigma}_i$ $(i = 1, \dots s)$ have at most weakly non-minimum-phase constraints.

(iv) The systems $\widetilde{\Sigma}_i$ $(i = 1, \dots s)$ with realization (8.56) *satisfy,*

$$\ker \widetilde{C}_i \subset \ker \widetilde{C}_i \widetilde{A}_i . \tag{8.57}$$

Moreover, the global stabilization problem as formulated in Problem 7.9 is solvable only if the above conditions (i)–(iv) and the following condition are satisfied:

(*v*) *The given system* Σ *with input* u *and output* z*, characterized by* (A, B, C_z, D_z)*, has no constraint infinite zeros of order greater than one.*

Proof : The necessity of conditions (*i*) and (*ii*) is obvious. The condition (*v*) is also a direct consequence of earlier arguments.

In order to show item (*iii*), consider one of the systems $\widetilde{\Sigma}_i$. This system has input constraints and output constraints in the sense that $\widetilde{v}_i$ and $\widetilde{z}_i$ must both be bounded, i.e., $\widetilde{v}_i \in \mathcal{V}_i$ and $\widetilde{z}_i \in \mathcal{S}_i$ for some bounded sets $\mathcal{V}_i$ and $\mathcal{S}_i$. Based on earlier theorems, the necessity of condition (*iii*) is then obvious.

To show the necessity of condition (*iv*), we proceed as follows. Consider the system $\widetilde{\Sigma}_i$ whose input $\widetilde{v}_i$ and output $\widetilde{z}_i$ are bounded. Assume $\widetilde{z}_i$ is constrained to be in the set $\mathcal{S}_i$. We will prove this implication by contradiction. Assume that there exists a vector $\check{x}$ such that $\widetilde{C}_i\check{x} = 0$ but $\widetilde{C}_i\widetilde{A}_i\check{x} \neq 0$. For any $\varepsilon > 0$, there exists a vector $\widehat{x}$ such that $\widetilde{C}_i\widehat{x}$ is in the interior of $\mathcal{S}_i$ but $\widetilde{C}_i\widehat{x} + \varepsilon\widetilde{C}_i\widetilde{A}_i\check{x} \notin \mathcal{S}_i$. Consider for any scalar λ the initial condition $x_i(0) = \widehat{x} + \lambda\check{x}$. It is easily verified that this initial condition is admissible. We have

$$\dot{z}_1(0) = \widetilde{C}_i\widetilde{A}_i\widehat{x} + \widetilde{C}_i\widetilde{B}_i\widehat{u}_i(0) + \lambda\widetilde{C}_i\widetilde{A}_i\check{x}. \tag{8.58}$$

For large enough λ, the last term in this derivative will dominate the first two. Recall in that respect that $\widehat{x}(0)$ is fixed and we can choose $\widehat{u}_i$ but it is constrained to be in a fixed bounded set. However, if we move in the direction $\widetilde{C}_i\widetilde{A}_i\check{x}$, then we will be outside the set $\mathcal{S}_i$ very quickly when we choose ε sufficiently small combined with the fact that

$$z_1(0) + \varepsilon\widetilde{C}_i\widetilde{A}_i\check{x} = \widetilde{C}_i\widehat{x} + \varepsilon\widetilde{C}_i\widetilde{A}_i\check{x} \notin \mathcal{S}_i.$$

Note that, since the complement of $\mathcal{S}_i$ is open, the small perturbation caused by the first two terms in the derivative of $\dot{z}_1(0)$ in (8.58) cannot avoid that we will leave $\mathcal{S}_i$ since they only cause a minor perturbation compared to the dominant third term. The initial condition $x_i(0)$ is in the interior of the admissible set of initial conditions for the system $\widetilde{\Sigma}_i$, but we cannot avoid constraint violation with this initial condition. This yields a contradiction to the claim that this system was semi-globally stabilizable. Therefore, such a $\check{x}$ for which $\widetilde{C}_i\check{x} = 0$ but $\widetilde{C}_i\widetilde{A}_i\check{x} \neq 0$ does not exist, and this yields the fourth condition. ∎

Remark 8.33 *The condition* (8.57) *immediately implies that the order of infinite zeros of each system* $\widetilde{\Sigma}_i$, $i = 1, \ldots s$, *is less than or equal to one.*

The following example indicates that the conditions given in Theorem 8.32 are just necessary conditions and are not sufficient to solve the constrained stabilization problems. Also, this example shows that the solvability conditions for global and semi-global stabilization in the case of non-right-invertible constraints (unlike the case of right-invertible constraints) in general depend on the *particular* choice of constraint sets $\mathcal{S}$ and $\mathcal{T}$.

Example 8.34 Consider the following system taken from [61]:

$$\begin{aligned}
\dot{x}_1 &= x_2,\\
\dot{x}_2 &= -a_1 x_1 - a_2 x_2 - b_1 u,\\
z &= \begin{pmatrix} u \\ x, \end{pmatrix}
\end{aligned}$$

where z is required to be constrained in hypercubes and

$$a_1 = 3575, \quad a_2 = 333, \quad b_1 = 305555.$$

Note that the transfer matrix from u to z in this example is non-right invertible. We obtain $\widetilde{\Sigma}_1$:

$$\widetilde{\Sigma}_1 : \begin{cases} \dot{\widetilde{x}}_1 = \begin{pmatrix} 0 & 1 \\ -a_1 & -a_2 \end{pmatrix} \widetilde{x}_1 + \begin{pmatrix} 0 & 0 \\ 0 & -b_1 \end{pmatrix} \widetilde{v}_1 \\ \widetilde{z}_1 = \widetilde{x}_1 \end{cases}$$

and

$$\widetilde{\Sigma}_2 : \begin{cases} \dot{\widetilde{x}}_2 = \begin{pmatrix} 0 & 1 \end{pmatrix} \widetilde{v}_2 \\ \widetilde{z}_2 = \widetilde{x}_2. \end{cases}$$

Note that to construct $\widetilde{\Sigma}_2$ from $\widetilde{\Sigma}_1$, we have removed the redundancy (a column equal to 0) in $\widetilde{B}_1$. This example satisfies the necessary conditions in Theorem 8.32. On the other hand, suppose we require that

$$u \in [c_1, d_1], \quad x_1 \in [c_2, d_2], \quad x_2 \in [c_3, d_3], \tag{8.59}$$

where $0 \in (c_i, d_i)$. We have 0 as an interior point of the constraints set, and hence, we have $c_2 < 0$ and $c_3 < 0$. Therefore, if

$$x_1(0) = c_2, \quad x_2(0) = c_3,$$

we see that x_1 will leave $[c_2, d_2]$, which shows that global constrained stabilization is not possible. Moreover, since $\dot{x}_2$ is bounded, an initial condition very close to the boundary will still violate the constraint conditions, and hence, semi-global stabilization is not possible either. Hence, for the given constraint sets (8.59), we cannot achieve semi-global or global constrained stabilization. However, it is trivial to show that there exist other constraint sets for x (for instance, ellipsoidal sets) such that we can achieve semi-global or global constrained stabilization. This implies that the solvability conditions depend on the particular choice of the constraint sets unlike in the case of right-invertible constraints.

8.4 Semi-global stabilization in admissible set for right-invertible constraints: discrete time

In Sect. 8.2, we considered continuous-time systems with right-invertible constraints and developed several results regarding semi-global stabilization in admissible set as formulated in Problem 7.8 for state feedback in Problem 7.10 for measurement feedback, and global stabilization in admissible set as formulated in Problem 7.9 for state feedback in Problem 7.11 for measurement feedback. In this section, we proceed to develop similar results, however, for discrete-time systems. Although our development here parallels the one in Sect. 8.2, as will be evident throughout this section, there exist several fundamental differences in every respect, that is, in the solvability conditions, in designing the controllers, as well as in constructing the proofs.

At this time, it is worth pointing out that, in contrast to the continuous-time case, for the discrete-time systems being considered in this section, the solvability conditions for the semi-global and global stabilization in the admissible set are the same. This is true for both state and measurement feedback.

We have the following theorem which is concerned with both semi-global and global stabilization Problems 7.8 and 7.9 via state feedback.

Theorem 8.35 *Consider the discrete-time system Σ as given by* (7.1) *and the constraint sets $\mathcal{S}$ and $\mathcal{T}$ that satisfy Assumption 7.1. Assume that the set $\mathcal{S}$ is bounded. Also, assume that the constraints are right invertible. Then, the semi-global stabilization problem in the admissible set as defined in Problem 7.8 or the global stabilization problem in the admissible set as defined in Problem 7.9 is solvable if and only if the following conditions hold:*

(i) (A, B) *is stabilizable.*

(ii) *The constraint invariant zeros of the given system Σ are all in the closed unit disc, i.e., the system Σ has at most weakly non-minimum-phase constraints.*

(iii) *The constraints are of type one, i.e., the given system Σ has no constraint infinite zeros of order greater than one.*

We point out that the controllers that solve both semi-global and global stabilization problems, in general, need to be nonlinear. However, in the semi-global case, the controller can be chosen either as a time-invariant nonlinear controller or as a time-varying linear controller as will be evident shortly when we present proofs.

The following remark points out that the specific features of the given constraint sets $\mathcal{S}$ and $\mathcal{T}$ do not have any role in the solvability of semi-global and global stabilization in the admissible set for right-invertible constraints.

Remark 8.36 *This remark is analogous to Remark 8.6 that concerns with the continuous-time case. We emphasize here a fundamental aspect of solvability conditions as given by Theorem 8.35, namely, that they are independent of any specific features of the given constraint sets.* That is, for the case of a right-invertible system Σ, if the semi-global and global stabilization problems in the admissible set via state feedback are solvable for some given constraint sets $\mathcal{S}$ and $\mathcal{T}$ satisfying Assumption 7.1, then these problems are also solvable for all constraint sets satisfying Assumption 7.1.

We now move on to the case of measurement feedback. We have the following theorem.

Theorem 8.37 *Consider the discrete-time system Σ as given by* (7.1) *and the constraint sets $\mathcal{S}$ and $\mathcal{T}$ that satisfy Assumption 7.1. Assume that the set $\mathcal{S}$ is bounded. Also, assume that the constraints are right invertible. Then, the semi-global stabilization problem in the admissible set via measurement feedback as defined in Problem 7.10 or global stabilization problem in the admissible set via measurement feedback as defined in Problem 7.11 is solvable if the following conditions hold:*

(*i*) (A, B) *is stabilizable.*

(*ii*) *The constraint invariant zeros of the given system Σ are all in the closed unit disc, i.e., the system Σ has at most weakly non-minimum-phase constraints.*

(*iii*) *The constraints are of type one, i.e., the given system Σ has no constraint infinite zeros of order greater than one.*

(*iv*) *The pair* (C_y, A) *is observable.*

(*v*) $\ker C_z \subset \ker C_z A$.

(*vi*) $\ker \begin{pmatrix} C_y & D_y \end{pmatrix} \subset \ker \begin{pmatrix} C_z & D_z \end{pmatrix}$.

Moreover, conditions (*i*)–(*iii*) *are necessary for the solvability of the semi-global stabilization problem in the admissible set via measurement feedback.*

The following remark addresses the need for condition (*iv*).

Remark 8.38 *This remark is analogous to Remark 8.5 that concerns with the continuous-time case. Some discussion regarding the condition* (*iv*) *in Theorem 8.37 is in order. Clearly, condition* (*iv*) *is sufficient but not necessary. Also, obviously, the detectability of the pair* (C_y, A) *is necessary to solve the posed problem.*

On the other hand, if the pair (C_y, A) is not observable, we can have a problem with unobservable dynamics which we cannot observe but which might result in violation of our constraints. Relaxing the condition (iv), and developing a set of necessary and sufficient conditions are not very difficult but highly technical and not that interesting.

The following remark addresses the conditions (*v*) and (*vi*) of Theorem 8.37.

Remark 8.39 *Note that condition* (vi) *states that the output is part of the measurements. The following example shows that conditions* (v) *and* (vi) *in Theorem 8.37 are natural, and hence are needed for discrete-time systems. Consider the system*

$$\begin{aligned} x_1(k+1) &= x_1(k) + x_2(k), \\ x_2(k+1) &= u(k) + x_1(k), \\ y(k) &= x_1(k) + 2x_2(k), \\ z(k) &= x_2(k). \end{aligned}$$

It is easy to see that all the conditions except (v) *and* (vi) *in the theorem are satisfied. Suppose we have a constraint $z(k) \in [-1, 1]$ for all $k \geqslant 0$. There exists a deadbeat observer which gives an exact state estimate for $x_1(k)$ and $x_2(k)$ for $k \geqslant 1$.*

The set of admissible initial conditions is characterized by the set of all initial conditions satisfying $|x_2(0)| \leqslant 1$. Using this information together with the first measurement $y(0)$, we can only conclude that

$$x_1(0) \in [y_1(0) - 2,\ y_1(0) + 2]. \tag{8.60}$$

But then, there is no choice for $u(0)$ to ensure that

$$x_2(1) = [x_1(0) + u(0)] \in [-1, 1]$$

for all possible values of $x_1(0)$ satisfying (8.60)*, i.e., there is no guarantee that at time $k = 1$, the constraint is not violated. Hence, the semi-global stabilization via measurement feedback is in general not possible without conditions* (v) *and* (vi)*.*

The following remark addresses the effect of the constraint sets $\mathcal{S}$ and $\mathcal{T}$ on the solvability of semi-global or stabilization problem via measurement feedback.

Remark 8.40 *The sufficient conditions for the solvability of semi-global as well as global stabilization via measurement feedback as given by Theorem 8.37 are independent of any specific features of the given constraint sets. But, in general, in the measurement feedback case, the solvability of the semi-global stabilization problem or global stabilization problem is dependent on the shape of the* *constraint sets even for the case of right-invertible constraints being considered in this section. But, this is not in contradiction with Theorem 8.37 since we only obtained sufficient conditions for solvability there. For example, consider the system,*

$$\begin{aligned} x_1(k+1) &= u_1(k), \\ x_2(k+1) &= u_2(k) + x_1(k), \\ y(k) &= x_2(k), \\ z_1(k) &= x_1(k), \\ z_2(k) &= x_2(k). \end{aligned}$$

Let the constrained output $[z_1 \;\; z_2]'$ *be in some set* $\mathcal{S}$ *that satisfies Assumption 7.1. Then, the posed semi-global and global stabilization problems can trivially be solved by state feedback* $u_1 = 0$ *and* $u_2 = -x_1$. *But in the measurement case, we can only implement this feedback for* $k \geqslant 1$. *At time* $k = 0$, *we have no information available about* $x_1(0)$ *except for the fact that the state must be in the admissible set of initial conditions:*

- *For the constraint set* $|z_1| \leqslant 1$ *and* $|z_2| \leqslant 2$, *the controller* $u_1 = 0$ *and* $u_2 = 0$ *trivially solves the global stabilization problem.*
- *For the constraint set* $|z_1| \leqslant 1$ *and* $|z_2| \leqslant 1/2$, *no measurement-based feedback can guarantee that the constraint is not violated at time* $k = 1$, *because the controller lacks information about* $x_1(0)$.

Note that the above is in contrast with continuous time where the solvability is always independent of the constraint set for right-invertible systems.

Remark 7.14 dictates that the constraints are right invertible whenever we have constraints only on actuators. The following remark exemplifies several aspects of having constraints only on actuators.

Remark 8.41 *This remark is analogous to Remark 8.7 that concerns with the continuous-time case. Consider the case when we have constraints only on actuator magnitude and rate, i.e., the case when* $C_z = 0$. *In other words, there are no state constraints, and only a subset of the input channels is subject to magnitude and rate constraints. Then, as Remark 7.14 points out, the constraints are right invertible. Also, in this case, it is straightforward to show that the constraint invariant zeros of* Σ *(i.e., the invariant zeros of the system* Σ *characterized by* $(A, B, 0, D_z)$*) coincide with a subset of the eigenvalues of* A. *This observation implies that the requirement of at most weakly non-minimum-phase constraints in Theorems 8.35 and 8.37 is equivalent to requiring that a particular subset of eigenvalues of* A *lies in the closed unit disc. Obviously, such a condition is always satisfied if we are dealing with critically unstable systems. Let us next assume that* *the given system Σ is controllable via the unconstrained input channels; it is then straightforward to see that the subsystem characterized by $(A, B, 0, D_z)$ does not have any invariant zeros, i.e., Σ does not have any constraint invariant zeros. Thus, for this special case, there will not be any constraints on the eigenvalues of A as obviously can be expected.*

It is worthwhile to consider another special case of the above when all the input channels are subject to magnitude and rate constraints which is the case considered in Chap. 6. In this case, $C_z = 0$ and $D_z = I_m$. For this special case, we observe that the admissible set of initial conditions $\mathcal{V}(\mathcal{S}, \mathcal{T})$ is indeed $\mathbb{R}^n$. Moreover, the constraint invariant zeros of Σ coincide with all the eigenvalues of A. As such, the requirement of at most weakly non-minimum-phase constraints in Theorems 8.35 and 8.37 is equivalent to requiring that the given system be critically unstable. Furthermore, for this special case, it is easy to see that there are no constraint infinite zeros of order greater than 1 and hence the condition (iii) of these Theorems 8.35 and 8.37 is automatically satisfied.

As in the continuous-time case, for discrete-time systems as well, whenever there is only input saturation, one of the important facts that emerged in Chap. 4 is that in general global stabilization requires nonlinear feedback, while semi-global stabilization can be achieved whenever it can be done by utilizing simply linear time-invariant feedback laws. Thus, as in the continuous-time case, a question that arises naturally is whether an analogous result is valid under a broad framework of state as well as input constraints as we are considering in this chapter. The following theorem answers this question:

Theorem 8.42 *Consider the discrete-time system Σ as given by* (7.1) *and constraint sets $\mathcal{S}$ and $\mathcal{T}$ that satisfy Assumption 7.1. Assume that the constraints are right invertible. Then the following hold:*

(*i*) *Under the condition that* $\operatorname{im} C_z \subset \mathcal{T}$ *(i.e., no rate constraints on states), if a semi-global stabilization problem in the admissible set via state feedback as defined in Problem 7.8 is solvable, then it is also solvable via a linear time-invariant state feedback law.*

(*ii*) *If* $\operatorname{im} C_z \not\subset \mathcal{T}$ *(i.e., rate constraints on states are present), whenever a semi-global stabilization problem in the admissible set via state feedback as defined in Problem 7.8 is solvable, in general it might not be solvable via a linear time-invariant state feedback law. That is, there exist a system Σ as given by* (7.1) *and constraint sets $\mathcal{S}$ and $\mathcal{T} \not\supset \operatorname{im} C_z$ that satisfy Assumption 7.1 for which the semi-global stabilization problem is solvable via a nonlinear feedback law but for which there exists no linear feedback law that solves the problem.*

Proof : The first statement of Theorem 8.42 follows from Theorem 8.35. The second statement follows from the following counter example. Consider the system,

$$\rho x = \begin{pmatrix} 1 & 0 & 0 & 0 & 0 & 0 & 0 \\ 0 & 1 & 0 & 0 & 0 & 0 & 0 \\ 0 & 0 & 1 & 0 & 0 & 0 & 0 \\ 1 & 0 & 0 & 1 & 3 & 0 & 0 \\ 0 & 1 & 0 & -3/2 & 5/2 & -3 & 0 \\ 0 & 0 & 1 & -3 & 0 & 1 & 0 \\ 0 & 0 & 0 & 0 & 0 & -3 & 1 \end{pmatrix} x + \begin{pmatrix} 1 & 0 & 0 & 0 \\ 0 & 1 & 0 & 0 \\ 0 & 0 & 1 & 0 \\ 0 & 0 & 0 & 1 \\ 0 & 0 & 0 & 1 \\ 0 & 0 & 0 & 1 \\ 0 & 0 & 0 & 1 \end{pmatrix} u$$

$$z = \begin{pmatrix} 0 & 0 & 0 & 1 & 0 & 0 & 0 \\ 0 & 0 & 0 & 0 & 1 & 0 & 0 \\ 0 & 0 & 0 & 0 & 0 & 1 & 0 \\ 0 & 0 & 0 & 0 & 0 & 0 & 1 \end{pmatrix} x$$

with no amplitude constraints and the rate constraint $\mathcal{T}$ given by three times the unit cube in $\mathbb{R}^4$:

$$\mathcal{T} = \{ z \in \mathbb{R}^4 \mid \|z\|_\infty < 3 \}.$$

Consider the following four initial conditions:

$$x_1 = \begin{pmatrix} 0 \\ 0 \\ 0 \\ 1 \\ 1 \\ 1 \\ 0 \end{pmatrix}, x_2 = \begin{pmatrix} 0 \\ 0 \\ 0 \\ 1 \\ 1 \\ -1 \\ 0 \end{pmatrix}, x_3 = \begin{pmatrix} 0 \\ 0 \\ 0 \\ -1 \\ 1 \\ 1 \\ 0 \end{pmatrix}, x_4 = \begin{pmatrix} 0 \\ 0 \\ 0 \\ 1 \\ -1 \\ 1 \\ 0 \end{pmatrix}.$$

These initial conditions are all feasible. But in order to satisfy the rate constraint at time 0 we see that $u_4(0)$ must be equal to 0, 0, 0, and 3 for $x(0) = x_1, x(0) = x_2, x(0) = x_3$ and $x(0) = x_4$, respectively. This can, however, never be done via a linear feedback. Note that this also excludes solvability of the semi-global problem by linear feedback because that still requires that for initial conditions close to x_1, we must have $u_4(0)$ close to 0, and similarly for the other three initial conditions. This is not possible with arbitrary accuracy with a linear feedback.

It remains to show that we can solve the semi-global stabilization problem for this system. This is, however, a direct consequence of Theorem 8.35. Note that the constructed system has *no* finite constraint invariant zeros.

Finally, that we can find a linear feedback in case rate constraints are not present is clear from the construction of a controller in the proof of Theorem 8.35. ■

8.4.1 Proofs and construction of controllers

As in the continuous-time case, before we proceed to the proofs and to construct appropriate controllers, we need to decompose the given system in a suitable way. For the right-invertible constraints considered in this section, the SCB of the system Σ_{uz} characterized by the quadruple (A, B, C_z, D_z) can be written in the following form:

$$\begin{cases} \rho x_a(k+1) = A_{aa}x_a + K_a\tilde{z} \\ \rho x_c(k+1) = A_{cc}x_c + B_c[u_c + J_a x_a] + K_c\tilde{z} \\ \rho x_d(k+1) = A_{dd}x_d + B_d[u_d + E_a x_a + E_c x_c + E_d x_d] + K_d\tilde{z} \\ y \quad = C_{ya}x_a + C_{yc}x_c + C_{yd}x_d + \widetilde{D}_y\tilde{u} \\ z_0 \quad = u_0 \\ z_d \quad = C_d x_d, \end{cases} \tag{8.61}$$

where

$$\begin{pmatrix} x_a \\ x_c \\ x_d \end{pmatrix} = \tilde{x} = \Gamma_s^{-1}x, \quad \begin{pmatrix} u_0 \\ u_c \\ u_d \end{pmatrix} = \tilde{u} = \Gamma_u^{-1}u, \quad \begin{pmatrix} z_0 \\ z_d \end{pmatrix} = \tilde{z} = \Gamma_z^{-1}z,$$

and x_a, x_c, x_d, u_0, u_c, u_d are of appropriate dimensions, while Γ_s , Γ_u, and Γ_z are transformation matrices. This decomposition renders the subsystem characterized by the state variables x_c and x_d strongly controllable[3] without finite zeros. Moreover, the pair (A_{aa}, K_a) is stabilizable.

We can extract from (8.61) two subsystems. The first subsystem is given by

$$\begin{cases} \rho x_a = A_{aa}x_a + K_a\tilde{z} \\ \quad\;\; = A_{aa}x_a + K_{a0}z_0 + K_{ad}z_d, \end{cases} \tag{8.62}$$

where $K_a = \begin{pmatrix} K_{a0}, & K_{ad} \end{pmatrix}$. The second subsystem extracted from (8.61) is given by

$$\begin{cases} \rho x_c = A_{cc}x_c + B_c[u_c + J_a x_a] + K_c\tilde{z} \\ \rho x_d = A_{dd}x_d + B_d[u_d + E_a x_a + E_c x_c \\ \qquad\qquad + E_d x_d] + K_d\tilde{z} \\ z_0 = u_0 \\ z_d = C_d x_d. \end{cases} \tag{8.63}$$

The solvability conditions in Theorems 8.35 and 8.37 both require that the constraints be at most weakly nonminimum phase and of type one. Once the constraints are of type one, the SCB representation of system Σ in (8.61) can be

[3] Definition 3.28 discusses strong controllability.

simplfied. More specifically, the equations for x_d and z_d have a simpler structure because of the first-order relative degree. To facilitate the proofs of Theorems 8.35 and 8.37, we rewrite (8.61) after simplification as

$$\begin{cases} \rho x_a &= A_{aa}x_a + K_a\tilde{z} \\ \rho x_c &= A_{cc}x_c + B_c[u_c + J_a x_a] + K_c\tilde{z} \\ \rho x_d &= u_d + G_a x_a + G_c x_c + G_d x_d \\ y &= C_{ya}x_a + C_{yc}x_c + C_{yd}x_d + \widetilde{D}_y\tilde{u} \\ z_0 &= u_0 \\ z_d &= x_d, \end{cases} \tag{8.64}$$

where G_a, G_c, and G_d are matrices with appropriate dimensions. Also, the second subsystem (8.63) can be rewritten as

$$\begin{cases} \rho x_c(k+1) &= A_{cc}x_c + B_c[u_c + J_a x_a] + K_c\tilde{z} \\ \rho x_d(k+1) &= u_d + G_a x_a + G_c x_c + G_d x_d \\ z_0 &= u_0 \\ z_d &= C_d x_d. \end{cases} \tag{8.65}$$

The two subsystems (8.62) and (8.63) form the system Σ_{uz}. We emphasize that the subsystem (8.62) represents the zero dynamics of the system Σ_{uz}. Moreover, the eigenvalues of A_{aa} are equal to the constraint invariant zeros of the given system Σ. By viewing $\tilde{z}$ as the input to this subsystem, we have a system with input constraints in the sense that $z(k) \in \mathcal{S}$ and $(z(k+1) - z(k)) \in \mathcal{T}$ for all $k \geqslant 0$. Since $\mathcal{S}$ and $\mathcal{T}$ satisfy Assumption 7.1, there exist appropriate sets $\mathcal{S}_0$, $\mathcal{S}_d$, $\mathcal{T}_0$, and $\mathcal{T}_d$ such that

$$z \in \mathcal{S} \text{ if and only if } z_0 \in \mathcal{S}_0 \text{ and } z_d \in \mathcal{S}_d, \tag{8.66a}$$

$$\rho z \in \mathcal{T} \text{ if and only if } \rho z_0 \in \mathcal{T}_0 \text{ and } \rho z_d \in \mathcal{T}_d. \tag{8.66b}$$

Now, we are at a position to define the admissible set for the subsystem Σ_2 as

$$\mathcal{V}_2(\mathcal{S}, \mathcal{T}) = \left\{ x_2 = \begin{pmatrix} x_c \\ x_d \end{pmatrix} \in \mathbb{R}^{n_2} \,\middle|\, C_d x_d \in \mathcal{S}_d \right\}, \tag{8.67}$$

where $n_2 = n_c + n_d$ is the dimension of x_2. Note that $\mathcal{T}$ does not affect the admissible set in discrete time for right-invertible systems because we know that the constraints must be of type one. x_a has no effect on $\mathcal{V}_2(\mathcal{S}, \mathcal{T})$.

As in the continuous-time case, both for construction of controllers and the proofs of the stated results, we use the decomposition in the two subsystems (8.62) and (8.63) as defined above. We observe clearly that we can control the first subsystem (8.62) only through $\tilde{z}$. Also, from the SCB decomposition, it follows that the second subsystem has no finite invariant zeros and is right invertible. This implies that we can guarantee by suitable choice of $\tilde{u}$ that $\tilde{z}$ is arbitrary close to any desired signal as will be evident soon. Therefore, we basically design a controller in two phases:

- First, design a desired feedback for the first subsystem (8.62) using $\tilde{z}$ as the (constrained) input signal such that the first subsystem exhibits a desired closed-loop behavior.
- Second, design a feedback for the second subsystem (8.63) with state x_2, input $\tilde{u}$ and output $\tilde{z}$ such that

  (*i*) The output $\tilde{z}$ is close to the desired feedback for the first subsystem.

  (*ii*) The output satisfies the constraints.

  (*iii*) The state x_2 of the second subsystem exhibits a desirable behavior.

All feedback designs here are constructed in accordance with this two-phase design.

We need to discuss next what kind of initial conditions can be considered for the first subsystem (8.62). In fact, since we have no state constraints on this subsystem, we can have arbitrary initial conditions for it. Hence, we consider arbitrary initial conditions in $\mathcal{W}_1 = \mathbb{R}^{n_a}$ in the global case, while in the semi-global case, we consider initial conditions in some arbitrary compact set $\mathcal{W}_1$.

Similarly, the initial conditions for the second subsystem must be in some set $\mathcal{W}_2$. In the global case, we have $\mathcal{W}_2 = \mathcal{V}_2(\mathcal{S}, \mathcal{T})$ using the definition in (8.67), while in the semi-global case, we have $\mathcal{W}_2$ as an arbitrary compact set contained in the interior of $\mathcal{V}_2(\mathcal{S}, \mathcal{T})$.

Proof of Theorem 8.35 : *Necessity:* The necessity of conditions (*i*) and (*ii*) is obvious. By the decomposition obtained above, the constrained variable $\tilde{z}$ becomes the input to the zero dynamics (8.62); hence, the system has to be at most weakly non-minimum phase, i.e., the poles of the zero dynamics must be in the closed unit disc. Next, we show the necessity of condition (*iii*).

We consider the global case first. Since we have right-invertible constraints, having no infinite zeros of order greater than one is equivalent to $(C_z B \quad D_z)$ being surjective. Therefore, if the system has infinite zeros of order greater than one, then there exists a vector $c \neq 0$ such that

$$c' D_z = 0 \quad \text{and} \quad c' C_z B = 0. \tag{8.68}$$

Moreover, since $\mathcal{T}$ contains zero in its interior, we can guarantee that $c \in \mathcal{T}$. Let $\zeta_0 \in \mathcal{S}$ be such that

$$\langle z, c \rangle \leqslant \langle \zeta_0, c \rangle$$

for all $z \in \mathcal{S}$. Since $\mathcal{S}$ is a compact and convex set, such a ζ_0 always exists at the boundary of $\mathcal{S}$. We have right-invertible constraints, or equivalently Σ_{uz} characterized by the quadruple (A, B, C_z, D_z) is right invertible. This implies that there exist an initial condition $x(0) = \xi_0$ and an input $u(0) = \mu_0$ such that the output z satisfies $z(0) = \zeta_0$ and $z(1) - z(0) = c$. Clearly, $\xi_0 \in \mathcal{V}(\mathcal{S}, \mathcal{T})$. But if the system starts at time 0 from ξ_0 then we have

$$\langle c, z(0) \rangle = \langle c, C_z \xi_0 \rangle = \langle c, \zeta_0 \rangle \quad \text{and} \quad \langle c, z(1) - z(0) \rangle = \langle c, c \rangle > 0$$

for any input signal u because of property (8.68). Hence, $\langle c, z(1)\rangle > \langle c, \zeta_0\rangle$ for any input u. By definition of ζ_0, this implies that $z(1) \notin \mathcal{S}$ for any input u. Therefore, there exist initial conditions in $\mathcal{V}(\mathcal{S}, \mathcal{T})$ which cannot be stabilized without violating the constraints. This yields a contradiction.

The necessity of condition (*iii*) for the semi-global case follows by a mild modification of the above argument. Choose a λ close to 1 from below such that $\langle c, c\rangle > (1-\lambda)\langle c, \zeta_0\rangle$, where ζ_0 is chosen as before. Let $z(0) = \lambda\zeta_0$. By the right invertibility of Σ_{uz} as above, there exist an initial condition ξ_0 and an input $u(0) = \mu_0$ such that the output z satisfies $z(0) = \lambda\zeta_0$ and $z(1) - z(0) = c$. Then, we can choose a compact set $\mathcal{A}_0$ in the interior of $\mathcal{V}(\mathcal{S}, \mathcal{T})$ so that $\xi_0 \in \mathcal{A}_0$. Since $\langle c, z(1) - z(0)\rangle = \langle c, c\rangle > 0$, we get $\langle c, z(1)\rangle = \langle c, z(0)\rangle + \langle c, c\rangle > \langle c, \zeta_0\rangle$. By the same argument as in the global case, this implies that $z(1) \notin \mathcal{S}$ for any input u, which is a contradiction.

The proof of sufficiency is by explicitly constructing a state feedback controller with the specified properties.

State feedback controller design for semi-global and global stabilization

Step 1 (Controller design for the zero dynamics)

We first design for the first subsystem given in (8.62) while viewing $\tilde{z}$ as an input variable. Let $v = \tilde{z} - \phi$, where the functions v and ϕ will become clear shortly. Then, (8.62) becomes

$$x_a(k+1) = A_{aa}x_a(k) + K_a\phi(k) + K_a v(k). \tag{8.69}$$

Note that the conditions of the theorem require that all eigenvalues of A_{aa} be in the closed unit disc. Viewing ϕ as an input to this subsystem, we can construct a state feedback law $\phi(k) = f(x_a(k))$ for the system (8.69), which has the following properties:

(a) It satisfies the constraints,

$$f(x_a(k)) \in \mathcal{S}, \quad (f(x_a(k+1)) - f(x_a(k))) \in \mathcal{T}, \qquad k \geqslant 0.$$

(b) It renders the zero equilibrium point of the closed-loop system of (8.69) semi-globally or globally attractive in the presence of any signal satisfying

$$\|v(k)\| \leqslant M\lambda^k, \qquad \lambda \in (0,1) \tag{8.70}$$

for some $M > 0$, i.e., $x_a(k) \to 0$ as $k \to \infty$.

(c) It renders the zero equilibrium point of the closed-loop system with $v = 0$ locally exponentially stable.

Note that the two parameters M and λ in (8.70) only depend on the size of the constraint sets $\mathcal{S}$ and $\mathcal{T}$. Whenever $\mathcal{S}$ and $\mathcal{T}$ are known, M and λ can be chosen a priori following the way specified in the design of *Step 2*. Knowing these facts, we are assured that the ℓ_2 norm of v signal is uniformly upper bounded. For completeness, the details of designing such a state feedback for the first subsystem in the semi-global or global sense are presented in Appendix 8.A.

Step 2 (Controller design for the second subsystem)

In this step, we design a control law for the second subsystem given in (8.63) so that the closed-loop system of the interconnection of the two subsystems with the control law is asymptotically stable and without constraint violation.

Choose $\lambda \in (0, 1)$ such that

$$(1-\lambda)\bar{\mathcal{S}}_d \subset \mathcal{T}_d, \tag{8.71}$$

where $\bar{\mathcal{S}}_d := \{\xi - \eta : \xi \in \mathcal{S}_d, \ \eta \in \mathcal{S}_d\}$. The control law is designed as follows. Partition f and v compatibly with the decomposition of z as

$$f(x_a) = \begin{pmatrix} f_0(x_a) \\ f_d(x_a) \end{pmatrix} \quad \text{and} \quad v = \begin{pmatrix} v_0 \\ v_d \end{pmatrix}.$$

Then choose

$$u_c(k) = F_c x_c(k) - J_a x_a(k), \tag{8.72}$$

where F_c is such that $A_{cc} + B_c F_c$ is Schur stable. Choose

$$u_0(k) = f_0(x_a(k)) \tag{8.73}$$

$$\begin{aligned} u_d(k) = {} & \lambda[x_d(k) - f_d(x_a(k))] + f_d(x_a(k+1)) \\ & - \lambda^{k+1}[f_d(x_a(k+1)) - f_d(x_a(k))] \\ & - G_a x_a(k) - G_c x_c(k) - G_d x_d(k), \end{aligned} \tag{8.74}$$

where $x_a(k+1) = A_{aa}x_a(k) + K_a z(k)$. Note that the control law for u_d is time varying, and is nonlinear in the global case and linear in the semi-global case. It remains to show that for the control law given above, we have $z(k) \in \mathcal{S}$ and $(z(k+1) - z(k)) \in \mathcal{T}$ for all $k \geqslant 0$; moreover, $v(k) = z(k) - f(x_a(k))$ satisfies (8.70) for a suitably chosen $M > 0$.

Given the feedback for u_d, we obtain

$$\begin{aligned} x_d(k+1) - f_d(x_a(k+1)) = {} & \lambda[x_d(k) - f_d(x_a(k))] \\ & - \lambda^{k+1}[f_d(x_a(k+1)) - f_d(x_a(k))]. \end{aligned} \tag{8.75}$$

Solving this difference equation yields that

$$x_d(k) = \lambda^k x_d(0) + (1-\lambda^k) f_d(x_a(k)). \tag{8.76}$$

Since both $x_d(0)$ and $f_d(x_a(k))$ are in the convex set $\mathcal{S}_d$, we have $z_d(k) = x_d(k) \in \mathcal{S}_d$. On the other hand,

$$\begin{aligned} x_d(k+1) - x_d(k) = {} & \lambda^k \{(1-\lambda)[f_d(x_a(k+1)) - x_d(0)]\} \\ & + (1-\lambda^k)[f_d(x_a(k+1)) - f_d(x_a(k))]. \end{aligned}$$

Hence, by (8.71), we get

$$z_d(k+1) - z_d(k) = x_d(k+1) - x_d(k) \in \mathcal{T}_d.$$

From (8.76), we see that

$$v_d(k) = x_d(k) - f_d(x_a(k)) = \lambda^k[x_d(0) - f_d(x_a(k))]. \tag{8.77}$$

Clearly, $x_d(0)$ must be in the admissible set. We can then conclude that both $x_d(0)$ and $f_d(x_a(k))$ are in the bounded set $\mathcal{S}_d$ and that $z_0 = u_0 = f_0(x_a)$; we find that there exists a $M > 0$ such that (8.70) holds.

So far, we have shown that the equilibrium point $x = 0$ of the overall closed-loop system is globally attractive. Since we have used a time-varying control law and the control law is nonlinear in the global case, the asymptotical stability of the equilibrium point $x = 0$ needs a careful verification. First, note that, according to the design of $f(x_a)$ presented in Appendix 8.A, the feedback $f(x_a)$ is globally Lipschitz and locally linear in terms of x_a. Then, it can be shown that for sufficiently small initial conditions $x_{a0} = x_a(0)$, $x_{c0} = x_c(0)$, and $x_{d0} = x_d(0)$, we have $\|x_a(k)\| \leqslant \kappa_1(\|x_{a0}\| + \|x_{d0}\|)$ for some constant $\kappa_1 > 0$ and all $k \geqslant 0$. This part of proof is presented in Appendix 8.B. From (8.77), we see that $\|v(k)\| \leqslant \kappa_2(\|x_{a0}\| + \|x_{d0}\|)$ for some constant $\kappa_2 > 0$ and all $k \geqslant 0$. From (8.76), it is straightforward that $\|x_d(k)\| \leqslant \kappa_3(\|x_{a0}\| + \|x_{d0}\|)$ for some constant $\kappa_3 > 0$ and all $k \geqslant 0$. Finally, viewing the dynamics of x_c as a Schur-stable system with disturbance $K_c z(k) = K_c[f(x_a(k)) + v(k)]$, we obtain that $\|x_c(k)\| \leqslant \kappa_4(\|x_{a0}\| + \|x_{c0}\| + \|x_{d0}\|)$ for some constant $\kappa_4 > 0$ and all $k \geqslant 0$. In conclusion, we have shown the local stability of the equilibrium point $x = 0$. This completes the proof. ∎

Proof of Theorem 8.37 : Note that condition (*v*) $\ker C_z \subset \ker C_z A$ implies that $G_a = G_c = 0$ in (8.64) and (8.65). Moreover, condition (*vi*) $\ker\begin{pmatrix} C_y & D_y \end{pmatrix} \subset \ker\begin{pmatrix} C_z & D_z \end{pmatrix}$ ensures that we can decompose y in a suitable basis such that

$$y = \begin{pmatrix} \tilde{y} \\ x_d \end{pmatrix} = \begin{pmatrix} \tilde{C}_{ya} & \tilde{C}_{yc} & 0 \\ 0 & 0 & I \end{pmatrix} \begin{pmatrix} x_a \\ x_c \\ x_d \end{pmatrix} + \begin{pmatrix} \tilde{D}_{yu} \\ 0 \end{pmatrix} \tilde{u},$$

which clearly indicates that the state x_d is directly determined by y. We get the following system:

$$\begin{aligned} x_a(k+1) &= A_{aa}x_a(k) + K_a z(k), \\ x_c(k+1) &= A_{cc}x_c(k) + K_c z(k), + B_c[u_c(k) + J_a x_a(k)], \\ x_d(k+1) &= u_d(k) + G_d x_d(k), \\ \tilde{y} \quad &= \tilde{C}_{ya}x_a + \tilde{C}_{yc}x_c + \tilde{D}_{yu}\tilde{u}, \\ z_0(k) &= u_0(k), \\ z_d(k) &= x_d(k). \end{aligned} \tag{8.78}$$

Since (C_y, A) is an observable pair, the SCB decomposition guarantees that the pair

$$\left(\begin{pmatrix} A_{aa} & 0 \\ B_c J_a & A_{cc} \end{pmatrix}, \begin{pmatrix} \widetilde{C}_{ya} & \widetilde{C}_{yc} \end{pmatrix}\right)$$

is also observable. That is, there exist matrices L_a and L_c such that

$$\widetilde{A} = \begin{pmatrix} A_{aa} - L_a\widetilde{C}_{ya} & -L_a\widetilde{C}_{yc} \\ B_c J_a - L_c\widetilde{C}_{ya} & A_{cc} - L_c\widetilde{C}_{yc} \end{pmatrix}$$

is Schur stable. For the above system, we use a reduced-order observer for the state variables (x_a, x_c):

$$\begin{aligned} \widehat{x}_a(k+1) &= A_{aa}\widehat{x}_a(k) + K_a z(k) + \\ &\qquad L_a[\widetilde{y}(k) - \widetilde{C}_{ya}\widehat{x}_a(k) - \widetilde{C}_{yc}\widehat{x}_c(k) - \widetilde{D}_{yu}\widetilde{u}(k)] \\ \widehat{x}_c(k+1) &= A_{cc}\widehat{x}_c(k) + K_c z(k) + B_c[u_c(k) + J_a\widehat{x}_a(k)] + \\ &\qquad L_c[\widetilde{y}(k) - \widetilde{C}_{ya}\widehat{x}_a(k) - \widetilde{C}_{yc}\widehat{x}_c(k) - \bar{D}_{yu}\widetilde{u}(k)]. \end{aligned}$$

Note that the measurement error is exponentially decaying.

The remaining design procedure follows the state feedback controller design presented in the proof of Theorem 8.35 with (x_a, x_c) replaced by $(\widehat{x}_a, \widehat{x}_c)$ in the controller, except that we have an additional exponentially decaying error perturbation as a result of the replacement. Note that this additional error disturbance can be accommodated in the error signal v in the state feedback design, which is taken care of by a properly designed feedback $z_0 = f(\widehat{x}_a)$ for the first subsystem. From the construction of state feedback, it can be verified that with the states (x_a, x_c) replaced by their measurements $(\widehat{x}_a, \widehat{x}_c)$, the constraints remain not violated. This completes the proof. ■

8.5 Semi-global and global stabilization in admissible set for non-right-invertible constraints: discrete time

For continuous-time systems, Sect. 8.3 considers both semi-global and global stabilization in the admissible set subject to non-right-invertible constraints. At first, for a given system Σ with non-right-invertible constraints and for a given pair of constraint sets $\mathcal{S}$ and $\mathcal{T}$ that satisfy Assumption 7.1, it develops necessary and sufficient conditions under which semi-global stabilization is solvable for non-right-invertible constraints. Such conditions depend on the shape of the *given* constraint set $\mathcal{S}$. Because of this, it resolves two fundamental questions which are as follows:

1. Given a system Σ with non-right-invertible constraints, for such a Σ, what are the solvability conditions for the Problems 7.8–7.11 when they are enhanced in the sense that they consider all possible pairs of constraint sets $\mathcal{S}$ and $\mathcal{T}$ that satisfy Assumption 7.1 instead of just one given pair of constraint sets $\mathcal{S}$ and $\mathcal{T}$?

2. Given a system Σ with non-right-invertible constraints, for such a Σ, does there exist a pair of constraint sets $\mathcal{S}$ and $\mathcal{T}$ that satisfy Assumption 7.1 for which semi-global and global stabilization problems as formally formulated in Problems 7.8–7.11 are solvable?

In this section, we proceed to develop similar results, however, for discrete-time systems. As said earlier, although our development here parallels the one in Sect. 8.3, as will be evident throughout this section, there exist several fundamental differences in every respect, that is, in the solvability conditions, in designing the controllers, as well as in constructing the proofs.

Our main focus in this section is often on amplitude constraints on the constrained output as expressed by $z(k) \in \mathcal{S}$ for all $k \geqslant 0$. The issues involved when both the amplitude and rate constraints exist on the constrained output $z(k)$ will only be briefly dealt with in Sect. 8.5.3.

We have the following theorem which is concerned with the semi-global stabilization Problem 7.8 via state feedback.

Theorem 8.43 *Consider the discrete-time system Σ as given by* (7.1) *and constraint sets $\mathcal{S}$ and $\mathcal{T} = \mathbb{R}^n$ that satisfy Assumption 7.1. Then, the semi-global stabilization problem in the admissible set via state feedback as defined in Problem 7.8 is solvable if and only if the following conditions hold:*

(i) (A, B) *is stabilizable.*

(ii) The constraints are at most weakly non-minimum phase.

(iii) For any $x \in \mathcal{V}(\mathcal{S}, \mathbb{R}^n)$, there exists a u such that $Ax + Bu \in \mathcal{V}(\mathcal{S}, \mathbb{R}^n)$ while $C_z x + D_z u \in \mathcal{S}$.

We consider next the case of measurement feedback that concerns with the semi-global stabilization. We have the following theorem.

Theorem 8.44 *Consider the discrete-time system Σ as given by* (7.1) *and constraint set $\mathcal{S}$ and $\mathcal{T} = \mathbb{R}^n$ that satisfies Assumption 7.1. Then, the semi-global stabilization problem in the admissible set via measurement feedback as defined in Problem 7.10 is solvable if the following conditions hold:*

(*i*) *(A, B) is stabilizable.*

(*ii*) *The constraints are at most weakly non-minimum phase.*

(*iii*) *For any $x \in \mathcal{V}(\mathcal{S}, \mathbb{R}^n)$, there exists a u such that $Ax + Bu \in \mathcal{V}(\mathcal{S}, \mathbb{R}^n)$ while $C_z x + D_z u \in \mathcal{S}$.*

(*iv*) *The pair (C_y, A) is observable.*

(*v*) *We have*

$$\ker C_y \subseteq \ker C_z A.$$

(*vi*) *We have*

$$\ker \begin{pmatrix} C_y & D_y \end{pmatrix} \subseteq \ker \begin{pmatrix} C_z & D_z \end{pmatrix}.$$

Note that in the above theorem, conditions (*i*)–(*iii*) are necessary. Condition (*iv*) can be weakened by assuming only detectability. However, clearly some additional assumptions would then be needed if unobservable states can affect the constrained output z. However, this is excluded by condition (*vi*). Regarding condition (*v*), we know that a necessary condition for solvability equals

$$\ker \begin{pmatrix} C_y \\ C_z \end{pmatrix} \subseteq \ker C_z A,$$

which is equal to condition (*v*) given condition (*vi*). Condition (*vi*) is not necessary but it is a natural condition to impose that the constrained variables z are part of the observations variables y, which is another way to express condition (*vi*).

We now proceed to prove Theorems 8.43 and 8.44. Expectedly, as in the continuous time, the proofs depend heavily on SCB of the system Σ_{uz} characterized by the quadruple (A, B, C_z, D_z). Consider the state, input, and constrained output transformation matrices, Γ_s, Γ_u, and Γ_z, and let

$$x = \Gamma_s \bar{x}, \quad u = \Gamma_u \bar{u}, \quad \text{and} \quad z = \Gamma_z \bar{z}$$

so that Σ_{uz} is in its SCB as given by Theorem 3.1. The given system (7.1) can then be written in SCB form as

$$\bar{\Sigma} : \begin{cases} \rho x_a &= A_{aa} x_a + K_a \bar{z} \\ \rho x_b &= A_{bb} x_b + K_b \bar{z} \\ \rho x_c &= A_{cc} x_c + K_c \bar{z} + B_c [u_c + J_a x_a] \\ \rho x_d &= u_d + G_a x_a + G_b x_b + G_c x_c + G_d x_d \\ y &= C_{ya} x_a + C_{yb} x_b + C_{yc} x_c + C_{yd} x_d + \bar{D}_y \bar{u} \\ z_0 &= u_0 \\ z_b &= C_b x_b \\ z_d &= C_d x_d, \end{cases} \tag{8.79}$$

where

$$\bar{x} = \begin{pmatrix} x_a \\ x_b \\ x_c \\ x_d \end{pmatrix}, \quad \bar{u} = \begin{pmatrix} u_0 \\ u_c \\ u_d \end{pmatrix}, \quad \bar{z} = \begin{pmatrix} z_0 \\ z_b \\ z_d \end{pmatrix},$$

and $G\bar{x} = G_a x_a + G_b x_b + G_c x_c + G_d x_d$.

We have $\bar{z}$ subject to the constraint $\bar{z}(k) \in \bar{\mathcal{S}}$ for all $k \geqslant 0$, where $\bar{\mathcal{S}} = \Gamma_z^{-1}\mathcal{S}$. Since $C_z' D_z = 0$, it is guaranteed that the new constraint set still satisfies Assumption 7.1. The admissible set in the new basis is defined as

$$\mathcal{V}(\bar{\mathcal{S}}) := \{\, \bar{x} \in \mathbb{R}^n \mid \exists u_0 \text{ such that } \begin{pmatrix} C_b x_b \\ u_0 \\ C_d x_d \end{pmatrix} \in \bar{\mathcal{S}}\,\}.$$

Proof of Theorem 8.43

We first establish the necessity of the conditions in Theorem 8.43. Using the SCB as introduced above, we can decompose the original system into two subsystems:

$$\Sigma_1 : \begin{cases} \rho x_a = A_{aa} x_a + K_{ab} C_b x_b + K_{a2}\zeta \\ \rho x_b = A_{bb} x_b + K_{bb} C_b x_b + K_{b2}\zeta \\ \rho x_d = A_{dd} x_d + B_d[u_d + G\bar{x}] + K_d \bar{z} \\ \zeta = \begin{pmatrix} 0 \\ C_d \end{pmatrix} x_d + \begin{pmatrix} I \\ 0 \end{pmatrix} u_0 \\ \bar{z} = \begin{pmatrix} C_b \\ 0 \end{pmatrix} x_b + \begin{pmatrix} 0 \\ I \end{pmatrix} \zeta, \end{cases} \tag{8.80}$$

and

$$\Sigma_2 : \left\{ \rho x_c = A_{cc} x_c + B_c[u_c + J_a x_a] + K_c \bar{z}. \right. \tag{8.81}$$

We have the following lemma regarding the necessary conditions for semi-global stabilization.

Lemma 8.45 *Consider the constrained system* (7.1). *The problem of semi-global stabilization in the admissible set via state feedback is solvable only if:*

(*i*) *The constraints are weakly non-minimum phase.*

(*ii*) *For any $x \in \mathcal{V}(\mathcal{S}, \mathbb{R}^n)$, there exists a u such that $Ax + Bu \in \mathcal{V}(\mathcal{S}, \mathbb{R}^n)$ while $C_z x + D_z u \in \mathcal{S}$.*

Moreover, condition (*ii*) *implies the following:*

(*iii*) *The constraints are weakly non-right invertible, i.e., the matrix C_b is injective.*

(iv) *The constraints are of type one, i.e., the matrix C_d is injective.*

Proof : We first note that the first equation of Σ_1 reads as

$$x_a(k+1) = A_{aa}x_a(k) + K_{ab}C_b x_b(k) + K_{a2}\zeta(k).$$

Clearly, $x(0) \in \mathcal{V}(\mathcal{S}, \mathbb{R}^n)$ still allows for an arbitrary initial condition $x_a(0)$ for this system (provided we choose the other initial conditions for x_b, x_c and x_d appropriately). On the other hand, the inputs $C_b x_b$ and ζ to this system are bounded. It is well known from the theory for linear systems subject to input constraints that if the system is exponentially unstable, there exist initial conditions $x_a(0)$ for which there exist no $C_b x_b$ and ζ such that x_a converges to zero. This is clearly in contradiction with the requirements for the semi-global stabilization in the admissible set. Therefore, A_{aa} must have its eigenvalues in the closed unit disc, or equivalently, the constraints are at most weakly non-minimum phase.

Condition (*ii*) is clearly necessary since the existence of $x_0 \in \mathcal{V}(\mathcal{S}, \mathbb{R}^n)$ for which there does not exist any u such that $Ax_0 + Bu \in \mathcal{V}(\mathcal{S}, \mathbb{R}^n)$ implies the existence of a $\tilde{x}_0 \in \operatorname{int}\mathcal{V}(\mathcal{S}, \mathbb{R}^n)$ for which there does not exist any u such that $A\tilde{x}_0 + Bu \in \mathcal{V}(\mathcal{S}, \mathbb{R}^n)$ because the set $\mathcal{V}(\mathcal{S}, \mathbb{R}^n)$ is closed. But clearly, semi-global stabilization in the admissible set requires that for all initial conditions in the interior of $\mathcal{V}(\mathcal{S}, \mathbb{R}^n)$, we must be able to avoid constraint violation. The fact that we cannot guarantee for some initial condition $x(0) = \tilde{x}_0$ that $x(1) \in \mathcal{V}(\mathcal{S}, \mathbb{R}^n)$ implies that we will get a constraint violation which yields a contradiction.

Assume that the constraints are not weakly non-right invertible or, in other words, the matrix C_b is not injective. In that case, we can find a x_b such that $C_b x_b = 0$. However, since the pair (C_b, A_{bb}) is observable, there always exists some $k < n$ such that $C_b A_{bb}^k x_b \neq 0$. Without loss of generality, choose k such that $C_b A_{bb}^{k-1} x_b = 0$. But then the initial condition,

$$\bar{x}(0) = \begin{pmatrix} 0 \\ \lambda A_{bb}^{k-1} x_b \\ 0 \\ 0 \end{pmatrix},$$

is in $\mathcal{V}(\bar{\mathcal{S}})$ for all λ. However,

$$\bar{z}(1) = \begin{pmatrix} \lambda C_b A_{bb}^k x_b + C_b K_{bb} C_b x_b + C_b K_{b2}\zeta(0) \\ \zeta(1) \end{pmatrix}$$

will not be in $\bar{\mathcal{S}}$ for sufficiently large λ since $\bar{\mathcal{S}}$ is bounded. After all, $\lambda C_b A_{bb}^k x_b$ can be made arbitrarily large by choosing λ large while all other terms are bounded since we know that $\bar{z}(0) \in \bar{\mathcal{S}}$. Therefore, $\bar{z}(1)$ is not in $\bar{\mathcal{S}}$ for any input even though $\bar{x}(0) \in \mathcal{V}(\bar{\mathcal{S}})$. This violates condition (*ii*).

In order to establish *(iv)*, we first assume that there exists an x_d such that $C_d x_d = 0$ while $C_d A_{dd} x_d + C_d B_d u_d \neq 0$ for any u_d. For any λ, there exists a $\bar{x}(0)$ in $\mathcal{V}(\bar{\mathcal{S}})$ which yields initial condition $x_d(0) = \lambda x_d$ for this system (provided we choose the other initial conditions for x_a, x_b and x_c appropriately). But then,

$$C_d x_d(1) = \lambda C_d A_{dd} x_d + C_d B_d[u_d(0) + G\bar{x}(0)] + C_d K_d \bar{z}(0)$$

can be made arbitrary large independent of our choice for $u_d(0)$ since the second term on the right cannot cancel the first term while the third term on the right is bounded. This violates condition *(ii)*. On other hand, if for all x_d satisfying $C_d x_d = 0$ there exists a u_d such that $C_d A_{dd} x_d + C_d B_d u_d = 0$, then there exists a matrix F such that for all x_d such that $C_d x_d = 0$ we have $C_d(A_{dd} + B_d F)x_d = 0$. But this in turn implies that $C_d B_d v = 0$ for some $v \neq 0$ yields that $C_d(A_{dd} + BF)^k Bv = 0$ for all k which is in contradiction with the left invertibility of $(A_{dd}, B_d, C_d, 0)$ as required by the properties of SCB. Hence, $C_d B_d$ is injective (which implies the infinite zeros are at most of order 1). The structure of the SCB then also guarantees that C_d is injective. ■

In order to establish sufficiency of the conditions of Theorem 8.43, we will construct an appropriate controller for the system Σ. Note that finding a controller for Σ_2 does not affect Σ_1 nor its constraints. Our design methodology will amount to designing a controller for the system Σ_1. However, since the eigenvalues of A_{aa} must be in the closed unit disc, it can be established that the critical part of the system is actually the x_b and x_d dynamics presented in the following subsystem:

$$\Sigma_{bd}: \begin{cases} x_b(k+1) = A_{bb}x_b(k) + K_{bb}C_b x_b(k) + K_{b2}\zeta(k) \\ x_d(k+1) = A_{dd}x_d(k) + B_d\tilde{u}_d(k) + K_d\bar{z}(k) \\ \quad\zeta(k) \quad = \begin{pmatrix} 0 \\ C_d \end{pmatrix} x_d(k) + \begin{pmatrix} I \\ 0 \end{pmatrix} u_0(k) \\ \quad\bar{z}(k) \quad = \begin{pmatrix} C_b \\ 0 \end{pmatrix} x_b(k) + \begin{pmatrix} 0 \\ I \end{pmatrix} \zeta(k), \end{cases} \tag{8.82}$$

where $\tilde{u}_d = u_d + G\bar{x}$. Also, we denote $x_{bd} = (x_b', \ x_d')'$.

We define the admissible set for subsystem Σ_{bd} as

$$\mathcal{V}_{bd}(\bar{\mathcal{S}}) := \left\{ x_b \in \mathbb{R}^{n_b}, x_d \in \mathbb{R}^{n_d} \;\middle|\; \exists u_0 \text{ such that } \begin{pmatrix} C_b x_b \\ u_0 \\ C_d x_d \end{pmatrix} \in \bar{\mathcal{S}} \right\}. \tag{8.83}$$

The proof of sufficiency can be sketched as follows. We will start the construction in Lemma 8.46 by determining an appropriate controller for the system Σ_{bd}. Then based on the result of Lemma 8.46, we construct a controller for Σ_1. Finally, after choosing a proper controller for Σ_2, we complete the proof.

The following lemma is concerned with semi-global stabilization for Σ_{bd}.

Lemma 8.46 *The problem of semi-global stabilization in the admissible set for Σ_{bd} is solvable by a static state feedback,*

$$u_0 = \overline{f}_1(x_b, x_d) \ \textit{and} \ \widetilde{u}_d = \overline{f}_2(x_b, x_d),$$

if the conditions of Theorem 8.43 are satisfied.

Proof : Using condition (*iii*), it is not difficult to construct a controller such that the state cannot leave the set $\mathcal{V}_{bd}(\overline{\mathcal{S}})$. However, to establish the convergence to the origin, we need to do some extra work.

We first transform Σ_{bd} into its controllable canonical form. That is, there is a nonsingular state transformation T such that with

$$\widetilde{x} = \begin{pmatrix} x_1(k) \\ x_2(k) \end{pmatrix} = T x_{bd},$$

the system Σ_{bd} given by (8.82) is transformed to the form

$$\widetilde{\Sigma}_{bd} : \begin{cases} \begin{pmatrix} x_1(k+1) \\ x_2(k+1) \end{pmatrix} = \begin{pmatrix} A_{11} & A_{12} \\ 0 & A_{22} \end{pmatrix} \begin{pmatrix} x_1(k) \\ x_2(k) \end{pmatrix} + \begin{pmatrix} B_1 \\ 0 \end{pmatrix} \widetilde{u}(k), \\ \overline{z}(k) = \begin{pmatrix} C_1 & C_2 \\ 0 & 0 \end{pmatrix} \begin{pmatrix} x_1(k) \\ x_2(k) \end{pmatrix} + \begin{pmatrix} 0 & 0 \\ I & 0 \end{pmatrix} \widetilde{u}(k), \end{cases}$$

where the dynamics of x_1 is controllable, the dynamics of x_2 is uncontrollable, and

$$\widetilde{u}(k) = \begin{pmatrix} u_0(k) \\ \widetilde{u}_d(k) \end{pmatrix}.$$

The admissible set of $\widetilde{\Sigma}_{bd}$ is given by

$$\widetilde{\mathcal{V}}_{bd}(\overline{\mathcal{S}}) = T\mathcal{V}_{bd}(\overline{\mathcal{S}}).$$

In order to construct a controller for $\widetilde{\Sigma}_{bd}$, define a modified system

$$\widetilde{\Sigma}^{\ell}_{bd} : \begin{cases} \rho \begin{pmatrix} x_1^{\ell} \\ x_2^{\ell} \end{pmatrix} = (1+\ell) \begin{pmatrix} A_{11} & A_{12} \\ 0 & A_{22} \end{pmatrix} \begin{pmatrix} x_1^{\ell} \\ x_2^{\ell} \end{pmatrix} + (1+\ell) \begin{pmatrix} B_1 \\ 0 \end{pmatrix} \widetilde{u}^{\ell}, \\ \overline{z}^{\ell} = \begin{pmatrix} C_1 & C_2 \\ 0 & 0 \end{pmatrix} \begin{pmatrix} x_1^{\ell} \\ x_2^{\ell} \end{pmatrix} + \begin{pmatrix} 0 & 0 \\ I & 0 \end{pmatrix} \widetilde{u}^{\ell}, \end{cases}$$

where $\ell > 0$ is small enough so that $\widetilde{\Sigma}^{\ell}_{bd}$ is still stabilizable. Let $\widetilde{\mathcal{R}}^{\ell}_{bd}(\overline{\mathcal{S}})$ be the largest set of initial conditions for the system $\widetilde{\Sigma}^{\ell}_{bd}$ for which there exists an input such that the constraints are satisfied while we stay inside the set for all k (note that we do **NOT** impose convergence to zero). We do claim that for any $\tilde{\rho}$, there exists an $\ell > 0$ sufficiently small such that

$$\tilde{\rho}\tilde{\mathcal{V}}_{bd}(\bar{\mathcal{S}}) \subset \tilde{\mathcal{R}}^{\ell}_{bd}(\bar{\mathcal{S}}) \subset \tilde{\mathcal{V}}_{bd}(\bar{\mathcal{S}}). \tag{8.84}$$

It is trivial to see that

$$\tilde{\mathcal{R}}^{\ell}_{bd}(\bar{\mathcal{S}}) \subset \tilde{\mathcal{V}}_{bd}(\bar{\mathcal{S}}).$$

It remains to establish that

$$\tilde{\rho}\tilde{\mathcal{V}}_{bd}(\bar{\mathcal{S}}) \subset \tilde{\mathcal{R}}^{\ell}_{bd}(\bar{\mathcal{S}}).$$

For any time $r > n$ for the system $\tilde{\Sigma}^{\ell}_{bd}$, since the dynamics of x_1 is controllable, there exists a $\delta^* > 0$ such that for any $\delta \in (0, \delta^*)$ and for any x_1 for which there exists a x_2 such that

$$\begin{pmatrix} x_1 \\ x_2 \end{pmatrix} \in \delta\tilde{\mathcal{V}}_{bd}(\bar{\mathcal{S}}),$$

there exists an input $\tilde{u}^{\ell} := (u^{\ell}_0, \tilde{u}^{\ell}_d)$ such that $x^{\ell}_1(r) = -x_1$ and $x^{\ell}_2(r) = 0$ with initial condition $\tilde{x}^{\ell}(0) = 0$, while

$$\bar{z}^{\ell}(k) \in \tfrac{1-\tilde{\rho}}{2}\mathcal{S}, \qquad k = 0, 1, \ldots, r-1.$$

Moreover, δ^* is independent of ℓ and r provided ℓ is small enough.

Let $r > n$ be such that for any $\tilde{x}^{\ell}(0) \in \tilde{\mathcal{V}}_{bd}(\bar{\mathcal{S}})$, we have

$$\begin{pmatrix} 0 \\ x^{\ell}_2(r) \end{pmatrix} \in \delta\tilde{\rho}\tilde{\mathcal{V}}_{bd}(\bar{\mathcal{S}})$$

for all ℓ sufficiently small. This is clearly possible due to the fact that the system is stabilizable and hence the uncontrollable dynamics of x^{ℓ}_2 must be asymptotically stable.

Consider any initial condition $\tilde{x}(0) \in \tilde{\mathcal{V}}_{bd}(\bar{\mathcal{S}})$. We have an input $\tilde{u}$ for the system $\tilde{\Sigma}_{bd}$ such that $\bar{z}(k) \in \bar{\mathcal{S}}$. Hence, for any $\tilde{\rho} < 1$, we can find, for any initial condition $\tilde{\rho}\tilde{x}(0) \in \tilde{\rho}\tilde{\mathcal{V}}_{bd}(\bar{\mathcal{S}})$, an input $\tilde{\rho}\tilde{u}$ for the system $\tilde{\Sigma}_{bd}$ such that $\bar{z}(k) \in \tilde{\rho}\bar{\mathcal{S}}$ for all k. But then, for ℓ small enough, we find that there exists an input, say $\tilde{u}^{\ell}_1$ for which we have $\tilde{x}^{\ell}(k) \in (1+\delta)\tilde{\rho}\tilde{\mathcal{V}}_{bd}(\bar{\mathcal{S}})$ and $\bar{z}^{\ell}(k) \in (1+\delta)\tilde{\rho}\bar{\mathcal{S}}$ for $k = 0, \ldots, r$. Also, we observe that if we choose $\delta < \frac{1-\tilde{\rho}}{2}$, we have

$$\delta\tilde{x}^{\ell}(r) = \begin{pmatrix} \delta x^{\ell}_1(r) \\ \delta x^{\ell}_2(r) \end{pmatrix} \in \delta\tilde{\mathcal{V}}_{bd}(\bar{\mathcal{S}}).$$

Choose

$$x_1 = \delta x^{\ell}_1(r).$$

Choose input, say $\tilde{u}_2^\ell$ such that, for $\tilde{x}^\ell(0) = 0$, we have $x_1^\ell(r) = -x_1$ and $x_2^\ell(r) = 0$ while $\bar{z}^\ell(k) \in \delta\mathcal{S}$. But then for initial condition $\tilde{x}^\ell(0) \in \tilde{\rho}\tilde{\mathcal{V}}_{bd}(\bar{\mathcal{S}})$, the input $\tilde{u}_1^\ell + \tilde{u}_2^\ell$ and $\delta < \min\{\delta^*, \frac{1-\tilde{\rho}}{2}\}$ we obtain that

$$\bar{z}^\ell(k) \in \bar{\mathcal{S}} \text{ for } k = 0, \dots, r-1,$$

and

$$\begin{aligned} x_{bd}^\ell(r) = (1-\delta)\begin{pmatrix} x_1^\ell(r) \\ x_2^\ell(r) \end{pmatrix} + \delta \begin{pmatrix} 0 \\ x_2^\ell(r) \end{pmatrix} &\in \left[(1-\delta)(1+\delta)\tilde{\rho} + \delta^2\tilde{\rho}\right] \tilde{\mathcal{V}}_{bd}(\bar{\mathcal{S}}) \\ &\in \tilde{\rho}\tilde{\mathcal{V}}_{bd}(\bar{\mathcal{S}}). \end{aligned}$$

If we repeat this construction between $k = r$ and $k = 2r$, and so forth, it becomes clear that we can find for any initial condition

$$\tilde{x}^\ell(0) \in \tilde{\rho}\tilde{\mathcal{V}}_{bd}(\bar{\mathcal{S}}),$$

an input such that

$$\bar{z}^\ell(k) \in \mathcal{S}$$

for all k. Hence $\tilde{x}^\ell(0) \in \tilde{\mathcal{R}}_{bd}^\ell(\bar{\mathcal{S}})$. This clearly implies that (8.84) is satisfied.

For semi-global stabilization in the admissible set, we take any compact set $\tilde{\mathcal{H}}_{bd}$ contained in the interior of $\tilde{\mathcal{V}}_{bd}(\bar{\mathcal{S}})$, and we construct a static controller which will stabilize the system and the domain of attraction contains $\tilde{\mathcal{H}}_{bd}$. But then clearly, using (8.84), we can find ℓ such that $\tilde{\mathcal{H}}_{bd} \subset \tilde{\mathcal{R}}_{bd}^\ell(\bar{\mathcal{S}})$. Next, we choose a feedback f on the boundary of $\tilde{\mathcal{R}}_{bd}^\ell(\bar{\mathcal{S}})$ such that, for any $\tilde{x}^\ell(k) \in \partial\tilde{\mathcal{R}}_{bd}^\ell(\bar{\mathcal{S}})$, we have $\tilde{x}^\ell(k+1) \in \tilde{\mathcal{R}}_{bd}^\ell(\bar{\mathcal{S}})$. We expand this feedback f to the whole state space. Define $g : \mathbb{R}^n \to \mathbb{R}^+$ such that for any x,

$$g(x)x \in \partial\tilde{\mathcal{R}}_{bd}^\ell(\bar{\mathcal{S}}).$$

Since $\tilde{\mathcal{R}}_{bd}^\ell(\bar{\mathcal{S}})$ is a convex set containing 0 in its interior, this mapping is well defined. Then we expand f to the whole state space by

$$\tilde{f}(x) = \frac{f(g(x)x)}{g(x)}.$$

This expansion has the property that for any $\eta > 0$, we have $\tilde{x}^\ell(k+1) \in \eta\tilde{\mathcal{R}}_{bd}^\ell(\bar{\mathcal{S}})$ for all $\tilde{x}^\ell(k) \in \eta\tilde{\mathcal{R}}_{bd}^\ell(\bar{\mathcal{S}})$. Note that $\tilde{f}$ is positively homogeneous, that is,

$$\tilde{f}(\tilde{\alpha}x) = \tilde{\alpha}\tilde{f}(x),$$

for any $\tilde{\alpha} > 0$.

Clearly, for the system $\tilde{\Sigma}_{bd}^\ell$ with the feedback

$$u_0^\ell(k) = \tilde{f}_1(x_{bd}^\ell(k)) \text{ and } \tilde{u}_d^\ell(k) = \tilde{f}_2(x_{bd}^\ell(k)),$$

and for all initial conditions in the set $\widetilde{\mathcal{R}}^{\ell}_{bd}(\overline{\mathcal{S}})$, we have $\widetilde{x}^{\ell}(k) \in \widetilde{\mathcal{R}}^{\ell}_{bd}(\overline{\mathcal{S}})$ for all k.

But then the feedback,

$$u_0(k) = \widetilde{f}_1(\widetilde{x}(k)) \text{ and } \widetilde{u}_d(k) = \widetilde{f}_2(\widetilde{x}(k)),$$

for the original system with $\widetilde{x}(0) = \widetilde{x}^{\ell}(0)$, results in a state

$$\widetilde{x}(k) = \frac{1}{(1+\ell)^k}\widetilde{x}^{\ell}(k).$$

Hence, we obviously have $x_{bd}(k) \in \mathcal{R}^{\ell}_{bd}(\overline{\mathcal{S}})$ for all k but also $x_{bd}(k) \to 0$ as $k \to \infty$.

Finally, the controller for the original system Σ_{bd} is of the form,

$$u_0(k) = \widetilde{f}_1(T^{-1}\widetilde{x}(k)) = \overline{f}_1(x_b, x_d)\widetilde{u}_d(k) \quad = \widetilde{f}_2(T^{-1}\widetilde{x}(k)) = \overline{f}_2(x_b, x_d).$$

■

We construct next a controller for Σ_1 based on the results of Lemma 8.46.

Lemma 8.47 *The problem of semi-global stabilization in the admissible set for Σ_1 is solvable by a static state feedback if the conditions of Theorem 8.43 are satisfied.*

Proof : It is easy to verify that the admissible set of initial conditions $\mathcal{V}_1(\overline{\mathcal{S}})$ and $\mathcal{V}_{bd}(\overline{\mathcal{S}})$ for Σ_1 and Σ_{bd}, respectively, have the relationship

$$\mathcal{V}_1(\overline{\mathcal{S}}) = \mathbb{R}^{n_a} \oplus \mathcal{V}_{bd}(\overline{\mathcal{S}}).$$

For any compact set $\mathcal{H}$ in $\mathcal{V}_1(\overline{\mathcal{S}})$, we choose a compact set $\mathcal{H}_a$ and $\widetilde{\rho} < 1$ such that

$$\mathcal{H} \subset \mathcal{H}_a \oplus \widetilde{\rho}\mathcal{V}_{bd}(\overline{\mathcal{S}}).$$

The controller constructed in Lemma 8.46, $u_0 = \overline{f}_1(x_{bd})$ and $u_d = \widetilde{u}_d - G\overline{x} = \overline{f}_2(x_{bd}) - G\overline{x}$, is such that for all initial conditions in $\widetilde{\rho}\mathcal{V}_{bd}(\overline{\mathcal{S}})$, the origin of the closed-loop system is exponentially stable. Hence, there exist $M > 0$ and λ with $|\lambda| < 1$ such that

$$\|x_{bd}(k)\| \leqslant M\lambda^k \tag{8.85}$$

for all k and for all $x_{bd}(0) \in \widetilde{\rho}\mathcal{V}_{bd}(\overline{\mathcal{S}})$.

Next, let P_0 be the semi-stabilizing solution of the discrete-time algebraic Riccati equation

$$P_0 = A_0'P_0A_0 + C_0'C_0 - A_0'P_0B_0(B_0'P_0B_0 + D_0'D_0)^{\dagger}B_0'P_0A_0,$$

where

$$A_0 = \begin{pmatrix} A_{aa} & K_{ab}C_b & K_{ad}C_d \\ 0 & A_{bb} + K_{bb}C_b & K_{bd}C_d \\ 0 & K_{db}C_b & A_{dd} \end{pmatrix}, \quad B_0 = \begin{pmatrix} B_{a0} & 0 \\ B_{b0} & 0 \\ B_{d0} & B_d \end{pmatrix},$$

$$C_0 = \begin{pmatrix} 0 & C_b & 0 \\ 0 & 0 & C_d \\ 0 & 0 & 0 \end{pmatrix}, \quad \text{and} \quad D_0 = \begin{pmatrix} 0 & 0 \\ 0 & 0 \\ I & 0 \end{pmatrix}.$$

We have

$$P_0 \begin{pmatrix} x_a \\ 0 \\ 0 \end{pmatrix} = 0 \tag{8.86}$$

for all $x_a \in R_{na}$ since the eigenvalue of A_{aa} are in the closed unit disc. Choose a level set,

$$V_0(c) := \{ \xi \in \mathbb{R}^{n_1} \mid \xi(k)' P_0 \xi(k) \leqslant c \},$$

such that we have

$$\big(C_0 - D_0(B_0' P_0 B_0 + D_0' D_0)^{\dagger} B_0' P_0 A_0\big)\xi \in \bar{\mathcal{S}}/3 \tag{8.87}$$

for all $\xi \in V_0(c)$. Then with the controller,

$$u_0 = \bar{f}_1(x_{bd}) \text{ and } u_d = \bar{f}_2(x_{bd}) - G\bar{x},$$

which we can abbreviate as

$$\begin{pmatrix} u_0 \\ u_d \end{pmatrix} = \bar{f}(\bar{x}),$$

there exists a T such that, for any initial state in

$$\mathcal{H}_a \oplus \tilde{\rho} \mathcal{V}_{bd}(\bar{\mathcal{S}}),$$

we have

$$\begin{pmatrix} x_a(T) \\ x_{bd}(T) \end{pmatrix} \in V_0(c). \tag{8.88}$$

Let P_ε be the stabilizing solution of the algebraic equation,

$$P_\varepsilon = A_0' P_\varepsilon A_0 + C_0' C_0 + \varepsilon I - A_0' P_\varepsilon B_0 (B_0' P_\varepsilon B_0 + I)^{-1} B_0' P_\varepsilon A_0.$$

We have $P_\varepsilon \to P_0$ as ε approach zero. Define the level set,

$$V_\varepsilon(c) := \{ \xi \in \mathbb{R}^{n_1} \mid \xi' P_\varepsilon \xi \leqslant c \},$$

such that for ε small enough, we have

$$\begin{pmatrix} x_a(T) \\ x_{bd}(T) \end{pmatrix} \in 2V_\varepsilon(c),$$

and

$$(C_0 - D_0(B_0' P_\epsilon B_0 + D_0' D_0)^\dagger B_0' P_\epsilon A_0)\xi \in \bar{\mathcal{S}}$$

for any initial condition $\xi \in 2V_\varepsilon(c)$. Hence, the feedback

$$\begin{pmatrix} u_0 \\ u_d \end{pmatrix} = -(B_0' P_\varepsilon B_0 + D_0' D_0)^\dagger B_0' P_\varepsilon A_0 \begin{pmatrix} x_a \\ x_{bd} \end{pmatrix}$$

is an asymptotically stabilizing controller for Σ_1 and achieves a domain of attraction containing $2V_\varepsilon(c)$. Next, consider the controller

$$\begin{pmatrix} u_0 \\ u_d \end{pmatrix} = \begin{cases} \bar{f}(\bar{x}), & x_{abd} \notin 2V_\varepsilon(c), . \\ -(B_0' P_\varepsilon B_0 + D_0' D_0)^\dagger B_0' P_\varepsilon A_0 x_{abd}, & x_{abd} \in 2V_\varepsilon(c). \end{cases}$$

It is easily verified that this controller asymptotically stabilizes the system. ■

The above lemma yields an appropriate controller for the subsystem Σ_1. Finally, we need to construct a controller for the original system Σ which will complete our proof of sufficiency for Theorem 8.43.

We are now ready to discuss the proof of Theorem 8.43. The necessity was already established in Lemma 8.45. For sufficiency, it is easily seen that the controllers designed in Lemma 8.47 combined with a controller,

$$u_c(k) = -J_a x_a(k) + F_c x_c(k),$$

where F_c is such that $A_{cc} + B_c F_c$ is asymptotically stable, solves the problem of semi-global stabilization in the admissible set via state feedback for the given system Σ.

Proof of Theorem 8.44 : At first, we note that the conditions of Theorem 8.44 imply that x_b and x_d can be directly deduced from the measurements. In other words, we have (in a suitable basis)

$$y = \begin{pmatrix} y_1 \\ y_2 \\ y_3 \end{pmatrix} = \begin{pmatrix} C_{ya} & 0 & C_{yc} & 0 \\ 0 & C_b & 0 & 0 \\ 0 & 0 & 0 & C_d \end{pmatrix} \begin{pmatrix} x_a \\ x_b \\ x_c \\ x_d \end{pmatrix} + \begin{pmatrix} D_y \\ 0 \\ 0 \end{pmatrix} u_0,$$

where C_b and C_d are injective. However, we need an observer to estimate x_a and x_c:

$$\rho\begin{pmatrix}\widehat{x}_a\\ \widehat{x}_c\end{pmatrix}=\begin{pmatrix}A_{aa} & 0\\ B_cJ_a & A_{cc}\end{pmatrix}\begin{pmatrix}\widehat{x}_a\\ \widehat{x}_c\end{pmatrix}+\begin{pmatrix}K_{ab} & K_{a2} & 0\\ K_{c1} & K_{c2} & B_c\end{pmatrix}\begin{pmatrix}C_bx_b\\ \zeta\\ u_c\end{pmatrix}$$
$$+\begin{pmatrix}L_a\\ L_c\end{pmatrix}\left[y_1-C_{ya}\widehat{x}_a-C_{yc}\widehat{x}_b-D_yu_0\right].$$

Clearly,

$$\rho\begin{pmatrix}\widehat{x}_a-x_a\\ \widehat{x}_c-x_c\end{pmatrix}=\begin{pmatrix}A_{aa}-L_aC_{ya} & -L_aC_{yc}\\ B_cJ_a-L_cC_{ya} & A_{cc}-L_cC_{yc}\end{pmatrix}\begin{pmatrix}\widehat{x}_a-x_a\\ \widehat{x}_c-x_c\end{pmatrix}$$
$$=\widetilde{A}_{ac}\begin{pmatrix}\widehat{x}_a-x_a\\ \widehat{x}_c-x_c\end{pmatrix},$$

where L_a and L_c are chosen such that $\widetilde{A}_{ac}$ is asymptotically stable.

The feedback $\bar{f}$ can be directly implemented even in the measurement feedback case since the condition (v) of Theorem 8.44 guarantees that

$$G=\begin{pmatrix}0 & G_b & 0 & G_d\end{pmatrix}+L_G\begin{pmatrix}C_{ya} & 0 & C_{yc} & 0\end{pmatrix}.$$

Hence, $u_d=\bar{f}_2(x_{bd})-G\bar{x}$ is equivalent to

$$u_d=\bar{f}_2(x_{bd})-G_bx_b-G_bx_d-L_Gy_1+L_GD_yu_0.$$

Next, we follow the same arguments as in Lemma 8.47 with small modifications such as the inclusion of x_c since the observer does not allow a separate controller design for x_c and x_a.

It is easy to verify that the admissible set of initial conditions $\mathcal{V}(\bar{\mathcal{S}})$ and $\mathcal{V}_{bd}(\bar{\mathcal{S}})$ for $\bar{\Sigma}$ and Σ_{bd}, respectively, have the relationship

$$\mathcal{V}(\bar{\mathcal{S}})=\mathbb{R}^{n_a}\oplus\mathbb{R}^{n_c}\oplus\mathcal{V}_{bd}(\bar{\mathcal{S}}).$$

For any compact set $\mathcal{H}$ in $\mathcal{V}(\bar{\mathcal{S}})$, we choose a compact set $\mathcal{H}_{ac}$ and $\widetilde{\rho}<1$ such that

$$\mathcal{H}\subset\mathcal{H}_{ac}\oplus\widetilde{\rho}\mathcal{V}_{bd}(\bar{\mathcal{S}}).$$

The controller $u_0=\bar{f}_1(x_{bd})$ and $u_d=\bar{f}_2(x_{bd})-G\bar{x}$ is such that for all initial conditions in $\widetilde{\rho}\mathcal{V}_{bd}(\bar{\mathcal{S}})$, the origin of the closed-loop system is exponentially stable. Hence, there exist $M>0$ and λ with $|\lambda|<1$ such that

$$\|x_{bd}(k)\|\leqslant M\lambda^k \tag{8.89}$$

for all k and for all $x_{bd}(0)\in\widetilde{\rho}\mathcal{V}_{bd}(\bar{\mathcal{S}})$.

Next, let P_0 be the semi-stabilizing solution of the discrete-time algebraic Riccati equation

$$P_0 = A_0' P_0 A_0 + C_0' C_0 - A_0' P_0 B_0 (B_0' P_0 B_0 + D_0' D_0)^{\dagger} B_0' P_0 A_0,$$

where

$$A_0 = \begin{pmatrix} A_{aa} & K_{ab}C_b & 0 & K_{ad}C_d \\ 0 & A_{bb} + K_{bb}C_b & 0 & K_{bd}C_d \\ B_c J_a & K_{cb}C_b & A_{cc} & K_{cd}C_d \\ 0 & K_{db}C_b & 0 & A_{dd} \end{pmatrix}, \quad B_0 = \begin{pmatrix} B_{a0} & 0 & 0 \\ B_{b0} & 0 & 0 \\ B_{c0} & B_c & 0 \\ B_{d0} & 0 & B_d \end{pmatrix},$$

$$C_0 = \begin{pmatrix} 0 & C_b & 0 & 0 \\ 0 & 0 & C_d & 0 \\ 0 & 0 & 0 & 0 \end{pmatrix}, \qquad \text{and} \qquad D_0 = \begin{pmatrix} 0 & 0 & 0 \\ 0 & 0 & 0 \\ I & 0 & 0 \end{pmatrix}.$$

We have

$$P_0 \begin{pmatrix} x_a \\ 0 \\ x_c \\ 0 \end{pmatrix} = 0 \tag{8.90}$$

for all $x_a \in \mathbb{R}^{n_a}$ and $x_c \in \mathbb{R}^{n_c}$ since the eigenvalue of A_{aa} are in the closed unit disc while u_c can stabilize the x_c dynamics without incurring cost. Choose a level set

$$\mathcal{V}_0(c) := \{\, \bar{x} \in \mathbb{R}^n \mid \bar{x}(k)' P_0 \bar{x}(k) \leqslant c \,\},$$

such that we have

$$(C_0 - D_0(B_0' P_0 B_0 + D_0' D_0)^{\dagger} B_0' P_0 A_0)\bar{x} \in \bar{\mathcal{S}}/3 \tag{8.91}$$

for all $\bar{x} \in \mathcal{V}_0(c)$. Then with the controller,

$$u_0 = \bar{f}_1(x_{bd}), \quad u_c = 0, \quad u_d = \bar{f}_2(x_{bd}) - G\bar{x},$$

which we can abbreviate as

$$\begin{pmatrix} u_0 \\ u_c \\ u_d \end{pmatrix} = \bar{f}(\bar{x}),$$

there exists a T such that, for any initial state in

$$\mathcal{H}_{ac} \oplus \tilde{\rho} \mathcal{V}_{bd}(\bar{\mathcal{S}}),$$

we have

$$\bar{x}(T) \in \mathcal{V}_0(c). \tag{8.92}$$

Let P_ε be the stabilizing solution of the algebraic equation

$$P_\varepsilon = A_0' P_\varepsilon A_0 + C_0' C_0 + \varepsilon I - A_0' P_\varepsilon B_0 (B_0' P_\varepsilon B_0 + D_0' D_0)^\dagger B_0' P_\varepsilon A_0.$$

We have $P_\varepsilon \to P_0$ as ε approach zero. Defining the level set,

$$V_\varepsilon(c) := \{\, \bar{x} \in \mathbb{R}^n \mid \bar{x}' P_\varepsilon \bar{x} \leqslant c \,\},$$

there exists an ε such that

$$\bar{x}(T) \in \tfrac{4}{3} V_\varepsilon(c).$$

Moreover, we can guarantee that for ε small enough,

$$\begin{pmatrix} x_a - \hat{x}_a \\ 0 \\ x_c - \hat{x}_c \\ 0 \end{pmatrix}(k) \in \tfrac{1}{3} V_\varepsilon(c)$$

for all $k > 0$ given initial conditions for the system and the observer in the compact sets $\mathcal{H}$ and $\mathcal{H}_{\text{obs}}$, respectively. For ε small enough, we have

$$\big(C_0 - D_0 (B_0' P_\epsilon B_0 + D_0' D_0)^\dagger B_0' P_\epsilon A_0\big)\bar{x} \in \bar{\mathcal{S}}$$

for any initial condition $\bar{x} \in 2V_\varepsilon(c)$. Note that

$$\hat{\bar{x}} = \begin{pmatrix} \hat{x}_a \\ x_b \\ \hat{x}_c \\ x_d \end{pmatrix} \in \frac{4}{3} V_\varepsilon(c)$$

implies that

$$\bar{x} = \begin{pmatrix} \hat{x}_a \\ x_b \\ \hat{x}_c \\ x_d \end{pmatrix} + \begin{pmatrix} x_a - \hat{x}_a \\ 0 \\ x_c - \hat{x}_c \\ 0 \end{pmatrix} \in 2V_\varepsilon(c),$$

and hence the feedback,

$$\begin{aligned} \begin{pmatrix} u_0 \\ u_c \\ u_d \end{pmatrix} &= -(B_0' P_\varepsilon B_0 + D_0' D_0)^\dagger B_0' P_\varepsilon A_0 \hat{\bar{x}} - \begin{pmatrix} 0 \\ 0 \\ G \end{pmatrix} \bar{x} \\ &= F_\varepsilon \hat{\bar{x}} + N y, \end{aligned}$$

with the associated observer is an asymptotically stabilizing controller for $\bar{\Sigma}$ and achieves a domain of attraction containing $\mathcal{H}_{\text{obs}} \oplus 2V_\varepsilon(c)$. Next, consider the following controller,

$$\begin{pmatrix} u_0 \\ u_c \\ u_d \end{pmatrix} = \begin{cases} \bar{f}(\hat{\bar{x}}), & \hat{\bar{x}} \notin \frac{4}{3}V_\varepsilon(c), \\ F_\varepsilon \hat{\bar{x}} + Ny, & \hat{\bar{x}} \in \frac{4}{3}V_\varepsilon(c), \end{cases}$$

together with our observer. It is easily verified that this controller asymptotically stabilizes the given system. ■

8.5.1 Exploration of complexity of non-right-invertible constraints

In this subsection, we proceed to answer the questions we posed earlier in order to shed some light on the complexities inherent in dealing with systems having non-right-invertible constraints. We first answer the first question: "*Given a system Σ with non-right invertible constraints, for such a Σ, what are the solvability conditions for the Problems 7.8–7.11 when they are enhanced in the sense that they consider all possible pairs of constraint sets $\mathcal{S}$ and $\mathcal{T}$ that satisfy Assumption 7.1 instead of just one given pair of constraint sets $\mathcal{S}$ and $\mathcal{T}$?*"

To proceed further, we need to extract a subsystem from the SCB decomposition as given in (8.79) of the system Σ_{uz}. Let

$$\begin{aligned} \rho\begin{pmatrix} x_a \\ x_b \end{pmatrix} &= \begin{pmatrix} A_{aa} & A_{ab} \\ 0 & \bar{A}_{bb} \end{pmatrix}\begin{pmatrix} x_a \\ x_b \end{pmatrix} + \begin{pmatrix} \bar{K}_a \\ \bar{K}_b \end{pmatrix} \bar{z}_{0d}, \\ z_b &= \begin{pmatrix} 0 & C_b \end{pmatrix}\begin{pmatrix} x_a \\ x_b, \end{pmatrix} \end{aligned} \tag{8.93}$$

where $\bar{z}'_{0d} = (z'_0, z'_d)'$ and

$$\begin{aligned} A_{ab} &= K_{ab} C_b, \\ \bar{A}_{bb} &= A_{bb} + K_{bb} C_b, \\ \bar{K}_a &= \begin{pmatrix} K_{a0} & K_{ad} \end{pmatrix}, \\ \bar{K}_b &= \begin{pmatrix} K_{b0} & K_{bd} \end{pmatrix}, \end{aligned}$$

when K_a and K_b are decomposed as

$$K_a = \begin{pmatrix} K_{a0} & K_{ab} & K_{ad} \end{pmatrix}, \quad K_b = \begin{pmatrix} K_{b0} & K_{bb} & K_{bd} \end{pmatrix}.$$

We have the following result.

Theorem 8.48 *Consider the system* (7.1)*. The following two statements are equivalent:*

(*i*) *The semi-global or global stabilization in the admissible set is possible for all constraint sets $\mathcal{S}$ and $\mathcal{T}$ satisfying Assumption 7.1.*

(*ii*) *The constraints of system* Σ *are at most weakly non-minimum phase and of type one. Moreover, the subsystem defined in* (8.93) *takes the following form:*

$$\begin{aligned} \rho \begin{pmatrix} x_a \\ x_b \end{pmatrix} &= \begin{pmatrix} A_{aa} & A_{ab} \\ 0 & \alpha I \end{pmatrix} \begin{pmatrix} x_a \\ x_b \end{pmatrix} + \begin{pmatrix} \bar{K}_a \\ 0 \end{pmatrix} \bar{z}_{0d}, \\ z_b &= \begin{pmatrix} 0 & C_b \end{pmatrix} \begin{pmatrix} x_a \\ x_b, \end{pmatrix} \end{aligned} \tag{8.94}$$

where the matirx C_b *is injective and* $\alpha \in [0, 1)$.

Proof : The proof of (*ii*) $\Rightarrow$ (*i*) is obvious. It remains to prove (*i*) $\Rightarrow$ (*ii*).

Consider the subsystem defined in (8.93). Note that the state x_a represents the finite zero dynamics of Σ_{uz}. Viewing $\bar{z}_{0d}$ as input to the zero dynamics and noting that z_b is constrained, the necessary condition for semi-global or global stabilization as stated in condition (*v*) of Theorem 8.49 (which will be presented shortly) requires that

$$\ker C_b \subset \ker C_b \bar{A}_{bb}.$$

This means that $\ker C_b$ is part of the zero dynamics. But all of the zero dynamics of the original system has been included in the dynamics of x_a. Hence, $\ker C_b = \{0\}$, i.e., C_b is injective.

Knowing that C_b is injective, we can choose a constraint set on z_b so that x_b is constrained to be arbitrarily small. However, $\bar{z}_{0d}(0)$ can be anywhere in the constraint set for $\bar{z}_{0d}$ which can be arbitrarily large. If $K_b \neq 0$ in (8.79), then we cannot guarantee that x_b is small enough to be in its constraint set and we get a constraint violation. Hence, we must have $K_b = 0$.

With $K_b = 0$, the subsystem of x_b becomes completely uncontrollable.

The fact that the dynamics of x_b is described by a state matrix αI follows using the same arguments as in Theorem 8.30 for the continuous time. For asymptotic stabilization of the whole system, we need $|\alpha| < 1$. However, if the constraint set on x_b is not symmetric, to avoid constraint violation, we must have $\alpha \in [0, 1)$. ■

We proceed next to answer the second question: "*Given a system* Σ *with non-right-invertible constraints, for such a* Σ, *does there exist a pair of constraint sets* $\mathcal{S}$ *and* $\mathcal{T}$ *that satisfy Assumption 7.1 for which semi-global and global stabilization problems as formally formulated in Problems 7.8–7.11 are solvable?*"

Once again, we extract some subsystems from the SCB decomposition as given in (8.79) of the system Σ_{uz}. Let

$$\tilde{A}_1 = \begin{pmatrix} A_{aa} & 0 \\ 0 & A_{bb} \end{pmatrix}, \quad \tilde{B}_1 = \begin{pmatrix} K_a \\ K_b \end{pmatrix}, \quad \tilde{C}_1 = \begin{pmatrix} 0 & C_b \end{pmatrix}, \quad \tilde{x}_1 = \begin{pmatrix} x_a \\ x_b \end{pmatrix},$$

$\widetilde{v}_1 = \bar{z}$, and $\widetilde{z}_1 = z_b$. We obtain for $i = 1$ the subsystem,

$$\widetilde{\Sigma}_i : \begin{cases} \widetilde{x}_i(k+1) = \widetilde{A}_i \widetilde{x}_i(k) + \widetilde{B}_i \widetilde{v}_i(k), \\ \widetilde{z}_i(k) \quad = \widetilde{C}_i \widetilde{x}_i(k), \end{cases} \tag{8.95}$$

where both $\widetilde{v}_1$ and $\widetilde{z}_1$ are constrained. If we ignore the constraint on $\widetilde{v}_1$, we can repeat the same procedure, used to obtain $\widetilde{\Sigma}_1$ from Σ_{uz}, to obtain $\widetilde{\Sigma}_2$ from $\widetilde{\Sigma}_1$ and so on. At each step of the construction, we should make sure that the matrix $\widetilde{B}_i$ has full column rank and the matrix $\widetilde{C}_i$ has full row rank. This can be done without loss of generality. This chain ends if we obtain a subsystem $\widetilde{\Sigma}_i$ which is right invertible in the sense that $\widetilde{\Sigma}_{i+1}$ satisfies $\widetilde{C}_{i+1} = 0$. Another possibility of termination is that at some step we get $\widetilde{B}_i = 0$, which obviously implies that we can end the chain. It can be shown easily that if the pair (A, B) of the given system Σ is stabilizable, then all the systems $\widetilde{\Sigma}_i$ as defined in (8.95) are stabilizable.

The following theorem contains certain inherent necessary conditions for semi-global or global stabilization in the admissible set whenever we have non-right-invertible constraints.

Theorem 8.49 *Consider the system Σ as given by* (7.1). *Let the sets $\mathcal{S}$ and $\mathcal{T}$ satisfy Assumption 7.1. Moreover, let the chain of systems $\widetilde{\Sigma}_i$ $(i = 1, \dots s)$ be as described above. Then the semi-global and global stabilization problems formulated in Problems 7.8 and 7.9 are solvable only if the following conditions are satisfied:*

(i) (A, B) is stabilizable.

(ii) The constraints of system Σ are at most weakly non-minimum phase.

(iii) The constraints of system Σ are of type one.

(iv) All the subsystems $\widetilde{\Sigma}_i$ $(i = 1, \dots s)$ have at most weakly non-minimum-phase constraints.

(v) The subsystems $\widetilde{\Sigma}_i$ $(i = 1, \dots s)$ with realization (8.95) *satisfy:*

$$\ker \widetilde{C}_i \subset \ker \widetilde{C}_i \widetilde{A}_i. \tag{8.96}$$

Proof : The necessity of these conditions except (v) is self-evident by considering each subsystem as an independent system with input and output constraints and recalling the necessary conditions in Theorem 8.35 for systems with output constraints. To see that the condition (v) is also necessary, we go back to the SCB decomposition used earlier in the proof of Theorem 8.35. As an illustration, let us look at the x_d equation in (8.63) at time 0. We must have

$$x_d(1) = u_d(0) + G_a x_a(0) + G_c x_c(0) + G_d x_d(0) \in \mathcal{S}_d \tag{8.97}$$

for all possible initial conditions, but keep in mind that now u_d is constrained following the way we obtain the decomposition of $\widetilde{\Sigma}_i$. Since x_a and x_c are completely unconstrained whereas $u_d(0)$ and $x_d(0)$ are constrained, condition (8.97) can be guaranteed only if G_a and G_c both equal 0. This is a condition equivalent to condition (v). ■

The following example indicates that the conditions given in Theorem 8.49 are just necessary but not sufficient conditions for solving the constrained stabilization problems. Also, this example shows that the solvability conditions for global and semi-global stabilization in the case of non-right-invertible constraints in general depend on the *particular* choice of constraint sets $\mathcal{S}$ and $\mathcal{T}$, unlike the case of right-invertible constraints.

Example 8.50 Consider the system:

$$\begin{aligned} x_1(k+1) &= x_2(k), \\ x_2(k+1) &= u(k), \\ z_1(k) &= x_1(k), \\ z_2(k) &= x_2(k). \end{aligned} \tag{8.98}$$

Note that the transfer matrix from u to z is non-right invertible and all the conditions in Theorem 8.49 are satisfied. If the constraint set is defined as

$$\mathcal{S} = \{z : |z_1| \leqslant 1, |z_2| \leqslant 2\} \quad \text{and} \quad \mathcal{T} = \mathbb{R}^2,$$

then for any initial condition with $x_1(0) = 0$ and $x_2(0) > 1$, we find that $x_1(1)$ will violate the constraints. Therefore constrained stabilization is not possible.

However, for the constraint set defined by

$$\mathcal{S} = \{z : |z_1| \leqslant 1, |z_2| \leqslant 1\} \quad \text{and} \quad \mathcal{T} = \mathbb{R}^2,$$

it is easily seen that the feedback $u = 0$ achieves constrained stabilization.

8.5.2 *Illustrative example*

Consider the following system:

$$\Sigma : \begin{cases} x(k+1) = Ax(k) + Bu(k), \\ y(k) = C_y x(k) + D_y u(k), \\ z(k) = C_z x(k) + D_z u(k), \end{cases} \tag{8.99}$$

where $x(k) \in \mathbb{R}^4$, $u(k) \in \mathbb{R}^3$, $y(k) \in \mathbb{R}^3$, $z(k) \in \mathbb{R}^3$ and

$$A = \begin{pmatrix} -1 & -1 & 1 & 3 \\ -2 & -1 & 1 & 2 \\ -5.5 & -3 & 2.5 & 5.5 \\ -4.5 & -2 & 1.5 & 6.5 \end{pmatrix}, \quad B = \begin{pmatrix} 3 & 5 & 3 \\ 2 & 3 & 3 \\ 7 & 8 & 8 \\ 8 & 11 & 9 \end{pmatrix},$$

$$C_y = \begin{pmatrix} 1 & 0 & 0 & 0 \\ -0.5 & -1 & 0.5 & 0.5 \\ -0.5 & 1 & -0.5 & 0.5 \end{pmatrix}, \quad D_y = \begin{pmatrix} 1 & 1 & 1 \\ 0 & 0 & 0 \\ 0 & 0 & 0 \end{pmatrix},$$

$$C_z = \begin{pmatrix} -0.5 & 0 & 0 & 0.5 \\ 0 & 1 & -0.5 & 0 \\ 0 & -1 & 0.5 & 0 \end{pmatrix}, \quad D_z = \begin{pmatrix} -0.5 & -0.5 & -0.5 \\ 0.5 & 0.5 & 0.5 \\ 0.5 & 0.5 & 0.5 \end{pmatrix}.$$

The system is subject to the constraints $z(k) \in \mathcal{S}$ where $\mathcal{S}$ is given by

$$\mathcal{S} = \{\gamma \in \mathbb{R}^3 \mid \begin{pmatrix} 0.5 & 0.5 & -0.5 \\ -0.5 & 0.5 & 0.5 \\ 0.5 & -0.5 & 0.5 \end{pmatrix} \gamma \in [-1,1] \times [-1,1] \times [-1,1]\}.$$

The problem is to stabilize the system with a priori given set $\mathcal{W}$ contained in its domain of attraction, where

$$\mathcal{W} = \{\gamma \in \mathbb{R}^4 \mid \begin{pmatrix} 0.5 & 0 & 0.5 & -0.5 \\ -0.5 & -1.0 & 0.5 & 0.5 \\ 0.5 & 0 & -0.5 & 0.5 \\ -0.5 & 1.0 & -0.5 & 0.5 \end{pmatrix} \gamma \in [-10,10] \times [-1,1] \times [-10,10] \times [-1,1]\}.$$

We first solve the semi-global stabilization in the admissible set via state feedback problem as follows:

Step 1: It is easy to verify that (A, B) is stabilizable.

Step 2: There exist a state transformation $\bar{x} = \Gamma_s^{-1} x$, an input basis transformation $\bar{u} = \Gamma_u^{-1} u$, and an output basis transformation $\bar{z} = \Gamma_z^{-1} z$ that converts the original system into its SCB form. These transformations are given by:

$$\bar{x} = \begin{pmatrix} x_a \\ x_b \\ x_c \\ x_d \end{pmatrix} = \begin{pmatrix} 0.5 & 0 & 0.5 & -0.5 \\ -0.5 & -1.0 & 0.5 & 0.5 \\ 0.5 & 0 & -0.5 & 0.5 \\ -0.5 & 1.0 & -0.5 & 0.5 \end{pmatrix} \begin{pmatrix} x_1 \\ x_2 \\ x_3 \\ x_4 \end{pmatrix},$$

$$\bar{u} = \begin{pmatrix} u_0 \\ u_c \\ u_d \end{pmatrix} = \begin{pmatrix} 1 & 1 & 1 \\ 0 & 1 & 0 \\ 0 & 1 & 1 \end{pmatrix} \begin{pmatrix} u_1 \\ u_2 \\ u_3 \end{pmatrix},$$

$$\bar{z} = \begin{pmatrix} z_b \\ z_d \\ z_0 \end{pmatrix} = \begin{pmatrix} 1 & 0 & 1 \\ 1 & 1 & 0 \\ 0 & 1 & 1 \end{pmatrix} \begin{pmatrix} z_1 \\ z_2 \\ z_2 \end{pmatrix}.$$

The transformed system is as follows:

$$\tilde{\Sigma} : \begin{cases} \bar{x}(k+1) = \bar{A}\bar{x}(k) + \bar{B}\bar{u}(k), \\ \bar{y}(k) = \bar{C}_y\bar{x}(k) + \bar{D}_y\bar{u}(k), \\ \bar{z}(k) = \bar{C}_z x(k) + \bar{D}_z\bar{u}(k), \end{cases} \tag{8.100}$$

where

$$\bar{A} = \begin{pmatrix} 1 & 2 & 0 & 1 \\ 0 & 3 & 0 & 2 \\ 2 & 2 & 2 & 2 \\ 0 & 1 & 0 & 1 \end{pmatrix}, \quad \bar{B} = \begin{pmatrix} 1 & 0 & 0 \\ 4 & 0 & 0 \\ 2 & 2 & 0 \\ 1 & 0 & 1 \end{pmatrix},$$

$$\bar{C}_y = \begin{pmatrix} 1 & 0 & 1 & 0 \\ 0 & 1 & 0 & 0 \\ 0 & 0 & 0 & 1 \end{pmatrix}, \quad \bar{D}_y = \begin{pmatrix} 1 & 0 & 0 \\ 0 & 0 & 0 \\ 0 & 0 & 0 \end{pmatrix},$$

$$\bar{C}_z = \begin{pmatrix} 0 & 1 & 0 & 0 \\ 0 & 0 & 0 & 1 \\ 0 & 0 & 0 & 0 \end{pmatrix}, \quad \bar{D}_z = \begin{pmatrix} 0 & 0 & 0 \\ 0 & 0 & 0 \\ 1 & 0 & 0 \end{pmatrix},$$

and the system is subject to the constraints $\bar{z}(k) \in \bar{\mathcal{S}}$, where $\bar{\mathcal{S}}$ is given by

$$\bar{\mathcal{S}} = [-1, 1] \times [-1, 1] \times [-1, 1],$$

and $\bar{\mathcal{W}}$ is given by

$$\bar{\mathcal{W}} = [-10, 10] \times [-1, 1] \times [-10, 10] \times [-1, 1].$$

Extract subsystem Σ_1 composed of x_a, x_b, and x_d dynamics:

$$\Sigma_1 : \begin{cases} x_{abd}(k+1) = A_0 x_{abd}(k) + B_0\bar{u}(k), \\ \bar{z}(k) \quad = C_0 x_{abd}(k) + D_0\bar{u}(k), \end{cases} \tag{8.101}$$

where

$$A_0 = \begin{pmatrix} 1 & 2 & 1 \\ 0 & 3 & 2 \\ 0 & 1 & 1 \end{pmatrix}, \quad B_0 = \begin{pmatrix} 1 & 0 \\ 4 & 0 \\ 1 & 1 \end{pmatrix},$$

$$C_0 = \begin{pmatrix} 0 & 1 & 0 \\ 0 & 0 & 1 \\ 0 & 0 & 0 \end{pmatrix}, \quad D_0 = \begin{pmatrix} 0 & 0 & 0 \\ 0 & 0 & 0 \\ 1 & 0 & 0 \end{pmatrix}.$$

Then extract x_d and x_d dynamics from Σ_1 and form Σ_{bd},

$$\Sigma_{bd} : \begin{cases} x_b(k+1) &= 3x_b(k) + 2x_d(k) + 4u_0(k), \\ x_d(k+1) &= x_b(k) + x_d(k) + u_0(k) + u_d(k), \\ \bar{z}(k) &= \begin{pmatrix} x_b(k) \\ x_d(k) \\ u_0(k) \end{pmatrix}. \end{cases} \tag{8.102}$$

Step 3: Design a state feedback for subsystem Σ_{bd}. A suitable controller is given by

$$\begin{pmatrix} u_0(k) \\ u_d(k) \end{pmatrix} = \bar{f}(x_{bd}(k)) = \begin{pmatrix} -0.625x_b(x) - 0.375x_d(k) \\ -0.375x_b(k) \end{pmatrix}.$$

Step 4: Design a state feedback controller for subsystem Σ_{abd}. Let P_0 be the semi-stabilizing solution of the discrete-time algebraic Riccati equation

$$P_0 = A_0' P_0 A_0 + C_0' C_0 - A_0' P_0 B_0 (B_0' P_0 B_0 + D_0' D_0)^{\dagger} B_0' P_0 A_0;$$

we have

$$P_0 = \begin{pmatrix} 0.0000 & 0.0000 & 0.0000 \\ 0.0000 & 1.5390 & 0.3593 \\ 0.0000 & 0.3593 & 1.2396 \end{pmatrix}.$$

Then choose $c = 0.2$ and $\epsilon = 0.000004$. P_ϵ is the stabilizing solution of the algebraic Riccati equation

$$P_\epsilon = A_0' P_\epsilon A_0 + C_0' C_0 + \epsilon I - A_0' P_\epsilon B_0 (B_0' P_\epsilon B_0 + D_0' D_0)^{\dagger} B_0' P_\epsilon A_0.$$

We have

$$P_\epsilon = \begin{pmatrix} 0.0015 & 0.0021 & 0.0009 \\ 0.0021 & 1.5421 & 0.3607 \\ 0.0009 & 0.3607 & 1.2401 \end{pmatrix}.$$

Choose the level set

$$V_\epsilon(c) = \left\{ \xi \in \mathbb{R}^3 \mid \xi' P_\epsilon \xi < c \right\}.$$

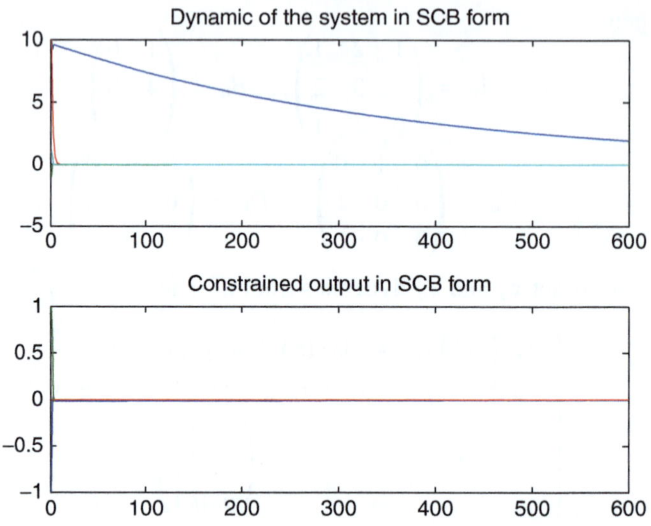

Figure 8.1: State feedback case

Design the low-gain feedback $F_\epsilon = -(B_0' P_\epsilon B_0 + D_0' D_0)^\dagger B_0' P_\epsilon A_0$, that is

$$F_\epsilon = \begin{pmatrix} 0.0004 & 0.7192 & 0.4794 \\ -0.0001 & 0.3175 & 0.5450 \end{pmatrix}.$$

Hence, the state feedback for Σ_1 is given by

$$\begin{pmatrix} u_0 \\ u_d \end{pmatrix} = \begin{cases} \bar{f}(x_{bd}), & x_{abd} \notin 2V_\epsilon(c), \\ F_\epsilon x_{abd}, & x_{abd} \in 2V_\epsilon(c). \end{cases}$$

Step 5: Design a state feedback controller for the entire system. Let

$$u_c(k) = -u_0(k) - x_a(k) - x_b(k) - \tfrac{3}{2} x_c(k) - x_d(k).$$

Then the controller designed in step 4 combined with this controller solves the problem of semi-global stabilization in the admissible set for the entire system, and we denote this controller as

$$\begin{pmatrix} u_0 \\ u_c \\ u_d \end{pmatrix} = \hat{f}(\bar{x}).$$

The simulation data for state feedback case is shown in Fig. 8.1.

Next, we solve the problem of semi-global stabilization in the admissible set via measurement feedback as follows:

Steps 1 and 2 are identical to state feedback case.

Step 3: We have

$$\bar{y}(k) = \bar{C}_y \bar{x} + \bar{D}_y \bar{u}.$$

Replace the x_b and x_d in the controller designed for state feedback with the observation $\widehat{x}_b$ and $\widehat{x}_d$ which can be directly determined from the measurement.

Step 4: Design an observer for x_a and x_c. The observer is given by

$$\rho \begin{pmatrix} \widehat{x}_a \\ \widehat{x}_c \end{pmatrix} = \begin{pmatrix} 1 & 0 \\ 2 & 2 \end{pmatrix} \begin{pmatrix} \widehat{x}_a \\ \widehat{x}_c \end{pmatrix} + \begin{pmatrix} 2 & 1 & 1 & 0 \\ 2 & 2 & 2 & 1 \end{pmatrix} \begin{pmatrix} x_b \\ x_d \\ u_0 \\ u_c \end{pmatrix} + \begin{pmatrix} 0.75 \\ 2.25 \end{pmatrix} [\bar{y}_1 - \widehat{x}_c - u_0].$$

Step 5: Design a measurement feedback for the entire system. Denote

$$\widehat{\bar{x}} = \begin{pmatrix} \widehat{x}_a \\ x_b \\ \widehat{x}_c \\ x_d \end{pmatrix}.$$

Let P_0 be the semi-stabilizing solution of algebraic Riccati equation,

$$P_0 = \bar{A}' P_0 \bar{A} + \bar{C}_z' \bar{C}_z - \bar{A}' P_0 \bar{B} (\bar{B}' P_0 \bar{B} + \bar{D}_z' \bar{D}_z)^{\dagger} \bar{B}' P_0 \bar{A}.$$

We have

$$P_0 = \begin{pmatrix} 0.0000 & 0.0000 & 0.0000 & 0.0000 \\ 0.0000 & 1.5390 & 0.0000 & 0.3593 \\ 0.0000 & 0.0000 & 0.0000 & 0.0000 \\ 0.0000 & 0.3593 & 0.0000 & 1.2396 \end{pmatrix}.$$

Choose $c = 0.2$ and $\epsilon = 0.000004$. P_ϵ is stabilizing solution of the algebraic Riccati equation,

$$P_\epsilon = \bar{A}' P_\epsilon \bar{A} + \bar{C}_z' \bar{C}_z + \epsilon I - \bar{A}' P_\epsilon \bar{B} (\bar{B}' P_\epsilon \bar{B} + \bar{D}_z' \bar{D}_z)^{\dagger} \bar{B}' P_\epsilon \bar{A}.$$

We have

$$P_\epsilon = \begin{pmatrix} 0.0015 & 0.0021 & 0.0000 & 0.0009 \\ 0.0021 & 1.5421 & 0.0000 & 0.3608 \\ 0.0000 & 0.0000 & 0.0000 & 0.0000 \\ 0.0009 & 0.3608 & 0.0000 & 1.2401 \end{pmatrix}.$$

Choose the level set

$$V_\epsilon(c) = \{\xi \in \mathbb{R}^4 \mid \xi' P_\epsilon \xi < c\}.$$

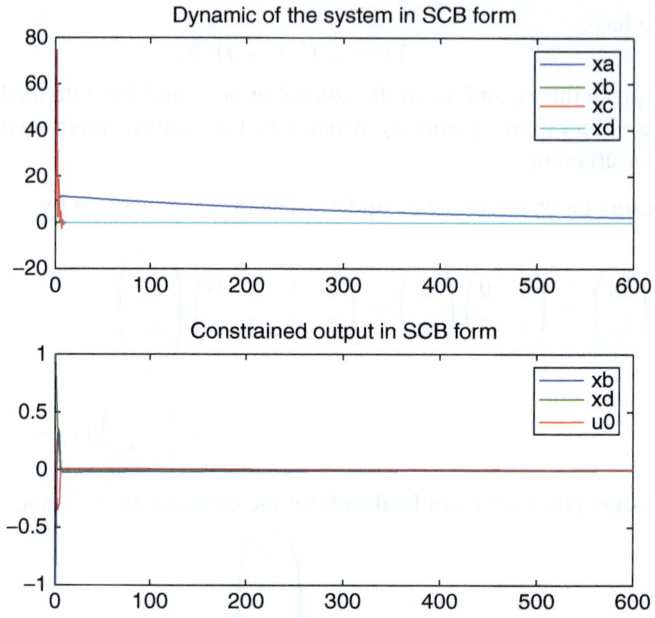

Figure 8.2: Measurement feedback case

Design the low-gain feedback $F_\epsilon = -(\bar{B}' P_\epsilon \bar{B} + \bar{D}_z' \bar{D}_z)^\dagger \bar{B}' P_\epsilon \bar{A}$, that is,

$$F_\epsilon = \begin{pmatrix} 0.0004 & 0.7192 & 0.0000 & 0.4794 \\ 0.9996 & 0.2808 & 1.0000 & 0.5206 \\ -0.0001 & 0.3175 & 0.0000 & 0.5450 \end{pmatrix}.$$

Then the following measurement feedback controller solves the problem of semi-global stabilization in the admissible set

$$\begin{pmatrix} u_0 \\ u_c \\ u_d \end{pmatrix} = \begin{cases} \hat{f}(\hat{\bar{x}}), & \hat{\bar{x}} \notin \frac{4}{3} V_\epsilon(c) \\ F_\epsilon \hat{\bar{x}}, & \hat{\bar{x}} \in \frac{4}{3} V_\epsilon(c), \end{cases}$$

where $\hat{f}$ is the controller designed in state feedback case. The simulation data for measurement feedback case is shown in Fig. 8.2.

8.5.3 *Discussion on semi-global stabilization in the presence of both amplitude and rate constraints*

In Sects. 8.2 (continuous-time) and 8.4 (discrete-time), we considered systems where, in addition to amplitude constraints, we have rate constraints as well on the constrained output. These sections considered right-invertible systems and dealt with semi-global stabilization in the admissible set. In the context of this section as well, we can consider systems which are subject to both rate and amplitude constraints on the constrained output. The issues related to this are discussed in this subsection.

In our main results of Theorems 8.43 and 8.44, we established that guaranteeing the fact that, for all initial conditions in the admissible set, we will still be in the admissible set *one* time step later, implies that we can stay in the admissible set forever without any constraint violations. The following example shows that this property does not hold when we consider rate constraints.

Example 8.51 Consider the system

$$x(k+1) = \begin{pmatrix} 0 & 1 & 0 \\ 0 & \widetilde{\alpha} & -\widetilde{\beta} \\ 0 & \widetilde{\beta} & \widetilde{\alpha} \end{pmatrix} x(k) + \begin{pmatrix} 1 & 0 \\ 0 & 0 \\ 0 & 1 \end{pmatrix} u,$$

where $\widetilde{\beta} = \sqrt{1-\widetilde{\alpha}^2}$ with the constraints:

$$x_1(k) \in [-1, 1],$$

$$u_2(k) \in [-\varepsilon, \varepsilon],$$

$$(u_1(k+1) - u_1(k)) \in [-\delta, \delta],$$

and

$$x_2^2(k) + x_3^2(k) \leqslant 49.$$

We claim that for any $r > 0$, there exist $\varepsilon, \widetilde{\alpha}$, and δ such that for all initial conditions in the admissible set, there exists an input which does not violate constraints in the first r steps even though there does exist an initial condition in the admissible set for which we will get a constraint violation at some time $k > r$ for any input.

Choose $\widetilde{\alpha}$ so close to 1 that $(x_2(k) - x_2(0)) \in [-1, 1]$ for all initial conditions with $k < r$. On the other hand, choose $\widetilde{\alpha}$ such that for $x_2(0) = 7$ and $x_3(0) = 0$, we have $x_2(5r) = 3$ for $u_2(k) = 0$. This is clearly possible.

If $u_1(k) = -x_2(0)$ for $k = 0, \dots, r$ and $u_2(0) = 0$, then we will not have any constraint violation in the first r steps.

On the other hand, for initial condition $x_2(0) = 7$ and $x_3(0) = 0$, we need to choose $u_1(0) \in [-8, -6]$. By choosing δ small enough, we must have

$$u_1(5r) \in [-\tfrac{17}{2}, -\tfrac{11}{2}].$$

On the other hand, for ε small enough, we will have

$$x_2(5r) \in [\tfrac{5}{2}, \tfrac{7}{2}]$$

for any choice of u_2 satisfying the constraints. But then

$$x_1(5r+1) \in [-6, -2],$$

and hence, we have a constraint violation.

The above example actually illustrates that acceptable initial conditions for the state are directly connected to acceptable initial conditions for the input. This connection cannot be ignored in the analysis, and hence, the approach of Theorems 8.43 and 8.44 cannot be directly applied.

One specific method to overcome this problem is to define an expanded system. Consider the system

$$\Sigma : \begin{cases} x(k+1) = Ax(k) + Bu(k), \\ y(k) \quad = C_y x(k) + D_y u(k), \\ z(k) \quad = C_z x(k) + D_z u(k), \end{cases} \tag{8.103}$$

subject to the constraint $z(k) \in \mathcal{S}$ and $z(k+1) - z(k) \in \mathcal{T}$ for all $k \geqslant 0$. This is clearly equivalent to

$$\Sigma : \begin{cases} x(k+1) \; = Ax(k) + Bx_1(k), \\ x_1(k+1) = x_1(k) + v(k), \\ y(k) \quad = C_y x(k) + D_y x_1(k), \\ z(k) \quad = C_z x(k) + D_z x_1(k), \\ z_1(k) \quad = C_z(A-I)x(k) + C_z B x_1(k) + D_z v(k), \end{cases} \tag{8.104}$$

provided $x_1(0) = u(0)$ and $v(k) = u(k+1) - u(k)$. Moreover, the constraints of the original system now convert to amplitude constraints: $z(k) \in \mathcal{S}$ and $z_1(k) \in \mathcal{T}$ for all $k \geqslant 0$.

There are, however, two main drawbacks to this approach. First of all, if we have right-invertible constraints, then this expansion will result in a system with non-right-invertible constraints. Hence, for systems with right-invertible constraints, the direct approach of Sects. 8.2 and 8.4 is preferable.

The second drawback of this expansion is the choice of the initial condition. If we can avoid constraint violations for a certain initial condition $x(0)$ for the system (8.103), then there exists an initial condition $x_1(0) = u(0)$ for the system (8.104) such that constraint violations can be avoided. It yields conservative results if we impose semi-global stabilization in the admissible set for the expanded system since we then suddenly need to guarantee that we can avoid constraint violations for *all* initial conditions for $x_1(0)$ in the admissible set. The above example illustrates that this connection between the initial conditions for state and input is intrinsic and cannot be ignored.

8.A Global and semi-global stabilization with constraints and ℓ_2 disturbances

In this section, we first develop a nonlinear control law satisfying the amplitude and rate constraints that achieves globally asymptotic stabilization for an asymptotically null controllable system without disturbance; meanwhile, it achieves global attractivity of the origin when an ℓ_2 disturbance is in presence. Then we develop a linear control for the semi-global case which achieves a similar result.

Theorem 8.52 *Consider the system*

$$x(k+1) = Ax(k) + Bu(k) + Bw(k) \tag{8.105}$$

with input subject to the amplitude and rate constraints,

$$\|u(k)\|_\infty \leqslant \tilde{\alpha}, \qquad \|u(k+1) - u(k)\|_\infty \leqslant \tilde{\beta}, \qquad \forall\, k \geqslant 0 \tag{8.106}$$

for some $\tilde{\alpha} > 0$ and $\tilde{\beta} > 0$. The sequence $w(k)$ is any disturbance in ℓ_2. Assume that (A, B) is stabilizable with all eigenvalues of A in the closed unit disc. Then, there exists a static nonlinear state feedback which has the following properties:

- *The constraints in* (8.106) *are not violated.*
- *In the absence of disturbance, the equilibrium point $x = 0$ of the closed-loop system is globally asymptotically stable and locally exponentially stable.*
- *In the presence of any ℓ_2, disturbance the state $x = 0$ remains globally attractive.*

Proof : We first recall Lemma 4.24. For simplicity, we choose $Q_\varepsilon = \varepsilon I$ and we define P_ε according to Lemma 4.24. We will use an adaptive-low-gain feedback with $\varepsilon(x)$ defined by (4.244). To simplify notation, we denote $\varepsilon_k = \varepsilon(x(k))$, $Q_k = Q_{\varepsilon(x(k))} = \varepsilon_k I$, and $P_k = P_{\varepsilon(x(k))}$. Following this, we define an adaptive-low-gain control law as

$$u(k) = -(B'P_kB + I)^{-1}B'P_kAx(k) \tag{8.107}$$

and show that there exists a sufficiently small $\delta^* > 0$ such that the control law satisfies the amplitude and rate constraints (8.106) and achieves global asymptotic stabilization of system (8.105) when $w = 0$.

Let $\tilde{\rho} = \min\{\tilde{\alpha},\ \tilde{\beta}/2\}$ and choose $\delta^* > 0$ small enough so that

$$2\lambda_{\max}(BB')\delta^{*2} \leqslant \tilde{\rho}^2$$

Then,

$$\begin{aligned}\|u(k)\|^2 &= x'(k)A'P_kB(B'P_kB+I)^{-2}B'P_kAx(k)\\ &\leqslant \left\|P_k^{-1/2}AP_k^{1/2}\right\| \lambda_{\max}(BB')\{x'(k)P_kx(k)\ \text{trace}\ P_k\}\\ &\leqslant 2\lambda_{\max}(BB')\delta^{*2}\\ &\leqslant \widetilde{\rho}^2 \leqslant \widetilde{\alpha}^2.\end{aligned}$$

This implies that $\|u(k)\|_\infty \leqslant \|u(k)\| \leqslant \widetilde{\alpha}$ for all k, i.e., the control law (8.107) does not violate the amplitude constraint. On the other hand, the above also yields that $\|u(k)\|_\infty \leqslant \widetilde{\beta}/2$ for all k. Hence,

$$\|u(k+1)-u(k)\|_\infty \leqslant \|u(k+1)\|_\infty + \|u(k)\|_\infty \leqslant \widetilde{\beta},$$

for all k which shows that the control law also does not violate the rate constraint either.

Next we show that the closed-loop system is globally asymptotically stable when $w \equiv 0$. Choose a Lyapunov function

$$V_k := V(x(k)) = x'(k)P_kx(k).$$

The variation of V_k along the state trajectory of the closed-loop system is

$$\begin{aligned}&V_{k+1}-V_k\\ &= x'(k+1)[P_{k+1}-P_k]x(k+1) - \varepsilon_k x'(k)x(k) - u'(k)u(k)\\ &\qquad\qquad + w'(k)B'P_kBw(k) - 2u'(k)w(k)\\ &= x'(k+1)[P_{k+1}-P_k]x(k+1) - \varepsilon_k\|x(k)\|^2 - \|u(k)+w(k)\|^2\\ &\qquad\qquad + w'(k)(B'P_kB+I)w(k).\end{aligned} \tag{8.108}$$

When $w \equiv 0$, we get

$$V(x(k+1)) - V(x(k)) \leqslant -\varepsilon_k\|x(k)\|^2 + x'(k+1)[P_{k+1}-P_k]x(k+1). \tag{8.109}$$

Consider the following two cases:

Case 1: If $\varepsilon(k+1) \leqslant \varepsilon_k$, we find by the monotonicity of P_ε that $P_{k+1} \leqslant P_k$ and using (8.109) that $V(x(k+1)) - V(x(k)) < 0$ for $x(k) \neq 0$.

Case 2: If $1 \geqslant \varepsilon(k+1) > \varepsilon_k$, then $P_{k+1} > P_k$ and

$$V(x(k))\ \text{trace}\ P_k = \delta^{*2} \geqslant V(x(k+1))\ \text{trace}\ P_{k+1},$$

which yields $V(x(k+1)) - V(x(k)) < 0$ for $x(k) \neq 0$.

In conclusion, the control law (8.107) guarantees that $V(x(k+1)) - V(x(k)) < 0$ for $x(k) \neq 0$, which implies the global asymptotic stability. The local exponential stability follows easily by noting that $\varepsilon(x(k)) \equiv 1$ if the system starts sufficiently close to the origin and the control law is linear and the input saturation is never overloaded.

The global attractivity of the origin in the presence of ℓ_2 disturbance follows from the following argument. From (8.108), we claim that if $V_{k+1} \geqslant V_k$, then

$$V_{k+1} - V_k \leqslant -\varepsilon_k \|x(k)\|^2 + \eta \|w(k)\|^2, \tag{8.110}$$

where $\eta = \lambda_{\max}[B' P_1 B + I]$ is associated with $\varepsilon = 1$. In deriving (8.110), we use that the scheduling defined in (4.244) guarantees that $P_{k+1} \leqslant P_k$ whenever $V_{k+1} \geqslant V_k$. This yields

$$V_{k+1} - V_k \leqslant \eta \|w(k)\|^2, \tag{8.111}$$

for all $k \geqslant 0$. This inequality guarantees that, given an ℓ_2 disturbance w, the state starting from anywhere in $\mathbb{R}^n$ is bounded. This implies that ε_k has a lower bound $\varepsilon_{\min} > 0$. It remains to show that the state x starting from any point in $\mathbb{R}^n$ is also in ℓ_2; hence, it approaches to the origin.

First, note that if the initial state is sufficiently close to the origin, say $\|x(0)\| \leqslant r_0$ for some $r_0 > 0$ small enough, and the disturbance is bounded by $\|w(k)\| \leqslant d_0$, then for sufficiently small d_0 the amplitude and rate constraints (8.106) will not be violated, and the closed-loop system is linear and exponentially stable. Hence, $x(k) \in \ell_2$.

Now, let $d_0^2 < \varepsilon_{\min} r_0 / \eta$. We show that for any initial state $x(0) \in \mathbb{R}^n$ and any disturbance $w \in \ell_2$, there exists a $K > 0$ such that $\|x(K)\| \leqslant r_0$ and $\|w(k)\| \leqslant d_0$ for all $k > K$. Since w is vanishing, there exists a $K_1 > 0$ such that $\|w(k)\| \leqslant d_0$ for all $k \geqslant K_1$. On the other hand, if $\|x(k)\| > r_0$ and $V_{k+1} \geqslant V_k$ for some $k \geqslant K_1$, then from (8.110), we have

$$V_{k+1} - V_k \leqslant -\varepsilon_k \|x(k)\|^2 + \eta \|w(k)\|^2 < -\varepsilon_{\min} r_0 + \eta d_0 < 0$$

for $k \geqslant K_1$. This contradiction yields that either $\|x(k)\| \leqslant r_0$ or $V_{k+1} < V_k$. For the former case, we are done. For the latter case, there exists a $K > K_1$ such that $\|x(K)\| \leqslant r_0$. In conclusion, there exists a $K > 0$ such that $\|x(K)\| \leqslant r_0$. This shows the global attractivity. ■

Theorem 8.53 *Consider the system* (8.105) *with input subject to the amplitude and rate constraints* (8.106)*. Assume the same condition as stated in Theorem* 8.52*. Then, given any compact set* $\mathcal{K}$ *in the state space and any* $D > 0$ *there exists a linear state feedback which has the following properties:*

- *The constraints in* (8.106) *are not violated.*
- *In the absence of disturbance, the equilibrium point* $x = 0$ *of the closed-loop system is asymptotically stable with* $\mathcal{K}$ *contained in the region of attraction.*

- *In the presence of any ℓ_2 disturbance satisfying $\|w\|_{\ell_2} \leqslant D$, the state $x = 0$ remains attractive.*

Proof : The proof of this theorem is easily adapted from the proof of Theorem 8.52. Since we are dealing with semi-global stabilization, we can fix ε to be a constant, instead of being state dependent. Let $V(x) = x' P_\varepsilon x$ be the Lyapunov function and $L_V(c) := \{x : x' P_\varepsilon x \leqslant c\}$ be the level set. Choose a sufficiently small $\varepsilon \in (0, 1]$ so that

$$2\eta D^2 \operatorname{trace} P_\varepsilon \leqslant \delta^{*2} \quad \text{and} \quad \mathcal{K} \subset L_V(\eta D^2)$$

where the constants η and δ^* are defined in the proof of Theorem 8.52. Following this choice of ε, we claim that the level set $L_V(2\eta D^2)$ is an invariant set for trajectories starting from any point in $\mathcal{K}$ and any disturbance w satisfying $\|w\|_{\ell_2} \leqslant D$. This claim follows easily from the inequality (8.110) which holds for all $k \geqslant 0$ when ε is fixed. The rest of the proof follows similar to the global case. ∎

8.B Completion to the proof of Theorem 8.35

Lemma 8.54 *Consider the following system*

$$x(k+1) = Ax(k) + \lambda^k G x(k)$$

where A is Schur stable and $|\lambda| < 1$. Then, for all $x(0) \in \mathbb{R}^n$ there exists a $\kappa > 0$ such that

$$\|x(k)\| \leqslant \kappa \|x(0)\|$$

for all $k \geqslant 0$.

Proof : Since A is Schur stable, there exists a positive definite matrix $P > 0$ such that $A' P A - P = -I$. Let $V(x) = x' P x$ and denote $V_k = V(x(k))$. Then

$$\begin{aligned} V_{k+1} - V_k &= -x'(k)x(k) + 2\lambda^k x'(k) G' P A x(k) + \lambda^{2k} x'(k) G' P G x(k) \\ &\leqslant 2|\lambda|^k \left(x'(k) A' P A x(k)\right)^{1/2} \left(x'(k) G' P G x(k)\right)^{1/2} \\ &\qquad + |\lambda|^{2k} x'(k) G' P G x(k) \\ &\leqslant (2\tilde{\beta}^{1/2} |\lambda|^k + \tilde{\beta} |\lambda|^{2k}) V_k \\ &\leqslant c_0 |\lambda|^k V_k, \end{aligned}$$

where we have used

$$x'(k) A' P A x(k) \leqslant x'(k) P x(k), \quad \tilde{\beta} = \lambda_{\max}(G' P G)/\lambda_{\min}(P),$$

and $c_0 = 2\widetilde{\beta}^{1/2} + \widetilde{\beta}$. It follows that

$$V_{k+1} \leqslant (1 + c_0|\lambda|^k)V_k \leqslant \exp\left\{c_0|\lambda|^k\right\} V_k.$$

Thus,

$$\prod_{i=0}^{k-1} \frac{V_{i+1}}{V_i} \leqslant \exp\left\{\sum_{k=0}^{\infty} c_0|\lambda|^k\right\},$$

i.e.,

$$V_k \leqslant \exp\left\{\sum_{k=0}^{\infty} c_0|\lambda|^k\right\} V_0$$

for all $k \geqslant 0$. Hence the lemma follows. ■

Completion to the Proof of Theorem 8.35 : Note that the feedback $f(x_a)$ as constructed in Appendix 8.A is globally Lipschitz and locally linear. Let $r > 0$ be sufficiently small and let $f(x_a) = -F_a x_a$ for $\|x_a\| \leqslant r$. As we shall see later, we can choose the initial conditions sufficiently small to guarantee that the trajectory of x_a remains in this ball. The construction of $f(x_a)$ guarantees that $\widetilde{A}_{aa} := A_{aa} - K_a F_a$ is Schur stable. We decompose $F_a = \left(F_{a0}', F_{ad}'\right)'$ and continue by writing out the first subsystem:

$$\begin{aligned}
x_a(k+1) &= A_{aa}x_a(k) + K_a f(x_a(k)) + K_a v(k) \\
&= (A_{aa} - K_a F_a)x_a(k) + K_a \begin{pmatrix} v_0 \\ v_d \end{pmatrix} \\
&= \widetilde{A}_{aa}x_a(k) + K_a \begin{pmatrix} 0 \\ \lambda^k x_d(0) - \lambda^k f_d(x_a(k)) \end{pmatrix} \\
&= \widetilde{A}_{aa}x_a(k) + \lambda^k K_a \begin{pmatrix} 0 \\ x_d(0) \end{pmatrix} + \lambda^k K_a \begin{pmatrix} 0 \\ F_{ad} \end{pmatrix} x_a(k).
\end{aligned}$$

This system is equivalent to the following dynamics:

$$\begin{pmatrix} x_a(k+1) \\ \xi(k+1) \end{pmatrix} = \begin{pmatrix} \widetilde{A}_{aa} & I \\ 0 & \lambda I \end{pmatrix} \begin{pmatrix} x_a(k) \\ \xi(k) \end{pmatrix} + \lambda^k G \begin{pmatrix} x_a(k) \\ \xi(k) \end{pmatrix},$$

where

$$\xi(0) = K_a \begin{pmatrix} 0 \\ x_d(0) \end{pmatrix} \quad \text{and} \quad G = \begin{pmatrix} K_a \begin{pmatrix} 0 \\ F_{ad} \end{pmatrix} & 0 \\ 0 & 0 \end{pmatrix}.$$

Applying Lemma 8.54, there exist $\kappa > 0$ and $\kappa_1 > 0$ such that

$$\begin{aligned}
\|x_a(k)\| &\leqslant \left\|\begin{pmatrix} x_a(k)' & \xi(k)' \end{pmatrix}'\right\| \leqslant \kappa \left\|\begin{pmatrix} x_a(0)' & \xi(0)' \end{pmatrix}'\right\| \\
&\leqslant \kappa_1(\|x_a(0)\| + \|x_d(0)\|.
\end{aligned}$$

■

9

Semi-global stabilization in the recoverable region: properties and computation of recoverable regions

9.1 Introduction

As in Chap. 8, we consider in this chapter constraints on state as well as input variables. As discussed in detail in Chap. 8, if the given system has at least one of the constraint invariant zeros in the open right-half plane (continuous time) or outside the unit disc (discrete time), that is, if it has non-minimum-phase constraints, then neither semi-global nor global stabilization in the admissible set is possible. Thus, whenever we have non-minimum-phase constraints, the semi-global stabilization is possible only in a certain proper subset of the admissible set. This gives rise to the notion of a *recoverable region* (set), sometimes also called the domain of null controllability or null controllable region. Generally speaking, for a system with constraints, an initial state is said to be *recoverable* if it can be driven to zero by some control without violating the constraints on the state and input. We can appropriately term the set of all recoverable initial conditions as the recoverable region. The recoverable region is thus indeed the maximum achievable domain of attraction in stabilizing linear systems subject to non-minimum-phase constraints. The goal of stabilization is to design a feedback, say $u = f(x)$, such that the constraints are not violated and moreover the region of attraction of the equilibrium point of the closed-loop system is equal to the recoverable region or an arbitrarily large subset contained within the recoverable region. Such a stabilization is termed as the semi-global stabilization in the recoverable region, and this is what we pursue in this chapter.

Our goals in this chapter are twofold. At first, we explore the properties and computational issues in constructing the recoverable region for a given constraint set by exploiting the structure of the given system. In this regard, we develop a method that lessens the computational complexity involved in obtaining the recoverable region. Our next goal, in accordance with the theme of this book, is to solve the problem of semi-global stabilization in the recoverable region via state feedback. Finally, we will briefly comment on the issues related to measurement feedback.

This chapter is based on our work in [167, 197].

A. Saberi et al., *Internal and External Stabilization of Linear Systems with Constraints*, Systems & Control: Foundations & Applications, DOI 10.1007/978-0-8176-4787-2_9,

9.2 Preliminaries

We start with a description of our system model Σ as given in (7.1) and an amplitude constraint set $\mathcal{S}$ that satisfies Assumption 7.4. Throughout this chapter, we assume that the rate constraint set $\mathcal{T} = \mathbb{R}^p$, i.e., there are no rate constraints. We define the admissible set of initial conditions as in Definition 7.5, however, with a slightly different notation to indicate its dependence on the given system Σ and the constraint set $\mathcal{S}$. We repeat this definition as follows:

Definition 9.1 *Consider the system Σ in* (7.1) *along with a constraint set $\mathcal{S}$ that satisfies Assumption 7.4. Then, the set*

$$\mathcal{V}(\Sigma, \mathcal{S}) := \{ x \in \mathbb{R}^n \mid \exists u \in \mathbb{R}^m \text{ such that } C_z x + D_z u \in \mathcal{S} \}$$

is said to be the ***admissible set*** *of initial conditions.*

Remark 9.2 *In view of Remark 7.7, we observe that the admissible set $\mathcal{V}(\Sigma, \mathcal{S})$ can be equivalently written as*

$$\mathcal{V}(\Sigma, \mathcal{S}) := \{ x \in \mathbb{R}^n \mid C_z x \in \mathcal{S} \}.$$

In Chap. 8, we looked at the semi-global case for conditions when for all compact sets $\mathcal{W}$ in the interior of admissible set $\mathcal{V}(\Sigma, \mathcal{S})$, there exists a controller which avoids constraint violation for all time and for all initial conditions in $\mathcal{W}$ while, additionally, guaranteeing that the state converges to zero. This is clearly not always possible when we have non-minimum-phase constraints. Hence, we define next what is known as a recoverable region or set $\mathcal{R}(\Sigma, \mathcal{S})$ as the largest set of initial conditions for which we can avoid constraint violation while steering the state to the origin.

Definition 9.3 *Consider the system Σ in* (7.1) *along with a constraint set $\mathcal{S}$ that satisfies Assumption 7.4. Then, the* ***recoverable region*** *$\mathcal{R}(\Sigma, \mathcal{S})$ of this system is the set of all initial states $x(0) \in \mathcal{V}(\Sigma, \mathcal{S})$ for which there exists a control input u such that $x(t) \to 0$ as $t \to \infty$ while $z(t) \in \mathcal{S}$ for all $t \geqslant 0$ (continuous time) or such that $x(k) \to 0$ as $k \to \infty$ while $z(k) \in \mathcal{S}$ for all $k \geqslant 0$ (discrete time).*

As we said in introduction, our goals in this chapter are twofold. At first, we explore the properties and computational issues in constructing the recoverable region $\mathcal{R}(\Sigma, \mathcal{S})$ for a given system Σ and for a given constraint set $\mathcal{S}$ by exploiting the structure of Σ. Then, we solve the problem of semi-global stabilization in the recoverable region via state feedback.

We have the following problem statement which expresses formally our goal of examining the properties and computational issues associated with $\mathcal{R}(\Sigma, \mathcal{S})$.

Problem 9.4 For a given system Σ as in (7.1) along with the constraint set $\mathcal{S}$ satisfying Assumption 7.4, examine the properties of $\mathcal{R}(\Sigma, \mathcal{S})$ and then explore the computational issues in constructing the recoverable region $\mathcal{R}(\Sigma, \mathcal{S})$.

We have the following problem statement which expresses formally our goal of solving via state feedback the semi-global stabilization problem in the recoverable region $\mathcal{R}(\Sigma, \mathcal{S})$.

Problem 9.5 Consider the system (7.1) with the constraint set $\mathcal{S}$ satisfying Assumption 7.4. The **semi-global stabilization problem in the recoverable region** $\mathcal{R}(\Sigma, \mathcal{S})$ is to find a family of static, possibly time varying, state feedbacks such that for any a priori given compact set $\mathcal{W}$ contained in the recoverable region $\mathcal{R}(\Sigma, \mathcal{S})$, there exists a feedback in this family such that the closed-loop system is asymptotically stable with a domain of attraction containing $\mathcal{W}$ in its interior and such that all the constraints are satisfied, i.e., $z(\nu) \in \mathcal{S}$ for all $\nu \geqslant 0$ provided $x(0) \in \mathcal{W}$.

9.3 Properties and computational issues of the recoverable region

Before we proceed to discuss the properties and computational issues of the recoverable region, it is appropriate to review briefly the existing literature. The earliest literature on recoverable regions can be traced back to the 1960s. For the case of input constraints, J. L. LeMay in 1964 first studied the conditions for characterizing the *maximal region of recoverability* and the *maximal region of reachability* [66]. LeMay also derived a method for calculation of a recoverable region $\mathcal{R}(\Sigma, \mathcal{S})$ based on optimal control techniques. It is known that for any state in the recoverable region, there exists a time-optimal control law that drives the state to zero. This fact builds a direct connection between the characterization of the recoverable region and time-optimal control. There exists a vast literature in the 1960s and 1970s that were devoted to time-optimal control, among them we mention [36, 39, 65, 114]. Ryan [121] presented a set of very detailed results of time-optimal control of systems with input constraints whose number of unstable eigenvalues is between one and four. He also provided some results for explicit characterization of the recoverable region $\mathcal{R}(\Sigma, \mathcal{S})$, including:

- Systems with one or two unstable real eigenvalues.
- Systems with two unstable complex eigenvalues.
- Systems with three unstable eigenvalues which are proportional, $(\lambda, 2\lambda, 3\lambda)$, where $\lambda > 0$.

- Some systems with four unstable poles can be reduced to systems with lower order unstable dynamics.

Note that the above results depend crucially on the fact that, in the case of only input constraints, the recoverable region $\mathcal{R}(\Sigma, \mathcal{S})$ is completely determined by the unstable dynamics. Stephan et al. extended some of LeMay's results to systems with input and state constraints [158, 159]. They examined computational issues of the recoverable regions for planar systems with state and input constraints. However, their work does not exploit the structure of a given system in order to reduce the computational burden. Recently, we studied the properties of the recoverable regions and then eased the computational burden in constructing them whenever there are constraints both on state and control variables for systems subject to non-minimum-phase constraints. We did so first for continuous-time systems in [167] and then for discrete-time systems in [197]. Unlike previous work, we explicitly exploit the structure of the given system to develop our results both for constructing the recoverable regions and for semi-global stabilization in the recoverable region. The results presented in this section are based on our work in [167] and [197].

Having reviewed the literature briefly, let us next observe an important point regarding our goal of exploring the properties and computational issues in constructing the recoverable region $\mathcal{R}(\Sigma, \mathcal{S})$. Since the computation of the admissible set of initial conditions $\mathcal{V}(\Sigma, \mathcal{S})$ is relatively trivial, we can enquire under what conditions the recoverable region $\mathcal{R}(\Sigma, \mathcal{S})$ coincides with the admissible set of initial conditions $\mathcal{V}(\Sigma, \mathcal{S})$. It is transparent from Chap. 8 that $\mathcal{R}(\Sigma, \mathcal{S})$ coincides with $\mathcal{V}(\Sigma, \mathcal{S})$ whenever the constraints are at most weakly non-minimum phase and right invertible. This property also holds for non-right-invertible constraints under certain conditions. However, whenever the constraints are strongly non-minimum phase, irrespective of whether they are right or non-right-invertible constraints, the recoverable region $\mathcal{R}(\Sigma, \mathcal{S})$ is always a proper subset of the admissible set of initial conditions $\mathcal{V}(\Sigma, \mathcal{S})$. In this section, our main interest is indeed to reduce the complexity in computing the recoverable region $\mathcal{R}(\Sigma, \mathcal{S})$ whenever the constraints are strongly non-minimum phase. We emphasize that such a construction is very involved, whereas the construction of the admissible set of initial conditions $\mathcal{V}(\Sigma, \mathcal{S})$ is quite straightforward. In order to reduce the complexities involved in the computation of $\mathcal{R}(\Sigma, \mathcal{S})$, as we said earlier, we exploit here the structural properties of the given system. In fact, by exploiting the structural properties, the recoverable region $\mathcal{R}(\Sigma, \mathcal{S})$ for a given system is constructed by constructing the same, however, for a reduced order subsystem of the given system. Such a reduction in the order or dimension of the system generally leads to a considerable simplification in the computational effort. One appreciates the reduction in the order of a system, when we note that in the literature so far, the recoverable region $\mathcal{R}(\Sigma, \mathcal{S})$ is constructed at the most for fourth-order systems.

Following the papers [167] and [197], we present below our results in two subsections, one for continuous-time systems and the other for discrete-time systems.

9.3.1 Continuous-time systems

For continuous-time systems, our goal in this subsection is to show how to reduce the complexities involved in computing the recoverable region $\mathcal{R}(\Sigma, \mathcal{S})$ by utilizing the structural properties of the given system.

The first set of properties of the recoverable region $\mathcal{R}(\Sigma, \mathcal{S})$ is more or less well known. They are compiled in the following lemma for easy reference.

Lemma 9.6 *Consider the system Σ in (7.1) and a compact, convex constraint set $\mathcal{S}$ containing 0 in the interior. The recoverable region $\mathcal{R}(\Sigma, \mathcal{S})$ for this system has the following properties:*

(i) If (A, B) is controllable, then for any initial $x(0) \in \mathcal{R}(\Sigma, \mathcal{S})$, there exist a $T > 0$ and an input signal u such that $x(T) = 0$ while $z(t) \in \mathcal{S}$ for all $t \in [0, T]$.

(ii) The set $\mathcal{R}(\Sigma, \mathcal{S})$ is convex and contains the origin as an interior point.

(iii) If (A, B) is stabilizable, then the set $\mathcal{R}(\Sigma, \mathcal{S})$ is open in case we have only input constraints, i.e., $C_z = 0$, but in general, this need not be true.

(iv) The set $\mathcal{R}(\Sigma, \mathcal{S})$ is bounded if all the invariant zeros of the system (7.1) are in the open right half plane, the system is left invertible, and the constraints are of type one.

Proof : See Appendix 9.A. ■

Remark 9.7 *Note that item (i) of the above lemma states that infinite-time recoverability is equivalent to finite-time recoverability, i.e., if we can bring the state to zero asymptotically without violating our constraints, then we can also bring the state to zero in finite time again without violating our constraints.*

Remark 9.8 *As is clear from the example to be presented later on, the recoverable region is in general not a polytope. Of course, like any set, it can be arbitrarily well approximated by a polytope.*

Remark 9.9 *Assume that $C_z = 0$ and $D_z = I$ in (7.1), i.e., the system is only subject to input constraints and does not have any state constraints. Then, under* *a suitable coordinate system in the state space, the system can be split into two subsystems:*

$$\begin{aligned} \Sigma_s : \ \dot{x}_s &= A_s x_s + B_s u, \\ \Sigma_u : \ \dot{x}_u &= A_u x_u + B_u u, \end{aligned}$$

where the eigenvalues of A_s are in the closed left half plane (at most critically unstable) and those of A_u are in the open right half plane (antistable). Then it was already established in [66] that:

(*i*) $\mathcal{R}(\Sigma_u, \mathcal{S})$ *is bounded.*

(*ii*) $x \in \mathcal{R}(\Sigma, \mathcal{S})$ *if and only if* $x_u \in \mathcal{R}(\Sigma_u, \mathcal{S})$.

The fact stated above tells us that, without state constraints, the recoverable region is completely determined by the exponentially unstable part of the system. On the other hand, for the case of state constraints, this decomposition is no longer possible. Later in this subsection, we show that, in general, a different type of order reduction is possible of which the above is actually a special case.

Next, we present our first reduction result for the set $\mathcal{R}(\Sigma, \mathcal{S})$. For this purpose, we first express the system in terms of the SCB which explicitly depicts both the finite and infinite zero structure of a given system. The SCB given in Chap. 3 is rewritten here, indicating the exact notation that is used in this subsection. As usual, choose appropriate coordinates in the state, input, and output spaces

$$x = T_x \bar{x}, \quad u = T_u \bar{u} + \bar{F} x, \quad z = T_z \bar{z},$$

where T_x, T_u, and T_z are transformation matrices and $\bar{F}$ is a preliminary feedback which will make the structure of the system more visible. With this objective, we also use the following decomposition for the state, output, and input of the system:

$$\bar{x} = \begin{pmatrix} x_a \\ x_b \\ x_c \\ x_d \end{pmatrix}, \quad \bar{z} = \begin{pmatrix} z_b \\ \zeta \end{pmatrix}, \quad \zeta = \begin{pmatrix} z_0 \\ z_d \end{pmatrix} \text{ and } \bar{u} = \begin{pmatrix} u_0 \\ u_c \\ u_d \end{pmatrix},$$

after which the system (7.1) takes the form

$$\bar{\Sigma} : \begin{cases} \dot{x}_a = A_{aa} x_a + K_{ab} z_b + K_{a2} \zeta, \\ \dot{x}_b = A_{bb} x_b + K_{bb} z_b + K_{b2} \zeta, \\ \dot{x}_c = A_{cc} x_c + B_c u_c + K_{cb} z_b + K_{c2} \zeta, \\ \dot{x}_d = A_{dd} x_d + B_d u_d + K_{db} z_b + K_{d2} \zeta, \\ \bar{z} = \begin{pmatrix} z_b \\ \zeta \end{pmatrix} = \begin{pmatrix} C_b x_b \\ u_0 \\ C_d x_d \end{pmatrix}. \end{cases} \tag{9.1}$$

Furthermore, the x_a equation can be decomposed as

$$\begin{aligned}\dot{x}_a^{-0} &= A_{aa}^{-0}x_a^{-0} + K_{ab}^{-0}C_b x_b + K_{a2}^{-0}\zeta,\\ \dot{x}_a^{+} &= A_{aa}^{+}x_a^{+} + K_{ab}^{+}C_b x_b + K_{a2}^{+}\zeta,\end{aligned}$$

where

$$x_a = \begin{pmatrix} x_a^{-0} \\ x_a^{+}\end{pmatrix}, \quad A_{aa} = \begin{pmatrix} A_{aa}^{-0} & 0 \\ 0 & A_{aa}^{+}\end{pmatrix},$$
$$K_{ab} = \begin{pmatrix} K_{ab}^{-0} \\ K_{ab}^{+}\end{pmatrix}, \quad K_{a2} = \begin{pmatrix} K_{a2}^{-0} \\ K_{a2}^{+}\end{pmatrix},$$

with all the eigenvalues of A_{aa}^{-0} in the closed left-half plane and all the eigenvalues of A_{aa}^{+} in the open right-half plane.

From the full system in the SCB form, we can extract a subsystem consisting of the state variables x_a and x_b, input variable ζ consisting of z_0 and z_d, and output $\bar{z}$:

$$\Sigma_1 : \begin{cases} \dot{x}_a = A_{aa}x_a + K_{ab}C_b x_b + K_{a2}\zeta, \\ \dot{x}_b = (A_{bb} + K_{bb}C_b)\, x_b + K_{b2}\zeta, \\ \bar{z} = \begin{pmatrix} C_b \\ 0 \end{pmatrix} x_b + \begin{pmatrix} 0 \\ I \end{pmatrix}\zeta. \end{cases} \tag{9.2}$$

The state dimension of this system equals $n_a + n_b$. Obviously, ζ is not an input of the original system. However, for the moment we view ζ as the input to this subsystem while $\bar{z}$ is a constrained output for this subsystem. A transformation of the system into the SCB form clearly affects the constraint set, and we obtain a new constraint set $\bar{\mathcal{S}} = T_z^{-1}\mathcal{S}$. Thus, the constraint on $\bar{z}$ becomes

$$\bar{z}(t) \in \bar{\mathcal{S}} \quad \text{for all } t \geqslant 0.$$

Let $\mathcal{R}(\Sigma_1, \bar{\mathcal{S}})$ be the recoverable region of subsystem Σ_1 with the constraint set $\bar{\mathcal{S}}$. The following theorem shows the relationship between the recoverable region of the full system Σ and the recoverable region of the subsystem Σ_1.

Theorem 9.10 *Consider the system Σ in* (7.1) *along with a constraint set $\mathcal{S}$ that satisfies Assumption 7.4. Assume that we have extracted the subsystem Σ_1 in* (9.2) *from Σ as described above. Then the closure of the set $\mathcal{R}(\Sigma, \mathcal{S})$ is equal to*

$$T_x \left\{ \begin{pmatrix} x_1 \\ x_2 \end{pmatrix} \,\middle|\, x_1 \in \overline{\mathcal{R}(\Sigma_1, \bar{\mathcal{S}})} \right\} \cap \mathcal{V}(\Sigma, \mathcal{S}), \tag{9.3}$$

where x_2 is of compatible dimension, i.e., $x_2 \in \mathbb{R}^{n_c+n_d}$, and the bar denotes the closure of the set.

In (9.3), x_1 is the state of Σ_1 consisting of x_a and x_b, while x_2 is the rest of the state variables of the system Σ which, in the SCB form, is composed of x_c and x_d.

Remark 9.11 *If $\mathcal{R}(\Sigma_1, \bar{\mathcal{S}})$ is approximated by a polytope consisting of all $x \in \mathbb{R}^{n_a+n_b}$ for which $R_1 x \leqslant q_1$ while $\mathcal{S}$ is described by all $z \in \mathbb{R}^p$ for which $Rz \leqslant q$, then $\mathcal{R}(\Sigma, \mathcal{S})$ is approximated by the set of all $x \in \mathbb{R}^n$ for which*

$$\begin{pmatrix} RC_z \\ \begin{pmatrix} R_1 & 0 \end{pmatrix} T_x^{-1} \end{pmatrix} x \leqslant \begin{pmatrix} q \\ q_1 \end{pmatrix}.$$

By improving the approximation of $\mathcal{R}(\Sigma_1, \bar{\mathcal{S}})$, we can approximate $\mathcal{R}(\Sigma, \mathcal{S})$ arbitrarily well in this way.

The decomposition of the recoverable region as presented in Theorem 9.10 is therefore very important from a computational point of view. Although it does not capture which boundary points of the recoverable region are actually part of the recoverable region itself, by approximating or exactly computing the set $\mathcal{R}(\Sigma_1, \bar{\mathcal{S}})$, we immediately obtain with arbitrary accuracy the set $\mathcal{R}(\Sigma, \mathcal{S})$.

As pointed out in [159], numerical computation of recoverable regions suffers from dimension growth. Papers such as [211] try to improve the gridding methods, but the exponential growth with dimension is not avoided. In this sense, any reduction of dimension in the computation of the recoverable region is crucial for improvement of computation efficiency. The above method allows us to obtain the recoverable set for the system Σ from the recoverable set of a lower-dimensional system in a transparent way. Note that the transformation into the SCB and the computations of the transformation matrices (in particular T_x) have been implemented in Matlab and Maple and work very well on numerous examples.

Proof : It is obvious that $\mathcal{R}(\Sigma, \mathcal{S})$ is contained in $\mathcal{V}(\Sigma, \mathcal{S})$. Moreover, assuming that in the first subsystem we have ζ as a free input, we clearly enlarge the recoverable set. The reverse inclusion follows from the proof of Theorem 9.18 (developed later on) since there we prove that for any compact set contained in the interior of (9.3), we can find a controller which contains this compact set in its constrained domain of attraction. ■

Let us next have a different look at the structure of the system Σ which will provide some interesting results for special cases. To do so, we need to define another subsystem. Consider the remaining dynamics in the system Σ besides the subsystem Σ_1. We consider the system in its SCB form, and we get the following description for the dynamics which together with Σ_1 describes the full system:

$$\Sigma_2 : \begin{cases} \dot{x}_c = A_{cc} x_c + K_{c2} \zeta + B_c u_c + K_{cb} z_b, \\ \dot{x}_d = A_{dd} x_d + K_{d2} \zeta + B_d u_d + K_{db} z_b, \\ \zeta = \begin{pmatrix} 0 \\ C_d \end{pmatrix} x_d + \begin{pmatrix} I \\ 0 \end{pmatrix} u_0. \end{cases} \tag{9.4}$$

Note that Σ_2 is only affected by Σ_1 via the signal z_b. When we set $z_b = 0$, then we decouple Σ_2 from Σ_1, and when we also ignore the constraints on z_b by setting

$$\mathcal{S}_2 := \left\{ \zeta \in \mathbb{R}^{n_2} \mid \exists\, z_b \text{ such that } \begin{pmatrix} z_b \\ \zeta \end{pmatrix} \in \bar{\mathcal{S}} \right\}$$

and view $\mathcal{S}_2$ as the constraint set for Σ_2, we obtain an independent system Σ_2. In this way, we define the recoverable region $\mathcal{R}(\Sigma_2, \mathcal{S}_2)$ for the second subsystem.

The following theorem establishes the recoverable set of the second subsystem Σ_2 and shows conditions under which we can completely characterize the recoverable set of the original system from the recoverable set of the subsystem Σ_1. Theorem 9.10 did not capture which boundary points belong to the recoverable set and the following theorem does this explicitly for a special case.

Theorem 9.12 *Consider the system Σ in* (7.1) *along with a constraint set $\mathcal{S}$ that satisfies Assumption 7.4. Assume that the system Σ has been decomposed into two subsystems in SCB as described by* (9.2) *and* (9.4)*. Then we have the following properties:*

(i) *It holds that $\overline{\mathcal{R}(\Sigma_2, \mathcal{S}_2)} = \mathcal{V}(\Sigma_2, \mathcal{S}_2)$.*

(ii) *If the constraints are right invertible, then $\mathcal{S}_2 = \bar{\mathcal{S}}$.*

(iii) *If the constraints are right invertible and of type one, then $\mathcal{R}(\Sigma_2, \mathcal{S}_2) = \mathcal{V}(\Sigma_2, \mathcal{S}_2)$, and*

$$\mathcal{R}(\Sigma, \mathcal{S}) = T_x \left[\mathcal{R}(\Sigma_1, \bar{\mathcal{S}}) \times \mathcal{V}(\Sigma_2, \mathcal{S}_2) \right]. \tag{9.5}$$

Proof : The first property is evident from the fact that the system Σ_2 has a special structure as constructed within the SCB. It is strongly controllable which yields that we can make ζ follow any trajectory with arbitrary accuracy and therefore any initial state that is admissible at time 0 can be steered to zero without violating any constraints. For details, we refer to the semi-global stabilization results in Chap. 8. The last two properties also follow easily from Chap. 8. ∎

We have achieved a reduction from computing the recoverable region for the system Σ to the computation of the recoverable region for the subsystem Σ_1. As noted before, a reduction in system order is crucial in making the computation of the recoverable region feasible. The question remains whether we can achieve further reductions. In SCB, the matrix A_{aa} is in fact a block diagonal matrix. With this one more step of refining, subsystem Σ_1 becomes:

$$\Sigma_1 : \begin{cases} \dot{x}_a^{-0} = A_{aa}^{-0} x_a^{-0} + K_{ab}^{-0} C_b x_b + K_{a2}^{-0} \zeta, \\ \dot{x}_a^{+} = A_{aa}^{+} x_a^{+} + K_{ab}^{+} C_b x_b + K_{a2}^{+} \zeta, \\ \dot{x}_b = (A_{bb} + K_{bb} C_b) x_b + K_{b2} \zeta, \\ \overline{z} = \begin{pmatrix} C_b \\ 0 \end{pmatrix} x_b + \begin{pmatrix} 0 \\ I \end{pmatrix} \zeta. \end{cases} \tag{9.6}$$

Note that the eigenvalues of A_{aa}^{-0} and A_{aa}^{+} are in the closed left-half plane and open right-half plane, respectively. We extract a subsystem from Σ_1:

$$\overline{\Sigma}_1 : \begin{cases} \dot{x}_a^{+} = A_{aa}^{+} x_a^{+} + K_{ab}^{+} C_b x_b + K_{a2}^{+} \zeta, \\ \dot{x}_b = (A_{bb} + K_{bb} C_b) x_b + K_{b2} \zeta, \\ \overline{z} = \begin{pmatrix} C_b \\ 0 \end{pmatrix} x_b + \begin{pmatrix} 0 \\ I \end{pmatrix} \zeta, \end{cases} \tag{9.7}$$

with state dimension $n_a^+ + n_b$. We can relate the recoverable region of Σ_1 to the recoverable region of $\overline{\Sigma}_1$, and then, using Theorem 9.10, we can relate the recoverable region of Σ to the recoverable region of $\overline{\Sigma}_1$.

Theorem 9.13 *Consider the system Σ in* (7.1) *along with a constraint set $\mathcal{S}$ that satisfies Assumption 7.4. Define Σ_1 by* (9.2) *and $\overline{\Sigma}_1$ by* (9.7). *We have:*

$$\mathcal{R}(\Sigma_1, \overline{\mathcal{S}}) = \mathbb{R}^{n_a^{-0}} \times \mathcal{R}(\overline{\Sigma}_1, \overline{\mathcal{S}}), \tag{9.8}$$

and the closure of $\mathcal{R}(\Sigma, \mathcal{S})$ is given by

$$T_x \left\{ \begin{pmatrix} x_1 \\ x_2 \\ x_3 \end{pmatrix} \,\middle|\, x_2 \in \overline{\mathcal{R}(\overline{\Sigma}_1, \overline{\mathcal{S}})} \right\} \cap \mathcal{V}(\Sigma, \mathcal{S}), \tag{9.9}$$

where x_1 and x_3 are of compatible dimension, i.e., $x_1 \in \mathbb{R}^{n_a^{-0}}$ and $x_3 \in \mathbb{R}^{n_c + n_d}$.

Using the decompositions from the SCB we have in the above that x_1 is equal to x_a^{-0}, x_2 denotes the variables of $\overline{\Sigma}_1$ consisting of x_a^+ and x_b, while x_3 is composed of x_c and x_d.

Proof : See Appendix 9.B. ■

Remark 9.14 *Again, as with Theorem 9.10, the above theorem does not capture the boundary points of the recoverable set. However, if $\mathcal{R}(\overline{\Sigma}_1, \overline{\mathcal{S}})$ is approximated* *by a polytope* $R_2 x \leqslant q_2$ *while* $\mathcal{S}$ *is described by* $Rz \leqslant q$*, then* $\mathcal{R}(\Sigma, \mathcal{S})$ *is approximated by*

$$\begin{pmatrix} RC_z \\ \begin{pmatrix} 0 & R_2 & 0 \end{pmatrix} T_x^{-1} \end{pmatrix} x \leqslant \begin{pmatrix} q \\ q_2 \end{pmatrix}.$$

By improving the approximation of $\mathcal{R}(\bar{\Sigma}_1, \bar{\mathcal{S}})$*, we can approximate* $\mathcal{R}(\Sigma, \mathcal{S})$ *arbitrarily well in this way.*

Since the transformation into SCB and the associated computation of T_x *is already implemented in Matlab and Maple, the remaining problem is the computation or approximation of* $\mathcal{R}(\bar{\Sigma}_1, \bar{\mathcal{S}})$.

If the constraint is right invertible and at most weakly non-minimum phase, then $\bar{\Sigma}_1$ *is actually an empty (zero-dimensional) system, and we have*

$$\mathcal{R}(\Sigma_1, \bar{\mathcal{S}}) = \mathbb{R}^{n_a},$$

and the closure of $\mathcal{R}(\Sigma, \mathcal{S})$ *is equal to the admissible set. If this subsystem* $\bar{\Sigma}_1$ *has dimension two or less, the tools from the book by Ryan [121] can be used. Otherwise, gridding tools are needed as mentioned in Remark 9.11.*

Note that the reduction of the computation of the recoverable region from Σ to the computation of the recoverable region for the lower order system $\bar{\Sigma}_1$ actually yields the result in Remark 9.9 as a special case.

9.3.2 Discrete-time systems

As in the previous subsection that pertains to continuous-time systems, our goal in this subsection, while considering discrete-time systems, is to show how to reduce the complexities involved in computing the recoverable region $\mathcal{R}(\Sigma, \mathcal{S})$ by utilizing the structural properties of the given system. For this purpose, as in the case of continuous-time systems, we first express the given system in terms of its SCB as given in (9.1).

As before, such a SCB allows us to decompose the system $\bar{\Sigma}$ into certain subsystems. In order to characterize the recoverable region efficiently, we can extract the first subsystem Σ_{a+b} from $\bar{\Sigma}$ as

$$\Sigma_{a+b} : \begin{cases} \rho x_a^+ &= A_{aa}^+ x_a^+ + K_{ab}^+ C_b x_b + K_{a2}^+ \zeta \\ \rho x_b &= (A_{bb} + K_{bb} C_b) x_b + K_{b2} \zeta \\ \bar{z} &= \begin{pmatrix} C_b \\ 0 \end{pmatrix} x_b + \begin{pmatrix} 0 \\ I \end{pmatrix} \zeta, \end{cases} \tag{9.10}$$

where A_{aa}^+ has all its eigenvalues outside the closed unit disc and ζ is as defined in (9.2).

We also make extensive use of the following subsystem Σ_{a+bd} of the original system:

$$\Sigma_{a+bd}: \begin{cases} \rho x_a^+ &= A_{aa}^+ x_a^+ + K_{ab}^+ C_b x_b + K_{a2}^+ \zeta \\ \rho x_b &= (A_{bb} + K_{bb} C_b) x_b + K_{b2} \zeta \\ \rho x_d &= A_{dd} x_d + B_d [u_d + G\overline{x}] + K_d \overline{z} \\ \zeta &= \begin{pmatrix} 0 \\ C_d \end{pmatrix} x_d + \begin{pmatrix} I \\ 0 \end{pmatrix} u_0 \\ \overline{z} &= \begin{pmatrix} C_b \\ 0 \end{pmatrix} x_b + \begin{pmatrix} 0 \\ I \end{pmatrix} \zeta. \end{cases} \tag{9.11}$$

We note that G and K_d can be appropriately defined from (9.1). In Σ_{a+bd}, we define

$$\widetilde{u}_d = u_d + G\overline{x} \text{ and } \widetilde{u} = \begin{pmatrix} u_0 \\ \widetilde{u}_d \end{pmatrix}.$$

Let $\mathcal{R}(\Sigma_{a+bd}, \overline{\mathcal{S}})$ denote the recoverable region of subsystem Σ_{a+bd} with input $\widetilde{u}$ where, as in the continuous-time case, $\overline{\mathcal{S}} = T_z^{-1}\mathcal{S}$ is the new constraint set when the given system is in the SCB form.

We claim that we can compute the recoverable region for the full system from the recoverable region for the subsystem Σ_{a+b}. We define

$$\mathcal{V}_q(\overline{\Sigma}, \overline{\mathcal{S}}) = \left\{ \overline{x} \in \mathbb{R}^n \mid x_{a+b} \in \mathcal{R}(\Sigma_{a+b}, \overline{\mathcal{S}}) \right\},$$

where q is an integer chosen larger than the maximum order of infinite zeros of the system. Next, we define the following recursion:

$$\mathcal{V}_k(\overline{\Sigma}, \overline{\mathcal{S}}) = \left\{ \overline{x} \in \mathbb{R}^n \mid \exists \overline{u} \text{ such that } \overline{A}\overline{x} + \overline{B}\overline{u} \in \mathcal{V}_{k+1}(\overline{\Sigma}, \overline{\mathcal{S}}) \text{ and } \overline{C}_z \overline{x} + \overline{D}_z \overline{u} \in \overline{\mathcal{S}} \right\} \tag{9.12}$$

for $k = q - 1, \ldots, 0$. In the above equation, the matrices $(\overline{A}, \overline{B}, \overline{C}_z, \overline{D}_z)$ are the system matrices of the SCB of the given system. Our first main result claims that $\mathcal{V}_0$ leads to the recoverable region for the original system.

Theorem 9.15 *Consider the system Σ as given in* (7.1) *along with a constraint set $\mathcal{S}$ that satisfies Assumption 7.4. We have*

$$\mathcal{R}(\Sigma, \mathcal{S}) = T_x^{-1} \mathcal{R}(\overline{\Sigma}, \mathcal{S}) = T_x^{-1} \mathcal{V}_0(\overline{\Sigma}, \mathcal{S}).$$

Remark 9.16 *A special case of the above theorem is obtained when the system Σ is right invertible and at most weakly non-minimum phase since in that case the system Σ_{a+b} is empty and we obtain*

$$\mathcal{V}_q(\overline{\Sigma}, \overline{\mathcal{S}}) = \mathbb{R}^n,$$

in which case we can obtain the recoverable region through a finite recursion. In general, however, the above theorem results only in a reduction of complexity since we need to obtain the recoverable region only for a system of lower dimension. However, note this is crucial since the classical results for computation of the recoverable region, such as the results in Ryan's book [121], consider only the cases $n = 2$ and $n = 3$. This is primarily because with growing dimension the required computational effort grows dramatically.

Proof : In order to prove Theorem 9.15, we need some preparatory work. Consider the recursive definition of $\mathcal{V}_k(\Sigma, \bar{\mathcal{S}})$ for $0 \leqslant k \leqslant q$. We define

$$\tilde{\mathcal{V}}_q(\Sigma_{a+bd}, \bar{\mathcal{S}}) := \left\{ x_{a+bd} \mid x_{a+b} \in \mathcal{R}(\Sigma_{a+b}, \bar{\mathcal{S}}) \right\},$$

and

$$\tilde{\mathcal{V}}_k(\Sigma_{a+bd}, \bar{\mathcal{S}}) = \Big\{ x_{a+bd} \mid \exists \tilde{u} \text{ such that } \tilde{C}_0 x_{a+bd} + \tilde{D}_0 \tilde{u} \in \bar{\mathcal{S}} \\ \text{and } \tilde{A}_0 x_{a+bd} + \tilde{B}_0 \tilde{u} \in \mathcal{V}_{k+1}(\Sigma_{a+bd}, \bar{\mathcal{S}}) \Big\},$$

where

$$\tilde{A}_0 = \begin{pmatrix} A_{aa}^+ & K_{ab}^+ C_b & K_{ad}^+ C_d \\ 0 & A_{bb} + K_{bb} C_b & K_{bd} C_d \\ 0 & K_{db} C_b & A_{dd} \end{pmatrix}, \quad \tilde{B}_0 = \begin{pmatrix} B_{a0}^+ & 0 \\ B_{b0} & 0 \\ B_{d0} & B_d \end{pmatrix},$$

$$\tilde{C}_0 = \begin{pmatrix} 0 & C_b & 0 \\ 0 & 0 & C_d \\ 0 & 0 & 0 \end{pmatrix}, \quad \tilde{D}_0 = \begin{pmatrix} 0 & 0 \\ 0 & 0 \\ I & 0 \end{pmatrix}, \quad \tilde{u} = \begin{pmatrix} u_0 \\ \tilde{u}_d \end{pmatrix}.$$

We have the relationship

$$\mathcal{V}_k(\bar{\Sigma}, \bar{\mathcal{S}}) = \mathbb{R}^{n_a^{0-}} \times \tilde{\mathcal{V}}_k(\Sigma_{a+bd}, \bar{\mathcal{S}}) \times \mathbb{R}^{n_c} \tag{9.13}$$

for $k = 0, \ldots, q$. This relationship between $\mathcal{V}_k(\bar{\Sigma}, \bar{\mathcal{S}})$ and $\tilde{\mathcal{V}}_k(\Sigma_{a+bd}, \bar{\mathcal{S}})$ for $0 \leqslant k \leqslant q$ results from the fact that the relationship is obviously true for $k = q$, while for $k < q$, we note that the dynamics of x_a^{0-} and x_c do not impact the constraints directly as shown by the structure of the SCB.

It can be easily verified that $\mathcal{V}_k(\bar{\Sigma}, \bar{\mathcal{S}})$ defined in (9.12) can be characterized as

$$\mathcal{V}_k(\bar{\Sigma}, \bar{\mathcal{S}}) = \left\{ \bar{x}(k) \mid \exists \bar{u} \text{ such that } \bar{z}(i) \in \bar{\mathcal{S}} \text{ for } i = k, \ldots, q \\ \text{and } x(q) \in \mathcal{V}_q(\bar{\Sigma}, \bar{\mathcal{S}}) \right\}. \tag{9.14}$$

Similarly, $\widetilde{\mathcal{V}}_k(\Sigma_{a+bd}, \overline{\mathcal{S}})$ is given by

$$\widetilde{\mathcal{V}}_k(\Sigma_{a+bd}, \overline{\mathcal{S}}) := \big\{ x_{a+bd}(k) \mid \exists \widetilde{u} \text{ such that } \overline{z}(i) \in \overline{\mathcal{S}} \\ \text{for } i = k, \dots, q \text{ and } x_{a+bd}(q) \in \widetilde{\mathcal{V}}_q(\Sigma_{a+bd}, \overline{\mathcal{S}}) \big\}. \quad (9.15)$$

As a first step in the proof of Theorem 9.15, the next lemma shows that the set $\widetilde{\mathcal{V}}_0(\overline{\Sigma}_{a+bd}, \overline{\mathcal{S}})$ is the recoverable region of the system Σ_{a+bd}.

Lemma 9.17 *Consider the system given by* (9.11) *with a constraint set* $\overline{\mathcal{S}}$ *satisfying Assumption 7.4. We have*

$$\mathcal{R}(\Sigma_{a+bd}, \overline{\mathcal{S}}) = \widetilde{\mathcal{V}}_0(\Sigma_{a+bd}, \overline{\mathcal{S}}).$$

Proof : We will show at first that

$$\mathcal{R}(\Sigma_{a+bd}, \overline{\mathcal{S}}) \subseteq \widetilde{\mathcal{V}}_0(\Sigma_{a+bd}, \overline{\mathcal{S}}). \quad (9.16)$$

Suppose $x_{a+bd}(0) \in \mathcal{R}(\Sigma_{a+bd}, \overline{\mathcal{S}})$. By definition, we know that there exists a $\widetilde{u}$ such that $x_{a+bd}(k) \to 0$ as $k \to \infty$ and $\overline{z}(k) \in \overline{\mathcal{S}}$ for all $k \geqslant 0$, which implies that, for subsystem Σ_{a+b}, there exists a sequence ζ such that $x_{a+b}(k)$ approaches zero as k goes to infinity. This means that $x_{a+b}(q) \in \mathcal{R}(\Sigma_{a+b}, \overline{\mathcal{S}})$. Hence, we have $x_{a+bd}(q) \in \widetilde{\mathcal{V}}_q(\Sigma_{a+bd}, \overline{\mathcal{S}})$. The fact that $x_{a+bd}(k) \in \widetilde{\mathcal{V}}_k(\Sigma_{a+bd}, \overline{\mathcal{S}})$ for $0 \leqslant k \leqslant q-1$ follows then directly from (9.15) since we have no constraint violation in the interval $[0, q]$. This concludes $x_{a+bd}(0) \in \widetilde{\mathcal{V}}_0(\Sigma_{a+bd}, \overline{\mathcal{S}})$.

The next step is to show that

$$\widetilde{\mathcal{V}}_0(\Sigma_{a+bd}, \overline{\mathcal{S}}) \subseteq \mathcal{R}(\Sigma_{a+bd}, \overline{\mathcal{S}}). \quad (9.17)$$

Suppose $x_{a+bd}(0) \in \widetilde{\mathcal{V}}_0(\Sigma_{a+bd}, \overline{\mathcal{S}})$ and define $x_{d0} = x_d(0)$. From the definition of $\widetilde{\mathcal{V}}_0(\Sigma_{a+bd}, \overline{\mathcal{S}})$, there exists a control input $\widetilde{u}(k)$, say, $\widetilde{u}_1(k)$, for $0 \leqslant k \leqslant q$, such that $x_{a+bd}(q) \in \widetilde{\mathcal{V}}_q(\Sigma_{a+bd}, \overline{\mathcal{S}})$ and no constraint violation occurs within the first $q-1$ steps. This implies that there exists a $\zeta(k)$ from 0 to q such that $x_{a+b}(q) \in \mathcal{R}(\Sigma_{a+b}, \overline{\mathcal{S}})$ and the constraints are not violated. From time q onward, since $x_{a+b}(q) \in \mathcal{R}(\Sigma_{a+b}, \overline{\mathcal{S}})$, there exists an input $\zeta(k)$ for $k \geqslant q$ that steers $x_{a+b}(k)$ to zero and causes no constraint violation.

In this way, we find a signal $\zeta(k)$ for all k. Clearly, we can generate this $\zeta(k)$ via a suitable input $\widetilde{u}(k)$, say, $\widetilde{u}_2(k)$, for $k \geqslant 0$ together with an appropriate initial condition $\widetilde{x}_{d0}$ because the x_d dynamics with inputs u_0 and $\widetilde{u}_d$ and output ζ is right invertible by construction.

Next, we note that x_{d0} and inputs $\widetilde{u}_1(k)$ also generate the same $\zeta(k)$ for $k = 0, \dots, q$. The special structure of x_d dynamics guarantees that the initial conditions x_{d0} and $\widetilde{x}_{d0}$ must be the same since they result in the same output $\zeta(k)$ for an interval at least as long as the order of the infinite zeros. The structure guarantees this even though the associated inputs might be different. We conclude that for our initial conditions, there exist inputs which generate the signal $\zeta(k)$ for all k.

We have noted before that this signal ζ is such that no constraint violations will occur. It remains to show that $x_{a+bd}(k) \to 0$ from time q onward. As noted earlier, this signal $\zeta(k)$ is such that $x_{a+b}(k) \to 0$ as $k \to \infty$, which also implies that $\zeta(k) \to 0$ as $k \to \infty$. Again, the structure of x_d dynamics guarantees that x_d also approaches zero as $k \to \infty$.

Hence, we can find an input $\tilde{u}(k)$ for all $k \geqslant 0$ such that

$$\lim_{k\to\infty} x_{a+bd}(k) = 0,$$

and no constraint violation occurs. ■

Now we proceed to prove Theorem 9.15. It is obvious from Lemma 9.17 and (9.13) that

$$\mathcal{V}_0(\bar{\Sigma}, \bar{\mathcal{S}}) = \mathbb{R}^{n_a^{0-}} \times \mathcal{R}(\Sigma_{a+bd}, \bar{\mathcal{S}}) \times \mathbb{R}^{n_c}.$$

Then, clearly, this implies that

$$\mathcal{R}(\bar{\Sigma}, \bar{\mathcal{S}}) \subseteq \mathcal{V}_0(\bar{\Sigma}, \bar{\mathcal{S}}).$$

It remains to prove the converse inclusion. This will be shown through an explicit controller design as presented later in the proof of Theorem 9.24. ■

9.4 Semi-global stabilization in the recoverable region

The first objective of this chapter is the reduction in the computation of the recoverable region as outlined in the previous section. The second objective of this chapter is to solve the semi-global stabilization problem (as stated in Problem 9.5) in the recoverable region by state feedback controllers. That is, our intention here is to show that semi-global stabilization can be achieved by a state feedback controller without violating the constraints for any compact subset $\mathcal{W}$ contained in the interior of $\mathcal{R}(\Sigma, \mathcal{S})$.

Before we proceed further, it is appropriate to mention that there are two lines of research in the literature on stabilization problems in the presence of non-minimum-phase constraints:

- A traditional line employs the construction of invariant sets. A common denominator in the stream of literature taking this approach is the idea of seeking a control law that does not violate the constraints posed on actuators and at the same time makes a subset of the admissible set invariant. Subsets of the admissible set which can be made invariant in this way are called positive invariant sets. Two candidate positively invariant sets widely used in the literature are *ellipsoidal sets* and *polyhedral sets*. Ellipsoidal sets are classical in control theory; however, they suffer from conservativeness in the approximation of the recoverable region. More recently, polyhedral sets have received great attention (see, e.g., [9, 11, 12, 30, 187], among others). In principle, polyhedral sets are not intrinsically conservative but this might require an exponential growth in the number of edges with the related exponential growth in the required numerical effort. For a detailed perspective in this line of research, the reader should consult the excellent review in [10]. Further information in this regard can be found in two survey papers [33, 41].

- The second line of research takes a fundamental view of global and semi-global stabilization relative to the recoverable region and follows the line of thought that has been described in previous chapters. Thus, in global stabilization problem, one would seek a stabilizing feedback law that does not violate the constraints posed and achieves a domain of attraction for the equilibrium point of the closed-loop system that is equal to the recoverable region. The semi-global stabilization problem deals with the issue of designing a family of stabilizing feedback laws such that, for any a priori given set, a member among the family of stabilizing feedback laws achieves a domain of attraction for the equilibrium point of the closed-loop system that is inside the priori given set and, moreover, does not violate the constraints posed. For semi-global stabilization problem, Choi [26] showed that for exponentially unstable *discrete time* linear systems subject to input constraints, any compact subset of the maximal recoverable region can be exponentially stabilized via periodic linear variable structure controllers. Moreover, Choi [27] showed that, in general, linear feedback cannot achieve global stabilization for discrete-time unstable systems. Also, Hu et al. [49] studied the possibility of semi-global stabilization of continuous-time systems with two unstable open-loop poles. It should be emphasized that all these works deal only with the case when the constraints are posed on the inputs. On the other hand, we provided in Chap. 8 the solvability conditions for semi-global stabilization in the admissible set whenever there are constraints both on state and control variables. Also, we studied recently semi-global stabilization in the recoverable region whenever there are constraints both on state and control variables for systems subject to non-minimum-phase constraints. We did so first for continuous-time systems in [167] and then for discrete-time systems in [197]. To distinguish our work with that of Choi [26] and Cwikel and Gutman [30], let us emphasize that [26] only considers input constraints and [30] uses a simple algorithm without exploiting structure. We exploit the structure of a given system to develop our results both on constructing the recoverable sets as we did in the previous section but also for semi-global stabilization in the recoverable region as we pursue in this section.

Following [167] and [197], we present below our results on semi-global stabilization in the recoverable region in two subsections, one for continuous-time systems and the other for discrete-time systems.

9.4.1 Continuous-time systems

For continuous-time systems, our goal in this subsection is to develop state feedback controllers that semi-globally stabilize in the recoverable region a given system with non-minimum-phase constraints. That is, we pursue here the possibility of stabilizing without violating the constraints for any compact subset $\mathcal{W}$ contained in the interior of $\mathcal{R}(\Sigma, \mathcal{S})$ by a continuous feedback. Regarding the existence of Lipschitz-continuous controllers, our main result is summarized in the following theorem.

Theorem 9.18 *Consider the system Σ in* (7.1) *along with a constraint set $\mathcal{S}$ that satisfies Assumption 7.4. Assume that (A, B) is stabilizable. Then, for any compact subset $\mathcal{W}$ contained in the interior of $\mathcal{R}(\Sigma, \mathcal{S})$, there exists a Lipschitz-continuous (in general nonlinear) feedback $u = f(x)$ such that the zero equilibrium point of the closed-loop system is asymptotically stable with a domain of attraction containing $\mathcal{W}$ and moreover $z(t) \in \mathcal{S}$ for all $t \geqslant 0$ when $x(0) \in \mathcal{W}$.*

Moreover, for all initial conditions inside $\mathcal{W}$, the state converges to the origin exponentially fast.

Remark 9.19 *Note that although this theorem is a pure existence result, we will also establish that we only need to design a controller for a subsystem which can have considerably lower dimension, and in this way, it does reduce the complexity of computational tools that are available for actually designing the controllers; note that, in all the available tools, complexity of computations grows exponentially with the dimension.*

In the previous section, we connected the recoverable set of the original system Σ to that of the reduced system $\bar{\Sigma}_1$ through the intermediate system Σ_1. We will use these three layers to look also into the design of controllers. We first look in the next subsection at the design of controllers for systems of the form $\bar{\Sigma}_1$. This will turn out to be the most involved design step. In the second subsection, we will extend a controller for this subsystem to come up with a controller for the original system Σ. Note that we assume in the proof that the system has no invariant zeros on the imaginary axis. This is without loss of generality since changing A to $A_\kappa = A + \kappa I$ with κ arbitrarily small would remove the zeros on the imaginary axis. Clearly, a controller for this system with a certain constrained domain of attraction would, when applied to the original system, always yield a larger domain of attraction, and according to the following lemma, the recoverable set of this new system would be only marginally smaller than the recoverable set of the original system.

Lemma 9.20 *Consider the system* Σ *in* (7.1) *and a convex, compact constraint set* $\mathcal{S}$ *containing* 0 *in the interior. Let* Σ^κ *be the system obtained from* Σ *by replacing* A *by* $A + \kappa I$ *with* $\kappa > 0$*. Assume that* (A, B) *is stabilizable. Then for any compact subset* $\mathcal{W} \subset \operatorname{int} \mathcal{R}(\Sigma, \mathcal{S})$*, there exists a* $\kappa^* > 0$ *such that*

$$\mathcal{W} \subset \mathcal{R}(\Sigma^\kappa, \mathcal{S}) \subset \mathcal{R}(\Sigma, \mathcal{S}),$$

for any $\kappa \in [0, \kappa^*]$.

Proof : See Appendix 9.C. ■

We now proceed to prove Theorem 9.18 for the subsystem $\bar{\Sigma}_1$.

Proof of Theorem 9.18 for the subsystem $\bar{\Sigma}_1$

As we mentioned before, the subsystem $\bar{\Sigma}_1$ in (9.7) is the core of the original system Σ, which causes most of the design difficulties under constraints. Therefore, we first prove Theorem 9.18 for systems which are left invertible, have relative degree zero, and have only antistable invariant zeros. Obviously, subsystem $\bar{\Sigma}_1$ is one of such systems. To simplify notation, we assume the system is in the following form

$$\Sigma_0 : \begin{cases} \dot{\xi} = & A_0\,\xi + \quad B_0\zeta, \\ \bar{z} = \begin{pmatrix} C_0 \\ 0 \end{pmatrix}\xi + \begin{pmatrix} 0 \\ I \end{pmatrix}\zeta, \end{cases} \tag{9.18}$$

with constraint $\bar{z}(t) \in \bar{\mathcal{S}}$ for all $t \geqslant 0$, where the unobservable eigenvalues of the pair (C_0, A_0), i.e., the invariant zeros, are in the open right-half plane (antistable) and (A_0, B_0) is stabilizable.

Consider the set $\mathcal{R}(\Sigma_0, \bar{\mathcal{S}})$. Our first objective is to choose the input in such a way that we stay inside this set. If this is possible, then we call the set positive invariant. In order to do this, we can try to choose at each boundary point of the set, an input such that the derivative of the state points inside or tangent to the set and then expand this feedback to the full set. We will show that this basic idea works, although we need to spend quite some effort on avoiding technical difficulties:

- We need the set $\mathcal{R}(\Sigma_0, \bar{\mathcal{S}})$ to be bounded and closed, since the suggested design is based on designing the feedback on the boundary.
- If the derivative does not point inside but tangent to the set, then we are not guaranteed that the state stays in the set.
- Our aim is to achieve asymptotic stability and the above idea only looks at achieving positive invariance and this is clearly not the same.

- The feedback that we choose in this way might not even be continuous, and therefore, we are not sure that the closed-loop system has a (unique) solution.

The first technical issue mentioned above can actually be resolved due to the extra structure of the system (9.18).

Lemma 9.21 *Consider the stabilizable linear system Σ_0 in* (9.18) *whose invariant zeros are antistable and a convex, compact constraint set $\overline{\mathcal{S}}$ containing* 0 *in the interior. The recoverable region $\mathcal{R}(\Sigma_0, \overline{\mathcal{S}})$ for this system has the following properties:*

(i) The set $\mathcal{R}(\Sigma_0, \overline{\mathcal{S}})$ is bounded.

(ii) For any initial condition $\xi_0 \in \partial\mathcal{R}(\Sigma_0, \overline{\mathcal{S}})$, there exists an input u such that the state of the system remains in $\overline{\mathcal{R}(\Sigma_0, \overline{\mathcal{S}})}$, while the constraint $\overline{z}(t) \in \overline{\mathcal{S}}$ is satisfied for all $t \geqslant 0$.

Proof : See Appendix 9.D. ■

By property *(ii)* of the above lemma, it seems feasible to find a feedback such that the compact set $\mathcal{R}(\Sigma_0, \overline{\mathcal{S}})$ becomes invariant. However, in order to avoid the technical difficulties mentioned before, it turns out that it is desirable to start working with an auxiliary system:

$$\Sigma_0^\kappa : \begin{cases} \dot{\xi} = & A_\kappa \xi + \quad B_0 \tilde{\zeta}, \\ \tilde{z} = \begin{pmatrix} C_0 \\ 0 \end{pmatrix} \xi + \begin{pmatrix} 0 \\ I \end{pmatrix} \tilde{\zeta}, \end{cases} \tag{9.19}$$

with constraint $\tilde{z}(t) \in \overline{\mathcal{S}}$ for all $t \geqslant 0$, where $A_\kappa = A + \kappa I$ for $\kappa \geqslant 0$. It is more difficult to keep the state inside a convex set $\mathcal{V}$ (containing 0) for this system due to the fact that the extra term $\kappa\xi$ always points outside the set $\mathcal{V}$.

All the technical difficulties mentioned before are resolved in this way. If we choose a direction for the derivative to point tangent or inside the set for this auxiliary system, then by reducing κ, we can guarantee that, for a slightly smaller κ, we can make the set positive invariant. Moreover, by reducing κ, we obtain some flexibility which enables us to make the feedback continuous and even Lipschitz continuous. Finally, if the state stays in the set for some positive κ, then for the original system, the state converges to zero exponentially.

Note that the recoverable set of this auxiliary system is close to the recoverable set of the original system by Lemma 9.20. The technical details of the above are in Appendix 9.E and yield the proof of Theorem 9.18 for the special case of system Σ_0 given in (9.18). However, in order to expand a controller of the subsystem

$\bar{\Sigma}_1$ to form a controller for the system Σ, we need a strengthened version of Theorem 9.18 which can handle small exponentially decaying disturbances. The details of this expansion from $\bar{\Sigma}_1$ to Σ are provided shortly.

Theorem 9.22 (Special case with disturbance) *Consider the system*

$$\Sigma_0^d : \begin{cases} \dot{\xi} = & A_0\xi + \quad B_0\zeta + d, \\ \eta = & \begin{pmatrix} C_0 \\ 0 \end{pmatrix}\xi + \begin{pmatrix} 0 \\ I \end{pmatrix}\zeta, \end{cases} \tag{9.20}$$

with the unobservable modes of (C_0, A_0) antistable and (A_0, B_0) stabilizable. Given $M > 0$ and a compact subset $\mathcal{W} \subset \operatorname{int} \mathcal{R}(\Sigma_0^d, \bar{\mathcal{S}})$, there exist a $\delta > 0$ and a Lipschitz-continuous feedback $\zeta = f(\xi)$ such that the equilibrium point 0 is asymptotically stable for all initial conditions in $\mathcal{W}$, and for any disturbance d satisfying

$$\|d(t)\| \leqslant M e^{-\delta t}, \tag{9.21}$$

the closed-loop system satisfies $\xi(t) \to 0$ as $t \to \infty$ and $\eta(t) \in \bar{\mathcal{S}}$ for all $t \geqslant 0$.

Proof : See Appendix 9.E. ■

We now proceed to prove Theorem 9.18.

Proof of Theorem 9.18 : Earlier, we have decomposed the original system Σ into two subsystems Σ_1 and Σ_2, and then, we established that the computational effort for determining the recoverable set is concentrated in system Σ_1. If we look more closely at the system Σ_1, we can extract another subsystem $\bar{\Sigma}_1$, and the recoverable set of this last subsystem is the core of the computational effort needed in determining the recoverable set.

This time, we want to establish a suitable controller with a domain of attraction containing an arbitrarily chosen compact set $\mathcal{W}$ which is itself contained in the interior of $\mathcal{R}(\Sigma, \mathcal{S})$. First, note that the recoverable set of the full system satisfies the structure established in Theorem 9.10. Therefore, we can find a compact set $\mathcal{W}_1$ such that $\mathcal{W}_1$ is contained in the interior of $\mathcal{R}(\Sigma_1, \bar{\mathcal{S}})$ and

$$\mathcal{W} \subset T_x \left\{ \begin{pmatrix} x_{ab} \\ x_{cd} \end{pmatrix} \middle| \, x_1 \in \mathcal{W}_1 \right\} \cap \mathcal{V}(\Sigma, \mathcal{S}),$$

where x_{ab} and x_{cd} denote the initial conditions of Σ_1 and Σ_2 respectively. We assume without loss of generality that we have no zeros on the imaginary axis which implies in the SCB structure that the dynamics of x_a^{-0} is asymptotically stable and hence can be exempted from stabilization. Consider the system,

$$\bar{\Sigma}_{1\kappa} : \begin{cases} \dot{x}_a^+ = A_{a\kappa}^+ x_a^+ + K_{ab}^+ + C_b x_b + K_{a2}^+ \zeta, \\ \dot{x}_b = (A_{b\kappa} + K_{bb} C_b)\, x_b + K_{b2} \zeta, \\ \bar{z} = \begin{pmatrix} C_b \\ 0 \end{pmatrix} x_b + \begin{pmatrix} 0 \\ I \end{pmatrix} \zeta, \end{cases} \tag{9.22}$$

with $A_{a\kappa}^+ = A_{aa}^+ + \kappa I$ and $A_{b\kappa} = A_{bb} + \kappa I$. The associated recoverable set $\mathcal{R}(\bar{\Sigma}_{1\kappa}, \mathcal{S})$ has the following property:

$$\left\{ \begin{pmatrix} x_1 \\ x_2 \end{pmatrix} \,\middle|\, x_1 \in \mathcal{R}(\bar{\Sigma}_{1\kappa}, \bar{\mathcal{S}}) \right\} \subset \mathcal{R}(\Sigma_1, \bar{\mathcal{S}}),$$

where x_1 denotes the initial condition for x_a^+ and x_b, while x_2 denotes the initial condition for x_a^{-0}. Moreover, similar to Lemma 9.20, it is easy to verify that for κ small enough,

$$\mathcal{W}_1 \subset T_x \left\{ \begin{pmatrix} x_1 \\ x_2 \end{pmatrix} \,\middle|\, x_1 \in \mathcal{R}(\bar{\Sigma}_{1\kappa}, \bar{\mathcal{S}}) \right\}.$$

Choose κ small enough such that this latter inclusion is satisfied. Then we can design a controller f for $\bar{\Sigma}_{1\kappa}$ according to Theorem 9.22, and it is easily verified that this controller when applied to Σ_1 creates an exponentially stable system with $\mathcal{W}_1$ contained in its domain of attraction which can handle exponentially decaying disturbances satisfying (9.21).

Next, we consider the second subsystem Σ_2 given by (9.4). This system has the nice structure that the mapping from $(\tilde{u}_d, u_0)$ to ζ is strongly controllable. Assume that the initial state of Σ is in the interior of the set

$$T_x \left\{ \begin{pmatrix} x_{ab} \\ x_{cd} \end{pmatrix} \,\middle|\, x_1 \in \mathcal{W}_1 \right\} \cap \mathcal{V}(\Sigma, \mathcal{S}). \tag{9.23}$$

Following the design methodology of Chap. 8, we can then design a feedback for inputs $(\tilde{u}_d, u_0)$ which stabilizes Σ_2 and such that $\zeta = f(x_1) + d$ with d satisfying (9.21) while satisfying the constraints. This controller is then easily seen to satisfy the conditions of Theorem 9.18. ■

Let us next briefly comment on measurement feedback controllers. If we design a family of static state feedback controllers that achieve semi-global stabilization in the recoverable region, then we can combine this with a high-gain observer to obtain semi-global stabilization in the recoverable region by measurement feedback. These high-gain observers have been designed in Chap. 4 either through the direct method or via a Riccati-based design. We can then rely on Lemma 4.62 to establish that this combination of state feedback controller with an appropriate observer has the required observer. Note that the use of Lemma 4.62 requires that the measurement error exhibits no peaking. But we can always choose an arbitrarily small $\tau > 0$ such that for $t > \tau$, the measurement error is bounded. Since we can easily ensure that our static state feedback is bounded, we can choose τ sufficiently small such that the state at time τ is ensured to lie in a compact set inside the recoverable region.

We illustrate below the results developed above by an example.

Example 9.23 We consider the system,

$$\dot{x} = \begin{pmatrix} 0 & 1 & 0 & 0 \\ -1 & 0 & 0 & 1 \\ 1 & -2 & 0 & 0 \\ 1 & 0 & -1 & 0 \end{pmatrix} x + \begin{pmatrix} 0 \\ 0 \\ 0 \\ 1 \end{pmatrix} u,$$

$$z = \begin{pmatrix} 0 & 1 & 0 & 0 \\ 0 & 0 & 0 & 1 \end{pmatrix} x,$$

with $\mathcal{S}$ given by

$$\mathcal{S} = \left\{ z \in \mathbb{R}^2 \mid -1 \leqslant z_1 \leqslant 4,\ -1 \leqslant z_2 \leqslant 1 \right\}.$$

In this case, we note that in the SCB context, x_a^0 corresponds to x_3 while x_d corresponds to x_4. Moreover, x_c is not present since the system is left invertible. Note that the system is not right invertible, and hence, we cannot rely on the relatively easy structure we obtained for right-invertible systems in Chap. 8.

In order to obtain the recoverable set, we first compute the system Σ_1 which is given by

$$\Sigma_1 : \begin{cases} \dot{\tilde{x}} = \begin{pmatrix} 0 & 1 & 0 \\ -1 & 0 & 0 \\ 1 & -2 & 0 \end{pmatrix} \tilde{x} + \begin{pmatrix} 0 \\ 1 \\ 0 \end{pmatrix} \zeta, \\ z = \begin{pmatrix} 0 & 1 & 0 \\ 0 & 0 & 0 \end{pmatrix} \tilde{x} + \begin{pmatrix} 0 \\ 1 \end{pmatrix} \zeta, \end{cases}$$

and then the system $\bar{\Sigma}_1$ is given by

$$\bar{\Sigma}_1 : \begin{cases} \dot{\bar{x}} = \begin{pmatrix} 0 & 1 \\ -1 & 0 \end{pmatrix} \bar{x} + \begin{pmatrix} 0 \\ 1 \\ 0 \end{pmatrix} \zeta, \\ z = \begin{pmatrix} 0 & 1 \\ 0 & 0 \end{pmatrix} \bar{x} + \begin{pmatrix} 0 \\ 1 \end{pmatrix} \zeta. \end{cases}$$

The recoverable region for this system can be computed using the techniques available from the work of Ryan [121]. We obtain the recoverable set $\mathcal{R}(\bar{\Sigma}_1, \mathcal{S})$

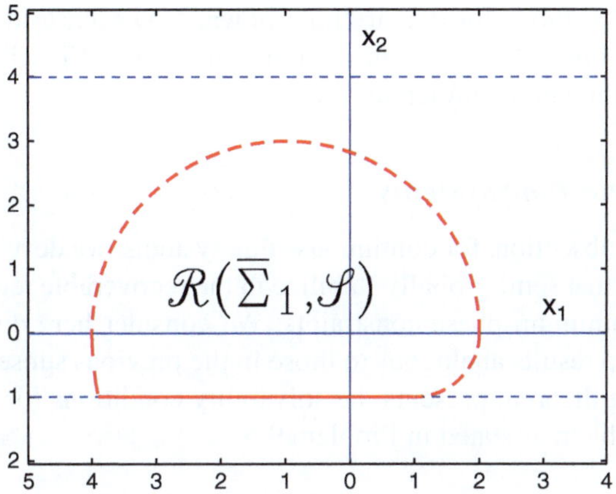

Figure 9.1: Recoverable region $\mathcal{R}(\bar{\Sigma}_1, \mathcal{S})$

given in Fig. 9.1. Next, consider the boundary. The dashed line does not belong to the recoverable set, while the solid line is part of the recoverable set. The theory developed in this subsection then tells us that

$$\overline{\mathcal{R}(\Sigma, \mathcal{S})} = \left\{ x \in \mathbb{R}^4 \mid \begin{pmatrix} x_1 \\ x_2 \end{pmatrix} \in \overline{\mathcal{R}(\bar{\Sigma}_1, \mathcal{S})}, -1 \leqslant x_4 \leqslant 1 \right\}.$$

Assume that we have a compact set $\mathcal{W}$ contained in the interior of $\mathcal{R}(\Sigma, \mathcal{S})$ and we want to obtain a controller which stabilizes the system and contains $\mathcal{W}$ in its domain of attraction while avoiding constraint violation when starting in the set $\mathcal{W}$.

The theory developed in this subsection tells us that we need to look at a modification of the system Σ_1:

$$\Sigma_{1,\varepsilon} : \begin{cases} \dot{\tilde{x}} = \begin{pmatrix} \varepsilon & 1 & 0 \\ -1 & \varepsilon & 0 \\ 1 & -2 & \varepsilon \end{pmatrix} \tilde{x} + \begin{pmatrix} 0 \\ 1 \\ 0 \end{pmatrix} \zeta, \\ z = \begin{pmatrix} 0 & 1 & 0 \\ 0 & 0 & 0 \end{pmatrix} \tilde{x} + \begin{pmatrix} 0 \\ 1 \end{pmatrix} \zeta, \end{cases}$$

for some $\varepsilon > 0$ small enough. We first need to design a controller which stabilizes this system and contains $\mathcal{W}_1$ in its domain of attraction while avoiding constraint violation when the initial condition is in the set $\mathcal{W}_1$ where

$$\mathcal{W}_1 = \begin{pmatrix} 1 & 0 & 0 & 0 \\ 0 & 1 & 0 & 0 \\ 0 & 0 & 1 & 0 \end{pmatrix} \mathcal{W}.$$

Efficient design methods for this are not known. However, gridding can be one option, and working with a lower dimensional subsystem Σ_1 will definitely make the gridding method more attractive.

9.4.2 Discrete-time systems

In the previous subsection, for continuous-time systems, we developed state feedback controllers that semi-globally stabilize in the recoverable region a given system with non-minimum-phase constraints. We consider here discrete-time systems and develop results analogous to those in the previous subsection.

The following theorem presents the solvability conditions for the semi-global stabilization problem as stated in Problem 9.5.

Theorem 9.24 : *Consider the system Σ as given in* (7.1) *along with a constraint set $\mathcal{S}$ that satisfies Assumption 7.4. The semi-global stabilization problem as defined in Problem 9.5 is solvable. More specifically, for any a priory given compact set $\mathcal{W}$ contained in the recoverable region $\mathcal{R}(\Sigma, \mathcal{S})$, there exists a time-invariant static state feedback $u(k) = f(x(k))$ such that the closed-loop system is asymptotically stable with a domain of attraction containing $\mathcal{W}$ in its interior and that all the constraints are satisfied, i.e., $z(k) \in \mathcal{S}$ for all $k \geqslant 0$.*

Before we start proving this theorem, it is necessary to define the following subsystem:

$$\Sigma_{abd} : \begin{cases} x_a(k+1) = A_{aa}x_a(k) + K_{ab}C_b x_b(k) + K_{a2}\zeta(k) \\ x_b(k+1) = (A_{bb} + K_{bb}C_b)x_b(k) + K_{b2}\zeta(k) \\ x_d(k+1) = A_{dd}x_d(k) + B_d[u_d(k) + G\overline{x}(k)] + K_d\overline{z}(k) \\ \quad\zeta(k) \quad = \begin{pmatrix} 0 \\ C_d \end{pmatrix} x_d(k) + \begin{pmatrix} I \\ 0 \end{pmatrix} u_0(k) \\ \quad\overline{z}(k) \quad = \begin{pmatrix} C_b \\ 0 \end{pmatrix} x_b(k) + \begin{pmatrix} 0 \\ I \end{pmatrix} \zeta(k). \end{cases} \tag{9.24}$$

Similarly,

$$\widetilde{u}_d(k) = u_d(k) + G\overline{x}(k) \text{ and } \widetilde{u}(k) = \begin{pmatrix} u_0(k) \\ \widetilde{u}_d(k) \end{pmatrix}.$$

Let $\mathcal{R}(\Sigma_{abd}, \overline{\mathcal{S}})$ denote the recoverable region of system Σ_{abd}.

Let us next give a brief road map of how we prove Theorem 9.24, that is, how we construct a semi-globally stabilizing controller for the given system Σ. Lemma 9.25 that follows considers the semi-global stabilization of the subsystem Σ_{a+bd} as given by (9.11). Based on the result of Lemma 9.25, we proceed to construct in Lemma 9.26 a semi-globally stabilizing controller for the newly defined subsystem Σ_{abd}. Finally, the controller constructed in Lemma 9.26 for the subsystem Σ_{abd} is augmented to form a semi-globally stabilizing controller for the given system Σ.

We proceed now to construct a semi-globally stabilizing controller for the subsystem Σ_{a+bd}.

Lemma 9.25 *The semi-global stabilization problem in recoverable region for Σ_{a+bd} is solvable by a nonlinear static state feedback of the form,*

$$u_0 = f_1(x_{a+bd}) \text{ and } \tilde{u}_d = f_2(x_{a+bd}).$$

Proof : To start with, we transform the subsystem Σ_{a+bd} into its controllable canonical form. That is, we utilize a nonsingular state transformation T,

$$\tilde{x} = \begin{pmatrix} x_1(k) \\ x_2(k) \end{pmatrix} = T x_{a+bd},$$

such that the system Σ_{a+bd} given by (9.11) is transformed to the form,

$$\tilde{\Sigma}_{a+bd} : \begin{cases} \begin{pmatrix} x_1(k+1) \\ x_2(k+1) \end{pmatrix} = \begin{pmatrix} A_{11} & A_{12} \\ 0 & A_{22} \end{pmatrix} \begin{pmatrix} x_1(k) \\ x_2(k) \end{pmatrix} + \begin{pmatrix} B_1 \\ 0 \end{pmatrix} \tilde{u}(k) \\ \bar{z}(k) = \begin{pmatrix} C_1 & C_2 \\ 0 & 0 \end{pmatrix} \begin{pmatrix} x_1(k) \\ x_2(k) \end{pmatrix} + \begin{pmatrix} 0 & 0 \\ I & 0 \end{pmatrix} \tilde{u}(k), \end{cases}$$

where the dynamics of x_1 is controllable, the dynamics of x_2 is uncontrollable, and

$$\tilde{u}(k) = \begin{pmatrix} u_0(k) \\ \tilde{u}_d(k) \end{pmatrix}.$$

We observe that the recoverable region of system $\tilde{\Sigma}_{a+bd}$ is given by

$$\mathcal{R}(\tilde{\Sigma}_{a+bd}, \mathcal{S}) = T\mathcal{R}(\Sigma_{a+bd}, \mathcal{S}).$$

In order to construct a controller for $\tilde{\Sigma}_{a+bd}$, we define a slightly modified form of $\tilde{\Sigma}_{a+bd}$. That is, we define the modified system

$$\tilde{\Sigma}^{\ell}_{a+bd} : \begin{cases} \rho \begin{pmatrix} x_1^{\ell} \\ x_2^{\ell} \end{pmatrix} = (1+\ell) \begin{pmatrix} A_{11} & A_{12} \\ 0 & A_{22} \end{pmatrix} \begin{pmatrix} x_1^{\ell} \\ x_2^{\ell} \end{pmatrix} + (1+\ell) \begin{pmatrix} B_1 \\ 0 \end{pmatrix} \tilde{u}^{\ell} \\ z = \begin{pmatrix} C_1 & C_2 \\ 0 & 0 \end{pmatrix} \begin{pmatrix} x_1^{\ell} \\ x_2^{\ell} \end{pmatrix} + \begin{pmatrix} 0 & 0 \\ I & 0 \end{pmatrix} \tilde{u}^{\ell}, \end{cases}$$

where $\ell > 0$ is small enough that $\tilde{\Sigma}^{\ell}_{a+bd}$ is still stabilizable.

Let $\widetilde{\mathcal{R}}(\widetilde{\Sigma}^{\ell}_{a+bd}, \overline{\mathcal{S}})$ be the largest set of initial conditions for the system $\widetilde{\Sigma}^{\ell}_{a+bd}$ for which there exists an input such that the constraints are satisfied while we stay inside the set for all k (and where we do **not** impose convergence to zero). We do claim that for any $\widetilde{\rho} \in (0,1)$, there exists an $\ell > 0$ sufficiently small such that

$$\widetilde{\rho}\mathcal{R}(\widetilde{\Sigma}_{a+bd}, \overline{\mathcal{S}}) \subset \widetilde{\mathcal{R}}(\widetilde{\Sigma}^{\ell}_{a+bd}, \overline{\mathcal{S}}) \subset \mathcal{R}(\widetilde{\Sigma}_{a+bd}, \overline{\mathcal{S}}). \tag{9.25}$$

It is trivial to see that

$$\widetilde{\mathcal{R}}(\widetilde{\Sigma}^{\ell}_{a+bd}, \overline{\mathcal{S}}) \subset \mathcal{R}(\widetilde{\Sigma}_{a+bd}, \overline{\mathcal{S}}).$$

It remains to establish that

$$\widetilde{\rho}\mathcal{R}(\widetilde{\Sigma}_{a+bd}, \overline{\mathcal{S}}) \subset \widetilde{\mathcal{R}}(\widetilde{\Sigma}^{\ell}_{a+bd}, \overline{\mathcal{S}}).$$

Consider any $r > n$ for the system $\widetilde{\Sigma}^{\ell}_{a+bd}$. Since the x_1^{ℓ} dynamics is controllable, there exists a $\delta^* > 0$ such that for any $\delta \in (0, \delta^*)$ and for any

$$\begin{pmatrix} x_1 \\ x_2 \end{pmatrix} \in \delta\mathcal{R}(\widetilde{\Sigma}_{a+bd}, \overline{\mathcal{S}}),$$

there exist an input $\widetilde{u}^{\ell} := \begin{pmatrix} u_0^{\ell} \\ u_d^{\ell} \end{pmatrix}$ and initial condition $\widetilde{x}^{\ell}(0) = 0$ such that $x_1^{\ell}(r) = -x_1$ and $x_2^{\ell}(r) = 0$ while

$$\overline{z}^{\ell}(k) \in \tfrac{1-\widetilde{\rho}}{2}\mathcal{S}, \qquad k = 0, 1, \dots, r-1.$$

Moreover, δ^* is independent of ℓ and r provided ℓ is small enough.

Let $r > n$ be such that for any $\widetilde{x}^{\ell}(0) \in \mathcal{R}(\widetilde{\Sigma}_{a+bd}, \overline{\mathcal{S}})$, we have

$$\begin{pmatrix} 0 \\ x_2^{\ell}(r) \end{pmatrix} \in \delta\widetilde{\rho}\mathcal{R}(\widetilde{\Sigma}_{a+bd}, \overline{\mathcal{S}})$$

for all ℓ sufficiently small. This is clearly possible due to the fact that the system is stabilizable, and hence, the uncontrollable dynamics of x_2^{ℓ} must be asymptotically stable.

Consider any initial condition $\widetilde{x}(0) \in \mathcal{R}(\widetilde{\Sigma}_{a+bd}, \overline{\mathcal{S}})$. We have an input $\widetilde{u}$ for the system $\widetilde{\Sigma}_{a+bd}$ such that $\overline{z}(k) \in \overline{\mathcal{S}}$. Hence, for any $\widetilde{\rho} < 1$, we can find, for any initial condition $\widetilde{x}(0) \in \widetilde{\rho}\mathcal{R}(\widetilde{\Sigma}_{a+bd}, \overline{\mathcal{S}})$, an input $\widetilde{\rho}\widetilde{u}$ for the system $\widetilde{\Sigma}_{a+bd}$ such that $\overline{z}(k) \in \widetilde{\rho}\overline{\mathcal{S}}$ for all k. But then for $\widetilde{x}^{\ell}(0) \in \widetilde{\rho}\mathcal{R}(\widetilde{\Sigma}_{a+bd}, \overline{\mathcal{S}})$ and ℓ small enough, we find that there exists a control input, say, $\widetilde{u}_1^{\ell}$, for which we have $\widetilde{x}^{\ell}(k) \in (1+\delta)\widetilde{\rho}\mathcal{R}(\widetilde{\Sigma}_{a+bd}, \overline{\mathcal{S}})$ for $k = 0, \dots, r$ and $\overline{z}^{\ell}(k) \in (1+\delta)\widetilde{\rho}\overline{\mathcal{S}}$ for $k = 0, \dots, r-1$. Also, we observe that if we choose $\delta < \frac{1-\widetilde{\rho}}{2}$, we have

$$\delta\widetilde{x}^{\ell}(r) = \begin{pmatrix} \delta x_1^{\ell}(r) \\ \delta x_2^{\ell}(r) \end{pmatrix} \in \delta\mathcal{R}(\widetilde{\Sigma}_{a+bd}, \overline{\mathcal{S}}).$$

Then let

$$x_1 = \delta x_1^\ell(r).$$

Hence, we can choose an input, say, $\tilde{u}_2^\ell$ such that for $\tilde{x}^\ell(0) = 0$, we have $x_1^\ell(r) = -x_1$ and $x_2^\ell(r) = 0$ while $\bar{z}^\ell(k) \in \frac{1-\tilde{\rho}}{2}\bar{\mathcal{S}}$. But then for initial condition $\tilde{x}^\ell(0) \in \tilde{\rho}\mathcal{R}(\tilde{\Sigma}_{a+bd}, \bar{\mathcal{S}})$, the input $\tilde{u}_1^\ell + \tilde{u}_2^\ell$ and $\delta < \min\{\delta^*, \frac{1-\tilde{\rho}}{2}\}$, we obtain that

$$\bar{z}^\ell(k) \in (1+\delta)\tilde{\rho}\bar{\mathcal{S}} + \tfrac{1-\tilde{\rho}}{2}\bar{\mathcal{S}} \subset \bar{\mathcal{S}} \text{ for } k = 0, \dots, r-1,$$

and

$$\begin{aligned}\tilde{x}^\ell(k) &= (1-\delta)\begin{pmatrix} x_1^\ell(k) \\ x_2^\ell(k) \end{pmatrix} + \delta \begin{pmatrix} 0 \\ x_2^\ell(k) \end{pmatrix} \\ &\in \left[(1-\delta)(1+\delta)\tilde{\rho} + \delta^2\tilde{\rho}\right] \mathcal{R}(\tilde{\Sigma}_{a+bd}, \bar{\mathcal{S}}) \\ &\in \tilde{\rho}\mathcal{R}(\tilde{\Sigma}_{a+bd}, \bar{\mathcal{S}}) \text{ for } k = 0, \dots, r.\end{aligned}$$

This is true due to the fact that $\bar{\mathcal{S}}$ and $\mathcal{R}(\tilde{\Sigma}_{a+bd}, \bar{\mathcal{S}})$ are convex and contain 0 as interior point.

If we repeat this construction between $k = r$ and $k = 2r$ and so forth, it becomes clear that we can find for any initial condition,

$$\tilde{x}^\ell(0) \in \tilde{\rho}\mathcal{R}(\tilde{\Sigma}_{a+bd}, \bar{\mathcal{S}}),$$

an input such that

$$\bar{z}^\ell(k) \in \bar{\mathcal{S}}$$

for all k. Hence, $\tilde{x}^\ell(0) \in \tilde{\mathcal{R}}(\tilde{\Sigma}_{a+bd}^\ell, \bar{\mathcal{S}})$. This clearly implies that (9.25) is satisfied.

For semi-global stabilization, we take any compact set $\tilde{\mathcal{H}}_{a+bd}$ contained in the interior of $\mathcal{R}(\tilde{\Sigma}_{a+bd}, \bar{\mathcal{S}})$, and we construct a static controller which will stabilize the system and the constrained domain of attraction contains $\tilde{\mathcal{H}}_{a+bd}$. But then clearly, using (9.25), we can find ℓ such that $\tilde{\mathcal{H}}_{a+bd} \subset \tilde{\mathcal{R}}(\tilde{\Sigma}_{a+bd}^\ell, \bar{\mathcal{S}})$. Next, we choose a feedback $\tilde{f}$ on the boundary of $\tilde{\mathcal{R}}(\tilde{\Sigma}_{a+bd}^\ell, \bar{\mathcal{S}})$ such that for any $\tilde{x}^\ell(k) \in \partial\tilde{\mathcal{R}}(\tilde{\Sigma}_{a+bd}^\ell, \bar{\mathcal{S}})$, we have $\tilde{x}^\ell(k+1) \in \tilde{\mathcal{R}}(\tilde{\Sigma}_{a+bd}^\ell, \bar{\mathcal{S}})$. We expand this feedback $\tilde{f}$ to the whole state space. To do so, define $g : \mathbb{R}^n \to \mathbb{R}^+$ such that for any x,

$$g(x)x \in \partial\tilde{\mathcal{R}}(\tilde{\Sigma}_{a+bd}^\ell, \bar{\mathcal{S}}).$$

Since $\tilde{\mathcal{R}}(\tilde{\Sigma}_{a+bd}^\ell, \bar{\mathcal{S}})$ is a convex set containing 0 in its interior, this mapping is well defined. Then we expand $\tilde{f}$ to the whole state space by

$$\bar{f}(x) = \frac{\tilde{f}(g(x)x)}{g(x)}.$$

This expansion has the property that for any $\eta > 0$, we have $x^\ell_{a+bd}(k+1) \in \eta\widetilde{\mathcal{R}}(\widetilde{\Sigma}^\ell_{a+bd}, \overline{\mathcal{S}})$ for all $\widetilde{x}^\ell(k) \in \eta\widetilde{\mathcal{R}}(\widetilde{\Sigma}^\ell_{a+bd}, \overline{\mathcal{S}})$. Note that $\overline{f}$ is positively homogeneous, that is,

$$\overline{f}(\alpha x) = \alpha \overline{f}(x)$$

for any $\alpha > 0$.

Clearly, for the system $\widetilde{\Sigma}^\ell_{a+bd}$ with the feedback

$$\begin{pmatrix} u^\ell_0 \\ \widetilde{u}^\ell_d \end{pmatrix} = \overline{f}(\widetilde{x}^\ell),$$

for all initial conditions in the set $\widetilde{\mathcal{R}}(\widetilde{\Sigma}^\ell_{a+bd}, \overline{\mathcal{S}})$, we have $\widetilde{x}^\ell(k) \in \widetilde{\mathcal{R}}(\widetilde{\Sigma}^\ell_{a+bd}, \overline{\mathcal{S}})$ for all k.

But then the feedback

$$\begin{pmatrix} u_0 \\ \widetilde{u}_d \end{pmatrix} = \overline{f}(\widetilde{x})$$

for the system $\widetilde{\Sigma}_{a+bd}$ with $\widetilde{x}(0) = \widetilde{x}^\ell(0)$ results in a state

$$\widetilde{x}(k) = \frac{1}{(1+\ell)^k}\widetilde{x}^\ell(k).$$

Hence, we obviously have $\widetilde{x} \in \widetilde{\mathcal{R}}(\widetilde{\Sigma}^\ell_{a+bd}, \overline{\mathcal{S}})$ for all k but also $\widetilde{x}(k) \to 0$ as $k \to \infty$.

Finally, the controller for the original system Σ_{a+bd} is given by

$$\begin{pmatrix} u_0 \\ \widetilde{u}_d \end{pmatrix} = \overline{f}(T^{-1}\widetilde{x}) = f(x_{a+}, x_b, x_d).$$ ■

Next, we design a controller for the subsystem Σ_{abd} based on the state feedback established for Σ_{a+bd}. We have the following lemma.

Lemma 9.26 *We have*

$$\mathcal{R}(\Sigma_{abd}, \overline{\mathcal{S}}) = \mathbb{R}^{n^{0-}_a} \times \mathcal{R}(\Sigma_{a+bd}, \overline{\mathcal{S}}), \tag{9.26}$$

where $\mathcal{R}(\Sigma_{abd}, \overline{\mathcal{S}})$ and $\mathcal{R}(\Sigma_{a+bd}, \overline{\mathcal{S}})$ are the recoverable regions of Σ_{abd} and Σ_{a+bd}, respectively. Moreover, the semi-global stabilization problem for Σ_{abd} is solvable.

Proof : It is easy to verify that $\mathcal{R}(\Sigma_{abd}, \overline{\mathcal{S}})$ and $\mathcal{R}(\Sigma_{a+bd}, \overline{\mathcal{S}})$ have the relationship,

$$\mathcal{R}_{abd}(\Sigma, \overline{\mathcal{S}}) \subseteq \mathbb{R}^{n^{0-}_a} \times \mathcal{R}(\Sigma_{a+bd}, \overline{\mathcal{S}}).$$

We will prove the reverse implication and establish the solvability of the stabilization problem for system Σ_{abd} by constructing a state feedback controller for it.

For any compact set $\mathcal{H}$ in $\mathbb{R}^{n_a^{0-}} \times \mathcal{R}(\Sigma_{a+bd}, \overline{\mathcal{S}})$, we choose a compact set $\mathcal{H}_1$ and $\widetilde{\rho} < 1$ such that

$$\mathcal{H} \subset \mathcal{H}_1 \times \widetilde{\rho}\mathcal{R}(\Sigma_{a+bd}, \overline{\mathcal{S}}).$$

The controllers $u_0 = f_1(x_{a+bd})$ and $\widetilde{u}_d = f_2(x_{a+bd})$ are such that, for all initial conditions in $\widetilde{\rho}\mathcal{R}(\Sigma_{a+bd}, \overline{\mathcal{S}})$, the origin of the closed-loop system is exponentially stable. Hence, there exist a $M > 0$ and a λ with $|\lambda| < 1$ such that

$$\|x_{a+bd}(k)\| \leqslant M\lambda^k \tag{9.27}$$

for all k and for all $x_{a+bd}(0) \in \widetilde{\rho}\mathcal{R}(\Sigma_{a+bd}, \overline{\mathcal{S}})$.

Next, let P_0 be the semi-stabilizing solution of the discrete-time algebraic Riccati equation,

$$P_0 = A_0' P_0 A_0 + C_0' C_0 - A_0' P_0 B_0 (B_0' P_0 B_0 + D_0' D_0)^{\dagger} B_0' P_0 A_0,$$

where

$$A_0 = \begin{pmatrix} A_{aa} & K_{ab}C_b & K_{ad}C_d \\ 0 & A_{bb} + K_{bb}C_b & K_{bd}C_d \\ 0 & K_{db}C_b & A_{dd} \end{pmatrix}, \quad B_0 = \begin{pmatrix} B_{a0} & 0 \\ B_{b0} & 0 \\ B_{d0} & B_d \end{pmatrix},$$

$$C_0 = \begin{pmatrix} 0 & C_b & 0 \\ 0 & 0 & C_d \\ 0 & 0 & 0 \end{pmatrix}, \quad \text{and} \quad D_0 = \begin{pmatrix} 0 & 0 \\ 0 & 0 \\ I & 0 \end{pmatrix}.$$

We have

$$P_0 \begin{pmatrix} x_a^{0-} \\ 0 \\ 0 \end{pmatrix} = 0 \tag{9.28}$$

for all $x_a^{0-} \in \mathbb{R}^{n_a^{0-}}$ since the eigenvalues of A_{aa}^{0-} are in the closed unit disc. Choose a level set,

$$V_0(c) := \{\xi \in \mathbb{R}^{n_{abd}} \mid \xi' P_0 \xi \leqslant c\}$$

such that we have

$$\left(C_0 + D_0(B_0' P_0 B_0 + D_0' D_0)^{\dagger} B_0' P_0 A_0\right)\xi \in \overline{\mathcal{S}}/3 \tag{9.29}$$

for all $\xi \in V_0(c)$. Then, with the controllers

$$u_0 = f_1(x_{a+bd}) \text{ and } \widetilde{u}_d = f_2(x_{a+bd}),$$

there exists a T such that, for any initial state in

$$\mathcal{H}_1 \times \tilde{\rho}\mathcal{R}(\Sigma_{a+bd}, \bar{\mathcal{S}}),$$

we have

$$\begin{pmatrix} x_{a^{0-}}(T) \\ x_{a+bd}(T) \end{pmatrix} \in V_0(c). \tag{9.30}$$

Let P_ε be the stabilizing solution of the algebraic equation

$$P_\varepsilon = A_0' P_\varepsilon A_0 + C_0' C_0 + \varepsilon I - A_0' P_\varepsilon B_0 (B_0' P_\varepsilon B_0 + D_0' D_0)^\dagger B_0' P_\varepsilon A_0.$$

We have $P_\varepsilon \to P_0$ as ε approaches zero. Define the level set

$$V_\varepsilon(c) := \{\, \xi \in R^{n_{abd}} \mid \xi' P_\varepsilon \xi \leqslant c \,\}.$$

Then, there exists an ε such that

$$\begin{pmatrix} x_{a^{0-}}(T) \\ x_{a+bd}(T) \end{pmatrix} \in 2V_\varepsilon(c).$$

For ε small enough, we have

$$\left[C_0 - D_0 (B_0' P_\varepsilon B_0 + D_0' D_0)^\dagger B_0' P_\varepsilon A_0 \right] \xi \in \bar{\mathcal{S}}$$

for any $\xi \in 2V_\varepsilon(c)$. Hence, the feedback

$$\begin{pmatrix} u_0 \\ \tilde{u}_d \end{pmatrix} = -(B_0' P_\varepsilon B_0 + D_0' D_0)^\dagger B_0' P_\varepsilon A_0 \begin{pmatrix} x_a^{0-} \\ x_{a+bd} \end{pmatrix}$$

is an asymptotically stabilizing controller for Σ_{abd} and achieves a domain of attraction containing $2V_\varepsilon(c)$. Next, consider the controller

$$\begin{pmatrix} u_0 \\ \tilde{u}_d \end{pmatrix} = \begin{cases} f(x_{a+bd}), & x_{abd} \notin 2V_\varepsilon(c) \\ -(B_0' P_\varepsilon B_0 + D_0' D_0)^\dagger B_0' P_\varepsilon A_0 x_{abd}, & x_{abd} \in 2V_\varepsilon(c). \end{cases}$$

It is easily verified that this controller asymptotically stabilizes the system. Hence we have

$$\mathbb{R}^{n_a^{0-}} \times \mathcal{R}(\Sigma_{a+bd}, \bar{\mathcal{S}}) \subset \mathcal{R}(\Sigma_{abd}, \bar{\mathcal{S}}). \qquad \blacksquare$$

The above lemma yields an appropriate controller for the subsystem Σ_{abd}. Finally, we need to construct a controller for the original system Σ which will complete our proof of Theorem 9.24 and will also complete our proof of Theorem 9.15.

Proof of Theorem 9.24 : After transforming the system into the form (9.1), it is easily seen that the controllers designed in Lemma 9.26 combined with a controller

$$u_c(k) = F_c x_c(k)$$

solve the semi-global constraint stabilization problem in recoverable region via state feedback, where F_c is such that $A_{cc} + B_c F_c$ is asymptotically stable. ■

Let us next comment on measurement feedback controllers. For discrete-time system with measurement feedback, it is not sensible to consider the recoverable region. After all, the recoverable region is intrinsically an open-loop concept relying on our knowledge of the state which, in the measurement feedback case, is clearly not available. Moreover, in the discrete time, in contrast with the continuous time, semi-global stabilization with measurement feedback is in general not possible. In continuous time a fast observer could guarantee a highly accurate estimate of the state in a short period of time in which we do not leave the recoverable region. However, in discrete time, it might take up to n time steps before we get a good estimate of the state, and in this period of time, we might leave the recoverable region.

An alternative is to use a concept such as maximum domain of attraction. Basically, we look for a measurement feedback controller with the largest constrained domain of attraction, i.e., the largest set of initial conditions for which we can guarantee convergence to the origin without constraint violation. However, this concept is also problematic in discrete-time systems. If two controllers achieve constrained domain of attractions $\mathcal{R}_1$ and $\mathcal{R}_2$ respectively, then there might not exist a controller for which $\mathcal{R}_1 \cup \mathcal{R}_2$ is contained in its constrained domain of attraction. This is established in the following example. The fact that this is not possible makes it impossible to decide which measurement feedback controller we should use.

Example 9.27 We consider the system

$$\Sigma_l : \begin{cases} x(k+1) = \begin{pmatrix} 0 & 1 \\ 0 & 2 \end{pmatrix} x(k) + \begin{pmatrix} 0 \\ 1 \end{pmatrix} u(k) \\ y(k) \quad = \begin{pmatrix} 1 & 0 \end{pmatrix} x(k) \\ z(k) \quad = x(k), \end{cases}$$

with a constraint set $\mathcal{S} = \{x \in \mathbb{R}^2 \mid x \in [-1,1] \times [-1,1]\}$. Note that there is one step delay from the input to the output. Consider the time $k = 0$. Suppose $x(0) \in [-1,1] \times \left[-\frac{3}{4}, \frac{1}{4}\right]$, we can choose $u(0) = \frac{1}{2}$ so that no constraint violation occurs at $k = 1$. Similarly, if $x(0) \in [-1,1] \times \left[-\frac{1}{4}, \frac{3}{4}\right]$, we can choose $u(0) = -\frac{1}{2}$ to avoid constraint violation. However, if $x(0) \in [-1,1] \times \left[-\frac{3}{4}, \frac{3}{4}\right]$ which is the union of these two regions, it is impossible to find a $u(0)$ that guarantees no constraint violation. After all, the measurement at time 0 does not yield any information about $x_2(0)$, and hence, we can never guarantee that the state $x_2(1)$ will be in $[-1, 1]$. Hence, we cannot avoid constraint violation.

9.A Proof of Lemma 9.6

We start with showing property (*i*). Since $0 \in \operatorname{int} \mathcal{S}$ and the system is linear and controllable, there exists a ball $\mathcal{B}(0, \varepsilon)$ around the origin with radius ε and time $T > 0$ such that for any $x(0) \in \mathcal{B}(0, \varepsilon)$, there exists a control u which steers the state to the origin in time T without violating the constraint. By definition, for any $x(0) \in \mathcal{R}(\Sigma, \mathcal{S})$, there exists an input u such that $x(t) \to 0$ as $t \to \infty$ while satisfying the constraints. Hence, there exists a time $T_1 > 0$ so that $x(t) \in \mathcal{B}(0, \varepsilon)$ for $t \geqslant T_1$. Therefore, it is possible to drive any initial state in $\mathcal{R}(\Sigma, \mathcal{S})$ to the origin in time $T + T_1$.

Property (*ii*) follows from the assumption that $\mathcal{S}$ is convex and $0 \in \mathcal{S}$.

To show property (*iii*), we note that already in [66] it was established that in the case of only input constraints, the recoverable set is open. In the case of general state and input constraints, the set $\mathcal{R}(\Sigma, \mathcal{S})$ need not be open. This is seen from the simple example $\dot{x} = u$ with $z = x$ and constraint set $\mathcal{S} = \{z \mid z \in [-1, 1]\}$ which yields $\mathcal{R}(\Sigma, \mathcal{S}) = \mathcal{S}$ which is obviously closed.

Finally, we consider property (*iv*). Under the conditions that the system Σ has a relative degree at most, one is left invertible, and with all invariant zeros antistable, the system Σ in the SCB takes the following form:

$$\begin{cases} \dot{x}_a = A_{aa}x_a + K_{ab}z_b + K_{a2}\zeta, \\ \dot{x}_b = A_{bb}x_b + K_{bb}z_b + K_{b2}\zeta, \\ \dot{x}_d = A_{dd}x_d + B_d u_d + K_{dd}z_b + K_{d2}\zeta, \\ \bar{z} = \begin{pmatrix} z_b \\ \zeta \end{pmatrix} = \begin{pmatrix} C_b x_b \\ u_0 \\ x_d \end{pmatrix}. \end{cases} \tag{9.31}$$

Firstly, since $\mathcal{S}$ is bounded, we find that $x_d(t)$ must be bounded. Secondly, the x_a^+ dynamics is antistable and controlled by the virtual input z_b and ζ which are bounded. It is a classical result that the recoverable region for this subsystem must be bounded. It remains to show that the recoverable region for $x_b(t)$ is also bounded. Consider the following subsystem:

$$\begin{aligned} \dot{x}_b &= (A_{bb} + K_{bb}C_b)x_b + K_{b2}\zeta, \\ z_b &= C_b x_b. \end{aligned}$$

It is known that (C_b, A_{bb}) is observable. Clearly, $(C_b, A_{bb} + K_{bb}C_b)$ is also observable. This system has input ζ and output z_b. Since both input ζ and output z_b are bounded, we can conclude that x_b is bounded.

9.B Proof of Theorem 9.13

We note that Theorem 9.10 helps us to relate the recoverable regions of $\bar{\Sigma}_1$ and Σ and we obtain (9.9). In order to obtain (9.8), we need to do a bit more work. Our proof is strongly motivated by the results of [66]. One inclusion is basically obvious:

$$\mathcal{R}(\Sigma_1, \bar{\mathcal{S}}) \subset \mathbb{R}^{n_a^{-0}} \times \mathcal{R}(\bar{\Sigma}_1, \bar{\mathcal{S}}). \tag{9.32}$$

For notational ease, we will denote an initial condition of Σ_1 by (x_1, x_2) with x_1 equal to the initial condition for x_a^{-0} and x_2 a vector consisting of initial conditions for x_a^+ and x_b.

We first note that for any $x_2 \in \mathcal{R}(\bar{\Sigma}_1, \bar{\mathcal{S}})$, we can find x_1 such that $(x_1, x_2) \in \mathcal{R}(\Sigma_1, \bar{\mathcal{S}})$. After all, if we choose an input u for $\bar{\Sigma}_1$ which steers x_2 to 0 at time T without violating constraints, then for this same input u, we can always choose x_1 such that for the initial condition (x_1, x_2), the system Σ_1 reaches the origin at time T. Moreover, since x_1 does not affect the constraints, the initial condition (x_1, x_2) is steered to zero without constraint violations.

Next, we note that for any $x_1 \in \mathbb{R}^{n_a^{-0}}$ we have $(x_1, 0) \in \mathcal{R}(\Sigma_1, \bar{\mathcal{S}})$. It is well known that we can locally stabilize a system using a linear feedback $u = -B'Px$ with P a solution of an algebraic Riccati equation and such that an ellipsoid of the form $x'Px \leqslant c$ is invariant for the closed-loop system while constraint violations are avoided. If we apply this to the system Σ_1, we find that P restricted to the part of the system composed of x_1 can be made arbitrarily small and this yields that we can guarantee that for any x_1, there exists a solution of the Riccati equation P such that $(x_1, 0)$ is contained in this invariant ellipsoidal set and for which we can hence avoid constraint violation. This clearly implies that $(x_1, 0) \in \mathcal{R}(\Sigma_1, \bar{\mathcal{S}})$. For further details regarding this type of arguments, we refer to [49].

We claim that for all (x_1, x_2) with $x_2 \in \mathcal{R}(\bar{\Sigma}_1, \bar{\mathcal{S}})$, we have $(x_1, x_2) \in \mathcal{R}(\Sigma_1, \bar{\mathcal{S}})$. In other words,

$$\mathcal{R}(\Sigma_1, \bar{\mathcal{S}}) \supset \mathbb{R}^{n_a^{-0}} \times \mathcal{R}(\bar{\Sigma}_1, \bar{\mathcal{S}}),$$

and combined with (9.32) the proof of (9.8) would be complete.

Let $\varepsilon > 0$ be given. Choose any (x_1, x_2) with $x_2 \in \mathcal{R}(\bar{\Sigma}_1, \bar{\mathcal{S}})$. We know that there exists a $\tilde{x}_1$ such that $(\tilde{x}_1, x_2)$ is in $\mathcal{R}(\Sigma_1, \bar{\mathcal{S}})$.

Choose $\lambda \in (0,1)$. We have

$$\left(\tfrac{x_1 - \lambda \tilde{x}_1}{1-\lambda}, 0 \right) \in \mathcal{R}(\Sigma_1, \bar{\mathcal{S}}).$$

This implies that

$$(x_1, x_2) = \lambda(\tilde{x}_1, x_2) + (1-\lambda) \left(\tfrac{x_1 - \lambda \tilde{x}_1}{1-\lambda}, 0 \right),$$

is an element of the set $\mathcal{R}(\Sigma_1, \bar{\mathcal{S}})$ due to convexity.

9.C Proof of Lemma 9.20

The proof of Lemma 9.20 requires the following lemma which can be proven easily.

Lemma 9.28 *Consider the system Σ in (7.1) and a convex, compact constraint set $\mathcal{S}$ containing 0 in the interior. Assume that (A, B) is stabilizable. For any compact subset $\mathcal{W} \subset \operatorname{int} \mathcal{R}(\Sigma, \mathcal{S})$ and any ball $\mathcal{B}(0, \varepsilon) \subset \mathcal{W}$ with $\varepsilon > 0$ sufficiently small, there exists a time $T > 0$ such that for any initial condition in $\mathcal{W}$, there exists an input u for which $x(T) \in \mathcal{B}(0, \varepsilon)$ and $z(t) \in \mathcal{S}$, $\forall\, t \in [0, T]$.*

Proof of Lemma 9.20 : We first note that for any $\kappa > 0$, we have

$$\mathcal{R}(\Sigma^\kappa, \mathcal{S}) \subset \mathcal{R}(\Sigma, \mathcal{S}). \tag{9.33}$$

This follows from a simple observation that if $x(t)$ and $u(t)$ satisfy system Σ^κ with initial condition $x(0)$ and the constraint, then $e^{-\kappa t} x(t)$ and $e^{-\kappa t} u(t)$ satisfy system Σ with the same initial condition and the constraint. Also, it is clear from the same observation that

$$\mathcal{R}(\Sigma^\ell, \mathcal{S}) \subset \mathcal{R}(\Sigma^\kappa, \mathcal{S}), \tag{9.34}$$

for $\ell > \kappa \geqslant 0$.

Next, we show that for any compact set $\mathcal{W}$ satisfying

$$\mathcal{W} \subset \tilde{\alpha} \mathcal{R}(\Sigma, \mathcal{S}), \qquad \tilde{\alpha} \in (0, 1),$$

there exists a $\kappa > 0$ such that

$$\mathcal{W} \subset \mathcal{R}(\Sigma^\kappa, \mathcal{S}). \tag{9.35}$$

Note that if (A, B) is stabilizable, then there exists a sufficiently small $\kappa_0 > 0$ such that for all $0 \leqslant \kappa \leqslant \kappa_0$ the pair (A_κ, B) is also stabilizable. Also note that by stabilizability, there exists an $\varepsilon > 0$ sufficiently small such that one can find for any point in $\mathcal{B}(0, \varepsilon)$, a control u such that the resulting trajectory goes to zero asymptotically without violating the constraints. We choose such a small $\varepsilon > 0$ for which this property holds for the system Σ^{κ_0}.

Next, we consider the system Σ. By Lemma 9.28, there exists a uniform $T > 0$ such that any initial state in $\mathcal{W}$ can be driven to the ball $\mathcal{B}(0, \tilde{\alpha}\varepsilon)$ in time T by a suitable u while respecting the constraints. Let $\tilde{x}_0 \in \mathcal{W} \subset \tilde{\alpha} \mathcal{R}(\Sigma, \mathcal{S})$. Then, there exist $\breve{x}(t)$ and $\breve{u}(t)$ satisfying

$$\begin{cases} \dot{\breve{x}}(t) = A\breve{x}(t) + B\breve{u}(t), \quad \breve{x}(0) = \tilde{x}_0, \\ \breve{z}(t) = C_z\breve{x}(t) + D_z\breve{u}(t) \in \tilde{\alpha}\mathcal{S}, \end{cases}$$

for $t \in [0, T]$ and $\breve{x}(T) \in \mathcal{B}(0, \tilde{\alpha}\varepsilon)$, where T does not depend on $\tilde{x}_0$. By choosing $\kappa > 0$ small enough so that $\tilde{\alpha} e^{\kappa T} < 1$, it is straightforward that $\tilde{x}(t) = e^{\kappa t}\breve{x}(t)$ and $\tilde{u}(t) = e^{\kappa t}\breve{u}(t)$ satisfy for $t \in [0, T]$,

$$\begin{cases} \dot{\tilde{x}}(t) = (A + \kappa I)\tilde{x}(t) + B\tilde{u}(t), & \tilde{x}(0) = \tilde{x}_0, \\ \tilde{z}(t) = C_z\tilde{x}(t) + D_z\tilde{u}(t) \in \mathcal{S}, \end{cases}$$

while $\tilde{x}(T) = x(T) \in \mathcal{B}(0, \varepsilon)$. Hence, $\tilde{x}_0 \in \mathcal{R}(\Sigma^\kappa, \mathcal{S})$. ■

9.D Proof of Lemma 9.21

In general, the recoverable region is not closed, and its closure is not easily characterized. However, because of the special structure of the system (9.18), the closure of the recoverable region, $\overline{\mathcal{R}(\Sigma_0, \mathcal{S})}$, can easily be characterized by the set of initial conditions of the system (9.18) for which there exists an input that keeps the state bounded while avoiding constraint violation. This result is stated in the following lemma, which leads to the proof of Lemma 9.21. Similar results have, for instance, been obtained in [66].

Lemma 9.29 *Consider the system Σ_0 in (9.18) with compact, convex constraint set $\mathcal{S}$ containing 0 in the interior. Assume that the unobservable modes of the pair (C_0, A_0) are antistable and the pair (A_0, B_0) is stabilizable. Then we have*

$$\overline{\mathcal{R}(\Sigma_0, \mathcal{S})} = \Omega(\Sigma_0, \mathcal{S}),$$

where

$$\Omega(\Sigma_0, \mathcal{S}) := \{\, \xi_0 \in \mathbb{R}^n \mid \exists \zeta \text{ such that the solution of } \Sigma_0 \text{ with } \xi(0) = \xi_0 \text{ satisfies } \xi \in L_\infty \text{ and } \overline{z}(t) \in \mathcal{S},\ \forall t \geqslant 0 \}.$$

Proof : We first show that $\Omega(\Sigma_0, \mathcal{S})$ is closed. We know that the unobservable modes of (C_0, A_0) are antistable. This implies that there exists a matrix K_0 such that all the eigenvalues of $A_0 - K_0C_0$ are in the open right half plane. Then for any $\xi_0 \in \Omega(\Sigma_0, \mathcal{S})$, we can write

$$e^{-(A_0 - K_0C_0)t}\xi(t) = \xi_0 + \int_0^t e^{-(A_0 - K_0C_0)\tau}[B_0\zeta(\tau) + K_0\overline{z}(\tau)]\mathrm{d}\tau.$$

Since ξ is bounded, we can take the limit and we obtain

$$\xi_0 = -\int_0^{\infty} e^{-(A_0 - K_0 C_0)\tau}[B_0 \zeta(\tau) + K_0 \overline{z}(\tau)]\mathrm{d}\tau.$$

This can be interpreted as the state at time 0, which starts from $\xi(-\infty) = 0$ under the control $\zeta(-t)$ while $\overline{z}(-t) \in \mathcal{S}$ for all $t \in (-\infty, 0]$.

Let $\{\xi_n\} \in \Omega(\Sigma_0, \mathcal{S})$ be a sequence such that $\lim_{n\to\infty} \xi_n = \overline{\xi}$. Then, for each ξ_n, there exists an associated ζ_n such that $\xi_n \in L_\infty$ and $\overline{z}_n(t) \in \mathcal{S}$ for all $t \geqslant 0$, where $\overline{z}_n$ is the associated constrained output of system Σ_0^- corresponding to the state ξ_n at time 0. Clearly,

$$\xi_n = -\int_0^{\infty} e^{(A_0 - K_0 C_0)\tau}[B_0 \zeta_n(\tau) + K_0 \overline{z}_n(\tau)]\mathrm{d}\tau.$$

Since both $\{\zeta_n\}$ and $\{\overline{z}_n\}$ are bounded subsets in $L_\infty = L_1^*$, the conjugate space of L_1, according to Alaoglu's theorem [3], there exist two subsequences ζ_{n_m} and $\overline{z}_{n_m}$ which converge in the *weak** sense to $\overline{\zeta}$ and $\overline{\overline{z}}$, respectively. Thus,

$$\overline{\xi} = -\int_0^{\infty} e^{(A_0 - K_0 C_0)\tau}[B_0 \overline{\zeta}(\tau) + K_0 \overline{\overline{z}}(\tau)]\mathrm{d}\tau. \tag{9.36}$$

Next, we show that the state $\xi(0) = \overline{\xi}$ and the input $\overline{\zeta}$ in fact determine the output $\overline{\overline{z}}$ in the *weak** sense. Since for any $t > 0$

$$\overline{z}_{n_m}(t) = C_0 e^{A_0 t} \xi_{n_m} + \int_0^{t} C_0 e^{A_0 (t-\tau)} B_0 \zeta_{n_m}(\tau)\mathrm{d}\tau,$$

where ξ_{n_m} converges to $\overline{\xi}$ and ζ_{n_m} converges to $\overline{\zeta}$ in the *weak** sense, the sequence $\overline{z}_{n_m}$ is pointwise convergent with a limit, say, $\tilde{\overline{z}}$. Since $\overline{z}_{n_m}$ is a bounded function for all n_m, according to Lebesgue's dominated convergence theorem, $\overline{z}_{n_m}$ converges to $\tilde{\overline{z}}$ in the *weak** sense. But we already knew that $\overline{z}_{n_m}$ converges in the *weak** sense to $\overline{\overline{z}}$. By the uniqueness of the *weak** limit, we obtain $\tilde{\overline{z}} = \overline{\overline{z}}$ and

$$\overline{\overline{z}}(t) = C_0 e^{A_0 t} \overline{\xi} + \int_0^{t} C_0 e^{A_0 (t-\tau)} B_0 \overline{\zeta}(\tau)\mathrm{d}\tau \tag{9.37}$$

for $t \geqslant 0$. Combining (9.36) and (9.37) shows that $\overline{\xi} \in \Omega(\Sigma_0, \mathcal{S})$ and hence $\Omega(\Sigma_0, \mathcal{S})$ is closed.

Clearly, for any $\xi_0 \in \mathcal{R}(\Sigma_0, \mathcal{S})$, there exist ζ and $T > 0$ such that given the system (9.18) with $\xi(0) = \xi_0$, we have $\xi(T) = 0$. Then, by defining $\zeta(t) = 0$

for $t > T$, it is trivial to see that $\xi_0 \in \Omega(\Sigma_0, \mathcal{S})$. This implies that $\mathcal{R}(\Sigma_0, \mathcal{S}) \subset \Omega(\Sigma_0, \mathcal{S})$. Therefore, we obtain that $\overline{\mathcal{R}}(\Sigma_0, \mathcal{S}) \subset \Omega(\Sigma_0, \mathcal{S})$. It remains to show that $\overline{\mathcal{R}}(\Sigma_0, \mathcal{S}) = \Omega(\Sigma_0, \mathcal{S})$.

Similar to part (*iv*) of Lemma 9.6, it is easy to verify that $\Omega(\Sigma_0, \mathcal{S})$ is bounded. Next, we show that given any $\xi_0 \in \tilde{\rho}\Omega(\Sigma_0, \mathcal{S})$ for $\tilde{\rho} \in (0, 1)$, there exists an input ζ such that it drives the state of the system Σ_0 with initial condition ξ_0 to zero without violating the constraint; that means $\xi_0 \in \mathcal{R}(\Sigma_0, \mathcal{S})$.

Note that the set $\Omega(\Sigma_0, \mathcal{S})$ is bounded since the invariant zeros are antistable and the system is left invertible. We decompose the system in a controllable and an uncontrollable part, i.e., $\xi(t) = (\xi_1(t), \xi_2(t))$ with ξ_1 controllable. Assume that we have an input ζ such that $\bar{z}(t) \in \mathcal{S}$ for all $t \geqslant 0$ with a bounded state ξ. Obviously, the uncontrollable part must be stable and hence $\xi_2(t) \to 0$ as $t \to \infty$. For any $\delta \in (0, 1)$, choose $T > 0$ such that $(0, \xi_2(t)) \in \delta\Omega(\Sigma_0, \mathcal{S})$ for all $t > T$ and any initial condition in $\Omega(\Sigma_0, \mathcal{S})$. Choose $\varepsilon > 0$ such that given any state of the form $(\xi_{1,0}, \xi_{2,0})$ in the set $\varepsilon\Omega(\Sigma_0, \mathcal{S})$, we can, starting at the origin at time 0, reach $(-\xi_{1,0}, 0)$ at time T_1 while guaranteeing $\bar{z}(t) \in (1 - \tilde{\rho})\mathcal{S}$ for all $t \in [0, T_1]$.

Choose any initial condition in $\tilde{\rho}\Omega(\Sigma_0, \mathcal{S})$. We have an input $\bar{\zeta}$ such that the state remains in $\tilde{\rho}\Omega(\Sigma_0, \mathcal{S})$ while $\bar{z}(t) \in \tilde{\rho}\mathcal{S}$ for all $t > 0$. Assume that we reach the state $(\bar{\xi}_1(T_1), \bar{\xi}_2(T_1))$ at time T_1 with this input. Clearly, $(\bar{\xi}_1(T_1), \bar{\xi}_2(T_1)) \in \tilde{\rho}\Omega(\Sigma_0, \mathcal{S})$.

Next, choose an input $\tilde{\zeta}$ such that we reach, starting at the origin, the state $(-\varepsilon\bar{\xi}_1(T_1), 0)$ while guaranteeing $\bar{z}(t) \in (1 - \tilde{\rho})\mathcal{S}$ for all $t \in [0, T_1]$. Then, the input $\bar{\zeta} + \tilde{\zeta}$ yields a state

$$((1-\varepsilon)\bar{\xi}_1(T_1), \bar{\xi}_2(T_1)) \in (1-\varepsilon)\tilde{\rho}\Omega(\Sigma_0, \mathcal{S}) + \delta\varepsilon\Omega(\Sigma_0, \mathcal{S}) \subset (1-\varepsilon/2)\tilde{\rho}\Omega(\Sigma_0, \mathcal{S})$$

when we choose δ small enough.

Putting together, we find that for initial condition $\xi_0 \in \tilde{\rho}\Omega(\Sigma_0, \mathcal{S})$, the system Σ_0 with the input $\zeta = \bar{\zeta} + \tilde{\zeta}$ yields that $\xi(T_1) \in (1 - \varepsilon/2)\tilde{\rho}\Omega(\Sigma_0, \mathcal{S})$ and the constraints are satisfied for $t \in [0, T_1]$. Repeatedly applying this argument, yields that $\xi(kT_1) \in (1 - \varepsilon/2)^k\tilde{\rho}\Omega(\Sigma_0, \mathcal{S})$ for $\xi_0 \in \tilde{\rho}\Omega(\Sigma_0, \mathcal{S})$ while the constraint is satisfied for all $t \in [0, kT_1]$ for any $k \in \mathbb{N}$. Since the state clearly converges to zero, we have shown that $\xi_0 \in \mathcal{R}(\Sigma_0, \mathcal{S})$. Consequently, $\overline{\mathcal{R}}(\Sigma_0, \mathcal{S}) = \Omega(\Sigma_0, \mathcal{S})$. ■

We now proceed to prove Lemma 9.21.

Proof of Lemma 9.21 : Property (*i*) is a direct consequence of property (*iv*) of Lemma 9.6 applied to the system (9.18). Property (*ii*) is a direct consequence of Lemma 9.29. ■

9.E Proof of Theorem 9.22

Some preliminary work is needed before we proceed to the proof of this theorem. Define

$$\mathcal{C}_0(\kappa) := \mathcal{R}(\Sigma_0^\kappa, \mathcal{S}).$$

From Lemmas 9.6 and 9.21, we know that $\mathcal{C}_0(\kappa)$ is convex and bounded. Also, by Lemma 9.20, for any compact subset $\mathcal{W} \subset \operatorname{int} \mathcal{R}(\Sigma_0, \mathcal{S})$, there exists an $\ell > \kappa > 0$ such that

$$\mathcal{W} \subset \mathcal{C}_0(\ell) \subset \mathcal{C}_0(\kappa) \subset \mathcal{R}(\Sigma_0, \mathcal{S}). \tag{9.38}$$

Next, we show the following:

(*i*) There exists a continuous feedback for the system (9.19) so that the closure of $\mathcal{C}_0(\ell)$ is an invariant set.

(*ii*) This feedback can be slightly modified to be Lipschitz, and when applied to the original system (9.18), $\overline{\mathcal{C}}_0(\ell)$ is again an invariant set while the state of the system with any initial condition in $\overline{\mathcal{C}}_0(\ell)$ will converge to the origin.

From these, we conclude that $\overline{\mathcal{C}}_0(\ell)$, and hence $\mathcal{W}$, is contained in the domain of attraction.

As stated before, we will try to achieve this by trying to guarantee that the trajectory points inward or tangent to this set in every boundary point by an appropriate choice of the input. In order to formalize this, we need the following set:

$$N_{\mathcal{V}}(\xi) := \{\, \eta \in \mathbb{R}^n \mid \|\eta\| = 1 \text{ and } \langle \xi' - \xi, \eta \rangle \leqslant 0,\ \forall\, \xi' \in \mathcal{V} \,\}.$$

Note that $N_{\mathcal{V}}(\xi)$ is the set of normals in the point ξ to the set $\mathcal{V}$ (as studied in, for instance, [119]). It is also shown in [119] that for a convex set $\mathcal{V}$, the set of normals is nonempty whenever ξ is a boundary point of $\mathcal{V}$. Its importance is due to the fact that if we start in ξ in the direction v, then this direction is tangent to or pointing inside $\mathcal{V}$ if and only if $\langle v, \eta \rangle \leqslant 0$ for all $\eta \in N_{\mathcal{V}}(\xi)$.

Let the relation (9.38) hold for $\ell > \kappa > 0$. Define $T_\kappa : \partial\mathcal{C}_0(\ell) \to \mathcal{P}(\mathbb{R}^n)$, where $\mathcal{P}(\mathbb{R}^n)$ denotes the collection of all subsets of $\mathbb{R}^n$, by

$$T_\kappa(\xi) := \left\{ A_\kappa \xi + B_0 \zeta \;\middle|\; \begin{pmatrix} C_0 \xi \\ \zeta \end{pmatrix} \in \mathcal{S} \text{ and } \langle A_\kappa \xi + B_0 \zeta, \eta \rangle \leqslant 0,\ \forall\, \eta \in N_{\mathcal{C}_0(\ell)}(\xi) \right\}.$$

The next lemma states some properties of $T_\kappa(\xi)$.

Lemma 9.30 *Assume that $\kappa < \ell$. Then we have:*

(*i*) *$T_\kappa(\xi)$ is convex and closed for every $\xi \in \partial\mathcal{C}_0(\ell)$.*

(*ii*) *For any point $\xi \in \partial\mathcal{C}_0(\ell)$, the sets $T_\kappa(\xi)$ and $T_\ell(\xi)$ are nonempty.*

Proof : (*i*): The convexity of $T_\kappa(\xi_0)$ is obvious by definition. The closedness is proven as follows. Let $\{A_\kappa\xi_0 + B_0\zeta_n\}_{n\in\mathbb{N}}$ be a convergent sequence in $T_\kappa(\xi_0)$ with limit p. For every $\eta \in N_\ell(\xi_0)$, we have $\langle A_\kappa\xi_0 + B_0\zeta_n, \eta\rangle \leqslant 0$ for all n. So $\langle p, \eta\rangle \leqslant 0$ for every $\eta \in N_\ell(\xi_0)$. Since $\{\zeta_n\}$ is in a compact set, it has a convergent subsequence $\{\zeta_{n_k}\}$ with a limit, say, ζ_0. Because

$$p = \lim_{k\to\infty} A_\kappa\xi_0 + B_0\zeta_{n_k} = A_\kappa\xi_0 + B_0\zeta_0 \leqslant 0$$

and $\langle p, \eta\rangle \leqslant 0$ for every $\eta \in N_\ell(\xi_0)$, we conclude that $p \in T_\kappa(\xi_0)$.

(*ii*): Since $0 \in \mathcal{C}_0(\ell)$, we have $\langle \xi, \eta\rangle \geqslant 0$ for any $\xi \in \partial\mathcal{C}_0(\ell)$ and any $\eta \in N_\ell(\xi)$. Hence, if $\langle A_\ell\xi + B_0\zeta, \eta\rangle \leqslant 0$, we get

$$\langle A_\kappa\xi + B_0\zeta, \eta\rangle = \langle A_\ell\xi + B_0\zeta, \eta\rangle + (\kappa - \ell)\langle \xi, \eta\rangle \leqslant 0$$

for $\kappa < \ell$ and any $\xi \in \partial\mathcal{C}_0(\ell)$, $\eta \in N_\ell(\xi)$. Therefore, if $T_\ell(\xi_0)$ is nonempty, so is the set $T_\kappa(\xi_0)$.

Assume that on the contrary, there exists a $\xi_0 \in \partial C_0(\ell)$ such that $T_\ell(\xi_0)$ is empty. Define

$$r(\xi_0) = \min_{\zeta} \left\{ \max_{\eta\in N_\ell(\xi_0)} \langle A_\ell\xi_0 + B_0\zeta, \eta\rangle \,\middle|\, \begin{pmatrix} C_0\xi_0 \\ \zeta \end{pmatrix} \in \mathcal{S} \right\}.$$

Since $N_\ell(\xi_0)$ is compact, the maximum exists. To justify the existence of a minimum, we show that the maximum exists and depends continuously on ζ. Let $m > 0$ be such that $\|B_0^{\mathrm{T}}\eta\| \leqslant m$ for all $\eta \in N_\ell(\xi_0)$. Then,

$$\begin{aligned} \left| \max_{\eta\in N_\ell(\xi_0)} \langle A_\ell\xi_0 + B_0\tilde{\zeta}, \eta\rangle - \max_{\eta\in N_\ell(\xi_0)} \langle A_\ell\xi_0 + B_0\zeta, \eta\rangle \right| &\leqslant \max_{\eta\in N_\ell(\xi_0)} \langle B(\tilde{\zeta} - \zeta), \eta\rangle \\ &\leqslant m\|\tilde{\zeta} - \zeta\|. \end{aligned}$$

Therefore, we are looking for the minimum of a continuous function over a compact set, which clearly always exists. Thus, $r(\xi_0)$ is well defined. If $r(\xi_0) \leqslant 0$, then the minimizing ζ^* would make $T_\ell(\xi_0)$ nonempty; hence, we have nothing to prove. Therefore, we focus on the case $r(\xi_0) > 0$. By property (*ii*) of Lemma 9.21, for $\xi(0) = \xi_0$ there exists an input ζ such that $\xi(t) \in \overline{\mathcal{C}}_0(\ell)$ for $t \in [0, \delta]$. This implies that

$$\max_{\eta\in N_\ell(\xi_0)} \langle \xi(t) - \xi(0), \eta\rangle \leqslant 0. \tag{9.39}$$

Since $N_\ell(\xi_0)$ is bounded, it follows that

$$\begin{aligned}\lim_{t\to 0+}\max_{\eta\in N_\ell(\xi_0)}\left\langle \tfrac{\xi(t)-\xi_0}{t},\eta\right\rangle &= \max_{\eta\in N_\ell(\xi_0)}\lim_{t\to 0+}\left\langle \tfrac{\xi(t)-\xi_0}{t},\eta\right\rangle \\ &= \max_{\eta\in N_\ell(\xi_0)}\langle A_\ell\xi(0)+B_0\zeta(0),\eta\rangle \geqslant r(\xi_0) > 0\end{aligned}$$

which yields a contradiction. This shows that $T_\ell(\xi)$ is nonempty for all $\xi \in \partial C_0(\ell)$. ■

Next, we ask the following: Will the state trajectory stay in $\overline{\mathcal{C}}(\kappa)$ for all $t \geqslant 0$ if we choose a feedback such that $\dot{\xi}(0) \in T_\kappa(x_0)$ for all initial conditions $\xi(0) \in \partial\mathcal{C}(\kappa)$? This can be addressed using Nagumo's theorem (see [3, 105]).

Theorem 9.31 (Nagumo) *Consider the system* (9.19)*. Assume that* (9.38) *is satisfied for* $\ell > \kappa > 0$*. Moreover, assume that there is a Lipschitz-continuous feedback* $\zeta = f(\xi)$ *such that* $A_\kappa\xi + B_0 f(\xi) \in T_\kappa(\xi)$ *for all* $\xi \in \partial\mathcal{C}_0(\ell)$*. Then, for any initial condition inside* $\mathcal{C}_0(\ell)$*, the solution of the differential equation remains in* $\overline{\mathcal{C}}_0(\ell)$.

Since $T_\kappa(\xi(t))$ is nonempty for $\xi(t) \in \partial\mathcal{C}_0(\ell)$, there exists a $\zeta(t)$ such that $\dot{\xi}(t) \in T_\kappa(x(t))$. In order to apply Nagumo's theorem, we need a continuous $\zeta(t)$ for feedback. The existence of a continuous feedback is assured by Michael's theorem. We first recall the formal definition of upper and lower semicontinuity of set valued functions (see, for instance, Aubin [3, Sects. 2.1.2 and 6.5.3]).

Definition 9.32 *Let* X *and* Y *be normed spaces,* $D \subset X$ *and* $F(\cdot)$ *a set-valued function from* D *to subsets of* Y *such that* $F(x)$ *is nonempty for all* $x \in D$.

F *is called* **upper semicontinuous** *at* $x_0 \in D$ *if, for any neighborhood* U *of* $F(x_0)$*, there exists an* $\varepsilon > 0$ *such that for all* $x' \in D$ *with* $\|x' - x\| < \varepsilon$*, we have* $F(x') \subset U$*.* F *is called upper semicontinuous if* F *is upper semicontinuous at every point of* D.

F *is called* **lower semicontinuous** *at* $x_0 \in D$ *if, for any* $y \in F(x_0)$ *and for any sequence* $\{x_n\} \in D$ *that converges to* x_0*, there exists a sequence* $\{y_n\}$ *with* $y_n \in F(x_n)$ *that converges to* y*.* F *is lower semicontinuous if* F *is lower semicontinuous at every point of* D.

Using the above, we can formulate Michael's theorem.

Theorem 9.33 (Michael [4, 99]) *Let* D *be a compact metric space and* Y *a Banach space. Every lower semicontinuous function* $F(\cdot)$ *from* D *to the nonempty, closed, and convex subsets of* Y *admits a continuous selection.*

In our case, $D = \partial\mathcal{C}_0(\ell)$ which is clearly a compact metric space and $Y = \mathbb{R}^n$ is clearly a Banach space. Lemma 9.30 assures that $T_\kappa(\xi)$ is not empty for all $\xi \in \partial\mathcal{C}_0(\ell)$; that is, $F = T_\kappa$ maps into nonempty subsets of Y. A continuous selection means that we can find a continuous function $h : \partial\mathcal{C}_0(\ell) \to \mathbb{R}^n$ such that $h(\xi) \in T_\kappa(\xi)$ for all $\xi \in \partial\mathcal{C}_0(\ell)$, which is the result we need. But, to apply Michael's theorem we need to establish that T_κ is lower semicontinuous and $T_\kappa(\xi)$ is closed and convex for all $\xi \in \partial\mathcal{C}_0(\ell)$. The set is closed and convex by Lemma 9.30 and lower semicontinuity is the content of next lemma.

Lemma 9.34 *Let the relation* (9.38) *hold for* $\ell > \kappa > 0$. *Then,* T_κ *is lower semicontinuous on* $\partial\mathcal{C}_0(\ell)$.

Michael's theorem leads to the existence of a continuous function h such that $h(\xi) \in T_\kappa(\xi)$ for all $\xi \in \partial\mathcal{C}_0(\ell)$. Let

$$\zeta = f(\xi) = B_0^\dagger[h(\xi) - A_\kappa\xi], \tag{9.40}$$

where $B_0^\dagger$ is the Moore-Penrose generalized inverse of B_0. Clearly, this $\zeta(\xi)$ is a continuous feedback on $\partial\mathcal{C}_0(\ell)$.

The control law in this proposition does not guarantee asymptotic stability. After a slight modification, we obtain a stabilizing continuous control law that achieves our goal.

Proof of Theorem 9.22 : Given that the system has a bounded input, there exists a t_1 such that at time t_1 for all initial conditions in $\mathcal{W}_1$ and any input satisfying the constraint, we are guaranteed to be inside the set $\mathcal{R}(\Sigma_0, \mathcal{S})$. We consider the system from time t_1 onward.

Let f be the continuous controller given by (9.40) whose existence followed from Michael's theorem. Let $\tilde{\rho} > 0$ be such that $\mathcal{B}(0, \tilde{\rho}) \subset \mathcal{C}_0(\ell)$. Then it is readily verified that

$$\langle \xi, \eta\rangle \geqslant \tilde{\rho}, \ \ \forall\, \xi \in \partial\mathcal{C}_0(\ell), \ \forall\, \eta \in N_{\mathcal{C}_0(\ell)}(\xi).$$

For any $M > 0$, choose δ such that $d(t)$ satisfies $\|d(t)\| \leqslant \tilde{\rho}\kappa/4$ for all $t > t_1$. It also follows that for all $\xi \in \partial\mathcal{C}_0(\ell)$ and $\eta \in N_{\mathcal{C}_0(\ell)}(\xi)$, we have

$$\langle A_{\kappa/2}\xi + B_0 f(\xi), \eta\rangle < -\frac{\tilde{\rho}\kappa}{2}. \tag{9.41}$$

Since f is a continuous function defined on the compact set $\partial\mathcal{C}_0(\ell)$, there exists a differentiable function f_0 on $\partial\mathcal{C}_0(\ell)$ such that $\|B[f(x) - f_0(x)]\| < \frac{\tilde{\rho}\kappa}{2}$. Thus, by (9.41), f_0 satisfies for all $\xi \in \partial\mathcal{C}_0(\ell)$ and for all $\eta \in N_{\mathcal{C}_0(\ell)}(\xi)$,

$$\langle A_{\kappa/2}\xi + B_0 f_0(\xi), \eta\rangle \leqslant 0. \tag{9.42}$$

Next, we extend the differentiable feedback f_0 defined on $\partial\mathcal{C}_0(\kappa)$ to a globally Lipschitz feedback f_1 defined on $\mathcal{C}_0(\ell)$. Define $\beta : \mathbb{R}^d \to \mathbb{R}_0^+$, where $d = \dim \xi$, as

$$\beta(\xi) := \inf\{\beta \geqslant 0 \mid \xi \in \beta\mathcal{C}_0(\ell)\}.$$

Clearly, $\xi \in \beta(\xi)\partial\mathcal{C}_0(\ell)$ for all $\xi \neq 0$. It is easily seen that the function β is Lipschitz and there exists a $M > 0$ such that

$$|\beta(\xi) - \beta(\xi')| \leqslant \beta(\xi - \xi') \leqslant M\|\xi - \xi'\|.$$

Define $f_1 : \mathbb{R}^d \to \mathbb{R}^m$ by

$$f_1(\xi) := \begin{cases} \beta(\xi) f_0\left(\frac{\xi}{\beta(\xi)}\right), & \xi \neq 0, \\ 0, & \xi = 0. \end{cases}$$

Since β is globally Lipschitz and f_0 differentiable, it is easily verified that f_1 is globally Lipschitz. Moreover, f_1 is positively homogeneous, i.e., $f_1(\gamma\xi) = \gamma f_1(\xi)$ for all $\xi \in \mathbb{R}^n$ and $\gamma > 0$, because $\beta(\gamma\xi) = \gamma(\beta\xi)$.

Noting that $f_1(\xi) = f_0(\xi)$ for $\xi \in \partial\mathcal{C}_0(\ell)$, utilizing (9.42), we find for all $\xi \in \partial\mathcal{C}_0(\ell)$ and $\eta \in N_{\mathcal{C}_0(\ell)}(\xi)$ that,

$$\langle A_{\kappa/4}\xi + B_0 f(\xi) + d, \eta\rangle \leqslant 0,$$

for all $t \geqslant t_1$. Then from Nagumo's theorem, we conclude that for all $\xi \in \mathcal{C}_0(\ell)$, the state $\xi(t)$ remains in $\overline{\mathcal{C}}_0(\ell)$ for all $t \geqslant t_1$. But if we apply the feedback $u = f_1(\xi)$ to system (9.18) with the same initial condition $\xi(0) \in \mathcal{C}_0(\ell)$ and let $\tilde{\xi}(t)$ be the solution of system (9.18), it is easy to see that

$$\tilde{\xi}(t) = e^{-\kappa t/4}\xi(t),$$

where we used the property that f_1 is positive homogeneous. Since $\xi(t)$ remains in $\overline{\mathcal{C}}_0(\ell)$, a bounded set, we conclude that $\tilde{\xi}(t)$ converges to zero exponentially, which shows that $\mathcal{W}$ is a subset of the domain of attraction of system Σ_0. ■

10

Sandwich systems: state feedback

10.1 Introduction

We studied internal stabilization of linear systems subject to actuator magnitude saturation in Chap. 4, and the same in Chap. 6, however, when the actuator is subject to both magnitude and rate saturation. The block diagram of Fig. 10.1 depicts the setup. Although such actuator saturation occurs ubiquitously, in this chapter we consider a broader class of nonlinear systems than that depicted in Fig. 10.1. As pointed out in Chap. 1, an important and common paradigm of nonlinear systems is that they are indeed linear systems in which *nonlinear elements are sandwiched or embedded* as shown in Fig. 10.2. A model of a common nonlinear element is a static nonlinearity followed by a linear system or vice versa. In either case, the block diagram of Fig. 10.2 depicts a commonly prevailing situation besides linear systems subject to merely actuator saturation.

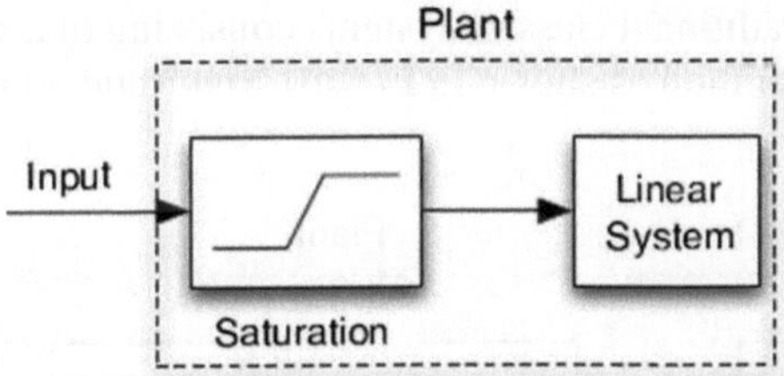

Figure 10.1: Linear system subject to actuator saturation

In view of the above comments, in this chapter we focus on the study of semi-global and global stabilization of the type of systems depicted by the block diagram of Fig. 10.2 and its generalizations, where the static nonlinear element is a *saturation* function as portrayed in Fig. 10.3. We refer to such systems as in Fig. 10.3 as *sandwich systems* because the saturation nonlinearity is *sandwiched* between two linear systems. We call the configuration of Fig. 10.3 as a single-layer sandwich system, where the single layer refers to a single saturation element that is sandwiched between two linear systems. A natural extension of this class of systems is depicted in Fig. 10.4, which shows a single-layer sandwich system

A. Saberi et al., *Internal and External Stabilization of Linear Systems with Constraints*, Systems & Control: Foundations & Applications, DOI 10.1007/978-0-8176-4787-2_10,

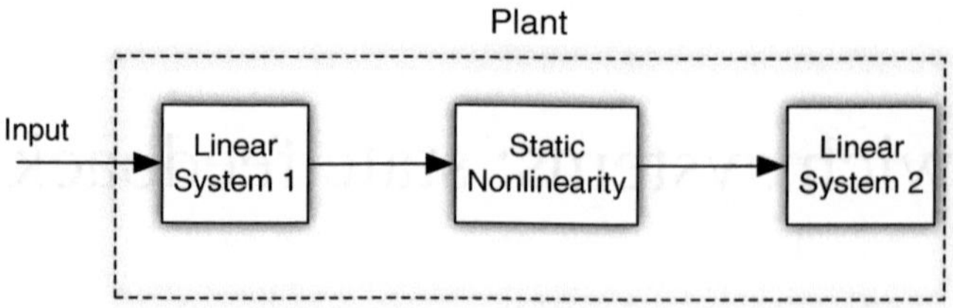

Figure 10.2: Static nonlinearity sandwiched between two linear systems

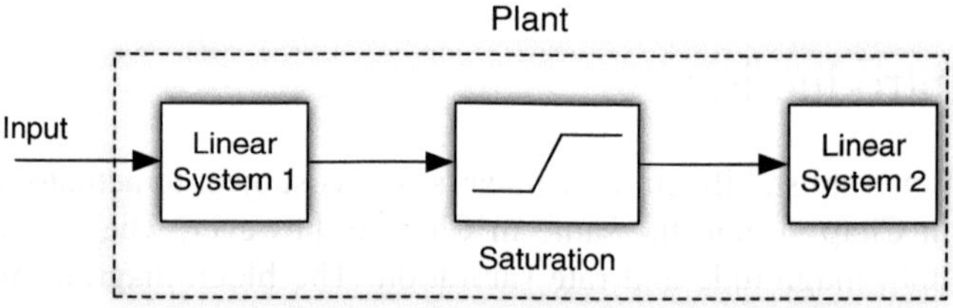

Figure 10.3: Single-layer sandwich system

which is also subject to actuator saturation. These types of systems can be further extended to multilayer sandwich systems and multilayer sandwich systems subject to actuator saturation, shown respectively in Figs. 10.5 and 10.6. Thus, the structures of cascaded systems illustrated by Figs. 10.3–10.6 are progressive generalizations of the traditional class of systems consisting of a single linear system with an actuator saturation as shown in Fig. 10.1 and studied in Chaps. 4 and 6.

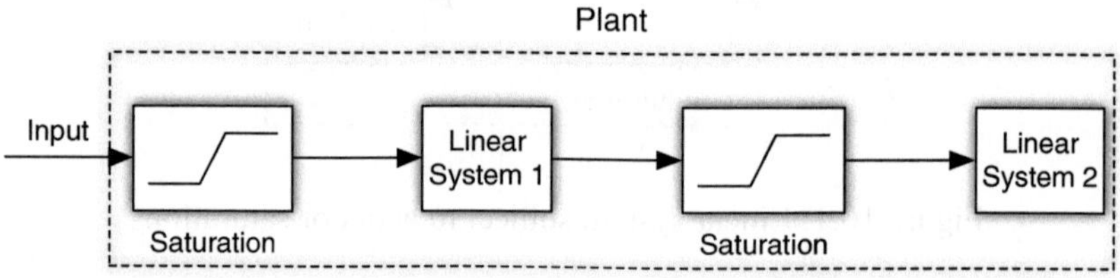

Figure 10.4: Single-layer sandwich system subject to actuator saturation

In this chapter, we first establish conditions for semi-global and global stabilizability of single-layer sandwich systems, portrayed in Fig. 10.3, and we construct appropriate control laws by state feedback. We then extend these stabilization results to single-layer sandwich systems subject to actuator saturation, portrayed in Fig. 10.4. The design methodologies that emerge from this extension are

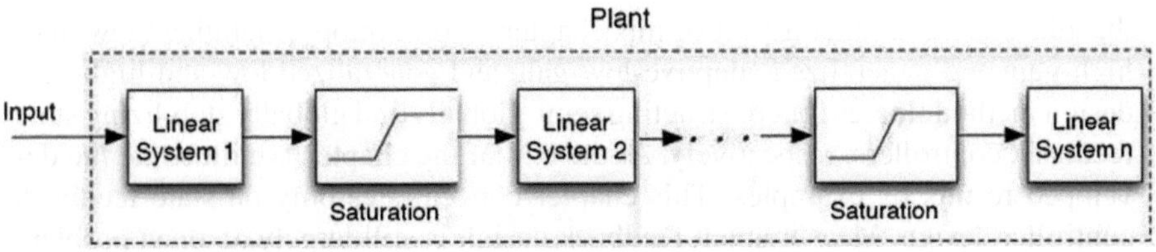

Figure 10.5: Multilayer sandwich system

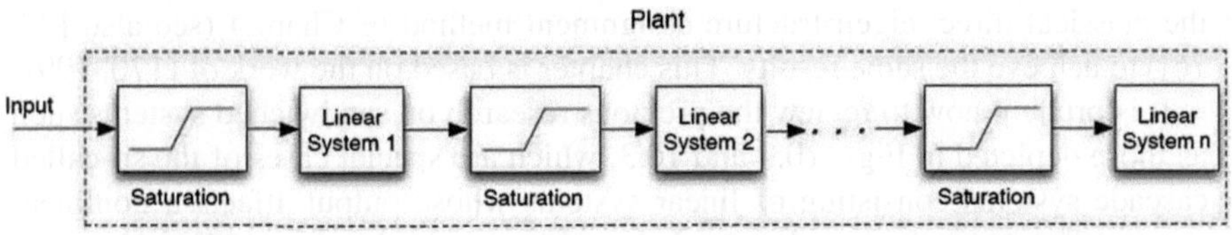

Figure 10.6: Multilayer sandwich system subject to actuator saturation

generalizations of the classical low-gain and adaptive-low-gain design methodologies, developed for semi-global and global stabilization of linear systems subject to only actuator saturation (Chap. 4). Indeed, when the first linear system is equal to the identity, the new design methodologies are reduced to their classical counterparts, and we therefore refer to the new design methodologies as *generalized low-gain* design (for semi-global stabilization) and *generalized adaptive-low-gain* design (for global stabilization). We furthermore discuss the natural extension of the results to the multilayer sandwich systems portrayed in Figs. 10.5 and 10.6.

The philosophy of generalized low-gain and generalized adaptive-low-gain design methodologies for constructing appropriate stabilizing controllers can be briefly sketched as follows: basically, these generalized low-gain methods seek to design controllers such that the saturation does not get activated after some finite time; thereafter, the design reduces to a simple low-gain or adaptive-low-gain design. However, as discussed in Chap. 4, such design methods based on simple low-gain or adaptive-low-gain design methods are conservative as they are constructed in such a way that the control forces do not exceed a certain level in an arbitrary a priori given region of the state space in the semi-global case or in the whole state space in the global case. The key in generalized low-gain design methods is to render the saturation inactive. Therefore, such generalized low-gain design methods do not allow full utilization of the available control capacity. Again, as discussed in Chap. 4, design methods based on low-and-high-gain feedback design are conceived to rectify the drawbacks of low-gain design methods and can utilize the available control capacity fully. As such, they have been successfully used for control problems beyond stabilization to enhance transient performance and to achieve robust stability and disturbance rejection in Chaps. 4–6. In view of this discussion, another aspect of this chapter is to develop generalized low-and-high-gain and generalized adaptive-low-gain and generalized low-and-high-gain design methodologies for constructing semi-globally and globally stabilizing state feedback controllers, respectively. At the end of the chapter, we illustrate the developed results by examples. This chapter concentrates only on state feedback controller design. Measurement feedback design is still largely an open problem for this class of systems.

For ease of presentation, we have chosen to base the design methodologies in this chapter on algebraic Riccati equations (AREs). It is also possible to generalize the classical direct eigenstructure assignment method of Chap. 4 (see also [75, 76]) to achieve the same results. This chapter is based on the work of [170, 199].

It is prudent now to review the previous research on sandwiched systems such as those depicted in Figs. 10.2 and 10.3, which are special cases of the so-called cascade systems consisting of linear systems whose output affects a nonlinear system. The research on such cascaded systems was initiated in [127] but has also been studied, for instance, in [147, 148]. Note that in our case the nonlinear system has a very special structure of an interconnection of a static nonlinearity with a linear system. Moreover, in these references the nonlinear system is assumed to be stable and the goal was to see whether the output of a stable linear system can affect the stability of the cascade system. The goal of this chapter, being focused on developing the conditions for stabilization as well as the methods of designing stabilizing controllers, is inherently different.

Also, some other researchers have previously studied linear systems with sandwiched nonlinearities. The most recent activity in this area is the work of Tao and his coworkers [175–178]. The main technique used in these papers is based on approximate inversion of nonlinearities. An example studied in these references is a dead zone, which is a right-invertible nonlinearity. By contrast, a saturation has a very limited range and cannot be inverted even approximately, except in a local region. The work of Tao et al. is therefore not applicable to the case of a saturation nonlinearity. To achieve our goal of semi-global and global stabilization, we need to face the saturation directly by exploiting the structural properties of the given linear systems.

It is also prudent to compare our study in this chapter to that in Chaps. 7–9. In particular, in Chap. 8, we considered linear systems subject to control input and state constraints as modeled by a constrained output. In this context, the controller is required to guarantee that the constrained output of a linear system remain in a given set. Clearly, a controller designed in this specific way can be used to guarantee that the saturation in the interconnection of Fig. 10.3 never gets activated, albeit with some drawbacks. As pointed out in Chap. 8, we cannot solve semi-global or global stabilization problems for arbitrarily large initial conditions since we cannot guarantee then that the saturation element is never activated. Because of this very reason, we defined there a set of admissible initial conditions and then studied semi-global and global stabilization in the admissible set. Even for the admissible set of initial conditions, utilizing the design philosophy presented in Chap. 8 is in fact conservative as it avoids saturation. Furthermore, the methods of design developed there require the structural condition that the linear system 1 as portrayed in Fig. 10.3 be weakly minimum phase. In contrast, the design methodologies developed in this chapter do not require this weakly minimum phase condition. Moreover, unlike in Chap. 8, activating the saturation element is not an issue. To illustrate the necessity not to avoid activating the saturation consider a car where an engine is modeled by linear dynamics followed by a saturation nonlinearity. In turn, the car dynamics is influenced by the saturated output of the engine dynamics. In this case, there is no reason to avoid saturation and hence a design which attempts to avoid saturation is inherently conservative.

10.2 Preliminaries and problem formulations

In this section, we describe the dynamic equations of three classes of sandwich systems, the class of single-layer sandwich systems (portrayed in Fig. 10.3), the class of single-layer sandwich systems subject to also actuator saturation (portrayed in Fig. 10.4), and its generalization, namely, the class of multilayer sandwich systems (portrayed in Fig. 10.6). We then formulate the semi-global and global stabilization problems for each class of these systems. Regarding different saturation elements present in sandwich systems, as will become clear in the design procedures, different saturation levels do not cause any intrinsic differences in controller design methodology except for some changes on ranges of certain design parameters. Therefore, without loss of generality, we assume that all the saturation elements studied in this chapter are indeed the same and equal to the standard saturation function as defined in (2.19) with the saturation level $\Delta = 1$.

Single-layer sandwich systems consisting of two interconnected systems, L_1 and L_2, are given by

$$L_1 : \begin{cases} \rho x = Ax + Bu, \\ z \ = Cx, \end{cases} \tag{10.1}$$

and

$$L_2 : \quad \rho\omega = M\omega + N\sigma(z), \tag{10.2}$$

where $x \in \mathbb{R}^{n_1}$, $\omega \in \mathbb{R}^{n_2}$, $u \in \mathbb{R}^{m_1}$, and $z \in \mathbb{R}^{m_2}$. The function $\sigma(\cdot)$ is the standard saturation function.

The semi-global and global stabilization problems for single-layer sandwich systems are formulated as follows:

Problem 10.1 Consider the systems given by (10.1) and (10.2). The *semi-global stabilization problem for single-layer sandwich systems* is said to be solvable if for any compact set $\mathcal{W} \subset \mathbb{R}^{n_1+n_2}$, there exists a state feedback control law $u = f(x, \omega)$ such that the origin of the closed-loop system is asymptotically stable with $\mathcal{W}$ contained in its domain of attraction.

Problem 10.2 Consider the systems given by (10.1) and (10.2). The *global stabilization problem for single-layer sandwich systems* is said to be solvable if there exists a state feedback control law $u = f(x, \omega)$ such that the origin of the closed-loop system is globally asymptotically stable.

The dynamics of system L_1 can be modified to include an actuator saturation, and we refer to the resulting system as $\overline{L}_1$. Single-layer sandwich systems subject to actuator saturation, as portrayed in Fig. 10.4, therefore consist of two systems, $\overline{L}_1$ and L_2, given by

$$\overline{L}_1 : \begin{cases} \rho x = Ax + B\sigma(u), \\ z = Cx, \end{cases} \tag{10.3}$$

and

$$L_2 : \quad \rho\omega = M\omega + N\sigma(z), \tag{10.4}$$

where as before $x \in \mathbb{R}^{n_1}$, $\omega \in \mathbb{R}^{n_2}$, $u \in \mathbb{R}^{m_1}$, and $z \in \mathbb{R}^{m_2}$.

The semi-global and global stabilization problems for single-layer sandwich systems subject to actuator saturation are now formulated as follows:

Problem 10.3 Consider the systems given by (10.3) and (10.4). The *semi-global stabilization problem for single-layer sandwich systems subject to actuator saturation* is said to be solvable if for any compact set $\mathcal{W} \subset \mathbb{R}^{n_1+n_2}$, there exists a state feedback control law $u = f(x, \omega)$ such that the origin of the closed-loop system is asymptotically stable with $\mathcal{W}$ contained in its domain of attraction.

Problem 10.4 Consider the systems given by (10.3) and (10.4). The *global stabilization problem for single-layer sandwich systems subject to actuator saturation* is said to be solvable if there exists a state feedback control law $u = f(x, \omega)$ such that the origin of the closed-loop system is globally asymptotically stable.

The above type of system configuration as in (10.3) and (10.4) can be generalized further to an interconnection of ν linear systems, namely, the multilayer nonlinear sandwich systems. Without loss of generality, we assume that the actuator is subject to saturation. Consider the following interconnection of ν systems:

$$L_i : \begin{cases} \rho x_i = A_i x_i + B_i \sigma(u_i), & i = 1, \dots, \nu \\ z_i = C_i x_i, & i = 1, \dots, \nu - 1 \\ u_i = z_{i-1}, & i = 2, \dots, \nu \end{cases} \tag{10.5}$$

where $x_i \in \mathbb{R}^{n_i}$, $u_i \in \mathbb{R}^{m_i}$ for $i = 1, \dots, \nu$, and $z_i \in \mathbb{R}^{m_{i+1}}$ for $i = 1, \dots, \nu - 1$.

The semi-global and global stabilization problems for the multilayer system can be defined as follows:

Problem 10.5 Consider the interconnected system as given by (10.5). The *semi-global stabilization problem for multilayer sandwich systems* is said to be solvable if for any compact set $\mathcal{W} \subset \mathbb{R}^{n_1+\cdots+n_\nu}$, there exists a state feedback control law $u_1 = f(x_1, \ldots, x_\nu)$ such that the origin of the closed-loop system is asymptotically stable with $\mathcal{W}$ contained in its domain of attraction.

Problem 10.6 Consider the interconnected system as given by (10.5). The *global stabilization problem for multilayer sandwich systems subject to actuator saturation* is said to be solvable if there exists a state feedback control law $u_1 = f(x_1, \ldots, x_\nu)$ such that the origin of the closed-loop system is globally asymptotically stable.

In what follows, at first we consider single-layer sandwich systems, then single-layer sandwich systems subject to actuator saturation, and finally general multi-layer sandwich systems. In this way, we progressively add different levels of complexity to our development.

10.3 Necessary and sufficient conditions for stabilization

In this section, we present the necessary and sufficient conditions for all the semi-global and global stabilization problems in the order they were formulated in Sect. 10.2.

10.3.1 Single-layer sandwich systems

In this subsection, we present two theorems that give the necessary and sufficient conditions for solving Problems 10.1 and 10.2.

Theorem 10.7 *Consider the systems given by* (10.1) *and* (10.2). *The* semi-global stabilization problem for single-layer sandwich systems, *as formulated in Problem 10.1, is solvable if and only if:*

(i) *The linearized cascade system is stabilizable, that is, $(\mathcal{A}, \mathcal{B})$ is stabilizable, where*

$$\mathcal{A} = \begin{pmatrix} A & 0 \\ NC & M \end{pmatrix} \text{ and } \mathcal{B} = \begin{pmatrix} B \\ 0 \end{pmatrix}. \tag{10.6}$$

(ii) *All the eigenvalues of M are in the closed left-half plane for continuous-time systems and in the closed unit disc for discrete-time systems.*

Moreover, semi-global stabilization can be achieved by a linear control law of the form $u = Fx + G\omega$.

Proof : The system L_2 must be stabilized through a saturated signal, and it is clear from Chap. 4 that this is only possible if the eigenvalues of M are in the closed left-half plane for continuous-time systems, and in the closed unit disc for discrete-time systems. The cascade system is linear in a neighborhood around the origin, and stabilizability of the nonlinear cascade system therefore requires stabilizability of the local linear system, which is equivalent to stabilizability of $(\mathcal{A}, \mathcal{B})$. We have thus established the necessity of the two conditions in Theorem 10.7. Sufficiency is established in Sect. 10.4 by explicit construction of a stabilizing controller. ■

Theorem 10.8 *Consider the systems given by* (10.1) *and* (10.2). *The* global stabilization problem for single-layer sandwich systems, *as formulated in Problem 10.2, is solvable if and only if:*

(*i*) *The linearized cascade system is stabilizable, that is,* $(\mathcal{A}, \mathcal{B})$ *is stabilizable, where* $\mathcal{A}$ *and* $\mathcal{B}$ *are given by* (10.6).

(*ii*) *The eigenvalues of* M *are in the closed left-half plane for continuous-time systems and in the closed unit disc for discrete-time systems.*

Proof : The proof of necessity follows along the lines of the proof of Theorem 10.7. Sufficiency is established in Sect. 10.4 by explicit construction of a stabilizing controller. ■

10.3.2 Single-layer sandwich systems subject to actuator saturation

In this subsection, we present two theorems that give the necessary and sufficient conditions for solving Problems 10.3 and 10.4 that pertain to single-layer sandwich systems subject to actuator saturation.

Theorem 10.9 *Consider the two systems given by* (10.3) *and* (10.4). *The* semi-global stabilization problem for single-layer sandwich systems subject to actuator saturation, *as formulated in Problem 10.3, is solvable if and only if*

(*i*) *The linearized cascade system is stabilizable, that is,* $(\mathcal{A}, \mathcal{B})$ *is stabilizable, where* $\mathcal{A}$ *and* $\mathcal{B}$ *are as defined in* (10.6).

(ii) The eigenvalues of M are in the closed left-half plane for continuous-time systems and in the closed unit disc for discrete-time systems.

(iii) The eigenvalues of A are in the closed left-half plane for continuous-time systems and in the closed unit disc for discrete-time systems.

Moreover, semi-global stabilization can be achieved by a linear control law of the form $u = Fx + G\omega$.

Proof : The systems $\bar{L}_1$ and L_2 must be stabilized through a saturated signal, and it is clear from Chap. 4 that this is only possible if the eigenvalues of A and M are all in the closed left-half plane for continuous-time systems and in the closed unit disc for discrete-time systems. The cascade system is linear in a neighborhood around the origin, and stabilizability of the nonlinear cascade system therefore requires stabilizability of the local linear system, which is equivalent to stabilizability of $(\mathcal{A}, \mathcal{B})$. We have thus established the necessity of the three conditions in Theorem 10.9. Sufficiency is established in Sect. 10.5 by explicit construction of a stabilizing controller. ■

Theorem 10.10 *Consider the two systems given by* (10.3) *and* (10.4). *The* global stabilization problem for single-layer sandwich systems subject to actuator saturation, *as formulated in Problem 10.4, is solvable if and only if:*

(i) The linearized cascade system is stabilizable, that is, $(\mathcal{A}, \mathcal{B})$ is stabilizable, where $\mathcal{A}$ and $\mathcal{B}$ are given by (10.6).

(ii) The eigenvalues of M are in the closed left-half plane for continuous-time systems and in the closed unit disc for discrete-time systems.

(iii) The eigenvalues of A are in the closed left-half plane for continuous-time systems and in the closed unit disc for discrete-time systems.

Proof : The proof of necessity follows along the lines of the proof of Theorem 10.9. Sufficiency is established in Sect. 10.5 by explicit construction of a stabilizing controller. ■

10.3.3 Multilayer sandwich systems

In this subsection, we establish the necessary and sufficient conditions for solving Problems 10.5 and 10.6 that pertain to multilayer sandwich system as given by (10.5).

Theorem 10.11 *Consider the interconnection of systems L_i, $i = 1, \ldots, \nu$ as given by* (10.5)*. Then, the semi-global stabilization problem, as formulated in Problem 10.5, is solvable if and only if*

(i) $(\mathcal{A}_0, \mathcal{B}_0)$ *is stabilizable, where*

$$\mathcal{A}_0 = \begin{pmatrix} A_1 & 0 & \cdots & \cdots & 0 \\ B_2C_1 & A_2 & \ddots & & \vdots \\ 0 & \ddots & \ddots & \ddots & \vdots \\ \vdots & \ddots & \ddots & A_{\nu-1} & 0 \\ 0 & \cdots & 0 & B_\nu C_{\nu-1} & A_\nu \end{pmatrix}, \quad \mathcal{B}_0 = \begin{pmatrix} B_1 \\ 0 \\ \vdots \\ 0 \end{pmatrix}. \tag{10.7}$$

(ii) *All A_i, $i = 1, \ldots, \nu$ have their eigenvalues in the closed left-half plane for continuous-time systems and in the closed unit disc for discrete-time systems.*

Moreover, the solution to the semi-global stabilization problem can be achieved by a linear state feedback law of the form $u = \sum_{i=1}^{\nu} F_i x_i$.

Proof : The necessity of condition (*i*) and (*ii*) can be proved following the same lines as in previous theorems. Sufficiency is proved in Sect. 10.6 by explicit construction of a semi-globally stabilizing controller. ■

Theorem 10.12 *Consider the interconnection of systems L_i, $i = 1, \ldots, \nu$ as given by* (10.5)*. Then, the global stabilization problem, as formulated in Problem 10.6, is solvable if and only if*

(i) $(\mathcal{A}_0, \mathcal{B}_0)$ *is stabilizable, where $\mathcal{A}_0$ and $\mathcal{B}_0$ are given in* (10.7)*.*

(ii) *All A_i, $i = 1, \ldots, \nu$ have their eigenvalues in the closed left-half plane for continuous-time systems and in the closed unit disc for discrete-time systems.*

Proof : Necessity of conditions (*i*) and (*ii*) can be shown following the same lines as in previous theorems. Sufficiency will be shown by explicitly constructing a globally stabilizing controller in Sect. 10.6. ■

Remark 10.13 *We remark that the solvability conditions for any of the three types of sandwich systems considered above are the same for either semi-global or global stabilization. However, there is still an intrinsic difference between the two cases, in that semi-global stabilization can be accomplished with a linear control law, whereas global stabilization generally requires a nonlinear control law.*

10.4 Generalized and adaptive-low-gain design for single-layer systems

In this section, we explicitly construct controllers to solve the semi-global and global stabilization problems described in Sect. 10.2 for single-layer systems where the actuator is not subject to saturation. In doing so, we prove sufficiency of the conditions in Theorems 10.7 and 10.8. We divide our development into two subsections, one for continuous-time systems and the other for discrete-time systems.

10.4.1 Continuous-time systems

We first present a generalized low-gain design for solving Problem 10.1 which concerns with semi-global stabilization of the origin of the single-layer sandwich system described by (10.1) and (10.2).

Generalized low-gain design for semi-global stabilization: We start by choosing an arbitrary F such that $A + BF$ is Hurwitz, and then let $u = Fx + v$, where v is a time-varying control input yet to be determined. Consider the resulting L_1 system,

$$\begin{aligned} \dot{x} &= (A + BF)x + Bv, \\ z &= Cx. \end{aligned} \tag{10.8}$$

We have

$$\begin{aligned} z(t) &= Ce^{(A+BF)t}x(0) + \int_0^t Ce^{(A+BF)(t-\tau)}Bv(\tau)\,\mathrm{d}\tau \\ &= Ce^{(A+BF)t}x(0) + z_0(t). \end{aligned} \tag{10.9}$$

Since $A + BF$ is Hurwitz, we know that there exists a $\delta > 0$ such that whenever

$$\|v(\tau)\| < \delta \quad \forall \tau > 0, \tag{10.10}$$

we have $\|z_0(t)\| < \frac{1}{2}$ for all $t > 0$.

Next we consider the linear system,

$$\dot{\bar{x}} = \tilde{\mathcal{A}}\bar{x} + \mathcal{B}v, \tag{10.11}$$

where

$$\tilde{\mathcal{A}} = \begin{pmatrix} A + BF & 0 \\ NC & M \end{pmatrix}, \quad \mathcal{B} = \begin{pmatrix} B \\ 0 \end{pmatrix}, \quad \bar{x} = \begin{pmatrix} x \\ \omega \end{pmatrix}. \tag{10.12}$$

Note that conditions (*i*) and (*ii*) of Theorem 10.7 and asymptotic stability of $A + BF$ together imply that $(\tilde{\mathcal{A}}, \mathcal{B})$ is stabilizable and $\tilde{\mathcal{A}}$ has all its eigenvalues in the closed left-half plane.

Our initial objective is, for any a priori given compact set $\bar{\mathcal{W}}$, to find a stabilizing controller for the system (10.11) such that $\bar{\mathcal{W}}$ is contained in its domain of attraction and $\|v(t)\| < \delta$ for all $t > 0$. This is accomplished in the following lemma whose proof follows easily from Theorem 4.21.

Lemma 10.14 *Consider the linear system* (10.11) *with constraint* $\|v(t)\| < \delta$, *and assume that* $(\mathcal{A}, \mathcal{B})$ *as given by* (10.6) *is stabilizable and that the eigenvalues of* M *are in the closed left-half plane. Let* $Q_\varepsilon > 0$ *be a parameterized family of matrices which is increasing in* $\varepsilon > 0$ *with* $\lim_{\varepsilon \to 0} Q_\varepsilon = 0$. *Then, for any a priori given compact set* $\bar{\mathcal{W}} \in \mathbb{R}^{n_1+n_2}$, *there exists an* ε^* *such that for any* $0 < \varepsilon < \varepsilon^*$, *the control law*

$$v = -\begin{pmatrix} B' & 0 \end{pmatrix} P_\varepsilon \begin{pmatrix} x_1 \\ x_2 \end{pmatrix} = -\mathcal{B}' P_\varepsilon \bar{x}, \tag{10.13}$$

where $P_\varepsilon = P_\varepsilon' > 0$ *satisfies the continuous-time algebraic Riccati equation (CARE),*

$$\tilde{\mathcal{A}}' P_\varepsilon + P_\varepsilon \tilde{\mathcal{A}} - P_\varepsilon \mathcal{B}\mathcal{B}' P_\varepsilon + Q_\varepsilon = 0, \tag{10.14}$$

achieves asymptotic stability of the equilibrium point $\bar{x} = 0$. *Moreover, for any initial condition in* $\bar{\mathcal{W}}$, $\|v(t)\| < \delta$ *for all* $t > 0$.

We can now use Lemma 10.14 to prove that a particular family of control laws achieves semi-global stability of the single-layer nonlinear sandwich system.

Theorem 10.15 *Consider the systems given by* (10.1) *and* (10.2) *that satisfy Conditions* (*i*) *and* (*ii*) *of Theorem 10.7. Let* F *be chosen such that* $A+BF$ *is Hurwitz,* *and let $P_\varepsilon = P_\varepsilon' > 0$ be defined by the CARE* (10.14). *Define what can be termed as generalized low-gain state feedback control law by*

$$u = Fx - \begin{pmatrix} B' & 0 \end{pmatrix} P_\varepsilon \begin{pmatrix} x \\ \omega \end{pmatrix} = F_{1,\varepsilon}x + F_{2,\varepsilon}\omega. \tag{10.15}$$

Then, for any compact set of initial conditions $\mathcal{W} \in \mathbb{R}^{n_1+n_2}$, there exists an $\varepsilon^* > 0$ such that for all ε with $0 < \varepsilon < \varepsilon^*$, the controller (10.15) *asymptotically stabilizes the origin with a domain of attraction containing $\mathcal{W}$.*

Proof : For simplicity and without loss of generality, let us choose $Q_\varepsilon = \varepsilon I$. Condition (*i*) of Theorem 10.7 implies the existence of $P_\varepsilon = P_\varepsilon' > 0$ satisfying (10.14). Moreover, Condition (*ii*) implies that $P_\varepsilon \to 0$ as $\varepsilon \to 0$, which in turn implies that $F_{1,\varepsilon} \to F$ and $F_{2,\varepsilon} \to 0$. The initial conditions belong to some compact set $\mathcal{W}$, and hence there exist compact sets $\mathcal{X} \subset \mathbb{R}^{n_1}$ and $\Omega \subset \mathbb{R}^{n_2}$ such that $x(0) \in \mathcal{X}$ and $\omega(0) \in \Omega$.

For $u = Fx$, there exists a $T > 0$ such that for any $x(0) \in \mathcal{X}$, we have $\|Ce^{(A+BF)t}x(0)\| < \frac{1}{2}$ for all $t > T$, and there exists a compact set $\bar{\mathcal{X}} \subset \mathbb{R}^{n_1}$ such that $x(t) \in \bar{\mathcal{X}}$ for all $t \in [0, T]$. This follows from the Hurwitz property of $A + BF$.

Since $\omega(0) \in \Omega$, where Ω is a compact set, and $\sigma(z(t))$ is bounded, we find that, independent of ε, there exists a compact set $\bar{\Omega}$ such that $\omega(t) \in \bar{\Omega}$ for all $t \in [0, T]$.

Next, there exists an $\varepsilon^* > 0$ such that for $u = F_{1,\varepsilon}x + F_{2,\varepsilon}\omega$ and $0 < \varepsilon < \varepsilon^*$, we have $x(t) \in 2\bar{\mathcal{X}}$ for all $t \in [0, T]$. This follows from the fact that $F_{1,\varepsilon} \to F$ and $F_{2,\varepsilon} \to 0$, while $\omega(t)$ is bounded.

We also note that, from Lemma 10.14, there exists an ε_2^* with $0 < \varepsilon_2^* < \varepsilon^*$ such that, for all $0 < \varepsilon < \varepsilon_2^*$, the controller

$$v = -\begin{pmatrix} B' & 0 \end{pmatrix} P_\varepsilon \begin{pmatrix} x \\ \omega \end{pmatrix} = -\mathcal{B}' P_\varepsilon \bar{x}$$

stabilizes system (10.11) and satisfies $\|v(t)\| < \delta$ for all $t > T$, given $x(T) \in 2\bar{\mathcal{X}}$ and $\omega(T) \in \bar{\Omega}$. This implies that $z(t)$ generated by (10.8) satisfies $\|z(t)\| < 1$ for all $t > T$. Then the interconnection of (10.1) and (10.2) with controller (10.15) is equivalent to the interconnection of (10.11) with controller (10.13) for $t > T$. The asymptotic stability of (10.11) with controller (10.13) implies that $x(t) \to 0$ and $\omega(t) \to 0$. Since this holds for any $(x(0), \omega(0)) \in \mathcal{W}$, it follows that $\mathcal{W}$ is contained in the domain of attraction. ■

Remark 10.16 *For semi-global stabilization, we can enlarge the domain of attraction by choosing a sufficiently small low-gain parameter. However, this incurs* *a deterioration of closed-loop performance near the origin since a small low-gain parameter results in conservativeness in feedback gain and hence does not allow full utilization of control capacity when the state is close to the origin. In order to rectify this problem, a generalized low-and-high-gain feedback design methodology is recently introduced in [170] and will be discussed subsequently in Sect. 10.7. It is shown there that a refined performance can be achieved with the so-called low-and-high-gain feedback controller.*

Remark 10.17 *To implement the semi-globally stabilizing controller, it is necessary to find appropriate low-gain parameters ε. It is difficult to derive tight upper bounds on ε analytically, and thus the parameters are typically found experimentally by gradually decreasing them until the desired stability is achieved.*

Generalized adaptive-low-gain design for global stabilization:

We present now a generalized scheduled or adaptive low-gain design for solving Problem 10.2 which concerns with global stabilization of the origin of the single-layer sandwich system described by (10.1) and (10.2). In the following, we show that the family of controllers defined by (10.15), with ε replaced by an adaptive-low-gain parameter $\varepsilon_a(\bar{x})$, solves Problem 10.2. Therefore, let F be such that $A + BF$ is Hurwitz, and let $P_{\varepsilon_a(\bar{x})}$ be defined by the CARE (10.14), with ε replaced by an adaptive parameter $\varepsilon_a(\bar{x})$. Choose δ such that the fact $\|v(t)\| < \delta$ for all $t > 0$ guarantees the fact that $\|z_0(t)\| < \frac{1}{2}$ for all $t > 0$, where z_0 is defined by (10.9).

Consider the adaptive parameter $\varepsilon_a(\bar{x})$ satisfying the following properties:

(*i*) $\varepsilon_a(\bar{x}) : \mathbb{R}^{n_1+n_2} \to (0, 1]$ is continuous and piecewise continuously differentiable.

(*ii*) There exists an open neighborhood $\mathcal{O}$ of the origin such that $\varepsilon_a(\bar{x}) = 1$ for all $\bar{x} \in \mathcal{O}$.

(*iii*) For any $\bar{x} \in \mathbb{R}^{n_1+n_2}$, we have $\|F_{\varepsilon_a(\bar{x})}\bar{x}\| \leqslant \delta$.

(*iv*) $\varepsilon_a(\bar{x}) \to 0$ as $\|\bar{x}\| \to \infty$.

(*v*) $\{\, \bar{x} \in \mathbb{R}^{n_1+n_2} \mid \bar{x}' P_{\varepsilon_a(\bar{x})}\bar{x} \leqslant c \,\}$ is a bounded set for all $c > 0$.

(*vi*) For any $\bar{x}_1, \bar{x}_2 \in \mathbb{R}^{n_1+n_2}$,

$$\bar{x}_1' P_{\varepsilon_a(\bar{x}_1)}\bar{x}_1 \leqslant \bar{x}_2' P_{\varepsilon_a(\bar{x}_2)}\bar{x}_2$$

implies $\varepsilon_a(x_1) \geqslant \varepsilon_a(x_2)$.

Note that the above properties yield that

$$\lim_{t\to\infty} \bar{x}'(t) P_{\varepsilon_a(\bar{x}(t))}\bar{x}(t) = 0 \implies \lim_{t\to\infty} \bar{x}(t) = 0.$$

Furthermore, we have

$$\frac{\mathrm{d}}{\mathrm{d}t} P_{\varepsilon_a(\bar{x}(t))} = \left(\frac{\mathrm{d}}{\mathrm{d}\varepsilon_a} P_{\varepsilon_a(\bar{x}(t))} \right) \left(\frac{\mathrm{d}}{\mathrm{d}t} \varepsilon_a(\bar{x}(t)) \right).$$

Since P_ε is increasing as a function of ε, this implies that, if $\varepsilon_a(\bar{x}(t))$ is increasing as a function of time, then $P_{\varepsilon_a(\bar{x}(t))}$ is also increasing as a function of time.

A particular choice satisfying the above criteria is

$$\varepsilon_a(\bar{x}) = \max \left\{ r \in (0,1] \mid (\bar{x}' P_r \bar{x}) \operatorname{trace} \mathcal{B}' P_r \mathcal{B} \leqslant \delta^2 \right\}, \tag{10.16}$$

where P_r is the unique positive definite solution of CARE (10.14) with r replacing ε.

Before we proceed further, we need the following lemma which is a slightly modified version of Lemma 4.51 (see also [98]) and which defines a control law that stabilizes the linear system (10.11).

Lemma 10.18 *Consider the linear system* (10.11) *and assume that* $(\mathcal{A}, \mathcal{B})$ *as given by* (10.6) *is stabilizable and that the eigenvalues of* M *are in the closed left-half plane. Define the control law,*

$$v = -\mathcal{B}' P_{\varepsilon_a(\bar{x})} \bar{x} \tag{10.17}$$

where $\bar{x}$ *is as defined in* (10.12). *Then, this control law achieves global stability of the equilibrium* $\bar{x} = 0$. *Moreover, the constraint* $\|v(t)\| \leq \delta$ *does not get violated for any* $t \geq 0$.

We can now use Lemma 10.18 to prove that a particular family of control laws achieves global stability of the single-layer nonlinear sandwich system.

Theorem 10.19 *Consider the systems given by* (10.1) *and* (10.2) *that satisfies Conditions* (i) *and* (ii) *of Theorem 10.8. Choose* F *such that* $A + BF$ *is Hurwitz. Let* $P_{\varepsilon_a(\bar{x})}$ *be defined by the CARE* (10.14), *with* ε *replaced by the adaptive-low-gain parameter* $\varepsilon_a(\bar{x})$ *defined by* (10.16). *Define the generalized adaptive-low-gain state feedback control law as*

$$u = Fx - \mathcal{B}' P_{\varepsilon_a(\bar{x})} \bar{x}, \tag{10.18}$$

where $\bar{x}$ *is as defined in* (10.12). *Then, the control law* (10.18) *achieves global asymptotic stability of the origin.*

Proof : Considering the interconnection of (10.1) and (10.2), we note that the saturation is not activated near the origin. Moreover, near the origin the control law (10.18) is given by $u = Fx - \mathcal{B}' P_1 \bar{x}$, which implies that the origin of the interconnection of (10.1), (10.2), and (10.18) is locally asymptotically stable. It remains to show that it is globally asymptotically stable.

Consider arbitrary initial conditions $x(0)$ and $\omega(0)$. There exists a $T > 0$ such that

$$\|Ce^{(A+BF)t} x(0)\| < \tfrac{1}{2}$$

for all $t > T$. Moreover, by construction $v = -\mathcal{B}' P_{\varepsilon_a(\bar{x})} \bar{x}$ yields $\|v(t)\| \leqslant \delta$ for all $t > 0$. This implies that $z(t)$ generated by (10.8) satisfies $\|z(t)\| < 1$ for all $t > T$. The interconnection of (10.1) and (10.2) with controller (10.18) therefore behaves like the interconnection of (10.11) with controller (10.17), for all $t > T$. Global asymptotic stability of (10.11) with controller (10.17) therefore implies that $\bar{x}(t) \to 0$ as $t \to \infty$. Since this property holds for any initial condition, the origin is globally asymptotically stable. ■

10.4.2 Discrete-time systems

This subsection pertaining to discrete-time systems is conceptually similar to the previous subsection that deals with continuous-time systems. For discrete-time systems, we explicitly present here a generalized low-gain design for solving Problem 10.1, concerning semi-global stabilization of the origin of the single-layer sandwich system described by (10.1) and (10.2).

Generalized low-gain design for semi-global stabilization: As we did before in the continuous-time, we start by choosing an arbitrary F such that $A + BF$ is asymptotically stable, and then let $u = Fx + v$, where v is a time-varying control input yet to be determined. Consider the resulting L_1 system,

$$\begin{aligned} x(k+1) &= (A+BF)x(k) + Bv(k) \\ z(k) &= Cx(k). \end{aligned} \tag{10.19}$$

We have

$$z(k) = C(A+BF)^k x(0) + \sum_{i=0}^{k-1} C(A+BF)^{k-i-1} Bv(i) \tag{10.20}$$

$$= C(A+BF)^k x(0) + z_0(k). \tag{10.21}$$

Define

$$\delta_1 = \frac{1}{2\sum_{k=0}^{\infty} \left\| C(A+BF)^k B \right\|}. \tag{10.22}$$

Since $A + BF$ is asymptotically stable, the above summation is well defined. We know that if

$$\|v(k)\| < \delta_1 \quad \forall k > 0, \tag{10.23}$$

then $\|z_0(k)\| < \frac{1}{2}$. Next we consider the system,

$$\bar{x}(k+1) = \tilde{\mathcal{A}}\bar{x}(k) + \mathcal{B}v(k), \tag{10.24}$$

where $\tilde{\mathcal{A}}$, $\mathcal{B}$, and $\bar{x}$ are as defined in (10.12). Note that condition i and ii of Theorem 10.7 and asymptotic stability of $A + BF$ implies that $(\tilde{\mathcal{A}}, \mathcal{B})$ is stabilizable and $\tilde{\mathcal{A}}$ has all its eigenvalues in the closed unit disc.

Our next objective is, for any a priori given compact set $\bar{\mathcal{W}}$, to find a stabilizing controller for the system (10.24) such that $\bar{\mathcal{W}}$ is contained in its domain of attraction and $\|v(k)\| < \delta_1$ for all $k > 0$.

We note that there exists a unique $P_\varepsilon > 0$ satisfying the DARE,

$$P_\varepsilon = \tilde{\mathcal{A}}' P_\varepsilon \tilde{\mathcal{A}} + \varepsilon I - \tilde{\mathcal{A}}' P_\varepsilon \mathcal{B}(\mathcal{B}' P_\varepsilon \mathcal{B} + I)^{-1} \mathcal{B}' P_\varepsilon \tilde{\mathcal{A}}. \tag{10.25}$$

The following lemma summarizes a result obtained in Chap. 4.

Lemma 10.20 *Consider the system* (10.24) *with constraint* $\|v(k)\| < \delta_1$, *and assume that* $(\tilde{\mathcal{A}}, \mathcal{B})$ *is stabilizable and* $\tilde{\mathcal{A}}$ *has its eigenvalues in the closed unit disc. For any a priori given compact set* $\bar{\mathcal{W}} \in \mathbb{R}^{n_1+n_2}$, *there exists an* ε^* *such that for any* $0 < \varepsilon < \varepsilon^*$, *the feedback control,*

$$v = -(\mathcal{B}' P_\varepsilon \mathcal{B} + I)^{-1} \mathcal{B}' P_\varepsilon \tilde{\mathcal{A}} \bar{x}, \tag{10.26}$$

achieves asymptotic stability of the equilibrium $\bar{x} = 0$ *with* $\bar{\mathcal{W}}$ *contained in its domain of attraction. Moreover, for any initial condition in* $\bar{\mathcal{W}}$, *the constraint* $\|v(k)\| < \delta_1$ *does not get violated for any* $k > 0$.

We can now use Lemma 10.20 to prove that a particular family of control laws achieves semi-global stability of the single-layer nonlinear sandwich system.

Theorem 10.21 *Consider the interconnection of the two systems given by* (10.1) *and* (10.2) *satisfying conditions (i) and (ii) of Theorem 10.7. Let* F *be an arbitrary matrix such that* $A + BF$ *is asymptotically stable. Let* $P_\varepsilon > 0$ *be the solution of DARE* (10.25). *Define what can be termed as generalized low-gain state feedback law by*

$$u = Fx - (\mathcal{B}' P_\varepsilon \mathcal{B} + I)^{-1} \mathcal{B}' P_\varepsilon \tilde{\mathcal{A}} \bar{x} = F_{1,\varepsilon} x + F_{2,\varepsilon} \omega. \tag{10.27}$$

For any compact set of initial conditions $\mathcal{W} \in \mathbb{R}^{n_1+n_2}$, *there exists an* $\varepsilon^* > 0$ *such that for all* ε *with* $0 < \varepsilon < \varepsilon^*$ *the controller* (10.27) *asymptotically stabilizes the equilibrium* $(0, 0)$ *with a domain of attraction containing* $\mathcal{W}$.

Proof : Condition (*i*) of Theorem 10.7 immediately implies the existence and uniqueness of $P_\varepsilon > 0$ satisfying the DARE (10.25). Moreover, condition (*ii*) immediately implies that

$$P_\varepsilon \to 0 \tag{10.28}$$

as $\varepsilon \to 0$. This immediately implies that

$$F_{1,\varepsilon} \to F, \qquad F_{2,\varepsilon} \to 0. \tag{10.29}$$

Note that the initial conditions are in some compact set $\mathcal{W}$ and hence there exist compact sets $\mathcal{X}$ and Ω such that $x(0) \in \mathcal{X}$ and $\omega(0) \in \Omega$.

Note that if we apply $u = Fx$, there exists a $K > 0$ such that for any $x(0) \in \mathcal{X}$ we have

$$\|C(A + BF)^k x(0)\| < \tfrac{1}{2}$$

for all $k > K$ and there exists a compact set $\bar{\mathcal{X}}$ such that $x(k) \in \bar{\mathcal{X}}$ for all $0 \leqslant k \leqslant K$. This immediately follows from the asymptotic stability of $A + BF$.

Since $\omega(0) \in \Omega$ which is a compact set and $\sigma(z(k))$ is bounded, we find that, independent of ε, there exists a compact set $\bar{\Omega}$ such that $\omega(k) \in \bar{\Omega}$ for all $0 \leqslant k \leqslant K$.

Next, there exists an $\varepsilon^\# > 0$ such that for

$$u(k) = F_{1,\varepsilon} x(k) + F_{2,\varepsilon}\omega(k)$$

and for $\varepsilon < \varepsilon^\#$ we have

$$x(k) \in 2\bar{\mathcal{X}}$$

for all $0 \leqslant k \leqslant K$. This follows from (10.29) while $\omega(k)$ is bounded in $\bar{\Omega}$.

From Lemma 10.20, we note that there exists an $\varepsilon^* < \varepsilon^\#$ such that for $\varepsilon < \varepsilon^*$, the controller,

$$v = -(\mathcal{B}' P_\varepsilon \mathcal{B} + I)^{-1} \mathcal{B}' P_\varepsilon \tilde{\mathcal{A}} \bar{x},$$

stabilizes system (10.24) and satisfies $\|v\| < \delta_1$ for all $k > 0$ given $\bar{x}(K) \in 2\bar{\mathcal{X}} \times \bar{\Omega}$. This implies that $z(k)$ generated by (10.19) satisfies $\|z(k)\| < 1$ for $k > K$. Then the interconnection of (10.1) and (10.2) with controller (10.27) for $k > K$ is equivalent to the interconnection of (10.24) with controller (10.26) for $k > K$. The asymptotic stability of the latter system follows from Lemma 10.20. Hence, we have

$$x(k) \to 0, \quad \omega(k) \to 0.$$

Since this follows for any $(x(0), \omega(0)) \in \mathcal{W}$, we find that $\mathcal{W}$ is contained in the domain of attraction as required. ■

Remark 10.22 *Remarks 10.16 and 10.17 apply to discrete-time systems as well. However, because of some inherent differences between continuous- and discrete-time systems, development of a low-and-high-gain design for discrete-time systems is not fully feasible as discussed at the end of Sect. 10.7 in Remark 10.41.*

We show here that the family of controllers defined by (10.27), with ε replaced by an adaptive-low-gain parameter $\varepsilon_a(\overline{x})$, solves Problem 10.2.

Generalized adaptive-low-gain design for global stabilization: We consider the adaptive parameter $\varepsilon_a(\overline{x})$ introduced earlier for continuous-time case and having the same properties as given there, but now selected as

$$\varepsilon_a(\overline{x}) = \max\{\, r \in (0, 1] \mid (\overline{x}' P_r \overline{x}) \operatorname{trace} \mathcal{B}' P_r \mathcal{B} \leqslant \tfrac{\delta_1^2}{M_p} \,\} \tag{10.30}$$

where P_r is the unique positive definite solution of DARE (10.25) with $\varepsilon = r$, δ_1 is defined by (10.22) and

$$M_p = \sigma_{\max}(P_1^{\frac{1}{2}} \mathcal{B}\mathcal{B}' P_1^{\frac{1}{2}}) + 1.$$

Here P_1 is the solution of DARE (10.25) with $\varepsilon = 1$. It can be shown easily that this adaptation guarantees that

$$\|(\mathcal{B}' P_{\varepsilon_a(\overline{x})} \mathcal{B} + I)^{-1} \mathcal{B}' P_{\varepsilon_a(\overline{x})} \widetilde{\mathcal{A}} \overline{x}\| \leqslant \delta_1,$$

where $P_{\varepsilon_a(\overline{x})}$ is the same as P_ε, which is the solution of DARE (10.25) with ε replaced by $\varepsilon_a(\overline{x})$.

To prove Theorem 10.8, we need the following lemma which is analogous to Lemma 10.18 and which defines a control law that stabilizes the linear system (10.24).

Lemma 10.23 *Consider the system* (10.24) *and assume that* $(\widetilde{\mathcal{A}}, \mathcal{B})$ *as given by* (10.12) *is stabilizable and that the eigenvalues of* $\widetilde{\mathcal{A}}$ *are within the closed unit disc. The control law*

$$v = -(\mathcal{B}' P_{\varepsilon_a(\overline{x})} \mathcal{B} + I)^{-1} \mathcal{B}' P_{\varepsilon_a(\overline{x})} \widetilde{\mathcal{A}} \overline{x} \tag{10.31}$$

achieves global stability of the equilibrium $\overline{x} = 0$. *Moreover, the constraint* $\|v(k)\| \leq \delta_1$ *does not get violated for any* $k \geq 0$.

We can now use Lemma 10.23 to prove that a particular family of control laws achieves global stability of the single-layer nonlinear sandwich system.

Theorem 10.24 *Consider the systems given by* (10.1) *and* (10.2), *satisfying Conditions i and ii of Theorem 10.8. Choose an arbitrary matrix* F *such that* $A + BF$ *is asymptotically stable. Let* $P_{\varepsilon_a(\overline{x})}$ *be the unique positive definite solution of DARE* (10.25), *with* ε *replaced by the adaptive-low-gain parameter* $\varepsilon_a(\overline{x})$ *defined by* (10.30). *Then, the generalized adaptive-low-gain state feedback control law*

$$u = Fx - (\mathcal{B}' P_{\varepsilon_a(\overline{x})} \mathcal{B} + I)^{-1} \mathcal{B}' P_{\varepsilon_a(\overline{x})} \widetilde{\mathcal{A}} \overline{x} \tag{10.32}$$

achieves global asymptotic stability of the origin where $\widetilde{\mathcal{A}}$ *and* $\mathcal{B}$ *are given by* (10.12).

Proof : Considering the interconnection of (10.1) and (10.2), we note that close to the origin the saturation does not get activated. Moreover, close to the origin the feedback (10.32) is given by

$$u = Fx - (\mathcal{B}'P_1\mathcal{B} + I)^{-1}\mathcal{B}'P_1\tilde{\mathcal{A}}\bar{x},$$

where P_1 is the solution of (10.25) with $\varepsilon = 1$. This immediately yields that the interconnection of (10.1), (10.2), and (10.32) is locally asymptotically stable. It remains to show that we have global asymptotic stability.

Consider an arbitrary initial condition $x(0)$ and $\omega(0)$. Then there exists a $K > 0$ such that

$$\|C(A + BF)^k x(0)\| < \tfrac{1}{2}$$

for $k > K$. Moreover, by construction

$$v = -(\mathcal{B}'P_{\varepsilon_a(\bar{x})}\mathcal{B} + I)^{-1}\mathcal{B}'P_{\varepsilon_a(\bar{x})}\tilde{\mathcal{A}}\bar{x}$$

yields $\|v(k)\| \leqslant \delta_1$ for all $k > 0$. However, this implies that $z(k)$ generated by (10.19) satisfies $\|z(k)\| < 1$ for all $k > K$. But this yields that the interconnection of (10.1) and (10.2) with controller (10.32) behaves for $k > K$ like the interconnection of (10.24) with controller (10.31). From Lemma 10.23, global asymptotic stability of the latter system then implies that $\bar{x}(k) \to 0$ as $k \to \infty$. Since this property holds for any initial condition and since we have local asymptotic stability, we can conclude that the controller yields global asymptotic stability. This completes the proof. ■

10.5 Low-gain design for single-layer systems with actuator saturation

In this section, we explicitly construct controllers to solve the semi-global and global stabilization problems described in Sect. 10.2 for single-layer systems where the actuator is subject to saturation. In doing so, we prove sufficiency of the conditions in Theorems 10.9 and 10.10. We divide our development into two subsections, one for continuous-time systems and the other for discrete-time systems.

10.5.1 Continuous-time systems

We first present a generalized low-gain design for solving Problem 10.3 which concerns with semi-global stabilization of the origin of the single-layer sandwich system subject to actuator saturation described by (10.3) and (10.4).

Generalized low-gain design for semi-global stabilization: We define the following family of generalized low-gain state feedback control laws:

$$u = -\mathcal{B}' P_{\varepsilon_1,\varepsilon_2} \begin{pmatrix} x \\ \omega \end{pmatrix} = F_{1,\varepsilon_1,\varepsilon_2} x + F_{2,\varepsilon_1,\varepsilon_2} \omega, \tag{10.33}$$

where $P_{\varepsilon_1,\varepsilon_2} = P'_{\varepsilon_1,\varepsilon_2} > 0$ satisfies the CARE,

$$\mathcal{A}' P_{\varepsilon_1,\varepsilon_2} + P_{\varepsilon_1,\varepsilon_2} \mathcal{A} - P_{\varepsilon_1,\varepsilon_2} \mathcal{B}\mathcal{B}' P_{\varepsilon_1,\varepsilon_2} + \begin{pmatrix} \varepsilon_1 I & 0 \\ 0 & \varepsilon_2 I \end{pmatrix} = 0. \tag{10.34}$$

The family of control laws is parameterized by the parameters $\varepsilon_1, \varepsilon_2 > 0$, and in the following, we show that semi-global stabilization is achieved for suitably chosen values of these parameters.

By Conditions (*ii*) and (*iii*) of Theorem 10.9, we know that the eigenvalues of M and A are in the closed left-half plane, and this implies that

$$\lim_{\varepsilon_2 \to 0} P_{\varepsilon_1,\varepsilon_2} = \begin{pmatrix} P_{\varepsilon_1} & 0 \\ 0 & 0 \end{pmatrix}, \qquad \lim_{\substack{\varepsilon_1 \to 0 \\ \varepsilon_2 \to 0}} P_{\varepsilon_1,\varepsilon_2} = 0,$$

where $P_{\varepsilon_1} = P'_{\varepsilon_1} > 0$ is the unique positive definite solution of the CARE,

$$A' P_{\varepsilon_1} + P_{\varepsilon_1} A - P_{\varepsilon_1} B B' P_{\varepsilon_1} + \varepsilon_1 I = 0. \tag{10.35}$$

This in turn implies that

$$F_{1,\varepsilon_1,\varepsilon_2} \to F_{\varepsilon_1} := -B' P_{\varepsilon_1}, \qquad F_{2,\varepsilon_1,\varepsilon_2} \to 0 \tag{10.36}$$

as $\varepsilon_2 \to 0$. Note that the initial conditions belong to some compact set $\mathcal{W}$, and hence there exist compact sets $\mathcal{X} \subset \mathbb{R}^{n_1}$ and $\Omega \subset \mathbb{R}^{n_2}$ such that $x(0) \in \mathcal{X}$ and $\omega(0) \in \Omega$.

For $u = F_{\varepsilon_1} x$, there exists an $\varepsilon_1^* > 0$ such that for all $0 < \varepsilon_1 < \varepsilon_1^*$ and for all $x(0) \in \mathcal{X}$,

$$\|F_{\varepsilon_1} x(t)\| = \|F_{\varepsilon_1} e^{(A+BF_{\varepsilon_1})t} x(0)\| \leqslant \tfrac{1}{4}. \tag{10.37}$$

Moreover, there exists a $T > 0$, dependent on ε_1, such that

$$x(T) \in \mathcal{V}_{\varepsilon_1} := \left\{ x \in \mathbb{R}^{n_1} \mid x' P_{\varepsilon_1} x < \delta \right\}$$

for all $x(0) \in \mathcal{X}$. Here $\delta > 0$ is such that $x \in \mathcal{V}_{\varepsilon_1}$ implies that $\|Cx\| \leqslant \frac{1}{4}$ and $\|F_{\varepsilon_1} x\| \leqslant \frac{1}{4}$. Since $\omega(0) \in \Omega$, where Ω is a compact set, and $\sigma(z(t))$ is bounded, it follows that there exists a compact set $\bar{\Omega}$, independent of ε_2, such that $\omega(t) \in \bar{\Omega}$ for all $t \in [0, T]$.

Next, we note that for $u = F_{\varepsilon_1} x$, we have $x(T) \in \mathcal{V}_{\varepsilon_1}$ and hence there exists an ε_2^*, dependent on ε_1, such that for all $0 < \varepsilon_2 < \varepsilon_2^*$, we have $x(T) \in 2\mathcal{V}_{\varepsilon_1}$ for

$u = F_{1,\varepsilon_1,\varepsilon_2}x + F_{2,\varepsilon_1,\varepsilon_2}\omega$. Here we use (10.36) and our earlier conclusion that $\omega(t)$ is bounded on $[0, T]$.

Define

$$\mathcal{V}_{\varepsilon_1,\varepsilon_2} = \left\{ \bar{x} \in \mathbb{R}^{n_1+n_2} \mid \bar{x}' P_{\varepsilon_1,\varepsilon_2}\bar{x} < 9\delta \right\}.$$

There exists an $\varepsilon_2^\#$ with $0 < \varepsilon_2^\# < \varepsilon_2^*$ such that, for $0 < \varepsilon_2 < \varepsilon_2^\#$, the following holds:

- If $x \in 2\mathcal{V}_{\varepsilon_1}$ and $\omega \in \bar{\Omega}$, then $(x, \omega) \in \mathcal{V}_{\varepsilon_1,\varepsilon_2}$.
- For any (x, ω) such that $(x, \omega) \in \mathcal{V}_{\varepsilon_1,\varepsilon_2}$, we have $x \in 3\mathcal{V}_{\varepsilon_1}$ and

$$\|F_{1,\varepsilon_1,\varepsilon_2}x + F_{2,\varepsilon_1,\varepsilon_2}\omega\| < 1.$$

To see this, note that $x \in 3\mathcal{V}_{\varepsilon_1}$ implies that $\|Cx\| \leqslant \frac{3}{4}$.

Consider the controller,

$$v = -\begin{pmatrix} B' & 0 \end{pmatrix} P_{\varepsilon_1,\varepsilon_2} \begin{pmatrix} x \\ \omega \end{pmatrix} = -\mathcal{B}' P_{\varepsilon_1,\varepsilon_2}\bar{x}, \tag{10.38}$$

for $0 < \varepsilon_2 < \varepsilon_2^\#$ where $\bar{x}$ is as defined in (10.12). This controller stabilizes the linear system

$$\dot{\bar{x}} = \mathcal{A}\bar{x} + \mathcal{B}v, \tag{10.39}$$

where $\mathcal{A}$ and $\mathcal{B}$ are as in (10.6). Given $x(T) \in 2\mathcal{V}_{\varepsilon_1}$ and $\omega(T) \in \bar{\Omega}$, we know that $(x(T), \omega(T)) \in \mathcal{V}_{\varepsilon_1,\varepsilon_2}$. Then the controller (10.38) applied to the system (10.39) guarantees that $(x(t), \omega(t)) \in \mathcal{V}_{\varepsilon_1,\varepsilon_2}$ for all $t > T$, which, given the properties of the set, implies that $\|v(t)\| < 1$ and $\|Cx(t)\| < 1$ for all $t > T$. Then the interconnection of (10.1) and (10.2) with controller (10.33) for $t > T$ is equivalent to the interconnection of (10.39) with controller (10.38) for $t > T$. The asymptotic stability of (10.39) with controller (10.38) implies that $x(t) \to 0$ and $\omega(t) \to 0$. Since this holds for any $(x(0), \omega(0)) \in \mathcal{W}$, it follows that $\mathcal{W}$ is contained in the domain of attraction. We have therefore established Theorem 10.25 given below, which proves that the family of control laws defined by (10.33) achieves semi-global stability.

Theorem 10.25 *Consider the systems given by* (10.3) *and* (10.4) *that satisfy Conditions (i), (ii), and (iii) of Theorem 10.9. For any compact set of initial conditions* $\mathcal{W} \in \mathbb{R}^{n_1+n_2}$, *there exists an* $\varepsilon_1^* > 0$ *such that for all* ε_1 *with* $0 < \varepsilon_1 < \varepsilon_1^*$, *there exists an* $\varepsilon_2^*(\varepsilon_1)$ *such that for all* $0 < \varepsilon_2 < \varepsilon_2^*(\varepsilon_1)$, *the controller defined by* (10.33) *asymptotically stabilizes the origin with a domain of attraction containing* $\mathcal{W}$.

We now present a generalized adaptive-low-gain design for solving Problem 10.4 which concerns with global stabilization of the single-layer sandwich system subject to actuator saturation described by (10.3) and (10.4).

Generalized adaptive-low-gain design for global stabilization: As in the previous section, we design a controller based on appropriate adaptation of low-gain parameter.

Let $P_{1,\varepsilon_1} = P'_{1,\varepsilon_1} > 0$ be the unique positive definite solution of the CARE,

$$A'P_{1,\varepsilon_1} + P_{1,\varepsilon_1}A - P_{1,\varepsilon_1}BB'P_{1,\varepsilon_1} + \varepsilon_1 I = 0. \tag{10.40}$$

We consider the adaptive parameter $\varepsilon_{a,1}(x)$ having the same properties as given in Sect. 10.4.1 on page 552. A particular choice satisfying such properties is given by

$$\varepsilon_{a,1}(x) = \max\left\{ r \in (0,1] \mid (x'P_{1,r}x)\,\mathrm{trace}\, B'P_{1,r}B \leqslant \frac{1}{4} \right\}. \tag{10.41}$$

Let $\ell > 0$ be such that $\varepsilon_{a,1}(x) < 1$ implies that $\|x\| > \ell$.

Next, consider $P_{2,\varepsilon_2} = P'_{2,\varepsilon_2} > 0$ satisfying

$$\begin{pmatrix} A+BF & 0 \\ NC & M \end{pmatrix}' P_{2,\varepsilon_2} + P_{2,\varepsilon_2}\begin{pmatrix} A+BF & 0 \\ NC & M \end{pmatrix} - P_{2,\varepsilon_2}\begin{pmatrix} BB' & 0 \\ 0 & 0 \end{pmatrix} P_{2,\varepsilon_2} + \varepsilon_2 I = 0, \tag{10.42}$$

where $F = -B'P_{1,1}$. Choose

$$\delta \leqslant \max\left\{ \frac{1}{2}, \frac{\ell}{4\|B'P_{1,1}\|} \right\},$$

such that the fact $\|v(t)\| < \delta$ for all $t > 0$ guarantees the fact that $\|z_0(t)\| < \frac{1}{2}$ for all $t > 0$, where z_0 is defined by (10.9). Consider an associated adaptive parameter $\varepsilon_{a,2}(\bar{x})$ that satisfies the same properties as given in Sect. 10.4.1 on page 552. A particular choice satisfying such properties is given by

$$\varepsilon_{a,2}(\bar{x}) = \max\left\{ r \in (0,1] \mid (\bar{x}'P_{2,r}\bar{x}) \left\| \begin{pmatrix} B \\ 0 \end{pmatrix}' P_{2,r} \begin{pmatrix} B \\ 0 \end{pmatrix} \right\| \leqslant \delta^2 \right\}. \tag{10.43}$$

The following theorem shows that a specified control law achieves global stability of the single-layer nonlinear sandwich system subject to actuator saturation.

Theorem 10.26 *Consider the two systems given by* (10.3) *and* (10.4) *that satisfy Conditions (i)–(iii) of Theorem 10.10. Let* $P_{1,\varepsilon_{a,1}(x)}$ *be defined by* (10.40) *with* ε_1 *replaced by the adaptive-low-gain parameter* $\varepsilon_{a,1}(x)$ *defined by* (10.41). *Let* $P_{2,\varepsilon_{a,2}(\bar{x})}$ *be defined by* (10.42) *with* ε_2 *replaced by the adaptive-low-gain parameter* $\varepsilon_{a,2}(\bar{x})$ *defined by* (10.43)*. Define the generalized adaptive-low-gain state feedback control law as*

$$u = -B'P_{1,\varepsilon_{a,1}(x)}x - \varepsilon_{a,1}(x)\begin{pmatrix} B' & 0 \end{pmatrix} P_{2,\varepsilon_{a,2}(\bar{x})}\bar{x}, \tag{10.44}$$

where $\bar{x}$ *is as defined in* (10.12)*. Then, the control law* (10.44) *achieves global asymptotic stability of the origin.*

Proof : Considering the interconnection of (10.3) and (10.4), we note that the saturation is not activated near the origin. Moreover, near the origin the control law (10.44) is given by $u = Fx - \mathcal{B} P_{2,1}\bar{x}$ where $F = -B'P_{1,1}$. This means that state matrix of the interconnection of (10.3), (10.4), and (10.44) equals

$$\tilde{A} - \mathcal{B}\mathcal{B}' P_{2,1},$$

which is Hurwitz by the properties of the ARE. We have therefore established local asymptotic stability. It remains to show that we have global asymptotic stability.

Define $\mathcal{V} = \{ x \in \mathbb{R}^{n_1} \mid \varepsilon_{a,1}(x) = 1 \}$. We want to establish that $\varepsilon_{a,1}(x)$ is strictly increasing in time when $x(t) \notin \mathcal{V}$.

Assume that this is not the case and we can find some $x(t) \notin \mathcal{V}$ such that $\varepsilon_{a,1}(x(t))$ is nonincreasing, that is, the derivative with respect to time is less than or equal to zero. We obtain

$$\begin{aligned}\frac{\mathrm{d}}{\mathrm{d}t} x'(t) P_{1,\varepsilon_{a,1}(x(t))} x(t) &= -\varepsilon_{a,1}(x(t)) x'(t) x(t) + x'(t) \left[\frac{\mathrm{d}}{\mathrm{d}t} P_{1,\varepsilon_{a,1}(x(t))} \right] x(t) \\ &\quad - x'(t) P_{1,\varepsilon_{a,1}(x(t))} BB' P_{1,\varepsilon_{a,1}(x(t))} x(t) \\ &\quad - 2\varepsilon_{a,1}(x(t)) x'(t) P_{1,\varepsilon_{a,1}(x(t))} B \begin{pmatrix} B \\ 0 \end{pmatrix}' P_{2,\varepsilon_{a,2}(\bar{x}(t))} \bar{x}(t).\end{aligned}$$

Since the derivative of $\varepsilon_{a,1}(x(t))$ with respect to time is less than or equal to zero, the properties of our adaptation imply that

$$\frac{\mathrm{d}}{\mathrm{d}t} P_{1,\varepsilon_{a,1}(x(t))} \leqslant 0.$$

Next, our adaptation guarantees that

$$\begin{aligned}\left\| x'(t) P_{1,\varepsilon_{a,1}(x(t))} B \begin{pmatrix} B' & 0 \end{pmatrix} P_{2,\varepsilon_{a,2}(\bar{x}(t))} \bar{x}(t) \right\| &\leqslant \delta \| B' P_{1,\varepsilon_{a,1}(x(t))} \| \| x(t) \| \\ &\leqslant \frac{\delta}{\ell} \| B' P_{1,1} \| \| x(t) \|^2 \\ &\leqslant \tfrac{1}{4} x'(t) x(t).\end{aligned}$$

Combining the above expressions, we obtain

$$\frac{\mathrm{d}}{\mathrm{d}t} x'(t) P_{1,\varepsilon_{a,1}(x(t))} x(t) \leqslant -\tfrac{1}{2} \varepsilon_{a,1}(x(t)) x'(t) x(t).$$

However, if $x'(t) P_{1,\varepsilon_{a,1}(x(t))} x(t)$ is decreasing, then the properties of our adaptation guarantee that $\varepsilon_{a,1}(x(t))$ is strictly increasing, which yields a contradiction.

Hence, if $x(t) \notin \mathcal{V}$, we find that $\varepsilon_{a,1}(x(t))$ is strictly increasing, and it follows that $x(t)$ converges to $\mathcal{V}$ and that it cannot escape from $\mathcal{V}$. On $\mathcal{V}$ we have $\varepsilon_{a,1} = 1$. Moreover,

$$u = Fx - \begin{pmatrix} B' & 0 \end{pmatrix} P_{\varepsilon_{a,2}(\bar{x})} \bar{x}$$

satisfies $\sigma(u) = u$. We can then apply Lemma 10.18 as in the previous subsection, and we conclude that the system therefore behaves like a stable system after a finite amount of time, and it follows that $x(t) \to 0$ and $\omega(t) \to 0$ as $t \to \infty$. ■

Remark 10.27 *Theorems 10.19 and 10.26 are stated with respect to the adaptive parameters described by* (10.16), (10.41), (10.43). *We remark, however, that these theorems are valid for all choices of adaptive parameters that satisfy the conditions enumerated earlier.*

10.5.2 Discrete-time systems

This subsection pertaining to discrete-time systems is conceptually similar to the previous subsection that deals with continuous-time systems.

We now present a generalized low-gain design for solving Problem 10.3, concerning semi-global stabilization of the origin of the single-layer sandwich system subject to input saturation described by (10.3) and (10.4).

Generalized low-gain design for semi-global stabilization: Let the matrix P_{1,ε_1} be the unique positive definite solution of the DARE,

$$P_{1,\varepsilon_1} = A'P_{1,\varepsilon_1}A + \varepsilon_1 I - A'P_{1,\varepsilon_1}B(B'P_{1,\varepsilon_1}B + I)^{-1}B'P_{1,\varepsilon_1}A, \tag{10.45}$$

and define

$$F_{1,\varepsilon_1} = -(B'P_{1,\varepsilon_1}B + I)^{-1}B'P_{1,\varepsilon_1}A. \tag{10.46}$$

Next, let $P_{2,\varepsilon_2} = P'_{2,\varepsilon_2} > 0$ be the unique positive definite solution of the DARE,

$$P_{2,\varepsilon_2} = \tilde{\mathcal{A}}'P_{2,\varepsilon_2}\tilde{\mathcal{A}} + \varepsilon_2 I - \tilde{\mathcal{A}}'P_{2,\varepsilon_2}\mathcal{B}(\mathcal{B}'P_{2,\varepsilon_2}\mathcal{B} + I)^{-1}\mathcal{B}'P_{2,\varepsilon_2}\tilde{\mathcal{A}}, \tag{10.47}$$

where $\tilde{\mathcal{A}}$ and $\mathcal{B}$ are given by

$$\tilde{\mathcal{A}} = \begin{pmatrix} A + BF_{1,\varepsilon_1} & 0 \\ NC & M \end{pmatrix}, \quad \mathcal{B} = \begin{pmatrix} B \\ 0 \end{pmatrix}.$$

Also define

$$F_{2,\varepsilon_2} = -(\mathcal{B}'P_{2,\varepsilon_2}\mathcal{B} + I)^{-1}\mathcal{B}'P_{2,\varepsilon_2}\tilde{\mathcal{A}}. \tag{10.48}$$

We define next the family of control laws,

$$u = F_{1,\varepsilon_1}x + F_{2,\varepsilon_2}\bar{x}. \tag{10.49}$$

This family of control laws is parameterized by the parameters $\varepsilon_1, \varepsilon_2 > 0$, and we show next that the semi-global stabilization is achieved for suitably chosen values of these parameters.

Theorem 10.28 *Consider the systems given by* (10.3) *and* (10.4)*, satisfying conditions (i)–(iii) of Theorem 10.9. For any compact set of initial conditions* $\mathcal{W} \in \mathbb{R}^{n_1+n_2}$*, there exists an* $\varepsilon_1^* > 0$ *such that for any* ε_1 *with* $0 < \varepsilon_1 < \varepsilon_1^*$*, there exists an* $\varepsilon_2^*(\varepsilon_1)$ *such that for all* $0 < \varepsilon_2 < \varepsilon_2^*(\varepsilon_1)$*, the controller defined by* (10.49) *asymptotically stabilizes the origin with a domain of attraction containing* $\mathcal{W}$*.*

Proof : By conditions (*ii*) and (*iii*) of Theorem 10.9, we know that the eigenvalues of A and M are in the closed unit disc. This implies that

$$\lim_{\varepsilon_1 \to 0} P_{1,\varepsilon_1} = 0, \qquad \lim_{\varepsilon_2 \to 0} P_{2,\varepsilon_2} = 0,$$

and hence implies that

$$\lim_{\varepsilon_1 \to 0} F_{1,\varepsilon_1} = 0, \qquad \lim_{\varepsilon_2 \to 0} F_{2,\varepsilon_2} \to 0. \tag{10.50}$$

Note that the initial conditions belong to some compact set $\mathcal{W}$, and hence there exist compact sets $\mathcal{X} \subset \mathbb{R}^{n_1}$ and $\Omega \subset \mathbb{R}^{n_2}$ such that $x(0) \in \mathcal{X}$ and $\omega(0) \in \Omega$.

Define a family of sets

$$\mathcal{V}_1(c) = \left\{ x \in \mathbb{R}^{n_1} \mid x' P_{1,\varepsilon_1} x \leqslant c \right\}.$$

If we apply $u = F_{1,\varepsilon_1} x$, there exists an $\varepsilon_1^* > 0$ such that for all $0 < \varepsilon_1 < \varepsilon_1^*$ and for all $x(0) \in \mathcal{X}$,

$$\| F_{1,\varepsilon_1} (A + B F_{1,\varepsilon_1})^k x(0) \| \leqslant \tfrac{1}{4}. \tag{10.51}$$

Moreover, there exists a $K > 0$, dependent on ε_1, such that $x(K) \in \mathcal{V}_1(c_1)$ for all $x(0) \in \mathcal{X}$. Here c_1 is such that $x \in \mathcal{V}_1(c_1)$ implies that $\|Cx\| \leqslant \frac{1}{4}$ and $\|F_{1,\varepsilon_1} x\| \leqslant \frac{1}{4}$. Since $\omega(0) \in \Omega$, where Ω is a compact set, and $\sigma(z(k))$ is bounded, it follows that there exists a compact set $\bar{\Omega}$, independent of ε_2, such that $\omega(k) \in \bar{\Omega}$ for all $0 \leqslant k \leqslant K$.

Define a family of level sets

$$\mathcal{V}_2(c) = \left\{ \bar{x} \in \mathbb{R}^{n_1+n_2} \mid x' P_{1,\varepsilon_1} x + \bar{x}' P_{2,\varepsilon_2} \bar{x} \leqslant c \right\}.$$

Next, we note that for $u = F_{1,\varepsilon_1} x$, we have $x(K) \in \mathcal{V}_1(c_1)$. If we apply $u = F_{1,\varepsilon_1} x + F_{2,\varepsilon_2} \bar{x}$, from (10.50) and our earlier conclusion that $\omega(k)$ is bounded for $0 \leqslant k \leqslant K$, we see that there exists an ε_2^*, dependent on ε_1, such that for all $0 < \varepsilon_2 < \varepsilon_2^*$, the following properties hold:

- $x(K) \in 2\mathcal{V}_1(c_1)$.
- If $x \in 2\mathcal{V}_1(c_1)$ and $\omega \in \bar{\Omega}$, then $\bar{x} \in 3\mathcal{V}_2(c_1)$.
- For any $\bar{x}$ such that $\bar{x} \in 3\mathcal{V}_2(c_1)$, we have $\|F_{2,\varepsilon_2} \bar{x}\| < \frac{1}{4}$.

At time $k = K$, we have $\bar{x} \in 3\mathcal{V}_2(c_1)$. This immediately implies that $\|F_{2,\varepsilon_2}\bar{x}\| \leqslant \frac{1}{4}$.

Note that for any $\bar{x} \in 3\mathcal{V}_2(c_1)$, we have

$$x' P_{1,\varepsilon} x \leqslant x' P_{1,\varepsilon} x + \bar{x}' P_{2,\varepsilon_2} \bar{x} \leqslant 9c_1,$$

and hence $x \in 3\mathcal{V}_1(c_1)$. But this implies that $\|F_{1,\varepsilon_1}x\| \leqslant \frac{3}{4}$. Therefore, we have

$$\|F_{1,\varepsilon_1}x + F_{2,\varepsilon_2}\bar{x}\| \leqslant 1.$$

Similarly, for any $\bar{x} \in 3\mathcal{V}_2(c_1)$, we have $x \in 3\mathcal{V}_1(c_1)$, and this implies that $\|Cx\| \leqslant \frac{3}{4}$. Therefore, for any $\bar{x} \in 3\mathcal{V}_2(c_1)$, both saturations are inactive.

We know at time K, the closed-loop system is linear and can be written as

$$\bar{x}(k+1) = (\tilde{\mathcal{A}} + \mathcal{B}F_{2,\varepsilon_2})\bar{x}(k). \tag{10.52}$$

It is straightforward to see that (10.52) is asymptotically stable and $3\mathcal{V}(c_1)$ is invariant. We know that both the saturations will remain inactive for all $k \geqslant K$. The asymptotic stability of (10.52) implies that $\bar{x}(k) \to 0$ as $k \to \infty$.

Since this holds for any $\bar{x}(0) \in \mathcal{W}$, it follows that $\mathcal{W}$ is contained in the domain of attraction. This completes the proof. ■

We now present a generalized adaptive-low-gain design methodology for solving Problem 10.4, concerning global stabilization of the single-layer sandwich system subject to input saturation described by (10.3), (10.4).

Generalized adaptive-low-gain design for global stabilization: As in previous section, this controller is formed from the semi-global controller (10.49) while using appropriate adaptation of low-gain parameters.

Let $P_{1,\varepsilon_1} = P_{1,\varepsilon_1}' > 0$ be the unique positive definite solution of the DARE (10.45) and F_{1,ε_1} be defined as (10.46) with adaptive parameter $\varepsilon_1 = \varepsilon_1(x)$.

As in Sect. 10.4.2, a particular choice of adaptation is given by

$$\varepsilon_1(x) = \max\{\, r \in (0,1] \mid (x' P_{1,r} x) \operatorname{trace} B' P_{1,r} B \leqslant \tfrac{1}{4M_2}\,\} \tag{10.53}$$

where $P_{1,r}$ is the solution of DARE (10.45) with $\varepsilon_1 = r$ and

$$M_2 = \sigma_{\max}(P_{1,1}^{\frac{1}{2}} BB' P_{1,1}^{\frac{1}{2}}) + 1.$$

Here $P_{1,1}$ is the solution of DARE (10.45) with $\varepsilon_1 = 1$.

It can be shown that the above adaptation guarantees that

$$\left\|(B' P_{1,\varepsilon_1(x)} B + I)^{-1} B' P_{1,\varepsilon_1(x)} x\right\| \leqslant \tfrac{1}{2}.$$

Let $\ell > 0$ be such that

$$(\lambda_{\max}(P_{1,1}) + \tfrac{1}{2})\ell^2 \leqslant \frac{1}{4M_2\|B' P_{1,1} B\|}.$$

Next, let $P_{2,\varepsilon_2} = P_{2,\varepsilon_2}' > 0$ be the unique positive definite solution of the DARE (10.47) and F_{2,ε_2} be defined by (10.48) where $\varepsilon_2 = \varepsilon_2(\overline{x})$ is an adaptive parameter and where, in both (10.47) and (10.48), we use

$$\widetilde{\mathcal{A}} = \begin{pmatrix} A + BF_{1,1} & 0 \\ NC & M \end{pmatrix}, \quad \mathcal{B} = \begin{pmatrix} B \\ 0 \end{pmatrix}. \tag{10.54}$$

Choose

$$\delta_2 = \min\left\{ \frac{1}{2}, \frac{\ell^2}{2(3\,\|B'P_{1,1}B\| + 1)}, \frac{1}{2\widetilde{\rho}} \right\}, \tag{10.55}$$

where $\widetilde{\rho} = \sum_{k=0}^{\infty} \|C(A+BF_{1,1})^k B\|$. Consider an associated adaptive parameter given by

$$\varepsilon_2(\overline{x}) = \max\{\, r \in (0,1] \mid (\overline{x}'P_{2,r}\overline{x}) \operatorname{trace} \mathcal{B}'P_{2,r}\mathcal{B} \leqslant \tfrac{\delta_2^2}{M_3} \,\} \tag{10.56}$$

where $P_{2,r}$ is the solution of DARE (10.47) with $\varepsilon_2 = r$ and

$$M_3 = \sigma_{\max}(P_{2,1}^{\frac{1}{2}} \mathcal{B}\mathcal{B}' P_{2,1}^{\frac{1}{2}}) + 1.$$

Here $P_{2,1}$ is the solution of DARE (10.47) with $\varepsilon_2 = 1$. We have $\|F_{2,\varepsilon}\| \leqslant \delta_2$.

The following theorem shows that a particular family of control laws achieves global stability of the single-layer nonlinear sandwich system subject to input saturation.

Theorem 10.29 *Consider the two systems given by* (10.3) *and* (10.4)*, satisfying conditions (i)–(iii) of Theorem 10.10. Let $P_{1,\varepsilon_1(x)}$ be the solution of DARE* (10.45) *with ε_1 replaced by the adaptive-low-gain parameter $\varepsilon_1(x)$ defined by* (10.53)*. Let $P_{2,\varepsilon_2(\overline{x})}$ be the solution of DARE* (10.47) *with ε_2 replaced by the adaptive-low-gain parameter $\varepsilon_2(\overline{x})$ defined by* (10.56)*. Then, the generalized adaptive-low-gain state feedback control law,*

$$u = F_{1,\varepsilon_1(x)}x + \varepsilon_1(x)F_{2,\varepsilon_2(\overline{x})}\overline{x}, \tag{10.57}$$

achieves global asymptotic stability of the origin where $F_{1,\varepsilon_1(x)}$ and $F_{2,\varepsilon_2(\overline{x})}$ are respectively defined by (10.46) *and* (10.48) *with ε_1 and ε_2 replaced by $\varepsilon_1(x)$ and $\varepsilon_2(\overline{x})$.*

Proof : Note that our adaptive parameter guarantees that $\|u(k)\| \leqslant 1$ for all $k \geqslant 0$. The input saturation is always inactive.

Considering the interconnection of (10.3) and (10.4), we note that the sandwiched saturation is not activated near the origin. Moreover, near the origin, the control law (10.57) is given by $u = F_{1,1}x + F_{2,1}\bar{x}$ where $F_{1,1}$ is $F_{1,\varepsilon_1(x)}$ with $\varepsilon_1(x) = 1$ and $F_{2,1}$ is $F_{2,\varepsilon_2(\bar{x})}$ with $\varepsilon_2(\bar{x}) = 1$. This means that state matrix of the interconnection of (10.3), (10.4), and (10.57) equals $\tilde{\mathcal{A}} + \mathcal{B}F_{2,1}$ which is aymptotically stable by the properties of the DARE (here $\tilde{\mathcal{A}}$ is as in (10.54)). We have therefore established local asymptotic stability. It remains to show that we have global asymptotic stability.

Define $V = x'P_{\varepsilon_1(x)}x$ and $\mathcal{V}_1 = \{x \in \mathbb{R}^{n_1} \mid \|x\| \leqslant \ell\}$ and $\mathcal{V}_2 = \{x \in \mathbb{R}^{n_1} \mid V(x) \leqslant (\lambda_{\max}(P_{1,1}) + 1/2)\ell^2\}$. Since $\|x(k)\| \leqslant \ell$ implies that $V(x) \leqslant \lambda_{\max}(P_{\varepsilon_1(x(k))})\|x(k)\|^2 \leqslant \lambda_{\max}(P_{1,1})\ell^2$, we have $\mathcal{V}_2 \supset \mathcal{V}_1$. Moreover, from the definition of ℓ, we have $\varepsilon_1(x) = 1$ for $x \in \mathcal{V}_2$. We first want to establish that $V(k)$ is strictly decreasing in time when $x \notin \mathcal{V}_1$.

Assume that this is not the case and we can find a $x(k) \notin \mathcal{V}_1$ such that $V(k+1) - V(k) \geqslant 0$. Denote $\varepsilon_1(x(k))$ and $P_{1,\varepsilon_1(x(k))}$ by $\varepsilon_1(k)$ and $P_1(k)$, respectively. We obtain

$$\begin{aligned}
&V(k+1) - V(k) \leqslant \\
&-\varepsilon_1(k)x(k)'x(k) - x(k+1)'P_1(k)x(k+1) + x(k+1)'P_1(k+1)x(k+1) \\
&-2x(k)'A'P_1(k)Bv_2(k) - 2v_1(k)'B'P_1(k)Bv_2(k) + v_2(k)'B'P_1(k)Bv_2(k),
\end{aligned}$$

where $v_1(k) = F_{1,\varepsilon_1(k)}x(k)$ and $v_2(k) = -\varepsilon_1(k)F_{2,\varepsilon_2(k)}\bar{x}(k)$.

Our adaptation guarantees that $\|v_1(k)\| \leqslant \frac{1}{2}$ and $\|v_2(k)\| \leqslant \varepsilon_1(k)\delta_2$ and hence

$$\begin{aligned}
\|x(k)'A'P_1(k)Bv_2(k)\| &= \|v_1(k)'(B'P_1(k)B + I)v_2(k)\| \\
&\leqslant \tfrac{1}{2}\varepsilon_1(k)(\|B'P_{1,1}B\| + 1)\delta_2 \\
\|v_1(k)'B'P_1(k)B'v_2(k)\| &\leqslant \tfrac{1}{2}\varepsilon_1(k)\,\|B'P_{1,1}B\|\,\delta_2 \\
\|v_2(k)'B'P_1(k)Bv_2(k)\| &\leqslant \varepsilon_1(k)^2\,\|B'P_{1,1}B\|\,\delta_2^2 \leqslant \varepsilon_1(k)\,\|B'P_{1,1}B\|\,\delta_2.
\end{aligned}$$

Therefore,

$$\begin{aligned}
&V(k+1) - V(k) \\
&\quad\leqslant -\varepsilon_1(k)x'(k)x(k) + x(k+1)'(P_1(k+1) - P_1(k))x(k+1) \\
&\qquad\qquad + \varepsilon_1(k)(3\,\|B'P_{1,1}B\| + 1)\delta_2 \\
&\quad\leqslant -\varepsilon_1(k)x'(k)x(k) + x(k+1)'(P_1(k+1) - P_1(k))x(k+1) + \tfrac{1}{2}\varepsilon_1(k)\ell^2 \qquad (10.58) \\
&\quad\leqslant -\tfrac{1}{2}\varepsilon_1(k)\|x(k)\|^2 + x(k+1)'(P_1(k+1) - P_1(k))x(k+1),
\end{aligned}$$

where we use, in the last inequality, the fact that $x(k) \notin \mathcal{V}_1$ and hence $\|x(k)\| \geqslant \ell$. Since $V(k+1) - V(k) \geqslant 0$, the properties of our scheduling imply that

$$x(k+1)'(P_1(k+1) - P_1(k))x(k+1) \leqslant 0. \tag{10.59}$$

We get

$$V(k+1) - V(k) \leqslant -\tfrac{1}{2}\varepsilon_1(k)\|x(k)\|^2 < 0.$$

This yields a contradiction. Hence, when $x(k) \notin \mathcal{V}_1$, we have $V(k)$ strictly decreasing, and it follows that $x(k)$ enters $\mathcal{V}_1$ within finite time, say K_1. When $x(k) \in \mathcal{V}_1$, we have either $V(k+1) - V(k) \leqslant 0$ or (10.59) and (10.58) yield that

$$V(k+1) - V(k) \leqslant \tfrac{1}{2}\varepsilon_1(k)\ell^2 \leqslant \tfrac{1}{2}\ell^2.$$

This implies that $V(k+1) \leqslant \lambda_{\max}(P_{1,1})\ell^2 + \frac{1}{2}\ell^2$ and hence $x(k+1) \in \mathcal{V}_2$. We find that if $x(k) \in \mathcal{V}_1$, then $x(k+1) \in \mathcal{V}_2$. On the other hand, if $x(k) \in \mathcal{V}_2 \backslash \mathcal{V}_1$, then $V(k)$ is strictly decreasing and hence $x(k+1) \in \mathcal{V}_2$. Therefore, $x(k)$ will enter $\mathcal{V}_2$ and it cannot escape from $\mathcal{V}_2$. On $\mathcal{V}_2$, we have $\varepsilon_1(k) = 1$. The L_1 system then becomes

$$\begin{aligned} x(k+1) &= (A + BF_{1,1})x(k) + Bv_2(k) \\ z(k) &= Cx(k), \end{aligned} \tag{10.60}$$

where $\|v_2(k)\| \leqslant \delta_2$. We have for any $k > K_1$

$$z(k) = C(A + BF_{1,1})^{k-K_1}x(K_1) + z_0(k)$$

where

$$z_0(k) = \sum_{i=K_1}^{k-1} C(A + BF_{1,1})^{k-i-1}Bv_2(i). \tag{10.61}$$

Given that $\delta_2 \leqslant \frac{1}{2\rho}$ as given by (10.55), we have $\|v(k)\| < \frac{1}{2\rho}$ for all $k > K_1$. But this guarantees that $\|z_0(k)\| < \frac{1}{2}$ for all $k > K_1$, where $z_0(k)$ is defined by (10.61). Therefore, there exists a K_2 such that for $k \geqslant K_2$,

$$\|C(A + BF_{1,1})^{k-K_1}x(K_1)\| \leqslant \tfrac{1}{2}$$

and hence $\|z(k)\| \leqslant 1$ for $k \geqslant K_2$. We can then apply Lemma 10.23 as in the previous subsection, and we conclude that the system therefore behaves like a stable system after a finite amount of time, and it follows that $x(k) \to 0$ and $\omega(k) \to 0$ as $k \to \infty$. ■

10.6 Low-gain design for multilayer systems with actuator saturation

As mentioned in the introduction, the results presented in this chapter are generalizations of classical low-gain design methodologies for linear systems subject to only actuator saturation. The principle behind classical low-gain design is to create a control law with a sufficiently low gain to avoid saturating the actuator. In the semi-global case, the gain is fixed, based on an a priori given set of possible initial conditions; in the global case, the gain is adapted to be sufficiently low regardless of the initial conditions.

For the single-layer sandwich system shown in Fig. 10.3, the principle is similar. However, the problem is more complex because the sandwiched saturation cannot be made inactive from the start by using low gain. Instead, the sandwiched saturation must be deactivated by controlling the states of the L_1 subsystem toward the origin. Conceptually, the control task can therefore be viewed as consisting of two subtasks. The first subtask is to control the states of the L_1 subsystem toward the origin in order to deactivate the sandwiched saturation after a finite time. Once the sandwiched saturation has been deactivated, the second subtask consists of controlling the state of the whole system to the origin without reactivating the sandwiched saturation.

To accomplish the two subtasks, the control law is divided into two terms: a term Fx that depends only on the state x of the L_1 system, referred to as the *L_1 term*, and a low-gain term that depends on ω as well, referred to as the *L_1/L_2 term*. The L_1 term ensures that x becomes small after an initial transient, so that after a finite time T, the sandwiched saturation is inactive. A sufficiently low gain for the L_1/L_2 term ensures that the sandwiched saturation remains inactive while the states are brought to the origin. This philosophy is what is pursued in Sect. 10.4.

When the single-layer sandwich system is extended with an actuator saturation, as in Fig. 10.4, we cannot freely select the L_1 term to make x small; we must also take care to avoid activation of the actuator saturation. The L_1 term is therefore also designed using a low-gain methodology, to ensure that the sandwiched saturation is inactive after a time T, without activating the actuator saturation in the mean time. The gain for the L_1/L_2 term is chosen sufficiently low, depending on the gain for the L_1 term, to ensure that both the sandwiched saturation and the actuator saturation remain inactive while the states are brought to the origin. This philosophy is what is pursued in Sect. 10.5.

As the above discussion makes clear, the design methodology for single-layer sandwich systems subject to actuator saturation is *nested*: first, the gain for the L_1 term is chosen sufficiently low, and then the gain for the L_1/L_2 term is chosen sufficiently low, depending on the gain of the L_1 term. This methodology can be naturally extended to multilayer sandwich systems of the types depicted in Figs. 10.5 and 10.6. Consider, for example, a double-layer sandwich system subject to input saturation. The control law is split into three terms, referred to as the L_1 term, the L_1/L_2 term, and the $L_1/L_2/L_3$ term. The gain of the L_1 term is chosen sufficiently low to make the first sandwiched saturation inactive after a finite time T_1, without activating the actuator saturation in the mean time. The gain of the L_1/L_2 term is then chosen sufficiently low, depending on the gain of the L_1 term, to ensure that the second sandwiched saturation is inactive after another finite time $T_2 > T_1$, without activating the first sandwiched saturation or the actuator saturation in the mean time. Finally, the gain of the $L_1/L_2/L_3$ term is chosen sufficiently low, depending on the gains of the L_1 term and the L_1/L_2 term, to ensure that the two sandwiched saturations and the actuator saturation remain inactive while the states are brought to the origin.

As in the case of single-layer sandwich systems, semi-global stabilization of multilayer sandwich systems can be achieved using fixed gains, and global stabilization can be achieved by appropriately adapting the gains. In either case, necessary and sufficient conditions for semi-global and global stabilization correspond to the conditions for single-layer sandwich systems. When there is no actuator saturation, the conditions are that (1) the linearized cascade system is stabilizable and (2) the eigenvalues of systems $L_2, L_3, \ldots$ are in the closed left-half plane(continuous-time systems) or within the closed unit disc (discrete-time systems). When the actuator is subject to saturation, the eigenvalues of the L_1 system must also be in the closed left-half plane.

In what follows, we develop the above design philosophy for multilayer sandwich systems by considering discrete-time systems. Although we focus here only on discrete-time systems, the methodology applies equally well for continuous-time systems.

10.6.1 Generalized low-gain design for semi-global stabilization

We now proceed to construct a linear semi-globally stabilizing controller for multilayer sandwich system (10.62) which solves the Problem 10.5 for discrete-time systems.

For discrete-time systems, we rewrite the multilayer nonlinear sandwich systems defined in Sect. 10.2 as

$$L_i : \begin{cases} x_i(k+1) = A_i x_i(k) + B_i \sigma(u_i(k)), \quad i = 1, \ldots, \nu \\ z_i(k) \quad = C_i x_i(k), \quad i = 1, \ldots, \nu - 1 \\ u_i(k) \quad = z_{i-1}(k), \quad i = 2, \ldots, \nu. \end{cases} \tag{10.62}$$

Let P_{ε_i} be the positive definite solution of DARE,

$$P_{\varepsilon_i} = \mathcal{A}_i' P_{\varepsilon_i} \mathcal{A}_i + \varepsilon_i I - \mathcal{A}_i' P_{\varepsilon_i} \mathcal{B}_i (\mathcal{B}_i' P_{\varepsilon_i} \mathcal{B}_i + I)^{-1} \mathcal{B}_i' P_{\varepsilon_i} \mathcal{A}_i, \tag{10.63}$$

and define

$$F_{\varepsilon_i} = -(\mathcal{B}_i' P_{\varepsilon_i} \mathcal{B}_i + I)^{-1} \mathcal{B}_i' P_{\varepsilon_i} \mathcal{A}_i, \tag{10.64}$$

where

$$\mathcal{A}_1 = A_1, \quad \mathcal{A}_i = \begin{pmatrix} \mathcal{A}_{i-1} + \mathcal{B}_{i-1} F_{\varepsilon_{i-1}} & 0 \\ B_i \mathcal{C}_{i-1} & A_i \end{pmatrix}, \quad i = 2, \ldots, \nu, \tag{10.65}$$

and

$$\mathcal{B}_i = \begin{pmatrix} B_1' & 0 & \cdots & 0 \end{pmatrix}', \quad \mathcal{C}_i = \begin{pmatrix} 0 & \ldots & 0 & C_i \end{pmatrix} \tag{10.66}$$

are of appropriate dimensions. The parameters ε_i, $i = 1, \ldots, \nu$ are to be determined subsequently.

We have the following theorem.

Theorem 10.30 *Consider the interconnection of* ν *systems as given by* (10.62), *satisfying conditions (i), (ii) of Theorem 10.11. Let* P_{ε_i} *be the solution of DARE in* (10.63) *with* $\varepsilon_i \in (0, 1]$, $i = 1, \dots, \nu$. *For any compact set* $\mathcal{W} \subset \mathbb{R}^{n_1+\cdots+n_\nu}$, *we can determine* ε_i, $i = 1, \dots, \nu$ *such that the controller*

$$u = \sum_{i=1}^{\nu} F_{\varepsilon_i} \chi_i \tag{10.67}$$

renders the origin asymptotically stable with a domain of attraction containing $\mathcal{W}$ *where*

$$\chi_i = \begin{pmatrix} x_1' & \cdots & x_i' \end{pmatrix}'.$$

Proof : For simplicity of presentation, denote P_{ε_i} and F_{ε_i} by P_i and F_i.

Conditions *(i)* and *(ii)* of Theorem 10.11 and the fact that all $\mathcal{A}_i + \mathcal{B}_i F_i$ are asymptotically stable imply that

$$\lim_{\varepsilon_i \to 0} P_i = 0, \quad \lim_{\varepsilon_i \to 0} F_i = 0. \tag{10.68}$$

Define the function

$$V_i(\chi_i) = \sum_{j=1}^{i} \chi_j' P_j \chi_j,$$

and the sets

$$\mathcal{V}_i(c) = \left\{ \chi_i \in \mathbb{R}^{\sum_{j=1}^{i} n_j} \mid V_i(\chi_i) \leqslant c \right\}.$$

Since $\mathcal{W}$ is compact, there exist for $i = 1, \dots, \nu$, compact sets $\mathcal{W}_i$ such that $\chi_\nu(0) \in \mathcal{W}$ implies that $x_i(0) \in \mathcal{W}_i$. We proceed next to determine ε_i recursively.

Determine ε_1

Let us consider applying a controller $v_1 = F_1 \chi_1 = F_1 x_1$.

Note that (10.68) implies that there exists an ε_1^* such that for any $\varepsilon \in (0, \varepsilon_1^*]$ and $x_1(0) \in \mathcal{W}_1$,

$$\| F_1 (A_1 + B_1 F_1)^k x_1(0) \| \leqslant \tfrac{1}{4^{\nu-1}}$$

for all $k \geqslant 0$.

Let c_1 be such that $x_1 \in \mathcal{V}_1(c_1)$ implies that $\|F_1 x_1\| \leqslant \frac{1}{4^{\nu-1}}$ and $\|C_1 x_1\| \leqslant \frac{1}{3^{\nu-1}}$. Since $A_1 + B_1 F_1$ is asymptotically stable, there exists a K_1 such that for all $x_1 \in \mathcal{W}_1$, we have $x_1(K_1) \in \mathcal{V}_1(c_1)$.

Determine ε_2

Since $x_2(0) \in \mathcal{W}_2$ and the input to L_2 is bounded, there exists a $\bar{\mathcal{W}}_2$ such that

$$x_2(k) \in \bar{\mathcal{W}}_2, \quad \text{for } k \leqslant K_1.$$

Let ε_1 be fixed. Consider applying the controller $v_2 = F_1 x_1 + F_2 \chi_2$. Due to (10.68), given $x_2 \in \mathcal{W}_2$, there exists an $\varepsilon_2^*(\varepsilon_1)$ such that the following properties hold:

(*i*) For any $\varepsilon_2 \in (0, \varepsilon_2^*(\varepsilon_1)]$, $x_1(K_1) \in 2\mathcal{V}_1(c_1)$.

(*ii*) For any $\varepsilon_2 \in (0, \varepsilon_2^*(\varepsilon_1)]$, $x_1 \in 2\mathcal{V}_1(c_1)$ and $x_2 \in \overline{\mathcal{W}}_2$ impliy that $\chi_2 \in 3\mathcal{V}_2(c_1)$.

(*iii*) For any $\varepsilon_2 \in (0, \varepsilon_2^*(\varepsilon_1)]$, $\chi_2 \in 3\mathcal{V}_2(c_1)$ implies that $\|F_2\chi_2\| \leqslant \frac{1}{4^{\nu-1}}$.

At time $k = K_1$, we know that $\chi_2 \in 3\mathcal{V}_2(c_1)$. For any $\chi_2 \in 3\mathcal{V}_2(c_1)$, we have $\|F_2\chi_2\| \leqslant \frac{1}{4^{\nu-1}}$. Also note that $\chi_2 \in 3\mathcal{V}_2(c_1)$ implies that $V_1(x_1) \leqslant 9c_1$. But this implies that $F_1x_1 \leqslant \frac{3}{4^{\nu-1}}$ and $\|C_1x_1\| \leqslant \frac{1}{3^{\nu-2}}$. We have

$$\|u\| = \|F_1x_1 + F_2\chi_2\| \leqslant \frac{1}{4^{\nu-2}}.$$

Therefore, both the first two saturation elements are inactive in $3\mathcal{V}_2(c_1)$, then it is straightforward to see that with controller v_2, we have $\chi_2(k) \in 3\mathcal{V}_2(c_1)$ for all $k \geqslant K_1$ and moreover $\chi_2(k) \to 0$ as $k \to \infty$.

Let c_2 be such that $\chi_2' P_2 \chi_2 \leqslant c_2$ implies that $\|C_1x_1\| \leqslant \frac{1}{3^{\nu-2}}$, and $\|\mathcal{C}_2\chi_2\| \leqslant \frac{1}{3^{\nu-2}}$. There exists then a K_2 such that for all $\chi_2(K_1) \in 3\mathcal{V}_2(c_1)$, we have $\chi_2(K_2) \in \mathcal{V}_2(c_2)$.

At time K_2, we get

(*i*) $\chi_2(K_2) \in \mathcal{V}_2(c_2)$.

(*ii*) $\|C_1x_1(K_2)\| \leqslant \frac{1}{3^{\nu-2}}$ and $\|\mathcal{C}_2\chi_2(K_2)\| \leqslant \frac{1}{3^{\nu-2}}$.

(*iii*) $\|F_1x_1 + F_2\chi_2\| \leqslant \frac{1}{4^{\nu-2}}$ for all $\chi_2 \in \mathcal{V}_2(c_2)$ and $k \leqslant K_2$.

Determine $\varepsilon_3, \dots, \varepsilon_n$

Consider the systems L_i, $i \geqslant 3$. At this moment, ε_j, c_j, and K_j for $j \leqslant i-1$ have been determined in previous $i-1$ steps. The resulting controller $v_{i-1} = \sum_{j=1}^{i-1} F_j\chi_j$ yields

(*i*) $\chi_{i-1}(K_{i-1}) \in \mathcal{V}_{i-1}(c_{i-1})$.

(*ii*) $\|\mathcal{C}_j\chi_j(K_{i-1})\| \leqslant \frac{1}{3^{\nu-i+1}}$ for all $j \leqslant i-1$.

(*iii*) $\|\sum_{j=1}^{i-1} F_j\chi_j\| \leqslant \frac{1}{4^{\nu-i+1}}$ for all $\chi_{i-1} \in \mathcal{V}_{i-1}(c_{i-1})$.

Since the input to L_i is bounded and $x_i(0) \in \mathcal{W}_i$, we know that there exists a $\overline{\mathcal{W}}_i$ such that $x_i(k) \in \overline{\mathcal{W}}_i$ for all $k \leqslant K_{i-1}$.

Consider the controller $v_i = \sum_{j=1}^{i} F_j\chi_j$. Equation (10.68) implies then that there exists an $\varepsilon_i^*(\varepsilon_1, \dots, \varepsilon_{i-1})$ such that the following properties hold:

(*i*) $\chi_{i-1}(K_{i-1}) \in 2\mathcal{V}_{i-1}(c_{i-1})$.

(*ii*) $\chi_{i-1} \in 2\mathcal{V}_{i-1}(c_{i-2})$ and $x_i \in \overline{\mathcal{W}}_i$ imply that $\chi_i \in 3\mathcal{V}_i(c_{i-1})$.

(*iii*) $\chi_i \in 3\mathcal{V}_i(c_{i-1})$ implies that $F_i\chi_i \leqslant \frac{1}{4^{\nu-i+1}}$.

Therefore, we get at $k = K_{i-1}$, $\chi_i(K_{i-1}) \in 3\mathcal{V}_i(c_{i-1})$, that is, $V_i(\chi) \leqslant 9c_{i-1}$. But this implies that $V_{i-1}(\chi) \leqslant 9c_{i-1}$. Hence, we get $\|\mathcal{C}_j\chi_j\| \leqslant \frac{1}{3^{\nu-i}}$ for all $j = 1, \dots, i-1$, and $\|F_i\chi_i\| \leqslant \frac{3}{4^{\nu-i+1}}$. Moreover,

$$\|v_i\| = \|F_j\chi_i\| + \|\sum_{j=1}^{i-1} F_j\chi_j\| \leqslant \frac{1}{4^{\nu-i+1}} + \frac{3}{4^{\nu-i+1}} = \frac{1}{4^{\nu-i}}.$$

Therefore, the first i saturation elements are inactive for any $\chi_i \in 3\mathcal{V}_i(c_{i-1})$. It is easy to see then that, with the controller v_i, $\chi(k) \in 3\mathcal{V}_i(c_{i-1})$ for all $k \geqslant K_{i-1}$ and moreover $\chi_i(k) \to 0$ as $k \to \infty$.

Let c_i be such that $V_i(\chi_i) \leqslant c_i$ implies that $\|\mathcal{C}_j\chi_j\| \leqslant \frac{1}{3^{\nu-i}}$ for all $j \leqslant i$. There exists a K_i such that $\chi_i(K_i) \in \mathcal{V}(c_i)$ for all $\chi_i(K_{i-1}) \in 3\mathcal{V}_i(c_{i-1})$.

At time K_i, we get

(*i*) $\chi_i(K_i) \in \mathcal{V}_i(c_i)$.

(*ii*) $\|\mathcal{C}_j\chi_j(K_{i-1})\| \leqslant \frac{1}{3^{\nu-i}}$ for all $j \leqslant i$.

(*iii*) $\|\sum_{j=1}^{i} F_j\chi_j\| \leqslant \frac{1}{4^{\nu-i}}$ for all $\chi_i \in \mathcal{V}_i(c_i)$.

Repeating this procedure, we can determine $\varepsilon_1, \dots, \varepsilon_\nu$, c_ν, K_ν and a controller $u(\chi_\nu) = v_\nu(\chi_\nu) = \sum_{i=1}^{\nu} F_i\chi_i$ such that for $k \geqslant K_\nu$ we have

(*i*) $\chi_\nu(K_\nu) \in \mathcal{V}_\nu(c_\nu)$.

(*ii*) $\|\mathcal{C}_j\chi_j(K_\nu)\| \leqslant 1$ for all $j \leqslant \nu$.

(*iii*) $\|\sum_{j=1}^{\nu} F_j\chi_j\| \leqslant 1$ for all $\chi_{i-1} \in 3\mathcal{V}_{i-1}(c_0)$.

Then the interconnection of ν systems (10.62) is equivalent to

$$\rho\chi = (\mathcal{A}_\nu + \mathcal{B}_\nu F_\nu)\chi_\nu.$$

The stability of this system implies that $\chi_\nu(k) \to 0$ as $k \to \infty$. This completes the proof. ∎

10.6.2 Generalized adaptive-low-gain design for global stabilization

In this section we construct a global stabilizing controller for a multilayer system to prove sufficiency of conditions (*i*) and (*ii*) in Theorem 10.12. This controller is formed by assembling semi-global stabilizing controller (10.67) with adaptive parameters.

Let $P_{\varepsilon_i(\chi_i)}$ be the positive definite solution of DARE (10.63) and $F_{\varepsilon_i(\chi_i)}$ be defined by (10.64) where $\varepsilon_i = \varepsilon_i(\chi_i)$ is an adaptive parameter, $\mathcal{B}$, is given by (10.66) and $\mathcal{A}$ is given by (10.65) with $F_{\varepsilon_{i-1}}$ replaced by $F_{i-1,1}$. Here we denote $F_{i,1} = F_{\varepsilon_i(\chi_i)}$ with $\varepsilon_i(\chi_i) = 1$.

We need ν adaptive parameters which satisfy similar properties as given in Sect. 10.4.1 on page 552. Choose

$$\delta_1 = \tfrac{1}{\nu},$$

$$\delta_i = \min\left\{\frac{1}{\nu}, \delta_{i-1}, \frac{\ell_{i-1}^2}{2(\nu-i+1)^2(\frac{\nu+2}{\nu}\left\|\mathcal{B}_{i-1}' P_{i-1,1}\mathcal{B}_{i-1}\right\| + \frac{2}{\nu})}, \frac{1}{2(\nu-i+1)\widetilde{\rho}_{i-1}}\right\}, \tag{10.69}$$

for $i = 2, \dots, \nu$, where ℓ_i is such that

$$(\lambda_{\max}(P_{i,1}) + \tfrac{1}{2})\ell_i^2 \leqslant \frac{\delta_i^2}{M_i\|B' P_{i,1} B\|}$$

and

$$\widetilde{\rho}_i = \sum_{k=0}^{\infty} \|\mathcal{C}_i(\mathcal{A}_i + \mathcal{B}_i F_{i,1})^k \mathcal{B}_i\|.$$

Consider the following adaptive parameters:

$$\varepsilon_i(\chi_i) = \max\{\, r \in (0,1] \mid (\chi_i' P_r \chi_i)\operatorname{trace} \mathcal{B}_i' P_r \mathcal{B}_i \leqslant \tfrac{\delta_i^2}{M_i}\}, \tag{10.70}$$

where P_r is the solution of the DARE (10.63) with

$$\varepsilon_i = r, \quad M_i = \sigma_{\max}(P_{i,1}^{\frac{1}{2}} \mathcal{B}_i \mathcal{B}_i' P_{i,1}^{\frac{1}{2}}) + 1$$

where $P_{i,1} = P_{\varepsilon_i(\chi_i)}$ with $\varepsilon_i(\chi_i) = 1$.

Consider the controller,

$$u_1 = \sum_{i=1}^{\nu}\Big(\prod_{j=0}^{i-1} \varepsilon_j(\chi_j)\Big) F_{\varepsilon_i(\chi_i)}\chi_i \tag{10.71}$$

with $\varepsilon_0 = 1$. It can be shown that our adaptation guarantees that

$$\|F_{\varepsilon_i(\chi_i)}\chi_i\| \leqslant \tfrac{1}{\nu},$$

and hence $\|u_1\| \leqslant 1$. This implies that the input saturation to the first system never gets activated.

The following theorem shows that the controller (10.71) with adaptive parameters defined by (10.70) achieves global asymptotic stability of the origin for multi-layer nonlinear sandwich system described by (10.62).

Theorem 10.31 *Consider the interconnection of systems L_i as given in* (10.5)*, satisfying conditions* (*i*) *and* (*ii*) *of Theorem 10.12. Then, the generalized adaptive-low-gain state feedback control law* (10.71) *achieves global asymptotic stability of the origin.*

Proof : For simplicity of presentation, we will denote $\varepsilon_i(\chi_i(k))$, $P_{\varepsilon_i(\chi_i(k))}$, and $F_{\varepsilon_i(\chi_i(k))}$ by $\varepsilon_i(k)$, $P_i(k)$, and $F_i(k)$, respectively. But we emphasize that they always depend on χ_i.

When the state is sufficiently close to the origin, all saturation elements are inactive and hence $\varepsilon_i(\chi_i) = 1$ for all $i = 1, \dots, \nu$. The state matrix of closed-loop system is given by $\mathcal{A}_\nu + \mathcal{B}_\nu F_{\nu,1}$. From the property of DARE, we know that the above matrix is asymptotically stable. Then local stability follows.

We shall prove global attractivity using induction. We have argued that for all $k \geqslant 0$, the input saturation on L_1 remains inactive and, by construction, $\varepsilon_0 = 1$. Suppose that there exists a K_i where $1 \leqslant i \leqslant \nu - 1$ such that $\varepsilon_j = 1$ for $j \leqslant i - 1$ and the first i saturation elements are inactive for all $k \geqslant K_i$. We show that there exists a K_{i+1} such that $\varepsilon_i = 1$ and saturation on L_{i+1} will be inactive for all $k \geqslant K_{i+1}$.

Since the first i saturation elements are inactive and all $\varepsilon_j = 1$ for $j \leqslant i - 1$, the interconnection of the first i systems is equivalent to the following system,

$$\dot{\chi}_i = \mathcal{A}_i \chi_i + \mathcal{B}_i v_1 \tag{10.72}$$

where $\mathcal{A}_i$ is given by (10.65) and v_1 is given by

$$v_1 = v_{1,1} + v_{1,2} = F_i \chi_i + \sum_{j=i+1}^{\nu} \Big(\prod_{t=i}^{j-1} \varepsilon_t\Big) F_j \chi_j.$$

Define $V_i(k) = \chi_i' P_i \chi_i$ and the family of sets $\mathcal{V}_{i,1} = \{\chi_i \in \mathbb{R}^{\sum_{j=1}^{i} n_j} \mid \|\chi_i\| \leqslant \ell_i\}$ and $\mathcal{V}_{i,2} = \{\chi_i \in \mathbb{R}^{\sum_{j=1}^{i} n_j} \mid V_i \leqslant (\lambda_{\max}(P_{i,1}) + 1/2)\ell_i^2\}$. Since $x(k) \in \mathcal{V}_{i,1}$ implies that

$$V_i(k) \leqslant \lambda_{\max}(P_i(k))\|\chi_i(k)\|^2 \leqslant \lambda_{\max}(P_{i,1})\ell_i^2,$$

we find that $\mathcal{V}_{i,1} \subset \mathcal{V}_{i,2}$. Moreover, the definition of ℓ_i implies that $\varepsilon_i(k) = 1$ for $\chi_i(k) \in \mathcal{V}_{i,2}$.

Evaluating $V_i(k+1) - V_i(k)$ along the trajectories yields

$$\begin{aligned} V_i(k+1) - V_i(k) \leqslant{}& -\varepsilon_i(k)\chi_i(k)'\chi_i(k) - \chi_i(k+1)'P_i(k)\chi_i(k+1) \\ &+ \chi_i(k+1)'P_i(k+1)\chi_i(k+1) - 2\chi_i(k)'\mathcal{A}_i' P_i(k)\mathcal{B}_i v_{1,2}(k) \\ &- 2v_{1,1}(k)'\mathcal{B}_i' P_i(k)\mathcal{B}_i v_{1,2}(k) + v_{1,2}(k)'\mathcal{B}_i' P_i(k)\mathcal{B}_i v_{1,2}(k), \end{aligned}$$

where

$$\begin{aligned} v_{1,1}(k) &= F_i(k)\chi_i(k), \\ v_{1,2}(k) &= \sum_{j=i+1}^{\nu} \Big(\prod_{t=i}^{j-1} \varepsilon_t(k)\Big) F_j(k)\chi_j(k). \end{aligned}$$

Our adaptation guarantees that

$$\|v_{1,1}(k)\| \leqslant \tfrac{1}{\nu},$$

and
$$\|v_{1,2}(k)\| \leqslant \varepsilon_i(k)(\nu - i)\delta_{i+1}.$$

Hence,
$$\begin{aligned}
\|\chi_i(k)'\mathcal{A}_i' P_i(k)\mathcal{B}_i v_{1,2}(k)\| &= \|v_{1,1}(k)'(\mathcal{B}_i' P_i(k)\mathcal{B}_i + I)v_{1,2}(k)\| \\
&\leqslant \varepsilon_i(k)\tfrac{(\nu-i)^2}{\nu}(\|\mathcal{B}_i' P_i(k)\mathcal{B}_i\| + 1)\delta_{i+1}, \\
\|v_{1,1}(k)'\mathcal{B}_i' P_i(k)\mathcal{B}_i v_{1,2}(k)\| &\leqslant \varepsilon_i(k)\tfrac{(\nu-i)^2}{\nu}\,\|\mathcal{B}_i' P_i(k)\mathcal{B}_i\|\,\delta_{i+1}, \\
\|v_{1,2}(k)'\mathcal{B}_i' P_i(k)\mathcal{B}_i v_{1,2}(k)\| &\leqslant \varepsilon_i(k)(\nu - i)^2\,\|\mathcal{B}_i' P_i(k)\mathcal{B}_i\|\,\delta_{i+1}.
\end{aligned}$$

With the above inequalities, we have

$$\begin{aligned}
V_i(k+1)-V_i(k) &\leqslant -\varepsilon_i(k)\chi_i'(k)\chi_i(k)+\chi_i(k+1)'(P_i(k+1)-P_i(k))\chi_i(k+1) \\
&\quad + \varepsilon_i(k)(\nu - i)^2(\tfrac{\nu+4}{\nu}\,\|\mathcal{B}_i' P_i(k)\mathcal{B}_i\| + \tfrac{2}{\nu})\delta_{i+1} \\
&\leqslant -\varepsilon_i(k)\|\chi_i(k)\|^2 + \chi_i(k+1)'(P_i(k+1) - P_i(k))\chi_i(k+1) + \tfrac{1}{2}\varepsilon_i(k)\ell_i^2.
\end{aligned}$$

Using the same argument as in the proof of Theorem 10.29, we can show that if $\chi_i(k) \notin \mathcal{V}_{i,1}$, then $V_i(k)$ is strictly decreasing and hence χ_i will enter $\mathcal{V}_{i,1}$ within finite time. On the other hand, if $\chi_i(k) \in \mathcal{V}_{i,1}$, then $\chi_i(k+1) \in \mathcal{V}_{i,2}$. Since $\mathcal{V}_{i,1} \subset \mathcal{V}_{i,2}$, we conclude that χ_i will enter $\mathcal{V}_{i,2}$ within finite time, say $K_{i,1}$, and cannot escape from it. On $\mathcal{V}_{i,2}$ we have $\varepsilon_i(k) = 1$.

Consider $z_i(k) = C_i x_i(k) = \mathcal{C}_i\chi_i(k)$ for $k \geqslant K_{i,1}$. Since $\varepsilon_i(k) = 1$, we have

$$z_i(k) = \mathcal{C}_i(\mathcal{A}_i + \mathcal{B}_i F_{i,1})^{k-K_{i,1}}\chi_i(K_{i,1}) + z_{i,0}(k),$$

where

$$z_{i,0}(k) = \sum_{j=K_{i,1}}^{k-1} \mathcal{C}_i(\mathcal{A}_i + \mathcal{B}_i F_{i,1})^{k-j-1}\mathcal{B}_i v_{1,2}(j). \tag{10.73}$$

Our adaptation guarantees that

$$v_{1,2} \leqslant (n - i)\delta_{i+1} \leqslant \frac{1}{2\bar{\rho}_i} = \frac{1}{2\sum_{k=0}^{\infty}\|\mathcal{C}_i(\mathcal{A}_i + \mathcal{B}_i F_{i,1})^k\mathcal{B}_i\|}.$$

This implies that $\|z_{i,0}(k)\| \leqslant \frac{1}{2}$ for all $k \geqslant K_{i,1}$. Since $\mathcal{A}_i + \mathcal{B}_i F_{i,1}$ is asymptotically stable, there exists a $K_{i+1} > K_{i,1}$ such that for all $k \geqslant K_i$, we have

$$\|z_i(k)\| \leqslant 1.$$

Therefore, the saturation on ν systems L_{i+1} will be inactive and $\varepsilon_i = 1$ for all $k \geqslant K_{i+1}$. By induction, there exists a K_ν such that all the saturations are inactive for $k \geqslant K_\nu$ and $\varepsilon_i = 1$ for all $i = 0, \dots, \nu - 1$.

Then the interconnection of ν systems (10.62) and controller (10.71) is equivalent to the interconnection of linear system,

$$\chi_\nu(k+1) = \mathcal{A}_\nu\chi_\nu(k) + \mathcal{B}_\nu v_1,$$

with controller

$$v_1 = F_{\varepsilon_\nu(\chi_\nu)}\chi_\nu = -(\mathcal{B}_\nu' P_{\varepsilon_\nu(\chi_\nu)}\mathcal{B}_\nu + I)^{-1}\mathcal{B}_\nu' P_{\varepsilon_\nu(\chi_\nu)}\mathcal{A}_\nu \chi_\nu.$$

It follows from Lemma 10.23 that the closed-loop system is globally asymptotically stable. This implies that $\chi_\nu(k) \to 0$ as $k \to \infty$. This completes the proof. ∎

10.7 Low-and-high-gain design for single-layer systems

In previous sections, we have developed generalized low-gain feedback design methodology and its adaptive version, respectively, for semi-global and global stabilization of sandwich systems. As we remarked in introduction, such generalized low-gain design methods do not allow full utilization of the available control capacity. In this section, we introduce a different strategy for stabilization of sandwich nonlinear systems, namely, a generalized low-and-high-gain feedback design methodology (for semi-global stabilization) and its adaptive version (for global stabilization) which allow full utilization of the available control capacity, and hence are capable of enhancing the system performance such as robust stability and disturbance rejection.

We consider continuous-time systems in one subsection and discrete-time systems in another.

10.7.1 Continuous-time systems

We first pursue semi-global stabilization. Let us consider the single-layer sandwich system as defined in (10.1) and (10.2).

Generalized low-and-high-gain design methodology for semi-global stabilization: As in Sect. 10.4, we first choose F such that $A+BF$ is asymptotically stable and consider the system,

$$\begin{aligned} \dot{x} &= (A+BF)x + Bv \\ z &= Cx, \end{aligned} \tag{10.74}$$

where $u = Fx + v$. We have

$$\begin{aligned} z(t) &= Ce^{(A+BF)t}x(0) + \int_0^t Ce^{(A+BF)(t-\tau)}Bv(\tau)\,\mathrm{d}\tau \\ &= Ce^{(A+BF)t}x(0) + z_0(t). \end{aligned}$$

Since $A + BF$ is asymptotically stable, we know that there exists a $\delta > 0$ such that whenever

$$\|v(\tau)\| < \delta \quad \forall \tau > 0, \tag{10.75}$$

we have $\|z_0(t)\| < \frac{1}{2}$ for all $t > 0$.

Next, we consider the system,

$$\begin{pmatrix} \dot{x} \\ \dot{\omega} \end{pmatrix} = \begin{pmatrix} A + BF & 0 \\ NC & M \end{pmatrix} \begin{pmatrix} x \\ \omega \end{pmatrix} + \begin{pmatrix} B \\ 0 \end{pmatrix} v, \tag{10.76}$$

which can be rewritten as

$$\dot{\bar{x}} = \widetilde{\mathcal{A}}\bar{x} + \mathcal{B}v, \tag{10.77}$$

where $\widetilde{\mathcal{A}}$, $\mathcal{B}$, and $\bar{x}$ are as defined in (10.12). Our initial objective is, for any a priori given compact set $\mathcal{W}$, to find a stabilizing controller for the system (10.76) such that $\mathcal{W}$ is contained in its domain of attraction and $\|v(\tau)\| < \delta$ for all $\tau > 0$.

Let $Q_\varepsilon > 0$ be a parameterized family of matrices which is increasing in $\varepsilon > 0$ with $\lim_{\varepsilon \to 0} Q_\varepsilon = 0$. In that case, for any $\varepsilon > 0$, there exists a $P_\varepsilon > 0$ satisfying the CARE,

$$\widetilde{\mathcal{A}}' P_\varepsilon + P_\varepsilon \widetilde{\mathcal{A}} - P_\varepsilon \mathcal{B}\mathcal{B}' P_\varepsilon + Q_\varepsilon = 0. \tag{10.78}$$

The above development is exactly the same as that in Sect. 10.4 leading to Lemma 10.14 but repeated here for continuity of presentation. The following lemma is a modified version of Lemma 10.14.

Lemma 10.32 *Consider the linear system given in* (10.76) *and assume that the pair* $(\mathcal{A}, \mathcal{B})$*, defined by* (10.6)*, is stabilizable and all the eigenvalues of* M *are in the closed left-half plane. Then, for any a priori given compact set* $\mathcal{W} \in \mathbb{R}^{n_1+n_2}$*, there exists an* ε^* *such that for any* $0 < \varepsilon < \varepsilon^*$ *and for any* $\alpha > 0$*, the generalized low-and-high-gain state feedback law,*

$$v = -\delta\sigma\left(\tfrac{1+\alpha}{\delta}\mathcal{B}' P_\varepsilon \bar{x}\right) = -\delta\sigma\left(\frac{1+\alpha}{\delta}\begin{pmatrix} B \\ 0 \end{pmatrix}' P_\varepsilon \begin{pmatrix} x \\ \omega \end{pmatrix}\right), \tag{10.79}$$

achieves asymptotic stability of the equilibrium point $\bar{x} = 0$*. Moreover, for any initial condition in* $\mathcal{W}$*, the constraint* $\|v(t)\| < \delta$ *does not get violated for any* $t > 0$*.*

Proof : Note that the condition that the system (10.76) is stabilizable immediately implies the existence of a $P_\varepsilon > 0$ satisfying the CARE (10.78), while the condition that the eigenvalues of M are in the closed left-half plane immediately implies that

$$P_\varepsilon \to 0 \tag{10.80}$$

as $\varepsilon \to 0$. Obviously, controller (10.79) satisfies $\|v\| < \delta$. It remains to show that such a controller achieves semi-global stabilization. Define $V(\bar{x}) = \bar{x}' P_\varepsilon \bar{x}$. Let c be defined as

$$c = \sup_{\substack{\varepsilon \in (0,1] \\ \bar{x} \in \mathcal{W}}} \{\bar{x}' P_\varepsilon \bar{x}\}.$$

There exists an ε^* such that for any $\varepsilon \in (0, \varepsilon^*]$, we have $\bar{x} \in \mathcal{L}_v(c) = \{\bar{x} \mid \bar{x}' P_\varepsilon \bar{x} \leqslant c\}$. This implies that

$$\|\tilde{v}\| \leqslant \delta$$

where we denote

$$\tilde{v} = -\begin{pmatrix} B \\ 0 \end{pmatrix}' P_\varepsilon \begin{pmatrix} x \\ \omega \end{pmatrix}.$$

Consider $\dot{V}$ along any trajectory,

$$\dot{V} \leqslant -\bar{x}' Q_\varepsilon \bar{x} - 2\delta \tilde{v}'[\sigma(\tfrac{1+\alpha}{\delta}\tilde{v}) - \tfrac{1}{\delta}\tilde{v}] = -\bar{x}' Q_\varepsilon \bar{x} - 2\delta \tilde{v}'[\sigma(\tfrac{1+\alpha}{\delta}\tilde{v}) - \sigma(\tfrac{1}{\delta}\tilde{v})].$$

We have $\dot{V} < 0$ for any $\alpha > 0$. This completes the proof. ■

This leads to the following result.

Theorem 10.33 *Consider the interconnection of the two systems given by* (10.1) *and* (10.2) *satisfying conditions (i) and (ii) of Theorem 10.7. Let F be such that $A + BF$ is asymptotically stable while $P_\varepsilon > 0$ is defined by the CARE* (10.78). *Define a low-and-high-gain state feedback law as,*

$$u = Fx - \delta\sigma(\tfrac{1+\alpha}{\delta} \begin{pmatrix} B \\ 0 \end{pmatrix}' P_\varepsilon \begin{pmatrix} x \\ \omega \end{pmatrix}). \tag{10.81}$$

Then, for any compact set of initial conditions $\mathcal{W} \in \mathbb{R}^{n_1+n_2}$, there exists an $\varepsilon^* > 0$ such that, for all ε with $0 < \varepsilon < \varepsilon^*$ and for any $\alpha > 0$, the controller (10.81) *asymptotically stabilizes the equilibrium point* $(0, 0)$ *with a domain of attraction containing $\mathcal{W}$.*

Proof : Consider any $(x(0), \omega(0)) \in \mathcal{W}$. Then there exists a $T > 0$ such that

$$\|Ce^{(A+BF)t}x(0)\| < \tfrac{1}{2}$$

for $t > T$. Denote

$$v(t) = -\delta\sigma(\tfrac{1+\alpha}{\delta} \begin{pmatrix} B \\ 0 \end{pmatrix}' P_\varepsilon \begin{pmatrix} x \\ \omega \end{pmatrix}).$$

By construction, we have $\|v(t)\| \leqslant \delta$ for $t > 0$. This implies that $\|z(t)\| \leqslant 1$ for $t \geqslant T$.

Since $A + BF$ is Hurwitz stable and the input to the second system is bounded, there exists a $\overline{\mathcal{W}}$ such that for any $(x(0), \omega(0)) \in \mathcal{W}$, we have $(x(T), \omega(T)) \in \overline{\mathcal{W}}$.

Then the interconnection of (10.1) and (10.2) with controller (10.81) for $t > T$ is equivalent to the interconnection of (10.11) with controller (10.79) for $t > T$. From Lemma 10.32, there exists an ε^* such that for any $\varepsilon \in (0, \varepsilon^*]$ and for any $\alpha > 0$, the closed-loop system of (10.11) and controller (10.79) is asymptotically stable with $(x(T), \omega(T)) \in \overline{\mathcal{W}}$. Therefore, we have

$$x(t) \to 0, \quad \omega(t) \to 0.$$

Since this follows for any $(x(0), \omega(0)) \in \mathcal{W}$, we find that $\mathcal{W}$ is contained in the domain of attraction as required. ∎

Remark 10.34 *If we simply set $\alpha = 0$ in the generalized low-and-high-gain state feedback controller as developed in* (10.81)*, what we get is a low-gain controller with a part of it, the part associated with the solution of the CARE* (10.78)*, in the argument of saturation nonlinearity $\sigma(.)$. However, we note that in this case we choose the low-gain parameter ε in such a way that the part associated with the solution of CARE* (10.78) *is less than δ in some large enough compact set. Then the saturation in the controller never gets activated; moreover, the saturation in the saturation element of sandwich system never gets activated either, and hence the controller reduces exactly to the generalized low-gain controller which we have constructed in* (10.15)*.*

We claim that the same controller given in (10.81) with ε being replaced by the adaptive-low-gain parameter $\varepsilon_a(\overline{x})$ as defined below solves the global stabilization problem.

Generalized adaptive-low-gain and generalized high-gain design methodology for global stabilization: As usual, we first look for an adaptive parameter having the same properties as given in Sect. 10.4.1 on page 552. A particular choice satisfying those properties is given by

$$\varepsilon_a(\overline{x}) = \max\{\, r \in (0, 1] \mid (\overline{x}' P_r \overline{x}) \operatorname{trace} \mathcal{B}' P_r \mathcal{B} \leqslant \delta^2 \,\}, \tag{10.82}$$

where P_r is the unique positive definite solution of CARE (10.78) with $\varepsilon = r$. Then, following the same procedure as before in connection with generalized adaptive-low-gain design, we first show the following result:

Lemma 10.35 *Consider the linear system given in* (10.76) *and assume that the pair $(\mathcal{A}, \mathcal{B})$, as defined by* (10.6)*, is stabilizable and the eigenvalues of M are in the closed left-half plane. Then, for any $\alpha > 0$, the feedback,*

$$v = -\delta\sigma\left(\tfrac{1+\alpha}{\delta}\mathcal{B}' P_{\varepsilon_a(\overline{x})}\overline{x}\right) = -\delta\sigma\left(\tfrac{1+\alpha}{\delta}\begin{pmatrix} B \\ 0 \end{pmatrix}' P_{\varepsilon_a(\overline{x})}\begin{pmatrix} x \\ \omega \end{pmatrix}\right), \tag{10.83}$$

achieves global stability of the equilibrium point $\overline{x} = 0$.

Proof : Obviously, controller (10.83) satisfies $\|v\| < \delta$. It remains to show that such a controller achieves global stabilization. Define $V(\overline{x}) = \overline{x}' P_{\varepsilon_a(\overline{x})} \overline{x}$, and denote

$$\widetilde{v} = -\begin{pmatrix} B \\ 0 \end{pmatrix}' P_{\varepsilon_a(\overline{x})} \overline{x}.$$

Consider $\dot{V}$ along any trajectory,

$$\dot{V} \leqslant -\overline{x}' Q_{\varepsilon_a(\overline{x})} \overline{x} - 2\delta \widetilde{v}' [\sigma(\tfrac{1+\alpha}{\delta} \widetilde{v}) - \tfrac{1}{\delta} \widetilde{v}] + \overline{x}' \frac{dP_{\varepsilon_a(\overline{x})}}{dt} \overline{x}.$$

By construction, $\|\frac{1}{\delta} \widetilde{v}\| < 1$. We get

$$\dot{V} \leqslant -\overline{x}' Q_{\varepsilon_a(\overline{x})} \overline{x} - 2\delta \widetilde{v}' [\sigma(\tfrac{1+\alpha}{\delta} \widetilde{v}) - \sigma(\tfrac{1}{\delta} \widetilde{v})] + \overline{x}' \frac{dP_{\varepsilon_a(\overline{x})}}{dt} \overline{x}.$$

If $\alpha > 0$, we have

$$\dot{V} < -\overline{x}' Q_{\varepsilon_a(\overline{x})} \overline{x} + \overline{x}' \frac{dP_{\varepsilon_a(\overline{x})}}{dt} \overline{x}.$$

The adaptive law (10.82) implies that

$$V(x) \left\| \mathcal{B}' P_{\varepsilon_a(\overline{x})} \mathcal{B} \right\| = \delta^2,$$

whenever $\varepsilon_a(\overline{x}) \neq 1$ or equivalently $P_{\varepsilon_a(\overline{x})}$ is not a constant locally. This implies that $\dot{V}$ and $\overline{x}' \frac{dP_{\varepsilon_a(\overline{x})}}{dt} \overline{x}$ are either both zero or of opposite signs. Hence, for $x \neq 0$, we have

$$\dot{V} < 0.$$

If not, we know that $\overline{x}' \frac{dP_{\varepsilon_a(\overline{x})}}{dt} \overline{x} \leqslant 0$. But this implies that $\dot{V} < -\overline{x}' Q_{\varepsilon_a(\overline{x})} \overline{x}$ which yields a contradiction. Therefore, the global asymptotic stability follows. ■

This leads to the following result.

Theorem 10.36 *Consider the interconnection of the two systems given by* (10.1) *and* (10.2) *satisfying conditions* (*i*) *and* (*ii*) *of Theorem 10.7. Choose* F *such that* $A + BF$ *is asymptotically stable. Let* P_ε *and* $\varepsilon_a(\overline{x})$ *be as defined by the CARE* (10.78) *and* (10.82), *respectively. In that case, for any* $\alpha > 0$, *the generalized adaptive-low-gain and generalized high-gain state feedback law,*

$$u = Fx - \delta\sigma\left(\tfrac{1+\alpha}{\delta} \mathcal{B}' P_{\varepsilon_a(\overline{x})} \overline{x}\right) = Fx - \delta\sigma\left(\tfrac{1+\alpha}{\delta} \begin{pmatrix} B \\ 0 \end{pmatrix}' P_{\varepsilon_a(\overline{x})} \begin{pmatrix} x \\ \omega \end{pmatrix}\right) \quad (10.84)$$

achieves global asymptotic stability.

Proof : If we consider the interconnection of (10.1) and (10.2), then we note that close to the origin the saturation does not get activated. Moreover, close to the origin the feedback (10.84) is given by

$$u = Fx - (1+\alpha)\begin{pmatrix} B \\ 0 \end{pmatrix}' P_1 \bar{x},$$

which immediately yields that the interconnection of (10.1), (10.2), and (10.84) is locally asymptotically stable. It remains to show that we have global asymptotic stability.

Consider an arbitrary initial condition $x(0)$ and $\omega(0)$. Then there exists a $T > 0$ such that

$$\|Ce^{(A+BF)t}x(0)\| < \tfrac{1}{2}$$

for $t > T$. Moreover, by construction, the control

$$v = -\delta\sigma\left(\tfrac{1+\alpha}{\delta}\mathcal{B}' P_{\varepsilon_a(\bar{x})}\bar{x}\right) = -\delta\sigma\left(\tfrac{1+\alpha}{\delta}\begin{pmatrix} B \\ 0 \end{pmatrix}' P_{\varepsilon_a(\bar{x})}\begin{pmatrix} x \\ \omega \end{pmatrix}\right)$$

yields $\|v(t)\| \leqslant \delta$ for all $t > 0$. However, this implies that $z(t)$ generated by (10.74) satisfies $\|z(t)\| < 1$ for all $t > T$. But this yields that the interconnection of (10.1) and (10.2) with controller (10.84) behaves for $t > T$ like the interconnection of (10.76) with controller (10.83). Global asymptotic stability of the latter system then implies that $\bar{x}(t) \to 0$ as $t \to \infty$. Since this property holds for any initial condition and since we have local asymptotic stability, we can conclude that the controller yields global asymptotic stability. This completes the proof. ∎

Similar to earlier chapters that utilize generalized low-and-high gain, the construction of our controller guarantees the saturation does not get activated after some finite time T and the stabilization of sandwich nonlinear systems becomes stabilization of a linear system subject to input saturation. It is clear from the proof that T is determined by the initial condition of L_1. Since $A + BF$ is Hurwitz stable with the preliminary feedback, this T can be fairly small. However, after time T, the design methodology presented above yields a regular low-and-high-gain feedback controller, while in the case of earlier chapters that utilize generalized low-gain, it reduces to the classical low-gain feedback controller. Therefore, we expect an enhanced system performance from our design technique. A numerical example given shortly in Sect. 10.8 illustrates this result. We like to emphasize that an appropriate selection of the matrix Q_ε plays an important role in the design process. A judicious choice of Q_ε can tremendously improve the performance.

10.7.2 Discrete-time systems

As in continuous-time case, we first pursue semi-global stabilization. We proceed to design here a controller which solves a semi-global stabilization problem for discrete-time sandwich systems. For ease of presentation, as before we denote $\bar{x} = (x, \omega)$.

Generalized low-and-high-gain design methodology for semi-global stabilization: Our design philosophy follows that in continuous-time systems and progresses in three steps:

Step 1

Choose a preliminary feedback F such that $A + BF$ is Schur stable.

Step 2

Define a δ as

$$\delta = \frac{\beta}{\sum_{k=0}^{\infty} \|C(A + BF)^k B\|} \tag{10.85}$$

for an arbitrary $\beta \in (0, 1)$. Such a δ is well defined since $A + BF$ is Schur stable. Note that in continuous-time case we used $\beta = 0.5$.

Step 3

Let $Q_\varepsilon > 0$ be a matrix function: $(0, 1] \to \mathbb{R}^{(n_1+n_2)\times(n_1+n_2)}$ which satisfies $\frac{\mathrm{d}Q_\varepsilon}{\mathrm{d}\varepsilon} > 0$ for $\varepsilon > 0$ and $\lim_{\varepsilon\to 0^+} Q_\varepsilon = 0$. Solve the following DARE (same as (10.25) except that εI is replaced with Q_ε),

$$P_\varepsilon = \tilde{\mathcal{A}}' P_\varepsilon \tilde{\mathcal{A}} - \tilde{\mathcal{A}}' P_\varepsilon \mathcal{B}(I + \mathcal{B}' P_\varepsilon \mathcal{B})^{-1} \mathcal{B}' P_\varepsilon \tilde{\mathcal{A}} + Q_\varepsilon = 0, \tag{10.86}$$

where $\tilde{\mathcal{A}}$ and $\mathcal{B}$ are as defined in (10.12), and ε as usual is a low-gain parameter to be chosen appropriately. Note that the condition (*i*) of Theorem 10.7 guarantees the existence of the positive definite solution P_ε. The generalized low-and-high-gain feedback can then be constructed as

$$u = Fx - \delta\sigma\left(\tfrac{1+\alpha}{\delta}(I + \mathcal{B}' P_\varepsilon \mathcal{B})^{-1} \mathcal{B}' P_\varepsilon \tilde{\mathcal{A}} \bar{x}\right). \tag{10.87}$$

We show in the next theorem that the above low-and-high-gain controller solves the semi-global stabilization problem:

Theorem 10.37 *Consider the interconnection of the two systems given by* (10.1) *and* (10.2) *satisfying conditions i and ii of Theorem 10.7. Define a state feedback control law as in* (10.87)*. Then, for any compact set of initial conditions $\mathcal{W} \in \mathbb{R}^{n_1+n_2}$, there exists an $\varepsilon^* > 0$ such that for all ε with $\varepsilon \in (0, \varepsilon^*]$ and for any*

$$\alpha \in \left[0, \frac{2}{\|\mathcal{B}' P_\varepsilon \mathcal{B}\|}\right],$$

the controller (10.87) *asymptotically stabilizes the equilibrium point* (0, 0) *with a domain of attraction containing $\mathcal{W}$.*

Proof : Define

$$v = -\delta\sigma\left(\tfrac{1+\alpha}{\delta}(I + \mathcal{B}' P_\varepsilon \mathcal{B})^{-1}\mathcal{B}' P_\varepsilon \widetilde{\mathcal{A}}\bar{x}\right). \tag{10.88}$$

Then the system L_1 given by (10.1) along with the preliminary feedback $u = Fx + v$ can be written as

$$\begin{aligned} x(k+1) &= (A + BF)x(k) + Bv(k) \\ z(k) &= Cx(k). \end{aligned} \tag{10.89}$$

We have

$$\begin{aligned} z(k) &= C(A + BF)^k x(0) + \sum_{i=0}^{k-1} C(A + BF)^{k-i-1} Bv(i) \\ &= C(A + BF)^k x(0) + z_0(k), \end{aligned}$$

where

$$z_0(k) = \sum_{i=0}^{k-1} C(A + BF)^{k-i-1} Bv(i). \tag{10.90}$$

For any a priori given set of initial conditions $\mathcal{W}$, there exists a $K > 0$ such that

$$\|C(A + BF)^k x(0)\| < 1 - \beta$$

for $k > K$ and any $x(0) \in \mathcal{W}$.

By construction, $\|v\| \leqslant \delta$ for all $k \geqslant 0$. From the definition of δ, we get

$$\|z_0(k)\| = \sum_{i=0}^{k-1} \|C(A + BF)^{k-i-1} B\| \|v(i)\| \leqslant \beta.$$

This implies that for all $k \geqslant K$ we have $\|z(k)\| \leqslant 1$, that is, the sandwiched saturation remains inactive after time K. Therefore, for all $k \geqslant K$, the closed-loop system is equivalent to the interconnection of the linear cascaded system

$$\bar{x}(k+1) = \widetilde{\mathcal{A}}\bar{x}(k) + \mathcal{B}v(k), \tag{10.91}$$

with the control v given by (10.88).

There exists a compact set $\overline{\mathcal{W}}$ such that for any $\bar{x}(0) \in \mathcal{W}$, we have $\bar{x}(K) \in \overline{\mathcal{W}}$. This is due to the fact that $\mathcal{W}$ is compact, $A + BF$ is Schur stable, and the input to L_2 is bounded.

In the next lemma which is akin to Lemma 10.20, we show that the interconnection of (10.91) and (10.88) is asymptotically stable with $\overline{\mathcal{W}}$ contained in its domain of attraction.

Lemma 10.38 *Consider the system* (10.91) *and assume that the pair* $(\mathcal{A}, \mathcal{B})$ *as given by* (10.6) *is stabilizable and all the eigenvalues of* M *are in the closed unit disc. Then, for any a priori given compact set* $\overline{\mathcal{W}} \in \mathbb{R}^{n_1+n_2}$, *there exists an* ε^* *such that for any* $0 < \varepsilon < \varepsilon^*$ *and for any*

$$\alpha \in \left[0, \frac{2}{\|\mathcal{B}' P_\varepsilon \mathcal{B}\|}\right], \tag{10.92}$$

the state feedback,

$$v = -\delta\sigma\left(\tfrac{1+\alpha}{\delta}(I + \mathcal{B}' P_\varepsilon \mathcal{B})^{-1} \mathcal{B}' P_\varepsilon \widetilde{\mathcal{A}} \overline{x}\right), \tag{10.93}$$

achieves asymptotic stability of the equilibrium point $\overline{x} = 0$ *with a domain of attraction containing* $\overline{\mathcal{W}}$.

Proof of Lemma 10.38 : First we introduce the following notation:

$$\widetilde{v} = (I + \mathcal{B}' P_\varepsilon \mathcal{B})^{-1} \mathcal{B}' P_\varepsilon \mathcal{A} \overline{x} \quad \text{and} \quad \mu = \|\mathcal{B}' P_\varepsilon \mathcal{B}\|.$$

Define $V(\overline{x}) = \overline{x}' P_\varepsilon \overline{x}$. Note that condition *(ii)* of Theorem 10.7 immediately implies that

$$P_\varepsilon \to 0 \tag{10.94}$$

as $\varepsilon \to 0$. Hence, there exists an ε^* such that for any $\varepsilon \in (0, \varepsilon^*]$ and $\overline{x} \in \mathcal{W}$, we have

$$|\widetilde{v}(k)| \leqslant \delta,$$

componentwise, and hence, given that α satisfies (10.92), we have

$$|\widetilde{v}(k)| \leqslant |v(k)| \leqslant \left(\tfrac{2}{\delta} + 1\right) |\widetilde{v}(k)| \tag{10.95}$$

componentwise, and since $v(k)$ and $\widetilde{v}(k)$, also componentwise, have the same sign, it follows that

$$\|v(k) - \tfrac{1+\mu}{\mu} \widetilde{v}(k)\| \leqslant \tfrac{1}{\mu} \|\widetilde{v}(k)\|. \tag{10.96}$$

Next, consider $V(k+1) - V(k)$ along any trajectory,

$$\begin{aligned}
V(k+1) - V(k)& \\
=&\widetilde{v}(k)'(I + \mathcal{B}' P_\varepsilon \mathcal{B})\widetilde{v}(k) - \overline{x}(k)' Q_\varepsilon \overline{x}(k) \\
&- 2\widetilde{v}(k)'(I + \mathcal{B}' P_\varepsilon \mathcal{B}) v(k) + v(k)' \mathcal{B}' P_\varepsilon \mathcal{B} v(k) \\
=&-\overline{x}(k)' Q_\varepsilon \overline{x}(k) - v(k)' v(k) \\
&+ [v(k) - \widetilde{v}(k)]'(I + \mathcal{B}' P_\varepsilon \mathcal{B})[v(k) - \widetilde{v}(k)] \\
\leqslant&(1+\mu)[v(k) - \widetilde{v}(k)]'[v(k) - \widetilde{v}(k)] - \overline{x}(k)' Q_\varepsilon \overline{x}(k) - v(k)' v(k) \\
\leqslant&-\overline{x}(k)' Q_\varepsilon \overline{x}(k) - \tfrac{1}{\mu} \widetilde{v}(k)' \widetilde{v}(k) \\
&+ \mu[v(k) - \tfrac{1+\mu}{\mu} \widetilde{v}(k)]'[v(k) - \tfrac{1+\mu}{\mu} \widetilde{v}(k)].
\end{aligned}$$

Using (10.96) we find,

$$\mu[v(k) - \tfrac{1+\mu}{\mu}\widetilde{v}(k)]'[v(k) - \tfrac{1+\mu}{\mu}\widetilde{v}(k)] - \tfrac{1}{\mu}\widetilde{v}(k)'\widetilde{v}(k) \leqslant 0.$$

We conclude that $V(k+1) - V(k) < 0$ for any $\varepsilon \in (0, \varepsilon^*]$, α satisfying (10.92) and $x \neq 0$. This completes the proof. ■

The above lemma indicates that

$$x(k) \to 0, \quad \omega(k) \to 0.$$

Since this follows for any $(x(0), \omega(0)) \in \mathcal{W}$, we find that $\mathcal{W}$ is contained in the domain of attraction as required. ■

Generalized adaptive-low-gain and generalized high-gain design methodology for global stabilization:

We claim that the same controller given in (10.87) with ε replaced by the scheduled low-gain parameter $\varepsilon_a(\overline{x})$ as defined below solves the global stabilization problem. Specifically, the design methodology follows in four steps:

Steps 1, 2, and 3

The above steps are exactly the same as for semi-global stabilization.

Step 4

As usual, we first look for an adaptive parameter $\varepsilon = \varepsilon_a(\overline{x})$ having the same properties as given in Sect. 10.4.1 on page 552. A particular choice satisfying such properties is given by

$$\varepsilon_a(\overline{x}) = \max\left\{ r \in (0,1] \mid (\overline{x}' P_r \overline{x}) \operatorname{trace} \mathcal{B}' P_r \mathcal{B} \leqslant \delta^2 / M_p \right\}, \tag{10.97}$$

where $M_p = \sigma_{\max}(P_1^{\frac{1}{2}} \mathcal{B}\mathcal{B}' P_1^{\frac{1}{2}}) + 1$ and P_r is the solution of (10.86) with $\varepsilon = r$.

We have the following result.

Theorem 10.39 *Consider the interconnection of the two systems given by* (10.1) *and* (10.2) *satisfying the conditions* (i) *and* (ii) *of Theorem 10.7. Choose F such that $A + BF$ is Schur stable. Let P_ε and ε_a be as defined by* (10.86) *and* (10.97), *respectively. In that case, for any*

$$\alpha \in \left[0, \frac{2}{\|\mathcal{B}' P_1 \mathcal{B}\|}\right],$$

the state feedback,

$$u = Fx - \delta\sigma\left(\tfrac{1+\alpha}{\delta}(I + \mathcal{B}' P_{\varepsilon_a(\overline{x})}\mathcal{B})^{-1}\mathcal{B}' P_{\varepsilon_a(\overline{x})}\widetilde{\mathcal{A}}\overline{x}\right), \tag{10.98}$$

achieves global asymptotic stability of the origin, where F, δ, $P_{\varepsilon_a(\overline{x})}$, and $\varepsilon_a(\overline{x})$ are obtained in Steps 1,2, 3, and 4, respectively.

Proof : Define

$$v = \delta\sigma\left(\tfrac{1+\alpha}{\delta}(I + \mathcal{B}' P_{\varepsilon_a(\bar{x})}\mathcal{B})^{-1}\mathcal{B}' P_{\varepsilon_a(\bar{x})}\tilde{\mathcal{A}}\bar{x}\right). \tag{10.99}$$

If we consider the interconnection of (10.1) and (10.2), then we note that close to the origin the saturation does not get activated. Moreover, close to the origin, the feedback (10.98) is given by

$$u = Fx - (1+\alpha)(I + \mathcal{B}' P_1 \mathcal{B})^{-1}\mathcal{B}' P_1 \tilde{\mathcal{A}}\bar{x},$$

which immediately shows that the interconnection of (10.1), (10.2), and (10.98) is locally asymptotically stable. It remains to show that we have global asymptotic stability.

Consider an arbitrary initial condition $x(0)$ and $\omega(0)$. Then there exists a $K > 0$ such that

$$\|C(A + BF)^k x(0)\| < 1 - \beta$$

for $k > K$. The definition of δ implies that $\|z_0(k)\| \leqslant \beta$ for all $k \geqslant 0$ where z_0 is given by (10.90). However, this implies that $z(k)$ generated by (10.89) satisfies $\|z(k)\| < 1$ for all $k > K$. Then the closed-loop system becomes the interconnection of the linear cascade system (10.91) with controller (10.99).

We next show the following lemma.

Lemma 10.40 *Consider the system* (10.91) *and assume that the pair* $(\mathcal{A}, \mathcal{B})$ *as given by* (10.6) *is stabilizable and all the eigenvalues of* $\tilde{\mathcal{A}}$ *are in the closed unit disc. Then, for any*

$$\alpha \in \left[0, \frac{2}{\|\mathcal{B}' P_1 \mathcal{B}\|}\right], \tag{10.100}$$

the feedback (10.99) *achieves global stability of the equilibrium point* $\bar{x} = 0$.

Proof of Lemma 10.40 : Define $V(\bar{x}) = \bar{x}' P_{\varepsilon_a(\bar{x})}\bar{x}$. Denote

$$\tilde{v}(k) = (I + \mathcal{B}' P_{\varepsilon_a(\bar{x}(k))}\mathcal{B})^{-1}\mathcal{B}' P_{\varepsilon_a(\bar{x}(k))}\tilde{\mathcal{A}}\bar{x},$$

and

$$\mu = \|\mathcal{B}' P_1 \mathcal{B}\|.$$

Consider $V(k+1) - V(k)$ along any trajectory,

$$\begin{aligned}
V(k+1) - V(k) &= -\bar{x}(k+1)'\left[P_{\varepsilon_a(\bar{x}(k))} - P_{\varepsilon_a(\bar{x}(k+1))}\right]\bar{x}(k+1) \\
&\quad - \bar{x}(k)' Q_{\varepsilon_a(\bar{x}(k))}\bar{x}(k) - 2(v(k) - \tilde{v}(k))'\tilde{v}(k) \\
&\quad + (v(k) - \tilde{v}(k))'\mathcal{B}' P_{\varepsilon_a(\bar{x}(k))}\mathcal{B}(v(k) - \tilde{v}(k)) \\
&\leqslant -\bar{x}(k+1)'\left[P_{\varepsilon_a(\bar{x}(k))} - P_{\varepsilon_a(\bar{x}(k+1))}\right]\bar{x}(k+1) \\
&\quad - \bar{x}(k)' Q_{\varepsilon_a(\bar{x}(k))}\bar{x}(k) - 2(v(k) - \tilde{v}(k))'\tilde{v}(k) \\
&\quad + \mu(v(k) - \tilde{v}(k))'(v(k) - \tilde{v}(k)).
\end{aligned}$$

Given that α satisfies (10.100), we have

$$|\widetilde{v}(k)| \leqslant |v(k)| \leqslant \left(\tfrac{2}{\delta}+1\right)|\widetilde{v}(k)|$$

componentwise, and since $v(k)$ and $\widetilde{v}(k)$, also componentwise, have the same sign, it follows that

$$\begin{aligned}V(k+1)-V(k) \leqslant & -\bar{x}(k)'Q_{\varepsilon_a(x(\bar{k}))}\bar{x}(k)\\ & - \bar{x}(k+1)'\left[P_{\varepsilon_a(\bar{x}(k))} - P_{\varepsilon_a(\bar{x}(k+1))}\right]\bar{x}(k+1).\end{aligned}$$

The scheduling law (10.97) guarantees that

$$V(k+1)-V(k) \text{ and } \bar{x}(k+1)'\left[P_{\varepsilon_a(\bar{x}(k+1))} - P_{\varepsilon_a(\bar{x}(k))}\right]\bar{x}(k+1)$$

cannot have the same signs. This implies that for $x \neq 0$ and α satisfying (10.100), we have

$$V(k+1)-V(k) < 0.$$

The global asymptotic stability follows. ■

The above lemma implies that $\bar{x}(k) \to 0$ as $k \to \infty$. Since this property holds for any initial condition and since we have local asymptotic stability, we can conclude that the controller achieves global asymptotic stability. This completes the proof. ■

Remark 10.41 *Comparing the low-and-high-gain controllers for discrete- and continuous-time sandwich systems, we can see that in the discrete-time case, the parameter α performs the role of high gain, which in the continuous-time case is just an arbitrary positive real number α. This shows an intrinsic difference between discrete-time and continuous-time high-gain techniques, namely, that in the continuous-time case, the high-gain parameter is completely free ($\alpha \in [0,\infty)$), while in the discrete-time case, the high-gain parameter has only a limited freedom as characterized by* (10.100). *It can actually be shown that by also scheduling α such that*

$$\alpha \in \left[0, \frac{2}{\|\mathcal{B}'P_{\varepsilon_a(\bar{x}(k))}\mathcal{B}\|}\right],$$

we can actually choose α larger without compromising global stability.

Note that in both the continuous- and discrete-time cases, the high-gain part does not affect the domain of attraction, but as demonstrated in an example shortly in the next section, the high-gain greatly improves the transient performance.

For stabilization of sandwich systems, despite the inherent difference between discrete- and continuous-time high-gain techniques, we still find a strict parallelism in choosing other design parameters, namely, the pre-feedback gain F and β. Once again as shown in the next section, these parameters also have a great impact on closed-loop performance.

10.8 Numerical examples

We illustrate here the results developed in previous sections by means of numerical examples.

10.8.1 Continuous-time systems

Example 1: (Semi-global stabilization of a single-layer system) Consider the two systems L_1 and L_2 given in (10.1) and (10.2),

$$L_1 : \begin{cases} \dot{x}(t) = \begin{pmatrix} 1 & 1 & 1 \\ 0 & 0 & 1 \\ 0 & 0 & 0 \end{pmatrix} x(t) + \begin{pmatrix} 0 \\ 0 \\ 1 \end{pmatrix} u(t) \\ z(t) = \begin{pmatrix} 0 & 1 & 0 \\ 0 & 0 & 1 \end{pmatrix} x(t), \end{cases}$$

and

$$L_2 : \quad \dot{\omega}(t) = \begin{pmatrix} 0 & 1 & 0 \\ 0 & 0 & 1 \\ 0 & 0 & 0 \end{pmatrix} \omega(t) + \begin{pmatrix} 0 & 0 \\ 0 & 1 \\ 1 & 1 \end{pmatrix} \sigma(z(t)).$$

For semi-global stabilization, we design below a low-gain as well as a low-and-high-gain state feedback controller that stabilizes the cascaded system of L_1 and L_2 with an a priori given compact set $\mathcal{W}$ to be contained in the domain of attraction of the closed-loop system, where

$$\mathcal{W} = \{\gamma \in \mathbb{R}^6 \mid \gamma \in [-10, 10]^6 \}.$$

Step 1. We start by choosing

$$F = \begin{pmatrix} -12 & -6 & -7 \end{pmatrix},$$

which makes $A + BF$ Hurwitz.
Step 2. We choose $\delta = 2.28$, which ensures that, if $\|v(t)\| < \delta$ for all $t > 0$, then $\|z_0(t)\| < \frac{1}{2}$ for all $t > 0$. We set the low-gain parameter $\varepsilon = 10^{-10}$ and choose $Q_\varepsilon = \varepsilon I$. After solving the associated CARE (10.14), we obtain the following generalized low-gain state feedback control law,

$$u = Fx - \begin{pmatrix} B \\ 0 \end{pmatrix}' P_\varepsilon \begin{pmatrix} x \\ \omega \end{pmatrix},$$

which is given explicitly as

$$u = \begin{pmatrix} -12.2879 & -6.0254 & -7.0237 \end{pmatrix} x + \begin{pmatrix} 0.00001 & 0.0017 & 0.1454 \end{pmatrix} \omega.$$

The simulation data of the closed-loop system under this generalized low-gain state feedback controller is shown in Fig. 10.7.
Step 3. As in Step 2, we set the low-gain parameter $\varepsilon = 10^{-10}$ and choose $Q_\varepsilon = \varepsilon I$. Also, we choose the high-gain parameter $\alpha = 1,000$. After solving the associated CARE (10.14), we obtain the following generalized low-and-high-gain state feedback controller,

$$\begin{aligned} u = \begin{pmatrix} -12 & -6 & -7 \end{pmatrix} x \\ - 2.28\sigma\{\begin{pmatrix} 126.4175 & 11.1549 & 10.4106 \end{pmatrix} x + \\ \begin{pmatrix} -0.0044 & -0.7486 & -63.8259 \end{pmatrix} \omega\}. \end{aligned}$$

The simulation data closed-loop system under this controller is shown in Fig. 10.8.

By comparing Figs. 10.7 and 10.8, we note that the generalized low-and-high-gain design enhances the performance by incurring much lower overshoot and undershoot.

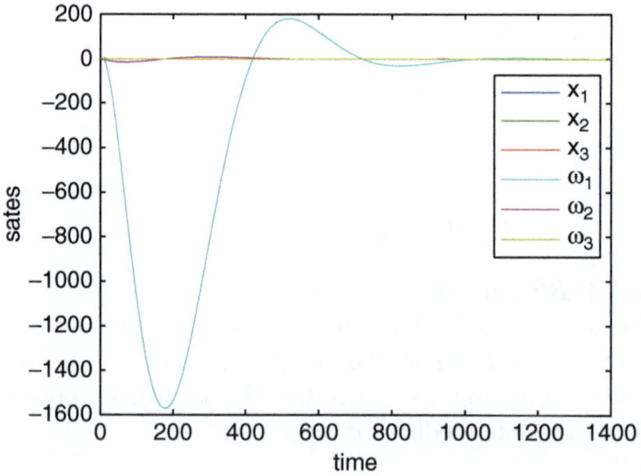

Figure 10.7: Semi-global stabilization by generalized low-gain state feedback—continuous-time system

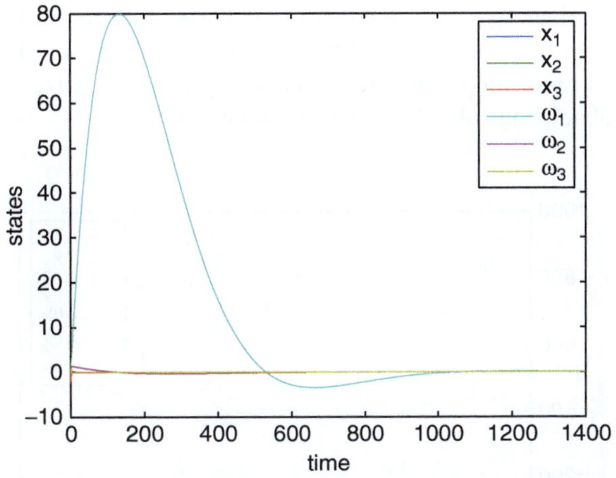

Figure 10.8: Semi-global stabilization by generalized low-and-high-gain state feedback—continuous-time system

Example 2: (Global stabilization of a single-layer system)

The two systems L_1 and L_2 in (10.1) and (10.2) are the same as in the preceding example. For global stabilization, we develop below an adaptive-low-gain as well as an adaptive-low-gain and generalized high-gain state feedback controller:
Step 1. Choose

$$F = \begin{pmatrix} -12 & -6 & -7 \end{pmatrix}$$

such that $A + BF$ is Hurwitz stable.
Step 2. Choose the same $\delta = 2.28$ as in the preceding example, and design a generalized adaptive-low-gain state feedback controller as

$$u = Fx - \begin{pmatrix} B \\ 0 \end{pmatrix}' P_{\varepsilon_a(\bar{x})}\bar{x},$$

where $P_{\varepsilon_a(\bar{x})}$ is given by the CARE (10.14) and (10.16). The simulation data of closed-loop system under this controller is shown in Fig. 10.9.
Step 3. We choose the same $\delta = 2.28$, and set $\alpha = 1,000$, and design a generalized low-and-high-gain state feedback controller as

$$u = Fx - \delta\sigma(\tfrac{1+\alpha}{\delta}\begin{pmatrix} B \\ 0 \end{pmatrix}' P_{\varepsilon_a(\bar{x})}\bar{x}),$$

where $P_{\varepsilon_a(\bar{x})}$ is given by the CARE (10.14) and (10.82). The resulting simulation data of the closed-loop system under this controller is shown in Fig. 10.10.

Clearly, the closed-loop dynamics achieved by the generalized low-and-high-gain state feedback controller has a lower overshoot.

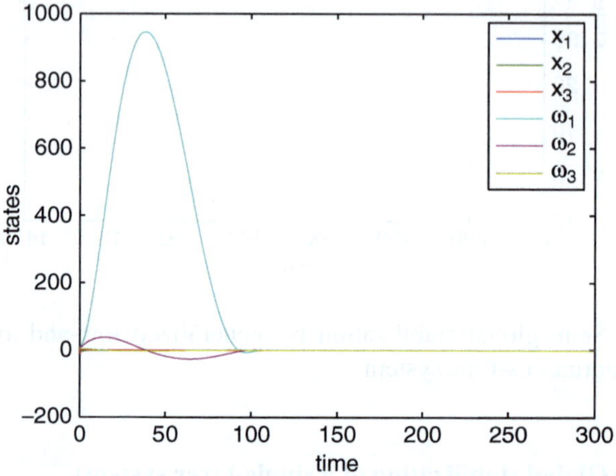

Figure 10.9: Global stabilization by generalized adaptive-low-gain state feedback—continuous-time system

Example 3: (The impact of Q_ε on performance)

Consider the same system as in Examples 1 and 2. Choose the same F and hence we have the same δ.

We observe that in the above examples, the first state element of system L_2 has the worst performance. Therefore, instead of $Q_\varepsilon = \varepsilon I$, choose

$$Q_\varepsilon = \begin{pmatrix} \varepsilon I & 0 \\ 0 & \tilde{Q}_\varepsilon \end{pmatrix},$$

where

$$\tilde{Q}_\varepsilon = \begin{pmatrix} 200\varepsilon & 0 & 0 \\ 0 & \varepsilon & 0 \\ 0 & 0 & \varepsilon \end{pmatrix}.$$

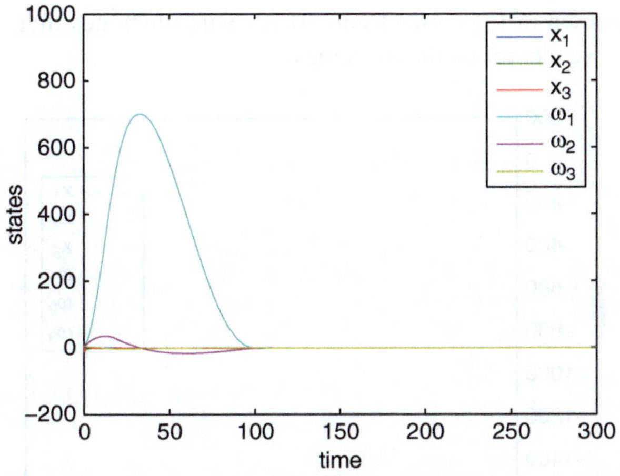

Figure 10.10: Global stabilization by generalized adaptive-low-gain and generalized high-gain state feedback—continuous-time system

However, with this Q_ε, we need to choose a relatively smaller ε. Set $\varepsilon = 2\times 10^{-12}$ and keep $\alpha = 1,000$. After solving the CARE (10.14), we obtain the following generalized low-and-high-gain state feedback controller:

$$\begin{aligned} u = &\begin{pmatrix} -12 & -6 & -7 \end{pmatrix} x \\ &- 2.28\upsilon\{\begin{pmatrix} 159.7642 & 14.2992 & 13.1163 \end{pmatrix} x + \\ &\qquad \begin{pmatrix} -0.0088 & -1.1917 & -80.8617 \end{pmatrix} \omega\}. \end{aligned}$$

Using the same ε and Q_ε but $\alpha = 0$, the generalized low-gain state feedback controller can be obtained as

$$\begin{aligned} u = &\begin{pmatrix} -12.36390 & -6.03257 & -7.02988 \end{pmatrix} x + \\ &\qquad \begin{pmatrix} 0.00002 & 0.00271 & 0.18418 \end{pmatrix} \omega. \end{aligned}$$

We reexamine then the semi-global stabilization of the interconnection of L_1 and L_2 by generalized low-and-high-gain state feedback and generalized low-gain state feedback, respectively. The simulation data of respective closed-loop systems are shown in Figs. 10.11 and 10.12. This illustrates that, with a proper choice of Q_ε, we can refine the dynamics.

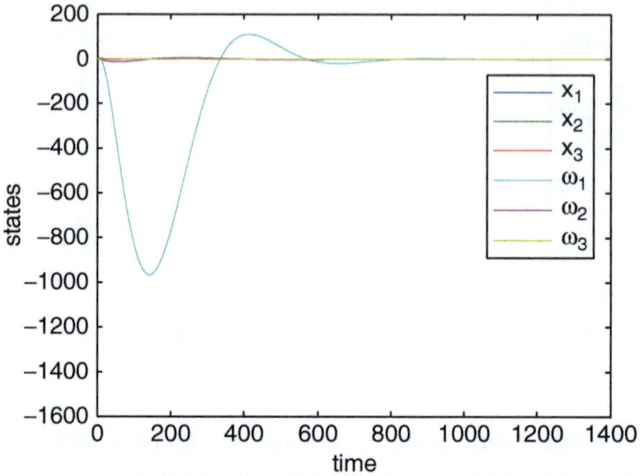

Figure 10.11: Semi-global stabilization by generalized low-gain state feedback with modified Q_ε—continuous-time system

10.8.2 Discrete-time systems

Example 4: (Semi-global and global stabilization of a single-layer system)

We consider the following two systems that constitute a single-layer sandwich system:

$$L_1 : \begin{cases} x(k+1) = \begin{pmatrix} 0 & -1 \\ 1 & 0 \end{pmatrix} x(k) + \begin{pmatrix} 0 \\ 1 \end{pmatrix} u(k), \\ z(k) = \begin{pmatrix} 1 & 0 \end{pmatrix} x(k), \end{cases} \tag{10.101}$$

and

$$L_2 : \quad \omega(k+1) = \begin{pmatrix} 0.8 & 0.6 \\ -0.6 & 0.8 \end{pmatrix} \omega(k) + \begin{pmatrix} 0 \\ 1 \end{pmatrix} \sigma(z(k)). \tag{10.102}$$

Given that $\mathcal{W} = [-1, 1]^4$, we design controllers for both semi-global and global stabilization of the origin of (10.101), (10.102).

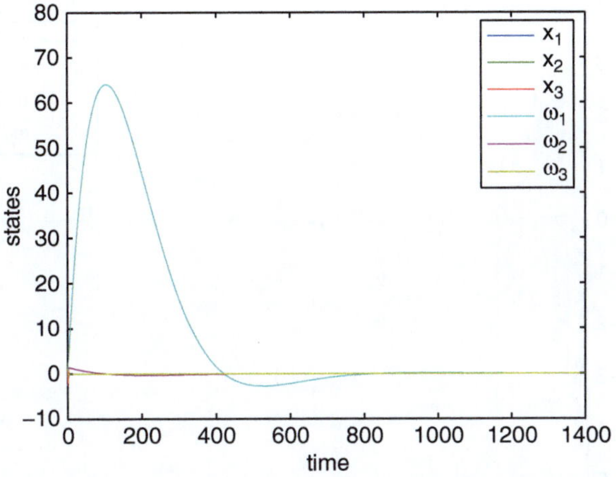

Figure 10.12: Semi-global stabilization by generalized low-and-high-gain state feedback with modified Q_ε—continuous-time system

We proceed now with generalized low-gain design for semi-global stabilization:

- Choose
$$F = \begin{pmatrix} 0.6500 & 1.2000 \end{pmatrix}.$$
- From (10.22), $\delta_1 = 0.0750$.
- Determine ε according to $\mathcal{W}$ and δ_1. We choose $\varepsilon = 10^{-3}$.
- The feedback controller is given by
$$u = \begin{pmatrix} 0.7216 & 1.0945 & 0.0060 & 0.0415 \end{pmatrix} \bar{x}.$$

The simulation data is shown in Fig. 10.13.

We proceed now with the generalized adaptive-low-gain design for global stabilization:

- Choose
$$F = \begin{pmatrix} 0.6500 & 1.2000 \end{pmatrix}.$$

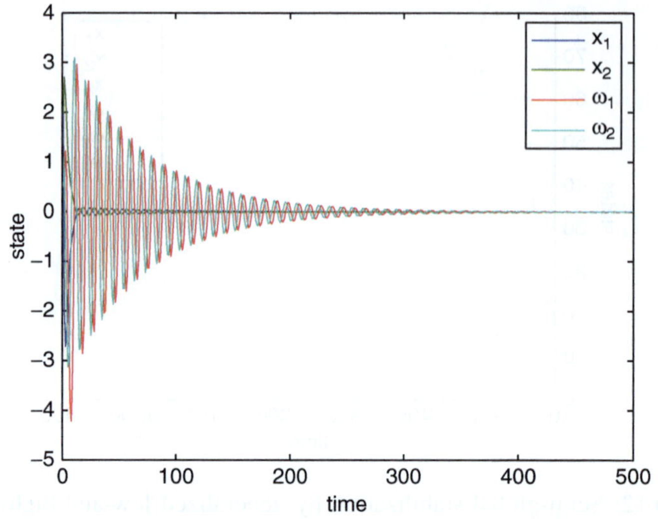

Figure 10.13: Semi-global Stabilization of a single layer system—Discrete-time system

- From (10.22), $\delta_1 = 0.0750$.
- The global stabilizing controller is formed by semi-global controller together with adaptive parameter.

The simulation data is shown in Fig. 10.14.

Example 5 (Semi-global and global stabilization of a single-layer system with input saturation): We consider the same systems as given by (10.101) and 10.102 and $\mathcal{W} = [-1, 1]^4$, however, with the actuator subject to saturation.

We proceed now with generalized low-gain design for semi-global stabilization:

- In accordance of the given $\mathcal{W}$, we choose $\varepsilon_1 = 0.0120$.
- In accordance of the given $\mathcal{W}$ and ε_1, we choose $\varepsilon_2 = 5 \times 10^{-4}$.
- The controller is given by

$$u = \begin{pmatrix} 0.1419 & -0.0170 & 0.0304 & -0.0069 \end{pmatrix} \bar{x}.$$

The simulation data is shown in Fig. 10.15.

We proceed now with generalized adaptive-low-gain design for global stabilization:

- In view of (10.55), we use $\delta = 0.0036$.

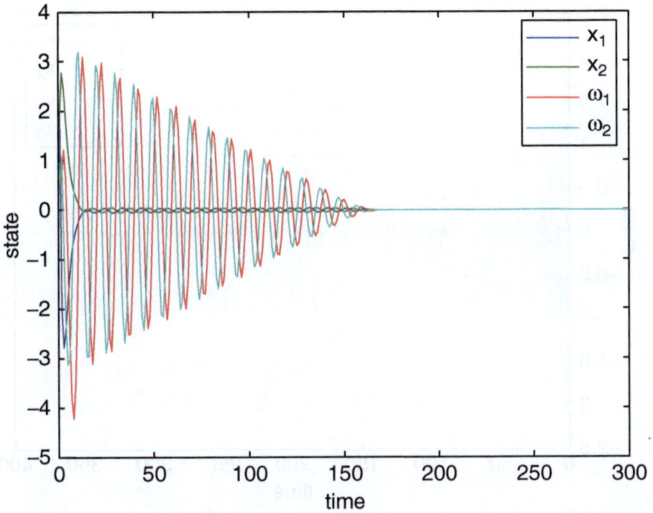

Figure 10.14: Global Stabilization of a single-layer system—discrete-time system

- $M_2 = 3.7321$ and $M_3 = 14.7373$.
- The controller is formed by semi-global stabilizing controller together with adaptation as given in (10.53) and (10.56).

The simulation data is shown in Fig. 10.16.

Example 6 (Semi-global and global stabilization of a multilayer system with input saturation): We now consider a multilayer sandwich system with the following three systems:

$$L_1 : \begin{cases} x_1(k+1) = \begin{pmatrix} 0 & -1 \\ 1 & 0 \end{pmatrix} x_1(k) + \begin{pmatrix} 0 \\ 1 \end{pmatrix} \sigma(u_1(k)), \\ z_1(k) \quad = \begin{pmatrix} 1 & 0 \end{pmatrix} x_1(k), \end{cases} \tag{10.103}$$

and

$$L_2 : \begin{cases} x_2(k+1) = \begin{pmatrix} 0.8 & 0.6 \\ -0.6 & 0.8 \end{pmatrix} x_2(k) + \begin{pmatrix} 0 \\ 1 \end{pmatrix} \sigma(z_1(k)), \\ z_2(k) \quad = \begin{pmatrix} 1 & 0 \end{pmatrix} x_2(k), \end{cases} \tag{10.104}$$

and

$$L_3 : \quad \omega(k+1) = \begin{pmatrix} 0.8 & 0.6 \\ -0.6 & 0.8 \end{pmatrix} \omega(k) + \begin{pmatrix} 0 \\ 1 \end{pmatrix} \sigma(z(k)). \tag{10.105}$$

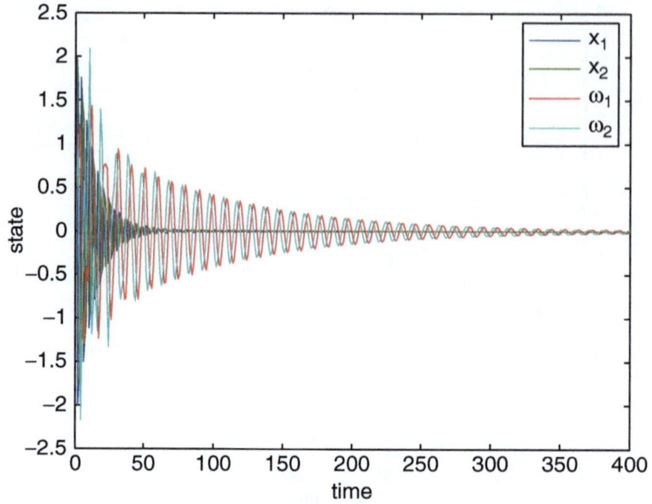

Figure 10.15: Semi-global stabilization of a single-layer system with Input saturation—Discrete-time

Given that $\mathcal{W} = [-1, 1]^6$, we design controllers for both semi-global and global stabilization of the origin of (10.103), (10.104) and (10.105).

We proceed next with generalized low-gain design for semi-global stabilization.

- In accordance of the given $\mathcal{W}$, we choose $\varepsilon_1 = 1.2 \times 10^{-4}$, $\varepsilon_2 = 10^{-6}$, and $\varepsilon_3 = 10^{-9}$.
- The controller is given by

$$u = \begin{pmatrix} 0.0133 & -0.0060 & 0.0116 & -0.0033 & -0.000015 & 0.000042 \end{pmatrix} \chi_3$$

  where $\chi_3 = (x_1, x_2, x_3)'$.

The simulation data is shown in Fig. 10.17.

We proceed next with generalized adaptive-low-gain design for global stabilization.

- In view of (10.69), we use $\delta_1 = \frac{1}{3}$, $\delta_2 = 0.0011$, and $\delta_3 = 1.5983 \times 10^{-6}$.
- We compute $M_1 = 3.7321$, $M_2 = 14.7373$, and $M_3 = 53.2378$.
- The global stabilizing controller consists of semi-global stabilizing controller and adaptation defined by (10.70).

The simulation data is shown in Fig. 10.18.

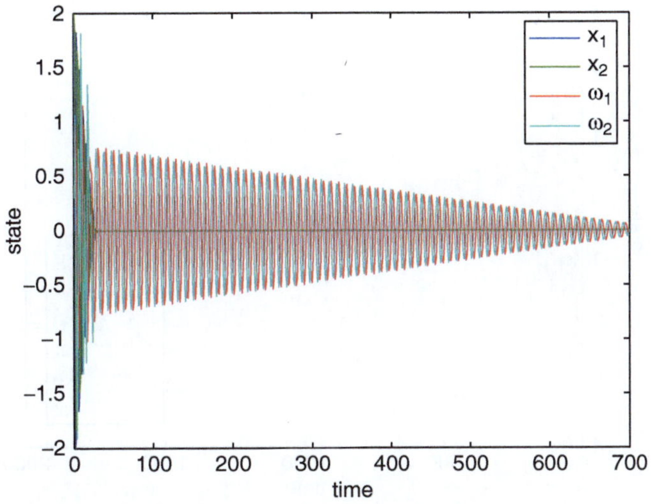

Figure 10.16: Global stabilization of a single-layer system with input saturation—discrete-time

Example 7: (Semi-global stabilization of a single-layer system—low-and-high-gain controllers) We consider, in this example, the low-and-high-gain design methodology and address the performance issues with regard to the choice of the design parameters: high-gain parameter α, magnitude of z_0 as represented by β, and pre-feedback gain F.

We consider semi-global stabilization of the following sandwich systems:

$$L_1 : \begin{cases} x(k+1) = \begin{pmatrix} 2 & 1 \\ 0 & 1 \end{pmatrix} x(k) + \begin{pmatrix} 0 \\ 1 \end{pmatrix} u(k) \\ z(k) \;\; = \begin{pmatrix} 1 & 0 \end{pmatrix} x(k), \end{cases} \tag{10.106}$$

and

$$L_2 : \quad \omega(k+1) = \begin{pmatrix} 0.8 & 0.6 \\ -0.6 & 0.8 \end{pmatrix} \omega(k) + \begin{pmatrix} 0 \\ 1 \end{pmatrix} \sigma(z(k)), \tag{10.107}$$

with initial conditions in $\mathcal{W} = \{\xi \in \mathbb{R}^4 \mid, \; \xi \in [-3, 3]^4\}$. We choose in particular $\bar{x}(0) = (-3, -3, 3, 3)'$ for all the simulations.

Impact of α: As we discussed in previous sections, the high-gain parameter does not affect the domain of attraction. However, it plays a crucial role in improving the transient performance. Note that, in the continuous-time case, the high-gain parameter can be any positive real number. Although for discrete-time systems, the high-gain parameter cannot be arbitrarily large, it can still enhance the transient performance greatly.

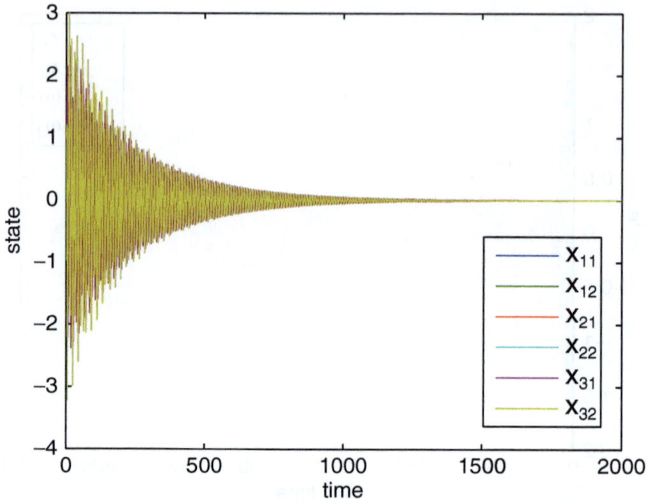

Figure 10.17: Semi-global stabilization of a multilayer system—discrete-time

In what follows, we shall compare the results for four different values of α, namely, $\alpha \in \{0, 0.1, 1, 2\}$. We proceed in three steps as described on page 583:

Step 1

Choose $F = (-2.1 \quad -1.9)$ such that the eigenvalues of $A + BF$ are at $\lambda_1 = 0.5$ and $\lambda_2 = 0.6$.

Step 2

Choose $\beta = 0.5$ and compute $\delta = 0.1$. The saturation time K is approximately 11.

Step 3

Choose $Q_\varepsilon = \varepsilon I_4$. Based on K, δ, and $\mathcal{W}$, we choose $\varepsilon = 10^{-4}$. For each α, solve DARE (10.86) with $Q_\varepsilon = \varepsilon I_4$ and $\varepsilon = 10^{-4}$.

The simulation data is shown in Figs. 10.19–10.22. Generally, the low-and-high-gain feedback greatly improves the transient performance, but this does not necessarily mean α should be chosen as the maximum value of 2. We find a small variation of performance among different choices of α. From $\alpha = 0.1$ to $\alpha = 1$, a small improvement can be observed. However, from $\alpha = 1$ to $\alpha = 2$, we get a bit of deterioration when state is close to the origin. The proper value of α may depend on particular circumstances.

Impact of β: In the generalized low-and-high-gain design methodology proposed in Sect. 10.7.2, we choose δ so that the zero state response of L_1 system, namely, z_0, remains less than β, where $\beta \in (0, 1)$. By choosing different β or equivalently z_0, we can tune the total control capacity allowed for the linear cas-

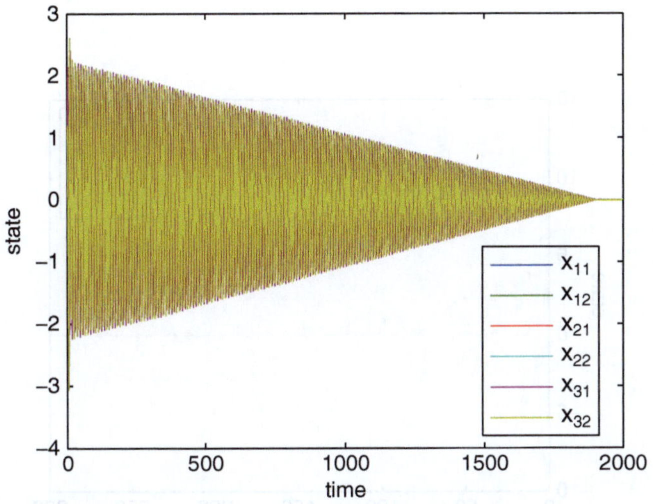

Figure 10.18: Global stabilization of a multilayer system—discrete-time

cade systems when saturation is inactive. We next explore the impact of this parameter β. (Note that, in the continuous-time case, β is fixed at 0.5; however, the following discussion is also applicable to continuous-time systems.)

The definition of δ indicates that a larger β will result in a larger δ and thus allows for better utilization of control forces when saturation remains inactive. We expect that this will refine the closed-loop performance. On the other hand, saturation is active until the zero input response of L_1 falls below $\frac{1}{\beta}$. Thus, a larger β may keep the saturation active for a longer period of time. The overall performance depends on the trade-off between these two effects.

We consider the same systems as in (10.106) and (10.107) and use the same value for F as chosen earlier.

If we choose $\beta = 0.1$, then $\delta = 0.02$ as given by (10.85) and the approximate saturation time $K = 9$. For K, δ, and $\mathcal{W}$ as given, a proper choice for ε is 10^{-5}. The simulation data is shown in Fig. 10.23.

If we choose $\beta = 0.9$, then we have $\delta = 0.18$ and $K = 14$. Based on K, δ, and $\mathcal{W}$, a proper choice for ε is ε is 10^{-3}. The simulation data is shown in Fig. 10.24.

In this case, the performance is improved by choosing a larger level of β or equivalently z_0.

Impact of F: We discuss here the choice of the pre-feedback gain matrix F. We note that F will affect the effective control capacity δ, saturation time K, and the solution P_ε of DARE (10.86). This impact is generally difficult to analyze quantitatively. We illustrate this by choosing a different value of F than what we did earlier.

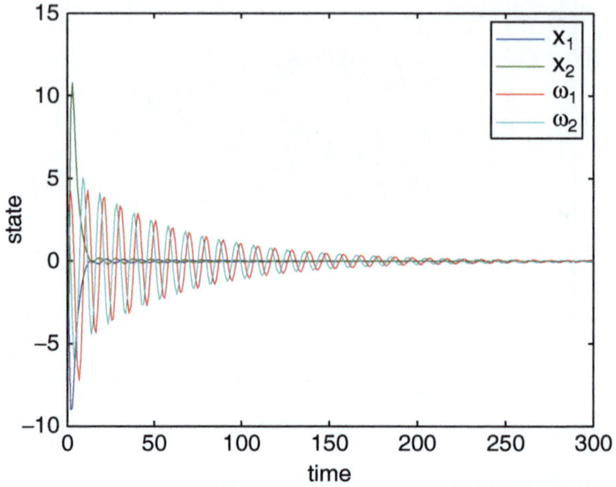

Figure 10.19: Semi-global stabilization via low-gain feedback

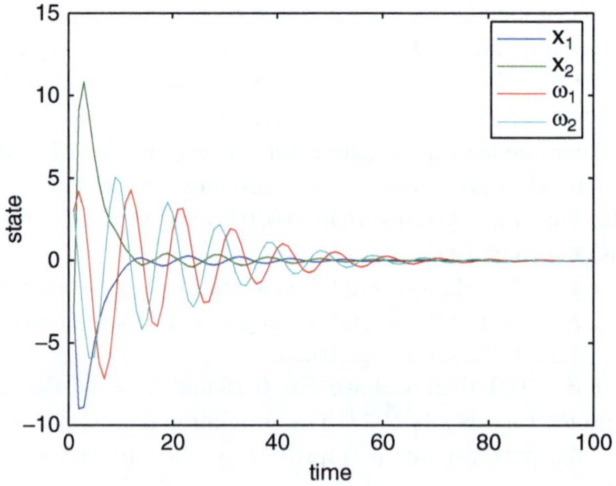

Figure 10.20: Semi-global stabilization via low-and-high-gain feedback with $\alpha = 0.1$

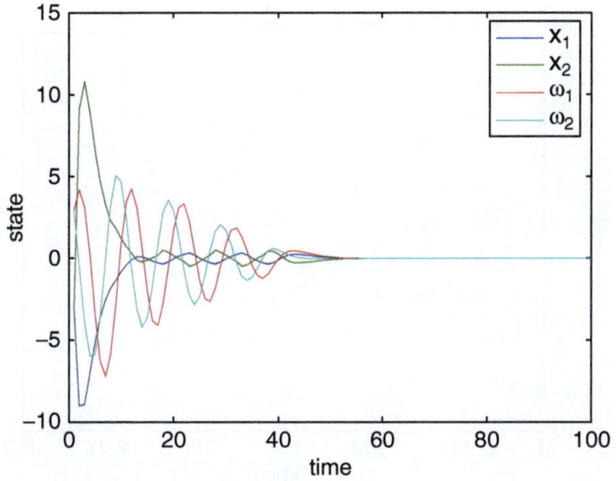

Figure 10.21: Semi-global stabilization via low-and-high-gain feedback with $\alpha = 1$

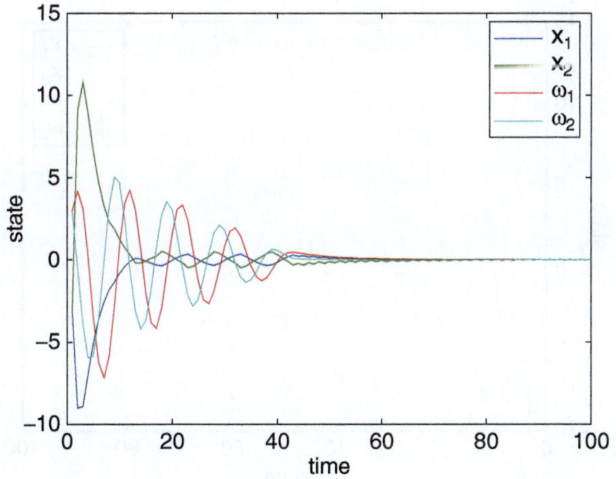

Figure 10.22: Semi-global stabilization via low-and-high-gain feedback with $\alpha = 2$

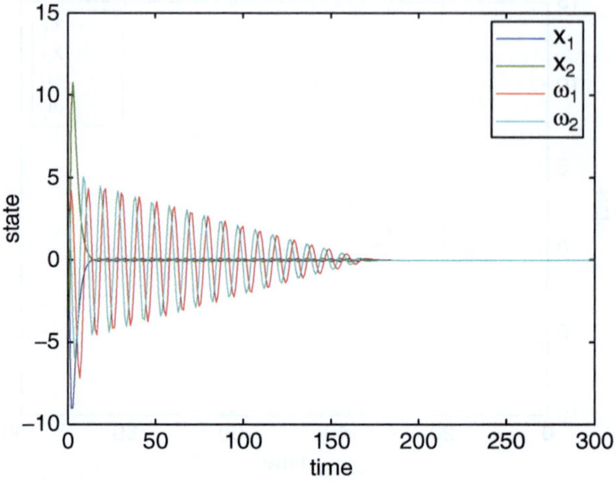

Figure 10.23: Semi-global stabilization $\|z_0\| = 0.1$

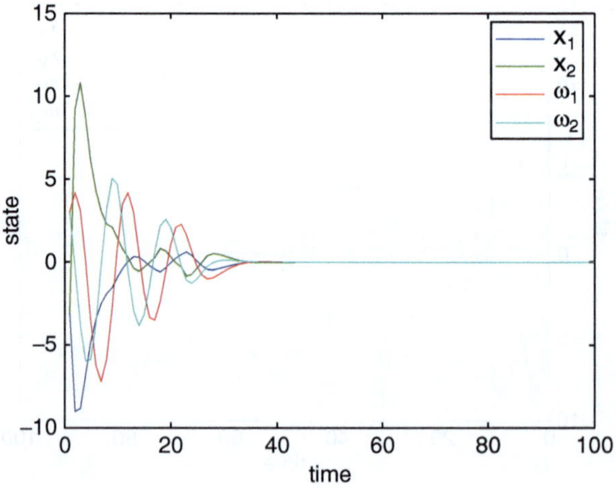

Figure 10.24: Semi-global stabilization $\|z_0\| = 0.9$

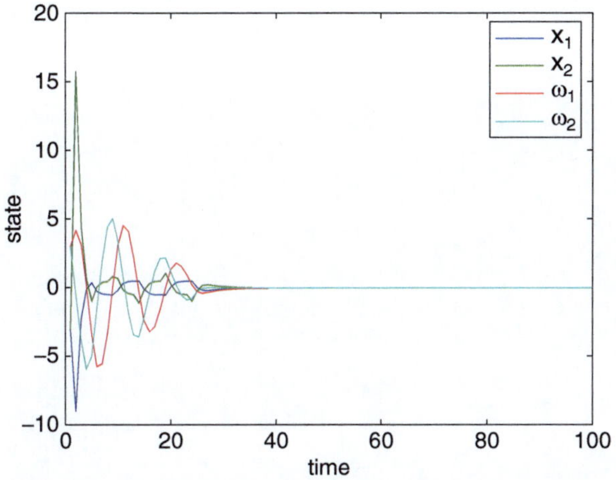

Figure 10.25: Semi-global stabilization via low-and-high-gain with different F

Let $\beta = 0.5$ and $\alpha = 1$ be fixed. We choose F such that the eigenvalues of $A + BF$ are at $\lambda_1 = 0.1$ and $\lambda_2 = 0.2$. As a result, we get

$$F = \begin{pmatrix} -3.4200 & -2.7000 \end{pmatrix}, \quad \delta = 0.36, \quad \varepsilon = 10^{-3}.$$

The simulation data is shown in Fig. 10.25. Comparing with Fig. 10.20, where we chose F such that the eigenvalues of $A + BF$ are at $\lambda_1 = 0.5$ and $\lambda_2 = 0.6$, we find a better convergence rate but a larger overshoot.

11

Simultaneous external and internal stabilization

11.1 Introduction

So far, we developed various design methodologies that attain internal stabilization in different contexts for linear systems with constraints, and in particular for linear systems subject to actuator saturation. We recall internal stability protects against a single impulse-like disturbance, whereas input–output stability or otherwise called external stability or L_p (ℓ_p) stability protects against external inputs or noise disturbances. In general, internal stability alone does not necessarily imply any aspects of external stability, although external stability in some cases implies some aspects of internal stability. As such, our focus now is on developing feedback control strategies for simultaneous internal and external stabilization. There exist various definitions of external stability as discussed in Sect. 2.8. Then, in view of such definitions, for simultaneous stabilization, we seek here besides internal stabilization different types of external or L_p (ℓ_p) stabilization, namely, (1) L_p (ℓ_p) stabilization with fixed initial conditions and without finite gain, (2) L_p (ℓ_p) stabilization with fixed initial conditions and with finite gain, (3) L_p (ℓ_p) stabilization with arbitrary initial conditions and without finite gain, and (4) L_p (ℓ_p) stabilization with arbitrary initial conditions and with finite gain and bias.

All the aforementioned simultaneous stabilization problems can be formulated in either global framework or semi-global framework. The global framework (respectively semi-global framework) implies that we seek both internal and external stabilization in a global sense (respectively in a semi-global sense). Another issue in the global framework is the input to state stability (ISS) problem discussed in Sect. 2.8. As said there, the notion of ISS makes an attempt to marry both the notions of internal stability and the L_∞-stability or ℓ_∞-stability. In fact, as pointed out in Remark 2.68, when the input d is identically zero, ISS implies the global asymptotic stability of the zero equilibrium point. In this sense, ISS is indeed a simultaneous stabilization concept.

Thus, we formulate here five problems in a global framework. We observe that the first two problems among the above mentioned problems are traditional simultaneous stabilization problems as they deal with fixed initial conditions, typically zero. On the other hand, the later three problems deal with arbitrary initial

A. Saberi et al., *Internal and External Stabilization of Linear Systems with Constraints*, Systems & Control: Foundations & Applications, DOI 10.1007/978-0-8176-4787-2_11,

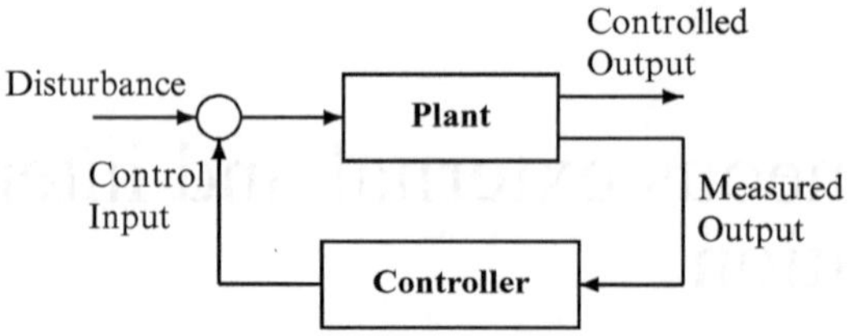

Figure 11.1: Closed-loop system with input-additive disturbance

conditions, and in these problems, the notions of external and internal stability are somewhat intermingled, i.e., guaranteeing external stability in the intended sense guarantees the internal stability in some sense or other.

Unlike the global framework, our focus in the semi-global framework is only on the first two problems, one that does not seek finite gain while the other seeks finite gain. Both of the problems are traditional simultaneous stabilization problems that deal with fixed initial conditions, typically zero.

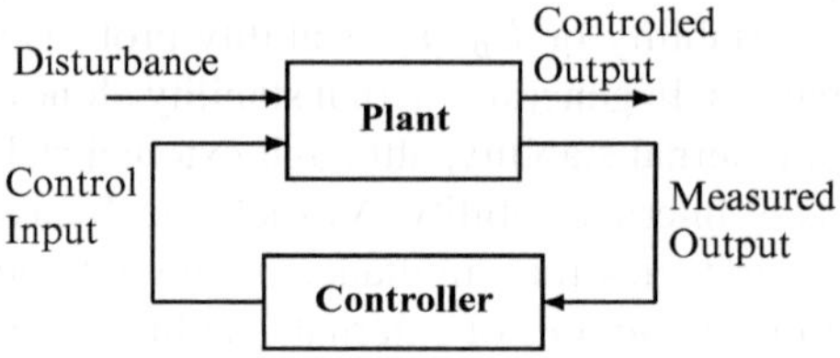

Figure 11.2: Closed-loop system with non-input-additive disturbance

Two fundamental questions that arise at this stage are these: What conditions do we need to impose on a given system in order to achieve simultaneous internal and external stabilization? Under such conditions, do we need linear or nonlinear feedback controllers? In view of earlier chapters dealing with internal stabilization alone, in general for simultaneous stabilization, the given system must necessarily be asymptotically null controllable with bounded control (ANCBC), or equivalently, it must be stabilizable and at most open-loop critically unstable.[1] Is ANCBC condition itself sufficient to achieve simultaneous stabilization as required in the problems discussed above? It turns out that the answer to this question depends intrinsically on how the disturbance signals act on the given system. Two distinct cases arise, the first case as illustrated in Fig. 11.1 corresponds to the situation when the control inputs and the disturbance signals are *additive*, and the second case as illustrated in Fig. 11.2 corresponds to the situation when

[1]We recall that a linear system is said to be open-loop critically unstable if all its open-loop poles are in the closed left-half plane for continuous-time systems or within or on the unit circle for discrete-time systems.

such signals are *nonadditive*. The achievable simultaneous stabilization results are profoundly different for these two cases. In fact, the level of complexity involved in the non-input-additive case compared to the input-additive case demands that we treat the two cases separately. As such, we treat each of these problems in different chapters that follow this introductory chapter which is basically devoted to introducing universal definitions, all pertaining to simultaneous stabilization.

11.2 Simultaneous stabilization in global framework: problem statements

The beforementioned simultaneous external and internal stabilization problems in a global framework are formulated formally in this section. Depending upon the type of external stabilization that is sought (see Sect. 2.8), there exist different types of such problems. These problems can be distinctly divided into two groups. The first group consists of those problems in which external stabilization is sought in the classical sense with fixed initial conditions normally set at the origin. On the other hand, the second group consists of those problems which utilize the recently introduced notions of external stability with arbitrary initial conditions and input to state stability (ISS). As we said in the introduction, in the problems of the second group, the notions of external and internal stability are somewhat intermingled, i.e., the notion of external stability imbeds in some sense or other the notion of internal stability. Also, in order to achieve stabilization in the sense defined in all the problems of this section, one would need the given linear system subject to saturation be ANCBC. Moreover, as can be expected, global simultaneous stabilization in general requires nonlinear feedback laws.

Our emphasis throughout this book is only on linear systems subject to actuator saturation. Also, as mentioned earlier, external disturbance can be additive (as depicted in Fig. 11.1) or nonadditive (as depicted in Fig. 11.2) to the control input. Consider a continuous-time linear system subject to actuator saturation where the disturbance is input-additive,

$$\Sigma_1^c : \begin{cases} \dot{x}(t) = Ax(t) + B\sigma(u(t) + d(t)), \\ z(t) = x(t), \quad t \geq 0, \\ y(t) = Cx(t), \quad t \geq 0, \end{cases} \tag{11.1a}$$

or similarly consider a discrete-time system,

$$\Sigma_1^d : \begin{cases} x(k+1) = Ax(k) + B\sigma(u(k) + d(k)), \\ \quad z(k) \quad = x(k), \quad k \geq 0, \\ \quad y(k) \quad = Cx(k), \quad k = 1, 2, \ldots, \end{cases} \tag{11.1b}$$

where as usual $x \in \mathbb{R}^n$ is the state, $u \in \mathbb{R}^m$ is the input, $z \in \mathbb{R}^q$ is the controlled variable, and $y \in \mathbb{R}^p$ is the measured output. Here σ denotes the standard saturation function defined as

$$\sigma(u) = [\text{sat}_\Delta(u_1), \ldots, \text{sat}_\Delta(u_m)],$$

where $\text{sat}_\Delta(s) = \text{sgn}(s) \min\{|s|, \Delta\}$ for some $\Delta > 0$.

Along the same lines, when the disturbance signals are non-input-additive, consider a continuous-time system of the form

$$\Sigma_2^c : \begin{cases} \dot{x}(t) = Ax(t) + B\sigma(u(t)) + Ed(t), \\ z(t) = x(t), \quad t \geq 0, \\ y(t) = Cx(t), \quad t \geq 0 \end{cases} \tag{11.2a}$$

or similarly consider a discrete-time system of the form

$$\Sigma_2^d : \begin{cases} x(k+1) = Ax(k) + B\sigma(u(k)) + Ed(k), \\ z(k) \quad = x(k), \quad k \geq 0, \\ y(k) \quad = Cx(k), \quad k = 1, 2, \ldots \end{cases} \tag{11.2b}$$

with $x \in \mathbb{R}^n$, $u \in \mathbb{R}^m$, $z \in \mathbb{R}^q$, and $y \in \mathbb{R}^p$. We have the following problem formulations valid for either or both the cases when the disturbance signal is additive or nonadditive to control input.

Problem 11.1 For any $p \in [1, \infty]$, the system as in (11.1) or as in (11.2) is said to be **simultaneously globally L_p (or ℓ_p) stabilizable with fixed initial conditions and without finite gain and globally asymptotically stabilizable** via static state (or dynamic measurement) feedback if there exists a static state (or dynamic measurement) feedback law such that the following conditions hold:

(*i*) The closed-loop system is L_p (or ℓ_p) stable with fixed initial conditions and without finite gain.

(*ii*) In the absence of any disturbance signal d, the equilibrium point of the closed-loop system is globally asymptotically stable.

The above problem is coined as the (G_p/G) problem with fixed initial conditions.

Problem 11.2 For any $p \in [1, \infty]$, the system as in (11.1) or as in (11.2) is said to be **simultaneously globally L_p (or ℓ_p) stabilizable with fixed initial conditions with finite gain and with zero bias and globally asymptotically stabilizable** via static state (or dynamic measurement) feedback if there exists a static state (or dynamic measurement) feedback law such that the following conditions hold:

(*i*) The closed-loop system is L_p (or ℓ_p) stable with fixed initial conditions with finite gain and with zero bias.

(*ii*) In the absence of any disturbance signal d, the equilibrium point of the closed-loop system is globally asymptotically stable.

The above problem is coined as the $(G_p/G)_{fg}$ problem with fixed initial conditions and with zero bias.

Problem 11.3 For any $p \in [1, \infty]$, the system as in (11.1) or as in (11.2) is said to be **simultaneously globally L_p (or ℓ_p) stabilizable with fixed initial conditions with finite gain and with bias and globally asymptotically stabilizable** via static state feedback (or dynamic measurement feedback) if there exists a static state (or dynamic measurement) feedback law such that the following conditions hold:

(*i*) The closed-loop system is L_p (or ℓ_p) stable with fixed initial conditions with finite gain and with bias.

(*ii*) In the absence of any disturbance signal d, the equilibrium point $x = 0$ of the closed-loop system is globally asymptotically stable.

The simultaneous global external and global internal stabilization problems formulated above utilize the classical concept of L_p (or ℓ_p) stability with fixed initial conditions while explicitly requiring internal stabilization. We formulate below external stabilization problems based on recently introduced notions of L_p (or ℓ_p) stability with arbitrary initial conditions and ISS stability.

Problem 11.4 For any $p \in [1, \infty]$, the system as in (11.1) or as in (11.2) is said to be **simultaneously globally L_p (or ℓ_p) stabilizable with arbitrary initial conditions and without finite gain and globally asymptotically stabilizable** via static state (or dynamic measurement) feedback if there exists a static state (or dynamic measurement) feedback law such that the following conditions hold:

(*i*) The closed-loop system is L_p (or ℓ_p) stable with arbitrary initial conditions and without finite gain.

(*ii*) In the absence of any disturbance signal d, the equilibrium point of the closed-loop system is globally asymptotically stable.

Problem 11.5 For any $p \in [1, \infty]$, the system as in (11.1) or as in (11.2) is said to be **simultaneously globally L_p (or ℓ_p) stabilizable with arbitrary initial conditions and with finite gain and bias and globally asymptotically stabilizable**

via static state (or dynamic measurement) if there exists a static state (or dynamic measurement) feedback law such that the following conditions hold:

(*i*) The closed-loop system is L_p (or ℓ_p) stable with arbitrary initial conditions and with finite gain and bias.

(*ii*) In the absence of any disturbance signal d, the equilibrium point of the closed-loop system is globally asymptotically stable.

As discussed in Sect. 2.8, there exists a connection between the notions of L_p stability with arbitrary initial conditions and internal stability. This connection plays a vivid and transparent role in the case of Problem 11.5. In fact, as the following lemma formalizes, the second condition of Problem 11.5 is imbedded in its first condition.

Lemma 11.6 *Let $p \in [1, \infty)$. For the system as in* (11.1) *with disturbance additive to the control input, and for Problem 11.5, we have*

$$\textit{Condition (i)} \Rightarrow \textit{Condition (ii)}.$$

Proof : The proof basically follows the same line as in Theorems 2.83 and 2.84. Let us first consider the continuous-time system. Suppose, with a feedback controller $u = f(x)$, condition (*i*) is attained, i.e., the closed-loop system

$$\dot{x} = Ax + B\sigma(f(x) + d)$$

satisfies

$$\|x\|_p \leqslant \gamma_p \|d\|_p + b_p(\|x_0\|)$$

for any $x_0 \in \mathbb{R}^n$ where $b_p(\cdot)$ is a class $\mathcal{K}$ function. This immediately implies that for $d = 0$, $x \in L_p$ for any $x(0)$. Suppose, for the sake of establishing a contradiction, that $x(t) \to 0$ does not hold. Then there exists a $\delta > 0$ such that, for any arbitrarily large $T \geqslant 0$, there is a $\tau \geqslant T$ such that $\|x(\tau)\| \geqslant 2\delta$. Let m be an upper bound on $\|Ax + B\sigma(f(x))\|$ for x in the closed ball $B(2\delta)$. Due to saturation, this bound is finite for finite δ.

For some τ such that $\|x(\tau)\| \geqslant 2\delta$, let $t_2 \geqslant \tau$ be the smallest value such that $\|x(t_2)\| = \delta$ and let t_1 be the largest value such that $\tau \leqslant t_1 < t_2$ and $\|x(t_2)\| = 2\delta$. Such t_1 and t_2 exist because $x \in L_p$ and x is absolutely continuous. Since $\|x(t)\| \leqslant 2\delta$ for all $t \in [t_1, t_2]$, we have, due to the absolute continuity of the solution,

$$\|x(t_1)\| - \|x(t_2)\| \leqslant \|x(t_2) - x(t_1)\| = \left\| \int_{t_1}^{t_2} [Ax(\tau) + B\sigma(f(x(\tau)))]\, d\tau \right\|$$

$$\leqslant \int_{t_1}^{t_2} \|Ax(\tau) + B\sigma(f(x(\tau)))\|\, d\tau \leqslant (t_2 - t_1)m.$$

Hence, $t_2 - t_1 \geqslant (\|x(t_1)\| - \|x(t_2)\|)/m = \delta/m$. Clearly, $\|x(t)\| \geqslant \delta$ for all $t \in [\tau, t_2]$, and furthermore $t_2 - \tau \geqslant t_2 - t_1 \geqslant \delta/m$. It follows that for each τ such that $\|x(\tau)\| \geqslant 2\delta$, we have $\|x(t)\| \geqslant \delta$ for all $t \in [\tau, \tau + \delta/m]$.

Let T be chosen large enough that

$$\int_T^\infty \|x(t)\|^p\, d\tau < \frac{\delta^{p+1}}{m}. \tag{11.3}$$

Such a T must exist, since $x(t) \in L_p$. Let $\tau \geqslant T$ be chosen such that $\|x(\tau)\| \geqslant 2\delta$. We have

$$\int_T^\infty \|x(t)\|^p\, d\tau \geqslant \int_\tau^{\tau+\delta/m} \|x(t)\|^p\, d\tau \geqslant \frac{\delta^{p+1}}{m}.$$

This contradicts (11.3), which proves that $x(t) \to 0$.

We proceed now to show local stability. For $p \in [1, \infty)$, and for any $\varepsilon > 0$, let M be such that $\|Ax + B\sigma(f(x))\| \leqslant M$ for $\|x\| \leqslant \varepsilon$, and $\delta \leqslant \frac{\varepsilon}{2}$ be such that

$$b_p(\delta)^p \leqslant \tfrac{\varepsilon^{p+1}}{2^{p+1}M}.$$

Therefore, for $d = 0$ and $\|x_0\| \leqslant \delta$, we have

$$\|x\|_p^p \leqslant \tfrac{\varepsilon^{p+1}}{2^{p+1}M}. \tag{11.4}$$

It remains to show that $\|x(t)\| < \varepsilon$ for all $t > 0$. We establish this by contradiction. Choose t_1 as the smallest t for which we have $\|x(t)\| = \varepsilon$. Let $t_2 < t_1$ be the largest value such that $\|x(t_2)\| = \varepsilon/2$. We have, owing to the absolute continuity of $x(\cdot)$,

$$\|x(t_1)\| - \|x(t_2)\| \leqslant \|x(t_1) - x(t_2)\|$$

$$\leqslant \| \int_{t_2}^{t_1} Ax(t) + B\sigma(f(x(t)))\mathrm{d}t \|$$

$$\leqslant \int_{t_2}^{t_1} M\,\mathrm{d}t \leqslant M(t_2 - t_1).$$

This yields that $t_1 - t_2 \geqslant \frac{\varepsilon}{2M}$ and hence we have for $p \in [1, \infty)$,

$$\int_0^\infty \|x(t)\|^p \mathrm{d}t \geqslant \int_{t_2}^{t_1} \|x(t)\|^p \mathrm{d}t > \frac{\varepsilon^{p+1}}{2^{p+1}M},$$

which contradicts (11.4). Therefore, such a t_1 does not exist. This, together with the fact that $\|x_0\| \leqslant \delta \leqslant \frac{\varepsilon}{2}$, implies that $\|x(t)\| < \varepsilon$ for all $t > 0$. The local stability follows.

We consider next the proof for discrete-time systems. In this case, it is rather simple owing to the nature of ℓ_p space. Suppose for $p \in [1, \infty)$, under a feedback controller $f(x)$, condition (*i*) is attained, i.e., the closed-loop system

$$\rho x = Ax + B\sigma(f(x) + d)$$

satisfies

$$\|x\|_p \leqslant \gamma_p \|d\|_p + b_p(\|x_0\|)$$

for any $x_0 \in \mathbb{R}^n$ where b_p is a class $\mathcal{K}$ function. By setting $d = 0$, we immediately have $x \in \ell_p$ for any x_0. This implies that $x(k) \to 0$ as $k \to \infty$ and hence global attractivity follows.

For any $\varepsilon > 0$, let $\delta \leqslant \varepsilon$ be such that $b_p(\delta) \leqslant \varepsilon$. Then for any $\|x_0\| \leqslant \delta$ and $d = 0$, we have for $p \in [1, \infty)$,

$$\|x\|_p \leqslant b_p(\|x_0\|) \leqslant b_p(\delta) \leqslant \varepsilon.$$

But then,

$$\|x\|_\infty \leqslant \|x\|_p \leqslant \varepsilon.$$

Therefore, we conclude local stability. ■

Remark 11.7 *The above lemma also holds when the disturbance is non-input-additive. However, in that case, as will be shown in a later chapter, the $L_p \backslash \ell_p$ stabilization with finite gain is in general not possible unless the open-loop system is already stable.*

Lemma 11.6 can be weakened if we do not require finite gain as discussed in the following lemma.

Lemma 11.8 *Let $p \in [1, \infty)$. For the system as in* (11.1) *with disturbance additive to the control input, and for Problem 11.4, we have*

$$\textit{Condition (i)} \Rightarrow \textit{Global attractivity.}$$

Proof : The proof follows exactly by the same argument as used in the proof of Lemma 11.6. ∎

Although the next problem does not explicitly seek internal stability, it imbeds the notion of internal stability within itself as we shall prove in the next chapter, Sect. 12.4.

Problem 11.9 The system as in (11.1) or as in (11.2) is said to be **globally ISS stabilizable** via static state feedback (or dynamical measurement feedback) if there exists a static state (or a dynamical measurement) feedback law such that the closed-loop system is ISS stable.

11.3 Simultaneous stabilization in semi-global framework: problem statements

As discussed in the introduction, our focus in the semi-global framework is only on two problems, one that does not seek finite gain while the other seeks finite gain. Both of the problems are traditional simultaneous stabilization problems that deal with fixed initial conditions, typically zero.

Before we proceed to define formally the problems we solve in the semi-global framework, we need to recall some notation.

Definition 11.10 *For a signal z with $z(t) \in \mathbb{R}^s$, and for any $p, q \in [1,\infty]$ and any $D > 0$, $L_{p,q}(D)$ denotes the set of all $z \in L_p$ such that $\|z\|_{L_q} \leqslant D$. Similarly, for discrete-time signals z with $z(t) \in \mathbb{R}^s$, and for any $p, q \in [1,\infty]$ and any $D > 0$, $\ell_{p,q}(D)$ denotes the set of all $z \in \ell_p$ such that $\|z\|_{\ell_q} \leqslant D$.*

Problem 11.11 For any $p, q \in [1,\infty]$, the system as in (11.1) or as in (11.2) is said to be **simultaneously L_q (or ℓ_q) semi-globally L_p-stabilizable (or ℓ_p-stabilizable) and semi-globally asymptotically stabilizable** via static state feedback (or dynamic measurement feedback with dynamic order n_c) if, for any a priori given (arbitrarily large) bounded set $\mathcal{X} \subset \mathbb{R}^n$ (or respectively, $\mathcal{X} \subset \mathbb{R}^{n+n_c}$) and any $D > 0$, there exists a static state feedback (or respectively, dynamic measurement feedback of order n_c), possibly depending on $\mathcal{X}$ and D, such that the following conditions hold:

(*i*) For continuous-time case, the closed-loop system is L_p-stable over the set $L_{p,q}(D)$. That is, $z \in L_p$ for all $d \in L_p$ with $\|d\|_{L_q} \leqslant D$ and $x(0) = 0$. Similarly, for discrete-time case, the closed-loop system is ℓ_p-stable over the set $\ell_{p,q}(D)$. That is, $z \in \ell_p$ for all $d \in \ell_p$ with $\|d\|_{\ell_q} \leqslant D$ and $x(0) = 0$.

(*ii*) In the absence of any disturbance signal d, the equilibrium point $x = 0$ of the closed-loop system is asymptotically stable with $\mathcal{X} \subset \mathbb{R}^n$ (or respectively, $\mathcal{X} \subset \mathbb{R}^{n+n_c}$) contained in its domain of attraction.

The above problem is coined as the $(SG_{p,q}/SG)$ problem.

Problem 11.12 For any $p, q \in [1, \infty]$, the system as in (11.1) or as in (11.2) is said to be **simultaneously L_q (or ℓ_q) semi-globally finite gain L_p-stabilizable (or ℓ_p-stabilizable) and semi-globally asymptotically stabilizable** via static state feedback (or dynamic measurement feedback with dynamic order n_c) if, for any a priori given (arbitrarily large) bounded set $\mathcal{X} \subset \mathbb{R}^n$ (or respectively, $\mathcal{X} \subset \mathbb{R}^{n+n_c}$) and any $D > 0$, there exists a static state feedback (or respectively, dynamic measurement feedback of order n_c), possibly depending on $\mathcal{X}$ and D, such that the following conditions hold:

(*i*) For continuous-time case, the closed-loop system is finite gain L_p-stable over the set $L_{p,q}(D)$. That is, there exists a positive constant γ_p such that with $x(0) = 0$, the following holds:

$$\|z\|_{L_p} \leq \gamma_p \|d\|_{L_p}, \quad \text{for all } d \in L_p \mid \|d\|_{L_q} \leqslant D.$$

Similarly, for the discrete-time case, the closed-loop system is finite gain ℓ_p-stable over the set $\ell_{p,q}(D)$. That is, there exists a positive constant γ_p such that with $x(0) = 0$, the following holds:

$$\|z\|_{\ell_p} \leq \gamma_p \|d\|_{\ell_p}, \quad \text{for all } d \in \ell_p \mid \|d\|_{\ell_q} \leqslant D.$$

(*ii*) In the absence of any disturbance signal d, the equilibrium point $x = 0$ of the closed-loop system is asymptotically stable with $\mathcal{X} \subset \mathbb{R}^n$ (or respectively, $\mathcal{X} \subset \mathbb{R}^{n+n_c}$) contained in its domain of attraction.

The above problem is coined as $(SG_{p,q}/SG)_{fg}$ problem.

Remark 11.13 *In the above two problem statements that deal with semi-global internal stabilization, we prespecified explicitly the dynamical order of the measurement feedback controller. This is necessitated because of the need to specify a priori bounded set $\mathcal{X}$ that has to be included in the region of attraction. However, such a pre-specification of the dynamical order of the measurement feedback controller is not necessary when we deal with global internal stabilization. This is so because global internal stabilization concerns the whole state space, i.e., the region of attraction is the whole state space.*

All the problems formulated in this chapter are studied in subsequent chapters.

12

Simultaneous external and internal stabilization: input-additive case

12.1 Introduction

For linear systems subject to saturation, Chap. 11 formulates five simultaneous global external and global internal stabilization problems and two simultaneous semi-global external and semi-global internal stabilization problems. As discussed there, two distinct cases arise, the first case corresponds to the situation when the control inputs and the disturbance signals are *additive*, while in the second case, such signals are *nonadditive*. In this chapter, under the condition that the given system is asymptotically null controllable with bounded control (ANCBC), we present control strategies that solve all the five global problems and the two semi-global problems for both continuous- and discrete-time systems. For state feedback, we will solve these problems completely. However, for the measurement feedback case, only partial results are available..

In general, all the simultaneous stabilization problems utilize nonlinear feedback controllers in a global framework and linear feedback controllers in a semi-global framework. However, for open-loop critically stable or neutrally stable[1] systems, even in a global framework, simultaneous stabilization can be achieved with linear feedback controllers.

As introduced in (11.1), a continuous-time linear system subject to actuator saturation where the disturbance is input-additive can be described by

$$\Sigma^c : \begin{cases} \dot{x}(t) = Ax(t) + B\sigma(u(t) + d(t)), \\ z(t) = x(t), \quad t \geq 0, \\ y(t) = Cx(t), \quad t \geq 0, \end{cases} \tag{12.1a}$$

where, as usual, $x \in \mathbb{R}^n$ is the state, $u \in \mathbb{R}^m$ is the input, $z \in \mathbb{R}^q$ is the controlled variable, and $y \in \mathbb{R}^p$ is the measured output. Similarly a discrete-time system is described by

[1] We recall that a linear system is said to be open-loop critically stable or neutrally stable if, for a continuous-time system, all its open-loop poles are in the closed left-half plane with those on the imaginary axis being simple or, for a discrete-time system, all its open-loop poles are within or on the unit circle with those on the unit circle being simple.

A. Saberi et al., *Internal and External Stabilization of Linear Systems with Constraints*, Systems & Control: Foundations & Applications, DOI 10.1007/978-0-8176-4787-2_12,

$$\Sigma^d : \begin{cases} x(k+1) = Ax(k) + B\sigma(u(k) + d(k)), \\ z(k) \quad = x(k), \quad k \geq 0, \\ y(k) \quad = Cx(k), \ k = 1, 2, \ldots. \end{cases} \tag{12.1b}$$

Here, σ denotes the standard saturation function defined as

$$\sigma(u) = [\text{sat}_\Delta(u_1), \ldots, \text{sat}_\Delta(u_m)],$$

where $\text{sat}_\Delta(s) = \text{sgn}(s) \min\{|s|, \Delta\}$ for some $\Delta > 0$.

For the above two systems where the disturbance d is additive to the control input, the goal of this chapter is to construct appropriate controllers that solve the simultaneous stabilization problems formulated in Chap. 11. Since all the problems require internal stabilization, it is obvious that the following two assumptions are *necessary* to solve these problems:

Assumption 12.1 *The pair (A, B) is stabilizable.*

Assumption 12.2 *The eigenvalues of A are in the closed left-half complex plane for continuous-time systems and are inside and/or on the unit circle for discrete-time systems.*

Also, we make the following additional assumption whenever a measurement feedback controller has to be used:

Assumption 12.3 *The pair (C, A) is observable.*

12.2 Simultaneous stabilization in a global framework: continuous time

In this section, we consider continuous-time systems. The objective will be to design appropriate controllers which solve the problem of simultaneous global external and global internal stabilization. To be specific, we consider in this section the first four problems, namely,

(*i*) Simultaneous global L_p stabilization with fixed initial conditions and without finite gain and global asymptotic stabilization, as defined by Problem 11.1 (the (G_p/G) problem)

(*ii*) Simultaneous global L_p stabilization with fixed initial conditions with finite gain and with zero bias and global asymptotic stabilization, as defined by Problem 11.2 (the $(G_p/G)_{fg}$ problem) (*iii*) Simultaneous global L_p stabilization with arbitrary initial conditions and without finite gain and bias and global asymptotic stabilization, as defined by Problem 11.4

(*iv*) Simultaneous global L_p stabilization with arbitrary initial conditions and with finite gain and bias and global asymptotic stabilization, as defined by Problem 11.5

We note that the solvability of $(G_p/G)_{fg}$ problem (Problem 11.2) implies the solvability of a simultaneous global L_p stabilization problem with fixed initial conditions with finite gain and with bias and global asymptotical stabilization (Problem 11.3).

In the following two sections, we consider state feedback and measurement feedback controllers, respectively.

12.2.1 State feedback

We consider state feedback controllers in this subsection. A natural question that arises first is, what are our design methodologies, and what kind of feedback control laws do we utilize? In tune with our development in earlier chapters related to internal stability, we utilize low-and-high-gain feedback control methodologies, that is, we utilize state feedback controllers of the form

$$\Sigma_{con}: \quad u = -(1+\alpha)B'P_\varepsilon x.$$

As usual, P_ε is the positive definite solution of the continuous-time algebraic Riccati equation (CARE) given in (4.42). Moreover, ε is a low-gain parameter and α is a high-gain parameter. Since we are aiming at global results for both external stabilization and internal asymptotic stabilization, based on our experience in dealing with internal stabilization problems, we must necessarily adapt (schedule) the low-gain and high-gain parameters ε and α to depend on the state x. Note that adaptation of the low-gain is discussed in Chap. 4 and is essential to achieve global internal stability. Low-gain adaptation is recalled below. It turns out that adaptation of the high-gain is crucial for achieving external stability. For adapting both low-gain and high-gain parameters, different adaptation functions can be used; however, they must satisfy certain properties, as described below.

Adaptation of the low-gain parameter: We need to adapt the low-gain parameter ε based on the state x. That is, one needs to use a function $\varepsilon_a(x)$ from $\mathbb{R}^n \to (0,1]$ in place of ε. Different properties which are sought in adapting the low-gain parameter $\varepsilon_a(x)$ are discussed earlier in Chap. 4. We recall them below with slight modifications to suit our present development:

(*i*) $\varepsilon_a(x) \in C^1$.

(*ii*) $\varepsilon_a(x) = 1$ for all x in an open neighborhood of the origin.

(*iii*) For any $x \in \mathbb{R}^n$, we have

$$\|B' P_{\varepsilon_a(x)} x\|_\infty \leqslant \Delta.$$

(*iv*) $\varepsilon_a(x) \to 0$ as $\|x\|_\infty \to \infty$.

(*v*) $\{ x \in \mathbb{R}^n \mid x' P_{\varepsilon_a(x)} x \leqslant c \}$ is a bounded set for all $c > 0$.

(*vi*) For any $x_1, x_2 \in \mathbb{R}^n$,

$$x_1' P_{\varepsilon_a(x_1)} x_1 \leqslant x_2' P_{\varepsilon_a(x_2)} x_2$$

implies that $\varepsilon_a(x_1) \geqslant \varepsilon_a(x_2)$.

In item (*iii*), $P_{\varepsilon_a(x_2)}$ is the positive definite solution of the CARE (4.42) when ε is replaced by $\varepsilon_a(x_2)$. Moreover, Δ is a design parameter to be chosen (often, it is the saturation level). A particular choice for the adaptive-low-gain parameter $\varepsilon_a(x)$ having all the above properties has been introduced by Megretski [98] in connection with the continuous-time systems. This is what we use throughout this section and it is given by

$$\varepsilon_a(x) = \max\{, r \in (0,1] \; : \; (x' P_r x) \operatorname{trace} P_r \leqslant \frac{\Delta^2}{\|BB'\|} \}. \tag{12.2}$$

Note that this is a nontrivial choice since Property (*iii*) requires that $\varepsilon_a(x)$ converges to 0 fast enough, yet Properties (*v*) and (*vi*) restrict the speed with which $\varepsilon_a(x)$ can converge to 0. We note that in this chapter, Lemma 12.54 given in "Appendix" plays a crucial role, and its proof relies on the explicit choice for the scheduling chosen in (12.2).

Adaptation of the high-gain parameter: We need to adapt the high-gain parameter α based on the state x. That is, one needs to use a function $\alpha_a(x)$ from $\mathbb{R}^n \to \mathbb{R}^+$ in place of α. The function $\alpha_a(x)$ has to be selected appropriately so that it does not affect the internal stability of the closed-loop system, that is, it does not affect the region of attraction, as is the case in adapting the low-gain parameter ε. On the other hand, the dependency of α on x can be advantageous to establish global L_p stability of the closed-loop system. For a globally defined adaptive controller, it turns out that the high-gain parameter $\alpha_a(x)$ has to go to ∞ as the norm of x goes to ∞, but the growth of $\alpha_a(x)$ that is required turns out to be problem dependent. That is, the adaptation of the $\alpha_a(x)$ depends on the choice of p in L_p stability whether the L_p stability of the closed-loop system is required to be with finite gain or without finite gain. Because of this, a strategic method of adapting the high-gain parameter was introduced in [126]. The adaptation of the $\alpha_a(x)$ in [126] directly links to the adaptation of the low-gain parameter ε and is given by

$$\alpha_a(x) = \alpha_o \frac{\lambda_{\max}(P_{\varepsilon_a})}{\lambda_{\min}(Q_{\varepsilon_a})\lambda_{\min}(P_{\varepsilon_a})}, \tag{12.3}$$

where α_o is a constant if we do not impose a finite gain or if we have $p = \infty$. Otherwise we choose

$$\alpha_o = \alpha_{oo}\left[\frac{\lambda_{\max}(P_{\varepsilon_a})}{\lambda_{\min}(Q_{\varepsilon_a})\lambda_{\min}(P_{\varepsilon_a})} \times \frac{1}{\|(sI - A + BB'P_{\varepsilon_a})^{-1}\|_{H_2}^2} \frac{\lambda_{\max}(\frac{dQ_\varepsilon}{d\varepsilon})}{\lambda_{\min}(\frac{dQ_\varepsilon}{d\varepsilon})}\right], \tag{12.4}$$

where α_{oo} is a design parameter.

Remark 12.4 *A typical choice of Q_ε is εI. In this case, obviously, the expression for α_o reduces to*

$$\alpha_o = \alpha_{oo}\left[\frac{\lambda_{max}(P_{\varepsilon_a})}{\varepsilon_a \lambda_{min}(P_{\varepsilon_a})} \frac{1}{\|(sI - A + BB'P_{\varepsilon_a})^{-1}\|_{H_2}^2}\right].$$

We proceed now with the construction of controllers and the analysis of the (G_p/G) and $(G_p/G)_{fg}$ problems when state feedback is considered. It turns out that the analysis of these problems for $p = \infty$ is radically different from the case where $p \in [1, \infty)$. As such, we separate out our development of the results related to these problems into two different cases, $p = \infty$ and $p \in [1, \infty)$.

For the system Σ^c given in (12.1a), the theorem given below deals with the $(G_p/G)_{fg}$ problem for $p = \infty$ while using state feedback.

Theorem 12.5 *Consider the system Σ^c of* (12.1a) *while using state feedback. For this system, under Assumptions 12.1 and 12.2, the problem of simultaneous global L_∞ stabilization with fixed initial conditions with finite gain and with zero bias and global asymptotic stabilization, that is, the $(G_\infty/G)_{fg}$ problem, is solvable, and the L_∞ gain can be made arbitrarily small. Moreover, the adaptive-low-and-high-gain design methodology can yield a control law that solves the problem. More specifically, for any specified gain $\gamma > 0$, there exists an $\alpha_o^* > 0$ such that the adaptive-low-and-high-gain design feedback law,*

$$\Sigma_{con}: \quad u = -(1 + \alpha_a(x))B'P_{\varepsilon_a(x)}x, \tag{12.5}$$

with ε_a defined by (12.2) *and α_a defined by* (12.3)*, where $\alpha_o > \alpha_o^*$ is a positive constant, has the following properties:*

(i) In the presence of a disturbance signal d, the closed-loop system is L_∞ stable with fixed initial conditions with finite gain and with zero bias. Moreover, the L_∞ gain is less than the specified γ.

(ii) In the absence of any disturbance signal d, the equilibrium point of the closed-loop system is globally asymptotically stable.

Remark 12.6 *Note that α_o can be chosen as a function of x or can be chosen as a constant as long as one makes sure that $\alpha_o > \alpha_o^*$.*

Proof : Property (*ii*) has been proven in [67] where we note that this proof is not affected if we use an adaptation of the high gain. Hence, we focus on the proof of Property (*i*).

Before we proceed, we recall the important Property (*iii*) on page 620 of the adaptive low-gain parameter ε_a. Choose a Lyapunov function $V(x) = x'P_{\varepsilon_a}x$ and let $L_V(c)$ be a level set defined as $L_V(c) = \{x \in \mathbb{R}^n : V(x) \leqslant c\}$.

For the derivative of $V(x)$ along the trajectory of the system (12.1a) with the feedback (12.5), we obtain the following inequality:

$$\begin{aligned}\frac{\mathrm{d}V}{\mathrm{d}t} &\leqslant -x'Qx + 2x'PB[\sigma(-B'Px - \alpha B'Px + d) + B'Px] \\ &\quad - x'PBB'Px + x'\frac{\mathrm{d}P}{\mathrm{d}t}x \\ &\leqslant -x'Qx - 2\sum_{i=1}^{m} v_i[\mathrm{sat}_\Delta(v_i + \alpha v_i + d_i) - \mathrm{sat}_\Delta(v_i)] + x'\frac{\mathrm{d}P}{\mathrm{d}t}x, \quad (12.6)\end{aligned}$$

where $v = -B'Px$, d_i is the ith component of d, v_i is the ith component of v, and we use $v_i = \mathrm{sat}_\Delta(v_i)$. Here, for simplicity, we have omitted the subscript ε_a, that is, $P = P_{\varepsilon_a}$, and $Q = Q_{\varepsilon_a}$, and $\alpha = \alpha_a(x)$.

If $|\alpha v_i| < |d_i|$, then we obtain the following inequality from (12.6):

$$\frac{\mathrm{d}V}{\mathrm{d}t} \leqslant -x'Qx + 2\sum_{i=1}^{m} \frac{|d_i|}{\alpha} \cdot |d_i| + x'\frac{\mathrm{d}P}{\mathrm{d}t}x. \tag{12.7}$$

On the other hand, if $|\alpha v_i| \geqslant |d_i|$, then

$$-v_i[\mathrm{sat}_\Delta(v_i + \alpha v_i + d_i) - \mathrm{sat}_\Delta(v_i)] \leqslant 0, \tag{12.8}$$

and hence, also in this case, (12.7) is satisfied. We then obtain,

$$\begin{aligned}\frac{\mathrm{d}V}{\mathrm{d}t} &\leqslant -x'Qx + 2\sum_{i=1}^{m} \frac{|d_i|^2}{\alpha} + x'\frac{\mathrm{d}P}{\mathrm{d}t}x \\ &\leqslant -\frac{\lambda_{\min}(Q)}{\lambda_{\max}(P)}V + 2\frac{\|d\|^2}{\alpha} + x'\frac{\mathrm{d}P}{\mathrm{d}t}x. \quad (12.9)\end{aligned}$$

We claim that for α equal to α_a, as defined in (12.3), we have

$$V(x) \leqslant \frac{2}{\alpha_o}\lambda_{\min}(P)\|d\|_\infty^2, \tag{12.10}$$

and

$$\|x\|^2 \leqslant \frac{2}{\alpha_o} \|d\|_\infty^2. \tag{12.11}$$

It is obvious that (12.11) is a consequence of (12.10), and therefore, it only remains to prove (12.10).

We know that inequality (12.10) is satisfied for $t = 0$. Assume that for some t_0, the inequality is satisfied for $t \leqslant t_0$, but the inequality is not satisfied for $t \in (t_0, t_0 + \nu]$ for some small $\nu > 0$. We consider three cases:

Case 1: If $\varepsilon_a(x(t_0)) = 1$, then $\frac{\mathrm{d}P}{\mathrm{d}t}(t_0) = 0$, and hence, (12.7) for $t = t_0$ reduces to

$$\frac{\mathrm{d}V}{\mathrm{d}t} \leqslant -\frac{\lambda_{\min}(Q)}{\lambda_{\max}(P)} V + \frac{2}{\alpha_o} \frac{\lambda_{\min}(Q)\lambda_{\min}(P)}{\lambda_{\max}(P)} \|d\|_\infty^2. \tag{12.12}$$

Since (12.10) is satisfied for $t = t_0$, we obtain that $\frac{\mathrm{d}V}{\mathrm{d}t}(t_0) \leqslant 0$. On the other hand, the right-hand side of (12.10) is 0. This contradicts that for $t > t_0$, at least one of the inequalities will be falsified.

Case 2: If $\frac{\mathrm{d}P}{\mathrm{d}t}(t_0) < 0$ and $\varepsilon_a(x(t_0)) \neq 1$, then there exists a small neighborhood of t_0 such that $\frac{\mathrm{d}P}{\mathrm{d}t}(t) < 0$ for all t in this neighborhood. We know that for $t \in (t_0, t_0 + \nu]$, the inequality (12.10) is not satisfied. But that implies by using (12.9) that $\frac{\mathrm{d}P}{\mathrm{d}t} < 0$ and $\frac{\mathrm{d}V}{\mathrm{d}t} < 0$. Since $\varepsilon_a(x(t_0)) \neq 1$, we have $\varepsilon_a(x(t)) \neq 1$ in a neighborhood of t, but this implies that

$$V \cdot \operatorname{trace} P = \frac{\Delta^2}{\|BB'\|} \tag{12.13}$$

in this neighborhood of t_0 which contradicts with V and P decreasing simultaneously.

Case 3: If $\frac{\mathrm{d}P}{\mathrm{d}t}(t_0) \geqslant 0$ and $\varepsilon_a \neq 1$, then we have that $\varepsilon_a(x(t)) \neq 1$ in a neighborhood of t, but this implies that (12.13) in this neighborhood of t_0, and we conclude that $\frac{\mathrm{d}V}{\mathrm{d}t} \leqslant 0$. But this implies that the right-hand side of (12.10) is increasing and the left-hand side is decreasing which contradicts the assumption made earlier.

Clearly, (12.11) then shows that the closed-loop system is L_∞ stable with fixed initial conditions with finite gain and with zero bias. Note that to achieve a finite gain less than γ, we should choose $\alpha_o > \frac{2}{\gamma}$. ■

We proceed now to the case of $p \in [1, \infty)$. We first consider the (G_p/G) problem, that is, the problem which does not seek finite gain.

Theorem 12.7 *Consider the system* Σ^c *of* (12.1a) *while using state feedback. For this system, under Assumptions 12.1 and 12.2, the problem of simultaneous global* L_p *stabilization with fixed initial conditions and without finite gain and global asymptotic stabilization, namely, the* (G_p/G) *problem, is solvable for any* $p \in [1, \infty)$. *Moreover, the adaptive-low-and-high-gain design methodology can* *yield a control law that solves the problem. In particular, for any $p \in [1,\infty)$, the adaptive-low-and-high-gain state feedback law,*

$$\Sigma_{con}: \quad u = -(1+\alpha_a(x))B'P_{\varepsilon_a(x)}x, \tag{12.14}$$

with ε_a defined by (12.2) and α_a defined by (12.3) with α_o any fixed positive constant, has the following properties:

(i) In the presence of a disturbance signal d, for any $p \in [1,\infty)$, the closed-loop system is L_p stable with fixed initial conditions and without finite gain.

(ii) In the absence of any disturbance signal d, the equilibrium point of the closed-loop system is globally asymptotically stable.

Remark 12.8 *Note that our method of adapting the high-gain parameter for solving the (G_p/G) problem requires only α_0 to be a positive constant for any $p \in [1,\infty)$. This, as we can see in the subsequent Theorem 12.9, is not the case if we additionally require a finite gain.*

Proof : We begin from (12.6). Here, for simplicity, we have again omitted the subscript ε_a, that is, $P = P_{\varepsilon_a}$ and $Q = Q_{\varepsilon_a}$, and $\alpha = \alpha_a(x)$.

If $|\alpha v_i| \geqslant |d_i|$, then we have (12.8). On the other hand, if $|\alpha v_i| > |d_i|$, then

$$-v_i[\operatorname{sat}_\Delta(v_i+\alpha v_i+d_i)-\operatorname{sat}_\Delta(v_i)] \leqslant \sum_{i=1}^m 2\frac{|d_i|}{\alpha}\Delta.$$

If we define the function V by $V(x) = x'P_{\varepsilon_a(x)}x$, then combining the above with (12.6) yields

$$\frac{\mathrm{d}V}{\mathrm{d}t} \leqslant -\frac{\lambda_{\min}(Q)}{\lambda_{\max}(P)}V + \frac{4m\Delta}{\alpha}\|d\| + x'\frac{\mathrm{d}P}{\mathrm{d}t}x. \tag{12.15}$$

We would like to show the following inequality:

$$V^{p-1}\frac{\mathrm{d}V}{\mathrm{d}t} < \left[\frac{\lambda_{\max}(P)}{\lambda_{\min}(Q)}\right]^{p-1}\left[\frac{4m\Delta}{\alpha}\right]^p \|d\|^p. \tag{12.16}$$

In order to do so, we consider two cases:

Case 1: If $\varepsilon_a = 1$, then $\frac{\mathrm{d}P}{\mathrm{d}t} = 0$. Hence, using (12.15), we find that $\frac{\mathrm{d}V}{\mathrm{d}t} > 0$ only if

$$V(x) \leqslant \frac{\lambda_{\max}(P)}{\lambda_{\min}(Q)}\frac{4m\Delta}{\alpha}\|d\|. \tag{12.17}$$

Since $\frac{\mathrm{d}P_{\varepsilon_a}}{\mathrm{d}t} = 0$, (12.15) also yields

$$\frac{\mathrm{d}V}{\mathrm{d}t} \leqslant \frac{4m\Delta}{\alpha}\|d\|. \tag{12.18}$$

Multiplying this inequality with V^{p-1} on the left-hand side and utilizing (12.17) on the right-hand side then yields (12.16) if $\frac{dV}{dt} > 0$.

On the other hand, if $\frac{dV}{dt} \leqslant 0$, then (12.16) is trivially satisfied.

Case 2: If $\varepsilon_a \neq 1$, then (12.13) is satisfied. We first assume that (12.17) is not satisfied. In that case, if $\frac{dP}{dt} < 0$, then (12.15) implies that $\frac{dV}{dt} \leqslant 0$ which contradicts that (12.13) is satisfied for a neighborhood around the current time. Therefore, if (12.17) is not satisfied, then $\frac{dP}{dt} \geqslant 0$ and $\frac{dV}{dt} \leqslant 0$. Obviously, in this case, (12.16) is satisfied.

On the other hand, if (12.17) is satisfied, then from (12.15), we obtain

$$\frac{dV}{dt} \leqslant \frac{4m\Delta}{\alpha}\|d\| + x'\frac{dP}{dt}x.$$

Therefore, $\frac{dV}{dt} \leqslant 0$ in which case (12.16) is obviously satisfied or $\frac{dP}{dt} \leqslant 0$ in which case combining (12.17) and (12.18) yields (12.16).

Therefore, the state trajectory of the closed-loop system satisfies (12.16). Integrating that inequality and using the definition of α together with the comparison theorem then implies that $V(x)$ is bounded from the above, that is, there exists a c such that the trajectory of the closed-loop system starting from $x(0) = 0$ will remain inside $L_V(c)$.

In summary, for any $p \in [1, \infty)$, we have already shown that in the presence of $d \in L_p$, the feedback (12.14) guarantees (12.16). This implies that the trajectory of the closed-loop system starting from $x(0) = 0$ remains inside $L_V(c)$, though c is dependent on the individual d.

Based on the definition of the adaptive low-gain parameter ε_a, it is then obvious that there exists an ε_a^* such that $\varepsilon_a^* \leqslant \varepsilon_a(x(t))$ for all t. We will use this to show that the closed-loop system is L_p stable.

From (12.6), there exist $\nu_1(\varepsilon_a) = \frac{\lambda_{\min} Q}{\lambda_{\max} P}$ and $\beta_1(\varepsilon_a) = 4m\|B'P^{\frac{1}{2}}\|$ such that

$$\frac{dV}{dt} \leqslant -\nu_1(\varepsilon_a)V + \beta_1(\varepsilon_a)V^{\frac{1}{2}}\|d\| + x'\frac{dP}{dt}x. \tag{12.19}$$

As demonstrated earlier, if $\varepsilon_a = 1$, then $\frac{dP}{dt} = 0$. We next concentrate on the case when $\varepsilon_a = 1$ does not hold. If so, we have (12.13); therefore, $\frac{dV}{dt}$ and $\frac{dP}{dt}$ have different signs, or both are zero. Hence, if $\frac{dV}{dt} \geqslant 0$, we have

$$\frac{dV}{dt} \leqslant -\nu_1(\varepsilon_a)V + \beta_1(\varepsilon_a)V^{\frac{1}{2}}\|d\|. \tag{12.20}$$

Next, we will discuss the case $\frac{dV}{dt} < 0$. Before that, we would like to point out some properties. Since we know that ε_a is bounded away from 0, we know that there exists a constant K (independent of x) such that (see Remark 12.55 in the "Appendix")

$$|x'\frac{dP}{dt}x| \leqslant K|\frac{dV}{dt}|. \tag{12.21}$$

If $\frac{dV}{dt} < 0$, the preceding inequality shows that

$$\frac{\mathrm{d}V}{\mathrm{d}t} \leqslant -\nu_1(\varepsilon_a)V + \beta_1(\varepsilon_a)V^{\frac{1}{2}}\|d\| - K\frac{\mathrm{d}V}{\mathrm{d}t}.$$

Thus, we have

$$(1+K)\frac{\mathrm{d}V}{\mathrm{d}t} \leqslant -\nu_1(\varepsilon_a)V + \beta_1(\varepsilon_a)V^{\frac{1}{2}}\|d\|. \tag{12.22}$$

Therefore, if $\varepsilon_a = 1$ or $\frac{dV}{dt} \geqslant 0$, we have (12.20), while if $\frac{dV}{dt} < 0$, we have (12.22). Therefore, there exist constants $\nu > 0$ and $\beta > 0$ with $\nu < \nu_1(\varepsilon_a)$ and $\beta > \beta_1(\varepsilon_a)$ for all $\varepsilon_a \in [\varepsilon_a^*, 1]$ such that

$$\frac{\mathrm{d}V}{\mathrm{d}t} \leqslant -\nu V + \beta V^{\frac{1}{2}}\|d\|. \tag{12.23}$$

Then, by using the standard comparison theorem, we can see easily that the closed-loop system is L_p stable with fixed initial conditions and without finite gain. This concludes our proof. ∎

The next theorem deals with $(G_p/G)_{fg}$ problem, that is, with the simultaneous global external stabilization problem with fixed initial conditions with finite gain and with zero bias and global asymptotic stabilization for $p \in [1, \infty)$.

Theorem 12.9 *Consider the system Σ^c of (12.1a) while using state feedback. For this system, under Assumptions 12.1 and 12.2, the problem of simultaneous global L_p stabilization with fixed initial conditions with finite gain and with zero bias and global asymptotic stabilization, namely, the $(G_p/G)_{fg}$ problem, is solvable for any $p \in [1, \infty)$. Also, the L_p gain can be made arbitrarily small. Moreover, the adaptive-low-and-high-gain design methodology can yield a control law that solves the problem. In particular, for any $p \in [1, \infty)$ and for any gain $\gamma > 0$, there exists an $\alpha_{oo}^* > 0$ such that the adaptive-low-and-high-gain feedback law,*

$$\Sigma_{con}: \quad u = -(1+\alpha_a(x))B'P_{\varepsilon_a(x)}x, \tag{12.24}$$

with ε_a defined by (12.2) and α_a defined by (12.3), where α_o defined by (12.4) with $\alpha_{oo} > \alpha_{oo}^*$, has the following properties:

(i) For any $p \in [1, \infty)$, in the presence of a disturbance signal d, the closed-loop system is L_p stable with fixed initial conditions with finite gain and with zero bias. Moreover, the L_p gain is less than or equal to γ.

(ii) In the absence of any disturbance signal d, the equilibrium point of the closed-loop system is globally asymptotically stable.

Proof: It is immediate from Theorem 12.7 and its proof that the system is globally asymptotically stable and is globally L_p stable with fixed initial conditions and without finite gain (the fact that we make α_o a function of x is easily seen not to change these properties). It remains to show that we have a finite gain.

Let γ be a positive constant which can be chosen arbitrarily small. We will derive a bound on

$$\int_{t_1}^{t_2} (\|x\|^p - \gamma \|d\|^p) \mathrm{d}t$$

when the system is moving from $c_1 = V(x(t_1))$ to $c_2 = V(x(t_2))$, where V is defined by $V(x) = x' P_{\varepsilon_a(x)} x$ and the bound will depend only on c_1 and c_2. We consider two cases: $\frac{\mathrm{d}V}{\mathrm{d}t} \geqslant 0$ on the interval $[t_1, t_2]$ and $\frac{\mathrm{d}V}{\mathrm{d}t} < 0$ on the interval $[t_1, t_2]$.

Case 1: ($\frac{\mathrm{d}V}{\mathrm{d}t} \geqslant 0$).

Using similar arguments as in the proofs of the previous theorems, we find that if V is increasing, we either have $\varepsilon_a = 1$ in which case $\frac{\mathrm{d}P}{\mathrm{d}t} = 0$ or $\varepsilon < 1$ in which case $\frac{\mathrm{d}P}{\mathrm{d}t} \leqslant 0$. By using that the saturation function is globally Lipschitz and bounded, we obtain

$$\frac{\mathrm{d}V}{\mathrm{d}t} \leqslant -x'Qx + 2 \sum_{i:v_i e_i < 0} |v_i| \operatorname{sat}_\Delta(|e_i|) \tag{12.25}$$

$$\leqslant 2\|B'P^{1/2}\| V^{1/2} \sum_{i:v_i e_i < 0} \operatorname{sat}_\Delta(|e_i|)$$

$$\leqslant N_1 V^{1/2} \sum_{i:v_i e_i < 0} \operatorname{sat}_\Delta(|e_i|), \tag{12.26}$$

for some constant $N_1 > 0$ where $e_i = d_i + \alpha v_i$. Note that if $e_i v_i \geqslant 0$ for all i then we have (12.8) for all i, and hence, $\frac{\mathrm{d}V}{\mathrm{d}t} < 0$ which yields a contradiction.

First, consider the case where

$$\sum_{i:v_i e_i < 0} \operatorname{sat}_\Delta(|e_i|) \leqslant V^{1/2}.$$

In that case,

$$\frac{\mathrm{d}V}{\mathrm{d}t} \leqslant N_1 V,$$

and we find that

$$\sum_{i:v_i e_i < 0} |v_i|^p \geqslant \frac{1}{pm^{p-1}} \Big(\sum_{i:v_i e_i < 0} |v_i| \Big)^p \geqslant \frac{1}{pm^{p-1}} \frac{(x'Qx)^p}{(2V^{1/2})^p}$$

$$\geqslant N_2 \frac{\lambda_{\min}^p(Q)}{\lambda_{\max}^p(P)} V^{p/2},$$

for some constant $N_2 > 0$ where the second inequality is based on $\frac{dV}{dt} \geqslant 0$ combined with (12.25) and $\sum_{i:v_i e_i<0} \mathrm{sat}_\Delta(|e_i|) \leqslant V^{1/2}$. Using the definition of α with α_{oo} large enough such that $N_2\gamma\alpha_o^p \geqslant 2\max\{\lambda_{\min}^{p/2}(P), 1\}$, which is obviously possible since $\lambda_{\min}(P)$ is bounded and α_o/α_{oo} has a lower bound larger than 0, we have

$$
\begin{aligned}
\frac{\|x\|^p - \gamma\|d\|^p}{\dot V} &\leqslant \frac{\|x\|^p - \gamma\alpha^p \sum_{i:v_i e_i<0} |v_i|^p}{\dot V} \\
&\leqslant \frac{1}{N_1}\left(\frac{V^{p/2-1}}{\lambda_{\min}^{p/2}(P)} - N_2\gamma\alpha^p \frac{\lambda_{\min}^p(Q)}{\lambda_{\max}^p(P)} V^{p/2-1}\right) \\
&\leqslant \frac{1 - N_2\gamma\alpha_o^p}{N_1}\frac{V^{p/2-1}}{\lambda_{\min}^{p/2}(P)}.
\end{aligned}
\tag{12.27}
$$

On the other hand, if

$$
\sum_{i:v_i e_i<0} \mathrm{sat}_\Delta(|e_i|) \geqslant V^{1/2}, \tag{12.28}
$$

then obviously V is bounded. Note that (12.25) implies that

$$
2\sum_{i:v_i e_i<0} |v_i e_i| \geqslant 2\sum_{i:v_i e_i<0} |v_i|\,\mathrm{sat}_\Delta(|e_i|) \geqslant x'Qx,
$$

and hence, we have the following reductions:

$$
\begin{aligned}
\sum_{i:v_i e_i<0} |d_i|^p &\geqslant \frac{1}{pm^{p-1}}\left(\sum_{i:v_i e_i<0} |d_i|\right)^p \\
&\geqslant F\left(\sum_{i:v_i e_i<0} |\frac{1}{\sqrt{\alpha}}e_i| + |\sqrt{\alpha}v_i|\right)^{p-1} \\
&\geqslant F\left(\sum_{i:v_i e_i<0} \left(|\frac{1}{\sqrt{\alpha}}e_i|^2 + |\sqrt{\alpha}v_i|^2\right)^{1/2}\right)^{p-1} \\
&\geqslant F\left(\sum_{i:v_i e_i<0} |\frac{1}{\sqrt{\alpha}}e_i|^2 + |\sqrt{\alpha}v_i|^2\right)^{p/2-1/2} \\
&\geqslant F\left(\sum_{i:v_i e_i<0} |e_i v_i|\right)^{p/2-1/2}
\end{aligned}
$$

$$
\begin{aligned}
&\geqslant \frac{\alpha^{p/2-1/2}(x'Qx)^{p/2-1/2}}{(2)^{p/2-1/2}pm^{p-1}}\left(\sum_{i:v_ie_i<0}|e_i|\right)\\
&\geqslant N_3\alpha_o^{p/2-1/2}\frac{V^{p/2-1/2}}{\lambda_{\min}^{p/2}(P)}\left(\sum_{i:v_ie_i<0}|e_i|\right), \qquad (12.29)
\end{aligned}
$$

where

$$
F=\frac{\alpha^{p/2-1/2}}{pm^{p-1}}\left(\sum_{i:v_ie_i<0}|e_i|\right),
$$

for some $N_3>0$ using that P is lower bounded and V is upper bounded using (12.28). Using (12.26) and (12.29) and choosing α_{oo} large enough such that $N_3\gamma\alpha_o^{p/2-1/2}\geqslant 2$, we have

$$
\begin{aligned}
\frac{\|x\|^p-\gamma\|d\|^p}{\dot V}&\leqslant\left(\frac{V^{p/2}}{\lambda_{\min}^{p/2}(P)}-\gamma\sum_{i:v_ie_i<0}|d_i|^p\right)\\
&\quad\times\frac{1}{N_1V^{1/2}\sum_{i:v_ie_i<0}\mathrm{sat}_\Delta(|e_i|)}\\
&\leqslant\frac{1}{N_1}\left(\frac{V^{p/2-1}}{\lambda_{\min}^{p/2}(P)}-N_3\gamma\alpha_o^{p/2-1/2}\frac{V^{p/2-1}}{\lambda_{\min}^{p/2}(P)}\right)\\
&\leqslant\frac{1-N_3\gamma\alpha_o^{p/2-1/2}}{N_1}\frac{V^{p/2-1}}{\lambda_{\min}^{p/2}(P)}. \qquad (12.30)
\end{aligned}
$$

Combining (12.30) and (12.27), we obtain

$$
\begin{aligned}
&\int_{t_1}^{t_2}(\|x\|^p-\gamma\|d\|^p)\mathrm{d}t\\
&\quad\leqslant\frac{1}{N_1}\int_{c_1}^{c_2}(1-\gamma\alpha_o^{p/2-1/2}\min\{N_2,N_3\})\frac{V^{p/2-1}}{\lambda_{\min}^{p/2}(P)}dV\\
&\quad\leqslant -N_4\int_{c_1}^{c_2}\gamma\alpha_o^{p/2-1/2}\frac{V^{p/2-1}}{\lambda_{\min}^{p/2}(P)}dV, \qquad (12.31)
\end{aligned}
$$

for some positive constant N_4 where we used that $\gamma\alpha_o^p\min\{N_2,N_3\}>2$.

Case 2: ($\frac{\mathrm{d}V}{\mathrm{d}t} < 0$).

Obviously a worst-case disturbance signal d wants to maximize

$$\int_{t_1}^{t_2} (\|x\|^p - \gamma \|d\|^p)\mathrm{d}t,$$

while V is decreasing from c_2 to c_1. For $d = 0$, the term is positive, and it is obvious that it is unfavorable for d to make $\|x\|^p - \gamma\|d\|^p \leqslant 0$ for any given time.

We define

$$\nu = \frac{\lambda_{\min}(Q)}{(1+K)\lambda_{\max}(P)},$$

where K is a function of V that should be such that (12.21) is satisfied. One K satisfying (12.21) is given by Lemma 12.54 in "Appendix",

$$K = \frac{N_5}{\lambda_{\min}(P)\|(sI - A + BB'P)^{-1}\|_{H_2}^2} \frac{\lambda_{\max}(\frac{\mathrm{d}Q_\varepsilon}{\mathrm{d}\varepsilon})}{\lambda_{\min}(\frac{\mathrm{d}Q_\varepsilon}{\mathrm{d}\varepsilon})},$$

for some positive constant N_5 which can be chosen such that additionally $K > 2$ for all V. Since d will not make $\|x\|^p - \gamma\|d\|^p$ negative, we find that

$$\|d\| \leqslant \gamma^{1/p}\|x\| \leqslant \frac{\gamma^{1/p}V^{1/2}}{\lambda_{\min}^{1/2}(P)}.$$

We consider two cases. If

$$|v_i| < \frac{\nu\lambda_{\min}^{1/2}(P)V^{1/2}}{2m\gamma^{1/p}}, \tag{12.32}$$

then

$$\begin{aligned} -v_i[\mathrm{sat}_\Delta(v_i + \alpha v_i + d_i) - \mathrm{sat}_\Delta(v_i)] &\leqslant \frac{\nu V^{1/2}\lambda_{\min}^{1/2}(P)}{2m\gamma^{1/p}}\|d\| \\ &\leqslant \frac{\nu}{2m}V. \end{aligned}$$

On the other hand, if (12.32) is not satisfied, then, for $v_i(d_i + \alpha v_i) < 0$, we find that

$$\begin{aligned} &-v_i[\mathrm{sat}_\Delta(v_i + \alpha v_i + d_i) - \mathrm{sat}_\Delta(v_i)] \\ &\quad \leqslant -v_i(d_i + \alpha v_i) \\ &\quad \leqslant |v_i|\frac{\gamma^{1/p}V^{1/2}}{\lambda_{\min}^{1/2}(P)} - \alpha v_i^2 \end{aligned}$$

$$\leqslant \left[\frac{\gamma^{1/p} V^{1/2}}{\lambda_{\min}^{1/2}(P)} - \frac{\alpha v}{2m\gamma^{1/p}} V^{1/2} \lambda_{\min}^{1/2}(P) \right] |v_i|,$$

which is negative for α_{oo} large enough, while for $v_i(d_i + \alpha v_i) \geqslant 0$, we get

$$-v_i[\operatorname{sat}_\Delta(v_i + \alpha v_i + d_i) - \operatorname{sat}_\Delta(v_i)] \leqslant 0.$$

Therefore, we can strengthen the argument used in the proof of the previous theorem to obtain, instead of (12.23),

$$\frac{\mathrm{d}V}{\mathrm{d}t} \leqslant -\frac{v}{2} V. \tag{12.33}$$

We have

$$\int_{t_1}^{t_2} (\|x\|^p - \gamma \|d\|^p) \mathrm{d}t \leqslant \int_{c_2}^{c_1} \frac{V^{p/2}}{\lambda_{\min}^{p/2}(P)} \frac{\mathrm{d}V}{\dot{V}}.$$

Note that since $c_2 > c_1$, we have, dV is negative, and hence, when combined with (12.33), it is obvious (where we use $1 + K < 2K$) that there exists a negative constant N_6 such that

$$\int_{t_1}^{t_2} (\|x\|^p - \gamma \|d\|^p) \mathrm{d}t \leqslant N_6 \int_{c_1}^{c_2} \frac{V^{p/2}}{\lambda_{\min}^{p/2+1}(P)} \frac{\lambda_{\max}(P)}{\lambda_{\min}(Q)} \times \frac{V^{-1}}{\|(sI - A + BB'P)^{-1}\|_{H_2}^2} \frac{\lambda_{\max}(\frac{\mathrm{d}Q_\varepsilon}{\mathrm{d}\varepsilon})}{\lambda_{\min}(\frac{\mathrm{d}Q_\varepsilon}{\mathrm{d}\varepsilon})} dV. \tag{12.34}$$

Conclusion of Case 1 and Case 2.

Therefore, if we consider a trajectory on the interval $[t_1, t_3]$ where we first have $\frac{\mathrm{d}V}{\mathrm{d}t} \geqslant 0$ while moving from $V = c_1$ to $V = c_2$ and then $\frac{\mathrm{d}V}{\mathrm{d}t} < 0$ while moving from $V = c_2$ back to $V = c_1$, then we have

$$\begin{aligned}\int_{t_1}^{t_3} (\|x\|^p - \gamma \|d\|^p) \mathrm{d}t \leqslant{}& -N_4 \int_{c_1}^{c_2} \gamma \alpha_o^{p/2-1/2} \frac{V^{p/2-1}}{\lambda_{\min}^{p/2}(P)} dV \\ &+ \int_{c_2}^{c_1} N_6 \frac{V^{p/2-1}}{\lambda_{\min}^{p/2+1}(P)} \frac{\lambda_{\max}(P)}{\lambda_{\min}(Q)} \\ &\times \frac{1}{\|(sI - A + BB'P)^{-1}\|_{H_2}^2} \frac{\lambda_{\max}(\frac{\mathrm{d}Q_\varepsilon}{\mathrm{d}\varepsilon})}{\lambda_{\min}(\frac{\mathrm{d}Q_\varepsilon}{\mathrm{d}\varepsilon})} dV\end{aligned}$$

using (12.31) and (12.34). The righthand side is negative if for all ε

$$\gamma\alpha_o^{p/2-1/2} \geq N_7 \frac{\lambda_{\max}(P)}{\lambda_{\min}(Q)\lambda_{\min}(P)} \frac{1}{\|(sI - A + BB'P)^{-1}\|_{H_2}^2} \frac{\lambda_{\max}(\frac{\mathrm{d}Q_\varepsilon}{\mathrm{d}\varepsilon})}{\lambda_{\min}(\frac{\mathrm{d}Q_\varepsilon}{\mathrm{d}\varepsilon})}$$

for a suitable constant N_7 which given our choice for α_o will always be satisfied if $\gamma\alpha_{oo}$ is large enough.

Given the fact that

$$\int_{t_1}^{t_3} (\|x\|^p - \gamma\|d\|^p)\mathrm{d}t \tag{12.35}$$

is always negative for any trajectory first moving up from c_1 to c_2 and then down from c_2 to c_1 can be used together with the fact that we rely on pointwise bounds that only rely on the value of $V(x)$ and not on x can be used to show that any trajectory which starts in c_1 and ends in c_1 will also yield that (12.35) is negative. This is related to the classical theory of dissipative systems (see, e.g., the overview article [185]). In particular, we obtain

$$\int_0^\infty (\|x\|^p - \gamma\|d\|^p)\mathrm{d}t \leqslant 0$$

since we know that any $d \in L_p$ yields a $x \in L_p$ where we use that the previous theorem guarantees that the system is L_p stable. Moreover, we can achieve this for any $\gamma > 0$, and hence, we can make the L_p gain arbitrarily small. ■

Remark 12.10 *Theorems 12.5 and 12.9 pronounce that the achievable L_p gain can be rendered arbitrarily small, that is, as small as required. That is, we can achieve almost disturbance decoupling (ADDP) under the standard Assumptions 12.1 and 12.2.*

In the previous chapter, Sect. 11.2 defines the notion of simultaneous L_p stabilization with arbitrary initial conditions with finite gain and with bias and global asymptotic stabilization. Proofs of Theorems 12.5 and 12.9 lead to the achievability of such a stabilization as the following theorem states:

Theorem 12.11 *Consider the system Σ^c of* (12.1a) *while using state feedback. Also, consider the control law*

$$\Sigma_{con}: \quad u = -(1 + \alpha_a(x))B'P_{\varepsilon_a(x)}x,$$

with ε_a *defined by* (12.2) *and* α_a *defined by* (12.3) *with* α_o *as in* (12.4)*. Then, under Assumptions 12.1 and 12.2, the controller* Σ_{con} *defined above achieves simultaneous* L_p *stabilization with arbitrary initial conditions with finite gain and with bias and global asymptotic stabilization (see Problem 11.5). Moreover, one can tune the adaptive-high-gain parameter to render the* L_p*-gain arbitrarily small.*

Proof : From the standard comparison principle, (12.23) in the proof of Theorem 12.7 immediately yields that there exist a $\gamma > 0$ and a function $\widetilde{b}_p : \mathbb{R}^n \to \mathbb{R}^+$ such that

$$\|x\|_{L_p} \leqslant \gamma_p \|d\|_{L_p} + \widetilde{b}_p(x_0).$$

Define a class $\mathcal{K}$ function b_p as

$$b_p(r) = \sup_{\|x_0\|=r} \widetilde{b}_p(x_0).$$

We have

$$\|x\|_{L_p} \leqslant \gamma_p \|d\|_{L_p} + b_p(\|x_0\|).$$

■

12.2.2 *Measurement feedback*

Our primary focus so far has been in constructing state feedback controllers. That is, we assumed so far that the complete state of the given system is available for feedback, and it is not corrupted with any disturbance signal. Once a state feedback controller is constructed, one often tries to construct a measurement feedback controller having an observer-based architecture. Accordingly, we construct here such measurement feedback controllers. By doing so, we solve here two problems both without finite gain, namely, (1) simultaneous global L_p stabilization with fixed initial conditions and without finite gain and global asymptotic stabilization, as formulated in Problem 11.1, and (2) global L_p stabilization with arbitrary initial conditions and without finite gain and bias and global asymptotic stabilization, as formulated in Problem 11.4. The development of a controller for both these problems is essentially the same. As such, we do not explicitly refer to initial conditions in our development. The corresponding finite gain problems are still open as they require complex high-gain observer.

Consider a traditional observer for the continuous-time system Σ^c given in (12.1a),

$$\dot{\widehat{x}} = A\widehat{x} + B\sigma(u) + K(y - C\widehat{x}),$$

where the gain K is such that $A - KC$ is Hurwitz stable. Also, consider a dynamic system,

$$\dot{\omega} = (A + BF)\omega + K(y - C\widehat{x}),$$

where the gain F is such that $A + BF$ is Hurwitz stable. Also, let P_ε be the solution of the CARE,

$$A'P_\varepsilon + P_\varepsilon A - P_\varepsilon BB'P_\varepsilon + Q_\varepsilon = 0, \tag{12.36}$$

where for $\varepsilon \in [0, 1]$,

$$Q_\varepsilon > 0, \quad \lim_{\varepsilon \to 0} Q_\varepsilon = 0, \quad \tfrac{\mathrm{d}Q_\varepsilon}{\mathrm{d}\varepsilon} > 0.$$

Let $z = \hat{x} - \omega$. Choose $u = -B'P_{\varepsilon_a(z)}z + F\omega$ where $\varepsilon_a(z)$ is determined by

$$\varepsilon_a(z) = \max\{r \in (0, 1] \mid z'P_r z \operatorname{trace} P_r \leqslant \tfrac{\Delta^2}{4\|BB'\|}\}, \tag{12.37}$$

and $P_{\varepsilon_a(z)}$ is the solution of the CARE (12.36) with ε replaced by $\varepsilon_a(z)$.

The above discussion leads to the dynamic observer-based controller as

$$\begin{cases} \dot{\hat{x}} = A\hat{x} + B\sigma(u) + K(y - C\hat{x}) \\ \dot{\omega} = (A + BF)\omega + K(y - C\hat{x}) \\ u = -B'P_{\varepsilon_a(z)}z + F\omega, \end{cases} \tag{12.38}$$

where K and F are such that $A - KC$ and $A + BF$ are Hurwitz stable, and $z = \hat{x} - \omega$.

We have the following theorem:

Theorem 12.12 *Consider the system Σ^c given in* (12.1a). *Let Assumptions 12.1, 12.2, and 12.3 be valid; also let σ be the standard saturation function as in Definition 2.19. Then, the nonlinear dynamic measurement feedback controller* (12.38) *solves the following two problems:*

(i) Simultaneous global L_p stabilization with fixed initial conditions and without finite gain and global asymptotic stabilization, as defined by Problem 11.1 (the (G_p/G) problem)

(ii) Simultaneous global L_p stabilization with arbitrary initial conditions and without finite gain and bias and global asymptotic stabilization, as defined by Problem 11.4

Proof : Define $e = x - \hat{x}$. The closed-loop system in terms of e, z, and ω is given by

$$\begin{cases} \dot{e} = (A - KC)e + B\big[\sigma(-B'P_{\varepsilon_a(z)}z + F\omega + d) \\ \qquad\qquad - \sigma(-B'P_{\varepsilon_a(z)}z + F\omega)\big] \\ \dot{z} = Az + B\sigma(-B'P_{\varepsilon_a(z)}z + F\omega) - BF\omega \\ \dot{\omega} = (A + BF)\omega + KCe. \end{cases}$$

In the absence of d, the above becomes

$$\begin{cases} \dot{e} = (A - KC)e \\ \dot{z} = Az + B\sigma(-B'P_{\varepsilon_a(z)}z + F\omega) - BF\omega \\ \dot{\omega} = (A + BF)\omega + KCe. \end{cases}$$

Clearly, the origin is locally exponentially stable. To see global attractivity, we note that $e \to 0$ and $\omega \to 0$ as time tends to infinity due to the fact that both $A + BF$ and $A - KC$ are Hurwitz stable. Then, there exists a T such that $\|F\omega(t)\| \leqslant \frac{\Delta}{2}$ for all $t > T$. This and the adaptation law (12.37) together imply that the saturation will be inactive for all $t > T$. Since σ is a standard saturation function, then the z dynamics becomes

$$\dot{z} = Az - BB'P_{\varepsilon_a(z)}z$$

which is globally attractive. This concludes global asymptotic stability. In the presence of d, for a standard saturation function, we have

$$\|\sigma(u + d) - \sigma(u)\| \leqslant \|d\|.$$

Therefore, $d \in L_p$ implies that $(\sigma(u + d) - \sigma(u)) \in L_p$. Since $A - KC$ and $A + BF$ are both Hurwitz stable, it follows that $e \in L_p$, $\omega \in L_p$, and $\omega \to 0$. Then, as before, there exists a T such that $\|F\omega(t)\| \leqslant \frac{\Delta}{2}$ for all $t > T$. Therefore, as before, we can conclude that the saturation is inactive for all $t > T$, and hence, z dynamics becomes

$$\dot{z} = Az - BB'P_{\varepsilon_a(z)}z.$$

This system is known to be globally asymptotically stable and locally exponentially stable. Hence, $z \in L_p$ and $x = e + \hat{x} = (e + z + \omega) \in L_p$. This completes the proof. ■

Remark 12.13 *As Theorem 12.12 clearly indicates, a $2n$-dimensional dynamic adaptive-low-gain feedback controller can solve simultaneous global L_p stabilization without finite gain and global asymptotic stabilization irrespective of the nature of initial conditions. This is remarkable especially when we note that to solve the same problems by a static state feedback controller, one needs an adaptive low-and-high-gain feedback controller.*

A similar result as in Theorem 12.12, however, with finite gain is challenging and is an open research problem.

12.3 Simultaneous stabilization in a global framework: discrete time

In this section, we consider discrete-time linear systems subject to actuator saturation when the disturbance is additive to the control input and design appropriate controllers for such systems in order to solve simultaneous global external and global internal stabilization. As such, this section is a counterpart of Sect. 12.2 that considers continuous-time systems. To be specific, we consider here the first four problems, namely,

(*i*) Simultaneous global ℓ_p stabilization with fixed initial conditions and without finite gain and global asymptotic stabilization, as defined by Problem 11.1 (the (G_p/G) problem)

(*ii*) Simultaneous global ℓ_p stabilization with fixed initial conditions with finite gain and with zero bias and global asymptotic stabilization, as defined by Problem 11.2 (the (G_p/G) problem)

(*iii*) Simultaneous global ℓ_p stabilization with arbitrary initial conditions and without finite gain and bias and global asymptotic stabilization, as defined by Problem 11.4

(*iv*) Simultaneous global ℓ_p stabilization with arbitrary initial conditions and with finite gain and bias and global asymptotic stabilization, as defined by Problem 11.5

We consider both state feedback and measurement feedback controllers one in each separate subsection.

12.3.1 State feedback

We first consider state feedback controllers based on the work of [198].

Under Assumptions 12.1 and 12.2, the system (12.1b) can be transformed into the form,

$$\begin{pmatrix} x_s(k+1) \\ x_u(k+1) \end{pmatrix} = \begin{pmatrix} A_s & 0 \\ 0 & A_u \end{pmatrix} \begin{pmatrix} x_s(k) \\ x_u(k) \end{pmatrix} + \begin{pmatrix} B_s \\ B_u \end{pmatrix} \sigma(u(k) + d(k)), \tag{12.39}$$

where A_s is Schur stable, A_u has all its eigenvalues on the unit circle, and (A_u, B_u) is controllable.

Suppose (G_p/G) and/or $(G_p/G)_{fg}$ of the x_u dynamics can be achieved by a feedback controller $u = f(x_u)$. If B_u has full column rank, it is straightforward to show that $u = f(x_u)$ also achieves (G_p/G) and/or $(G_p/G)_{fg}$ of the overall system (see below). However, it takes some effort to reach the same conclusion in the general case. We show this in Appendix 12.B under a generic assumption on controller structure. Therefore, without loss of generality, the following assumption is made throughout this section:

Assumption 12.14 *(i)* (A, B) *is controllable.*

(ii) A *has all its eigenvalues on the unit circle.*

Controller design

As in the previous section, the controller design in this section is based on the classical low-gain and low-and-high-gain feedback design methodologies and their adapted or scheduled versions. As discussed earlier, the low-gain feedback can be constructed using different approaches such as direct eigenstructure assignment and H_2 and H_∞ ARE based methods. However, in this section, we choose a recently developed parametric Lyapunov equation method [215, 216] to build the low-gain feedback because of its special properties. As will become clear later on, it greatly simplifies the expressions for our controllers and the subsequent analysis.

As we said often, the low-gain feedback does not allow complete utilization of our control capacities. On the other hand, the low-and-high-gain feedback does so. The low-and-high-gain feedback is composed of a low-gain and a high-gain feedback. As in the continuous-time case, the solvability of simultaneous global external and internal stabilization problems critically rely on a proper choice of high gain. At first, we recall the low-gain feedback design and then propose a new high-gain design methodology well suited to discrete-time systems.

Low-gain state feedback

In this subsection, we review the low-gain feedback design methodology recently introduced in [215, 216] which is based on the solution of a parametric Lyapunov equation. Five key properties of the parametric Lyapunov equation are summarized in the next lemma, where the first three properties are adopted from [216].

Lemma 12.15 *Assume that* (A, B) *is controllable and* A *has all its eigenvalues on the unit circle. For any* $\varepsilon \in (0, 1)$, *the parametric ARE,*

$$(1-\varepsilon)P_\varepsilon = A'P_\varepsilon A - A'P_\varepsilon B(I + B'P_\varepsilon B)^{-1}B'P_\varepsilon A, \tag{12.40}$$

has a unique positive definite solution $P_\varepsilon = W_\varepsilon^{-1}$ *where* W_ε *is the solution for* W *of*

$$W - \frac{1}{1-\varepsilon}AWA' = -BB'.$$

Moreover, the following properties hold:

(i) $A_c(\varepsilon) = A - B(I + B'P_\varepsilon B)^{-1}B'P_\varepsilon A$ *is Schur stable for any* $\varepsilon \in (0, 1)$.

(ii) $\frac{dP_\varepsilon}{d\varepsilon} > 0$ *for any* $\varepsilon \in (0, 1)$.

(iii) $\lim_{\varepsilon \downarrow 0} P_\varepsilon = 0$.

(*iv*) *There exists an* ε^* *such that for any* $\varepsilon \in (0, \varepsilon^*]$,

$$\|[P_\varepsilon^{\frac{1}{2}} A P_\varepsilon^{-\frac{1}{2}}\| \leqslant \sqrt{2}.$$

(*v*) *Let* ε^* *be given by Property (iv). There exists a* M_{ε^*} *such that*

$$\|\tfrac{B' P_\varepsilon B}{\varepsilon}\| \leqslant M_{\varepsilon^*}$$

for all $\varepsilon \in (0, \varepsilon^*]$.

Proof : The existence of the positive definite solution $P_\varepsilon = W_\varepsilon^{-1}$ and Properties (*i*), (*ii*) and (*iii*) were shown in [216]. Regarding Property (*iv*), multiplying by $P_\varepsilon^{-1/2}$ on both sides of (12.40) gives

$$V_\varepsilon'[I - P_\varepsilon^{\frac{1}{2}} B(I + B' P_\varepsilon B)^{-1} B' P_\varepsilon^{\frac{1}{2}}] V_\varepsilon = (1 - \varepsilon) I,$$

where $V_\varepsilon = P_\varepsilon^{1/2} A P_\varepsilon^{-1/2}$. Since $P_\varepsilon \to 0$ as $\varepsilon \to 0$, there exists an ε^* such that for any $\varepsilon \in (0, \varepsilon^*]$,

$$I - P_\varepsilon^{\frac{1}{2}} B(I + B' P_\varepsilon B)^{-1} B' P_\varepsilon^{\frac{1}{2}} \geqslant \tfrac{1}{2} I.$$

Hence,

$$V_\varepsilon' V_\varepsilon < 2I,$$

or equivalently

$$\|V_\varepsilon\| \leqslant \sqrt{2}.$$

It remains to show Property (*v*). Note that W_ε is a rational matrix in ε, and thus, P_ε is a rational matrix in ε. Property (*iii*) implies that $P = \varepsilon \overline{P}_\varepsilon$ where $\overline{P}_\varepsilon$ is rational in ε and $\overline{P}_\varepsilon$ is bounded for $\varepsilon \in (0, \varepsilon^*]$. This clearly implies that there exists a M_{ε^*} such that Property (*v*) holds. This concludes the proof of Lemma 12.15. ∎

We define the low-gain state feedback which is a family of parameterized state feedback laws given by

$$u_L(x) = F_L x = -(I + B' P_\varepsilon B)^{-1} B' P_\varepsilon A x, \tag{12.41}$$

where P_ε is the solution of (12.40). Here, as usual, ε is called the low-gain parameter. From the properties given by Lemma 12.15, it can be seen that the magnitude of the control input can be made arbitrarily small by choosing ε sufficiently small so that the input never saturates for any, a priori given, set of initial conditions.

Low-and-high-gain feedback: We proceed now to design a low-and-high-gain state feedback which is composed of a low-gain state feedback and a high-gain state feedback as

$$u_{LH}(x) = F_{LH}x = F_L x + F_H x, \tag{12.42}$$

where $F_L x$ is given by (12.41). The high-gain feedback is of the form

$$F_H x = \alpha F_L x,$$

where α, as before, is called the high-gain parameter.

For continuous-time systems and for semi-global stabilization, the high-gain parameter α can be any positive real number in principle. However, this is not the case for discrete-time systems. In order to preserve local asymptotic stability, this high gain has to be bounded at least near the origin. The existing results in literature on the choice of high-gain parameter for discrete-time system are really sparse. To the best of our knowledge, the only available result is in [84, 85] where the high-gain parameter is a nonlinear function of x. This nonlinear high gain always yields a control smaller than Δ in magnitude, which lacks the capability of dealing with disturbances. Furthermore, to solve the global external and internal stabilization problem, we need to schedule the high-gain parameter with respect to x. Moreover, this nonlinear high-gain parameter is also not suitable for adaptation since it will make the analysis extremely complicated. Instead, we need a constant high-gain parameter so that the controller (12.42) remains linear. A suitable choice of such a high-gain parameter satisfies

$$\alpha \in \left[0, \frac{2}{\|B'P_\varepsilon B\|}\right], \tag{12.43}$$

where P_ε is the solution of parametric Lyapunov equation (12.40).

The following lemma justifies the above selection of the high-gain parameter by considering the local stabilization of system (12.1b) over a given compact set $\mathcal{X}$:

Lemma 12.16 *Consider the system* (12.1b) *satisfying Assumption 12.14. Let P_ε be the solution of* (12.40)*. For any a priori given compact set $\mathcal{X}$, there exists an ε^* such that for any $\varepsilon \in [0, \varepsilon^*]$ and α satisfying* (12.43)*, the origin of the interconnection of* (12.1b) *with the low-and-high-gain feedback*

$$u_{LH} = -(1+\alpha)(I + B'P_\varepsilon B)^{-1}B'P_\varepsilon Ax$$

is locally asymptotically stable with domain of attraction containing $\mathcal{X}$.

Proof : Let c be such that

$$c = \sup_{\substack{\varepsilon \in (0,\varepsilon^*] \\ x \in \mathcal{X}}} x'P_\varepsilon x,$$

where ε^* is given by property (*iv*) in Lemma 12.15. Define a Lyapunov function $V(x) = x' P_\varepsilon x$ and a level set $\mathcal{V}(c) = \{x \mid V(x) \leqslant c\}$. We have $\mathcal{X} \subset \mathcal{V}_c$. From Lemma 12.15, there exists an $\varepsilon_1 \leqslant \varepsilon^*$ such that for any $\varepsilon \in (0, \varepsilon_1]$ and $x \in \mathcal{V}_c$,

$$\|(I + B' P_\varepsilon B)^{-1} B' P_\varepsilon A x\| \leqslant \Delta.$$

Define $\mu = \|B' P_\varepsilon B\|$. We evaluate $V(k+1) - V(k)$ along the trajectories as

$$\begin{aligned}
&V(k+1) - V(k)\\
&= -\varepsilon V(k) - \sigma(u_{LH})'\sigma(u_{LH})\\
&\qquad + [\sigma(u_{LH}) - u_L]'(I + B'PB)[\sigma(u_{LH}) - u_L]\\
&\leqslant -\varepsilon V(k) - \sigma(u_{LH})'\sigma(u_{LH})\\
&\qquad + (1+\mu)[\sigma(u_{LH}) - u_L]'[\sigma(u_{LH}) - u_L)]\\
&= -\varepsilon V(k) - \tfrac{1+\mu}{\mu}\|u_L\|^2 + \mu\|\sigma(u_{LH}) - \tfrac{1+\mu}{\mu} u_L\|^2,
\end{aligned}$$

where we abbreviated $u_{LH}(k)$ and $u_L(k)$ by u_{LH} and u_L, respectively. Since $\|u_L\| \leqslant \Delta$ and α satisfies (12.43), we have

$$\|u_L\| \leqslant \|\sigma(u_{LH})\| \leqslant (1 + \tfrac{2}{\mu})\|u_L\|.$$

This implies that

$$\|\sigma(u_{LH}) - \tfrac{1+\mu}{\mu} u_L\| \leqslant \tfrac{1}{\mu}\|u_L\|,$$

and thus,

$$\mu\|\sigma(u_{LH}) - \tfrac{1+\mu}{\mu} u_L\|^2 - \tfrac{1}{\mu}\|u_L\|^2 \leqslant 0.$$

Combining the above, we get for any $x(k) \in \mathcal{V}(c)$,

$$V(k+1) - V(k) \leqslant -\varepsilon V(k).$$

We conclude local asymptotic stability of the origin with a domain of attraction containing $\mathcal{X}$. ∎

Remark 12.17 *We would like to explain the role played by the high-gain parameter α in the controller design. For semi-global asymptotic stabilization, the domain of attraction is basically determined by the low-gain parameter ε provided that α lies in a proper range. When α is too large, stabilization is not possible. This is different from continuous-time systems for which the high-gain parameter α does not have any impact on internal stability. But like continuous-time systems, α plays a dominant role in issues other than internal stability such as external stabilization, robust stabilization, and disturbance rejection.*

Scheduling of the low-gain parameter

In the semi-global framework, with controller (12.41), the domain of attraction of the closed-loop system is determined by the low-gain parameter ε. In order to solve the global stabilization problem, this ε can be adapted with respect to the state. For discrete-time systems, this has been done in the literature; see, for instance, [48].

An adaptive low-gain feedback controller for global stabilization is given by

$$u_L(x) = F_{\varepsilon(x)}x = -(B'P_{\varepsilon(x)}B + I)^{-1}B'P_{\varepsilon(x)}Ax, \tag{12.44}$$

where $P_{\varepsilon(x)}$ is the solution of (12.40) with ε replaced by $\varepsilon(x)$. The adapted parameter $\varepsilon(x)$ should satisfy the following properties which are slight modifications of those we used in continuous-time case:

(*i*) $\varepsilon_a(x) \in C^1$.

(*ii*) $\varepsilon_a(x) = 1$ for all x in an open neighborhood of the origin.

(*iii*) For any $x \in \mathbb{R}^n$, we have

$$\|(I + B'P_{\varepsilon_a(x)}B)^{-1}B'P_{\varepsilon_a(x)}Ax\|_\infty \leqslant \Delta.$$

(*iv*) $\varepsilon_a(x) \to 0$ as $\|x\|_\infty \to \infty$.

(*v*) $\{\, x \in \mathbb{R}^n \mid x'P_{\varepsilon_a(x)}x \leqslant c \,\}$ is a bounded set for all $c > 0$.

(*vi*) For any $x_1, x_2 \in \mathbb{R}^n$,

$$x_1'P_{\varepsilon_a(x_1)}x_1 \leqslant x_2'P_{\varepsilon_a(x_2)}x_2$$

implies that $\varepsilon_a(x_1) \geqslant \varepsilon_a(x_2)$.

A particular choice of scheduling satisfying the above conditions is given in [48],

$$\varepsilon_a(x) = \max\left\{ r \in (0, \varepsilon^*] \mid (x'P_r x) \operatorname{trace} P_r \leqslant \frac{\Delta^2}{2\|BB'\|} \right\}, \tag{12.45}$$

where $\varepsilon^* \in (0,1)$ is any a priori given constant, while P_r is the unique positive definite solution of parametric Lyapunov equation (12.40) with $\varepsilon = r$. We should note that in this section, we rely on a key technical bound (12.61) whose proof explicitly relies on (12.45).

Note that the adaptive-low-gain controller (12.44) with (12.45) satisfies

$$\|(B'P_{\varepsilon_a(x)}B + I)^{-1}B'P_{\varepsilon_a(x)}Ax\| \leqslant \Delta.$$

To see this, observe that

$$\begin{aligned}
&\|(B'P_{\varepsilon_a(x)}B+I)^{-1}B'P_{\varepsilon_a(x)}Ax\|^2\\
&\quad\leqslant\|B'P_{\varepsilon_a(x)}Ax\|^2\\
&\quad\leqslant\|B'P_{\varepsilon_a(x)}^{1/2}\|^2\|P_{\varepsilon_a(x)}^{1/2}AP_{\varepsilon_a(x)}^{-1/2}\|^2\|P_{\varepsilon_a(x)}^{1/2}x\|^2\\
&\quad\leqslant 2\|BB'\|\|P_{\varepsilon_a(x)}\|x'P_{\varepsilon_a(x)}x\\
&\qquad\text{(where we use Property } (iv) \text{ of Lemma 12.15)}\\
&\quad\leqslant\Delta^2.
\end{aligned}$$

Scheduling of high-gain parameter

As emphasized earlier, the high-gain parameter plays a crucial role in dealing with external inputs/disturbances. In order to solve the simultaneous external and internal stabilization problems for continuous-time systems, different methods for scheduling the high-gain parameter have been developed earlier in this chapter and in the literature [48, 67, 126]. Unfortunately, none of them carry over to the discrete-time case because the high gain has to be restricted near the origin. In this subsection, we introduce a new scheduling of the high-gain parameter with which we shall solve the (G_p/G) and $(G_p/G)_{fg}$ problems, as formulated in the previous chapter (Sect. 11.2).

Our scheduling depends on the specific control objective. If one is not interested in finite gain, the following adaptive high gain suffices to solve (G_p/G) problem:

$$\alpha_0(x)=\frac{1}{\|B'P_{\varepsilon_a(x)}B\|}. \tag{12.46}$$

Clearly, this high gain satisfies the constraint (12.43) mentioned earlier. We observe that this high-gain parameter is radially unbounded. However, if we further pursue finite gain ℓ_p stabilization, the rate of growth of $\alpha(x)$ with respect to $\|x\|$ as given in (12.46) is not sufficient for us. The scheduled high-gain parameter must rise quickly enough to overwhelm any disturbances in ℓ_p before the state is steered so large that it actually prevents finite gain. Therefore, we shall introduce a different scheduling of high-gain parameter. In order to do so, we need the following lemma:

Lemma 12.18 *Assume that* $2p\geqslant 1$. *For any* $\eta>1$, *there exists a* $\beta>0$ *such that*

$$(u+v)^p\leqslant u^p+\eta u^p+\beta v^p \tag{12.47}$$

for all $u,v\geqslant 0$.

Proof : The lemma is a known result for $p\geqslant 1$; see, for instance, [151]. For $p\in[\frac{1}{2},1)$, we have $2p\geqslant 1$ and then

$$(\sqrt{u+v})^{2p} \leqslant (\sqrt{u}+\sqrt{v})^{2p} \leqslant u^{2p} + \eta u^{2p} + \beta v^{2p},$$

where we use the lemma with p replaced by $2p$ which is the known case. ∎

We proceed now to present a scheduled high-gain parameter. Let ε^* and M_{ε^*} be given by Lemma 12.15 and let P^* be the solution of (12.40) with $\varepsilon = \varepsilon^*$. The adapted high-gain parameter is then given by

$$\alpha_a(x) = \begin{cases} \alpha_0(x) = \frac{1}{\|B' P_{\varepsilon_a(x)} B\|}, & x' P_{\varepsilon_a}(x) x \leqslant c \\ \frac{8\alpha_1(x)}{\varepsilon_a(x)\lambda_{\min} P_{\varepsilon_a(x)}}, & \text{otherwise} \end{cases} \tag{12.48}$$

with

$$\alpha_1(x) = \frac{\lambda_{\max} P_{\varepsilon_a(x)}}{\lambda_{\min} P_{\varepsilon_1(x)}} \alpha_2(x), \tag{12.49}$$

where

$$\alpha_2(x) = \begin{cases} 1 & p = \infty \\ \left[\frac{\alpha_p \beta(\varepsilon_a(x))}{1 - \left(1 - \frac{\varepsilon_1(x)}{4(1+L_{\varepsilon_1(x)})}\right)^{p/2}} + 1 \right]^{2/p}, & p \in [1, \infty), \end{cases}$$

and where α_p is a positive constant to be determined later and c, $\varepsilon_1(x)$, and L_s are given by

$$\begin{aligned} c &= \Delta^2 \max\{4M_{\varepsilon^*}, 4(1 + \|B' P^* B\|)\}, \qquad (12.50) \\ \varepsilon_1(x) &= \max \left\{ r \in (0, \varepsilon^*] \mid x' P_r x \operatorname{trace} P_r \leqslant \tfrac{\Delta^2}{2\|BD'\|} \right\}, \\ L_s &= \tfrac{\operatorname{trace} P^*}{\lambda_{\min} P_s}. \end{aligned}$$

Finally, in order to define $\beta(\varepsilon) > 1$, we first define $\eta(\varepsilon)$ satisfying

$$\left[1 - \tfrac{\varepsilon}{4(1+L_\varepsilon)}\right]^{p/2} \leqslant (1 + \eta(\varepsilon)) \left[1 - \tfrac{\varepsilon}{2(1+L_\varepsilon)}\right]^{p/2} < 1.$$

Next, we choose $\beta(\varepsilon) > 1$ such that Lemma 12.18 holds for $\eta = \eta(\varepsilon)$. In other words, $\beta(\varepsilon)$ is such that for a given $p > 1/2$, ε and $\eta(\varepsilon)$,

$$(u+v)^p \leqslant (1 + \eta(\varepsilon)) u^p + \beta(\varepsilon) v^p$$

for all $u > 0, v > 0$.

We are now ready to solve the simultaneous external and internal stabilization problems, as formulated in the previous chapter (Sect. 11.2). We first study the simultaneous stabilization without finite gain, as formulated in Problems 11.1 and 11.4. Then we will solve Problems 11.2 and 11.5 which seek finite gain.

The theorem given below solves the global ℓ_p stabilization with arbitrary initial conditions and without finite gain, as formulated in Problem 11.4.

Theorem 12.19 *Consider the system* Σ^d *of* (12.1b) *satisfying Assumption 12.14. For any* $p \in [1,\infty]$, *the* ℓ_p *stabilization with arbitrary initial conditions and without finite gain, as formulated in Problem 11.4, can be solved by the adaptive-low-gain and high-gain controller,*

$$u = -(1+\alpha_0(x))(I + B'P_{\varepsilon_a(x)}B)^{-1}B'P_{\varepsilon_a(x)}Ax, \tag{12.51}$$

where $P_{\varepsilon_a(x)}$ *is the solution of* (12.40), $\varepsilon_a(x)$ *is determined by the scheduling* (12.45), *and* $\alpha_0(x)$ *is determined by* (12.46).

Theorem 12.19 immediately yields the following result:

Corollary 12.20 *Consider a system* Σ^d *of the form* (12.1b) *satisfying Assumption 12.14. For any* $p \in [1,\infty]$, *the* (G_P/G), *as formulated in Problem 11.1, can be solved by the same adaptive-low-gain and high-gain controller as given in* (12.51).

Proof of Theorem 12.19 : In this proof, we denote $\varepsilon_a(x(k))$, $\alpha_0(x(k))$, and $P_{\varepsilon_a(x(k))}$ by $\varepsilon_a(k)$, $\alpha_0(k)$, and $P(k)$, respectively. This abbreviation should not cause any notational confusions.

Define

$$\begin{aligned}
v(k) &= -(I + B'P(k)B)^{-1}B'P(k)Ax(k),\\
u(k) &= v(k) + \alpha_0(k)v(k),\\
\mu(k) &= \|B'P(k)B\|.
\end{aligned}$$

We have shown that (12.45) implies that $\|v(k)\|_\infty < \Delta$.

We proceed now to show global asymptotic stability. In the absence of d, we can evaluate the increment of $V(k)$ along the trajectory as

$$\begin{aligned}
&V(k+1) - V(k)\\
&\quad = x(k+1)'[P(k+1) - P(k)]x(k+1) - \varepsilon_a(k)V(k) - \|\sigma(u(k))\|^2\\
&\qquad + [\sigma(u(k)) - v(k)]'(I + B'P(k)B)[\sigma(u(k)) - v(k)]\\
&\quad \leqslant x(k+1)'[P(k+1) - P(k)]x(k+1) - \varepsilon_a(k)V(k) - \|\sigma(u(k))\|^2\\
&\qquad + (1+\mu(k))[\sigma(u(k)) - v(k)]'[\sigma(u(k)) - v(k)]\\
&\quad = x(k+1)'[P(k+1) - P(k)]x(k+1) - \varepsilon_a(k)V(k)\\
&\qquad - \tfrac{1+\mu(k)}{\mu(k)}\|v(k)\|^2 + \mu(k)\|\sigma(u(k)) - \tfrac{1+\mu(k)}{\mu(k)}v(k)\|^2.
\end{aligned}$$

As noted before, $\|v(k)\| \leqslant \Delta$ for all $k \geqslant 0$, and therefore,

$$\|v(k)\| \leqslant \|\sigma(u(k))\| \leqslant (1 + \tfrac{1}{\mu(k)})\|v(k)\|.$$

This implies that

$$\|\sigma(u(k)) - \tfrac{1+\mu(k)}{\mu(k)} v(k)\| \leqslant \tfrac{1}{\mu(k)} \|v(k)\|,$$

and thus,

$$\mu(k)\|\sigma(u(k)) - \tfrac{1+\mu(k)}{\mu(k)} v(k)\|^2 - \tfrac{1+\mu(k)}{\mu(k)} \|v(k)\|^2 \leqslant -\|v(k)\|^2.$$

Finally, we get

$$V(k+1) - V(k) \leqslant -\varepsilon_a(k)V(k) + x(k+1)'[P(k+1) - P(k)]x(k+1). \quad (12.52)$$

Our scheduling (12.45) implies that

$$V(k+1) - V(k) \text{ and } x(k+1)'[P(k+1) - P(k)]x(k+1)$$

cannot have the same sign. To see this, assume that

$$V(k+1) > V(k) \text{ and } P(k+1) > P(k). \quad (12.53)$$

This implies that $\varepsilon_a(k) < \varepsilon^*$. If

$$V(k) \operatorname{trace} P(k) < \tfrac{\Delta^2}{2\|BB'\|},$$

then (12.45) implies that $\varepsilon_a(k) = \varepsilon^*$, which yields a contradiction. If

$$V(k) \operatorname{trace}(P(k)) = \tfrac{\Delta^2}{2\|BB'\|},$$

then

$$V(k+1) \operatorname{trace} P(k+1) > \tfrac{\Delta^2}{2\|BB'\|}$$

since by assumption (12.53). But this is impossible by our scheduling (12.45). A similar argument can be used to establish that

$$V(k+1) - V(k) < 0 \text{ and } P(k+1) - P(k) < 0$$

cannot happen simultaneously either. Using this property, (12.52) then implies that for all $x \neq 0$,

$$V(k+1) - V(k) < 0.$$

This concludes the global asymptotic stability.

What remains is to show ℓ_p stability. Similar to our earlier development, we have

$$\begin{aligned} V(k+1) - V(k) \leqslant & - x(k+1)'[P(k) - P(k+1)]x(k+1) \\ & - \varepsilon_a(k)V(k) - \tfrac{1}{\mu(k)}\|v(k)\|^2 \\ & + \mu(k)\|\sigma(u(k) + d(k)) - \tfrac{1+\mu(k)}{\mu(k)} v(k)\|^2. \end{aligned}$$

Let $d_i(k)$, $v_i(k)$, and $u_i(k)$ denote the ith element of $d(k)$, $v(k)$, and $u(k)$, respectively.

If $|d_i(k)| \leqslant \frac{1}{\mu(k)}|v_i(k)|$, recalling that $|v_i(k)| \leqslant \Delta$, we have

$$|v_i(k)| \leqslant |\operatorname{sat}_\Delta(u_i(k)+d_i(k))| \leqslant (1+\tfrac{2}{\mu(k)})|v_i(k)|.$$

Hence,

$$\mu(k)|\operatorname{sat}_\Delta(u_i(k)+d_i(k)) - \tfrac{1+\mu(k)}{\mu(k)}v_i(k)|^2 - \tfrac{1}{\mu(k)}|v_i(k)|^2 \leqslant 0.$$

If $|d_i(k)| \geqslant \frac{1}{\mu(k)}|v_i(k)|$, we have

$$\begin{aligned}
&-\tfrac{1}{\mu(k)}|v_i(k)|^2 + \mu(k)|\operatorname{sat}_\Delta(u_i(k)+d_i(k)) - \tfrac{1+\mu(k)}{\mu(k)}v_i(k)|^2\\
&\leqslant \mu(k)\left[(1+\mu(k))d_i(k) + d_i(k) + (1+\mu(k))d_i(k)\right]^2 + \mu(k)|d_i(k)|^2\\
&\leqslant a\mu(k)|d_i(k)|^2,
\end{aligned}$$

where $a = (2\mu^*+3)^2+1$, $\mu^* = \|B'P^*B\|$, and P^* is the solution of (12.40) with $\varepsilon = \varepsilon^*$. Therefore, we conclude that

$$\begin{aligned}
V(k+1)-V(k) \leqslant x(k+1)'[P(k+1)-P(k)]x(k+1)&\\
-\varepsilon_a(k)V(k) + a\mu(k)\|d(k)\|^2.& \qquad (12.54)
\end{aligned}$$

Note that this implies that

$$V(k+1)-V(k) \leqslant \max\{-\varepsilon_a(k)V(k) + a\mu(k)\|d(k)\|^2, 0\} \qquad (12.55)$$

since $V(k+1)-V(k)$ and $x(k+1)'[P(k+1)-P(k)]x(k+1)$ cannot have the same sign. Let us first address the case of $p=\infty$. We will show that there exists a c_1 such that $V(k) \leqslant c_1$ for all $k \geqslant 0$ with $V(0)=0$.

If

$$V(k) \geqslant a\tfrac{\mu(k)}{\varepsilon_a(k)}\|d(k)\|^2, \qquad (12.56)$$

we have

$$V(k+1)-V(k) \leqslant 0. \qquad (12.57)$$

Property (v) of Lemma 12.15 then yields that there exists a M_{ε^*} independent of k and d such that $V(k) \geqslant aM_{\varepsilon^*}\|d\|_\infty^2$ implies that (12.56) is satisfied and therefore $V(k+1)-V(k) \leqslant 0$.

On the other hand, if (12.56) is not satisfied, we have

$$V(k+1)-V(k) \leqslant a\mu(k)\|d(k)\|^2 \leqslant a\mu^*\|d\|_\infty^2.$$

We conclude that

$$V(k) \leqslant V(0) + aM_{\varepsilon^*}\|d\|_\infty^2 + a\mu^*\|d\|_\infty^2. \qquad (12.58)$$

Property (v) of our scheduling therefore implies that $x(k)$ is bounded for all $k \geqslant 0$. This shows ℓ_∞ stability of the closed-loop system with arbitrary initial conditions.

We proceed now with the case of $p \in [1, \infty)$. First of all, due to the fact that $\|d\|_\infty \leqslant \|d\|_p$, (12.58) implies that $V(k)$ is bounded for all $k \geqslant 0$. Hence, by our scheduling, there exists an ε_0 such that $\varepsilon_a(k) \geqslant \varepsilon_0$ for all $k \geqslant 0$.

Next, we consider two possible cases:

Case 1: For $V(k+1) - V(k) \geqslant 0$, (12.55) implies that

$$V(k+1) - V(k) \leqslant -\varepsilon_a(k)V(k) + a\mu(k)\|d(k)\|^2. \tag{12.59}$$

Case 2: For $V(k+1) - V(k) \leqslant 0$, our scheduling implies that

$$x(k+1)'[P(k+1) - P(k)]x(k+1) \geqslant 0.$$

But this implies that $\varepsilon_a(k) \leqslant \varepsilon_a(k+1) \leqslant \varepsilon^*$, and thus,

$$V(k+1) \operatorname{trace} P(k+1) \leqslant V(k) \operatorname{trace} P(k).$$

Hence,

$$[V(k+1) - V(k)] \operatorname{trace} P(k+1) \leqslant -V(k) \operatorname{trace}[P(k+1) - P(k)].$$

Then we have

$$\begin{aligned}
&|x(k+1)'[P(k+1) - P(k)]x(k+1)| && (12.60)\\
&\quad \leqslant |\operatorname{trace}(P(k+1) - P(k))| \cdot \|x(k+1)\|^2\\
&\quad \leqslant \frac{\operatorname{trace}(P(k+1))}{V(k)} \cdot |V(k+1) - V(k)| \cdot \|x(k+1)\|^2\\
&\quad \leqslant \frac{V(k+1)\operatorname{trace}(P(k+1))}{V(k)\lambda_{\min} P(k+1)} \cdot |V(k+1) - V(k)|\\
&\quad \leqslant \frac{\operatorname{trace}(P^*)}{\lambda_{\min} P(k)} \cdot |V(k+1) - V(k)|\\
&\quad \leqslant L(k) \cdot |V(k+1) - V(k)|, && (12.61)
\end{aligned}$$

where $L(k) = \frac{\operatorname{trace}(P^*)}{\lambda_{\min}(P(k))}$. We have

$$V(k+1) - V(k) \leqslant \tfrac{-\varepsilon_a(k)}{1+L(k)} V(k) + a\mu(k)\|d(k)\|^2. \tag{12.62}$$

Given $\varepsilon_a(k) \in [\varepsilon_0, \varepsilon^*]$ for all $k \geqslant 0$, (12.59) in case 1 and (12.62) in case 2 ensure that

$$V(k+1) - V(k) \leqslant -\tfrac{\varepsilon_0}{1+L} V(k) + a\mu^* \|d(k)\|^2, \tag{12.63}$$

where $L = \frac{\operatorname{trace}(P^*)}{\lambda_{\min} P_0}$ and P_0 is the solution of (12.40) with $\varepsilon = \varepsilon_0$. Also, $\varepsilon_0 < 1$ implies that $\varepsilon_0/(1+L) < 1$.

Applying Lemma 12.18 with η such that

$$(1+\eta)(1-\tfrac{\varepsilon_0}{1+L})^{p/2} < 1,$$

we find that there exists a β such that

$$V(k+1)^{p/2} \leqslant (1+\eta)(1-\tfrac{\varepsilon_0}{1+L})^{p/2} V(k)^{p/2} + \beta(a\mu^*)^{p/2}\|d(k)\|^p.$$

This yields

$$\left[1-(1+\eta)(1-\tfrac{\varepsilon_0}{1+L})^{p/2}\right] \|V^{p/2}\|_1 \leqslant \beta(a\mu^*)^{p/2}\|d\|_p^p + V(0)^{p/2}.$$

Since $\varepsilon_a(k) \geqslant \varepsilon_0$ for all k,

$$\begin{aligned}\|x\|_p^p &\leqslant \frac{\|V^{p/2}\|_1}{(\lambda_{\min} P_0)^{p/2}} \\ &\leqslant \frac{\beta(a\mu^*)^{p/2}}{(\lambda_{\min} P_0)^{p/2}\left[1-(1+\eta)(1-\frac{\varepsilon_0}{1+L})^{p/2}\right]}\|d\|_p^p \\ &\quad + \frac{V(0)^{p/2}}{(\lambda_{\min} P_0)^{p/2}\left[1-(1+\eta)(1-\frac{\varepsilon_0}{1+L})^{p/2}\right]},\end{aligned} \tag{12.64}$$

we conclude that $d \in \ell_p$ implies that $x \in \ell_p$ for any $x(0) \in \mathbb{R}^n$. This concludes the proof of Theorem 12.19. ■

We observe from (12.58) and (12.64) that as $\|d\|_p$ and $x(0)$ become larger, the ε_0 becomes smaller, and the ℓ_p gain becomes larger. In order to pursue finite gain ℓ_p stabilization, it is necessary to modify the high-gain parameter. We first consider the case $p = \infty$.

Theorem 12.21 *Consider the system Σ^d of* (12.1b) *satisfying Assumption 12.14. For $p = \infty$, ℓ_p stabilization with arbitrary initial conditions with finite gain and with bias, as formulated in Problem 11.5, can be achieved by the adaptive-low-gain and high-gain controller,*

$$u = -(1+\alpha_a(x))(I + B'P_{\varepsilon_a(x)}B)^{-1}B'P_{\varepsilon_a(x)}Ax, \tag{12.65}$$

where $P_{\varepsilon_a(x)}$ is the solution of (12.40) *with $\varepsilon = \varepsilon_a(x)$, $\varepsilon_a(x)$ is determined adaptively by* (12.45)*, and $\alpha_a(x)$ is determined by* (12.48) *and* (12.49)*.*

Theorem 12.21 readily yields the following corollary:

Corollary 12.22 *Consider a system Σ^d of the form* (12.1b) *which satisfies Assumption 12.14. For $p = \infty$, the $(G_p/G)_{fg}$, as formulated in Problem 11.2, can be solved by the same adaptive-low-gain and high-gain controller as given in* (12.65)*.*

Proof of Theorem 12.21 : For simplicity, we denote $P_{\varepsilon_a(x(k))}$ and $P_{\varepsilon_1(x(k))}$, respectively, by $P(k)$ and $P_1(k)$ whenever this does not cause any notational confusions.

Define

$$\begin{aligned} v(k) &= -(I + B'P(k)B)^{-1}B'P(k)Ax(k), \\ u(k) &= v(k) + \alpha_a(k)v(k). \end{aligned}$$

We have already shown that the controller (12.65) along with (12.45) satisfies $\|v\|_\infty < \Delta$.

Define a Lyapunov function $V(k) = x(k)'P(k)x(k)$ and a set

$$\mathcal{V}(c) = \{ x \mid V(x) \leqslant c \}$$

with c defined by (12.50). Owing to Property (*v*) of Lemma 12.15, it is easy to verify that for $x(k) \in \mathcal{V}(c)^c$, the following inequality holds:

$$\varepsilon_a(k)V(k) \geqslant 4\varepsilon_a(k)M_{\varepsilon^*}\Delta^2 \geqslant 8\|B'P(k)B\|\Delta^2. \tag{12.66}$$

In the absence of d, we can evaluate the increment of V along the trajectory as

$$\begin{aligned} V(k+1) - V(k) = x(k+1)'\,&[P(k+1) - P(k)]\,x(k+1) - \varepsilon_a(k)V(k) \\ &- 2v(k)'[\sigma(u(k)) - v(k)] \\ &+ [\sigma(u(k)) - v(k)]'\,B'P(k)B\,[\sigma(u(k)) - v(k)]\,. \end{aligned}$$

Also, $\|v(k)\| \leqslant \Delta$ implies that $-2v(k)'[\sigma(u(k)) - v(k)] \leqslant 0$ for any $\alpha_a(k) > 0$. Using this property, we find that for $x(k) \in \mathcal{V}(c)^c$,

$$\begin{aligned} V(k+1) - V(k) &\leqslant x(k+1)'\,[P(k+1) - P(k)]\,x(k+1) - \varepsilon_a(k)V(k) \\ &\quad - 2v(k)'[\sigma(u(k)) - v(k)] + 4\|B'P(k)B\|\Delta^2 \\ &\leqslant x(k+1)'\,[P(k+1) - P(k)]\,x(k+1) - \tfrac{\varepsilon_a(k)}{2}V(k). \end{aligned}$$

The last inequality is owing to (12.66). If

$$x(k+1)'\,[P(k+1) - P(k)]\,x(k+1) < 0, \tag{12.67}$$

the last inequality implies that $V(k+1) - V(k) < 0$. But we have argued earlier that (12.67) and $V(k+1) - V(k) < 0$ cannot happen simultaneously by our scheduling (12.45). Therefore, $x(k+1)'\,[P(k+1) - P(k)]\,x(k+1) \geqslant 0$. From the proof of Theorem 12.19,

$$x(k+1)'\,[P(k+1) - P(k)]\,x(k+1) \leqslant L(k)[V(k) - V(k+1)].$$

Hence, for $x(k) \in \mathcal{V}(c)^c$,

$$V(k+1) - V(k) < -\tfrac{\varepsilon_a(k)}{2(1+L(k))}V(k).$$

The trajectory will enter $\mathcal{V}(c)$ within finite time. However, for $x(k) \in \mathcal{V}(c)$, we have already proved in the proof of Theorem 12.19 that

$$V(k+1) - V(k) < 0$$

since in $\mathcal{V}(c)$, $\alpha_a(k) = \alpha_0(k)$. This proves global asymptotic stability of the origin.

We proceed now to show ℓ_∞ stability with arbitrary initial conditions with finite gain with bias. In order to do so, we first find an upper bound of $\frac{V(k)}{\lambda_{\min} P(k)}$ in terms of $\|d\|_\infty$ and then conclude ℓ_∞ stability by observing that

$$\|x\|_\infty^2 \leqslant \|\tfrac{V}{\lambda_{\min} P}\|_\infty.$$

To this end, we note that the case $V(k+1) - V(k) \leqslant 0$ is not interesting since it is equivalent with

$$\tfrac{V(k+1)}{\lambda_{\min} P(k+1)} - \tfrac{V(k)}{\lambda_{\min} P(k)} \leqslant 0$$

due to the fact that $V(k+1) \leqslant V(k)$ implies $\lambda_{\min} P(k+1) \geqslant \lambda_{\min} P(k)$. Therefore, it will not affect the upper bound of $\frac{V(k)}{\lambda_{\min} P(k)}$. In view of this, we only consider the case $V(k+1) - V(k) > 0$ throughout the rest of the proof.

Suppose $V(k+1) - V(k) > 0$, scheduling (12.45) implies that

$$x(k+1)' \left[P(k+1) - P(k)\right] x(k+1) \leqslant 0.$$

By construction, $\|v(k)\| \leqslant \Delta$. We get

$$\begin{aligned} V(k+1)-&V(k) \\ &\leqslant -\varepsilon_a(k)V(k) - 2v(k)'[\sigma(u(k)+d(k)) - v(k)] + 4\|B'P^*B\|\Delta^2 \\ &\leqslant\ 4(1+\|B'P^*B\|)\Delta^2. \end{aligned}$$

Since $c > 4(1+\|B'P^*B\|)\Delta^2$, we have

$$V(k+1) - V(k) \leqslant c. \tag{12.68}$$

The above inequality holds for any $x(k) \in \mathbb{R}^n$. Since different high gains are applied in different regions, we have two possible cases:
Case 1: $x(k) \in \mathcal{V}(c)^c$. Then (12.68) implies that $V(k+1) \leqslant 2V(k)$. But this implies that $\varepsilon_1(k) \leqslant \varepsilon_a(k+1)$ and $P_1(k) \leqslant P(k+1)$. Let $v_i(k)$ and $d_i(k)$ denote the ith element of $v(k)$ and $d(k)$.
If $|d_i(k)| < \alpha_a(k)|v_i(k)|$, then

$$-v_i(k)\left[\operatorname{sat}_\Delta(v_i(k) + \alpha_a(k)v_i(k) + d_i(k)) - v_i(k)\right] \leqslant 0.$$

If $|d_i(k)| \leqslant |\alpha_a(k)v_i(k)|$, we have

$$\begin{aligned}
&- v_i(k)\,[\mathrm{sat}_\Delta(v_i(k) + \alpha_a(k)v_i(k) + d_i(k)) - v_i(k)]\\
&\quad = -v_i(k)\,[\mathrm{sat}_\Delta(v_i(k) + \alpha_a(k)v_i(k) + d_i(k)) - \sigma(v_i(k))]\\
&\quad \leqslant \tfrac{|d_i(k)|}{\alpha_a(k)} \cdot |2d_i(k)|\\
&\quad = \tfrac{2d_i(k)^2}{\alpha_a(k)}.
\end{aligned}$$

In summary, we find that

$$-2v(k)'[\sigma(u(k) + d(k)) - v(k)] \leqslant \tfrac{4\|d(k)\|^2}{\alpha_a(k)}.$$

This yields

$$\begin{aligned}
V(k+1) - V(k) &\leqslant -\tfrac{\varepsilon_a(k)}{2} V(k) - 2v(k)'[\sigma(u(k) + d(k)) - v(k)]\\
&\leqslant -\tfrac{\varepsilon_a(k)}{2} V(k) + 4\tfrac{\|d(k)\|^2}{\alpha_a(k)}\\
&\leqslant -\tfrac{\varepsilon_a(k)\lambda_{\min} P(k)}{2}(\|x(k)\|^2 - \tfrac{\|d(k)\|^2}{\alpha_1(k)}).
\end{aligned}$$

Clearly, $V(k+1) - V(k) \geqslant 0$ requires that

$$\|x(k)\|^2 \leqslant \tfrac{\|d(k)\|^2}{\alpha_1(k)}.$$

Then

$$\begin{aligned}
\tfrac{V(k+1)}{\lambda_{\min} P(k+1)} \leqslant \tfrac{2V(k)}{\lambda_{\min} P_1(k)} &\leqslant \tfrac{2\lambda_{\max} P(k)}{\lambda_{\min} P_1(k)} \|x(k)\|^2\\
&= 2\alpha_1 \|x(k)\|^2 \leqslant 2\|d(k)\|^2. \qquad (12.69)
\end{aligned}$$

Case 2: $x(k) \in \mathcal{V}(c)$. We have $\alpha_a(k) = \alpha_0(k)$ and hence the same controller as in Theorem 12.19. In the proof of Theorem 12.19, the following two properties have already been shown:

(*i*) If $V(k) \geqslant aM_{\varepsilon^*}\|d(k)\|^2$, we have $V(k+1) - V(k) \leqslant 0$.

(*ii*) $V(k+1) - V(k) \leqslant a\mu^*\|d(k)\|^2$.

We can immediately draw the conclusion that for $V(k+1) - V(k) > 0$ and $x(k) \in \mathcal{V}(c)$,

$$V(k+1) \leqslant (aM_{\varepsilon^*} + a\mu^*)\|d(k)\|^2.$$

On the other hand, (12.68) and the fact $V(k) \leqslant c$ imply that $V(k+1) \leqslant 2c$. But this implies that there exists a λ_1 independent of d such that

$$\tfrac{V(k+1)}{\lambda_{\min} P(k+1)} \leqslant \tfrac{aM_{\varepsilon^*} + a\mu^*}{\lambda_1} \|d\|_\infty^2. \qquad (12.70)$$

In summary, whenever $V(k)$ or, equivalently, $\frac{V(k)}{\lambda_{\min} P(k)}$ is increasing, we have either (12.70) or (12.69) holds depending on $x(k) \in \mathcal{V}(c)$ or not. Therefore,

$$\|\frac{V}{\lambda_{\min} P}\|_\infty \leqslant \frac{V(0)}{\lambda_{\min} P(0)} + \max\{2, \frac{aM_{\varepsilon*}+a\mu^*}{\lambda_1}\}\|d\|_\infty.$$

Using the fact that $\|x\|_\infty^2 \leqslant \|\frac{V}{\lambda_{\min} P}\|_\infty$, we have

$$\|x\|_\infty \leqslant \sqrt{\|\frac{V}{\lambda_{\min} P}\|_\infty} \leqslant \sqrt{\frac{V(0)}{\lambda_{\min} P(0)}} + \max\{\sqrt{2}, \sqrt{\frac{aM_{\varepsilon*}+a\mu^*}{\lambda_1}}\}\|d\|_\infty. \quad (12.71)$$

Note that $\sqrt{\frac{V(0)}{\lambda_{\min} P(0)}}$ is clearly a class $\mathcal{K}$ function of $\|x(0)\|$. The finite gain ℓ_∞ stability of closed-loop system with arbitrary initial conditions and bias follows. ■

In Theorem 12.21, we only need to consider the case that $V(x(k))$ is increasing. However, this does not work when the external input d is in ℓ_p with $p \in [1, \infty)$. The decay rate of $V(x(k))$ when $V(x(k))$ is decreasing definitely has an impact on the ℓ_p norm of x. Therefore, we have to consider both cases and obtain bounds on $\|x\|_p$ in terms of $\|d\|_p$. As will be seen in the next theorem, it requires even more complicated high-gain design and involved analysis.

Theorem 12.23 *Consider the system Σ^d of* (12.1b) *satisfying Assumption 12.14. For any $p \in [1, \infty)$, the ℓ_p stabilization with arbitrary initial conditions with finite gain with bias problem, as formulated in Problem 11.5, can be solved by the adaptive-low-gain and high-gain controller,*

$$u = -(1 + \alpha_a(x))(I + B'P_{\varepsilon_a(x)}B)^{-1}B'P_{\varepsilon_a(x)}Ax, \quad (12.72)$$

where $P_{\varepsilon_a(x)}$ is the solution of (12.40) *with $\varepsilon = \varepsilon_a(x)$, $\varepsilon_a(x)$ is determined adaptively by* (12.45), *and $\alpha_a(x)$ is determined by* (12.48), (12.49) *with α_p sufficiently large.*

Theorem 12.23 also produces as a special case the solution to $(G_p/G)_{fg}$. This is stated in the following corollary:

Corollary 12.24 *Consider a system Σ^d of the form* (12.1b) *satisfying Assumption 12.14. For any $p \in [1, \infty)$, the $(G_p/G)_{fg}$, as formulated in Problem 11.2, can be solved by the adaptive-low-gain and high-gain controller as given in* (12.72).

Proof of Theorem 12.23 : For simplicity, we denote

$$\varepsilon_a(x(k)), \varepsilon_1(x(k)), \alpha_1(x(k)), \alpha_a(x(k)), \text{ and } \beta(\varepsilon_a(x(k)))$$

by $\varepsilon_a(k)$, $\varepsilon_1(k)$, $\alpha_1(k)$, $\alpha_a(k)$, and $\beta(k)$, respectively, and we denote $P_{\varepsilon_a(x(k))}$, $P_{\varepsilon_1(x(k))}$, and $L_{\varepsilon_1(x(k))}$, respectively, by $P(k)$, $P_1(k)$, and $L_1(k)$. This does not cause any notational confusions.

Define

$$\begin{aligned} v(k) &= -(I + B'P(k)B)^{-1}B'P(k)Ax(k) \\ u(k) &= v(k) + \alpha_f(k)v(k). \end{aligned}$$

We have already shown that $v(k)$ along with (12.45) satisfies $\|v\|_\infty < \Delta$.

Define a Lyapunov function $V(k) = x(k)'P(k)x(k)$ and a set

$$\mathcal{V}(c) = \{x \mid V(x) \leqslant c\},$$

where c is as defined in (12.50). As in the proof of Theorem 12.21, for $x \in \mathcal{V}(c)^c$, the following inequality holds:

$$\varepsilon_a(k)V(k) \geqslant 4\varepsilon_a(k)M_{\varepsilon^*}\Delta^2 \geqslant 8\|B'P(k)B\|\Delta^2. \tag{12.73}$$

Using exactly the same argument as used in Theorem 12.21, we conclude the global asymptotic stability of the origin of the closed-loop system.

It remains to prove global ℓ_p stability with finite gain. The proof proceeds in several steps:

Step 1. Define a function

$$\nu(s) = \frac{s^{p/2}}{(\lambda_{\min} P_s)^{p/2}\left[1 - \left(1 - \frac{\varepsilon_s}{4(1+L_s)}\right)\right]^{p/2}},$$

where ε_s is a function of s as given by

$$\varepsilon_s = \max\{r \in [0, \varepsilon^*] \mid s\|P_r\| \leqslant \tfrac{\Delta^2}{2\|BB'\|}\},$$

and P_s is the solution of (12.40) with $\varepsilon = \varepsilon_s$, $L_s = \frac{\operatorname{trace}(P^*)}{\lambda_{\min} P_s}$. Note that, if s is strictly increasing then, by the property of our scheduling, ε_s is decreasing, and hence, $\lambda_{\min} P_s$ is decreasing, and L_s is increasing. This implies that $\nu(s)$ is strictly increasing and is a class $\mathcal{K}$ function.

Define

$$\kappa = \frac{(\lambda_{\min} P^*)^{p/2}\left[1-\left(1-\frac{\varepsilon^*}{4(1+L^*)}\right)^{p/2}\right]}{(\lambda_{\min} P_{2c})^{p/2}\left[1-\left(1-\frac{\varepsilon_{2c}}{4(1+L_{2c})}\right)^{p/2}\right]},$$

where P^* is the solution of (12.40) with $\varepsilon = \varepsilon^*$ and

$$L^* = \frac{\text{trace}(P^*)}{\lambda_{\min} P^*}.$$

Since c is given, ε_{2c}, P_{2c}, L_{2c}, and κ are fixed constants. Choose

$$\alpha_p > \max\left\{1 + \kappa, \left(\lambda_{\min} P^*\right)^{p/2}\right\}.$$

We have $\alpha_a(k) \geqslant 1$ for any $x(k)$.

In what follows, for $k_1, k_2 \in \mathbb{Z}^+$ such that $k_1 \leqslant k_2$, we denote by $\overline{k_1, k_2}$ the set of integers $\{k_1, k_1 + 1, \ldots, k_2\}$. We can always divide the whole time horizon into a sequence of successive intervals $\{I_i\}_{i \geqslant 1}$ with $I_i = \overline{k_i, k_{i+1} - 1}$ such that for each I_i, one of the following cases holds:

(*i*) For any $k \in I_i$, $x(k) \in \mathcal{V}(2c)^c$ and $V(k + 1) - V(k) > 0$.

(*ii*) For any $k \in I_i$, $x(k) \in \mathcal{V}(2c)^c$ and $V(k + 1) - V(k) \leqslant 0$.

(*iii*) For any $k \in I_i$, $x(k) \in \mathcal{V}(2c)$ with $k_{i+1} < \infty$.

(*iv*) For any $k \in I_i$, $x(k) \in \mathcal{V}(2c)$ with $k_{i+1} = \infty$.

Step 2. For case (*i*), since $V(k+1) - V(k) > 0$, the adaptation (12.45) implies that $x(k + 1)' [P(k + 1) - P(k)] x(k + 1) \leqslant 0$. As in the proof of Theorem 12.21, we find

$$V(k + 1) - V(k) \leqslant -\frac{\varepsilon_a(k)\lambda_{\min} P(k)}{2} \left[\|x(k)\|^2 - \frac{\|d(k)\|^2}{\alpha_1(k)}\right].$$

Then, $V(k + 1) - V(k) > 0$ implies that

$$\|d(k)\|^2 \geqslant \alpha_1(k)\|x(k)\|^2 \geqslant \|x(k)\|^2 \tag{12.74}$$

since $\alpha_1(k) \geqslant 1$ by construction.

Furthermore, we have already shown that for all $x(k)$, $V(k + 1) - V(k) \leqslant c$. Hence,

$$V(k + 1) \leqslant 2V(k).$$

From the definition of $\varepsilon_1(k)$ and $L_1(k)$, this implies that

$$\begin{aligned} &\varepsilon_1(k) \leqslant \varepsilon_a(k + 1),\; L_1(k) \geqslant L(k + 1), \text{ and} \\ &\qquad\qquad \lambda_{\min} P(\varepsilon_1(k)) \leqslant \lambda_{\min} P(\varepsilon_a(k + 1)). \end{aligned} \tag{12.75}$$

Consider specifically $k = k_{i+1} - 1$. We have

$$\begin{aligned} &\|d(k_{i+1} - 1)\|^p - \|x(k_{i+1} - 1)\|^p \\ &\quad \geqslant \left(\alpha_1(k_{i+1} - 1)^{p/2} - 1\right) \|x(k_{i+1} - 1)\|^p \end{aligned}$$

$$
\begin{aligned}
&\geqslant \frac{\alpha_p \lambda_{\max} P(k_{i+1}-1)^{p/2} \|x(k_{i+1}-1)\|^p}{\lambda_{\min} P_1(k_{i+1}-1)^{p/2}\left[1-\left(1-\frac{\varepsilon_1(k_{i+1}-1)}{4(1+L_1(k_{i+1}-1))}\right)^{p/2}\right]} \\
&\geqslant \frac{\alpha_p V(k_{i+1}-1)^{p/2}}{\lambda_{\min} P(k_{i+1}-1)^{p/2}\left[1-\left(1-\frac{\varepsilon(k_{i+1}-1)}{4(1+L(k_{i+1}-1))}\right)^{p/2}\right]} \\
&\geqslant \frac{(1+\kappa) V(k_{i+1})^{p/2}}{\lambda_{\min} P(k_{i+1})^{p/2}\left[1-\left(1-\frac{\varepsilon(k_{i+1})}{4(1+L(k_{i+1}))}\right)^{p/2}\right]},
\end{aligned}
$$

where we use (12.75), $\alpha_p > 1+\kappa$, and $V(k_{i+1}-1) > V(k_{i+1})$ in the derivation of the last inequality. We get

$$
\|d(k_{i+1}-1)\|^p \geqslant \|x(k_{i+1}-1)\|^p + (1+\kappa)\nu(V(k_{i+1})). \tag{12.76}
$$

Then (12.74) and (12.76) yield

$$
\sum_{k=k_i}^{k_{i+1}-1} \|x(k)\|^p \leqslant \sum_{k=k_i}^{k_{i+1}-1} \|d(k)\|^p - (1+\kappa)\nu(V(k_{i+1})).
$$

Step 3. For case (*ii*), the following relationship has been established in the proof of Theorem 12.19:

$$
\begin{aligned}
0 &\leqslant x(k+1)'[P(k+1) - P(k))x(k+1)] \\
&\leqslant L(k)(V(k) - V(k+1)),
\end{aligned}
$$

where $L(k) = \frac{\operatorname{trace}(P^*)}{\lambda_{\min} P(k)}$. Therefore,

$$
\begin{aligned}
V(k+1) - V(k) &\leqslant -\frac{\varepsilon_a(k)}{2(1+L(k))} V(k) + \frac{\varepsilon_a(k)\lambda_{\min} P(k)}{\alpha_1(k)(1+L(k))} \|d(k)\|^2 \\
&\leqslant -\frac{\varepsilon_a(k)}{2(1+L(k))} V(k) + \frac{\lambda_{\min} P(k)}{\alpha_1(k)} \|d(k)\|^2,
\end{aligned}
$$

and hence,

$$
V(k+1) \leqslant \left[1 - \frac{\varepsilon_a(k)}{2(1+L(k))}\right] V(k) + \frac{\lambda_{\min} P(k)}{\alpha_1(k)} \|d(k)\|^2.
$$

Since $V(k)$ is decreasing, we have $\lambda_{\min} P(k+1) \geqslant \lambda_{\min} P(k)$, and

$$
\frac{V(k+1)}{\lambda_{\min} P(k+1)} \leqslant \left[1 - \frac{\varepsilon_a(k)}{2(1+L(k))}\right] \frac{V(k)}{\lambda_{\min} P(k)} + \frac{1}{\alpha_1(k)} \|d(k)\|^2.
$$

By definition of $\beta(k)$,

$$
\left(\frac{V(k+1)}{\lambda_{\min} P(k+1)}\right)^{p/2} \leqslant \left[1 - \frac{\varepsilon_a(k)}{4(1+L(k))}\right]^{p/2} \left(\frac{V(k)}{\lambda_{\min} P(k)}\right)^{p/2} + \beta(k) \frac{\|d(k)\|^p}{\alpha_1(k)^{p/2}}.
$$

Using standard comparison principle, we get for $k \geqslant k_i$,

$$\left(\frac{V(k)}{\lambda_{\min}P(k)}\right)^{p/2} \leqslant \prod_{j=k_i}^{k}\left[1-\frac{\varepsilon(j)}{4(1+L(j))}\right]^{p/2}\left(\frac{V(k_i)}{\lambda_{\min}P(k_i)}\right)^{p/2} + \sum_{j=k_i}^{k-1}\left(\prod_{s=j}^{k-1}\left[1-\frac{\varepsilon(s)}{4(1+L(s))}\right]^{p/2}\right)\frac{\beta(j)}{\alpha_1(j)^{p/2}}\|d(j)\|^p.$$

Since $V(k)$ is decreasing, $\left[1-\frac{\varepsilon_a(k)}{4(1+L(k))}\right]^{p/2}$ is decreasing. Hence,

$$\left(\frac{V(k)}{\lambda_{\min}P(k)}\right)^{p/2} \leqslant \left\{\left[1-\frac{\varepsilon_a(k_i)}{4(1+L(k_i))}\right]^{p/2}\right\}^{k-k_i}\left(\frac{V(k_i)}{\lambda_{\min}P(k_i)}\right)^{p/2} + \sum_{j=k_i}^{k-1}\left\{\left[1-\frac{\varepsilon(j)}{4(1+L(j))}\right]^{p/2}\right\}^{k-1-j}\frac{\beta(j)}{\alpha_1(j)^{p/2}}\|d(j)\|^p.$$

We have

$$\sum_{k=k_i}^{k_{i+1}-1}\left(\frac{V(k)}{\lambda_{\min}P(k)}\right)^{p/2} \leqslant \frac{1}{1-\left[1-\frac{\varepsilon_a(k_i)}{4(1+L(k_i))}\right]^{p/2}}\left(\frac{V(k_i)}{\lambda_{\min}P(k_i)}\right)^{p/2} + \sum_{j=k_i}^{k_{i+1}-2}\frac{\beta(j)}{1-\left[1-\frac{\varepsilon(j)}{4(1+L(j))}\right]^{p/2}}\frac{\|d(j)\|^p}{\alpha_1(j)^{p/2}}.$$

By definition, for any $x(k)$,

$$\varepsilon_1(k) \leqslant \varepsilon_a(k) \text{ and } L_1(k) \geqslant L(k),$$

and from (12.49),

$$\alpha_1(j)^{p/2} \geqslant \frac{\beta(j)}{1-\left[1-\frac{\varepsilon_1(j)}{4(1+L_1(j))}\right]^{p/2}} \geqslant \frac{\beta(j)}{1-\left[1-\frac{\varepsilon(j)}{4(1+L(j))}\right]^{p/2}}.$$

We conclude that

$$\sum_{k=k_i}^{k_{i+1}-1}\|x(k)\|^p \leqslant \sum_{k=k_i}^{k_{i+1}-1}\|d(j)\|^p + \nu(V(k_i)).$$

Note that $\nu(V(k_i))$ is increasing. Therefore, $\nu(V(k_i)) \geqslant \nu(V(k_{i+1}))$. We can rewrite the above inequality as

$$\sum_{k=k_i}^{k_{i+1}-1}\|x(k)\|^p \leqslant \sum_{k=k_i}^{k_{i+1}-1}\|d(j)\|^p + (1+\kappa)\nu(V(k_i)) - \kappa\nu(V(k_{i+1})).$$

Step 4. For case (*iii*) and (*iv*), if $x(k) \in \mathcal{V}(c)$, from (12.59) and (12.62), we have

$$V(k+1) - V(k) \leqslant -\tfrac{\varepsilon_a(k)}{1+L(k)} V(k) + a\mu^* \|d(k)\|^2.$$

If $x(k) \in \mathcal{V}(c)^c \cap \mathcal{V}(2c)$ and $V(k+1) - V(k) > 0$, then

$$x(k+1)'[P(k+1) - P(k)]x(k+1) \leqslant 0.$$

We obtain

$$\begin{aligned} V(k+1) - V(k) &\leqslant -\tfrac{\varepsilon_a(k)}{2} V(k) - 2v(k)'[\sigma(u(k)+d(k)) - v(k)] \\ &\leqslant -\tfrac{\varepsilon_a(k)}{2} V(k) + 4\tfrac{\|d(k)\|^2}{\alpha_a(k)} \\ &\leqslant -\tfrac{\varepsilon_a(k)}{2} V(k) + 4\|d(k)\|^2. \end{aligned}$$

If $x(k) \in \mathcal{V}(c)^c \cap \mathcal{V}(2c)$ and $V(k+1) - V(k) \leqslant 0$, then

$$x(k+1)'[P(k+1) - P(k)]x(k+1) \leqslant L(k)(V(k) - V(k+1)).$$

This yields

$$\begin{aligned} V(k+1) - V(k) &\leqslant -\tfrac{\varepsilon_a(k)}{2(1+L(k))} V(k) + 4\tfrac{\|d(k)\|^2}{\alpha_a(k)(1+L(k))} \\ &\leqslant -\tfrac{\varepsilon_a(k)}{2(1+L(k))} V(k) + 4\|d(k)\|^2. \end{aligned}$$

Hence, there exists a $\zeta = \max\{4, a\mu^*\}$ such that for all $x(k) \in \mathcal{V}(2c)$, we have

$$V(k+1) - V(k) \leqslant -\tfrac{\varepsilon_a(k)}{2(1+L(k))} V(k) + \zeta\|d(k)\|^2.$$

Note that our adaptation (12.45) and the fact that $V(x) \leq 2c$ imply that $\varepsilon_a(k) \geqslant \varepsilon_{2c}$ for $k = k_i, \dots, k_{i+1} - 1$, and hence,

$$-\tfrac{\varepsilon_a(k)}{2(1+L(k))} \leqslant -\tfrac{\varepsilon_{2c}}{2(1+L_{2c})}, \qquad \lambda_{\min} P(k) \geqslant \lambda_{\min} P_{2c}.$$

Choose η_{2c} such that

$$\left[1 - \tfrac{\varepsilon_{2c}}{4(1+L_{2c})}\right]^{p/2} \leqslant (1+\eta_{2c})\left[1 - \tfrac{\varepsilon_{2c}}{2(1+L_{2c})}\right]^{p/2} < 1.$$

Applying Lemma 12.18, there exists a β_{2c} independent of d and k such that

$$V(k+1)^{p/2} \leqslant \left[1 - \tfrac{\varepsilon_{2c}}{4(1+L_{2c})}\right]^{p/2} V(k)^{p/2} + \beta_{2c}\zeta^{p/2}\|d(k)\|^p.$$

Using the same comparison principle as used in case (*ii*), we can find a constant γ_1 solely dependent on β_{2c} and ζ such that

$$\sum_{k=k_i}^{k_{i+1}-1} \|x(k)\|^p \leqslant \sum_{k=k_i}^{k_{i+1}-1} \frac{V(k)^{p/2}}{(\lambda_{\min} P_{2c})^{p/2}}$$

$$\leqslant \gamma_1 \sum_{k=k_i}^{k_{i+1}-2} \|d(k)\|^p + \frac{V(k_i)^{p/2}}{(\lambda_{\min} P_{2c})^{p/2}\left[1-\left(1-\frac{\varepsilon_{2c}}{4(1+L_{2c})}\right)^{p/2}\right]}$$

$$\leqslant \gamma_1 \sum_{k=k_i}^{k_{i+1}-2} \|d(k)\|^p + \frac{\kappa V(k_i)^{p/2}}{(\lambda_{\min} P(k_i))^{p/2}\left[1-\left(1-\frac{\varepsilon_a(k_i)}{4(1+L(k_i))}\right)^{p/2}\right]}$$

$$\leqslant \gamma_1 \sum_{k=k_i}^{k_{i+1}-2} \|d(k)\|^p + \kappa\nu(V(k_i)).$$

For case (*iii*) where $k_{i+1} < \infty$, consider specifically $k = k_{i+1} - 1$. Since the states are leaving $\mathcal{V}(2c)$, we have $V(k_{i+1}) - V(k_{i+1} - 1) > 0$. Moreover, we have argued that the increment of $V(k)$ for any $x(k)$ is at most c. This implies that $x(k_{i+1} - 1) \in \mathcal{V}(c)^c \cap \mathcal{V}(2c)$. Following the same argument as used in case (*i*), we have

$$\|d(k_{i+1} - 1)\|^p \geqslant \|x(k_{i+1} - 1)\|^p + (1 + \kappa)\nu(V(k_{i+1})).$$

Finally, we conclude for $k \in \overline{k_i, k_{i+1} - 1}$,

$$\sum_{k=k_i}^{k_{i+1}-1} \|x(k)\|^p \leqslant \gamma_1 \sum_{k=k_i}^{k_{i+1}-1} \|d(k)\|^p + \kappa\nu(V(k_i)) - (1 + \kappa)\nu(V(k_{i+1})).$$

For case (*iv*) where $k_{i+1} = \infty$, we only have

$$\sum_{k=k_i}^{k_{i+1}} \|x(k)\|^p \leqslant \gamma_1 \sum_{k=k_i}^{k_{i+1}} \|d(k)\|^p + \kappa\nu(V(k_i)).$$

Step 5. In summary of previous steps, we find the following results:

- If I_i belongs to case (*i*) ,

$$\sum_{k_i}^{k_{i+1}-1} \|x(k)\|^p \leqslant \sum_{k=k_i}^{k_{i+1}-1} \|d(k)\|^p - (1 + \kappa)\nu(V(k_{i+1})).$$

- If I_i belongs to case (*ii*),

$$\sum_{k=k_i}^{k_{i+1}-1} \|x(k)\|^p \leqslant \sum_{k=k_i}^{k_{i+1}-1} \|d(j)\|^p + (1 + \kappa)\nu(V(k_i)) - \kappa\nu(V(k_{i+1})).$$

- If I_i belongs to case (*iii*),

$$\sum_{k=k_i}^{k_{i+1}-1} \|x(k)\|^p \leqslant \gamma_1 \sum_{k=k_i}^{k_{i+1}-1} \|d(k)\|^p + \kappa\nu(V(k_i)) - (1 + \kappa)\nu(V(k_{i+1})).$$

- If I_i belongs to case (*iv*),

$$\sum_{k=k_i}^{k_{i+1}} \|x(k)\|^p \leqslant \gamma_1 \sum_{k=k_i}^{k_{i+1}} \|d(k)\|^p + \kappa\nu(V(k_i)).$$

Note that if I_i belongs to cases (*i*), (*iii*), and (*iv*), we have either $i = 1$ or I_{i-1} belongs to cases (*i*), (*ii*) or (*iii*). Then the positive term $\kappa\nu(V(k_i))$ of I_i can always be canceled by the corresponding negative term of I_{i-1} for $i > 1$.

Similarly, if I_i belongs to case (*ii*), we have either $i = 1$ or I_{i-1} belongs to case (*i*) or (*iii*). The positive term $(1+\kappa)\nu(V(k_i))$ can also be canceled by the negative term of I_{i-1} for $i > 1$.

In conclusion, we find that for any $x(0)$ and k,

$$\sum_{k=0}^{k} \|x(k)\|^p \leqslant \max\{1, \gamma_1\} \sum_{k=0}^{k} \|d(k)\|^p + (1+\kappa)\nu(V(0)).$$

This completes the proof. ■

12.3.2 *Measurement feedback*

Our primary focus so far has been in constructing state feedback controllers. As usual, once a state feedback controller is constructed, we can construct a measurement feedback controller having an observer-based architecture. By doing so, we solve here two problems both without finite gain, namely, (1) simultaneous global ℓ_p stabilization with fixed initial conditions and without finite gain and global asymptotic stabilization, as formulated in Problem 11.1, and (2) global ℓ_p stabilization with arbitrary initial conditions and without finite gain and bias and global asymptotic stabilization, as formulated in Problem 11.4. As in the continuous time, studied in Sect. 12.2.2, our development here for both of the above problems is essentially the same. As such, we do not explicitly refer to initial conditions in our development. The corresponding finite gain problems are still open as they require complex high-gain observer.

Consider a standard observer for the discrete-time system Σ^d given in (12.1b),

$$\rho\hat{x} = A\hat{x} + B\sigma(u) + K(y - C\hat{x}),$$

where the gain K is such that $A - KC$ is Schur stable. Also, consider a dynamic system,

$$\rho\omega = (A + BF)\omega + K(y - C\hat{x}),$$

where the gain F is such that $A + BF$ is Schur stable. Next, let P_ε be the solution of the DARE,

$$P_\varepsilon = A'P_\varepsilon A - AP_\varepsilon B(I + B'P_\varepsilon B)^{-1}B'P_\varepsilon A + Q_\varepsilon, \tag{12.77}$$

where for $\varepsilon \in [0, 1]$,

$$Q_\varepsilon > 0, \quad \lim_{\varepsilon \to 0} Q_\varepsilon = 0, \quad \tfrac{\mathrm{d}Q_\varepsilon}{\mathrm{d}\varepsilon} > 0.$$

Let $z = \widehat{x} - \omega$. Choose $u = -(I + B'P_{\varepsilon_a(z)}B)^{-1}B'P_{\varepsilon_a(z)}Az + F\omega$ where $\varepsilon_a(z)$ is determined by

$$\varepsilon_a(z) = \max\{r \in (0, 1] \mid z'P_r z \operatorname{trace} B'P_r B\| \leqslant \tfrac{\Delta^2}{8}\} \tag{12.78}$$

and $P_{\varepsilon_a(z)}$ is the solution of the DARE (12.77) with ε replaced by $\varepsilon_a(z)$.

The above discussion leads to the dynamic observer-based controller as

$$\begin{cases} \rho\widehat{x} = A\widehat{x} + B\sigma(u) + K(y - C\widehat{x}) \\ \rho\omega = (A + BF)\omega + K(y - C\widehat{x}) \\ u \;= -(I + B'P_{\varepsilon_a(z)}B)^{-1}B'P_{\varepsilon_a(z)}Az + F\omega, \end{cases} \tag{12.79}$$

where K and F are such that $A-KC$ and $A+BF$ are Schur stable, and $z = \widehat{x}-\omega$.

We have the following theorem:

Theorem 12.25 *Consider the discrete-time system Σ^d given in* (12.1b)*. Let Assumptions 12.1, 12.2, and 12.3 be valid, also let σ be the standard saturation function as in Definition 2.19. Then, the nonlinear dynamic measurement feedback controller* (12.79) *solves the following two problems:*

(i) *Simultaneous global ℓ_p stabilization with fixed initial conditions and without finite gain and global asymptotic stabilization, as defined by Problem 11.1 (the (G_p/G) problem)*

(ii) *Simultaneous global ℓ_p stabilization with arbitrary initial conditions and without finite gain and bias and global asymptotic stabilization, as defined by Problem 11.4.*

Proof : Define $e = x - \widehat{x}$. The closed-loop system in terms of e, z, and ω is given by

$$\begin{cases} \rho e = (A - KC)e + B\big[\sigma(-(I + B'P_{\varepsilon_a(z)}B)^{-1}B'P_{\varepsilon_a(z)}Az + F\omega + d) \\ \qquad\qquad -\sigma(-(I + B'P_{\varepsilon_a(z)}B)^{-1}B'P_{\varepsilon_a(z)}Az + F\omega)\big] \\ \rho z = Az + B\sigma(-(I + B'P_{\varepsilon_a(z)}B)^{-1}B'P_{\varepsilon_a(z)}Az + F\omega) - BF\omega \\ \rho\omega = (A + BF)\omega + KCe. \end{cases}$$

In the absence of d, the above becomes

$$\begin{cases} \rho e \;= (A - KC)e \\ \rho z \;= Az + B\sigma(-(I + B'P_{\varepsilon(z)}B)^{-1}B'P_{\varepsilon(z)}Az + F\omega) - BF\omega \\ \rho\omega = (A + BF)\omega + KCe. \end{cases}$$

Clearly, the origin is locally exponentially stable. To see global attractivity, we note that $e \to 0$ and $\omega \to 0$ as time tends to infinity due to the fact that both $A+BF$ and $A-KC$ are Schur stable. Then, there exists a N such that $\|F\omega(k)\| \leqslant \frac{\Delta}{2}$ for all $k > N$. This and the adaptation law (12.78) together imply that the saturation will be inactive for all $k > N$ (see also (4.244) and the inequalities below it). Since σ is a standard saturation function, the z dynamics then becomes

$$\rho z = Az - B(I + B'P_{\varepsilon_a(z)}B)^{-1}B'P_{\varepsilon_a(z)}Az$$

which is globally attractive. This concludes global asymptotic stability.

In the presence of d, for a standard saturation function, we have

$$\|\sigma(u+d) - \sigma(u)\| \leqslant \|d\|.$$

Therefore, $d \in \ell_p$ implies that $(\sigma(u+d) - \sigma(u)) \in \ell_p$. Since $A - KC$ and $A + BF$ are both Schur stable, it follows that $e \in \ell_p$, $\omega \in \ell_p$ and $\omega \to 0$. Then, as before, there exists a N such that $\|F\omega(k)\| \leqslant \frac{\Delta}{2}$ for all $k > N$. Therefore, as before, we can conclude that the saturation is inactive for all $k > N$, and hence, z dynamics becomes

$$\rho z = Az - B(I + B'P_{\varepsilon_a(z)}B)^{-1}B'P_{\varepsilon_a(z)}Az.$$

This system is known to be globally asymptotically stable and locally exponentially stable. Hence, $z \in \ell_p$ and $x = e + \hat{x} = (e + z + \omega) \in \ell_p$. This completes the proof. ∎

Remark 12.26 *As Theorem 12.25 clearly indicates, a $2n$-dimensional dynamic adaptive-low-gain feedback controller can solve simultaneous global ℓ_p stabilization without finite gain and global asymptotic stabilization irrespective of the nature of initial conditions. As in the continuous-time case, this is remarkable especially when we note that to solve the same problems by a static state feedback controller, one needs an adaptive-low-and-high-gain feedback controller.*

A similar result as in Theorem 12.25, however, with finite gain is challenging and is an open research problem as in the case of continuous-time systems.

12.4 ISS stabilization with state feedback: continuous time

In a global framework, a recent notion, for a combined notion of external and internal stability, is global input-to-state stability (ISS) defined earlier in Sects. 2.8 and (11.2).

The following theorem states that the same controller defined in Theorem 12.11 achieves ISS stability:

Theorem 12.27 *Consider the system* Σ^c *of* (12.1a) *while using state feedback. Also, consider the controller* Σ_{con} *defined in Theorem 12.11. Then, under Assumptions 12.1 and 12.2, this controller* Σ_{con} *also solves the global ISS stabilization problem.*

Proof : Regarding ISS, Theorem 12.5 has shown that if x_0 satisfies

$$V(x_0) \leqslant \tfrac{2}{\alpha_o}\lambda_{\min}(P_{\varepsilon_a})\|d\|_\infty,$$

we have

$$V(x(t)) \leqslant \tfrac{2}{\alpha_o}\lambda_{\min}(P_{\varepsilon_a})\|d\|_\infty \tag{12.80}$$

for all $t > 0$.

We next consider the case when the initial condition is such that

$$V(x_0) \geqslant \tfrac{2}{\alpha_o}\lambda_{\min}(P_{\varepsilon_a})\}\|d\|_\infty. \tag{12.81}$$

We note that

$$V(x) \geqslant \tfrac{2}{\alpha_o}\lambda_{\min}(P_{\varepsilon_a})\}\|d\|_\infty \tag{12.82}$$

implies $\dot{V} < 0$; we find that if x_0 is such that (12.81) is satisfied, then $V(x(t)) \leqslant V(x_0)$ for all $t > 0$.

Since $\dot{V}(x) < 0$ for any x such that (12.82) is satisfied, there exists an $\varepsilon^*(x_0)$ such that $\varepsilon_a \geqslant \varepsilon^*(x_0)$ for all $t > 0$. Next, we define γ by

$$\gamma(c) = \inf_{\|\xi\|\leqslant c} \inf_{\varepsilon_a\in[\varepsilon^*(\xi),1]} \tfrac{\lambda_{\min} Q_{\varepsilon_a}}{\lambda_{\max} P_{\varepsilon_a}}.$$

We know γ is nonincreasing. Then

$$\begin{aligned}\dot{V} &\leqslant \tfrac{\lambda_{\min} Q_{\varepsilon_a}}{\lambda_{\max} P_{\varepsilon_a}}(-V + \tfrac{2}{\alpha_o}\lambda_{\min}(P_{\varepsilon_a})\|d\|_\infty^2) + x'\tfrac{P_{\varepsilon_a}}{\mathrm{d}t}x \\ &\leqslant \gamma(\|x_0\|)(-V + \tfrac{2}{\alpha_o}\lambda_{\min}(P_{\varepsilon_a})\|d\|_\infty^2) + x'\tfrac{P_{\varepsilon_a}}{\mathrm{d}t}x.\end{aligned}$$

Since $V(x) \leqslant V(x_0)$ for all $t > 0$, ε_a is bounded away from zero. We have proved that there exists a constant $k(\varepsilon^*(x_0)) > 0$ such that

$$|x'\tfrac{\mathrm{d}P_{\varepsilon_a}}{\mathrm{d}t}x| \leqslant k(\varepsilon^*(x_0))|\dot{V}|.$$

Moreover, $x'\frac{\mathrm{d}P_{\varepsilon_a}}{\mathrm{d}t}x$ and $\dot{V}$ cannot have the same signs. Define a class $\mathcal{K}$ function K as

$$K(c) = \sup_{\|\xi\|\leqslant c} k(\varepsilon^*(\xi)).$$

Therefore, we have

$$\dot{V} \leqslant -\tfrac{\gamma(\|x_0\|)}{1+K(\|x_0\|)}(V - \tfrac{2}{\alpha_o}\lambda_{\min}(P_{\varepsilon_a})\|d\|^2).$$

It follows from the standard comparison principle that

$$V(x(t)) \leqslant V(x_0)e^{-\frac{\gamma(\|x_0\|)}{1+K(\|x_0\|)}t} + \frac{2}{\alpha_o}\lambda_{\min}(P_{\varepsilon_a}) \sup_{\tau\in[0,t]} \|d(\tau)\|^2. \tag{12.83}$$

From (12.80) and (12.80), we conclude that for any $x_0 \in \mathbb{R}^n$,

$$V(x(t)) \leqslant V(x_0)e^{-\frac{\gamma(\|x_0\|)}{1+K(\|x_0\|)}t} + \frac{2}{\alpha_o}\lambda_{\min}(P_{\varepsilon_a})\|d\|_\infty^2, \tag{12.84}$$

that is,

$$\|x(t)\|^2 \leqslant \frac{V(x_0)}{\lambda_{\min}P_{\varepsilon a}}e^{-\frac{\gamma(\|x_0\|)}{1+K(\|x_0\|)}t} + \frac{2}{\alpha_o}\|d\|_\infty^2.$$

Note that γ is nonincreasing and K is a class $\mathcal{K}$ function. We can define a class $\mathcal{KL}$ function as

$$\beta(\|x_0\|,t) = \sqrt{\frac{\sup_{\|x\|\leqslant\|x_0\|} V(x)}{\inf_{\|x\|\leqslant\|x_0\|} \lambda_{\min}(P_{\varepsilon^*(x)})}}e^{-\frac{\gamma(\|x_0\|)}{2(1+K(\|x_0\|))}t}.$$

We have that when trajectories are such that

$$\{x \mid V(x) \geqslant \frac{2}{\alpha_o}\lambda_{\min}(P_{\varepsilon_a})\|d\|_\infty^2\},$$

then

$$\|x(t)\| \leqslant \beta(\|x_0\|,t) + \sqrt{\frac{2}{\alpha_o}\|d\|_\infty^2}.$$

However, when trajectories enter and remain in

$$\{x \mid V(x) \leqslant \frac{2}{\alpha_o}\lambda_{\min}(P_{\varepsilon_a})\|d\|_\infty\},$$

we obviously have

$$\|x(t)\| \leqslant \sqrt{\frac{2}{\alpha_o}\|d\|_\infty^2}.$$

We conclude that for any initial condition $x_0 \in \mathbb{R}^n$, we have

$$\|x(t)\| \leqslant \beta(\|x_0\|,t) + \sqrt{\frac{2}{\alpha_o}}\|d\|_\infty. \tag{12.85}$$

This completes the proof. ■

Remark 12.28 *As we discussed in an earlier chapter in Sect. 2.8, the notion of ISS makes an attempt to marry both the notions of internal stability and the L_∞ stability or ℓ_∞ stability. In fact, as pointed out in Remark 2.68, when the input d is identically zero, the ISS implies the global asymptotic stability of the zero equilibrium point. In this sense, ISS is indeed a simultaneous stabilization concept.*

12.5 ISS stabilization with state feedback: discrete time

The following theorem states that the same controller defined in Theorem 12.23 achieves ISS stability:

Theorem 12.29 *Consider the system* Σ^d *of* (12.1b) *while using state feedback. Let Assumption 12.14 hold. Also, consider the same controller used in Theorem 12.21, namely,*

$$u = -(1+\alpha_a(x))(I + B'P_{\varepsilon_a(x)}B)^{-1}B'P_{\varepsilon_a(x)}Ax,$$

where $P_{\varepsilon_a(x)}$ *is the solution of* (12.40) *and where*

$$\alpha_a(x) = \begin{cases} \alpha_0(x) = \frac{1}{\|B'P_{\varepsilon_a(x)}B\|}, & x'P_{\varepsilon_a}(x)x \leqslant c \\ \frac{8\alpha_1(x)}{\varepsilon_a(x)\lambda_{\min}P_{\varepsilon_a(x)}}, & \text{otherwise} \end{cases} \tag{12.86}$$

with

$$\alpha_1(x) = \frac{\lambda_{\max}P_{\varepsilon_a(x)}}{\lambda_{\min}P_{\varepsilon_1(x)}}.$$

Here, c, $\varepsilon_1(x)$ *are given by*

$$c = \Delta^2 \max\{4M_{\varepsilon^*}, 4(1 + \|B'P^*B\|)\},$$
$$\varepsilon_1(x) = \max\{r \in (0,\varepsilon^*] \mid x'P_rx \cdot \operatorname{trace} P_r \leqslant \tfrac{\Delta^2}{2\|BB'\|}\},$$

where P^* *is the solution of* (12.40) *when* ε *takes the value* ε^*. *Then, this controller solves the global ISS stabilization problem.*

We have already proven in Theorem 12.21 that whenever $V(k+1) - V(k) \geqslant 0$ or equivalently,

$$\frac{V(k+1)}{\lambda_{\min}P(k+1)} - \frac{V(k)}{\lambda_{\min}P(k)} \geqslant 0,$$

we have

$$\frac{V(k+1)}{\lambda_{\min}P(k+1)} \leqslant \max\{2, \frac{aM^* + a\mu^*}{\lambda_1}\}\|d\|_\infty^2.$$

The definitions of a and M^* are the same as in Theorems 12.19 and 12.23. Also, λ_1 is such that $\lambda_{\min}P \geqslant \lambda_1$ for $x \in \mathcal{V}(c)$ where the set $\mathcal{V}(c)$ is also the same as in these theorems.

This implies that

$$\|x(k)\|^2 \leqslant \frac{V(k)}{\lambda_{\min} P(k)} \leqslant \max\left\{\frac{V(0)}{\lambda_{\min} P(0)}, \max\{2, \frac{aM_{\varepsilon^*}+a\mu^*}{\lambda_1}\}\|d\|_\infty^2\right\}, \quad \forall k \geqslant 0. \tag{12.87}$$

If we can show that when

$$\frac{V(k)}{\lambda_{\min} P(k)} > 2\max\{2, \frac{aM_{\varepsilon^*}+a\mu^*}{\lambda_1}\}\|d\|_\infty^2,$$

$\frac{V(k)}{\lambda_{\min} P(k)}$ can be bounded by a class $\mathcal{KL}$ function $\beta(\|x(0)\|, k)$, then we shall have

$$\|x(k)\|^2 \leqslant \frac{V(k)}{\lambda_{\min} P(k)} \leqslant \beta(\|x(0)\|, k) + 2\max\{2, \frac{aM_{\varepsilon^*}+a\mu^*}{\lambda_1}\}\|d\|_\infty^2.$$

Note that if $\frac{V(0)}{\lambda_{\min} P(0)} \leqslant 2\max\{2, \frac{aM_{\varepsilon^*}+a\mu^*}{\lambda_1}\}\|d\|_\infty^2$, (12.87) implies that

$$\|x(k)\|^2 \leqslant \frac{V(k)}{\lambda_{\min} P(k)} \leqslant 2\max\{2, \frac{aM_{\varepsilon^*}+a\mu^*}{\lambda_1}\}\|d\|_\infty^2, \quad \forall k \geqslant 0. \tag{12.88}$$

In what follows, we only consider the case

$$\frac{V(0)}{\lambda_{\min} P(0)} > 2\max\{2, \frac{aM_{\varepsilon^*}+a\mu^*}{\lambda_1}\}\|d\|_\infty^2.$$

In this case, we have

$$\|x(k)\|^2 \leqslant \frac{V(k)}{\lambda_{\min} P(k)} \leqslant \frac{V(0)}{\lambda_{\min} P(0)}$$

for any $k \geqslant 0$. This implies that $\varepsilon_a(k) \geqslant \varepsilon(0)$ for all $k \geqslant 0$. We also have

$$V(k+1) - V(k) \leqslant 0$$

until

$$\frac{V(k)}{\lambda_{\min} P(k)} \leqslant \max\{2, \frac{aM_{\varepsilon^*}+a\mu^*}{\lambda_1}\}\|d\|_\infty^2.$$

Then we have two cases:

1. Let $x(k) \in \mathcal{V}(c)^c$. From the proof of Theorem 12.21, we have

$$\begin{aligned}
&V(k+1) - V(k)\\
&\leqslant -\frac{\varepsilon_a(k)}{2} V(k) + 4\frac{\|d(k)\|^2}{\alpha_a(k)} + x(k+1)'[P(k+1) - P(k)]x(k+1)\\
&\leqslant -\frac{\varepsilon_a(k)}{4} V(k) - \frac{\varepsilon_a(k)\lambda_{\min} P(k)}{4}\left(\frac{V(k)}{\lambda_{\min} P(k)} - 2\frac{\|d(k)\|^2}{\alpha_1(k)}\right)\\
&\qquad + x(k+1)'[P(k+1) - P(k)]x(k+1).
\end{aligned}$$

Note that $\alpha_1(k) > 1$ for all k. When

$$\frac{V(k)}{\lambda_{\min} P(k)} \geqslant 2\max\{2, \frac{aM_{\varepsilon^*}+a\mu^*}{\lambda_1}\}\|d\|_\infty,$$

we have

$$\frac{V(k)}{\lambda_{\min} P(k)} - 2\frac{\|d(k)\|^2}{\alpha_1(k)} \geqslant 0,$$

and

$$V(k+1) - V(k) \leqslant -\tfrac{\varepsilon_a(k)}{4} V(k) + x(k+1)'[P(k+1) - P(k)]x(k+1).$$

We have also shown in the proof of Theorem 12.19 that $x(k+1)'[P(k+1) - P(k)]x(k+1)$ and $V(k+1) - V(k)$ cannot have the same signs and that when $V(k+1) - V(k) \leqslant 0$,

$$|x(k+1)'[P(k+1) - P(k)]x(k+1)| \leqslant L(k) \cdot |V(k+1) - V(k)|,$$

where $L(k) = \frac{\text{trace}(P^*)}{\lambda_{\min}(P(k))}$. Hence,

$$V(k+1) - V(k) \leqslant \tfrac{-\varepsilon_a(k)}{4(1+L(k))} V(k).$$

Given $\varepsilon_a(k) \in [\varepsilon(0), \varepsilon^*]$ for all $k \geqslant 0$, we get

$$V(k+1) - V(k) \leqslant -\tfrac{\varepsilon(0)}{4(1+L(0))} V(k), \tag{12.89}$$

where $L(0) = \frac{\text{trace}(P^*)}{\lambda_{\min} P(0)}$. Also, $\varepsilon(0) < 1$ implies that $\varepsilon(0)/(1 + L(0)) < 1$.

2. Let $x(k) \in \mathcal{V}(c)$. In that case, $\alpha_a = \alpha_0$, and we have proven in Theorem 12.19 that

$$\begin{aligned} V(k+1) - V(k) & \\ &\leqslant -\varepsilon_a(k)V(k) + a\mu(k)\|d(k)\|^2 + x(k+1)'[P(k+1) - P(k)]x(k+1) \\ &\leqslant -\tfrac{\varepsilon_a(k)}{2} V(k) - \tfrac{\varepsilon_a(k)}{2}\left(V(k) - a\tfrac{2\mu(k)}{\varepsilon_a(k)}\|d\|_\infty^2\right) \\ &\qquad + x(k+1)'[P(k+1) - P(k)]x(k+1). \end{aligned}$$

Note that for $x(k) \in \mathcal{V}(c)$, $\lambda_{\min} P(k) \geqslant \lambda_1$, and hence, $\frac{V(k)}{\lambda_{\min} P(k)} \geqslant \frac{aM^*}{\lambda_1}\|d\|_\infty^2$ implies that

$$V(k) \geqslant aM^*\|d\|_\infty^2 \geqslant a\tfrac{\mu(k)}{\varepsilon_a(k)}\|d\|_\infty^2.$$

The last inequality is due to Lemma 12.15. Therefore, we get

$$V(k+1) - V(k) \leqslant -\tfrac{\varepsilon_a(k)}{2} V(k) + x(k+1)'[P(k+1) - P(k)]x(k+1),$$

and since $\varepsilon_a(k) \geqslant \varepsilon(0)$ for all $k \geqslant 0$, similar to the previous case, we get

$$V(k+1) - V(k) \leqslant -\tfrac{\varepsilon(0)}{2(1+L(0))} V(k). \tag{12.90}$$

Combining (12.89) and (12.90), if $\frac{V(0)}{\lambda_{\min} P(0)} \geqslant 2 \max\{2, \frac{aM_{\varepsilon *}+a\mu^*}{\lambda_1}\}\|d\|_\infty^2$, we have

$$V(k) \leqslant (1 - \tfrac{\varepsilon(0)}{4(1+L(0))})^k V(0)$$

until

$$\frac{V(k)}{\lambda_{\min} P(0)} \leqslant 2 \max\{2, \frac{aM_{\varepsilon *}+a\mu^*}{\lambda_1}\}\|d\|_\infty^2.$$

Note that $\varepsilon_a(k) \geqslant \bar{\varepsilon}$ for all $k \geqslant 0$ in this case. Hence,

$$\|x(k)\| \leqslant (1 - \tfrac{\varepsilon(0)}{4(1+L(0))})^k \frac{V(0)}{\lambda_{\min} P(0)}. \tag{12.91}$$

Let us summarize the results up to this point.

(*i*) When $\frac{V(0)}{\lambda_{\min} P(0)} \leqslant 2 \max\{2, \frac{aM_{\varepsilon *}+a\mu^*}{\lambda_1}\}\|d\|_\infty^2$, we have (12.88).

(*ii*) When $\frac{V(0)}{\lambda_{\min} P(0)} \geqslant 2 \max\{2, \frac{aM_{\varepsilon *}+a\mu^*}{\lambda_1}\}\|d\|_\infty^2$, we have (12.91).

Therefore, we have

$$\|x(k)\|^2 \leqslant (1 - \tfrac{\varepsilon(0)}{4(1+L)})^k \frac{V(0)}{\lambda_{\min} P(0)} + 2 \max\{2, \frac{aM_{\varepsilon *}+a\mu^*}{\lambda_1}\}\|d\|_\infty^2.$$

This concludes ISS of the closed-loop system.

12.6 Achieving (G_p/G) and $(G_p/G)_{fg}$ with a linear control law

So far, we considered general linear systems subject to actuator saturation which are *asymptotically null controllable with bounded control* (ANCBC), that is, we considered linear systems which, in the absence of saturation, have all their open-loop poles in the closed left-half plane besides being stabilizable. Global internal stabilization of such systems in general requires nonlinear feedback control laws. Consequently, simultaneous external and internal stabilization of such systems requires necessarily nonlinear feedback control laws as well. This brings forth an important question:

For what class of linear systems can a saturated linear feedback control law achieve simultaneous global internal stability as well as external stability?

12.6.1 $(G_p/G)_{fg}$ problem with state feedback for neutrally stable systems: continuous time

We answer next the above question for continuous-time neutrally stable systems. In particular, we show here that, for neutrally stable linear systems, a saturated linear feedback control law can achieve simultaneous global internal stability as well as external stability. Moreover, this can be done for any fixed initial conditions including the origin.

If the given system Σ^c in (12.1a) is open-loop neutrally stable and (A, B) is stabilizable, then in a suitable basis, we have

$$A = \begin{pmatrix} A_c & 0 \\ 0 & A_s \end{pmatrix}, \qquad B = \begin{pmatrix} B_c \\ B_s \end{pmatrix} \tag{12.92}$$

where A_c satisfies $A_c + A_c' = 0$, A_s is asymptotically stable, and (A_c, B_c) is controllable. For the moment. without much loss of generality, we ignore the asymptotically stable subsystem and assume that the system satisfies the condition $A + A' = 0$ and that (A, B) is controllable.

We first need to recall a result of [154]. Consider a nonlinear system,

$$\dot{x} = f(x, u), \tag{12.93}$$

where $x \in \mathbb{R}^n$ and $u \in \mathbb{R}^m$. The following definitions and results are extracted from [154]:

Definition 12.30 *A continuous function $V : \mathbb{R}^n \to \mathbb{R}_{\geqslant 0}$ is called an ISS Lyapunov function for system* (12.93) *if the following hold:*

(i) There exist a class $\mathcal{K}_\infty$ function, α_1, and α_2 such that

$$\alpha_1(\|\xi\|) \leqslant V(\xi) \leqslant \alpha_2(\|\xi\|), \quad \forall \xi \in \mathbb{R}^n.$$

(ii) There exist a class $\mathcal{K}_\infty$ function, α_3, and $\widetilde{\sigma}$ such that

$$\dot{V}(\xi, u) \leqslant -\alpha_3(\|\xi\|) + \widetilde{\sigma}(\|\mu\|)$$

for all $\xi \in \mathbb{R}^n$ and $\mu \in \mathbb{R}^m$.

Remark 12.31 *Note that (i) is equivalent to the condition that $V(0) = 0$ and that V is a radially unbounded, positive definite function.*

Lemma 12.32 *System* (12.93) *is ISS if it admits a smooth ISS Lyapunov function.*

We have the following theorem whose proof is adapted from [151] and [89]:

Theorem 12.33 *Consider the system,*

$$\dot{x} = Ax + B\sigma(u + d), \qquad x(0) = x_0, \tag{12.94}$$

where (A, B) *is controllable and* $A + A' = 0$*. Also, consider the control law* $u = -\kappa B'x$*. Then, the following hold:*

(i) *For a given* $p \in [1, \infty]$ *and for* $\kappa > 0$*, the closed-loop system comprising of* (12.94) *and the control law* $u = -\kappa B'x$ *is* L_p *stable with arbitrary initial conditions with finite gain and with bias, as defined in Definition* 2.63.

(ii) *For* $\kappa > 0$*, the closed-loop system comprising of* (12.94) *and the control law* $u = -\kappa B'x$ *is input-to-state stable (ISS) as defined in Definition* 2.65.

Proof : Proof of global internal stability follows from Sect. 4.6.1. Choose $V_1 = \frac{\|x\|^{p+1}}{p+1}$. Differentiating along the trajectories yields

$$\begin{aligned}\dot{V}_1 &= -\|x\|^{p-1}x'B\sigma(\kappa B'x - d)\\ &\leqslant -\|x\|^{p-1}(B'x - \tfrac{d}{\kappa})'\sigma(\kappa B'x - d) + \|x\|^{p-1}\tfrac{\|d\|}{\kappa}.\end{aligned}$$

Let P be the solution of Lyapunov equation

$$P(A - \kappa BB') + (A - \kappa BB')'P = -I.$$

Define $V_2 = \frac{1}{p}(x'Px)^{p/2}$. There exist α and β such that

$$\begin{aligned}\dot{V}_2 &\leqslant (x'Px)^{(p-2)/2}x'P[(A - \kappa BB')x + B(\kappa B'x - \sigma(\kappa B'x + d))]\\ &\leqslant -\alpha\|x\|^p + \beta\|x\|^{p-1}\|\kappa B'x - d - \sigma(\kappa B'x - d)\| + \beta\|x\|^{p-1}\|d\|\\ &\leqslant -\alpha\|x\|^p + \beta\|x\|^{p-1}(\kappa B'x - d)'\sigma(\kappa B'x - d) + \beta\|x\|^{p-1}\|d\|.\end{aligned}$$

Define $V = \beta\kappa V_1 + V_2$. We get

$$\dot{V} \leqslant -\alpha\|x\|^p + 2\beta\|x\|^{p-1}\|d\|.$$

From Young's inequality, there exists a ν_p such that

$$2\beta\|x\|^{p-1}\|d\| \leqslant \tfrac{\alpha}{2}\|x\|^p + \nu_p\|d\|^p.$$

Therefore, we have

$$\dot{V} \leqslant -\tfrac{\alpha}{2}\|x\|^p + \nu_p\|d\|^p. \tag{12.95}$$

This implies that

$$\begin{aligned}\|x\|_{L_p}^p &\leqslant \tfrac{2}{\alpha}(V(0) - V(\infty)) + \tfrac{2\nu_p}{\alpha}\|d\|_{L_p}^p\\ &\leqslant \tfrac{2}{\alpha}V(0) + \tfrac{2\nu_p}{\alpha}\|d\|_{L_p}^p,\end{aligned}$$

that is,

$$\|x\|_{L_p} \leqslant (\tfrac{2}{\alpha}V(0))^{1/p} + (\tfrac{2\nu_p}{\alpha})^{1/p}\|d\|_{L_p}.$$

Define a class $\mathcal{K}$ function b_p as

$$b_p(c) = \sup_{\|x_0\| \leqslant c} (\tfrac{2}{\alpha}V(x_0))^{1/p},$$

and a γ_p as

$$\gamma_p = (\tfrac{2\nu_p}{\alpha})^{1/p}.$$

Finally, we conclude that

$$\|x\|_{L_p} \leqslant b_p(\|x_0\|) + \gamma_p\|d\|_{L_p}.$$

Note that this L_p gain γ_p is independent of controller gain κ.

Note that (10.58) for $p = 1$ immediately yields

$$\dot{V}(t) \leqslant -\frac{\alpha}{2}V^{1/2}(t) + \nu_1\|d(t)\|.$$

Hence, if

$$V^{1/2}(0) \leqslant \frac{4\nu_1}{\alpha}\|d\|_\infty,$$

then

$$V^{1/2}(t) \leqslant \frac{4\nu_1}{\alpha}\|d\|_\infty$$

for all $t > 0$. On the other hand, if

$$V^{1/2}(t) \geqslant \frac{4\nu_1}{\alpha}\|d\|_\infty,$$

then we have

$$\dot{V} \leqslant -\frac{\alpha}{4}V^{1/2},$$

and therefore, there exists a class $\mathcal{KL}$ function $\widetilde{\beta}$ such that

$$V(t) \leqslant \widetilde{\beta}(V(0), t).$$

This implies that there exists a class $\mathcal{KL}$ function β such that

$$\|x(t)\| \leqslant \beta(\|x_0\|, t) + \gamma_p\|d\|_\infty, \tag{12.96}$$

which, by definition, guarantees ISS. ■

12.6.2 $(G_p/G)_{fg}$ problem with state feedback for neutrally stable systems: discrete time

We consider now discrete-time neutrally stable systems and show that a saturated linear feedback control law can achieve simultaneous global internal stability as well as external stability. Moreover, this can be done for any fixed initial conditions including the origin.

Without much loss of generality, we ignore the asymptotically stable poles of the given discrete-time neutral system. That is, we assume that the given discrete-time neutral system is of the form,

$$x(k+1) = Ax(k) + B\sigma(u(k) + d(k)), \qquad x(0) = x_0, \tag{12.97}$$

where (A, B) is controllable and $A'A = I$.

We first recall certain results from [55]. Consider a discrete-time non-linear system,

$$x(k+1) = f(x(k), u(k)), \tag{12.98}$$

where $x \in \mathbb{R}^n$ and $u \in \mathbb{R}^m$. The following definitions and results are extracted from [55]:

Definition 12.34 *A continuous function $V : \mathbb{R}^n \to \mathbb{R}_{\geqslant 0}$ is called an ISS Lyapunov function for system* (12.98) *if the following holds:*

(i) There exist a class $\mathcal{K}_\infty$ function, $\tilde{\alpha}_1$, and $\tilde{\alpha}_2$ such that

$$\tilde{\alpha}_1(\|\xi\|) \leqslant V(\xi) \leqslant \tilde{\alpha}_2(\|\xi\|), \quad \forall \xi \in \mathbb{R}^n.$$

(ii) There exist a $\mathcal{K}_\infty$ function, $\tilde{\alpha}_3$, and $\mathcal{K}$ function $\tilde{\sigma}$ such that

$$V(f(\xi, \mu)) - V(\xi) \leqslant -\tilde{\alpha}_3(\|\xi\|) + \tilde{\sigma}(\|\mu\|)$$

for all $\xi \in \mathbb{R}^n$ and $\mu \in \mathbb{R}^m$.

Lemma 12.35 *System* (12.98) *is ISS if it admits a continuous ISS Lyapunov function.*

We have the following result:

Theorem 12.36 *Consider the discrete-time neutral system of the form,* (12.97) *system, where (A, B) is controllable, $A'A = I$, and $\sigma(\cdot)$ is a standard saturation* *function. Also, consider the control law $u = -\kappa B'Ax$ where $\kappa > 0$ is a design parameter. Then, the following hold:*

(i) *For a given $p \in [1, \infty]$, there exists a $\kappa^* > 0$ such that for $\kappa \in (0, \kappa^*]$, the closed-loop system comprising of (12.97) and control law $u = -\kappa B'Ax$ is ℓ_p stable with arbitrary initial conditions with finite gain and with bias, as defined in Definition 2.63.*

(ii) *There exists a $\kappa^* > 0$ such that for $\kappa \in (0, \kappa^*]$, the closed-loop system comprising of (12.97) and the control law $u = -\kappa B'Ax$ is input-to-state stable (ISS), as defined in Definition 2.65.*

Proof : Proof of global internal stability follows from Sect. 4.6.1. Item (*i*) is proven in [5] and [23], respectively, for $p \in (1, \infty]$ and $p = 1$.

Item (*ii*) follows from [5] and [55]. The proof can be sketched as follows. Let $P(\kappa)$ be the positive definite solution of Lyapunov equation

$$\bar{A}(\kappa)'P(\kappa)\bar{A}(\kappa) - P(\kappa) = -I,$$

where $\bar{A}(\kappa) = A - \kappa BB'A$. It is proven in [5] that there exist $\kappa^* > 0$ and functions $\theta, \alpha, \beta : \mathbb{R}^+ \to \mathbb{R}^+$ such that $V(x) = x'P(\kappa)x + \theta(\kappa)\|x\|^3$ satisfies

$$V(x(k+1)) - V(x(k)) \leqslant -\alpha(\kappa)\|x\|^2 + \beta(\kappa)\|u\|^2, \quad \forall \kappa \in (0, \kappa^*].$$

According to Definition 12.34, $V(x)$ is an ISS Lyapunov function. The conclusion of ISS stability of closed-loop system then follows from Lemma 12.35 straightforwardly. ∎

12.7 Simultaneous stabilization in a semi-global framework: continuous time

In this section, we consider simultaneous stabilization in semi-global framework for continuous-time systems. Our interest is to solve $(SG_{p,q}/SG)$ problem and $(SG_{p,q}/SG)_{fg}$ problem, as defined respectively in Problems 11.11 and 11.12. We consider both state and measurement feedback, each in one subsection.

12.7.1 State feedback

We consider first state feedback controllers and show that, under Assumptions 12.1 and 12.2, the $(SG_{p,q}/SG)_{fg}$ problem is solvable via linear state feedback laws. Clearly, the $(SG_{p,q}/SG)$ problem is solvable whenever the $(SG_{p,q}/SG)_{fg}$ problem is solvable.

Theorem 12.37 *Consider the system* Σ^c *of* (12.1a). *For this system, under Assumptions 12.1 and 12.2, the* $(SG_{p,q}/SG)_{fg}$ *problem is solvable for any* $p, q \in [1,\infty]$. *Moreover, low-and-high-gain design technique* (4.193) *can yield a control law that solves the problem. More specifically, for any* $p, q \in [1,\infty]$, *any a priori given (arbitrarily large) bounded set* $\mathcal{X} \subset \mathbb{R}^n$, *and any* $D > 0$, *there exists an* ε^*, *and, for any* $\varepsilon \in (0, \varepsilon^*]$, *there exists a* α_q^* *such that the low-and-high-gain feedback,*

$$u = -(1+\alpha)B'P_\varepsilon x, \quad \varepsilon \in (0,\varepsilon^*], \ \alpha \geqslant \alpha_q^*, \tag{12.99}$$

has the following properties:

(*i*) *In the presence of a disturbance signal d, the closed-loop system is finite gain* L_p *stable over the set* $L_{p,q}(D)$.

(*ii*) *In the absence of any disturbance signal d, the equilibrium point* $x = 0$ *of the closed-loop system is asymptotically stable with* $\mathcal{X}$ *contained in its domain of attraction.*

Proof : In [86], the result for $p \in [1,\infty]$ and $q = \infty$ was given. Hence, we only need to discuss the case $p \in [1,\infty]$ and $q = [1,\infty)$. In our proof, we first show Property (*ii*) and then Property (*i*).

Let $D > 0$ and $\mathcal{X}$ be given. Consider a Lyapunov function $V(x) = x'P_\varepsilon x$. Let c be a positive number such that

$$c \geqslant \sup_{x\in\mathcal{X},\varepsilon\in(0,1]} x'P_\varepsilon x.$$

Such a c exists since $\lim_{\varepsilon\to 0} P_\varepsilon = 0$ and $\mathcal{X}$ is bounded. Also, for any Lyapunov function $V(x)$ and any $c > 0$, let us define a level set $L_V(c)$ as

$$L_V(c) = \{x \in \mathbb{R}^n \mid V(x) < c\}.$$

We note that there exists an $\varepsilon_1^* > 0$ such that for any $\varepsilon \in (0,\varepsilon_1^*]$, $x \in L_V(c)$ implies that $\|B'P_\varepsilon x\|_\infty < \Delta$. It is noted that ε_1^* depends only on $\mathcal{X}$ and Δ. In Theorem 4.41, it was shown that, in the absence of d, for any fixed $\varepsilon \in (0,\varepsilon_1^*]$ and $\alpha \geqslant 0$, the equilibrium point $x = 0$ of the closed-loop system is locally asymptotically stable with $\mathcal{X} \subset L_V(c)$ contained in its domain of attraction. This proves Property (*ii*).

In order to show Property (*i*), a major step is to show that there exist an $\varepsilon^* \in (0,\varepsilon_1^*]$ and a $\alpha_q^* \geqslant 0$ such that for any $\varepsilon \in (0,\varepsilon^*]$ and any $\alpha \geqslant \alpha_q^*$, the trajectory of the closed-loop system comprising of (12.1a) and (12.99) with $x(0) = 0$ remains in $L_V(c)$ for $t \geqslant 0$ even in the presence of $d \in L_{p,q}(D)$. This is what we do next.

Let $d \in L_{p,q}(D)$. Then, for all $x \in L_V(c)$,

$$\begin{aligned}\frac{\mathrm{d}V}{\mathrm{d}t} &\leqslant -x'Qx + 2x'PB[\sigma(-(1+\alpha)B'Px + d) + B'Px] - x'PBB'Px \\ &\leqslant -x'Qx - 2\sum_{i=1}^{m} v_i[\mathrm{sat}_\Delta((1+\alpha)v_i + d_i) - \mathrm{sat}_\Delta(v_i)],\end{aligned} \tag{12.100}$$

where $v = -B'Px$, d_i is the ith component of d, while v_i is the ith component of v.

Observe that

$$|\alpha v_i| \geqslant |d_i| \Longrightarrow -v_i[\text{sat}_\Delta((1+\alpha)v_i + d_i) - \text{sat}_\Delta(v_i)] \leqslant 0,$$

and

$$\begin{aligned}|\alpha v_i| < |d_i| \Longrightarrow & - v_i[\text{sat}_\Delta((1+\alpha)v_i + d_i) - \text{sat}_\Delta(v_i)] \\ & \leqslant \frac{|d_i|^\theta}{\alpha^\theta}|v_i|^{1-\theta}|\,\text{sat}_\Delta((1+\alpha)v_i + d_i) - \text{sat}_\Delta(v_i)| \\ & \leqslant \frac{2}{\alpha^\theta}\|B'P_\varepsilon^{\frac{1}{2}}\|^{1-\theta}V^{\frac{1-\theta}{2}}|d_i|^{1+\theta},\end{aligned}$$

where $\theta \in [0,1]$ is such that $1+\theta \leqslant q$ and will be specified later. Hence, combining these two observations, for any $x \in L_V(c)$, (12.100) can be rewritten as

$$\frac{\mathrm{d}V}{\mathrm{d}t} \leqslant -\nu_1(\varepsilon)V + \frac{1}{\alpha^\theta}\beta_1(\varepsilon)V^{\frac{1-\theta}{2}}\|d\|^{1+\theta}, \tag{12.101}$$

where $\beta_1(\varepsilon) = 4m\|B'P_\varepsilon^{\frac{1}{2}}\|^{1-\theta}$ and $\nu_1(\varepsilon) = \lambda_{\min}(Q_\varepsilon)/\lambda_{\max}(P_\varepsilon) > 0$. In the following, we divide our development into different cases depending on the value of q.

CASE 1: Here we assume that $p \in [1,\infty]$ and $q = 1$. Thus, $\theta = 0$, and (12.101) can be simplified as

$$\frac{\mathrm{d}V}{\mathrm{d}t} \leqslant -\nu_1(\varepsilon)V + \beta_1(\varepsilon)V^{\frac{1}{2}}\|d\|.$$

Dividing both sides with $V^{\frac{1}{2}}$ when $x \neq 0$ gives

$$\frac{\mathrm{d}V^{\frac{1}{2}}}{\mathrm{d}t} \leqslant -\frac{1}{2}\nu_1(\varepsilon)V^{\frac{1}{2}} + \frac{1}{2}\beta_1(\varepsilon)\|d\|.$$

Taking the integral on both sides, and in view of $\|d\|_{L_1} \leqslant D$ and $x(0) = 0$, we get that $V^{\frac{1}{2}}(x) \leqslant \frac{1}{2}\beta_1(\varepsilon)D$. Recalling the Properties 2 and 3 of the CARE (4.42), and using the definition of $\beta_1(\varepsilon)$, we know that there exists an $\varepsilon^* \in (0, \varepsilon_1^*]$ such that for any $\varepsilon \in (0,\varepsilon^*]$, we have $\frac{1}{2}\beta_1(\varepsilon)D < c^{\frac{1}{2}}$. This shows that, for any $\varepsilon \in (0,\varepsilon^*]$, the trajectory of the closed-loop system comprising of (12.1a) and (12.99) starting from $x(0) = 0$ will stay inside $L_V(c)$ for all t even in the presence of a disturbance signal $d \in L_{p,q}(D)$.

CASE 2: Here we assume that $p \in [1,\infty]$ and $q \in (1,2]$. Denoting $\theta = q-1$, when $x \neq 0$, (12.101) yields

$$V^{-\frac{1-\theta}{2}}\frac{\mathrm{d}V}{\mathrm{d}t} \leqslant \frac{1}{\alpha^\theta}\beta_1(\varepsilon)\|d\|^{1+\theta}.$$

Also, $d \in L_q$ implies that $\|d\|^{1+\theta} \in L_1$. Moreover, $\| \|d\|^{1+\theta} \|_1 \leqslant D^q$. Integrating the above inequality with $x(0) = 0$, we get

$$V^{\frac{1+\theta}{2}}(x) \leqslant \frac{\beta_1(\varepsilon)}{\alpha^\theta} \| \|d\|^{1+\theta} \|_1 \leqslant \frac{\beta_1(\varepsilon)}{\alpha^\theta} D^q.$$

Choose any $\alpha_q^*(D,\varepsilon)$ such that $\alpha_q^*(D,\varepsilon) > [\frac{\beta_1(\varepsilon)}{(\sqrt{c})^{1+\theta}} D^q]^{\frac{1}{\theta}}$. Then for any $\alpha \geqslant \alpha_q^*(D,\varepsilon)$, we have $V(x) < c$. In other words, we have shown that for any given D, and $q \in (1,2]$, there exists an $\varepsilon^* = \varepsilon_1^*$, and for any $\varepsilon \in (0, \varepsilon^*]$, there exists a α_q^*, and for any $\alpha \geqslant \alpha_q^*$, the trajectory of the closed-loop system comprising of (12.1a) and (12.99) starting from $x(0) = 0$ will stay within $L_V(c)$ for all t even in the presence of $d \in L_{p,q}(D)$.
CASE 3: Here we assume that $p \in [1,\infty]$ and $q = (2,\infty)$. For this case, $1+\theta < q$ for any $\theta \in [0,1]$. Choose any $\theta \in (0,1]$. Let $W = V^{\frac{1+\theta}{2}}$. For any $\varepsilon \in (0,\varepsilon^*]$ where $\varepsilon^* = \varepsilon_1^*$, (12.101) can be rewritten as

$$\frac{\mathrm{d}W}{\mathrm{d}t} \leqslant -\nu(\varepsilon)W + \frac{1}{\alpha^\theta}\beta(\varepsilon)\|d\|^{1+\theta},$$

where $\nu(\varepsilon) > 0$ and $\beta(\varepsilon) > 0$. We need to show that there exists a $\alpha_q^* > 0$ such that for any $\alpha \geqslant \alpha_q^*$, we have $W(t) < c^{\frac{1+\theta}{2}}$ for any $t \in \mathbb{R}$. To achieve this, we introduce the scalar equation

$$\frac{\mathrm{d}\widehat{W}}{\mathrm{d}t} = -\nu(\varepsilon)\widehat{W} + \frac{1}{\alpha^\theta}\beta(\varepsilon)\|d\|^{1+\theta} \tag{12.102}$$

with $\widehat{W}(0) = 0$. The solution of the above equation is

$$\widehat{W}(t) = \int_0^t e^{-\nu(\varepsilon)(t-\tau)} \frac{1}{\alpha^\theta}\beta(\varepsilon)\|d(\tau)\|^{1+\theta}\mathrm{d}\tau.$$

We can give a bound for $\widehat{W}(t)$ as

$$\begin{aligned}\widehat{W}(t) &\leqslant \frac{\beta(\varepsilon)}{\alpha^\theta}(\int_0^t [e^{-\nu(\varepsilon)\tau}]^{\bar{p}}\mathrm{d}\tau)^{1/\bar{p}} \cdot (\int_0^t [\|d(\tau)\|^{1+\theta}]^{\bar{q}}\mathrm{d}\tau)^{1/\bar{q}} \\ &\leqslant \frac{\beta(\varepsilon)}{\alpha^\theta}(\int_0^t [e^{-\nu(\varepsilon)\tau}]^{\bar{p}}\mathrm{d}\tau)^{1/\bar{p}} \cdot D^{1/\bar{q}},\end{aligned} \tag{12.103}$$

where $(1+\theta)\bar{q} = q$ and $\frac{1}{\bar{p}} + \frac{1}{\bar{q}} = 1$, and the Hölder inequality is used. Hence,

$$\widehat{W}(t) \leqslant \frac{\beta(\varepsilon)}{\alpha^\theta} D^{\frac{1+\theta}{q}} (\frac{1}{\nu(\varepsilon)\bar{p}})^{\frac{1}{\bar{p}}}.$$

For simplicity, fix θ as $\theta = 1$, choose any α_q^* such that

$$\alpha_q^* > \frac{\beta(\varepsilon)}{c} D^{\frac{2}{q}} (\nu(\varepsilon)\overline{p})^{-\frac{1}{p}}.$$

Thus, for $\alpha \geqslant \alpha_q^*$, $\widehat{W}(t) < c$ for any $t \in \mathbb{R}^n$.

Using the standard comparison theorem, we note that $W(t) \leqslant \widehat{W}(t)$ for any $t \in \mathbb{R}$. This shows that any trajectory of the closed-loop system comprising of (12.1a) and (12.99) starting from $x = 0$ will remain within the level set $L_V(c)$ for any $\alpha \geqslant \alpha_q^*(D, \varepsilon)$ for all t even in the presence of a disturbance signal $d \in L_{p,q}(D)$.

In conclusion, so far, we have shown that for any $\varepsilon \in (0, \varepsilon^*]$ and $\alpha \geqslant \alpha_q^*$, the feedback law (12.99) satisfies (a) Property 2 in the absence of d and (b) the trajectory of the closed-loop system comprising of (12.1a) and (12.99) starting from $x(0) = 0$ remains in $L_V(c)$ for all t even in the presence of any $d \in L_{p,q}(D)$.

From the above conclusion and in view of (12.100), it follows that for any fixed $\varepsilon \in (0, \varepsilon^*]$ and any $\alpha \geqslant \alpha_q^*$, the trajectory of the closed-loop system starting from $x(0) = 0$ satisfies the inequality (12.101) for $\theta = 0$. Let $V = W^2$, then there exist an $\nu_2(\varepsilon) > 0$ and a $\beta_2(\varepsilon) > 0$ such that

$$\frac{\mathrm{d}W}{\mathrm{d}t} \leqslant -\nu_2(\varepsilon) W + \beta_2(\varepsilon)\|d\|. \tag{12.104}$$

Using the standard comparison theorem, with $x(0) = 0$, $d \in L_p$ implies that $W \in L_p$, and hence, $x \in L_p$. Moreover,

$$\|W\|_p \leqslant \frac{\beta_2(\varepsilon)}{\nu_2(\varepsilon)} \|d\|_p,$$

and this implies that

$$\|x\|_p \leqslant \gamma_p(\varepsilon)\|d\|_p \tag{12.105}$$

for some $\gamma_p(\varepsilon) > 0$ independent of d. This completes the proof. ∎

An interesting question arises at this stage. That is, whether we can solve the $(SG_{p,q}/SG)$ problem via only low-gain state feedback. More generally, the question can be posed as follows: for what value of q can one have $\alpha_q^* = 0$. Already, the proof of Theorem 12.37 alludes that for $q = 1$ and $p \in [1, \infty]$, we have $\alpha = 0$. In the following lemma, we will show that, for $q \in [1, 2]$ and $p \in [1, \infty]$, one can always choose $\alpha_q^* = 0$:

Lemma 12.38 *In Theorem 12.37, for any $p \in [1, \infty]$, and any $q \in [1, 2]$, one can take α_q^* as zero. More specifically, for any $p \in [1, \infty]$, any $q \in [1, 2]$, any a priori given (arbitrarily large) bounded set $\mathcal{X} \subset \mathbb{R}^n$, and any $D > 0$, there exists an ε^* such that for any $\varepsilon \in (0, \varepsilon^*]$ the low-and-high-gain feedback,*

$$u = -B' P_\varepsilon x - \alpha B' P_\varepsilon x, \tag{12.106}$$

has the following properties for any $\alpha \geqslant 0$:

(i) *In the presence of a disturbance signal d, the closed-loop system is finite gain L_p stable over the set $L_{p,q}(D)$.*

(ii) *In the absence of any disturbance signal d, the equilibrium point $x = 0$ of the closed-loop system is asymptotically stable with $\mathcal{X}$ contained in its domain of attraction.*

Proof : Since Property (*ii*) has been shown in the proof of Theorem 12.99, we concentrate on the proof of Property (*i*).

Let $p \in [1, \infty]$, $q \in [1, 2]$, $D > 0$, and $\mathcal{X}$ be given. Consider a Lyapunov function $V(x) = x' P_\varepsilon x$. Let $c > 0$ be a constant to be chosen shortly. Choose an $\varepsilon^* > 0$ such that for any $\varepsilon \in (0, \varepsilon^*]$ and for any $x \in L_V(c)$, we have $\|B' P_\varepsilon x\|_\infty < \Delta$.

To prove Property (*i*), we first need to show that a disturbance $d \in L_{p,q}(D)$ results in a trajectory of the closed-loop system comprising of (12.1a) and (12.106) which, starting from $x(0) = 0$, will stay inside $L_V(c)$ for all $t > 0$.

To this end, by evaluating the derivative of V along the trajectory of the closed-loop system (12.1a) and (12.106) starting from $x(0) = 0$, we obtain for all $x \in L_V(c)$,

$$\begin{aligned}\frac{\mathrm{d}V}{\mathrm{d}t} &\leqslant -x'Qx + 2x'PB[\sigma(-B'Px - \alpha B'Px + d) + \tfrac{1}{2}B'Px] \\ &\leqslant -x'Qx - 2\sum_{i=1}^{m} v_i[\mathrm{sat}_\Delta(v_i + \alpha v_i + d_i) - \mathrm{sat}_\Delta(\frac{v_i}{2})], \qquad (12.107)\end{aligned}$$

where $v = B'Px$, d_i is the ith component of d, while v_i is the ith component of v.

Let $\tilde{\alpha} = \alpha + 1/2$. Then, we observe that

$$|\tilde{\alpha} v_i| \geqslant |d_i| \Longrightarrow -v_i[\mathrm{sat}_\Delta(\frac{v_i}{2} + \tilde{\alpha} v_i + d_i) - \mathrm{sat}_\Delta(\frac{v_i}{2})] \leqslant 0.$$

Similarly, we note that

$$\begin{aligned}|\tilde{\alpha} v_i| < |d_i| \Longrightarrow &- v_i[\mathrm{sat}_\Delta(\frac{v_i}{2} + \tilde{\alpha} v_i + d_i) - \mathrm{sat}_\Delta(\frac{v_i}{2})] \\ &\leqslant (\frac{d_i}{\tilde{\alpha}})^\theta |v_i|^{1-\theta} |\mathrm{sat}_\Delta(v_i + d_i) - \mathrm{sat}_\Delta(\frac{v_i}{2})| \\ &\leqslant \frac{2}{\tilde{\alpha}^\theta} \|B' P_\varepsilon^{\frac{1}{2}}\|^{1-\theta} V^{\frac{1-\theta}{2}} |d_i|^{1+\theta},\end{aligned}$$

where $\theta \in [0, 1]$. Hence, for any $x \in L_V(c)$, (12.107) can be written as

$$\frac{\mathrm{d}V}{\mathrm{d}t} \leqslant -\nu_1(\varepsilon)V + \frac{\beta_1(\varepsilon)}{\tilde{\alpha}^\theta} V^{\frac{1-\theta}{2}} \|d\|^{1+\theta}, \qquad (12.108)$$

where $\nu_1(\varepsilon) = \lambda_{\min}(Q_\varepsilon)/\lambda_{\max}(P_\varepsilon)$ and $\beta_1(\varepsilon) = 2m\|B'P_\varepsilon^{\frac{1}{2}}\|^{1-\theta}$. Let $\theta = q-1$. We note that $\theta \in [0,1]$ for $q \in [1,2]$. Dividing both sides of (12.108) with $V^{\frac{1-\theta}{2}}$ when $x \neq 0$ gives

$$\frac{\mathrm{d}V^{\frac{1+\theta}{2}}}{\mathrm{d}t} \leqslant -\frac{1}{2}\nu_1(\varepsilon)V^{\frac{1+\theta}{2}} + \frac{1}{(\tilde{\alpha})^\theta}\beta_1(\varepsilon)\|d\|^q.$$

With the assumption that $d \in L_{p,q}(D)$, taking the integral on both sides with $x(0) = 0$ yields

$$V^{\frac{1+\theta}{2}}(x) \leqslant \frac{1}{(\tilde{\alpha})^\theta}\beta_1(\varepsilon)D^q.$$

Note that

$$\tilde{\alpha}^\theta \geqslant \left(\tfrac{1}{2}\right)^\theta.$$

Recalling the Properties (*ii*) and (*iii*) of the CARE (4.42), in Lemma 4.20 and using the definition of $\beta_1(\varepsilon)$, we know that

$$\left(\frac{\beta_1(\varepsilon)D^q}{2^{-\theta}}\right)^{-\frac{1+\theta}{2}} \leqslant \left(\frac{\beta_1(1)D^q}{2^{-\theta}}\right)^{-\frac{1+\theta}{2}} \quad \text{for } \varepsilon \in (0,1].$$

Following the proof of the internal stabilization result of Theorem 12.38, we can choose c such that

$$c > \max\left\{\left(\frac{\beta_1(1)D^q}{2^{-\theta}}\right)^{-\frac{1+\theta}{2}}, \sup_{x\in\mathcal{X},\varepsilon\in(0,1]} x'P_\varepsilon x\right\}.$$

Such a choice of c leads to the fact that $\mathcal{X} \in L_V(c)$. From the above proof and the proof of Theorem 12.37, we have, in the absence of d, that the closed-loop system comprising of (12.1a) and (12.106) is locally asymptotically stable with $\mathcal{X}$ contained in its domain of attraction, and in the presence of $d \in L_{p,q}(D)$, the trajectory of the closed-loop systems (12.1a) and (12.106) starting from $x(0) = 0$ will remain within $L_V(c)$ for any $\alpha \geqslant 0$ and for all t.

Consequently, (12.108) yields for $\theta = 0$:

$$\frac{\mathrm{d}V}{\mathrm{d}t} \leqslant -\nu_1(\varepsilon)V + 4m\|B'P_\varepsilon^{1/2}\|V^{1/2}\|d\|. \tag{12.109}$$

Next, we use the above inequality to prove the L_p stability with finite gain over the set $L_{p,q}(D)$, and in fact, this follows from the application of the standard comparison theorem to (12.109). The rest of the proof is similar to that at the end of the proof of Theorem 12.37. This completes the proof. ■

Remark 12.39 *From Lemma 12.38, since we can choose any $\alpha \geqslant 0$, we conclude that the low-gain feedback (which means $\alpha = 0$) can solve the $(SG_{p,q}/SG)$ problem for $q \in [1,2]$ and $p \in [1,\infty]$.*

Remark 12.40 *It is straightforward to show that the result of Theorem 12.37 still holds if we relax in our controller design methodology the positive definiteness requirement on the matrix Q_ε to semi-positive definiteness. An interesting choice for Q_ε would be $Q_\varepsilon = \varepsilon^2 BB'$. This yields a solution P_ε of CARE* (4.42) *as εI. With this choice of Q_ε, our low-and-high-gain state feedback law reduces to $u = -\mu B'x$ where $\mu = (1 + \alpha)\varepsilon$. This feedback law is exactly the one that was used in [89]. Thus, in view of the results of [89], one can conclude that if the open-loop system is critically stable (and hence without loss of generality assuming that $A' + A = 0$), our low-and-high-gain state feedback law solves the (G_p/G) problem.*

12.7.2 Measurement feedback

Our primary focus in the previous subsection has been in constructing state feedback controllers. That is, we assumed so far that the complete state of the given system is available for feedback and it is not corrupted with any disturbance signal. Once a state feedback controller is constructed, one often tries to construct a measurement feedback controller having a linear observer-based architecture. Accordingly, we construct here such measurement feedback controllers for certain $(SG_{p,q}/SG)$ problems which we utilize either a low-gain or an adaptive-low-gain design to construct the corresponding state feedback controllers. In this regard, let us recall from Lemma 12.38 (see also Remark 12.39) that a low-gain state feedback controller can solve the $(SG_{p,q}/SG)$ problem for $p \in [1, \infty]$ and $q \in [1, 2]$. This easily leads us to the construction of a low-gain measurement feedback controller having a linear observer-based architecture.

Consider the observer

$$\dot{\hat{x}} = A\hat{x} + B\sigma(u) + K(y - C\hat{x}), \tag{12.110}$$

where K is chosen such that $A - KC$ is Hurwitz stable, and the control law

$$u = -B' P_\varepsilon \hat{x}, \tag{12.111}$$

where as usual P_ε is the positive definite solution of the CARE (4.42) parameterized in a low-gain parameter ε.

We have the following theorem:

Theorem 12.41 *Consider the system Σ^c of* (12.1a). *For this system, under Assumptions 12.1, 12.2, and 12.3, the $(SG_{p,q}/SG)_{fg}$ problem via measurement feedback is solvable for any $p \in [1, \infty]$ and any $q \in [1, 2]$. In fact, there exists an ε^* such that for any $\varepsilon \in (0, \varepsilon^*]$, the low-gain measurement feedback controller comprising of the observer* (12.110) *and the control law* (12.111) *solves the $(SG_{p,q}/SG)_{fg}$ problem for any $p \in [1, \infty]$ and any $q \in [1, 2]$.*

Proof : Let $p \in [1,\infty]$, $q \in [1,2]$, $D > 0$, and $\mathcal{X} \subset \mathbb{R}^{2n}$ be given. It is shown in Chap. 4 that there exists an ε_1^* such that for $\varepsilon \in (0,\varepsilon_1^*]$, in the absence of any disturbance signal, the closed-loop system comprising of (12.1a), (12.110), and (12.111) is asymptotically stable, and the set $\mathcal{X}$ is included in the domain of attraction of the equilibrium $x = 0$ of the closed-loop system. To prove the L_p stability of the closed-loop system in the presence of $d \in L_{p,q}(D)$, we observe that the closed-loop system can be written as

$$\begin{cases} \dot{e} = (A - KC)e + B(\sigma(u) - \sigma(u+d)), \\ \dot{x} = Ax + B\sigma(F_\varepsilon x + F_\varepsilon e + d), \end{cases} \tag{12.112}$$

where $e := \hat{x} - x$ and $F_\varepsilon = -B'P_\varepsilon$. We note that $d \in L_{p,q}(D)$ implies that $e \in L_{p,q}(D_e)$ where D_e is a fixed positive number independent of ε. Defining $\bar{d} := F_\varepsilon e + d$, we can see that $\bar{d} \in L_{p,q}(\bar{D})$ where $\bar{D}$ is independent of ε because e is independent of ε and F_ε is uniformly bounded for $\varepsilon \in (0,1]$. Applying the result we derived earlier for the state feedback case, we note that there exists an $\varepsilon^* \leqslant \varepsilon_1^*$ such that for $\varepsilon \in (0,\varepsilon^*]$, the system (12.112) is L_p stable with finite gain over the set $L_{p,q}(\bar{D})$. This completes our proof. ∎

Remark 12.42 *We would like to emphasize an aspect of the measurement feedback controller given in Theorem 12.41. It has a linear observer-based architecture. However, it does not constitute to a mere implementation of a state feedback law via an observer. This is because, as we note, both the state feedback controller and the measurement feedback controller depend on a parameter ε; but the upper bound ε^* on ε that is required for the state feedback controller can however be different to that of the measurement feedback controller. This remark, or a similar one, is applicable to certain other measurement feedback controllers developed in this chapter.*

12.8 Simultaneous stabilization in a semi-global framework: discrete time

In this section, we consider simultaneous stabilization in semi-global framework for discrete-time systems. As in the continuous-time case, our interest is to solve the $(SG_{p,q}/SG)$ problem and the $(SG_{p,q}/SG)_{fg}$ problem, as defined respectively in Problems 11.11 and 11.12. We consider both state and measurement feedback, each in one subsection.

12.8.1 State feedback

The following theorem solves the $(SG_{p,q}/SG)_{fg}$ problem via state feedback. Clearly, the $(SG_{p,q}/SG)$ problem is solvable whenever the $(SG_{p,q}/SG)_{fg}$ problem is solvable.

Theorem 12.43 *Consider the system* Σ^d *of* (12.1b). *For this system, under Assumptions 12.1 and 12.2, the* $(SG_{p,q}/SG)_{fg}$ *problem is solvable for any* $p \in [1, \infty]$ *and any* $q \in [1, 2]$. *Moreover, the low-gain design technique* (4.70) *can yield a control law that solves the problem. More specifically, for any* $p \in [1, \infty]$, *and any* $q \in [1, 2]$, *any a priori given (arbitrarily large) bounded set* $\mathcal{X} \subset \mathbb{R}^n$, *and any* $D > 0$, *there exists an* ε^*, *such that the low-gain feedback,*

$$u(k) := -(B'P_\varepsilon B + I)^{-1} B'P_\varepsilon Ax(k), \quad \varepsilon \in (0, \varepsilon^*], \tag{12.113}$$

where P_ε *is the solution of parameterized discrete-time algebraic Riccati equation (DARE) as given in* (4.66) *has the following properties:*

(*i*) *In the presence of a disturbance signal d, the closed-loop system is finite gain* ℓ_p *stable over the set* $d \in \ell_{p,q}(D)$.

(*ii*) *In the absence of any disturbance signal d, the equilibrium point* $x = 0$ *of the closed-loop system is asymptotically stable with* $\mathcal{X}$ *contained in its domain of attraction.*

Proof : Let $p \in [1, \infty]$, $q \in [1, 2]$, $D > 0$, and $\mathcal{X}$ be given. It is shown in Chap. 4 that, for any a priori given (arbitrarily large) bounded set $\mathcal{X} \subset R^n$, in the absence of any disturbance signal d, there exists an $\varepsilon_1^* > 0$ such that, for each $\varepsilon \in (0, \varepsilon_1^*]$, the equilibrium point $x = 0$ of the closed-loop system comprising (12.1b) and (12.113) is locally exponentially stable and $\mathcal{X}$ is contained in its domain of attraction. Thus, we need to focus only on the proof of Property 1. That is, we need to show that there exists an $\varepsilon^* \in (0, \varepsilon_1^*]$ such that for any $\varepsilon \in (0, \varepsilon^*]$, in the presence of $d \in \ell_{p,q}(D)$, the closed-loop system comprising of (12.1b) and (12.113) satisfies Property (*i*).

To show Property (*i*), we first need to find a level set such that in the presence of $d \in \ell_{p,q}(D)$, the trajectory of the closed-loop system comprising of (12.1b) and (12.113) starting from $x(0) = 0$ will remain inside it for all k.

Consider a Lyapunov function $V(x) = x'P_\varepsilon x$. Let $c > 0$ be a constant to be chosen shortly. Choose ε^* where $\varepsilon^* \leqslant \varepsilon_1^*$ such that for $x(k) \in L_V(c)$ and $\|u_L(k)\|_\infty < \Delta$. By evaluating $V(x(k+1)) - V(x(k))$ along the trajectory of the closed-loop system comprising of (12.1b) and (12.113), we obtain that for any $x(k) \in L_V(c)$,

$$\begin{aligned} V(x(k+1)) - V(x(k)) &\\ = -x(k)'Qx(k) - u(k)'u(k) &\\ + 2x(k)'A'P_\varepsilon B(B'P_\varepsilon B + I)^{-1}(\overline{\sigma}(k) - u(k)) &\\ + (\overline{\sigma}(k) - u(k))'B'P_\varepsilon B(\overline{\sigma}(k) - u(k)) &\\ \leqslant -x(k)'Qx(k) &\\ + (\overline{\sigma}(k) - \sigma(u(k)))'(I + B'P_\varepsilon B)(\overline{\sigma}(k) - \sigma(u(k))) &\\ \leqslant -x(k)'Qx(k) + a_1^2\|d(k)\|^2, & \end{aligned} \tag{12.114}$$

where for brevity, we denote $\overline{\sigma}(k) = \sigma(u(k) + d(k))$. Also,

$$a_1^2 := \lambda_{\max}(I + B'P_1B)$$

is a constant independent of ε.

We note that d being an element of $\ell_{p,q}(D)$ implies that $\|d(k)\|_\infty \leqslant D$ and

$$\sum_{k=1}^{\infty} \|d(k)\|^2 \leqslant \sum_{k=1}^{\infty} \|d(k)\|^q D^{2-q} \leqslant D^2.$$

Let $c := a_1^2 D^2$. Taking sum from both sides of (12.114) yields

$$V(x(k)) \leqslant c$$

for any $k \geqslant 0$. Hence, in the presence of $d \in \ell_{p,q}(D)$, the trajectory of the closed-loop systems (12.1b) and (12.113) starting from $x(0) = 0$ will stay inside $L_V(c)$ for any $k \geqslant 0$, and moreover, the trajectory satisfies (12.114).

Next, we can rewrite (12.114) as

$$V(x(k+1)) - V(x(k)) \leqslant -\nu V(x(k)) + a_1^2 \|d(k)\|^2 \tag{12.115}$$

where $0 < \nu < 1$, and this follows from the property of the DARE.

Define $W(k) := V(x(k))^{\frac{1}{2}}$; it is straightforward to show that (12.115) implies that

$$W(k+1) \leqslant (1-\nu)^{\frac{1}{2}} W(k) + a_1 \|d(k)\| \tag{12.116}$$

with $W(0) = 0$. Applying comparison theorem, the preceding inequality with initial condition $W(0) = 0$ yields that $d \in \ell_p$ implies that $W \in \ell_p$ for any $p \in [1, \infty]$ and hence implies $x \in \ell_p$. Moreover, there exists a constant $\gamma_p > 0$ such that

$$\|x\|_{\ell_p} \leqslant \gamma_p \|d\|_{\ell_p}.$$

This completes the proof. ■

12.8.2 *Measurement feedback*

Here we construct measurement feedback controllers for discrete-time systems. As in Sect. 12.2.2, our controllers have a linear observer-based architecture. Consider the observer-based controller,

$$\begin{aligned} \hat{x}(k+1) &= A\hat{x}(k) + B\sigma(u(k)) + K(y(k) - C\hat{x}(k)), \\ u(k) &= -(B'P_\varepsilon B + I)^{-1} B'P_\varepsilon A\hat{x}(k), \end{aligned} \tag{12.117}$$

where K is chosen such that $A - KC$ is Schur stable.

We have the following theorem:

Theorem 12.44 *Consider the system* Σ^d *of* (12.1b). *For this system, under Assumptions 12.1, 12.2, and 12.3, the* $(SG_{p,q}/SG)_{fg}$ *problem is solvable for any* $p \in [1,\infty]$ *and any* $q \in [1,2]$. *In fact, there exists an* ε^* *such that for an* $\varepsilon \in (0,\varepsilon^*]$, *the low-gain measurement feedback controller given by* (12.117) *solves the* $(SG_{p,q}/SG)_{fg}$ *problem for any* $p \in [1,\infty]$ *and any* $q \in [1,2]$.

Proof : The proof of internal stability was shown in Chap. 4. Since $(B'P_\varepsilon B + I)^{-1}B'P_\varepsilon A$ is uniformly bounded, the proof of external stabilization is similar to the continuous-time counterpart given in the proof of Theorem 12.41. This completes the proof. ∎

12.9 Role of the location of the open-loop poles of the system

For linear systems subject to actuator saturation, earlier sections explored in depth two broad and important simultaneous stabilization problems, (1) simultaneous global external stabilization and global internal stabilization problem and (2) simultaneous semi-global external stabilization and semi-global internal stabilization problem. As we have seen, global framework requires nonlinear feedback control laws while linear feedback control laws suffice in semi-global framework. Appropriate feedback control laws are constructed for these problems under the assumption that the given linear system is ANCBC. This begs the question how important is this assumption. In this regard, Chap. 4 dictates that global internal stabilization, as well as semi-global internal stabilization, is possible only if the given system satisfies the ANCBC assumption. However, local internal stabilization can be achieved without the ANCBC assumption. That is, the location of the poles does not play any role at all in achieving local internal stabilization. This is true whether we deal with a continuous- or a discrete-time system. Prompted by this, it is worthwhile to examine (1) the simultaneous global external stabilization and local internal stabilization problem and (2) the simultaneous semi-global external stabilization and local internal stabilization problem, both with or without finite gain. We have the following formal problem statements:

Problem 12.45 Consider the system Σ^c given in (12.1a) or the system Σ^d given in (12.1b). For any $p \in [1,\infty]$, this system is said to be **simultaneously globally L_p stabilizable (or ℓ_p stabilizable) and locally asymptotically stabilizable** via static state feedback (or dynamic measurement feedback) if there exists a static state (or respectively, dynamic measurement) feedback law, such that the following conditions hold:

(*i*) For continuous-time case, the closed-loop system with $x(0) = 0$ is L_p-stable, that is, $\|z\| \in L_p$ for all $d \in L_p$. Similarly, for discrete-time case, the closed-loop system with $x(0) = 0$ is ℓ_p stable, that is, $\|z\| \in \ell_p$ for all $d \in \ell_p$.

(*ii*) In the absence of any disturbance signal d, the equilibrium point $x = 0$ of the closed-loop system is locally asymptotically stable.

The above problem is coined as (G_p/L) problem.

Problem 12.46 Consider the system Σ^c given in (12.1a) or the system Σ^d given in (12.1b). For any $p \in [1, \infty]$, this system is said to be **simultaneously globally finite gain L_p-stabilizable (or ℓ_p stabilizable) and locally asymptotically stabilizable** via static state feedback (or dynamic measurement feedback) if there exists a static state (or respectively, dynamic measurement) feedback law such that the following conditions hold:

(*i*) For continuous-time case, the closed-loop system is finite gain L_p stable, that is, there exists a positive constant γ_p such that with $x(0) = 0$, the following holds:

$$\|z\|_p \leq \gamma_p \|d\|_p, \quad \text{for all } d \in L_p.$$

Similarly, for the discrete-time case, the closed-loop system is finite gain ℓ_p stable, that is, there exists a positive constant γ_p such that with $x(0) = 0$, the following holds:

$$\|z\|_p \leq \gamma_p \|d\|_p, \quad \text{for all } d \in \ell_p.$$

(*ii*) In the absence of any disturbance signal d, the equilibrium point $x = 0$ of the closed-loop system is locally asymptotically stable.

The above problem is coined as $(G_p/L)_{fg}$ problem.

Problem 12.47 Consider the system Σ^c given in (12.1a) or the system Σ^d given in (12.1b). For any $p, q \in [1, \infty]$, this system is said to be **simultaneously L_q (or ℓ_q) semi-globally L_p stabilizable (or ℓ_p-stabilizable) and locally asymptotically stabilizable** via static state feedback (or dynamic measurement feedback with dynamical order n_c) if, for any a priori given (arbitrarily large) finite number $D > 0$, there exists a static state (or respectively, dynamic measurement) feedback law, possibly depending on D (or respectively, n_c and D), such that the following conditions hold:

(*i*) For continuous-time case, the closed-loop system is L_p stable over the set $L_{p,q}(D)$. That is, $z \in L_p$ for all $d \in L_p$ with $\|d\|_{L_q} \leqslant D$ and $x(0) = 0$. Similarly, for discrete-time case, the closed-loop system is ℓ_p stable over the set $\ell_{p,q}(D)$. That is, $z \in \ell_p$ for all $d \in \ell_p$ with $\|d\|_q \leqslant D$ and $x(0) = 0$.

(*ii*) In the absence of any disturbance signal d, the equilibrium point $x = 0$ of the closed-loop system is locally asymptotically stable.

The above problem is coined as $(SG_{p,q}/L)$ problem.

Problem 12.48 Consider the system Σ^c given in (12.1a) or the system Σ^d given in (12.1b). For any $p, q \in [1, \infty]$, this sytem is said to be **simultaneously L_q (or ℓ_q) semi-globally finite gain L_p-stabilizable (or ℓ_p-stabilizable) and locally asymptotically stabilizable** via static state feedback (or dynamic measurement feedback with dynamical order n_c) if, for any a priori given (arbitrarily large) finite number $D > 0$, there exists a static state (or respectively, dynamic measurement) feedback law, possibly depending on D (or respectively, n_c, and D), such that the following conditions hold:

(*i*) For continuous-time case, the closed-loop system is finite gain L_p stable over the set $L_{p,q}(D)$. That is, there exists a positive constant γ_p such that with $x(0) = 0$, the following holds:

$$\|z\|_p \leq \gamma_p \|d\|_p, \quad \text{for all } d \in L_p \text{ with } \|d\|_{L_q} \leqslant D.$$

Similarly, for the discrete-time case, the closed-loop system is finite gain ℓ_p stable over the set $\ell_{p,q}(D)$. That is, there exists a positive constant γ_p such that with $x(0) = 0$, the following holds:

$$\|z\|_p \leq \gamma_p \|d\|_p, \quad \text{for all } d \in \ell_p \text{ with } \|d\|_q \leqslant D.$$

(*ii*) In the absence of any disturbance signal d, the equilibrium point $x = 0$ of the closed-loop system is locally asymptotically stable.

The above problem is coined as $(SG_{p,q}/L)_{fg}$ problem.

Some of the above four problems have been examined in the literature. It is shown in [86] that, *for continuous-time linear systems* with saturating actuators, bounded external ($q = \infty$) finite gain L_p stabilization and local asymptotic stabilization can always be achieved simultaneously irrespective of the location of the open-loop poles of the given system. The location of open-loop poles, as we know already, affects only the solvability of global and semi-global internal stabilization.

On the other hand, *for discrete-time linear systems* with saturating actuators, it is shown in [47] that neither global external stabilization nor semi-global external stabilization is possible whenever there exists one or more controllable poles located strictly outside the unit circle. This negative result is possibly owing to the lack of high-gain feedback in discrete-time systems.

Based on the results of [86] and [47], our interest in this section is to examine some of the above problems when the entire state is available for feedback. This is done just to show the role of the open-loop poles on achieving external stability of linear systems with saturating actuators. We assume throughout this section that the pair (A, B) is stabilizable.

12.9.1 Continuous-time systems

In this subsection, for continuous-time linear systems subject to actuator saturation, we consider only $(SG_{p,q}/L)_{fg}$ problem just as a contrast to the discrete-time case to be considered in the next subsection. One can pursue other problems in a similar manner. For clarity, we assume that the saturation function σ is the one given in Definition 2.19.

We have the following result based on [86]:

Theorem 12.49 *Consider the continuous-time system Σ^c given in* (12.1a). *For any $p \in [1,\infty]$, $q \in [1,\infty]$ and for any $D > 0$, the sytem Σ^c is simultaneously L_q semi-globally finite-gain L_p stabilizable and locally asymptotically stabilizable via static linear state feedback, that is, the problem $(SG_{p,q}/L)_{fg}$ is solvable via static linear state feedback.*

Proof : Let us consider a family of high-gain state feedback control laws,

$$u = -(1+\alpha)B'Px,$$

where P is the solution of CARE,

$$A'P + PA - 2PBB'P + Q = 0,$$

and where Q is any positive definite matrix. Then, the closed-loop system takes the form

$$\dot{x} = Ax + B\sigma(-(1+\alpha)B'Px + d). \tag{12.118}$$

As we did often, let us pick a Lyapunov function $V(x) = x'Px$ and let c be such that $x \in L_V(c)$ implies that $\|B'Px\| \leqslant \Delta$, where the level set $L_V(c)$ is defined as $L_V(c) = \{x \in \mathbb{R}^n : V(x) \leqslant c\}$. The evaluation of $\dot{V}$ along the trajectories of the closed-loop system in the absence of d shows that for all $x \in L_V(c)$,

$$\begin{aligned}\dot{V} &= -x'Qx + 2x'PB[\sigma(-(1+\alpha)B'Px) + B'Px] \\ &= -x'Qx - 2\sum_{i=1}^{m} v_i[\mathrm{sat}_\Delta((1+\alpha)v_i) - \mathrm{sat}_\Delta(v_i)] \\ &\leqslant -x'Qx,\end{aligned} \tag{12.119}$$

where we have defined $v \in \mathbb{R}^m$ by $v = -B'Px$.

Equation (12.119) shows that the equilibrium point $x = 0$ of the closed-loop system (12.118) is locally asymptotically stable in the absence of d. It remains to show that the closed-loop system (12.118) is also L_q semi-globally finite gain L_p stable. In this regard, it can be shown easily that in the presence of d and for all $x \in L_V(c)$,

$$\dot{V} = -x'Qx + \tfrac{4m\delta D^2}{\alpha}.$$

Let

$$\alpha^* = \frac{4m\delta D^2 \lambda_{\max}(P)}{\lambda_{\min}(Q)c}.$$

Then, $L_V(c)$ is an invariant set for all $\alpha \geqslant \alpha^*$. We next rewrite the closed-loop system (12.118) as

$$\begin{aligned}\dot{x} = Ax &+ B\sigma(-(1+\alpha)B'Px) \\ &+ B[\sigma(-(1+\alpha)B'Px + d) - \sigma(-(1+\alpha)B'Px)]. \quad (12.120)\end{aligned}$$

Using (12.120), we obtain that, for all $x \in L_V(c)$,

$$\dot{V} \leqslant -x'Qx + 2V^{1/2}\|B'P^{1/2}\|\|d\|, \quad \text{for } \alpha \geqslant \alpha^*.$$

The rest of the proof follows by utilizing the standard comparison theorems as we often did (see for details [86]). ■

12.9.2 Discrete-time systems

This subsection is devoted to discrete-time linear systems. Unlike in continuous-time case, our results here are negative and are based on [47].

Theorem 12.50 *Consider the discrete-time system Σ^d given in* (12.1b). *Suppose that A has at least one controllable eigenvalue strictly outside the unit circle. Then, for Σ^d and for any $p \in [1, \infty]$, the global ℓ_p stabilization cannot be achieved via any state feedback law. That is, there does not exist any static state feedback law that renders $z \in \ell_p$ when $d \in \ell_p$.*

Proof : We prove this theorem by contradiction. Suppose that there exists a feedback law $u = F(x)$ that achieves global ℓ_p stabilization, we construct an external input $d \in \ell_p$ such that the state of the system diverges to infinity. Without loss of generality, we assume that the pair (A, B) is in the form (see, e.g., [21]),

$$\Gamma_x^{-1} A \Gamma_x = \begin{pmatrix} A_1 & A_{1,2} & \cdots & A_{1,r} & A_{1\bar{c}} \\ 0 & A_2 & \cdots & A_{2,r} & A_{2\bar{c}} \\ \vdots & \vdots & \ddots & \vdots & \vdots \\ 0 & 0 & \cdots & A_r & A_{r\bar{c}} \\ 0 & 0 & \cdots & 0 & A_{\bar{c}} \end{pmatrix}, \qquad B = \begin{pmatrix} B_1 \\ B_2 \\ \vdots \\ B_r \\ 0 \end{pmatrix}, \quad (12.121)$$

where $A_{\bar{c}}$ contains all the uncontrollable poles of the open-loop system, and all the remaining diagonal blocks are either 1×1 matrices or 2×2 matrices of the form

$$\bar{\rho} \begin{pmatrix} \bar{\alpha} & \bar{\beta} \\ -\bar{\beta} & \bar{\alpha} \end{pmatrix}$$

with $\bar{\rho} > 0$, $\bar{\beta} \neq 0$, and $\bar{\alpha}^2 + \bar{\beta}^2 = 1$, having complex conjugate eigenvalues $\bar{\rho}(\bar{\alpha} \pm j\bar{\beta})$. Also, assume that the diagonal blocks are arranged in such a way that A_r is either a real number λ with $|\lambda| > 1$ or the said 2×2 matrix with $\bar{\rho} > 1$. In constructing the "destabilizing input" $d(k)$, we consider two separate cases, A_r is a real number or it is a 2×2 matrix.

Case 1: Let A_r be λ with $|\lambda| > 1$. Without loss of generality, let this eigenvalue be positive and hence $\lambda > 1$. The dynamics of x_r (the state associated with A_r) is then given by (assuming $x_{\bar{c}}(0) = 0$)

$$x_r(k+1) = \lambda x_r(k) + \sum_{i=1}^{m} B_{ri} \operatorname{sat}_\Delta(F_i(x(k)) + d_i(k)),$$

where B_{ri} is the ith element of $B_r \in \mathbb{R}^{1\times m}$ and $F_i(x(k))$ is the ith element of the feedback function $F(x(k)) \in \mathbb{R}^{m\times 1}$. The stabilizability of the pair (A, B) implies that the matrix $B_r \in \mathbb{R}^{1\times m}$ has a nonzero element, say $B_{ri_o} > 0$.

We now proceed recursively to construct a $d(k) \in \ell_p$ such that the state x_r diverges to infinity and thus contradicts the assumption that $x \in \ell_p$. We first choose $d_{i_o}(0) = -F_{i_o}(x(0)) + 1$ and $d_j(0) = -F_j(x(0))$ for $j \neq i_o$. With such a choice of $d(0)$, we have $x_r(1) = B_{ri_o} \operatorname{sat}_\Delta(1)$. We next choose $d(1) = -F(x(1))$ and obtain $x_r(2) = \lambda x_r(1) = \lambda B_{ri_o}$. Continuing the same way, we have $d(k) = -F(x(k))$ and $x_r(k+1) = \lambda x_r(k) = \lambda^k B_{ri_o} \operatorname{sat}_\Delta(1)$. Recalling that $\lambda > 1$, there exists a finite k_o such that

$$x_r(k_o+1) = \lambda^{k_o} B_{ri_o} \operatorname{sat}_\Delta(1) \geqslant \left(\sum_{i=1}^{m} |B_{ri} \operatorname{sat}_\Delta(1)|\right) / (\lambda - 1) + 1.$$

We then choose $d(k) \equiv 0$ for all $k \geqslant k_o+1$. Clearly, $d(k) \in \ell_p$ for all $p \in [1, \infty]$. It is then straightforward to verify that for $k \geqslant k_o + 2$,

$$x_r(k) \geqslant \left(\sum_{i=1}^{m} |B_{ri} \operatorname{sat}_\Delta(1)|\right) / (\lambda - 1) + \lambda^{k-k_o-1}.$$

Hence, $x_r(k) \to \infty$ as $k \to \infty$, and thus, $x(k) \notin \ell_p$ for any $p \in [1, \infty]$.

Case 2: Let

$$A_r = \bar{\rho}\begin{pmatrix} \bar{\alpha} & \bar{\beta} \\ -\bar{\beta} & \bar{\alpha} \end{pmatrix} \quad \text{with } \bar{\rho} > 1.$$

The dynamics of x_r (the state associated with A_r) is then given by

$$x_r(k+1) = A_r x_r(k) + \sum_{i=1}^{m} B_{ri} \operatorname{sat}_\Delta(F_i(x(k)) + d_i(k)),$$

where $B_{ri} \in \mathbb{R}^{2\times 1}$ is the ith column of $B_r \in \mathbb{R}^{2\times m}$ and $F_i(x(k))$ is the ith element of the feedback function $F(x(k))$. The stabilizability of the pair (A, B) implies that at least one column of the matrix B_r, say B_{ri_o}, is nonzero.

We now proceed recursively to construct a $d(k) \in \ell_p$ such that the state x_r diverges to infinity and thus contradicts the assumption that $x \in \ell_p$. We first choose $d_{i_o}(0) = -F_{io}(x(0)) + 1$ and $d_j(0) = -F_j(x(0))$ for $j \neq i_o$. With such a choice of $d(0)$, we have $x_r(1) = B_{ri_o}\sigma_{io}(1) = B_{ri_o}$. We next choose $d(1) = -F(x(1))$ and obtain $x_r(2) = A_r B_{ri_o}$. Continuing the same way, we have $d(k) = -F(x(k))$ and $x_r(k+1) = A_r^k B_{ri_o}$. Observing that the matrix operator A_r consists of a rotation and scaling up by a factor $\overline{\rho} > 1$, we have $\|x_r(k+1)\| = \overline{\rho}^k \|B_{ri_o}\|$. Recalling that $\overline{\rho} > 1$, there exists a finite k_o such that

$$\|x_r(k_o+1)\| = \overline{\rho}^{k_o}\|B_{ri_o}\| \geqslant \left(\sum_{i=1}^{m} |B_{ri}|\right) / (\overline{\rho}-1) + 1.$$

We then choose $d(k) \equiv 0$ for all $k \geqslant k_o+1$. Clearly, $d(k) \in \ell_p$ for all $p \in [1,\infty]$. It is then straightforward to verify that for $k \geqslant k_o + 2$, we have

$$\|x_r(k)\| \geqslant \left(\sum_{i=1}^{m} \|B_{ri}\|\right) / (\overline{\rho}-1) + \overline{\rho}^{k-k_o-1}.$$

Hence, $\|x_r(k)\| \to \infty$ as $k \to \infty$, and thus, $x(k) \notin \ell_p$ for any $p \in [1,\infty]$. ■

Remark 12.51 *An obvious consequence of Theorem 12.50 is that neither the problem (G/L) nor $(G/L)_{fg}$ is solvable whenever A has at least one controllable pole strictly outside the unit circle.*

Theorem 12.52 *Consider the discrete-time system Σ^d given in* (12.1b). *We have the following:*

(i) For the single input case, that is, for m=1, suppose that A has at least one controllable eigenvalue strictly outside the unit circle. Then, for Σ^d and for any $p, q \in [1,\infty]$, the ℓ_q semi-global ℓ_p stabilization cannot be achieved by any static linear state feedback law. That is, there exists a D^* such that for any $D > D^*$, there does not exist any static linear state feedback law such that the following holds:

$$z \in p \quad \text{for all } d \in \ell_p \text{ with } \|d\|_q \leqslant D,$$

for any a priori given (arbitrarily large) finite number D.

(ii) For general multi-input case (i.e., for $m \geqslant 1$), suppose that A has a controllable eigenvalue whose magnitude is greater than 2. Then, for Σ^d and for any $p, q \in [1,\infty]$, the ℓ_q semi-global ℓ_p stabilization cannot be achieved by any nonlinear static state feedback law $u = F(x)$ with $F(0) = 0$. That is, there exists a D^* such that for any $D > D^*$, there does not exist any

static linear or nonlinear state feedback law $u = F(x)$ *with* $F(0) = 0$ *such that the following holds:*

$$z \in p \quad \text{for all } d \in \ell_p \text{ with } \|d\|_q \leqslant D,$$

for any a priori given (arbitrarily large) finite number D.

Proof : We first consider the proof of part (*i*). We prove this part of the theorem by showing that there exists a $D(q) > 0$ such that, for any given linear feedback law $u = Fx$, there exists an external input $d(k) \in \ell_{p,q}(D(q))$ such that the state $\|x(k)\|$ diverges to infinity and hence $x(k) \notin \ell_p$ for any $p \in [1, \infty]$.

Following the arguments in [13], it can be shown that if a feedback law $u = Fx$ achieves ℓ_q semi-global ℓ_p stabilization, then all the eigenvalues of $A + BF$ must lie inside or on the unit circle. Hence, without loss of generality, we assume that F is such that $A + BF$ has all its eigenvalues within or on the unit circle. It then follows from [56] (pages 198–199), that in the single-input case for all F such that the eigenvalues of $A + BF$ are within or on the unit circle, $\|F\| \leqslant \phi$, for some positive constant ϕ independent of F. Let d be chosen in the same way as in the proof of Theorem 12.50.

Since the closed-loop system is a linear system driven by a bounded input signal σ, there exists a $M > 0$, independent of the control input, such that $\|x\| \leqslant M$ for all $k \in \{0, 1, 2, \ldots, k_o\}$.

If we define

$$D(q) = \begin{cases} k^{1/q}(\phi M + 1) & \text{if } q \in [1, \infty) \\ \phi M + 1 & \text{if } q = \infty, \end{cases}$$

then $d(k) \in \ell_{p,q}(D(q))$ for all $p,\ q \in [1, \infty]$. However, as shown in the proof of Theorem 12.50, after a time k_o, the state reaches to "the point of no return", after which the state will approach infinity no matter what value the actuator output $\sigma(Fx)$ is.

Along the same lines as above and those in the proof of Theorem 12.50, we prove part (*ii*) of the theorem by showing that there exists a $d \in \ell_p$ with $\|d\|_q \leqslant \sqrt{m}$ such that the state diverges to infinity for any nonlinear state feedback law $u = F(x)$ with $F(0) = 0$. Without loss of generality, we assume that the pair (A, B) is in the form given in (12.121) where $A_{\bar{c}}$ contains all the uncontrollable poles of the open-loop system, and all the remaining diagonal blocks are either 1×1 matrices or 2×2 matrices of the form

$$\bar{\rho}\begin{pmatrix} \bar{\alpha} & \bar{\beta} \\ -\bar{\beta} & \bar{\alpha} \end{pmatrix}$$

with $\bar{\rho} > 0$, $\bar{\beta} \neq 0$, and $\bar{\alpha}^2 + \bar{\beta}^2 = 1$, having complex conjugate eigenvalues $\bar{\rho}(\bar{\alpha} \pm j\bar{\beta})$. Also, assume that the diagonal blocks are arranged in such a way that

A_r is either a real number λ with $|\lambda| > 2$ or the said 2×2 matrix with $\bar{\rho} > 2$. In constructing the "destabilizing input" $d(k)$, we consider two separate cases, A_r is a real number or it is a 2×2 matrix.

Case 1: Let A_r be λ with $|\lambda| > 2$. Without loss of generality, let this eigenvalue be positive and hence $\lambda > 2$. The dynamics of x_r (the state associated with A_r) is then given by

$$x_r(k+1) = \lambda x_r(k) + \sum_{i=1}^{m} B_{ri} \operatorname{sat}_\Delta(F_i(x(k)) + d_i(k))$$

where B_{ri} is the ith element of $B_r \in \mathbb{R}^{1\times m}$ and $F_i(x(k))$ is the ith element of the feedback function $F(x(k)) \in \mathbb{R}^{m\times 1}$. The stabilizability of the pair (A, B) implies that the matrix $B_r \in \mathbb{R}^{1\times m}$ has a nonzero element. Let us choose

$$d(0) = \begin{pmatrix} \operatorname{sgn}(B_{r1}) & \operatorname{sgn}(B_{r2}) & \cdots & \operatorname{sgn}(B_{rm}) \end{pmatrix}',$$

and $d(k) \equiv 0$ for all $k \geqslant 1$. Obviously, $d(k) \in \ell_{p,q}(D(q))$ for any $p, q \in [1, \infty]$. Then, it is easily verified that, for $k \geqslant 1$, we have

$$x_r(k) \geqslant \lambda^{k-1} \sum_{i=1}^{m} |B_{ri}| - \sum_{j=0}^{k-2} \lambda^j \sum_{i=1}^{m} |B_{ri}| = \sum_{i=1}^{m} |B_{ri}| \left(\lambda^{k-1} - \sum_{j=0}^{k-2} \lambda^j \right)$$

$$= \sum_{i=1}^{m} |B_{ri}| \left[(\lambda - 2) \sum_{j=0}^{k-2} \lambda^j + 1 \right].$$

Recalling that $\lambda > 2$, it is clear that $x_r(k)$ diverges to infinity as k tends to infinity. Hence, $x(k) \notin \ell_p$ for any $p \in [1, \infty]$.

Case 2: Let

$$A_r = \bar{\rho} \begin{pmatrix} \bar{\alpha} & \bar{\beta} \\ -\bar{\beta} & \bar{\alpha} \end{pmatrix} \quad \text{with } \bar{\rho} > 2.$$

The dynamics of x_r (the state associated with A_r) is then given by

$$x_r(k+1) = A_r x_r(k) + \sum_{i=1}^{m} B_{ri} \operatorname{sat}_\Delta(F_i(x(k)) + d_i(k))$$

where $B_{ri} \in \mathbb{R}^{2\times 1}$ is the ith column of $B_r \in \mathbb{R}^{2\times m}$ and $F_i(x(k))$ is the ith element of the feedback function $F(x(k))$. The stabilizability of the pair (A, B) implies that at least one column of the matrix B_r is nonzero. Let $\eta^* \in \mathbb{R}^m$ maximize $\|B_r \eta\|$ subject to the constraint $\eta_i \leqslant 1$ for all $i = 1$ to m. Let $b_r^* = \|B_r \eta^*\|$. Clearly, $b_r^* \neq 0$. We now choose $d(0) = \eta^*$, $d(k) \equiv 0$ for all $k \geqslant 1$. Obviously, $d(k) \in \ell_{p,q}(D(q))$ for any $p, q \in [1, \infty]$. It is then straightforward to verify that for $k \geqslant 1$,

$$\|x_r(k)\| \geqslant \bar{\rho}^{k-1} b_r^* - \sum_{j=0}^{k-2} \bar{\rho}^j b_r^*$$

$$= b_r^* \left(\bar{\rho}^{k-1} - \sum_{j=0}^{k-2} \bar{\rho}^j \right) = b_r^* \left[(\bar{\rho} - 2) \sum_{j=0}^{k-2} \bar{\rho}^j + 1 \right].$$

Recalling that $\bar{\rho} > 2$, it is clear that $\|x_r(k)\| \to \infty$ as $k \to \infty$, and thus, $x(k) \notin \ell_p$ for any $p \in [1, \infty]$. ■

Remark 12.53 *An obvious consequence of Theorem 12.52 is that neither the problem* (SG/L) *nor* $(SG/L)_{fg}$ *is solvable whenever* A *has at least one controllable pole whose magnitude is greater than 2.*

12.A Appendix: a preliminary lemma

Lemma 12.54 *Consider* ε_a *as defined by* (12.2) *and let* P_{ε_a} *be the positive definite solution of the algebraic Riccati equation* (4.42) *when* ε *is replaced by* ε_a. *Moreover, define* V *by* $V(x) = x' P_{\varepsilon_a(x)} x$. *Then, there exists a function* K *of* ε_a *such that for any function* x *from* $\mathbb{R}^+$ *to* $\mathbb{R}^n$, *we have*

$$|x' \frac{dP_{\varepsilon_a}}{dt} x| \leqslant K |\frac{dV(x)}{dt}|. \tag{12.122}$$

Moreover, one particular choice for K *is given by*

$$K = \frac{N_5}{\lambda_{min}(P_{\varepsilon_a}) \|(sI - A + BB'P_{\varepsilon_a})^{-1}\|_{H_2}^2} \frac{\lambda_{max}(\frac{dQ_{\varepsilon_a}}{d\varepsilon_a})}{\lambda_{min}(\frac{dQ_{\varepsilon_a}}{d\varepsilon_a})} \tag{12.123}$$

for some positive constant N_5.

Remark 12.55 *It is obvious that the* K *given in the previous lemma is upper bounded as soon as we know that* ε_a *is bounded away from* 0. *Therefore, for any* ε_a^*, *there exists a constant* K *such that* (12.122) *is satisfied for any function* x *that guarantees that* $\varepsilon_a(x) > \varepsilon_a^*$ *for all* $t > 0$.

Proof : First note that if $\varepsilon_a = 1$, then $\frac{dP}{dt} = 0$, and the inequality is always satisfied for any $K > 0$. Therefore, we can concentrate on the case $\varepsilon_a \neq 1$.

From the definition of ε_a, using that $\varepsilon_a \neq 1$, we have

$$V(x)\,\mathrm{trace}(B'\frac{\mathrm{d}P_{\varepsilon_a}}{\mathrm{d}t}B) + \frac{\mathrm{d}V(x)}{\mathrm{d}t}\,\mathrm{trace}(B'PB) = 0,$$

which yields

$$|\,\mathrm{trace}(B'\frac{\mathrm{d}P_{\varepsilon_a}}{\mathrm{d}t}B)| = \frac{\mathrm{trace}(B'P_{\varepsilon_a}B)}{V(x)}\frac{\mathrm{d}V(x)}{\mathrm{d}t} \leqslant \frac{K_1}{V(x)}\frac{\mathrm{d}V(x)}{\mathrm{d}t} \tag{12.124}$$

for some constant K_1 since P_{ε_a} is upper bounded. On the other hand,

$$\begin{aligned}|\,\mathrm{trace}(x\frac{\mathrm{d}P_{\varepsilon_a}}{\mathrm{d}t}x)| &\leqslant \frac{\lambda_{\max}(\frac{\mathrm{d}P_{\varepsilon_a}}{\mathrm{d}\varepsilon_a})\|x\|^2}{\lambda_{\min}(\frac{\mathrm{d}P_{\varepsilon_a}}{\mathrm{d}\varepsilon_a})\|B\|^2}|\,\mathrm{trace}(B'\frac{\mathrm{d}P_{\varepsilon_a}}{\mathrm{d}t}B)| \\ &\leqslant K_2\frac{V}{\lambda_{\min}(P_{\varepsilon_a})}\frac{\lambda_{\max}(\frac{\mathrm{d}P_{\varepsilon_a}}{\mathrm{d}\varepsilon_a})}{\lambda_{\min}(\frac{\mathrm{d}P_{\varepsilon_a}}{\mathrm{d}\varepsilon_a})}\end{aligned} \tag{12.125}$$

for some constant $K_2 > 0$. It remains to derive a relation between the derivatives of Q and the derivatives of P_{ε_a}. We will find that the Riccati equation immediately yields

$$(A - BB'P_{\varepsilon_a})'\frac{\mathrm{d}P_{\varepsilon_a}}{\mathrm{d}\varepsilon_a} + \frac{\mathrm{d}P_{\varepsilon_a}}{\mathrm{d}\varepsilon_a}(A - BB'P_{\varepsilon_a}) + \frac{\mathrm{d}Q_{\varepsilon_a}}{\mathrm{d}\varepsilon_a} = 0.$$

We recall that $A - BB'P_{\varepsilon_a}$ is stable, and therefore, we find that the solution of the preceding *Lyapunov* equation is given by

$$\frac{\mathrm{d}P_{\varepsilon_a}}{\mathrm{d}\varepsilon_a} = \int_0^\infty e^{(A-BB'P_{\varepsilon_a})'t}\frac{\mathrm{d}Q_{\varepsilon_a}}{\mathrm{d}\varepsilon_a}e^{(A-BB'P_{\varepsilon_a})t}\mathrm{d}t,$$

and this leads to

$$\lambda_{\max}(\frac{\mathrm{d}P_{\varepsilon_a}}{\mathrm{d}\varepsilon_a}) \leqslant \left(\int_0^\infty e^{(A-BB'P_{\varepsilon_a})'t}e^{(A-BB'P_{\varepsilon_a})t}\mathrm{d}t\right)\lambda_{\max}(\frac{\mathrm{d}Q_{\varepsilon_a}}{\mathrm{d}\varepsilon_a}),$$

or, in other words,

$$\lambda_{\max}(\frac{\mathrm{d}P_{\varepsilon_a}}{\mathrm{d}\varepsilon_a}) \leqslant \frac{1}{\|(sI - A + BB'P_{\varepsilon_a})^{-1}\|_{H_2}^2}\lambda_{\max}\left(\frac{\mathrm{d}Q_{\varepsilon_a}}{\mathrm{d}\varepsilon_a}\right).$$

On the other hand, from [62], we obtain

$$\begin{aligned}\lambda_{\min}(\frac{\mathrm{d}P_{\varepsilon_a}}{\mathrm{d}\varepsilon_a}) &\geqslant \frac{1}{\|A - BB'P)\|}\lambda_{\max}\left(\frac{\mathrm{d}Q_{\varepsilon_a}}{\mathrm{d}\varepsilon_a}\right) \\ &\geqslant K_3\lambda_{\max}\left(\frac{\mathrm{d}Q_{\varepsilon_a}}{\mathrm{d}\varepsilon_a}\right)\end{aligned}$$

for some constant K_3 since $A - BB'P$ is bounded. Combining the last two inequalities with (12.122) and (12.125) yields that there exists a constant $N_5 > 0$ such that K as defined by (12.123) satisfies (12.122). ■

12.B Appendix: controller of the form $u = B'f(x_u)$ for discrete-time systems

We show in this section that for system (12.39) if a feedback controller of the form $u = B'f(x_u)$ achieves (G/G_p) and/or $(G/G_p)_{fg}$ for the unstable dynamics x_u, it also achieves (G/G_p) and/or $(G/G_p)_{fg}$ for the overall system.

Let us consider the unstable part of the input-additive case,

$$\rho x_u = Ax_u + B_u\sigma(u + d).$$

Assume that we have a feedback $u = B_u' f(x_u)$ such that $x_u \in \ell_p$ and, if possible, with finite gain,

$$\|x_u\|_p \leqslant c_1\|d\|_p. \tag{12.126}$$

Note that we impose a bit of special structure on the feedback. Namely, $u = B_u' f(x_u)$ instead of $u = f(x_u)$, but all our standard controllers satisfy this property which is easily seen if we recall that

$$-(I + B'P_\varepsilon B)^{-1}B'P_\varepsilon A = B'P_\varepsilon(I + BB'P_\varepsilon)^{-1}A.$$

If we achieve (G/G_p) for the unstable dynamics, then it is easily verified that we must have

$$B_u\sigma(B_u' f(x_u) + d) \in \ell_p$$

while achieving $(G/G_p)_{fg}$ for the unstable dynamics. This implies that

$$\|B_u\sigma(B_u' f(x_u) + d)\|_p \leqslant c_2\|d\|_p. \tag{12.127}$$

Now, in order to incorporate the stable dynamics, we want to establish that

$$\sigma(B_u' f(x_u) + d) \in \ell_p$$

and ideally with a finite gain

$$\|\sigma(B_u' f(x_u) + d)\|_p \leqslant c_3\|d\|_p. \tag{12.128}$$

This implies that for the stable dynamics, we will have

$$\|x_s\|_p \leqslant \gamma\|\sigma(B_u' f(x_u) + d)\|_p \leqslant \gamma c_3\|d\|_p,$$

where γ is the ℓ_p gain of the stable dynamics characterized by the pair (A_s, B_s).

From (12.128) and the fact that A_s is Schur stable, we conclude that there exists a c_4 such that

$$\|x_s\|_p \leqslant c_4\|\sigma(f(x_u)+d)\|_p \leqslant c_3c_4\|d\|_p.$$

This, together with (12.126), concludes the finite gain ℓ_p stability.

To establish (12.128), we first note that

$$B_u\sigma(B_u' f(x_u)+d) = B_u\sigma(B_u' f(x_u)) + B_u d_1$$

with $\|d_1\|_p \leqslant \|d\|_p$. But this implies that

$$\|B_u\sigma(B_u' f(x_u))\|_p \leqslant \|B_u\sigma(B_u' f(x_u)+d)\|_p + \|B_u\|\|d\|_p.$$

In other words, it is sufficient to prove that

$$\|\sigma(B_u' f(x_u))\|_p \leqslant c_4\|B_u\sigma(B_u' f(x_u))\|_p \tag{12.129}$$

in order to obtain

$$\begin{aligned}\|\sigma(B_u' f(x_u)+d)\|_p &\leqslant \|\sigma(B_u' f(x_u))\|_p + \|d\|_p\\ &\leqslant c_4\|B_u\sigma(B_u' f(x_u))\|_p + \|d\|_p\\ &\leqslant c_4\|B_u\sigma(B_u' f(x_u)+d)\|_p\\ &\qquad + (1+c_4\|B_u\|)\,\|d\|_p\\ &\leqslant (c_4c_2+1+c_4\|B_u\|)\,\|d\|_p,\end{aligned}$$

where we used (12.127).

It remains to verify (12.129) which is implied by the following static inequality:

$$\|\sigma(B_u' v)\| \leqslant c_4\|B_u\sigma(B_u' v)\|. \tag{12.130}$$

Since this is a static problem, we are using vector norms.

Note that we can find a matrix S such that

$$B_u = S\begin{pmatrix} B_{u1}\\ 0\end{pmatrix}$$

with B_{u1} surjective. Next, we note that it is sufficient to prove that

$$\|\sigma(B_{u1}' w)\| \leqslant c_5\|B_{u1}\sigma(B_{u1}' w)\| \tag{12.131}$$

for some suitably chosen c_5 since for $w = Sv$, we get

$$\begin{aligned}\|\sigma(B_u' v)\| &\leqslant c_5\|B_{u1}\sigma(B_u' v)\|\\ &\leqslant \frac{c_5}{\sigma_{\min}(S)}\|SB_{u1}\sigma(B_u' v)\|\\ &\leqslant \frac{c_5}{\sigma_{\min}(S)}\|B_u\sigma(B_u' v)\|,\end{aligned}$$

which yields (12.130) for a suitably chosen c_4. It remains to show (12.131). We consider two cases. If $B'_{u1}w$ saturates at least one channel, then

$$\begin{aligned}\|B_{u1}\sigma(B'_{u1}w)\| &\geqslant \langle B'_{u1}w_n, \sigma(B'_{u1}w)\rangle \\ &\geqslant \|B'_{u1}w_n\|_\infty \\ &\geqslant \frac{1}{\sqrt{m}}\sigma_{\min}(B'_{u1}),\end{aligned}$$

where

$$w_n = \frac{w}{\|w\|}.$$

In that case,

$$\begin{aligned}\|\sigma(B'_{u1}w)\| &\leqslant \sqrt{m}\,\|\sigma(B'_{u1}w)\|_\infty \\ &= \sqrt{m} \\ &\leqslant \frac{m}{\sigma_{\min}(B'_{u1})}\|B_{u1}\sigma(B'_{u1}w)\|.\end{aligned}$$

On the other hand without saturation,

$$\begin{aligned}\|B'_{u1}w\| &\leqslant \|B'_{u1}(B'_{u1}B'_{u1})^{-1}B_{u1}B'_{u1}w\|_2 \\ &\leqslant \|B'_{u1}(B'_{u1}B'_{u1})^{-1}\|\;\|B_{u1}B'_{u1}w\|.\end{aligned}$$

Combining the two cases with and without saturation yields (12.131) for a suitably chosen c_5, that is,

$$c_5 \geqslant \max\left\{\frac{m}{\sigma_{\min}(B'_{u1})},\; \|B'_{u1}(B'_{u1}B'_{u1})^{-1}\|\right\}.$$

13

Simultaneous external and internal stabilization: non-input-additive case

13.1 Introduction

For the case when the external signals appear additive to the control input, Chap. 12 develops control strategies for several simultaneous stabilization problems in both global and semi-global setting for both continuous- and discrete-time systems. In this chapter, we tackle the same problems, however, for the case of non-input-additive external signals. Unlike in the input-additive case where our results are more or less complete, we present here only some limited results.

As introduced in (11.1), a continuous-time linear system subject to actuator saturation where the disturbance is non-input additive can be described by

$$\Sigma^c : \begin{cases} \dot{x}(t) = Ax(t) + B\sigma(u(t)) + Ed(t), \\ z(t) = x(t), \quad t \geq 0, \\ y(t) = Cx(t), \quad t \geq 0 \end{cases} \tag{13.1a}$$

and similarly consider a discrete-time system of the form

$$\Sigma^d : \begin{cases} x(k+1) = Ax(k) + B\sigma(u(k)) + Ed(k), \\ \quad z(k) \quad = x(k), \quad k \geq 0, \\ \quad y(k) \quad = Cx(k), \quad k = 1, 2, \dots, \end{cases} \tag{13.1b}$$

where, as usual, $x \in \mathbb{R}^n$ is the state, $u \in \mathbb{R}^m$ is the input, $z \in \mathbb{R}^q$ is the controlled variable, and $y \in \mathbb{R}^p$ is the measured output. For simplicity, throughout this chapter, we consider a standard saturation function σ.

13.2 Simultaneous stabilization in global framework

We are concerned here with simultaneous external and internal stabilization problems in a global setting for the case of non-input-additive disturbance, that is, we consider systems of the form (13.1a). We consider here only two classical problems, namely, (G_p/G) and $(G_p/G)_{fg}$ problems. This is because our results, especially for $(G_p/G)_{fg}$ problem, are more or less negative and are vastly different

A. Saberi et al., *Internal and External Stabilization of Linear Systems with Constraints*, Systems & Control: Foundations & Applications, DOI 10.1007/978-0-8176-4787-2_13,

from the input-additive case discussed in the previous chapter. In fact, unlike the input-additive case, we start seeing a complex picture emerging here, and such a complex picture more or less persists in subsequent chapters as well, although the exact nature of the complexity varies from chapter to chapter. The fact that the results for the $(G_p/G)_{fg}$ problem are negative simply precludes considering the other global problems defined in Chap. 11.

Without much fanfare, let us first consider the problem $(G_p/G)_{fg}$, i.e., the problem of simultaneous global asymptotic stabilization and global L_p stabilization with finite-gain.

As said above, it turns out that an extremely restrictive necessary condition must be satisfied to achieve $(G_p/G)_{fg}$. In fact, the result given below indicates that if the disturbance excites those marginally unstable modes of the open-loop system, then $(G_p/G)_{fg}$ is not solvable. We state and prove this result only for continuous-time systems. The same result holds analogously for discrete-time systems as well.

Theorem 13.1 *Consider the system Σ in* (13.1a)*. The problem of simultaneous global asymptotic stabilization and global L_p stabilization with finite gain (the $(G_p/G)_{fg}$ problem) as defined in Problem 11.2 is solvable by any feedback control law* only if *the auxiliary system Σ_{aux} defined as,*

$$\Sigma_{\text{aux}} : \left\{ \begin{aligned} \dot{x} &= Ax + Ed \\ z &= C_z x \end{aligned} \right. \tag{13.2}$$

is globally L_p stable with finite gain, i.e., there exists a $\gamma'_p > 0$ such that

$$\|z\|_p \leqslant \gamma'_p \|d\|_p$$

for zero initial conditions.

Remark 13.2 *If (C_z, A) is detectable and (A, B) is ANCBC, then the necessary condition in the above theorem is also sufficient for state feedback. This is seen as follows: Since Σ_{aux} is L_p stable and detectable, the unstable internal dynamics of Σ_{aux} must be uncontrollable from d. Thus, for system Σ, we only need to design a globally stabilizing controller for the unstable part and choose zero control for the stable part. Then, we obtain global asymptotic stabilization and global L_p stabilization with finite gain for system Σ. However, this argument is not valid for measurement feedback because in that case the disturbance can affect the unstable dynamics via the estimated state of the unstable dynamics.*

Remark 13.3 *Note that even when the open-loop system is critically stable, the $(G_p/G)_{fg}$ problem is not solvable in general. Another interesting point of*

Theorem 13.1 is that the lowest gain γ_p' set by the auxiliary system Σ_{aux} provides a lower bound for the possible L_p gains achievable by any saturated control law for the original system. This indicates that, whenever the disturbance is not-input-additive and the control is saturated, the global framework for L_p stabilization with finite gain is an over-requirement. Hence, it becomes more meaningful to work with the notion of L_p stabilization for some restricted set of disturbances. One way to restrict the disturbance is by imposing a uniform bound on the L_p norm, which is the so-called semi-global L_p stabilization. But this might not be the best choice. In this sense, it becomes an important issue to identify those non-conservative subsets of L_p disturbances for which the finite-gain L_p stabilization is possible.

Proof : We assume zero initial conditions from the beginning, and let $d \in L_p$. Then we have $z = z_{u,0} + z_{0,d}$, where $z_{u,0}$ is the output of the system for input u and zero disturbance, while $z_{0,d}$ is the output of the system for zero input and disturbance d. Since the saturation function is bounded, for any given $T > 0$, there exists a constant $M > 0$ such that $\|z_{u,0}\|_{L_p[0,T]} < M$ for any input u and zero disturbance. Obviously,

$$\|z_{u,d}\|_{L_p[0,T]} \geqslant \|z_{0,d}\|_{L_p[0,T]} - \|z_{u,0}\|_{L_p[0,T]}.$$

Hence, for any $\lambda > 0$,

$$\|z_{u,\lambda d}\|_{L_p[0,T]} \geqslant \lambda\|z_{0,d}\|_{L_p[0,T]} - M,$$

because the controlled output z is linear in terms of d. This yields

$$\frac{\|z_{u,\lambda d}\|_{L_p[0,T]}}{\|\lambda d\|_{L_p[0,T]}} \geqslant \frac{\|z_{0,d}\|_{L_p[0,T]}}{\|d\|_{L_p[0,T]}} - \frac{M}{\lambda\|d\|_{L_p[0,T]}},$$

which holds for any nonzero disturbance $d \in L_p$. Letting $\lambda \to \infty$, we find that

$$\sup_d \frac{\|z_{u,d}\|_{L_p[0,T]}}{\|d\|_{L_p[0,T]}} \geqslant \sup_d \frac{\|z_{0,d}\|_{L_p[0,T]}}{\|d\|_{L_p[0,T]}}, \tag{13.3}$$

and hence

$$\inf_u \sup_d \frac{\|z_{u,d}\|_{L_p[0,T]}}{\|d\|_{L_p[0,T]}} \geqslant \sup_d \frac{\|z_{0,d}\|_{L_p[0,T]}}{\|d\|_{L_p[0,T]}}, \tag{13.4}$$

where $\inf_u$ indicates the infimum over all possible controls.

Suppose that Σ_{aux} is not finite-gain L_p stable. Then, for any $\gamma > 0$, there exist $d^* \in L_p$, $(d^* \neq 0)$, and $T > 0$ such that

$$\|z_{0,d^*}\|_{L_p[0,T]} > \gamma\|d^*\|_{L_p[0,T]}.$$

Thus, it follows from (13.4) that

$$\inf_u \sup_d \frac{\|z_{u,d}\|_p}{\|d\|_p} \geqslant \inf_u \sup_d \frac{\|z_{u,d}\|_{L_p[0,T]}}{\|d\|_{L_p[0,T]}} \geqslant \frac{\|z_{0,d^*}\|_{L_p[0,T]}}{\|d^*\|_{L_p[0,T]}} > \gamma.$$

Hence, it is impossible to achieve a finite gain for system Σ by any control law. ■

Theorem 13.1 implies that in general the simultaneous global L_p (or ℓ_p) stabilization and global asymptotic stabilization *with finite gain* is not achievable. This leads us to study the same problem however without the requirement of finite gain. In this case, unlike the results of Theorem 13.1, next theorem provides a positive result. That is, *whenever we do not require finite gain*, the simultaneous global L_p (or ℓ_p) stabilization and global asymptotic stabilization with arbitrary initial conditions is achievable for all $p \in [1, \infty)$.[1] We emphasize that the initial conditions of the given system need not be fixed at $x(0) = 0$, they can be arbitrary.

Theorem 13.4 *Consider the continuous- or discrete-time system Σ in* (13.1). *Assume that the pair (A, B) is asymptotically null controllable with bounded control (ANCBC), and (C, A) is detectable. Then, the problem of simultaneous global L_p (or ℓ_p) stabilization and global asymptotic stabilization with arbitrary initial conditions and without finite gain, as defined in Problem 11.4, is solvable for any $p \in [1, \infty)$ via measurement feedback controllers.*

Remark 13.5 *We observe that in the discrete-time case, Theorem 13.4 implies that for any $d \in \ell_p$ with $p \in [1, \infty)$, one can design a measurement feedback law such that the controlled output z goes to zero asymptotically because $z \in \ell_p$. In other words, the closed-loop system remains globally attractive in the presence of any $d \in \ell_p$. A similar conclusion can be made for continuous-time systems, due to Lemma 2.4.*

Proof : We first consider continuous-time systems. Let P_ε be the solution of the CARE,

$$A'P_\varepsilon + P_\varepsilon A - P_\varepsilon BB'P_\varepsilon + Q_\varepsilon = 0, \tag{13.5}$$

where for $\varepsilon \in [0, 1]$,

$$Q_\varepsilon > 0, \quad \lim_{\varepsilon \to 0} Q_\varepsilon = 0, \quad \frac{\mathrm{d}Q_\varepsilon}{\mathrm{d}\varepsilon} > 0.$$

---

[1] Note that it is impossible to achieve external stability for disturbance signals belonging to L_∞ (ℓ_∞); however, a class of such signals are identified in Chap. 15 for which saturated linear feedback control laws exist that preserve the boundedness of states.

Consider a dynamic $2n$-dimensional observer-based measurement feedback controller,

$$\begin{cases} \dot{\hat{x}} = A\hat{x} + B\sigma(u) + K(y - C\hat{x}) \\ \dot{\omega} = (A + BF)\omega + K(y - C\hat{x}) \\ u = -B'P_{\varepsilon_a(z)}z + F\omega, \end{cases} \tag{13.6}$$

where K and F are such that $A - KC$ and $A + BF$ are Hurwitz stable, $z = \hat{x} - \omega$, and $\varepsilon_a(z)$ is determined by

$$\varepsilon_a(z) = \max\{r \in (0,1] \mid z'P_r z \operatorname{trace}(B'P_r B) \leqslant \tfrac{\Delta^2}{4}\}, \tag{13.7}$$

and where $P_{\varepsilon_a(z)}$ is the solution of CARE (13.5) with ε replaced by $\varepsilon_a(z)$.

Define $e = x - \hat{x}$. The closed-loop system in terms of e, z, and ω is given by

$$\begin{cases} \dot{e} = (A - KC)e + Ed \\ \dot{z} = Az + B\sigma(-B'P_{\varepsilon_a(z)}z + F\omega) - BF\omega \\ \dot{\omega} = (A + BF)\omega + KCe. \end{cases}$$

In the absence of d, the above becomes:

$$\begin{cases} \dot{e} = (A - KC)e \\ \dot{z} = Az + B\sigma(-B'P_{\varepsilon_a(z)}z + F\omega) - BF\omega \\ \dot{\omega} = (A + BF)\omega + KCe. \end{cases}$$

Clearly, the origin is locally exponentially stable. To see global attractivity, we note that $e \to 0$ and $\omega \to 0$ as time tends to infinity due to the fact that both $A + BF$ and $A - KC$ are Hurwitz stable. Then, there exists a T such that $\|F\omega(t)\| \leqslant \frac{\Delta}{2}$ for all $t > T$. This and the adapting law (13.7) together imply that the saturation will be inactive for all $t > T$. Since σ is a standard saturation function, then the z dynamics becomes,

$$\dot{z} = Az - BB'P_{\varepsilon_a(z)}z$$

which is globally attractive. This concludes global asymptotic stability.

In the presence of d, since $A - KC$ and $A + BF$ are both Hurwitz stable, it follows that $e \in L_p$, $\omega \in L_p$ and $\omega \to 0$ (see Sect. 2.9). Then, as before, there exists a T such that $\|F\omega(t)\| \leqslant \frac{\Delta}{2}$ for all $t > T$. Therefore, as before, we can conclude that the saturation is inactive for all $t > T$, and hence, z dynamics becomes

$$\dot{z} = Az - BB'P_{\varepsilon_a(z)}z.$$

This system is known to be globally asymptotically stable and locally exponentially stable. Hence, $z \in L_p$ and $x = e + \hat{x} = (e + z + \omega) \in L_p$.

We consider next discrete-time systems. Consider a discrete-time linear system Σ^d given in (13.1b), and let P_ε be the solution of the DARE,

$$P_\varepsilon = A' P_\varepsilon A - A P_\varepsilon B(I + B' P_\varepsilon B)^{-1} B' P_\varepsilon A + Q_\varepsilon, \tag{13.8}$$

where for $\varepsilon \in [0, 1]$,

$$Q_\varepsilon > 0, \quad \lim_{\varepsilon \to 0} Q_\varepsilon = 0, \quad \tfrac{\mathrm{d}Q_\varepsilon}{\mathrm{d}\varepsilon} > 0.$$

Consider a dynamic $2n$-dimensional observer-based measurement feedback controller

$$\begin{cases} \rho\hat{x} = A\hat{x} + B\sigma(u) + K(y - C\hat{x}) \\ \rho\omega = (A + BF)\omega + K(y - C\hat{x}) \\ u \;\,= -(I + B' P_{\varepsilon_a(z)} B)^{-1} B' P_{\varepsilon_a(z)} Az + F\omega, \end{cases} \tag{13.9}$$

where K and F are such that $A - KC$, and $A + BF$ are Schur stable, $z = \hat{x} - \omega$, and $\varepsilon_a(z)$ is determined by

$$\varepsilon_a(z) = \max\{r \in (0, 1] \mid z' P_r z \operatorname{trace}(B' P_r B) \leqslant \tfrac{\Delta^2}{8}\}, \tag{13.10}$$

and where $P_{\varepsilon_a(z)}$ is the solution of the DARE (13.8) with ε replaced by $\varepsilon_a(z)$.

Define $e = x - \hat{x}$. The closed-loop system in terms of e, z, and ω is given by

$$\begin{cases} \rho e \;= (A - KC)e + Ed \\ \rho z \;= Az + B\sigma(-(I + B' P_{\varepsilon_a(z)} B)^{-1} B' P_{\varepsilon_a(z)} Az + F\omega) - BF\omega \\ \rho\omega = (A + BF)\omega + KCe. \end{cases} \tag{13.11}$$

In the absence of d, the above becomes:

$$\begin{cases} \rho e \;= (A - KC)e \\ \rho z \;= Az + B\sigma(-(I + B' P_{\varepsilon(z)} B)^{-1} B' P_{\varepsilon(z)} Az + F\omega) - BF\omega \\ \rho\omega = (A + BF)\omega + KCe. \end{cases}$$

Clearly, the origin is locally exponentially stable. To see global attractivity, we note that $e \to 0$ and $\omega \to 0$ as time tends to infinity due to the fact that both $A+BF$ and $A-KC$ are Schur stable. Then, there exists a N such that $\| F\omega(k)\| \leqslant \frac{\Delta}{2}$ for all $k > N$. This and the adapting law (13.10) together imply that the saturation will be inactive for all $k > N$ (see also (4.244) and the inequalities below it). Since σ is a standard saturation function, the z dynamics then becomes,

$$\rho z = Az - B(I + B' P_{\varepsilon_a(z)} B)^{-1} B' P_{\varepsilon_a(z)} Az$$

which is globally attractive. This concludes global asymptotic stability.

In the presence of d, for a standard saturation function, since $A - KC$ and $A + BF$ are both Schur stable, it follows that $e \in \ell_p$, $\omega \in \ell_p$, and $\omega \to 0$

Then, as before, there exists a N such that $\|F\omega(k)\| \leqslant \frac{\Delta}{2}$ for all $k > N$. Therefore, as before, we can conclude that the saturation is inactive for all $k > N$, and hence, z dynamics becomes,

$$\rho z = Az - B(I + B'P_{\varepsilon_a(z)}B)^{-1}B'P_{\varepsilon_a(z)}Az.$$

This system is known to be globally asymptotically stable and locally exponentially stable (see Chap. 4). Hence, $z \in \ell_p$ and $x = e + \hat{x} = (e + z + \omega) \in \ell_p$. ■

Remark 13.6 *Admittedly, the results of Theorem 13.4 which utilizes a 2n-dimensional dynamic measurement feedback pertain only to L_p (ℓ_p) stabilization without finite gain. Even then, the results are remarkable especially when we note that similar results by static state feedback are sparse. Furthermore, these results are true irrespective of the nature of initial conditions.*

13.3 Semi-global external stabilization and global asymptotic stabilization

As said earlier, Theorem 13.1 in general precludes achieving simultaneous global L_p (or ℓ_p) stabilization and global asymptotic stabilization *with finite gain*. This begs the question whether any positive results are feasible if the requirement of *global L_p* (or ℓ_p) stabilization is relaxed to *semi-global L_p* (or ℓ_p) stabilization. To pursue this, let us first formally define what can be termed as $(SG_p/G)_{fg}$ problem.

Problem 13.7 Consider the continuous- or discrete-time system Σ in (13.1). For any $p \in [1, \infty]$, this system is said to be **simultaneously semi-globally finite gain L_p stabilizable (or ℓ_p stabilizable) and globally asymptotically stabilizable** via static state feedback (or dynamic measurement feedback with dynamical order n_c) if, for any a priori given (arbitrarily large) $D > 0$, there exists a static state (or respectively, dynamic measurement) feedback law, possibly depending on D (or respectively, n_c, and D), such that the following conditions hold:

(*i*) For continuous-time case, the closed-loop system is finite gain L_p stable over the set $L_{p,p}(D)$. That is, there exists a positive constant γ_p such that with $x(0) = 0$, the following holds:

$$\|z\|_p \leq \gamma_p \|d\|_p, \quad \text{for all } d \in L_p \text{ with } \|d\|_p \leqslant D.$$

Similarly, for the discrete-time case, the closed-loop system is finite gain ℓ_p stable over the set $\ell_{p,p}(D)$. That is, there exists a positive constant γ_p such that with $x(0) = 0$, the following holds:

$$\|z\|_p \leq \gamma_p \|d\|_p, \quad \text{for all } d \in \ell_p \text{ with } \|d\|_p \leqslant D.$$

(*ii*) In the absence of any disturbance signal d, the equilibrium point $x = 0$ of the closed-loop system is globally asymptotically stable .

The above problem is coined as $(SG_{p,q}/G)_{fg}$ problem.

We have the following result:

Theorem 13.8 *Consider the continuous or discrete-time system Σ in* (13.1)*. Assume that the pair (A, B) is asymptotically null controllable with bounded control (ANCBC), and (C, A) is detectable. Then the problem of simultaneous global asymptotic stabilization and semi-global L_p (ℓ_p) stabilization with finite gain, as defined in Problem 13.7, is solvable for $p \in [1, 2]$ via either state or measurement feedback controllers.*

Remark 13.9 *A result for the continuous-time systems with $p = 2$ was reported in [98].*

Proof : We only provide a proof for the continuous-time case. The proof for discrete-time case follows along the same lines. Consider the H_∞ algebraic Riccati equation,

$$A'P + PA - PBB'P + \gamma^{-2}PEE'P + \varepsilon I = 0. \tag{13.12}$$

It can be shown that there exists a $\gamma > 0$ such that for all $\varepsilon \in (0, 1]$, there exists a unique positive definite solution $P_\varepsilon > 0$ to (13.12) such that $A - BB'P_\varepsilon + \gamma^{-2}EE'P_\varepsilon$ is asymptotically stable. Moreover, $P_\varepsilon \to 0$ as $\varepsilon \to 0$ and $\frac{\mathrm{d}P_\varepsilon}{\mathrm{d}\varepsilon} > 0$ for $\varepsilon \in (0, 1]$. We define the adapting scheme as

$$\varepsilon(x) := \max\left\{\varepsilon \in (0, 1] : x'P_\varepsilon x \operatorname{trace} P_\varepsilon \leqslant \frac{\nu^2}{\|BB'\|}\right\}. \tag{13.13}$$

Let the control law be $u(x) = -B'P_{\varepsilon(x)}x$. For simplicity, we use the shorthand $P = P_{\varepsilon(x)}$. Then, clearly,

$$\|u\|^2 = x'PBB'Px \leqslant \|BB'\|\lambda_{\max}(P)x'Px \leqslant \|BB'\|x'Px \operatorname{trace}(P) \leqslant \nu^2.$$

Hence, the adaptive control law never activates the saturation.

Define $V(x) = x'P_{\varepsilon(x)}x$. The derivative of V along the system trajectory is

$$\dot{V} = -\varepsilon\|x\|^2 - \|u\|^2 + \left[2d'E'Px - \gamma^{-2}x'PEE'Px\right] + x'\dot{P}x. \tag{13.14}$$

Let $h = E'Px$. We claim that for $p \geqslant 1$, there exists a $c_1 > 0$ such that

$$2\|d\|\|h\| - \gamma^{-2}\|h\|^2 \leqslant c_1\|h\|^{2-p}\|d\|^p. \tag{13.15}$$

This is seen as follows. First, (13.15) holds if $\|h\| = 0$. For $\|h\| \neq 0$,

$$2\|d\|\|h\| - \gamma^{-2}\|h\|^2 = \gamma^{-2}\|h\|^2 \left(\frac{2\gamma^2\|d\|}{\|h\|} - 1 \right)$$
$$\leqslant \gamma^{-2}\|h\|^2 \left(\frac{2\gamma^2\|d\|}{\|h\|} \right)^p = \gamma^{-2}(2\gamma^2)^p \|h\|^{2-p}\|d\|^p.$$

Further noting that $P = P_{\varepsilon(x)} \leqslant P_1$, we get

$$\|h\|^2 = x'PEE'Px \leqslant \|EE'\|\lambda_{\max}[P_1]V(x).$$

Then it follows from (13.15) that there exists a constant $c_2 > 0$ such that for $p \in [1, 2]$,

$$2\|d\|\|h\| - \gamma^{-2}\|h\|^2 \leqslant c_2 V^{1-p/2}\|d\|^p. \tag{13.16}$$

Continuing from (13.14), we get

$$\dot{V} + \varepsilon\|x\|^2 + \|u\|^2 \leqslant c_2 V^{1-p/2}\|d\|^p + x'\dot{P}x. \tag{13.17}$$

Let $d = 0$. We claim that $\dot{V} < 0$ for $x \neq 0$. In fact, when x is sufficiently large, the adaptation scheme (13.13) implies that $0 < \varepsilon < 1$ and

$$V(x)\,\mathrm{trace}(P) = \nu^2/\|BB'\|.$$

Hence, if $\dot{V} \geqslant 0$ for $x \neq 0$, then $\dot{P} \leqslant 0$, which contradicts (13.17) with $d = 0$. Thus, ultimately, the state trajectory enters a neighborhood of the origin where $\varepsilon(x) \equiv 1$ and $\dot{P} = 0$. It is easy to see that in this region, we have local exponential stability. This shows global asymptotic stabilization.

It remains to derive the finite-gain stability by assuming zero initial condition and $d \in L_p(D)$ for some $D > 0$ and $p \in [1, 2]$. Noting that $\dot{V}$ and $\dot{P}$ always have opposite signs for $\dot{P} \neq 0$, we obtain from (13.17) that

$$\dot{V} \leqslant c_2 V^{1-p/2}\|d\|^p.$$

This implies that for any given $d \in L_p(D)$, we have

$$V^{p/2}(x) \leqslant \frac{p}{2} c_2 D^p, \qquad \forall t \geqslant 0,$$

which means that the state x starting from the origin remains bounded. Then the adaptation scheme guarantees that there is a lower bound for $\varepsilon(x)$, i.e., $\exists\, \varepsilon^* > 0$ such that $\varepsilon(x) > \varepsilon^*$, given a specific disturbance $d \in L_p(D)$.

For $\dot{P} \neq 0$, we have

$$V(x)\,\mathrm{trace}(P) = \nu^2/\|BB'\|,$$

which implies

$$\dot{V}\,\mathrm{trace}(P) + V\,\mathrm{trace}(\dot{P}) = 0$$

and

$$V(x) \geqslant \frac{\nu^2}{\|BB'\|\,\mathrm{trace}\,P_1}.$$

Hence, for $\dot{V} < 0$ and $\dot{P} > 0$,

$$x'\dot{P}x \leqslant \|x\|^2\,\mathrm{trace}(\dot{P}) = \|x\|^2 \frac{\mathrm{trace}(P)}{V}(-\dot{V}) \leqslant -M\dot{V},$$

for some constant $M > 0$, where we have used the facts that x is bounded, $\mathrm{trace}(P) \leqslant \mathrm{trace}(P_1)$, and

$$\frac{\|x\|^2}{V}$$

has a positive lower bound. Then, it follows from (13.17) that for $\dot{V} < 0$,

$$\dot{V} + \frac{\varepsilon}{1+M}\|x\|^2 \leqslant \frac{c_2}{1+M} V^{1-p/2}\|d\|^p$$

which implies that for $p \in [1,2]$,

$$\dot{V} + \frac{\varepsilon}{1+M}\left(\frac{V}{\lambda_{\max}(P_1)}\right)^{1-p/2}\|x\|^p \leqslant \frac{c_2}{1+M} V^{1-p/2}\|d\|^p.$$

That is, there exist $\beta_1 > 0$ and $\beta_2 > 0$ such that

$$\frac{\mathrm{d}V^{p/2}}{\mathrm{d}t} + \beta_1\|x\|^p \leqslant \beta_2\|d\|^p. \tag{13.18}$$

On the other hand, for $\dot{V} \geqslant 0$ (thus $\dot{P} \leqslant 0$), it follows similarly from (13.17) that there exist $\beta_3 > 0$ and $\beta_4 > 0$ such that

$$\frac{\mathrm{d}V^{p/2}}{\mathrm{d}t} + \beta_3\|x\|^p \leqslant \beta_4\|d\|^p. \tag{13.19}$$

Combining (13.18) and (13.19), we obtain

$$\frac{\mathrm{d}V^{p/2}}{\mathrm{d}t} + \eta_1\|x\|^p \leqslant \eta_2\|d\|^p, \tag{13.20}$$

where $\eta_1 = \min\{\beta_1, \beta_3\}$ and $\eta_2 = \max\{\beta_2, \beta_4\}$. This dissipation inequality shows that

$$\|x\|_p \leqslant \gamma\|d\|_p,$$

where $\gamma = (\eta_2/\eta_1)^{1/p}$.

So far, we have shown the finite-gain L_p stability with state feedback. For the measurement feedback case, we use the controller,

$$\begin{cases} \dot{\hat{x}} = A\hat{x} + B\sigma(u) + K(y - C\hat{x}) \\ u = f(\hat{x}), \end{cases} \tag{13.21}$$

where K is chosen such that $A-KC$ is Hurwitz and f is the nonlinear state feedback we constructed in the state feedback but with E replaced by KC. Defining $e = x - \hat{x}$, we obtain

$$\dot{e} = (A - KC)e + Ed. \tag{13.22}$$

Since the error dynamics is exponentially stable, for all $d \in L_p(D)$, we have $e \in L_p$ and

$$\|e\|_p \leqslant \gamma_1 \|d\|_p,$$

for some $\gamma_1 > 0$. Therefore, there exists a $D' > 0$ such that $(e', d')' \in L_p(D')$. Then, from the result for state feedback and (13.21), we conclude that

$$\|\hat{x}\|_p \leqslant \gamma_2(\|e\|_p + \|d\|_p)$$

for some $\gamma_2 > 0$. Consequently, we obtain that

$$\begin{aligned} \|x\|_p &\leqslant \|\hat{x}\|_p + \|e\|_p, \\ &\leqslant \gamma_2\|d\|_p + (\gamma_2 + 1)\|e\|_p, \\ &\leqslant \{\gamma_2 + \gamma_1(\gamma_2 + 1)\}\, \|d\|_p. \end{aligned}$$

The finite gain from d to z follows immediately. ∎

13.4 L_p (ℓ_p) stabilization for $p \in [1,\infty)$ of open-loop neutrally stable linear systems with saturated linear control laws

We considered so far general linear systems with saturated feedback control laws when the given system is open-loop asymptotically null controllable with bounded control (ANCBC). When the external disturbances are non-input-additive, it is clear that in general there is *no* control law that can achieve L_p or ℓ_p stability *with finite gain*, unless the disturbance affects only the asymptotically stable modes of the system. However, *nonlinear feedback control laws* are constructed that can achieve simultaneous global asymptotic stability as well as L_p stability or ℓ_p stability *without finite gain*. Then a question arises: *Can a linear feedback control law achieve L_p stability without finite gain if it achieves global asymptotic stability*? The answer to this question in general is of course negative. However, as shown in [169] and as will be discussed in Chap. 14, for a double integrator which belongs to the class of marginally unstable systems, a saturated linear internally stabilizing feedback control law can achieve L_p stability without finite gain for $p \in [1, 2]$, but not for $p > 2$.

The above discussion brings forth another important question: *For what class of linear systems subject to non-input-additive disturbances can a saturated linear feedback control law achieve global asymptotic stability as well as L_p or ℓ_p stability with possibly arbitrary initial conditions*?

For the case of input-additive disturbance, Chap. 12 shows that for open-loop neutrally stable systems with input saturation, global asymptotic stability and the traditional L_p stability or ℓ_p stability with zero initial conditions can be achieved by a saturated linear feedback control law for all $p \in [1, \infty]$. In this section, we generalize these results in two ways by considering *arbitrary initial conditions as well as non-input-additive disturbances*. More specifically, we show that for open-loop neutrally stable systems with input saturation and for non-input-additive disturbance, there exist linear feedback control laws that achieve global asymptotic stability and external L_p (ℓ_p) stability for all $p \in [1, \infty)$ and for *arbitrary* initial conditions. In the course of our development here, for the same class of systems, we also show that the same saturated linear feedback control laws also achieve another type of external stability in the sense that any vanishing disturbance produces a vanishing state for arbitrary initial conditions. Such a property is usually possessed by a linear exponentially stable system.

Before we present a general result, we first work out two special cases, one for continuous-time and another for discrete-time systems. That is, in Sects. 13.4.1 and 13.4.2, we derive, respectively, the L_p and ℓ_p stability results by state feedback for open-loop neutrally stable systems with all their open-loop poles on the imaginary axis (continuous time) or on the unit circle (discrete time). These results are then extended in Sect. 13.4.3 to general open-loop neutrally stable systems by using measurement feedback controllers. For simplicity, we consider throughout standard saturation functions.

13.4.1 Continuous-time systems

If a continuous-time system (13.1a) is open-loop neutrally stable and (A, B) is stabilizable, then in a suitable basis, we have

$$A = \begin{pmatrix} A_c & 0 \\ 0 & A_s \end{pmatrix}, \qquad B = \begin{pmatrix} B_c \\ B_s \end{pmatrix}, \tag{13.23}$$

where A_c satisfies $A_c + A_c' = 0$, A_s is asymptotically stable, and (A_c, B_c) is controllable. For the moment, we ignore the asymptotically stable subsystem and assume that the system satisfies the condition $A + A' = 0$ and that (A, B) is controllable. Also, we assume that $y = z = x$, that is, we only consider state feedback and L_p stability with arbitrary initial conditions with the output equal to the state. A general result is presented in Sect. 13.4.3.

Theorem 13.10 *Consider the system*

$$\dot{x} = Ax + B\sigma(u) + d, \qquad x(0) = x_0, \tag{13.24}$$

where (A, B) *is controllable and* $A' + A = 0$. *Then for any* $\kappa > 0$, *the linear feedback control law* $u = -\kappa B'x$ *achieves simultaneous global* L_p *stability with arbitrary initial conditions for all* $p \in [1, \infty)$ *and global asymptotic stability.*

Remark 13.11 *The system in the form of* (13.24) *is in general not finite-gain* L_p *stabilizable as was shown in [166]. Also, note that the bounded disturbance* $d = -B\sigma(u) + x_0$ *with* x_0 *an unstable eigenvector of* A *yields an unbounded state* x *for zero initial condition. Hence, the system is not* L_∞ *stable for any controller. This idea can be easily extended to establish that the system is not* L_∞ *stable for any fixed initial condition and any controller.*

Proof : The result of global asymptotic stability follows from Chap. 4. We proceed to show global external stability. According to Lemma 13.21 in the appendix, we only need to show that the state of the following system:

$$\dot{x} = Ax - B\sigma(\kappa B'x) + Bd \tag{13.25}$$

is in L_p for all $d \in L_p \cap L_\infty \cap \mathcal{C}_0$.

Since d is vanishing, there exists a $T_0 > 0$ such that $\|d(t)\|_\infty < 1/2$ for all $t > T_0$. Let

$$V_1(x) = \frac{\|x\|^{p+1}}{p+1}. \tag{13.26}$$

Then, for $t > T_0$, we have $d = \sigma(d)$ and

$$\begin{aligned}
\dot{V}_1 &= \|x\|^{p-1}x'[Ax - B\sigma(\kappa B'x) + Bd] \\
&= \frac{1}{\kappa}\|x\|^{p-1}[-u'\sigma(u) - u'd] \\
&= \frac{1}{\kappa}\|x\|^{p-1}\{-u'[\sigma(d) - \sigma(d-u)] + u'[\sigma(u-d) - \sigma(u)]\} \\
&\leqslant -\frac{1}{2\kappa}\|x\|^{p-1}u'\sigma(u) + \frac{2\sqrt{m}}{\kappa}\|x\|^{p-1}\|d\|,
\end{aligned}$$

where in the last inequality, we have used the inequalities (13.51) and (13.52) in the appendix.

Next, since $A - \kappa BB'$ is Hurwitz, there exists a matrix $P > 0$ such that

$$(A - \kappa BB')P + P(A - \kappa BB') = -I.$$

We define

$$V_2(x) = \frac{1}{p}(x'Px)^{p/2}.$$

Clearly, there exist $\alpha > 0$ and $\beta > 0$ such that

$$\begin{aligned}\dot{V}_2 &= (x'Px)^{(p-2)/2}\{x'P\left[(A - \kappa BB')x + B(\sigma(u) - u) + Bd\right]\} \\ &\leqslant -\alpha\|x\|^p + \beta\|x\|^{p-1}\|u - \sigma(u)\| + \beta\|x\|^{p-1}\|d\| \\ &\leqslant -\alpha\|x\|^p + \beta\|x\|^{p-1}u'\sigma(u) + \beta\|x\|^{p-1}\|d\|,\end{aligned} \tag{13.27}$$

where we have used the inequality (13.53).

Now, letting $V(x) = 2\beta\kappa V_1(x) + V_2(x)$, we obtain

$$\dot{V} \leqslant -\alpha\|x\|^p + \gamma_1\|x\|^{p-1}\|d\| = -\alpha\|x\|^{p-1}\left(\|x\| - \frac{\gamma_1}{\alpha}\|d\|\right) \tag{13.28}$$

for $t > T_0$, where $\gamma_1 = \beta(1 + 4\sqrt{m})$. Now since d is vanishing, for any $\varepsilon_1 > 0$ and $\varepsilon_2 > 0$, there exists a $T_1 > T_0$ such that

$$\|d(t)\| < \min\left\{\varepsilon_2, \frac{\varepsilon_1}{2}\left(\frac{\alpha}{\gamma_1}\right)\right\}$$

for all $t > T_1$. Hence, if $\|x(t)\| > \varepsilon_1$ for some $t > T_1$, then

$$\dot{V}(t) < -\tfrac{\varepsilon_1\alpha}{2}\|x(t)\|^{p-1} < -\tfrac{\alpha}{2}\varepsilon_1^p.$$

Therefore, there exists a $T > T_1$ such that $\|x(T)\| < \varepsilon_1$. Since the closed-loop system without saturation is exponentially stable, we can choose ε_1 and ε_2 sufficiently small so that if $\|x(T)\| < \varepsilon_1$ and $\|d(t)\| < \varepsilon_2$ for all $t > T$, then the saturation of $u = -\kappa B'x$ will never be activated again. That is, the closed-loop system is ultimately linear for $t > T$ and exponentially stable. Hence, the L_p stability follows immediately. ■

As a by-product of the above proof, we obtain that the same linear feedback control law achieves CICS stability as defined in Definitions 2.70.

Theorem 13.12 *Consider the system* (13.24) *with the same conditions as in Theorem 13.10. Then for any $\kappa > 0$, the linear feedback control law $u = -\kappa B'x$ achieves CICS stability.*

Proof : According to Lemma 13.21, we only need to prove this theorem for a matched disturbance. From the proof of Theorem 13.10, we see that, given any vanishing disturbance and any initial condition, the saturation element will not be activated again after some finite time. Hence, by the fact that the closed-loop system is exponentially stable, the CICS stability follows easily. ■

13.4.2 Discrete-time systems

This subsection is parallel to Sect. 13.4.1 but deals with discrete-time systems. We assume in this subsection that (A, B) is controllable, A satisfies $A'A = I$, and $y = z = x$. A more general result is stated in Sect. 13.4.3.

Theorem 13.13 *Consider the system*

$$\rho x = Ax + B\sigma(u) + d, \qquad x(0) = x_0, \tag{13.29}$$

where (A, B) is controllable and $A'A = I$. Then there exists a $\kappa^* > 0$ such that for any $\kappa \in (0, \kappa^*]$, the linear feedback control law $u = -\kappa B'Ax$ achieves simultaneous global ℓ_p stability with arbitrary initial conditions for all $p \in [1, \infty)$ and global asymptotic stability.

Remark 13.14 *The control law $u = -\kappa B'Ax$ is also used in [5, 26].*

We first establish local exponential stability when using the feedback law given in Theorem 13.13.

Lemma 13.15 *Assume that $A'A = I$ and $\kappa B'B < 2I$ for some $\kappa > 0$. Then $A_\kappa := (I - \kappa BB')A$ is asymptotically stable if and only if (A, B) is controllable.*

Proof : Since $A'A = I$, all eigenvalues of the matrix have norm 1 and are hence unstable. Hence, stabilizability is equivalent to controllability, from which we have the necessity.

Note that for any $x \in \mathbb{C}^n$, we have

$$x^* A_\kappa' A_\kappa x = \|x\|^2 - \kappa x^* A' B(2I - \kappa B'B)B'Ax \leqslant \|x\|^2, \tag{13.30}$$

where x^* denotes the conjugate transpose of x. Assume we have an eigenvector x of A_κ, i.e.,

$$A_\kappa x = (A - \kappa BB'A)x = \lambda x.$$

Then from (13.30), we immediately obtain that $|\lambda| \leqslant 1$. If $|\lambda| = 1$, then from (13.30), we must have that $B'Ax = 0$. It follows that $Ax = \lambda x$, and using $A'A = I$, we obtain $x^* A = \lambda^* x^*$. This yields that $x \neq 0$ satisfies $x^* A = \lambda^* x^*$ and $x^* B = 0$, which is contradictory to the controllability of (A, B). ∎

We need also two inequalities as presented in the following lemma:

Lemma 13.16 *Let $\kappa B'B < I$, $A'A = I$, and $u = -\kappa B'Ax$, where $\kappa > 0$. Then the following two inequalities hold:*

$$u'\sigma(u) \leqslant \kappa\|x\|^2, \tag{13.31}$$
$$u'\sigma(u) \leqslant \sqrt{\kappa m}\|x\|. \tag{13.32}$$

Proof : Since $\kappa B'B < I$ is obviously equivalent to $\kappa BB' < I$, inequality (13.31) follows from

$$u'\sigma(u) \leqslant u'u = \kappa^2 x'A'BB'Ax \leqslant \kappa\|x\|^2.$$

Inequality (13.32) follows from

$$\begin{aligned} u'\sigma(u) &\leqslant \sum_{i=1}^{m} |\kappa(B'Ax)_i| \\ &\leqslant \kappa\sqrt{m}\|B'Ax\| \\ &= \sqrt{m}(\kappa^2 x'A'BB'Ax)^{1/2} \\ &\leqslant \sqrt{\kappa m}\|x\|, \end{aligned}$$

where $(\cdot)_i$ indicates the ith component. ■

Proof of Theorem 13.13 : The result of global asymptotic stability follows from Chap. 4. We proceed to show global external stability. According to Lemma 13.22, we only need to prove the theorem for matched disturbances, that is, it suffices to establish the ℓ_p stability of the following closed-loop system

$$\rho x = Ax - B\sigma(\kappa B'Ax) + Bd, \qquad x(0) = x_0 \tag{13.33}$$

for all $x_0 \in \mathbb{R}^n$ and $d \in \ell_p$. We choose in advance a small $\kappa^* > 0$ so that

$$2\kappa B'B < I \tag{13.34}$$

for all $\kappa \in (0, \kappa^*]$.

Since d is vanishing, there exists a $K_0 > 0$ such that $\|d(k)\| < 1/4$ for all $k \geqslant K_0$. Using the inequality

$$2\sigma'(u)B'Bd \leqslant \sigma'(u)B'B\sigma(u) + d'B'Bd$$

and inequalities (13.34) and (13.54), we get for $k \geqslant K_0$,

$$\begin{aligned}
&\|\rho x\|^2 - \|x\|^2 \\
&= \|Ax + B\sigma(u) + Bd\|^2 - \|x\|^2 \\
&= \sigma'(u)B'B\sigma(u) + d'B'Bd + 2\sigma'(u)B'Bd + 2\sigma'(u)B'Ax + 2d'B'Ax \\
&\leqslant -\frac{2}{\kappa}\sigma'(u)u - \frac{2}{\kappa}u'd + 2\sigma'(u)B'B\sigma(u) + 2d'B'Bd \\
&\leqslant -\frac{1}{\kappa}u'\sigma(u) - \frac{1}{\kappa}u'(2d) + \frac{1}{\kappa}\|d\|^2 \\
&\leqslant -\frac{1}{2\kappa}u'\sigma(u) - \frac{1}{2\kappa}u'[\sigma(u) + 4d] + \frac{1}{\kappa}\|d\|^2 \\
&\leqslant -\lambda_1 u'\sigma(u) + \lambda_2\|d\|^2,
\end{aligned} \tag{13.35}$$

with $\lambda_1 = \frac{1}{2\kappa}, \lambda_2 = \frac{3}{\kappa}$.

Note that by inequality (13.31), we have:

$$\lambda_1 u'\sigma(u) \leqslant \frac{1}{2}\|x\|^2 \leqslant \|x\|^2 + \lambda_2\|d\|^2. \tag{13.36}$$

Now, let $V_1(x) = \|x\|^3$. Denoting ρV_1 by $V_1(\rho x)$, by inequality (13.35), we have

$$\begin{aligned}
\rho V_1 &\leqslant [\|x\|^2 + \lambda_2\|d\|^2 - \lambda_1 u'\sigma(u)]^{\frac{3}{2}} \\
&\leqslant [\|x\|^2 + \lambda_2\|d\|^2]^{\frac{3}{2}} - \lambda_1[\|x\|^2 + \lambda_2\|d\|^2]^{\frac{1}{2}} u'\sigma(u) \\
&\leqslant [\|x\|^2 + \lambda_2\|d\|^2]^{\frac{3}{2}} - \lambda_1\|x\|u'\sigma(u) \\
&\leqslant \|x\|^3 + \varepsilon_1\|x\|^2 + \lambda_2\beta_1\|d\|^2 - \lambda_3\left[u'\sigma(u)\right]^2,
\end{aligned}$$

where $\lambda_3 = \frac{\lambda_1}{\sqrt{\kappa m}}$, and we used (13.55) and (13.36) in the second inequality, (13.59) and (13.32) in the last inequality. Recall from Lemma 13.28 that we can make ε_1 arbitrary small at the expense of a larger β_1. Consequently, we obtain

$$\rho V_1 - V_1 \leqslant -\lambda_3[u'\sigma(u)]^2 + \varepsilon_1\|x\|^2 + \lambda_2\beta_1\|d\|^2, \tag{13.37}$$

for all $k \geqslant K_0$.

Let $A_\kappa = A - \kappa BB'A$, which is Schur stable by Lemma 13.15. Then there exists a $P > 0$ such that

$$A_\kappa' P A_\kappa - P = -I.$$

We have

$$x'A_\kappa' P A_\kappa x - x'Px = -\|x\|^2. \tag{13.38}$$

Let $V_2(x) = x'Px$ and $v = B[\sigma(u) - u] + Bd$. Then, applying inequality (13.58), (13.38), and (13.53), we have

$$\begin{aligned}
\rho V_2 - V_2 &= (\rho x)'P(\rho x) - x'Px \\
&= (x'A_\kappa' + v')P(A_\kappa x + v) - x'Px \\
&= \|P^{1/2}A_\kappa x + P^{1/2}v\|^2 - x'Px \\
&\leqslant (\|P^{1/2}A_\kappa x\| + \|P^{1/2}v\|)^2 - x'Px \\
&\leqslant x'A_\kappa' PA_\kappa x - x'Px + \varepsilon_2\|P^{1/2}A_\kappa x\|^2 + \beta_2\|P^{1/2}v\|^2 \\
&\leqslant -\|x\|^2 + \varepsilon_3\|x\|^2 + \beta_3\|v\|^2 \\
&\leqslant -(1-\varepsilon_3)\|x\|^2 + \beta_4\|\sigma(u) - u\|^2 + \beta_5\|d\|^2 \\
&\leqslant -(1-\varepsilon_3)\|x\|^2 + \beta_4\left[u'\sigma(u)\right]^2 + \beta_5\|d\|^2,
\end{aligned} \tag{13.39}$$

where $\varepsilon_3 = \varepsilon_2\|P^{1/2}A_\kappa\|^2$, $\beta_3 = \beta_2\|P\|$, and $\beta_4,\ \beta_5 > 0$ are appropriately defined constants. According to Lemma 13.28, ε_2 (and hence ε_3) can be made arbitrarily small by choosing a larger $\beta_2 > 0$ and hence larger β_4 and β_5. Here, we take $\varepsilon_3 < 1/2$. Now, define

$$V = \frac{\beta_4}{\lambda_3}V_1 + V_2.$$

It follows from (13.37) and (13.39) that for $k \geqslant K_0$,

$$\rho V - V \leqslant -(1 - \varepsilon_3 - \varepsilon_4)\|x\|^2 + \beta\|d\|^2, \tag{13.40}$$

where $\varepsilon_4 = \frac{\varepsilon_1\beta_4}{\lambda_3}$ and $\beta = \frac{\beta_4}{\lambda_3}\lambda_2\beta_1 + \beta_5$. As we pointed out before, ε_4 can be chosen arbitrarily small by choosing a small ε_1. Hence, we assume that ε_1 has been chosen so that $\varepsilon_4 < 1/2$. Thus, $\eta := (1 - \varepsilon_3 - \varepsilon_4) > 0$. Analogously to the proof of Theorem 13.10 in continuous time, the inequality (13.40) can be used to show that the saturation at input will not be activated again after some finite time because the disturbance d is vanishing. Since the closed-loop system without saturation is exponentially stable and the disturbance is in ℓ_p, it easily follows that the state $x \in \ell_p$. This completes the proof. ■

Note that the proof of Theorem 13.13 is similar to the proof of Theorem 13.10. Hence, along the same lines as in the proof of Theorem 13.12, we can achieve CICS stabilization.

Theorem 13.17 *Consider the system* (13.29) *with the same condition as in Theorem 13.13. Then there exists a* $\kappa^* > 0$ *such that for any* $\kappa \in (0, \kappa^*]$, *the linear feedback control law* $u = -\kappa B'Ax$ *achieves CICS stabilization.*

Proof : Similar to the proof of Theorem 13.12. ■

Corollary 13.18 *Consider the system*

$$\rho x = Ax + B\sigma(u + d), \qquad x(0) = x_0, \tag{13.41}$$

where (A, B) is controllable and $A'A = I$. Then, for sufficiently small $\kappa>0$, the linear control law $u = -\kappa B'Ax$ achieves converging input converging state (CICS) stability with arbitrary initial condition as well as ℓ_p stability with arbitrary initial condition for any $p \in [1, \infty)$.

Proof : We have

$$\rho x = Ax + B\sigma(u + d) = Ax + B\sigma(u) + Bv$$

with $v = \sigma(u + d) - \sigma(u)$. Since

$$\|v(k)\| = \|\sigma(u(k) + d(k)) - \sigma(u(k))\| \leqslant \|d(k)\|,$$

we have $v \in c_0$ if $d \in c_0$ and $v \in \ell_p$ if $d \in \ell_p$. Then, the CICS stability and ℓ_p stability are direct consequences of Theorems 13.13 and 13.17. ■

Remark 13.19 *Finite-gain ℓ_p stability for the controller $u = -\kappa B'Ax$ and the system* (13.41) *has been established recently for $p \in (1, \infty]$ in [5]. In [5] the case $p = 1$ was excluded. The above corollary shows that for the case $p = 1$, we can at least achieve ℓ_p stability, although the result for finite gain is still not available.*

13.4.3 Generalization of L_p (ℓ_p) stability results

We consider in this section continuous- and discrete-time systems of the following form:

$$\begin{cases} \rho x = Ax + B\sigma(u) + d_1, & x(0) = x_0 \\ y = Cx + d_2 \\ z = C_z x + D_z\sigma(u) + d_3, \end{cases} \tag{13.42}$$

where ρx, as usual, represents either time derivative in continuous time or time shift in discrete time, x with $x(t) \in \mathbb{R}^n$ is the state, $\sigma(u)$ represents the standard vector saturation element on the input u with $u(t) \in \mathbb{R}^m$, $y(t) \in \mathbb{R}^r$ is the measurement output, where z is the controlled output with $z(t) \in \mathbb{R}^q$. The signals d_1, d_2, and d_3 represent external disturbances to the system. We assume that $d_i \in L_p$ $(i = 1, 2, 3)$ in continuous time and $d_i \in \ell_p$ $(i = 1, 2, 3)$ in discrete time. As in the previous subsection, we assume that A is open-loop neutrally stable, (A, B) is stabilizable, and (C, A) is detectable.

We extend here the results developed in the previous two subsections to the general system (13.42). Also, instead of using state feedback, we implement the state feedback laws via measurement. These extensions are direct consequences of the previous results. For completeness, we state the general results in this section, sketch the ideas, but omit the proofs.

Under our general assumptions, we know that in a suitable basis we have

$$A = \begin{pmatrix} A_c & 0 \\ 0 & A_s \end{pmatrix}, \qquad B = \begin{pmatrix} B_c \\ B_s \end{pmatrix}, \tag{13.43}$$

where A_c satisfies $A_c + A_c' = 0$ for continuous-time systems and $A_c' A_c = I$ for discrete-time systems, A_s is asymptotically stable, and (A_c, B_c) is controllable. We have the following theorem:

Theorem 13.20 *Consider the system* (13.42) *which satisfies the conditions stated above. Then there exists an observer-based linear feedback control law which achieves simultaneous global L_p stability with arbitrary initial conditions (continuous time) or global ℓ_p stability with arbitrary initial conditions (discrete time) for all $p \in [1,\infty)$ and global asymptotic stability. That is, given any $d_i \in L_p$ $(\ell_p)(i = 1, 2, 3)$ and an arbitrary initial conditions, we have $z \in L_p(\ell_p)$, and in the absence of any disturbances, we have global asymptotic stability. A suitable controller is given by*

$$\begin{cases} \rho\hat{x} = A\hat{x} + B\sigma(F\hat{x}) + L(C\hat{x} - y), & \hat{x}(0) = \hat{x}_0, \\ u = F\hat{x}, \end{cases} \tag{13.44}$$

where L is chosen so that $(A + LC)$ is asymptotically stable,

$$F = \begin{pmatrix} -\kappa B_c' & 0 \end{pmatrix}$$

for any $\kappa > 0$ in continuous time, and

$$F = \begin{pmatrix} -\kappa B_c' A_c & 0 \end{pmatrix}$$

with $\kappa > 0$ small enough in discrete time.

Proof : Global asymptotic stability result follows from Chap. 4. Regarding global external stability, we sketch here only the main idea behind the proof for the continuous-time case and omit the details. Given the suitable basis for which we obtained (13.43), the system can be put in the following diagonal form:

$$\begin{cases} \dot{x}_c = A_c x_c + B_c \sigma(u) + d_c \\ \dot{x}_s = A_s x_s + B_s \sigma(u) + d_s, \end{cases}$$

where the state has been decomposed into $x' = [x_c',\ x_s']$. We define $e = x - \widehat{x}$ and we get

$$\dot{e} = (A + LC)e + (d_1 + Ld_2).$$

Since $(d_1 + Ld_2) \in L_p$, obviously, $e \in L_p$. Given the controller as presented in the theorem, we obtain

$$\begin{cases} \dot{x}_c = A_c x_c + B_c \sigma(-B_c' x_c + B_c' e) + d_c \\ \dot{x}_s = A_s x_s + B_s \sigma(-B_c' x_c + B_c' e) + d_s. \end{cases}$$

Applying Theorem 13.10 to the first subsystem, we have $x_c \in L_p$, since $e \in L_p$ and $d_c \in L_p$. The second subsystem is exponentially stable with inputs x_c, e, and d_s in L_p. It follows that $x_s \in L_p$. Thus, we have $z \in L_p$. ∎

13.A Appendix: Some Preliminary Lemmas

This appendix contains several lemmas that are used in the proof of the main results.

In functional analysis, a general L_p signal for any $p \in [1, \infty)$ may be highly complicated; it may not be bounded or vanishing, which is not convenient in analysis. Fortunately, the following lemma is helpful in that if the control law is globally Lipschitz in state, we only need to consider those L_p signals that are bounded, vanishing, and matched with the control for the purpose of proving general L_p stability. This is adapted from [89].

Lemma 13.21 *Let $p \in [1, \infty)$ and (A, B) be stabilizable. Let F be Lipschitz continuous, i.e., there exists a constant $L > 0$ such that*

$$\|F(\xi) - F(\eta)\| \leqslant L\|\xi - \eta\| \tag{13.45}$$

for all $\xi,\ \eta \in \mathbb{R}^n$. Then, we have the following equivalences:

(i) The system

$$\dot{x} = Ax - B\sigma(F(x)) + d, \qquad x(0) = x_0 \tag{13.46}$$

is L_p stable for all $d \in L_p$ and for all $x_0 \in \mathbb{R}^n$ if and only if the state of system

$$\dot{\xi} = A\xi - B\sigma(F(\xi)) + B\widetilde{d}, \qquad \xi(0) = \xi_0 \tag{13.47}$$

is in L_p for all $\widetilde{d} \in L_p \cap L_\infty \cap \mathcal{C}_0$ and for all $\xi_0 \in \mathbb{R}^n$.

(*ii*) *The system* (13.46) *is finite-gain* L_p *stable if and only if system* (13.47) *is finite-gain* L_p *stable for all disturbances* $\widetilde{d} \in L_p \cap L_\infty \cap \mathcal{C}_0$.

(*iii*) *The system* (13.46) *is* $\mathcal{C}_0$ *stable for arbitrary initial conditions if and only if the system* (13.47) *has the same property.*

Proof : The necessity for each equivalence is self-evident. For sufficiency of (*i*), we introduce an auxiliary linear system

$$\dot{y} = (A - BK)y + d, \tag{13.48}$$

where K is such that $A - BK$ is Hurwitz. Define $z = x - y$. Then

$$\dot{z} = Az - B\sigma(F(z)) + B\widetilde{d},$$

where

$$\widetilde{d} = -[\sigma(F(z + y)) - \sigma(F(z))] + Ky.$$

Since the system (13.48) is L_p stable, by Lemma 2.4, the signal y is vanishing and bounded. Since the functions σ and F are both Lipschitz continuous, we have

$$\|\sigma(F(z + y)) - \sigma(F(z))\| \leqslant \|F(z + y) - F(z)\| \leqslant L\|y\|.$$

Thus, $\widetilde{d}$ is also vanishing and bounded. If $z \in L_p$ for all such $\widetilde{d} \in L_p \cap L_\infty \cap \mathcal{C}_0$, then $x = z + y \in L_p$.

From the above argument, it is easy to see that the equivalence also holds for L_p stability with finite gain by observing that system (13.48) is also finite-gain L_p stable. Also, it is easy to obtain the equivalence of $\mathcal{C}_0$ stability between the systems (13.46) and (13.47) because the system (13.48) is converging input converging state (CICS) stable. ■

A similar result holds for discrete-time systems. Since any ℓ_p signal ($1 \leqslant p < \infty$) is automatically vanishing and bounded, the discrete version of Lemma 13.21 can be simplified.

Lemma 13.22 *Let* $p \in [1, \infty)$ *and* (A, B) *be stabilizable. Let* F *be Lipschitz continuous, i.e., there exists a constant* $L > 0$ *such that* (13.45) *is satisfied for all* $\xi,\ \eta \in \mathbb{R}^n$. *Then, the system*

$$\rho x = Ax - B\sigma(F(x)) + d, \qquad x(0) = x_0 \tag{13.49}$$

is ℓ_p *stable for arbitrary initial conditions if and only if the system*

$$\rho z = Az - B\sigma(F(z)) + B\widetilde{d}, \qquad z(0) = z_0 \tag{13.50}$$

is ℓ_p *stable for arbitrary initial conditions. Moreover, the equivalence also holds for the other cases: finite-gain* ℓ_p *stability and converging input converging state (CICS) stability both with fixed or arbitrary initial conditions.*

13.B Some inequalities

This appendix contains some inequalities we used in the text.

Lemma 13.23 *For arbitrary vectors* $u, d \in \mathbb{R}^m$, *we have*

$$|u'[\sigma(u+d)-\sigma(u)]| \leqslant (2\sqrt{m})\|d\|. \tag{13.51}$$

Proof : We first establish (13.51) for scalars u and d. Because of symmetry, we only need to prove this inequality for all $d > 0$. Since $d > 0$, it simply follows that the inequality holds for all $u \geqslant 1$. It also holds for all $u + d \leqslant -1$ because $u \leqslant -1 - d < -1$. Since the case for $|u| \leqslant 1$ is easy, it remains to verify the following two cases:

Case 1 $u < -1,\ u + d > 1$ in which case, $-1 > u > 1 - d > -d$. Hence,

$$|u[\sigma(u+d)-\sigma(u)]| = |u(1+1)| = 2|u| < 2|d|.$$

Case 2 $u < -1,\ |u+d| \leqslant 1$. If $u < -1,\ -1 \leqslant u + d \leqslant 0$, then

$$\begin{aligned}|u[\sigma(u+d)-\sigma(u)]| &= |u(u+d+1)| = (-u)(u+d+1)\\ &= -u(1+u) - ud - d + d\\ &= -(1+u)(u+d) + d < d = |d|\end{aligned}$$

while, if $u < -1,\ 0 \leqslant u + d \leqslant 1$, then

$$\begin{aligned}&|u[\sigma(u+d)-\sigma(u)]|\\ &\quad = |u(u+d+1)| = (-u)(u+d+1)\\ &\quad = -(1+u)(u+d) - d + 2d\\ &\quad \leqslant -1 - u - d + 2d = -1 - (u+d) + 2d < 2d = 2|d|.\end{aligned}$$

The general vector case then follows from

$$|u'[\sigma(u-d)-\sigma(u)]| \leqslant \sum_{i=1}^{m} |u_i[\sigma(u_i-d_i)-\sigma(u_i)]| \leqslant 2\sum_{i=1}^{m} |d_i| \leqslant (2\sqrt{m})\|d\|.$$

■

Lemma 13.24 *Consider two arbitrary vectors* $u, d \in \mathbb{R}^m$. *Then, for any* $d \in \mathbb{R}^m$ *satisfying* $\|d\| < \frac{1}{2}$, *we have*

$$2u'[\sigma(d)-\sigma(d-u)] \geqslant u'\sigma(u). \tag{13.52}$$

Proof : We first establish (13.52) for scalars u and d with $|d| < \frac{1}{2}$. We consider the following three cases:

- If $d - u > 1$, then $-1/2 > d - 1 > u$ and

$$u[\sigma(d) - \sigma(d-u)] = |u||d-1| > \frac{1}{2}|u| \geqslant \frac{1}{2}u\sigma(u).$$

- If $d - u < -1$, then $1/2 < d + 1 < u$ and

$$u[\sigma(d) - \sigma(d-u)] = u(d+1) > \frac{1}{2}u \geqslant \frac{1}{2}u\sigma(u).$$

- If $|d - u| \leqslant 1$, then

$$u[\sigma(d) - \sigma(d-u)] = u^2 \geqslant u\sigma(u) \geqslant \frac{1}{2}u\sigma(u).$$

The general case with u and d vectors then immediately follows from

$$u'[\sigma(d) - \sigma(d-u)] = \sum_{i=1}^{m} u_i[\sigma(d_i) - \sigma(d_i - u_i)] \geqslant \frac{1}{2}\sum_{i=1}^{m} u_i\sigma(u_i) = \frac{1}{2}u'\sigma(u),$$

■

Lemma 13.25 *We have the following inequality:*

$$\|u - \sigma(u)\| \leqslant u'\sigma(u). \tag{13.53}$$

Moreover, for any $d \in \mathbb{R}^m$ satisfying $\|d\| < 1$, we have

$$-u'[\sigma(u) + d] \leqslant \frac{\|d\|^2}{4}. \tag{13.54}$$

Proof : Inequality (13.53) follows from

$$\|u - \sigma(u)\| \leqslant \sum_{i=1}^{m} |u_i - \sigma(u_i)| \leqslant \sum_{i=1}^{m} u_i\sigma(u_i) = u'\sigma(u).$$

To show inequality (13.54), we consider two cases. If $|u_i| > |d_i|$, then

$$-u_i[\sigma(u_i) + d_i] \leqslant 0.$$

If $|u_i| \leqslant |d_i| < 1$, then

$$-u_i[\sigma(u_i) + d_i] = -u_i(u_i + d_i) \leqslant \frac{d_i^2}{4}.$$

Hence, inequality (13.54) follows. ■

Lemma 13.26 *Assume $b \geqslant a \geqslant 0$ and $q \geqslant 1$. Then,*

$$(b-a)^q - b^q \leqslant -b^{q-1}a. \tag{13.55}$$

Proof : This follows from direct verification. ∎

Lemma 13.27 *Assume $q > r > 0$. For any $\varepsilon > 0$, there exists a $\beta > 0$ such that*

$$u^{q-r}v^r \leqslant \varepsilon u^q + \beta v^q \tag{13.56}$$

for all $u,\ v \geqslant 0$.

Proof : If $u = 0$, the inequality holds trivially. If $u \neq 0$, let $x = \frac{v}{u} \geqslant 0$. Then, it suffices to prove that

$$x^r - \beta x^q \leqslant \varepsilon. \tag{13.57}$$

It is easy to verify that $g(x) = x^r - \beta x^q$ attains its maximum at

$$\left(\frac{r}{\beta q}\right)^{\frac{1}{q-r}},$$

while the maximum is equal to

$$g_{\max} = \left(\frac{r}{\beta q}\right)^{\frac{r}{q-r}} \left(1 - \frac{r}{q}\right).$$

Clearly, $g_{\max}$ converges to zero as $\beta \to \infty$ and hence for a suitable choice of β, we know that (13.57) is satisfied. ∎

Lemma 13.28 *Assume $q > 1$. For any $\varepsilon > 0$, there exists a $\beta > 0$ such that*

$$(u+v)^q \leqslant u^q + \varepsilon u^q + \beta v^q \tag{13.58}$$

for all $u,\ v \geqslant 0$. On the other hand, if v is bounded by $M > 0$, then for any $\varepsilon > 0$, there exists a β (depending on M) such that

$$(u+v)^q \leqslant u^q + \varepsilon u^{q-1/2} + \beta v^{q-1/2} \tag{13.59}$$

for all $u \geqslant 0$ and all v satisfying $0 \leqslant v \leqslant M$.

Proof : The first part follows from Lemma 12.18.

For the second inequality, we first note for any $q \geqslant 1$, there exists a $N > 0$ such that

$$(1+z)^q - 1 \leqslant Nz \tag{13.60}$$

for $z \in [0,1]$. Then we obtain for $0 \leqslant v \leqslant u$ that

$$\begin{aligned}(u+v)^q - u^q &= u^q \left[(1+\tfrac{v}{u})^q - 1 \right] \\ &\leqslant N u^{q-1} v \\ &\leqslant N M^{1/2} u^{q-1} v^{1/2} \\ &\leqslant \varepsilon u^{q-1/2} + \beta_0 v^{q-1/2} \end{aligned} \tag{13.61}$$

where the first inequality is a consequence of (13.60) and the final inequality is a consequence of Lemma 13.27 which guarantees that for any $\varepsilon > 0$, there exists a suitable β_0 satisfying the last inequality.

On the other hand, for $v > u$, we have

$$(u+v)^q - u^q \leqslant 2^q v^q \leqslant M^{1/2} 2^q v^{q-1/2}. \tag{13.62}$$

Combining (13.62) and (13.61) yields the required result. ■

14

The double integrator with linear control laws subject to saturation

14.1 Introduction

As discussed in previous chapters, for linear systems subject to actuator saturation, simultaneous external and internal stabilization is possible only for systems which are asymptotically null controllable with bounded control (ANCBC). Even then, in general, nonlinear feedback control laws are required. Of particular interest is the use of linear feedback control laws. Clearly, as shown in Chaps. 12 and 13, for open-loop neutrally stable systems, the global asymptotic stability and external L_p (ℓ_p) stability for $p \in [1, \infty)$ with arbitrary initial conditions can be achieved by linear feedback control laws. We continue in this chapter the theme of pursuing systems other than open-loop neutrally stable ones for which such a simultaneous stabilization is feasible. We focus here on a canonical class of ANCBC systems, namely, a double integrator which is ubiquitously used. In fact, the double-integrator system is commonly seen in control applications including low friction, free rigid-body motion, such as single-axis spacecraft rotation and rotary crane motion (see [117], and the references therein). All these issues are the motivating factors why we choose in this chapter to reexamine the notions of external and internal stabilization via linear feedback control laws for a double integrator with linear feedback control laws subject to saturation.

As is obvious from Chap. 4, double integrator can be asymptotically stabilized globally with linear saturated feedback control laws. In fact, any internally stabilizing controller for the system without input saturation also stabilizes the system when the saturation element is present. Unlike the case of internal stability, the picture that emerges regarding the external stability as portrayed in this chapter is complex and challenging.

A double-integrator system subject to actuator saturation is described by

$$\begin{cases} \dot{x}_1 = x_2 \\ \dot{x}_2 = \sigma(u). \end{cases} \tag{14.1}$$

Let us utilize a linear state feedback control law,

$$u = -k_1 x_1 - k_2 x_2, \tag{14.2}$$

A. Saberi et al., *Internal and External Stabilization of Linear Systems with Constraints*, Systems & Control: Foundations & Applications, DOI 10.1007/978-0-8176-4787-2_14,

where $k_1 > 0$ and $k_2 > 0$. Then, a closed-loop system in the presence of a non-input-additive external disturbance d can be written as

$$\begin{cases} \dot{x}_1 = x_2 \\ \dot{x}_2 = \sigma(-k_1 x_1 - k_2 x_2) + d. \end{cases} \tag{14.3}$$

Whenever the external disturbance d is input additive, the above closed-loop system can be rewritten as

$$\begin{cases} \dot{x}_1 = x_2 \\ \dot{x}_2 = \sigma(-k_1 x_1 - k_2 x_2 + d), \end{cases} \tag{14.4}$$

where, clearly, the disturbance d directly affects the system through the input channel.

Our goals in this section are multifold; examination of (1) simultaneous global L_p stability and global internal stability of closed-loop double-integrator system in the presence of non-input-additive external disturbance d (system (14.3)) with arbitrary initial conditions, (2) simultaneous global L_p stability and global internal stability of closed-loop double-integrator system in the presence of input-additive external disturbance d (system (14.4)) with arbitrary initial conditions, (3) ISS stability of closed-loop double-integrator system in the presence of non-input-additive external disturbance d (system (14.3)). As we shall see, this examination portrays a complex and intricate picture regarding the external stability of double-integrator system. This motivates us to consider an additional goal, (4) construction of a class of sustained external disturbance signals under which some external stability properties of the system (14.3) are preserved.

In the absence of external disturbance d, both the closed-loop systems (14.3) and (14.4) coalesce are globally asymptotically stable for $k_1 > 0$ and $k_2 > 0$. Hence, we concentrate throughout the chapter in examining various L_p stability properties. This chapter is based on the results of [169], [150], [203], and [196].

14.2 L_p stability: non-input-additive case

In this section, we consider the double-integrator system (14.3) subject to actuator saturation and having a linear static state feedback law and non-input-additive disturbances in order to examine its L_p stability. The main result of this section is stated in the following theorem.

Theorem 14.1 *For any $k_1 > 0$ and $k_2 > 0$, the system*

$$\begin{cases} \dot{x}_1 = x_2 \\ \dot{x}_2 = \sigma(-k_1 x_1 - k_2 x_2) + d \end{cases} \tag{14.5}$$

is L_p stable for $p \in [1,2]$. In fact, for this double integrator system with non-input-additive disturbances, the simultaneous global L_p stabilization with arbitrary initial conditions and without finite-gain and globally asymptotic stabilization (as defined in Problem 11.4) is achievable for all $p \in [1,2]$ by any linear static state feedback. Furthermore, the above system (14.5) is not L_p stable for all $p \in (2,\infty]$.

Remark 14.2 *It is worth noting that the L_p stability stated in the above theorem is independent of the specific linear feedback controller. If $p \in [1,2]$, then any linear feedback controller that achieves global asymptotic stability also yields L_p stability, whereas for $p > 2$, there is no linear feedback controller that achieves L_p stability.*

Before proceeding to the proof of this theorem, we make some observations. By a simple scaling of the state and time, system (14.3) is equivalent to the new system:

$$\begin{cases} \dot{x}_1 = x_2 \\ \dot{x}_2 = \lambda\sigma(-x_1 - x_2) + \lambda d, \end{cases} \tag{14.6}$$

where $\lambda > 0$. We then introduce new coordinates:

$$y_1 = x_1 + x_2, \quad y_2 = x_2, \quad v = \lambda d \tag{14.7}$$

to obtain the system,

$$\begin{cases} \dot{y}_1 = y_2 - \lambda\sigma(y_1) + v \\ \dot{y}_2 = \quad\; - \lambda\sigma(y_1) + v. \end{cases} \tag{14.8}$$

Introducing a Lyapunov function $V(y_1, y_2)$ for this system,

$$V(y_1, y_2) = \lambda \int_0^{y_1} \sigma(\eta)\mathrm{d}\eta + \frac{1}{2}y_2^2, \tag{14.9}$$

we get

$$\tfrac{\mathrm{d}V}{\mathrm{d}t} = -\lambda^2\sigma^2(y_1) + \lambda\sigma(y_1)v + y_2 v. \tag{14.10}$$

If there is no disturbance at all, i.e., $v(t) \equiv 0$, then global asymptotic stability follows from LaSalle's invariance principle.

Note that because of saturation, L_∞ stability is impossible since

$$v = \lambda\sigma(y_1) + 1$$

is bounded and yields an unbounded state (we will elaborate more on this in the next section). Hence, in the proof of Theorem 14.1, we will focus on the L_p

stability for $p < \infty$. By Lemma 13.21 of Chap. 13, it suffices to consider only those L_p disturbance signals v which are vanishing, i.e., with the property that $\lim_{t \to \infty} v(t) = 0$. Theorem 14.1 follows after we establish three lemmas.

Before we state the following lemma, let us generalize slightly the notion of L_p stability as stated in Definition 2.55.

Definition 14.3 *The system* (14.3) *or the system* (14.4) *is said to be* L_p/L_q *stable if the state* $x \in L_p$ *given* $d \in L_q$ *and zero initial conditions. If* $p = q$, *we simply refer to this property as* L_p *stability as was done in earlier chapters.*

Lemma 14.4 L_p *stability of system* (14.8) *is equivalent to* L_∞/L_p *stability of system* (14.5) *for* $p \in [1, \infty)$.

Proof : If system (14.5) is L_p stable, then the state x and the derivative $\dot{x}$ are in L_p which implies that the state is bounded.

Conversely, we claim that if the state is bounded given any L_p disturbance, then the state is in L_p. Using Lemma 13.21 in Appendix of Chap. 13, it is sufficient to consider disturbances v with the property that $\lim_{t \to \infty} v(t) = 0$. Consider an arbitrary disturbance v; then because the state x is bounded, there exists a convergent subsequence with some limit $\bar{x}$, i.e., there exists an unbounded sequence t_i $(i \in \mathbb{N})$ with $t_i > t_{i-1}$ such that $\lim_{i \to \infty} x(t_i) = \bar{x}$.

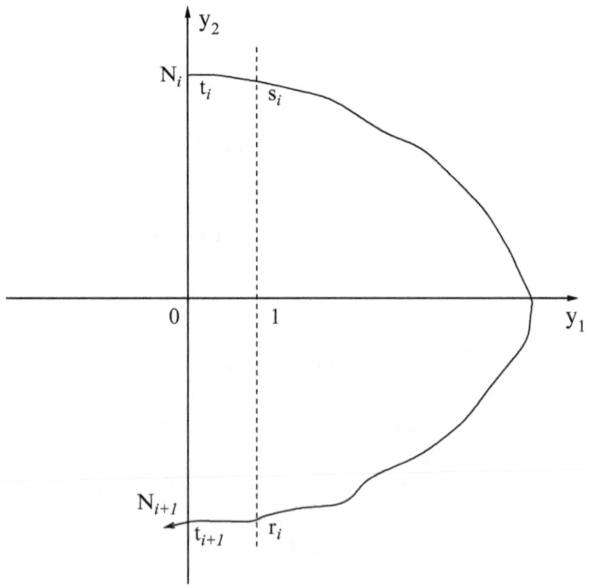

Figure 14.1: Proof of Lemma 14.5

Let ε and ρ be such that for any initial condition x_0 satisfying $\|x_0\| < \varepsilon$ and $\|v\|_\infty < \rho$, the saturation will never get activated. This is clearly possible. Since the system is globally asymptotically stable, the system converges to zero for initial condition $\bar{x}$ with disturbance $v = 0$, and hence, there exists a $T > 0$ such that starting in $\bar{x}$, we are in a ball of size $\varepsilon/2$ around the origin after T seconds. But then, it is easy to see that given $\varepsilon > 0$, there exists a $\delta < \rho$ such that for any initial condition $x(0) = x_0$ satisfying $\|\bar{x} - x_0\| < \delta$ and $\|v\|_\infty < \delta$, we have $\|x(T)\| < \varepsilon$.

But then, choose t_i such that $\|v(t)\| < \delta$ for all $t > t_i$ and $\|x(t_i) - \bar{x}\| < \delta$. Then we know that for $s > t_i + T$, we have $\|x(s)\| < \varepsilon$ and $\|v(t)\|_\infty < \rho$ for $t > s$. But then for $t > T + t_i$, the saturation will never get activated, and we have a stable linear system with disturbance $v \in L_p$. But this implies that $x \in L_p$. ■

In the next lemma and elsewhere, for any $D > 0$ and $1 \leqslant p \leqslant \infty$, we shall use the notation

$$L_p(D) := \{x \in L_p : \|x\|_p \leqslant D\}.$$

Lemma 14.5 *There exists a $\delta > 0$ such that, if $v \in L_2(\delta) \cap L_\infty(\delta)$, then the state y of system* (14.8) *is bounded for all initial conditions.*

Proof : We want to establish that the state remains bounded. We will investigate a period of time that the state is sufficiently large starting at time $t = t_0$ and show that the state can only grow a limited amount for $t > t_0$ and hence will remain bounded. For initial conditions sufficiently large, it is easily verified that the state will circle around the origin, and hence t_i $(i = 1, \ldots)$ as given by

$$\begin{aligned} t_1 &= \min\{\, t > t_0 \mid y_1(t) = 0\,\} \\ t_{i+1} &= \min\{\, t > t_i \mid y_1(t) = 0\,\} \end{aligned}$$

are well-defined. We define $N_i := y_2(t_i)$. We also define,

$$\begin{aligned} s_i &:= \min\{t > t_i : |y_1(t)| = 1\} \\ r_i &:= \min\{t > s_i : |y_1(t)| = 1\}. \end{aligned}$$

From the above, it is obvious that when the state is large enough or, equivalently, for N_i large enough, s_i and t_i are well-defined and $t_i < s_i < r_i < t_{i+1}$.

We will study the trajectory on the interval $[t_i, t_{i+1}]$ in detail. For ease of exposition, we consider the case $N_i > 0$. The case $N_i < 0$ can be shown using similar arguments. In the above, we have shown that the trajectory on the interval $[t_i, t_{i+1}]$ can be depicted as in Fig. 14.1. It is also clear that $N_{i+1} < 0$.

We will analyze the three time intervals $[t_i, s_i]$, $[s_i, r_i]$, and $[r_i, t_{i+1}]$ separately.

The interval $[t_i, s_i]$. The derivative of y_2 is bounded from above by $\lambda + \delta$ and hence

$$y_2(t) > N_i - (\lambda + \delta)(t - t_i)$$

for $t \in [t_i, s_i]$. But then, given that $\dot{y}_1 > y_2 - \lambda - \delta$, we find that

$$s_i - t_i < \frac{2}{N_i}, \tag{14.11}$$

assuming N_i is sufficiently large. It is also clear that $y_2(s_i) = \widetilde{N}_i$ satisfies $N_i/2 < \widetilde{N}_i < 2N_i$. We obtain that $\lambda\sigma(y_1) + y_2 < 2N_i$ holds for $t \in [t_i, s_i]$ if N_i is large. This implies that

$$\int_{t_i}^{s_i} [\lambda\sigma(y_1) + y_2]^2 \mathrm{d}t \leqslant \int_{t_i}^{s_i} (2N_i)^2 \mathrm{d}t \leqslant 8N_i,$$

and using (14.10),

$$\begin{aligned} V(s_i) &\leqslant V(t_i) + \int_{t_i}^{s_i} [\lambda\sigma(y_1) + y_2] v \mathrm{d}t \\ &\leqslant V(t_i) + \left(\int_{t_i}^{s_i} [\lambda\sigma(y_1) + y_2]^2 \mathrm{d}t \right)^{1/2} \left(\int_{t_i}^{s_i} v^2(t) \mathrm{d}t \right)^{1/2} \\ &\leqslant V(t_i) + \sqrt{8N_i} \left(\int_{t_i}^{s_i} v^2(t) \mathrm{d}t \right)^{1/2}. \end{aligned} \tag{14.12}$$

The interval $[r_i, t_{i+1}]$. Applying similar arguments, we obtain that $t_{i+1} - r_i < -2/N_{i+1}$, and if we define $y_2(r_i) = \widetilde{N}_{i+1}$, then $-2N_{i+1} < \widetilde{N}_{i+1} < -N_{i+1}/2$. Finally,

$$V(t_{i+1}) \leqslant V(r_i) + \sqrt{8|N_{i+1}|} \left(\int_{r_i}^{t_{i+1}} v^2(t) \mathrm{d}t \right)^{1/2}. \tag{14.13}$$

The interval $[s_i, r_i]$. Note that for the trajectory from s_i to r_i, we have $-\lambda - \delta \leqslant \dot{y}_2 \leqslant -\lambda + \delta$. It follows that

$$-(\lambda + \delta)(r_i - s_i) \leqslant y_2(r_i) - y_2(s_i) \leqslant -(\lambda - \delta)(r_i - s_i),$$

and if we use that $y_2(s_i) > N_i/2$ and $y_2(r_i) < N_{i+1}/2$, we get

$$(r_i - s_i) \geqslant \frac{1}{\lambda+\delta}[y_2(s_i) - y_2(r_i)] \geqslant \frac{1}{2(\lambda+\delta)}(N_i - N_{i+1}). \quad (14.14)$$

From the expression of $\dot{V}$ in (14.10), the inequality (14.14), and

$$\frac{1}{\lambda+\delta} \leqslant \frac{-1}{\dot{y}_2} \leqslant \frac{1}{\lambda-\delta}, \quad (14.15)$$

we get

$$\begin{aligned} V(r_i) &\leqslant V(s_i) - \lambda^2(r_i - s_i) + \int_{s_i}^{r_i} [\lambda\sigma(y_1) + y_2]v\,\mathrm{d}t \\ &\leqslant V(s_i) - \frac{\lambda^2(N_i - N_{i+1})}{2(\lambda+\delta)} + \left(\int_{s_i}^{r_i} [\lambda\sigma(y_1) + y_2]^2\mathrm{d}t\right)^{1/2} \left(\int_{s_i}^{r_i} v^2(t)\mathrm{d}t\right)^{1/2} \\ &\leqslant V(s_i) - \frac{\lambda}{4}N_i + \left(\int_{s_i}^{r_i} [\lambda\sigma(y_1) + y_2]^2\mathrm{d}t\right)^{1/2} \left(\int_{s_i}^{r_i} v^2(t)\mathrm{d}t\right)^{1/2}. \end{aligned}$$

Then,

$$\begin{aligned} \int_{s_i}^{r_i} [\lambda\sigma(y_1) + y_2]^2\mathrm{d}t &= \int_{\widetilde{N}_{i+1}}^{\widetilde{N}_i} (\lambda + y_2)^2 \frac{1}{(-\dot{y}_2)}\mathrm{d}y_2 \\ &\leqslant \frac{1}{\lambda-\delta}\int_{\widetilde{N}_{i+1}}^{\widetilde{N}_i} (\lambda + y_2)^2\mathrm{d}y_2 \\ &\leqslant \frac{1}{3(\lambda-\delta)}\left[(\widetilde{N}_i + \lambda)^3 - (\widetilde{N}_{i+1} + \lambda)^3\right]. \end{aligned}$$

This gives rise to

$$V(r_i) \leqslant V(s_i) - \frac{\lambda}{4}N_i + \alpha N_i^{3/2}A_i, \quad (14.16)$$

for some constant $\alpha > 0$ where

$$A_i = \left(\int_{t_i}^{t_{i+1}} v^2\mathrm{d}t\right)^{1/2},$$

and we have used $-\widetilde{N}_{i+1} \leqslant 6\widetilde{N}_i \leqslant 12N_i$, which is to be proved later.

Combining (14.12), (14.13), and (14.16), we finally obtain that

$$V(t_{i+1}) \leqslant V(t_i) - \frac{\lambda}{4} N_i + \alpha N_i^{3/2} A_i. \tag{14.17}$$

From (14.17), we claim that

$$N_{i+1}^2 \leqslant N_i^2 \left(1 + \tfrac{4}{\lambda}\alpha^2 A_i^2\right). \tag{14.18}$$

First, note that $V(t_i) = \frac{1}{2} N_i^2$ and then is easily seen from the fact that if $4\alpha A_i \leqslant \lambda N_i^{-1/2}$, then

$$N_{i+1}^2 \leqslant N_i^2 \leqslant N_i^2 \left(1 + \tfrac{8}{\lambda}\alpha^2 A_i^2\right),$$

while if $4\alpha A_i \geqslant \lambda N_i^{-1/2}$, we have

$$N_{i+1}^2 \leqslant N_i^2 + 2\alpha N_i^{3/2} A_i \leqslant N_i^2 \left(1 + \tfrac{8}{\lambda}\alpha^2 A_i^2\right).$$

We have only shown inequality (14.18) for $N_i > 0$. But a similar argument can be used for $N_i < 0$ to yield the same inequality, and hence, (14.18) holds for all i. Recall that $\prod_{i=1}^{\infty}(1 + b_i)$ with $b_i > 0$ is convergent if and only if $\sum_{i=1}^{\infty} b_i$ is convergent. We find that

$$N_i^2 \leqslant N_1^2 \prod_{i=1}^{\infty} \left(1 + \tfrac{8}{\lambda}\alpha^2 A_i^2\right) < \infty$$

because $v \in L_2$ implies that $\sum_{i=1}^{\infty} A_i^2 < \infty$.

It remains to show that $-\tilde{N}_{i+1} \leqslant 6\tilde{N}_i$. Let $y_2(p_i) = 0$ where $s_i < p_i < r_i$. Then, noticing $\dot{y}_1 = y_2 - \lambda + v \leqslant y_2 + \delta$ with $y_2 \geqslant 0$ from s_i to p_i and using (14.15), we get

$$\begin{aligned} y_1(p_i) - 1 = \int_{s_i}^{p_i} \dot{y}_1(t)\mathrm{d}t &\leqslant \int_{\tilde{N}_i}^{0} (y_2 + \delta)\frac{1}{\dot{y}_2}\mathrm{d}y_2 \\ &\leqslant \frac{1}{\lambda - \delta}\int_{0}^{\tilde{N}_i} (y_2 + \delta)\mathrm{d}y_2 \\ &\leqslant \frac{1}{\lambda - \delta}\left(\frac{1}{2}\tilde{N}_i^2 + \delta\tilde{N}_i\right). \end{aligned}$$

Also noticing that $\dot{y}_1 = y_2 - \lambda + v \leqslant y_2 - (\lambda - \delta)$ with $y_2 \leqslant 0$ from p_i to r_i, we get

$$\begin{aligned}
1 - y_1(p_i) &= \int_{p_i}^{r_i} \dot{y}_1 \mathrm{d}t \leqslant \int_{p_i}^{r_i} [y_2 - (\lambda - \delta)] \mathrm{d}t \\
&= \int_{0}^{\widetilde{N}_{i+1}} [y_2 - (\lambda - \delta)] \frac{1}{\dot{y}_2} \mathrm{d}y_2 \\
&\leqslant \frac{1}{\lambda + \delta} \int_{\widetilde{N}_{i+1}}^{0} [y_2 - (\lambda - \delta)] \mathrm{d}y_2 \\
&= \frac{1}{\lambda + \delta} \left[-\frac{1}{2} \widetilde{N}_{i+1}^2 + (\lambda - \delta) \widetilde{N}_{i+1} \right].
\end{aligned}$$

It follows that

$$\frac{1}{\lambda + \delta} \left[\frac{1}{2} \widetilde{N}_{i+1}^2 - (\lambda - \delta) \widetilde{N}_{i+1} \right] \leqslant \frac{1}{\lambda - \delta} \left[\frac{1}{2} \widetilde{N}_i^2 + \delta \widetilde{N}_i \right].$$

This yields

$$\frac{1}{2(\lambda + \delta)} \widetilde{N}_{i+1}^2 \leqslant \frac{1}{\lambda - \delta} \widetilde{N}_i^2,$$

which in turn yields $|\widetilde{N}_{i+1}| \leqslant 6 |\widetilde{N}_i|$. This completes the proof. ■

Let us next recall $\mathcal{C}_0$, as defined in Definition 2.69, denotes the set of all vanishing functions. We have the following lemma:

Lemma 14.6 *For $p > 2$ and any $\delta > 0$, there exists for any large enough initial condition, a disturbance $d \in L_p(\delta) \cap L_\infty(\delta) \cap \mathcal{C}_0$ such that the state y of system* (14.8) *is unbounded.*

Proof : We will use the same notation as in the proof of Lemma 14.5. Again, provided the initial conditions are large enough, we know that the state will basically circle around the origin, and the time instants t_i are well-defined for $i = 1, 2, \ldots$, provided we make sure that the state does not become too small. It will be an obvious consequence from our construction that the state does not get small. We again define r_i and s_i which are well-defined since if N_i is large enough, we always reach the area with $|y_1| > 1$. We set our disturbance equal to zero for $t \in [0, t_1]$. As before, we do our analysis for the case $y(t_i) = N_i > 0$ but the case when $y(t_i) < 0$ is identical with some obvious modifications.

The interval $[t_i, s_i]$ Choose $v(t) = 0$ for $t \in [t_i, s_i]$. Then from (14.10) and (14.11), we get

$$V(s_i) \geqslant V(t_i) - (s_i - t_i) \geqslant V(t_i) - \frac{2}{N_i}. \tag{14.19}$$

The interval $[r_i, t_{i+1}]$ Similarly, choosing $v(t) = 0$ for $t \in [r_i, t_{i+1}]$, we have

$$V(t_{i+1}) \geqslant V(r_i) - (s_i - t_i) \geqslant V(r_i) + \frac{2}{N_{i+1}}. \tag{14.20}$$

The interval $[s_i, r_i]$ For the trajectory from s_i to r_i, we choose

$$v = \alpha_i \operatorname{sgn}(\lambda + y_2)|\lambda + y_2|^{1/(p-1)},$$

where $\alpha_i > 0$ is a constant to be chosen later. Then, using

$$-(r_i - s_i) \geqslant \frac{1}{\lambda - \delta}[y_2(r_i) - y_2(s_i)] \geqslant \frac{2}{(\lambda - \delta)}(N_{i+1} - N_i), \tag{14.21}$$

we obtain

$$\begin{aligned} V(r_i) &= V(s_i) - \lambda^2(r_i - s_i) + \int_{s_i}^{r_i} (\lambda + y_2) v \mathrm{d}t \\ &\geqslant V(s_i) + \frac{2\lambda^2}{\lambda - \delta}(N_{i+1} - N_i) + \alpha_i \int_{s_i}^{r_i} |\lambda + y_2|^{\frac{p}{p-1}} \mathrm{d}t. \end{aligned}$$

Note that

$$\begin{aligned} \int_{s_i}^{r_i} |\lambda + y_2|^{\frac{p}{p-1}} \mathrm{d}t \geqslant \int_{s_i}^{p_i} |\lambda + y_2|^{\frac{p}{p-1}} \mathrm{d}t &\geqslant \int_{0}^{\tilde{N}_i} (\lambda + y_2)^{\frac{p}{p-1}} \frac{1}{(-\dot{y}_2)} \mathrm{d}y_2 \\ &\geqslant \frac{1}{\lambda + \delta} \int_{0}^{\tilde{N}_i} (y_2)^{\frac{p}{p-1}} \mathrm{d}y_2 \\ &= \frac{1}{\lambda + \delta} \left(\frac{p-1}{2p-1} \right) \tilde{N}_i^{\frac{2p-1}{p-1}} \\ &\geqslant b^{\frac{p}{p-1}} N_i^{\frac{2p-1}{p-1}} \end{aligned}$$

for some constant $b > 0$ and where we have used that $\widetilde{N}_i \geqslant N_i/2$. Letting

$$A_i = \left(\int_{s_i}^{r_i} |v|^p \mathrm{d}t \right)^{1/p} = \alpha_i \left(\int_{s_i}^{r_i} |\lambda + y_2|^{\frac{p}{p-1}} \mathrm{d}t \right)^{1/p},$$

we finally obtain that

$$V(r_i) \geqslant V(s_i) - aN_i + bA_i N_i^{2-\frac{1}{p}} \tag{14.22}$$

for some constant $a > 0$, where we have used the linear bound $-N_{i+1} \leqslant 12N_i$ obtained before.

Putting together (14.19), (14.20), and (14.22) yields

$$V(t_{i+1}) \geqslant V(t_i) - aN_i + bA_i N_i^{2-\frac{1}{p}}. \tag{14.23}$$

Now, choose α_i so that

$$A_i = b^{-1} N_i^{\frac{1}{p}-2} \left(aN_i + N_i + \tfrac{1}{2} \right).$$

Then

$$N_{i+1}^2 \geqslant N_i^2 + 2N_i + 1 = (N_i + 1)^2.$$

Hence, $|N_i| \geqslant |N_1| + i$. But then, it is easy to verify that

$$\sum_{i=1}^{\infty} A_i^p \leqslant [2(a+2)b^{-1}]^p \sum_{i=1}^{\infty} \frac{1}{(|N_1| + i)^{p-1}} < \infty$$

for $p > 2$ and hence $v \in L_p$. Note that a and b do not depend on N_1 and hence for N_1 large enough, we find that $d \in L_p(\delta)$. Next, we note that

$$\alpha_i \leqslant c N_i^{-1-\frac{1}{p-1}}$$

for some constant c independent of N_i. But then, in terms of the disturbance $d(t)$ (note that $v(t) = \lambda d(t)$), we obtain that

$$\|d(t)\| \leqslant d_1 N_i^{-1}, \quad t \in [t_i, t_{i+1}]$$

for a constant d_1 independent of N_i. Since N_i is increasing and converges to infinity, this clearly implies that for N_1 large enough, we have $d \in L_\infty(\delta)$ and moreover $d \in \mathcal{C}_0$. This completes the proof. ■

Proof of Theorem 14.1 : According to Lemma 13.21 of Appendix of Chap. 13, it is sufficient to consider those L_p disturbances that converge to zero as time tends to infinity. This implies that for any $\delta > 0$, there exists a time $\tilde{t}$ such that $v = \lambda d$ restricted to $[\tilde{t}, \infty)$ is in $L_2(\delta) \cap L_\infty(\delta)$. But then, according to Lemma 14.5 for the case $p = 2$, we have the state remaining bounded. This implies that the system is L_∞/L_2 stable, and hence, L_2 stability follows directly from Lemmas 14.4.

Since $L_p \cap L_\infty \subset L_2$ for all $p \in [1, 2)$, we can use the L_2 stability to conclude that the state is converging to zero as time tends to infinity. Thus, ultimately, the saturation has no effect, i.e., the system is linear. But this immediately implies L_p stability for all $p \in [1, 2)$.

We still need to establish that the system is not L_p stable for $p > 2$. We know from Lemma 14.6 that if the initial condition is large enough, then we can find a disturbance such that the state becomes unbounded and therefore the state will definitely not be in L_p. If we start at time $t = 0$ and we choose some time $\tilde{t} > 0$, then we can clearly choose a disturbance $\bar{d}$ which is constant and large on the interval $[0, \tilde{t}]$ such that the initial state at time $t = \tilde{t}$ is sufficiently large to apply Lemma 14.6 and hence the existence of a disturbance $\tilde{d} \in L_p$ on the interval $[\tilde{t}, \infty)$ such that the state is not in L_p. Note that the disturbance signal d which is equal to $\bar{d}$ on $[0, \tilde{t}]$ and equal to $\tilde{d} \in L_p$ on the interval $[\tilde{t}, \infty)$ is clearly in L_p. ∎

14.3 Further examination of L_∞ stability for non-input-additive disturbances

It is clear from Theorem 14.1 that L_p stability for $p > 2$ is not feasible whenever the external signals or disturbances are non-input additive. As such, L_∞ stability is not feasible. In fact, if the amplitude of disturbance is greater than one, it overwhelms the control input, and thus L_∞ stability is impossible. Nevertheless, it is fruitful to reexamine L_∞ stability in order to learn clearly different aspects of it. This is our goal in this section. Toward this goal, we consider the system given in (14.8) which is a rewritten version of double integrator with external disturbances not additive to control input. The arguments we use in this section do not depend on the value of λ, and hence, without any loss of generality, we set $\lambda = 1$ and rewrite (14.8) as

$$\begin{cases} \dot{y}_1 = y_2 - \sigma(y_1) + d \\ \dot{y}_2 = \quad\;\; - \sigma(y_1) + d. \end{cases} \tag{14.24}$$

As said earlier, it is obvious that any constant disturbance $d(t) \equiv c$ with magnitude $|c| > 1$ drives the state of system (14.24) to infinity. If a constant disturbance $d(t) \equiv c$ has magnitude $|c| = 1$, the state can still diverge. To see this, let $c = 1$ in (14.24). Then, starting from $(0, 0)$, direct integration of (14.24) shows that $(y_1(t), y_2(t)) = (1, 0.546)$ at $t = 1.21$. For all $t > 1.21$, $y_2(t) \equiv 0.546$ and $y_1(t) = 0.546(t - 1.21) + 1 \;\to \infty$ as $t \to \infty$.

In this section, we show that if the *constant* disturbance is of magnitude $|c| < 1$, then the state remains bounded. However, we will construct also an example to show that there is a vanishing disturbance with magnitude less than one which drives the state from the origin to infinity. We first have the following result.

Theorem 14.7 *Consider the system* (14.24) *with constant disturbance* $d(t) \equiv c$ *where* $|c| < 1$. *Then the trajectory starting from the origin is bounded and finally approaches the equilibrium* $(c, 0)$.

Proof : It is easy to see that the point $(c, 0)$ is an isolated equilibrium for $|c| < 1$. It follows from (14.24) that

$$y_2 \dot{y}_2 = -[\sigma(y_1) - d]\dot{y}_1 - [\sigma(y_1) - d]^2.$$

Taking the integral and noting that $d(t) \equiv c$ is a constant, we obtain

$$\int_0^t [\sigma(y_1) - c]^2 d\tau + \frac{1}{2} y_2^2(t) + \int_0^{y_1(t)} \sigma(\eta) \mathrm{d}\eta = c y_1(t). \tag{14.25}$$

Thus, $c y_1(t) > 0$ for $t > 0$. Assume for the moment that $0 < c < 1$. Then, we have $y_1(t) > 0$ for all $t > 0$.

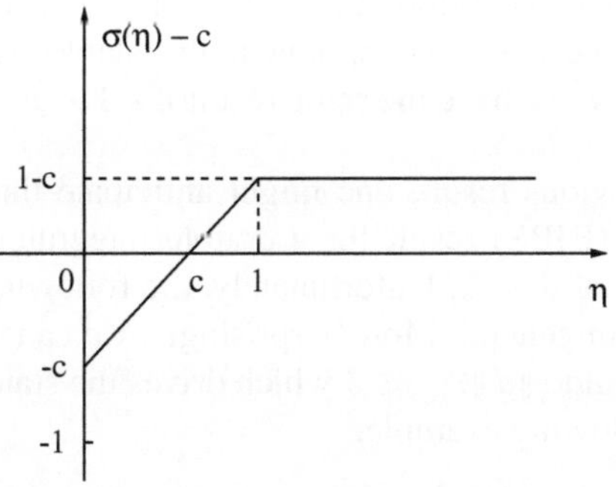

Figure 14.2: Plot of $\sigma(\eta) - c$

Rewrite (14.25) as

$$\int_0^t [\sigma(y_1) - c]^2 \mathrm{d}\tau + \frac{1}{2} y_2^2(t) + \int_0^{y_1(t)} [\sigma(\eta) - c] \mathrm{d}\eta = 0. \tag{14.26}$$

The plot of $[\sigma(\eta) - c]$ is shown in Fig. 14.2, from which we see that

$$\int_0^{y_1} [\sigma(\eta) - c] \mathrm{d}\eta$$

goes to infinity if $y_1 \to +\infty$ and reaches the minimum $-\frac{c^2}{2}$ at $y_1 = c$. Thus, it follows from the identity (14.26) that both $y_1(t)$ and $y_2(t)$ are bounded for all $t > 0$. Furthermore, we have $|y_2(t)| \leqslant c$ for all $t \geqslant 0$ since

$$\int_0^{y_1(t)} [\sigma(\eta) - c]\mathrm{d}\eta \geqslant -\frac{c^2}{2}$$

(see Fig. 14.2).

Next, we show that $(y_1, y_2) \to (c, 0)$ as $t \to \infty$. Note that the identity (14.26) also implies that

$$\int_0^t [\sigma(y_1) - c]^2 \mathrm{d}\tau \leqslant \frac{c^2}{2}.$$

That is, $[\sigma(y_1) - c] \in L_2$. The Lipschitz property of σ implies that

$$|\sigma(y_1(t_1)) - \sigma(y_1(t_2))| \leqslant |y_1(t_1) - y_1(t_2)| = |\dot{y}_1(t^*)| \cdot |t_1 - t_2|,$$

where t^* is between t_1 and t_2. It is seen from the first equation in (14.24) that $\dot{y}_1(t)$ is bounded for all $t > 0$. Thus, we have the uniform continuity of $[\sigma(y_1) - c]$. Then Barbalat's Lemma implies that $\sigma(y_1(t)) \to c$ as $t \to \infty$. Since $|c| < 1$, in fact we have shown $y_1(t) \to c$. Therefore, the system (14.24) is ultimately unsaturated. Without saturation, the system is exponentially stable. Hence, $y_2(t) \to 0$ as $t \to \infty$. By symmetry, we have the same result if $-1 < c < 0$. ■

Following the previous result, one might anticipate that there is a bounded-input-bounded-state (BIBS) result for a double integrator if the disturbance is restricted to $\|d\|_{L_\infty} \leqslant \delta < 1$. Unfortunately, the following example shows that even this is not true in general. More surprisingly, we can create a vanishing disturbance with magnitude $\|d\|_{L_\infty} \leqslant \delta$ which drives the state from zero to infinity. We first have the following example.

Example 14.8 The state of the following system

$$\begin{cases} \dot{y}_1 = y_2 - \sigma(y_1) + d \\ \dot{y}_2 = \quad\;\; - \sigma(y_1) + d, \quad y_1(0) = y_2(0) = 0, \end{cases} \tag{14.27}$$

with

$$d(t) = \begin{cases} 0.9, & \text{if } \; y_2(t) \geqslant 0 \\ -0.9, & \text{if } \; y_2(t) < 0 \end{cases}$$

is unbounded.

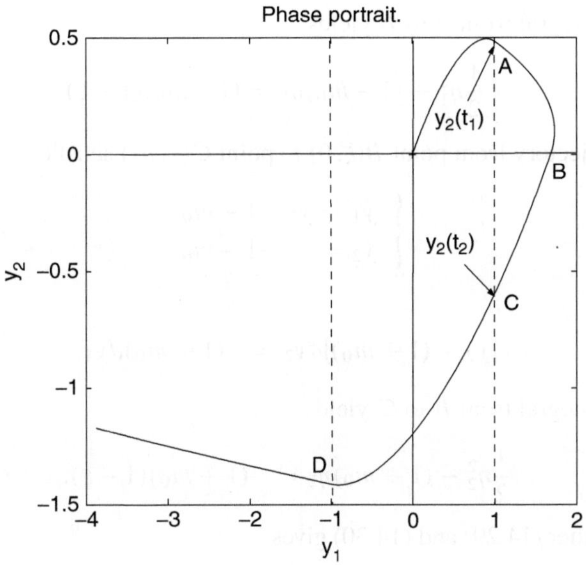

Figure 14.3: Plot of trajectory

Proof : For visualization purpose, some part of the trajectory starting from the origin is shown in Fig. 14.3.

We label the y_2 coordinates where the trajectory crosses the line $y_1 = 1$ as $y_2(t_1),\ y_2(t_2), y_2(t_3), \ldots$ in the order of increasing time $0 < t_1 < t_2 < t_3 < \cdots$ and show that $y_2(t_{2k-1}) \to +\infty$ as $k \to \infty$. First, directly solving the differential equation yields $y_2(t_1) = 0.48$ at $t_1 = 1.44$ (point A in Fig. 14.3). Then, $y_2(t_2) = -0.5827$ at some $t_2 > t_1$ (point C in Fig. 14.3). Next, we show that $|y_2(t_{2k})| > |y_2(t_{2k-1})|$ for all k.

To present a general analysis, we let

$$d(t) = \begin{cases} m_0, & \text{if } y_2(t) \geqslant 0 \\ -m_0, & \text{if } y_2(t) < 0, \end{cases} \tag{14.28}$$

where $m_0 \in (0, 1)$, and we denote three generic points: $A(1, \eta_1)$, $B(\xi, 0)$, and $C(1, \eta_2)$, as shown in Fig. 14.3. We will show that point C is always below the mirror image of point A when η_1 is large.

First, the trajectory from point $A(1, \eta_1)$ to point $B(\xi, 0)$ satisfies

$$\begin{cases} \dot{y}_1 = y_2 - 1 + m_0 \\ \dot{y}_2 = \quad -1 + m_0, \end{cases}$$

i.e.,

$$[y_2 - (1 - m_0)]dy_2 = -(1 - m_0)dy_1.$$

Taking the integral from A to B yields

$$-\frac{1}{2}\eta_1^2 + (1-m_0)\eta_1 = (1-m_0)(1-\xi). \tag{14.29}$$

Next, the trajectory from point $B(\xi, 0)$ to point $C(1, \eta_2)$ satisfies

$$\begin{cases} \dot{y}_1 = y_2 - 1 - m_0 \\ \dot{y}_2 = \quad\;\; -1 - m_0, \end{cases}$$

i.e.,

$$[y_2 - (1+m_0)]dy_2 = -(1+m_0)dy_1.$$

Taking the integral from B to C yields

$$\frac{1}{2}\eta_2^2 - (1+m_0)\eta_2 = -(1+m_0)(1-\xi). \tag{14.30}$$

Putting together (14.29) and (14.30) gives

$$\eta_2^2 - 2(1+m_0)\eta_2 = \frac{1+m_0}{1-m_0}\eta_1^2 - 2(1+m_0)\eta_1.$$

It follows that

$$[\eta_2 - (1+m_0)]^2 = [\eta_1 + (1+m_0)]^2 + \left[\frac{2m_0}{1-m_0}\eta_1^2 - 4(1+m_0)\eta_1\right].$$

Since $\eta_2 < 0$, we finally obtain

$$[|\eta_2| + (1+m_0)]^2 - [\eta_1 + (1+m_0)]^2 = \frac{2m_0}{1-m_0}\eta_1^2 - 4(1+m_0)\eta_1. \tag{14.31}$$

This implies that, if $\eta_1 > 2(1-m_0^2)/m_0$, we always have $|\eta_2| > \eta_1$, and the distance between $|\eta_2|$ and η_1 becomes arbitrarily large for any fixed $m_0 \in (0,1)$ as $\eta_1 \to \infty$.

Taking $m_0 = 0.9$, we see that if $\eta_1 > 0.42$, we have $|\eta_2| > \eta_1$. That is, the negative crossing points on the line $y_1 = 1$ is lower than the mirror image of the previous positive crossing point. Observe that when $\eta_1 \to \infty$, we have $|\eta_2| \approx \sqrt{19}\,\eta_1$, which indicates the order of $|y_2(t)|$ diverging to ∞ along the $y_1 = 1$ line.

The next thing is to ensure that the y_2 coordinate at point D (crossing the line $y_1 = -1$) is always lower than the point C. Note that from point C to D, $\dot{y}_2 = -y_1 - 0.9$ and y_1 decreases from 1 to -1. This means that y_2 decreases before y_1 hits -0.9. Because -0.9 is close to -1, we are almost sure that y_2 at D is lower than y_2 at C. To justify this, let $V = \frac{1}{2}[y_1^2(t) + y_2^2(t)]$. The trajectory from C to D satisfies

$$\begin{aligned} \dot{y}_1 &= y_2 - y_1 - m_0 \\ \dot{y}_2 &= -y_1 - m_0. \end{aligned}$$

Hence,

$$\dot{V} = -y_1^2 - y_1 m_0 + |y_2| m_0. \tag{14.32}$$

Note that the first two terms on the right-hand side are bounded. Hence, as $|y_2|$ gets large, it is guaranteed that $\dot{V} > 0$ for $|y_1| < 1$. For $m_0 = 0.9$, we have $\dot{V} > 0$ if $y_2 < -2.1$. In other words, for y_2 at C lower than -2.1, we always have y_2 at D lower than y_2 at C. For $0.2 < |y_2| < 2.1$, by estimating $|dy_2/dy_1|$, it can be verified that $|\Delta y_2| > 0.3$ for $y_1 \in [-0.9, 1]$, but $|\Delta y_2| < 0.1$ for $y_1 \in [-1, -0.9]$. This means that point D is lower than point C for $-2.1 < y_2 < -0.2$, as can be seen from the plot in Fig. 14.3.

The remaining argument relies on the symmetry; that is, the crossing points on the line $y_1 = -1$ have the similar behavior to those on the line $y_1 = 1$. Therefore, we can conclude the divergence of the crossing points along the two lines. This shows that the trajectory is unbounded for the fixed $m_0 = 0.9$. The trajectory over a longer period of time for $m_0 = 0.9$ is shown in Fig. 14.4. ∎

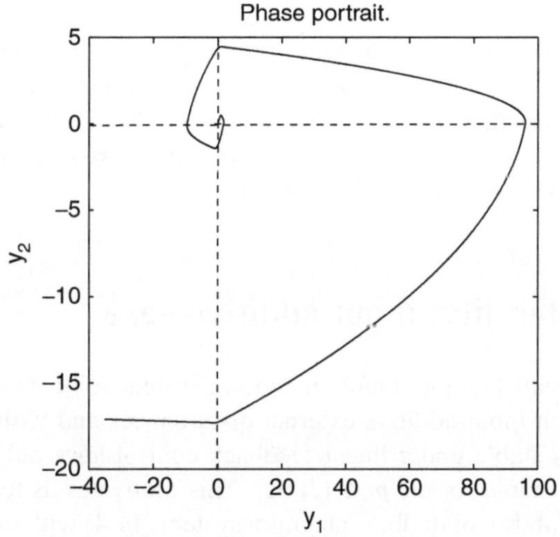

Figure 14.4: Plot of trajectory

From (14.31) in the proof, we see that, as $\eta_1 \to \infty$, η_2 is always lower than $-\eta_1$ and the distance between them can be arbitrarily large even for small m_0. This together with (14.32) shows that we can gradually reduce m_0 when η_1 gets large so that $d(t) \to 0$ while the trajectory diverges. Such a disturbance signal is constructed below.

We denote by $\{\eta_n\}$ the y-coordinates of the sequence of points where the trajectory crosses the line $\{(1, y_2) : y_2 > 0\}$ (such as point A in Fig. 14.3) for odd n then the line $\{(-1, y_2) : y_2 < 0\}$ (such as point D in Fig. 14.3) for even n, alternatively. Choose $m_0 = 0.9$ before the trajectory hits $(1, \eta_1)$, and then

$m_n = 2/|\eta_n|$ for the trajectory from η_n to η_{n+1}. Here, we use the sequence $\{\eta_n\}$ to represent the crossing points. We emphasize that the disturbance $d(t)$ is still defined by (14.28) with m_0 replaced by m_n selected above. For such a piecewise constant disturbance signal $d(t)$, it is easy to verify, using the argument before, that

$$(|\eta_{n+1}| + a_n)^2 - (|\eta_n| + a_n)^2 \geqslant \frac{16}{|\eta_n|},$$

where $a_n = 1 + m_n = 1 + 2/|\eta_n|$. Since $|\eta_n|$ is increasing, if it is upper bounded, it has a limit, say $\eta^* > 0$. But such an η^* would satisfy $0 \geqslant 16/\eta^*$, which is a contradiction. Hence, $|\eta_n| \to \infty$. This argument leads to the following result.

Theorem 14.9 *There exists a vanishing disturbance with magnitude less than one which drives the state of the system* (14.24) *from zero to infinity.*

Remark 14.10 *As was shown in Sect. 12.6.1 of Chap. 12, for neutrally stable systems, there exists a linear control law that achieves L_p stability for all $p \in [1, \infty)$ and ISS stability, i.e., any vanishing disturbance produces a vanishing state. Clearly, a single integrator is a neutrally stable system. However, the double integrator* (14.24) *does not have ISS stability.*

14.4 L_p stability: input-additive case

Section 14.2 reveals that the double-integrator system subject to actuator saturation and with non-input-additive external disturbances and with arbitrary initial conditions is L_p stable under linear feedback control laws only for $p \in [1, 2]$, and it is not L_p stable for all $p \in (2, \infty]$. This motivates us to consider in this section the L_p stability of double-integrator system (14.4) with input-additive disturbances and having a linear static saturated state feedback law with $k_1 > 0$ and $k_2 > 0$.

We note that a direct consequence of Theorem 14.1 in Sect. 14.2 is that such a system is L_p stable for all $p \in [1, 2]$ by simply rewriting the second equation in the system (14.4) as $\dot{x}_2 = \sigma(-k_1 x_1 - k_2 x_2) + v$ with a new disturbance v as

$$v = \sigma(k_1 x_1 + k_2 x_2) - \sigma(k_1 x_1 + k_2 x_2 - d).$$

It is easily verified that $\|v\|_p \leqslant \|d\|_p$. However, for $p \in (2, \infty]$, the result regarding L_p stability of the system (14.4) having input-additive disturbance is quite different from the non-input-additive case studied in the previous section. In fact, we have a positive result this time as given below.

Theorem 14.11 *The system given in* (14.4) *with* $k_1 > 0$ *and* $k_2 > 0$ *is* L_p *stable for all* $p \in [1,\infty]$. *In fact, for this double integrator system with input-additive disturbances, the simultaneous global* L_p *stabilization with arbitrary initial conditions and without finite-gain and globally asymptotic stabilization (as defined in Problem 11.4) is achieved for all* $p \in [1,\infty]$ *by any linear static state feedback.*

Proof : We prove that for arbitrary initial conditions and disturbance d satisfying $\|d\|_p < \delta$, the state remains bounded. This is clearly sufficient. Firstly, there obviously exists a T such that the tail of $d(t)$ with $t > T$ has L_p norm less than δ, and we can take $x(T)$ as initial condition. Secondly, if the system is L_∞/L_p stable, then we can derive an equivalent of Lemma 14.4 for the input-additive case to show that the system is L_p stable.

We again use the y-coordinates and study the following system:

$$\left\{\begin{array}{l} \dot{y}_1 = y_2 - \lambda\sigma(y_1 + d) \\ \dot{y}_2 = \qquad - \lambda\sigma(y_1 + d). \end{array}\right. \tag{14.33}$$

Assume that $|y_1(t)| < 3$ and y_2 large. With a bounded derivative of y_2, we obtain that y_1 is increasing and will become larger than 3, provided y_2 started large enough. While y_1 is larger than 3, we have

$$\begin{aligned} y_2(r) &= y_2(s) - \lambda(r-s) + \lambda \int_s^r v(\tau)d\tau \\ &\leqslant y_2(s) - \lambda(r-s) + \lambda \int_s^r |d(\tau)|^p d\tau \\ &\leqslant y_2(s) - \lambda(r-s) + \lambda\delta^p \end{aligned}$$

with

$$v = \sigma(y) - \sigma(y+d). \tag{14.34}$$

Since $v(t)$ is such that if $|d(t)| < 2$, then $v(t) = 0$ and otherwise

$$|v(t)| < |d(t)| < |d(t)|^p. \tag{14.35}$$

If we know that the trajectory is not getting close to zero and hence y_1 and y_2 are not simultaneously small, then eventually y_2 is negative and very small, and we have $|y_1| < 3$. Then it is easy to see that eventually we end up with $y_1(t) < -3$. Therefore, we see that we make the same loop as depicted in Fig. 14.1. We will investigate a period of time that we stay away from the ball with radius 10 around the origin. Hence, we move between three different regions:

- $\mathcal{D}_1 := \{y \in \mathbb{R}^2 \mid \|y\| > 10, y_1 > 3\}$
- $\mathcal{D}_2 := \{y \in \mathbb{R}^2 \mid \|y\| > 10, |y_1| < 3\}$
- $\mathcal{D}_3 := \{y \in \mathbb{R}^2 \mid \|y\| > 10, y_1 < -3\}$.

We define t_1, s_i, r_i, and t_{i+1} as

$$\begin{aligned} t_1 &= \min\{t > 0 \mid y_1(t) = 0\} \\ s_i &= \min\{t > t_i \mid |y_1(t)| = 3\} \\ r_i &= \min\{t > s_i \mid |y_1(t)| = 3\} \\ t_{i+1} &= \min\{t > r_i \mid y_1(t) = 0\} \end{aligned}$$

which, as argued before, are well defined.

The interval $[t_i, s_i]$. It is easy to check that

$$s_i - t_i \leqslant \frac{2}{N_i},$$

where $y_2(t_i) = N_i$. But then, we have

$$\begin{aligned} V(s_i) &\leqslant V(t_i) + 2N_i(s_i - t_i)^{\frac{p-1}{p}} \left(\int_{t_i}^{s_i} |v(t)|^p \mathrm{d}t \right)^{\frac{1}{p}} \\ &\leqslant V(t_i) + 4N_i^{\frac{1}{p}} \left(\int_{t_i}^{s_i} |v(t)|^p \mathrm{d}t \right)^{\frac{1}{p}} \\ &\leqslant V(t_i) + 4N_i^{\frac{1}{p}} \left[1 + \int_{t_i}^{s_i} |v(t)|^p \mathrm{d}t \right] \\ &\leqslant V(t_i) + 4N_i^{\frac{1}{p}} \left[1 + \int_{t_i}^{s_i} |d(t)|^p \mathrm{d}t \right] \end{aligned}$$

using Hölder's inequality, where v is defined by (14.34).

The interval $[r_i, t_{i+1}]$. Applying similar arguments, we obtain

$$V(t_{i+1}) \leqslant V(r_i) + 4|N_{i+1}|^{\frac{1}{p}} \left[1 + \int_{r_i}^{t_{i+1}} |d(t)|^p \mathrm{d}t \right].$$

The interval $[s_i, r_i]$. First, we note that similar as before, we have

$$\frac{N_i}{2} - \frac{N_{i+1}}{2} \leqslant r_i - s_i \leqslant 2N_i - 2N_{i+1}.$$

Hence,

$$\begin{aligned} V(r_i) &\leqslant V(s_i) - (r_i - s_i) + \int_{s_i}^{r_i} (\lambda + y_2) v \mathrm{d}t \\ &\leqslant V(s_i) - \frac{N_i - N_{i+1}}{2} + 2N_i \int_{s_i}^{r_i} |v(t)| \mathrm{d}t \\ &\leqslant V(s_i) - \frac{N_i - N_{i+1}}{2} + 2N_i \int_{s_i}^{r_i} |w(t)|^p \mathrm{d}t, \end{aligned}$$

where the last step is due to (14.35).

Putting everything together, we obtain

$$\begin{aligned} V(t_{i+1}) &\leqslant V(t_i) - \frac{N_i - N_{i+1}}{2} + 4N_i^{\frac{1}{p}} + 4|N_i|^{\frac{1}{p}} + 2N_i A_i \\ &\leqslant V(t_i) - \frac{1}{4} N_i + 2N_i A_i \end{aligned}$$

with

$$A_i = \int_{t_i}^{t_{i+1}} |w(t)|^p \mathrm{d}t < \delta$$

or in other words

$$N_{i+1}^2 \leqslant N_i^2 - \frac{1}{4} N_i$$

which clearly implies that N_i is bounded and hence the state is bounded. From Lemma 14.4, we then conclude that the state is in L_p. ∎

14.5 Input-to-state stability

Results in the previous two sections have shown that L_p stability of a double integrator with a saturating linear controller has a drastically different nature compared to its linear counterpart. In this section, our interest is to examine the input-to-state stability (ISS) as stated in Definition 2.65 for a double integrator with a saturating linear controller and with non-input-additive disturbances. By the definition, an immediate consequence of an ISS system is that for any fixed initial state $x_0 \in \mathbb{R}^n$, any bounded disturbance d must produce a bounded state. However, in this section, we show that the double-integrator system (14.3) which is controlled by a linear stabilizing control (14.2) is not ISS from d to x even for disturbances restricted to $d \in L_\infty(\delta)$ with $\delta > 0$. In fact, we show that for any arbitrarily small $\delta > 0$, we can find an external signal $d \in L_\infty(\delta)$ and an initial state so that the state trajectory of the closed-loop system is unbounded.

The main result of this section is summarized in the following theorem:

Theorem 14.12 *Consider the system*

$$\begin{cases} \dot{x}_1 &= x_2 \\ \dot{x}_2 &= \sigma(-k_1 x_1 - k_2 x_2) + d, \end{cases} \tag{14.36}$$

where $k_1,\ k_2$ are constants and d is an external disturbance. For any $k_1,\ k_2 > 0$, this system is globally asymptotically stable in the absence of disturbance. However, in the presence of disturbance d, this system is not ISS even for external disturbances that are restricted to the set $L_\infty(\delta)$ with $\delta > 0$ arbitrarily small.

Moreover, for any $\delta > 0$ and for any initial condition large enough, there exists a vanishing disturbance d with magnitude less than δ, i.e., $d \in \mathcal{C}_0 \cap L_\infty(\delta)$ which drives the state to infinity.

Proof : Note that this result follows directly from Lemma 14.6. However, we can construct a disturbance with a simpler structure to establish that the system is not ISS.

We again transform the system into the following form:

$$\begin{cases} \dot{y}_1 = y_2 - \lambda\,[\sigma(y_1) - d]\,, \\ \dot{y}_2 = \qquad - \lambda\,[\sigma(y_1) - d]\,. \end{cases} \tag{14.37}$$

Using the same Lyapunov function as before

$$V(y_1, y_2) = \lambda \int_0^{y_1} \sigma(\eta)d\eta + \frac{1}{2}y_2^2, \tag{14.38}$$

we obtain

$$\dot{V} = -\lambda^2\sigma^2(y_1) + \lambda^2\sigma(y_1)d + \lambda y_2 d, \tag{14.39}$$

which simply leads to global asymptotic stability if $d = 0$ by LaSalle's invariance principle.

Next, given any $\delta > 0$, we will construct a specific disturbance signal d with amplitude $|d(t)| \leqslant \delta$ for all $t \geqslant 0$ and show that there is an initial state such that the trajectory of system (14.37) starting from the initial state with the constructed disturbance diverges to infinity.

Assume that system (14.37) starts from some point on the ray $\{(1, y_2) : y_2 > 0\}$, say point A in Fig. 14.5. Define a piecewise constant disturbance as follows:

$$d(t) := \begin{cases} d_0, & \text{if } y_2(t) \geqslant 0, \\ -d_0, & \text{if } y_2(t) < 0, \end{cases} \tag{14.40}$$

where d_0 is any constant satisfying $0 < d_0 \leqslant \delta$. Some analysis is needed to see the behavior of points A, B, C, and D in Fig. 14.5, where in terms of coordinates, $A = (1, y_2(0))$, $B = (y_1(t_1), 0)$, $C = (1, y_2(t_2))$, and $D = (-1, y_2(t_2))$ with $0 < t_1 < t_2 < t_3$. To simplify notation, we let $\eta_1 = y_2(0)$, $\xi = y_1(t_1)$, and $\eta_2 = y_2(t_2)$. Thus, we have $A = (1, \eta_1)$, $B = (\xi, 0)$, and $C = (1, \eta_2)$.

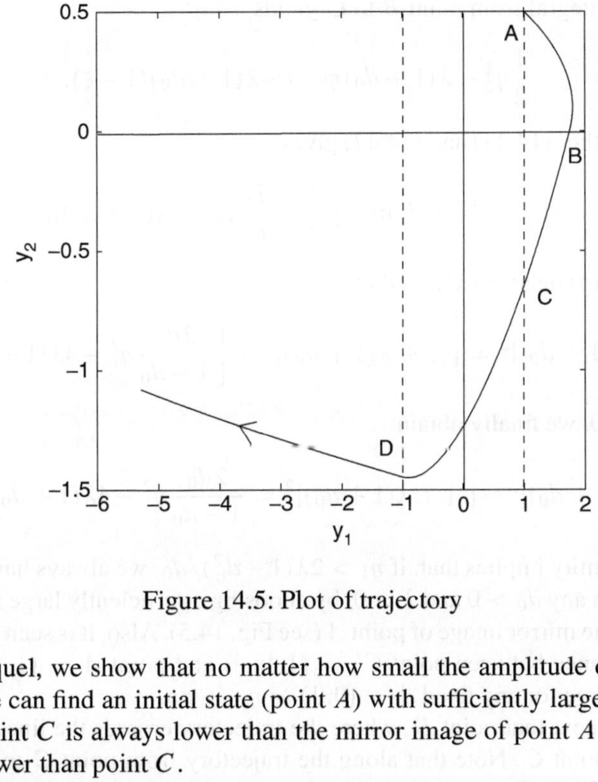

Figure 14.5: Plot of trajectory

In the sequel, we show that no matter how small the amplitude of the disturbance is, we can find an initial state (point A) with sufficiently large $\eta_1 = y_2(0)$ such that point C is always lower than the mirror image of point A and point D is further lower than point C.

First, the trajectory from point $A = (1, \eta_1)$ to point $B = (\xi, 0)$ satisfies

$$\begin{cases} \dot{y}_1 = y_2 - \lambda(1 - d_0), \\ \dot{y}_2 = -\lambda(1 - d_0), \end{cases}$$

i.e.,

$$[y_2 - \lambda(1 - d_0)]dy_2 = -\lambda(1 - d_0)dy_1.$$

Taking the integral from point A to B yields

$$-\frac{1}{2}\eta_1^2 + \lambda(1-d_0)\eta_1 = \lambda(1-d_0)(1-\xi). \tag{14.41}$$

Next, the trajectory from point $B = (\xi, 0)$ to point $C = (1, \eta_2)$ satisfies

$$\begin{cases} \dot{y}_1 = y_2 - \lambda(1+d_0), \\ \dot{y}_2 = -\lambda(1+d_0), \end{cases}$$

i.e.,

$$[y_2 - \lambda(1+d_0)]dy_2 = -\lambda(1+d_0)dy_1.$$

Taking the integral from point B to C yields

$$\frac{1}{2}\eta_2^2 - \lambda(1+d_0)\eta_2 = -\lambda(1+d_0)(1-\xi). \tag{14.42}$$

Putting together (14.41) and (14.42) gives

$$\eta_2^2 - 2\lambda(1+d_0)\eta_2 = \frac{1+d_0}{1-d_0}\eta_1^2 - 2\lambda(1+d_0)\eta_1.$$

A simple rearrangement yields that

$$[\eta_2 - \lambda(1+d_0)]^2 = [\eta_1 + \lambda(1+d_0)]^2 + \left[\frac{2d_0}{1-d_0}\eta_1^2 - 4\lambda(1+d_0)\eta_1\right].$$

Since $\eta_2 < 0$, we finally obtain

$$[|\eta_2| + \lambda(1+d_0)]^2 - [\eta_1 + \lambda(1+d_0)]^2 = \frac{2d_0}{1-d_0}\eta_1^2 - 4\lambda(1+d_0)\eta_1. \tag{14.43}$$

This key identity implies that, if $\eta_1 > 2\lambda(1-d_0^2)/d_0$, we always have $|\eta_2| > \eta_1$. Hence, given any $d_0 > 0$ and $\lambda > 0$, by choosing a sufficiently large η_1 point C is lower than the mirror image of point A (see Fig. 14.5). Also, it is seen from (14.43) that the distance between point $C' = (1, |\eta_2|)$ and point $A = (1, \eta_1)$ becomes arbitrarily large for any fixed $d_0 \in (0, 1)$ as $\eta_1 \to \infty$.

We show next that point D, where the trajectory crosses the line $y_1 = -1$, is lower than point C. Note that along the trajectory from point C to D, we have $|y_1(t)| \leqslant 1$. For this part of trajectory, V_λ defined in (14.38) becomes

$$V(y_1, y_2) = \frac{1}{2}\left[\lambda y_1^2 + y_2^2\right].$$

Now, given the disturbance of fixed magnitude $d = -d_0$ (since during this period $y_2(t) < 0$), it follows from (14.39) that

$$\dot{V} = -\lambda^2 y_1^2 - \lambda^2 y_1 d_0 + \lambda |y_2| d_0. \tag{14.44}$$

So, if $|y_2|$ is large enough, say $|y_2| > 2\lambda/d_0$ (assume $0 < d_0 < 1$), then $\dot{V} > 0$ for the trajectory from point C to D. This implies that point D is lower than C when $|\eta_2| = |y_2(t_2)|$ is large, which is possible by choosing η_1 large.

By considering the points A and C as two points of the trajectory when it crosses the line $y_1 = 1$ and noting the symmetry, it is easy to see from (14.43) and (14.44) that the trajectory of system (14.37) starting from point A diverges to infinity, given the disturbance defined in (14.40). Furthermore, as the trajectory diverges, the difference between η_1 and $|\eta_2|$ increases. The proof is now complete. ∎

The above results point out that there exist bounded disturbances with arbitrarily small L_∞ norm that cause the states to grow unbounded from certain initial conditions. Even more dramatically, they point out that unbounded growth can be achieved by *vanishing* disturbances with arbitrarily small L_∞ norm.

The work of [203] extends the above negative results to classes of small L_∞ signals with further restrictions. In what follows, we show that certain typical disturbances belonging to even more restricted classes than the ones considered above can drive the state unbounded. We omit the proofs of these results because they are rather straightforward variations of the proof of Theorem 14.12.

We first consider disturbances that are not only bounded but have one or more derivatives that are bounded. Such derivative bounds are often appropriate in modeling real disturbances, and hence, there has been some discussion on whether disturbances with limited derivatives (i.e., sluggish disturbances) have different rejection properties. In fact, as formalized below, we find that even such derivative-bounded signals can drive the state unbounded.

Theorem 14.13 *Consider the system as given in* (14.36) *which is repeated below:*

$$\begin{cases} \dot{x}_1 = x_2 \\ \dot{x}_2 = \sigma(-k_1 x_1 - k_2 x_2) + d, \end{cases} \tag{14.45}$$

where k_1, k_2 are constants and d is an external disturbance that is not only in $L_\infty(\delta)$ but also satisfies $|\frac{d^i w}{dt^i}| < \delta_i$ for $i = 1, \ldots, k$, for some finite k. Then, for any $\delta > 0$, $\delta_1 > 0, \ldots, \delta_k > 0$, there is such a disturbance d that drives the state of the system in (14.45) *unbounded for some initial condition x_0.*

Theorem 14.13 indicates that even when the disturbance is constrained to be small and have small derivatives, a feedback-controlled double-integrator with actuator saturation is not ISS. We note briefly that this theorem can be proved by smoothing out the disturbance d of Sect. 14.5 that drives the state unbounded. In a recent paper by Sontag [153], some new definitions besides ISS are given, one is differentiable k-ISS (D^kISS) stability. A system is D^kISS if there exist $\beta \in KL$ and $\gamma_i \in K_\infty$, $i = 0, 1, 2 \dots, k$, such that the inequality

$$|x(t, x_0, u)| \leq \beta(|x_0|, t) + \sum_{i=0}^{k} \gamma_i(\|u^{(i)}\|_\infty)$$

holds for all initial values and inputs. From Theorem 14.13, we know that the given system in (14.45) is not D^kISS, which indicates that perhaps some refinement of this definition is needed to appropriately capture external-stability notions.

Let us next consider the response of the system in (14.45) to positive-valued disturbances, which are appropriate models for disturbances in, e.g., industrial or chemical processes and communication networks. In some contexts, it is known that the responses of systems to strictly positive disturbances are qualitatively different from the response to general disturbances (e.g., [108]). However, as formalized next, even small positive disturbances may drive the state of the saturating double-integrator unbounded.

Theorem 14.14 *Consider the system as given in* (14.45) *and disturbances d that are in $L_\infty(\delta)$ and are also positive valued. For any $\delta > 0$, there is such a disturbance that drives the state unbounded for some initial condition x_0.*

We also note that the notions in Theorems 14.13 and 14.14 can be combined, i.e., even small positive-valued disturbances with bounded derivatives can drive the state unbounded. It is worth noting that, when disturbances are constrained as in Theorems 14.13 and 14.14, the initial condition must be further from the origin than in the unconstrained case, for a disturbance to drive the system unbounded.

14.6 Stable response under integral-bounded non-input-additive disturbances

Theorems 14.12 and 14.13 state negative results, namely, for a given linear static feedback law, there always exist external disturbance signals that cause the trajectories of the double-integrator system subject to actuator saturation to grow unbounded from some initial conditions. These results reveal that external stability of nonlinear systems is essentially different from that of linear systems. An important revelation is that the external stability of nonlinear systems cannot be separated from the internal state behavior. Also, the above results clearly point out that, for sustained disturbances, one cannot always have suitable L_p stability problem formulations even for innocently simple systems such as a double integrator, and thus lead to fundamental questions about the classical external stability notions currently in use for nonlinear control systems. The trouble is that sustained disturbances contain mathematical functions that are not reasonable models for common disturbances, and thus it is impossible to get a good "stable" response if we use these functions as models for disturbances. In other words, one would need to examine carefully the signal space of external disturbances to come up from an engineering point of view a class of sensible sustained disturbances. In this regard, [203] and [196] identify classes of small L_∞ signals with certain restrictions under which the boundedness of the state of the system (14.3) is preserved whatever might be the initial conditions of the system are. Our intention in this section is to report these results here. Since the results of [196] include those of [203], we base this section on [196].

At first, we recall the set of integral-bounded disturbance signals Ω_M:

$$\Omega_M = \left\{ d \in L_\infty \mid \forall\, t_1, t_2 \geqslant 0, \left| \int_{t_1}^{t_2} d(t)\, dt \right| \leqslant M \right\}. \tag{14.46}$$

We have the following results.

Theorem 14.15 *Consider the system given in* (14.45)*. Let M be given. If k_1 and k_2 satisfy $\frac{k_2}{k_1} > 16M$, then for any $d \in \Omega_M$ and any initial condition, we have $x_1, x_2 \in L_\infty$.*

The proof of Theorem 14.15 is a consequence of Lemmas 14.16 and 14.17 which are stated and proved below.

Lemma 14.16 *Consider the system*

$$\begin{aligned} \dot{x}_1 &= x_2 + y, \\ \dot{x}_2 &= \sigma(-k_1 x_1 - k_2 x_2), \end{aligned} \tag{14.47}$$

where $\|y\|_\infty < 2M$ and $\frac{k_2}{k_1} > 16M$. In that case, we have $x_1, x_2 \in L_\infty$ for any initial condition.

Proof : Define a positive definite function V as

$$V = \int_0^{k_1 x_1} \sigma(s)\, ds + \int_0^{k_1 x_1 + k_2 x_2} \sigma(s)\, ds + k_1 x_2^2.$$

This function was first introduced in [22]. Differentiating V along the trajectories yields

$$\begin{aligned}\dot{V} &= (k_1x_2 + k_1y)\sigma(k_1x_1) - 2k_1x_2\sigma(k_1x_1 + k_2x_2)\\ &\quad + [k_1x_2 + k_1y - k_2\sigma(k_1x_1 + k_2x_2)]\,\sigma(k_1x_1 + k_2x_2)\\ &= k_1x_2\,[\sigma(k_1x_1) - \sigma(k_1x_1 + k_2x_2)] - k_2\sigma^2(k_1x_1 + k_2x_2)\\ &\quad + k_1y[\sigma(k_1x_1 + k_2x_2) + \sigma(k_1x_1)]\\ &\leqslant k_1x_2\,[\sigma(k_1x_1) - \sigma(k_1x_1 + k_2x_2)] - k_2\sigma^2(k_1x_1 + k_2x_2) + 2k_1|y|.\end{aligned}$$

If $|k_1x_1 + k_2x_2| > \frac{1}{2}$, then

$$-k_2\sigma^2(k_1x_1 + k_2x_2) + 2k_1|y| \leqslant -16Mk_1 \times \tfrac{1}{4} + 4k_1M \leqslant 0.$$

Hence,

$$\dot{V} \leqslant k_1x_2\,[\sigma(k_1x_1) - \sigma(k_1x_1 + k_2x_2)] \leqslant 0.$$

If $|k_1x_1 + k_2x_2| \leqslant \frac{1}{2}$, then using Lemma 13.24, we get

$$k_1x_2\,[\sigma(k_1x_1) - \sigma(k_1x_1 + k_2x_2)] \leqslant -\tfrac{k_1}{2}x_2\sigma(k_2x_2).$$

If we also have that $|x_2| \geqslant \max\{8M, \frac{1}{k_2}\}$, then

$$k_1x_2\,[\sigma(k_1x_1) - \sigma(k_1x_1 + k_2x_2)] \leqslant -\tfrac{k_1}{2}x_2\sigma(k_2x_2) \leqslant -4k_1M,$$

which yields $\dot{V} \leqslant 0$. We therefore conclude that $\dot{V} \leqslant 0$ outside the region defined by $|k_1x_1 + k_2x_2| \leqslant \frac{1}{2}$ and $|x_2| \leqslant \max\{8M, \frac{1}{k_2}\}$. Hence, V remains bounded, which implies that $x_1, x_2 \in L_\infty$. ■

Now consider the double-integrator system (14.45). We construct a fictitious state

$$\begin{aligned}\dot{y} &= \sigma(-k_1x_1 - k_2x_2) - \sigma(-k_1x_1 - k_2x_2 + k_2y) + d,\\ y(0) &= 0.\end{aligned}$$

By defining $z = x_2 - y$, we obtain the augmented system

$$\begin{aligned}\dot{x}_1 &= z + y,\\ \dot{z} &= \sigma(-k_1x_1 - k_2z),\\ \dot{y} &= \sigma(-k_1x_1 - k_2x_2) - \sigma(-k_1x_1 - k_2x_2 + k_2y) + d,\end{aligned}$$

with $y(0) = 0$, $z(0) = x_2(0)$. From Lemma 14.16, we know that given $\frac{k_2}{k_1} > 16M$, x_1 and z remain bounded provided $|y| \leqslant 2M$. The latter statement is proven by the following lemma.

Lemma 14.17 *Consider the system*

$$\dot{y} = \sigma(v) - \sigma(v + k_2 y) + d, \quad y(0) = 0, \tag{14.48}$$

where $k_2 > 0$*,* $d \in \Omega_M$*, and* v *is continuous. We have* $|y(t)| \leqslant 2M$ *for all* $t \geqslant 0$.

Proof : Define

$$\dot{\bar{y}} = d, \quad \bar{y}(0) = 0.$$

Since $d \in \Omega_M$, the solution satisfies $|\bar{y}| \leqslant M$. Define $\tilde{y} = y - \bar{y}$. We have

$$\dot{\tilde{y}} = \sigma(v) - \sigma(v + k_2(\tilde{y} + \bar{y})), \quad \tilde{y} = 0.$$

Define a positive definite function $\widetilde{V} = \tilde{y}^2$. Taking the derivative of $\widetilde{V}$ with respect to t, we get

$$\dot{\widetilde{V}} = \tilde{y}\left[\sigma(v) - \sigma(v + k_2(\tilde{y} + \bar{y}))\right].$$

If $\widetilde{V} \geqslant M^2$, then $|\tilde{y}| \geqslant M \geqslant |\bar{y}|$, which implies that $k_2(\tilde{y} + \bar{y})$ has the same sign as $\tilde{y}$. It then follows that $\dot{\widetilde{V}} \leqslant 0$. Since $\widetilde{V}(0) = 0$, we can conclude that $\widetilde{V} \leqslant M^2$ and $|\tilde{y}| \leqslant M$ for all $t \geqslant 0$, and it follows that $|y| \leqslant |\bar{y}| + |\tilde{y}| \leqslant 2M$. ■

Proof of Theorem 14.15 : Lemma 14.17 implies that y is bounded. From Lemma 14.16, we then know that x_1 and z are bounded. Clearly then also $x_2 = z + y$ is bounded, and the proof is complete. ■

An immediate consequence of Theorem 14.15 is that if k_1 and k_2 are arbitrary positive real numbers, then boundedness is guaranteed if the integral bound M is sufficiently small. This is formally stated in the following corollary.

Corollary 14.18 *For any given* $k_1 > 0$ *and* $k_2 > 0$*, we have* $x_1, x_2 \in L_\infty$ *if* $d \in \Omega_M$ *with* $M \leqslant \frac{k_2}{16k_1}$.

14.6.1 Integral-bounded disturbances with DC bias

We consider here integral-bounded disturbances that are biased by a DC signal. The next theorem shows that, if the magnitude of the bias is less than 1 by a known margin, and an integral bound M is known a priori, then k_1, k_2 can be chosen to ensure boundedness of x_1, x_2.

Theorem 14.19 *Let $M > 0$ and $\delta \in (0, 1]$ be given, and suppose that $d = d_1 + d_2$ where d_1 is a constant with $|d_1| \leqslant 1 - \delta$ and $d_2 \in \Omega_M$. If k_1, k_2 satisfy $k_2 \geqslant \max\{\frac{1-\delta}{M}, \frac{48k_1M}{\delta^2}\}$, then $x_1, x_2 \in L_\infty$.*

Proof : The closed-loop system is given by

$$\begin{aligned} \dot{x}_1 &= x_2, \\ \dot{x}_2 &= \sigma(-k_1x_1 - k_2x_2) + d_1 + d_2. \end{aligned}$$

We construct a fictitious state

$$\begin{aligned} \dot{y} &= \sigma(-k_1x_1 - k_2x_2) - \sigma(-k_1x_1 - k_2x_2 + k_2y) + d_2, \\ y(0) &= 0. \end{aligned}$$

Lemma 14.16 shows that $|y| \leqslant 2M$. Similar to the proof of Theorem 14.15, we define $z = x_2 - y$ and convert the closed-loop system to the form

$$\begin{aligned} \dot{x}_1 &= z + y, \\ \dot{z} &= \sigma(-k_1x_1 - k_2z) + d_1, \end{aligned}$$

with $z(0) = x_2(0)$ and $|y| \leqslant 2M$.

We also introduce another fictitious state

$$\dot{w} = \sigma(-k_1x_1 - k_2z) - \sigma(-k_1x_1 - k_2z + k_2w - d_1)$$

with $w(0) = 0$. Following the same argument as in the proof of Lemma 14.17, we can show that $|w| \leqslant \frac{1-\delta}{k_2} \leqslant M$. Define $\xi_1 = x_1$, $\xi_2 = z - w = x_2 - y - w$. Then (14.45) can be transformed into

$$\begin{aligned} \dot{\xi}_1 &= \xi_2 + w + y, \\ \dot{\xi}_2 &= \sigma(-k_1\xi_1 - k_2\xi_2 - d_1) + d_1, \end{aligned}$$

where $\xi_1(0) = x_1(0)$, $\xi_2(0) = x_2(0)$, and $|w + y| \leqslant M + 2M = 3M$. Since w and y are bounded, we know that x_1 and x_2 are bounded if ξ_1 and ξ_2 are bounded.

Define $\tilde{\sigma}_{d_1}(s) = \sigma(s - d_1) + d_1$ with $|d_1| \leqslant 1 - \delta$. Then

$$\tilde{\sigma}_{d_1}(s) = \begin{cases} 1 + d_1, & s \geqslant 1 + d_1, \\ s, & -1 + d_1 \leqslant s < 1 + d_1, \\ -1 + d_1, & s \leqslant -1 + d_1. \end{cases} \tag{14.49}$$

This function can be viewed as a generalized saturation function, which is visualized in Fig. 14.6. It is easy to verify that $\tilde{\sigma}_{d_1}$ satisfies the following properties:

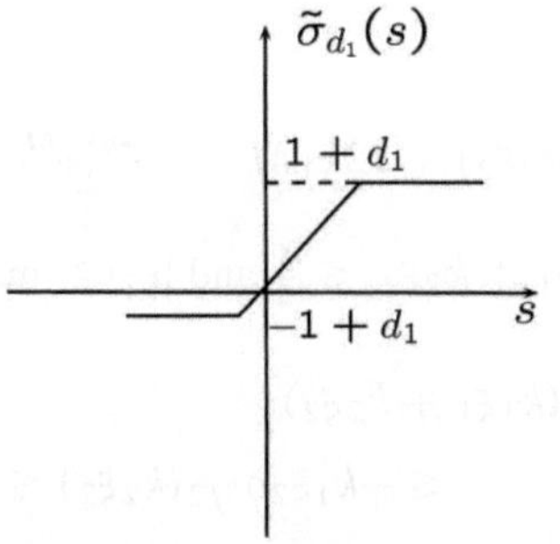

Figure 14.6: Generalized saturation function $\widetilde{\sigma}_{d_1}(s)$

(i) $|\widetilde{\sigma}_{d_1}(s)| \leqslant 2$

(ii) $s\widetilde{\sigma}_{d_1}(s) \geqslant 0$ and $s\widetilde{\sigma}_{d_1}(s) = 0$ iff $s = 0$

(iii) $s\left[\widetilde{\sigma}_{d_1}(v+s) - \widetilde{\sigma}_{d_1}(v)\right] \geqslant 0$

Moreover, it is shown in Lemma 14.22 in Appendix that if $|v| \leqslant \frac{\delta}{2}$, then

$$s\left[\widetilde{\sigma}_{d_1}(v+s) - \widetilde{\sigma}_{d_1}(v)\right] \geqslant s\sigma_{\delta/2}(s),$$

where $\sigma_{\delta/2}(s)$ is the standard saturation function with saturation level $\delta/2$, which is defined by $\sigma_{\delta/2}(s) = \operatorname{sgn}(s)\min\{\delta/2, |s|\}$.

With this generalized saturation function, the closed-loop system can be rewritten as

$$\begin{aligned}\dot{\xi}_1 &= \xi_2 + w + y,\\ \dot{\xi}_2 &= \widetilde{\sigma}_{d_1}(-k_1\xi_1 - k_2\xi_2).\end{aligned}$$

Define a positive definite function

$$V = \int_0^{k_1\xi_1} \widetilde{\sigma}_{d_1}(s)\,ds + \int_0^{k_1\xi_1+k_2\xi_2} \widetilde{\sigma}_{d_1}(s)\,ds + k_1\xi_2^2.$$

Differentiating V along the trajectory yields

$$\begin{aligned}\dot{V} &= (k_1\xi_2 + k_1w + k_1y)\widetilde{\sigma}_{d_1}(k_1\xi_1) - 2k_1\xi_2\widetilde{\sigma}_{d_1}(k_1\xi_1 + k_2\xi_2)\\ &\quad + \left[k_1\xi_2 + k_1w + k_1y - k_2\widetilde{\sigma}_{d_1}(k_1\xi_1 + k_2\xi_2)\right]\widetilde{\sigma}_{d_1}(k_1\xi_1 + k_2\xi_2)\\ &= k_1\xi_2\left[\widetilde{\sigma}_{d_1}(k_1\xi_1) - \widetilde{\sigma}_{d_1}(k_1\xi_1 + k_2\xi_2)\right] - k_2\widetilde{\sigma}_{d_1}^2(k_1\xi_1 + k_2\xi_2)\\ &\quad + k_1(w+y)\left[\widetilde{\sigma}_{d_1}(k_1\xi_1) + \widetilde{\sigma}_{d_1}(k_1\xi_1 + k_2\xi_2)\right]\\ &\leqslant k_1\xi_2\left[\widetilde{\sigma}_{d_1}(k_1\xi_1) - \widetilde{\sigma}_{d_1}(k_1\xi_1 + k_2\xi_2)\right] - k_2\widetilde{\sigma}_{d_1}^2(k_1\xi_1 + k_2\xi_2)\\ &\quad + 12k_1M.\end{aligned}$$

If $|k_1\xi_1 + k_2\xi_2| \geqslant \frac{\delta}{2}$, then

$$-k_2\tilde{\sigma}_{d_1}^2(k_1\xi_1 + k_2\xi_2) + 12k_1M \leqslant -\frac{48k_1M}{\delta^2}\frac{\delta^2}{4} + 12k_1M = 0,$$

and hence, $\dot{V} \leqslant 0$. If $|k_1\xi_1 + k_2\xi_2| \leqslant \frac{\delta}{2}$ and $|\xi_2| \geqslant \max\{\frac{\delta}{2k_2}, \frac{24M}{\delta}\}$, then

$$\begin{aligned} k_1\xi_2 & \left[\tilde{\sigma}_{d_1}(k_1\xi_1) - \tilde{\sigma}_{d_1}(k_1\xi_1 + k_2\xi_2)\right] \\ & \leqslant -k_1\xi_2\sigma_{\delta/2}(k_2\xi_2) \leqslant -k_1\frac{24M}{\delta}\frac{\delta}{2} \leqslant -12k_1M, \end{aligned}$$

and hence, $\dot{V} \leqslant 0$. We therefore find that $\dot{V} \leqslant 0$ outside the region defined by $|k_1\xi_1 + k_2\xi_2| \leqslant \frac{1}{2}$ and $|\xi_2| \leqslant \max\{\frac{\delta}{2k_2}, \frac{24M}{\delta}\}$. It follows that V remains bounded, which implies that ξ_1 and ξ_2 remain bounded. ■

14.6.2 Sinusoidal disturbances with DC bias

Our final result concerns a special case where the disturbance consists of a finite number of sinusoids together with a DC bias of magnitude less than 1. In this case, as shown in the next theorem, any internally stabilizing linear static feedback controller guarantees that the states of the system (14.45) remain bounded.

Theorem 14.20 *Consider the system* (14.45) *with $k_1 > 0$ and $k_2 > 0$. Suppose that $d = d_1 + d_2$, where d_1 is a constant satisfying $|d_1| < 1$ and d_2 is generated by an exogenous system*

$$\begin{aligned} \dot{w} &= Aw, \quad w(0) = w_0, \\ d &= Cw, \end{aligned}$$

where A is non-singular and satisfies $A + A' = 0$. We have $x_1, x_2 \in L_\infty$ for any initial condition.

Proof : We can rewrite the closed-loop system in a compact form:

$$\begin{pmatrix} \dot{w} \\ \dot{x}_1 \\ \dot{x}_2 \end{pmatrix} = \begin{pmatrix} A & 0 & 0 \\ 0 & 0 & 1 \\ C & 0 & 0 \end{pmatrix} \begin{pmatrix} w \\ x_1 \\ x_2 \end{pmatrix} + \begin{pmatrix} 0 \\ 0 \\ 1 \end{pmatrix} \left[\sigma(-k_1x_1 - k_2x_2) + d_1\right].$$

Consider the state transformation

$$\begin{pmatrix} w \\ \bar{x}_1 \\ \bar{x}_2 \end{pmatrix} = \begin{pmatrix} I & 0 & 0 \\ -CA^{-2} & 1 & 0 \\ -CA^{-1} & 0 & 1 \end{pmatrix} \begin{pmatrix} w \\ x_1 \\ x_2 \end{pmatrix}.$$

This transformation results in the system

$$\begin{pmatrix}\dot{w}\\ \dot{\overline{x}}_1\\ \dot{\overline{x}}_2\end{pmatrix}=\begin{pmatrix}A&0&0\\0&0&1\\0&0&0\end{pmatrix}\begin{pmatrix}w\\ \overline{x}_1\\ \overline{x}_2\end{pmatrix}+\begin{pmatrix}0\\0\\1\end{pmatrix}$$
$$\times\left[\sigma\left(-(k_1CA^{-2}+k_2CA^{-1})w-k_1\overline{x}_1-k_2\overline{x}_2\right)+d_1\right].$$

Define $v=-(k_1CA^{-2}+k_2CA^{-1})w+d_1$. Then

$$\begin{aligned}\sigma(-(k_1CA^{-2}+k_2CA^{-1})w-k_1\overline{x}_1-k_2\overline{x}_2)+d_1&\\ &=\sigma(v-k_1\overline{x}_1-k_2\overline{x}_2-d_1)+d_1\\ &=\widetilde{\sigma}_{d_1}(-k_1\overline{x}_1-k_2\overline{x}_2+v),\end{aligned}$$

where $\widetilde{\sigma}_{d_1}$ is the generalized saturation function defined in the proof of Theorem 14.19. The dynamics of $\overline{x}_1$ and $\overline{x}_2$ can now be written as

$$\begin{aligned}\dot{\overline{x}}_1&=\overline{x}_2,\\ \dot{\overline{x}}_2&=\widetilde{\sigma}_{d_1}(-k_1\overline{x}_1-k_2\overline{x}_2+v).\end{aligned}$$

Clearly $v\in L_\infty$. It was shown by [22] that the $(\overline{x}_1,\overline{x}_2)$ dynamics is L_∞ stable from v to $\overline{x}_1$ and $\overline{x}_2$ for any $k_1>0$ and $k_2>0$. ∎

Remark 14.21 *For the ease of presentation, we used a standard saturation function with saturation level* 1*, but all the results obtained above can be easily extended to the case where a saturation function with arbitrary saturation level Δ is used.*

14.A Appendix

Lemma 14.22 *The generalized saturation function $\widetilde{\sigma}_{d_1}$ defined in* (14.49) *with $|d_1|\leqslant 1-\delta$ satisfies*

$$s\left[\widetilde{\sigma}_{d_1}(s+v)-\widetilde{\sigma}_{d_1}(v)\right]\geqslant s\sigma_{\delta/2}(s)$$

for $|v|\leqslant\frac{\delta}{2}$, where $\sigma_{\delta/2}(s)$ denotes the standard saturation function with saturation level $\delta/2$ defined as $\sigma_{\delta/2}(s)=\operatorname{sgn}(s)\min\{\delta/2,|s|\}$.

Proof : If $|s|<\frac{\delta}{2}$, we have $|v+s|\leqslant\delta\leqslant 1-|d_1|$. By definition (14.49), we have $\widetilde{\sigma}_{d_1}(s+v)=s+v$. Hence,

$$\widetilde{\sigma}_{d_1}(s+v)-\widetilde{\sigma}_{d_1}(v)=s+v-v=s.$$

If $|s| \geqslant \frac{\delta}{2}$, it can be seen from Fig. 14.6 that

$$|\tilde{\sigma}_{d_1}(s+v) - \tilde{\sigma}_{d_1}(v)| \geqslant |\operatorname{sgn}(s)\tfrac{\delta}{2} + v - v| = \tfrac{\delta}{2}.$$

Hence, $s\left[\tilde{\sigma}_{d_1}(s+v) - \tilde{\sigma}_{d_1}(v)\right] \geqslant s\sigma_{\delta/2}(s)$. ■

15

Simultaneous internal and external stabilization in the presence of a class of non-input-additive sustained disturbances: continuous time

15.1 Introduction

We continue here with the theme of simultaneous internal and external stabilization in the presence of non-input-additive disturbances. As discussed in Chap. 13, for such non-input-additive disturbances, L_p stabilization with finite gain is impossible, but L_p stabilization without finite gain is always attainable via a nonlinear dynamic low-gain feedback law for all $p \in [1, \infty)$ (i.e., for all disturbances whose "energy" vanishes asymptotically). In the case of open-loop neutrally stable system, this can be done via a linear state feedback law. Nevertheless, all these results apply only to L_p disturbances for $p \in [1, \infty)$, and not to sustained signals belonging to L_∞. One can argue easily that L_∞ stability in general is impossible in the presence of non-input-additive sustained disturbance signals as they can dominate the saturated control signal. To exemplify this further, we considered in Chap. 14 the canonical case of a double-integrator system. Among other results, we showed there that whatever might be a linear feedback law, there exist bounded disturbances with *arbitrarily small* L_∞ norm that cause the states of it to grow unbounded from certain initial conditions. Even more dramatically, it turns out that unbounded growth can be achieved by *vanishing* disturbances with arbitrarily small L_∞ norm. On the positive side, we also identified there a class of integral-bounded non-input-additive sustained disturbances for which L_∞ stabilization can be attained. Thus, a fundamental question arises whether for general systems which are asymptotically null controllable with bounded control (ANCBC) one can identify a set of non-input-additive sustained disturbances for which a feedback control law can be determined such that

1. In the absence of disturbances, the origin of the closed-loop system is globally asymptotically stable.

2. If the disturbances belong to the given set, the states of the closed-loop system are bounded for any arbitrarily specified initial conditions.

In this chapter, our focus is to answer this fundamental question positively by identifying such a set of sustained disturbances. We consider here only continuous time. Discussions on discrete-time systems follow in the next chapter.

A. Saberi et al., *Internal and External Stabilization of Linear Systems with Constraints*, Systems & Control: Foundations & Applications, DOI 10.1007/978-0-8176-4787-2_15,

Regarding the architecture of this chapter, after certain preliminary discussions in Sect. 15.2, we consider in Sect. 15.3 neutrally stable systems which allow a linear feedback law. We move on then to Sect. 15.4 to consider general critically unstable systems (or equivalently general ANCBC systems).

This chapter is based on our work [126, 192, 195, 196].

15.2 Preliminaries

Consider the system

$$\dot{x} = Ax + B\sigma(u) + Ed, \quad x(0) = x_0, \tag{15.1}$$

where $x \in \mathbb{R}^n$, $u \in \mathbb{R}^m$, and $d \in \mathbb{R}^p$. The pair (A, B) is stabilizable, and A has all its eigenvalues in the closed left-half plane. That is, the given system is asymptotically null controllable with bounded control (ANCBC).

As pointed out in the introduction, for systems of the form (15.1), our goal in this chapter is to identify a class of L_∞ disturbance signals d such that for appropriately designed feedback control, laws the closed-loop system exhibits global asymptotic stability in the absence of disturbances d, and in the presence of such disturbances d, the state variables are bounded. Toward this goal, we first define the following set of *integral-bounded* disturbance signals,

$$\mathcal{S}_\infty = \left\{ d \in L_\infty \mid \exists M \ s.t. \forall t_2 > t_1 > 0, \left\| \int_{t_1}^{t_2} d(t)\,\mathrm{d}t \right\| < M \right\}.$$

The set $\mathcal{S}_\infty$ represents signals that have a uniformly bounded integral over every time interval, that is, signals that have no sustained DC bias. We next introduce another set of disturbance signals,

$$\Omega_\infty = \{d \in L_\infty \mid \forall i \in 1,\dots,q,\ d(t)\sin\omega_i t \in \mathcal{S}_\infty \text{ and } d(t)\cos\omega_i t \in \mathcal{S}_\infty\}, \tag{15.2}$$

where $\pm j\omega_i$, $i \in 1,\dots,q$ represent the imaginary-axis eigenvalues of A. The set Ω_∞ consists of those signals that remain integral-bounded when multiplied by $\sin\omega_i t$ and $\cos\omega_i t$. This definition is a natural generalization of $\mathcal{S}_\infty$, since $\Omega_\infty = \mathcal{S}_\infty$ for $\omega_i = 0$.

In practical terms, a signal that belongs to Ω_∞ is a signal that has no sustained frequency component at any of the frequencies ω_i, $i \in 1,\dots,q$. To see this, note that we can equivalently write

$$\Omega_\infty = \left\{ d \in L_\infty \mid \exists M \ s.\,t.\ \forall i \in 1,\dots,q, \forall t_2 > t_1 > 0, \left\| \int_{t_1}^{t_2} d(t)e^{j\omega_i t}\,\mathrm{d}t \right\| < M \right\}. \tag{15.3}$$

The integral

$$\int_{t_1}^{t_2} d(t)e^{j\omega_i t}\,\mathrm{d}t$$

is easily recognized as the value at ω_i of the Fourier transform of the signal $d(t)$ truncated to the interval $[t_1, t_2]$. The definition of Ω_∞ implies that this value must be uniformly bounded regardless of the choice of t_1 and t_2.

15.3 Neutrally stable systems

We first consider neutrally stable systems where A has only semi-simple eigenvalues on the imaginary axis. Then, the system (15.1) that is neutrally stable can be decomposed into the following form:

$$\begin{pmatrix} \dot{x}_s \\ \dot{x}_u \end{pmatrix} = \begin{pmatrix} A_s & 0 \\ 0 & A_u \end{pmatrix}\begin{pmatrix} x_s \\ x_u \end{pmatrix} + \begin{pmatrix} B_s \\ B_u \end{pmatrix}\sigma(u) + \begin{pmatrix} E_s \\ E_u \end{pmatrix}d,$$

where A_s is Hurwitz stable, A_u has all its eigenvalues on the imaginary axis, and (A_u, B_u) is controllable. Since A_s is Hurwitz stable and σ and d are bounded, it follows that the x_s dynamics will remain bounded no matter what controller is used. Therefore, without loss of generality, we can ignore the asymptotically stable dynamics and assume in (15.1) that (A, B) is controllable and all the eigenvalues of A are on the imaginary axis. Without loss of generality, we can then assume that $A + A' = 0$.

Guided by Sect. 4.6.1 that considered global asymptotic stabilization of neutrally stable systems via linear state feedback laws, we employ here once again a linear static state feedback law $u = -B'x$, which results in a closed-loop system,

$$\dot{x} = Ax - B\sigma(B'x) + Ed, \quad x(0) = x_0. \tag{15.4}$$

It is clear from the results of Sect. 4.6.1 that, in the absence of d, the origin of (15.4) is globally asymptotically stable. Therefore, in the presence of the disturbance d, we focus next on the boundedness of the closed-loop states.

In what follows, we show that the trajectories of the controlled system (15.4) remain bounded for all disturbances that belong to Ω_∞. We also show that this result holds if we add a sufficiently small signal that does not belong to Ω_∞.

Second-order single frequency system:
We start by considering an example system with a pair of complex eigenvalues at $\pm j$:

$$\begin{pmatrix} \dot{x}_1 \\ \dot{x}_2 \end{pmatrix} = \begin{pmatrix} 0 & 1 \\ -1 & 0 \end{pmatrix}\begin{pmatrix} x_1 \\ x_2 \end{pmatrix} - \begin{pmatrix} 0 \\ 1 \end{pmatrix}\sigma(x_2) + \begin{pmatrix} e_1 \\ e_2 \end{pmatrix}d, \quad \begin{pmatrix} x_1(0) \\ x_2(0) \end{pmatrix} = x_0. \tag{15.5}$$

We have the following result:

Theorem 15.1 *Let $d \in \Omega_\infty$. Then, the trajectories of* (15.5) *remain bounded for any initial condition.*

Proof : To analyze the system, we start by introducing a rotation matrix

$$R = \begin{pmatrix} \cos t & -\sin t \\ \sin t & \cos t \end{pmatrix},$$

which represents a counterclockwise rotation by an angle t. The dynamics of the rotation matrix is given by

$$\dot{R} = -R \begin{pmatrix} 0 & 1 \\ -1 & 0 \end{pmatrix}.$$

We shall study the dynamics of x from a rotated coordinate frame, and toward this end, we define the rotated state $y = Rx$. The dynamics of y is given by

$$\begin{aligned} \dot{y} &= \dot{R}x + R\dot{x} \\ &= R\left(\begin{pmatrix} e_1 \\ e_2 \end{pmatrix} d - \begin{pmatrix} 0 \\ 1 \end{pmatrix} \sigma(x_2) \right) \\ &= R\left(\begin{pmatrix} e_1 \\ e_2 \end{pmatrix} d - \begin{pmatrix} 0 \\ 1 \end{pmatrix} \sigma\left(\begin{pmatrix} 0 & 1 \end{pmatrix} R'y \right) \right), \quad y(0) = x(0) = x_0. \end{aligned}$$

We define next a fictitious system,

$$\dot{\tilde{y}} = R \begin{pmatrix} e_1 \\ e_2 \end{pmatrix} d, \quad \tilde{y}(0) = x_0. \tag{15.6}$$

We know from the definition of Ω_∞ that the signal $d(t)$ is integral-bounded when multiplied by $\sin t$ and $\cos t$. It therefore follows that the right-hand side of (15.6) is integral-bounded, and hence $\tilde{y} \in L_\infty$.

Consider the difference between y and the fictitious state $\tilde{y}$, given by $z = y - \tilde{y}$, with dynamics

$$\begin{aligned} \dot{z} &= -R \begin{pmatrix} 0 \\ 1 \end{pmatrix} \sigma\left(\begin{pmatrix} 0 & 1 \end{pmatrix} R'y \right) \\ &= -R \begin{pmatrix} 0 \\ 1 \end{pmatrix} \sigma\left(\begin{pmatrix} 0 & 1 \end{pmatrix} R'z + \delta \right), \quad z(0) = 0, \end{aligned}$$

where

$$\delta = \begin{pmatrix} 0 & 1 \end{pmatrix} R' \widetilde{y} \in L_\infty.$$

We rotate z back to the original coordinate frame by introducing $w = R'z$, thereby obtaining the dynamics,

$$\begin{aligned} \dot{w} &= \dot{R}' z + R' \dot{z} \\ &= \begin{pmatrix} 0 & 1 \\ -1 & 0 \end{pmatrix} w - \begin{pmatrix} 0 \\ 1 \end{pmatrix} \sigma\big(\begin{pmatrix} 0 & 1 \end{pmatrix} w + \delta\big), \quad w(0) = 0. \end{aligned}$$

It is shown in Theorem (12.33) that the above system is L_∞ stable with respect to the input δ, and hence $w \in L_\infty$. Finally, we have $x = w + R'\widetilde{y}$, and hence, $x \in L_\infty$. ■

To demonstrate the importance of the disturbance belonging to Ω_∞, we shall now show that if d contains a large frequency component at $\pm j$, the states of (15.5) will diverge toward infinity for any initial condition. Suppose therefore that $d(t) = a \sin(t + \theta)$, where a is an amplitude yet to be chosen. For ease of presentation, we assume that $e_1 = 0, e_2 = 0$. Consider the dynamics of the rotated state y from the proof of Theorem 15.1. We have

$$\begin{aligned} \dot{y} &= R \begin{pmatrix} 0 \\ 1 \end{pmatrix} \Big(d - \sigma\big(\begin{pmatrix} 0 & 1 \end{pmatrix} R' y\big)\Big) \\ &= a \begin{pmatrix} -\sin t \sin(t + \theta) \\ \cos t \sin(t + \theta) \end{pmatrix} - R \begin{pmatrix} 0 \\ 1 \end{pmatrix} \sigma\big(\begin{pmatrix} 0 & 1 \end{pmatrix} R' y\big). \end{aligned}$$

Using appropriate trigonometric identities, the dynamics can be rewritten as

$$\dot{y} = \frac{a}{2} \begin{pmatrix} \cos(2t + \theta) - \cos(-\theta) \\ \sin(2t + \theta) - \sin(-\theta) \end{pmatrix} - R \begin{pmatrix} 0 \\ 1 \end{pmatrix} \sigma\big(\begin{pmatrix} 0 & 1 \end{pmatrix} R' y\big).$$

We have either $|\sin(-\theta)| \geqslant \sqrt{2}/2$ or $|\cos(-\theta)| \geqslant \sqrt{2}/2$. Without loss of generality, we assume that $|\sin(-\theta)| \geqslant \sqrt{2}/2$. Let a be chosen such that $a \geqslant 4(1 + \varepsilon)/\sqrt{2}$, where ε is a positive number. For the trajectory $y_2(t)$, we have

$$\begin{aligned} |y_2(t)| = \Big| y_2(0) + \int_0^t & \frac{a}{2} (\sin(2\tau + \theta) - \sin(-\theta)) \\ & - \begin{pmatrix} 0 & 1 \end{pmatrix} R(\tau) \begin{pmatrix} 0 \\ 1 \end{pmatrix} \sigma\big(\begin{pmatrix} 0 & 1 \end{pmatrix} R'(\tau) y(\tau)\big) \, d\tau \Big|. \end{aligned}$$

Noting that the last term of the integrand is bounded by ± 1, and using the bound $|a/2\sin(-\theta)| \geqslant \sqrt{2}a/4 \geqslant 1+\varepsilon$, we therefore have

$$\begin{aligned} |y_2(t)| &\geqslant -|y_2(0)| - \frac{a}{2}\left|\int_0^t \sin(2\tau+\theta)\,\mathrm{d}\tau\right| + \int_0^t \varepsilon\,\mathrm{d}\tau \\ &\geqslant -|y_2(0)| - \frac{a}{2} + \varepsilon t. \end{aligned}$$

This shows that $y_2(t)$ diverges toward infinity.

Connection to a single integrator system:

Before moving on to the case of general multifrequency systems, it is instructive to compare some aspects of the above example with similar results for a single integrator system. A single integrator system with a saturated control input and an external disturbance has the form

$$\dot{x} = \sigma(u) + ed.$$

In the absence of disturbances, the open-loop response of this system is stationary. It is intuitively easy to see that a large DC bias in d would constitute a problem because it would tend to dominate the bounded control term $\sigma(\cdot)$, thus leading to unboundedness. The absence of such a DC bias is guaranteed by d belonging to $\mathcal{S}_\infty$.

The system with eigenvalues at $\pm j$ has the form

$$\dot{x} = \begin{pmatrix} 0 & 1 \\ -1 & 0 \end{pmatrix} x + \begin{pmatrix} 0 \\ 1 \end{pmatrix} \sigma(u) + \begin{pmatrix} e_1 \\ e_2 \end{pmatrix} d.$$

In the absence of disturbances, the open-loop response of this system is oscillatory rather than stationary, and it is less obvious why a disturbance that does not belong to Ω_∞ could be problematic. By introducing a rotated state $y = Rx$, however, we obtain the dynamics

$$\dot{y} = R\begin{pmatrix} 0 \\ 1 \end{pmatrix} \sigma(u) + R\begin{pmatrix} 0 \\ 1 \end{pmatrix} d.$$

In the absence of disturbances, the open-loop response of y is stationary, and the dynamics of y are strikingly similar to the single-integrator case. In particular, it is easy to see that a large DC bias in the term

$$R\begin{pmatrix} 0 \\ 1 \end{pmatrix} d \tag{15.7}$$

would constitute a problem, because it would tend to dominate the bounded control term. Analogous to the single-integrator case, the absence of such a bias is guaranteed if (15.7) belongs to $\mathcal{S}_\infty$, which is equivalent to d belonging to Ω_∞.

In the single-integrator case, a DC bias in d can be tolerated if it is sufficiently small. Similarly, a small signal that does not belong to Ω_∞ can be tolerated for systems with complex eigenvalues. This is demonstrated shortly by considering general multifrequency systems.

Multifrequency systems:
We now extend Theorem 15.1 to multifrequency neutrally stable systems. Consider

$$\dot{x} = Ax - B\sigma(B'x) + Ed, \quad x(0) = x_0, \tag{15.8}$$

where $A + A' = 0$ and (A, B) is controllable. Without loss of generality, we assume that

$$A = \begin{pmatrix} A_1 & 0 & \cdots & 0 \\ 0 & \ddots & \ddots & \vdots \\ \vdots & \ddots & A_q & 0 \\ 0 & \cdots & 0 & 0 \end{pmatrix}, \quad x = \begin{pmatrix} x_1 \\ \vdots \\ x_q \\ x_0 \end{pmatrix}, \tag{15.9}$$

where $x_i \in \mathbb{R}^2$, $i = 1, \dots, q$, $x_0 \in \mathbb{R}^{n-2q}$ and

$$A_i = \begin{pmatrix} 0 & \omega_i \\ -\omega_i & 0 \end{pmatrix}, \quad i = 1, \dots, q \tag{15.10}$$

with $2q \leqslant n$. We have the following theorem:

Theorem 15.2 *Let $d \in \Omega_\infty$. Then, the states of* (15.8) *remain bounded for any initial condition.*

Proof : Consider the rotation matrix

$$R = \begin{pmatrix} R_1 & & & \\ & \ddots & & \\ & & R_q & \\ & & & I \end{pmatrix}, \tag{15.11}$$

where

$$R_i = \begin{pmatrix} \cos\omega_i t & -\sin\omega_i t \\ \sin\omega_i t & \cos\omega_i t \end{pmatrix}. \tag{15.12}$$

Note that R is unitary, i.e., $RR' = I$, and moreover,

$$\dot{R} = -RA.$$

Define a transformed state $y = Rx$. As a result,

$$\dot{y} = -RB'\sigma(B'R'y) + REd, \quad y(0) = x_0.$$

Introduce a fictitious system

$$\dot{\tilde{y}} = REd, \quad \tilde{y}(0) = x_0.$$

It follows from the definition of Ω_∞ that $\tilde{y} \in L_\infty$. Next, define the difference between y and $\tilde{y}$ by $z = y - \tilde{y}$. We get

$$\dot{z} = -RB\sigma(B'R'z + B'R'\tilde{y}), \quad z(0) = 0.$$

Finally, transform z back to the original coordinates by defining $w = R'z$. Note that

$$\dot{R}' = AR'.$$

Hence,

$$\dot{w} = Aw - B\sigma(B'w + B'R'\tilde{y}), \quad w(0) = 0.$$

This system is L_∞ stable. Thus, we have $w \in L_\infty$. Since $x = w + R'\tilde{y}$ and $\tilde{y} \in L_\infty$ is bounded for all t, we conclude that $x \in L_\infty$. ■

We shall prove next that the states of (15.8) also remain bounded if a small signal that does not belong to Ω_∞ is added on top of the original signal in Ω_∞. Consider the system,

$$\dot{x} = Ax - B\sigma(B'x) + E_1 d_1 + E_2 d_2, \quad x(0) = x_0, \tag{15.13}$$

where $A + A' = 0$ and (A, B) is controllable, and without loss of generality, we assume that A is in the form of (15.9). We have the following result:

Theorem 15.3 *Let $d_1 \in \Omega_\infty$ and $d_2 \in L_\infty(\delta)$. Then, for δ sufficiently small, the states of system* (15.13) *remain bounded for all initial conditions.*

Proof : Using the same sequence of transformations as introduced in the proof of Theorem 15.2, we get the following transformed system:

$$\dot{w} = Aw - B\sigma(B'w + B'R'\tilde{y}) + E_2 d_2, \quad w(0) = 0,$$

where $w = x - R'\tilde{y}$ and

$$\dot{\tilde{y}} = RE_1 d_1, \quad \tilde{y} = x_0.$$

The fact that $d_1 \in \Omega_\infty$ implies that $\tilde{y} \in L_\infty$. Introduce another fictitious system,

$$\dot{\bar{w}} = (A - BB')\bar{w} + E_2 d_2, \quad \bar{w}(0) = 0.$$

Since $A - BB'$ is Hurwitz stable and $d_2 \in L_\infty(\delta)$, we have $\overline{w} \in L_\infty$, and moreover, $\|B'\overline{w}\|_\infty \leqslant \frac{1}{2}$ for a sufficiently small δ.

Define $\zeta = w - \overline{w}$. Then ζ has the dynamics

$$\dot{\zeta} = A\zeta - B\sigma(B'\zeta + B'\overline{w} + B'R'\widetilde{y}) + BB'\overline{w}.$$

Let $B'\zeta$, $B'\overline{w} + B'R'\widetilde{y}$, and $B'\overline{w}$ be denoted, respectively, by u, v, μ. Define $V_1 = \frac{1}{3}\|\zeta\|^3$. Differentiating V_1 along the trajectories yields,

$$\begin{aligned}
\dot{V}_1 &= \|\zeta\|u'\left[\sigma(-u+v)+\mu\right] \\
&\leqslant \|\zeta\|(u-v)'\left[-\sigma(u-v)+\mu\right] + 2\|\zeta\|\|v\|_\infty \\
&= \|\zeta\|\left\{(u-v)'[-\sigma(u-v)+\sigma(u-v+\mu)]\right. \\
&\qquad \left.+(u-v)'[-\sigma(u-v+\mu)+\sigma(\mu)]\right\} + 2\|\zeta\|\|v\|_\infty \\
&\leqslant -\tfrac{1}{2}\|\zeta\|(u-v)\sigma(u-v) + \sqrt{m}\|\zeta\| + 2\|\zeta\|\|v\|_\infty.
\end{aligned}$$

The last inequality in the above is obtained by using Lemmas 13.23 and 13.24. Next, since $A - BB'$ is Hurwitz stable, there exists a $P > 0$ satisfying

$$(A - BB')'P + P(A - BB') = -I.$$

Define $V_2 = \zeta' P\zeta$. There exists an $\widetilde{\alpha}$ such that

$$\begin{aligned}
\dot{V}_2 &= -\|\zeta\|^2 + 2\zeta'P[B\sigma(-u+v) + Bu + B\mu] \\
&= -\|\zeta\|^2 + 2\zeta'P[B(\sigma(-u+v)+u-v) + B\mu + Bv] \\
&\leqslant -\|\zeta\|^2 + 2\widetilde{\alpha}\|\zeta\|(u-v)\sigma(u-v) + \widetilde{\alpha}\|\zeta\| + 2\widetilde{\alpha}\|\zeta\|\|v\|_\infty.
\end{aligned}$$

The last inequality in the above is obtained by using Lemma 13.25.

Finally, define a Lyapunov candidate $V = 4\widetilde{\alpha}V_1 + V_2$. We find that

$$\begin{aligned}
\dot{V} &\leqslant -\|\zeta\|^2 + (4\widetilde{\alpha}\sqrt{m} + \widetilde{\alpha})\|\zeta\| + 10\widetilde{\alpha}\|\zeta\|\|v\|_\infty \\
&= -\|\zeta\|\left[\|\zeta\| - 4\widetilde{\alpha}\sqrt{m} - \widetilde{\alpha} - 10\widetilde{\alpha}\|v\|_\infty\right].
\end{aligned}$$

Lemma 13.23 is used in arriving at the above inequality. Thus, $\dot{V} \leqslant 0$ for $\|\zeta\| \geqslant 4\widetilde{\alpha}\sqrt{m} + \widetilde{\alpha} + 10\widetilde{\alpha}\|v\|_\infty$. This implies that $\zeta \in L_\infty$. Since $x = \overline{w} + \widetilde{w}\zeta + R'\widetilde{y}$, we conclude that $x \in L_\infty$. ■

15.4 Critically unstable systems

We now consider a general critically unstable linear system with input saturation and disturbances,

$$\dot{x} = Ax + B\sigma(u) + \mathcal{E}d, \quad x(0) = x_0, \tag{15.14}$$

where (A, B) is controllable and A has all its eigenvalues in the closed-left-half plane. Without loss of generality, we assume that $x = (x_1', x_2', \ldots, x_q')'$, and A, B are in the following Jordan canonical form:

$$A = \begin{pmatrix} \bar{A}_1 & 0 & 0 & 0 \\ 0 & \bar{A}_2 & \ldots & 0 \\ \vdots & \vdots & \ddots & \vdots \\ 0 & 0 & \ldots & \bar{A}_q \end{pmatrix}, \quad B = \begin{pmatrix} B_1 \\ B_2 \\ \vdots \\ B_q \end{pmatrix}, \quad E = \begin{pmatrix} E_1 \\ E_2 \\ \vdots \\ E_q \end{pmatrix}, \tag{15.15}$$

where

$$x_i = \begin{pmatrix} x_{i,1} \\ x_{i,2} \\ \vdots \\ x_{i,n_i-1} \\ x_{i,n_i} \end{pmatrix}, \quad \bar{A}_i = \underbrace{\begin{pmatrix} A_i & I & 0 & \ldots & 0 \\ 0 & A_i & I & \ldots & 0 \\ \vdots & \vdots & \ddots & \ddots & \vdots \\ 0 & 0 & \ldots & A_i & I \\ 0 & 0 & 0 & 0 & A_i \end{pmatrix}}_{n_i \times n_i \text{ blocks}},$$
$$B_i = \begin{pmatrix} B_{i,1} \\ B_{i,2} \\ \vdots \\ B_{i,n_i-1} \\ B_{i,n_i} \end{pmatrix}, \quad E_i = \begin{pmatrix} E_{i,1} \\ E_{i,2} \\ \vdots \\ E_{i,n_i-1} \\ E_{i,n_i} \end{pmatrix}, \tag{15.16}$$

$x_{i,j} \in \mathbb{R}^{p_i}$ and $A_i + A_i' = 0$. We can further assume that A_i is in the form of (15.9) and (15.10). Note that the above form can be obtained by assembling together those blocks corresponding to the eigenvalues with the same Jordan block size in the real Jordan canonical form.

Based on the above Jordan structure of matrix A, we say the disturbance d is *aligned* if $E_{i,n_i} \neq 0$ for any $i = 1, \ldots, q$ and *misaligned* if $E_{i,n_i} = 0$ for all $i = 1, \ldots, q$. We use here the words "aligned" and "misaligned" because of a lack of better terminology. To explore further while using this terminology, we rewrite the given system (15.14) as

$$\dot{x} = Ax + B\sigma(u) + \bar{E}_1 d_1 + \bar{E}_2 d_2, \quad x(0) = x_0, \tag{15.17}$$

with

$$\overline{E}_1 = \begin{pmatrix} \overline{E}_{1,1} \\ \overline{E}_{1,2} \\ \vdots \\ \overline{E}_{1,q} \end{pmatrix}, \quad \overline{E}_{1,i} = \begin{pmatrix} E_{i,1} \\ E_{i,2} \\ \vdots \\ E_{i,n_i-1} \\ 0 \end{pmatrix} \tag{15.18}$$

and

$$\overline{E}_2 = \begin{pmatrix} \overline{E}_{2,1} \\ \overline{E}_{2,2} \\ \vdots \\ \overline{E}_{2,q} \end{pmatrix}, \quad \overline{E}_{2,i} = \begin{pmatrix} 0 \\ 0 \\ \vdots \\ 0 \\ E_{i,n_i} \end{pmatrix}. \tag{15.19}$$

Note that the above representation delineates aligned and misaligned disturbances.

We first present the result for the above system when $d_1 \in L_\infty$ and $d_2 \in \Omega_\infty$. As will become clear later, the result derived based on this can be straightforwardly extended to the general case. We shall show that if the disturbances are aligned with input and belong to Ω_∞ or misaligned and belong to L_∞, a controller can be designed such that the states of the closed-loop system remain bounded for any initial condition while yielding a globally asymptotically stable equilibrium. For neutrally stable system, we have shown that this can be achieved by a linear state feedback. However, a nonlinear feedback controller is generally needed for critically unstable systems. In what follows, we present a nonlinear dynamic low-gain state feedback design methodology which solves the said problem.

Let P_ε be the solution to a parametric Lyapunov equation (PLE),

$$A'P_\varepsilon + P_\varepsilon A - P_\varepsilon BB'P_\varepsilon + \varepsilon P_\varepsilon = 0. \tag{15.20}$$

Lemma 15.9 in the appendix tells us that for any given matrix $\overline{E}_1$ in the form of (15.18), there exists a M such that for $\varepsilon \in (0, 1]$, we have

$$\overline{E}_1' P_\varepsilon \overline{E}_1 \leqslant M\varepsilon^2 I.$$

Consider the dynamical state feedback controller,

$$\begin{cases} \dot{\hat{x}}_i = A_i \hat{x}_i + B_{i,n_i}\sigma(-B'P_{\varepsilon_a(\overline{x})}\overline{x}), & i = 1, \dots, q \\ u = -\widehat{B}'(x_b - \hat{x}) - B'P_{\varepsilon_a(\overline{x})}\overline{x}, \end{cases} \tag{15.21}$$

where

$$\widehat{B} = \begin{pmatrix} B_{1,n_1} \\ B_{2,n_2} \\ \vdots \\ B_{q,n_q} \end{pmatrix}, \quad x_b = \begin{pmatrix} x_{1,n_1} \\ x_{2,n_2} \\ \vdots \\ x_{q,n_q} \end{pmatrix}, \quad \widehat{x} = \begin{pmatrix} \widehat{x}_1 \\ \widehat{x}_2 \\ \vdots \\ \widehat{x}_q \end{pmatrix};$$

$P_{\varepsilon_a(\overline{x})}$ is the solution to PLE (15.20) with $\varepsilon = \varepsilon_a(\overline{x})$, and $\varepsilon_a(\overline{x})$ is determined by

$$\varepsilon = \varepsilon_a(\overline{x}) := \max\{r \in [0,1] \mid (\overline{x}' P_r \overline{x}) \times \operatorname{trace}(B' P_r B) \leqslant \delta^2\}, \quad (15.22)$$

where $\overline{x}$ is x with x_{i,n_i} replaced by $\widehat{x}_i$, and $\delta \in (0, \frac{1}{2}]$ is a design parameter to be determined later. The scheduling (15.22) guarantees that $\|B' P_{\varepsilon_a(\overline{x})} \overline{x}\| \leqslant \delta$ for all $\overline{x}$.

We have the following result in view of properties of scheduling as enumerated in (12.2.1).

Theorem 15.4 *Consider the system* (15.17) *with controller* (15.21). *The following hold:*

(i) *In the absence of d_1 and d_2, the origin of the closed-loop system is globally asymptotically stable.*

(ii) *$x \in L_\infty$ for any $x(0)$, $d_1 \in L_\infty$, and $d_2 \in \Omega_\infty$.*

Proof : For simplicity of our presentation, we omit the parameter ε_a in the expression of $P_{\varepsilon_a(\overline{x})}$. Define

$$\widetilde{x} = x_b - \widehat{x} = \begin{pmatrix} x_{1,n_1} - \widehat{x}_1 \\ x_{2,n_2} - \widehat{x}_2 \\ \vdots \\ x_{q,n_q} - \widehat{x}_q \end{pmatrix}.$$

We have

$$\dot{\widetilde{x}} = \widehat{A}\widetilde{x} + \widehat{B}\sigma(-\widehat{B}'\widetilde{x} - B' P_{\varepsilon_a(\overline{x})}\overline{x}) - \widehat{B}\sigma(-B' P_{\varepsilon_a(\overline{x})}\overline{x}) + \widehat{E}_2 d_2,$$

where

$$\widehat{A} = \begin{pmatrix} A_1 & 0 & 0 & 0 \\ 0 & A_2 & \dots & 0 \\ \vdots & \vdots & \ddots & \vdots \\ 0 & 0 & \dots & A_q \end{pmatrix}, \quad \widehat{E}_2 = \begin{pmatrix} E_{1,n_1} \\ E_{2,n_2} \\ \vdots \\ E_{q,n_q} \end{pmatrix}. \quad (15.23)$$

Note that (A, B) is controllable, implies that $(\widehat{A}, \widehat{B})$ is controllable. Moreover, $\widehat{A} + \widehat{A}' = 0$. The closed-loop system can be written in terms of $\widetilde{x}$ and $\overline{x}$ as

$$\left\{\begin{aligned} \dot{\overline{x}} &= A\overline{x} + B\sigma(-B'P\overline{x}) + \overline{E}_1 d_1 + \mathcal{J}\widetilde{x} \\ &\quad + \overline{B}\left[\sigma(-\widehat{B}'\widetilde{x} - B'P\overline{x}) - \sigma(-B'P\overline{x})\right] \\ \dot{\widetilde{x}} &= \widehat{A}\widetilde{x} + \widehat{B}\sigma(-\widehat{B}'\widetilde{x} - B'P_{\varepsilon_a(\overline{x})}\overline{x}) \\ &\quad - \widehat{B}\sigma(-B'P_{\varepsilon_a(\overline{x})}\overline{x}) + \widehat{E}_2 d_2, \end{aligned}\right. \tag{15.24}$$

where $\overline{B}$ is B with B_{i,n_i} blocks set to zero and

$$\mathcal{J} = \begin{pmatrix} \mathcal{J}_1 \\ \mathcal{J}_2 \\ \vdots \\ \mathcal{J}_q \end{pmatrix}, \quad \mathcal{J}_i = \begin{pmatrix} 0 \\ \vdots \\ 0 \\ I \\ 0 \end{pmatrix}, \quad \overline{I}_i = (0 \quad \cdots \quad \underset{\substack{\uparrow \\ i\text{th block}}}{I} \quad \cdots \quad 0).$$

It should be noted that $\overline{B}$, $\overline{E}_1$, and $\mathcal{J}$ are all in the form of (15.18).

We first prove global asymptotic stability in the autonomous case. Let $v = -B'P\overline{x}$. Our scheduling (15.22) guarantees that $\|v\| \leqslant \delta \leqslant \frac{1}{2}$ for any $\overline{x}$. Consider the dynamics of $\widetilde{x}$,

$$\dot{\widetilde{x}} = \widehat{A}\widetilde{x} + \widehat{B}\sigma(-\widehat{B}'\widetilde{x} + v) - \widehat{B}\sigma(v).$$

Define a Lyapunov function as $V_1 = \widetilde{x}^2$. Differentiating V_1 along the trajectories yields

$$\dot{V}_1 = 2\widetilde{x}'\widehat{B}[\sigma(-\widehat{B}'\widetilde{x} + v) - \sigma(v)].$$

Since $\|v\| \leqslant \frac{1}{2}$, Lemma 13.24 yields that

$$\dot{V}_1 \leqslant -\widetilde{x}'B\sigma(B'\widetilde{x}).$$

Since $\widetilde{x}$ has a bounded derivative, this implies that we must have that

$$\lim_{t\to\infty} \widehat{B}'\widetilde{x}(t) = 0,$$

which implies that there exists a T_0 such that

$$\|\widehat{B}'\widetilde{x}(t)\| \leqslant \tfrac{1}{2} \quad \text{for} \quad t \geqslant T_0,$$

and hence,

$$\dot{\widetilde{x}} = (\widehat{A} - \widehat{B}\widehat{B}')\widetilde{x},$$

and since this system matrix is Hurwitz stable, we have $\widetilde{x}(t) \to 0$ as $t \to \infty$.

For $t > T_0$, we have for $\overline{x}$ dynamics that

$$\dot{\overline{x}} = A\overline{x} + B\sigma(-B'P\overline{x}) + \overline{\mathcal{J}}\widetilde{x},$$

where $\overline{\mathcal{J}} = \mathcal{J}fbar\mathcal{B}\widehat{\mathcal{B}}'$. Define $V_2 = \overline{x}'P\overline{x}$ and a set

$$\mathcal{K} = \{\overline{x} \mid V_2(\overline{x}) \leqslant \tfrac{\delta^2}{\mathrm{trace}(B'P_1B)}\},$$

where P_1 is the solution of PLE (15.20) with $\varepsilon = 1$. It can be easily seen from (15.22) that $\varepsilon_a(\overline{x}) = 1$ for $\overline{x} \in \mathcal{K}$. Next, consider the derivative of V_2,

$$\begin{aligned}\dot{V}_2 &\leqslant -\varepsilon V_2 + 2\overline{x}'P\mathcal{J}\widetilde{x} + 2\overline{x}'P\overline{\mathcal{J}}\widetilde{x} + \overline{x}'\tfrac{\mathrm{d}P}{\mathrm{d}t}\overline{x}\\ &\leqslant -\varepsilon V_2 + 2\sqrt{V_2}\|P^{1/2}\mathcal{J}\widetilde{x}\| + \overline{x}'\tfrac{\mathrm{d}P}{\mathrm{d}t}\overline{x}.\end{aligned}$$

Note that $\mathcal{J}$ and $\overline{B}$ and hence $\overline{\mathcal{J}}$ are in the form of (15.18). Lemma 15.9 shows that there exists a M such that

$$\|P^{1/2}\overline{\mathcal{J}}\widetilde{x}\| = \sqrt{\widetilde{x}'\overline{\mathcal{J}}'P\mathcal{J}\widetilde{x}} \leqslant \varepsilon\sqrt{M}\|\widetilde{x}\|.$$

Hence,

$$\begin{aligned}\dot{V}_2 &\leqslant -\varepsilon V_2 + 2\varepsilon\sqrt{M}\|\widetilde{x}\|\sqrt{V_2} + \overline{x}'\tfrac{\mathrm{d}P}{\mathrm{d}t}\overline{x}\\ &\leqslant -\varepsilon\sqrt{V_2}\left[\sqrt{V_2} - 2\sqrt{M}\|\widetilde{x}\|\right] + \overline{x}'\tfrac{\mathrm{d}P}{\mathrm{d}t}\overline{x}.\end{aligned}$$

Since $\widetilde{x} \to 0$, there exists a $T_1 > T_0$ such that for $t \geqslant T_1$,

$$\|\widetilde{x}\| \leqslant \frac{\delta^2}{4\sqrt{M}\sqrt{\mathrm{trace}(B'P_1B)}}.$$

Therefore, for $t \geqslant T_1$ and $\overline{x} \notin \mathcal{K}$, $\sqrt{V_2} - 2\sqrt{M}\|\widetilde{x}\| \geqslant \frac{\sqrt{V_2}}{2}$, and thus,

$$\dot{V}_2 \leqslant -\tfrac{\varepsilon}{2}V_2 + \overline{x}'\tfrac{\mathrm{d}P}{\mathrm{d}t}\overline{x}.$$

Since $\dot{V}_2$ cannot have the same sign as $\overline{x}'\frac{dP}{dt}\overline{x}$ (see [48]), we conclude that $\dot{V}_2 < 0$ for $\overline{x} \notin \mathcal{K}$ and $t > T_1$. This implies that $\overline{x}$ will enter $\mathcal{K}$ within finite time, say $T_2 > T_1$, and remain in $\mathcal{K}$ thereafter. For $t > T_2$ and $\overline{x} \in \mathcal{K}$, we have $\varepsilon_a(\overline{x}) = 1$ and $\|\widehat{B}'\widetilde{x}\| \leqslant 1 - \delta$. All saturations are inactive, and the system becomes

$$\begin{cases}\dot{\overline{x}} = A\overline{x} - BB'P_1\overline{x} + \overline{\mathcal{J}}\widetilde{x},\\ \dot{\widetilde{x}} = \widehat{A}\widetilde{x} - \widehat{B}\widehat{B}'\widetilde{x}.\end{cases}$$

The global asymptotic stability follows from the properties that $\widehat{A} - \widehat{B}\widehat{B}'$ and $A - BB'P_1$ are Hurwitz stable.

We proceed next to show the boundedness of trajectories in the presence of d_1 and d_2. Define

$$R = e^{\widehat{A}'t}$$

and $y = R\widetilde{x}$. Note that the above R is a compact form of the rotation defined in (15.11) and (15.12). We have $\dot{R} = -R\widehat{A}$. Then

$$\dot{y} = R\widehat{B}\sigma(-\widehat{B}'R'y + v) - R\widehat{B}\sigma(v) + R\widehat{E}_2 d_2$$

with $y(0) = \widetilde{x}_0$, where $v = -B'P\overline{x}$. Let $\overline{y}$ satisfy

$$\dot{\overline{y}} = R\widehat{E}_2 d_2, \quad \overline{y}(0) = \widetilde{x}_0.$$

The solution is given by

$$\overline{y}(t) = \overline{y}(0) + \int_0^t R\widehat{E}_2 d_2 \mathrm{d}t.$$

Note that R comprises elements of the form $\sin(\omega_i t)$ and $\cos(\omega_i t)$, where $j\omega_i$ is the eigenvalue of A. By the definition of Ω_∞ in (15.2), we find that $\overline{y} \in L_\infty$. Define $\widetilde{y} = y - \overline{y}$. Then

$$\dot{\widetilde{y}} = R\widehat{B}\sigma(-\widehat{B}'R'\widetilde{y} - \widehat{B}'R'\overline{y} + v) - R\widehat{B}\sigma(v), \quad \widetilde{y}(0) = 0.$$

Again, define $z = R'\widetilde{y}$. We get

$$\dot{z} = \widehat{A}z + \widehat{B}\sigma(-\widehat{B}'z - u) - \widehat{B}\sigma(v), \quad z(0) = 0,$$

where

$$u = \widehat{B}'R'\overline{y} - v.$$

Since $u \in L_\infty$ and $\|v\| \leqslant \delta < \frac{1}{2}$, it follows from Lemma 15.7 that $z \in L_\infty$. This implies that $\widetilde{x} \in L_\infty$.

Consider the dynamics of $\overline{x}$,

$$\dot{\overline{x}} = A\overline{x} + B\sigma(-B'P\overline{x}) + \overline{E}_1 d_1 + \mathcal{J}\widetilde{x} + \overline{B}\zeta,$$

where $\zeta = \sigma(-\widehat{B}'\widetilde{x} - B'P\overline{x}) - \sigma(-B'P\overline{x})$. Since $\sigma(\cdot)$ is globally Lipschitz with Lipschitz constant 1, we have that $\|\zeta\| \leqslant \|\widehat{B}'\widetilde{x}\|$, and thus, $\zeta \in L_\infty$.

By differentiating $V_2 = \overline{x}P\overline{x}$, we obtain

$$\begin{aligned}
\dot{V}_2 &\leqslant -\varepsilon V_2 + 2\overline{x}'P\overline{E}_1 d_1 + 2\overline{x}'P\mathcal{J}\widetilde{x} + 2\overline{x}'P\overline{B}\zeta + 2\overline{x}'\tfrac{\mathrm{d}P}{\mathrm{d}t}\overline{x} \\
&\leqslant -\varepsilon V_2 + 2\sqrt{V_2}\|P^{1/2}\overline{E}_1 d_1\| + 2\sqrt{V_2}\|P^{1/2}\overline{B}\zeta\| \\
&\quad + 2\sqrt{V_2}\|P^{1/2}\mathcal{J}\widetilde{x}\| + \overline{x}'\tfrac{\mathrm{d}P}{\mathrm{d}t}\overline{x}.
\end{aligned}$$

We have shown that according to Lemma 15.9, there exist M, M_1, and M_2 such that

$$\|P^{1/2}\mathcal{J}\widetilde{x}\| \leqslant \varepsilon M\|\widetilde{x}\|, \quad \|P^{1/2}\overline{E}_1 d_1\| \leqslant \varepsilon M_1\|d_1\|, \quad \text{and} \quad \|P^{1/2}\overline{\zeta}\| \leqslant \varepsilon M\|\zeta\|.$$

Therefore,

$$\dot{V}_2 \leqslant -\varepsilon\sqrt{V_2}\Big[\sqrt{V}_2 - 2M_1\|d_1\| - 2M\|\tilde{x}\| - 2M_2\|\zeta\|\Big] + \bar{x}'\tfrac{dP}{dt}\bar{x}.$$

If $\sqrt{V_2} \geqslant 2M_1\|d_1\|_\infty + 2M\|\tilde{x}\|_\infty$, we have

$$\dot{V}_2 \leqslant \bar{x}'\tfrac{dP}{dt}\bar{x}.$$

Since $\dot{V}_2$ and $\bar{x}'\frac{dP}{dt}\bar{x}$ cannot have the same sign, we find that

$$\dot{V}_2 \leqslant 0$$

for $\bar{x} \in \{\sqrt{V_2} \geqslant 2M_1\|d_1\|_\infty + 2M\|\tilde{x}\|_\infty\}$, which implies that $\bar{x} \in L_\infty$, and hence, $x \in L_\infty$ in view of the properties of scheduling as enumerated in (12.2.1). ■

The combination of aligned and misaligned disturbances could appear in a more general fashion than those in Theorem 15.4. In general, for any critically unstable system with input saturation and nonadditive disturbances, we can always assume the following system configuration with the aligned as well as misaligned disturbances,

$$\dot{x} = Ax + B\sigma(u) + \bar{E}_1\begin{pmatrix} d_1 \\ d_2 \end{pmatrix} + \bar{E}_2 d_2, \quad x(0) = x_0, \tag{15.25}$$

where A and B are given by (15.15) and (15.16), $\bar{E}_1$ and $\bar{E}_2$ are in the form of (15.18) and (15.19) but with appropriate dimensions.

Based on Theorem 15.4, we can straightforwardly draw the following conclusion:

Theorem 15.5 *Consider the system* (15.25) *with controller* (15.21). *The following hold:*

(i) *In the absence of d_1 and d_2, the origin of the closed-loop system is globally asymptotically stable.*

(ii) *$x \in L_\infty$ for any $x(0)$, $d_1 \in L_\infty$, and $d_2 \in \Omega_\infty$.*

Proof : Note that by definition, $d_2 \in L_\infty$ if $d_2 \in \Omega_\infty$. We already know that $d_1 \in L_\infty$. Then the result is a direct consequence of Theorem 15.4. ■

We can extend the results of Theorem 15.5 by adding another sustained disturbance with a sufficiently small magnitude. To do so, consider the following system configuration:

$$\dot{x} = Ax + B\sigma(u) + \bar{E}_1\begin{pmatrix} d_1 \\ d_2 \end{pmatrix} + \bar{E}_2 d_2 + \bar{E}_3 d_3, \quad x(0) = x_0, \tag{15.26}$$

where A and B are given by (15.15) and (15.16), $\bar{E}_1$ and $\bar{E}_2$ are in the form of (15.18) and (15.19) but with appropriate, dimensions and $\bar{E}_3$ is arbitrary.

Theorem 15.6 *Consider the system* (15.26) *with controller* (15.21). *The following hold:*

(*i*) *In the absence of d_1, d_2, and d_3, the origin of the closed-loop system is globally asymptotically stable.*

(*ii*) *$x \in L_\infty$ for any $x(0)$, $d_1 \in L_\infty$, $d_2 \in \Omega_\infty$, and $d_3 \in L(\delta_1)$ with δ_1 sufficiently small.*

Proof : The global asymptotic stability of the origin in the absence of disturbances has already been shown in the proof of Theorem 15.4.

For any matrix $\bar{E}_3$, we can always write that $\bar{E}_3 = \bar{E}_{3,m} + \bar{E}_{3,a}$, where $\bar{E}_{3,m}$ is in the misaligned form of (15.18) and $\bar{E}_{3,a}$ is in the aligned form of (15.19). To be precise, we can write

$$E_{3,m} = \begin{pmatrix} \bar{E}_{m,1} \\ \bar{E}_{m,2} \\ \vdots \\ \bar{E}_{m,q} \end{pmatrix}, \quad \bar{E}_{m,i} = \begin{pmatrix} E_{3,i,1} \\ E_{3,i,2} \\ \vdots \\ E_{3,i,n_i-1} \\ 0 \end{pmatrix}$$

and

$$\bar{E}_{3,a} = \begin{pmatrix} \bar{E}_{a,1} \\ \bar{E}_{a,2} \\ \vdots \\ \bar{E}_{a,q} \end{pmatrix}, \quad \bar{E}_{a,i} = \begin{pmatrix} 0 \\ 0 \\ \vdots \\ 0 \\ E_{3,i,n_i} \end{pmatrix}.$$

The system (15.26) can be rewritten in the following form:

$$\dot{x} = Ax + B\sigma(u) + \tilde{E}_1 \begin{bmatrix} d_1 \\ d_2 \\ d_3 \end{bmatrix} + \tilde{E}_2 \begin{bmatrix} d_2 \\ d_3 \end{bmatrix}, \quad x(0) = x_0,$$

where

$$\tilde{E}_1 = \begin{bmatrix} \bar{E}_1 & \bar{E}_{3,m} \end{bmatrix}, \quad \tilde{E}_3 = \begin{bmatrix} \bar{E}_2 & \bar{E}_{3,a} \end{bmatrix}.$$

Define $\tilde{x} = x - \bar{x}$ and let $\tilde{d}_1 = (d_1', d_2', d_3')'$. Consider the closed-loop system in terms of $\tilde{x}$ and $\bar{x}$ as in the proof of Theorem 15.4,

$$\begin{cases} \dot{\bar{x}} = A\bar{x} + B\sigma(-B'P\bar{x}) + \tilde{E}_1\tilde{d}_1 + \mathcal{J}\tilde{x} \\ \qquad\qquad + \bar{B}\left[\sigma(-\hat{B}'\tilde{x} - B'P\bar{x}) - \sigma(-B'P\bar{x})\right] \\ \dot{\tilde{x}} = \hat{A}\tilde{x} + \hat{B}\sigma(-\hat{B}'\tilde{x} - B'P\bar{x}) - \hat{B}\sigma(-B'P\bar{x}) + \hat{E}_2 d_2 + \hat{E}_3 d_3, \end{cases}$$

where $\hat{E}_2$ is given by (15.23) and

$$\hat{E}_3 = \begin{pmatrix} E_{3,1,n_1} \\ E_{3,2,n_2} \\ \vdots \\ E_{3,q,n_q} \end{pmatrix}.$$

Define $R = e^{\hat{A}'t}$ and $y = R\tilde{x}$. Then

$$\dot{y} = R\hat{B}\sigma(-\hat{B}'R'y + v) - R\hat{B}\sigma(v) + R\hat{E}_2 d_2 + \hat{E}_3 d_3$$

with $y(0) = \tilde{x}_0$, where $v = -B'P\bar{x}$. Let $\bar{y}$ satisfy

$$\dot{\bar{y}} = R\hat{E}_2 d_2, \quad \bar{y}(0) = \tilde{x}_0.$$

Since $d_2 \in \Omega_\infty$, we find that $\bar{y} \in L_\infty$. Define $\tilde{y} = y - \bar{y}$. Then

$$\dot{\tilde{y}} = R\hat{B}\sigma(-\hat{B}'R'\tilde{y} - \hat{B}'R'\bar{y} + v) - R\hat{B}\sigma(v) + R\hat{E}_3 d_3$$

with $\tilde{y}(0) = 0$. Again define $z = R'\tilde{y}$. We get

$$\dot{z} = \hat{A}z + \hat{B}\sigma(-\hat{B}'z - u) - \hat{B}\sigma(v) + \hat{E}_3 d_3, \qquad z(0) = 0,$$

where $u = \hat{B}'R'\bar{y} - v$. Consider an auxiliary system

$$\dot{w} = (\hat{A} + \hat{B}\hat{B}')w + \hat{E}_3 d_3, \quad w(0) = 0.$$

Let $\delta_1 \leqslant \frac{1}{4\gamma\|\hat{B}\|}$, where γ is the L_∞ gain of pair $(\hat{A} - \hat{B}\hat{B}', E_3)$. Then we have that $\|Fw\|_\infty \leqslant 1/4$. Let $\xi = z - w$. We have that

$$\dot{\xi} = \hat{A}\xi + \hat{B}\sigma(-\hat{B}'\xi + u) - \hat{B}\sigma(v) - \hat{B}\hat{F}w, \quad \xi(0) = 0,$$

where $u = -\hat{B}'w - \hat{B}'R'\bar{y} + v$. Since $u \in L_\infty$ and $\|\sigma(v) + \hat{F}w\|_\infty \leqslant 1/4 + 1/4 = 1/2$, it follows from Lemma 15.7 in the appendix that $\xi \in L_\infty$. This implies that $\tilde{x} \in L_\infty$.

Consider the dynamics of $\bar{x}$,

$$\dot{\bar{x}} = A\bar{x} + B\sigma(-B'P\bar{x}) + \tilde{E}_1\tilde{d}_1 + \mathcal{J}\tilde{x} + \bar{B}\zeta,$$

where $\zeta = \sigma(-\widehat{B}'\widetilde{x} - B'P\overline{x}) - \sigma(-B'P\overline{x})$, and $\widetilde{x} \in L_\infty$ implies that $\zeta \in L_\infty$. Note that $\widetilde{d}_1 \in L_\infty$ and $\widetilde{E}_1$ are also in the form of (15.18). Following the same argument as in the proof of Theorem 15.4, we can show that $\overline{x} \in L_\infty$. ∎

Appendix 15.A

The following lemma for a neutrally stable system is adapted from [89].

Lemma 15.7 *Assume that (A, B) is controllable and $A + A' = 0$. The system,*

$$\dot{x} = Ax - B\sigma(B'x + v_1) - v_2, \quad x(0) = 0,$$

satisfies $x \in L_\infty$ whenever $v_1 \in L_\infty$ and $v_2 \in L_\infty(1/2)$.

Proof : Denote $u = B'x$ and define $V_1 = \frac{1}{3}\|x\|^3$. Differentiating V_1 along the trajectories yields

$$\begin{aligned}
\dot{V}_1 &= \|x\|u'\left[\sigma(-u + v_1) + v_2\right] \\
&\leqslant \|x\|(u - v_1)'\left[-\sigma(u - v_1) + v_2\right] + 2\|x\|\|v_1\|_\infty \\
&= \|x\| \left\{(u - v_1)'[-\sigma(u - v_1) + \sigma(u - v_1 + v_2)]\right. \\
&\quad \left. +(u - v_1)'[-\sigma(u - v_1 + v_2) + \sigma(v_2)]\right\} + 2\|x\|\|v_1\|_\infty \\
&\leqslant -\tfrac{1}{2}\|x\|(u - v_1)\sigma(u - v_1) + 2\sqrt{m}\|x\|\|v_2\|_\infty + 2\|x\|\|v_1\|_\infty.
\end{aligned}$$

The last inequality results from Lemmas 13.23 and 13.24 and the condition that $\|v_2\| \leqslant \frac{1}{2}$.

Next, since $A - BB'$ is Hurwitz stable, there exists a $P > 0$ satisfying

$$(A - BB')'P + P(A - BB') = -I.$$

Define $V_2 = x'Px$. There exists an α such that

$$\begin{aligned}
\dot{V}_2 &= -\|x\|^2 + 2x'P[B\sigma(-u + v_1) + Bu + Bv_2] \\
&= -\|x\|^2 + 2x'P[B(\sigma(-u + v_1) + u - v_1) + Bv_2 + Bv] \\
&\leqslant -\|x\|^2 + 2\alpha\|x\|(u - v_1)\sigma(u - v_1) + 2\alpha\|x\|\|v_2\|_\infty + 2\alpha\|x\|\|v_1\|_\infty,
\end{aligned}$$

where the inequality in Lemma 13.25 is used to derive the last inequality.

Finally, define a Lyapunov candidate $V = 4\alpha V_1 + V_2$. We find that

$$\begin{aligned}
\dot{V} &\leqslant -\|x\|^2 + (8\alpha\sqrt{m} + 2\alpha)\|x\|\|v_2\|_\infty + 10\alpha\|x\|\|v_1\|_\infty \\
&= -\|x\|\left[\|x\| - (8\alpha\sqrt{m} + 2\alpha)\|v_2\|_\infty - 10\alpha\|v_1\|_\infty\right].
\end{aligned}$$

Hence, $\dot{V} \leqslant 0$ for $\|x\| \geqslant (8\alpha\sqrt{m} + 2\alpha)\|v_2\|_\infty + 10\alpha\|v_1\|_\infty$. Let c be such that

$$\{x \mid V(x) \leqslant c\} \supset \{x \mid \|x\| \leqslant (8\alpha\sqrt{m} + 2\alpha)\|v_2\|_\infty + 10\alpha\|v_1\|_\infty\}.$$

We have $\dot{V} \leqslant 0$ for $x \notin \{x \mid V(x) \leqslant c\}$. This implies that $x(t) \in \{x \mid V(x) \leqslant c\}$ for all $t \geqslant 0$. ■

Lemma 15.8 *Consider the system*

$$\dot{y}(t) = \sigma(v(t)) - \sigma(v(t) + k(t)y) + d, \tag{15.27}$$

where $d \in \Omega_\infty$ and $k(t) > 0$ and $v(t)$ are continuous. In that case, we have $y \in L_\infty$ for all $y(0)$.

Proof : Define

$$\dot{\bar{y}} = d, \quad \bar{y}(0) = y(0).$$

Since $d \in \Omega_\infty$, there exists a $M > 0$ such that $|\bar{y}(t)| \leqslant |y(0)| + M$ for all $t > 0$. Define $\tilde{y} = y - \bar{y}$. We have

$$\dot{\tilde{y}} = \sigma(v) - \sigma(v + k(\tilde{y} + \bar{y})), \quad \tilde{y} = 0.$$

Let $\tilde{V} = \tilde{y}^2$. Taking the derivative of $\tilde{V}$ with respect to t, we get

$$\dot{\tilde{V}} = \tilde{y}\left[\sigma(v) - \sigma(v + k(\tilde{y} + \bar{y}))\right].$$

If $\tilde{V} \geqslant (|y(0)| + M)^2$, then $|\tilde{y}| \geqslant M + |y(0)| \geqslant |\bar{y}|$. But this implies that $k(\tilde{y} + \bar{y})$ has the same sign as $\tilde{y}$. Thus,

$$\dot{\tilde{V}} = \tilde{y}\left[\sigma(v) - \sigma(v + k(\tilde{y} + \bar{y}))\right] \leqslant 0.$$

Since $\tilde{V}(0) = 0$, we have $\tilde{V} \leqslant (|y(0)| + M)^2$ and $|\tilde{y}| \leqslant |y(0)| + M$ for all $t > 0$. Therefore, $|y| \leqslant |\tilde{y}| + |\bar{y}| \leqslant 2M + 2y(0)$. ■

Lemma 15.9 *Let P_ε be the solution to PLE* (15.20)*. For any given matrix $\bar{E}_1$ in the form of* (15.18)*, there exists a M such that for $\varepsilon \in (0, 1]$, we have*

$$\bar{E}_1' P_\varepsilon \bar{E}_1 \leqslant M\varepsilon^2 I.$$

Proof : It is shown in [215] that $P_\varepsilon \to 0$ as $\varepsilon \to 0$. Let

$$P_\varepsilon = \varepsilon P_1 + \varepsilon^2 P_2 + \ldots + \varepsilon^i P_i + \ldots.$$

Substituting P_ε in (15.20), we find that P_1 satisfies

$$P_1 A + A' P_1 = 0, \tag{15.28}$$

where A is given by (15.15). Consider the diagonal block of P_1, say $P_{1,i}$ corresponding to $\overline{A}_i$ block. $P_{1,i}$ must satisfy

$$\overline{A}_i' P_{1,i} + P_{1,i} \overline{A}_i = 0, \tag{15.29}$$

where $\overline{A}_i$ is given by (15.16). Suppose

$$P_{1,i} = \begin{pmatrix} \overline{P}_{11} & \overline{P}_{12}' \\ \overline{P}_{12} & \overline{P}_{22} \end{pmatrix},$$

where $\overline{P}_{11} \in \mathbb{R}^{p_i \times p_i}$, $\overline{P}_{12}$ and $\overline{P}_{22}$ are of appropriate dimension.

Define

$$\xi_{i,j} = \begin{pmatrix} x_{i,j} \\ 0 \\ \vdots \\ 0 \end{pmatrix},$$

where $x_{i,j}$, $j = 1, \ldots, p_i$ are the eigenvectors of A_i associated with the eigenvalues λ_j, $j = 1, \ldots, p_l$. Clearly, we have $\overline{A}_i \xi_{i,j} = \lambda_j \xi_{i,j}$, and thus, $\xi_{i,j}$ is an eigenvector of $\overline{A}_i$. Note that $\overline{A}_i$ has p_i linearly independent eigenvectors.

We shall have

$$(\overline{A}_i' P_{1,i} + \overline{P}_{1,i} \overline{A}_i)\xi_{i,j} = 0.$$

This implies that

$$\overline{A}_i' P_{1,i}\xi_{i,j} = -\lambda_j P_{1,i}\xi_{i,j}.$$

In words, $P_{1,i}\xi_{i,j}$, $j = 1, \ldots, p_i$ are the eigenvectors of $\overline{A}_i'$ associated with the eigenvalue $-\lambda_j$, $j = 1, \ldots, p_i$.

On the other hand, we have a set of eigenvectors of $\widehat{A}'$ in the form of

$$\nu_{i,j} = \begin{pmatrix} 0 \\ \vdots \\ 0 \\ v_{i,j} \end{pmatrix}, i = 1, \ldots, p_i,$$

where $v_{i,j}$ are the eigenvectors of A_i' associated with the eigenvalue λ_j.

Note that $\bar{A}'$ has only p_i linearly independent eigenvectors. Therefore,

$$P_1\xi_{i,j} = \begin{pmatrix} \bar{P}_{11}x_i \\ \bar{P}_{12}x_i \end{pmatrix} \in \operatorname{span}\{v_{i,1},\ldots,v_{i,p_i}\}.$$

This implies that $\bar{P}_{11}x_{i,j} = 0$, $j = 1,\ldots,p_i$. Since $x_{i,j}$ forms a basis of $\mathbb{R}^{p_i}$, we must have $\bar{P}_{11} = 0$, and hence, $\bar{P}_{12} = 0$ due to the fact that $P_{1,i}$ is positive semi-definite.

Applying the above argument to $\bar{P}_{22}$ and recursively, we shall eventually find that

$$P_{1,i} = \begin{pmatrix} 0 & 0 & 0 & \ldots & 0 \\ 0 & 0 & 0 & \ldots & 0 \\ \vdots & \vdots & \ddots & \ddots & \vdots \\ 0 & 0 & \ldots & 0 & 0 \\ 0 & 0 & 0 & 0 & \bar{P}_{n_i n_i} \end{pmatrix}.$$

Due to positive semi-definiteness of P_1, for any given matrix $\bar{E}_1$ in the form of (15.18), we must have $P_1\bar{E}_1 = 0$. This implies that $\bar{E}_1' P_\varepsilon \bar{E}_1$ must be of order ε^2. This completes the proof. ■

16

Simultaneous internal and external stabilization in the presence of a class of non-input-additive sustained disturbances: discrete time

16.1 Introduction

For discrete-time general critically unstable linear systems subject to actuator saturation, this chapter is a counterpart of Chap. 15 which pertains to continuous-time systems. That is, our goal here for discrete-time systems is to identify a set of non-input-additive sustained disturbances for which a feedback control law can be determined such that

1. In the absence of disturbances, the origin of the closed-loop system is globally asymptotically stable.

2. If the disturbances belong to the given set, the states of the closed-loop system are bounded for any arbitrarily specified initial conditions.

Regarding the architecture of this chapter, after certain preliminary discussions in Sect. 16.2, we consider in Sect. 16.3 neutrally stable systems which allow a linear feedback law. We then move on to Sect. 16.4 to consider general critically unstable systems (or equivalently general ANCBC systems).

This chapter is based on our work [48, 193].

16.2 Preliminaries

Consider the system,

$$x(k+1) = Ax(k) + B\sigma(u(k)) + Ed(k), \quad x(0) = x_0, \tag{16.1}$$

where $x \in \mathbb{R}^n$, $u \in \mathbb{R}^m$ and $d \in \mathbb{R}^p$. The pair (A, B) is stabilizable, and A has all its eigenvalues in the closed unit disc. That is, the given system is asymptotically null controllable with bounded control (ANCBC). As pointed out in the introduction, for systems of the form (16.1), our goal in this chapter is to identify a class of ℓ_∞ disturbance signals d such that for appropriately designed feedback control laws, the closed-loop system exhibits global asymptotic stability in the absence

A. Saberi et al., *Internal and External Stabilization of Linear Systems with Constraints*, Systems & Control: Foundations & Applications, DOI 10.1007/978-0-8176-4787-2_16,

of disturbances d, and in the presence of such disturbances d, the state variables of it are bounded. Consistent with previous chapters, we use $\ell_\infty(\delta)$ to denote the set of ℓ_∞ signals whose ℓ_∞ norm is less than δ.

We define a set of discrete disturbances

$$\Omega_\infty = \Bigg\{ d \in \ell_\infty \mid \exists M > 0, \text{ such that } \forall i \in 1,\ldots,q,\ \forall k_2 \geqslant k_1 \geqslant 0,$$
$$\left\| \sum_{k=k_1}^{k_2} d(k)\cos(\theta_i k) \right\| \leqslant M, \quad \left\| \sum_{k=k_1}^{k_2} d(k)\sin(\theta_i k) \right\| \leqslant M \Bigg\}, \tag{16.2}$$

where $e^{j\theta_i}$, $i \in 1,\ldots,q$, represent the eigenvalues of A on the unit circle.

Note that Ω_∞ contains signals which do not have sustained component at the discrete frequency θ_i. Like in the continuous-time case, we can also rewrite the above definition as

$$\Omega_\infty = \Bigg\{ d \in \ell_\infty \mid \exists M > 0, \text{ such that } \forall i \in 1,\ldots,q,$$
$$\forall k_2 \geqslant k_1 \geqslant 0, \quad \left\| \sum_{k=k_1}^{k_2} d(k) z_i^k \right\| \leqslant M \Bigg\}, \tag{16.3}$$

where $z_i = e^{j\theta_i}$, $i = 1,\ldots,q$. Since $d \in \ell_\infty$, the power series $\sum_0^\infty d(z)z^k$, i.e., the z-transform of d, always has a radius of convergence 1. For $|z| = 1$, Definition (16.3) implies that all the partial sums of the power series are bounded at $z = z_i$.

We shall need the following inequality, the proof of which is given in the appendix:

Lemma 16.1 *For two vectors $s, t \in \mathbb{R}^m$, and for $\|t\| \leqslant 1$, we have*

$$\|\sigma(s+t) - \sigma(t)\| \leqslant 2\|\sigma(\tfrac{1}{2}s)\|.$$

16.3 Neutrally stable systems

In this section, we deal with neutrally stable systems. Consider the following system:

$$x(k+1) = Ax(k) + B\sigma(u(k)) + Ed(k), \quad x(0) = x_0.$$

We assume that (A, B) is controllable, $A'A = I$, and $d \in \ell_\infty$.

We use a linear state feedback controller $u = -\kappa B'Ax$ which gives a closed-loop system as

$$x(k+1) = Ax(k) + B\sigma(-\kappa B'Ax(k)) + Ed(k), \quad x(0) = x_0.$$

In view of the development in Sect. 4.6.1, it is easy to see that for κ such that $4\kappa B'B \leqslant 1$, the origin of the closed-loop system in the absence of disturbances is globally asymptotically stable. As such, we focus here only on the boundedness of closed-loop states with disturbances. The development given below basically parallels the same as in the continuous-time case discussed in the previous chapter. We apply a sequence of successive rotations to state coordinates and eventually convert the non-input-additive disturbances to input-additive disturbances using the property of Ω_∞.

16.3.1 *Single-frequency systems*

We start by considering an example system with a pair of complex eigenvalues at $\pm j$:

$$\rho\begin{pmatrix} x_1 \\ x_2 \end{pmatrix} = \begin{pmatrix} 0 & 1 \\ -1 & 0 \end{pmatrix}\begin{pmatrix} x_1 \\ x_2 \end{pmatrix} + \begin{pmatrix} 0 \\ 1 \end{pmatrix}\sigma(\kappa x_1) + \begin{pmatrix} e_1 \\ e_2 \end{pmatrix} d, \tag{16.4}$$

where $d \in \Omega_\infty$.

Theorem 16.2 *For κ such that $4\kappa B'B \leqslant 1$, the trajectories of* (16.4) *remain bounded for any initial condition.*

Proof : To analyze the system, we introduce a rotation matrix

$$R(k) = \begin{pmatrix} 0 & -1 \\ 1 & 0 \end{pmatrix}^k,$$

which represents a counterclockwise rotation by an angle $\frac{k}{2}\pi$. The dynamics of the rotation matrix are given by

$$R(k+1) = R(k)\begin{pmatrix} 0 & -1 \\ 1 & 0 \end{pmatrix}.$$

We shall study the dynamics of x from a rotated coordinate frame, and toward this end, we define the rotated state $y = Rx$. The dynamics of y is given by

$$\begin{aligned} y(k+1) &= y(k) + R(k+1)\left[\begin{pmatrix} 0 \\ 1 \end{pmatrix}\sigma(\kappa x_1(k)) + \begin{pmatrix} e_1 \\ e_2 \end{pmatrix} d(k)\right] \\ &= y(k) + R(k+1)\left[\begin{pmatrix} 0 \\ 1 \end{pmatrix}\sigma\left(\begin{pmatrix} \kappa & 0 \end{pmatrix} R'(k)y(k)\right) + \begin{pmatrix} e_1 \\ e_2 \end{pmatrix} d(k)\right], \end{aligned}$$

with $y(0) = x(0) = x_0$. Next, define a fictitious system

$$\tilde{y}(k+1) = \tilde{y}(k) + R(k+1)\begin{pmatrix} e_1 \\ e_2 \end{pmatrix} d(k), \quad \tilde{y}(0) = x_0. \tag{16.5}$$

The solution to this dynamic system is

$$\tilde{y}(k) = \tilde{y}(0) + \sum_{i=0}^{k-1} \begin{pmatrix} 0 & -1 \\ 1 & 0 \end{pmatrix}^{i+1} \begin{pmatrix} e_1 \\ e_2 \end{pmatrix} d(i).$$

It follows from the definition of Ω_∞ that the sum in the right-hand side is bounded for all k and hence $\tilde{y} \in \ell_\infty$.

Consider the difference between y and the fictitious state $\tilde{y}$, given by $z = y - \tilde{y}$, with dynamics

$$\begin{aligned} z(k+1) &= z(k) + R(k+1)\begin{pmatrix} 0 \\ 1 \end{pmatrix} \sigma\left(\begin{pmatrix} \kappa & 0 \end{pmatrix} R'(k) y(k)\right) \\ &= z(k) + R(k+1)\begin{pmatrix} 0 \\ 1 \end{pmatrix} \sigma\left(\begin{pmatrix} \kappa & 0 \end{pmatrix} R'(k) z(k) + \delta(k)\right), \end{aligned}$$

with $z(0) = 0$, where

$$\delta = \begin{pmatrix} \kappa & 0 \end{pmatrix} R' \tilde{y} \in \ell_\infty.$$

We rotate z back to the original coordinate frame by introducing $w = R'z$, thereby obtaining the dynamics

$$w(k+1) = \begin{pmatrix} 0 & 1 \\ -1 & 0 \end{pmatrix} w(k) + \begin{pmatrix} 0 \\ 1 \end{pmatrix} \sigma\left(\begin{pmatrix} \kappa & 0 \end{pmatrix} w(k) + \delta(k)\right), \quad w(0) = 0.$$

It is shown in Lemma 16.8 given in Appendix that the above system is ℓ_∞ stable with respect to the input δ for κ such that $4\kappa B'B \leqslant 1$ and hence $w \in \ell_\infty$. Finally, we have that $x = w + R'\tilde{y}$ and hence $x \in \ell_\infty$. ■

To demonstrate the importance of the disturbance belonging to Ω_∞, we shall now show that if d contains a large component at discrete frequency $\pm\frac{\pi}{2}$, the states of (16.4) could diverge toward infinity for any initial condition. Suppose that

$$d(k) = a \sin\left(\tfrac{k\pi}{2} + \theta\right),$$

where a is an amplitude yet to be chosen. For ease of presentation, we assume that $e_1 = 0$ and $e_2 = 1$. Consider the dynamics of the rotated state y from the proof of Theorem 16.2. We have

$$y(k+1) = y(k) + R(k+1)\left[\begin{pmatrix}0\\1\end{pmatrix}\sigma(u(k)) + \begin{pmatrix}0\\1\end{pmatrix} a\sin(\tfrac{k\pi}{2}+\theta)\right],$$
$$= y(k) - a\begin{pmatrix}\cos(\frac{k\pi}{2})\sin(\frac{k\pi}{2}+\theta)\\ \sin(\frac{k\pi}{2})\sin(\frac{k\pi}{2}+\theta)\end{pmatrix} - \begin{pmatrix}\cos(\frac{k\pi}{2})\\ \sin(\frac{k\pi}{2})\end{pmatrix}\sigma(u(k)),$$

where for ease of notation, we have used

$$u = \kappa x_1 = \begin{pmatrix}\kappa & 0\end{pmatrix} R' y.$$

Using appropriate trigonometric identities, the dynamics can be rewritten as

$$y(k+1) = y(k) - \tfrac{a}{2}\begin{pmatrix}\sin(\theta) + \sin(k\pi+\theta)\\ \cos(\theta) - \cos(k\pi+\theta)\end{pmatrix} - \begin{pmatrix}\cos(\frac{k\pi}{2})\\ \sin(\frac{k\pi}{2})\end{pmatrix}\sigma(u(k)).$$

We have either $|\sin(\theta)| \geqslant \sqrt{2}/2$ or $|\cos(\theta)| \geqslant \sqrt{2}/2$. Without loss of generality, we assume that $|\cos(\theta)| \geqslant \sqrt{2}/2$. Let a be chosen such that $a \geqslant 4(1+\varepsilon)/\sqrt{2}$, where ε is a positive number. For the trajectory $y_2(k)$, we have

$$|y_2(k)| = \left| y_2(0) - \sum_{i=0}^{k-1} \tfrac{a}{2}\big(\cos(\theta) - \cos(i\pi+\theta)\big) - \sum_{i=0}^{k-1} \sin(\tfrac{i\pi}{2})\sigma(u(i)) \right|.$$

Noting that $\sin(\frac{i\pi}{2})\sigma(u(i))$ is bounded by ± 1 and using the bound

$$|\tfrac{a}{2}\cos(\theta)| \geqslant \tfrac{a}{4}\sqrt{2} \geqslant 1+\varepsilon,$$

we therefore have

$$|y_2(k)| \geqslant -|y_2(0)| - \tfrac{a}{2}\left|\sum_{i=0}^{k-1}\cos(i\pi+\theta)\right| + \sum_{i=0}^{k-1}\varepsilon$$
$$\geqslant -|y_2(0)| - \tfrac{a}{2} + \varepsilon k.$$

This shows that $y_2(k)$ diverges toward infinity.

16.3.2 Multifrequency systems

Theorem 16.3 *Consider the system*

$$x(k+1) = Ax(k) - B\sigma(\kappa B'Ax(k)) + Ed(k), \quad x(0) = x_0 \qquad (16.6)$$

where (A, B) is controllable, $A'A = I$, and $d \in \Omega_\infty$. Then for κ such that $4\kappa B'B \leqslant I$, we have for any initial condition the state $x(k)$ bounded for all $k \geqslant 0$.

Proof : Define $R(k) = (A')^k$. Since $A'A = I$, R represents a time-varying rotation matrix with difference equation $R(k+1) = R(k)A'$. Also, define $y = Rx$. The transformed system becomes

$$y(k+1) = y(k) - R(k)A'B\sigma(\kappa B'AR'(k)y(k)) + R(k)A'Ed(k),$$

with $y(0) = x_0$. Introduce a fictitious system

$$\tilde{y}(k+1) = \tilde{y}(k) + R(k)A'Ed(k), \quad \tilde{y}(0) = x_0.$$

Note that $d \in \Omega_\infty$ implies that there exists a $M > 0$ such that

$$\forall k_2 > k_1 > 0, \quad \left\| \sum_{k=k_1}^{k_2} (A')^{k+1} Ed(k) \right\| \leqslant M.$$

Therefore, we observe that $\tilde{y} \in \ell_\infty$. Let $z = y - \tilde{y}$. We get

$$z(k+1) = z(k) - R(k)A'B\sigma\left(\kappa B'AR'(k)z(k) + \kappa B'AR'(k)\tilde{y}(k)\right),$$

with $z(0) = 0$. Finally, define $w = R'z$. The dynamics of w is given by

$$w(k+1) = Aw(k) - B\sigma(\kappa B'Aw(k) + v(k)), \quad w(0) = 0,$$

where $v(k) = \kappa B'A^{k+1}\tilde{y}(k)$. For $4\kappa B'B \leqslant I$, the above system is ℓ_∞ stable with respect to v. Thus, $\tilde{y} \in \ell_\infty$ implies that $w \in \ell_\infty$. Note that $x(k) = w(k) + R'(k)\tilde{y}(k)$. Therefore, we conclude $x \in \ell_\infty$. ■

Next theorem shows that a small disturbance that does not belong to Ω_∞ can also be tolerated.

Theorem 16.4 *Consider the discrete-time system*

$$x(k+1) = Ax(k) - B\sigma(\kappa B'Ax(k)) + E_1d_1(k) + E_2d_2(k) \tag{16.7}$$

with $x(0) = x_0$, where $d_1 \in \Omega_\infty$ and $d_2 \in \ell_\infty(\delta)$. Then, for κ such that $4\kappa B'B \leqslant I$ and with δ sufficiently small, we have for any initial condition the state $x(k)$ bounded for $k \geqslant 0$.

Proof : Following the same lines as in the proof of Theorem 16.3, we shall get a transformed system

$$w(k+1) = Aw(k) - B\sigma(\kappa B'Aw(k) + \kappa B'AR'(k)\tilde{y}(k)) + E_2d_2(k),$$

with $w(0) = 0$, where $w = x - R'\tilde{y}$ and $\tilde{y}$ satisfies

$$\tilde{y}(k+1) = \tilde{y}(k) + R(k+1)A'E_1d_1(k), \quad \tilde{y}(0) = x_0,$$

and hence $\widetilde{y} \in \ell_\infty$. Introduce an auxiliary system

$$\bar{w}(k+1) = (A+BF)\bar{w}(k) + E_2 d_2(k), \quad \bar{w}(0) = 0,$$

where F is such that $A+BF$ is asymptotically stable. Since $d_2 \in \ell_\infty$, we find that $F\bar{w} \in \ell_\infty$. Moreover, we have $\|F\bar{w}\|_\infty \leqslant \frac{1}{2}$ provided that δ is small enough.

Define $\widetilde{w} = w - \bar{w}$. Then we get

$$\widetilde{w}(k+1) = A\widetilde{w}(k) - B\sigma(\kappa B'A\widetilde{w}(k) + v_1(k)) + Bv_2(k), \quad \widetilde{w}(0) = 0,$$

where $v_1 = \kappa B'A\bar{w} + \kappa B'AR'\widetilde{y}$ and $v_2 = F\bar{w}$. Note that $\|v_2\|_\infty \leqslant \frac{1}{2}$. Then, Lemma 16.8 given in Appendix shows that $\widetilde{w} \in \ell_\infty$. Since $x = \widetilde{w} + \bar{w} + R'\widetilde{y}$, we conclude that $x \in \ell_\infty$ for any initial condition. ■

16.4 Critically unstable systems

16.4.1 Formulation

We consider next general critically unstable systems. The system given in (16.1) with possibly different type of disturbances can be rewritten as

$$x(k+1) = Ax(k) + B\sigma(u(k)) + Ed(k), \quad x(0) = x_0 \tag{16.8}$$

where, without loss of generality, we assume that (A, B) is controllable and A has all its eigenvalues on the unit circle. Again, without loss of generality, we assume that $x = (x_1', x_2', \dots, x_q')'$ and A, B, and E have the following structure:

$$A = \begin{pmatrix} \bar{A}_1 & 0 & \cdots & 0 \\ 0 & \bar{A}_2 & \ddots & \vdots \\ \vdots & \ddots & \ddots & 0 \\ 0 & \cdots & 0 & \bar{A}_q \end{pmatrix}, B = \begin{pmatrix} B_1 \\ B_2 \\ \vdots \\ B_q \end{pmatrix}, \quad E = \begin{pmatrix} E_1 \\ E_2 \\ \vdots \\ E_q \end{pmatrix}, \tag{16.9}$$

where

$$x_i = \begin{pmatrix} x_{i,1} \\ x_{i,2} \\ \vdots \\ x_{i,n_i-1} \\ x_{i,n_i} \end{pmatrix}, \quad \bar{A}_i = \underbrace{\begin{pmatrix} A_i & I & 0 & \cdots & 0 \\ 0 & A_i & I & \ddots & \vdots \\ \vdots & \ddots & \ddots & \ddots & 0 \\ \vdots & & \ddots & A_i & I \\ 0 & \cdots & \cdots & 0 & A_i \end{pmatrix}}_{n_i \times n_i \text{ blocks}}. \tag{16.10}$$

Here, $x_{i,j} \in \mathbb{R}^{p_i}$ and $A_i' A_i = I$. Finally,

$$B_i = \begin{pmatrix} B_{i,1} \\ B_{i,2} \\ \vdots \\ B_{i,n_i-1} \\ B_{i,n_i} \end{pmatrix}, \quad E_i = \begin{pmatrix} E_{i,1} \\ E_{i,2} \\ \vdots \\ E_{i,n_i-1} \\ E_{i,n_i} \end{pmatrix}. \tag{16.11}$$

Note that the above form can be obtained by assembling together in the real Jordan canonical form those blocks corresponding to the eigenvalues with the same Jordan block size.

As in the continuous time which was discussed in the previous chapter, based on the above Jordan structure of matrix A, we say the disturbance d is *aligned* if $E_{i,n_i} \neq 0$ for some $i = 1, \dots, q$ and *misaligned* if $E_{i,n_i} = 0$ for all $i = 1, \dots, q$. Again, we use here the words "aligned" and "misaligned" because of a lack of better terminology. To explore further while using this terminology, we rewrite the given system (16.9) as

$$x(k+1) = Ax(k) + B\sigma(u(k)) + \bar{E}_1 d_1(k) + \bar{E}_2 d_2(k), \tag{16.12}$$

with $x(0) = x_0$, where

$$\bar{E}_1 = \begin{pmatrix} \bar{E}_{1,1} \\ \bar{E}_{1,2} \\ \vdots \\ \bar{E}_{1,q} \end{pmatrix}, \quad \bar{E}_{1,i} = \begin{pmatrix} E_{i,1} \\ E_{i,2} \\ \vdots \\ E_{i,n_i-1} \\ 0 \end{pmatrix} \tag{16.13}$$

and

$$\bar{E}_2 = \begin{pmatrix} \bar{E}_{2,1} \\ \bar{E}_{2,2} \\ \vdots \\ \bar{E}_{2,q} \end{pmatrix}, \quad \bar{E}_{2,i} = \begin{pmatrix} 0 \\ 0 \\ \vdots \\ 0 \\ E_{i,n_i} \end{pmatrix}. \tag{16.14}$$

We first present the result for the above type of disturbances. As will become clear later, the result derived based on the above form can be straightforwardly extended to other type of disturbances.

We shall design a nonlinear dynamic state feedback controller that will enable us to solve our problem. For $\varepsilon \in (0, 0.9]$, let P_ε be the solution of the discrete parametric Lyapunov equation (DPLE):

$$(1-\varepsilon)P_\varepsilon = A'P_\varepsilon A - A'P_\varepsilon B(B'P_\varepsilon B + I)^{-1}B'P_\varepsilon A. \tag{16.15}$$

The existence of P_ε for $\varepsilon \in (0,1)$ has been established in Lemma 12.15. When A is given by (16.9) and (16.10), an important property of P_ε is shown in Lemma 16.12 given in Appendix.

Consider the following dynamic state feedback controller

$$\begin{cases} \widehat{x}_i(k+1) = A_i\widehat{x}_i(k) + B_{i,n_i}\sigma\left(-F_{\varepsilon_a(\bar{x}(k))}\bar{x}(k)\right), & i=1,\dots,q \\ u(k) \quad = -\kappa\widehat{B}'\widehat{A}(x_b(k)-\widehat{x}(k)) - F_{\varepsilon_a(\bar{x}(k))}\bar{x}(k), \end{cases} \tag{16.16}$$

where $\bar{x}$ is x with x_{i,n_i} replaced by $\widehat{x}_i$,

$$x_b = \begin{pmatrix} x_{1,n_1} \\ x_{2,n_2} \\ \vdots \\ x_{q,n_q} \end{pmatrix}, \quad \widehat{x} = \begin{pmatrix} \widehat{x}_1 \\ \widehat{x}_2 \\ \vdots \\ \widehat{x}_q \end{pmatrix}, \quad \widehat{A} = \begin{pmatrix} A_1 & 0 & \cdots & 0 \\ 0 & A_2 & \ddots & \vdots \\ \vdots & \ddots & \ddots & 0 \\ 0 & \cdots & 0 & A_q \end{pmatrix}, \quad \widehat{B} = \begin{pmatrix} B_{1,n_1} \\ B_{2,n_2} \\ \vdots \\ B_{q,n_q} \end{pmatrix},$$

$$F_{\varepsilon_a(\bar{x})} = (B'P_{\varepsilon_a(\bar{x})}B + I)^{-1}B'P_{\varepsilon_a(\bar{x})}A,$$

and where $P_{\varepsilon_a(\bar{x})}$ is the solution to the discrete parametric Lyapunov equation (16.15) with $\varepsilon = \varepsilon_a(\bar{x})$ and $\varepsilon_a(\bar{x})$ is determined by

$$\varepsilon = \varepsilon_a(\bar{x}) := \max\{r \in (0, 0.9] \mid (\bar{x}'P_r\bar{x}) \times \operatorname{trace}(P_r) \leqslant \tfrac{\delta^2}{2\operatorname{trace}(BB')}\}, \tag{16.17}$$

where $\delta = \frac{1}{4}$ and P_r is the unique positive definite solution of the discrete parametric Lyapunov equation (16.15) with $\varepsilon = r$ (see the properties of scheduling as enumerated in (12.2.1)).

Note that we have

$$\|F_{\varepsilon_a(\bar{x})}\bar{x}\| \leqslant \tfrac{1}{4},$$

for any $\bar{x}$. We have the following result:

Theorem 16.5 *Consider the system* (16.12) *with controller* (16.16) *for κ such that $8\kappa B'B \leqslant I$. Then the following hold:*

(*i*) *In the absence of d_1 and d_2, the origin is global asymptotically stable.*

(*ii*) *In the presence of d_1 and d_2, $x \in \ell_\infty$ for any $x(0)$, $d_1 \in \ell_\infty$, and $d_2 \in \Omega_\infty$.*

Proof : In the proof, we omit the parameter ε_a in the expressions for P_{ε_a} and F_{ε_a} for ease of notation and use $\varepsilon(k)$, $P(k)$ and $F(k)$ to express their dependence on time. Define

$$\widetilde{x} = x_b - \widehat{x} = \begin{pmatrix} x_{1,n_1} - \widehat{x}_1 \\ x_{2,n_2} - \widehat{x}_2 \\ \vdots \\ x_{q,n_q} - \widehat{x}_q \end{pmatrix}.$$

We have

$$\rho\widetilde{x} = \widehat{A}\widetilde{x} + \widehat{B}\sigma(-\kappa\widehat{B}'\widehat{A}\widetilde{x} - F\overline{x}) - \widehat{B}\sigma(-F\overline{x}) + \widehat{E}_2 d_2,$$

where

$$\widehat{E}_2 = \begin{pmatrix} E_{1,n_1} \\ E_{2,n_2} \\ \vdots \\ E_{q,n_q} \end{pmatrix}. \tag{16.18}$$

Note that the controllability of (A, B) implies the controllability of $(\widehat{A}, \widehat{B})$. Moreover, $\widehat{A}\widehat{A}' = I$. The closed-loop system can then be written in terms of $\widetilde{x}, \overline{x}$ as

$$\begin{cases} \rho\overline{x} = A\overline{x} + B\sigma(-F\overline{x}) + \overline{B}\left[\sigma(-\kappa\widehat{B}'\widehat{A}\widetilde{x} - Fx) - \sigma(-Fx)\right] + \overline{E}_1 d_1 + \mathcal{I}\widetilde{x} \\ \rho\widetilde{x} = \widehat{A}\widetilde{x} + \widehat{B}\sigma(-\kappa\widehat{B}'\widehat{A}\widetilde{x} - F\overline{x}) - \widehat{B}\sigma(-F\overline{x}) + \widehat{E}_2 d_2, \end{cases} \tag{16.19}$$

where $\overline{B}$ is B with B_{i,n_i} blocks set to zero and

$$\mathcal{I} = \begin{pmatrix} \mathcal{I}_1 \\ \mathcal{I}_2 \\ \vdots \\ \mathcal{I}_q \end{pmatrix}, \quad \mathcal{I}_i = \begin{pmatrix} 0 \\ \vdots \\ 0 \\ \overline{I}_i \\ 0 \end{pmatrix}, \quad \overline{I}_i = [0 \quad \cdots \quad \underset{\substack{\uparrow \\ i\text{th block}}}{I} \quad \cdots \quad 0].$$

We first prove global asymptotic stability in the autonomous case. Let $v = -F\overline{x}$. Our scheduling (16.17) guarantees that $\|v\| \leqslant \delta = \frac{1}{4}$ for any $\overline{x}$. Consider the dynamics of $\widetilde{x}$:

$$\rho\widetilde{x} = \widehat{A}\widetilde{x} + \widehat{B}\sigma(-\kappa\widehat{B}'\widehat{A}\widetilde{x} + v) - \widehat{B}\sigma(v).$$

Note that $8\kappa B'B \leqslant I$ implies that $8\kappa \widehat{B}'\widehat{B} \leqslant I$. Define a Lyapunov function as $V_1 = \widetilde{x}^2$. Let $\widetilde{u} = \kappa \widehat{B}'\widehat{A}\widetilde{x}$. The increment of V_1 along the trajectories is given by

$$\begin{aligned}\rho V_1 - V_1 &= [\sigma(-\widetilde{u} + v) - \sigma(v)]' \, \widehat{B}'\widehat{B} \, [\sigma(-\widetilde{u} + v) - \sigma(v)] \\ &\qquad + \tfrac{2}{\kappa}\widetilde{u}' \, [\sigma(-\widetilde{u} + v) - \sigma(v)] \\ &\leqslant -\tfrac{1}{\kappa}\widetilde{u}\sigma(\widetilde{u}) + \tfrac{1}{2\kappa}\|\sigma(\tfrac{1}{2}\widetilde{u})\|^2 \\ &\leqslant -\tfrac{1}{\kappa}\widetilde{u}\sigma(\widetilde{u}) + \tfrac{1}{2\kappa}\|\sigma(\widetilde{u})\|^2 \\ &\leqslant -\tfrac{1}{2\kappa}\widetilde{u}\sigma(\widetilde{u}) \\ &= -\tfrac{1}{2}\widetilde{x}'\widehat{A}'\widehat{B}\sigma(\kappa \widehat{B}'\widehat{A}\widetilde{x}),\end{aligned}$$

where we use $8\kappa \widehat{B}'\widehat{B} \leqslant I$, Lemma 13.23, and Lemma 16.1.

This clearly implies that $\widehat{B}'\widehat{A}\widetilde{x}(k) \to 0$ as $k \to \infty$, and hence there exists a K_0 such that we have

$$\|\kappa \widehat{B}'\widehat{A}\widetilde{x}(k)\| \leqslant \tfrac{1}{2}$$

for $k \geqslant K_0$, and hence we obtain

$$\widetilde{x}(k+1) = (\widehat{A} - \kappa \widehat{B}\widehat{B}'\widehat{A})\widetilde{x},$$

and therefore, $\widetilde{x}(k) \to 0$ as $k \to \infty$ because the matrix $\widehat{A} - \kappa \widehat{B}\widehat{B}'\widehat{A}$ is Schur stable.

For $k \geqslant K_0$, we have for $\overline{x}$ dynamics,

$$\overline{x}(k+1) = A\overline{x} + B\sigma(-F\overline{x}) + \overline{\mathcal{J}}\widetilde{x},$$

where $\overline{\mathcal{J}} = \mathcal{J} - \kappa \overline{B}\widehat{B}'\widehat{A}$. Define $V_2 = \overline{x}'P\overline{x}$ and a set

$$\mathcal{K} = \left\{ \overline{x} \mid V_2(\overline{x}) \leqslant \alpha^2 \triangleq \frac{\delta^2}{2\,\mathrm{trace}(BB')\,\mathrm{trace}(P^*)} \right\},$$

where P^* is the solution of (16.15) with $\varepsilon = 0.9$. It can be easily seen from (16.17) that for $\overline{x} \in \mathcal{K}$, $\varepsilon_a(\overline{x}) = 0.9$. Next, consider the increment of V_2 along the trajectory. There exists a β independent of d such that

$$\begin{aligned}&V_2(k+1) - V_2(k) \\ &\quad \leqslant -\varepsilon V_2(k) - 2\sigma(F\overline{x}(k))'B'P(k)\overline{\mathcal{J}}\widetilde{x}(k) + 2\overline{x}'(k)AP(k)\overline{\mathcal{J}}\widetilde{x}(k) \\ &\qquad + \widetilde{x}'(k)\overline{\mathcal{J}}'P(k)\overline{\mathcal{J}}\widetilde{x}(k) + \overline{x}'(k+1)'[P(k+1) - P(k)]\overline{x}(k+1) \\ &\quad \leqslant -\varepsilon V_2(k) + 2\|A\|\sqrt{V_2(k)}\|P^{1/2}(k)\overline{\mathcal{J}}\widetilde{x}(k)\| + \beta\|P^{1/2}(k)\overline{\mathcal{J}}\widetilde{x}(k)\| \\ &\qquad + \|P^{1/2}(k)\overline{\mathcal{J}}\widetilde{x}(k)\|^2 + \overline{x}'(k+1)[P(k+1) - P(k)]\overline{x}(k+1).\end{aligned}$$

Note that $\mathcal{J}$, $\overline{B}$, and hence $\overline{\mathcal{J}}$ are all in the form of (16.13). Lemma 16.12 given in Appendix shows that there exists a M such that

$$\|P^{1/2}\overline{\mathcal{J}}\widetilde{x}\| = \sqrt{\widetilde{x}'\overline{\mathcal{J}}'P\overline{\mathcal{J}}\widetilde{x}} \leqslant \varepsilon\sqrt{M}\|\widetilde{x}\|.$$

Moreover, for $\overline{x} \notin \mathcal{K}$, $V_2 \geqslant \alpha^2$. Hence, we have for $\overline{x} \notin \mathcal{K}$,

$$\begin{aligned}
&V_2(k+1) - V_2(k)\\
&\quad \leqslant -\varepsilon V_2(k) + 2\varepsilon\sqrt{MV_2(k)}\|A\|\|\widetilde{x}(k)\| + \tfrac{\beta}{\alpha}\varepsilon\sqrt{MV_2(k)}\|\widetilde{x}(k)\|\\
&\qquad + \varepsilon\tfrac{1}{\alpha}M\sqrt{V_2(k)}\|\widetilde{x}(k)\|^2 + \overline{x}'(k+1)[P(k+1) - P(k)]\overline{x}(k+1)\\
&\quad \leqslant -\varepsilon\sqrt{V_2(k)}\Big[\sqrt{V_2(k)} - 2\sqrt{M}\|A\|\|\widetilde{x}(k)\| - \tfrac{\beta}{\alpha}\sqrt{M}\|\widetilde{x}(k)\|\\
&\qquad - \tfrac{1}{\alpha}M\|\widetilde{x}(k)\|^2\Big] + \overline{x}'(k+1)[P(k+1) - P(k)]\overline{x}(k+1).
\end{aligned}$$

Since $\widetilde{x} \to 0$, there exists a $K_1 > K_0$ such that for $k \geqslant K_1$,

$$\|\widetilde{x}\| \leqslant \min\left\{1, \frac{\alpha^2}{4\sqrt{M}\|A\|\alpha + 2\beta\sqrt{M} + 2M}\right\}.$$

Therefore, for $k \geqslant K_1$ and $\overline{x} \notin \mathcal{K}$,

$$\sqrt{V_2} - (2\sqrt{M}\|A\| - \tfrac{\beta}{\alpha}\sqrt{M})\|\widetilde{x}\| - \tfrac{1}{\alpha}M\|\widetilde{x}\|^2 \geqslant \sqrt{V_2} - \tfrac{\alpha}{2} \geqslant \tfrac{\sqrt{V_2}}{2},$$

and thus

$$V_2(k+1) - V_2(k) \leqslant -\tfrac{\varepsilon}{2}V_2(k) + \overline{x}'(k+1)[P(k+1) - P(k)]\overline{x}(k+1).$$

Since $V_2(k+1) - V_2(k)$ cannot have the same sign as

$$\overline{x}'(k+1)'[P(k+1) - P(k)]\overline{x}(k+1)$$

(which can be established similar to the argument in the proof of Theorem 12.19), we conclude that for $\overline{x} \notin \mathcal{K}$ and $k > K_1$,

$$V_2(k+1) - V_2(k) < 0.$$

This implies that $\overline{x}$ will enter $\mathcal{K}$ within finite time, say $K_2 > K_1$. For $k > K_2$ and $\overline{x} \in \mathcal{K}$, we have $\varepsilon = 0.9$ and $\|\widehat{B}'\widehat{A}\widetilde{x} + F\overline{x}\| \leqslant \frac{1}{2} + \frac{1}{4} = \frac{3}{4}$. All saturations are inactive, and the system becomes

$$\begin{cases}
\overline{x}(k+1) = [A - BF^*]\overline{x}(k) + \mathcal{J}\widetilde{x}(k),\\
\widetilde{x}(k+1) = [\widehat{A} - \kappa\widehat{B}\widehat{B}'\widehat{A}]\widetilde{x}(k),
\end{cases}$$

where

$$F^* = (B'P^*B + I)^{-1}B'P^*A$$

and P^* is the solution of (16.15) with $\varepsilon = 0.9$. It is clear that we shall also have $V_2(k+1) - V_2(k) < 0$ for $\bar{x} \in \mathcal{K}$. Therefore, $\bar{x}$ will remain in $\mathcal{K}$ for $k > K_2$. The global asymptotic stability follows from the properties that $A - BF^*$ and $\widehat{A} - \kappa\widehat{B}\widehat{B}'\widehat{A}$ are Schur stable with $8\kappa\widehat{B}\widehat{B}' \leqslant I$.

We proceed next to show the boundedness of trajectories in the presence of d_1 and d_2. Define

$$R(k) = (\widehat{A}')^k$$

and $y = R\widetilde{x}$. Note that since $\widehat{A}'\widehat{A} = I$, R defines a discrete-time rotation matrix. We have that $R(k+1) = R(k)\widehat{A}'$. Then

$$y(k+1) = y(k) + R\widehat{A}'\widehat{B}\sigma(-\kappa\widehat{B}'\widehat{A}R'y + v) - R\widehat{A}'\widehat{B}\sigma(v) + R\widehat{A}'\widehat{E}_2 d_2$$

with $y(0) = \widetilde{x}_0$, where $v = -F\bar{x}$. Let $\bar{y}$ satisfy

$$\bar{y}(k+1) = \bar{y}(k) + R(k)\widehat{A}'\widehat{E}_2 d_2, \quad \bar{y}(0) = \widetilde{x}_0.$$

Since $d_2 \in \Omega_\infty$, we find that $\bar{y} \in \ell_\infty$. Define $\widetilde{y} = y - \bar{y}$. Then

$$\begin{aligned}\widetilde{y}(k+1) = \widetilde{y}(k) + R(k)\widehat{A}'\widehat{B}\sigma(-\kappa\widehat{B}'\widehat{A}R'(k)\widetilde{y}(k) - \kappa\widehat{B}'\widehat{A}R'(k)\bar{y}(k) + v(k))\\ - R(k)\widehat{A}'\widehat{B}\sigma(v(k)),\end{aligned}$$

with $\widetilde{y}(0) = 0$. Again, define $z = R'\widetilde{y}$. We get

$$z(k+1) = \widehat{A}z(k) + \widehat{B}\sigma(-\kappa\widehat{B}'\widehat{A}z(k) - \widetilde{u}(k)) - \widehat{B}\sigma(v(k)), \quad z(0) = 0,$$

where

$$\widetilde{u} = \kappa\widehat{B}'\widehat{A}R'\bar{y} - v.$$

Since $\widetilde{u} \in \ell_\infty$ and $\|v\| \leqslant \delta = \frac{1}{4}$, it follows from Lemma 16.8 that $z \in \ell_\infty$. This implies that $\widetilde{x} \in \ell_\infty$.

Consider the dynamics of $\bar{x}$

$$\bar{x}(k+1) = A\bar{x}(k) + B\sigma(-F\bar{x}(k)) + \bar{B}\zeta(k) + \bar{E}_1 d_1(k) + \mathcal{J}\widetilde{x}(k),$$

where

$$\zeta(k) = \sigma(-\kappa\widehat{B}'\widehat{A}\widetilde{x}(k) - F\bar{x}(k)) - \sigma(-F\bar{x}(k)).$$

Because σ is globally Lipschitz with Lipschitz constant 1, we have

$$\|\zeta(k)\| \leqslant \|\kappa\widehat{B}'\widehat{A}\widetilde{x}(k)\|.$$

Hence, $\zeta \in \ell_\infty$. There exists a β such that

$$\begin{aligned}
&V_2(k+1) - V_2(k) \\
&\leqslant -\varepsilon V_2(k) - 2\sigma(F\bar{x}(k))' B' P(k) \mathcal{J}\tilde{x}(k) - 2\sigma(F\bar{x}(k))' B' P(k) \bar{E}_1 d_1(k) \\
&\quad - 2\sigma(F\bar{x}(k))' BP\bar{B}\zeta(k) + 2\bar{x}(k)' A' P \bar{B}\zeta(k) \\
&\quad + 2\bar{x}'(k) AP(k) \mathcal{J}\tilde{x}(k) + 2\bar{x}'(k) AP(k) \bar{E}_1 d_1(k) \\
&\quad + (d_1'(k)\bar{E}_1' + \tilde{x}'(k)\mathcal{J}' + \zeta(k)'\bar{B}') P(k) (\bar{E}_1 d_1(k) + \mathcal{J}\tilde{x}(k) + \bar{B}\zeta(k)) \\
&\quad + \bar{x}'(k+1)[P(k+1) - P(k)]\bar{x}(k+1) \\
&\leqslant -\varepsilon V_2(k) + \left(\|P^{1/2}(k)\mathcal{J}\tilde{x}(k)\| + \|P^{1/2}(k)\bar{E}_1 d_1(k)\| + \|P^{1/2}\bar{B}\zeta(k)\| \right)^2 \\
&\quad + \left(2\|A\|\sqrt{V_2(k)} + \beta \right) \\
&\qquad\qquad \cdot \left(\|P^{1/2}(k)\mathcal{J}\tilde{x}(k)\| + \|P^{1/2}(k)\bar{E}_1 d_1(k)\| + \|P^{1/2}\bar{B}\zeta(k)\| \right) \\
&\quad + \bar{x}'(k+1)[P(k+1) - P(k)]\bar{x}(k+1).
\end{aligned}$$

We have already shown that according to Lemma 16.12, there exist M, M_1, and M_2 such that

$$\begin{aligned}
\|P^{1/2}\mathcal{J}\tilde{x}\| &\leqslant \varepsilon\sqrt{M}\|\tilde{x}\|, \\
\|P^{1/2}\bar{E}_1 d_1\| &\leqslant \varepsilon\sqrt{M_1}\|d_1\|, \\
\|P^{1/2}\bar{B}\zeta\| &\leqslant \varepsilon\sqrt{M_2}\|\zeta\|.
\end{aligned}$$

Define a set $\mathcal{V}$ as

$$\mathcal{V} = \{\bar{x} \mid V_2(\bar{x}) \leqslant c\},$$

where c is such that $V_2 \geqslant c$ implies that

$$\begin{aligned}
\tfrac{1}{2} V_2 \geqslant (2\|A\|\sqrt{V_2} + \beta)(\sqrt{M}\|\tilde{x}\|_{\ell_\infty} + \sqrt{M_1}\|d_1\|_{\ell_\infty} + \sqrt{M_2}\|\zeta\|_{\ell_\infty}) \\
+ 2M\|\tilde{x}\|^2_{\ell_\infty} + 2M_1\|d_1\|^2_{\ell_\infty} + 2M_2\|\zeta\|^2_{\ell_\infty}.
\end{aligned}$$

Therefore, for $\bar{x} \notin \mathcal{V}$, we have

$$V_2(k+1) - V_2(k) \leqslant -\tfrac{1}{2}\varepsilon V_2(k) + \bar{x}'(k+1)[P(k+1) - P(k)]\bar{x}(k+1).$$

Since $V_2(k+1) - V_2(k)$ and $\bar{x}'(k+1)[P(k+1) - P(k)]\bar{x}(k+1)$ cannot have the same sign, we find that

$$V_2(k+1) - V_2(k) < 0, \quad \bar{x} \notin \mathcal{V}. \tag{16.20}$$

On the other hand, for $\bar{x} \in \mathcal{V}$, assume that $V_2(k+1) - V_2(k) \geqslant 0$. In that case, we have

$$\bar{x}'(k+1)[P(k+1) - P(k)]\bar{x}(k+1) \leqslant 0.$$

Then

$$
\begin{aligned}
V_2(k+1) - V_2(k) \leqslant \Delta \triangleq\ & 2M\|\tilde{x}\|_{\ell_\infty}^2 + 2M_1\|d_1\|_{\ell_\infty}^2 + 2M_2\|\zeta\|_{\ell_\infty}^2 \\
& + (2\|A\|\sqrt{c} + \beta)(\sqrt{M}\|\tilde{x}\|_{\ell_\infty} + \sqrt{M_1}\|d_1\|_{\ell_\infty} + \sqrt{M_2}\|\zeta\|_{\ell_\infty}).
\end{aligned}
$$

Hence, the maximum increment of V_2 inside $\mathcal{V}$ is Δ. In view of this, (16.20) and definition of $\mathcal{V}$, we conclude that $V_2 \leqslant \max\{c + \Delta, V_2(0)\}$, which implies that $\bar{x} \in \ell_\infty$ and hence $x \in \ell_\infty$. ■

The combination of aligned and misaligned disturbances could appear in a more general fashion than those in Theorem 16.5. Without loss of generality, for any critically unstable system with input saturation and nonadditive disturbances, we can always assume the following system configuration:

$$
\rho x = Ax + B\sigma(u) + \bar{E}_1\begin{pmatrix} d_1 \\ d_2 \end{pmatrix} + \bar{E}_2 d_2, \quad x(0) = x_0, \tag{16.21}
$$

where A and B are given by (16.9)–(16.11), $\bar{E}_1$; and $\bar{E}_2$ are in the form of (16.13) and (16.14) but with appropriate dimensions.

Based on Theorem 16.5, we can immediately draw the following conclusion.

Theorem 16.6 *Consider the system* (16.21) *with controller* (16.16) *for* κ *such that* $8\kappa B'B \leqslant I$. *Then we have:*

(i) In the absence of d_1 *and* d_2, *the origin is global asymptotically stable.*

(ii) In the presence of d_1 *and* d_2, $x \in \ell_\infty$ *for any* $x(0)$, $d_1 \in \ell_\infty$, *and* $d_2 \in \Omega_\infty$.

Proof : Note that by definition, $d_2 \in \ell_\infty$ if $d_2 \in \Omega_\infty$. Therefore, $d_1, d_2 \in \ell_\infty$. Then the result is a direct consequence of Theorem 16.5. ■

We shall prove next that a small aligned disturbance which does not belong to Ω_∞ is permitted. Moreover, it turns out that a combination of aligned Ω_∞ disturbance, misaligned ℓ_∞ disturbance, and an arbitrary disturbance which belongs to $\ell_\infty(\delta_1)$ can be handled by the same controller (16.16) given δ_1 sufficiently small. Such a case is formulated as follows:

$$
\rho x = Ax + B\sigma(u) + \bar{E}_1\begin{pmatrix} d_1 \\ d_2 \end{pmatrix} + \bar{E}_2 d_2 + \bar{E}_3 d_3, \quad x(0) = x_0, \tag{16.22}
$$

where A and B are given by (16.9) and (16.10), $\bar{E}_1$ and $\bar{E}_2$ are in the form of (16.13) and (16.14) but with appropriate dimensions, and $\bar{E}_3$ is arbitrary.

Theorem 16.7 *Consider the system* (16.22) *with controller* (16.16) *for* κ *such that* $8\kappa B'B \leqslant I$. *Then we have:*

(*i*) *In the absence of* d_1, d_2, *and* d_3, *the origin is global asymptotically stable.*

(*ii*) *In the presence of* d_1, d_2, *and* d_3, $x \in \ell_\infty$ *for any* $x(0)$, $d_1 \in \ell_\infty$, $d_2 \in \Omega_\infty$, *and* $d_3 \in \ell(\delta_1)$ *with* δ_1 *sufficiently small.*

Proof : As in the proof of Theorem 16.5, we omit the parameter $\varepsilon_a(\bar{x})$ in the expressions $P_{\varepsilon_a(\bar{x})}$ and $F_{\varepsilon_a(\bar{x})}$. The global asymptotic stability of the origin in the absence of disturbances has already been shown in the proof of Theorem 16.5.

For any matrix $\bar{E}_3$, we can always write that $\bar{E}_3 = \bar{E}_{3,m} + \bar{E}_{3,a}$, where $\bar{E}_{3,m}$ is in the misaligned form of (16.13) and $\bar{E}_{3,a}$ is in the aligned form of (16.14). To be precise, we can write

$$\bar{E}_{3,m} = \begin{pmatrix} \bar{E}_{m,1} \\ \bar{E}_{m,2} \\ \vdots \\ \bar{E}_{m,q} \end{pmatrix}, \quad \bar{E}_{m,i} = \begin{pmatrix} E_{3,i,1} \\ E_{3,i,2} \\ \vdots \\ E_{3,i,n_i-1} \\ 0 \end{pmatrix}$$

and

$$\bar{E}_{3,a} = \begin{pmatrix} \bar{E}_{a,1} \\ \bar{E}_{a,2} \\ \vdots \\ \bar{E}_{a,q} \end{pmatrix}, \quad \bar{E}_{a,i} = \begin{pmatrix} 0 \\ 0 \\ \vdots \\ 0 \\ E_{3,i,n_i} \end{pmatrix}.$$

The system (16.22) can then be written in the form

$$\rho x = Ax + B\sigma(u) + \tilde{E}_1 \begin{pmatrix} d_1 \\ d_2 \\ d_3 \end{pmatrix} + \tilde{E}_2 \begin{pmatrix} d_2 \\ d_3 \end{pmatrix}, \quad x(0) = x_0,$$

where

$$\tilde{E}_1 = \begin{pmatrix} \bar{E}_1 & \bar{E}_{3,m} \end{pmatrix}, \quad \tilde{E}_3 = \begin{pmatrix} \bar{E}_2 & \bar{E}_{3,a} \end{pmatrix}.$$

Define $\tilde{x} = x - \bar{x}$ and let $\tilde{d}_1 = (d_1', d_2', d_3')$. Consider the closed-loop system in terms of $\tilde{x}$ and $\bar{x}$ as in the proof of Theorem 16.5:

$$\begin{cases} \rho\bar{x} = A\bar{x} + B\sigma(-F\bar{x}) + \bar{B}\left[\sigma(-\kappa\hat{B}'\hat{A}\tilde{x} - F\bar{x}) - \sigma(-F\bar{x})\right] + \tilde{E}_1\tilde{d}_1 + \mathcal{J}\tilde{x} \\ \rho\tilde{x} = \hat{A}\tilde{x} + \hat{B}\sigma(-\kappa\hat{B}'\hat{A}\tilde{x} - F\bar{x}) - \hat{B}\sigma(-F\bar{x}) + \hat{E}_2 d_2 + \hat{E}_3 d_3, \end{cases}$$

where $\widehat{E}_2$ is given by (16.18) and

$$\widehat{E}_3 = \begin{pmatrix} E_{3,1,n_1} \\ E_{3,2,n_2} \\ \vdots \\ E_{3,q,n_q} \end{pmatrix}.$$

Similarly, define $R = (\widehat{A}')^k$ and $y = R\widetilde{x}$. Then

$$\begin{aligned} y(k+1) = y(k) + R(k+1)\widehat{B}\sigma(-\kappa\widehat{B}'\widehat{A}R'(k)y(k)+v(k)) - R(k+1)\widehat{B}\sigma(v(k)) \\ + R(k+1)\widehat{E}_2 d_2(k) + \widehat{E}_3 d_3(k), \end{aligned}$$

with $y(0) = \widetilde{x}(0)$, where $v = -F\overline{x}$. Let $\overline{y}$ satisfy

$$\overline{y}(k+1) = \overline{y}(k) + R(k+1)\widehat{E}_2 d_2(k), \quad \overline{y}(0) = \widetilde{x}_0.$$

Since $d_2 \in \Omega_\infty$, we find that $\overline{y} \in \ell_\infty$. Define $\widetilde{y} = y - \overline{y}$. Then

$$\begin{aligned} \widetilde{y}(k+1) = \widetilde{y}(k) + R(k+1)\widehat{B}\sigma(-\kappa\widehat{B}'\widehat{A}R'(k)\widetilde{y}(k) - \kappa\widehat{B}'\widehat{A}R'\overline{y}(k) + v(k)) \\ - R(k+1)\widehat{B}\sigma(v(k)) + R(k+1)\widehat{E}_3 d_3(k), \end{aligned}$$

with $\widetilde{y}(0) = 0$. Again, define $z = R'\widetilde{y}$. We get

$$z(k+1) = \widehat{A}z(k) + \widehat{B}\sigma(-\kappa\widehat{B}'\widehat{A}z(k) - \widetilde{u}(k)) - \widehat{B}\sigma(v) + \widehat{E}_3 d_3(k), \quad z(0) = 0,$$

where

$$\widetilde{u} = \kappa\widehat{B}'\widehat{A}R'\overline{y} - v.$$

Consider an auxiliary system

$$w(k+1) = (\widehat{A} - \kappa\widehat{B}\widehat{B}'\widehat{A})w(k) + \widehat{E}_3 d_3(k), \quad w(0) = 0.$$

Let δ_1 be small enough such that $\delta_1 \leqslant \frac{1}{4\kappa\|\widehat{B}'\widehat{A}\|\gamma}$, where γ is the ℓ_∞ gain of pair $(\widehat{A} - \kappa\widehat{B}\widehat{B}'\widehat{A}, E_3)$. Hence, $d_3 \in \ell_\infty(\delta_1)$ implies that $\|\kappa\widehat{B}'\widehat{A}w\| \leqslant \frac{1}{4}$. Consider $\xi = z - w$. We get

$$\rho\xi = \widehat{A}\xi + \widehat{B}\sigma(-\kappa\widehat{B}'\widehat{A}\xi - \widehat{u}) - \widehat{B}\sigma(v) + \kappa\widehat{B}\widehat{B}'\widehat{A}w,$$

where $\widehat{u} = \kappa\widehat{B}'\widehat{A}w + \kappa\widehat{B}'\widehat{A}R'\overline{y} - v$. Since $\widehat{u} \in \ell_\infty$ and $\|v\| + \kappa\|\widehat{B}'\widehat{A}w\| \leqslant \frac{1}{4} + \frac{1}{4} = \frac{1}{2}$, it follows from Lemma 16.8 that $\xi \in \ell_\infty$. This finally implies that $\widetilde{x} \in \ell_\infty$.

Consider the dynamics of $\overline{x}$

$$\overline{x}(k+1) = A\overline{x} + B\sigma(-F\overline{x}) + \widetilde{\overline{E}}_1\widetilde{d}_1 + \mathcal{J}\widetilde{\overline{x}} + \overline{B}\zeta,$$

where $\zeta = \sigma(-\kappa\widehat{B}'\widehat{A}\widetilde{\overline{x}} - F\overline{x}) - \sigma(-F\overline{x})$. Note that $\zeta \in \ell_\infty$, $\widetilde{d}_1 \in \ell_\infty$, and $\widetilde{E}_1$ is also in the form of (16.13). Following the same argument as in the proof of Theorem 16.5, we can show that $\overline{x} \in \ell_\infty$. ■

16.A Proofs of some lemmas

We first prove the following lemma that was used in the proof of one of our main theorems.

Lemma 16.8 *Suppose $A'A = I$ and (A, B) is controllable. Consider*

$$\rho x(k) = Ax - B\sigma(-\kappa B'Ax + v_1) + Bv_2, \quad x(0) = x_0.$$

Then for κ satisfying $4\kappa B'B \leqslant I$, we have:

(i) *In the absence of v_1 and v_2, the origin is globally asymptotically stable.*

(ii) *In the presence of v_1 and v_2, $x \in \ell_\infty$ for all $v_1 \in \ell_\infty$ and $v_2 \in \ell_\infty(1/2)$ and any initial condition.*

In order to prove the above result, we first need the following lemmas. The first lemma has already been established in Sect. (4.6.1).

Lemma 16.9 *Assume that $A'A = I$ and $\kappa B'B \leqslant 2I$ for some $\kappa > 0$. Then, $\widetilde{A} = A - \kappa BB'A$ is Schur stable if and only if (A, B) is controllable.*

The next two lemmas can be easily verified:

Lemma 16.10 *For any $t \in \mathbb{R}^m$ satisfying $\|t\| \leqslant 1$, we have*

$$-s'[\sigma(s) + t] \leqslant \tfrac{\|t\|^2}{4}. \tag{16.23}$$

Lemma 16.11 *For any $s \in \mathbb{R}^m$,*

$$\|s - \sigma(s)\| \leqslant s'\sigma(s). \tag{16.24}$$

Proof of Lemma 16.8 : Denote $\kappa B'Ax$ by y. Define $V_1 = \|x\|^2$. We have

$$\begin{aligned}\rho V_1 - V_1 &= \|Ax + B\sigma(-y - v_1) + Bv_2\|^2 - \|x\|^2 \\ &= \tfrac{2}{\kappa} y'[\sigma(-y - v_1) + v_2] \\ &\qquad + \left[\sigma(-y - v_1)' + v_2'\right] B'B \left[\sigma(-y - v_1) + v_2\right] \\ &\leqslant \tfrac{2}{\kappa}[y' + v_1'][\sigma(-y - v_1) + v_2] - \tfrac{2}{\kappa} v_1'[\sigma(-y - v_1) + v_2] \\ &\qquad + \tfrac{1}{4\kappa}\|\sigma(-y - v_1) + v_2\|^2,\end{aligned}$$

where we use condition $4\kappa B'B \leqslant I$. We can assume that $\|v_2\| \leqslant \frac{1}{2}$, and since σ is bounded by ± 1, we find that $-v_1'[\sigma(-y - v_1) + v_2] \leqslant 2\|v_1\|$. This yields

$$\begin{aligned}\rho V_1 - V_1 &\leqslant \tfrac{2}{\kappa}[y' - v_1'][\sigma(-y - v_1) + v_2] + \tfrac{1}{2\kappa}\|\sigma(y + v_1)\|^2 \\ &\qquad + \tfrac{1}{2\kappa}\|v_2\|^2 + \tfrac{4}{\kappa}\|v_1\| \\ &\leqslant \tfrac{2}{\kappa}[y + v_1]'[\sigma(-y - v_1) + v_2] + \tfrac{1}{2\kappa}\|\sigma(y + v_1)\|^2 \\ &\qquad + \tfrac{1}{2\kappa}\|v_2\| + \tfrac{4}{\kappa}\|v_1\|.\end{aligned}$$

Note that

$$\begin{aligned}&\tfrac{2}{\kappa}[y' + v_1'][\sigma(-y - v_1) + v_2] \\ &\qquad = \tfrac{1}{\kappa}[y' + v_1']\sigma(-y - v_1) + \tfrac{1}{\kappa}[y + v_1]'[\sigma(-y - v_1) + 2v_2] \\ &\qquad \leqslant \tfrac{1}{\kappa}[y' + v_1']\sigma(-y - v_1) + \tfrac{1}{\kappa}\|v_2\|^2 \\ &\qquad \leqslant \tfrac{1}{\kappa}[y' + v_1]'\sigma(-y - v_1) + \tfrac{1}{\kappa}\|v_2\|,\end{aligned}$$

where we use (16.23) and $\|v\| \leqslant \frac{1}{2}$. Therefore,

$$\begin{aligned}\rho V_1 - V_1 &\leqslant \tfrac{1}{\kappa}[y' + v_1']\sigma(-y - v_1) + \tfrac{1}{2\kappa}\|\sigma(y + v_1)\|^2 \\ &\qquad + \tfrac{3}{2\kappa}\|v_2\| + \tfrac{4}{\kappa}\|v_1\| \\ &\leqslant \tfrac{1}{2\kappa}[y' + v_1']\sigma(-y - v_1) + \tfrac{3}{2\kappa}\|v_2\| + \tfrac{4}{\kappa}\|v_1\|.\end{aligned} \tag{16.25}$$

Since $4\kappa B'B \leqslant I$, $\widetilde{A} = A - \kappa BB'A$ is Schur stable. Let P be the solution to the Lyapunov equation:

$$\widetilde{A}' P \widetilde{A} - P + I = 0.$$

Define $V_2 = \|P^{1/2}x\|$. We have

$$\begin{aligned}\rho V_2 - V_2 &= \|P^{1/2}\widetilde{A}x + P^{1/2}B[y + v_1 - \sigma(y + v_1) + (v_2 - v_1)]\| \\ &\qquad - \|P^{1/2}x\| \\ &\leqslant \|P^{1/2}\widetilde{A}x\| + \|P^{1/2}B[y + v_1 - \sigma(y + v_1) + (v_2 - v_1)]\| \\ &\qquad - \|P^{1/2}x\|.\end{aligned}$$

For $x \neq 0$, there exists a $\beta > 0$ such that

$$\|P^{1/2}\widetilde{A}x\| - \|P^{1/2}x\| = \frac{\|P^{1/2}\widetilde{A}x\|^2 - \|P^{1/2}x\|^2}{\|P^{1/2}\widetilde{A}x\| + \|P^{1/2}x\|} = \frac{-\|x\|^2}{\|P^{1/2}\widetilde{A}x\| + \|P^{1/2}x\|}$$
$$\leqslant -\beta\|x\|.$$

Obviously, the above also holds for $x = 0$. Hence,

$$\begin{aligned}\rho V_2 - V_2 &\leqslant -\beta\|x\| + \|P^{1/2}B\|\|(y + v_1) - \sigma(y + v_1)\| \\ &\quad + \|P^{1/2}B\|(\|v\| + \|u\|) \\ &\leqslant -\beta\|x\| + \|P^{1/2}B\|(y + v_1)'\sigma(y + v_1) \\ &\quad + \|P^{1/2}B\|(\|v_2\| + \|v_1\|), \end{aligned} \tag{16.26}$$

where we use (16.24).

Define $V = 2\kappa\|P^{1/2}B\|V_1 + V_2$. We obtain from (16.25) and (16.26) that

$$\rho V - V \leqslant -\beta\|x\| + 9\|P^{1/2}B\|\|v_1\| + 4\|P^{1/2}B\|\|v_2\|. \tag{16.27}$$

This immediately implies that $x \in \ell_\infty$ for any initial condition. ∎

Proof of Lemma 16.1 : Let s_i and t_i denote each element of s and t.
Case 1: $s_i + t_i \geqslant 1$; we have that

$$|\sigma(s_i + t_i) - \sigma(t_i)| = 1 - t_i \leqslant 2.$$

Also,

$$|\sigma(s_i + t_i) - \sigma(t_i)| = 1 - t_i \leqslant |u_i|.$$

Hence,

$$|\sigma(s_i + t_i) - \sigma(t_i)| \leqslant 2|\sigma(\tfrac{1}{2}s_i)|.$$

Case 2: $|s_i + t_i| < 1$; this implies that $|s_i| \leqslant 2$,

$$|\sigma(s_i + t_i) - \sigma(t_i)| = |s_i| = 2|\sigma(\tfrac{1}{2}s_i)|.$$

Case 3: $s_i + t_i \leqslant -1$

$$|\sigma(s_i + t_i) - \sigma(t_i)| = |-1 - t_i| = 1 + t_i \leqslant |s_i|.$$

Also,

$$|\sigma(s_i + t_i) - \sigma(t_i)| = |-1 - t_i| = 1 + t_i \leqslant 2.$$

Hence,

$$|\sigma(s_i + t_i) - \sigma(t_i)| \leqslant 2|\sigma(\tfrac{1}{2}s_i)|.$$ ∎

Lemma 16.12 *Let P_ε be the solution of the discrete parametric Lyapunov equation* (16.15). *For any given matrix $\overline{E}_1$ of the form* (16.13), *there exists a M such that for $\varepsilon \in (0, 0.9]$,*

$$\overline{E}_1' P_\varepsilon \overline{E}_1 \leqslant M\varepsilon^2 I.$$

Proof : It is shown in [216] that $P_\varepsilon \to 0$ as $\varepsilon \to 0$. Let

$$P_\varepsilon = \varepsilon P_1 + \varepsilon^2 P_2 + \ldots + \varepsilon^i P_i + \ldots.$$

Substituting P_ε in (16.15), we find that P_1 satisfies

$$A' P_1 A - P_1 = 0, \tag{16.28}$$

where A is given by (16.9). In a block decomposition of P_1 compatible with the block decomposition of A, we denote the diagonal block matrices by $P_{1,i}$ with $i = 1, \ldots, q$. We note that $P_{1,i}$ must satisfy

$$\overline{A}_i' P_{1,i} \overline{A}_i - P_{1,i} = 0, \tag{16.29}$$

where $\overline{A}_i$ is given by (16.10). Suppose

$$P_{1,i} = \begin{pmatrix} \overline{P}_{11} & \overline{P}_{12}' \\ \overline{P}_{12} & \overline{P}_{22} \end{pmatrix},$$

where $\overline{P}_{11} \in \mathbb{R}^{p_i \times p_i}$ and $\overline{P}_{12}$ and $\overline{P}_{22}$ are of appropriate dimension.

Define

$$\xi_{i,j} = \begin{pmatrix} x_{i,j} \\ 0 \\ \vdots \\ 0 \end{pmatrix},$$

where $x_{i,j}$, $j = 1, \ldots, p_i$ are the eigenvectors of A_i associated with the eigenvalues λ_i, $j = 1, \ldots, p_i$. Clearly, we have $\overline{A}_i \xi_{i,j} = \lambda_j \xi_{i,j}$.

We shall have that

$$(\overline{A}_i' P_{1,i} \overline{A}_i - \overline{P}_{1,i})\xi_{i,j} = 0.$$

This implies that

$$\overline{A}_i' P_{1,i} \xi_{i,j} = \lambda_j^* P_{1,i} \xi_{i,j}.$$

In words, $P_{1,i}\xi_{i,j}$, $j = 1, \ldots, p_i$ are the eigenvectors of $\overline{A}'$ associated with the eigenvalue $-\lambda_j$, $j = 1, \ldots, p_i$. On the other hand, we have a set of eigenvectors of $\overline{A}_i'$ in the form of

$$\nu_{i,j} = \begin{pmatrix} 0 \\ \vdots \\ 0 \\ v_{i,j} \end{pmatrix}, i = 1, \dots, p_i,$$

where $v_{i,j}$ are the eigenvectors of A_i' associated with the eigenvalue λ_j, $j = 1, \dots, p_i$.

Note that $\bar{A}_i'$ also has only p_i linearly independent eigenvectors. Therefore,

$$P_1 \xi_{i,j} = \begin{pmatrix} \bar{P}_{11} x_{i,j} \\ \bar{P}_{12} x_{i,j} \end{pmatrix} \in \operatorname{span}\{\nu_{i,1}, \dots, \nu_{i,p_i}\}.$$

This implies that $\bar{P}_{11} x_{i,j} = 0$, $j = 1, \dots, p_i$. Since $x_{i,j}$ forms a basis of $\mathbb{R}^{p_i}$, we must have $\bar{P}_{11} = 0$ and hence $\bar{P}_{12} = 0$ due to the fact that $P_{1,i}$ is positive semi-definite.

Applying the above argument to $\bar{P}_{22}$ and recursively, we shall eventually find that

$$P_{1,i} = \begin{pmatrix} 0 & 0 & \cdots & 0 \\ 0 & \ddots & \ddots & \vdots \\ \vdots & \ddots & 0 & 0 \\ 0 & \cdots & 0 & \bar{P}_{n_i n_i} \end{pmatrix}.$$

Due to positive semi-definiteness of P_1, for any given matrix $\bar{E}_1$ in the form of (16.13), we must have $P_1 \bar{E}_1 = 0$. This implies that $\mathcal{E}_1' P_\varepsilon \bar{E}_1$ must be of order ε^2. This completes the proof. ∎

17

External and internal stabilization under the presence of stochastic disturbances

17.1 Introduction

So far, all the results discussed in this book are in deterministic setting. A new frontier for the next phase of research is in stochastic setting. That is, to consider disturbances which are modeled as colored noise which in turn can be modeled as white noise followed by a linear system. Then, the goal is to investigate simultaneous external and internal stabilization of linear systems subject to constraints when the disturbances are modeled stochastically. To be precise, the goal is to develop feedback controllers for such systems such that:

1. In the absence of disturbances, the origin of the closed-loop system is globally asymptotically stable.
2. In the presence of disturbances, the states of the closed-loop system have finite variance for random Gaussian distributed initial conditions, possibly independent of the external disturbances.

Because of the requirement of global internal stability, the above simultaneous external and internal stabilization problem necessitates that the given linear system is asymptotically null controllable with bounded control (ANCBC). In this chapter, the above problem is solved for two special classes of ANCBC systems, (1) both continuous- and discrete-time open-loop neutrally stable systems and (2) continuous-time double-integrator system. We utilize linear static state feedback control laws for both classes of systems.

The simultaneous global internal stabilization and semi-global external stochastic stabilization problem as mentioned above requires the state of the closed-loop system to have finite variance in the presence of disturbances. A broader goal could be to minimize the variance of the state of a closed-loop system. This is not pursued in this chapter.

This chapter is organized as follows. For both discrete- and continuous-time systems, a formal problem formulation is given in Sect. 17.2. The main results for open-loop neutrally stable systems are given in Sect. 17.3. The proofs of such

A. Saberi et al., *Internal and External Stabilization of Linear Systems with Constraints*, Systems & Control: Foundations & Applications, DOI 10.1007/978-0-8176-4787-2_17,

results are given in Sect. 17.3.1 for discrete-time case and in Sect. 17.3.2 for continuous-time case. The main results and their proofs for a double integrator with linear feedback are presented in Sect. 17.4.

This chapter is mainly based on our work [165, 168, 171].

17.2 Problem formulation

In this chapter, as before, σ denotes the standard saturation function introduced in Definition 2.19. In the case of discrete time, we consider systems of the form,

$$x(k+1) = Ax(k) + B\sigma(u(k)) + Ew(k), \tag{17.1}$$

where the state x, the control u, and the disturbance w are vector-valued signals of dimension n, m, and ℓ, respectively. Here, $k \in \mathbb{Z}_+$, w is a white-noise stochastic process with covariance matrix Q and mean 0; the initial condition x_0 of (17.1) is a Gaussian random vector independent of $w(k)$ for all $k \geqslant 0$.

In the case of continuous time, we consider the stochastic differential equation of the form

$$\mathrm{d}x(t) = Ax(t)\mathrm{d}t + B\sigma(u(t))\mathrm{d}t + E\mathrm{d}w(t), \tag{17.2}$$

where once again the state x, the control u, and the disturbance w are vector-valued signals of dimension n, m, and ℓ, respectively. Here, w is a Wiener process (a process of ℓ independent Brownian motions) with mean 0 and rate Q, that is, $\mathrm{Var}[w(t)] = Qt$ and the initial condition x_0 of (17.2) is a Gaussian random vector which is independent of w. Its solution x is rigorously defined through Wiener integrals and is a Gauss-Markov process. See, for instance, [109].

An *admissible feedback* is a nonlinear feedback of the form

$$u(\cdot) = f(x(\cdot)). \tag{17.3}$$

For discrete-time systems, we assume that $f : \mathbb{R}^n \to \mathbb{R}^m$ is a continuous map with $f(0) = 0$. On the other hand, for continuous-time systems, we assume that f is a Lipschitz-continuous mapping with $f(0) = 0$.

We can consider three versions of external stochastic stability. We define this for discrete-time systems. If a controller exists which achieves a bounded variance of the state for all possible covariance matrices Q of the white-noise stochastic process w, then we refer to this as global external stochastic stability. On the other hand, if for all $M > 0$ there exists a controller which achieves a bounded variance of the state for all possible covariance matrices Q with $\|Q\| < M$ of the white-noise stochastic process w, then we refer to this as semi-global external stochastic stability. Finally, if there exists a M and if there exists a controller which achieves a bounded variance of the state for all possible covariance matrices Q with $\|Q\| < M$ of the white-noise stochastic process w, then we refer to this as local external stochastic stability. In the continuous-time systems, we have the same definitions, but Q refers to the rate of the Wiener process instead of the covariance matrix.

We are interested in the following problems:

Problem 17.1 Consider the system (17.1) for discrete time, and the system (17.2) for continuous time. Then, the *simultaneous global internal stabilization and global external stochastic stabilization problem* is to find an admissible feedback (17.3) such that the following properties hold:

(*i*) In the absence of the external input w, the equilibrium point $x = 0$ of the controlled system (17.1)–(17.3) or (17.2) and (17.3) is globally asymptotically stable.

(*ii*) The variance $\mathrm{Var}(x(k))$ or $\mathrm{Var}(x(t))$ of the state of the controlled system (17.1)–(17.3) or (17.2) and (17.3) is bounded over $k \geqslant 0$ or $t \geqslant 0$ for any $Q > 0$.

Problem 17.2 Consider the system (17.1) for discrete time, and the system (17.2) for continuous time. Then, the *simultaneous global internal stabilization and semi-global external stochastic stabilization problem* is, for any given $M > 0$, to find an admissible feedback (17.3) such that the following properties hold:

(*i*) In the absence of the external input w, the equilibrium point $x = 0$ of the controlled system (17.1)–(17.3) or (17.2)–(17.3) is globally asymptotically stable.

(*ii*) The variance $\mathrm{Var}(x(k))$ or $\mathrm{Var}(x(t))$ of the state of the controlled system (17.1)–(17.3) or (17.2) and (17.3) is bounded over $k \geqslant 0$ or $t \geqslant 0$ provided $\|Q\| < M$.

Problem 17.3 Consider the system (17.1) for discrete time, and the system (17.2) for continuous time. Then, the *simultaneous global internal stabilization and local external stochastic stabilization problem* is to find an $M > 0$ and an admissible feedback (17.3) such that the following properties hold:

(*i*) In the absence of the external input w, the equilibrium point $x = 0$ of the controlled system (17.1)–(17.3) or (17.2) and (17.3) is globally asymptotically stable.

(*ii*) The variance $\mathrm{Var}(x(k))$ or $\mathrm{Var}(x(t))$ of the state of the controlled system (17.1)–(17.3) or (17.2) and (17.3) is bounded over $k \geqslant 0$ or $t \geqslant 0$ provided $\|Q\| < M$.

Remark 17.4 *Note that even for small Q, for all $N > 0$, the probability that $\|X(t)\| > N$ will be small but nonzero (except for the trivial case when $E = 0$). Since the state is not bounded, it therefore makes sense to require global internal stability even in the case of local external stochastic stability.*

The fact that controllers exist that achieve global asymptotic stability in the absence of disturbances as described in condition (*i*) is well known and is discussed in Chap. 4. The main objective then is to look at the additional requirement on the variance of the state.

17.3 Open-loop neutrally stable systems

The following result pertains to discrete-time open-loop neutrally stable systems:

Theorem 17.5 *Consider the system* (17.1) *and suppose that (A, B) is stabilizable while A is neutrally stable, i.e., the eigenvalues of A are in the closed unit disc and the eigenvalues on the unit circle have equal geometric and algebraic multiplicity. Then, there exists a linear feedback which solves the simultaneous global internal stabilization and global external stochastic stabilization problem as defined in Problem 17.1.*

The following theorem pertains to continuous-time systems:

Theorem 17.6 *Consider the system* (17.2) *and suppose that (A, B) is stabilizable while A is neutrally stable, i.e., the eigenvalues of A are in the closed left half plane and the eigenvalues on the imaginary axis have equal geometric and algebraic multiplicity. Then, there exists a linear feedback which solves the simultaneous global internal stabilization and global external stochastic stabilization problem as defined in Problem 17.1.*

17.3.1 Proofs for the discrete-time case

We first present a few little lemmas that we need later:

Lemma 17.7 *For any $a \in \mathbb{R}^m$ and $b \in \mathbb{R}^m$ and $\lambda > 1$, we have*

$$(a+b)'\sigma(a+b) \geqslant \left(1 - \tfrac{1}{\lambda}\right) a'\sigma(a) - (\lambda+1)\|b\|^2,$$

where σ is the standard saturation function.

Proof : It is easily verified that for any $\alpha \in \mathbb{R}$ and $\beta \in \mathbb{R}$ and $n > 1$, we have

$$(\alpha+\beta)\sigma(\alpha+\beta) \geqslant \left(1-\tfrac{1}{\lambda}\right)\alpha\sigma(\alpha) - (\lambda+1)\beta^2,$$

The vector case then follows immediately. ∎

Lemma 17.8 *For any convex function $f : \mathbb{R} \to \mathbb{R}$, we have*

$$f(x) + f(-x) \leqslant f(y) + f(-y)$$

if $|x| \leqslant |y|$.

Proof : Assume without loss of generality that x and y are positive. The lemma follows from the fact that

$$f(x) \leqslant \frac{y+x}{2y} f(y) + \frac{y-x}{2y} f(-y)$$
$$f(-x) \leqslant \frac{y-x}{2y} f(y) + \frac{y+x}{2y} f(-y),$$

and adding these two inequalities. ∎

For ease of notation, we introduce the function $F : \mathbb{R}^m \to \mathbb{R}$ defined by

$$F(v) = 2v'\sigma(\kappa v). \tag{17.4}$$

The following property of F will be useful later and can be easily verified.

Lemma 17.9 *For all $u \in \mathbb{R}^m$, we have*

$$F(v) \geqslant 2\|v\| - \frac{m}{2\kappa}.$$

Using a simple basis transformation, we can assume without loss of generality that

$$A = \begin{pmatrix} A_{11} & 0 \\ 0 & A_{22} \end{pmatrix}, \quad B = \begin{pmatrix} B_1 \\ B_2 \end{pmatrix}, \quad E = \begin{pmatrix} E_1 \\ E_2 \end{pmatrix}$$

with A_{11} asymptotically stable while $A_{22}'A_{22} = I$. This is guaranteed by the fact that A is neutrally stable. Next, we consider the subsystem:

$$x_2(t+1) = A_{22}x_2(t) + B_2u(t) + E_2w(t). \tag{17.5}$$

If we find an admissible feedback of the form $u(t) = f(x_2(t))$ which solves the simultaneous global internal stabilization and global external stochastic stabilization problem as defined in Problem 17.1 for the system (17.5), then it is easily verified that this controller also solves the same problem for the original system (17.1).

Proof of Theorem 17.5 : Without loss of generality, using the above argument, we assume that the original system satisfies $A'A = I$. Then, as in Sect. 4.6.1, Sect. 12.6, as well as Sect. 13.4, we consider a feedback of the form

$$u(k) = -\kappa B'Ax(k). \tag{17.6}$$

The general case of the theorem can then easily be established using the reduction to the system (17.5) as described earlier.

If the system is not affected by noise, it is well known that the feedback (17.6) achieves asymptotic stability which is also easily verified by noting that $V(x) = x'x$ is a suitable Lyapunov function.

In the derivation, we will make the role of the design parameter κ explicit. However, we will show that the interconnection of (17.1) and (17.6) results in a bounded variance of the state independent of the choice of Q for any positive value of κ, i.e., we solve the global external stochastic stabilization problem.

According to a crucial result of [113] presented in Lemma 17.16 and first used in the context of this problem in [116], it is sufficient to establish that there exist $a, b, J > 0$ such that for all $k \geqslant 0$:

$$\mathbb{E}\left[\, \|x(k+1)\| - \|x(k)\| \mid x(k) \,\right] \leqslant -a$$

provided $\|x(k)\| > J$ and

$$\mathbb{E}\left[\, \big|\|x(k+1)\| - \|x(k)\|\big|^4 \mid x(k) \,\right] \leqslant b.$$

However, we will actually establish that there exist $a, b, J > 0$ such that for all $k \geqslant 0$,

$$\mathbb{E}\left[\, \|x(k+n)\| - \|x(k)\| \mid x(k) \,\right] \leqslant -a \tag{17.7}$$

provided $\|x(k)\| > J$ and

$$\mathbb{E}\left[\, \big|\|x(k+n)\| - \|x(k)\|\big|^4 \mid x(k) \,\right] \leqslant b, \tag{17.8}$$

where n is the dimension of the state space. Since it is trivial to verify that $x(0), \ldots, x(n-1)$ have a bounded variance, it is easily seen that (17.7) and (17.8) also guarantee that the variance of the state is bounded.

In order to verify (17.8), we consider $z(k) = (A')^k x(k)$. It is easily verified (using that $A'A = I$) that we have

$$z(k+1) = z(k) + (A')^{k+1} B\sigma(u(k)) + (A')^{k+1} Ew(k).$$

This yields

$$\|z(k+1)\| \leqslant \|z(k)\| + \|B\| + \|E\|\|w(k)\|.$$

We note that $\|z(k)\| = \|x(k)\|$, and hence, it is easily seen that we have

$$\|x(k+n)\| \leqslant \|x(k)\| + n\|B\| + \|E\| \sum_{i=0}^{n-1} \|w(k+i)\|.$$

This yields

$$\mathbb{E}\left[\,|\|x(k+n)\| - \|x(k)\||^4 \mid x(k) \right] \leqslant \mathbb{E} \left| n\|B\| + \|E\| \sum_{i=0}^{n-1} \|w(k+i)\| \right|^4 \leqslant b$$

for a suitable b since all moments of $w(k)$ are bounded since it has a Gaussian distribution.

Using the feedback (17.6), the dynamics (17.1), and the fact that $A'A = I$, we note that there exists a $N > 0$ such that

$$\mathbb{E}\left[x(k+1)'x(k+1) \mid x(k) \right] \leqslant x(k)'x(k) - F\left(B'Ax(k)\right) + N \tag{17.9}$$

for all k using the notation (17.4) introduced before. Note that N depends on the covariance matrix Q. We find that

$$\begin{aligned} &\mathbb{E}\left[x(k+n)'x(k+n) \mid x(k) \right] \\ &\qquad \leqslant x(k)'x(k) - \sum_{i=0}^{n-1} \mathbb{E}\left[F\left(B'Ax(k+i)\right) \mid x(k) \right] + nN. \end{aligned} \tag{17.10}$$

The following lemma presents a crucial inequality.

Lemma 17.10 *We have for $i = 1, \dots, n$ that*

$$\mathbb{E}\left[F\left(B'Ax(k+i)\right) \mid x(k)\right] \geqslant \frac{1}{2^i} F\left(B'A^i x(k)\right) - \kappa M, \tag{17.11}$$

where M is a constant which depends on the covariance matrix Q but is independent of κ.

Proof : We first consider

$$F\left(B'A^t x(k+1)\right)$$

for some integer $t \geqslant 1$. We note that

$$B'A^t x(k+1) = \underbrace{B'A^{t+1}x(k)}_{a} + \underbrace{B'A^t Ew(k) - B'A^t B\sigma(\kappa B'Ax(k))}_{b}.$$

Using Lemma 17.7 with $\lambda = 2$, we get

$$\begin{aligned} &F\left(B'A^t x(k+1)\right) \\ &\qquad \geqslant \tfrac{1}{2} F\left(B'A^{t+1}x(k)\right) - 3\kappa \left\| B'A^t E w(k) - B'A^t B\sigma(\kappa B'Ax(k))\right\|^2. \end{aligned}$$

Next, we note that for suitable M_1 and M_2 independent of t and κ, we have

$$3\left\| B'A^t E w(k) - B'A^t B\sigma(\kappa B'Ax(k))\right\|^2 \leqslant M_1 + M_2\|w(k)\|^2$$

since the saturation function is bounded. Therefore,

$$F\left(B'A^t x(k+1)\right) \geqslant \tfrac{1}{2} F\left(B'A^{t+1}x(k)\right) - \kappa M_1 - \kappa M_2\|w(k)\|^2. \quad (17.12)$$

Since (17.12) is true for all k, we find that

$$\begin{aligned} F\left(B'A^j x(k+i-j+1)\right) \geqslant \tfrac{1}{2} F\left(B'A^{j+1}x(k+i-j)\right) \\ - \kappa M_1 - \kappa M_2\|w(k+i-j)\|^2. \end{aligned}$$

By repeatedly applying the above inequality starting with $j = 1$, we get

$$F\left(B'Ax(k+i)\right) \geqslant \frac{1}{2^i} F\left(B'A^i x(k)\right) - 2\kappa M_1 - \kappa M_2 \sum_{j=0}^{i-1} \|w(k+i-j)\|^2,$$

and taking the conditional expectation and defining

$$M = 2M_1 + M_2 \operatorname{trace} Q,$$

we find (17.11). ■

Using the last lemma and (17.10), we find that

$$\begin{aligned} &\mathbb{E}\left[x(k+n)'x(k+n) \mid x(k) \right] \\ &\qquad \leqslant x(k)'x(k) - \sum_{i=0}^{n-1} \frac{1}{2^i} F\left(B'A^{i+1}x(k)\right) + \widetilde{M} \quad (17.13) \end{aligned}$$

for a suitable $\widetilde{M}$ which is independent of x and w and can be chosen independent of κ provided we have a known upper bound for κ.

Next, we note that since (A, B) is controllable while A is invertible, the matrix

$$\begin{pmatrix} B'A \\ B'A^2 \\ \vdots \\ B'A^n \end{pmatrix}$$

is injective which implies that there exists an α (the smallest singular value of this matrix divided by n) such that for any x there exists a positive integer $t \leqslant n$ such that

$$\|B'A^t x\| \geqslant \alpha \|x\|$$

and then it is easily seen, using Lemma 17.9, that there exists a β such that for any $x \in \mathbb{R}^n$ we have

$$\begin{aligned}\sum_{i=0}^{n-1} \frac{1}{2^i} F(B'A^{i+1}x) &\geqslant \sum_{i=0}^{n-1} \frac{1}{2^{i-1}} \|B'A^{i+1}x\| - \frac{m}{2^{i+1}\kappa} \\ &\geqslant 2\beta\|x\| - \frac{m}{\kappa},\end{aligned}$$

where $\beta = \frac{\alpha}{2^n}$. This yields

$$\mathbb{E}\left[\, \|x(k+n)\|^2 \mid x(k) \,\right] \leqslant \|x(k)\|^2 - 2\beta\|x(k)\| + \frac{m}{\kappa} + \tilde{M}.$$

Using Jensen's inequality, we obtain

$$\begin{aligned}\mathbb{E}\left[\, \|x(k+n)\| \mid x(k) \,\right] &\leqslant \sqrt{\mathbb{E}\left[\, \|x(k+n)\|^2 \mid x(k) \,\right]} \\ &\leqslant \sqrt{\|x(k)\|^2 - 2\beta\|x(k)\| + \frac{m}{\kappa} + \tilde{M}} \\ &\leqslant \|x(k)\| - \tfrac{1}{2}\beta,\end{aligned}$$

provided

$$\|x(k)\| \geqslant J := \min\left\{ \frac{m + \kappa\tilde{M}}{\kappa\beta}, \frac{\beta}{2} \right\}.$$

Therefore, we note (17.8) is satisfied which completes the proof. ■

17.3.2 *Proofs for the continuous-time case*

Proof of Theorem 17.6 : Without loss of generality, we assume that the original system satisfies $A + A' = 0$. Then, as in Sect. 4.6.1, Sect. 12.6, as well as Sect. 13.4, we consider a feedback of the form

$$u(t) = -\kappa B'x(t). \tag{17.14}$$

The general case of the theorem can then easily be established using a reduction that ignores the asymptotically stable dynamics in a similar fashion as in the discrete-time case and as in Sect. 12.6.1.

If the system is not affected by noise, it is well known that the feedback (17.14) achieves asymptotic stability which is also easily verified by noting that $V(x) = x'x$ is a suitable Lyapunov function.

In order to prove external stochastic stability, we will show that the interconnection of (17.1) and (17.6) results in a bounded variance of the state independent of the rate Q of the Wiener process. We will make the role of the design parameter κ explicit in the bounds obtained below but the global external stochastic stability is a property that is obtained independent of our choice for κ.

We again rely on a crucial result of [113] as presented in Lemma 17.16. It is therefore sufficient to establish that there exist $a, b, J > 0$ such that for all $t \geqslant 0$,

$$\mathbb{E}\left[\, \|x(t+1)\| - \|x(t)\| \mid x(t) \,\right] \leqslant -a \tag{17.15}$$

provided $\|x(t)\| > J$ and

$$\mathbb{E}\Big[\, \big|\|x(t+1)\| - \|x(t)\|\big|^4 \mid x(k) \,\Big] \leqslant b. \tag{17.16}$$

Since it is trivial to verify that $x(t), t \in [0, 1)$ has a bounded variance, it is easily seen from the results in [113] that (17.15) and (17.16) also guarantee that the variance of the state is bounded.

In order to verify (17.16), we consider $z(t) = e^{-At}x(t)$. It is easily verified that we have

$$\mathrm{d}z(t) = e^{-At}B\sigma(u(k))\mathrm{d}t + e^{-At}E\mathrm{d}w(t)$$

which yields

$$z(t+1) = z(t) + \int_t^{t+1} e^{-A\tau}B\sigma(u(\tau))\mathrm{d}\tau + \int_t^{t+1} e^{-A\tau}E\mathrm{d}w(\tau).$$

We note that $\|z(t)\| = \|x(t)\|$ (using that $A + A' = 0$), and hence it is easily seen that we have

$$\|x(t+1)\| \leqslant \|x(t)\| + \|B\| + \|v_t\|,$$

where

$$v_t := \int_t^{t+1} e^{-A\tau}E\mathrm{d}w(\tau)$$

is a Gaussian stochastic variable. This yields

$$\mathbb{E}\Big[\, \big|\|x(t+1)\| - \|x(t)\|\big|^4 \mid x(t) \,\Big] \leqslant \mathbb{E}\Big[\, \big|\|B\| + \|v_t\|\big|^4 \mid x(t) \,\Big] \leqslant b$$

for a suitable constant b using the fact that v_t and $x(t)$ are independent. This establishes (17.15).

In order to establish (17.16), we will use the following lemma:

Lemma 17.11 *Assume that* (A, B) *is controllable. Then there exist a constant* β *such that for all* κ,

$$\tfrac{1}{2}\int_t^{t+1} F(B'e^{A(\tau-t)}x(t))\mathrm{d}\tau \geqslant 2\beta\|x(t)\| - \tfrac{m}{4\kappa}.$$

Proof : First, using Lemma 17.9, we obtain

$$\tfrac{1}{2}\int_t^{t+1} F(B'e^{A(\tau-t)}x(t))\mathrm{d}\tau \geqslant \int_t^{t+1} \|B'e^{A(\tau-t)}x(t)\|\mathrm{d}\tau - \tfrac{m}{4\kappa}.$$

If there is no $\beta > 0$ such that

$$\int_t^{t+1} \|B'e^{A(\tau-t)}x(t)\|\mathrm{d}\tau \geqslant 2\beta\|x(t)\|$$

for all $x(t) \in \mathbb{R}^n$, then there exists a sequence $\{x_n\} \subset \mathbb{R}^n$ with $\|x_n\| = 1$ such that

$$\int_t^{t+1} \|B'e^{A(\tau-t)}x_n\|\mathrm{d}\tau \to 0,$$

but then, let $\bar{x} \neq 0$ be the limit of a convergent subsequence of $\{x_n\}$. Then we have

$$\int_t^{t+1} \|B'e^{A(\tau-t)}\bar{x}\|\mathrm{d}\tau = 0$$

which implies that

$$B'e^{A's}\bar{x} = 0$$

for all $s \in [-1, 0]$. But controllability of (A, B) implies that (B', A') is observable which yields a contradiction. ∎

To study the effect of the noise, we first note that

$$\begin{aligned} x(\tau) &= e^{A(\tau-t)}x(t) + \int_t^{\tau} e^{A(\tau-s)}B\sigma(u(s))\mathrm{d}s + \int_t^{\tau} e^{A(\tau-s)}E\mathrm{d}w(s) \\ &=: e^{A(\tau-t)}x(t) + v_t^1(\tau) \end{aligned} \tag{17.17}$$

Moreover, we have

$$\mathrm{d}x'x = (2x'B\sigma(u) + \operatorname{trace} EQE')\mathrm{d}t + 2x'E\mathrm{d}w$$

using $A + A' = 0$ and Itô's lemma (see, for instance, [109]). This yields that

$$\mathbb{E}\left[\, x(t+1)'x(t+1) \mid x(t)\right] = x(t)'x(t) - \tfrac{2}{\kappa}\,\mathbb{E}\left[\int_t^{t+1} u(\tau)'\sigma(u(\tau))\mathrm{d}\tau \mid x(t)\right] + \operatorname{trace} EQE',$$

where we used (17.14). We have

$$\begin{aligned}\tfrac{2}{\kappa}\int_t^{t+1} u(\tau)'\sigma(u(\tau))\mathrm{d}\tau &= \int_t^{t+1} F(B'x(\tau))d\tau \\ &\geqslant \int_t^{t+1} \tfrac{1}{2}F(B'e^{A(\tau-t)}x(t)) - 3\kappa\|v_t^1(\tau)\|^2 d\tau,\end{aligned}$$

where we used Lemma 17.7 with $\lambda = 2$. This yields

$$\tfrac{2}{\kappa}\,\mathbb{E}\left[\int_t^{t+1} u(\tau)'\sigma(u(\tau))\mathrm{d}\tau \mid x(t)\right] \geqslant \int_t^{t+1} \tfrac{1}{2}F(B'e^{A(\tau-t)}x(t)) - N$$

for some $N > 0$ since v_t^1 is, according to (17.17), the sum of a bounded term and a term independent of $x(t)$ with bounded moments. Combining the above with Lemma 17.11, we get

$$\mathbb{E}\left[\, \|x(t+1)\|^2 \mid x(t)\right] = \|x(t)\|^2 - 2\beta\|x(t)\| + \frac{m}{4\kappa} + \tilde{N}$$

for some appropriate $\tilde{N} > 0$.

Using Jensen's inequality, we obtain

$$\begin{aligned}\mathbb{E}\left[\, \|x(t+1)\| \mid x(t)\right] &\leqslant \sqrt{\mathbb{E}\left[\, \|x(t+1)\|^2 \mid x(t)\right]} \\ &\leqslant \sqrt{\|x(t)\|^2 - 2\beta\|x(t)\| + \frac{m}{4\kappa} + \tilde{N}} \\ &\leqslant \|x(t)\| - \tfrac{1}{2}\beta\end{aligned}$$

provided

$$\|x(k)\| \geqslant J := \min\left\{\frac{m + 4\kappa\tilde{N}}{4\kappa\beta}, \frac{\beta}{2}\right\}.$$

Therefore, we note that (17.16) is satisfied which completes the proof. ∎

17.4 Double-integrator system

As a follow-up on our previous work in Chap. 14 as well as in [168], we consider here the canonical example of double integrator which has played an important role in obtaining a much needed insight as to what is possible and what is not in the presence of an input saturation:

$$\begin{aligned} \mathrm{d}x_1 &= x_2 \mathrm{d}t + e_1 \mathrm{d}w_t \\ \mathrm{d}x_2 &= \sigma(u(t))\mathrm{d}t + e_2 \mathrm{d}w_t, \end{aligned} \tag{17.18}$$

where w_t is a Wiener processes with mean 0 and rate Q. The initial condition $(x_1(0), x_2(0))'$ is a Gaussian random vector which is independent of w_t.

As discussed in Chap. 14, it is well known that any linear controller which asymptotically stabilizes the above linear system without the saturation element will still internally stabilize the system with the saturation function present, i.e., any controller of the form

$$u = -k_1 x_1 - k_2 x_2 \tag{17.19}$$

will achieve global internal stability (when $w \equiv 0$) provided $k_1 > 0$ and $k_2 > 0$. This can be easily verified using the Lyapunov function:

$$V(x_1, x_2) = \int_0^{-k_1 x_1 - k_2 x_2} \sigma(\xi)\mathrm{d}\xi + \tfrac{k_1}{2} x_2' x_2. \tag{17.20}$$

As such, we rewrite (17.18) with $u = -k_1 x_1 - k_2 x_2$ for positive k_1 and k_2 as

$$\begin{cases} \mathrm{d}x_1 = x_2 \mathrm{d}t + e_1 \mathrm{d}w_t \\ \mathrm{d}x_2 = \sigma(-k_1 x_1 - k_2 x_2)\mathrm{d}t + e_2 \mathrm{d}w_t. \end{cases} \tag{17.21}$$

At first, we consider the simultaneous global internal stabilization and global external stochastic stabilization as defined in Problem 17.1 for the double-integrator system as given in (17.18) while utilizing the linear state feedback law (17.19) that is for the system (17.21). In this regard, we obtain a negative result that says that global external stochastic stability cannot be achieved with any $k_1 > 0$ and $k_2 > 0$ whenever $e_2 \neq 0$.

Theorem 17.12 *Consider the double-integrator system* (17.21) *which utilizes linear state feedback. Then, any $k_1 > 0$ and $k_2 > 0$ will result in an unbounded variance of the state for Q large enough whenever $e_2 \neq 0$.*

Explained in simple words, the above theorem states that global internal stability does not imply the global external stability. Moreover, it shows that simultaneous global internal stabilization and global external stochastic stabilization as defined in Problem 17.1 is impossible to achieve for the double integrator. The above theorem does not, however, exclude the possibility of achieving local or semi-global external stochastic stabilization as we discuss shortly.

We proceed next to prove Theorem 17.12.

Proof : We first use a change of variables,

$$\begin{aligned} y_1(\tau) &= k_1 x_1(\tfrac{k_2}{k_1}\tau) + k_2 x_2(\tfrac{k_2}{k_1}\tau), \\ y_2(\tau) &= k_2 x_2(\tfrac{k_2}{k_1}\tau), \end{aligned}$$

while

$$\widetilde{w}(\tau) = \sqrt{\tfrac{k_1}{k_2}}\; w\left(\tfrac{k_2}{k_1}\tau\right).$$

We note that $\widetilde{w}$ (like w) is Wiener process with mean 0 and rate Q. This allows us to rewrite the system (17.21) as

$$\begin{aligned} \mathrm{d}y_1(\tau) &= y_2(\tau)\mathrm{d}\tau - \lambda\sigma(y_1(\tau))\mathrm{d}\tau + \widetilde{e}_1\mathrm{d}\widetilde{w}(\tau) \\ \mathrm{d}y_2(\tau) &= -\lambda\sigma(y_1(\tau))\mathrm{d}\tau + \widetilde{e}_2\mathrm{d}\widetilde{w}(\tau), \end{aligned} \tag{17.22}$$

where

$$\lambda = \tfrac{k_2^2}{k_1},$$

and

$$\widetilde{e}_1 = \sqrt{\tfrac{k_2}{k_1}}(k_1 e_1 + k_2 e_2), \quad \widetilde{e}_2 = \sqrt{\tfrac{k_2}{k_1}} k_2 e_2.$$

Next, we look at the Lyapunov function V defined in (17.20) which in our new coordinates reduces to the form

$$V(y_1, y_2) = \int_0^{y_1} \sigma(\xi)\mathrm{d}\xi + \tfrac{1}{2\lambda} y_2' y_2. \tag{17.23}$$

We define $y = (y_1, y_2)$.

We consider two cases. If $|y_1| \leqslant 1$, we find that

$$\begin{aligned} dV = -\lambda\sigma(y_1)^2\mathrm{d}\tau + \left[\sigma(y_1)\widetilde{e}_1 + \tfrac{1}{\lambda} y_2\widetilde{e}_2\right]\mathrm{d}\widetilde{w} \\ + \left[\widetilde{e}_1 Q\widetilde{e}_1' + \tfrac{1}{\lambda}\widetilde{e}_2 Q\widetilde{e}_2'\right]\mathrm{d}\tau, \end{aligned} \tag{17.24}$$

while for $|y_1| > 1$, we get

$$dV = -\lambda\sigma(y_1)^2\mathrm{d}\tau + \left[\sigma(y_1)\widetilde{e}_1 + \tfrac{1}{\lambda} y_2\widetilde{e}_2\right]\mathrm{d}\widetilde{w} + \tfrac{1}{\lambda}\widetilde{e}_2 Q\widetilde{e}_2'\mathrm{d}\tau. \tag{17.25}$$

Note that in the above two equations, the last term is the special correction term that comes from Itô's lemma. For details, we refer to [109].

Combining the above, we find the lower bound,

$$\mathbb{E}[V(y(t))] \geqslant \mathbb{E}[V(y(t_0))] - \int_{t_0}^{t} \lambda \mathrm{d}\tau + \int_{t_0}^{t} \tfrac{1}{\lambda}\tilde{e}_2 Q \tilde{e}_2' \mathrm{d}\tau.$$

It is clear that $V \leqslant \|y\|^2$. Using the above lower bound, we find $\mathrm{Var}[y]$ is unbounded whenever

$$\tilde{e}_2 Q \tilde{e}_2' \geqslant \lambda^2.$$

Here, we used that $\mathrm{Var}[y] = \mathbb{E}\,\|y\|^2$ given that y has mean zero. The above clearly means that for any given controller, there exist stochastic disturbances with sufficiently large variance that will result in an unbounded variance for the state. The only possible exception is when $e_2 = 0$. Finally, note that if the variance of y is unbounded, then we clearly also have the variance of x unbounded. ■

The above theorem clearly states that the simultaneous global internal stabilization and global external stochastic stabilization as defined in Problem 17.1 is impossible to achieve for the double-integrator system with linear feedback.

This prompts us to examine for the double-integrator system the simultaneous global internal stabilization and semi-global external stochastic stabilization problem as defined in Problem 17.2. In this connection, we have a positive result, namely if Q has a known upper bound, we can always find k_1 and k_2 for which the states will have a bounded variance.

This result has only been obtained for a particular case of double-integrator systems where $e_2 = 0$, i.e., we consider the system,

$$\begin{cases} \mathrm{d}x_1 = x_2 \mathrm{d}t + \mathrm{d}w_t \\ \mathrm{d}x_2 = \sigma(-k_1 x_1 - k_2 x_2)\mathrm{d}t, \end{cases} \tag{17.26}$$

where k_1 and k_2 are positive real numbers, w_t is a Wiener process with mean 0 and rate q, and the initial condition $(x_1(0), x_2(0))'$ is a Gaussian random vector which is independent of w_t.

We have the following results:

Theorem 17.13 *Consider the system* (17.26)*. For any a priori given $\bar{q} > 0$, there exist $k_1 > 0$ and $k_2 > 0$ such that the state has a bounded variance for any w_t with $q \leqslant \bar{q}$.*

Theorem 17.14 *Consider the system* (17.26)*. For any $k_1 > 0$ and $k_2 > 0$, there exists a $\bar{q} > 0$ such that for any w_t with $q < \bar{q}$, the closed-loop state has a bounded variance.*

We conjecture that the above two theorems are also correct for the more general case of system 17.21 with $e_2 \neq 0$. In other words, we claim that for any $k_1 > 0$ and $k_2 > 0$, we achieve local external stochastic stability while we can achieve semi-global external stochastic stability by suitable chosen linear feedbacks.

Theorems 17.13 and 17.14 are an immediate consequence of the following proposition:

Proposition 17.15 *Consider the system* (17.26). *For any given* $q > 0$, Var$[x]$ *is bounded if* $\frac{k_2}{k_1^2} > 16q$.

Proof : Define

$$V(x_1, x_2) = \int_0^{k_1 x_1} \sigma(s)\mathrm{d}s + \int_0^{k_1 x_1 + k_2 x_2} \sigma(s)\mathrm{d}s + k_1 x_2^2. \tag{17.27}$$

We have

$$\|x\|^2 \leqslant \frac{2(V+1)^2 + 2V}{k_1}.$$

Therefore, Var$[x]$ is bounded if $\mathbb{E}[V^r]$ is bounded for $r \in (0, 3)$. Thanks to Lemma 17.16 in Appendix, it suffices to show that there exist $a > 0$, $b > 0$, and $J > 0$ such that

(*i*) $\mathbb{E}[V(t+1) - V(t) \mid V(t)] \leqslant -a$ on the event that $V(t) \geqslant J$;

(*ii*) $\mathbb{E}[|V(t+1) - V(t)|^4 \mid V(t)] \leqslant b$.

From Itô's formula, we have

$$\begin{aligned}\mathrm{d}V &= [k_1 x_2 \mathrm{d}t + k_1 \mathrm{d}w_t]\sigma(k_1 x_1) \\ &\quad + [k_1 x_2 \mathrm{d}t + k_1 \mathrm{d}w_t - k_2 \sigma(k_1 x_1 + k_2 x_2)\mathrm{d}t]\sigma(k_1 x_1 + k_2 x_2) \\ &\quad - 2k_1 x_2 \sigma(k_1 x_1 + k_2 x_2)\mathrm{d}t + \sum_{i,j} \tfrac{\partial^2 V}{\partial x_i \partial x_j} \mathrm{d}x_i \mathrm{d}x_j.\end{aligned}$$

Then

$$\begin{aligned}\mathrm{d}V &= -k_2 \sigma^2(k_1 x_1 + k_2 x_2)^2 \mathrm{d}t + k_1 x_2[\sigma(k_1 x_1) - \sigma(k_1 x_1 + k_2 x_2)]\mathrm{d}t \\ &\quad + [k_1 \sigma(k_1 x_1) + k_1 \sigma(k_1 x_1 + k_2 x_2)]\mathrm{d}w_t + \sum_{i,j} \tfrac{\partial^2 V}{\partial x_i \partial x_j} \mathrm{d}x_i \mathrm{d}x_j.\end{aligned}$$

Consider the Itô correction term,

$$\sum_{i,j} \frac{\partial^2 V}{\partial x_i \partial x_j} \mathrm{d}x_i \mathrm{d}x_j = \frac{\partial}{\partial x_1}[k_1\sigma(k_1x_1) + k_1\sigma(k_1x_1 + k_2x_2)]\mathrm{d}x_1^2$$
$$+ 2\frac{\partial}{\partial x_2}[k_1\sigma(k_1x_1 + k_2x_2)]\mathrm{d}x_1\mathrm{d}x_2 + 2k_1\mathrm{d}x_2^2$$
$$= \begin{cases} 0, & |k_1x_1| > 1 \,\&\, |k_1x_1 + k_2x_2| > 1 \\ k_1^2, & |k_1x_1| \leqslant 1 \,\&\, |k_1x_1 + k_2x_2| > 1 \\ k_1^2, & |k_1x_1| > 1 \,\&\, |k_1x_1 + k_2x_2| \leqslant 1 \\ 2k_1^2, & |k_1x_1| \leqslant 1 \,\&\, |k_1x_1 + k_2x_2| \leqslant 1. \end{cases}$$

Therefore, we have

$$\mathrm{d}V \leqslant \phi_t \mathrm{d}t + v_t \mathrm{d}w_t + 2k_1^2 q \mathrm{d}t,$$

where

$$v_t := k_1\sigma(k_1x_1) + k_1\sigma(k_1x_1 + k_2x_2)$$

andv

$$\phi_t := -k_2\sigma^2(k_1x_1 + k_2x_2)^2 + k_1x_2[\sigma(k_1x_1) - \sigma(k_1x_1 + k_2x_2)].v$$

Note that $\phi_t \leqslant 0$. Then

$$|V(t+1) - V(t)| \leqslant \left| \int_t^{t+1} v_t \mathrm{d}w_t \right| + 2k_1^2 q.$$

We first prove item *(ii)*. It suffices to prove that there exists a J such that

$$\mathbb{E}\left[\left| \int_t^{t+1} v_t \mathrm{d}w_t \right|^4 \right] \leqslant J.$$

Define $W_s = \int_t^s v_\tau \mathrm{d}w_\tau$. We have $\mathrm{d}W_s = v_s \mathrm{d}w_s$. Define $Y_s = W_s^4$. Then

$$\mathrm{d}Y_s = 4W_s^3 v_s \mathrm{d}w_s + 12W_s^2 v_s^2 \mathrm{d}s.$$

Hence, we have

$$\mathbb{E}[Y_{t+1}] = 12\,\mathbb{E}\left[\int_t^{t+1} W_s^2 v_s^2 \mathrm{d}s \right] \leqslant 24k_1^2 \int_t^{t+1} \mathbb{E}\left[W_s^2\right] \mathrm{d}s.$$

From Itô isometry, we have, for $s \in [t, t+1]$,

$$\mathbb{E}\left[W_s^2\right] = \mathbb{E}\left[\left(\int_t^s v_t \mathrm{d}w_t\right)^2\right] = \mathbb{E}\left[\int_t^s v_t^2 \mathrm{d}t\right] \leqslant 4k_1^2.$$

Therefore,

$$\mathbb{E}\left[\left|\int_t^{t+1} v_t \mathrm{d}w_t\right|^4\right] = \mathbb{E}\left[Y_{t+1}\right] \leqslant 96k_1^4.v$$

It remains to show item (i). Define $c = \max\{8k_1 q, \frac{1}{k_2}\}$. We have two cases:

Case 1:

$x_2(t) \geqslant c+1$. This implies that $x_2 \geqslant c$ for $[t, t+1]$.

(i) If $|k_1x_1 + k_2x_2| \geqslant \frac{1}{2}$, we have

$$\begin{aligned} &-k_2\sigma^2(k_1x_1 + k_2x_2) + k_1x_2[\sigma(k_1x_1) - \sigma(k_1x_1 + k_2x_2)] \\ &\qquad \leqslant -k_2\sigma^2(k_1x_1 + k_2x_2) \leqslant -\frac{k_2}{4} \leqslant -4k_1^2 q; \end{aligned}$$

(ii) if $|k_1x_1 + k_2x_2| \leqslant \frac{1}{2}$, we have

$$\begin{aligned} &-k_2\sigma^2(k_1x_1 + k_2x_2) + k_1x_2[\sigma(k_1x_1) - \sigma(k_1x_1 + k_2x_2)] \\ &\leqslant k_1x_2[\sigma(k_1x_1) - \sigma(k_1x_1 + k_2x_2)] \leqslant -\frac{1}{2}k_1x_2\sigma(k_2x_2) \leqslant -4k_1^2 q. \end{aligned}$$

In either possibility, we shall have

$$\mathbb{E}[\phi_t] \leqslant -4k_1^2 q.$$

Case 2:

$|x_2(t)| \leqslant c+1$. This implies that $|x_2| \leqslant c+2$ in $[t, t+1]$. Let $J_0 > 0$ be such that for any $t > J_0$

$$\int_t^\infty \frac{1}{\sqrt{2\pi q}} e^{-\frac{s^2}{2q}} \mathrm{d}s \leqslant \frac{1}{2},$$

and let J be such that

$$\left.\begin{array}{r} V(t) \geqslant J \\ |x_2(t)| \leqslant c+1 \end{array}\right\} \Rightarrow |x_1(t)| \geqslant \tfrac{k_2(c+2)+J_0+1}{k_1} + c + 2.$$

Without loss of generality, we assume that $x_1(t) > \frac{k_2(c+2)+J_0+1}{k_1} + c + 2$. Then for $s \in [t, t+1]$,

$$k_1 x_1(s) + k_2 x_2(s) = M_1(s) + M_2(s),$$

where

$$M_1(s) = k_1 x_1(t) + k_2 x_2(s) + k_1 \int_t^s x_2(\tau)\mathrm{d}\tau, \quad M_2(s) = \int_t^s \mathrm{d}w_\tau.$$

Note that given $x_1(t) > \frac{k_2(c+2)+J_0+1}{k_1} + c + 2$ and $|x_2(s)| \leqslant c+2$, we have for $s \in [t, t+1]$,

$$M_1(s) \geqslant J_0 + 1.$$

Hence, $k_1 x_1(s) + k_2 x_2(s) \in [-1, 1]$ implies that $M_2(s) \leqslant -J_0$, which yields that

$$\mathbb{P}[k_1 x_1(s) + k_2 x_2(s) \in [-1, 1] \mid V(t) \geqslant J] \leqslant \mathbb{P}[M(s) \leqslant -J_0].$$

This implies that

$$\mathbb{E}[\sigma^2(k_1 x_1(s) + k_2 x_2(s)) \mid V(t) \geqslant J] \geqslant 1 - \int_{J_0}^{\infty} \tfrac{1}{\sqrt{2\pi q}} e^{-\frac{s^2}{2q}} \mathrm{d}s \geqslant \frac{1}{2}.$$

Therefore,

$$\mathbb{E}[\phi_t] \leqslant -k_2\, \mathbb{E}[\sigma^2(k_1 x_1(s) + k_2 x_2(s))] \leqslant -\tfrac{k_2}{2} \leqslant -8k_1^2 q.$$

In summary, for $\frac{k_2}{k_1^2} \geqslant 16q$, we find in both cases that

$$\mathbb{E}[\phi_t] \leqslant -4k_1^2 q.$$

Therefore,

$$\mathbb{E}[V(t+1) - V(t) \mid V(t) \geqslant J] = \phi_t + 2k_1^2 q \leqslant -2k_1^q.$$

This completes the proof. ■

17.A Appendix

We recall here a crucial lemma proved in [113].

Lemma 17.16 *Let X_n be random variables and $\{\mathcal{F}_n\}$ be the filtration to which $\{X_n\}$ is adapted and suppose that there exist constants $a > 0$, J and $V < \infty$, and $p > 2$ such that $X_0 < J$ and for all n*

$$\mathbb{E}(X_{n+1} - X_n \mid \mathcal{F}_n) < -a, \quad \text{on the event that } \{X_n > J\} \tag{17.28}$$

and

$$\mathbb{E}(|X_{n+1} - X_n|^p \mid \mathcal{F}_n) \leqslant V. \tag{17.29}$$

Then, for any $r \in (0, p-1)$, there is a $c = c(a, J, V, p, r)$ such that $\mathbb{E}(X_n^+)^r < c$ for all n.

References

[1] J. ACKERMANN AND T. BÜNTE, "Actuator rate limits in robust car steering control", in Proc. 36th CDC, San Diego, CA, 1997, pp. 4726–4731.

[2] B.D.O. ANDERSON AND J.B. MOORE, *Linear optimal control*, Prentice-Hall, Englewood Cliffs, NJ, 1971.

[3] J.P. AUBIN, *Viability theory*, Birkhäuser, 1991.

[4] J.P. AUBIN AND H. FRANKOWSKA, *Set-valued analysis*, Birkhäuser, 1990.

[5] X. BAO, Z. LIN, AND E. D. SONTAG, "Finite gain stabilization of discrete-time linear systems subject to actuator saturation", Automatica, 36(2), 2000, pp. 269–277.

[6] J.M. BERG, K.D. HAMMETT, C.A. SCHWARTZ, AND S.S. BANDA, "An analysis of the destabilizing effect of daisy chained rate-limited actuators", IEEE Trans. Control Systems Technology, 4(2), 1996, pp. 171–176.

[7] D.S. BERNSTEIN AND A.N. MICHEL, "A chronological bibliography on saturating actuators", Int. J. Robust & Nonlinear Control, 5(5), 1995, pp. 375–380.

[8] ——, Guest Eds., *Special Issue on saturating actuators*, Int. J. Robust & Nonlinear Control, 5(5), 1995, pp. 375–540.

[9] G. BITSORIS, "On the linear decentralized constrained regulation problem of discrete-time dynamical systems", Information and Decision Technologies, 14(3), 1988, pp. 229–239.

[10] F. BLANCHINI, "Set invariance in control", Automatica, 35(11), 1999, pp. 1747–1769.

[11] F. BLANCHINI AND S. MIANI, "Constrained stabilization of continuous-time linear systems", Syst. & Contr. Letters, 28(2), 1996, pp. 95–102.

[12] ——, "Constrained stabilization via smooth Lyapunov functions", Syst. & Contr. Letters, 35(3), 1998, pp. 155–163.

[13] H. BOURLÈS, "Local l_p-stability and local small gain theorem for discrete-time systems", IEEE Trans. Aut. Contr., 41(6), 1996, pp. 903–907.

[14] S. BOYD, V. BALAKRISHNAN, AND P. KABAMBA, "A bisection method for computing the H_∞-norm of a transfer matrix and related problems", Math. Contr. Sign. & Syst., 2(3), 1989, pp. 207–219.

[15] S.P. BOYD AND C.H. BARRATT, *Linear controller design: limits of performance*, Information and system sciences, Prentice Hall, Englewood Cliffs, NJ, 1991.

[16] N.A. BRUINSMA AND M. STEINBUCH, "A fast algorithm to compute the H_∞ norm of a transfer matrix", Syst. & Contr. Letters, 14(4), 1994, pp. 287–293.

[17] E. CAMACHO AND C. BORDONS, *Model predictive control*, Springer Verlag, 1998.

A. Saberi et al., *Internal and External Stabilization of Linear Systems with Constraints*, Systems & Control: Foundations & Applications, DOI 10.1007/978-0-8176-4787-2,

[18] B.M. CHEN, *Robust and H_∞ control*, Communication and Control Engineering Series, Springer Verlag, 2000.

[19] B.M. CHEN, Z. LIN, AND Y. SHAMASH, *Linear systems theory: a structural decomposition approach*, Birkhäuser, Boston, 2004.

[20] B. M. CHEN, A. SABERI, AND P. SANNUTI, "Explicit expressions for cascade factorization of general nonminimum phase systems", IEEE Trans. Aut. Contr., 37(3), 1992, pp. 358–363.

[21] C.-T. CHEN, *Linear system theory and design*, Holt, Rinehart and Winston, New York, 1984.

[22] Y. CHITOUR, "On the L^p stabilization of the double integrator subject to input saturation", ESAIM: Control, Optimization and Calculus of Variations, 6, 2001, pp. 291–331.

[23] Y. CHITOUR AND Z. LIN, "Finite gain ℓ_p stabilization of discrete-time linear systems subject to actuator saturation: the case of $p = 1$", IEEE Trans. Aut. Contr., 48(12), 2003, pp. 2196–2198.

[24] HO-LIM CHOI AND JONG-TAE LIM, "Stabilization of a chain of integrators with an unknown delay in the input by adaptive output feedback", IEEE Trans. Aut. Contr., 51(8), 2006, pp. 1359–1363.

[25] J. CHOI, "Connections between local stability in Lyapunov and input / output senses", IEEE Trans. Aut. Contr., 40(12), 1995, pp. 2139 –2143.

[26] ——, "On the stabilization of linear discrete-time systems subject to input saturation", Syst. & Contr. Letters, 36(3), 1999, pp. 241–244.

[27] ——, "On the constrained asymptotic stabilizability of unstable linear discrete time systems via linear feedback", in American Control Conference, Arlington, VA, 2001, pp. 4926–4929.

[28] D. CHU, X. LIU, AND C.E. TAN, "On the numerical computation of structural decomposition in systems and control", IEEE Trans. Aut. Contr., 47(11), 2002, pp. 1786–1799.

[29] C. COMMAULT AND J.M. DION, "Structure at infinity of linear multivariable systems: a geometric approach", IEEE Trans. Aut. Contr., 27(3), 1982, pp. 693–696.

[30] M. CWIKEL AND P. GUTMAN, "Convergence of an algorithm to find maximal state constraint sets for discrete-time linear dynamical systems with bounded controls and states", IEEE Trans. Aut. Contr., 31(5), 1986, pp. 457–459.

[31] R. DATKO, "A procedure for determination of the exponential stability of certain differential-difference equations", Quart. Appl. Math., 36(3), 1978, pp. 279–292.

[32] C. CANUDAS DE WIT AND P. LISCHINSKY, "Adaptive friction compensation with partially known dynamic friction model", Int. J. Adapt. Contr. and Sign. Proc., 11(1), 1997, pp. 65–80.

[33] A. DONTCHEV AND F. LEMPIO, "Difference methods for differential inclusions: a survey", SIAM Review, 34(2), 1992, pp. 263–294.

[34] J.C. DOYLE, K. GLOVER, P.P. KHARGONEKAR, AND B.A. FRANCIS, "State space solutions to standard H_2 and H_∞ control problems", IEEE Trans. Aut. Contr., 34(8), 1989, pp. 831–847.

[35] F. ESFANDIARI AND H.K. KHALIL, "Output feedback stabilization of fully linearizable systems", Int. J. Contr., 56(5), 1992, pp. 1007–1037.

[36] I. FLÜGGE-LOTZ, *Discontinuous and optimal control*, McGraw-Hill, New York, 1968.

[37] E. FRIDMAN, "New Lyapunov-Kasovskii functionals for stability of linear retarded and neutral type systems", Syst. & Contr. Letters, 43(4), 2001, pp. 309–319.

[38] A.T. FULLER, "In-the-large stability of relay and saturating control systems with linear controller", Int. J. Contr., 10(4), 1969, pp. 457–480.

[39] ——, Ed., *Nonlinear stochastic control systems*, Taylor and Francis, London, 1970.

[40] F.R. GANTMACHER, *The theory of matrices*, Chelsea, New York, 1959.

[41] J.E. GAYEK, "A survey of techniques for approximating reachable and controllable sets", in Proc. 30th CDC, Brighton, England, 1991, pp. 1724–1729.

[42] H.F. GRIP AND A. SABERI, "Structural decomposition of linear multivariable systems using symbolic computations", Int. J. Contr., 83(7), 2010, pp. 1414–1426.

[43] K. GU, V. L. KHARITONOV, AND J. CHEN, *Stability of time-delay systems*, Birkhäuser, Boston, MA, 2003.

[44] W. HAHN, *Stability of motion*, vol. 138 of Die Grundlehren der mathematischen Wissenschaften in Einzeldarstellungen, Springer Verlag, Berlin, 1967.

[45] M.L.J. HAUTUS, "Controllability and observability conditions of linear autonomous systems", Proc. Nederl. Akad. Wetensch., Ser. A, 72(5), 1969, pp. 443–448.

[46] D. HILL AND P. MOYLAN, "Connections between finite-gain and asymptotic stability", IEEE Trans. Aut. Contr., 25(5), 1980, pp. 931–936.

[47] P. HOU, A. SABERI, AND Z. LIN, "On ℓ_p-stabilization of strictly unstable discrete-time linear systems with saturating actuators", Int. J. Robust & Nonlinear Control, 8(14), 1998, pp. 1227–1236.

[48] P. HOU, A. SABERI, Z. LIN, AND P. SANNUTI, "Simultaneously external and internal stabilization for continuous and discrete-time critically unstable systems with saturating actuators", Automatica, 34(12), 1998, pp. 1547–1557.

[49] T. HU, Z. LIN, AND L. QIU, "Stabilization of exponentially unstable linear systems with saturating actuators", IEEE Trans. Aut. Contr., 45(6), 2001, pp. 973–979.

[50] Y.S. HUNG AND A.G.J. MACFARLANE, "On the relationships between the unbounded asymptote behavior of multivariable root loci, impulse response and infinite zeros", Int. J. Contr., 34(1), 1981, pp. 31–69.

[51] L. IANNELLI, K.H. JOHANSSON, U.T. JÖNSSON, AND F. VASCA, "Dither for smoothing relay feedback systems", IEEE Trans. Circ. & Syst.-I Fundamental theory and applications, 50(8), 2003, pp. 1025–1035.

[52] ——, "Averaging of nonsmooth systems using dither", Automatica, 42(4), 2006, pp. 669–676.

[53] R.N. IZMAĬLOV, "“Peak” effect in stationary linear systems in scalar inputs and outputs", Automation and Remote Control, 48(8, part 1), 1987, pp. 1018–1024. Translated from Avtomatika i Telemekhanika, 1987, No. 8, pp. 56–62.

[54] ——, "The peak effect in stationary linear systems with multivariate inputs and outputs", Automation and Remote Control, 49(1, part 1), 1988, pp. 40–47. Translated from Avtomatika i Telemekhanika, 1988, No. 1, pp. 52–60.

[55] Z.P. JIANG AND Y. WANG, "Input-to-state stability for discrete-time nonlinear systems", Automatica, 37(6), 2001, pp. 857–869.

[56] T. KAILATH, *Linear Systems*, Prentice-Hall, Englewood Cliffs, NJ, 1980.

[57] N. KARCANIAS AND B. KOUVARITAKIS, "The output zeroing problem and its relationship to the invariant zero structure: a matrix pencil approach", Int. J. Contr., 30(2), 1979, pp. 395–415.

[58] HASSAN K. KHALIL, *Nonlinear systems*, Prentice Hall, Upper Saddle River, 2nd Ed., 1996.

[59] V.L. KHARITONOV, S.-I. NICULESCU, J. MORENO, AND W. MICHIELS, "Static output feedback stabilization: necessary conditions for multiple delay controllers", IEEE Trans. Aut. Contr., 50(1), 2005, pp. 82–86.

[60] U. KIENCKE AND L. NIELSEN, *Automotive control systems: for engine, driveline and vehicle*, Springer Verlag, Berlin, 2000.

[61] R.L. KOSUT, "Adaptive robust control via transfer function uncertainty estimation", in American Control Conference, Atlanta, GA, 1988, pp. 349–354.

[62] W.H. KWON AND A.E. PEARSON, "A note on the algebraic Riccati equation", IEEE Trans. Aut. Contr., 22(1), 1977, pp. 143–144.

[63] P. LANCASTER, *Theory of matrices*, Academic Press, New York, 1969.

[64] P. LANCASTER AND L. RODMAN, *Algebraic Riccati equations*, Clarendon Press, Oxford, U.K., 1995.

[65] E.B. LEE AND L. MARKUS, *Foundations of optimal control theory*, John Wiley & Sons, New York, 1967.

[66] J.L. LEMAY, "Recoverable and reachable zones for control systems with linear plants and bounded controller outputs", IEEE Trans. Aut. Contr., 9(4), 1964, pp. 346–354.

[67] Z. LIN, "H_∞-almost disturbance decoupling with internal stability for linear systems subject to input saturation", IEEE Trans. Aut. Contr., 42(7), 1997, pp. 992–995.

[68] ——, "Robust semi-global stabilization of linear systems with imperfect actuators", Syst. & Contr. Letters, 29(4), 1997, pp. 215–221.

[69] ——, "Semi-global stabilization of linear systems with position and rate-limited actuators", Syst. & Contr. Letters, 30(1), 1997, pp. 1–11.

[70] ——, "Global control of linear systems with saturating actuators", Automatica, 34(7), 1998, pp. 897–905.

[71] ——, *Low gain feedback*, vol. 240 of Lecture Notes in Control and Inform. Sci., Springer Verlag, Berlin, 1998.

[72] ——, "Semi-global stabilization of discrete-time linear systems with position and rate limited actuators", in Proc. 37th CDC, Tampa, FL., 1998, pp. 395–400.

[73] Z. LIN, M. PACHTER, S. BANDA, AND Y. SHAMASH, "Stabilizing feedback design for linear systems with rate limited actuators", in Control of uncertain systems with bounded inputs, S. Tarbouriech and G. Garcia, eds., Springer Verlag, 1997, pp. 173–186.

[74] Z. LIN AND A. SABERI, "A low-and-high gain approach to semi-global stabilization and/or semi-global practical stabilization of a class of linear systems subject to input saturation via linear state and output feedback", in Proc. 32nd CDC, San Antonio, TX, 1993, pp. 1820–1821.

[75] ——, "Semi-global exponential stabilization of linear systems subject to "input saturation" via linear feedbacks", Syst. & Contr. Letters, 21(3), 1993, pp. 225–239.

[76] ——, "Semi-global exponential stabilization of linear discrete-time systems subject to 'input saturation' via linear feedbacks", Syst. & Contr. Letters, 24(2), 1995, pp. 125–132.

[77] ——, "A semi-global low-and-high gain design technique for linear systems with input saturation – stabilization and disturbance rejection", Int. J. Robust & Nonlinear Control, 5(5), 1995, pp. 381–398.

[78] ——, "Low-and-high gain design technique for linear systems subject to input saturation – a direct method", Int. J. Robust & Nonlinear Control, 7(12), 1997, pp. 1071–1101.

[79] Z. LIN, A. SABERI, AND B.M. CHEN, *Linear systems toolbox*, A.J. Controls Inc., Seattle, WA, 1991. Washington State Univ. Techn. Report No. EE/CS 0097.

[80] ——, "Linear systems toolbox: system analysis and control design in the Matlab environment", in Proc. 1st IEEE Conf. Control Appl., Dayton, OH, 1992, pp. 659–664.

[81] Z. LIN, A. SABERI, P. SANNUTI, AND Y. SHAMASH, "Perfect regulation of linear multivariable systems a low-and-high-gain design", in Proc. Workshop on Advances in Control and its Applications (Urbana Champaign, IL, 1994), Lecture Notes in Control and Inform. Sci., London, 1996, Springer Verlag, pp. 173–192.

[82] ——, "A direct method of constructing H_2 suboptimal controllers – continuous-time systems", J. Optim. Th. & Appl., 99(3), 1998, pp. 585–616.

[83] Z. LIN, A. SABERI, AND A.A. STOORVOGEL, "Semi-global stabilization of linear discrete-time systems subject to input saturation via linear feedback - an ARE-based approach", IEEE Trans. Aut. Contr., 41(8), 1996, pp. 1203–1207.

[84] Z. LIN, A. SABERI, A. STOORVOGEL, AND R. MANTRI, "An improvement to the low gain design for discrete-time linear systems subject to input saturation – solution of semi-global output regulation problems", in Proc. 13th IFAC world congress, vol. F, San Francisco, 1996, pp. 1–6.

[85] ——, “An improvement to the low gain design for discrete-time linear systems in the presence of actuator saturation nonlinearity”, Int. J. Robust & Nonlinear Control, 10(3), 2000, pp. 117–135.

[86] Z. LIN, A. SABERI, AND A.R. TEEL, “Simultanuous L_p-stabilization and internal stabilization of linear systems subject to input saturation — state feedback case”, Syst. & Contr. Letters, 25(3), 1995, pp. 219–226.

[87] ——, “The almost disturbance decoupling problem with internal stability for linear systems subject to input saturation – state feedback case”, Automatica, 32(4), 1996, pp. 619–624.

[88] Z. LIN, A.A. STOORVOGEL, AND A. SABERI, “Output regulation for linear systems subject to input saturation”, Automatica, 32(1), 1996, pp. 29–47.

[89] W. LIU, Y. CHITOUR, AND E.D. SONTAG, “On finite-gain stabilizability of linear systems subject to input saturation”, SIAM J. Contr. & Opt., 34(4), 1996, pp. 1190–1219.

[90] W. LIU, Y. CHITOUR, AND E. D. SONTAG, “Remarks on finite gain stabilizability of linear systems subject to input saturation”, in Proc. 32nd CDC, San Antonio, TX, 1993, pp. 1808–1813.

[91] X. LIU, B.M. CHEN, AND Z. LIN, “Linear systems toolkit in Matlab: structural decompositions and their applications”, Journal of Control Theory and Applications, 3(3), 2005, pp. 287–294.

[92] A.G.J. MACFARLANE AND N. KARCANIAS, “Poles and zeros of linear multivariable systems: a survey of the algebraic, geometric and complex-variable theory”, Int. J. Contr., 24(1), 1976, pp. 33–74.

[93] J.M. MACIEJOWSKI, *Predictive control with constraints*, Prentice Hall, 2002.

[94] J.L. MASSERA, “On Liapounoff’s Conditions of Stability”, The Annals of Mathematics, Second Series, 50(3), 1949, pp. 705–721.

[95] D.Q. MAYNE, J.B. RAWLINGS, C.V. RAO, AND P.O.M. SCOKAERT, “Constrained model predictive control: stability and optimality”, Automatica, 36(6), 2000, pp. 789–814.

[96] F. MAZENC, S. MONDIE, AND R. FRANCISCO, “Global asymptotic stabilization of feedforward systems with delay in the input”, IEEE Trans. Aut. Contr., 49(5), 2004, pp. 844–850.

[97] F. MAZENC, S. MONDIE, AND S.-I. NICULESCU, “Global asymptotic stabilization for chains of integrators with a delay in the input”, IEEE Trans. Aut. Contr., 48(1), 2003, pp. 57–63.

[98] A. MEGRETSKI, “L_2 BIBO output feedback stabilization with saturated control”, in Proc. 13th IFAC world congress, vol. D, San Francisco, 1996, pp. 435–440.

[99] E. MICHAEL, “Continuous selections I”, Annals of Mathematics, 63(2), 1956, pp. 361–381.

[100] W. MICHIELS AND D. ROOSE, “Global stabilization of multiple integrators with time delay and input constraints”, in Proceedings of the 3th Workshop on Time-Delay Systems (TDS2001), T. Abdallah, K. Gu, and S.-I. Niculescu, eds., Santa Fe, NM, 2001, Pergamon, pp. 243–248.

[101] T. MITA, "On maximal unobservable subspace, zeros, and their applications", Int. J. Contr., 25(6), 1977, pp. 885–899.

[102] ——, "On the synthesis of an unknown observer for a class of multi-input/output systems", Int. J. Contr., 26(6), 1977, pp. 841–851.

[103] A.S. MORSE, "Structural invariants of linear multivariable systems", SIAM J. Contr. & Opt., 11(3), 1973, pp. 446–465.

[104] P. MOYLAN, "Stable inversion of linear systems", IEEE Trans. Aut. Contr., 22(1), 1977, pp. 74–78.

[105] M. NAGUMO, "Über die Lage der Integralkurven gewöhnlicher Differentialgleichungen", Proc. Phys. Math. Soc. Japan, 24, 1942, pp. 551–559.

[106] S.I. NICULESCU AND W. MICHIELS, "Stabilizing a chain of integrators using multiple delays", IEEE Trans. Aut. Contr., 49(5), 2004, pp. 802–807.

[107] S.-I. NICULESCU, "On delay-dependent stability under model transformations of some neutral linear systems", Int. J. Contr., 74(6), 2001, pp. 609–617.

[108] M. NIKOLAOU AND V. MANOUSIOUTHAKIS, "A hybrid approach to nonlinear system stability and performance", AIChE Journal, 35(4), 1989, pp. 559–572.

[109] B. ØKSENDAL, *Stochastic differential equations: an introduction with applications*, Universitext, Springer-Verlag, Berlin, Sixth Ed., 2003.

[110] D.H. OWENS, "Invariant zeros of multivariable systems: a geometric analysis", Int. J. Contr., 26(4), 1977, pp. 537–548.

[111] H.K. OZCETIN, A. SABERI, AND P. SANNUTI, "Special coordinate basis for order reduction of linear multivariable systems", Int. J. Contr., 52(1), 1990, pp. 191–226.

[112] HANS B. PACEJKA, *Tire and vehicle dynamics*, Butterworth-Heinemann, Amsterdam, 2nd Ed., 2006.

[113] R. PEMANTLE AND J.S. ROSENTHAL, "Moment conditions for a sequence with negative drift to be uniformly bounded in L^r", Stochastic Processes and their Applications, 82(1), 1999, pp. 143–155.

[114] L.S. PONTRYAGIN, V.G. BOLTYANSKII, R.V. GAMKRELIDZE, AND E.R. MISHCHENKO, *The mathematical theory of optimal processes*, Wiley, New York, 1962.

[115] A. C. PUGH AND P. A. RATCLIFFE, "On the zeros and poles of a rational matrix", Int. J. Contr., 30(2), 1979, pp. 213–227.

[116] F. RAMPONI, D. CHATTERJEE, A. MILIAS-ARGEITIS, P. HOKAYEM, AND J. LYGEROS, "Attaining mean square boundedness of a neutrally stable noisy linear system with a bounded control input", IEEE Trans. Aut. Contr., 55(10), 2010, pp. 2414–2418.

[117] V.G. RAO AND D.S. BERNSTEIN, "Naive control of the double integrator", IEEE Control Systems Magazine, 21(5), 2001, pp. 86–97.

[118] J.P. RICHARD, "Time-delay systems: an overview of some recent advances and open problems", Automatica, 39(10), 2003, pp. 1667–1694.

[119] R.T. ROCKAFELLAR, *Convex analysis*, Princeton University Press, Princeton, 1970.

[120] H. H. ROSENBROCK, *State-space and multivariable theory*, John Wiley, New York, 1970.

[121] E.P. RYAN, *Optimal relay and saturating control system synthesis*, Peter Peregrinus Ltd., 1982.

[122] A. SABERI, B.M. CHEN, AND P. SANNUTI, "Theory of LTR for non-minimum phase systems, recoverable targets, and recovery in a subspace. Part 1: analysis", Int. J. Contr., 53(5), 1991, pp. 1067–1115.

[123] ——, *Loop transfer recovery : analysis and design*, Springer Verlag, Berlin, 1993.

[124] A. SABERI, J. HAN, AND A.A. STOORVOGEL, "Constrained stabilization problems for linear plants", Automatica, 38(4), 2002, pp. 639–654.

[125] A. SABERI, J. HAN, A.A. STOORVOGEL, AND G. SHI, "Constrained stabilization problems for discrete-time linear plants", Int. J. Robust & Nonlinear Control, 14(5), 2004, pp. 435–461.

[126] A. SABERI, P. HOU, AND A.A. STOORVOGEL, "On simultaneous global external and global internal stabilization of critically unstable linear systems with saturating actuators", IEEE Trans. Aut. Contr., 45(6), 2000, pp. 1042–1052.

[127] A. SABERI, P.V. KOKOTOVIC, AND H.J. SUSSMANN, "Global stabilization of partially linear composite systems", SIAM J. Contr. & Opt., 28(6), 1990, pp. 1491–1503.

[128] A. SABERI, Z. LIN, AND A. TEEL, "Control of linear systems with saturating actuators", IEEE Trans. Aut. Contr., 41(3), 1996, pp. 368–378.

[129] A. SABERI, H. OZCETIN, AND P. SANNUTI, "New structural invariants of linear multivariable systems", Int. J. Contr., 56(4), 1992, pp. 877–900.

[130] A. SABERI AND P. SANNUTI, "Squaring down by static and dynamic compensators", IEEE Trans. Aut. Contr., 33(4), 1988, pp. 358–365.

[131] ——, "Observer design for loop transfer recovery and for uncertain dynamical systems", IEEE Trans. Aut. Contr., 35(8), 1990, pp. 878–897.

[132] ——, "Squaring down of non-strictly proper systems", Int. J. Contr., 51(3), 1990, pp. 621–629.

[133] A. SABERI, P. SANNUTI, AND B. M. CHEN, *H_2 Optimal Control*, Prentice Hall, Englewood Cliffs, NJ, 1995.

[134] A. SABERI AND A.A. STOORVOGEL, Guest Eds., *Special issue on control problems with constraints*, Int. J. Robust & Nonlinear Control, 9(10), 1999, pp. 583–734.

[135] A. SABERI, A.A. STOORVOGEL, AND P. SANNUTI, *Control of linear systems with regulation and input constraints*, Communication and Control Engineering Series, Springer Verlag, Berlin, 2000.

[136] ——, *Filtering theory: with applications to fault detection, isolation, and estimation*, Birkhäuser, Boston, MA, 2007.

[137] A. SABERI, A.A. STOORVOGEL, G. SHI, AND P. SANNUTI, "Semi-global stabilization of linear systems subject to non-right invertible constraints", Int. J. Robust & Nonlinear Control, 14(13–14), 2004, pp. 1087–1103.

[138] P. SANNUTI, "A direct singular perturbation analysis of high-gain and cheap control problems", Automatica, 19(1), 1983, pp. 41–51.

[139] P. SANNUTI AND A. SABERI, "Special coordinate basis for multivariable linear systems – finite and infinite zero structure, squaring down and decoupling", Int. J. Contr., 45(5), 1987, pp. 1655–1704.

[140] P. SANNUTI AND H.S. WASON, "A singular perturbation canonical form of invertible systems: determination of multivariable root-Loci", Int. J. Contr., 37(6), 1983, pp. 1259–1286.

[141] ——, "Multiple time-scale decomposition in cheap control problems – singular control", IEEE Trans. Aut. Contr., 30(7), 1985, pp. 633–644.

[142] S. SASTRY, *Nonlinear systems: analysis, stability and control*, vol. 10 of Interdisciplinary Applied Mathematics, Springer Verlag, London, 1999.

[143] S. SASTRY, J. HAUSER, AND P. KOKOTOVIC, "Zero dynamics of regularly perturbed systems may be singularly perturbed", Syst. & Contr. Letters, 13(4), 1989, pp. 299–314.

[144] C. SCHERER, *The Riccati inequality and state-space H_∞-optimal control*, PhD thesis, Universität Würzburg, 1991.

[145] W.E. SCHMITENDORF AND B.R. BARMISH, "Null controllability of linear systems with constrained controls", SIAM J. Contr. & Opt., 18(4), 1980, pp. 327–345.

[146] L. SCHUCHMAN, "Dither signals and their effect on quantization errors", IEEE Trans. on Commun. Technol., 4(12), 1964, pp. 162–165.

[147] P. SEIBERT AND R. SUAREZ, "Global stabilization of nonlinear cascade systems", Syst. & Contr. Letters, 14(4), 1990, pp. 347–352.

[148] ——, "Global stabilization of a certain class of nonlinear systems", Syst. & Contr. Letters, 16(1), 1991, pp. 17–23.

[149] GUOYONG SHI, *Control of linear systems with constraints – internal and external stabilization and output regulation*, PhD thesis, School of Electrical Engineering and Computer Science, 2002.

[150] G. SHI AND A. SABERI, "On the input-to-state stability (ISS) of a double integrator with saturated linear control laws", in American Control Conference, Anchorage, Alaska, 2002, pp. 59–61.

[151] G. SHI, A. SABERI, AND A.A. STOORVOGEL, "On the L_p (ℓ_p) stabilization of open-loop neutrally stable linear plants with input subject to amplitude saturation", Int. J. Robust & Nonlinear Control, 13(8), 2003, pp. 735–754.

[152] L.M. SILVERMAN, "Inversion of multivariable linear systems", IEEE Trans. Aut. Contr., 14(3), 1969, pp. 270–276.

[153] E.D. SONTAG, "Comments on integral variants of ISS", Syst. & Contr. Letters, 34(1-2), 1998, pp. 93–100.

[154] ——, "Input to state stability: Basic concepts and results", in Nonlinear and Optimal Control Theory, P. Nistri and G. Stefani, eds., Springer-Verlag, Berlin, 2007, pp. 163–220.

[155] E.D. SONTAG AND H.J. SUSSMANN, "Nonlinear output feedback design for linear systems with saturating controls", in Proc. 29th CDC, Honolulu, 1990, pp. 3414–3416.

[156] E.D. SONTAG AND Y. WANG, "On characterization of the input-to-state stability", Syst. & Contr. Letters, 24(5), 1995, pp. 351–359.

[157] E. SOROKA AND U. SHAKED, "The properties of reduced order minimum variance filters for systems with partially perfect measurements", IEEE Trans. Aut. Contr., 33(11), 1988, pp. 1022–1034.

[158] J. STEPHAN, M. BODSON, AND J. LEHOCZKY, "Properties of recoverable sets for systems with input and state constraints", in American Control Conference, Seattle, WA, 1995, pp. 3912–3913.

[159] ——, "Calculation of recoverable sets for systems with input and state constraints", Opt. Control Appl. & Meth., 19(4), 1998, pp. 247–269.

[160] A.A. STOORVOGEL, "The discrete time H_∞ control problem with measurement feedback", SIAM J. Contr. & Opt., 30(1), 1992, pp. 182–202.

[161] ——, *The H_∞ control problem: a state space approach*, Prentice-Hall, Englewood Cliffs, NJ, 1992.

[162] ——, "The H_∞ control problem with zeros on the boundary of the stability domain", Int. J. Contr., 63(6), 1996, pp. 1029–1053.

[163] A.A. STOORVOGEL AND A. SABERI, "Continuity properties of solutions to H_2 and H_∞ Riccati equation", Syst. & Contr. Letters, 27(4), 1996, pp. 209–222.

[164] ——, "Output regulation of linear plants with actuators subject to amplitude and rate constraints", Int. J. Robust & Nonlinear Control, 9(10), 1999, pp. 631–657.

[165] A. STOORVOGEL AND A. SABERI, "On external semi-global stochastic stabilization of a double integrator with input saturation", in Proc. 47th CDC, Cancun, Mexico, 2008, pp. 3493–3497.

[166] A.A. STOORVOGEL, A. SABERI, AND G. SHI, "On achieving L_p (ℓ_p) performance with global internal stability for linear plants with saturating actuators", in Proc. 38th CDC, Phoenix, AZ, 1999, pp. 2762–2767.

[167] ——, "Properties of recoverable region and semi-global stabilization in recoverable region for linear systems subject to constraints", Automatica, 40(9), 2004, pp. 1481–1494.

[168] A.A. STOORVOGEL, A. SABERI, AND S. WEILAND, "On external semi-global stochastic stabilization of linear systems with input saturation", in American Control Conference, New York, 2007, pp. 5845–5850.

[169] A.A. STOORVOGEL, G. SHI, AND A. SABERI, "External stability of a double integrator with saturated linear control laws", Dynamics of Continuous Discrete and Impulsive Systems, Series B: Applications & Algorithms, 11(4–5), 2004, pp. 429–451.

[170] A.A. STOORVOGEL, X. WANG, A. SABERI, AND P. SANNUTI, "Stabilization of sandwich non-linear systems with low-and-high gain feedback design", in American Control Conference, Baltimore, MD, 2010, pp. 4217–4222.

[171] A.A. Stoorvogel, S. Weiland, and A. Saberi, "On stabilization of linear systems with stochastic disturbances and input saturation", in Proc. 43rd CDC, The Bahamas, 2004, pp. 3007–3012.

[172] H.J. Sussmann and P.V. Kokotovic, "The peaking phenomenon and the global stabilization of nonlinear systems", IEEE Trans. Aut. Contr., 36(4), 1991, pp. 424–440.

[173] H.J. Sussmann, E.D. Sontag, and Y. Yang, "A general result on the stabilization of linear systems using bounded controls", IEEE Trans. Aut. Contr., 39(12), 1994, pp. 2411–2425.

[174] H.J. Sussmann and Y. Yang, "On the stabilizability of multiple integrators by means of bounded feedback controls", in Proc. 30th CDC, Brighton, U.K., 1991, pp. 70–72.

[175] A. Taware and G. Tao, "Neural-hybrid control of systems with sandwiched dead-zones", Int. J. Adapt. Contr. and Sign. Proc., 16(7), 2002, pp. 473–496.

[176] ——, "An adaptive dead-zone inverse controller for systems with sandwiched dead-zones", Int. J. Contr., 76(8), 2003, pp. 755–769.

[177] ——, *Control of sandwich nonlinear systems*, vol. 288 of Lecture notes in control and information sciences, Springer Verlag, 2003.

[178] A. Taware, G. Tao, and C. Teolis, "Design and analysis of a hybrid control scheme for sandwich nonsmooth nonlinear systems", IEEE Trans. Aut. Contr., 47(1), 2002, pp. 145–150.

[179] A.R. Teel, *Feedback stabilization: nonlinear solutions to inherently nonlinear problems*, PhD thesis, Electronics Research Laboratory, College of Engineering, University of California, 1992.

[180] ——, "Global stabilization and restricted tracking for multiple integrators with bounded controls", Syst. & Contr. Letters, 18(3), 1992, pp. 165–171.

[181] ——, "Semi-global stabilization of linear null-controllable systems with input nonlinearities", IEEE Trans. Aut. Contr., 40(1), 1995, pp. 96–100.

[182] Andrew Teel, "Asymptotic convergence from $\mathcal{L}_p$ stability", IEEE Trans. Aut. Contr., 44(11), 1999, pp. 2169–2170.

[183] H.L. Trentelman, "The regular free-endpoint linear quadratic problem with indefinite cost", SIAM J. Contr. & Opt., 27(1), 1989, pp. 27–42.

[184] H.L. Trentelman, A.A. Stoorvogel, and M.L.J. Hautus, *Control theory for linear systems*, Communication and Control Engineering Series, Springer Verlag, London, 2001.

[185] H.L. Trentelman and J.C. Willems, "The dissipation inequality and the algebraic Riccati equation", in The Riccati Equation, S. Bittanti, A J. Laub, and J C. Willems, eds., Springer Verlag, sep 1991, pp. 197–242.

[186] F. Tyan and D.S. Bernstein, "Global stabilization of systems containing a double integrator using a saturated linear controller", Int. J. Robust & Nonlinear Control, 9(15), 1999, pp. 1143–1156.

[187] M. Vassilaki, J.C. Hennet, and G. Bitsoris, "Feedback control of linear discrete-time systems under state and control constraints", Int. J. Contr., 47(6), 1988, pp. 1727–1735.

[188] G. VERGHESE, *Infinite frequency behavior in generalized dynamical systems*, PhD thesis, Department of Electrical Engineering, Stanford University, 1978.

[189] M. VIDYASAGAR, *Nonlinear systems analysis*, Prentice Hall, London, 2nd Ed., 1992.

[190] M. VIDYASAGAR AND A. VANNELLI, "New relationships between input-output and Lyapunov stability", IEEE Trans. Aut. Contr., 27(2), 1982, pp. 481–483.

[191] X. WANG, H.F. GRIP, A. SABERI, A.A. STOORVOGEL, AND I. SABERI, "Remarks on the relationship between $\mathcal{L}_p$ stability and internal stability of nonlinear systems", in American Control Conference, San Francisco, CA, 2011, pp. 1969–1970.

[192] X. WANG, A. SABERI, H.F. GRIP, AND A.A. STOORVOGEL, "Control of linear systems with input saturation and disturbances – continuous-time system", Submitted for publication, 2011.

[193] ——, "Control of linear systems with input saturation and disturbances – discrete-time systems", Submitted for publication, 2011.

[194] X. WANG, A. SABERI, AND A.A. STOORVOGEL, "Stabilization of linear system with input saturation and unknown delay", Submitted for publication, 2012.

[195] X. WANG, A. SABERI, A.A. STOORVOGEL, AND H.F. GRIP, "Control of a chain of integrators subject to input saturation and disturbances", To appear in *Int. J. Robust & Nonlinear Control*. Full version can be found at http://onlinelibrary.wiley.com/doi/10.1002/rnc.1767/full.

[196] ——, "Further results on the disturbance response of a double integrator controlled by saturating linear static state feedback", Automatica, 48(2), 2012, pp. 430–435.

[197] X. WANG, A. SABERI, A.A. STOORVOGEL, S. ROY, AND P. SANNUTI, "Computation of the recoverable region and stabilization problem in the recoverable region for discrete-time systems", Int. J. Contr., 82(10), 2009, pp. 1870–1881.

[198] X. WANG, A. SABERI, A.A. STOORVOGEL, AND P. SANNUTI, "Simultaneous global external and internal stabilization of linear time-invariant discrete-time systems subject to actuator saturation", Automatica, 48(5), 2012, pp. 699–711.

[199] X. WANG, A.A. STOORVOGEL, A. SABERI, H.F. GRIP, S. ROY, AND P. SANNUTI, "Stabilization of a class of nonlinear sandwich systems via state feedback", IEEE Trans. Aut. Contr., 55(9), 2010, pp. 2156–2160.

[200] X. WANG, A.A. STOORVOGEL, A. SABERI, H.F. GRIP, AND P. SANNUTI, "H_2 and H_∞ low-gain theory", in American Control Conference, San Francisco, CA, 2011, pp. 4463–4469.

[201] X. WANG, A.A. STOORVOGEL, A. SABERI, AND P. SANNUTI, "H_2 and H_∞ low gain theory", Submitted for publication, 2010.

[202] ——, "Discrete-time H_2 and H_∞ low-gain theory", Int. J. Robust & Nonlinear Control, 22(7), 2012, pp. 743–762.

[203] Z. WEN, S. ROY, AND A. SABERI, "On the disturbance response and external stability of a saturating static-feedback-controlled double integrator", Automatica, 44(8), 2008, pp. 2191 – 2196.

[204] J.C. WILLEMS, *The analysis of feedback systems*, MIT Press, 1971.

[205] W.M. WONHAM, *Linear multivariable control: a geometric approach*, Springer Verlag, New York, Third Ed., 1985.

[206] T. YANG, A.A. STOORVOGEL, AND A. SABERI, "Issues on global stabilization of linear systems subject to actuator saturation", in Proceedings 4th IFAC Symposium on System, Structure and Control, L. Jetto, ed., Ancona, Italy, 2010, pp. 231–236.

[207] ——, "Global stabilization of the discrete-time double integrator using a saturated linear state feedback controller", in American Control Conference, San Francisco, CA, 2011, pp. 4440–4445.

[208] T. YANG, A.A. STOORVOGEL, X. WANG, AND A. SABERI, "Periodic behavior of locally stabilizing saturated linear controllers for the discrete-time double integrator", in Proceedings 4th IFAC Symposium on System, Structure and Control, L. Jetto, ed., Ancona, Italy, 2010, pp. 237–241.

[209] Y. YANG, *Global stabilization of linear systems with bounded feedback*, PhD thesis, Rutgers University, New Brunswick, 1993.

[210] Y. YANG, E.D. SONTAG, AND H.J. SUSSMANN, "Global stabilization of linear discrete-time systems with bounded feedback", Syst. & Contr. Letters, 30(5), 1997, pp. 273–281.

[211] C.A. YFOULIS, A. MUIR, AND P.E. WELLSTEAD, "A new approach for estimating controllable and recoverable regions with state and control constraints", Int. J. Robust & Nonlinear Control, 12(7), 2002, pp. 561–589.

[212] G. ZAMES AND N.A. SHNEYDOR, "Dither in non-linear systems", IEEE Trans. Aut. Contr., 21(5), 1976, pp. 660–667.

[213] ——, "Structural stabilization and quenching by dither in non-linear systems", IEEE Trans. Aut. Contr., 22(3), 1977, pp. 352–361.

[214] J.R. ZHANG, C.R. KNOSPE, AND P. TSIOTRAS, "Stability of linear time-delay systems: a delay-dependent criterion with a tight conservatism bound", in American Control Conference, Chicago, IL, 2000, pp. 1458–1462.

[215] B. ZHOU, G.R. DUAN, AND Z. LIN, "A parametric Lyapunov equation approach to the design of low gain feedback", IEEE Trans. Aut. Contr., 53(6), 2008, pp. 1548–1554.

[216] B. ZHOU, Z. LIN, AND G.R. DUAN, "A parametric Lyapunov equation approach to low gain feedback design for discrete time systems", Automatica, 45(1), 2009, pp. 238–244.

[217] ——, "Global and semi-global stabilization of linear systems with multiple delays and saturations in the input", SIAM J. Contr. & Opt., 53(8), 2010, pp. 5294–5332.

Index

A. Saberi et al., *Internal and External Stabilization of Linear Systems with Constraints*, Systems & Control: Foundations & Applications, DOI 10.1007/978-0-8176-4787-2,

MIX
Papier aus verantwortungsvollen Quellen
Paper from responsible sources
FSC® C105338

FSC
www.fsc.org

If you have any concerns about our products,
you can contact us on
ProductSafety@springernature.com

In case Publisher is established outside the EU,
the EU authorized representative is:
Springer Nature Customer Service Center GmbH
Europaplatz 3, 69115 Heidelberg, Germany

Printed by Libri Plureos GmbH
in Hamburg, Germany